CHEMISTRY
Structure and Dynamics
Fifth Edition

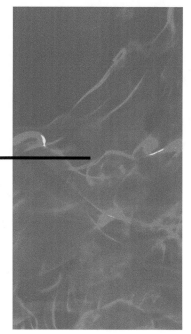

CHEMISTRY
Structure and Dynamics
Fifth Edition

James N. Spencer
Franklin and Marshall College

George M. Bodner
Purdue University

Lyman H. Rickard
Millersville University

WILEY

John Wiley & Sons, Inc.

VICE PRESIDENT AND EXECUTIVE PUBLISHER   Kaye Pace
ASSOCIATE PUBLISHER   Petra Recter
ACQUISITIONS EDITOR   Nicholas Ferrari
ASSOCIATE EDITOR   Aly Rentrop
MARKETING MANAGER   Kristine Ruff
EXECUTIVE MEDIA EDITOR   Thomas Kulesa
MEDIA EDITOR   Marc Wezdecki
SENIOR PHOTO EDITOR   Lisa Gee
SENIOR ILLUSTRATION EDITOR   Anna Melhorn
PRODUCTION MANAGER   Dorothy Sinclair
SENIOR PRODUCTION EDITOR   Trish McFadden
DESIGN DIRECTOR   Harry Nolan
SENIOR DESIGNER   Kevin Murphy
EDITORIAL ASSISTANT   Lauren Stauber
COVER IMAGE   © Aleksander Trankov/iStockphoto
PRODUCTION MANAGEMENT   Ingrao Associates

This book was set in 10.5/12.5 Times Roman by Aptara and printed and bound by Courier Kendalville. The cover was printed by Courier Kendalville.

This book is printed on acid free paper. ∞

To order books or for customer service, please call 1-800-CALL WILEY (225-5945).

ISBN-13 978-0-470-587119
ISBN-10 0-470-587113

Printed in the United States of America

10 9 8 7 6 5

Preface

The fifth edition of *Chemistry: Structure and Dynamics,* retains its hallmark features—brevity, flexibility, and currency—while increasing its ease of use in the classroom. This edition includes:

- Expanded discussions of Le Châtelier's Principle, solubility, buffers, indicators and acid-base titrations.
- A rearrangement of Chapters 1 and 2 including the integration of units, unit conversions, scientific notation and significant figures into the context of the text as they are needed in the discussion. Students are referred to Appendix A for a more in depth discussion of these topics.
- Expanded development of the use of Dalton's law in Chapter 6.
- A new section on work has been added to Chapter 7.
- A new section on buffers including the development and use of the Henderson-Hasselbalch equation has been added to Chapter 11. An added section discusses acid-base indicators and their color changes.
- The electrolysis of molten NaCl, aqueous water, and aqueous NaCl are now presented in Chapter 12.
- A new development of the Second Law in terms of changes that take place in the system and surroundings as water freezes or melts is presented in Chapter 13.
- Chapter 14 introduces the rate limiting step terminology for a reaction sequence.
- A new Chapter 16 provides an Introduction to Organic Chemistry.
- New problems, examples, and exercises have been added throughout the text.

Moog/Farrell's *Chemistry: A Guided Inquiry* is available in a new edition to extend the text instruction. Moog and Farrell provide a cooperative teaching pedagogy which can be used for recitation sections, as an in-class workbook or as a take-home study guide. The pedagogy used in *Chemistry: A Guided Inquiry* was developed as part of the Process Oriented Guided Inquiry Learning (POGIL) project. See http://pogil.org for a more detailed description of this approach to teaching Introductory Chemistry.

In 1989 the Division of Chemical Education of the American Chemical Society recognized the need to foster the development of alternative introductory chemistry curricula. The Task Force on the General Chemistry Curriculum was created to meet the need. We were members of the Task Force and our book, *Chemistry: Structure and Dynamics,* is one of the products of the work of the Task Force.

This text encourages innovation in the teaching of general chemistry by providing flexible materials that allow instructors to build a custom curriculum appropriate for their students. We achieved this flexibility through the use of a core/modular structure. The text contains the core and chapter appendices called Special Topics. The modules are furnished separately to supplement the core. The core provides the fundamentals that all students need to prepare them for the degree programs in which an introductory chemistry course is a prerequisite. By selecting topics from the chapter appendices (Special Topics) and modules, instructors can customize their course to fit the needs of their particular students.

The text is written for students taking introductory chemistry for science and math majors. Its flexibility, however, makes it appropriate for most introductory chemistry courses. The choice of concepts to include in the core is based on two primary criteria. First, the concepts should be the most fundamental building blocks for understanding chemistry—concepts that provide the basis on which remaining parts of the core and the modules are built. Second, these concepts should be perceived by *the students* as being directly applicable to their majors or careers.

Chemistry: Structure and Dynamics is characterized by the following key features:

- **Brevity.** The core contains approximately 700 pages with 50 pages of chapter appendices (Special Topics).
- **Flexibility.** Modules and chapter appendices can be added to the core to provide a curriculum that meets the needs of the students at a particular institution.
- **Currency.** New models and current methods of understanding chemical concepts not yet found in traditional texts are used to introduce material.
- **Unifying Themes.** Major themes are used to link the core material into a unified whole.
- **Balanced Coverage.** Organic and biochemical examples are used throughout the text to provide a more balanced coverage of all areas of chemistry.

  A new chapter 16 provides an introduction to organic chemistry.

- **Conceptual Nature.** There is an increased emphasis on conceptual questions and problems at the end of the chapters. A large selection of traditional problems are provided, but these are supplemented with conceptual problems and discussion questions that ask students to explain, describe, or suggest experiments.

THE CORE

Traditional texts present concepts and principles in isolation, with little, if any, connection between each concept or principle and the rest of the material. Unifying themes are used to integrate the core topics in *Chemistry: Structure and Dynamics*.

- The *process of science* theme incorporates experimental data in discussions that traditional texts present as the *product* of scientific process. Our goal is to provide students with the evidence that will allow them to understand why chemists believe what they do.
- A second theme is the interrelationships between chemistry on the macroscopic and microscopic scales. This theme is developed to help students understand both how and why chemists make observations on the macroscopic scale so as to comprehend the microscopic world of atoms and molecules, and

then use the resulting understanding of microscopic structure and properties to explain, predict, and—most importantly—control the macroscopic properties of matter.

- Atomic and molecular structure are developed early in the text and then repeatedly used to help students understand the physical and chemical properties of matter.

The fundamental concepts in the core portion of *Chemistry: Structure and Dynamics* can be found in current traditional texts. But these texts contain so much additional material that it is tempting to give too little time and emphasis to these fundamental concepts. As a result, the average student does not gain a full appreciation of the core concepts necessary to build a solid understanding of chemistry. *Chemistry: Structure and Dynamics* develops these core concepts using concrete models and then illustrates them with practical applications relevant to the students' experience.

THE CHAPTER APPENDICES (SPECIAL TOPICS) AND MODULES

The chapter appendices extend the core material. The core chapter on bonding, for example, covers Lewis structures, molecular geometry, and the concept of polarity. A chapter appendix is then available that includes hybridization, valence bond theory, and molecular orbital theory. The modules are designed to introduce new topics, such as biochemistry, polymer chemistry, and coordination chemistry.

The core/modular approach has advantages over the traditional 1000-page texts. When traditional texts are used, sections or even whole chapters are skipped. This can be frustrating to those students who depend heavily on the text for understanding new material because later topics in the text are often explained using concepts that have been omitted. *Chemistry: Structure and Dynamics* avoids this problem by building the modules on the concepts presented in the core. The core covers all concepts that are prerequisite to the modules.

NEW MODELS

The new models used in this text can be divided into three types: (1) data-driven models, (2) models that reflect current understanding of chemical theories, and (3) models that make it easier for students to understand traditional concepts. The development of electron configurations from experimental photoelectron spectral (PES) data is an example of a data-driven model that supports the unifying theme of the process of science by demonstrating to students how experimental data can be used to construct models. This approach gives students a more concrete, and still scientifically correct, foundation on which to base their understanding of electron configuration than does use of the more abstract quantum numbers. The use of experimental data and the graphical representation of that data to develop the gas laws is another example of this type of model. Once the kinetic molecular theory is developed, it is used throughout the text to provide a consistent background for understanding temperature, heat, and equilibrium processes.

Models that reflect current understanding of chemical theories include the replacement of the valence shell electron pair repulsion (VSEPR) theory for predicting molecular geometry with Gillespie's more recent electron domain (ED) model and the use of bond-type triangles to explain the interrelationship of covalent, ionic, and metallic bonding. Bond-type triangles are then used in the text to help explain physical and chemical properties of compounds and to predict properties of new materials.

New models in the text are also used to present familiar concepts in innovative ways that make it easier for students to understand. For example, enthalpies of atom combination replace enthalpies of formation. Because chemistry is concerned with the making and breaking of bonds, thermochemical calculations are first done by considering the energetics of breaking the bonds necessary to produce atoms in the gas phase and then allowing these atoms to combine to form molecules. In the enthalpy of formation approach, the standard states are the elements in their stable states of aggregation. This construct places another concept between the student and the idea that chemical reactions liberate or absorb heat through the making and breaking of bonds. The major strength of the atom combination method is the use of the powerful visual model of reactants being broken down into gaseous atoms and then recombining to form products. The advantage to this method is that bond breaking always requires an input of energy, whereas bond making produces energy. Instructors accustomed to thinking of enthalpy changes in terms of enthalpies of formation may at first find this method more awkward. However, students find this approach much easier to visualize than enthalpies of formation.

A major advantage of the atom combination approach is that the same standard states and reaction diagrams are used for all thermodynamic parameters. Traditionally, third-law entropies are used in conjunction with enthalpies based on elemental standard states to introduce free energy. The use of third-law entropies that are based on yet a different concept and standard state further confuses students. The atom combination approach clearly shows students the origin of the entropies of substances and, because of the direct relationship with enthalpies and free energies, makes all these concepts more accessible to students.

Another example of the use of models that reflect current understanding of chemical theories is the introduction of average valence electron energies (AVEE), calculated from PES data, which describe how tightly an atom holds on to its electrons. AVEE values are used to predict which elements will form cations and which will form anions. This value is then used to develop the concept of electronegativity, thus giving electronegativity a more concrete meaning for students. Not only do the students develop the electronegativity concept from the same data used to determine the electronic structure of atoms, but they also see how electronegativities relate to oxidation numbers, partial charges, and formal charges. This theme is carried through the text and provides a unifying thread for oxidation–reduction, resonance structures, and polarity.

USING THE CORE TEXT AND MODULES

The core text consists of 16 chapters. With the exception of Chapter 2, the core chapters are designed to be covered in order. Chapter 2 covers molarity and stoichiometry. This information is provided early in the text for those instructors who traditionally cover these topics at the beginning of the course. Mole calculations and stoichiometry are not required until Chapter 6. However, the mole concept is used qualitatively in Chapters 3–5. Therefore, instructors can choose to cover portions of Chapter 2 as it fits their curriculum.

The core may be supplemented using the chapter appendices, called Special Topics (found at the end of Chapters 4, 6, 8, 9, 11, 12, and 14), or modules (available online). The chapter appendices extend the core topics found in the chapters they accompany. The material in the chapter appendices may be used or skipped at the discretion of the instructor because subsequent core chapters depend only on the concepts covered in the core chapters. Topics covered in the core chapters and their accompanying appendices are listed in the table of contents.

The modules, available online at www.wiley.com/college/spencer, may be inserted at various points in the text depending on the core material prerequisite to the modules. The table below lists the core chapters that are prerequisite for each module.

Prerequisites for Modules

Module	Prerequisite Core Chapters
I. Chemistry of the Nonmetals	Chapter 12
II. Transition-Metal Chemistry	Chapter 5 & 4 Appendixes
III. Complex-Ion Equilibria	Chapter 11
IV. Organic Chemistry: Structure and Nomenclature of Hydrocarbons	Chapter 7 & 4 Appendixes
V. Organic Chemistry: Functional Groups	Structure and Nomenclature Module
VI. Organic Chemistry: Reaction Mechanisms	Functional Groups Module
VII. Polymer Chemistry	Chapter 7 & an Organic Module
VIII. Biochemistry	Chapter 11 and an Organic Module
IX. Chemical Analysis	Chapter 1–8

TO THE STUDENT

This text, *Chemistry: Structure and Dynamics,* differs from most texts currently available for introductory chemistry. The text contains a short core of material that should provide a basis for the understanding of the fundamental concepts of chemistry, and new models to teach familiar concepts in ways that should make the concepts easier to grasp. Consistent themes throughout the text tie together many seemingly isolated topics.

The chapters in this text are designed to be read linearly, from the beginning to the end, rather than reading individual topics within the chapter. It is important to answer the Checkpoints and to work through the Exercises as you encounter them in your reading. Important new concepts are sometimes introduced in the Checkpoints and Exercises. In addition, answering the Checkpoints and working the Exercises will change your reading of the chapter from passive into active reading. The checkpoints will assist you in determining if you have understood what you have just read. Answers to the checkpoints can be found in Appendix D. Solutions to selected end-of-chapter problems are found in Appendix C.

STUDENT AND INSTRUCTOR RESOURCES

Chemistry; A Guided Inquiry, written by Richard S. Moog and John J. Farrell, of Franklin and Marshall College. This supplement facilitates implementation of cooperative learning in general chemistry and can be used as part of a recitation section, as a workbook, or as a means of interacting with students in the classroom. This supporting book uses all of the new approaches featured in *Chemistry: Structure and Dynamics,* and features problem assignments from the Core text. The book uses guided inquiry in which data, written descriptions, models, and figures are used to develop chemical concepts.

INSTRUCTOR RESOURCES

Wiley Resource Kit: The **Wiley Resource Kit** provides a simple and hassle-free way to integrate the most sought after instructor and student tools into any Learning Management System (LMS).

With the Wiley Resource Kit, you'll have:

- FREE access to resources that complement your course & textbook
- Immediate availability in one convenient location

- No Cartridges, Plug-ins, or license fees!
- Compatibility with any Learning Management System (Blackboard, WebCT, Angel, and more!)

WILEY ● RESOURCE KIT   The **Wiley Resource** Delivery Kit consists of two main elements:

Student Link-ins

WILEY STUDENT LINK-INS

Deliver high-value resources instantly to students through a pedagogically sound, Wiley hosted page within your LMS course.

Respondus®

RESPONDUS TEST BANK NETWORK

Provides tests and quizzes for Wiley's leading titles for easy publication into your LMS course, as well as for printed tests. *For schools without a campus-wide license to Respondus, Wiley will provide one for no additional cost.

Instructor's Manual, prepared by James Spencer, George Bodner, and Lyman Rickard. The Instructor's Manual highlights material in the text, which differs from more traditional texts, focusing on how the text's method of presentation can best be utilized in the course. Further information is given on PES, atom combination parameters, bond-type triangles, AVEE, and partial charge calculations.

Instructor's Solutions Manual, prepared by Alex Grushow. The Instructor Solutions Manual provides complete solutions to all text problems.

Computerized Test Bank, prepared by Wimba. This Test Bank contains multiple-choice and short-answer questions.

Test Bank, prepared by Dan Freedman. This Test Bank contains multiple-choice and short-answer questions designed for an introductory chemistry course.

Classroom Response System Questions, prepared by David A. Cleary. This resource is a collection of interactive lecture questions.

Power Point Lecture Slides, prepared by David A. Cleary. These slides outline topics in the text, and are accompanied by illustrations in the text.

STUDENT RESOURCES

Student Solution Manual, prepared by Alex Grushow. The Student Solutions Manual provides complete solutions to all odd-numbered problems.

ACKNOWLEDGMENTS

There are many people who contributed to this textbook. First are all the members of the Task Force on the General Chemistry Curriculum, without whose discussions and ideas this project would never have been initiated. We are also indebted to many who reviewed all or part of the text.

Janice Alexander
Flathead Valley Community College

Linda Allen
Louisiana State University

Dennis M. Anjo
Cal State, Long Beach

Chris Bailey
Wells College

David W. Ball
Cleveland State University

Jay Bardole
Vincennes University

William Bare
Randolph-Macon Woman's College

Jack Barbera
Northern Arizona University

Elisabeth Bell-Loncella
University of Pittsburgh, Johnstown

Steven D. Bennett
Bloomsburg University

Debra Boehmler
University of Maryland

Chris Bowers
Ohio Northern University

Thomas R. Burkholder
Central Connecticut State University

Bruce Burnham
Rider University

Sheila Cancella
Raritan Valley Community College

Feng Chen
Rider University

David A. Cleary
Gonzaga University

Martin Cowie
University of Alberta

Paul H. Davis
Santa Clara University

Michael Doyle
University of Maryland

Robert Eierman
University of Wisconsin-Eau Claire

William Evans
University of California, Irvine

Deniel Freedman
Suny-New Paltz

Larry Gerdom
Trevecca Nazarene University

Tom Gilbert
Northeastern University

L. Peter Gold
Pennsylvania State University

Stan Grenda
University of Nevada, Las Vegas

Eugene Grimley III
Elon College

Thomas Grover
Gustavus Adolphus College

Alexander Grushow
Rider University

Keith Hansen
Lamar University

Lee Hansen
Brigham Young University

Paul Hanson
University of New Orleans

David Harvey
DePauw University

Craig Hoag
SUNY Pittsburgh

Mike Iannone
Millersville University

John Jefferson
Luther College

Pamela St. John
SUNY New Paltz

Keith Kester
Colorado College

Leslie Kinsland
University of Southwestern Louisiana

Nancy Konigsberg-Kerner
University of Michigan, Ann Arbor

George Kraus
College of Southern Maryland

David Lewis
Colgate University

Robert Loeschen
California State University, Long Beach

Baird Lloyd
Miami University of Ohio

David MacInnes Jr.
Guilford College

Asoka Marasinghe
Minnesota State University

Doug Martin
Sonoma State University

Claude Mertzenich
Luther College

Patricia Metz
Texas Tech University

David Millican
Guilford College

Susan Morante
Mont Royal College

Edward Paul
Stockton College

Michael Prushan
La Salle University

Olga Rinco
Luther College

Wayne E. Steinmetz
Pomona College

E. B. Robertson
University of Calgary

Robert Stewart, Jr.
Miami (OH) University

Carey Rosenthal
Drexel University

Wesley Stites
University of Arkansas

Doug Rustad
Sonoma State University

Duane Swank
Pacific Lutheran University

Patricia Schroeder
Johnson County Community College

Robert L. Swofford
Wake Forest University

Tim Schroeder
Southern Arkansas University

Sandra Turchi
Millersville University

Karl D. Sienerth
Elon University

John B. Vincent
University of Alabama

Karl Sohlberg
Drexel University

Gloria Brown Wright
Central Connecticut State University

Larry Spreer
University of the Pacific

John Woolcock
Indiana University of Pennsylvania

William Stanclift
*Northern Virginia Community College,
Annandale*

Andrew Zanella
Claremont College

Particular contributions were made by John Farrell and Rick Moog of Franklin and Marshall College; Ron Gillespie of McMaster University; Dudley Herschbach of Harvard University; Lee Allen of Princeton University; Gordon Sproul of the University of South Carolina at Beaufort; and Alex Grushow at Rider University.

It is a particular pleasure to acknowledge the support and assistance of the staff at John Wiley and Sons, Inc., Nick Ferrari, Karen Gulliver, Aly Rentrop, Catherine Donovan, Patricia McFadden, and Suzanne Ingrao who attended to the countless details associated with this project.

Finally, to Kathy, Christine, and Lynette we owe a debt for their patience and encouragement.

JAMES N. SPENCER
GEORGE M. BODNER
LYMAN H. RICKARD

Contents

Chapter 4
The Covalent Bond   123

Chapter 5
Ionic and Metallic Bonds   177

Chapter 6
Gases   221

Chapter 7
Making and Breaking of Bonds 264

Chapter 8
Liquids and Solutions 313

Chapter 9
Solids 367

Chapter 10
The Connection Between Kinetics and Equilibrium   408

Chapter 11
Acids and Bases   468

Modules
(available at www.wiley.com/college/spencer)

Module 1
Chemistry of the Nonmetals

Module 2
Transition Metal Chemistry

Module 3
Complex Ion Equilibria

Module 4
Organic Chemistry: Structure and Nomenclature
of Hydrocarbons

Module 5
Organic Chemistry: Functional Groups

Module 6
Organic Chemistry: Reaction Mechanisms

Module 7
Polymer Chemistry

Module 8
Biochemistry

Module 9
Chemical Analysis

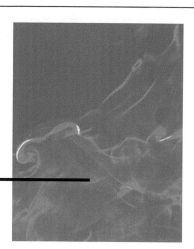

Chapter One
ELEMENTS AND COMPOUNDS

1.1 Chemistry: A Definition

It seems logical to start a book of this nature with the question: What is chemistry? Most dictionaries define chemistry *as the science that deals with the composition, structure, and properties of substances and the reactions by which one substance is converted into another.* Knowing the definition of chemistry, however, is not the same as understanding what it means.

One way to understand the nature of chemistry is to look at examples of what it isn't. In 1921, a group from the American Museum of Natural History began excavations at an archaeological site on Dragon-Bone Hill, near the town of Chou-k'outien, 34 miles southwest of Beijing, China. Fossils found at this site were assigned to a new species, *Homo erectus pekinensis,* commonly known as Peking man. These excavations suggest that for at least 500,000 years, people have known enough about the properties of stone to make tools, and they have been able to take advantage of the chemical reactions involved in combustion in order to cook food. But even the most liberal interpretation would not allow us to call this chemistry because of the absence of any evidence of control over these reactions or processes.

The ability to control the transformation of one substance into another can be traced back to the origin of two different technologies: brewing and metallurgy. People have been brewing beer for at least 12,000 years, since the time when the first cereal grains were cultivated, and the process of extracting metals from ores has been practiced for at least 6000 years, since copper was first produced by heating the ore malachite.

But brewing beer by burying barley until it germinates and then allowing the barley sprouts to ferment in the open air wasn't chemistry. Nor was extracting copper metal from one of its ores because this process was carried out without any understanding of what was happening or why. Even the discovery around 3500 B.C. that copper mixed with 10 to 12% tin gave a new metal that was harder than copper, and yet easier to melt and cast, was not chemistry. The preparation of bronze was a major breakthrough in metallurgy, but it didn't provide us with an understanding of how to make other metals.

Between the sixth and the third centuries B.C., the Greek philosophers tried to build a theoretical **model** for the behavior of the natural world. They argued that the world was made up of four primary, or *elementary,* substances: fire, air, earth, and water. These substances differed in two properties: hot versus cold, and dry versus wet. Fire was hot and dry; air was hot and wet; earth was cold and dry; water was cold and wet.

This model was the first step toward the goal of understanding the properties and compositions of different substances and the reactions that convert one substance to another. But some elements of modern chemistry were still missing. This model could explain certain observations of how the natural world behaved, but it couldn't predict new observations or behaviors. It was also based on pure speculation. In fact, its proponents weren't interested in using the results of experiments to test the model.

Modern chemistry is based on certain general principles.

- **One of the goals of chemistry is to recognize patterns in the way different substances behave.** An example might be the discovery in 1794 by the French chemist Antoine Lavoisier that many substances that burn in air gain weight.

- **Once a pattern is recognized, it should be possible to develop a model that explains these observations.** Lavoisier concluded that substances that burn in air combine with the oxygen in the air to form products that weigh more than the starting material.

- **These models should allow us to predict the behavior of other substances.** In 1869, Dmitri Mendeléeff[1] used his model of the behavior of the known elements to predict the properties of elements that had not yet been discovered.

- **When possible, the models should be quantitative.** They should not only predict what happens, but by how much.

- **The models should be able to make predictions that can be tested experimentally.** Mendeléeff's periodic table was accepted by other chemists because of the agreement between his predictions and the results of experiments based on these predictions.

The term *model* is defined as a noun or an adjective that describes a simplified or idealized description of an idea, object, event, process, or system that focuses attention on certain aspects of the system. Models are often expressed as equations that explain what has been observed in the past and make predictions about what might be observed in the future.

Chemists think in terms of constructing, evaluating, refining, adapting, modifying, and extending models that are based on their experiences with the world in which they work and live. Some have gone so far as to suggest that "modeling" is the essence of thinking and working scientifically. As you encounter various models in the course of reading this book, it is important to recognize that these models fit experimental data, more or less, under certain conditions and within certain limitations. They are not examples of "something that must be obeyed."

In essence, *chemistry is an experimental science.* Experiment serves two important roles. It forms the basis of observations that define the problems that models must explain, and it provides a way of checking the validity of new models. This text emphasizes an experimental approach to chemistry. As often as possible, it presents the experimental basis of chemistry before the theoretical explanations of these observations.

1.2 Elements, Compounds, and Mixtures

Matter is defined as anything that has mass and occupies space. All substances that we encounter—whether natural or synthetic—are matter. Matter can be divided into three general categories: elements, compounds, and mixtures.

Elements are substances that contain only one kind of atom. To date, 118 elements have been discovered. They include a number of substances with which you are familiar, such as the oxygen in the atmosphere, the aluminum in aluminum foil, the iron in nails, the copper in electrical wires, and so on. Elements are the fundamental building blocks from which all other substances are made.

---

[1]There are at least half a dozen ways of spelling Mendeléeff's name because of disagreements about transliterations from the Cyrillic alphabet. The version used here is the spelling that Mendeléeff himself used when he visited England in 1887.

Imagine cutting a piece of gold metal in half and then repeating this process again and again and again. In theory, we should eventually end up with a single gold atom. If we tried to split this atom in half, we would end up with something that no longer retains any of the characteristics of the element. An **atom** is therefore the smallest particle that can be used to identify an element.

Compounds are substances that contain more than one element combined in fixed proportions. Water, for example, is composed of the elements hydrogen and oxygen in the ratio of two atoms of hydrogen to one atom of oxygen. If we tried to divide a sample of water into infinitesimally small portions, we would eventually end up with a single molecule of water containing two hydrogen atoms and one oxygen atom. If we tried to break this molecule into its individual atoms, we would no longer have water. A **molecule** is therefore the smallest particle that can be used to identify a compound.

Both elements and compounds have a *constant composition*. Water, for example, is always 88.8% oxygen by weight, regardless of where it is found. When pure, the salt used to flavor food has exactly the same composition regardless of whether it was dug from mines beneath the surface of the earth or obtained by evaporating seawater. No matter where it comes from, salt always contains 1.54 times as much chlorine by weight as sodium. Pure substances also have constant chemical and physical properties. Pure water always freezes at 0°C and boils at 100°C at atmospheric pressure.

Mixtures, such as a cup of coffee, have different compositions from sample to sample, and therefore varying properties. If you are a coffee drinker, you will have noted that cups of coffee from your home, the college cafeteria, and a gourmet coffeehouse aren't the same. They vary in appearance, aroma, and flavor because of differences in the composition of this mixture. Mixtures can be classified as **homogeneous** or **heterogeneous.** A homogeneous mixture is uniform; the composition is the same throughout the mixture. An individual cup of coffee is a homogenous mixture because the composition throughout the cup is the same. A heterogeneous mixture does not have the same composition throughout. An example of a heterogeneous mixture is a shovelful of dirt that may have grass from the top of the soil, with rich top soil and then gravel on the bottom.

1.3 Atomic Symbols

When describing atoms, chemists use a shorthand notation to save both time and space. Each element is represented by a unique symbol. Most of these symbols make sense because they are derived from the name of the element.

H = hydrogen	B = boron
C = carbon	N = nitrogen
O = oxygen	P = phosphorus
Se = selenium	Si = silicon
Mg = magnesium	Br = bromine
Al = aluminum	Ca = calcium
Cr = chromium	Zn = zinc

Symbols that don't seem to make sense can be traced back to the Latin or German names of the elements. Fortunately, there are only a handful of elements in this category.

Ag = silver   Na = sodium
Au = gold   Pb = lead
Cu = copper   Sb = antimony
Fe = iron   Sn = tin
Hg = mercury   W = tungsten
K = potassium

1.4 Chemical Formulas

The composition of a compound can be represented by a **chemical formula** that represents the relative number of atoms of different elements in the compound, as shown in Figure 1.1. By convention, no subscript is written when a molecule contains only one atom or an element. Thus, water is H_2O and carbon dioxide is CO_2.

Compounds can be divided into two general categories: molecular and ionic. Water (H_2O), carbon dioxide (CO_2), and butane (C_4H_{10}) are examples of **molecular compounds.** The smallest particle in each of these compounds is a molecule that doesn't carry an electric charge. **Ionic compounds** contain both positive and negative particles that form an extended three-dimensional structure. The chemical formula of an ionic compound describes the overall ratio of positive and negative particles in this network. Sodium chloride (NaCl) is the best-known example of an ionic compound.

Elements can also exist in the form of molecules, but these molecules are composed of identical atoms (Figure 1.2). The oxygen we breathe, for example, consists of molecules that contain two oxygen atoms, O_2. Elemental phosphorus molecules are composed of four phosphorus atoms (P_4), and elemental sulfur contains molecules are composed of eight sulfur atoms (S_8).

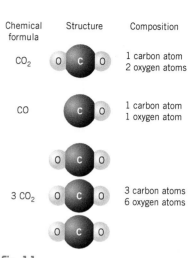

Fig. 1.1 The formula CO_2 describes a molecule that contains one carbon atom and two oxygen atoms. The formula CO tells us that this molecule consists of one carbon and one oxygen atom. A collection of three CO_2 molecules would be described by writing "3 CO_2."

Chemical formula	Structure	Composition
CO_2		1 carbon atom 2 oxygen atoms
CO		1 carbon atom 1 oxygen atom
3 CO_2		3 carbon atoms 6 oxygen atoms

■ Exercise 1.1

Describe the difference between the following pairs of symbols.
(a) Ni and NI_3
(b) 2 N and N_2
(c) Sn and S_2N_2

Solution

(a) Ni represents the element Ni. NI_3 represents a compound composed of the elements nitrogen and iodine in a one to three ratio.

(b) 2 N represents two individual atoms of nitrogen. N_2 represents two atoms of nitrogen bonded together to form one molecule.

(c) Sn represents the element tin. S_2N_2 represents a molecule formed from two atoms of sulfur and two atoms of nitrogen.

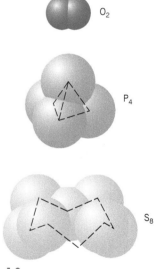

Fig. 1.2 At room temperature, oxygen exists as O_2 molecules, phosphorus forms P_4 molecules, and sulfur forms cyclic S_8 molecules.

The only way to determine whether a substance is an element or a compound is to try to break it down into simpler substances. Molecules of an element can be broken down into only one kind of atom. If a substance can be decomposed into more than one kind of atom, it is a compound. Water, for example,

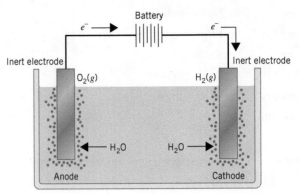

Fig. 1.3 Electrolysis of water results in the production of oxygen gas and hydrogen gas.

Table 1.1
Examples of Compounds, and Mixtures

Iron metal	Element
Carbon dioxide gas	Compound
Gasoline	Mixture
Distilled water	Compound
Tap water	Mixture
Sugar	Compound
Air	Mixture

can be decomposed into hydrogen and oxygen by passing an electric current through the liquid, as shown in Figure 1.3. In a similar fashion, salt can be decomposed into its elements—sodium and chlorine—by passing an electric current through a molten sample. Table 1.1 provides examples of common elements, compounds, and mixtures.

 ## Exercise 1.2

Classify each of the following as elements, compounds, or mixtures. Classify the mixtures as homogeneous or heterogeneous.

(a) helium gas   (d) lead sulfide

(b) Raisin Bran cereal   (e) a glass of milk

(c) lead metal   (f) potassium bromide

Solution

(a) Helium is an element.

(b) Raisin Bran cereal is a heterogeneous mixture. The cereal does not have a constant composition; the ratio of raisins to bran flakes varies from one box to another. The mixture is heterogeneous because the ratio of raisins to flakes also varies from the top of an individual box to the bottom.

(c) Lead is an element.

(d) Lead sulfide is a compound made up of the two elements lead and sulfur.

(e) Milk is a homogeneous mixture. The composition of one glass of milk differs from another (whole milk, skim milk), but an individual glass of milk should have the same composition throughout.

(f) Potassium bromide is a compound made up of the two elements potassium and bromine.

► **CHECKPOINT**

Describe the difference between the symbols 8 S and S_8.

1.5 Evidence for the Existence of Atoms

Most students believe in atoms. If asked to describe the evidence on which they base this belief, however, they hesitate.

Our senses argue against the existence of atoms.

- The atmosphere in which we live feels like a continuous fluid.
- We don't feel bombarded by collisions with individual particles in the air.
- The water we drink acts like a continuous fluid.
- We can take a glass of water, pour out half, divide the remaining water in half, and repeat this process again and again, seemingly forever.

Because our senses suggest that matter is continuous, it isn't surprising that the debate about the existence of atoms goes back to the ancient Greeks and continued well into the twentieth century.

Experiments with gases that first became possible at the turn of the nineteenth century led John Dalton in 1803 to propose a model for the atom based on the following assumptions:

- Matter is made up of atoms that are indivisible and indestructible.
- All atoms of an element are identical.
- Atoms of different elements have different weights and different chemical properties.
- Atoms of different elements combine in simple whole-number ratios to form compounds.
- Atoms cannot be created or destroyed. When a compound is decomposed, the atoms are recovered unchanged.

Dalton's assumptions form the basis of the modern atomic theory. However, modern experiments have shown that not all atoms of an element are exactly the same and that atoms can be broken down into subatomic particles. Only recently has direct evidence for the existence of atoms become available. Using the scanning tunneling microscope (STM) developed in the 1980s, scientists have finally been able to observe and even manipulate individual atoms. The "molecular man" shown in Figure 1.4 was formed by using an STM probe to move 28 CO molecules into position on a platinum surface.

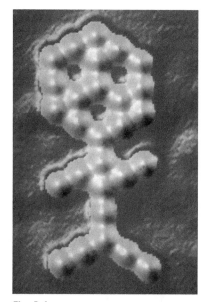

Fig. 1.4 This "molecular man" was formed by moving carbon monoxide molecules into position on a platinum surface.

1.6 The Role of Measurement in Chemistry

There are two kinds of scientific knowledge. One of them revolves around the *product* of science, noting the existence of models that describe or explain the results of experiments, such as the periodic table or Boyle's law. The other form of scientific knowledge focuses on the *process* by which science is done. Many students already know that water is represented by the formula H_2O, or that the salt used to improve the flavor of foods has the formula NaCl, when they take their first chemistry course. But they are less familiar with the process by which chemists have obtained this information. This section introduces one of the consequences of the fact that the models chemists construct to explain the results of experiments or observations of the world around us are often based on measurements of one or more quantities.

All measurements contain a number that indicates the magnitude of the quantity being measured and a set of units that provide a basis for comparing the quantity with a standard reference. The results of measurements can be reported

Table 1.2
Metric System Prefixes

Prefix	Symbol	Meaning
femto-	f	$\times$ 1/1,000,000,000,000,000 (10^{-15})
pico-	p	$\times$ 1/1,000,000,000,000 (10^{-12})
nano-	n	$\times$ 1/1,000,000,000 (10^{-9})
micro-	μ	$\times$ 1/1,000,000 (10^{-6})
milli-	m	$\times$ 1/1,000 (10^{-3})
centi-	c	$\times$ 1/100 (10^{-2})
deci-	d	$\times$ 1/10 (10^{-1})
kilo-	k	$\times$ 1,000 (10^{3})
mega-	M	$\times$ 1,000,000 (10^{6})
giga-	G	$\times$ 1,000,000,000 (10^{9})
tera-	T	1,000,000,000,000 (10^{12})

with several systems, each containing units for properties such as length, volume, weight, and time.

The irregular **English system of units** was replaced by a system based on decimals called the **metric system** for which the fundamental units of measurement for length, volume, and mass are meters, liters, and grams. An advantage of the metric system is the ease with which the base units can be converted into a unit that is more appropriate for the quantity measured. This is done be adding a prefix to the name of the base unit. These prefixes are given in Table 1.2. The prefix *kilo* (k), implies multiplication by a factor of 1000. Thus a kilometer is equal to 1000 meters.

$$1 \text{ km} = 1000 \text{ m} \quad \text{or} \quad 0.001 \text{ km} = 1 \text{ m}$$

To convert between kilometers and meters we use the conversion factor $\left(\frac{1000 \text{ m}}{\text{km}}\right)$.

The prefix *milli* (m), means division by a factor of 1000. A milliliter (mL) is therefore equal to 0.001 liters.

$$1 \text{ mL} = 0.001 \text{ L} \quad \text{or} \quad 1000 \text{ mL} = 1 \text{ L}$$

To convert between liters and milliliters we use the conversion factor $\left(\frac{1000 \text{ mL}}{\text{L}}\right)$.

Another advantage of the metric system is the link between the base units of length and volume. By definition, a liter is equal to the volume of a cube exactly 10 cm tall, 10 cm long, and 10 cm wide. Because the volume of a cube with these dimensions is 1000 cubic centimeters (1000 cm^3) and a liter contains 1000 milliliters, 1 mL is equivalent to 1 cm^3.

In 1960 the International System of Units, abbreviated **SI**, was adopted. The seven base units for the SI system are given in Table 1.3.

Conversion factors are used to convert between units. For example, to convert 0.248 kg to g, we use the conversion factor $\left(\frac{1000 \text{ g}}{\text{kg}}\right)$.

$$0.248 \text{ kg} \left(\frac{1000 \text{ g}}{\text{kg}}\right) = 248 \text{ g}$$

By definition there are exactly 12 inches in 1 foot. This can be expressed as

$$12 \text{ inches} \equiv 1 \text{ foot}$$

➤ **CHECKPOINT**

How many milligrams (mg) are in 0.529 gram?

How many centimeters (cm) are in 2.59 meters?

Table 1.3
SI Base Units

Physical Quantity	Name of Unit Symbol	
Length	meter	m
Mass	kilogram	kg
Time	second	s
Temperature	Kelvin	K
Electric current	ampere	A
Amount of substance	mole	mol
Luminous intensity	candela	cd

where $\equiv$ signifies that 12 inches is the physical equivalence of 1 foot. However, in an algebraic equation containing a number and units, both the number and the units must follow the rules of algebra. Thus it is not correct to write:

$$12 \text{ inches} = 1 \text{ foot} \quad \text{or} \quad 12 \text{ in} = 1 \text{ ft}$$

Both sides of the equation must have the same dimensional units. In addition, the units must be related linearly, $y = mx + b$, and pass through the origin of the plot of unit 1 versus unit 2. This requires the intercept b to be zero. A plot of inches versus feet gives a linear relationship:

$$y = mx + b$$

$$y \text{ in} = \left(\frac{12 \text{ in}}{1 \text{ ft}}\right)(x \text{ ft}) + 0$$

The slope of the plot, 12 in/ft, is a constant ratio between inches and feet. The slope is a proportionality constant usually called a **conversion factor** for converting inches to feet. A conversion factor is a fraction with its numerator and denominator expressed in different units. The conversion factor for inches and feet is 12 in/1 ft or 1 ft/12 in. If we wanted to determine the number of inches in 2.5 feet we would use the conversion factor to obtain the desired new unit of inches.

$$2.5 \text{ ft}\left(\frac{12 \text{ in}}{1 \text{ ft}}\right) = 30 \text{ in}$$

The units in the conversion factor cancel the units of feet and leave the units of inches. In general, if unit 1 is to be converted to unit 2, the conversion factor is used:

$$\text{Unit 1} \times \text{conversion factor} = \text{Unit 2}$$

A more in-depth discussion of unit conversions can be found in Appendix A.

1.7 The Structure of Atoms

We now know that atoms are not indivisible. They are composed of the three fundamental subatomic particles listed in Table 1.4: **electrons, protons,** and **neutrons,** which are in turn assumed to be composed of still smaller particles, the so-called

Table 1.4
Fundamental Subatomic Particles

Particle	Symbol	Absolute Charge (C)	Relative Charge	Absolute Mass (g)	Relative Mass
Electron	e^-	-1.60×10^{-19}	-1	9.11×10^{-28}	0
Proton	p^+	1.60×10^{-19}	$+1$	1.673×10^{-24}	1
Neutron	n^0	0	0	1.675×10^{-24}	1

up and down quarks. Chemists normally refer to electrons, protons and neutrons as fundamental particles because they are the building blocks of all atoms. Although gold atoms and oxygen atoms are quite different from one another, the electrons, protons, and neutrons found within a gold atom are indistinguishable from the electrons, protons, and neutrons found within an oxygen atom.

Chemists routinely work with numbers that are extremely small. The measured mass of an electron, for example, is 0.000,000,000,000,000,000,000,000,000, 000,911 kilogram. They also work with numbers that are extremely large. There are 10,030,000,000,000,000,000,000 carbon atoms in a 1-carat diamond. There isn't a calculator made that will accept either of these numbers as they are written here. Before these numbers can be used, it is necessary to convert them to **scientific notation,** that is to convert them to a number between 1 and 10 multiplied by 10 raised to some exponent. Referring back to the mass of an electron, we see that the very awkward number can be written in scientific notation as 9.11×10^{-31} kg, or by applying the prefix 1 kg is equivalent to 1000 g, the mass of an electron can be written as 9.11×10^{-28} g. Other examples of the use of scientific notation can be found in Appendix A.

> ➤ **CHECKPOINT**
>
> Convert the following decimal numbers into scientific notation.
>
> 0.000000472
>
> 10,030,000,000,000,000,000,000
>
> Convert the following scientific notation numbers into decimal form.
>
> 7.54×10^{-8}
>
> 3.668×10^6

▪ Exercise 1.3

Match the following items with their appropriate masses: a carbon atom, an *E. coli* bacterium, a penny, an automobile, and the earth.

2.9×10^{-13} g,   2.5 g, 6.0×10^{27} g, 2.0×10^{-23} g, 1.1×10^6 g

Solution

a carbon atom	2.0×10^{-23} g
an *E. coli* bacterium	2.9×10^{-13} g
a penny	2.5 g
an automobile	1.1×10^6 g
the earth	6.0×10^{27} g

The electrons, protons, and neutrons in an atom differ in terms of both the charge on the particle and its mass. The magnitude of **absolute** charge on an electron is equal to that of proton, but the sign of the charge on the two particles is different. The neutron carries no net electric charge. Because the magnitude of the charge on an electron and a proton is the same—they differ only in the sign of the charge—the **relative** charge on these particles is -1 or $+1$, as shown in Table 1.4.

Because the charge on a proton has the same magnitude as the charge on an electron, the charge on one proton exactly balances the charge on an electron, and vice versa. Thus, atoms are electrically neutral when they contain the same number of electrons and protons.

The absolute mass of each of the three subatomic particles is given in Table 1.4 in units of grams, the fundamental unit for measurement of mass. The last column in this table gives the relative mass of these particles. Because the mass of a proton is almost the same as that of a neutron, both particles are assigned a relative mass of 1. Because the ratio of the mass of an electron to that of a proton is so very small, it is considered negligible, and the electron is assigned a relative mass of zero.

The protons and neutrons in an atom are concentrated in the **nucleus,** which contains most of the mass of the atom. For example, 99.97% of the mass of a carbon atom can be found in the nucleus of that atom. The term *nucleus* comes from the Latin word meaning "little nut." This term was chosen to convey the image that the nucleus of an atom occupies an infinitesimally small fraction of the volume of an atom. The radius of an atom is approximately 10,000 times larger than its nucleus. To appreciate the relative size of an atom and its nucleus, imagine that we could expand an atom until it was the size of the Superdome. The nucleus would be the size of a small pea suspended above the 50-yard line, with electrons moving throughout the arena. Thus, most of the volume of an atom is empty space through which the electrons move.

It is impossible to determine the exact position or path of an electron. Because of this, chemists often visualize electrons as a cloud of negative charge spread throughout the volume of space surrounding the nucleus, as shown in Figure 1.5. The size of the atom is assumed to be equal to the volume occupied by this cloud of negative charge.

Fig. 1.5 The exact position of an electron in an atom cannot be determined. Electrons are therefore often described as a cloud of negative charge spread out in the space surrounding the nucleus. The boundary of the atom is not a physical boundary but instead is a volume that contains the electron density of the atom.

1.8 Atomic Number and Mass Number

The number of protons in the nucleus of an atom determines the identity of the atom. Every carbon atom ($Z = 6$) has 6 protons in the nucleus of the atom, whereas sodium atoms ($Z = 11$) have 11. Each element has therefore been assigned an **atomic number** (Z) between 1 and 118 that describes the number of protons in the nucleus of an atom of that element. Neutral atoms contain just enough electrons to balance the charge on the nucleus. The nucleus of a neutral carbon atom would be surrounded by 6 electrons; a neutral sodium atom would contain 11 electrons.

The nucleus of an atom is also described by a **mass number** (A), which is the sum of the number of protons and neutrons in the nucleus. The difference between the mass number and the atomic number of an atom is therefore equal to the number of neutrons in the nucleus of that atom. A carbon atom with a mass number of 12 would contain 6 protons and 6 neutrons. A sodium atom with a mass number of 23 would contain 11 protons and 12 neutrons.

A shorthand notation has been developed to describe the number of neutrons and protons in the nucleus of an atom. The atomic number is written in the bottom-left corner of the symbol for the element, and the mass number is written in the top-left corner $^{A}_{Z}X$. The atomic number of carbon is six because the nucleus of each carbon atom contains 6 protons. Because the nucleus of a typical carbon atom also contains 6 neutrons, the mass number of this atom would be 12 and the atom would be given the symbol $^{12}_{6}C$. The nucleus of a neutral $^{12}_{6}C$ atom would be surrounded by 6 electrons to balance the positive charge. A sodium atom with 11 protons and 12 neutrons in the nucleus would be given the symbol $^{23}_{11}Na$ to indicate that the atomic number is 11 and the mass number is 23.

Because each element has a unique atomic number and a unique symbol, it is redundant to give both the symbol for the element and its atomic number. Thus, the atoms discussed in this section are usually written as ^{12}C and ^{23}Na.

1.9 Isotopes

The number of protons in the nucleus of an atom determines the identity of the atom. As a result, all atoms of an element must have the same number of protons. But they don't have to contain the same number of neutrons.

Atoms with the same atomic number but different numbers of neutrons are called **isotopes.** Carbon, for example, has three naturally occurring isotopes: ^{12}C, ^{13}C, and ^{14}C. ^{12}C has 6 protons and 6 neutrons; ^{13}C has 6 protons and 7 neutrons; ^{14}C has 6 protons and 8 neutrons.

Each element occurs in nature as a mixture of its isotopes. Consider a "lead" pencil, for example. These pencils don't contain the element lead, which is fortunate because many people chew on pencils and lead can be very toxic. They contain a substance once known as "black lead" and now known as graphite that is mixed with clay; the more clay, the harder the pencil.

The graphite in a pencil contains a mixture of ^{12}C, ^{13}C, and ^{14}C atoms. The three isotopes, however, do not occur to the same extent. Most of the atoms (98.892%) are ^{12}C, a small percentage (1.108%) are ^{13}C, and only about 1 in about 10^{12} is the radioactive isotope of carbon, ^{14}C. The percentage of atoms occurring as a given isotope found in nature is referred to as the **natural abundance** of that isotope. Some elements, such as fluorine, have only one naturally occurring isotope, ^{19}F, whereas other elements have several, as shown in Table 1.5.

It should be noted that in Table 1.5 the natural abundances reported for the isotopes contain different numbers of digits. For example, the natural abundance of the ^{1}H isotope is given to five digits, while that of ^{6}Li contains only three. The number of digits reported in a measurement expresses the confidence in that meas-

Table 1.5
Common Isotopes of Some of the Lighter Elements

Isotope	Natural Abundance (%)	Mass (g)	Mass (amu)
^{1}H	99.985	1.6735×10^{-24}	1.0078
^{2}H	0.015	3.3443×10^{-24}	2.0141
^{6}Li	7.42	9.9883×10^{-24}	6.0151
^{7}Li	92.58	1.1650×10^{-23}	7.0160
^{10}B	19.7	1.6627×10^{-23}	10.012
^{11}B	80.3	1.8281×10^{-23}	11.009
^{12}C	98.892	1.9926×10^{-23}	12.000...
^{13}C	1.108	2.1592×10^{-23}	13.003
^{16}O	99.76	2.6560×10^{-23}	15.995
^{17}O	0.04	2.8228×10^{-23}	16.999
^{18}O	0.20	2.9888×10^{-23}	17.999
^{20}Ne	90.51	3.3198×10^{-23}	19.992
^{21}Ne	0.27	3.4861×10^{-23}	20.993
^{22}Ne	9.22	3.6518×10^{-23}	21.991

urement and is referred to as the number of significant figures for the measurement. Digits for which there is a high degree of confidence or certainty are referred to as *significant.*

Thus, it may be concluded from the table that some natural abundances are known to a greater degree of confidence than others. The natural abundance given for ^{1}H contains five digits, indicating that the first four digits, are known with a high degree of confidence but there is uncertainty in the last digit on the right. For ^{6}Li only three digits are given and so only three significant figures are known, the uncertainty being in the last digit on the right.

At first glance it might seem that the number of significant figures can be determined by counting the number of digits given. Unfortunately, zeros present a problem. Leading zeros are never significant. Zeros between two significant figures are always significant. Trailing zeros that are not needed to hold the decimal point are significant.

In a number such as 0.004050, the first three zeros are leading and not significant. The zero between the 4 and 5 is significant because it is between two significant figures. The trailing zero after the 5 is significant because this zero is not necessary to show the magnitude of the number. It is present to show that the uncertainty in this measurement is in the 6th digit after the decimal. Thus this measurement contains four significant digits. For further discussion and examples see Appendix A. In particular, use the worksheet in A.6 to practice your skills with significant figures. A good way to determine the number of significant figures in a measurement is to write the number in scientific notation. The preceding example becomes 4.050×10^{-13}. Converting a measurement to scientific notation or to another unit never changes the number of significant figures in the measurement.

It is important to recognize that some conversion factors are based on definitions, not measurements. For example, the mass of an atom of ^{12}C is defined to be exactly 12 amu. **A number based on a definition has an infinite number of significant figures.**

Because the mass of an atom is so very small—on the order of 10^{-23} grams—it is often more useful to know the relative mass of an atom than it is to know the absolute mass in grams (the fundamental unit for measurements of mass). The relative mass of each isotope in Table 1.5 is given in **atomic mass units (amu).** The unit of amu is defined such that the mass of an atom of ^{12}C is exactly 12 amu.

> ➤ **CHECKPOINT**
>
> In Table 1.5 determine the number of significant figures for the natural abundances of ^{1}H, ^{2}H, ^{10}B, ^{17}O, and ^{20}Ne.

> ➤ **CHECKPOINT**
>
> There are two naturally occurring isotopes of lithium, ^{6}Li and ^{7}Li. According to the data in Table 1.5, how many ^{6}Li atoms would be found in a sample of 10,000 lithium atoms selected at random? How many would be ^{7}Li?

Exercise 1.4

According to Table 1.5, the absolute mass of a ^{1}H atom is 1.6735×10^{-24} grams, whereas the absolute mass of a ^{12}C atom is 1.9926×10^{-23} grams. Calculate the ratio of the mass of a ^{12}C atom to that of a ^{1}H atom when the masses are measured in units of grams. Use this ratio to calculate the mass of a ^{1}H atom in units of amu if the mass of a ^{12}C atom is defined as exactly 12 amu.

Solution

The relative mass of ^{1}H and ^{12}C atoms can be calculated from their absolute masses in grams.

$$\frac{^{12}\text{C}}{^1\text{H}} = \frac{1.9926 \times 10^{-23}}{1.6735 \times 10^{-24}} = 11.907$$

We can set up this problem as follows:

$$\frac{^{12}C}{^{1}H} = \frac{12.000 \text{ amu}}{x} = 11.907$$

If the mass of a ^{12}C atom is defined as exactly 12 amu, then the mass of a ^{1}H atom to five significant figures must be 1.0078 amu.

$$x = \frac{12.000 \text{ amu}}{11.907} = 1.0078 \text{ amu}$$

This is the value of the mass in units of amu for a ^{1}H atom in Table 1.5.

There are five significant figures in the relative masses of ^{1}H and ^{12}C. There are five significant figures given in the ratio, 11.907. **When measurements are multiplied or divided, the answer can contain no more total significant figures than the measurement with the fewest number of significant figures.** In the division of 12 amu by 11.907, there are an infinite number of significant figures in the 12 amu because the mass of the ^{12}C isotope has been defined to be exactly 12 amu. There are five significant figures in the ratio 11.907 because the ratio is based on a calculation of measurements. Therefore, the result of the division is limited to five significant figures.

> **CHECKPOINT**

Calculate the ratio of the natural abundance of ^{1}H to ^{2}H. How many significant figures are allowed in the ratio?

1.10 The Difference between Atoms and Ions

Imagine that you had a small piece of sodium metal and a crystal of table salt. The sodium metal contains neutral Na atoms. If you dropped it into water, it would instantly react to give H_2 gas, which would burst into flame. The table salt contains positively charged Na^+ ions. When dropped into water, the salt would dissolve to give a solution with a characteristic salty flavor.

It is difficult, if not quite impossible, to change the number of protons in the nucleus of an atom. It takes much less energy, however, to add or remove electrons from an atom to form electrically charged particles known as **ions.**

Neutral atoms are turned into positively charged ions by removing one or more electrons, as shown in Figure 1.6. By convention, these positively charged ions are called **cations.** A Na^+ ion or cation that has 10 electrons and 11 protons is produced by removing one electron from a neutral sodium atom that contains 11 electrons and 11 protons. Ions with larger positive charges can be produced by removing more electrons. A neutral aluminum atom, for example, has 13 electrons and 13 protons. If we remove three electrons from this atom, we get a positively charged Al^{3+} ion that has 10 electrons and 13 protons, for a net charge of $+3$.

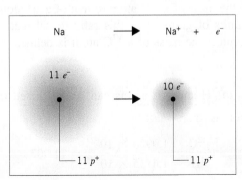

Fig. 1.6 Removing an electron from a neutral sodium atom produces a Na^+ ion that has a net charge of $+1$.

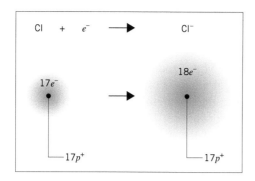

Fig. 1.7 Adding an extra electron to a neutral chlorine atom produces a Cl^- ion that has a net charge of -1.

Neutral atoms that gain extra electrons become negatively charged ions, or **anions,** as shown in Figure 1.7. A neutral chlorine atom, for example, has 17 protons and 17 electrons. By adding one more electron to this atom, a Cl^- ion is formed that has 18 electrons and 17 protons, for a net charge of -1.

As we have seen, the gain or loss of electrons by an atom to form negative or positive ions has an enormous impact on the chemical and physical properties of the atom. Sodium metal, which consists of neutral sodium atoms, reacts violently with water. But positively charged Na^+ ions are so unreactive with water they are essentially inert. Neutral chlorine atoms instantly combine to form Cl_2 molecules, which are so reactive that entire communities are evacuated when trains carrying chlorine gas derail. Negatively charged Cl^- ions are essentially inert to chemical reactions.

The enormous difference between the chemistry of neutral atoms and their ions means that it is necessary to pay close attention to the symbols to make sure that atoms and ions are not confused with one another.

Exercise 1.5

Find the number of protons, electrons, and neutrons in each of the following atoms and ions.

(a) $^{24}Mg^{2+}$

(b) $^{79}Br^-$

Solution

(a) The atomic number of magnesium is 12, which means that the nucleus of this ion contains 12 protons. Because the mass number of the ion is 24 and it contains 12 protons, the nucleus of the atom must contain 12 neutrons. Because the ion carries a charge of $+2$, there must be two more protons (positive charges) than electrons (negative charges). This ion therefore contains 10 electrons.

(b) Bromine has an atomic number of 35, and there are therefore 35 protons in the nucleus of this ion. Because the mass number of the ion is 79 and it contains 35 protons, there must be 44 neutrons in the nucleus of the atom. Because the ion has a -1 charge, the ion must have one more electron than a neutral atom. This ion therefore contains 36 electrons.

1.11 Polyatomic Ions

Simple ions, such as the Mg^{2+} and N^{3-} ions, are formed by adding or subtracting electrons from neutral atoms. **Polyatomic ions** are electrically charged molecules composed of more than one atom. You will commonly encounter only two polyatomic positive ions or cations. These are the ammonium and hydronium ions, NH_4^+ and H_3O^+. A few of the more common negative ions or anions are listed in Table 1.6.

Table 1.6
Common Polyatomic Negative Ions

	−1 ions		
HCO_3^-	Hydrogen carbonate (bicarbonate)	OH^-	Hydroxide
$CH_3CO_2^-$	Acetate	ClO_4^-	Perchlorate
NO_3^-	Nitrate	ClO_3^-	Chlorate
NO_2^-	Nitrite	ClO_2^-	Chlorite
MnO_4^-	Permanganate	ClO^-	Hypochlorite
CN^-	Cyanide		
	−2 ions		
CO_3^{2-}	Carbonate	O_2^{2-}	Peroxide
SO_4^{2-}	Sulfate	CrO_4^{2-}	Chromate
SO_3^{2-}	Sulfite	$Cr_2O_7^{2-}$	Dichromate
$S_2O_3^{2-}$	Thiosulfate		
	−3 ions		
PO_4^{3-}	Phosphate	AsO_4^{3-}	Arsenate
BO_3^{3-}	Borate		

➤ **CHECKPOINT**

What ions can be found in each of the following ionic compounds: NaOH, K_2SO_4, $BaSO_4$, and $Be_3(PO_4)_2$?

1.12 The Periodic Table

While trying to organize a discussion of the properties of the elements for a chemistry course at the Technological Institute in St. Petersburg, Dmitri Ivanovitch Mendeléeff listed the properties of each element on a different card. As he arranged the cards in different orders, he noticed that the properties of the elements repeated in a periodic fashion when the elements were listed more or less in order of increasing atomic weight. In 1869 Mendeléeff published the first of a series of papers outlining a **periodic table** of the elements in which the properties of the elements repeated in a periodic fashion.

More than 700 versions of the periodic table were proposed in the first 100 years after the publication of Mendeléeff's table. A modern version of the table is shown in Figure 1.8. In this version the elements are arranged in order of increasing atomic number, which is written above the symbol for the atom.

The vertical columns in the periodic table are known as **groups,** or families. Traditionally these groups have been distinguished by a **group number** consisting of a Roman numeral followed by either an A or a B. In the United States,

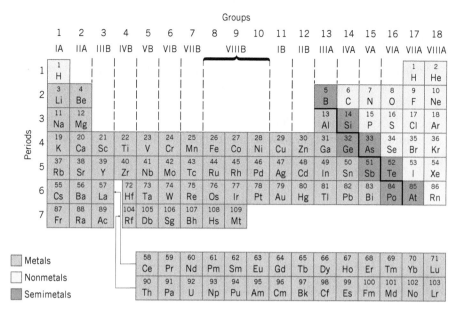

Fig. 1.8 A modern version of the periodic table.

the elements in the first column on the left-hand side of the table were historically known as Group IA. The next column was IIA, then IIIB, and so on across the periodic table to VIIIA.

Unfortunately, the same notation wasn't used in all countries. The elements known as Group VIA in the United States were Group VIB in Europe. A new convention for the periodic table has been proposed that numbers the columns from 1 to 18, reading from left to right. This convention has obvious advantages. It is perfectly regular and therefore unambiguous. The advantages of the old format are less obvious, but they are equally real. This book therefore introduces the new convention but retains the old.

The elements in a vertical column of the periodic table have similar chemical properties. Elements in the first column, for example, combine in similar ways with chlorine to form compounds with similar chemical formulas: HCl, LiCl, NaCl, KCl, and so on.

The horizontal rows in the periodic table are called **periods.** The first period contains only two elements: hydrogen (H) and helium (He). The second period contains eight elements (Li, Be, B, C, N, O, F, and Ne). Although there are nine horizontal rows in the periodic table in Figure 1.8, there are only seven periods. The two rows at the bottom of the table belong in the sixth and seventh periods. These rows are listed at the bottom to prevent the table from becoming so large that it becomes unwieldy.

The elements in the periodic table can be divided into three categories: **metals, nonmetals,** and **semimetals.** The dividing line between the metals and the nonmetals in Figure 1.8 is marked with a heavy stair-step line. As you can see from Figure 1.8, more than 75% of the elements are metals. These elements are found toward the bottom-left side of the table.

Only 17 elements are nonmetals. With only one exception—hydrogen, which appears on both sides of the table in Figure 1.8—these elements are clustered in the upper-right corner of the periodic table. A cluster of elements that are neither metals nor nonmetals can be found between the metals and nonmetals in Figure 1.8. These elements are called the semimetals, or metalloids.

➤ **CHECKPOINT**

What are the atomic numbers of the elements that have the atomic symbols F and Pb? What are the atomic symbols of the elements with atomic numbers 24 and 74?

Exercise 1.6

Classify each element in Group IVA as a metal, a nonmetal, or a semimetal.

Solution

Group IVA contains five elements: carbon, silicon, germanium, tin, and lead. According to Figure 1.8, these elements fall into the following categories.

Nonmetal:	C
Semimetal:	Si and Ge
Metal:	Sn and Pb

1.13 The Macroscopic, Atomic, and Symbolic Worlds of Chemistry

Chemists work in three very different worlds, represented by Figure 1.9. Most measurements are done in the **macroscopic world**—with objects visible to the naked eye. On the macroscopic scale, water is a liquid that freezes at 0°C and boils at 100°C at one atmosphere pressure. When you walk into a chemical laboratory, you'll find a variety of bottles, tubes, flasks, and beakers designed to study samples of liquids and solids large enough to be seen. You may also find sophisticated instruments that can be used to analyze very small quantities of materials, but even these samples are visible to the naked eye.

Although they perform experiments on the macroscopic scale, chemists think about the behavior of matter in terms of a world of atoms and molecules. In this **atomic world,** water is no longer a liquid that freezes at 0°C and boils at 100°C, but individual molecules that contain two hydrogen atoms and an oxygen atom.

One of the challenges students face is understanding the process by which chemists perform experiments on the macroscopic scale that can be interpreted in terms of the structure of matter on the atomic scale. The task of bridging the gap between the atomic and macroscopic worlds is made more difficult by the fact that chemists also work in a **symbolic world,** in which they represent water as H_2O and write equations such as the following to represent what happens when hydrogen and oxygen react to form water.

$$2\ H_2 + O_2 \longrightarrow 2\ H_2O$$

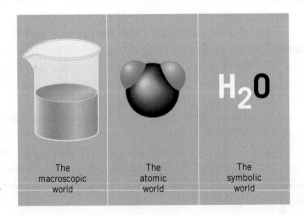

Fig. 1.9 Water on the scale of the macroscopic, atomic, and symbolic worlds.

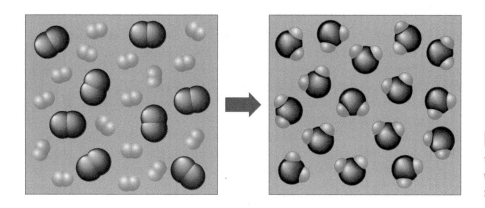

Fig. 1.10 A mechanical model for the reaction between H_2 and O_2 on the atomic scale to form water molecules.

Chemists use the same symbols to describe what happens on both the macroscopic and the atomic scales. The symbol H_2O, for example, may be used to represent both a single water molecule or the water in a beaker.

It is easy to forget the link between the symbols chemists use to represent reactions and the particles involved in these reactions. Figure 1.10 provides an example of how you might envision the reaction between hydrogen and oxygen on the atomic scale. The reaction starts with a mixture of H_2 and O_2 molecules, each containing a pair of atoms. It produces water molecules that contain two hydrogen atoms and an oxygen atom.

1.14 The Mass of an Atom

Atoms are so small that a sliver of copper metal just big enough to detect on a good analytical balance contains about 1×10^{17} atoms. As a result, it is impossible to measure the absolute mass of a single atom. We can measure the relative masses of different atoms, however, with an instrument known as a mass spectrometer.

Because the mass spectrometer can tell us only the relative mass of an atom, we need a standard with which our measurement can be compared. The standard used to calibrate these measurements is the ^{12}C isotope of carbon. The unit in which atomic mass measurements are reported is the atomic mass unit or amu (see Section 1.9). By definition, the mass of a single atom of the ^{12}C isotope is exactly 12 atomic mass units, or 12 amu.

Most elements exist in nature as mixtures of isotopes. As we have seen, the graphite in a lead pencil is composed of a mixture of ^{12}C (98.892%, 12.000 amu), ^{13}C (1.108%, 13.003 amu), and an infinitesimally small amount of ^{14}C atoms. It is therefore useful to calculate the average mass of a sample of carbon atoms. Because there is a large difference in the natural abundance of these isotopes, the average mass of a carbon atom must be a weighted average of the masses of the different isotopes. Because the amount of ^{14}C is so small, the average mass of a carbon atom is calculated using only the two most abundant isotopes of the element.

$$\left(12.000 \text{ amu} \times \frac{98.892}{100} \right) + \left(13.003 \text{ amu} \times \frac{1.108}{100} \right) = 12.011 \text{ amu}$$

$$(11.867 \text{ amu}) + (0.1441 \text{ amu}) = 12.011 \text{ amu}$$

The average mass of a carbon atom is much closer to the mass of a ^{12}C atom than a ^{13}C atom because the vast majority of the atoms in a sample of carbon are ^{12}C.

This weighted average of all the naturally occurring isotopes of an atom has traditionally been known as the **atomic weight** of the element. It is this value that is reported beneath the symbol of the element in the periodic table. It is important to remember that the atomic weight of carbon is 12.011 amu, even though no individual carbon atom actually has a mass of 12.011 amu.

When performing math calculations like the one shown above, it is necessary to perform the calculations in steps in order to maintain the correct number of significant figures. In the first multiplication, the atomic mass of ^{12}C and the 100 are definitions and therefore known to an infinite number of significant figures. The 100 is based on the definition of percent. The number of significant figures in the result is therefore limited by the five significant figures in the measurement of natural abundance, 98.892. In the second multiplication there are two measurements; 13.003 amu with five significant figures and the percent abundance of 1.108 with four significant figures. The result of 0.1441 amu is limited to just four significant figures. Addition and subtraction are treated differently than multiplication and division when counting significant figures. **When measurements are added or subtracted, the number of significant figures to the right of the decimal in the answer is determined by the measurement with the fewest digits to the right of the decimal.** In the calculation above there are three digits to the right of the decimal in 11.867 amu and there are four digits to the right of the decimal in 0.1441 amu. We are therefore limited to only three digits to the right of the decimal in the final answer of 12.011 amu. Refer to Appendix A for a more detailed discussion of maintaining the correct number of significant figures during calculations.

Exercise 1.7

Calculate the atomic weight of chlorine if 75.77% of the atoms have a mass of 34.97 amu and 24.23% have a mass of 36.97 amu.

Solution

Percent literally means "per hundred." Chlorine is therefore a mixture of atoms for which 75.77 parts per hundred have a mass of 34.97 amu and 24.23 parts per hundred have a mass of 36.97 amu. The atomic weight of chlorine is therefore 35.46 amu.

$$\left(34.97\,amu \times \frac{75.77}{100}\right) + \left(36.97\,amu \times \frac{24.23}{100}\right) = 35.46\,amu$$
$$26.50\,amu + 8.958\,amu = 35.46\,amu$$

No atom of chlorine has a mass of 35.46 amu. This is the average mass of a chlorine atom in a large group of naturally occurring chlorine atoms.

The atomic weight obtained from the above calculation (35.46 amu) is slightly different from the value found for chlorine in the periodic table (35.453 amu). Using the rules for significant figures yields a result with four significant figures. This means that the first three digits (35.4) are known with certainty but the last reported digit (6) has some degree of uncertainty, and this is indeed the digit that is in disagreement with the five significant figure atomic weight given in the periodic table.

Exercise 1.8

There are two naturally occurring isotopes of element *X*. One of these isotopes has a natural abundance of 70.5% and a relative mass of 204.97 amu. The second isotope is lighter. Identify element *X* and state your reasoning. Give your best estimate of the number of neutrons, protons, and electrons in each isotope.

Solution

Because the second isotope is lighter, the average atomic weight of the element must be less than 204.97 amu. The element with an atomic weight closest to this value is thallium, Tl. The atomic weight of Tl found in the periodic table is 204.38 amu.

The atomic weight of the heavier isotope is 204.97 amu. This would suggest that the mass number for this isotope is 205. The atomic number of Tl is 81. Tl therefore has 81 protons. An electrically neutral atom would have 81 electrons. The mass number of 205 minus the 81 protons gives 124 neutrons.

The lighter isotope must have an atomic weight less than 204.38 amu. It has a natural abundance of 29.5%; therefore, its atomic weight must be approximately 203 amu. A mass number of 203 would give 81 protons, 81 electrons and 122 neutrons.

1.15 Chemical Reactions and the Law of Conservation of Atoms

We have focused so far on individual compounds such as water (H_2O) and carbon dioxide (CO_2). Much of the fascination of chemistry, however, revolves around chemical reactions. The first breakthrough in the study of chemical reactions resulted from the work of the French chemist Antoine Lavoisier between 1772 and 1794. Lavoisier noted that the total mass of all of the products of a chemical reaction is always the same as the total mass of all of the starting materials consumed in the reaction. His results led to one of the fundamental laws of chemical behavior: the **law of conservation of mass,** which states that matter is conserved in a chemical reaction.

We now understand *why* matter is conserved—atoms are neither created nor destroyed in a chemical reaction. The hydrogen atoms in an H_2 molecule can combine with oxygen atoms in an O_2 molecule to form H_2O, as shown in Figure 1.11. But the number of hydrogen and oxygen atoms before and after the reaction must be the same. The total mass of the products of a reaction therefore must be the same as the total mass of the reactants that undergo reaction.

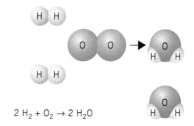

$$2\,H_2 + O_2 \rightarrow 2\,H_2O$$

Fig. 1.11 Mass is conserved in chemical reactions such as the reaction between hydrogen, H_2, and oxygen, O_2, to form water because atoms are neither created nor destroyed.

1.16 Chemical Equations as a Representation of Chemical Reactions

It is possible to describe a chemical reaction in words, but it is much easier to describe it with a **chemical equation.** The formulas of the starting materials, or **reactants,** are written on the left side of the equation, and the formulas of the **products** are written on the right. Instead of an equal sign, the reactants and products are separated by an arrow. The reaction between hydrogen and oxygen to form water shown in Figure 1.11 is represented by the following equation.

$$2\ H_2 + O_2 \longrightarrow 2\ H_2O$$

It is often useful to indicate whether the reactants or products are solids, liquids, or gases by writing an *s, l,* or *g* in parentheses after the symbol for the reactants or products.

$$2\ H_2(g) + O_2(g) \longrightarrow 2\ H_2O(l)$$

Many of the reactions you will encounter in this course occur when solutions of two substances dissolved in water are mixed. These **aqueous** solutions (from the Latin word *aqua* meaning "water") are so important we use the special symbol *aq* to describe them. In this way we can distinguish between glucose as a solid, $C_6H_{12}O_6(s)$, and solutions of this sugar dissolved in water, $C_6H_{12}O_6(aq)$, or between salt as an ionic solid, $NaCl(s)$, and solutions of salt dissolved in water, $NaCl(aq)$. The process by which a sample dissolves in water will be indicated by equations such as the following.

$$C_6H_{12}O_6(s) \xrightarrow{\ H_2O\ } C_6H_{12}O_6(aq)$$

Ionic compounds break up into their component ions when they dissolve in water. Therefore the aqueous forms of these compounds may be written as aqueous ions. Since salt is an ionic compound, the chemical equation describing the dissolution of salt can be written as an **ionic equation.**

$$NaCl(s) \xrightarrow{\ H_2O\ } Na^+(aq) + Cl^-(aq)$$

Some molecular compounds also form ions when they dissolve in water.

$$HCl(g) \xrightarrow{\ H_2O\ } H^+(aq) + Cl^-(aq)$$

Chemical equations are such a powerful shorthand for describing chemical reactions that we tend to think about reactions in terms of these equations. It is important to remember that a chemical equation is a statement of *what can happen,* not necessarily *what will happen.* The following equation, for example, does not guarantee that hydrogen will react with oxygen to form water.

$$2\ H_2(g) + O_2(g) \longrightarrow 2\ H_2O(l)$$

It is possible to fill a balloon with a mixture of hydrogen and oxygen and find that no reaction occurs until the balloon is touched with a flame. All the equation tells us is what would happen if, or when, the reaction occurs.

► CHECKPOINT

The overall reaction between $HCl(aq)$ and $NaOH(aq)$ can be described by the following equation:

$$HCl(aq) + NaOH(aq) \\ \longrightarrow NaCl(aq) + H_2O(l)$$

HCl, NaOH, and NaCl break up into their respective ions in aqueous solution. Write the ionic equation for this reaction.

1.17 Balancing Chemical Equations

There is no sequence of rules that can be blindly followed to generate a balanced chemical equation. All we can do is manipulate the coefficients written in front of the formulas of the various reactants and products until the number of atoms of each element on both sides of the equation is the same.

Remember that only the coefficients in front of the formulas of the components of the reaction can be changed when balancing an equation. The subscripts in the chemical formulas cannot be changed when balancing an equation because that would change the identity of the products and reactants.

Persistence is required to balance chemical equations; the equation must be explored until the number of atoms of each element is the same on both sides of the equation. When doing this, it is usually a good idea to tackle the easiest part of a problem first.

Consider, for example, the equation for the combustion of glucose ($C_6H_{12}O_6$). Everything that we digest, at one point or another, gets turned into a sugar that is oxidized to provide the energy that fuels our bodies. Although a variety of sugars can be used as fuels, the primary source of energy that drives our bodies is glucose, or *blood sugar* as it is also known. The bloodstream delivers both glucose and oxygen to tissues, where they react to give a mixture of carbon dioxide and water.

$$C_6H_{12}O_6(aq) + O_2(g) \longrightarrow CO_2(g) + H_2O(l)$$

If you look at this equation carefully, you will notice that all of the carbon atoms in glucose end up in CO_2 and all of the hydrogen atoms end up in H_2O, but there are two sources of oxygen among the starting materials and two compounds that contain oxygen among the products. This means that there is no way to predict the number of O_2 molecules consumed in this reaction until we know how many CO_2 and H_2O molecules are produced.

We might therefore start the process of balancing this equation by noting that there are 6 carbon atoms in each $C_6H_{12}O_6$ molecule. Thus 6 CO_2 molecules are formed for every $C_6H_{12}O_6$ molecule consumed.

$$\textit{1 } C_6H_{12}O_6 + O_2 \longrightarrow \textbf{6 } CO_2 + H_2O$$

There are 12 hydrogen atoms in each $C_6H_{12}O_6$ molecule, which means there must be 12 hydrogen atoms, or 6 H_2O molecules, on the right-hand side of the equation.

$$\textit{1 } C_6H_{12}O_6 + O_2 \longrightarrow 6\ CO_2 + \textbf{6 } H_2O$$

Now that the carbon and hydrogen atoms are balanced, we can try to balance the oxygen atoms. There are 12 oxygen atoms in 6 CO_2 molecules and 6 oxygen atoms in 6 H_2O molecules. To balance the 18 oxygen atoms in the products of this reaction, we need a total of 18 oxygen atoms in the starting materials. But each $C_6H_{12}O_6$ molecule already contains 6 oxygen atoms. We therefore need 6 O_2 molecules among the reactants.

$$\boxed{6\,C + 12\,H + 18\,O}$$

$$\boxed{C_6H_{12}O_6 + 6\,O_2} \qquad\longrightarrow\qquad \boxed{6\,CO_2 + 6\,H_2O}$$

The balanced equation for this reaction is therefore written as follows.

$$C_6H_{12}O_6(aq) + 6\ O_2(g) \longrightarrow 6\ CO_2(g) + 6\ H_2O(l)$$

There are now 6 carbon atoms, 12 hydrogen atoms, and 18 oxygen atoms on each side of the equation, as shown in Figure 1.12.

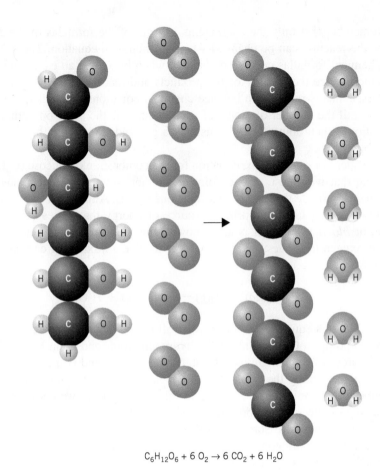

Fig. 1.12 A mechanical model on the atomic scale for the reaction between glucose ($C_6H_{12}O_6$) and O_2 to form CO_2 and H_2O. Note that the number of carbon, hydrogen, and oxygen atoms is the same in both the reactants and the products of the reaction.

$$C_6H_{12}O_6 + 6\ O_2 \rightarrow 6\ CO_2 + 6\ H_2O$$

Exercise 1.9

Write a balanced equation for the reaction that occurs when ammonia (NH_3) burns in air to form nitrogen oxide (NO) and water.

$$NH_3(g) + O_2(g) \longrightarrow NO(g) + H_2O(g)$$

Solution

We might start by balancing the nitrogen atoms because all of the nitrogen atoms in ammonia end up in nitrogen oxide. If we start with 1 molecule of ammonia and form 1 molecule of NO, the nitrogen atoms are balanced.

$$1\ NH_3 + O_2 \longrightarrow 1\ NO + H_2O$$

We can then turn to the hydrogen atoms. We have 3 hydrogen atoms on the left and 2 hydrogen atoms on the right in this equation. One way of balancing the hydrogen atoms is to look for the lowest common multiple: $2 \times 3 = 6$. We therefore set up the equation so that there are 6 hydrogen atoms on both sides. Doing this doubles the amount of NH_3 consumed in the reaction, so we have to double the amount of NO produced.

$$2\ NH_3 + O_2 \longrightarrow 2\ NO + 3\ H_2O$$

Because both the nitrogen and hydrogen atoms are balanced, the only task left is to balance the oxygen atoms. There are 5 oxygen atoms on the right side of this equation, so we need 5 oxygen atoms on the left.

$$2 \ NH_3 + 2\frac{1}{2} \ O_2 \longrightarrow 2 \ NO + 3 \ H_2O$$

There is no such thing, however, as a half of an oxygen molecule. If we insist that chemical equations must "work" on both the atomic and macroscopic scales, we must multiply the equation by 2. The balanced equation for the reaction is therefore written as follows.

$$4 \ NH_3(g) + 5 \ O_2(g) \longrightarrow 4 \ NO(g) + 6 \ H_2O(g)$$

All of the atoms in the reactants are now accounted for in the products. The 12 hydrogen atoms in 4 NH_3, for example, are found in the 6 water molecules, as shown in Figure 1.13.

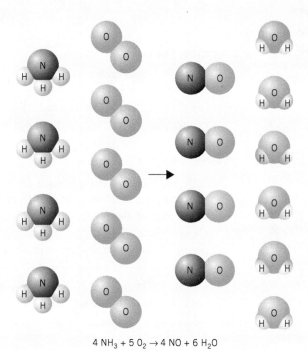

$$4 \ NH_3 + 5 \ O_2 \rightarrow 4 \ NO + 6 \ H_2O$$

Fig. 1.13 The reaction between ammonia and oxygen to form NO and water on the atomic scale. All atoms in the reactants must be accounted for in the products.

Key Terms

Absolute measurement	Cation	Group
Anion	Chemical equation	Group number
Aqueous	Chemical formula	Heterogeneous
Atom	Compound	Homogeneous
Atomic mass unit (amu)	Conversion Factor	Ion
Atomic number	Electron	Ionic compound
Atomic weight	Element	Ionic equation
Atomic world	English system of units	Isotope

Law of conservation of mass
Macroscopic world
Mass number
Matter
Metal
Metric system
Mixture
Model
Molecular compound

Molecule
Natural abundance
Neutron
Nonmetal
Nucleus
Period
Periodic table
Polyatomic ions
Products

Proton
Reactants
Relative measurement
Scientific notation
Semimetal
SI system
Significant figures
Symbolic world

Problems

Chemistry: A Definition

1. How would you describe the goals of modern chemistry?

2. It was known by the eleventh century that the addition of alum, prepared from a mineral, to animal skins aided in the tanning process. Could the practitioners of this tanning procedure be considered to be chemists?

3. The early Greek philosophers debated the idea of whether matter is continuous or consists of small indivisible particles. They performed no experiments. What role does experimentation play in chemistry?

Elements, Compounds, and Mixtures

4. Define the following terms: *element, compound,* and *mixture.* Give an example of each.

5. Describe the difference between elements and compounds on the macroscopic scale (objects are visible to the naked eye) and on the atomic scale.

6. Classify the following substances into the categories of elements, compounds, and mixtures. Use as many labels as necessary to classify each substance. Use whatever reference books you need to identify each substance.
   (a) diamond (b) brass (c) soil
   (d) glass (e) cotton (f) milk of magnesia
   (g) salt (h) iron (i) steel

7. Granite consists primarily of three minerals in varying composition: feldspar, plagioclase, and quartz. Is granite an element, a compound, or a mixture?

8. Describe what the formula P_4S_3 tells us about this compound.

9. What information does the formula SO_3 give us about this compound?

Atomic Symbols

10. List the symbols for the following elements.
    (a) antimony (b) gold (c) iron
    (d) mercury (e) potassium (f) silver
    (g) tin (h) tungsten

11. Name the elements with the following symbols.
    (a) Na (b) Mg (c) Al (d) Si
    (e) P (f) Cl (g) Ar

12. Name the elements with the following symbols.
    (a) Ti (b) V (c) Cr (d) Mn (e) Fe
    (f) Co (g) Ni (h) Cu (i) Zn

13. Name the elements with the following symbols.
    (a) Mo (b) W (c) Rh (d) Ir (e) Pd
    (f) Pt (j) Ag (h) Au (i) Hg

Chemical Formulas

14. Describe the difference between the following pairs of symbols.
    (a) Co and CO (b) Cs and CS_2
    (c) Ho and H_2O (d) 4 P and P_4

Evidence for the Existence of Atoms

15. Describe some of the evidence for the existence of atoms and some of the evidence from our senses that seems to deny the existence of atoms.

16. Choose one of Dalton's assumptions and design an experiment that would support or refute the assumption.

17. Why is the atomic theory so widely accepted?

18. Did any of Dalton's assumptions give any clues as to the structure of the atom?

19. According to Dalton, how do atoms of different elements differ?

20. One of Dalton's assumptions was that atoms cannot be created or destroyed. Does this mean that the number of atoms in the universe has remained unchanged?

The Role of Measurement in Chemistry

21. Calculate the number of seconds in a year.

22. Define the following prefixes from the metric system:
    (a) nano- (b) micro- (c) milli- (d) centi- (e) kilo-

23. Light is a small portion of the electromagnetic spectrum that is visible to the naked eye. It has wavelengths between about 4×10^{-5} and 7×10^{-5} centimeters. Calculate the range of wavelengths of light in units of micrometers and nanometers.

24. Liquor, which used to be sold in "fifths," is now sold in 750-mL bottles. If a fifth is one-fifth of a gallon, which

is the better buy: a fifth of scotch selling for $12.50 or a 750-mL bottle selling for the same price?

25. Air flow is measured in units of cubic feet per minute (CFM). Convert 100 CFM into units of cubic meters per second.

26. The LD_{50} for a drug is the dose that would be lethal for 50% of the population. LD_{50} for aspirin in rats is 1.75 grams per kilogram of body weight. Calculate the number of tablets containing 325 mg of aspirin a 70-kg human would have to consume to achieve this dose.

27. Determine the number of significant figures in the following numbers. See Appendix A.
    (a) 0.00641   (b) 0.07850
    (c) 500   (d) 50,003

28. Determine the number of significant figures in the following numbers. See Appendix A.
    (a) 3.4×10^{-2}
    (b) 5.98521×10^{3}
    (c) 8.709×10^{-6}
    (d) 7.00×10^{-5}

29. Round off the following numbers to three significant figures. See Appendix A.
    (a) 474.53   (b) 0.067981
    (c) 9.463×10^{10} (d) 30.0974

30. Convert the following numbers to scientific notation.
    (a) 11.98
    (b) 0.0046940
    (c) 4,679,000

31. Do the following calculations. (Keep track of significant figures.) See Appendix A.
    (a) $132.76 + 21.16071$
    (b) $32 + 0.9767$
    (c) $3.02 \times 10^{4} + 1.69 \times 10^{3}$
    (d) $4.18 \times 10^{-2} + 1.29 \times 10^{-3}$

The Structure of Atoms

32. Describe the differences between a proton, a neutron, and an electron.

33. One of Dalton's assumptions is now known to be in error. Which one is it?

34. What similarities are there between an atom of iron and an atom of mercury?

35. What are the three fundamental subatomic particles that make up an atom? Give the relative charge on each of these particles.

36. What is a neutral atom?

37. Which of the particles that make up an atom is lightest?

38. Where is the weight of the atom concentrated?

39. How does the radius of an atom compare to the size of the nucleus?

Atomic Number and Mass Number

40. Describe the relationship between the atomic number, mass number, number of protons, number of neutrons, and number of electrons in a calcium atom, ^{40}Ca.

41. Write the symbol for the atom that contains 24 protons, 24 electrons, and 28 neutrons.

42. Calculate the number of protons and neutrons in the nucleus and the number of electrons surrounding the nucleus of a ^{39}K atom. What are the atomic number and the mass number of this atom?

43. Calculate the number of protons and neutrons in the nucleus and the number of electrons surrounding the nucleus of an ^{127}I atom. What are the atomic number and the mass number of this atom?

44. Identify the element that has atoms with mass numbers of 20 that contain 11 neutrons.

45. Give the symbol for the atom that has 34 protons, 45 neutrons, and 34 electrons.

46. Calculate the number of electrons in a ^{134}Ba atom.

47. Complete the following table.

Isotope	Atomic Number (Z)	Mass Number (A)	Number of Electrons
^{31}P	15	—	—
^{18}O	—	—	8
—	19	39	19
^{58}Ni	—	58	—

Isotopes

48. What is the ratio of the mass of a ^{12}C atom to a ^{13}C atom?

49. How many times heavier is a ^{6}Li atom than a ^{1}H atom?

50. If you were to select one oxygen atom at random, what would its mass in grams most likely be? (One amu is equal to 1.66054×10^{-24} grams.)

51. The ratio of the mass of a ^{12}C atom to that of an unknown atom is 0.750239. Identify the unknown atom.

52. Divide the mass of a ^{1}H atom in atomic mass units by the mass of the atom in grams. Do the same for ^{2}H and ^{12}C. Does this suggest a relationship between the atomic mass in grams and amu?

53. Complete the following table. Table 1.5 may be useful.

Mass (grams)	Z	A	Number of Neutrons	Mass (amu)
1.6627×10^{-23}	—	—	—	10.0129
—	12	—	12	23.9850
—	8	18	—	—
1.7752×10^{-22}	—	107	60	—

54. Without referring to Table 1.5, which is heavier, an atom of ^{11}B or of ^{12}C? Justify your answer.

55. How many common isotopes of oxygen occur naturally on Earth?

56. What do all isotopes of oxygen have in common? In what ways are they different?

57. If you select one carbon atom at random, what is the mass of that atom likely to be (in grams and in amu)? (One amu is equal to 1.66054×10^{-24} grams.)

58. What is the mass (in amu) of 100 ^{12}C atoms? Of 100 ^{13}C atoms? (One amu is equal to 1.66054×10^{-24} grams.)

59. What would be the total mass of 100 carbon atoms selected at random?

    (a) 1200.00 amu

    (b) Slightly more than 1200.00 amu

    (c) Slightly less than 1200.00 amu

    (d) 1300.3 amu

    (e) Slightly less than 1300.3 amu

    Explain your reasoning.

The Difference between Atoms and Ions

60. Describe the difference between the following pairs of symbols.

    (a) H and H^+ (b) H and H^-

    (c) 2 H and H_2 (d) H^+ and H^-

61. Explain the difference between H^+ ions, H atoms, and H_2 molecules on the atomic scale.

62. Calculate the number of electrons, protons, and neutrons in a $^{134}Ba^{2+}$ ion.

63. Write the symbol for the atom or ion that contains 24 protons, 21 electrons, and 28 neutrons.

64. How many protons, neutrons, and electrons are in the $^{127}I^-$ ion?

65. Give the symbol for the atom or ion that has 34 protons, 45 neutrons, and 36 electrons.

66. Complete the following table.

Isotope	Atomic Number (Z)	Mass Number (A)	Number of Electrons
$^{31}P^{3-}$	—	—	—
$^{18}O^{2-}$	—	—	—
$^{58}Ni^{2+}$	—	—	—
—	12	24	10
—	13	27	10
—	35	80	36

Polyatomic Ions

67. What are polyatomic ions?

68. List three polyatomic ions by name and chemical formula for which the charges are -1, -2, and -3.

69. Give two common polyatomic ions that have positive charges.

The Periodic Table

70. Describe the differences between periods and groups of elements in the periodic table.

71. Mendeléeff placed both silver and copper in the same group as lithium and sodium. Look up the chemistry of these four elements in the *CRC Handbook of Chemistry and Physics*. Describe some of the similarities that allow these elements to be classified in a single group on the basis of their chemical properties.

72. Which of the following are nonmetals?

    (a) Li   (b) Be   (c) B   (d) C

    (e) N   (f) O

73. Place each of the following elements in the correct group on the periodic table.

    (a) K   (b) Si   (c) Ca   (d) S

    (e) Mg   (f) He   (g) I

74. Of the following sets of elements, which are in the same period of the periodic table? The same group?

    (a) Be, B, C   (b) Be, Mg, Ca   (c) P, S, Al

    (d) As, N, P   (e) Sb, Te, Xe   (f) K, Rb, Sr

75. How many elements are in Group IA?

76. How many elements are in the second period? The third period? The fourth period?

77. In which of the following sets of elements should all elements have similar chemical properties?

    (a) O, S, Se   (b) F, Cl, Te   (c) Al, Si, P

    (d) Ca, Sr, Ba   (e) K, Ca, Sc   (f) N, O, F

The Macroscopic, Atomic, and Symbolic Worlds of Chemistry

78. Which of the following samples exist on the macroscopic scale?

    (a) an atom of gold

    (b) a gold ring

    (c) a sample of gold ore

    (d) a sample of gold dust

79. How would a chemist symbolize gold on the atomic scale?

80. (a) How would a chemist symbolize gold on the macroscopic scale?

    (b) Give a symbolic representation that chemists would use for a bar of iron. What symbolic representation would the chemist use for an atom of iron?

The Mass of an Atom

81. Calculate the atomic weight of bromine if naturally occurring bromine is 50.69% ^{79}Br atoms with a mass of

78.9183 amu and 49.31% ^{81}Br atoms with a mass of 80.9163 amu.

82. Naturally occurring zinc is 48.6% ^{64}Zn atoms (63.9291 amu), 27.9% ^{66}Zn atoms (65.9260 amu), 4.1% ^{67}Zn atoms (66.9721 amu), 18.8% ^{68}Zn atoms (67.9249 amu), and 0.6% ^{70}Zn atoms (69.9253 amu). Calculate the atomic weight of zinc.

83. What is the total mass in amu of a sample of 100,000 carbon atoms selected at random? What is the average mass of a carbon atom? Does any carbon atom have this mass?

84. What is the average mass of an Mg atom in amu for a large collection of magnesium atoms?

85. What is the average mass in amu of an iodine atom?

86. Identify the element that contains atoms that have an average mass of 28.086 amu.

87. There are two naturally occurring isotopes of element X. One of these isotopes has a natural abundance of 80.3% and a relative mass of 11.00931 amu. The second isotope is lighter. Identify element X and state your reasoning. Give your best estimate of the number of neutrons, protons, and electrons in each isotope.

88. Element X has only two naturally occurring isotopes. The most abundant of these two isotopes has a mass of 7.01600 amu and accounts for more than 90% of the isotopic atoms.

(a) Identify element X. Explain your reasoning.

(b) Give the mass number and the number of protons, electrons, and neutrons for each of the two isotopes.

(c) The element X combines with various polyatomic anions to produce several compounds. The formulas of the compounds formed are XBr, X_2SO_4, and X_3PO_4. What is the charge on the ion formed by X? How many electrons does this ion have?

89. (a) There are two naturally occurring isotopes of silver; ^{107}Ag (106.90509 amu) is 51.84% and ^{109}Ag (108.90476 amu) is 48.16% abundant. Calculate the average atomic mass of silver. How will you know if your answer is correct?

(b) How many protons does a ^{107}Ag atom have? How many protons does a ^{109}Ag atom have?

(c) Give the number of neutrons and electrons in ^{107}Ag and ^{109}Ag atoms.

90. Element X has only two naturally occurring isotopes. One has a relative mass of 78.9183 amu, and the other has a relative mass of 80.9163 amu.

(a) Which element is this most likely to be? Explain.

(b) Without doing a calculation, estimate the percent abundance of these two isotopes. Explain how you arrived at your answer.

91. Complete the following table:

Isotope	Atomic Number	Mass Number	Number of Electrons	Number of Neutrons	% Abundance
^{6}Li	—	—	—	—	7.42
—	3	7	—	—	92.58
^{20}Ne	—	—	—	—	90.51
—	10	21	—	—	0.27
$^-$Ne	—	22	—	—	9.22

92. 100 Li atoms are selected at random. The total mass will be:

(a) more than 600 amu   (b) less than 600 amu

(c) 694.1 amu   (d) 700.0 amu

Explain your answer.

If 10,000 Ne atoms are selected at random, how many will have a mass number of 20?

93. When calculating the average atomic mass from percent abundance, you can always quickly check your answer. How?

94. There is only one naturally occurring isotope of this element. If that isotope has a mass of 26.982 amu, identify the element.

Chemical Reactions and the Law of Conservation of Atoms

95. If a candle is burned in a closed container filled with oxygen, will the mass of the container and contents be the same as, more than, or less than the original mass of the container, oxygen, and candle? Explain.

96. When gasoline is burned in air, are there more atoms, fewer atoms, or the same number as before burning? Explain.

97. Give an interpretation on a microscopic scale for why mass is conserved in a reaction.

98. What observation did Lavoisier make that led him to formulate the law of conservation?

99. What does the conservation of atoms in a chemical reaction tell us about what must happen to the atoms during the reaction?

Chemical Equations as a Representation of Chemical Reactions

100. State in a complete, grammatically correct sentence what the following symbolic equation represents.

$$2\,H_2(g) + O_2(g) \longrightarrow 2\,H_2O(g)$$

Do the same for this reaction.

$$2\,H_2(g) + O_2(g) \longrightarrow 2\,H_2O(l)$$

101. State in a complete, grammatically correct sentence what the following symbolic equation represents.

$$KI(s) \longrightarrow K^+(aq) + I^-(aq)$$

102. State in words what the following symbolic equation means.

$$CO_2(g) + H_2O(l) \longrightarrow H_2CO_3(aq)$$

Balancing Chemical Equations

103. Balance the following chemical equations.
    (a) $Cr(s) + O_2(g) \longrightarrow Cr_2O_3(s)$
    (b) $SiH_4(g) \longrightarrow Si(s) + H_2(g)$
    (c) $SO_3(g) \longrightarrow SO_2(g) + O_2(g)$

104. Balance the following chemical equations.
    (a) $Pb(NO_3)_2(s) \longrightarrow PbO(s) + NO_2(g) + O_2(g)$
    (b) $NH_4NO_2(s) \longrightarrow N_2(g) + H_2O(g)$
    (c) $(NH_4)_2Cr_2O_7(s)$
    $\qquad\qquad \longrightarrow N_2(g) + Cr_2O_3(s) + H_2O(g)$

105. Balance the following chemical equations.
    (a) $CH_4(g) + O_2(g) \longrightarrow CO_2(g) + H_2O(g)$
    (b) $H_2S(g) + O_2(g) \longrightarrow H_2O(g) + SO_2(g)$
    (c) $B_5H_9(g) + O_2(g) \longrightarrow B_2O_3(g) + H_2O(g)$

106. Balance the following chemical equations.
    (a) $PF_3(g) + H_2O(l) \longrightarrow H_3PO_3(aq) + HF(aq)$
    (b) $P_4O_{10}(s) + H_2O(l) \longrightarrow H_3PO_4(aq)$

107. Balance the following chemical equations.
    (a) $C_3H_8(g) + O_2(g) \longrightarrow CO_2(g) + H_2O(g)$
    (b) $C_2H_5OH(l) + O_2(g) \longrightarrow CO_2(g) + H_2O(g)$
    (c) $C_6H_{12}O_6(s) + O_2(g) \longrightarrow CO_2(g) + H_2O(l)$

Integrated Problems

108. A sealed bottle contains oxygen gas (O_2) and liquid butyl alcohol ($C_4H_{10}O$). There is enough oxygen in the bottle to react completely with the butyl alcohol to produce carbon dioxide (CO_2) and water (H_2O) gas. Write a chemical equation to describe this reaction. Assume that the bottle remains sealed during the reaction. Compare the number of molecules in the bottle before the reaction occurs ($C_4H_{10}O$ and O_2) with the number of molecules present in the bottle after the reaction (CO_2 and H_2O). Will the number of molecules in the bottle increase, decrease, or remain the same as the reaction takes place?

109. The mass number of the atom X in Group IIA from which an ion is formed is 40. The formula of the ionic compound formed with the carbonate ion is XCO_3. How many electrons, protons, and neutrons does the ion X have? What is the chemical symbol for X?

110. Element X is a metal whose chemical properties are similar to potassium. There is only one isotope of atom X. The mass of X in amu is 22.98976. Use the trends in masses in Table 1.5 to identify element X.

111. Complete the following table for uncharged atoms.

Classification	Group	Period	Number of Electrons	Atomic Symbol
Metal	—	—	11	—
—	IVA	—	—	Ge
—	—	—	—	B
Semimetal	—	3	—	—
—	VIIA	4	—	—

112. In 1999 a \$125 million Mars Climate Orbiter was destroyed when it flew too close to Mars. The loss was due to a failure to convert between English and metric units of measurement during the design of the navigation system. Data in units of pounds of force were used instead of the metric unit of newtons of force. A newton is defined as $1 \text{ kg} \cdot \text{m/sec}^2$. A pound is $1 \text{ slug} \cdot \text{ft/sec}^2$ where a mass of 1 slug is equal to 14.6 kg. Determine the conversion factor to convert pounds into newtons. Convert 5261 pounds of force into newtons.

Chapter Two

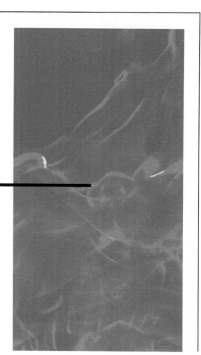

THE MOLE: THE LINK BETWEEN THE MACROSCOPIC AND THE ATOMIC WORLDS OF CHEMISTRY

2.1 The Mole as the Bridge between the Macroscopic and Atomic Scales

Imagine that you pick the following items off the shelves of a grocery store: a dozen eggs, a 1-lb bag of sugar, a 5-lb bag of flour, and a quart of milk. When you open the egg carton, you know exactly how many eggs it should contain—a dozen. But the same can't be said about either the sugar, the flour, or the milk. A recipe may call for 1 egg, but it never calls for 1 grain of sugar because a grain of sugar is too small to be useful. Recipes therefore tend to call for half a cup of sugar or two cups of flour or a cup of milk.

Chemists face a similar problem because it takes an enormous number of atoms to give a sample large enough to be seen with the naked eye. (A dot of graphite from a pencil just large enough to be weighed on an analytical balance contains approximately 5×10^{19} atoms.) Chemists therefore created a unit known as the **mole** (from Latin, meaning "a huge pile") that can serve as the bridge between chemistry on the macroscopic and atomic scales.

> **A mole is the amount of substance that contains as many elementary units as there are atoms in exactly 12 grams of the carbon-12 isotope.**

A single ^{12}C atom has a mass of exactly 12 amu, and a mole of ^{12}C atoms has a defined mass of exactly 12 grams.

> 1 ^{12}C atom has a mass of 12.00... amu
>
> 1 mole of ^{12}C atoms has a mass of 12.000... g

As noted in Section 1.14, the average mass of naturally occurring carbon atoms is 12.011 amu because of the presence of the ^{12}C, ^{13}C, and, to some extent, the ^{14}C isotopes. Thus, one mole of carbon atoms would have a mass of 12.011 grams.

The mole is the most fundamental unit of chemistry because it allows us to determine the number of elementary particles in a sample of a pure substance from measurements of the mass of the sample. Assume, for example, that we want a sample of aluminum metal that contains the same number of atoms as a mole of carbon atoms. We can start by looking up the atomic weight of aluminum in the periodic table.

> 1 Al atom has a mass of 26.982 amu

The atomic weight of aluminum is a little more than twice that of a ^{12}C atom. As a result, aluminum atoms are a little more than twice as heavy as a ^{12}C atom.

$$\frac{1 \text{ Al atom}}{1 \, ^{12}C \text{ atom}} = \frac{26.982 \text{ amu}}{12 \text{ amu}} = 2.2485$$

(The ratio of the mass of an aluminum atom to a ^{12}C atom in this calculation can be given to five significant figures because the mass of ^{12}C is based on a definition, not a measurement.)

If a mole of aluminum contains exactly the same number of atoms as a mole of ^{12}C, then a mole of aluminum must have a mass that is 2.2485 times the mass of a mole of ^{12}C atoms.

$$\frac{1 \text{ mol Al}}{1 \text{ mol } ^{12}C} = \frac{26.982 \text{ g}}{12 \text{ g}} = 2.2485$$

Thus, the mass of a mole of aluminum is 26.982 grams.

Figure 2.1 shows two beakers. The beaker on the left contains 12.011 grams of carbon, whereas the beaker on the right contains 26.982 grams of aluminum. Both beakers therefore contain 1 mole of the respective elements, and there are the same number of atoms in both beakers.

The concept of the mole serves as a bridge between the atomic and macroscopic worlds. By definition:

> **A mole of atoms of any element has a mass in grams equal to the atomic weight of the element.**

Because the atomic weight of Al is 26.982 amu, a mole of aluminum atoms would have a mass of 26.982 grams.

The mass of a mole of a substance is often called the **molar mass.** The molar mass of ^{12}C, for example, is 12 grams per mole (abbreviated "mol"). The molar mass of a sample of carbon that contains both ^{12}C and ^{13}C atoms in their natural abundances would be 12.011 grams/mol. The relationship between atomic weight and molar mass is illustrated by the following examples.

Element	Atomic Weight	Molar Mass
Carbon	12.011 amu	12.011 g/mol
Aluminum	26.982 amu	26.982 g/mol
Iron	55.847 amu	55.847 g/mol

The key to understanding the concept of the mole is recognizing that 12.011 grams of carbon contain the same number of atoms as 26.982 grams of aluminum or 55.847 grams of iron.

Fig. 2.1 Because each beaker contains 1 mole of the element, the two beakers contain the same number of atoms.

➤ **CHECKPOINT**

What is the atomic weight and molar mass of potassium? Of uranium?

2.2 The Mole as a Collection of Atoms

For many years, chemists used the concept of a mole without knowing exactly how many particles there were in a mole of elemental carbon or aluminum metal. All that matters is the fact that there are the same number of particles in a mole of each of these elements.

The only way to determine the number of particles in a mole is to measure the same quantity on both the atomic and the macroscopic scales. In 1910 Robert Millikan measured the charge in coulombs on a single electron: $1.6 \times 10^{-19}C$. Because the charge on a mole of electrons, 96,485.3415 C, was already known, it was possible to estimate the number of electrons in a mole for the first time. Using more recent data, we get the following results.

$$\frac{96,485.3415 \text{ C}}{1 \text{ mol}} \times \frac{1 \text{ electron}}{1.60217733 \times 10^{-19}\text{C}} = 6.02213873 \times 10^{23}\frac{\text{electrons}}{\text{mol}}$$

The number of particles in a mole is known as **Avogadro's number,** 6.0221×10^{23}, and is a pure number. The unit conversion factor, $6.0221 \times 10^{23} \text{ mol}^{-1}$, is known as **Avogadro's constant.** Avogadro's number is so large it is difficult to comprehend. It would take 6 million million galaxies the size of the Milky Way to yield 6.02×10^{23} stars. At the speed of light, it would take 102 billion years to travel 6.02×10^{23} miles. There are only about 40 times this number of drops of water in all the oceans on Earth.

In everyday life units such as dozen (12) are used to describe a collection of items. The mole is sometimes referred to as the "chemist's dozen." The concept of the mole can be applied to any particle. We can talk about a mole of Mg atoms, a mole of Na^+ ions, a mole of electrons, or a mole of glucose molecules ($C_6H_{12}O_6$) Each time we use the term, we refer to Avogadro's number of items.

1 mole of Mg atoms contains 6.02×10^{23} Mg atoms.

1 mole of Na^+ ions contains 6.02×10^{23} Na^+ ions.

1 mole of electrons contains 6.02×10^{23} electrons.

1 mole of $C_6H_{12}O_6$ molecules contains 6.02×10^{23} $C_6H_{12}O_6$ molecules.

Once we know the number of elementary particles in a mole, we can determine the number of particles in a sample of a pure substance by weighing the sample. To see how this is done, let's first consider a process by which objects of known mass are counted by weighing a sample.

Suppose that the mass of a dozen balls is found to be 107 grams, as shown in Figure 2.2. If a sample that contains an unknown number of balls has a mass of 178 grams. How many balls are in the unknown sample?

We can build two conversion factors from our knowledge of the mass of a dozen balls.

$$\frac{107 \text{ g}}{1 \text{ dozen balls}} \quad \text{or} \quad \frac{1 \text{ dozen balls}}{107 \text{ g}}$$

Which conversion factor should we use? A technique known as *dimensional analysis* can guide us to the correct conversion factor. All we have to do is keep track of what happens to the units during the calculation. If the units cancel as expected, the calculation has been set up properly.

In this case, we know the mass of the unknown sample and the mass of a dozen balls. We therefore set up the calculation as follows:

$$178 \text{ g} \times \frac{1 \text{ dozen balls}}{107 \text{ g}} = 1.66 \text{ dozen balls}$$

We can now calculate the number of balls in the sample from the fact that there are 12 balls in a dozen.

$$1.66 \text{ dozen} \times \frac{12 \text{ balls}}{1 \text{ dozen}} = 20 \text{ balls}$$

In this example a dozen is analogous to a mole, and 107 grams/dozen is analogous to the molar mass of an element.

We can use the logic developed in the example shown above to calculate the number of carbon atoms in a 1-carat diamond. All we need to know is that a

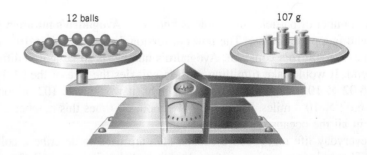

Fig. 2.2 A sample of a dozen balls that weigh 107 grams.

diamond can be thought of as a single crystal that contains only carbon atoms and that the mass of a carat is defined as 200.0 milligrams. Because 1 gram is equal to 1000 milligrams, a 1-carat diamond has a mass of 0.2000 grams.

$$200.0 \text{ mg} \times \frac{1 \text{ g}}{1000 \text{ mg}} = 0.2000 \text{ g}$$

The atomic weight of carbon is 12.011 amu, which means that the molar mass of carbon is 12.011 grams per mole.

1 mol of C has a mass of 12.011 g

We therefore start the calculation by determining the number of moles of carbon in the diamond.

$$0.2000 \text{ g C} \times \frac{1 \text{ mol C}}{12.011 \text{ g C}} = 0.01665 \text{ mol C}$$

We can then use Avogadro's constant to calculate the number of carbon atoms in the diamond.

$$0.01665 \text{ mol C} \times \frac{6.022 \times 10^{23} \text{ atoms}}{1 \text{ mol C}} = 1.003 \times 10^{22} \text{C atoms}$$

> ➤ CHECKPOINT
>
> How many aluminum atoms are there in 1.0 gram of pure aluminum?

2.3 Converting Grams into Moles and Number of Atoms

The mole is the bridge between chemistry on the macroscopic scale, where we do experiments, and the atomic scale, where we think about the implications of these experiments. As a result, one of the most common calculations in chemistry involves converting measurements of the mass of a sample into the number of moles of the substance it contains. We accomplish this by taking the mass of a compound and multiplying by the inverse of its molar mass (conversion factor) as shown below.

$$n = (mass) \times \left(\frac{1}{molar \ mass} \right)$$

where n = moles

Consider the following question: How many moles of sulfur atoms do 45.5 grams of sulfur contain?

The first step in calculations like the one above involves finding out the number of grams in one mole of sulfur. According to the periodic table, the atomic weight of sulfur is 32.066 amu. This means that a mole of sulfur atoms would have a mass of 32.066 grams.

1 mol of S has a mass of 32.066 g

We don't need five significant figures, however, because we know the mass of the sample to only three significant figures. Let's therefore carry four significant figures as we transform this equality into two conversion factors.

$$\frac{1 \text{ mol S}}{32.07 \text{ g S}} \quad \text{or} \quad \frac{32.07 \text{ g S}}{1 \text{ mol S}}$$

Careful analysis of the units involved in converting from grams to moles suggests that we should multiply the mass of the sample in grams by the conversion factor (molar mass) on the left to determine the moles of sulfur atoms in the sample.

$$n = (mass) \times \left(\frac{1}{molar\ mass} \right)$$

$$n = 45.5\ \text{g S} \times \frac{1\ \text{mol S}}{32.07\ \text{g S}} = 1.42\ \text{mol S}$$

If we want to know how many atoms are in the sample, we can multiply the number of moles of sulfur atoms by the number of atoms in one mole, Avogadro's constant.

$$1.42\ \text{mol S} \times \frac{6.022 \times 10^{23}\text{S atoms}}{1\ \text{mol S}} = 8.55 \times 10^{23}\text{S atoms}$$

In general, you need two pieces of information to do calculations of this nature. You need to know the mass of a mole of the substance and the number of particles in a mole.

We convert mass into moles or moles into mass by using the molar mass of a substance. We convert the moles *of an element* into atoms, or vice versa, by using Avogadro's constant. This is summarized in the following diagram.

$$\textbf{Mass} \quad \xleftarrow{\textit{Molar mass}} \quad \textbf{Moles} \quad \xleftarrow{\textit{Avogadro's constant}} \quad \textbf{Atoms}$$
Macroscopic Scale   *Atomic Scale*

Exercise 2.1

Calculate the mass of a sample of iron metal that contains 0.250 moles of iron atoms.

Solution

According to the periodic table, the atomic weight of iron to four significant figures is 55.85 amu. One mole of iron would therefore have a mass of 55.85 grams. The atomic weight or molar mass of iron can be represented in terms of either of the following conversion factors.

$$\frac{1\ \text{mol Fe}}{55.85\ \text{g Fe}} \quad \text{or} \quad \frac{55.85\ \text{g Fe}}{1\ \text{mol Fe}}$$

In order to convert from moles to grams, we need the conversion factor that allows the units of moles to cancel.

$$0.250\ \text{mol Fe} \times \frac{55.85\ \text{g FeS}}{1\ \text{mol Fe}} = 14.0\ \text{g Fe}$$

Exercise 2.2

Calculate the number of atoms in a 0.123-gram sample of aluminum foil.

Solution

Before we can do anything else, we need to know how many moles of aluminum metal are in the sample. This can be calculated from the mass of the sample and the molar mass of aluminum, which is 26.98 g/mol to four significant figures.

$$0.123 \text{ g Al} \times \frac{1 \text{ mol Al}}{26.98 \text{ g Al}} = 4.56 \times 10^{-3} \text{ mol Al}$$

We can then use Avogadro's constant to calculate the number of atoms in the sample.

$$4.56 \times 10^{-3} \text{ mol Al} \times \frac{6.022 \times 10^{23} \text{ Al atoms}}{1 \text{ mol Al}} = 2.75 \times 10^{21} \text{ Al atoms}$$

2.4 The Mole as a Collection of Molecules

Before we can apply the concept of the mole to compounds such as carbon dioxide (CO_2) or the sugar that flows through your bloodstream known as glucose ($C_6H_{12}O_6$), we have to be able to calculate the molecular weight of these compounds. As might be expected, the molecular weight of a compound is the sum of the atomic weights of the atoms in the formula of the compound.

 Exercise 2.3

Calculate both the average mass of a single molecule of carbon dioxide and glucose and the molecular weight of these compounds.

Solution

The average mass of a *single molecule of carbon dioxide* is the sum of the atomic weights of the three atoms in a CO_2 molecule.

Average mass of a CO_2 molecule:

Mass of 1 C atom is 1(12.011 amu) = 12.011 amu
Mass of 2 O atoms is 2(15.999 amu) = 31.998 amu
_____
44.009 amu

The mass of a *mole of carbon dioxide* therefore would be 44.009 grams.

The average mass of a molecule of glucose is the sum of the atomic weights of the 24 atoms in a $C_6H_{12}O_6$ molecule.

Average mass of a $C_6H_{12}O_6$ molecule:

mass of 6 C atoms is 6(12.011 amu) =   72.066 amu
mass of 12 H atoms is 12(1.0079 amu) =   12.095 amu
mass of 6 O atoms is 6(15.999 amu) =   95.994 amu
_____
180.155 amu

The molecular weight of this compound is therefore 180.155 grams/mol.

For many years, chemists referred to the results of the calculations in the previous exercise as the **molecular weight** of the compound. This term is somewhat misleading for several reasons. First, no $C_6H_{12}O_6$ molecule would have a mass equal to 180.155 amu. This is the *average mass* of the sugar molecules in a sample, which contain ^{12}C and ^{13}C atoms as well as a mixture of hydrogen and oxygen isotopes. Second, some compounds, as we shall see, don't exist as molecules, so it is misleading to talk about their "molecular" weight. Some chemists therefore recommend that we describe the results of these calculations as the *mass of a mole* or the *molar mass* of a compound. Because the term *molecular weight* has been so extensively used by chemists and is widely found in the chemical literature, we will use the terms *molar mass* and *molecular weight* interchangeably.

Exercise 2.4

Describe the difference between the mass of a mole of oxygen atoms (O) and the mass of a mole of oxygen molecules (O_2).

Solution

Because the atomic weight of oxygen is 15.999 amu, a mole of oxygen atoms has a mass of 15.999 grams. Each O_2 molecule has two atoms, however, so the molecular weight of O_2 molecules is twice as large as the atomic weight of the atom.

1 mol of O has a mass of 15.999 g   1 mol O_2 has a mass of 31.998 g

The diagram we used to summarize mass–mole conversions for elements can be used for chemical compounds. In this case, however, we can take the calculation one step further by using the formula of the compound to calculate the number of atoms of a given element in the sample.

$$\text{Mass} \xleftrightarrow{\textit{Molar mass}} \text{Moles} \xleftrightarrow{\textit{Avogadro's constant}} \text{Molecules} \xleftrightarrow{\textit{chemical formula}} \text{Atoms}$$
Macroscopic Scale   *Atomic Scale*

To illustrate the power of the mole concept, consider the following question: What is the formula of carbon dioxide if 2.73 grams of carbon combine with 7.27 grams of oxygen molecules (O_2) when the carbon burns?

The first step in answering an unfamiliar problem often involves drawing a diagram that helps us organize the information in the problem and visualize the process taking place. We could start, for example, with the diagram in Figure 2.3, which summarizes the relationship between the mass of carbon and the oxygen gas consumed in this reaction.

The next step in any problem of this kind is to convert grams into moles, when possible. To do this, we need to know the relationship between the number of grams and the number of moles of the substance. It doesn't matter which element we start with in this example because we eventually have to work with both, so let's arbitrarily start with carbon.

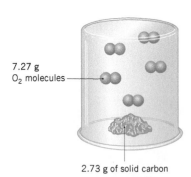

7.27 g
O_2 molecules

2.73 g of solid carbon

Fig. 2.3 2.73 grams of carbon react with 7.27 grams of O_2 molecules when the carbon burns to form carbon dioxide.

The atomic weight of carbon is 12.011 amu, which means that a mole of carbon has a mass of 12.011 grams. We can use this information to construct two conversion factors.

$$\frac{1 \text{ mol C}}{12.011 \text{ g C}} \quad \text{or} \quad \frac{12.011 \text{ g C}}{1 \text{ mol C}}$$

Converting grams of carbon into moles requires a conversion factor that has units of moles in the numerator and grams in the denominator. Analysis of the units involved in the calculation therefore suggests setting up the problem as follows.

$$2.73 \text{ g C} \times \frac{1 \text{ mol C}}{12.01 \text{ g C}} = 0.227 \text{ mol C}$$

A similar approach can be used to convert grams of oxygen into moles of oxygen molecules.

$$7.27 \text{ g O}_2 \times \frac{1 \text{ mol O}_2}{32.00 \text{ g O}_2} = 0.227 \text{ mol O}_2 \text{ molecules}$$

and then moles of oxygen atoms.

$$0.227 \text{ mol O}_2 \text{ molecules} \times \frac{2 \text{ O atoms}}{1 \text{ O}_2 \text{ molecules}} = 0.454 \text{ mol O atoms}$$

So far we have found that reacting 2.73 grams of carbon and 7.27 grams of oxygen corresponds to 0.227 mol of carbon reacting with 0.454 mol of oxygen atoms. Because atoms are neither created nor destroyed in a chemical reaction, the same number of atoms of each element must be found on both sides of the equation used to describe the reaction. The product of this reaction must therefore have a mass of 10.00 grams (2.73 grams + 7.27 grams) and must contain 0.227 mol of carbon atoms and 0.454 mol of oxygen atoms.

We can now reread the question and ask: Have we made any progress toward the answer? In this case, we are trying to find the chemical formula for carbon dioxide, which gives the ratio of carbon atoms to oxygen atoms. The next step in the problem therefore involves determining the relationship between the number of moles of carbon atoms and moles of oxygen atoms in the sample of carbon dioxide.

$$\frac{0.454 \text{ mol O}}{0.227 \text{ mol C}} = 2.00$$

There are twice as many moles of oxygen atoms as there are moles of carbon atoms in the sample. Because a mole of atoms always contains the same number of atoms, the only possible conclusion is that there are twice as many oxygen atoms as carbon atoms in the compound. In other words, the formula for carbon dioxide must be CO_2.

This calculation shows how the mole concept can be used as the bridge between macroscopic measurements (the mass of carbon and oxygen) and the microscopic atomic world (the number of carbon and oxygen atoms in a carbon dioxide molecule).

Exercise 2.5

Determine how many carbon atoms there are in an 0.500-gram sample of carbon dioxide, CO_2.

Solution

The first step in this calculation involves converting the mass of the sample into the moles of CO_2 using the molecular weight of CO_2 calculated in Exercise 2.3.

$$0.500 \text{ g } CO_2 \times \frac{1 \text{ mol } CO_2}{44.01 \text{ g } CO_2} = 1.14 \times 10^{-2} \text{ moles of } CO_2$$

Once we know the moles of CO_2 in the sample, we can use Avogadro's constant to calculate the number of CO_2 molecules in the sample.

$$1.14 \times 10^{-2} \text{ mol } CO_2 \times \frac{6.022 \times 10^{23} \text{ } CO_2 \text{ molecules}}{1 \text{ mol } CO_2}$$
$$= 6.86 \times 10^{21} \text{ } CO_2 \text{ molecules}$$

We can now use the chemical formula for carbon dioxide to determine the number of carbon atoms in the sample.

$$6.86 \times 10^{21} \text{ } CO_2 \text{ molecules} \times \frac{1 \text{ C atom}}{1 \text{ } CO_2 \text{ molecules}} = 6.86 \times 10^{21} \text{ C atoms}$$

> ➤ **CHECKPOINT**
>
> How many carbon atoms are there in one molecule of acetylene, C_2H_2? How many moles of carbon atoms are in one mole of C_2H_2? How many carbon atoms are in one mole of C_2H_2?

2.5 Percent by Mass

It is often useful to know the **percent by mass** of the different elements in a compound. Percent by mass can be determined experimentally or calculated from the chemical formula. When the formula of a compound is known, the first step in this calculation involves determining the molecular weight of the compound. The molecular weight of ethanol (CH_3CH_2OH), for example, would be calculated as follows.

Average mass of a CH_3CH_2OH molecule:

$$
\begin{array}{ll}
2 \text{ C atoms: } 2(12.011 \text{ amu}) = & 24.022 \text{ amu} \\
6 \text{ H atoms: } 6(1.0079 \text{ amu}) = & 6.0474 \text{ amu} \\
1 \text{ O atom: } 1(15.999 \text{ amu}) = & \underline{15.999 \text{ amu}} \\
& 46.068 \text{ amu}
\end{array}
$$

A mole of ethanol would therefore weigh 46.068 grams.

As we have noted, *percent* literally means "parts per hundred." The percent by mass of carbon in ethanol is therefore the mass of carbon in a mole of ethanol divided by the mass of a mole of ethanol, times 100.

$$\frac{24.022 \text{ g C}}{46.068 \text{ g } CH_3CH_2OH} \times 100 = 52.145\% \text{ C}$$

The percent by mass of hydrogen and oxygen in ethanol can be calculated in a similar fashion.

$$\frac{6.0474 \text{ g H}}{46.068 \text{ g CH}_3\text{CH}_2\text{OH}} \times 100 = 13.127\% \text{ H}$$

$$\frac{15.999 \text{ g O}}{46.068 \text{ g CH}_3\text{CH}_2\text{OH}} \times 100 = 34.729\% \text{ O}$$

2.6 Determining the Formula of a Compound

Section 2.4 showed one way to determine the formula of a compound. By carefully measuring the amount of carbon and oxygen that combined to form carbon dioxide, it was possible to show that the formula of this compound is CO_2. Let's look at another way to approach this problem using percent by mass data. This time we will examine the compound methane, once known as marsh gas because it was first collected above certain swamps, or marshes, in Britain.

Methane is 74.9% carbon and 25.1% hydrogen by mass. A 100-gram sample of the gas therefore contains 74.9 grams of carbon and 25.1 grams of hydrogen.

$$100 \text{ g methane} \times \frac{74.9 \text{ g C}}{100 \text{ g methane}} = 74.9 \text{ g C}$$

$$100 \text{ g methane} \times \frac{25.1 \text{ g H}}{100 \text{ g methane}} = 25.1 \text{ g H}$$

This is useful information because we can use the molar mass of these elements to convert the grams of carbon and hydrogen in this sample into moles of each element.

$$74.9 \text{ g C} \times \frac{1 \text{ mol C}}{12.01 \text{ g C}} = 6.24 \text{ mol C}$$

$$25.1 \text{ g H} \times \frac{1 \text{ mol H}}{1.008 \text{ g H}} = 24.9 \text{ mol H}$$

We now know the number of moles of carbon atoms and the number of moles of hydrogen atoms in a sample. We also know that there are always the same number of atoms in a mole of atoms of any element. It might therefore be useful to look at the ratio of the moles of these elements in the sample.

$$\frac{24.9 \text{ mol H}}{6.24 \text{ mol C}} = 3.99$$

A 100-gram sample of methane therefore contains four times as many moles of hydrogen atoms as moles of carbon atoms, within experimental error. This means that there are four times as many hydrogen atoms as carbon atoms in this sample.

This experiment tells us the simplest or **empirical formula** (smallest whole-number ratio of atoms) of the compound, but not necessarily the **molecular formula** (actual number of each kind of atom in a molecule). These results are consistent with molecules that contain one carbon atom and four hydrogen atoms: CH_4. But they are also consistent with formulas such as C_2H_8, C_3H_{12}, C_4H_{16}, and so on. All we know at this point is that the molecular formula for the molecule is some multiple of the empirical formula, CH_4. Using other experimental techniques, we can show that the molecular weight of methane is 16 grams/mol, which is consistent with a molecular formula of CH_4.

Natural and synthetic sources of vitamin C.

Exercise 2.6

Vitamin C (ascorbic acid) is found in citrus fruits, or it can be obtained from dietary supplements such as vitamin C tablets. Calculate the empirical formula for vitamin C, which is 40.9% C, 54.5% O, and 4.58% H by mass. Use the fact that the molar mass of vitamin C is 176 g/mol to determine the molecular formula for vitamin C.

Solution

We can start by calculating the number of grams of each element in a 100-gram sample of vitamin C.

$$100 \text{ g} \times 40.9\% \text{ C} = 40.9 \text{ g C}$$
$$100 \text{ g} \times 54.5\% \text{ O} = 54.5 \text{ g O}$$
$$100 \text{ g} \times 4.58\% \text{ H} = 4.58 \text{ g H}$$

We then convert the number of grams of each element into the number of moles of atoms of that element.

$$40.9 \text{ g C} \times \frac{1 \text{ mol C}}{12.01 \text{ g C}} = 3.41 \text{ mol C}$$

$$54.5 \text{ g O} \times \frac{1 \text{ mol O}}{16.00 \text{ g O}} = 3.41 \text{ mol O}$$

$$4.58 \text{ g H} \times \frac{1 \text{ mol H}}{1.008 \text{ g H}} = 4.54 \text{ mol H}$$

Because we are interested in the simplest whole-number ratio of these elements, we now divide through by the element with the smallest number of moles of atoms.

$$\frac{3.41 \text{ mol O}}{3.41 \text{ mol C}} = 1.00 \quad \text{and} \quad \frac{4.54 \text{ mol H}}{3.41 \text{ mol C}} = 1.33$$

The ratio of C to H to O atoms in vitamin C is therefore $1 : 1\frac{1}{3} : 1$. But it doesn't make sense to write the ratio of atoms as $CH_{1\frac{1}{3}}O$ because there is no such thing as one–third of a hydrogen atom. We therefore multiply this ratio by small whole numbers until we get a formula in which all of the coefficients are integers.

$$2(CH_{1\frac{1}{3}}O) = C_2H_{2\frac{2}{3}}O_2$$
$$3(CH_{1\frac{1}{3}}O) = C_3H_4O_3$$

The empirical formula of vitamin C is therefore $C_3H_4O_3$.

The molecular weight of $C_3H_4O_3$ would be 88.06 amu, to four significant figures. The mass of a mole of $C_3H_4O_3$ molecules is therefore 88.06 grams.

$$
\begin{array}{ll}
\text{3 C atoms: } 3(12.01 \text{ amu}) = & 36.03 \text{ amu} \\
\text{4 H atoms: } 4(1.008 \text{ amu}) = & 4.032 \text{ amu} \\
\text{3 O atoms: } 3(16.00 \text{ amu}) = & \underline{48.00 \text{ amu}} \\
& 88.06 \text{ amu}
\end{array}
$$

This is as far as we can go with percent by mass data. Since we also know the molar mass of vitamin C, however, we can compare the mass of a mole of this compound with the mass of a mole of $C_3H_4O_3$ molecules.

$$\frac{176 \text{ g/mol of vitamin C}}{88.06 \text{ g/mol of } C_3H_4O_3} = 2.00$$

The only logical conclusion is that a molecule of vitamin C is twice as large as the empirical formula for this compound. In other words, the molecular formula of vitamin C is $C_6H_8O_5$.

● ●

➤ **CHECKPOINT**

Calcium carbide was once used in miner's lamps. A sample of calcium carbide large enough to contain one mole of calcium atoms also contains 24 grams of carbon. What is the empirical formula of the compound? What additional information is necessary to determine the molecular formula?

● ●

 Exercise 2.7

A student is assigned to analyze a sample of ascorbic acid (vitamin C) and is told that ascorbic acid contains only C, H, and O. The student analyzed an 8.437 mg sample and recorded the following results.

	Mass (g)
C	3.46×10^{-3}
O	4.6×10^{-3}

(a) Determine the mass of H in grams in the compound to the correct number of significant figures.

The student then recorded the following atomic masses.

	Atomic mass (g/mol)
C	12.011
H	1.01
O	16

(b) Calculate the number of moles of C, H, and O in the sample to the correct number of significant figures.

(c) Use the student's recorded data to determine the mole ratio, C:H:O and the empirical formula of the compound.

Solution

(a) The mass of H in the sample is determined by subtracting the mass of C and O from the mass of the sample.

mass of C and O = $(3.46 \times 10^{-3} \text{ g}) + (4.6 \times 10^{-3} \text{ g}) = 8.1 \times 10^{-3} \text{ g}$

Because the mass of O is known to only one digit after the decimal, the result can have only one digit after the decimal.

Before finding the mass of H, the mass of the sample is converted from mg to g in order to have all of the masses in the same unit.

$$8.437 \text{ mg} \times \frac{1 \text{ g}}{1000 \text{ mg}} = 8.437 \times 10^{-3} \text{ g}$$

mass of H = $(8.437 \times 10^{-3} \text{ g}) - (8.1 \times 10^{-3} \text{ g}) = 0.3 \times 10^{-3} \text{ g} = 3 \times 10^{-4} \text{ g H}$

Once again, because the mass of C plus O is known to only one digit after the decimal, the result can have only one digit after the decimal.

(b) $\text{mol C} = (3.46 \times 10^{-3} \text{ g C}) \times \dfrac{1 \text{ mol C}}{12.011 \text{ g C}} = 2.88 \times 10^{-4} \text{ mol C}$

$\text{mol O} = (4.6 \times 10^{-3} \text{ g O}) \times \dfrac{1 \text{ mol O}}{16 \text{ g O}} = 2.9 \times 10^{-4} \text{ mol O}$

$\text{mol H} = (3 \times 10^{-4} \text{ g H}) \times \dfrac{1 \text{ mol H}}{1.01 \text{ g H}} = 3 \times 10^{-4} \text{ mol H}$

(c) The ratio of C:H:O is 2.88×10^{-4} mol C : 3×10^{-4} mol H : 2.9×10^{-4} mol O

Divide each value by the smallest number of moles (2.88×10^{-4} mol C) to find the smallest whole number ratio.

$$\frac{2.88 \times 10^{-4} \text{ mol C}}{2.88 \times 10^{-4} \text{ mol C}} = 1.00$$

$$\frac{3 \times 10^{-4} \text{ mol H}}{2.88 \times 10^{-4} \text{ mol C}} = 1$$

$$\frac{2.9 \times 10^{-4} \text{ mol O}}{2.88 \times 10^{-4} \text{ mol C}} = 1.0$$

This gives a C:H:O ratio of $1.00 : 1 : 1.0$. The empirical formula based on the above data is therefore CHO.

However, this is not the correct empirical formula for ascorbic acid. The empirical formula for ascorbic acid determined in the previous example is $C_3H_4O_3$. Why was the incorrect formula obtained? The student's data are correct, and the calculations are all done correctly. The problem lies in the number of significant figures recorded by the student. The one significant figure after the decimal for the mass of oxygen limits us to just one significant figure for the mass of C plus O. However, when this value with just one digit after the decimal is subtracted from the sample mass we get only one digit after the decimal for H (0.3×10^{-3} g $= 3 \times 10^{-4}$ g H).

In order to improve the reliability of the results, the student should have determined the mass of C and O to the same level of confidence as was obtained for the mass of the sample. Using the following data, we can repeat the above calculations.

Mass (g)	
C	3.458×10^{-3}
O	4.586×10^{-3}

We will also use a sufficient number of significant figures for the atomic masses from the periodic table such that the number of significant figures associated with the calculations is limited by the significant figures in the measurements rather than by the values obtained from the periodic table.

Atomic mass (g/mol)	
C	12.011
H	1.008
O	15.999

(a) mass of C and O = $(3.458 \times 10^{-3} \text{ g}) + (4.586 \times 10^{-3} \text{ g}) = 8.044 \times 10^{-3} \text{ g}$

mass H = $(8.437 \times 10^{-3} \text{ g}) - (8.044 \times 10^{-3} \text{ g}) = 0.393 \times 10^{-3} \text{ g}$
$\quad\quad = 3.93 \times 10^{-4} \text{ g H}$

(b) mol C = $(3.458 \times 10^{-3} \text{ g C}) \times \dfrac{1 \text{ mol C}}{12.011 \text{ g C}} = 2.879 \times 10^{-4} \text{ mol C}$

mol O = $(4.586 \times 10^{-3} \text{ g O}) \times \dfrac{1 \text{ mol O}}{15.999 \text{ g O}} = 2.866 \times 10^{-4} \text{ mol O}$

mol H = $(3.93 \times 10^{-4} \text{ g H}) \times \dfrac{1 \text{ mol H}}{1.008 \text{ g H}} = 3.90 \times 10^{-4} \text{ mol H}$

(c) $\dfrac{2.879 \times 10^{-4} \text{ mol C}}{2.866 \times 10^{-4} \text{ mol O}} = 1.005$

$\dfrac{2.866 \times 10^{-4} \text{ mol O}}{2.866 \times 10^{-4} \text{ mol O}} = 1.000$

$\dfrac{3.90 \times 10^{-4} \text{ mol H}}{2.866 \times 10^{-4} \text{ mol O}} = 1.36$

This gives a C:H:O ratio of 1.005 : 1.36 : 1.000. To achieve a whole-number ratio, we multiply each value by 3 to give a ratio of 3.015 : 4.07 : 3.000. This can be rounded to give an empirical formula of $C_3H_4O_3$.

• •

2.7 Two Views of Chemical Equations: Molecules versus Moles

Chemical equations can be used to represent reactions on either the atomic or macroscopic scale. As a result, the following equation can be read in either of two ways.

$$2 \text{ H}_2(g) + \text{O}_2(g) \longrightarrow 2 \text{ H}_2\text{O}(g)$$

• When hydrogen reacts with oxygen, 2 molecules of hydrogen and 1 molecule of oxygen are consumed for every 2 molecules of water produced.

• When hydrogen reacts with oxygen, 2 moles of hydrogen and 1 mole of oxygen are consumed for every 2 moles of water produced.

It doesn't matter whether we think of the reaction in terms of molecules or moles; chemical equations must be balanced. There must be the same number of atoms of each element on both sides of the equation. As a result, the total mass of the reactants will be equal to the total mass of the products of the reaction.

On the atomic scale, the following equation is balanced because the total mass of the reactants in atomic mass units is equal to the total mass of the products.

$$2 \text{ H}_2(g) + \text{O}_2(g) \longrightarrow 2 \text{ H}_2\text{O}(g)$$
$$2 \times 2 \text{ amu} + 32 \text{ amu} \quad\quad 2 \times 18 \text{ amu}$$
$$36 \text{ amu} \quad\quad\quad\quad 36 \text{ amu}$$

On the macroscopic scale, the equation is balanced because the mass of 2 moles of hydrogen and 1 mole of oxygen is equal to the mass of 2 moles of water.

$$2\,H_2(g) + O_2(g) \longrightarrow 2\,H_2O(g)$$
$$2 \times 2\,g + 32\,g \qquad\qquad 2 \times 18\,g$$
$$36\,g \qquad\qquad\qquad 36\,g$$

The following diagram is a useful way of visualizing the relationship between the mass of the starting materials and products of the reaction. The box on the left shows the reactants, and the box on the right shows the products. The box centered above the arrow for the reaction represents the atoms found in either the products or the reactants.

$$\boxed{4\,H + 2\,O}$$

$$\boxed{2\,H_2(g) + O_2(g)} \quad \longrightarrow \quad \boxed{2\,H_2O(g)}$$

If we think about this reaction in terms of H_2 and O_2 molecules combining to form H_2O molecules, the equation is balanced because we have 4 hydrogen atoms and 2 oxygen atoms on both sides of the equation. If we think about the reaction in terms of moles of starting materials and products, the equation must be balanced because we have 4 moles of hydrogen atoms and 2 moles of oxygen atoms on both sides of the arrow.

It is important to recognize that reactions seldom occur by passing through an intermediate stage in which they form isolated atoms. But this approach can be a useful way to emphasize the fact that atoms are conserved in a chemical reaction. Each and every atom among the starting materials must be found in one of the products of the reaction.

2.8   Mole Ratios and Chemical Equations

Science has two fundamental goals: explaining observations about the world around us and predicting what will happen under a particular set of conditions. Any chemical equation explains something about the world, but a balanced chemical equation has the added advantage of allowing us to predict what happens when the reaction takes place.

Consider the reaction that occurs when the rocket fuel known as hydrazine (N_2H_4) burns in air to form N_2 gas and water vapor, as shown in Figure 2.4.

$$N_2H_4(l) + O_2(g) \longrightarrow N_2(g) + 2\,H_2O(g)$$

According to this equation, one mole of oxygen is consumed for each mole of hydrazine that burns when this rocket fuel is ignited. How many moles of oxygen would be needed to consume a sample that contains 5.20 mol of hydrazine?

We can start by transforming the relationship between the moles of hydrazine and oxygen consumed in this reaction into the following **mole ratios.** The numerator and denominator in mole ratios are counting numbers taken from the balanced chemical equation and therefore are exact. In other words they are known to an infinite number of significant figures.

$$\frac{1\ \text{mol}\ O_2}{1\ \text{mol}\ N_2H_4} \quad \text{or} \quad \frac{1\ \text{mol}\ N_2H_4}{1\ \text{mol}\ O_2}$$

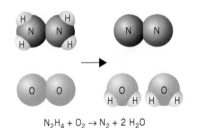

$$N_2H_4 + O_2 \rightarrow N_2 + 2\,H_2O$$

Fig. 2.4   The reaction between N_2H_4 and oxygen. Note that 1 mole of hydrazine reacts with 1 mole of O_2.

To determine the number of moles of oxygen needed to react with 5.20 mol of N_2H_4, we must decide which of these mole ratios to use. Careful attention to the units in which the information in this problem is expressed suggests that we should use the mole ratio on the left because it allows us to convert moles of hydrazine to moles of oxygen.

$$5.20 \text{ mol } N_2H_4 \times \frac{1 \text{ mol } O_2}{1 \text{ mol } N_2H_4} = 5.20 \text{ mol } O_2$$

Exercise 2.8

How many moles of water are formed when 5.20 mol of hydrazine react with excess oxygen?

Solution

Because the hydrazine is being burned in the presence of excess oxygen, this reaction should continue until all of the hydrazine is consumed. In order to predict how much water is produced in the reaction, we need to know the mole ratio of water produced to hydrazine consumed. According to the balanced equation, two moles of water are formed for every mole of hydrazine consumed in this reaction.

$$\frac{2 \text{ mol } H_2O_2}{1 \text{ mol } N_2H_4} \quad \text{or} \quad \frac{1 \text{ mol } N_2H_4}{2 \text{ mol } H_2O}$$

By carefully considering the units of the calculation, we can conclude that the number of moles of water produced in this reaction can be calculated from the following equation.

$$5.20 \text{ mol } N_2H_4 \times \frac{2 \text{ mol } H_2O}{1 \text{ mol } N_2H_4} = 10.4 \text{ mol } H_2O$$

Let's now use what we have learned to see how a balanced equation can be used to predict the amount of O_2 we would have to breathe to digest 10.00 grams of the sugar that flows through our bloodstream. We start with the balanced equation for the reaction between glucose and oxygen to form carbon dioxide and water.

$$C_6H_{12}O_6(aq) + 6\,O_2(g) \longrightarrow 6\,CO_2(g) + 6\,H_2O(l)$$

We then ask the fundamental question: How many moles of $C_6H_{12}O_6$ molecules would be present in a sample of this compound that has a mass of 10.00 grams?

The only way we can convert grams of a substance into moles is to know the ratio of grams per mole in a sample of this substance. In other words, we need to know the molecular weight of glucose. The molecular weight of glucose that was calculated in Exercise 2.3 can be used to construct a pair of unit factors.

$$\frac{1 \text{ mol } C_6H_{12}O_6}{180.16 \text{ g } C_6H_{12}O_6} \quad \text{or} \quad \frac{180.16 \text{ g } C_6H_{12}O_6}{1 \text{ mol } C_6H_{12}O_6}$$

By paying attention to the way the units cancel during the calculation, we can choose the correct conversion factor to convert grams of sugar into moles.

$$10.00 \text{ g C}_6\text{H}_{12}\text{O}_6 \times \frac{1 \text{ mol C}_6\text{H}_{12}\text{O}_6}{180.16 \text{ g C}_6\text{H}_{12}\text{O}_6} = 0.05551 \text{ mol C}_6\text{H}_{12}\text{O}_6$$

We now turn to the balanced equation for the reaction.

$$\text{C}_6\text{H}_{12}\text{O}_6(aq) + 6 \text{ O}_2(g) \longrightarrow 6 \text{ CO}_2(g) + 6 \text{ H}_2\text{O}(l)$$

This equation can be used to construct two mole ratios that describe the relationship between the moles of sugar and moles of oxygen consumed in the reaction.

$$\frac{6 \text{ mol O}_2}{1 \text{ mol C}_6\text{H}_{12}\text{O}_6} \quad \text{or} \quad \frac{1 \text{ mol C}_6\text{H}_{12}\text{O}_6}{6 \text{ mol O}_2}$$

By once again focusing on the units of this problem, we can select the correct mole ratio to convert moles of sugar into an equivalent number of moles of oxygen.

$$0.05551 \text{ mol C}_6\text{H}_{12}\text{O}_6 \times \frac{6 \text{ mol O}_2}{1 \text{ mol C}_6\text{H}_{12}\text{O}_6} = 0.3331 \text{ mol O}_2$$

We now need only one more step to complete our calculation. We need to convert the number of moles of O_2 consumed in the reaction into grams of oxygen. In Exercise 2.4 we concluded that the molecular weight of O_2 is exactly twice the atomic weight of the element. The next step in the calculation therefore involves multiplying the number of moles of O_2 consumed in the reaction by the molecular weight of oxygen.

$$0.3331 \text{ mol O}_2 \times \frac{32.00 \text{ g O}_2}{1 \text{ mol O}_2} = 10.66 \text{ g O}_2$$

We now have the answer to our original question. We need to breathe 10.66 grams of oxygen to digest 10.00 grams of the glucose that we carry through our bloodstream as the source of the energy needed to fuel our bodies.

➤ **CHECKPOINT**

The following reaction has been proposed as a way to produce methanol (CH_3OH) from gases produced by the controlled burning of coal. It is therefore the first step in the process of transforming coal into liquid fuels that could replace the gasoline that fuels today's cars.

$$\text{CO}(g) + 2 \text{ H}_2(g) \longrightarrow \text{CH}_3\text{OH}(g)$$

How many moles of H_2 are needed to consume 2 moles of CO according to the reaction described by the balanced equation? How many H_2 molecules would be needed to consume 2 moles of CO? How many H atoms would be needed?

2.9 Stoichiometry

By now, you have encountered all the steps necessary to do calculations of the sort that are grouped under the heading **stoichiometry.** The goal of these calculations is to use a balanced equation to predict the relationships between the amounts of the reactants and the products of a reaction. There are three steps in these calculations.

- Find the starting material or product of the reaction for which you know both the mass of the sample and the formula of the substance. Use the molecular weight of this substance to convert the number of grams in the sample into an equivalent number of moles.
- Use the balanced equation for the reaction to create a mole ratio that can convert the number of moles of this substance into moles of another component of the reaction.
- Use the molecular weight of the other component of the reaction to convert the number of moles involved in the reaction into grams of that substance.

Exercise 2.9

In theory, we could burn ammonia (NH_3) as a fuel. This is seldom done, however, because using ammonia as a source of nitrogen to fertilize crops is too important. Calculate the number of grams of ammonia (NH_3) needed to prepare 3.00 grams of nitrogen oxide (NO) when it burns in the presence of excess oxygen.

$$4\ NH_3(g) + 5\ O_2(g) \longrightarrow 4\ NO(g) + 6\ H_2O(g)$$

Solution

The only component of this reaction about which we know both the formula of the compound and the mass of the sample is nitrogen oxide. We therefore start by converting 3.00 grams of NO into an equivalent number of moles of the compound. To do this, we need to calculate the molecular weight of NO, which is 30.01 grams/mol, to four significant figures. The number of moles of NO formed in this reaction can therefore be calculated as follows.

$$3.00\ \text{g NO} \times \frac{1\ \text{mol NO}}{30.01\ \text{g NO}} = 0.100\ \text{mol NO}$$

We now use the balanced equation for the reaction to determine the mole ratio that allows us to calculate the number of moles of NH_3 needed to produce 0.100 mole of NO.

$$0.100\ \text{mol NO} \times \frac{4\ \text{mol NH}_3}{4\ \text{mol NO}} = 0.100\ \text{mol NH}_3$$

We then use the molecular weight of NH_3 to calculate the mass of ammonia consumed in the reaction.

$$0.100\ \text{mol NH}_3 \times \frac{17.03\ \text{g NH}_3}{1\ \text{mol NH}_3} = 1.70\ \text{g NH}_3$$

According to this calculation, we need to start with 1.70 grams of ammonia in excess oxygen to obtain 3.00 grams of nitrogen oxide.

➤ **CHECKPOINT**

How many grams of H_2 are required to consume 2 moles of CO in the reaction described by the following balanced equation?

$$CO(g) + 2\ H_2(g) \longrightarrow CH_3OH(g)$$

2.10 The Stoichiometry of the Breathalyzer

A patent was issued in 1958 for the Breathalyzer, which is one method for determining whether an individual is "driving under the influence" (DUI) or "driving while intoxicated" (DWI). The chemistry behind the Breathalyzer is described by the following equation.

$$3\ CH_3CH_3OH(g) + 2\ Cr_2O_7{}^{2-}(aq) + 16\ H^+(aq)$$
$$\longrightarrow 3\ CH_3CO_2H(aq) + 4\ Cr^{3+}(aq) + 11\ H_2O(l)$$

The dichromate ion has a yellow-orange color, whereas aqueous solutions of the Cr^{3+} ion produced in this reaction are green. The extent to which the color of the solution changes is therefore a direct measure of the amount of alcohol in the breath sample. Measurements of the alcohol on the breath are then converted into estimates of the concentration of alcohol in the blood by assuming that 2100 mL of air exhaled from the lungs contains the same amount of alcohol as 1 mL of blood.

An urban myth has been circulating in recent years that assumes someone can "cheat" on a Breathalyzer test by placing a copper penny in their mouth. (Modern folklore apparently suggests that this decreases the amount of alcohol on the breath.) Copper metal will, in fact, catalyze the following reaction, in which the ethanol in alcoholic beverages is oxidized to acetaldehyde.

$$CH_3CH_2OH \xrightarrow{Cu} CH_3CHO + H_2$$

There is only one minor problem—*the copper penny will only catalyze this reaction when it has been heated until it glows red-hot!*

2.11 The Nuts and Bolts of Limiting Reagents

According to Exercise 2.9, we need 1.70 grams of ammonia to make 3.00 grams of nitrogen oxide by the reaction described by the following equation.

$$4 NH_3(g) + 5 O_2(g) \longrightarrow 4 NO(g) + 6 H_2O(g)$$

But we also need something else—we need enough oxygen for the reaction to take place.

Figure 2.5 shows the amount of NO produced when 1.70 grams of ammonia are allowed to react with different amounts of oxygen. At first, the amount of NO produced is directly proportional to the amount of O_2 present when the reaction begins. At some point, however, the yield of the reaction reaches a maximum. No matter how much O_2 we add to the system, no more NO is produced.

We eventually reach a point at which the reaction runs out of NH_3 before all the O_2 is consumed. When this happens, the reaction must stop. No matter how much O_2 is added to the system, we can't get more than 3.00 grams of NO from 1.70 grams of NH_3.

When there isn't enough NH_3 to consume all the O_2 in the reaction, the amount of NH_3 limits the amount of NO that can be produced. Ammonia is therefore the **limiting reagent** in this reaction. Because there is more O_2 than we need, it is the **excess reagent.** The concept of limiting reagent is important because chemists frequently run reactions in which only a limited amount of one of the reactants is present. An analogy might help clarify what goes on in limiting reagent problems.

Let's start with exactly 10 nuts and 10 bolts, as shown in Figure 2.6. How many *NB* "molecules" can be made by screwing 1 nut (*N*) onto each bolt (*B*)?

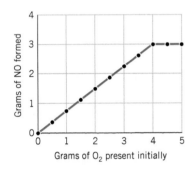

Fig. 2.5 A graph of the amount of NO that can be produced by adding different amounts of O_2 to 1.70 grams of NH_3. Addition of oxygen up to 4.0 grams produces more NO, but from this point on no matter how much O_2 is added, no more NO is produced.

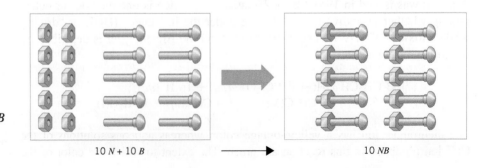

Fig. 2.6 Starting with 10 nuts (*N*) and 10 bolts (*B*), we can make 10 *NB* molecules, with no nuts or bolts left over.

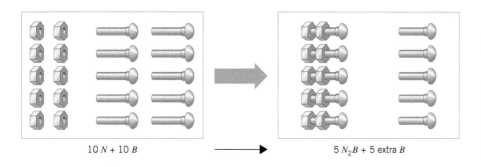

$10\ N + 10\ B$ $5\ N_2B + 5$ extra B

Fig. 2.7 Starting with 10 nuts and 10 bolts, we can make only 5 N_2B molecules, and we will have 5 bolts left over. Because the number of N_2B molecules is limited by the number of nuts, the nuts are the limiting reagent in this analogy.

The answer is obvious: 10. After that, we run out of both nuts and bolts. Because we run out of both nuts and bolts at the same time, neither is a limiting reagent.

Now let's assemble N_2B molecules by screwing 2 nuts onto each bolt. Starting with 10 nuts and 10 bolts, we can make only 5 N_2B molecules, as shown in Figure 2.7. Because we run out of nuts, they must be the limiting reagent. Because 5 bolts are left over, they are the excess reagent.

Let's now extend the analogy to a slightly more difficult problem in which we assemble as many N_2B molecules as possible from a collection of 30 nuts and 20 bolts. There are three possibilities: (1) We have too many nuts and not enough bolts, (2) we have too many bolts and not enough nuts, or (3) we have just the right number of both nuts and bolts.

One way to approach the problem is to pick one of these alternatives and test it. Because there are more nuts (30) than bolts (20), let's assume that we have too many nuts and not enough bolts. In other words, let's assume that bolts are the limiting reagent in this problem. Now let's test that assumption. According to the formula, N_2B, we need 2 nuts for every bolt. Thus, we need 40 nuts to consume 20 bolts.

$$20 \text{ bolts} \times \frac{2 \text{ nuts}}{1 \text{ bolt}} = 40 \text{ nuts}$$

According to this calculation, we need more nuts (40) than we have (30). Thus our original assumption is wrong. We don't run out of bolts; we run out of nuts.

Because our original assumption is wrong, let's turn it around and try the opposite assumption. Now let's assume that nuts are the limiting reagent and calculate the number of bolts we need.

$$30 \text{ nuts} \times \frac{1 \text{ bolt}}{2 \text{ nuts}} = 15 \text{ bolts}$$

Do we have enough bolts to use up all the nuts? Yes, we need only 15 bolts and we have 20 bolts to choose from.

Our second assumption is correct. The limiting reagent in this case is nuts, and the excess reagent is bolts. We can now calculate the number of N_2B molecules that can be assembled from 30 nuts and 20 bolts. Because the limiting reagent is nuts, the number of nuts limits the number of N_2B molecules we can make. Because we get one N_2B molecule for every 2 nuts, we can make a total of 15 of the N_2B molecules.

$$30 \text{ nuts} \times \frac{1 \text{ N}_2\text{B molecule}}{2 \text{ nuts}} = 15 \text{ N}_2\text{B molecules}$$

The following sequence of steps is helpful in working limiting reagent problems.

- Consider the possibility that there might be a limiting amount of one of the reactants.
- Assume that one of the reactants is the limiting reagent.
- See if you have enough of the other reactant to consume the material you have assumed to be the limiting reagent.
- If you do, your original assumption was correct.
- If you don't, assume that another reagent is the limiting reagent and test this assumption.
- Once you have identified the limiting reagent, calculate the amount of product formed.

The white light emitted during fireworks displays is produced by burning magnesium metal.

Exercise 2.10

Magnesium metal burns rapidly in air to form magnesium oxide. This reaction gives off an enormous amount of energy in the form of light and is used in both flares and fireworks. What mass of magnesium oxide (MgO) is formed when 10.0 grams of magnesium react with 10.0 grams of O_2?

Solution

The first step in solving the problem involves writing a balanced equation for the reaction.

$$2\ Mg(s) + O_2(g) \longrightarrow 2\ MgO(s)$$

We then pick one of the reactants and assume it is the limiting reagent. For the sake of argument, let's assume that magnesium is the limiting reagent and O_2 is present in excess. Our immediate goal is to test the validity of this assumption. If it is correct, we will have more O_2 than we need to burn 10.0 grams of magnesium. If it is wrong, O_2 is the limiting reagent.

We start by converting grams of magnesium into moles of magnesium.

$$10.0\ g\ Mg \times \frac{1\ mol\ Mg}{24.31\ g\ Mg} = 0.411\ mol\ Mg$$

We then use the balanced equation to predict the number of moles of O_2 needed to burn this much magnesium. According to the equation for the reaction, it takes 1 mole of O_2 to burn 2 moles of magnesium. We therefore need 0.206 mol of O_2 to consume all of the magnesium.

$$0.411\ mol\ Mg \times \frac{1\ mol\ O_2}{2\ mol\ Mg} = 0.206\ mol\ O_2$$

We now calculate the mass of the O_2 that would be present in 0.206 mol of O_2.

$$0.206\ mol\ O_2 \times \frac{32.00\ g\ O_2}{1\ mol\ O_2} = 6.59\ g\ O_2$$

According to this calculation, we need 6.59 grams of O_2 to burn all the magnesium. Because we have 10.0 grams of O_2, our original assumption was correct. We have more than enough O_2 and only a limited amount of magnesium.

We can now calculate the amount of magnesium oxide formed when all of the limiting reagent is consumed. The balanced equation suggests that 2 moles of MgO are produced for every 2 moles of magnesium consumed. Thus 0.411 mol of MgO can be formed in this reaction.

$$0.411 \text{ mol Mg} \times \frac{2 \text{ mol MgO}}{2 \text{ mol Mg}} = 0.411 \text{ mol MgO}$$

We can use the molar mass of MgO to calculate the number of grams of MgO that can be formed.

$$0.411 \text{ mol MgO} \times \frac{40.30 \text{ g MgO}}{1 \text{ mol MgO}} = 16.6 \text{ g MgO}$$

We can check the result of our calculations by noting that 10.0 grams of magnesium combine with roughly 6.6 grams of O_2 to form 16.6 grams of MgO (3.4 grams of O_2 remain unused). Mass is therefore conserved, and we can feel confident that our calculations are correct.

● ●

➤ **CHECKPOINT**

Assume that 4 grams of $I_2(s)$ are allowed to react with 4 grams of Mg(s) according to the following chemical equation.

$$Mg(s) + I_2(s) \longrightarrow MgI_2(s)$$

What is the limiting reagent in this reaction?

2.12 Density

Imagine that you were given samples of two metals: tin and zinc. At first glance, they look more or less the same. How could you decide which sample was tin and which was zinc? All you would need is a balance to measure the mass of each sample and a way of measuring its volume.

Neither measurement, by itself, would be useful. But the ratio of the mass of the sample to its volume is a characteristic property of the metal known as its **density** (*d*).

$$\text{density} = \frac{\text{mass}}{\text{volume}}$$

The mass of the sample (*m*) is usually expressed in grams and the volume (*V*) in milliliters or cubic centimeters. Because 1 mL is defined as exactly 1 cm^3, densities are reported in units of either g/mL or g/cm^3.

Density is a characteristic property of a substance, as shown by the data in Table 2.1. Tin has a density of 5.75 g/cm^3, whereas the density of zinc is 7.14 g/cm^3. Thus we could identify these metals by weighing a sample, determining its volume, and then calculating the ratio of these measurements in units of g/cm^3.

Alternatively, we could weigh out 1.0 gram of each metal and immerse it in water. The volume of 1.0 gram of zinc would be significantly smaller than 1.0 gram of tin because it is so much denser. The volume of water displaced by the metal would therefore be larger for the sample of tin.

Table 2.1
Densities of Common Metals or Alloys (grams/cm³)

Aluminum	2.70	Molybdenum	10.2
Brass	8.47	Nickel	8.90
Cadmium	8.64	Osmium	22.48
Calcium	1.54	Palladium	12.02
Chromium	7.20	Platinum	21.45
Cobalt	8.9	Potassium	0.86
Copper	8.6	Rhodium	12.4
Gold	19.3	Silver	10.5
Iridium	22.42	Sodium	0.97
Iron	7.86	Tin (gray)	5.75
Lead	11.34	Titanium	4.5
Lithium	0.534	Tungsten	19.3
Magnesium	1.74	Uranium	18.9
Manganese	7.20	Vanadium	6.1
Mercury	13.55	Zinc	7.14

Exercise 2.11

(a) What is the density of gold if a gold brick that has a volume of 10.6 cm³ weighs 205 grams?

(b) What would be the mass in grams of a gold rod that has a volume of 27.6 cm³?

Solution

(a) The density of a substance can be found by dividing the mass of a sample by the volume of the same sample.

$$d = \frac{m}{V}$$

The two pieces of information needed to calculate the density of the gold brick are therefore known.

$$d = \frac{205 \text{ g}}{10.6 \text{ cm}^3} = 19.3 \text{ g/cm}^3$$

(b) By analyzing the units of the calculation, we can see that it is possible to get the mass of the gold rod by multiplying the volume in cm³ by the density in g/cm³.

$$27.6 \text{ cm}^3 \times \frac{19.3 \text{ g}}{1 \text{ cm}^3} = 533 \text{ g}$$

2.13 Solute, Solvent, and Solution

Hydrogen chloride (HCl) and ammonia (NH_3) are gases at room temperature that are notoriously difficult to work with. Chemists have therefore traditionally found it easier to handle these compounds by dissolving them in water to form hydrochloric acid and aqueous ammonia, respectively.

$$HCl(g) \xrightarrow{H_2O} HCl(aq)$$

$$NH_3(g) \xrightarrow{H_2O} NH_3(aq)$$

The result of dissolving one of these gases in water is a **solution.** Solutions are uniform mixtures; the composition is the same throughout the mixture.

Solutions contain two components: a solute and a solvent. The substance that dissolves is the **solute.** The substance in which the solute dissolves is called the **solvent.** Two general rules can be used to decide which component of a solution is the solute and which is the solvent.

- Any reagent that undergoes a change in state when it forms a solution is the *solute*. Thus when gaseous HCl dissolves in water to form an aqueous solution, HCl is the solute.

- If neither component of the solution undergoes a change in state, the component present in the smallest quantity is the *solute*.

Figure 2.8 shows how a solid such as copper(II) sulfate pentahydrate [$Cu(SO_4) \cdot 5 H_2O$] can be dissolved in water to form an aqueous solution. $Cu(SO_4) \cdot 5 H_2O$ is the solute because it dissolves in water, which is the solvent, to form an aqueous solution.

Table 2.2 gives examples of different kinds of solutions. $CuSO_4 \cdot 5 H_2O$ is a dark-blue solid that forms a blue solution when dissolved in water. When H_2 gas dissolves in platinum metal to form a solid solution, it undergoes a change in state, so H_2 is the solute. When liquid mercury dissolves in sodium metal the solution is a solid, which means that the mercury undergoes a change in state and is the solute. Wine that is 12% ethanol (CH_3CH_2OH) by volume is a solution of a small quantity of liquid ethanol (the solute) in a larger volume of liquid water (the solvent). In a 50:50 mixture, either component of the mixture can be thought of as the solute.

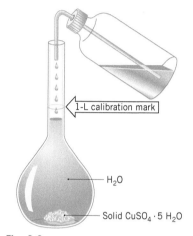

Fig. 2.8 A solution of $CuSO_4$ in water is made by dissolving $CuSO_4 \cdot 5 H_2O$ in a solvent (H_2O)

Table 2.2
Examples of Solutions

Solute	Solvent	Solution
$CuSO_4 \cdot 5 H_2O(s)$	$H_2O(l)$	$CuSO_4 \cdot 5 H_2O(aq)$
$H_2(g)$	$Pt(\underline{s})$	$H_2/Pt(s)$
$Hg(l)$	$Na(s)$	$Na/Hg(s)$
$CH_3CH_2OH(l)$	$H_2O(l)$	Wine

2.14 Concentration

The amount of either the solute or the solvent in a solution depends on the volume of the solution. A liter of concentrated hydrochloric acid obviously contains more HCl than a 1 mL sample. The ratio of the amount of solute to the amount of solvent or solution, however, is a characteristic property of the solution that does not depend on the size of the sample. This quantity is known as the **concentration** of the solution.

$$\text{concentration} = \frac{\text{amount of solute}}{\text{amount of solvent or solution}}$$

The concept of concentration is a common one. We talk about concentrated orange juice, which must be *diluted* with water. We even describe certain laundry products as *concentrated,* which means that we don't have to use as much of them.

2.15   Molarity as a Way to Count Particles in a Solution

Chemists use one concentration unit more than any other: **molarity (*M*).** The molarity of a solution is defined as the moles of solute per liter of solution. Molarity is therefore calculated by dividing the moles of solute in the solution by the volume of the solution in liters.

$$\text{molarity} = \frac{\text{moles of solute}}{\text{liters of solution}}$$

 Exercise 2.12

Copper sulfate is available as blue crystals that contain water molecules coordinated to the Cu^{2+} ions in the crystal. Because the crystals contain five water molecules per Cu^{2+} ion, the compound is called a *pentahydrate,* and the formula is written as $CuSO_4 \cdot 5\,H_2O$. Calculate the molarity of a solution prepared by dissolving 1.25 grams of this compound in enough water to give 50.0 mL of solution.

Solution

A useful strategy for solving problems involves looking at the goal of the problem and asking: What information do we need to have in order to answer the question? The molarity of a solution is calculated by dividing the moles of solute by the volume of the solution. We therefore need two pieces of information: the moles of solute and the volume of the solution in liters.

The volume of the solution in liters can be calculated by using the appropriate conversion factor.

$$50.0 \text{ mL} \times \frac{1 \text{ L}}{1000 \text{ mL}} = 0.0500 \text{ L}$$

The moles of solute can be calculated from the mass of solute used to prepare the solution and the mass of a mole of this compound.

$$1.25 \text{ g } CuSO_4 \cdot 5\,H_2O \times \frac{1 \text{ mol } CuSO_4 \cdot 5\,H_2O}{249.7 \text{ g } CuSO_4 \cdot 5\,H_2O} = 0.00501 \text{ mol } CuSO_4 \cdot 5\,H_2O$$

The molarity of the solution is then calculated by dividing the moles of solute in the solution by the volume of the solution.

$$\frac{0.00501 \text{ mol } CuSO_4 \cdot 5 \text{ H}_2O}{0.0500 \text{ L}} = 0.100 \text{ M } CuSO_4 \cdot 5 \text{ H}_2O$$

• •

Sections 2.2 and 2.3 described how measurements of the mass of a sample of pure substance can be used to determine the number of moles of the substance that are present in a sample. Because solutions are often liquids, it is much easier to measure the *volume of the solution* than the *mass of the solute*. Molarity can then be used to count the number of moles of solute in that volume of a solution.

Molarity has units of moles per liter. The product of the molarity of a solution times its volume in liters is therefore equal to the moles of solute dissolved in the solution.

$$\frac{\text{mol}}{\text{L}} \times \text{L} = \text{mol}$$

We can write this relationship in terms of the following generic equation, where M is molarity, V is volume in liters, and n is moles.

$$M \times V = n$$

Exercise 2.13

How many moles of sodium sulfate, Na_2SO_4, are present in 250 mL of a 0.150 M solution of sodium sulfate?

Solution

We start with the relationship between the molarity of the solution (M), the volume of the solution being studied (V), and the moles of solute in the sample (n).

$$M \times V = n$$

We then substitute the known values of the concentration and the volume of solution into this equation.

$$(0.150 \, M) \times (0.250 \, L) = n$$

Solving for the value of n gives us the number of moles of Na_2SO_4 in this volume of the solution.

$$n = 0.0375 \text{ mol of } Na_2SO_4$$

• •

It is possible to determine the number of atoms, ions, or molecules of solute in a solution using measurements of density and volume if the percent by mass of the solute in the solution is known.

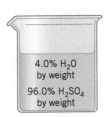

4.0% H₂O
by weight

96.0% H₂SO₄
by weight

Concentrated H₂SO₄
1.84 g/cm³

Fig. 2.9   The relevant information for Exercise 2.14.

Exercise 2.14

Concentrated sulfuric acid is 96.0% H_2SO_4 by mass. (The remaining 4.0% is water.) Calculate the moles of H_2SO_4 in 1.00 liter of concentrated sulfuric acid if the density of this solution is 1.84 grams/cm³.

Solution

The relevant information in this problem is summarized in Figure 2.9. We know the density of the solution, the volume of the solution, the percent by mass due to sulfuric acid, and the goal of the problem: to calculate the concentration in units of moles of sulfuric acid per liter of solution.

Because we know the density in units of grams per cubic centimeter, we might start by calculating the volume of the solution in cubic centimeters.

$$1.00 \text{ L} \times \frac{1000 \text{ mL}}{1 \text{ L}} \times \frac{1 \text{ cm}^3}{1 \text{ mL}} = 1.00 \times 10^3 \text{ cm}^3$$

We can then combine the volume of the solution with the density to calculate the mass of the solution.

$$1.00 \times 10^3 \text{ cm}^3 \times \frac{1.84 \text{ g}}{1 \text{ cm}^3} = 1.84 \times 10^3 \text{ g of } H_2SO_4 \text{ solution}$$

At this point it might be useful to ask: Are we making any progress toward the answer to this exercise? Our goal is to determine the moles of H_2SO_4 in this solution, and we know the total mass of the solution. We might therefore use the fact that this solution is 96.0% H_2SO_4 by mass to calculate the number of grams of H_2SO_4 in 1 liter of the solution.

$$1.84 \times 10^3 \text{ g soln} \times \frac{96.0 \text{ g } H_2SO_4}{100 \text{ g } H_2SO_4 \text{ soln}} = 1.77 \times 10^3 \text{ g of pure } H_2SO_4$$

We can then use the molecular weight of H_2SO_4 to calculate the number of moles of this compound in a liter of the solution.

$$1.77 \times 10^3 \text{ g } H_2SO_4 \times \frac{1.00 \text{ mol } H_2SO_4}{98.1 \text{ g } H_2SO_4} = 18.0 \text{ mol } H_2SO_4$$

According to this calculation, concentrated sulfuric acid contains 18.0 moles of H_2SO_4 per liter.

► **CHECKPOINT**

Explain why chemists might find it more useful to describe concentrated H_2SO_4 as a solution that contains 18.0 moles of solute per liter instead of as a solution that is 96.0% H_2SO_4 by mass.

2.16   Dilution Calculations

Anyone who has ever made a pitcher of orange juice by adding water to a can of frozen orange juice should appreciate that adding more solvent to a solution to decrease the solute concentration is known as **dilution.** Starting with a known volume of a solution of known molarity, we can prepare a more dilute solution of any desired concentration, as shown in Figure 2.10.

Exercise 2.15

Describe how you would prepare 2.50 L of a 0.360 M solution of sulfuric acid (H_2SO_4) starting with concentrated sulfuric acid that is 18.0 M.

Solution

We have one piece of information about the concentrated H_2SO_4 solution—the concentration is 18.0 mol per liter—and two pieces of information about the dilute solution—the volume is 2.50 liters and the concentration is 0.360 mol per liter, as shown in Figure 2.11.

It seems reasonable to start with the solution about which we know the most. If we know the concentration (0.360 M) and the volume (2.50 liters) of the dilute sulfuric acid solution we are trying to prepare, we can calculate the moles of H_2SO_4 it must contain.

$$\frac{0.360 \text{ mol sulfuric acid}}{1 \text{ L}} \times 2.50 \text{ L} = 0.900 \text{ mol sulfuric acid}$$

What volume of concentrated H_2SO_4 would contain the same moles of H_2SO_4? We can start with the equation that describes the relationship between the molarity (M) of a solution, the volume of the solution (V), and the moles of solute in the solution (n).

$$M \times V = n$$

We then substitute into this equation the molarity of the concentrated sulfuric acid solution and the moles of sulfuric acid molecules needed to prepare the dilute solution,

$$\frac{18.0 \text{ mol sulfuric acid}}{1 \text{ L}} \times V = 0.900 \text{ mol sulfuric acid}$$

And then solve this equation for the volume of the solution.

$$V = 0.0500 \text{ L}$$

According to this calculation, we can prepare 2.50 L of 0.360 M H_2SO_4 solution by adding 50.0 mL of concentrated sulfuric acid to enough water to give a total volume of 2.50 L.[1]

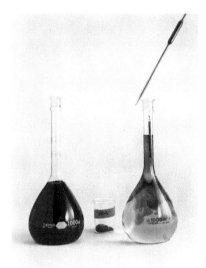

Fig. 2.10 The number of moles of MnO_4^- ion in a solution remains the same when the solution is diluted. The moles of solute removed from the initial solution with the pipet are the same moles of solute found in the final solution. The intensity of the color decreases, however, because the solution becomes more dilute.

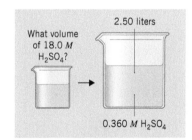

Fig. 2.11 The volume and concentration of the dilute H_2SO_4 solution and the concentration of the concentrated solution in Exercise 2.15.

2.17 Solution Stoichiometry

In Section 2.8 we saw how measurements of mass could be used to understand the relationship between the amounts of starting materials consumed and products generated in a chemical reaction between pure substances. In this section we will see how measurements of molarity can be used to obtain a similar goal for reactions that occur when solutions are mixed.

---

[1]Never add water to concentrated acid; always add the acid to water to avoid accidents.

Oxalic acid ($H_2C_2O_4$) is a natural product often found in plants. The roots and leaves of rhubarb are unusually rich in oxalic acid. The reaction between oxalic acid and sodium hydroxide (NaOH) can be described by the following equation.

$$H_2C_2O_4(aq) + 2\ NaOH(aq) \longrightarrow 2\ Na^+(aq) + C_2O_4^{2-}(aq) + 2\ H_2O(l)$$

Calculate the concentration of an oxalic acid solution if it takes 34.0 mL of an 0.200 M NaOH solution to consume the acid in 25.0 mL of the oxalic acid solution.

The above chemical equation can be rewritten to simplify it. We start by noting that NaOH dissociates into Na^+ and OH^- ions when it dissolves in water.

$$H_2C_2O_4(aq) + 2\ Na^+(aq) + 2\ OH^-(aq) \longrightarrow 2\ Na^+(aq) + C_2O_4^{2-}(aq) + 2\ H_2O(l)$$

Because the 2 $Na^+(aq)$ term appears on both sides of the equation, we can simplify it by writing the **net ionic equation** obtained by removing this term from both sides of the equation.

$$H_2C_2O_4(aq) + 2\ OH^-(aq) \longrightarrow C_2O_4^{2-}(aq) + 2\ H_2O(l)$$

We know only the volume of the oxalic acid solution, but we know both the volume and the concentration of the NaOH solution, as shown in Figure 2.12.

It therefore seems reasonable to start by calculating the number of moles of NaOH in this solution.

$$\frac{0.200\ mol\ NaOH}{1\ L} \times 0.0340\ L = 6.80 \times 10^{-3}\ mol\ NaOH$$

We now know the number of moles of NaOH consumed in the reaction, and we have a balanced chemical equation for the reaction that occurs when the two solutions are mixed. We can therefore calculate the number of moles of $H_2C_2O_4$ needed to consume this much NaOH.

$$6.80 \times 10^{-3}\ mol\ NaOH \times \frac{1\ mol\ H_2C_2O_4}{2\ mol\ NaOH} = 3.40 \times 10^{-3}\ mol\ H_2C_2O_4$$

This calculation gives us the number of moles of $H_2C_2O_4$ in the original oxalic acid solution. Using the volume of this solution given in the statement of the problem, we can calculate the number of moles of oxalic acid per liter of this solution, which is the molarity of the solution.

$$\frac{3.40 \times 10^{-3}\ mol\ H_2C_2O_4}{0.0250\ L} = 0.136\ M\ H_2C_2O_4$$

The oxalic acid solution therefore has a concentration of 0.136 mol per liter.

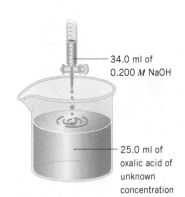

34.0 ml of 0.200 M NaOH

25.0 ml of oxalic acid of unknown concentration

Fig. 2.12 Addition of sodium hydroxide to oxalic acid.

Exercise 2.16

Calculate the volume of 1.50 M HCl that would consume 25.0 grams of $CaCO_3$ according to the following balanced equation.

$$CaCO_3(s) + 2\ HCl(aq) \longrightarrow CaCl_2(aq) + CO_2(g) + H_2O(l)$$

Solution

We have only one piece of information about the HCl solution—its concentration is 1.50 M—and only one piece of information about the CaCO$_3$—it has a mass of 25.0 grams, as shown in Figure 2.13.

Which of these numbers do we start with? There is nothing we can do with the concentration of the HCl solution unless we know either the number of moles of HCl consumed in the reaction or the volume of the solution. We can work with the mass of CaCO$_3$ consumed in the reaction, however. We can start by converting grams of CaCO$_3$ into moles of CaCO$_3$.

$$25.0 \text{ g CaCO}_3 \times \frac{1 \text{ mol CaCO}_3}{100.1 \text{ g CaCO}_3} = 0.250 \text{ mol CaCO}_3$$

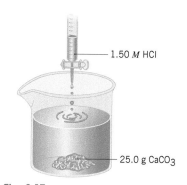

Fig. 2.13 The reaction between hydrochloric acid and calcium carbonate described in Exercise 2.16.

We then ask: Does this information get us any closer to our goal of calculating the volume of HCl consumed in this reaction? We know the number of moles of CaCO$_3$ present initially, and we have a balanced equation that states that 2 moles of HCl are consumed for every mole of CaCO$_3$. We can therefore calculate the number of moles of HCl consumed in the reaction.

$$0.250 \text{ mol CaCO}_3 \times \frac{2 \text{ mol HCl}}{1 \text{ mol CaCO}_3} = 0.500 \text{ mol HCl}$$

We now know the number of moles of HCl (0.500 mol) consumed in the reaction and the molarity of the solution (1.50 M). The number of moles and the molarity are related to volume by the following equation.

$$M \times V = n$$

Substituting the known values of the concentration of the HCl solution and the number of moles of HCl into this equation gives the following result.

$$\frac{1.50 \text{ mol HCl}}{1 \text{ L}} \times V = 0.500 \text{ mol HCl}$$

We then solve the equation for the volume of the solution that would contain this amount of HCl.

$$V = 0.333 \text{ L}$$

According to this calculation, we need 333 mL of 1.50 M HCl to consume 25.0 grams of CaCO$_3$.

• •

With a little imagination, the concept of concentration can be used to do far more interesting calculations.

 •

Exercise 2.17

Assume that a metal, M, reacts with hydrochloric acid according to the following balanced equation.

$$M(s) + 2 \text{ HCl}(aq) \longrightarrow M^{2+}(aq) + 2 \text{ Cl}^-(aq) + H_2(g)$$

Calculate the molar mass of the metal if 125 mL of 0.200 *M* HCl reacts with 0.304 gram of the metal.

Solution

At first glance, it seems that we don't have enough information to solve this problem. The only way to proceed with such a question is to start by identifying what we know, do what can be done, and see where this leads us. In other words, start by exploring the problem. What do we know?

- The metal reacts with hydrochloric acid according to the balanced equation given above.
- We start with 0.304 gram of the metal.
- It takes 125 mL of 0.200 *M* HCl to consume the metal.

What can we do with this information? We know the volume (125 mL) and the concentration (0.200 *M*) of a solution. We might therefore start by calculating the number of moles of solute in this solution.

$$\frac{0.200 \text{ mol HCl}}{1 \text{ L}} \times 0.125 \text{ L} = 0.0250 \text{ mol HCl}$$

Now what? We know the number of moles of HCl, and we have a balanced equation. Furthermore, we are interested in one of the properties of the metal. It seems reasonable to convert moles of HCl consumed in this reaction into moles of metal consumed.

$$0.0250 \text{ mol HCl} \times \frac{1 \text{ mol M}}{2 \text{ mol HCl}} = 0.0125 \text{ mol M}$$

It is important never to lose sight of the goal of the problem. In this case, the problem asks for the molar mass of the metal. It might be useful to go to the end of the problem and work backward. Molar mass has units of grams per mole. If we knew both the number of grams and the number of moles of metal in a sample, we could calculate the molar mass of the metal.

But we already have that information. We know the number of moles of metal (0.0125 mol) in a sample of known mass (0.304 gram). The ratio of these numbers is the molar mass of the metal.

$$\frac{0.304 \text{ g M}}{0.0125 \text{ mol M}} = 24.3 \text{ g/mol}$$

By looking at a table of atomic weights, we can deduce that the metal is magnesium.

Key Terms

Aqueous	Concentration	Excess reagent
Avogadro's constant	Density	Limiting reagent
Avogadro's number	Dilution	Molar mass
Chemical equation	Empirical formula	Molarity (*M*)

Mole
Mole ratio
Molecular formula
Molecular weight

Net ionic equation
Percent by mass
Solute

Solution
Solvent
Stoichiometry

Problems

The Mole as the Bridge between the Macroscopic and Atomic Scales

1. If a new scale of atomic weights was defined based on the assumption that the mass of a ^{12}C atom was exactly 1 amu, what would be the atomic weight of neon?

2. Identify the element that has an atomic weight 4.33 times as large as carbon.

3. Give the atomic weight and the molar mass of the following atoms.

    (a) Li   (b) C   (c) Mg   (d) Cu

4. Which of the following pairs of elements contains the same number of atoms?

    (a) 12.011 grams C, 12.011 grams Na

    (b) 22.99 grams Na, 12.011 grams C

    (c) 39.10 grams K, 9.012 grams Be

    (d) 85.47 grams Rb, 6.941 grams Li

5. Which would weigh the most, 1000 Al atoms or 1000 Si atoms?

6. What is wrong with the following question: Which contains the most atoms, a mole of Fe or a mole of Cu?

7. How many grams of each of the following would be required to make 1 mole?

    (a) Ca   (b) Sr   (c) Se   (d) Ge

8. Of each pair of the following atoms, pick the heavier atom.

    (a) Ni, Co   (b) Zn, Al   (c) Ga, Ge   (d) Pb, Sn

9. If the average mass of a chromium atom is 51.996 amu, what is the mass of a mole of chromium atoms?

10. If the average sulfur atom is approximately twice as heavy as the average oxygen atom, what is the ratio of the mass of a mole of sulfur atoms to the mass of a mole of oxygen atoms?

11. Calculate the mass in grams of a mole of atoms of the following elements.

    (a) C   (b) Ni   (c) Hg

12. If eggs sell for $0.90 a dozen, what does it cost to buy 2.5 dozen eggs? If the molar mass of carbon is 12.011 grams, what is the mass of 2.5 moles of carbon atoms?

The Mole as a Collection of Atoms

13. What would be the value of Avogadro's number if a mole were defined as the number of ^{12}C atoms in 12 lb of ^{12}C?

14. Calculate the number of atoms in 16 grams of O_2, 31 grams of P_4, and 32 grams of S_2.

15. Benzaldehyde has the pleasant, distinctive odor of almonds. What is the weight of a mole of benzaldehyde if a single molecule has a mass of 1.762×10^{-22} grams?

16. What is the mass in grams of one ^1H atom? Of one ^{12}C atom?

17. What is the mass in grams of 4.35×10^6 atoms of ^{12}C?

18. What is the mass in grams of 6.022×10^{23} atoms of ^{12}C?

19. What is the mass in grams of a molecule of carbon dioxide that has one ^{12}C atom and two ^{16}O atoms?

Converting Grams into Moles and Number of Atoms

20. How many atoms are there in 25.0 grams of Sn?

21. Calculate the mass in grams of a sample of copper metal that contains 1.65 mol of copper atoms.

22. How many grams of lithium contain 4.56×10^{23} atoms of lithium?

23. 2.0 mol of silver metal contains how many silver atoms?

24. How many atoms are there in

    (a) 1 mol of Si?   (b) 2 mol of Si?

    (c) 0.5 mol of Si   (c) 0.10 mol of Si

25. What is the mass of an atom of iron in grams?

26. If one atom of ^1H weighs 1.6735×10^{-24} grams, what is the mass of 6.022×10^{23} atoms of ^1H?

The Mole as a Collection of Molecules

27. How many carbon and hydrogen atoms could be found in a sample of one dozen methane, CH_4, molecules? In one mole of methane molecules?

28. What is the average mass in amu of one methane molecule? What is the mass in grams of one mole of methane?

29. How many hydrogen atoms are present in 1.00 mol of hydrogen gas, H_2? How many H_2 molecules? What is the mass of the sample?

30. Indicate whether each of the following statements is true or false, and explain your reasoning.

    (a) One mole of NH_3 weighs more than 1 mol of H_2O.

    (b) There are more carbon atoms in 48 grams of CO_2 than in 12 grams of diamond (a pure form of carbon).

(c) There are equal numbers of nitrogen atoms in 1 mol of NH_3 and 1 mol of N_2.

(d) The number of Cu atoms in 100 grams of Cu(s) is the same as the number of Cu atoms in 100 gram of copper(II) oxide, CuO.

(e) The number of Ni atoms in 1 mol of Ni(s) is the same as the number of Ni atoms in 1 mol of nickel(II) chloride, $NiCl_2$.

31. Which pair of samples contains the same number of hydrogen atoms?
    (a) 1 mol of NH_3 and 1 mol of N_2H_4
    (b) 2 mol of NH_3 and 1 mol of N_2H_4
    (c) 2 mol of NH_3 and 3 mol of N_2H_4
    (d) 4 mol of NH_3 and 3 mol of N_2H_4

32. Which of the following contains the largest number of carbon atoms?
    (a) 0.10 mol of acetic acid, CH_3CO_2H
    (b) 0.25 mol of carbon dioxide, CO_2
    (c) 0.050 mol of glucose, $C_6H_{12}O_6$
    (d) 0.0010 mol of sucrose, $C_{12}H_{22}O_{11}$

33. Calculate the molecular weights of formic acid, HCO_2H, and formaldehyde, H_2CO.

34. Calculate the molecular weight of the following compounds.
    (a) methane, CH_4
    (b) glucose, $C_6H_{12}O_6$
    (c) diethyl ether, $(CH_3CH_2)_2O$
    (d) thioacetamide, CH_3CSNH_2

35. Calculate the molecular weight of the following compounds.
    (a) tetraphosphorus decasulfide, P_4S_{10}
    (b) nitrogen dioxide, NO_2
    (c) zinc sulfide, ZnS
    (d) potassium permanganate, $KMnO_4$

36. Calculate the molecular weight of the following compounds.
    (a) chromium hexacarbonyl, $Cr(CO)_6$
    (b) iron(III) nitrate, $Fe(NO_3)_3$
    (c) potassium dichromate, $K_2Cr_2O_7$
    (d) calcium phosphate, $Ca_3(PO_4)_2$

37. Root beer hasn't tasted the same since the FDA outlawed the use of sassafras oil as a food additive because sassafras oil is 80% safrole, which has been shown to cause cancer in rats and mice. Calculate the molecular weight of safrole, $C_{10}H_{10}O_2$.

38. MSG ($C_5H_8NNaO_4$) is a spice used in Chinese cooking that causes some people to feel light-headed (a disorder known as *Chinese restaurant syndrome*). Calculate the molecular weight of MSG.

39. Calculate the molecular weight of the active ingredients in the following prescription drugs.
    (a) Darvon, $C_{22}H_{30}ClNO_2$
    (b) Valium, $C_{16}H_{13}ClN_2O$
    (c) Tetracycline, $C_{22}H_{24}N_2O_8$

40. Calculate the atomic weight of platinum if 0.8170 mol of the metal has a mass of 159.4 grams.

41. Calculate the mass of 0.0582 mol of carbon tetrachloride, CCl_4.

42. Calculate the moles of carbon tetrachloride, CCl_4, in 100 grams of CCl_4.

43. Calculate the number of moles in 5.72 grams of Al.

Percent by Mass

44. Calculate the percent by mass of chromium in each of the following oxides.
    (a) CrO   (b) Cr_2O_3 (c) CrO_3

45. Calculate the percent by mass of nitrogen in the following fertilizers.
    (a) $(NH_4)_2SO_4$ (b) KNO_3
    (c) $NaNO_3$ (d) $(H_2N)_2CO$

46. Calculate the percent by mass of carbon, hydrogen, and chlorine in DDT, $C_{14}H_9Cl_5$.

47. Emeralds are gem-quality forms of the mineral beryl, $Be_3Al_2(SiO_3)_6$. Calculate the percent by mass of silicon in beryl.

48. Osteoporosis is a disease common in older women who have not had enough calcium in their diets. Calcium can be added to the diet by tablets that contain either calcium carbonate ($CaCO_3$), calcium sulfate ($CaSO_4$), or calcium phosphate [$Ca_3(PO_4)_2$]. On a per-gram basis, which is the most efficient way of getting Ca^{2+} ions into the body?

Determining the Formula of a Compound

49. Calculate the number of moles of carbon atoms in 0.244 gram of calcium carbide, CaC_2.

50. Calculate the number of moles of phosphorus in 15.95 grams of tetraphosphorus decaoxide, P_4O_{10}.

51. Calculate the number of chlorine atoms in 0.756 gram of K_2PtCl_6.

52. Calculate the number of oxygen atoms in the following samples.
    (a) 0.100 mol of potassium permanganate, $KMnO_4$
    (b) 0.25 mol of dinitrogen pentoxide, N_2O_5
    (c) 0.45 mol of penicillin, $C_{16}H_{17}N_2O_5SK$

53. A molecule containing only nitrogen and oxygen contains 36.8% N by mass.
    (a) How many grams of N would be found in a 100-gram sample of the compound? How many grams of O would be found in the same sample?

(b) How many moles of N would be found in a 100-gram sample of the compound? How many moles of O would be found in the same sample?

(c) What is the ratio of the number of moles of O to the number of moles of N?

(d) What is the empirical formula of the compound?

54. Stannous fluoride, or "Fluoristan," is added to toothpaste to help prevent tooth decay. What is the empirical formula for stannous fluoride if the compound is 24.25% F and 75.75% Sn by mass?

55. Iron reacts with oxygen to form three compounds: FeO, Fe_2O_3, and Fe_3O_4. One of these compounds, known as magnetite, is 72.36% Fe and 27.64% O by mass. What is the formula of magnetite?

56. The most abundant ore of manganese is an oxide known as pyrolusite, which is 36.8% O and 63.2% Mn by mass. Which of the following oxides of manganese is pyrolusite?

(a) MnO   (b) MnO_2 (c) Mn_2O_3

(d) MnO_3 (e) Mn_2O_7

57. Nitrogen combines with oxygen to form a variety of compounds, including N_2O, NO, NO_2, N_2O_3, N_2O_4, and N_2O_5. One of these compounds is called nitrous oxide, or "laughing gas." What is the formula of nitrous oxide if this compound is 63.65% N and 36.35% O by mass?

58. Chalcopyrite is a bronze-colored mineral that is 34.59% Cu, 30.45% Fe, and 34.96% S by mass. Calculate the empirical formula for the mineral.

59. A compound of xenon and fluorine is found to be 53.5% xenon by mass. What is the empirical formula of the compound?

60. In 1914, E. Merck and Company synthesized and patented a compound known as MDMA as an appetite suppressant. Although it was never marketed, it has reappeared in recent years as a street drug known as ecstasy. What is the empirical formula of the compound if it contains 68.4% C, 7.8% H, 7.2% N, and 16.6% O by mass?

61. What is the empirical formula of the compound that contains 0.483 gram of nitrogen and 1.104 grams of oxygen?

(a) N_2O (b) NO   (c) NO_2

(d) N_2O_3 (e) N_2O_4

62. What is the empirical formula of the compound formed when 9.33 grams of copper metal react with excess chlorine to give 14.54 grams of the compound?

63. Is it possible to determine the molecular formula of a compound solely from its percent composition? Why or why not?

64. β-Carotene is the protovitamin from which nature builds vitamin A. It is widely distributed in the plant and animal kingdoms, always occurring in plants together with chlorophyll. Calculate the molecular formula for β-carotene if the compound is 89.49% C and 10.51% H by mass and its molecular weight is 536.89 grams/mol.

65. The phenolphthalein used as an indicator in acid–base titrations has also been used as the active ingredient in laxatives such as ExLax. Calculate the molecular formula for phenolphthalein if the compound is 75.46% C, 4.43% H, and 20.10% O by mass and has a molecular weight of 318.31 grams per mole.

66. Caffeine is a central nervous system stimulant found in coffee, tea, and cola nuts. Calculate the molecular formula of caffeine if the compound is 49.48% C, 5.19% H, 28.85% N, and 16.48% O by mass and has a molecular weight of 194.2 grams per mole.

67. Aspartame, also known as NutraSweet, is 160 times sweeter than sugar when dissolved in water. The true name for this artificial sweetener is N-L-α-aspartyl-L-phenylalanine methyl ester. Calculate the molecular formula of aspartame if the compound is 57.14% C, 6.16% H, 9.52% N, and 27.18% O by mass and has a molecular weight of 294.30 grams per mole.

Two Views of Chemical Equations: Molecules Versus Moles

68. Give two ways of interpreting the following equation.

$$H_2(g) + Cl_2(g) \longrightarrow 2\,HCl(g)$$

69. Show that the following equation is balanced by calculating the masses of the products and reactants in both amu and grams.

$$3Ca(s) + N_2(g) \longrightarrow Ca_3N_2(s)$$

70. 2.0 mol of $H_2(g)$ is mixed with 1.0 mol of $O_2(g)$ and allowed to react as shown in Sect. 2.7. How many atoms of H are initially present? How many atoms of O are initially present? How many atoms of H and O will there be in the product?

(a) How many moles of H_2O will be formed if all the H_2 and O_2 react?

(b) How many molecules of H_2 and O_2 were initially present?

(c) How many molecules of H_2O were formed?

Mole Ratios and Chemical Equations

71. How many moles of CO_2 are produced when 5 moles of O_2 are consumed in the following reaction?

$$2\,CO(g) + O_2(g) \longrightarrow 2\,CO_2(g)$$

72. Does the total number of moles of gas present increase, decrease, or remain the same when the following reaction occurs?

$$2\ CO(g) + O_2(g) \longrightarrow 2\ CO_2(g)$$

73. How many moles of CuO would be required to produce 12 mol of copper metal in the following reaction?

$$CuO(s) + H_2(g) \longrightarrow Cu(s) + H_2O(g)$$

74. Carbon disulfide burns in oxygen to form carbon dioxide and sulfur dioxide.

$$CS_2(l) + 3\ O_2(g) \longrightarrow CO_2(g) + 2\ SO_2(g)$$

Calculate the number of O_2 molecules it would take to consume 500 molecules of CS_2. Calculate the number of moles of O_2 it would take to consume 5.00 moles of CS_2.

75. Calculate the number of moles of oxygen produced when 6.75 moles of manganese (IV) oxide decompose to form Mn_3O_4 and O_2.

$$3\ MnO_2(s) \longrightarrow Mn_3O_4(s) + O_2(g)$$

76. Calculate the number of moles of carbon monoxide needed to reduce 3.00 mol of iron(III) oxide to iron metal.

$$Fe_2O_3(s) + 3\ CO(g) \longrightarrow 2\ Fe(s) + 3\ CO_2(g)$$

Stoichiometry

77. Describe the steps needed to calculate the number of grams of CO_2 produced in the following reaction if x grams of CO are consumed.

$$2\ CO(g) + O_2(g) \longrightarrow 2\ CO_2(g)$$

78. Calculate the mass of oxygen released when enough mercury(II) oxide decomposes to give 25 grams of liquid mercury.

$$2\ HgO(s) \longrightarrow 2\ Hg(l) + O_2(g)$$

79. Calculate the mass of CO_2 produced and the mass of oxygen consumed when 10.0 grams of methane (CH_4) are burned in oxygen to produce CO_2 and H_2O.

80. How many pounds of sulfur react with 10.0 pounds of zinc to form zinc sulfide, ZnS?

81. Calculate the mass of oxygen that can be prepared by decomposing 25.0 grams of potassium chlorate.

$$2\ KClO_3(s) \longrightarrow 2\ KCl(s) + 3\ O_2(g)$$

82. Predict the formula of the compound produced when 1.00 gram of chromium metal reacts with 0.923 grams of oxygen, O_2.

83. Ethanol, or ethyl alcohol, is produced by the fermentation of sugars such as glucose.

$$C_6H_{12}O_6(aq) \longrightarrow 2\ C_2H_5OH(aq) + 2\ CO_2(g)$$

Calculate the number of kilograms of alcohol that can be produced from 1.00 kilogram of glucose.

84. Calculate the number of pounds of aluminum metal that can be obtained from 1.000 ton of bauxite, $Al_2O_3 \cdot 2\ H_2O$.

85. Calculate the mass of phosphine, PH_3, that can be prepared when 10.0 grams of calcium phosphide, Ca_3P_2, react with excess water.

$$Ca_3P_2(s) + 6\ H_2O(l) \longrightarrow 3\ Ca(OH)_2(aq) + 2\ PH_3(g)$$

86. Hydrogen chloride can be obtained by reacting phosphorus trichloride with excess water and then boiling the HCl gas out of the solution.

$$PCl_3(g) + 3\ H_2O(l) \longrightarrow 3\ HCl(aq) + H_3PO_3(aq)$$

Calculate the mass of HCl gas that can be prepared from 15.0 grams of PCl_3.

87. Nitrogen reacts with hydrogen to form ammonia,

$$N_2(g) + 3\ H_2(g) \longrightarrow 2\ NH_3(g)$$

which burns in the presence of oxygen to form nitrogen oxide,

$$4\ NH_3(g) + 5\ O_2(g) \longrightarrow 4\ NO(g) + 6\ H_2O(l)$$

which reacts with excess oxygen to form nitrogen dioxide,

$$2\ NO(g) + O_2(g) \longrightarrow 2\ NO_2(g)$$

which dissolves in water to give nitric acid,

$$3\ NO_2(g) + H_2O(l) \longrightarrow 2\ HNO_3(aq) + NO(g)$$

Calculate the mass of nitrogen needed to make 150 grams of nitric acid, assuming an excess of all other reactants.

The Nuts and Bolts of Limiting Reagents

88. Calculate the number of water molecules that can be prepared from 500 H_2 molecules and 500 O_2 molecules.

$$2\ H_2(g) + O_2(g) \longrightarrow 2\ H_2O(l)$$

What would happen to the potential yield of water molecules if the amount of O_2 were doubled? What if the amount of H_2 were doubled?

89. Calculate the number of moles of P_4S_{10} that can be produced from 0.500 mol of P_4 and 0.500 mol of S_8.

$$4 P_4(s) + 5 S_8(s) \longrightarrow 4 P_4S_{10}(s)$$

What would happen to the potential yield of P_4S_{10} if the amount of P_4 were doubled? What if the amount of S_8 were doubled?

90. Calculate the number of moles of nitrogen dioxide, NO_2, that could be prepared from 0.35 mol of nitrogen oxide and 0.25 mol of oxygen.

$$2 NO(g) + O_2(g) \longrightarrow 2 NO_2(g)$$

Identify the limiting reagent and the excess reagent in the reaction. What would happen to the potential yield of NO_2 if the amount of NO were increased? What if the amount of O_2 were increased?

91. Calculate the mass of hydrogen chloride that can be produced from 10.0 grams of hydrogen and 10.0 grams of chlorine.

$$H_2(g) + Cl_2(g) \longrightarrow 2 HCl(g)$$

What would have to be done to increase the amount of hydrogen chloride produced in the reaction?

92. Calculate the mass of calcium nitride, Ca_3N_2, that can be prepared from 54.9 grams of calcium and 43.2 grams of nitrogen.

$$3 Ca(s) + N_2(g) \longrightarrow Ca_3N_2(s)$$

93. PF_3 reacts with XeF_4 to give PF_5.

$$2 PF_3(g) + XeF_4(s) \longrightarrow 2 PF_5(g) + Xe(g)$$

How many moles of PF_5 can be produced from 100.0 grams of PF_3 and 50.0 grams of XeF_4?

94. Trimethyl aluminum, $Al(CH_3)_3$, must be handled in an apparatus from which oxygen has been rigorously excluded because the compound bursts into flame in the presence of oxygen. Calculate the mass of trimethyl aluminum that can be prepared from 5.00 grams of aluminum metal and 25.0 grams of dimethyl mercury.

$$2 Al(s) + 3 Hg(CH_3)_2(l) \longrightarrow 2 Al(CH_3)_3(l) + 3 Hg(l)$$

95. The thermite reaction, used to weld rails together in the building of railroads, is described by the following equation.

$$Fe_2O_3(s) + 2 Al(s) \longrightarrow Al_2O_3(s) + 2 Fe(l)$$

Calculate the mass of iron metal that can be prepared from 150 grams of aluminum and 250 grams of iron(III) oxide.

Density

96. Which weighs the most, 10.0 cm^3 of iron or 5.0 cm^3 of silver? The density of iron is 7.9 grams/cm^3 and that of silver is 10.5 grams/cm^3.

97. Two strips of metal each weighing 100 grams are placed into a cylinder containing water. The volume of water displaced is 8.8 mL in one case and 37.0 mL in the other case. Identify the two metal strips.

Metal	Au	Fe	Al	Pb	Ag
Density (grams/mL)	19.3	7.9	2.7	11.3	10.5

98. What is the density of a metal that has a volume of 20.0 mL and weighs 271 grams?

99. Mercury is a liquid metal at room temperature with a density of 13.6 grams/mL. What is the mass in grams of 5.6 mL of mercury?

Solute, Solvent, and Solution

100. Define the terms *solution, solvent,* and *solute* and give an example of each.

101. Which of the following are solutions?
    (a) chicken noodle soup (b) air
    (c) wine (d) table salt (NaCl)

Concentration and Molarity as a Way of Counting Particles in Solution

102. Describe in detail the steps you would take to prepare 125 mL of 0.745 *M* oxalic acid, starting with solid oxalic acid dihydrate ($H_2C_2O_4 \cdot 2 H_2O$). Describe the glassware you would need, the chemicals, the amounts of each chemical, and the sequence of steps you would take.

103. Hydrochloric acid was once known as muriatic acid because it was the "marine" acid—it was made from seawater. Muriatic acid is still sold in many hardware stores for cleaning bricks and tile. What is the molarity of this solution if 125 mL contains 27.3 grams of HCl?

104. Silver chloride is only marginally soluble in water; only 0.00019 gram of AgCl dissolves in 100 mL of water. Calculate the molarity of this solution.

105. Ammonia (NH_3) is relatively soluble in water. Calculate the molarity of a solution that contains 252 grams of NH_3 per liter.

106. During a physical exam, one of the authors was found to have a cholesterol level of 1.60 milligrams per deciliter (0.100 L). If the molecular weight of cholesterol is 386.67 grams per mole, what is the cholesterol level in his blood in units of moles per liter?

107. At 25°C, 5.77 grams of chlorine gas dissolve in 1.00 liter of solution. Calculate the molarity of Cl_2 in this solution.

108. Calculate the mass of Na_2SO_4 needed to prepare 0.500 L of a 0.150 M solution.

109. When asked to prepare a liter of 1.00 M K_2CrO_4, a student weighed out exactly 1.00 mole of K_2CrO_4 and added this solid to 1.00 L of water in a volumetric flask. What did the student do wrong? Did the student get a solution that was more concentrated than 1.00 M or less concentrated than 1.00 M? How would you prepare the solution?

110. You can make the chromic acid bath commonly used to clean glassware in the lab by dissolving 92 grams of sodium dichromate ($Na_2Cr_2O_7 \cdot 2 H_2O$) in enough water to give 458 mL of solution and then adding 800 mL of concentrated sulfuric acid. Calculate the molarity of the $Cr_2O_7^{2-}$ ion in the solution.

111. People who smoke marijuana can be detected by looking for the tetrahydrocannabinols (THC) that are the active ingredient in marijuana. The present limit on detection of THC in urine is 20 nanograms of THC per milliliter of urine (20 ng/mL). Calculate the molarity of the solution at the limit of detection if the molecular weight of THC is 315 grams/mol.

112. What is the molarity of a solution formed by the dissolution of 1.25 grams of KCl in 500 mL of solution? How could you make a solution that is twice as concentrated?

113. If 2.75 grams of $AgNO_3$ are dissolved in 250 mL of solution, what is the molarity of the solution? How could you prepare a solution that is half as concentrated?

114. How many grams of NaOH would need to be dissolved in 250.0 mL of solution to produce a 1.25 M solution?

115. 500 mL of 0.50 M solution of NaOH contain how many moles of NaOH?

116. How many grams of NaOH are in 500 mL of a 0.50 M solution?

117. Which is more concentrated, 500 mL of a 0.20 M solution of NaCl or 250 mL of a 0.25 M solution of NaCl?

118. What is the molarity of a solution containing 0.25 gram of $CuSO_4$ in 125 mL of solution? Describe how you would prepare a solution of this molarity.

Dilution Calculations

119. 0.275 gram of $AgNO_3$ is dissolved in 500 mL of solution.
    (a) What is the molarity of this solution?
    (b) If 10.0 mL of this solution are transferred to a flask and diluted to 500 mL, what is the concentration of the resulting solution?
    (c) If 10.0 mL of the solution in (b) are transferred to a flask and diluted to 250 mL, what is the concentration of the resulting solution?

120. Describe how you would prepare 500 mL of a 0.10 M solution of HCl from a 12.0 M solution of HCl.

121. A 0.050 M solution of $CuSO_4$ is diluted to double the volume. What is the concentration of the new solution?

122. 100.0 mL of an 18.0 M solution of H_2SO_4 are transferred to a flask and diluted to 500.0 mL. What is the concentration of this new solution?

123. It is desired to prepare 250 mL of a 0.10 M solution of HCl from a 6.0 M solution of HCl in water. Describe how to do this.

124. 100.0 mL of a 0.050 M solution of NaCl in water are diluted to 250.0 mL. What is the final concentration of the solution?

125. To what final volume must 100 mL of a 1.20 M solution of KF be diluted to produce a 0.45 M solution?

126. 1.00 L of a 1.0 M solution of a sugar is diluted to 1.75 L. What is the final concentration?

127. Calculate the volume of 17.4 M acetic acid needed to prepare 1.00 L of 3.00 M acetic acid.

128. Calculate the concentration of the solution formed when 15.0 mL of 6.00 M HCl are diluted with 25.0 mL of water.

129. Describe how you would prepare 0.200 liter of 1.25 M nitric acid from a solution that is 5.94 M HNO_3.

Solution Stoichiometry

130. Calculate the concentration of an aqueous KCl solution if 25.00 mL of this solution give 0.430 grams of AgCl when treated with excess $AgNO_3$.

$$KCl(aq) + AgNO_3(aq) \longrightarrow AgCl(s) + KNO_3(aq)$$

131. Calculate the volume of 0.25 M NaI that would be needed to react with all of the Hg^{2+} ion from 45 mL of a 0.10 M $Hg(NO_3)_2$ solution.

$$2 NaI(aq) + Hg(NO_3)_2(aq)$$
$$\longrightarrow HgI_2(s) + 2 NaNO_3(aq)$$

132. Calculate the molarity of an acetic acid (CH_3CO_2H) solution if 34.57 mL of the solution are needed to react with 25.19 mL of 0.1025 M sodium hydroxide.

$$CH_3CO_2H(aq) + NaOH(aq)$$
$$\longrightarrow Na^+(aq) + CH_3CO_2^-(aq) + H_2O(l)$$

133. Calculate the molarity of a sodium hydroxide solution if 10.42 mL of this solution are needed to react with 25.00 mL of 0.2043 M oxalic acid ($H_2C_2O_4$).

$$H_2C_2O_4(aq) + 2 NaOH(aq)$$
$$\longrightarrow Na_2C_2O_4(aq) + 2 H_2O(l)$$

134. Calculate the volume of 0.0985 M sulfuric acid (H_2SO_4) that would be needed to react with 10.89 mL of a 0.01043 M aqueous ammonia (NH_3) solution.

$$H_2SO_4(aq) + 2\,NH_3(aq) \longrightarrow (NH_4)_2SO_4(aq)$$

135. α-D-Glucopyranose reacts with the periodate ion (IO_4^-) as follows.

$$C_6H_{12}O_6(aq) + 5\,IO_4^-(aq)$$
$$\longrightarrow 5\,IO_3^-(aq) + 5\,HCO_2H(aq) + H_2CO(aq)$$

Calculate the molarity of the glucopyranose solution if 25.0 mL of 0.750 M IO_4^- are required to consume 10.0 mL of the sugar solution.

136. Oxalic acid reacts with the chromate ion in acidic solution as follows.

$$3\,H_2C_6O_4(aq) + 2\,CrO_4^{2-}(aq) + 10\,H^+(aq)$$
$$\longrightarrow 6\,CO_2(g) + 2\,Cr^{3+}(aq) + 8\,H_2O(l)$$

Calculate the molarity of the oxalic acid ($H_2C_2O_4$) solution if 10.0 mL of the solution consume 40.0 mL of 0.0250 M CrO_4^{2-}.

Integrated Problems

137. A compound that combines in fixed amounts with one or more molecules of water is known as a hydrate. In the lab, a 5.00-gram sample of the hydrate of barium chloride, $BaCl_2 \cdot xH_2O$, is heated to drive off the water. After heating, 4.26 grams of anhydrous barium chloride, $BaCl_2$, remain. What is the value of x in the formula of the hydrate, $BaCl_2 \cdot xH_2O$?

138. Predict the formula of the compound produced when 1.00 gram of chromium metal reacts with 0.923 gram of oxygen atoms, O.

139. A 3.500-gram sample of an oxide of manganese contains 1.288 grams of oxygen. What is the empirical formula of the compound?

140. Cocaine is a naturally occurring substance that can be extracted from the leaves of the coca plant, which grows in South America (and is not to be confused with chocolate, or cocoa, which is extracted from the seeds of another South American plant). If the chemical formula for cocaine is $C_{17}H_{21}O_4N$, what is the percentage by mass of carbon, hydrogen, oxygen, and nitrogen in the compound? Comment on the ease with which elemental analysis of the carbon and hydrogen in a compound can be used to distinguish between the white, crystalline powder known as aspirin ($C_9H_8O_4$), which is used to cure headaches, and the white, crystalline powder known as cocaine, which is more likely to cause headaches.

141. A crucible and sample of $CaCO_3$ weighing 42.670 grams were heated until the compound decomposed to form CaO and CO_2.

$$CaCO_3(s) \longrightarrow CaO(s) + CO_2(g)$$

The crucible had a mass of 35.351 grams. What is the theoretical mass of the crucible and residue after the decomposition is complete?

142. Nitrogen reacts with red-hot magnesium to form magnesium nitride,

$$3\,Mg(s) + N_2(g) \longrightarrow Mg_3N_2(s)$$

which reacts with water to form magnesium hydroxide and ammonia.

$$Mg_3N_2(s) + 6\,H_2O \longrightarrow 3\,Mg(OH)_2(aq) + 2\,NH_3(aq)$$

Calculate the number of grams of magnesium that would be needed to prepare 15.0 grams of ammonia.

143. A 2.50-gram sample of bronze was dissolved in sulfuric acid. The copper in the alloy reacted with sulfuric acid as follows.

$$Cu(s) + 2\,H_2SO_4(aq) \longrightarrow CuSO_4(aq) + SO_2(g) + 2\,H_2O(l)$$

The $CuSO_4$ formed in the reaction was mixed with KI to form CuI.

$$2\,CuSO_4(aq) + 5\,I^-(aq) \longrightarrow 2\,CuI(s) + I_3^-(aq) + 2\,SO_4^{2-}(aq)$$

The I_3^- ion formed in this reaction was then titrated with $S_2O_3^{2-}$.

$$I_3^-(aq) + 2\,S_2O_3^{2-}(aq) \longrightarrow 3\,I^-(aq) + S_4O_6^{2-}(aq)$$

Calculate the percentage by mass of copper in the original sample if 31.5 mL of 1.00 M $S_2O_3^{2-}$ were consumed in the titration.

144. Assume that you start with a glass of water, a glass of methanol, and a teaspoon. Exactly one teaspoon of water is removed from the glass of water and added to the glass of methanol. The resulting mixture of methanol is stirred until the two liquids are thoroughly mixed. Exactly one teaspoon of this mixture is then transferred back to the water. Which of the following statements is true?

(a) The volume of water that ends up in the methanol is larger than the volume of methanol that ends up in the water.

(b) The net volume of water transferred to the methanol is smaller than the net volume of methanol transferred to the water.

(c) The net volume of water added to the methanol is exactly the same as the net volume of methanol added to the water.

145. Iron can react with O_2 to produce two different oxides, $Fe_2O_3(s)$ or $Fe_3O_4(s)$. Write the chemical equations that describe both reactions. If 167.6 grams of Fe react completely with excess $O_2(g)$ to produce 231.6 grams of product, which oxide was formed?

146. Assume that two experiments are performed on the following chemical reaction:

$$2\,Br^-(aq) \;+\; Cl_2(aq) \;\longrightarrow\; Br_2(aq) \;+\; 2\,Cl^-(aq)$$

Colorless   Colorless   Red   Colorless

• Experiment 1: 100 mL of a 0.0100 M solution of Br^- are added to 100 mL of a 0.0200 M solution of Cl_2.

• Experiment 2: 100 mL of a 0.0100 M solution of Br^- are added to 100 mL of a 0.0500 M solution of Cl_2.

If the reaction between aqueous solutions of the Br^- ion and Cl_2 goes to completion, which of the following would you expect to observe after mixing the two solutions? Explain your answer.

(a) The solution formed in experiment 1 will be a darker red.

(b) The solution formed in experiment 2 will be a darker red.

(c) The solutions formed in both experiments will be the same shade of red.

Chapter Three

THE STRUCTURE OF THE ATOM

J. J. Thomson (left), who showed that electrons were subatomic particles, and his student, Ernest Rutherford (right), who proposed that atoms were composed of negatively charged electrons orbiting an infinitesimally small positively charged nucleus.

3.1 Rutherford's Model of the Atom

Shortly after radioactivity was discovered at the turn of the twentieth century, Ernest Rutherford became interested in the alpha particles, or $^4_2He^{2+}$ ions emitted by uranium metal and its compounds. Rutherford found that α particles were absorbed by a thin sheet of metal, but they could pass through metal foil if it was thin enough.

Rutherford noticed that a narrow beam of α particles was broadened as it passed through the metal foil. Working with his assistant Hans Geiger, Rutherford measured the angle through which the α particles were scattered by a thin piece of gold foil. Because it is unusually ductile, gold can be made into a foil that is only 0.00004 cm—or about 2000 atoms—thick. When the foil was bombarded with α particles, Rutherford and Geiger found that the angle of scattering was small, on the order of 1°.

These results were consistent with Rutherford's expectations. He knew that the α particle had a considerable mass for a subatomic particle and that it moved quite rapidly. Although the α particles should be scattered slightly by collisions with the atoms through which they passed, Rutherford expected these particles to pass through the metal foil much the way a rifle bullet would pass through a bag of sand.

One day, Geiger suggested that a research project should be given to Ernest Marsden, who was working in Rutherford's laboratory. Rutherford responded, "Why not let him see whether any α particles can be scattered through a large angle?" When this experiment was done, Marsden found that a small fraction (perhaps 1 in 20,000) of the α particles were scattered through angles larger than 90°, as shown in Figure 3.1a. Many years later, reflecting on his reaction to these results, Rutherford said, "It was quite the most incredible event that has ever happened to me in my life. It was almost as incredible as if you fired a 15-inch shell at a piece of tissue paper and it came back and hit you."

Rutherford found that he could explain Marsden's results by assuming that the positive charge and most of the mass of an atom are concentrated in a small fraction of the total volume, which he called the **nucleus** (or "little nut"). Most of the α particles were able to pass through the gold foil without encountering

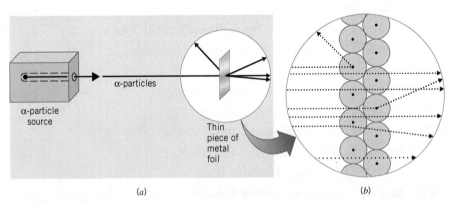

(a)   (b)

Fig. 3.1 (a) A block diagram of the Rutherford–Marsden–Geiger experiment. (b) Most α particles pass through empty space between the nuclei. A few come close enough to be repelled by the nucleus and are deflected through small angles. Occasionally, an α particle travels along a path that would lead to a direct hit with the nucleus. These particles are deflected through large angles by the force of repulsion between the particle and the positively charged nucleus of the atom.

anything large enough to significantly deflect their path. A small fraction of the α particles came close to the nucleus of a gold atom as they passed through the foil. When this happened, the force of repulsion between the positively charged α particle and the nucleus deflected the α particle by a small angle, as shown in Figure 3.1b. Occasionally, an α particle traveled along a path that would eventually lead to a direct collision with the nucleus of one of the 2000 or so atoms through which it had to pass. When this happened, repulsion between the nucleus and the α particle deflected the α particle through an angle of 90° or more.

By carefully measuring the fraction of the α particles deflected through large angles, Rutherford was able to estimate the size of the nucleus. According to his calculations, the radius of the nucleus is at least 10,000 times smaller than the radius of the atom. The vast majority of the volume of an atom is therefore empty space.

➤ **CHECKPOINT**

Assume that a circle 1 cm in diameter is used to represent the nucleus of an atom. Calculate the size of the circle that would have to be used to represent the diameter of the atom.

3.2 Particles and Waves

Rutherford's model assumes that most of the mass and all of the positive charge of an atom are concentrated in an infinitesimally small nucleus surrounded by a sea of lightweight, negatively charged electrons. Our next goal is to develop a picture of how these electrons are distributed around the nucleus. Much of what we know about the arrangement of electrons in an atom has been obtained by studying the interaction between matter and different forms of **electromagnetic radiation.** We therefore need to understand what we mean when we say that electromagnetic radiation has some of the properties of both a particle and a wave.

Scientists divide matter into two categories: particles and waves. **Particles** are easy to understand because they have a measurable mass and they occupy space. **Waves** are more challenging. They have no mass, and yet they carry energy as they travel through space. The best way to demonstrate that waves carry energy is to watch what happens when a pebble is tossed into a lake. As the waves produced by this action travel across the lake, they set in motion any leaves that lie on the surface.

In addition to their ability to carry energy, waves have four other characteristic properties: speed, frequency, wavelength, and amplitude. As we watch waves travel across the surface of a lake, we can see that they move at a certain **speed.** By watching waves strike a pier at the edge of the lake, we can see that they are also characterized by a **frequency** (v), which is the number of wave cycles that hit the pier per unit of time. The frequency of a wave is therefore reported in units of cycles per second (s^{-1}) or hertz (Hz). The idealized drawing of a wave in Figure 3.2 illustrates the definitions of amplitude and wavelength.

Waves travel through space, whereas particles occupy space.

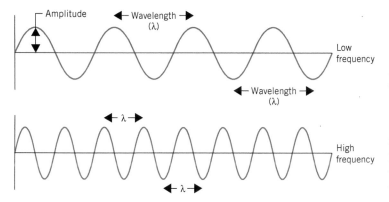

Fig. 3.2 The wavelength (λ) is the shortest distance between repeating points on a wave. The *amplitude* is the difference between the height of the highest or lowest points on the wave and its midline. The *frequency* (v) is the number of waves (or cycles) that pass a fixed point per unit time. The product of frequency times wavelength is the speed at which the wave moves through space.

The **wavelength** (λ) is the shortest distance between repeating points on the wave. The **amplitude** is the difference between highest or lowest points on the wave and the middle of the wave.

If we measure the frequency (v) of a wave in cycles per second and the wavelength (λ) in meters, the product of the two numbers has the units of meters per second. The product of the frequency times the wavelength of a wave is therefore the speed at which the wave travels through space.

$$v\lambda = speed$$

To understand the relationship among the speed, frequency, and wavelength of a wave, it may be useful to consider a concrete example. Imagine that you are at a railroad crossing, watching a train that consists of 45-ft-long boxcars go by. Assume that it takes 3.0 seconds for each boxcar to pass in front of your car. The "frequency" of the train is therefore 1 car every 3.0 seconds. How fast is the train moving? The speed at which the train moves through space is the product of the "frequency" of this phenomenon times its "wavelength."

$$\frac{1}{3.0\,\text{s}} \times 45\,\text{ft} = 15\,\frac{\text{ft}}{\text{s}} \approx 10\,\frac{\text{mi}}{\text{hr}}$$

Exercise 3.1

Orchestras in the United States tune their instruments to an "A" that has a frequency of 440 cycles per second, or 440 Hz. If the speed of sound is 1116 feet per second (1116 ft/s), what is the wavelength of this note?

Solution

The product of the frequency times the wavelength of any wave is equal to the speed with which the wave travels through space.

$$\nu\lambda = speed$$

Substituting the speed with which the note travels through space and the frequency of the note into this equation gives the following result:

$$(440\,\text{s}^{-1})(\lambda) = 1116\,\text{ft/s}$$

Solving for λ gives a wavelength of 2.54 ft for the note.

3.3 Light and Other Forms of Electromagnetic Radiation

In 1865 James Clerk Maxwell proposed that light is a wave with both *electric* and *magnetic* components. Light is therefore a form of electromagnetic radiation. Because it is a wave, light is bent when it enters a glass prism. When white light is focused on a prism, the light rays of different wavelengths are bent by differing amounts and the light is transformed into a band, or **spectrum,** of different colors. Starting from the side of the spectrum where the light is bent by the smallest angle, the colors are red, orange, yellow, green, blue, and violet.

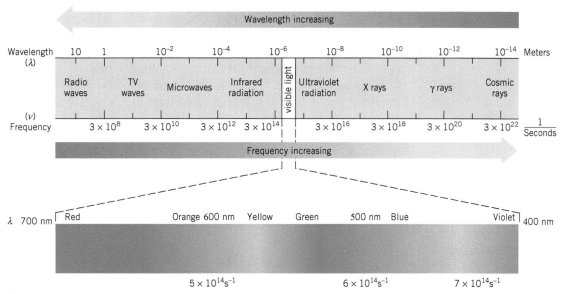

Fig. 3.3 The visible spectrum is the small portion of the total electromagnetic spectrum our eyes can see. Other forms of electromagnetic radiation include radio and TV waves, microwaves, infrared and ultraviolet radiation, as well as X rays, γ rays, and cosmic rays.

Visible light consists of a narrow band of frequencies and wavelengths in the small portion of the electromagnetic spectrum that our eyes can see. It includes radiation with wavelengths between about 400 nm (violet) and 800 nm (red). Because the wavelength of electromagnetic radiation can be as long as 40 m or as short as 10^{-5} nm, the visible spectrum is only a tiny portion of the total range of electromagnetic radiation.

The electromagnetic spectrum is divided into the categories of radio and TV waves, microwaves, infrared radiation, visible light, ultraviolet radiation, X rays, γ rays, and cosmic rays, as shown in Figure 3.3. These different forms of radiation all travel at the speed of light (c). They differ, however, in their frequencies and wavelengths. The product of the frequency times the wavelength of electromagnetic radiation is always equal to the speed of light.

$$\nu\lambda = c$$

As a result, electromagnetic radiation that has a long wavelength has a low frequency, and radiation with a high frequency has a short wavelength.

> ► CHECKPOINT
>
> What is the range of frequencies of the waves used in a microwave oven?

 Exercise 3.2

About half of the energy from the Sun that enters the Earth's atmosphere is absorbed by the surface of the planet. Because this heats the surface of the planet, the Earth emits low-energy, long-wavelength infrared (IR) radiation back into space. The majority of this outgoing IR radiation is captured by the greenhouse gases in the atmosphere, such as CO_2, H_2O, and CH_4. Without this greenhouse effect life on this planet would not exist because the Earth would have an average annual temperature of $-18°C$ (0°F) rather than the average annual temperature on Earth of 15°C (59°F). Thus, it is not the greenhouse effect (by itself) that concerns atmospheric scientists. It is the *enhanced* greenhouse effect that results from the increased levels of anthropogenic (human-made) CO_2 and other greenhouse gases in the atmosphere.

CO_2 molecules absorb IR radiation in narrow bands at two wavelengths: 4.26 μm and 15.0 μm. Calculate the frequency of the IR radiation that has a wavelength of 4.26 μm if the speed of light is 3.00×10^8 m/s, to three significant figures.

Solution

The product of the frequency times the wavelength of any wave is equal to the speed with which the wave travels through space. In this case, the wave travels at the speed of light: 3.00×10^8 m/s. Before we can calculate the frequency of the radiation, we have to convert the wavelength into units of meters.

$$4.26\,\mu\text{m} \times \frac{1\,\text{m}}{10^6\,\mu\text{m}} = 4.26 \times 10^{-6}\,\text{m}$$

We then substitute the wavelength of the IR radiation in units of meters and the speed of light in meters per second into the following equation: $\nu\lambda = c$.

$$\nu(4.26 \times 10^{-6}\,\text{m}) = 3.00 \times 10^8\,\text{m/s}$$

We can then solve for the frequency of the IR radiation in units of cycles per second.

$$\nu = 7.04 \times 10^{13}\,\text{s}^{-1}$$

3.4 Atomic Spectra

For more than 200 years, chemists have known that sodium salts emit a yellow color when added to a flame. Robert Bunsen, however, was the first to systematically study this phenomenon. (Bunsen went so far as to design a new burner that would produce a colorless flame for this work.) Between 1855 and 1860, Bunsen and his colleague Gustav Kirchhoff developed a spectroscope that focused the light from the burner flame onto a prism that separated the light into its spectrum. Using this device, Bunsen and Kirchhoff were able to show that the **emission spectrum** of sodium salts contains two narrow bands of radiation in the yellow portion of the spectrum.

Chemists and physicists soon began using the spectroscope to catalog the wavelengths of light emitted or absorbed by a variety of compounds. These data were then used to detect the presence of certain elements in everything from mineral water to sunlight. No obvious patterns were discovered in the data, however, until 1885, when Johann Jacob Balmer analyzed the emission spectrum of excited hydrogen atoms.

When an electric current is passed through a glass tube that contains hydrogen gas at low pressure, hydrogen atoms are produced and the tube gives off a pink-violet light. When the light is passed through a prism (as shown in Figure 3.4), four narrow lines of bright light are observed against a black background. These narrow bands have the characteristic wavelengths and colors

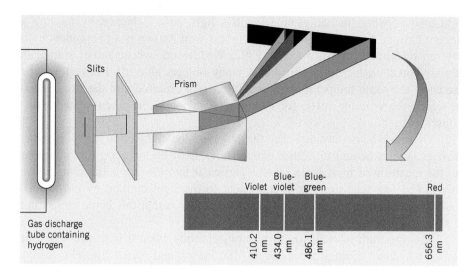

Violet Blue-violet Blue-green Red

410.2 nm 434.0 nm 486.1 nm 656.3 nm

Fig. 3.4 The light given off when a tube filled with H_2 gas is excited with an electric discharge can be separated into four narrow lines of visible light when it is passed through a prism.

shown in Table 3.1. Balmer found that these data fit the following equation to within ± 0.02%.

$$\frac{1}{\lambda} = R_H\left(\frac{1}{2^2} - \frac{1}{n^2}\right)$$

In this equation, R_H is a constant known as the Rydberg constant, which is equal to 1.09737×10^{-2} nm^{-1}, and n is an integer equal to either 3, 4, 5, or 6.

Between 1906 and 1924, four more series of lines were discovered in the emission spectrum of hydrogen by searching the infrared (IR) spectrum at longer wavelengths and the ultraviolet (UV) spectrum at shorter wavelengths. Theodore Lyman discovered one series of lines in the UV portion of the spectrum, and Friedrich Paschen, F.S. Brackett, and A.H. Pfund each discovered another series of lines in the IR spectrum. Each of these lines fits the same general equation, where n_1 and n_2 are integers and R_H is the Rydberg constant.

$$\frac{1}{\lambda} = R_H\left(\frac{1}{n_1^2} - \frac{1}{n_2^2}\right)$$

Table 3.1

Characteristic Lines in the Visible Spectrum of Hydrogen

Color	Wavelength (nm)
Red	656.3
Blue-green	486.1
Blue-violet	434.0
Violet	410.2

3.5 The Wave-Packet Model of Electromagnetic Radiation

The emission spectrum of the hydrogen atom raises several important questions.

- Why do hydrogen atoms give off only a handful of narrow lines of radiation when they emit light?
- Why do hydrogen atoms emit light when excited that has exactly the same wavelengths as the light they absorb when they are not excited?
- Why do the lines in the hydrogen spectrum depend on the integer relationship observed by Balmer and others?

Answers to these questions came from work on a related topic.

It is common knowledge that objects give off light when heated. Examples range from the gentle red glow of an electric burner on a stove to the bright

Both the bottom of the crucible and the wire gauze on which it rests will glow by giving off light of the same color when the crucible is heated with a Bunsen burner.

light emitted when the tungsten wire in a light bulb is heated by passing an electric current through the wire. What is less well known is a phenomenon discovered by Thomas Wedgwood in 1792. Wedgwood, whose father started the famous porcelain factory, noticed that many objects give off a red glow when heated to the same temperature. The bottom of a crucible and the iron triangle on which the crucible rests, for example, both glow red when heated with a Bunsen burner.

Wedgwood also noticed that the color of the light emitted by an object changes as it is heated to higher temperatures until the object glows white-hot, but the spectrum of light given off at a particular temperature is the same for any object. The fact that sunlight is equivalent to the light emitted by an object at about 5500°C, for example, has led to the assumption that this is the temperature of the surface of the Sun.

The spectrum of radiation that an object emits when it is heated was eventually understood through the work of Max Planck and Albert Einstein. According to the Planck–Einstein model, light and other forms of electromagnetic radiation behave to some extent as both a particle and a wave. As a wave, light carries an amount of energy that is proportional to the frequency of this wave-particle.

$$E = h\nu$$

In this equation, h is a proportionality constant known as Planck's constant, which is equal to 6.626×10^{-34} joule-seconds (J · s).

But electromagnetic radiation also behaves to some extent as a particle. Light isn't continuous; it is a stream of small bundles or packets of energy. These "energy packets" were given the name **photons.** When an atom or a molecule absorbs a photon, it therefore gains a finite amount of energy equal to the product of Planck's constant times the frequency of the photon.

We can now understand why an object gives off red light when heated until it just starts to glow. Red light has the longest wavelength, and therefore the smallest frequency, of any form of visible radiation. Because the energy of electromagnetic radiation is proportional to its frequency, red light carries the smallest amount of energy of any form of visible radiation. The light given off when an object just starts to glow, therefore, has the color characteristic of the lowest-energy form of electromagnetic radiation our eyes can see. As the object becomes hotter, it also emits light at higher frequencies, until it eventually appears to give off white light.

Exercise 3.3

Calculate both the energy of a single photon of IR radiation with a wavelength of 4.26 μm that could be absorbed by CO_2 in the atmosphere and the energy of a mole of such photons.

Solution

In Exercise 3.2 we found that IR radiation with a wavelength of 4.26 μm has a frequency of 7.04×10^{13} s^{-1}. Substituting this frequency into the Planck–Einstein equation gives the following result, in units of joules (J).

$$E = h\nu$$
$$E = (6.626 \times 10^{-34}\,\text{J} \cdot \text{s})(7.04 \times 10^{13}\,\text{s}^{-1}) = 4.66 \times 10^{-20}\,\text{J}$$

A single photon of IR radiation therefore carries an insignificant amount of energy. But a mole of these photons carries 28.1 kJ of energy.

$$\frac{4.66 \times 10^{-20}\,\text{J}}{1\,\text{photon}} \times \frac{6.02 \times 10^{23}\,\text{photons}}{1\,\text{mol}} = 2.81 \times 10^4\,\text{J/mol} = 28.1\,\text{kJ/mol}$$

This is enough energy to raise the temperature of 100 mL of water by almost 70°C.

● ●

➤ **CHECKPOINT**

Explain how the color of a heated metal bar could be used to determine its temperature.

3.6 The Bohr Model of the Atom

Although Rutherford was never able to incorporate electrons into his model of the atom, one of his students, Niels Bohr, proposed a model for the hydrogen atom that accounted for its spectrum. The **Bohr model** assumed that the negatively charged electron and the positively charged nucleus of a hydrogen atom were held together by the force of attraction between oppositely charged particles. According to Coulomb's law (shown below), this force is directly proportional to the charge on the electron (q_e) and the charge on the proton in the nucleus of the atom (q_p), and inversely proportional to the square of the distance between the particles (r^2).

$$F = \frac{q_e \times q_p}{r^2}$$

Bohr found that if he made the following assumptions, he could derive the equation that fit the experimental data obtained by Balmer, Lyman, Paschen, Brackett, and Pfund.

- The electron in a hydrogen atom travels around the nucleus of the atom in a circular orbit.
- The energy of the electron in a given orbit is proportional to its distance from the nucleus. (It takes energy to move an electron from a region close to the nucleus to one that is farther away.)
- Only orbits with certain energies are allowed.
- Light (or other forms of electromagnetic radiation) is absorbed when an electron moves from a lower-energy orbit into one that has a higher energy. The energy of the light that is absorbed when this happens is equal to the difference between the energies of the two orbits.
- Light is emitted when an electron falls from a higher-energy orbit into a lower-energy orbit. The energy of the light that is emitted when this happens is equal to the difference between the energies of the two orbits.

Although the Bohr model was remarkably successful for the hydrogen atom, it was unable to explain the properties of atoms with more than one electron. The Bohr model was therefore eventually replaced by a **quantum mechanical model** of the atom. The quantum mechanical model has the advantage that it is more powerful, but it achieves this power at a significant cost in terms of the ease with which it can be visualized.

We will retain four ideas from the Bohr model as we try to visualize the quantum mechanical model of the atom, although these ideas will be modified slightly.

- Electrons are attracted to the nucleus of an atom by the force of attraction between oppositely charged objects.
- Electrons reside in regions in space that are at different distances from the nucleus.
- There are only certain regions in space in which an electron can reside.
- Atoms emit or absorb radiation when an electron moves from one of these regions of space to another.

3.7 The Energy States of the Hydrogen Atom

As we have seen, atoms emit or absorb radiation at certain discrete frequencies or wavelengths. Because the energy of this radiation is directly proportional to its frequency, this means that the atoms emit or absorb radiation with only certain specific energies. This suggests that there are only certain specific stable *energy states* or *energy levels* within an atom. If this is true, these *energy states* are countable. In other words, the energy states of an atom are **quantized.** Because it only contains a single electron, we can conclude that the energy of the electron in a hydrogen atom is quantized.

Table 3.1 reported the wavelength and the color of the radiation for the four lines in the visible spectrum of the hydrogen atom. Table 3.2 also includes both the frequency and the energy of this radiation.

The data in Table 3.2 describe only the characteristic lines in the visible spectrum of the hydrogen atom analyzed by Balmer. Figure 3.5 also contains data for the lines in the IR and UV spectrum of the hydrogen atom discovered by Lyman, Paschen, Brackett, and Pfund. Each of these lines corresponds to a transition between a pair of energy levels labeled $n = 1, 2, 3, 4$, and so on. By convention, the lowest-energy level for the hydrogen atom is the $n = 1$ state. The highest-energy level would be the $n = \infty$ state. According to this diagram, it would take a photon with a wavelength of 121.57 nm to excite an electron from the $n = 1$ to the $n = 2$ state. Conversely, a photon with a wavelength of 121.57 nm would be emitted when the electron fell from the $n = 2$ to $n = 1$ state.

The relative energy for each level in Figure 3.5 is given on the left side of this diagram. These values are reported as negative numbers to convey the fact that energy is given off when an electron falls from one of the energy levels toward the top of this diagram into an energy level that lies farther down on the

Table 3.2

Characteristic Lines in the Visible Spectrum of the Hydrogen Atom

Color	Wavelength (nm)	Frequency (s^{-1})	Energy (kJ/mol)
Red	656.3	4.568×10^{14}	182.2
Blue-green	486.1	6.167×10^{14}	246.0
Blue-violet	434.0	6.908×10^{14}	275.5
Violet	410.2	7.309×10^{14}	291.6

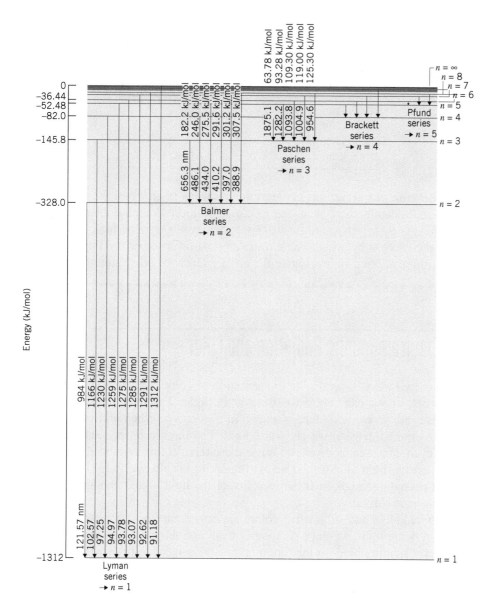

Fig. 3.5 According to the Bohr model, the wavelengths and corresponding energies in the ultraviolet spectrum of the hydrogen atom discovered by Lyman were the result of electrons dropping from higher-energy states into the $n = 1$ state. The Balmer, Paschen, Brackett, and Pfund series result from electrons falling from high-energy states into the $n = 2$, $n = 3$, $n = 4$, and $n = 5$ states, respectively.

diagram. Thus, moving an electron from the $n = 1$ to the $n = 2$ state would correspond to the absorption of a photon with an energy of 984 kJ/mol.

$$\Delta E = E_2 - E_1 = (-328 \text{ kJ/mol}) - (-1312 \text{ kJ/mol}) = 984 \text{ kJ/mol}$$

If the electron then falls back into the $n = 1$ state, a photon with an energy of 984 kJ/mol would be given off.

Exercise 3.4

Identify the transitions between different energy states that give rise to the four lines in the spectrum of the hydrogen atom given in Table 3.2.

Solution

Table 3.2 contains the following values for the energies of the four lines in the visible spectrum of the hydrogen atom: 182.2, 246.0, 275.5, and 291.6 kJ/mol.

Figure 3.5 suggests that each of these lines corresponds to a transition between the second energy state and a higher-energy state.

Photon Energy	Corresponding Transition
182.2 kJ/mol	$3 \rightarrow 2$
246.0 kJ/mol	$4 \rightarrow 2$
275.5 kJ/mol	$5 \rightarrow 2$
291.6 kJ/mol	$6 \rightarrow 2$

Because of interactions between the electrons in an atom, the spectra for atoms that contain more than one electron are more complex than the spectrum for the hydrogen atom. As we'll see, however, the same basic postulates about the process by which radiation is emitted or absorbed by the hydrogen atom can be applied to investigate the energy levels of other atoms.

3.8 Electromagnetic Radiation and Color

There are two ways of producing the sensation of color. We can add color where none exists, or we can subtract it from white light. The three primary *additive colors* are red, green, and blue, as shown in Figure 3.6. When all three are present at the same intensity, we get white light. The three primary *subtractive colors* are cyan, magenta, and yellow. When these three colors are absorbed with the same intensity, light is absorbed across the entire visible spectrum. If the object absorbs strongly enough, or if the intensity of the light is dim enough, all of the light can be absorbed.

The additive and subtractive colors are complementary. If we subtract yellow from white light, we get a mixture of cyan and magenta, and the light looks blue. If we subtract blue—a mixture of cyan and magenta—from white light, the light looks yellow. Aqueous solutions of the $Cu(NH_3)_4^{2+}$ complex ion have a deep blue color, for example, because this complex ion absorbs light in the yellow portion of the electromagnetic spectrum. Conversely, solutions that contain the CrO_4^{2-} ion appear yellow because they absorb blue light.

As we have seen, light is absorbed when it carries just enough energy to excite an electron from one energy level on an atom or molecule to another. Compounds therefore absorb light when the difference in energy between the lower energy level and the higher energy level corresponds to a wavelength in the narrow band of the electromagnetic spectrum that is visible to the naked eye.

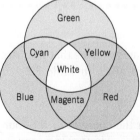

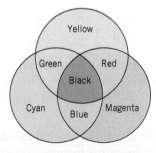

Primary additive colors   Primary subtractive colors

Fig. 3.6 The additive and subtractive colors.

The characteristic color of the light emitted by various elements can be demonstrated by filling a series of salt shakers with a variety of powdered metals (such as Mg or Fe) or ionic compounds that contain various positive or negative ions (such as the Na^+, K^+, Li^+, Sr^{2+}, Ba^{2+}, Cu^{2+}, and BO_3^{3-} ions). The salt shakers can then be used, one at a time, to sprinkle the metals or salts into the flame of a burner.

For the ionic compounds, the heat of the flame provides enough energy to excite an electron from a lower-energy level of one of the ions into a higher-energy level. Light is then emitted when this electron falls back to the lower-energy level. General rules for predicting the color of the light emitted by various salts are summarized as follows.

Na^+	Yellow
Li^+, Sr^{2+}	Red
K^+, Rb^+, Cs^+	Violet
Ba^{2+}, BO_3^{3-}	Yellow-green
Copper halides	Blue
Other copper salts	Green

When metals are placed in a flame, they react with oxygen in the atmosphere. The energy given off by this reaction provides the energy needed to excite an electron on atoms of metals such as Mg, which gives off the bright-white light characteristic of military flares, and Fe, which gives off the yellow light seen each Fourth of July when children light sparklers.

3.9 The First Ionization Energy

According to the data in Figure 3.5, the difference in energy between the $n = 1$ and $n = \infty$ levels is 1312 kJ/mol. This suggests that it should take 1312 kJ of energy to remove the electrons from a mole of isolated hydrogen atoms in the gas phase.

$$H(g) + h\nu \longrightarrow H^+(g) + e^-$$

This quantity is the first ionization energy of hydrogen. For elements that contain more than one electron, the first ionization energy can be defined as the smallest amount of energy needed to remove an electron from neutral atoms of the element in the gas phase to form positive ions with a +1 charge.

We can bring the magnitude of the first ionization energy for hydrogen into perspective by noting that it is more than one and a half times as large as the energy released when we burn a mole of the methane that fuels the Bunsen burners in a chemistry laboratory.

$$CH_4(g) + 2\,O_2(g) \longrightarrow CO_2(g) + 2\,H_2O(g)$$

So much energy is consumed when the electrons are removed from a mole of hydrogen atoms that an equivalent amount of energy in the form of heat would be able to raise the temperature of 4 L of water by more than 75°C!

If we divide the first ionization energy of hydrogen by Avogadro's constant, we can obtain the ionization energy of a single hydrogen atom.

$$\frac{1312 \text{ kJ}}{1 \text{ mol}} \times \frac{1 \text{ mol}}{6.022 \times 10^{23} \text{ atoms}} \times \frac{1000 \text{ J}}{1 \text{ kJ}} = 2.179 \times 10^{-18} \text{ J}$$

We can then use the Planck–Einstein equation to calculate the frequency of the radiation that has this much energy.

$$E = h\nu$$
$$2.179 \times 10^{-18} \, \text{J} = (6.626 \times 10^{-34} \, \text{J} \cdot \text{s})(\nu)$$
$$\nu = 3.288 \times 10^{15} \, s^{-1}$$

We can then use the relationship between the frequency and wavelength of electromagnetic radiation to calculate the wavelength of this radiation.

$$\nu\lambda = C$$
$$(3.288 \times 10^{15} \, s^{-1})(\lambda) = 2.998 \times 10^8 \, \text{m/s}$$
$$\lambda = 9.118 \times 10^{-8} \, \text{m} = 91.18 \, \text{nm}$$

This frequency and wavelength correspond to electromagnetic radiation in the ultraviolet region of the spectrum. If we need even more energy, we could turn to a source of X-ray radiation.

We can therefore measure the first ionization energy of an atom by shining ultraviolet light or X rays on a sample of neutral atoms in the gas phase. Some of the UV or X-ray photons will have enough energy to knock an electron out of the atom. The **first ionization energy (*IE*)** of an atom is measured by determining the radiation with the smallest amount of energy needed to remove an electron from the atom. The electron that is removed is the outermost or most loosely held electron on the atom.

Experimental values of the first ionization energies for the neutral atoms in the gas phase for the first 20 elements are given in Table 3.3.

A plot of the first ionization energy versus the atomic number of these elements is shown in Figure 3.7.

There are several clear patterns in these data.

- It doesn't matter whether we compare H and He, Li and Ne, or Na and Ar. The first ionization energy increases by a factor of between 2 and 4 as we go from left to right across a row of the periodic table.

- There is a dramatic drop in the first ionization energy as we go from the end of one row of the periodic table to the beginning of the next. It doesn't matter whether we compare helium and lithium, or neon and sodium: The first ionization energy decreases by more than a factor of 4.

Table 3.3
First Ionization Energies for Gas-Phase Atoms of the First 20 Elements

Symbol	Z	IE (kJ/mol)	Symbol	Z	IE (kJ/mol)
H	1	1312.0	Na	11	495.8
He	2	2372.3	Mg	12	737.7
Li	3	520.2	Al	13	577.6
Be	4	899.4	Si	14	786.4
B	5	800.6	P	15	1011.7
C	6	1086.4	S	16	999.6
N	7	1402.3	Cl	17	1251.1
O	8	1313.9	Ar	18	1520.5
F	9	1681.0	K	19	418.8
Ne	10	2080.6	Ca	20	589.8

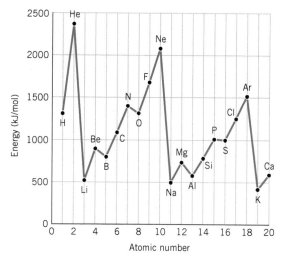

Fig. 3.7 Plot of the first ionization energies of the first 20 elements in the periodic table. There is a gradual increase in the first ionization energy across a row of the periodic table from H to He, from Li to Ne, and from Na to Ar. K and Ca appear to start a similar trend. There is a gradual decrease in the first ionization energy down a column of the periodic table, as can be seen by comparing the values for He, Ne, and Ar.

- There is a gradual decrease in the first ionization energy as we go down a column of the periodic table. Consider going from He (IE = 2372 kJ/mol) to Ne (IE = 2081 kJ/mol) to Ar (IE = 1521 kJ/mol), for example.

- There are minor exceptions to the gradual increase in the first ionization energy across a row of the periodic table. The increase from Li (IE = 520 kJ/mol) to Be (IE = 899 kJ/mol), for example, is followed by a small decrease as we continue to B (IE = 801 kJ/mol).

> ► **CHECKPOINT**
>
> Which would you expect to be larger, the first ionization energy of Rb or Sr? Explain why. Which would you expect to be larger, the first ionization energy of Cl or Br? Explain why.

3.10 The Shell Model

The data in Table 3.3 are consistent with the idea that the attraction between an electron and the nucleus of an atom becomes stronger as the charge on the nucleus increases and becomes weaker as the distance from the nucleus increases.

Consider the first and second elements in the periodic table, for example. The first ionization energy increases by a factor of about 2 as we go from H (IE = 1312 kJ/mol) to He (IE = 2372 kJ/mol). This is consistent with the hypothesis that the electrons on a helium atom feel roughly twice the force of attraction for the nucleus than the electron on a hydrogen atom because the positive charge on the helium nucleus is twice as large as the charge on the hydrogen nucleus.

If we extended this argument to lithium, we would predict that the first ionization energy for a lithium atom would be 1.5 times as large as the first ionization energy of helium and 3 times as large as the first ionization energy of hydrogen because the charge on the nucleus is now +3. According to the data in Table 3.3, however, nothing could be further from the truth. The first ionization energy of lithium (IE = 520 kJ/mol) is only about 20% as large as that of helium (IE = 2372 kJ/mol), and it is less than half as large as that of hydrogen (IE = 1312 kJ/mol).

The only way to explain these data is to assume that the electron removed when the first ionization energy of lithium is measured is farther away from the nucleus than the two electrons on a helium atom. Because this electron is farther from the nucleus, it is much easier to remove from the atom.

The data in Table 3.3 therefore suggest that the electrons in an atom are arranged in shells. The **shell model** of the atom assumes that the hydrogen and helium atoms consist of a nucleus surrounded by either one or two electrons in a single shell, relatively close to the nucleus, as shown in Figure 3.8. Lithium contains two electrons that lie relatively close to the nucleus of the atom, in the same shell

Fig. 3.8 The shell model assumes that the energy of an electron depends on the charge on the nucleus and the distance from the nucleus. The nucleus of a helium atom has twice the positive charge of the nucleus of a hydrogen atom. The first ionization energy of He is therefore about twice that of H.

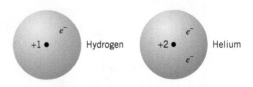

as the electrons in H and He, and one electron in a shell that is farther from the nucleus, as shown in Figure 3.9

The nucleus and the two inner electrons constitute the core of the lithium atom. The outermost electron in lithium doesn't experience the full +3 nuclear charge but rather a charge reduced by the underlying electrons. It is convenient to define a **core charge,** which, although it doesn't give the actual charge felt by an outer-shell electron, is useful for organizing atomic properties. The core charge of an element is equal to the sum of the positive charge on the nucleus of the atom and the negative charge on the electrons in all but the outermost shell of electrons on the atom. The core charge for elements in the second row of the periodic table, for example, is the sum of the positive charge on the nucleus and the negative charge on the two inner-shell core electrons. The core charge on a lithium atom is therefore +1.

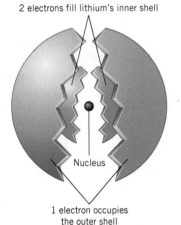

2 electrons fill lithium's inner shell

Nucleus

1 electron occupies
the outer shell

Fig. 3.9 The lithium atom consists of a nucleus with a positive charge of + 3, two electrons in a shell close to the nucleus, and one electron in a shell farther from the nucleus. [Reprinted from Carl Snyder, *The Extraordinary Chemistry of Ordinary Things,* New York, John Wiley & Sons, 1992, p. 24.]

Core charge

Li: (+3 nuclear charge) + (−2 inner electron charge) = +1 (core charge)

Because the outermost electron in lithium is at a larger distance from the nucleus and experiences a smaller attraction for the nucleus than the electrons in a He atom, it takes less energy to remove this electron from the atom. As a result, Li has a significantly smaller first ionization energy than helium.

Lithium marks the start of a series of eight elements characterized by a general increase in the first ionization energy with atomic number. Although there are minor variations in this trend, it is consistent with the idea that eight electrons are added to the atom in the second shell of electrons.

The shell that lies closest to the nucleus is described as the $n = 1$ shell of electrons. The data in Table 3.3 suggest that hydrogen has a single electron in the $n = 1$ shell and that helium has two electrons in this shell. Lithium has two electrons in the $n = 1$ shell and a third electron in the $n = 2$ shell. The next element, Be, has a first ionization energy (899 kJ/mol) that is larger than that for Li, which reflects the increased charge on the core:

Core charge

Be: (+4 nuclear charge) + (−2 inner electron charge) = +2 (core charge)

Note that for the elements in the first row of the periodic table, the core charge is the same as the atomic number Z, whereas the core charge is $Z − 2$ for elements in the second row. Beryllium has two electrons in both the $n = 1$ and $n = 2$ shells, and neon has two electrons in the $n = 1$ shell and eight electrons in the $n = 2$ shell, as shown in Figure 3.10.

➤ **CHECKPOINT**

What are the core charges for carbon and fluorine atoms?

Fig. 3.10 The shell model for four atoms from the second period of the periodic table. The core charge can be found from the shell model. The core charges are +1 for Li, +3 for B, + 6 for O, and + 8 for Ne.

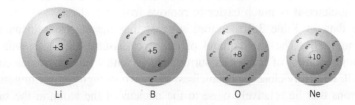

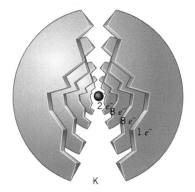

Fig. 3.11   The shell model of the sodium atom. There are three shells and a core charge of + 1. [Adapted from Carl Snyder, *The Extraordinary Chemistry of Ordinary Things,* New York, John Wiley & Sons, 1992.]

Fig. 3.12   The shell model of the potassium atom. There are now four shells. The first shell is complete with two electrons, the $n = 2$ and $n = 3$ shells hold eight electrons, and the fourth shell has only one electron. [Adapted from Carl Snyder, *The Extraordinary Chemistry of Ordinary Things,* New York, John Wiley & Sons, 1992.]

Once the eighth electron has been added to the $n = 2$ shell, there is a precipitous drop in the first ionization energy as we go from neon ($IE = 2081$ kJ/mol) to sodium ($IE = 496$ kJ/mol). This is similar to the drop observed from helium to lithium. It even has a similar magnitude: The first ionization energy drops by a factor of about 4 from neon to sodium. This suggests that the eleventh electron in the sodium atom is placed in a third shell ($n = 3$), at an even greater distance from the nucleus than the second shell of electrons used for the elements between Li and Ne, as shown in Figure 3.11.

The data in Table 3.3 for the elements between Na ($Z = 11$) and Ar ($Z = 18$) mirror the pattern observed between Li ($Z = 3$) and Ne ($Z = 10$). With slight variations, there is a gradual increase in the first ionization energy of these elements. This increase is then followed by another precipitous drop in the first ionization energy as we go from Ar ($IE = 1521$ kJ/mol) to potassium ($IE = 419$ kJ/mol). These data suggest that we leave the third shell at this point and enter a fourth shell ($n = 4$), as shown in Figure 3.12.

> ➤ **CHECKPOINT**
>
> Why aren't the first ionization data (Table 3.3) consistent with having nine electrons in the second shell of sodium?

3.11 The Shell Model and the Periodic Table

Many of the trends in the data in Table 3.3 are reflected by the structure of the periodic table. Note that there are two elements in the first row of the periodic table, and we concluded that two electrons occupy the first shell in our shell model. There are eight electrons in the second and third shells, and eight elements in the second and third rows of the periodic table.

Furthermore, the number of electrons in each of the outermost shells on an atom is consistent with the group in which the element is found in the periodic table. We have concluded that H, Li, Na, and K each have one electron in their outermost shell, and each of these elements is in Group IA of the periodic table. The decrease in the ionization energy as we go down the column of Group IA elements, from H to K, is also consistent with all of these atoms having a core charge of +1, with the outermost electron in a shell that is progressively farther and farther from the nucleus. In fact, the number of electrons in the outermost shell of each atom among the first 20 elements in the periodic table corresponds exactly with the group number for that element.

As we will see in the next chapter, the electrons in the outermost shell of an atom are involved in the formation of bonds between atoms. Because these bonds are relatively strong, the electrons that form them are often called the **valence electrons** (from the Latin stem *valens* meaning "to be strong").

The periodic table was originally created to group elements that had similar chemical and physical properties. The observation that the arrangement of electrons in the atom deduced from the shell model is reflected in the arrangement of elements in the periodic table suggests that many of the chemical and physical properties of the elements are related to the number of electrons in the outermost shell, the electrons that are the valence electrons in these atoms.

> **CHECKPOINT**
>
> Is the radius of the valence shell of Na larger, smaller, or the same as the radius of the valence shell of Li?

3.12 Photoelectron Spectroscopy and the Structure of Atoms

The shell model of the atom that we have deduced from first ionization energy data assumes that the electrons in an atom are arranged in shells about the nucleus, with successive shells being further and further from the nucleus of the atom. The data that were used to derive this model, however, were the first ionization energies of the elements, which reflect the ease with which the outermost electron, or highest-energy electron, can be removed from the atom. These data therefore represent the smallest or minimum energy needed to remove an electron from the atom.

For atoms that have many electrons, we would expect that it would take more energy to remove an electron from an inner shell than it does to remove the electron from the valence shell. It would take more energy to remove an electron from the $n = 2$ shell than from the $n = 3$ shell, for example, and even more energy to remove an electron from the $n = 1$ shell for a given atom.

This raises an interesting question: Do all of the electrons in a given shell have the same energy? This question can be answered by using a slightly different technique to measure the energy required to remove an electron from a neutral atom in the gas phase to form a positively charged ion.

This time, we will shine radiation on the sample that has enough energy to excite an atom to the point that an electron can be ejected from any shell of the atom to form a positively charged ion. The experiment, which is diagrammed in Figure 3.13, is known as **photoelectron spectroscopy (PES).**

Fig. 3.13 A block diagram of the photoelectron spectroscopy (PES) experiment. Absorption of a high-energy electron leads to the ejection of an electron from the atom. The kinetic energy (*KE*) of the ejected electron is measured, and the energy required to remove the electron from the atom is calculated from the difference between the energy of the photon (*hv*) and the kinetic energy of the photoelectron. Because electrons can be removed from any shell in the atom, this experiment is different from measurements of first ionization energies, which remove electrons from only the outermost shell. [Reprinted from R. J. Gillespie et al., *Atoms, Molecules and Reactions,* © 1994. Printed and electronically reproduced by permission of Pearson Education, Inc., Upper Saddle River, New Jersey.]

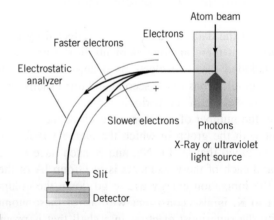

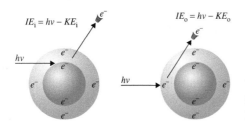

Fig. 3.14 Absorption of a high-energy photon can result in the ejection of an electron from any shell of an atom. The ionization energy of the inner-shell electrons, IE_i, will be larger than that of the outer-shell electrons, IE_o.

The PES experiment begins with the absorption of a high-energy UV or X-ray photon that carries more energy than the ionization energy (IE) of the atom. The excess energy is carried off by the electron ejected from the atom in the form of kinetic energy (KE). The energy of the photon ($h\nu$) absorbed by the atom is therefore equal to the sum of the ionization energy (IE) of the atom and the kinetic energy (KE) of the electron that is ejected from the atom.

$$h\nu = IE + KE$$

This means that the ionization energy is equal to the difference between the energy of the radiation absorbed by the atom ($h\nu$) and the kinetic energy of the photoelectron ejected (KE) when this radiation is absorbed.

$$IE = h\nu - KE$$

PES differs from the experiment used to obtain the first ionization energies given in Table 3.3 by its ability to remove electrons *from any shell in the atom,* as shown in Figure 3.14. Not only can an electron from the outermost shell be removed, but an electron from one of the shells deep within the core of electrons that surround the nucleus can be ejected. Only a single electron is removed from a given atom, but that electron can come from any energy level. As a result, PES allows us to measure the energy needed to remove any electron on an atom.

Data from PES experiments are obtained as peaks in a spectrum that plots the intensity of the observed signal on the vertical axis versus the energy needed to eject an electron (IE) on the horizontal axis, as shown in Figure 3.15. By convention, the spectrum is plotted so that energy *decreases* from left to right on the horizontal axis. The height of the peak in the simulated PES spectra in this book is directly proportional to the number of electrons of equivalent energy ejected during the experiment. If we see two peaks, for example, that have a relative height of 2:1, we can conclude that one of the energy levels from which electrons are removed in this experiment contains twice as many electrons as the other.

As we examine PES data in the next section, it is important to remember that the electromagnetic radiation energy supplied can not only remove an electron from the outermost shell; it can remove an electron from any one of the shells in the core of the atom.

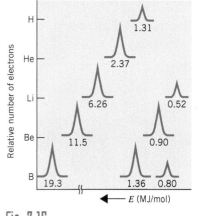

Fig. 3.15 Simulated photoelectron spectra of the first five elements in the periodic table. The energy needed to remove an electron from an atom increases from right to left. The energy required to remove an electron from the hydrogen atom is 1.312 MJ/mol, and that for the helium atom is 2.372 MJ/mol. The height of the He spectrum is twice that of the H spectrum because He has twice as many electrons as H. The spectra have been adjusted so that the peak heights of spectra of different atoms are directly comparable.

➤ **CHECKPOINT**

What factor determines the number of peaks in the photoelectron spectrum of an atom?

3.13 Electron Configurations from Photoelectron Spectroscopy

Hydrogen has only one peak in the photoelectron spectrum in Figure 3.15 because it contains only a single electron. As expected, this peak comes at an energy of 1312 kJ/mol, or, in units of megajoules, 1.312 MJ/mol. This is the energy required to remove the electrons from one mole of hydrogen atoms.

➤ **CHECKPOINT**

Why is there only one peak in the photoelectron spectrum of He even though there are two electrons in an He atom?

Helium also has only one peak in the PES experiment, which occurs at an energy of 2372 kJ/mol or 2.372 MJ/mol. The peak for helium is shifted to the left in Figure 3.15 when compared to the peak in the spectrum for hydrogen. This corresponds to a shift toward a larger ionization energy because it takes more energy to remove an electron from a helium atom than it does to remove an electron from a hydrogen atom. The height of the peak for helium in Figure 3.15 is also twice that for the peak in the spectrum for hydrogen. This is consistent with our hypothesis that the two electrons in a helium atom both occupy the $n = 1$ shell.

Our shell model of the atom leads us to expect two peaks in the PES spectrum for lithium, which is exactly what is observed. These peaks occur at ionization energies of 6.26 and 0.52 MJ/mol, and they have a relative intensity of 2:1. The outermost electron in the Li atom is relatively easy to remove ($IE = 0.52$ MJ/mol) because it is in the $n = 2$ shell. But it takes a great deal of energy ($IE = 6.26$ MJ/mol) to reach into the $n = 1$ shell because the electrons in that shell lie close to a nucleus that carries a charge of +3.

Two peaks are also observed in the PES spectrum of beryllium, with a relative intensity of 1:1. In this case, it takes an enormous amount of energy to reach into the $n = 1$ shell ($IE = 11.5$ MJ/mol) to remove one of the electrons that lie close to the nucleus with its charge of +4. It takes quite a bit less energy ($IE = 0.90$ MJ/mol) to remove one of the electrons in the $n = 2$ shell, as shown in Figure 3.15.

An interesting phenomenon occurs when we compare the photoelectron spectrum of boron with the spectra for the first four elements shown in Figure 3.15. There are now three distinct peaks in the spectrum, at energies of 19.3, 1.36, and 0.80 MJ/mol. There is a peak in the PES spectrum that corresponds to removing one of the electrons from the $n = 1$ shell ($IE = 19.3$ MJ/mol). But there are now two peaks, with a relative intensity of 2:1, that correspond to removing electrons from the $n = 2$ shell ($IE = 1.36$ and 0.80 MJ/mol).

The same phenomenon occurs in the PES spectra for carbon, nitrogen, oxygen, fluorine, and neon, as shown in Table 3.4. In each case, we see three peaks. As the charge on the nucleus increases, it takes more and more energy to remove an electron from the $n = 1$ shell, until, by the time we get to neon, it takes 84.0 MJ/mol to remove an electron from this shell. (This is more than 100 times the energy given off in a typical chemical reaction.)

Table 3.4

Ionization Energies for Gas-Phase Atoms of the First 10 Elements Obtained from Photoelectron Spectra

Element	IE (MJ/mol)		
	First Peak	Second Peak	Third Peak
H	1.31		
He	2.37		
Li	6.26	0.52	
Be	11.5	0.90	
B	19.3	1.36	0.80
C	28.6	1.72	1.09
N	39.6	2.45	1.40
O	52.6	3.12	1.31
F	67.2	3.88	1.68
Ne	84.0	4.68	2.08

Source: D. A. Shirley et al., *Physical Review* **B 1977**, *15*, 544–552.

The PES spectra for B, C, N, O, F, and Ne contain a second peak, of gradually increasing energy because of the increasing nuclear charge, that has the same intensity as the peak for the $n = 1$ shell. In each case, we get a third peak, of gradually increasing energy, that corresponds to the electrons that are the easiest to remove from the atoms. The intensity of the third peak increases from element to element, representing a single electron for boron up to a total of six electrons for neon, as shown in Figures 3.15 and 3.16.

The PES data for the first 10 elements reinforce our belief in the shell model of the atom. But they suggest that we have to refine this model to explain why the electrons in the $n = 2$ shell seem to occupy different energy levels. In other words, we have to introduce the concept of **subshells** within the shells of electrons into our model of the atom.

When the first evidence for subshells was discovered shortly after the turn of the twentieth century, a shorthand notation was introduced in which these subshells were described as either *s, p, d,* or *f.* Within any shell of electrons, it always takes the largest amount of energy to remove an electron from the *s* subshell.

The PES data in Table 3.4 suggest that there is only one subshell in the $n = 1$ shell. Chemists usually represent this as the $1s$ subshell, where the number represents the shell and the letter represents the subshell. The data in Table 3.4 also suggest that that the $1s$ subshell can hold a maximum of two electrons.

The PES spectra suggest that the third and fourth electrons on an atom are added to a $2s$ subshell. Once we have four electrons, however, the $2s$ subshell seems to be filled, and we have to add the fifth electron to the next subshell: $2p$.

By convention, the information we have deduced from the PES data is referred to as the atom's **electron configuration** and is written as follows.

$$
\begin{array}{ll}
\text{H}(Z = 1) & 1s^1 \\
\text{He}(Z = 2) & 1s^2 \\
\text{Li}(Z = 3) & 1s^2 2s^1 \\
\text{Be}(Z = 4) & 1s^2 2s^2 \\
\text{B}(Z = 5) & 1s^2 2s^2 2p^1
\end{array}
$$

The superscripts in the electron configurations designate the number of electrons in each subshell. The existence of two subshells within the $n = 2$ shell explains the minor inversion in the first ionization energies of boron and beryllium shown in Figure 3.7. It is slightly easier to remove the outermost electron from B ($IE = 0.80$ MJ/mol) than from Be ($IE = 0.90$ MJ/mol), in spite of the greater charge on the nucleus of the boron atom, because the outermost electron of B is in the $2p$ subshell, whereas the outermost electron in Be has to be removed from the $2s$ subshell.

As we continue across the second row of the periodic table, from B to Ne, the number of electrons in the $2p$ subshell gradually increases to six. The best evidence for this is the growth in the intensity of the peak corresponding to the $2p$ subshell. In boron, this peak has half the height of the $1s$ or $2s$ peaks. When we get to neon, we find it has three times the height of the $1s$ or $2s$ peaks. This suggests that the $2p$ subshell can hold a maximum of six electrons. We can therefore continue the process of translating PES data into the electron configurations for the atoms as follows.

$$
\begin{array}{ll}
\text{C}(Z = 6) & 1s^2 2s^2 2p^2 \\
\text{N}(Z = 7) & 1s^2 2s^2 2p^3 \\
\text{O}(Z = 8) & 1s^2 2s^2 2p^4 \\
\text{F}(Z = 9) & 1s^2 2s^2 2p^5 \\
\text{Ne}(Z = 10) & 1s^2 2s^2 2p^6
\end{array}
$$

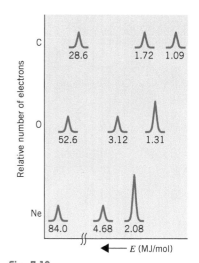

Fig. 3.16 Simulated photoelectron spectra of C, O, and Ne. The nuclear charge increases from carbon to oxygen to neon, and the peaks in the spectra shift to the left to higher energies. The peak farthest to the right shows the electrons most easily removed, and the peak heights (relative number of electrons) increase in a ratio of 2:4:6.

Table 3.5

Ionization Energies for Gas-Phase Atoms of Elements 11 through 21 Obtained from Photoelectron Spectra

Element	IE (MJ/mol)						
	1s	2s	2p	3s	3p	3d	4s
Na	104	6.84	3.67	0.50			
Mg	126	9.07	5.31	0.74			
Al	151	12.1	7.79	1.09	0.58		
Si	178	15.1	10.3	1.46	0.79		
P	208	18.7	13.5	1.95	1.01		
S	239	22.7	16.5	2.05	1.00		
Cl	273	26.8	20.2	2.44	1.25		
Ar	309	31.5	24.1	2.82	1.52		
K	347	37.1	29.1	3.93	2.38		0.42
Ca	390	42.7	34.0	4.65	2.90		0.59
Sc	433	48.5	39.2	5.44	3.24	0.77	0.63

Source: D. A. Shirley et al., *Physical Review* **B 1977**, *15*, 544–552.

PES data for the next 11 elements in the periodic table are given in Table 3.5, in which the columns are labeled in terms of the representations chemists use for the subshells: 1s, 2s, 2p, 3s, 3p, 3d, and 4s.

Sodium has four peaks in the photoelectron spectrum, corresponding to the loss of electrons from the 1s, 2s, 2p, and 3s subshells, as shown in Figure 3.17. It is easier to remove the electron in the $n = 3$ shell in sodium ($IE = 0.50$ MJ/mol) than the electrons in the $n = 2$ shell ($IE = 6.84$ and 3.67 MJ/mol), which in turn are easier to remove than the electrons in the $n = 1$ shell ($IE = 104$ MJ/mol).

$$\text{Na}(Z = 11) \qquad 1s^2 2s^2 2p^6 3s^1$$

Magnesium also gives four peaks in the PES experiment, which is consistent with the following electron configuration.

$$\text{Mg}(Z = 12) \qquad 1s^2 2s^2 2p^6 3s^2$$

Aluminum and each of the subsequent elements up to argon have five peaks in the PES spectrum. The new peak corresponds to the 3p subshell, and these elements have the following electron configurations.

$$\text{Al}(Z = 13) \qquad 1s^2 2s^2 2p^6 3s^2 3p^1$$
$$\text{Si}(Z = 14) \qquad 1s^2 2s^2 2p^6 3s^2 3p^2$$
$$\text{P}(Z = 15) \qquad 1s^2 2s^2 2p^6 3s^2 3p^3$$
$$\text{S}(Z = 16) \qquad 1s^2 2s^2 2p^6 3s^2 3p^4$$
$$\text{Cl}(Z = 17) \qquad 1s^2 2s^2 2p^6 3s^2 3p^5$$
$$\text{Ar}(Z = 18) \qquad 1s^2 2s^2 2p^6 3s^2 3p^6$$

The electron configurations of the elements in the third row of the periodic table therefore follow the same pattern as the corresponding elements in the second row.

By the time we get to potassium and calcium, we find six peaks in the PES spectrum, with the smallest ionization energy comparable to, but

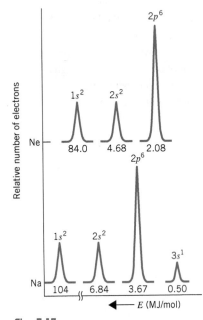

Fig. 3.17 Simulated photoelectron spectra of Ne and Na. Sodium has four peaks corresponding to loss of electrons from the *1s, 2s, 2p,* and *3s* subshells. The heights of the peaks correspond to the number of electrons in a subshell.

➤ **CHECKPOINT**

Explain the relative intensity of the three peaks in the PES spectrum of Ne (Figure 3.17), which is 2:2:6. Assign each of these peaks to a subshell.

slightly less than, the ionization energies of the 3s electrons on sodium and magnesium. We therefore write the electron configurations of potassium and calcium as follows.

$$K(Z = 19) \qquad 1s^2 2s^2 2p^6 3s^2 3p^6 4s^1$$
$$Ca(Z = 20) \qquad 1s^2 2s^2 2p^6 3s^2 3p^6 4s^2$$

Figure 3.18 shows the relative energies of the shells and subshells in the first six elements in the periodic table: H, He, Li, Be, B, and C. The same labels (1s, 2s, and 2p) are used to describe the different energy states in these atoms. The energy associated with a given label, however, changes significantly from one element to another. Once again, the energies are given as negative numbers to indicate that the atom must absorb energy to remove an electron from one of these subshells.

The energy of the 1s subshell gradually decreases from -1.31 MJ/mol in H to -2.37 MJ/mol in He, -6.26 MJ/mol in Li, and so on, until we reach an energy of -28.6 MJ/mol in C. This is consistent with the shell model of the atom, which assumes that these electrons are close to a nucleus with a charge that increases from $+1$ to $+6$. The energy of the 2s orbital also decreases as the charge on the nucleus increases, as we would expect. But the energy of the 1s orbital is always lower than that of the 2s, which is lower than that of the 2p.

An interesting phenomenon occurs in the next element in the periodic table, scandium (Sc, $Z = 21$). Our shell model predicts that the subshells used to hold electrons in calcium are all filled. The twenty-first electron therefore has to go into a new subshell. But the ionization energy for the new peak that appears in the PES spectrum doesn't occur at a lower energy than the subshells used previously, which has been observed for every other subshell as it has appeared. The new peak appears at a higher energy than the 4s subshell on scandium.

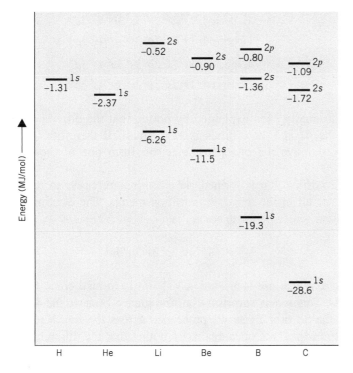

Fig. 3.18 Relative energies of the electrons within the shells and subshells in H, He, Li, Be, B, and C atoms in units of megajoules per mole (MJ/mol). The labels *1s, 2s,* and *2p* are used to describe the energy states. Note that the *1s* electrons are more tightly held as the nuclear charge increases from H through C.

Evidence from other forms of spectroscopy suggest that the twenty-first electron on the scandium atom goes into the $n = 3$ shell, not the $n = 4$ shell. The subshell used to hold this electron is known as the $3d$ subshell. As we can see from the data in Table 3.5, the $3d$ subshell is very close in energy to the $4s$ subshell. For scandium, it is easier to remove a $4s$ electron than a $3d$ electron. The electron configuration of scandium is usually written as follows:

$$\text{Sc} \, (Z = 21) \qquad 1s^2 2s^2 2p^6 3s^2 3p^6 4s^2 3d^1$$

This way of writing electron configurations is convenient because it follows the arrangement of the elements in the periodic table.

Scandium is followed by nine elements—known as the **transition metals**—that have no analogs in the second and third rows of the periodic table. In these elements, electrons continue to fill the $3d$ subshell until we reach zinc.

$$
\begin{array}{ll}
\text{Ti} & 1s^2 2s^2 2p^6 3s^2 3p^6 4s^2 3d^2 \\
\text{V} & 1s^2 2s^2 2p^6 3s^2 3p^6 4s^2 3d^3 \\
\quad \vdots & \\
\text{Zn} & 1s^2 2s^2 2p^6 3s^2 3p^6 4s^2 3d^{10}
\end{array}
$$

Thus, the $3d$ subshell can contain a maximum of 10 electrons, giving a total of up to 18 electrons in the $n = 3$ shell.

The next element in the periodic table is gallium, at which point the $4p$ level begins to fill, giving six elements (gallium to krypton) that have electron configurations analogous to those of the elements in the second and third rows of the periodic table.

An inspection of the electron configurations of the elements given in Appendix B.15 reveals electron configurations that differ from what we would predict using the periodic table. Two examples are chromium and copper.

$$
\begin{array}{ll}
\text{Cr} & \text{Expected: } 1s^2 2s^2 2p^6 3s^2 3p^6 4s^2 3d^4 \\
& \text{Observed: } 1s^2 2s^2 2p^6 3s^2 3p^6 4s^1 3d^5 \\
\text{Cu} & \text{Expected: } 1s^2 2s^2 2p^6 3s^2 3p^6 4s^2 3d^9 \\
& \text{Observed: } 1s^2 2s^2 2p^6 3s^2 3p^6 4s^1 3d^{10}
\end{array}
$$

This has traditionally been explained by noting that the difference between the energies of the $4s$ and $3d$ orbitals is very small. It therefore takes relatively little energy to move the outermost electron from one of these subshells to the other.

When a positive ion is formed, $4s$ electrons are easier to remove than the $3d$ electrons for all of the first-row transition metals. The electron configuration of the vanadium ion, V^{2+}, for example, is

$$V^{2+} \qquad 1s^2 2s^2 2p^6 3s^2 3p^6 3d^3$$

The two electrons that are lost when a V^{2+} ion is formed come from the outermost shell, $4s$. This seems surprising, at first glance because the $4s$ orbitals were filled before the $3d$ orbitals as we proceeded across the fourth row of the periodic table. It is explained, once again, by noting that the difference between the energies of $4s$ and $3d$ orbitals is very small and that the relative energy of these orbitals may be different in neutral transition-metal atoms and their ions.

➤ **CHECKPOINT**

Using data from Table 3.5, sketch the photoelectron spectrum of scandium, Sc. Give the relative intensities of the peaks and assign energies to each peak. Use the photoelectron spectrum to list the order in which electrons are removed from subshells of scandium. Use the electron configuration of Sc to list the order in which its subshells are filled with electrons. Is there a difference in the orders in which electrons are added to and taken away from Sc?

3.14 Allowed Combinations of Quantum Numbers

We have used modern experimental evidence to describe the way electrons are distributed around the nucleus of an atom. Experiments carried out near the beginning of the twentieth century were responsible for the development of the model that remains the underpinning of our understanding of the atomic world. Scientists became more and more perplexed as they attempted to explain the results of these early experiments. It was apparent that particles as small as atoms and electrons did not conform to the same physical laws as could be applied to everyday objects.

In 1926, the Austrian physicist Erwin Schrödinger worked out a mathematical way of dealing with this problem by assuming that electrons simultaneously have the properties of both a particle and a wave. The mathematical solutions he obtained were consistent with the assumption that electrons of different energies were found in different regions around the nucleus. These different regions of stability were the result of complex interactions between the electrostatic fields generated by the charged particles that form an atom. They suggested, however, that the stable energy states in which an electron could exist were quantized. Furthermore, Schrödinger showed that these stable states could be characterized by (1) the distance of the electron from the nucleus, (2) the momentum of the electron, and (3) the location of these regions of stability in space.

The new "wave mechanics" introduced certain complications that were beyond what we experience in ordinary life. The idea of quantization places restrictions on our ability to make precise measurements at the atomic level. In our everyday world we can describe the motion of a baseball or an airplane by giving their successive positions and velocities. The motion of atomic particles, however, cannot be described in this way.

To measure the position of an object on the macroscopic scale, the object must be illuminated with light so that we can see it. The photons of illuminating light are scattered by the object and strike our eye, allowing us to measure the position of the object. However, in the atomic world the energy of a photon of light can cause electrons to change energy states. This means that the electron is no longer in its original energy state, which means that either its velocity or its position is no longer the same. Therefore, if we precisely know the electron's position, we cannot know its velocity, and vice versa.

The German physicist Werner Heisenberg formulated this idea in an uncertainty principle, which states that the better the position of an electron is known, the less well its velocity is known. For this reason we often describe an electron as a cloud of electron density within the atom instead of describing it as a single particle.

Schrödinger's model describes the regions in space, or **orbitals,** where electrons are most likely to be found. Instead of trying to tell us where the electron is at any time, this model gives the probability that an electron can be found in a given region of space. The model no longer tells us where the electron is; it only tells us where it might be.

The Schrödinger model uses three coordinates, or three *quantum numbers,* to describe the orbitals in which electrons can be found. These coordinates are known as the principal (n), angular (l), and magnetic (m_l) quantum numbers. The quantum numbers describe the size, shape, and orientation in space of the orbitals on an atom.

The **principal quantum number** (n) describes the size of the orbital—orbitals for which $n = 2$ are larger than those for which $n = 1$, for example. As we have

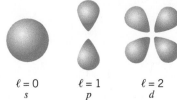

$\ell = 0$ $\ell = 1$ $\ell = 2$
  s   p   d

Fig. 3.19 The angular quantum number specifies the shape of the orbital. When $l = 0$, the orbital is spherical. When $l = 1$, it is polar. When $l = 2$, the orbital typically has the shape of a cloverleaf.

> ➤ **CHECKPOINT**
>
> Explain why three quantum numbers are required to describe an orbital. What feature of the three-dimensional structure of an orbital is specified by each of these quantum numbers?

seen, it takes energy to excite an electron from an orbital in which the electron is close to the nucleus ($n = 1$) into an orbital in which it is farther from the nucleus ($n = 2$ or higher). The principal quantum number therefore indirectly describes the energy of an orbital.

The **angular quantum number** (l) describes the shape of the orbital. Orbitals have shapes that are best described as spherical ($l = 0$), polar ($l = 1$), or cloverleaf ($l = 2$), as shown in Figure 3.19. They can take on even more complex shapes as the value of the angular quantum number becomes larger.

There is only one way in which a sphere ($l = 0$) can be oriented in space. Orbitals that have polar ($l = 1$) or cloverleaf ($l = 2$) shapes, however, can point in different directions. We therefore need a third quantum number, known as the **magnetic quantum number** (ml), to describe the orientation in space of a particular orbital. (It is called the *magnetic* quantum number because the effect of different orientations of orbitals was first observed in the presence of a magnetic field.)

Atomic spectra have shown that electrons in atoms occupy discrete energy states or energy levels. The quantum mechanical model allows only certain combinations of quantum numbers that can be used to describe these states. The allowed combinations of quantum numbers can be determined from the following set of rules.

Selection Rules Governing Allowed Combinations of Quantum Numbers

- The three quantum numbers (n, l, and m_l) are all integers.
- The principal quantum number (n) cannot be zero. The allowed values of n are therefore 1, 2, 3, 4, …
- The angular quantum number (l) can be any integer between 0 and $n - 1$. If $n = 3$, for example, l can be either 0, 1, or 2.
- The magnetic quantum number (m_l) can be any integer between $-l$ and $+l$. If $l = 2$, m_l can be -2, -1, 0, 1, or 2.

3.15 Shells and Subshells of Orbitals

Orbitals that have the same value of the principal quantum number form a **shell.** Orbitals within a shell are divided into *subshells* that are labeled with the same value of the angular quantum number. As we have seen, chemists describe the shell and subshell in which an orbital belongs with a two-character code such as $2p$ or $4f$. The first character indicates the shell ($n = 2$ or $n = 4$), and the second character identifies the subshell. By convention, the following lowercase letters are used to indicate different subshells.

s	$l = 0$
p	$l = 1$
d	$l = 2$
f	$l = 3$

Although there is no pattern in the first four letters (s, p, d, f) used to describe the different values of the angular quantum number, the letters progress alphabetically from that point (g, h, and so on). Some of the allowed combinations of the n and l quantum numbers are shown in Figure 3.20.

The third rule limiting allowed combinations of the n, l, and m_l quantum numbers has an important consequence. It says that the number of subshells in a

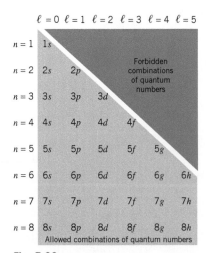

Fig. 3.20 Some allowed combinations of the principal (n) and angular (l) quantum numbers.

shell is equal to the principal quantum number for the shell. The $n = 3$ shell, for example, contains three subshells: the $3s$, $3p$, and $3d$ orbitals.

Let's look at some of the possible combinations of the n, l, and m_l quantum numbers. We'll start with the first shell, for which $n = 1$. According to the third selection rule, the angular quantum number (l) can be any integer between 0 and $n - 1$. Thus, if $n = 1$, l can only be 0. The fourth rule limits the magnetic quantum number (m_l) to integers between $-l$ and $+l$. Thus, when $l = 0$, m_l must be 0. There is only one orbital in the $n = 1$ shell because there is only one way in which a sphere can be oriented in space. The only allowed combination of quantum numbers for which $n = 1$ is the following, which is known as the $1s$ orbital

n	l	m_l	
1	0	0	$1s$

Let's now look at the orbitals in the second shell. When $n = 2$, l can be either 0 or 1. When $l = 0$, m_l must be zero. When $l = 1$, m_l can be either -1, 0, or 1. Thus, there are four orbitals in the $n = 2$ shell with the following combination of quantum numbers.

n	l	m_l	
2	0	0	$2s$
2	1	-1	
2	1	0	$2p$
2	1	1	

There is only one orbital in the $2s$ subshell. But there are three orbitals in the $2p$ subshell, each at a 90° angle to one another. One of these orbitals is oriented along the x axis, another along the y axis, and the third along the z axis of a coordinate system, as shown in Figure 3.21. These orbitals are therefore known as the $2p_x$, $2p_y$, and $2p_z$ orbitals.

There are nine orbitals in the $n = 3$ shell.

n	l	m_l	
3	0	0	$3s$
3	1	-1	
3	1	0	$3p$
3	1	1	
3	2	-2	
3	2	-1	
3	2	0	$3d$
3	2	1	
3	2	2	

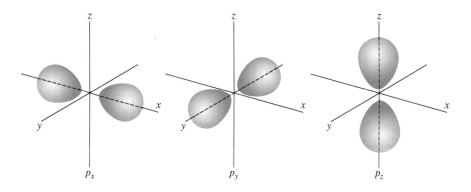

p_x p_y p_z

Fig. 3.21 There are three possible values of the magnetic quantum number (m_l) when the angular quantum number (l) is 1. The three values of m_l restrict the orientations of the orbitals in space.

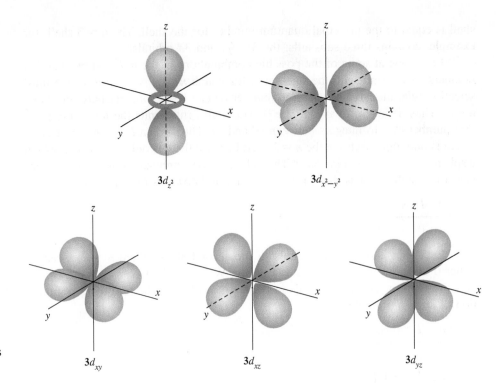

Fig. 3.22 There are five possible values of the magnetic quantum number (m_l) when the angular quantum number (l) is 2. The five values of m_l correspond to the five different orientations of the orbitals in space.

There is one orbital in the 3s subshell and there are three orbitals in the 3p subshell. The $n = 3$ shell, however, also includes five 3d orbitals.

The five different orientations of orbitals in the 3d subshell are shown in Figure 3.22. One of the orbitals lies in the xy plane of an xyz coordinate system and is called the $3d_{xy}$ orbital. The $3d_{xz}$ and $3d_{yz}$ orbitals have the same shape, but they lie between the axes of the coordinate system in the xz and yz planes. The fourth orbital in this subshell lies along the x and y axes and is called the $3d_{x^2-y^2}$ orbital. Most of the space occupied by the fifth orbital lies along the z axis, and this orbital is called the $3d_{z^2}$ orbital.

There is one orbital in the $n = 1$ shell, four orbitals in the $n = 2$ shell, and nine orbitals in the $n = 3$ shell. Thus, in general, the number of orbitals in a shell is equal to the square of the principal quantum number: $1^2 = 1$, $2^2 = 4$, $3^2 = 9$, $4^2 = 16$, and so on. There is one orbital in an s subshell ($l = 0$); there are three orbitals in a p subshell ($l = 1$) and five orbitals in a d subshell ($l = 2$). The number of orbitals in a subshell is therefore $2(l) + 1$.

3.16 Orbitals and the Pauli Exclusion Principle

In Chapter 1, we discussed the characteristic charge and mass of an electron. The Schrödinger model is based on the fact that electrons exhibit some of the behavior of a wave as well. In 1920, the German physicists Otto Stern and Walter Gerlach discovered a fourth property of the electron.

The Stern–Gerlach experiment began by heating a piece of silver metal until silver atoms evaporated from the surface. These atoms were passed through slits into an evacuated chamber, where they were allowed to pass through a magnetic field. Because the magnetic field was inhomogeneous, the atoms felt a net force pushing them in a direction perpendicular to their path as they passed through the magnetic field. As a result, the path of the silver atoms was bent or deflected by the magnetic field.

Once they passed through the magnetic field, the atoms struck a glass plate, where they were deposited as silver metal. When the experiment was done at very low residual pressures and long exposure times, Stern and Gerlach found that the silver metal was deposited in two closely spaced areas, suggesting that silver atoms ($Z = 47$) interact with a magnetic field in two distinctly different ways. Similar results were obtained with lithium ($Z = 3$), copper ($Z = 29$), and gold ($Z = 79$) but not with zinc ($Z = 30$), cadmium ($Z = 48$), or mercury ($Z = 80$).

This experiment suggests that there is a fundamental difference between atoms that contain an odd number of electrons and at least some of those that contain an even number of electrons. Atoms that have an odd number of electrons seem to behave as if there is at least one electron that spins in either a clockwise or counterclockwise direction on its axis. A negatively charged electron that spins on its axis would produce a tiny magnetic field that would interact with the magnetic field through which the beam of atoms passes. Atoms in which all of the electrons were paired with an electron of opposite **spin** wouldn't have magnetic properties and would pass through the magnetic field without being deflected in either direction.

Silver has an odd number of electrons ($Z = 47$). Therefore, even if most of the electrons pair so that the spins on these electrons cancel, there must be at least one odd electron on each silver atom that is unpaired. This unpaired electron interacts with an external magnetic field to cause the paths of moving silver atoms to be deflected in one direction if the electron spins clockwise on its axis and in the other direction if it spins counterclockwise.

A further consequence of an electron having a spin is that electrons that have the same spin (both electrons spinning in a clockwise direction, for example) have a low probability of being close together and a high probability of being far apart. There is no such restriction on electrons of opposite spin being close together.

Because electrons of the same spin must keep apart, an electron tends to exclude all other electrons of the same spin from the space that it occupies. However, an electron of opposite spin may enter this space. This space from which other electrons of the same spin tend to be excluded is called the *orbital* of the electron. Two electrons with opposite spins that occupy a given orbital are said to be "paired."

In 1924, the Austrian physicist Wolfgang Pauli proposed the following hypothesis, which has become known as the Pauli exclusion principle.

> **No more than two electrons can occupy an orbital, and, if there are two electrons in an orbital, the spins of the electrons must be paired.**

The concept of the spin of an electron and the Pauli exclusion principle provide the last step needed to complete our model of the structure of the atom.

The electrons in a shell or a subshell occupy three-dimensional orbitals in space. An orbital can be empty or it can contain either one or two electrons. If there are two electrons in the orbital, they must have different spins. The different spins are described in terms of a **spin quantum number** (m_s). The allowed values for the spin quantum number are $+1/2$ and $-1/2$.

The allowed combinations of n, l, and m_l quantum numbers for the orbitals in the first four shells are given in Table 3.6. For each of these orbitals, there are two allowed values of the spin quantum number m_s. As a result, 2 electrons can occupy either the $1s$, $2s$, $3s$, or $4s$ orbitals; 6 electrons can occupy the three orbitals in the $2p$, $3p$, or $4p$ subshells; 10 electrons can occupy the five orbitals in the $3d$ or $4d$ subshells; and 14 electrons can occupy the seven orbitals in the $4f$ subshell.

The quantum mechanical model that we have described agrees with the PES data that were introduced in Sections 3.12 and 3.13.

➤ **CHECKPOINT**

Predict the results of a Stern–Gerlach experiment on a beam of:
(a) Li atoms   (b) Be atoms
(c) B atoms   (d) N atoms

Table 3.6
Allowed Combinations of Quantum Numbers for the n = 1, 2, 3, or 4 Shells

n	l	m_l	Subshell Notation	Number of Orbitals in Subshell	Number of Electrons Needed to Fill Subshell
1	0	0	$1s$	1	2 total in 1st shell = 2
2	0	0	$2s$	1	2
2	1	1, 0, −1	$2p$	3	6 total in 2nd shell = 8
3	0	0	$3s$	1	2
3	1	1, 0, −1	$3p$	3	6
3	2	2, 1, 0, −1, −2	$3d$	5	10 total in 3rd shell = 18
4	0	0	$4s$	1	2
4	1	1, 0, −1	$4p$	3	6
4	2	2, 1, 0, −1, −2	$4d$	5	10
4	3	3, 2, 1, 0, −1, −2, −3	$4f$	7	14 total in 4th shell = 32

3.17 Predicting Electron Configurations

It is clear from the electron configurations of the first 36 elements in the periodic table that they follow a regular pattern. These configurations can be predicted from the diagram in Figure 3.23. The order in which atomic orbitals are filled can be obtained by following the arrows in this diagram, starting at the top of the first line and proceeding on to the second, third, and fourth lines, and so on.

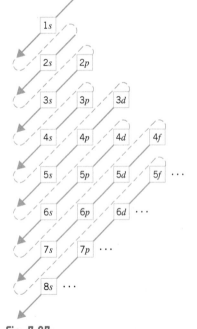

Fig. 3.23 The order in which atomic orbitals are filled can be predicted by following the arrows in this diagram.

Exercise 3.5

Predict the electron configuration for a neutral tin atom (Sn, Z = 50).

Solution

We start by predicting the order of atomic orbitals from the diagram in Figure 3.23. We then add electrons to these orbitals, starting with the $1s$ orbital, until all 50 electrons have been included.

The electron configuration for an atom can be written by remembering that each orbital can hold two electrons. Because an s subshell contains only one orbital, only two electrons can be added to this subshell. There are three orbitals in a p subshell, however, so a set of p orbitals can hold up to six electrons. There are five orbitals in a d subshell, so a set of d orbitals can hold up to ten electrons. The following is the complete electron configuration for tin.

$$\text{Sn}\,(Z = 50) \qquad 1s^2\,2s^2\,2p^6\,3s^2\,3p^6\,4s^2\,3d^{10}\,4p^6\,5s^2\,4d^{10}\,5p^2$$

A list of electron configurations of the elements can be found in Appendix B, Table B.15. As noted in Section 3.13, a handful of elements have electron configurations that differ slightly from the predictions of this model. What is impressive, however, is the number of elements for which the predictions of this model are correct.

3.18 Electron Configurations and the Periodic Table

Atoms in Group VIIIA, such as He, Ne, and Ar, have electron configurations with filled shells of orbitals. By convention, we therefore write abbreviated electron configurations in terms of the number of electrons beyond the previous element with a filled s and p subshell electron configuration. The electron configuration of lithium, for example, could be written as follows.

$$\text{Li}(Z = 3) \quad [\text{He}]2s^1$$

When the electron configurations of the elements are arranged so that we can compare elements in the horizontal rows of the periodic table, we find that these rows correspond to the filling of shells or subshells. The second row, for example, contains elements in which the $2s$ and $2p$ subshells in the $n = 2$ shell are filled.

$$
\begin{aligned}
\text{Li}(Z = 3) &\quad [\text{He}]2s^1 \\
\text{Be}(Z = 4) &\quad [\text{He}]2s^2 \\
\text{B}(Z = 5) &\quad [\text{He}]2s^2 2p^1 \\
\text{C}(Z = 6) &\quad [\text{He}]2s^2 2p^2 \\
\text{N}(Z = 7) &\quad [\text{He}]2s^2 2p^3 \\
\text{O}(Z = 8) &\quad [\text{He}]2s^2 2p^4 \\
\text{F}(Z = 9) &\quad [\text{He}]2s^2 2p^5 \\
\text{Ne}(Z = 10) &\quad [\text{He}]2s^2 2p^6
\end{aligned}
$$

There is an obvious pattern within the vertical columns, or groups, of the periodic table as well. The elements in groups often have similar configurations for their outermost electrons. This relationship can be seen by looking at the electron configurations of elements in columns on either side of the periodic table.

Group IA		Group VIIA	
H	$1s^1$		
Li	$[\text{He}]\, 2s^1$	F	$[\text{He}]\, 2s^2\, 2p^5$
Na	$[\text{Ne}]\, 3s^1$	Cl	$[\text{Ne}]\, 3s^2\, 3p^5$
K	$[\text{Ar}]\, 4s^1$	Br	$[\text{Ar}]\, 4s^2\, 3d^{10}\, 4p^5$
Rb	$[\text{Kr}]\, 5s^1$	I	$[\text{Kr}]\, 5s^2\, 4d^{10}\, 5p^5$
Cs	$[\text{Xe}]\, 6s^1$	At	$[\text{Xe}]\, 6s^2\, 4f^{14}\, 5d^{10}\, 6p^5$

Figure 3.24 shows the relationship between the periodic table and the orbitals being filled during the process by which the electron configuration is generated. The 2 columns on the left side of the periodic table correspond to the filling of an s orbital. The next 10 columns include elements in which the five orbitals in a d subshell are filled. The 6 columns on the right represent the filling of the three orbitals in a p subshell (except for He, which has only one occupied s orbital). Finally, the 14 columns at the bottom of the table correspond to the filling of the seven orbitals in an f subshell.

Elements are organized by groups in the periodic table because they have similar chemical and physical properties. As we have seen, elements in the same group have similar electron configurations for their outermost electrons. We can therefore conclude that many of the chemical and physical properties of an atom are a consequence of the electron configurations of their outermost electrons.

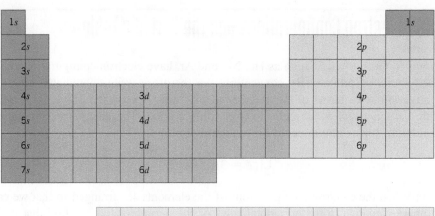

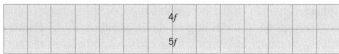

Fig. 3.24 The periodic table reflects the order in which atomic orbitals are filled. The *s* orbitals are filled in the two columns on the far left, and the *p* orbitals are filled in the six columns on the right. The *d* orbitals are filled along the transition between the *s* and *p* orbitals. The *f* orbitals are filled in the two long rows of elements at the bottom of the table.

 Exercise 3.6

Predict the electron configurations for calcium ($Z = 20$) and zinc ($Z = 30$) from their positions in the periodic table.

Solution

Calcium is in the second column and the fourth row of the table. The second column corresponds to the filling of an *s* orbital. The 1*s* orbital is filled in the first row, the 2*s* orbital in the second row, and so on. By the time we get to the fourth row, we are filling the 4*s* orbital. Calcium therefore has all of the electrons of argon, plus a filled 4*s* orbital.

$$Ca\,(Z = 20) \quad [Ar]\,4s^2$$

Zinc is the tenth element in the region of the periodic table where *d* orbitals are filled. Zinc therefore has a filled subshell of *d* orbitals. The only question is: Which set of *d* orbitals is filled? Although zinc is in the fourth row of the periodic table, the first time *d* orbitals occur is in the $n = 3$ shell. The following is therefore the abbreviated electron configuration for zinc.

$$Zn\,(Z = 30) \quad [Ar]\,4s^2\,3d^{10}$$

> ➤ **CHECKPOINT**
>
> Predict the number of unpaired electrons there are in each of the following atoms. Which of these atoms would have magnetic properties? (See Section 3.16.)
> (a) carbon   (b) nitrogen
> (c) neon   (d) fluorine

3.19 Electron Configurations and Hund's Rules

It is sometimes useful to illustrate the electron configuration of an atom in terms of an **orbital diagram** that shows the spins of the electrons. By convention, electron spins are represented by arrows pointing up or down. Each orbital is represented by a single line or box.

Consider the electrons in the $n = 2$ shell of a boron atom, for example.

B$(Z = 5)$ $[He]2s^2 2p^1$ $\underset{2s}{\uparrow\downarrow}$ $\underset{2p}{\uparrow\quad\_\_\quad\_\_}$

The $2s$ orbital is designated by a single line, whereas three lines are needed to represent the three $2p$ orbitals. There is only one electron in the $2p$ subshell, so we add a single electron to one of the three orbitals in this subshell, with the spin of that electron shown in an arbitrarily chosen direction.

A problem arises when we try to adapt the diagram to the next element, carbon. Where do we put the second electron? Do we put it in the same orbital, to form a pair of electrons of opposite spin? Do we put it in a different orbital in the subshell with the same spin as the first electron? Or do we put it in a different orbital, but with the spins of the two electrons paired? The German physicist Friedrich Hund found that the most stable arrangement of electrons can be predicted from the following rules.

- One electron is added to each orbital in a subshell before two electrons are added to any orbital in the subshell.
- Electrons are added to a subshell with the same spin until each orbital in the subshell has at least one electron.

Hund's rules are a consequence of the spin and repulsion of electrons discussed earlier. Electrons tend to avoid one another and can exist in the same orbital only if their spins are paired.

According to **Hund's rules,** the electrons in the $2p$ subshell on a carbon atom occupy two different orbitals and have the same spin. These electrons can be represented as follows:

C$(Z = 6)$ $[He]2s^2 2p^2$ $\underset{2p}{\uparrow\quad\uparrow\quad\_\_}$

When we get to N $(Z = 7)$, we have to put one electron into each of the three orbitals in the $2p$ subshell, all with the same spins.

N$(Z = 7)$ $[He]2s^2 2p^3$ $\underset{2p}{\uparrow\quad\uparrow\quad\uparrow}$

Because each orbital in the $2p$ subshell now contains one electron, the next electron added to the subshell must have the opposite spin, thereby filling one of the $2p$ orbitals.

O$(Z = 8)$ $[He]2s^2 2p^4$ $\underset{2p}{\uparrow\downarrow\quad\uparrow\quad\uparrow}$

The ninth electron fills a second orbital in the subshell.

F$(Z = 9)$ $[He]2s^2 2p^5$ $\underset{2p}{\uparrow\downarrow\quad\uparrow\downarrow\quad\uparrow}$

The tenth electron completes the $2p$ subshell.

Ne$(Z = 10)$ $[He]2s^2 2p^6$ $\underset{2p}{\uparrow\downarrow\quad\uparrow\downarrow\quad\uparrow\downarrow}$

3.20 The Sizes of Atoms: Metallic Radii

The size of an atom influences many of the chemical and physical properties of the atom. It therefore would be useful to have an accurate measurement of the size of an isolated atom. Unfortunately, the size of an isolated atom can't be measured because we can't determine the location of the electrons that surround the nucleus. We therefore estimate the size of an atom by assuming that the radius of the atom is equal to half the distance between equivalent nuclei of adjacent atoms in a solid. This technique is best suited to elements that are metals, which form solids composed of extended planes of atoms of that element. The results of these measurements are known as **metallic radii.**

Because more than 75% of the elements are metals, metallic radii are available for most elements in the periodic table. Figure 3.25 shows the relationship between the metallic radii for elements in Groups IA and IIA. There are two general trends in the data.

- Atoms become *larger* as we go down a column of the periodic table,
- Atoms become *smaller* as we go from left to right across a row of the periodic table.

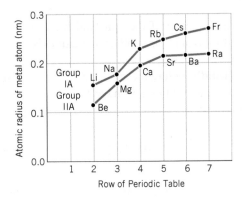

Fig. 3.25 Metallic radii increase as we go down a column of the periodic table. With rare exceptions, they decrease from left to right across a row of the table.

The first trend can be explained by looking at the electron configurations of the atoms. As we go down the periodic table, electrons are placed in larger and larger shells, but the core charge remains the same. When this happens, the size of the atom should increase.

The second trend is a bit surprising. We might expect atoms to become *larger* as we go across a row of the periodic table because each element has one more electron than the preceding element. But the additional electrons are added to the same shell. Because the number of protons in the nucleus increases as we go across a row of the table, the core charge increases and the force of attraction between the nucleus and the electrons that surround it also increases. The nucleus therefore tends to pull the electrons in a given shell closer, and the atoms become *smaller*.

3.21 The Sizes of Atoms: Covalent Radii

The size of an atom can also be estimated by measuring the distance between adjacent atoms in compounds that are neither ionic nor metallic. These compounds are called covalent compounds. The **covalent radius** of a chlorine atom,

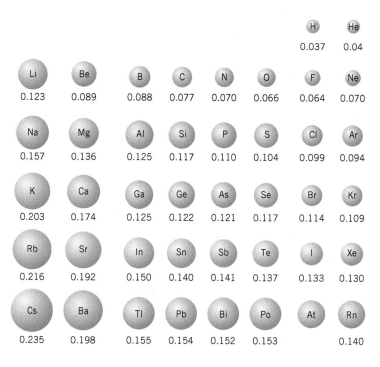

Fig. 3.26 Covalent radii, in units of nanometers, for the main-group elements. Atoms become larger as we go down a column of the periodic table and smaller as we go from left to right across a row of the table. The radii of the Noble gases are atomic radii.

for example, is assumed to be equal to one-half the distance between the nuclei of the atoms in a Cl_2 molecule.

The covalent radii of the main-group elements are given in Figure 3.26. These data confirm the trends observed for metallic radii. Atoms *become larger* as we go down a column of the periodic table, and they *become smaller* as we go across a row of the table.

Table B.4 in Appendix B contains covalent and metallic radii for a number of elements. The covalent radius for an element is usually a little smaller than the metallic radius. This can be explained by noting that covalent bonds tend to squeeze the atoms together, as shown in Figure 3.27.

> ➤ **CHECKPOINT**
>
> What is the difference between covalent and metallic radii?

Li   Li
Gaseous Li₂ molecule
r is the covalent radius
r = 0.123 nm

Li   Li
Lithium metal
r is the metallic radius
r = 0.152 nm

Fig. 3.27 The radii of atoms and ions can be determined by measuring the distance between adjacent nuclei. The covalent radius of lithium can be determined by measuring the internuclear distance in the gaseous Li_2 molecule, which exists only at very high temperatures. Li_2 is covalently bonded, and the Li atoms are pulled tightly together. The metallic radius of lithium can be determined from measurements of solid lithium metal. In lithium metal, each lithium atom is surrounded by other lithium atoms, all immersed in a sea of electrons.

3.22 The Relative Sizes of Atoms and Their Ions

The radii of ions can also be determined from measurements of the distance between adjacent nuclei. Consider LiCl, for example, in which Li^+ and Cl^- ions are held together by the strong coulombic force of attraction between particles of opposite charge. Measurements for ionic solids, however, must take into

Table 3.7

Covalent Radii of Neutral Group VIIA Atoms and Ionic Radii of Their Negative Ions

Element	Covalent Radius (nm)	Ion	Ionic Radius (nm)
F	0.064	F⁻	0.136
Cl	0.099	Cl⁻	0.181
Br	0.1142	Br⁻	0.196
I	0.1333	I⁻	0.216

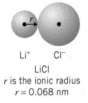

Li⁺   Cl⁻
LiCl
r is the ionic radius
r = 0.068 nm

Fig. 3.28 The ionic radii of the Li⁺ and Cl⁻ ions can be determined from measurements of lithium chloride.

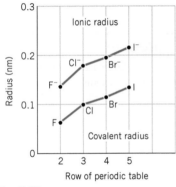

Fig. 3.29 A comparison of the radii of the F⁻, Cl⁻, Br⁻, and I⁻ ions with the covalent radii of the corresponding neutral atoms. In each case, the negatively charged ion is larger than the neutral atom.

account the fact that Li^+ ions differ significantly in size from Cl^- ions, as shown in Figure 3.28.

Table 3.7 and Figure 3.29 compare the covalent radii of neutral fluorine, chlorine, bromine, and iodine atoms with the radii of the corresponding F^-, Cl^-, Br^-, and I^- ions. In each case, the negative ion is much larger than the atom from which it was formed. In fact, the negative ion can be more than twice as large as the neutral atom.

The only difference between an atom and its ions is the number of electrons that surround the nucleus. A neutral chlorine atom, for example, contains 17 electrons, while a Cl^- ion contains 18 electrons.

$$Cl \quad [Ne]3s^2 3p^5 \qquad Cl^- \quad [Ne]3s^2 3p^6$$

Because the nucleus can't hold the 18 electrons in the Cl^- ion as tightly as the 17 electrons in the neutral Cl atom, the negative ion is significantly larger than the atom from which it forms.

Extending this line of reasoning suggests that positive ions should be smaller than the atoms from which they are formed. The 11 protons in the nucleus of an Na^+ ion, for example, should be able to hold the 10 electrons on the ion more tightly than the 11 electrons on a neutral sodium atom. The electrons removed from an atom to form a positive ion are the electrons that are most loosely held: the electrons in the outer shell. The Na^+ ion therefore should be much smaller than a sodium atom.

$$Na \qquad 1s^2 2s^2 2p^6 3s^1$$
$$Na^+ \qquad 1s^2 2s^2 2p^6$$

Table 3.8 and Figure 3.30 provide data to test this hypothesis. Here the covalent radii for neutral atoms of the Group IA elements are compared with the ionic radii

Table 3.8

Covalent Radii of Neutral Group IA Atoms and Ionic Radii of Their Positive Ions

Element	Covalent Radius (nm)	Ion	Ionic Radius (nm)
Li	0.123	Li⁺	0.068
Na	0.157	Na⁺	0.095
K	0.2025	K⁺	0.133
Rb	0.216	Rb⁺	0.148
Cs	0.235	Cs⁺	0.169

for the corresponding positive ions. In each case, the positive ion is much smaller than the atom from which it forms.

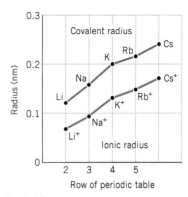

Fig. 3.30 A comparison of the radii of the Li^+, Na^+, K^+, Rb^+, and Cs^+ ions with the covalent radii of the corresponding neutral atoms. In each case, the positively charged ion is smaller than the neutral atom.

Exercise 3.7

Compare the sizes of neutral sodium and chlorine atoms and their Na^+ and Cl^- ions.

Solution

A neutral sodium atom is significantly larger than a neutral chlorine atom because chlorine has a larger core charge than sodium and their outer electrons are in the same shell.

$$Na \gg Cl$$
$$(0.157 \text{ nm}) \quad (0.099 \text{ nm})$$

But an Na^+ ion is only one-half the size of a Cl^- ion because the Na^+ ion is so much smaller than a neutral sodium atom, whereas the Cl^- ion is so much larger than a neutral chlorine atom.

$$Na^+ \ll Cl^-$$
$$(0.095 \text{ nm}) \quad (0.181 \text{ nm})$$

The size of these particles therefore increases in the following order.

$$Na^+ \approx Cl \ll Na \ll Cl^-$$

3.23 Patterns in Ionic Radii

The ionic radii in Tables 3.7 and 3.8 confirm one of the patterns observed for both metallic and covalent radii: Atoms become larger as we go down a column of the periodic table. We can examine trends in ionic radii across a row of the periodic table by comparing data for atoms and ions that are **isoelectronic.** By definition, isoelectronic atoms or ions have the same number of electrons. Table 3.9 summarizes data on the radii of a series of isoelectronic ions of second- and third-row elements.

The data in Table 3.9 can be explained by noting that these ions all have 10 electrons, but the number of protons in the nucleus increases from 6 in the C^{4-} ion to 13 in the Al^{3+} ion. As the charge on the nucleus becomes larger, the nucleus can hold the same number of electrons more tightly. As a result, the ions become significantly smaller—in this series, by a factor of 5 from C^{4-} to Al^{3+}.

▶ **CHECKPOINT**

Where would the radius of the Si^{4+} ion fit into Table 3.9?

Table 3.9
Radii for Isoelectronic Second-Row and Third-Row Atoms or Ions

Atom or Ion	Radius (nm)	Electron Configuration
C^{4-}	0.260	$1s^2\, 2s^2\, 2p^6$
N^{3-}	0.171	$1s^2\, 2s^2\, 2p^6$
O^{2-}	0.140	$1s^2\, 2s^2\, 2p^6$
F^-	0.136	$1s^2\, 2s^2\, 2p^6$
Na^+	0.095	$1s^2\, 2s^2\, 2p^6$
Mg^{2+}	0.065	$1s^2\, 2s^2\, 2p^6$
Al^{3+}	0.050	$1s^2\, 2s^2\, 2p^6$

3.24 Second, Third, Fourth, and Higher Ionization Energies

Sodium forms Na^+ ions, magnesium forms Mg^{2+} ions, and aluminum forms Al^{3+} ions. Why doesn't sodium form Na^{2+} ions, or even Na^{3+} ions? The answer can be obtained from data for the second, third, and higher ionization energies of the element.

As we have seen, the first ionization energy represents the energy it takes to remove the *outermost* electron from a neutral atom in the gas phase.

$$Na(g) + energy \longrightarrow Na^+(g) + e^-$$

For sodium, this is the energy required to remove the electron from the $3s$ orbital.

$$Na \qquad 1s^2 2s^2 2p^6 3s^1$$
$$Na^+ \qquad 1s^2 2s^2 2p^6$$

The *second ionization energy* of sodium is the energy it would take to remove another electron to form an Na^{2+} ion in the gas phase.

$$Na^+(g) + energy \longrightarrow Na^{2+}(g) + e^-$$
$$(1s^2\, 2s^2\, 2p^6) \qquad\qquad (1s^2\, 2s^2\, 2p^5)$$

The *third ionization energy* of sodium represents the process by which the Na^{2+} ion is converted to an Na^{3+} ion.

$$Na^{2+}(g) + energy \longrightarrow Na^{3+}(g) + e^-$$
$$(1s^2 2s^2 2p^5) \qquad\qquad (1s^2 2s^2 2p^4)$$

▶ **CHECKPOINT**

How do the second, third, and higher ionization energies differ from ionization energies obtained from PES?

The energy required to form an Na^{3+} ion from an Na atom in the gas phase is the sum of the first, second, and third ionization energies of the element.

A complete set of data for the ionization energies of the elements is given in Table B.5 in Appendix B. For the moment, let's look at the first, second, third, and fourth ionization energies of sodium, magnesium, and aluminum listed in Table 3.10.

It doesn't take much energy to remove one electron from a sodium atom to form an Na^+ ion with a filled-shell electron configuration. Once this happens, it takes almost 10 times as much energy to remove a second electron because the next available electrons are in the $n = 2$ shell, which is closer to the nucleus. Because it takes so much energy to remove the second electron, sodium generally forms compounds that contain Na^+ ions rather than Na^{2+} or Na^{3+} ions.

Table 3.10

First, Second, Third, and Fourth Ionization Energies (kJ/mol) of Gas-Phase Atoms of Sodium, Magnesium, and Aluminum

	First *IE*	Second *IE*	Third *IE*	Fourth *IE*
Na	495.8	4562.4	6912	9,543
Mg	737.7	1450.6	7732.6	10,540
Al	577.6	1816.6	2744.7	11,577

A similar pattern is observed when the ionization energies of magnesium are analyzed. The first ionization energy of magnesium is larger than that of sodium because magnesium has one more proton in its nucleus to hold onto the electrons in the 3s orbital.

$$\text{Mg} \qquad [\text{Ne}]3s^2$$

The second ionization energy of Mg is larger than the first because it always takes more energy to remove an electron from a positively charged ion than from a neutral atom. The third ionization energy of magnesium is enormous, however, because an electron would have to be removed from the $n = 2$ shell to form an Mg^{3+} ion.

The same pattern can be seen in the ionization energies of aluminum. The first ionization energy of aluminum is smaller than that of magnesium because it involves removing an electron from a 3p rather than a 3s orbital. The second ionization energy of aluminum is larger than the first, and the third ionization energy is even larger. Although it takes a considerable amount of energy to remove three electrons from an aluminum atom to form an Al^{3+} ion, the energy needed to remove a fourth electron is astronomical. Thus it would be a mistake to look for an Al^{4+} ion as the product of a chemical reaction.

Exercise 3.8

Predict the group in the periodic table in which an element with the following ionization energies would most likely be found.

$$
\begin{aligned}
1st\ IE &= \quad 786\ \text{KJ/mol} \\
2nd\ IE &= 1{,}577\ \text{KJ/mol} \\
3rd\ IE &= 3{,}232\ \text{KJ/mol} \\
4th\ IE &= 4{,}355\ \text{KJ/mol} \\
5th\ IE &= 16{,}091\ \text{KJ/mol}
\end{aligned}
$$

Solution

The gradual increase in the energy needed to remove the first, second, third, and fourth electrons from this element is followed by an abrupt increase in the energy required to remove one more electron. This is consistent with an element that has four electrons in the outermost shell, and we might expect the element to be in Group IVA of the periodic table. These data are in fact the ionization energies of silicon.

$$\text{Si} \qquad [\text{Ne}]3s^2 3p^2$$

The trends in the ionization energies of the elements can be used to explain why elements on the left side of the periodic table are more likely than those on the right to form positive ions. The ionization energies of elements on the left side of the table are much smaller than those of elements on the right. Consider the first ionization energies for sodium and chlorine, for example.

➤ **CHECKPOINT**

What is the maximum charge under normal conditions for an atom from Group IVA?

$$
\begin{aligned}
&\text{Na} \qquad \text{1st } IE = 495.8 \text{ KJ/mol} \\
&\text{Cl} \qquad \text{1st } IE = 1251.1 \text{ KJ/mol}
\end{aligned}
$$

Elements on the left side of the periodic table are therefore more likely to form positive ions.

Trends in ionization energies can also be used to explain why the maximum positive charge found on atoms under normal conditions for the main-group elements is equal to the group number of the element. The number of valence electrons is equal to the group number, so the maximum positive charge on an ion is also equal to the group number. Because aluminum is in Group IIIA, for example, it can lose only three electrons before an inner subshell is reached. Thus the maximum positive charge on an aluminum ion is +3.

3.25 Average Valence Electron Energy (AVEE)[1]

Ionization energies provide a measure of how tightly the electrons are held in an isolated atom. It would be useful to have a single quantity that reflects the *average* ionization energy of the valence electrons on an atom. This quantity, which is known as the **average valence electron energy (AVEE),** can be calculated from the ionization energies for the valence electrons on the atom obtained by photoelectron spectroscopy. Because there are different numbers of electrons in the various subshells, the AVEE is calculated as a weighted average.

Consider carbon, for example, which has two electrons in the valence $2s$ subshell and two electrons in the valence $2p$ subshell. The ionization energies of electrons from the $2s$ subshell ($IE_s = 1.72$ MJ/mol) and the $2p$ subshell ($IE_p = 1.09$ MJ/mol) are given in Table 3.4. The average valence electron energy for a carbon atom can therefore be calculated as follows.

$$
\begin{aligned}
\text{AVEE}_\text{C} &= \left[\frac{(2 \times IE_s) + (2 \times IE_p)}{2 + 2} \right] = \left[\frac{(2 \times 1.72 \text{ MJ/mol}) + (2 \times 1.09 \text{ MJ/mol})}{4} \right] \\
&= 1.41 \text{ MJ/mol}
\end{aligned}
$$

Fluorine, on the other hand, would have an AVEE that is significantly larger.

$$
\begin{aligned}
\text{AVEE}_\text{F} &= \left[\frac{(2 \times IE_s) + (5 \times IE_p)}{2 + 5} \right] = \left[\frac{(2 \times 3.88 \text{ MJ/mol}) + (5 \times 1.68 \text{ MJ/mol})}{7} \right] \\
&= 2.31 \text{ MJ/mol}
\end{aligned}
$$

The AVEE values for the main-group elements calculated from photoelectron spectroscopy data are given in Figure 3.31.

---

[1]L. C. Allen, *Journal of the American Chemical Society,* **111,** 9003 (1989).

H									He
1.31									2.37
Li	Be		B	C	N	O	F		Ne
0.52	0.90		1.17	1.41	1.82	1.91	2.31		2.73
Na	Mg		Al	Si	P	S	Cl		Ar
0.50	0.74		0.92	1.13	1.39	1.35	1.59		1.85
K	Ca	Sc	Ga	Ge	As	Se	Br		Kr
0.42	0.59	0.68	1.00	1.07	1.26	1.3	1.53		1.69
Rb	Sr	Y	In	Sn	Sb	Te	I		Xe
0.40	0.55	0.57	0.94	1.04	1.13	1.2	1.35		1.47

Fig. 3.31 Average valence electron energies (AVEE) for the main-group elements, in megajoules per mole (MJ/mol). [L. C. Allen, *Journal of the American Chemical Society*, **111,** 9003 (1989).]

With the exception of phosphorus, there is a systematic increase in AVEE from left to right across each row of the periodic table. Because the charge on the nucleus steadily increases as we go across each row, and the size of the atom gradually decreases, the valence electrons on each atom are held more tightly as we go across the row.

AVEE measures two important quantities: the attraction of an atom for its electrons and the difference between the energies of the subshells in the outermost or valence shell. Figure 3.32 shows the relative energies of the 2s and 2p electrons in B, C, N, O, and F. These data suggest that the difference between the energies of the subshells in the valence shell of an atom increases significantly as we go across a row of the periodic table. The opposite effect is seen as we proceed down a vertical column in the periodic table. As we go down a column in the periodic table, the difference between the energies of the valence subshells becomes smaller. As a result, AVEE values increase toward the upper-right corner of the periodic table and decrease toward the lower-left corner of the table.

The metals on the left side of the periodic table have AVEE values that are relatively small compared with the nonmetals on the right side of the table. Sodium and magnesium, for example, have AVEE values of 0.50 and 0.74 MJ/mol, respectively, whereas the AVEE values for O and Cl are 1.91 and 1.59 MJ/mol. Thus sodium and magnesium are more likely to lose electrons to form positive ions than O and Cl, and O and Cl are more likely to add electrons to form negative ions than Na and Mg.

3.26 AVEE and Metallicity

Because the average valence electron energy provides a measure of how tightly an atom holds onto its valence electrons and the difference between the energies of the valence electron subshells, it can be used to explore the dividing line between the metals and nonmetals in the periodic table. The semimetals or metalloids that lie along the stairstep line separating the metals from the nonmetals in the periodic table all have similar AVEE values. B, Si, Ge, As, Sb, and Te all have AVEE values that lie between 1.07 and 1.26 MJ/mol, and they are the only elements in the periodic table that have values within this range, as shown in Figure 3.33.

> **CHECKPOINT**

Why do AVEE values increase from left to right across a period and from bottom to top of a group? What other periodic properties follow these trends?

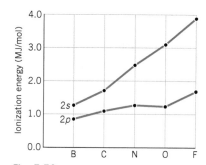

Fig. 3.32 Ionization energies of the electrons in the 2s and 2p subshells of B, C, N, O, and F. Two trends can be observed: The ionization energies of the subshells increase (the electrons are more tightly held) as we move across the period, and the difference in energy between the s and the p electrons also increases as we go across the period.

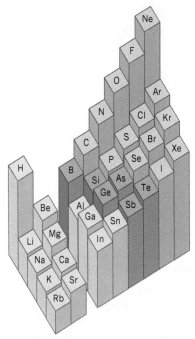

Fig. 3.33 Three-dimensional plot of the average valence electron energies (AVEE) of the main-group elements versus position in the periodic table. The AVEE is a measure of how tightly atoms hold onto their valence electrons and the energy gap between valence subshells.

The AVEE of an atom of a given element can therefore be used to decide whether the element is a metal or nonmetal. Atoms with an AVEE value below 1.07 MJ/mol are metals, whereas those with an AVEE greater than 1.26 MJ/mol are nonmetals. Elements with AVEE values in the range of 1.07 to 1.26 MJ/mol have properties intermediate between those of metals and nonmetals and are known as the *semimetals,* or *metalloids.*

Figure 3.33 raises an interesting point: The AVEE value for hydrogen (1.31 MJ/mol) is only slightly larger than the six elements that are labeled as semimetals (B, Si, Ge, As, Sb, and Te). This is consistent with the chemistry of hydrogen, which forms compounds such as HCl that are analogous to those formed by the metals in Group IA (NaCl, KCl, and so on), as well as compounds such as MgH_2 that are analogous to those formed by the nonmetals in Group VIIA (MgF_2, $MgCl_2$, $MgBr_2$, and so on). Because of its position in Figure 3.33, we would suspect that hydrogen might be difficult to classify unambiguously as a metal or nonmetal. The fact that hydrogen is found as H_2 molecules in the gas phase at room temperature and atmosphere pressure leads most students to consider it a nonmetal. It should be noted, however, that elemental hydrogen forms a metal at high pressures, such as the pressures in the center of the giant planets such as Jupiter and Saturn.

The power of AVEE data can be illustrated by considering the difference between the second and third versus the fourth and fifth rows of the periodic table. In the second and third rows, there is only one semimetal each (B and Si). In the fourth and fifth rows, there are two (Ge and As, and Sb and Te). If we tried to extend the pattern seen in the fourth and fifth rows of the periodic table to the second and third rows, we would predict that beryllium and aluminum would be semimetals. But the AVEE values for beryllium (0.90) and aluminum (0.92) clearly indicate that these elements should be metals. This is consistent with the observation that both elements exhibit all of the characteristic properties of a metal: They have a metallic luster, they are good conductors of heat and electricity, and they are both malleable and ductile.

In general, two factors contribute to metallic behavior: Valence electrons must be easy to remove, and the difference between the energies of valence subshells must be small. AVEE is a convenient measure of both effects. The increasing trend for metallicity down a group also follows the AVEE values. Carbon is a nonmetal; but as we go down Group IVA through Si, Ge, Sn, and Pb, metallic behavior increases. This reflects two things: The electrons are becoming easier to remove, and the valence subshells are becoming closer to each other in energy.

Key Terms

Amplitude
Angular quantum number
Average valence electron energy (AVEE)
Bohr model
Core charge
Covalent radius
Electromagnetic radiation
Electron configuration
Emission spectrum
First ionization energy
Frequency

Hund's rules
Isoelectronic
Magnetic quantum number
Metallic radii
Nucleus
Orbital diagram
Orbitals
Particles
Photoelectron spectroscopy (PES)
Photons
Principal quantum number
Quantized

Quantum mechanical model
Shell
Shell model
Spectrum
Speed
Spin
Spin quantum number
Subshell
Transition metals
Valence electrons
Wavelength
Waves

Problems

Rutherford's Model of the Atom

1. According to Rutherford's model of the atom, where is the positive charge concentrated?

2. What does Rutherford's model predict about the relative size of the nucleus and the radius of the atom?

3. What is the nucleus composed of according to Rutherford's model?

4. A chemistry text published in 1922—before the discovery of the neutron—proposed a model of the atom based on only two subatomic particles: electrons and protons. All of the protons and some of the electrons were assumed to be concentrated in the nucleus of the atom. The other electrons revolved around the nucleus. The number of protons in the nucleus was equal to the mass of the atom. The charge on the nucleus was equal to the number of protons minus the number of electrons in the nucleus. Enough electrons were then added to the atom to neutralize the charge on the nucleus. Use this model to calculate the number of protons and electrons in a neutral fluorine atom, F, and a fluoride ion, F⁻. Compare this calculation with results obtained by assuming that the nucleus is composed of protons and neutrons.

Particles and Waves

5. Describe the difference between a particle and a wave.

6. To examine the relationship among the frequency, wavelength, and speed of a wave, imagine that you are sitting at a railroad crossing, watching a train that consists of 45-ft boxcars go by. Furthermore assume it takes 1.0 s for each boxcar to pass in front of your car. Calculate the "frequency" of a boxcar and the product of the 45-ft "wavelength" times this frequency. Convert the final answer into units of miles per hour.

7. An octave in a musical scale corresponds to a change in the frequency of a note by a factor of 2. If a note with a frequency of 440 Hz is an "A," then the "A" one octave above this note has a frequency of 880 Hz. What happens to the wavelength of the sound as the frequency increases by a factor of 2? What happens to the speed at which the sound travels to your ear?

8. The human ear is capable of hearing sound waves with frequencies between about 20 and 20,000 Hz. If the speed of sound is 340.3 m/s at sea level and 20°C, what is the wavelength in meters of the longest wave the human ear can hear?

Light and Other Forms of Electromagnetic Radiation

9. Calculate the wavelength in meters of light that has a frequency of 5.0×10^{14} cycles per second.

10. Calculate the frequency of red light that has a wavelength of 700 nm.

11. Which has the longer wavelength, red light or blue light?

12. Which has the larger frequency, radio waves or microwaves?

13. In a magnetic field of 2.35 tesla, ^{13}C nuclei absorb electromagnetic radiation that has a frequency of 25.147 MHz. Calculate the wavelength of this radiation. In which region of the electromagnetic spectrum does the radiation fall?

14. The meter has been defined as 1,650,763.73 wavelengths of the orange-red line of the emission spectrum of ^{86}Kr. Calculate the frequency of this radiation. In what portion of the electromagnetic spectrum does the radiation fall?

15. Methylene blue, $C_{16}H_{18}ClN_3S$, absorbs light most intensely at wavelengths of 668 and 609 nm. What color is the light absorbed by the dye?

Atomic Spectra

16. A pink-violet light is observed when current is passed through a glass tube containing hydrogen gas. When this light emerges from a prism, four narrow lines of visible light are seen (Table 3.1). What are the frequencies of these four lines?

17. Sodium salts give off a characteristic yellow-orange light when added to the flame of a Bunsen burner. The yellow-orange color is due to two narrow bands of radiation with wavelengths of 588.9953 and 589.5923 nm. Calculate the frequencies of these emission lines.

The Wave Packet Model of Electromagnetic Radiation

18. Which has more energy, ultraviolet or infrared radiation?

19. Which has more energy, light with a wavelength of 580 nm or light with a wavelength of 660 nm?

20. List the four lines in the visible region of the emission spectrum of hydrogen in order of increasing energy. See Table 3.1.

21. Calculate the energy in joules of the radiation in the emission spectrum of the hydrogen atom that has a wavelength of 656.3 nm.

22. Calculate the energy in joules of a single particle of radiation broadcast by an amateur radio operator who transmits at a wavelength of 10 m.

23. Cl_2 molecules can dissociate to form chlorine atoms by absorbing electromagnetic radiation. It takes 243.4 kJ of energy to break the bonds in a mole of Cl_2 molecules. What is the wavelength of the radiation that has just enough energy to decompose a Cl_2 molecule to chlorine atoms? In what portion of the spectrum is this wavelength found?

The Bohr Model of the Atom

24. Why do hydrogen atoms emit or absorb discrete amounts of energy?

25. How does the Bohr model of the atom differ from Rutherford's model?

26. In the Bohr atom, what holds the electron and nucleus together?

27. As the distance between the electron and the nucleus increases, what happens to the force of attraction between them according to Bohr's model?

28. For two charged particles separated by a distance r, the force between the particles is given by Coulomb's law:

$$F = \frac{q_1 \times q_2}{r^2}$$

   (a) If the particles are separated by an infinite distance, what is the magnitude of the force of attraction between them?

   (b) If the signs of the charges on the particles are the same, is the sign of the force between them positive or negative?

   (c) If the signs of the charges on the particles are not the same, is the sign of the force between them positive or negative?

   (d) Is the sign of the force between the particles in the hydrogen atom positive or negative?

   (e) Which of the following has the greatest force between the particles?

   (i) An electron 1000 picometers (pm) from another particle of charge +1 or an electron 500 pm from another particle of charge +1?

   (ii) An electron 1000 pm from another particle of charge +1 or an electron 1000 pm from another particle of charge +2?

29. (a) Which of the following arrangements has the largest force of attraction between the charged species?

$$\oplus \!-\!\!\frac{r}{}\!\!-\! \ominus \qquad \textcircled{\scriptsize 2+} \!-\!\!\frac{r}{}\!\!-\! \textcircled{\scriptsize 2-} \qquad \textcircled{\scriptsize 2+} \!-\!\!\frac{r}{}\!\!-\! \textcircled{\scriptsize 2+}$$
$$\text{(i)} \qquad\qquad \text{(ii)} \qquad\qquad \text{(iii)}$$

   (b) Which of the following has the smallest force of attraction between charged species?

$$\oplus \!-\!\!\frac{r}{}\!\!-\! \ominus \qquad \oplus \!-\!\!\frac{2r}{}\!\!-\! \ominus \qquad \textcircled{\scriptsize 2+} \!-\!\!\frac{2r}{}\!\!-\! \textcircled{\scriptsize 2-}$$
$$\text{(i)} \qquad\qquad \text{(ii)} \qquad\qquad \text{(iii)}$$

   (c) Which of the following would require the smallest amount of energy to remove the negatively charged particle an infinite distance away from the positively charged particle?

$$\oplus \!-\!\!\frac{r}{}\!\!-\! \ominus \qquad \textcircled{\scriptsize 2+} \!-\!\!\frac{2r}{}\!\!-\! \ominus \qquad \textcircled{\scriptsize 2+} \!-\!\!\frac{2r}{}\!\!-\! \textcircled{\scriptsize 2-}$$
$$\text{(i)} \qquad\qquad \text{(ii)} \qquad\qquad \text{(iii)}$$

The Energy States of the Hydrogen Atom

30. Refer to Figure 3.5. How much energy is required to remove the electron completely from the hydrogen atom?

31. If the H-atom electron in the $n = 1$ state absorbs energy equivalent to 2.178×10^{-18} J, what energy state will it occupy?

32. How much energy in joules is required to change the energy state of the H-atom electron from $n = 2$ to $n = 3$? From $n = 2$ to $n = 4$? From $n = 2$ to infinity?

33. If the energy state of the H-atom electron is changed from $n = 3$ to $n = 2$, what is the energy change in joules and the frequency of the photon emitted?

34. From which energy state is it easier to remove the electron from a hydrogen atom, the $n = 1$ or $n = 2$ energy state? In which state is the electron most stable?

35. Suppose that 2.091×10^{-18} J is absorbed by the electron of a hydrogen atom in the $n = 1$ energy state. Describe the final energy state of the atom.

The First Ionization Energy

36. Explain why it takes energy to remove an electron from an isolated atom in the gas phase.

37. List the following elements in order of increasing first ionization energy.

   (a) Li   (b) Be   (c) F   (d) Na

The Shell Model

38. Use the data in Table 3.3 to construct a shell model of the atom similar to that in Figure 3.10 for Na through Ca.

39. Calculate the core charge for the atoms Na through Ca.

40. How would you expect the ionization energy of Na to compare with that of He? Explain.

41. How would you expect the ionization energy of Cl^- to compare with that of Ar? Explain.

42. Predict the order of the first ionization energies for I, Xe, and Cs. Explain your reasoning.

43. Which electron is harder to remove from an Li atom, the one in the outermost shell or the one in the innermost shell? Explain.

44. Static electricity results from the buildup of charge on a material. If wool is rubbed on a piece of rubber, the rubber becomes negatively charged and the wool becomes positively charged. Use the shell model for the structure of the atom to explain static electricity.

45. According to the shell model, why is the first ionization energy of Cl smaller than that of F?

46. If a single electron is removed from a Li atom, the resulting Li^+ ion has only two electrons, both in the $n = 1$ shell. In this respect it is very similar to an He atom. How would you expect the ionization energy of the Li^+ ion to compare with that of a He atom? Explain your reasoning.

47. If a single electron is added to an F atom, the resulting F⁻ ion has a total of eight valence electrons in the $n = 2$ shell. In this respect it is very similar to an Ne atom. How would you expect the first ionization energy of the F⁻ ion to compare with that of an Ne atom? Explain your reasoning.

48. Predict the order of the first ionization energies for the atoms Br, Kr, and Rb. Explain your reasoning.

49. Which has the lowest first ionization energy, He or Be? Explain.

50. Describe the general trend in first ionization energies from left to right across the second row of the periodic table.

51. Describe the general trend in first ionization energies from top to bottom of a column of the periodic table.

52. Explain why the first ionization energy of N is smaller than that of F.

53. Explain why the first ionization energy of hydrogen is so much larger than the first ionization energy of sodium.

The Shell Model and the Periodic Table

54. According to the shell model, why do first ionization energies increase across a row (period) of the periodic table?

55. According to the shell model, why do first ionization energies decrease down a column (group) of the periodic table?

56. Which of the following elements should have the largest first ionization energy?
   (a) B   (b) C   (c) N   (d) Mg   (e) Al

57. Which of the following elements should have the smallest ionization energy?
   (a) Mg   (b) Ca   (c) Si   (d) S   (e) Se

58. How many valence electrons does each of the atoms from H through Ne have?

59. What is the relationship among the core charge, number of valence electrons, and group number?

60. Why are valence electrons easier to remove from an atom than core electrons?

Photoelectron Spectroscopy and the Structure of Atoms

61. Do all of the electrons in a given shell have the same energy? Explain why or why not.

62. Refer to Table 3.3. If radiation with an energy of 520.2 kJ/mol strikes a lithium atom, can an electron be removed?

63. If radiation with energy of 483.6 kJ/mol strikes a sodium atom, could an electron be removed? See Table 3.5.

64. In a single PES experiment, how many electrons are removed from an individual atom?

65. If a photon with an energy of 920.6 kJ/mol strikes a B atom, and the kinetic energy of the electron that is ejected is 120.0 kJ/mol, what is the ionization energy of the ejected electron?

66. What determines the height of each peak in a photoelectron spectrum?

67. Why are the number of peaks in the PES for H and He the same?

68. Why does the PES of Li have two peaks? Why are the peaks of different heights?

69. In Figure 3.15, the energy associated with the peak representing the largest ionization energy increases from H to B. Why?

70. Why is the lowest-energy peak in Figure 3.15 assigned to Li?

71. Why are there three peaks in the PES of B?

72. Refer to Figure 3.15. If photons of energy 1.40 MJ/mol strike a B atom, what electrons could be removed? What electrons in a Be atom could be removed by photons of this energy?

Electron Configurations from Photoelectron Spectroscopy

73. Why are two of the three peaks in the PES spectrum of neon assigned to the $n = 2$ shell rather than to the $n = 1$ shell? What is the rationale for assuming that the peak at 84.0 MJ/mol corresponds to electrons in the $n = 1$ shell?

74. Roughly sketch the photoelectron spectra for Al and S. Give the relative intensities of the peaks.

75. What element do you think should give rise to the photoelectron spectrum shown in Figure 3P.1? Explain your reasoning.

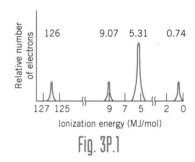

Fig. 3P.1

76. Use the ionization energies given below for Li, Na, and Ar to predict the photoelectron spectrum of K.

| Element | Ionization Energy (MJ/mol) | | | | |
	$1s$	$2s$	$2p$	$3s$	$3p$
Li	6.26	0.52			
Na	104	6.84	3.67	0.50	
Ar	309	31.5	24.1	2.82	1.52

(a) Consider the first three shells (18 electrons) of K. For these 18 electrons, indicate the relative energies of the peaks and their relative intensities in the PES spectrum of this element

(b) If the nineteenth electron of K is found in the $n = 4$ shell, would the ionization energy be closest to 0.42, 1.4, or 2.0 MJ/mol? Explain. (*Hint:* Compare to Na and Li.) Show a predicted photoelectron spectrum based on this assumption.

(c) If the nineteenth electron of K is found in the third subshell of the $n = 3$ shell, would the ionization energy be closest to 0.42, 1.4, or 2.0 MJ/mol? Explain. (*Hint:* Compare to Ar.) Show a predicted photoelectron spectrum based on this assumption.

(d) Given the photoelectron spectrum of K in Figure 3P.2, predict whether the nineteenth electron of K is found in the $n = 4$ or $n = 3$ shell. Explain your reasoning.

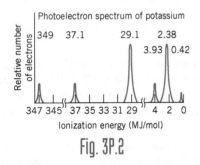

Fig. 3P.2

77. Identify the element whose photoelectron spectrum is shown in Figure 3P.3. (*Note:* In Figure 3.P3, the peak that arises from the 1s electrons has been omitted.)

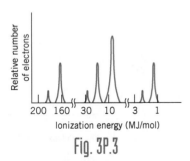

Fig. 3P.3

78. The PES for element *X* is shown in Figure 3P.4.

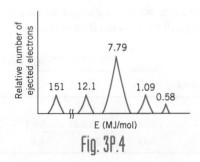

Fig. 3P.4

(a) What is the group number of this element?

(b) What is the maximum positive charge that this atom is likely to have under normal chemical conditions? Explain.

(c) What is this atom's core charge?

(d) Which element is this? Explain your reasoning.

(e) Assume that this atom is irradiated with light of 206-nm wavelength. Which, if any, of the atom's electrons could be removed by the photons of this light? (*Note:* 10^{-9} m = 1 nm.)

79. The PES spectrum for atom *X* is shown in Figure 3P.5.

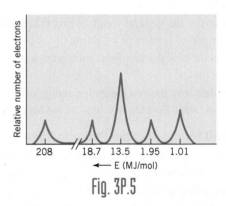

Fig. 3P.5

(a) Identify atom *X* and explain how you arrived at your conclusion.

(b) Sketch the shell model for atom *X* that corresponds to the PES spectrum.

(c) Sketch the PES spectrum for the atom having eight fewer protons than atom *X*. Label the axis clearly and show the approximate energies.

(d) If photons of wavelength 1.20×10^{-8} m are used to bombard atom *X*, which, if any, of the electrons of atom *X* could be completely removed from the atom?

Allowed Combinations of Quantum Numbers

80. Sketch the shapes of the orbitals for which $l = 0$, $l = 1$, and $l = 2$.

81. If $l = 1$, what values can m_l have? What values can l have if $n = 2$?

82. Identify the quantum number that specifies each of the following.

(a) The size of the orbital

(b) The shape of the orbital

(c) The way the orbital is oriented in space

Shells and Subshells of Orbitals

83. Determine the allowed values of the angular quantum number, l, when the principal quantum number is 4. Describe the difference between orbitals that have the

same principal quantum number and different angular quantum numbers.

84. Write the combination of quantum numbers for every electron in the $n = 1$ and $n = 2$ shells using the selection rules outlined in this chapter.

85. Identify the symbols used to describe orbitals for which $l = 0, 1, 2,$ and 3.

86. Determine the number of orbitals in the $n = 3$, $n = 4$, and $n = 5$ shells.

87. Which of the following orbitals cannot exist?

(a) 6s   (b) 3p   (c) 2d   (d) 4f   (e) 3f

Orbitals and the Pauli Exclusion Principle

88. What is meant by the term *paired electrons*?

89. What causes an atom to be magnetic?

90. Under what circumstances can an atom have no magnetic properties?

91. What would be the results of a Stern–Gerlach experiment on a beam of these atoms?

(a) H   (b) He   (c) Be   (d) F

92. Which of the following is a legitimate set of $n, l, m_l,$ and m_s quantum numbers?

(a) $4, -2, -1, \frac{1}{2}$ (b) $4, 2, 3, \frac{1}{2}$

(c) $4, 3, 0, 1$ (d) $4, 0, 0, -\frac{1}{2}$

93. Which of the following is a legitimate set of $n, l, m_l,$ and m_s quantum numbers?

(a) $0, 0, 0, \frac{1}{2}$ (b) $8, 4, -3, -\frac{1}{2}$

(c) $3, 3, 2, \frac{1}{2}$ (d) $2, 1, -2, -\frac{1}{2}$

(e) $5, 3, 3, -1$

94. Determine the number of allowed values of the magnetic quantum number when $n = 5$ and $l = 3$.

95. Calculate the maximum number of unpaired electrons that can be placed in a 5d subshell.

96. Explain why the difference between the atomic numbers of pairs of elements in a vertical column, or group, of the periodic table is either 8, 18, or 32.

97. Write the combination of quantum numbers for every electron in the $n = 1$ and $n = 2$ shells assuming that the selection rules for assigning quantum numbers are changed to the following.

(a) The principal quantum number can be any integer greater than or equal to 1.

(b) The angular quantum number can have any value between 0 and n.

(c) The magnetic quantum number can have any value between 0 and 1.

(d) The spin quantum number can have a value of either +1 or −1.

98. Which of the following sets of $n, l, m_l,$ and m_s quantum numbers can be used to describe an electron in a 2p orbital?

(a) $2, 1, 0, -\frac{1}{2}$ (b) $2, 0, 0, \frac{1}{2}$ (c) $2, 2, 1, \frac{1}{2}$

(d) $3, 2, 1, -\frac{1}{2}$ (e) $3, 1, 0, \frac{1}{2}$

99. Calculate the maximum number of electrons that can have the quantum numbers $n = 4$ and $l = 3$.

100. Calculate the number of electrons in an atom that can simultaneously possess the quantum numbers $n = 4$ and $m_s = \frac{1}{2}$.

101. Determine the allowed values of the spin quantum number, m_s, when $n = 5$, $l = 2$, and $m_l = -1$.

102. Determine the number of subshells in the $n = 3$ and $n = 4$ shells.

103. Calculate the maximum number of electrons in the $n = 1$, $n = 2$, $n = 3$, and $n = 4$ shells.

104. Calculate the maximum number of electrons that can fit into a 4d subshell.

105. Which of the following is a possible set of $n, l, m_l,$ and m_s quantum numbers for the last electron added to form a gallium atom $(Z = 31)$?

(a) $3, 1, 0, -\frac{1}{2}$ (b) $3, 2, 1, \frac{1}{2}$ (c) $4, 0, 0, \frac{1}{2}$

(d) $4, 1, 1, \frac{1}{2}$ (e) $4, 2, 2, \frac{1}{2}$

106. Which of the following is a possible set of $n, l, m_l,$ and ms quantum numbers for the last electron added to form an As^{3+} ion?

(a) $3, 1, -1, \frac{1}{2}$ (b) $4, 0, 0, -\frac{1}{2}$ (c) $3, 2, 0, \frac{1}{2}$

(d) $4, 1, -1, \frac{1}{2}$ (e) $5, 0, 0, \frac{1}{2}$

107. What requirement must two electrons meet if they are to occupy the same orbital?

Predicting Electron Configurations

108. Which of the following sets of subshells for yttrium is arranged in the correct order of filling?

(a) 3d, 4s, 4p, 5s, 4d   (b) 3d, 4s, 4p, 4d, 5s

(c) 4s, 3d, 4p, 5s, 4d   (d) 4s, 3d, 4p, 4d, 5s

109. Which of the following orders of filling of orbitals is incorrect?

(a) 3s, 4s, 5s   (b) 5s, 5p, 5d   (c) 5s, 4d, 5p

(d) 6s, 4f, 5d   (e) 6s, 5f, 6p

110. As atomic orbitals are filled, the 6p orbitals are filled immediately after which of the following orbitals?

(a) 4f   (b) 5d   (c) 6s   (d) 7s

Electron Configurations and the Periodic Table

111. Describe some of the evidence that could be used to justify the argument that the modern periodic table is based on similarities in the chemical properties of the elements. Describe evidence that could be used to justify the argument that the periodic table groups elements with similar electron configurations.

112. Determine the row and column of the periodic table in which you would expect to find the first element to have 3d electrons in its electron configuration.

113. Determine the row and column of the periodic table in which you would expect to find the element that has five more electrons than the rare gas krypton.

114. Determine the group of the periodic table in which an element with the following electron configuration belongs.

    $$[X] \quad 1s^2 2s^2 2p^6 3s^2 4s^2 3d^{10} 4p^6 5s^2 4d^{10} 5p^3$$

115. In which group of the periodic table should element 119 belong if and when it is discovered?

116. Describe the groups of the periodic table in which the s, p, d, and f subshells are filled.

117. Give the electron configuration of the nitrogen atom.

118. Give the electron configuration for nickel.

119. The electron configuration of Si is $1s^2 2s^2 2p^6 3s^2 3p^x$, where x is which of the following?
    (a) 1   (b) 2   (c) 3   (d) 4   (e) 6

120. Which of the following is the correct electron configuration for the P^{3-} ion?
    (a) [Ne]   (b) [Ne] $3s^2$
    (c) [Ne] $3s^2 3p^3$ (d) [Ne] $3s^2 3p^6$

121. Which of the following is the correct electron configuration for the bromide ion, Br^-?
    (a) [Ar] $4s^2 4p^5$ (b) [Ar] $4s^2 3d^{10} 4p^5$
    (c) [Ar] $4s^2 3d^{10} 4p^6$ (d) [Ar] $4s^2 3d^{10} 4p^6 5s^1$

122. Determine the number of electrons in the third shell of a vanadium atom.

123. Determine the number of electrons in s orbitals in the Ti^{2+} ion.

124. Theoreticians predict that the element with atomic number 114 will be more stable than the elements with atomic numbers between 103 and 114. On the basis of its electron configuration, in which group of the periodic table should element 114 be placed?

125. Which is the first element to have $4d$ electrons in its electron configuration?
    (a) Ca   (b) Sc   (c) Rb   (d) Y   (e) La

126. Which of the following contains sets of atoms or ions that have equivalent electron configurations?
    (a) B^{3+}, C^{4+}, H^+, He   (b) Na^+, Ne, N^{3+}, O^{2-}
    (c) Mg^{2+}, F^-, Na^+, O^{2-} (d) Ne, Ar, Xe, Kr
    (e) O^{2-}, S^{2-}, Se^{2-}, Te^{2-}

Electron Configurations and Hund's Rules

127. Which of the following electron configurations for carbon satisfies Hund's rules?

    (a) $1s^2 2s^2 2p^2$ $\dfrac{\uparrow}{} \dfrac{\downarrow}{} \dfrac{}{}$

    $2p$

    (b) $1s^2 2s^2 2p^2$ $\dfrac{\uparrow}{} \dfrac{\uparrow}{} \dfrac{}{}$

    $2p$

    (c) $1s^2 2s^2 2p^2$ $\dfrac{\uparrow}{} \dfrac{}{} \dfrac{\downarrow}{}$

    $2p$

    (d) $1s^2 2s^2 2p^2$ $\dfrac{\uparrow\downarrow}{} \dfrac{}{} \dfrac{}{}$

    $2p$

128. Draw orbital diagrams for the following atoms.
    (a) Si   (b) V   (c) Ga   (d) Cl   (e) Na

129. Which of the following orbital diagrams are incorrect for all electrons in the lowest-energy levels of an atom?

    (a) $\dfrac{\uparrow\downarrow}{s} \quad \dfrac{\uparrow\downarrow}{}\dfrac{\uparrow}{}\dfrac{}{}$

    p

    (b) $\dfrac{\uparrow}{s} \quad \dfrac{\uparrow\downarrow}{}\dfrac{\uparrow}{}\dfrac{\uparrow}{}$

    p

    (c) $\dfrac{\uparrow\downarrow}{s} \quad \dfrac{\uparrow}{}\dfrac{\uparrow}{}\dfrac{\uparrow}{}$

    p

    (d) $\dfrac{}{s} \quad \dfrac{\uparrow\downarrow}{}\dfrac{\uparrow\downarrow}{}\dfrac{\downarrow}{}$

    p

130. Which of the following neutral atoms has the largest number of unpaired electrons?
    (a) Na   (b) Al   (c) Si   (d) P   (e) S

131. Which of the following ions has five unpaired electrons?
    (a) Ti^{4+} (b) Co^{2+} (c) V^{3+} (d) Fe^{3+} (e) Zn^{2+}

132. Atoms that have unpaired electrons are magnetic. Those that have no unpaired electrons are not magnetic. Which of the following atoms or ions are magnetic? Show your work.
    (a) H   (b) He   (c) F^- (d) Na
    (e) Mg   (f) Si   (g) Cr^{3+}

The Sizes of Atoms: Metallic Radii

133. Describe what happens to the sizes of the atoms as we go down a column of the periodic table. Explain.

134. Describe what happens to the sizes of the atoms as we go across a row of the periodic table from left to right. Explain.

135. At one time, the size of an atom was given in units of angstroms because the radius of a typical atom was about 1 angstrom (Å). Now they are given in a variety of units. If the radius of a gold atom is 1.442 Å, and 1 Å is equal to 10^{-8} cm, what is the radius of this atom in nanometers and in picometers?

136. Which of the following atoms has the smallest radius?
    (a) Na   (b) Mg   (c) Al   (d) K   (e) Ca

The Sizes of Atoms: Covalent Radii

137. Explain why the covalent radius of an atom is smaller than the metallic radius of the atom.

138. Which of the following atoms has the largest covalent radius?
    (a) N   (b) O   (c) F   (d) P   (e) S

139. What happens to the covalent radii as the periodic table is traversed from left to right? From top to bottom? Explain.

140. Why does the covalent radius undergo a dramatic change from Xe to Cs?

The Relative Sizes of Atoms and Their Ions

141. Look up the covalent radii for magnesium and sulfur atoms and the ionic radii of Mg^{2+} and S^{2-} ions in Appendix B.4. Explain why Mg^{2+} ions are smaller than S^{2-} ions even though magnesium atoms are larger than sulfur atoms.

142. Explain why the radius of a Pb^{2+} ion (0.120 nm) is very much larger than that of a Pb^{4+} ion (0.084 nm).

143. Predict the relative sizes of the Fe^{2+} and Fe^{3+} ions, which can be found in a variety of proteins, including hemoglobin, myoglobin, and the cytochromes.

144. Describe what happens to the radius of an atom when electrons are removed to form a positive ion. Describe what happens to the radius of the atom when electrons are added to form a negative ion.

145. Predict the order of increasing ionic radius for the following ions: F^-, Cl^-, Br^-, and I^-. Compare your predictions with the data for the ions in Appendix B.4. Explain any differences between your predictions and the experiment.

Patterns in Ionic Radii

146. Predict whether the Al^{3+} or the Mg^{2+} ion is the smaller. Explain.

147. Which of the following ions has the largest radius? Explain.
(a) Na^+ (b) Mg^{2+} (c) S^{2-} (d) Cl^- (e) Se^{2-}

148. Which of the following atoms or ions is the smallest? Explain.
(a) Na (b) Mg (c) Na^+ (d) Mg^{2+} (e) O^{2-}

149. Which of the following ions has the smallest radius? Explain
(a) K^+ (b) Li^+ (c) Be^{2+} (d) O^{2-} (e) F^-

150. Sort the following atoms or ions into isoelectronic groups.
(a) N^{3-} (b) Ar (c) F^- (d) Ne
(e) P^{3-} (f) Ca^{2+} (g) Al^{3+} (h) Si^{4+}
(i) Na^+ (j) S^{2-} (k) Cl^- (l) O^{2-}
(m) K^+ (n) Mg^{2+}

151. Which of the following isoelectronic ions is the largest? Explain.
(a) Mn^{7+} (b) P^{3-} (c) S^{2-} (d) Sc^{3+} (e) Ti^{4+}

152. Arrange the following ions in order of increasing ionic radius.
(a) I^- (b) Cs^+ (c) Ba^{2+}

Second, Third, Fourth, and Higher Ionization Energies

153. Which of the following atoms or ions has the largest ionization energy?
(a) P (b) P^+ (c) P^{2+} (d) P^{3+} (e) P^{4+}

154. Explain why the second ionization energy of sodium is so much larger than the first ionization energy of the element.

155. What is the most probable electron configuration for the element that has the following ionization energies?

$$1st\ IE = 578\ KJ/mol$$
$$2nd\ IE = 1817$$
$$3rd\ IE = 2745$$
$$4th\ IE = 2745$$
$$5th\ IE = 14,831$$

(a) [Ne] (b) [Ne] $3s^1$
(c) [Ne] $3s^2$ (d) [Ne] $3s^2 3p^1$
(e) [Ne] $3s^2 3p^2$ (f) [Ne] $3s^2 3p^3$

156. Which of the following ionization energies is the largest?
(a) 1st IE of Ba
(b) 1st IE of Mg
(c) 2nd IE of Ba
(d) 2nd IE of Mg
(e) 3rd IE of Al
(f) 3rd IE of Mg

157. Which of the following elements should have the largest second ionization energy? Explain.
(a) Na (b) Mg (c) Al (d) Si (e) P

158. Which of the following elements should have the largest third ionization energy? Explain.
(a) B (b) C (c) N
(d) Mg (e) Al

159. List the following elements in order of increasing second ionization energy.
(a) Li (b) Be (c) Na
(d) Mg (e) Ne

160. Some elements, such as tin and lead, have more than one common ion. Use the electron configurations to predict the most likely ions of Sn and Pb.

Average Valence Electron Energy (AVEE)

161. Why is Be more likely to form the Be^{2+} ion than O is to form the O^{2+} ion?

162. Use Table 3.4 to calculate the AVEE of B and F. Compare your results to the values given in Figure 3.31.

163. Arrange the following in order of increasing AVEE without referring to Figure 3.31.
(a) P, Mg, Cl
(b) S, O, Se, F
(c) K, P, O

164. What two quantities does the AVEE measure?

165. How does the AVEE change from left to right across the periodic table? Explain.

166. How does the AVEE change from top to bottom down a group on the periodic table?

167. Describe what happens to the difference between the energies of subshells as the value of n becomes larger.

AVEE and Metallicity

168. Refer to Figure 1.8 and use AVEE data to explain why the elements N, P, As, Sb, and Bi range from nonmetals to semimetals to a metal.

169. Are large values of AVEE associated with metals or nonmetals? Explain.

170. Use AVEE to explain why in passing from the left-hand side of the periodic table to the right-hand side, metallic character decreases.

171. Explain why C is a nonmetal, Si is a semimetal, and Sn is a metal.

172. Arrange the following in order of increasing metallicity: Pb, Bi, Au, Ba.

173. Arrange the following in order of increasing nonmetallic behavior: B, Al, Ga, Tl.

174. Refer to Figure 3.31 and predict which is more metallic:
    (a) P or S   (b) As or Se
    (c) Sn or Sb   (d) Ga or Ge

175. As an atom gets larger, what happens to the energy required to remove valence electrons and to the distance in energy between shells?

Integrated Problems

176. The most recent estimates give values of about 10^{-10} m for the radius of an atom and 10^{-14} m for the radius of the nucleus of the atom. Calculate the fraction of the total volume of an atom that is essentially empty space.

177. Consider the following ions/atoms: O^{2-}, F^-, Ne, Na^+, and Mg^{2+}. Arrange them in order of increasing ionization energy. Also arrange them in order of increasing radius. Now consider the following atoms: O, F, Ne, Na, Mg. Arrange them in order of increasing ionization energy. Also arrange them in order of increasing radius.

178. Two *hypothetical* shell models of a lithium atom are shown in Figure 3P.6.

(a) Use Model 1 to predict the relationship (larger, smaller, or equal) between the first ionization energy of Li and the first ionization energy of He. Explain your reasoning.

(b) Use Model 2 to predict the relationship (larger, smaller, or equal) between the first ionization energy of Li and the first ionization energy of He. Explain your reasoning.

(c) Use Table 3.3 to decide which model is most consistent with the observed ionization energy of Li. Explain your reasoning.

179. The following atoms and ions all have the same electronic structure (they are isoelectronic): Ar, S^{2-}, K^+. Arrange them in order of increasing first ionization energy. Arrange them in order of increasing atomic radii. Explain your reasoning.

Questions 180–185 relate to the following data.

An atom with an equal number of spin-up and spin-down electrons is said to be *diamagnetic* because the atom is repelled by a magnetic field. In this case we say that all of the electrons are "paired." If this is not the case—if there are one or more unpaired electrons on an atom—the atom is attracted to a magnetic field, and it is said to be *paramagnetic*. The strength of the attraction is an experimentally measurable quantity known as the magnetic moment. The magnitude of the *magnetic moment* (measured in magnetons) is related to (but not proportional to) the number of unpaired electrons present. In other words, the larger the number of unpaired electrons, the larger the magnetic moment. Here are some experimental data collected by an investigator of this phenomenon.

Magnetic Moments of Several Elements

Element	Type	Magnetic Moments (magnetons)
H	Paramagnetic	1.7
He	Diamagnetic	0
B	Paramagnetic	1.7
C	Paramagnetic	2.8
N	Paramagnetic	3.9
O	Paramagnetic	2.8
Ne	Diamagnetic	0

180. Why is the situation of equal numbers of spin-up and spin-down electrons referred to as all electrons being "paired"?

181. How many unpaired electrons are in the following atoms?
    (a) C   (b) N   (c) O
    (d) Ne   (e) F

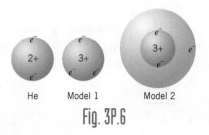

He   Model 1   Model 2

Fig. 3P.6

182. How many "pairs" of electrons are there in a filled p shell?

183. On the basis of information provided in the table of Magnetic Moments of Several Elements, predict the results of a Stern–Gerlach experiment on each of the atoms listed.

184. An ion, X^{2+}, is known to be from the first transition metal series. The ion is paramagnetic, with four unpaired electrons. What two possible elements could X be?

185. Element Z is diamagnetic. Its most common ion is Z^{2+}. Z has the next to the lowest first ionization energy in its group. The energy required to remove an electron from Z^{2+} is extremely high. Identify element Z. Give the chemical formula of the ionic compounds formed from Z^{2+} and the negative ions O^{2-} and Cl^{-}.

186. (a) Give complete electron configurations for Ca, In, Si

(b) The following list gives the first four ionization energies for one of these atoms. Identify the atom. Explain how the four ionization energies match the electron configuration of the atom.

	Energy (kJ/mol)
First Ionization Energy	558
$A \longrightarrow A^{+} + e$	
Second Ionization Energy	1820
$A^{+} \longrightarrow A^{2+} + e$	
Third Ionization Energy	2704
$A^{2+} \longrightarrow A^{3+} + e$	
Fourth Ionization Energy	5200
$A^{3+} \longrightarrow A^{4+} + e$	

187. Three atoms—X, Y, and Z—have the following relation to each other. Atom X has one less proton than atom Y, and atom Z has one more proton than atom Y. The following table lists additional characteristics.

(a) Identify atoms X, Y, and Z. Explain how you arrived at your identification. Enter the identity in the table under Chemical Symbol.

(b) Fill in the missing underlined entries in the following table. In some cases your best estimates are acceptable.

188. The following data are known for the oxygen atom.

Ionization Energies (MJ/mol)

52.6   3.12   1.31

(a) Sketch the shell model for the oxygen atom.

(b) If photons of wavelength 2.39×10^{-8} m strike an O atom, which, if any, of the electrons could be completely removed from the atom?

(c) What is the first ionization energy, IE, for O?

(d) Calculate the AVEE for O.

(e) Sketch the PES spectrum for O and F. Show clearly how the two spectra differ.

189. The following data are known for element X.

Number of Peaks in the PES	Covalent Radius of Neutral Atom	First IE	AVEE	Core Charge
5	0.099 nm	1.251 MJ/mol	1.59	+7

(a) Identify element X. Explain your reasoning.

(b) Give the complete electron configuration for element X.

(c) Give the electron configuration for X^{-1}. How would the radius of X^{-1} compare to the radius for X? Explain.

(d) How would IE for X compare to the energy required to remove the most loosely held electron from X^{-}? Explain.

(e) How would the AVEE for X compare to the AVEE for the element with one more proton? Explain.

(f) Qualitatively compare the covalent radius of X to that of the element with a core charge of $+7$ and a first IE of 1.681 MJ/mol.

(g) How many valence electrons does X have?

190. Three elements, X, Y, and Z, have the electron configurations:

$1s^2 2s^2 2p^6 3s^2 3p^6$
$1s^2 2s^2 2p^6 3s^2$
$1s^2 2s^2 2p^6 3s^2 3p^6 4s^1$

The first ionization energies are known to be (not in any order): 0.4188 MJ/mol, 0.7377 MJ/mol, and

Atom	Chemical Symbol	Behavior in a Magnetic Field	First Ionization Energy (IE) (MJ/mol)	Radius of Covalent Atom (nm)	Number of PES Peaks	Core Charge	AVEE	Number of Valence Electrons
X	—	not deflected	2.08	0.070	—	—	—	—
Y	—	—	—	0.16	4	—	0.50	—
Z	—	not deflected	0.74	—	—	—	—	—

1.5205 MJ/mol. The covalent radii of these elements are (not in any order): 0.094 nm, 0.136 nm, and 0.202 nm.

(a) Identify each element and match the appropriate values of ionization energy and atomic radius to each configuration.

(b) Which of X, Y, and Z has the smallest AVEE?

(c) Which is(are) paramagnetic?

191. (a) Which of the following is the most probable electron configuration for the atom that has these ionization energies? Explain.

	MJ/mol
1st IE	0.899
2nd IE	1.757
3rd IE	14.848
4th IE	21.006

(i) [He] $2s^1$

(ii) [He] $2s^2$

(iii) [He] $2s^2 2p^1$

(iv) [He] $2s^2 2p^2$

(v) [He] $3s^1$

(b) How many valence electrons does this atom have? Explain.

(c) What is the core charge of this atom?

(d) Why is the second electron more difficult to remove than the first electron?

Chapter Four

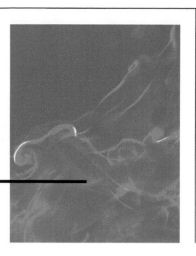

THE COVALENT BOND

Ever since John Dalton introduced his atomic theory in 1803, chemists have tried to understand the forces that hold atoms together in chemical compounds. The goal of this chapter is to build a model that explains the bonds between atoms in covalent molecules and to use this model to predict the structures of these molecules.

4.1 Valence Electrons

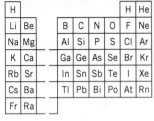

Fig. 4.1 Lewis structures were first applied to the atoms of the main-group elements, which are found on either side of the periodic table.

In 1902, while trying to find a way to explain the periodic table to a beginning chemistry class, G. N. Lewis discovered that the chemistry of the main-group elements shown in Figure 4.1 could be explained by assuming that atoms of these elements gain or lose electrons until they have eight electrons (a so-called octet of electrons) in the outermost shell of electrons of the atom. The magnitude of this achievement can be appreciated by noting that Lewis's **octet rule** was generated only five years after J. J. Thomson's discovery of the electron and nine years *before* Ernest Rutherford proposed that the atom consisted of an infinitesimally small nucleus surrounded by a sea of electrons.

The electrons in the outermost shell eventually became known as the **valence electrons.** This name reflects the fact that the number of bonds an element can form is called its *valence*. Because the number of electrons in the outermost shells in the Lewis theory controls the number of bonds the atom can form, these outermost electrons are the valence electrons.

Figure 4.2, which utilizes the PES data from Chapter 3, shows the relative energies of the $1s$, $2s$, $2p$, $3s$, and $3p$ orbitals as the atomic number increases from H ($Z = 1$) to Sc ($Z = 21$). By the time we get to lithium ($Z = 3$), we already begin to see a significant difference between the energy needed to remove an electron from the valence $2s$ orbital and the $1s$ core orbital. This gap continues to widen, and, by neon ($Z = 10$), the $1s$ electron is extremely difficult to remove. At argon ($Z = 18$), we see that the $2s$ and $2p$ core electrons are much more tightly held than the $3s$ and $3p$ valence electrons. There is not much difference, however, between the energy required to remove an electron from the $2s$ and $2p$ orbitals or between the $3s$ and $3p$ orbitals. As the atomic number increases, the electrons in the core shells or subshells become increasingly more difficult to remove from the atom. It is only the electrons in the valence orbitals that are relatively easy to remove from an atom.

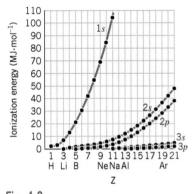

Fig. 4.2 Photoelectron ionization energies for the $1s$, $2s$, $2p$, $3s$, and $3p$ orbitals of the first 21 elements. The energy required to remove a $1s$ electron increases rapidly with increasing nuclear charge, Z. The difference in the energy required to remove an s electron and a p electron from the same shell is never large. This difference gets smaller in subsequent shells. [Reprinted from R. J. Gillespie et al., *Atoms, Molecules and Reactions,* © 1994. Printed and electronically reproduced by permission of Pearson Education, Inc., Upper Saddle River, New Jersey.]

The valence electrons are therefore those electrons on an atom that can be gained or lost in a chemical reaction. Consider fluorine, for example, which has the following electron configuration:

$$\text{F} \qquad 1s^2\,2s^2\,2p^5 = [\text{He}]\,2s^2\,2p^5$$

As we can see from Figure 4.2, there is a significant difference between the energy needed to remove an electron from the $1s$ core orbital and the $2s$ or $2p$ valence orbitals. As a result, only the electrons in the $2s$ and $2p$ orbitals on fluorine are valence electrons. Fluorine therefore has a total of seven valence electrons, the two electrons in the $2s$ orbital and the five electrons in the $2p$ orbitals.

Let's now consider gallium, which has the following electron configuration, in which the symbol [Ar] represents the 18 core electrons in the gallium atom.

$$\text{Ga} \qquad [\text{Ar}]\,4s^2\,3d^{10}\,4p^1$$

Filled d or f subshells are seldom involved in chemical reactions. As a result, the electrons in filled d or f subshells are not considered when counting the number of valence electrons. Gallium therefore has three valence electrons, the electrons in the 4s and 4p valence orbitals.

For the main-group elements in Figure 4.1, the number of valence electrons is equal to the group number. Fluorine, in Group VIIA, has seven valence electrons. Gallium, in Group IIIA, has three valence electrons.

Vanadium has the following electron configuration.

$$V \qquad [Ar]4s^23d^3$$

The core electrons represented by the symbol [Ar] don't count as valence electrons, but the 4s and 3d do. Vanadium therefore has five valence-shell electrons. As a general rule, we can define valence electrons as the electrons on an atom that are not present in the previous rare gas, ignoring filled d or f subsells.

> **CHECKPOINT**
>
> How many valence electrons does an atom in Group IVA have?

Exercise 4.1

Determine the number of valence electrons in neutral atoms of the following elements.

(a) Si

(b) Mn

(c) Sb

(d) Pb

Solution

We start by writing the electron configuration for each element.

Si	[Ne] $3s^2\,3p^2$
Mn	[Ar] $4s^2\,3d^5$
Sb	[Kr] $5s^2\,4d^{10}\,5p^3$
Pb	[Xe] $6s^2\,4f^{14}\,5d^{10}\,6p^2$

Ignoring filled d and f subshells, we conclude that neutral atoms of these elements contain the following numbers of valence electrons.

(a) Si = 4

(b) Mn = 7

(c) Sb = 5

(d) Pb = 4

4.2 The Covalent Bond

By 1916 Lewis realized that there is another way atoms can achieve an octet of valence electrons: They can share electrons with other atoms until each of their respective valence shells contains eight electrons. Two fluorine atoms, for example, can form a stable F_2 molecule in which each atom is surrounded by eight valence electrons by sharing a pair of electrons. A pair of oxygen atoms can form an O_2 molecule in which each atom has a total of eight valence electrons by sharing two pairs of electrons.

Whenever Lewis applied this model, he noted that the atoms seem to share pairs of electrons. He also observed that most molecules contain an even number of electrons, which suggests that the electrons exist in pairs. He therefore introduced a system of notation known as **Lewis structures** in which each atom is surrounded by up to four pairs of dots corresponding to the eight possible valence electrons. The Lewis structures of F_2 and O_2 were therefore written as shown on the left in Figure 4.3. This symbolism is still in use today. The only significant change is the use of lines to indicate bonds between atoms formed by the sharing of a pair of electrons, as shown on the right in Figure 4.3.

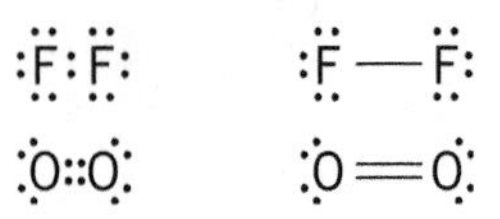

The prefix *co-* is used to indicate when things are joined or equal (for example, *coexist*, *cooperate*, and *coordinate*). It is therefore appropriate that the term **covalent bond** is used to describe the bonds in molecules that result from the sharing of one or more pairs of valence electrons. As might be expected, molecules held together by covalent bonds are called **covalent molecules.**

:F̈:F̈:   :F̈ — F̈:
:Ö::Ö: :Ö = Ö:

Fig. 4.3 Lewis structures of F_2 and O_2.

4.3 How Does the Sharing of Electrons Bond Atoms?

To understand how atoms can be held together by sharing a pair of electrons, let's look at the simplest covalent bond, the bond that forms when two isolated hydrogen atoms come together to form an H_2 molecule.

$$H \cdot + \cdot H \longrightarrow H{-}H$$

Each hydrogen atom contains a proton surrounded by a spherical cloud of electron density that corresponds to a single electron. The proton and electron on an isolated hydrogen atom are held together by the force of attraction between oppositely charged particles. The magnitude of this force is equal to the product of the charge on the electron (q_e) times the charge on the proton (q_p) divided by the square of the distance between these particles (r^2).

$$F = \frac{q_e q_p}{r^2}$$

When a pair of isolated hydrogen atoms are brought together, two new forces of attraction appear, as shown in Figure 4.4a, because of the attraction between the electron on one atom and the proton on the other. But two forces of repulsion are also created, as shown in Figure 4.4b, because the two negatively charged electrons repel each other, as do the two positively charged protons.

Fig. 4.4 (a) Two forces of attraction act to bring a pair of hydrogen atoms together—the forces of attraction between the electron of each atom and the proton of the other atom. (b) Two forces of repulsion drive a pair of hydrogen atoms apart—the repulsion between the two protons and the repulsion between the two electrons.

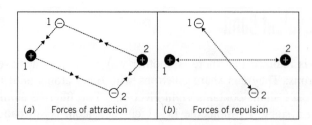
(a) Forces of attraction   (b) Forces of repulsion

At first glance, it might seem that the two new repulsive forces would balance the two new attractive forces. If this happened, the H_2 molecule would be no more stable than a pair of isolated hydrogen atoms. There is a way to maximize the new forces of attraction and minimize the new forces of repulsions, however.

The force of repulsion between the protons can be minimized if the pair of electrons is placed between the two nuclei. The distance between the electron in one atom and the nucleus of the other is now smaller than the distance between the two nuclei, as shown in Figure 4.5. As a result, the force of attraction between each electron and the nucleus of the other atom is larger than the force of repulsion between the two nuclei. Placing the two electrons between the two nuclei, however, implies that they have to occupy the same region of space. The only way for two electrons to occupy the same region of space is if they have opposite spins, so that each electron may enter the domain of the other. When this happens, the electrons are said to be "paired."

The net result of pairing the electrons is to create a region in space in which both electrons can reside that concentrates the electron density between the two nuclei. These two electrons are said to occupy the same *bonding domain*. Each hydrogen atom shares a pair of electrons with the other hydrogen atom, and it is the sharing of these electrons that leads to the formation of a covalent bond.

Pairing the electrons and placing them between the two nuclei creates a system that is more stable than a pair of isolated atoms if the nuclei are close enough to share the pair of electrons, but not so close that repulsion between the nuclei becomes too large. The hydrogen atoms in an H_2 molecule are therefore held together (or bonded) by the sharing of a pair of electrons, and this bond is the strongest when the distance between the two nuclei is 0.074 nm.

Fig. 4.5 If the electrons have opposite spins and are restricted to the region directly between the two nuclei, the attractive forces are larger than the repulsive forces in the hydrogen molecule. As a result, the H_2 molecule is more stable than a pair of isolated hydrogen atoms.

➤ **CHECKPOINT**

Explain why the distance between the nuclei of the two hydrogen atoms in the H_2^+ ion is longer (0.106 nm) than the distance between the hydrogen atoms in a neutral H_2 molecule (0.074 nm).

4.4 Using Lewis Structures to Understand the Formation of Bonds

It is so easy to get caught up in the process of generating Lewis structures that we lose sight of the role they played in the development of chemistry. For at least a century before the theory was proposed, chemists had evidence that hydrogen and oxygen existed as diatomic H_2 and O_2 molecules. Lewis provided the first model that could explain why hydrogen atoms come together to form H_2 molecules and why oxygen atoms spontaneously combine to form O_2 molecules. Although Lewis's octet rule works well for many molecules, there are other, more complex bonding models that are more powerful in certain cases. The Lewis model has the advantage of simplicity and is often the first model used by chemists to describe the bonding in a molecule.

Lewis structures can be used to explain why hydrogen and oxygen atoms combine to form H_2O molecules. We start by noting that the hydrogen atom has only one valence electron and that the oxygen atom has six valence electrons.

$$H \qquad 1s^1$$
$$O \qquad 1s^2\, 2s^2\, 2p^4$$

We then represent neutral atoms of these elements with the following symbols.

$$H\cdot \qquad \cdot \overset{\cdot\cdot}{\underset{\cdot\cdot}{O}}\cdot$$

As we bring the atoms together, the electron on each hydrogen atom combines with an electron on the oxygen atom of opposite spin to occupy a region of space between the two nuclei.

$$H\cdot \; + \; \cdot \ddot{O} \cdot \; + \; \cdot H \; \longrightarrow \; H\!-\!\ddot{O}\!-\!H$$

The result is an H_2O molecule in which each hydrogen atom shares a pair of electrons with the oxygen atom and the oxygen atom has access to an octet of valence electrons.

The Lewis structure of water is often written with a line representing each pair of electrons that is shared by two atoms.

$$H\!-\!\overset{\cdot\cdot}{\underset{\cdot\cdot}{O}}\!-\!H$$

There are four regions in space, or **domains,** where electrons can be found around the oxygen atom in the water molecule. Two of these domains are called **bonding domains** because they hold the pairs of electrons used for bonding. The other two domains contain pairs of electrons that are described as **nonbonding electrons.** These domains are called **nonbonding domains.** The nonbonding electrons are in the valence shell of oxygen and are considered to belong exclusively to the oxygen atom in the molecule. Thus there are eight total electrons in the nonbonding and bonding domains of oxygen, and there are two electrons in the bonding domain of each hydrogen atom.

Let's apply our technique for generating Lewis structures to carbon dioxide (CO_2). We start by determining the number of valence electrons on each atom. Carbon has four valence electrons, and each oxygen has six.

$$C \qquad [He]2s^2 2p^2$$
$$O \qquad [He]2s^2 2p^4$$

We can represent this information by the following symbols.

$$:\ddot{O}\cdot \qquad \cdot \dot{C}\cdot \qquad \cdot \ddot{O}:$$

We now combine one electron from each atom to form covalent bonds between the atoms.

$$:\ddot{O}\!-\!\dot{C}\!-\!\ddot{O}:$$

When this is done, each oxygen atom has a total of seven valence electrons, and the carbon atom has a total of six valence electrons. Because none of the atoms have a filled valence shell, we combine another electron on each atom to form two more bonds. The result is a Lewis structure in which each atom has eight electrons in its valence shell.

$$\ddot{O}\!=\!C\!=\!\ddot{O}$$

> ➤ **CHECKPOINT**
>
> In the CO_2 molecule, where are the nonbonding electrons? How many electrons are in bonding domains?

4.5 Drawing Skeleton Structures

The most difficult step in generating the Lewis structure of a molecule is the step in which the skeleton structure of the molecule is written. The first step in drawing the skeleton structure involves selecting the central atom.

As a general rule, the element with the smallest AVEE (see Figure 3.31) is at the center of the structure. Another hint as to the skeleton structure involves looking at the number of atoms of each element. When there is only one atom of an element in a formula, that atom is often the central atom. Thus the formulas of thionyl chloride ($SOCl_2$) and sulfuryl chloride (SO_2Cl_2) can be translated into the following skeleton structures.

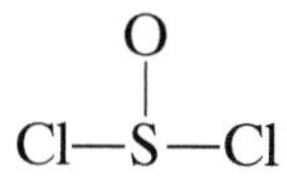

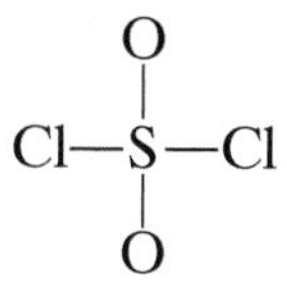

$$
\begin{array}{cc}
\overset{\textstyle O}{\underset{\textstyle |}{}} & \overset{\textstyle O}{\underset{\textstyle \|}{}} \\
Cl\!-\!S\!-\!Cl & Cl\!-\!S\!-\!Cl \\
 & \underset{\textstyle O}{\underset{\textstyle |}{}}
\end{array}
$$

Chemists often write the formulas for complex molecules in a way that provides hints about the skeleton structure of the molecule. Consider acetic acid, for example, for which the formula CH_3CO_2H is written to indicate that this molecule contains the following skeleton structure.

$$
\begin{array}{c}
H \\
| \quad\quad O \\
H\!-\!C\!-\!C \\
| \quad\quad\; \diagdown \\
H \quad\quad O\!-\!H
\end{array}
$$

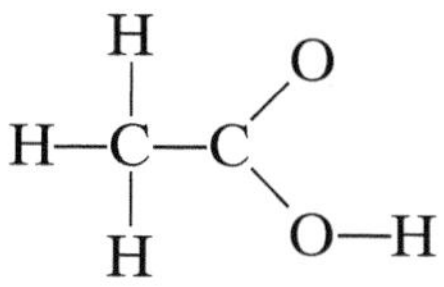

Sometimes more than one skeleton structure can be written for a given chemical formula. Consider dimethyl ether and ethanol, for example, which both have the formula C_2H_6O. The formula for dimethyl ether is often written as CH_3OCH_3, which translates into the following skeleton structure.

$$
\begin{array}{c}
H \quad\quad\; H \\
| \quad\quad\;\; | \\
H\!-\!C\!-\!O\!-\!C\!-\!H \\
| \quad\quad\;\; | \\
H \quad\quad\; H
\end{array}
$$

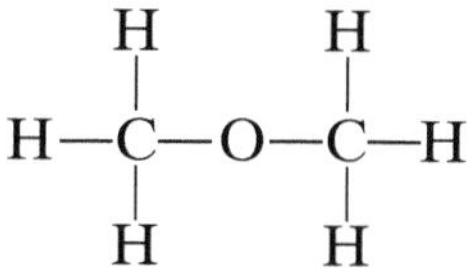

The formula for ethanol is written as CH_3CH_2OH because this compound has the following skeleton structure.

$$
\begin{array}{c}
H \;\; H \\
| \;\; | \\
H\!-\!C\!-\!C\!-\!O\!-\!H \\
| \;\; | \\
H \;\; H
\end{array}
$$

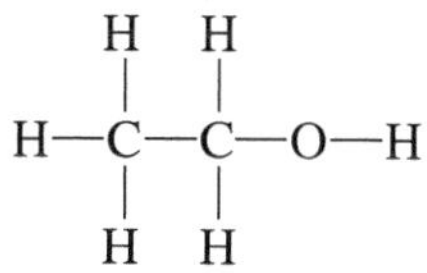

When two structures differ only in the arrangement of atoms, the compounds they represent are called **isomers.** Dimethyl ether and ethanol, shown above, are therefore isomers.

> ➤ **CHECKPOINT**
>
> Why can H never serve as the central atom in a Lewis structure?

> ➤ **CHECKPOINT**
>
> What is the difference between a skeleton structure and a Lewis structure?

4.6 A Step-by-Step Approach to Writing Lewis Structures

The method for writing Lewis structures used in Section 4.4 can be time consuming. For all but the simplest molecules, the following step-by-step process is faster.

- Write the skeleton structure of the molecule.
- Determine the number of valence electrons in the molecule.

- Use two valence electrons to form each bond in the skeleton structure.
- Try to place eight electrons in the valence shells of the atoms by distributing the remaining valence electrons as nonbonding electrons.

The first step in this process involves deciding which atoms in the molecule are connected by covalent bonds. As we have seen, the formula of the compound often provides a hint as to the skeleton structure. Consider PCl_3, for example. The formula for the molecule suggests the following skeleton structure.

$$Cl-\underset{\underset{\textstyle Cl}{|}}{P}-Cl$$

The second step in generating the Lewis structure of a molecule involves calculating the number of valence electrons in the molecule or ion. For a neutral molecule, this is the sum of the valence electrons on each atom. If the molecule has a positive charge or negative charge, we add one electron for each negative charge and subtract an electron for each positive charge.

Phosphorus is in Group VA of the periodic table. The phosphorus atom in PCl_3 therefore contributes 5 valence electrons to the total number of valence electrons. Chlorine is in Group VIIA, which means each neutral chlorine atom contains 7 valence electrons. The three chlorine atoms therefore contribute a total of 21 valence electrons. Because PCl_3 has no net charge, no additional electrons need to be added or subtracted. Thus, PCl_3 has a total of 26 valence electrons.

$$PCl_3 \qquad 5 + 3(7) = 26$$

The third step in this process assumes that the skeleton structure of the molecule is held together by covalent bonds. The valence electrons are therefore divided into two categories: *bonding electrons* and *nonbonding electrons.* Because it takes two electrons to form a covalent bond, we can calculate the number of nonbonding electrons in the molecule by subtracting two electrons for each bond in the skeleton structure from the total number of valence electrons.

There are three covalent bonds in the skeleton structure for PCl_3. As a result, 6 of the 26 valence electrons must be used as bonding electrons. This leaves 20 nonbonding electrons in the valence shell.

$$
\begin{array}{r}
26 \text{ valence electrons} \\
-6 \text{ bonding electrons} \\
\hline
20 \text{ nonbonding electrons}
\end{array}
$$

The last step in the process by which Lewis structures are generated involves using the nonbonding valence electrons to satisfy the octets of the atoms in the molecule. Each chlorine atom in PCl_3 already has 2 electrons—the electrons in the P—Cl covalent bond. Because each chlorine atom needs 6 nonbonding electrons to satisfy its octet, it takes 18 nonbonding electrons to satisfy the three chlorine atoms. This leaves one pair of nonbonding electrons, which can be used to fill the valence shell of the central atom.

$$:\!\ddot{C}\!l-\underset{\underset{\textstyle :\ddot{C}l:}{|}}{P}-\ddot{C}\!l\!:$$

Lewis structures can also be used to describe polyatomic ions such as the ammonium ion, NH_4^+. We begin by drawing the skeleton structure.

$$
\begin{array}{c}
H \\
| \\
H-N-H \\
| \\
H
\end{array}
$$

Nitrogen is in Group VA of the periodic table, which means that a neutral nitrogen atom contains five valence electrons. Each hydrogen atom has one valence electron. Thus the four hydrogen atoms can contribute a total of four electrons to the number of valence electrons in the molecule. Since the NH_4^+ ion is a polyatomic positive ion, or cation, one electron is removed from the total to account for the positive charge on the molecule, leaving a total of eight valence electrons.

$$ NH_4^+ \quad 5 + 4(1) - 1 = 8 $$

Each of the four covalent bonds in the NH_4^+ ion contains two electrons. Therefore, all eight valence electrons are used in the bonding, and no electrons remain to be distributed as nonbonding electrons.

$$
\begin{array}{r}
8 \text{ valence electrons} \\
-8 \text{ bonding electrons} \\
\hline
0 \text{ nonbonding electrons}
\end{array}
$$

This leaves us with eight electrons surrounding the nitrogen and two electrons shared with each of the four hydrogen atoms. Thus each atom is surrounded by the expected number of electrons. The final structure is drawn in brackets, and the charge on the polyatomic ion is shown at the upper-right corners.

$$
\left[
\begin{array}{c}
H \\
| \\
H-N-H \\
| \\
H
\end{array}
\right]^+
$$

The brackets are used to remind us that the charge doesn't reside on any particular atom in this polyatomic ion. It is spread over the five atoms that contribute to the skeleton structure.

When calculating the number of valence electrons on a polyatomic negative ion, or anion, enough electrons must be added to the total number of valence electrons to account for the overall negative charge. The NO_3^- ion, for example, contains a total of 24 valence electrons.

$$ NO_3^- \quad 5 + 3(6) + 1 = 24 $$

➤ **CHECKPOINT**

Draw Lewis structures for carbon tetrachloride, CCl_4, and the hydroxide anion, OH^-.

4.7 Molecules That Don't Seem to Satisfy the Octet Rule

NOT ENOUGH ELECTRONS

Occasionally we encounter a molecule that doesn't seem to have enough valence electrons. When this happens, we have to remember why atoms share electrons in the first place. If we can't get a satisfactory Lewis structure by sharing a single pair of electrons, it may be possible to achieve that goal by sharing two or even three pairs of electrons to form double or triple bonds.

Consider formaldehyde (H_2CO), for example, which is used as a preservative in biology labs because it is a disinfectant that kills most bacteria. The H_2CO molecule contains a total of 12 valence electrons.

$$H_2CO \qquad 2(1) + 4 + 6 = 12$$

The formula of the molecule suggests the following skeleton structure.

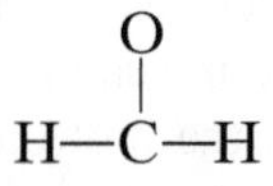

$$
\begin{array}{c}
O \\
| \\
H-C-H
\end{array}
$$

There are three covalent bonds in the skeleton structure, which means that six valence electrons must be used as bonding electrons. This leaves six nonbonding electrons. It is impossible, however, to satisfy the octets of the atoms in this molecule with only six nonbonding electrons. When the nonbonding electrons are used to satisfy the octet of the oxygen atom, the carbon atom has a total of only six valence electrons.

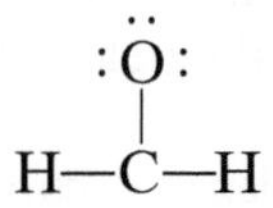

$$
\begin{array}{c}
:\overset{..}{O}: \\
| \\
H-C-H
\end{array}
$$

Let's assume that the carbon and oxygen atoms share two pairs of electrons to make a double bond, as shown below. There are now four bonds in the skeleton structure, which leaves only four nonbonding electrons. This is enough, however, to satisfy the octets of the carbon and oxygen atoms. The formaldehyde molecule can now be described in terms of two C—H single-bond domains, a C=O double bond domain, and two nonbonding domains localized on the oxygen atom.

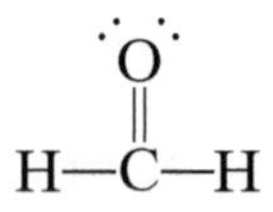

$$
\begin{array}{c}
\overset{..}{\underset{..}{O}} \\
\| \\
H-C-H
\end{array}
$$

Every once in a while, we encounter a molecule for which it seems impossible to write a satisfactory Lewis structure. Consider boron trifluoride (BF_3), for example, which contains 24 valence electrons.

$$BF_3 \qquad 3 + 3(7) = 24$$

There are three covalent bonds in the skeleton structure for the molecule. Because it takes 6 electrons to form the skeleton structure, there are 18 nonbonding valence electrons. But each fluorine atom needs 6 additional electrons to satisfy its octet. Thus all of the nonbonding electrons are used by the three fluorine atoms. As a result, we run out of electrons while the boron atom has only 6 valence electrons.

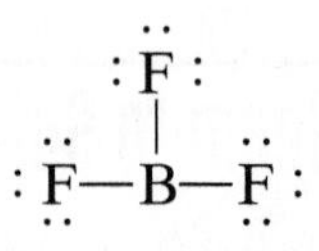

$$
\begin{array}{c}
:\overset{..}{F}: \\
| \\
:\overset{..}{F}-B-\overset{..}{F}:
\end{array}
$$

The next step would be to look for the possibility of forming a double or triple bond. However, chemists have learned that the atoms that form strong double or triple bonds are C, N, O, P, and S. Because neither boron nor fluorine belongs in that category, we have to stop with what appears to be an unsatisfactory Lewis structure.

TOO MANY ELECTRONS

It is also possible to encounter a molecule that seems to have too many valence electrons. When that happens, we expand the valence shell of the central atom. Consider the Lewis structure for sulfur tetrafluoride (SF_4), for example, which contains 34 valence electrons.

$$SF_4 \qquad 6 + 4(7) = 34$$

There are four covalent bonds in the skeleton structure for SF_4.

$$
\begin{array}{c}
F \\
| \\
F—S—F \\
| \\
F
\end{array}
$$

Because this requires using 8 valence electrons to form the covalent bonds that hold the molecule together, there are 26 nonbonding valence electrons.

Each fluorine atom needs 6 additional electrons to satisfy its octet. Because there are four F atoms, we need 24 nonbonding electrons for this purpose. But there are 26 nonbonding electrons in the molecule. In other words, after we have satisfied the octets for all five atoms, we still have one more pair of valence electrons. We therefore expand the valence shell of the central atom to hold more than 8 electrons.

$$
\begin{array}{cc}
:\ddot{F} & \ddot{F}: \\
\diagdown & \diagup \\
 & S \\
\diagup & \diagdown \\
:\ddot{F} & \ddot{F}:
\end{array}
$$

> ➤ **CHECKPOINT**
>
> Elements in the first and second rows of the periodic table do not have valence-shell d orbitals. Use the fact that nitrogen and oxygen do not contain $2d$ orbitals to explain why these elements cannot expand their valence shell.

This raises an interesting question: How does the sulfur atom in SF_4 hold 10 electrons in its valence shell? The electron configuration for a neutral sulfur atom seems to suggest that only 8 electrons will fit in the valence shell of the atom because it takes 8 electrons to fill the $3s$ and $3p$ orbitals. But sulfur also has valence-shell $3d$ orbitals.

$$S \qquad [Ne]3s^2 3p^4 3d^0$$

Because the $3d$ orbitals on a neutral sulfur atom are all empty, the traditional explanation for why sulfur seems able to hold 10 valence electrons is to assume that the extra pair of electrons on the sulfur atom is placed in one of these empty $3d$ orbitals.

Exercise 4.2

Write the Lewis structure for xenon tetrafluoride (XeF_4).

Solution

Xenon (Group VIIIA) has 8 valence electrons, and fluorine (Group VIIA) has 7. Thus, there are 36 valence electrons in the molecule.

$$XeF_4 \qquad 8 + 4(7) = 36$$

The skeleton structure for the molecule contains four covalent bonds.

$$
\begin{array}{c}
\text{F} \\
| \\
\text{F—Xe—F} \\
| \\
\text{F}
\end{array}
$$

Because 8 electrons are used to form the skeleton structure, there are 28 non-bonding valence electrons. If each fluorine atom needs 6 nonbonding electrons, a total of 24 nonbonding electrons are used to complete the octets of the F atoms. But this leaves 4 extra nonbonding electrons. Because the octet of each atom appears to be satisfied and we have electrons left over, we expand the valence shell of the central atom until it contains a total of 12 electrons.

An enormous number of molecules follow the octet rule because this rule is obeyed by elements such as C, N, and O that form so many molecules. There are exceptions to the octet rule, however. The most common exceptions are the elements in the first row of the periodic table (H and He), which fill their valence shell with only one pair of electrons. We've already seen other exceptions: BF_3, SF_4, and XeF_4. BF_3 doesn't have enough valence electrons to obey the octet rule, whereas SF_4 and XeF_4 have too many. Elements in the third or higher rows of the periodic table can exceed an octet but are rarely found to be electron deficient. Elements in the second row of the periodic table never exceed an octet but are sometimes electron deficient.

EXCEPTIONS TO THESE GENERAL RULES

There are a limited number of molecules that appear to be exceptions to the general rules outlined above. Consider NO_2, for example, which contains a total of 17 valence electrons.

$$NO_2 \qquad 5 + 2(6) = 17$$

Because there are an odd number of electrons in this molecule, it is impossible to write a Lewis structure in which all of the electrons are paired. The most common Lewis structure for NO_2 assumes that there is an unpaired electron in one of the domains on the nitrogen atom.

4.8 Bond Lengths

The **bond length** is defined as the distance between two atoms involved in a chemical bond. Bond lengths are usually given in units of nanometers (nm), where one nm is 10^{-9} meters, or picometers, where one pm is 10^{-12} meters.

Table 4.1
Carbon–Carbon and Carbon–Hydrogen Bond Lengths in Selected Molecules[a]

Molecule	Carbon–Carbon Bond Length (nm)	Carbon–Hydrogen Bond Length (nm)
Ethane	0.154	0.110
Graphite	0.142	
Benzene	0.139	0.110[b]
Ethylene	0.133	0.109
Acetylene	0.120	0.106[b]

[a]From Emil J. Margolis, *Bonding and Structure,* Appleton, Century, Croft, New York, 1968, with permission from Plenum Publishing Corp.
[b]David R. Lide, ed., *CRC Handbook of Chemistry and Physics*, 75th ed., CRC Press, Boca Raton, FL, 1994.

Carbon–carbon and carbon–hydrogen bond lengths for a number of compounds are given in Table 4.1.

The C—C and C—H bond lengths for three of these compounds—ethane (H_3C—CH_3), ethylene (H_2C=CH_2), and acetylene (HC≡CH)—are shown in Figure 4.6. The triple bond in acetylene is significantly shorter than the double bond in ethylene, which is shorter than the single bond in ethane. This suggests that the atoms in a triple bond are pulled closer to one another than the atoms in a double bond, which are pulled closer to one another than the atoms in a double bond.

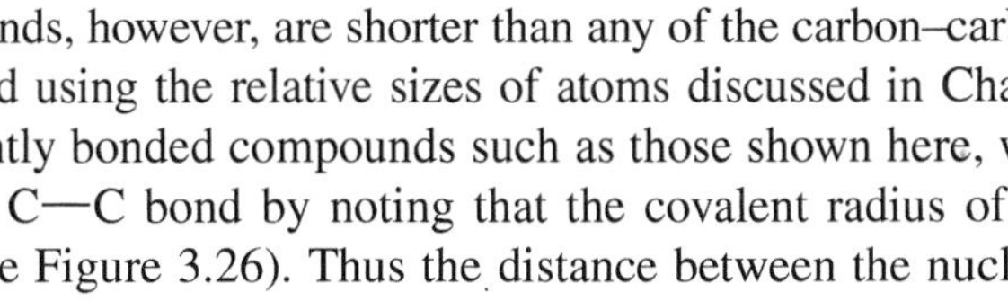

Fig. 4.6   Bond lengths (in nm) in ethane, ethylene, and acetylene. The carbon–carbon bond length becomes shorter as the bond changes from a single to a double to a triple bond. The carbon–hydrogen bond length is about the same in all three compounds.

Because each C—H bond is formed by sharing a single pair of electrons, the C—H bond length is approximately the same in the three compounds in Figure 4.6. These C—H bonds, however, are shorter than any of the carbon–carbon bonds. This can be explained using the relative sizes of atoms discussed in Chapter 3.

In covalently bonded compounds such as those shown here, we can estimate the length of a C—C bond by noting that the covalent radius of a carbon atom is 0.077 nm (see Figure 3.26). Thus the distance between the nuclei of two covalently bonded carbon atoms should be roughly twice the radius of a carbon atom, or 0.154 nm. The carbon–hydrogen bond length should be roughly equal to the sum of the radius of a carbon atom and a hydrogen atom, or 0.114 nm.

Exercise 4.3

Use the covalent radii given in Chapter 3 to estimate the bond lengths for all bonds in the following compounds.

(a) H_2O

(b) CH_3OH

(c) CH_3OCH_3

(d) CH_3SCH_3

Solution

Before any bond lengths can be estimated, we need to write the Lewis structures of these molecules.

The covalent radii of hydrocarbon, carbon, oxygen, and sulfur atoms can be obtained from the data in Figure 3.26.

H	0.037 nm
O	0.066 nm
C	0.077 nm
S	0.104 nm

The best estimates for the bond lengths in these molecules are therefore:

The data given in Figure 3.26 are average values taken from a variety of compounds. Therefore the bond lengths obtained by adding these radii are only approximations, but they are quite close to the actual values. It should also be noted that the covalent radii given in Figure 3.26 apply only to single bonds. The bond lengths in compounds that contain double or triple bonds are shorter than those estimated from the radii of the atoms.

● ●

➤ **CHECKPOINT**

In which compound would you expect the oxygen–oxygen bond length to be the longest, the O_2 molecule that contains an O=O double bond or hydrogen peroxide (H_2O_2) that contains an O—O single bond? Explain your answer.

4.9 Resonance Hybrids

Two equivalent Lewis structures can be written for some compounds, such as sulfur dioxide, as shown in Figure 4.7. The only difference between these structures is the identity of the oxygen atom to which the double bond is formed. As a result, they must be equally satisfactory representations of the molecule. This raises an important question: Which of the Lewis structures for SO_2 is correct?

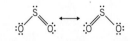

Fig. 4.7 Two equally satisfactory Lewis structures can be written for SO_2

Interestingly enough, neither of the structures is correct. The two Lewis structures suggest that one of the sulfur–oxygen bond lengths is shorter than the other. Every experiment that has been done to probe the structure of this molecule, however, suggests that the two sulfur–oxygen bonds have identical bond lengths. Moreover, the actual bond length is shorter than expected for a S—O single bond and longer than expected for a S=O double bond.

SO_2 is not unique. We often find that there are several equally satisfactory ways of writing the Lewis structures of compounds that contain double or triple bonds. When this happens, the best description of the structure of the molecule is a **resonance hybrid** of all the possible Lewis structures.

The meaning of the term *resonance* can be best understood by an analogy. In music, the notes in a chord are often said to resonate; that is, they mix to give something that is more than the sum of its parts. In a similar sense, the two Lewis structures for the SO_2 molecule are in resonance. They mix to give a hybrid in which each of the sulfur–oxygen bonds behaves as if it had a bond order of 1.5. The fact that SO_2 is a resonance hybrid of two Lewis structures is indicated by writing a double-headed arrow between the Lewis structures, as shown in Figure 4.7.

It is important to distinguish between resonance structures and isomers, such as dimethyl ether and ethanol, discussed in Section 4.5. Isomers have the same chemical formula but different arrangements of the atoms. Resonance structures have the same arrangement of atoms but a different arrangement of electrons, leading to multiple bonds in different positions in the resonance structures.

The relationship between the SO_2 molecule and its Lewis structures can be illustrated by an analogy. Suppose a knight of the Round Table returns to Camelot to describe a wondrous beast encountered during his search for the Holy Grail. He suggests that the animal looked something like a unicorn because it had a large horn in the center of its forehead. But it also looked something like a dragon because it was huge, ugly, and thick skinned. The beast was, in fact, a rhinoceros. The animal is real, but it was described as a hybrid of two mythical animals. The SO_2 molecule is also real, but we have to use two mythical Lewis structures to describe its bonding.

Exercise 4.4

The acetic acid found in vinegar dissociates to some extent in water to give the acetate ion, $CH_3CO_2^-$. Write two alternative Lewis structures for the acetate ion.

Solution

We can start by noting that the acetate ion contains two carbon atoms (Group IVA), three hydrogen atoms (Group IA), and two oxygen atoms (Group VIA). It also carries a negative charge, which means that the $CH_3CO_2^-$ ion contains a total of 24 valence electrons.

$$CH_3CO_2^- \qquad 2(4) + 3(1) + 2(6) + 1 = 24$$

The skeleton structure contains six covalent bonds, which leaves 12 nonbonding electrons. Unfortunately, it would take all 12 nonbonding electrons to satisfy the octets of the oxygen atoms, which leaves no nonbonding electrons for the carbon atom on the right in the following attempt at the Lewis structure of this molecule.

Because there aren't enough electrons to satisfy the octets of the atoms, we assume that there is at least one $C\!=\!O$ double bond. There are now seven covalent bonds in the skeleton structure, which leaves only 10 nonbonding electrons. Fortunately, this is enough.

This process gives us one satisfactory Lewis structure for the acetate ion. We can get another by changing the location of the $C\!=\!O$ double bond. As a result, the acetate ion is a resonance hybrid of the following Lewis structures.

> ➤ **CHECKPOINT**
>
> Explain why the two resonance structures of the acetate ion do not represent a pair of isomers.

> ➤ **CHECKPOINT**
>
> Whenever you pump your own gasoline, you are exposed to small quantities of benzene (C_6H_6) that evaporate from gasoline. The structure of benzene is based on a six-membered ring of carbon atoms in which all of the C—C bonds are the same length. Draw the two resonance structures for benzene. Explain why the carbon–carbon bond length in benzene is smaller than expected for a C—C single bond but longer than expected for a $C\!=\!C$ double bond.

Exercise 4.5

The molecule N_2O has the skeleton structure NNO. Which of the following Lewis structures are acceptable? For structures that are not acceptable explain why.

(a) $:\!\ddot{N}\!-\!\ddot{N}\!-\!\ddot{O}\!:$ (e) $:\!\ddot{N}\!=\!N\!-\!\ddot{O}\!:$

(b) $:\!\ddot{N}\!-\!N\!=\!\ddot{O}\!:$ (f) $:\!N\!\equiv\!\ddot{N}\!-\!\ddot{O}\!:$

(c) $:\!\ddot{N}\!=\!N\!=\!\ddot{O}\!:$ (g) $:\!\ddot{N}\!=\!\ddot{N}\!-\!\ddot{O}\!:$

(d) $:\!N\!\equiv\!N\!-\!\ddot{O}\!:$ (h) $:\!\ddot{N}\!-\!N\!\equiv\!O\!:$

Solution

The N_2O molecule has a total of 16 valence electrons, 5 from each nitrogen atom and 6 from the oxygen atom.

(a) This is not an acceptable Lewis structure because the valence shell of the central nitrogen is not filled. There are only four electrons around the central nitrogen atom.

(b) This is not an acceptable structure because there are only six electrons around the central nitrogen atom.

(c) This is an acceptable structure.

(d) This is also an acceptable structure.

(e) This is not an acceptable structure because there are only six electrons around the central nitrogen atom.

(f) This is not an acceptable structure because the central nitrogen atom has more than a filled valence shell. Because nitrogen is in the second row of the periodic table, it can accommodate only eight electrons.

(g) This structure contains a total of 18 electrons. Only 16 electrons are available in the valence shells of the atoms.

(h) This is an acceptable Lewis structure.

Structures (c), (d), and (h) are resonance structures of one another.

4.10 Electronegativity

A covalent bond results from the sharing of a pair of valence electrons by two atoms. When the atoms are identical, they must share the electrons equally. There is no difference between the electron density on the two oxygen atoms in an O_2 molecule, for example.

The same can't be said about molecules that contain different atoms. Consider the HCl molecule, for example. If there is any difference between the relative ability of hydrogen and chlorine to draw the electrons in a bond toward itself, the electrons in the H—Cl bond won't be shared equally. The electrons in the bond will be drawn closer to one atom or the other.

The relative ability of an atom to draw electrons in a bond toward itself is called the **electronegativity** (*EN*) of the atom. For many years, chemists have recognized that some atoms attract electrons in a bond better than other atoms. Thus F and O are more electronegative than Na and Mg. But there is no direct way to measure the electronegativity of an atom. Instead, properties that are assumed to depend on electronegativity are measured and then compared to one another in order to determine a relative scale of electronegativity. There are currently at least 15 electronegativity scales in use. Because there is a close agreement among electronegativity values of an atom on different scales, our confidence in these values is increased.

The first scale of electronegativities was created by Linus Pauling. He assigned fluorine an electronegativity of 4.0 and determined the electronegativities of the atoms of the other elements relative to fluorine. Every electronegativity scale that has been proposed since then has been adjusted so that the electronegativity of fluorine is about 4.

A new scale was proposed in 1989 that links the electronegativity of an atom to the average valence electron energy (AVEE) obtained from photoelectron spectroscopy.[1] This scale assumes that atoms that best resist the loss of valence electrons are the atoms that are most likely to draw electrons in a bond toward themselves. Thus the AVEE for an atom is a direct measure of the electronegativity of the atom. The AVEE data given in Figure 3.31 have been used to calculate the electronegativities shown in Figure 4.8*a* by multiplying by an appropriate factor to give fluorine an electronegativity value of about 4. (These data can also be found in Table B.7 in Appendix B and on the back cover of the text.)

Figure 4.8 shows two electronegativity scales. Figure 4.8*a* shows the scale based on AVEE, and Figure 4.8*b* shows the original Pauling scale. Because the AVEE scale is based on PES experimental measurements, discussed in Chapter 3, we will use the AVEE electronegativity scale exclusively in this text. This scale

---

[1] L. C. Allen, *Journal of the American Chemical Society,* **111,** 9003 (1989).

H 2.30																H 2.30	He 4.16
Li 0.91	Be 1.58											B 2.05	C 2.54	N 3.07	O 3.61	F 4.19	Ne 4.79
Na 0.87	Mg 1.29											Al 1.61	Si 1.92	P 2.25	S 2.59	Cl 2.87	Ar 3.24
K 0.73	Ca 1.03	Sc 1.2	Ti 1.3	V 1.4	Cr 1.5	Mn 1.6	Fe 1.7	Co 1.8	Ni 1.9	Cu 1.8	Zn 1.6	Ga 1.76	Ge 1.99	As 2.21	Se 2.42	Br 2.69	Kr 2.97
Rb 0.71	Sr 0.96	Y 1.0	Zr 1.1	Nb 1.3	Mo 1.4	Tc 1.5	Ru 1.7	Rh 1.8	Pd 1.9	Ag 2.0	Cd 1.5	In 1.66	Sn 1.82	Sb 1.98	Te 2.16	I 2.36	Xe 2.58
Cs 0.66	Ba 0.88										Hg 1.76						

(a) AVEE Scale

H 2.1																H 2.1	He –
Li 1.0	Be 1.5											B 2.0	C 2.5	N 3.0	O 3.5	F 4.0	Ne –
Na 0.9	Mg 1.2											Al 1.5	Si 1.8	P 2.1	S 2.5	Cl 3.0	Ar –
K 0.8	Ca 1.0	Sc 1.3	Ti 1.5	V 1.6	Cr 1.6	Mn 1.5	Fe 1.8	Co 1.9	Ni 1.9	Cu 1.9	Zn 1.6	Ga 1.6	Ge 1.8	As 2.0	Se 2.4	Br 2.8	Kr –
Rb 0.8	Sr 1.0	Y 1.2	Zr 1.4	Nb 1.6	Mo 1.8	Tc 1.9	Ru 2.2	Rh 2.2	Pd 2.2	Ag 1.9	Cd 1.7	In 1.7	Sn 1.8	Sb 1.9	Te 2.1	I 2.5	Xe –
Cs 0.7	Ba 0.9										Hg 1.9						

(b) Pauling Scale

Fig. 4.8   (a) Electronegativities of the elements calculated from photoelectron spectroscopy and refined AVEE. The AVEE data from Chapter 3 have been adjusted to give fluorine an electronegativity value close to 4. [Reprinted from L. C. Allen and E. T. Knight, *Journal of Molecular Structure,* **261,** 313 (1992).] (b) Electronegativities based on the Pauling scale.

offers the additional advantage of providing electronegativities for the noble gases, which will be useful when discussing the compounds formed by these elements.

Figure 4.9 shows the result of adding the electronegativities of the main-group elements to the periodic table as a third axis. Because the electronegativity of He (*EN* = 4.16) is smaller than that of Ne (*EN* = 4.79), the column for He cannot be seen in Figure 4.9. It is obscured by the taller column that represents the electronegativity of Ne.

The data in Figures 4.8 and 4.9 show several clear patterns.

- Electronegativity increases in a regular fashion from left to right across a row of the periodic table.

- Electronegativity decreases from top to bottom down a column of the periodic table.

When atoms with large differences in electronegativities combine, the electron density in the resulting bond is pulled toward the more electronegative atom. These compounds are called ionic compounds, which are discussed in Chapter 5. NaCl, for example, is an ionic compound in which most of the electron density in the bond is transferred from the sodium atom to the more electronegative chlorine atom.

In the covalent molecules discussed so far in this chapter, the electrons in the covalent bonds are shared more or less evenly between bonding atoms. If the atoms in a bond have significantly different electronegativities, the more electronegative atom will attract more of the electron density in the bond. This type of bond is referred to as a **polar covalent bond.** One end of the bond has a partial positive charge ($\delta+$), and the other end has a partial negative charge ($\delta-$).

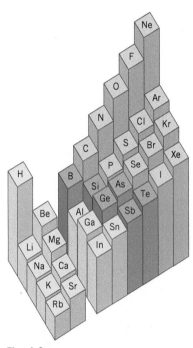

Fig. 4.9   Three-dimensional plot of the electronegativities of the main-group elements versus position in the periodic table. [Reprinted from L. C. Allen, *Journal of the American Chemical Society,* **111,** 9003 (1989).]

➤ **CHECKPOINT**

Classify the bonds in NO and O_2 as either covalent or polar covalent.

4.11 Partial Charge

It would be useful to have a quantitative measure of the extent to which the charge in an individual bond is located on one of the atoms that form the bond. This can be achieved by calculating the **partial charge** on the atom. The partial charge on an atom is proportional to the difference between the electronegativities of the atoms that form the bond. When the difference is relatively small, the electrons are shared more or less equally. When the difference is relatively large, there is a significant amount of charge separation in the bond.

To illustrate how the partial charge on an atom can be calculated, let's look at the HCl molecule.

$$H\!-\!\ddot{\underset{\cdot\cdot}{Cl}}:$$

According to the data in Figure 4.8a, the electronegativity of chlorine is 2.87 and that of hydrogen is 2.30. Thus the electrons of the H—Cl bond should be attracted to the chlorine slightly more than to the hydrogen. This would result in a partial charge on both atoms, with an excess of electron density on the chlorine atom. The symbol δ is used to indicate the partial charge on an atom that results from unequal sharing of electrons. A $\delta-$ is written above the symbol for the element that has the larger share of the electron density in the bond, whereas $\delta+$ is used to indicate a smaller share of electron density.

$$^{\delta+}H\!-\!Cl^{\delta-}$$

The partial charge on an atom in a molecule is determined by comparing the electron density associated with the free atom—one that is not involved in a bond—to the electron density associated with the atom when it is involved in a bond. Since only the outer-shell valence electrons are involved in bonding, we consider only those electrons. We begin the calculation by determining the number of valence electrons (V) in an isolated atom from the group number of the element in the periodic table.

The valence electron density associated with an atom in a molecule results from contributions from both the electrons in the covalent bond (B) and the nonbonding electrons (N) around the atom. The difference between the electronegativities of the atoms in a bond has no effect on the number of nonbonding electrons. But the difference in electronegativity does affect the number of bonding electrons that are assigned to each atom. The calculation of partial charge therefore involves the electronegativities of the two atoms (EN_a and EN_b) that form the bond.

The partial charge on an atom A in a covalent bond to atom B is given the symbol δ_a. The partial charge on this atom is calculated by subtracting the number of nonbonding electrons on the atom (N_a) from the number of valence electrons on the atom (V_a) and then subtracting the number of bonding electrons (B_a) multiplied by the fraction of the total electronegativity of the two atoms that can be assigned to that atom.[2]

$$\delta_a = V_a - N_a - B_a\left(\frac{EN_a}{EN_a + EN_b}\right)$$

To illustrate how the formula is used, let's calculate the partial charge on the chlorine atom in HCl. A neutral chlorine atom would have seven valence electrons,

---

[2]L. C. Allen, *Journal of the American Chemical Society,* **111,** 9115 (1989).

Table 4.2

Partial Charge on Group VIIA Atoms in Combination with Hydrogen

Molecule	δ
HF	-0.29
HCl	-0.11
HBr	-0.08
HI	-0.01

➤ **CHECKPOINT**

In a hypothetical AB molecule, if the electronegativity of B is significantly larger than that of A, which atom, A or B, would have the negative partial charge?

and there are six nonbonding electrons on the chlorine atom in the Lewis structure for HCl. We then substitute the electronegativities of chlorine (2.87) and hydrogen (2.30) into the above equation to get the following result.

$$\delta_{Cl} = 7 - 6 - 2\left(\frac{2.87}{2.87 + 2.30}\right) = -0.11$$

The partial charge on the chlorine is -0.11. The magnitude of the partial charge on the hydrogen atom must be the same as the magnitude of the partial charge on the chlorine atom, but the sign of the charge is different. We can demonstrate this by noting that a neutral hydrogen atom has one electron, there are no nonbonding valence electrons on the hydrogen atom in HCl, and the appropriate electronegativities of hydrogen and chlorine would be substituted into the equation that defines partial charge as follows.

$$\delta_{H} = 1 - 0 - 2\left(\frac{2.30}{2.30 + 2.87}\right) = 0.11$$

Because it is more electronegative than the hydrogen atom, the chlorine atom attracts the electrons in the H—Cl to itself until the partial charge on chlorine is -0.11 and the partial charge on the hydrogen atom is $+0.11$. This means that the chlorine atom in HCl has more electron density associated with it than a free chlorine atom. This molecule is therefore slightly polarized.

HCl is a neutral molecule, so the sum of the partial charges on the two atoms must be zero. The electron density that is lost by the hydrogen atom must be gained by the chlorine atom. Table 4.2 lists the partial charges on compounds formed between hydrogen and other Group VIIA atoms.

4.12 Formal Charge

The actual charge on an atom in a molecule is best represented by the partial charge. But partial charges are difficult to calculate by hand for complex molecules. Chemists therefore find it useful to calculate the **formal charge** on an atom, which helps us identify the atoms that are most likely to carry a significant amount of positive or negative charge. While formal charges are useful, it should be remembered that they do not represent the actual charges on the atoms.

The first step in the calculation of formal charge involves dividing the electrons in each covalent bond equally between the atoms that form the bond. The number of valence electrons formally assigned to each atom is then compared with the number of valence electrons on a neutral atom of the element. If the atom has more valence electrons than a neutral atom, it is assigned a formal negative charge. If it has fewer valence electrons, it carries a formal positive charge.

Consider the amino acid glycine, for example, which is the single most common amino acid found in the class of tough, insoluble proteins known as the α-keratins that form the skin, hair, fur, scales, nails, and horns of mammals, birds, and reptiles. The Lewis structure of glycine is written as follows.

The concept of formal charge can be used to understand the meaning of the positive and negative signs in the Lewis structure of this molecule.

We start by arbitrarily dividing pairs of bonding electrons in each covalent bond so that each atom in a bond is formally assigned half of the electrons, as shown in Figure 4.10. Once this is done, the nitrogen has four valence electrons, which is one fewer than a neutral nitrogen atom. The nitrogen therefore carries a formal charge of +1 in this Lewis structure. Both of the carbon atoms formally have four valence electrons, which is equal to the number of valence electrons on a neutral carbon atom. As a result, neither carbon atom has a formal charge. The oxygen atom in the C=O double bond formally has six valence electrons, which means it has no formal charge. The other oxygen atom, however, has seven valence electrons, which means that it formally has a charge of −1.

Although the glycine molecule has no net charge, one end of the molecule has a positive formal charge, and the other end has a negative formal charge. As a result, it isn't surprising that chemists write the formula for this amino acid as $H_3N^+CH_2CO_2^-$.

We can summarize the process by which the formal charge on an atom is calculated by noting that we start with the number of valence electrons (V_a) on a neutral atom of the element. We then subtract all of the nonbonding electrons on the atom (N_a) and half of the bonding electrons on the atom (B_a). The formal charge (FC_a) on atom A is therefore given by the following equation.

$$FC_a = V_a - N_a - \frac{B_a}{2}$$

To illustrate how this equation is used, let's apply the equation to calculating the formal charge on the nitrogen atom in glycine. Nitrogen has five valence electrons, there are no nonbonding electrons on the atom, and there are eight bonding electrons. As we have seen, the formal charge on the nitrogen atom is therefore +1.

$$FC_N = 5 - 0 - \frac{8}{2} = +1$$

The following example shows another reason why chemists find formal charge useful. According to Exercise 4.5, three equally valid Lewis structures can be drawn for dinitrogen oxide, N_2O, a gas commonly known as "laughing gas" that is used as an anesthetic by dentists.

$$:N{\equiv}N{-}\ddot{O}: \qquad \ddot{N}{=}N{=}\ddot{O} \qquad :\ddot{N}{-}N{\equiv}O:$$
$$\textit{I} \qquad\qquad \textit{II} \qquad\qquad \textit{III}$$

Which structure best represents the actual bonding in the molecule? Are the two nitrogen atoms connected by a single, a double, or a triple bond?

We begin by using what has come to be accepted as a general rule for choosing the Lewis structure for a molecule that best represents what we know about the properties of the compound. The best structure is considered to be the one in which the atoms have the smallest formal charges and the negative formal charges are on the more electronegative atoms. The formal charges on each atom in the three Lewis structures of N_2O are as follows.

$$\overset{\oplus 1 \quad \ominus 1}{:N{\equiv}N{-}\ddot{O}:} \qquad \overset{\ominus 1 \quad \oplus 1}{\ddot{N}{=}N{=}\ddot{O}} \qquad \overset{\ominus 2 \quad \oplus 1 \quad \oplus 1}{:\ddot{N}{-}N{\equiv}O:}$$
$$\textit{I} \qquad\qquad\qquad \textit{II} \qquad\qquad\qquad \textit{III}$$

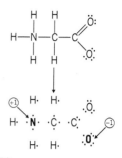

Fig. 4.10 The first step in calculating the formal charge on the atoms in glycine involves dividing pairs of bonding electrons equally between the atoms in each bond. Formal charges are shown in circles to distinguish them from actual charges.

➤ **CHECKPOINT**

Calculate the formal charge on each atom in SO_2.

Note that the sum of the formal charges must always equal the net charge on the molecule, which, in this case, is zero because N_2O is a neutral molecule.

Structure III has two serious flaws. First, it puts a negative formal charge on the *less* electronegative nitrogen atom, rather than the *more* electronegative oxygen atom. It also has the largest formal charges. Structure II also puts a negative formal charge on nitrogen, instead of oxygen, and therefore isn't as good as structure I.

Are there experimental data that could be used to support structure I? Because nitrogen forms single, double, and triple bonds in the proposed structures, bond length data should be helpful. We expect a N—N single bond to be longer than a N=N double bond, which should be longer than a N≡N triple bond. The following are typical nitrogen–nitrogen bond lengths for these three types of bonds.

$$
\begin{array}{ll}
\text{N—N} & 0.146\ \text{nm} \\
\text{N=N} & 0.125\ \text{nm} \\
\text{N≡N} & 0.110\ \text{nm}
\end{array}
$$

An experimental determination of the nitrogen–nitrogen bond length in N_2O gives a result of 0.113 nm, which is consistent with the triple bond suggested by structure I. The experimental data support the formal charge calculation that suggests structure I is the best Lewis structure for N_2O.

There is still some debate about the best Lewis structure for a compound. At one time, chemists preferred to arrange the electrons in a Lewis structure in order to minimize the formal charge on each atom. Consider SO_2, for example. In Figure 4.7, we wrote two Lewis structures for this molecule that each contain one S—O single bond and one S=O double bond, as shown by structure I on the left below. In theory, we could write a Lewis structure in which there were two S=O double bonds, as shown by structure II on the right.

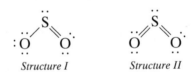

Structure I   Structure II

In structure I, sulfur has 8 valence electrons and would have a formal charge of +1, whereas in structure II sulfur has 10 valence electrons and the formal charge on the sulfur would be zero. Without additional information, it isn't possible to unambiguously conclude which is the best representation of the molecule. Recent advances in the theory of bonding have suggested that it is not appropriate to introduce more double bonds into a structure than needed to satisfy the octet of the central atom.[3] In this text we therefore expand the octets of the central atom when necessary, to form structures such as SF_4 and XeF_4, and introduce double (or triple) bonds as needed to satisfy the octet of the central atom, but we will not add additional multiple bonds to the Lewis structure to minimize the formal charges in the structure. Therefore, until conclusive evidence is available, we'll use structure I to describe SO_2 rather than structure II.

The equation used to determine the formal charge on an atom

$$
FC_a = V_a - N_a - \frac{B_a}{2}
$$

[3]L. Suidan, J. K. Badenhoop, E. D. Glendenins, and F. Weinhold, *Journal of Chemical Education*, **72**, 583 (1995).

is a special case of the more general equation used to calculate the partial charge on the atom.

$$\delta_a = V_a - N_a - B\left(\frac{EN_a}{EN_a + EN_b}\right)$$

The difference between the two expressions is that when formal charge is calculated, the electrons are assumed to be equally shared ($EN_a = EN_b$).

Calculations of formal charge are therefore a more approximate description of the distribution of electrons in a bond than calculations of the partial charge on the atoms. Consider HCl, for example. The formal charge on both the hydrogen and chlorine atoms is zero because the electronegativities of the hydrogen and chlorine atom are treated as if they were equal in formal charge calculations. As we have seen, however, the actual charge on the chlorine atom is -0.11, whereas the actual charge on the hydrogen atom is $+0.11$, because the chlorine atom is significantly more electronegative than the hydrogen atom.

4.13 The Shapes of Molecules

The shape of a molecule often plays a vital role in determining its chemistry. The changes in the three-dimensional structure of proteins that occur when an egg is heated, for example, are the primary source of the differences between a raw egg and a cooked one.

An illustration of how sensitive biomolecules are to changes in their three-dimensional structure is provided by the chemistry of hemoglobin, a protein with a molecular weight of 65,000 amu that carries oxygen through the body. This protein contains four chains of amino acids, two *a* chains and two *b* chains, as shown in Figure 4.11.

Sickle-cell anemia occurs when the identity of a single amino acid among the 146 amino acids on a *b* chain in hemoglobin is changed. The substitution of valine for glutamic acid at the sixth position on this chain produces a change in the structure of the hemoglobin, which interferes with its ability to pick up oxygen at low pressures. The result is so severe that children who inherit this disorder from both parents seldom live past age 2.

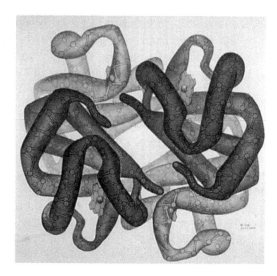

Fig. 4.11 Change in the identity of one of the 146 amino acids on one of the chains in hemoglobin causes a large enough change in the structure of hemoglobin to interfere with its ability to carry oxygen through the blood. [Illustration, Irving Geis. Image from Irving Geis Collection/ Howard Hughes Medical Institute. Rights owned by HHMI. Not to be reproduced without permission.]

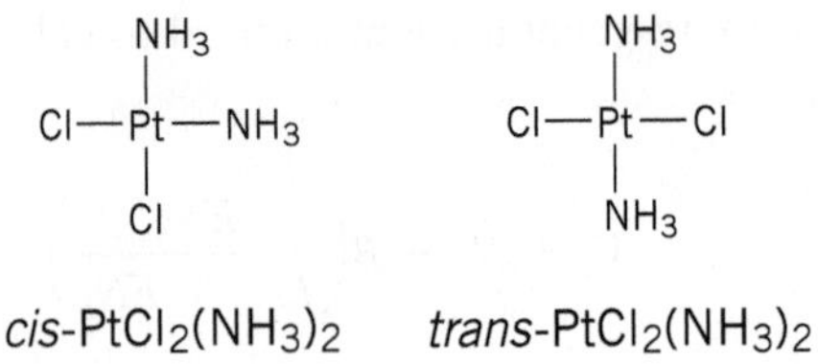

Fig. 4.12   The *cis* and *trans* isomer of $PtCl_2(NH_3)_2$.   *cis*-$PtCl_2(NH_3)_2$ *trans*-$PtCl_2(NH_3)_2$

A compound with the formula $PtCl_2(NH_3)_2$ provides another example of the importance of the three-dimensional shape of a molecule. For more than 100 years, chemists have known that this compound exists as a pair of isomers with the same chemical formula but slightly different structures, as shown in Figure 4.12. One of these compounds is known as the *cis* isomer (from the Latin prefix meaning "on this side") because the two chlorines can be thought of as being on the same side of the platinum atom. The other is the *trans* isomer (from the Latin prefix that means "across, or on the other side") because the chlorines are now across from each other.

While studying the effect of an electric current on *E. coli* bacteria, a group of researchers at Michigan State discovered that using platinum electrodes to carry the electric current inadvertently led to the formation of the square-planar *cis*-$PtCl_2(NH_3)_2$ complex. This result was totally unexpected, but it explained why the bacteria cells seemed to grow to as much as 300 times their normal length instead of reproducing to form new cells. They concluded that this platinum complex interfered with the process by which bacteria cells reproduce by cell mitosis. After extensive experiments with mammalian tumor cells, this compound was proposed for use as an anti-tumor agent that was given the trade name *cisplatin*. Approved for clinical use by the FDA in 1978, *cisplatin* revolutionized the treatment of certain cancers.

Because of its square-planar structure, *cisplatin* can slip into one of the grooves in the structure of DNA. Each of the chlorine atoms is gradually replaced by a bond to one of the strands of DNA, so that the drug effectively cross-links these strands. When this happens, the cell can no longer undergo the replication of DNA needed in order for cell mitosis to occur, and the natural repair mechanism of the body eventually causes the tumor cell to die.

Although the discovery of *cisplatin* represented a major step forward in the treatment of various forms of cancer, it had a number of serious side effects. As a result, it has been replaced by other drugs that also inhibit cell mitosis and therefore slow down the rate at which tumors grow. One of the most commonly used anti-tumor agents is now a drug that was given the common name *taxol* and is now known by the generic name *paclitaxel* ($C_{47}H_{51}NO_{14}$). The so-called line structure of this molecule as it would be drawn by an organic chemist is shown in Figure 4.13. The three-dimensional structure of this compound is shown in Figure 4.14.

Taxol was first isolated from the bark of a yew tree that grows in the Pacific northwest. Like so many other compounds isolated from natural sources, it was first screened for biological activity. When it was found to be unusually active as an anti-tumor agent, there was initially some concern about the environmental impact of harvesting this compound from trees. In 1994, Robert Holton, who was a professor at Florida State, worked out a synthesis of this complex molecule beginning with a starting material that had been used as a constituent of perfumes for at least 200 years. The unique structure of the *taxol* molecule allows it to bind to microtubules that are part of the cell's apparatus for dividing and replicating

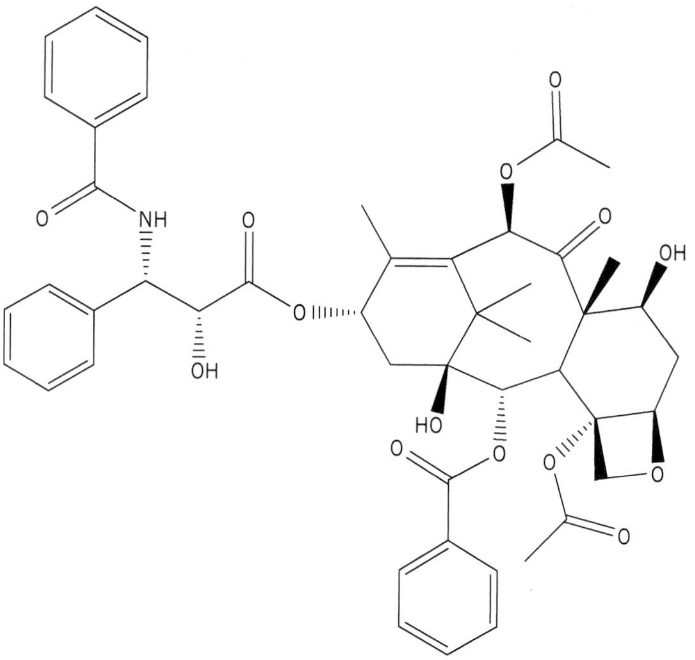

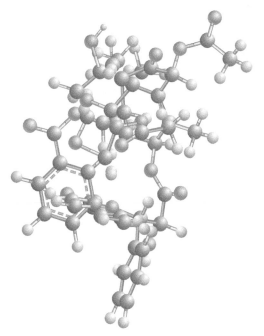

Fig. 4.13 A line structure drawing of the structure of anti-tumor drug *taxol*.

itself through cell mitosis, which inevitably leads to cell death. *Taxol* is therefore commonly used today to treat breast cancer, lung cancer, and ovarian cancel. It is also being used to treat Kaposi's sarcoma in AIDS patients.

The side effects of chemotherapy drugs such as *cisplatin* and *taxol* can be understood by noting that they slow down the rapid cell division that is the characteristic property of cancerous tumors. But these drugs cannot differentiate between tumor cells and normal cells. Fortunately, the normal cells eventually grow back.

Fig. 4.14 The three-dimensional structure of a *taxol* molecule.

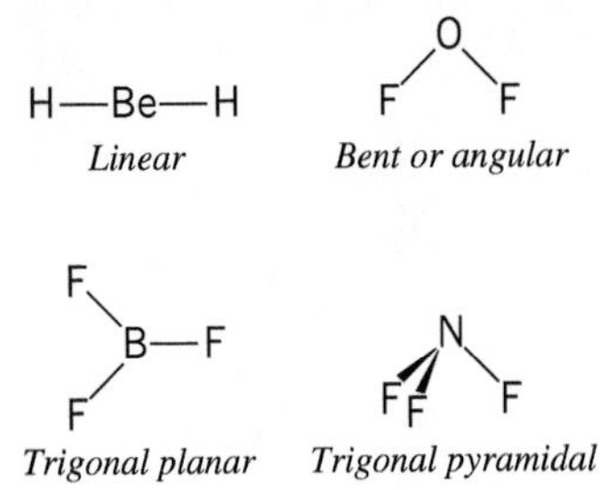

H—Be—H
Linear

F—O—F (shown as angular)
Bent or angular

F, B—F (trigonal)
Trigonal planar

N with F's
Trigonal pyramidal

Fig. 4.15 There is no obvious relationship between the chemical formula and the shape of a molecule. BeH$_2$ is linear, whereas OF$_2$ is bent. BF$_3$ is a planar molecule, whereas NF$_3$ is pyramidal.

4.14 Predicting the Shapes of Molecules (The Electron Domain Model)

There is no direct relationship between the formula of a compound and the shape of its molecules, as shown by the examples in Figure 4.15. The three-dimensional shapes of these molecules can be predicted from their Lewis structures, however, with a model first developed in the 1960s.[4,5] When it was introduced, it was called the *valence-shell electron pair repulsion* (*VSEPR*) *model*. The model has been modified, however, and is now referred to as the **electron domain (ED) model.**[6]

The spatial arrangement of atoms in molecules conforms to a relatively small number of types. Angles between selected atoms tend generally to be 90°, 109.5°, 120°, or 180°. The limited number of ways of arranging atoms and the arrangement of atoms in an enormous array of molecular structures can be explained by the electron domain model.

The electrons in the valence shell of an atom in a molecule are usually present as pairs of electrons with opposite spins. There could be a single pair of electrons, such as the bonding domain in HCl, or two or even three pairs of electrons, such as the bonding domains in ethylene (H$_2$C=CH$_2$) and acetylene (HC≡CH). They can also be present as a nonbonding domain, such as the three pairs of nonbonding electrons on each of the fluorine atoms in the F$_2$ molecule shown in Figure 4.3. In rare situations, there can even be a domain that contains only a single unpaired electron, such as the nonbonding domain on the nitrogen atom in the Lewis structure for NO$_2$ discussed at the end of Section 4.7.

The ED model assumes that the geometry around each atom in a molecule can be predicted by arranging these domains of electron density in space in a geometry that keeps the domains separated as far as possible. Thus two domains will be separated by an angle of 180°. For three domains the best arrangement is an angle of 120°, whereas four domains are oriented in space toward the corners of a tetrahedron with an angle between electrons domains of 109.5°. Figure 4.16 shows how the ED model predicts the way domains, treated as spheres, would be arranged about a central atom.

The electrons in a bonding domain are shared by two atoms, whereas the electrons in a nonbonding domain belong entirely to the valence shell of the atom on which they are placed. Therefore, a nonbonding domain tends to spread out and occupy a larger space than does a bonding domain.

We can see how the ED model is used by applying it to BeH$_2$. The Lewis structure of the molecule suggests that there are two pairs of bonding electrons, or two bonding domains, in the valence shell of the central atom. Note that this

Fig. 4.16 Arrangement of domains about a central atom. Two domains take up a position on opposite sides of a central atom, three domains are arranged at an angle of 120°, and four domains are arranged toward the corners of a tetrahedron at an angle of 109.5°.

Two domains 180°   Three domains 120°   Four domains 109.5°

[4]R. J. Gillespie and R. S. Nyholm, *Quarterly Review of the Chemical Society,* **11,** 339 (1957).
[5]R. J. Gillespie, *Journal of Chemical Education,* **40,** 295 (1963).
[6]R. J. Gillespie, *Journal of Chemical Education,* **69,** 116 (1992).

molecule is an exception to the octet rule because there are only two pairs of electrons in the valence shell of the central atom.

We can keep the two bonding domains as far apart as possible by arranging them on either side of the beryllium atom. The ED model therefore predicts that BeH_2 should be a **linear** molecule, with a 180° angle between the two Be—H bonds.

The ED model can also be applied to predict the shape or geometry of molecules with multiple bonds. Consider the Lewis structure of CO_2, for example.

$$\ddot{O}=C=\ddot{O}$$

There are four pairs of bonding electrons in the valence shell of the carbon atom but only two domains in which those electrons can be found. (These two domains correspond to the two pairs of electrons in the C=O double bond on the left and the two pairs in the double bond on the right.) As a result, the ED model predicts that this molecule will also have a linear geometry, with a 180° angle between the two double bonds.

The two electron pairs in the C=O double bond in CO_2 occupy more space than a single-bond domain. The three electron pairs in a triple bond would require even more space than a double-bond domain.

There are three domains in the valence shell of the central atom in boron trifluoride (BF_3) where electrons can be found. These domains correspond to the three pairs of bonding electrons in the B—F bonds.

$$:\ddot{F}—\ddot{B}—\ddot{F}:$$

The optimum geometry to keep the three domains in the valence shell of the boron atom as far apart as possible is an equilateral triangle. The ED model therefore predicts a **trigonal planar** geometry for the BF_3 molecule, with a F—B—F bond angle of 120°, as shown in Figure 4.17.

The advantage of counting domains of electron density rather than pairs of electrons can be illustrated by considering the geometry of formaldehyde, H_2CO.

$$H—C—H$$

The Lewis structure of the molecule suggests that there are four pairs of electrons in the valence shell of the central atom. Two pairs form a double-bond domain, and there are two single-bond domains. Because the four pairs of electrons can be found in three bonding domains, the ED model predicts that this molecule would have a trigonal planar geometry, just like BF_3.

At first glance, we might expect that the H—C—H bond angle in formaldehyde would be 120°, the internal angle in an equilateral triangle. Experiment suggests that it is a little smaller than this, only 118°. This can be explained by noting that the C=O double-bond domain occupies more space than the domains

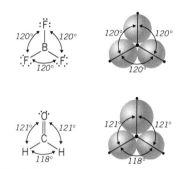

Fig. 4.17 Three equal bonding domains surround the central B atom in BF_3 giving an angle of 120°. The three bonding domains around the carbon atom in formaldehyde are not the same size, and the H—C—O angle increases to 121°.

► **CHECKPOINT**

The structure of benzene was discussed in the Checkpoint at the end of Section 4.9. Determine the number of electron domains around each carbon atom.

---

that contain the electrons in the C—H bonds. As a result, the H—C—O Bond angles are slightly larger than 120°, and the H—C—H bond angle is slightly smaller than 120°, as shown in Figure 4.17.

BeH_2, CO_2, BF_3, and H_2CO are all two-dimensional molecules in which the atoms lie in the same plane. If we place the same restriction on methane (CH_4), we would get a square-planar geometry, with an H—C—H angle of 90°.

$$H$$
$$|$$
$$H—C—H$$
$$|$$
$$H$$

There is a more efficient way of arranging the four domains in the valence shell of the central atom in CH_4, however. If the four domains are arranged toward the corners of a **tetrahedron,** the H—C—H angle increases to 109.5°. This structure is preferred by the ED model because it keeps the four domains of electron density further apart.

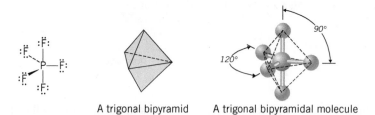

A tetrahedron   A tetrahedral molecule

PF_5 has five single-bond domains in the valence shell of the phosphorus atom.

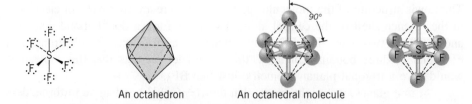

A trigonal bipyramid   A trigonal bipyramidal molecule

The best arrangement of the domains—to keep them separated as much as possible—is to orient them toward the corners of a **trigonal bipyramid.** Three of the positions in a trigonal bipyramid are labeled *equatorial* because they lie along the equator of the molecule. The other two are said to be *axial* because they lie along an axis perpendicular to the equatorial plane. The angle between the three equatorial positions is 120°, whereas the angle between an axial and an equatorial position is 90°.

There are six single-bond domains on the sulfur atom in SF_6.

An octahedron   An octahedral molecule

The optimum geometry for the molecule would involve arranging these six domains toward the corners of an **octahedron.** The term *octahedron* literally means "eight sides," but it is the six corners, or vertices, that interest us. To envision the geometry of an SF_6 molecule, imagine four fluorine atoms lying in a plane around the sulfur atom with one fluorine atom above the plane and another below. All F—S—F angles in SF_6 are 90°.

4.15 The Role of Nonbonding Electrons in the ED Model

All of our examples, so far, have contained only bonding electrons in the valence shell of the central atom. What happens when we apply the ED model to atoms that also contain nonbonding electrons? Consider ammonia (NH_3) and water (H_2O), for example.

$$H—\overset{\displaystyle ..}{\underset{\displaystyle |}{N}}—H \qquad H—\overset{\displaystyle ..}{\underset{\displaystyle ..}{O}}—H$$
$$H$$

In each case, there are four domains in the valence shell of the central atom where electrons can be found. The valence electrons on the central atom in both NH_3 and H_2O therefore should be distributed toward the corners of a tetrahedron, as shown in Figure 4.18. Our goal, however, isn't to predict the distribution of valence electrons. Rather, it is to use the distribution of electrons to predict the geometry of the molecule, which describes how the atoms are distributed in space. In previous examples, the way the valence electrons were distributed and the geometry of the molecule have been the same. Once we include nonbonding electrons, this is no longer true.

The ED model predicts that the valence electrons on the central atoms in ammonia and water will be oriented toward the corners of a tetrahedron. Because we can't locate the nonbonding electrons with any precision, this prediction can't be tested directly. But the results of the ED model can be used to predict the positions of the atoms in the molecules, which can be tested experimentally. If we focus on the positions of the atoms in ammonia, we predict that the NH_3 molecule should have a shape best described as **trigonal pyramidal,** with the nitrogen at the top of the pyramid. Water, on the other hand, should have a shape that can be described as **bent,** or **angular.** Both predictions have been shown to be correct, which reinforces our faith in the ED model.

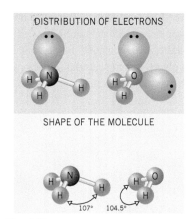

Fig. 4.18 The ED model predicts that the valence electrons on the central atom in NH_3 and H_2O will be oriented toward the corners of a tetrahedron. The shape of the molecules, however, is determined by the positions of the atoms. Ammonia is therefore described as trigonal pyramidal, and water is described as bent, or angular.

Exercise 4.6

Use the Lewis structure of the PF_3 molecule to predict the shape of this molecule.

$$:\overset{\displaystyle ..}{\underset{\displaystyle ..}{F}}—\overset{\displaystyle ..}{\underset{\displaystyle |}{P}}—\overset{\displaystyle ..}{\underset{\displaystyle ..}{F}}:$$
$$:\overset{\displaystyle ..}{\underset{\displaystyle ..}{F}}:$$

Solution

The Lewis structure suggests that there are four domains in the valence shell of the phosphorus atom where electrons can be found. There are electron pairs in the three P—F single bonds and a pair of nonbonding electrons. The geometry for the molecule is based on arranging these domains toward the corners of a tetrahedron. The shape of the molecule is therefore best described as *trigonal pyramidal,* like the geometry of ammonia.

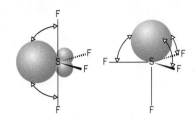

Fig. 4.19 Crowding of bonding and nonbonding domains in SF₄ is minimized if the nonbonding domain is placed in an equatorial position, as shown in the structure on the left.

When we extend the ED model to molecules in which the domains are distributed toward the corners of a trigonal bipyramid, we run into the question of whether nonbonding electrons should be placed in equatorial or axial positions. Experimentally, we find that nonbonding electrons usually occupy equatorial positions in a trigonal bipyramid.

To understand why, we have to recognize that nonbonding electron domains take up more space than bonding electron domains. Nonbonding domains only need to be close to one nucleus, and there is a considerable amount of space in which nonbonding electrons can reside and still be near the nucleus of the atom. Bonding electrons, however, must be simultaneously close to two nuclei, and only a small region of space between the nuclei satisfies this restriction.

Figure 4.19 can help us understand why nonbonding electrons are placed in equatorial positions in a trigonal bipyramid. If the nonbonding electron domain in SF₄ is placed in an axial position, it will be relatively close (90°) to *three* bonding-pair domains. But if the nonbonding domain is placed in an equatorial position, it will be 90° away from only *two* bonding domains. As a result, an equatorial position is less crowded than an axial position in a trigonal bipyramid geometry. The relatively large nonbonding domains therefore occupy the equatorial positions in this geometry.

The results of applying the ED model to SF₄, ClF₃, and the I₃⁻ ion are shown in Figure 4.20. When the nonbonding domain on the sulfur atom in SF₄ is placed in an equatorial position, the geometry around the sulfur atom in this molecule can be best described as having a **seesaw,** or *teeter-totter,* shape. The bonding and nonbonding domains in the valence shell of chlorine in ClF₃ can best be accommodated by placing both nonbonding domains in equatorial positions in a trigonal bipyramid. When this is done, we get a geometry that can be described as **T-shaped.** The Lewis structure of the triiodide (I₃⁻) ion suggests a trigonal bipyramidal distribution of valence electrons on the central atom. When the three nonbonding electron domains on the central I atom are placed in equatorial positions, the net result is a linear molecule.

The predictions of the ED model are summarized in Table 4.3. The number of bonding and nonbonding domains around the central atom determines the way these domains will be distributed in space around that atom. This, in turn, deter-

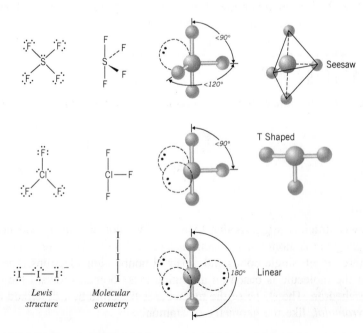

Lewis structure   *Molecular geometry*

Fig. 4.20 Structures of SF₄, ClF₃, and the I₃⁻ ion.

Table 4.3
Relationship between Number of Electron Domains and Geometry Around an Atom

Electron Domains	Bonding Domains	Nonbonding Domains	Distribution of Electron Domains	Molecular Geometry	Examples
2 (sp)	2	0	Linear	Linear	BeH_2, CO_2
	1	1		Linear	CO, N_2
3 (sp^2)	3	0	Trigonal planar	Trigonal planar	BF_3, CO_3^{2-}
	2	1		Bent	O_3, SO_2
	1	2		Linear	O_2
4 (sp^3)	4	0	Tetrahedral	Tetrahedral	CH_4, SO_4^{2-}
	3	1		Trigonal pyramidal	NH_3, H_3O^+
	2	2		Bent	H_2O, ICl_2^+
	1	3		Linear	HF, OH^-
5 (sp^3d)	5	0	Trigonal bipyramidal	Trigonal bipyramidal	PF_5
	4	1		Seesaw	SF_4, IF_4^+
	3	2		T-shaped	ClF_3
	2	3		Linear	I_3^-, XeF_2
6 (sp^3d^2)	6	0	Octahedral	Octahedral	SF_6, PF_6^-
	5	1		Square pyramidal	BrF_5, $SbCl_5^{2-}$
	4	2		Square planar	XeF_4, ICl_4^-

mines the geometry that would be described by the positions of the atoms to which the central atom is bound. For example, SO_2 has three electron domains around the central atom (two bonding domains and one nonbonding domain). We can determine the geometry around the central atom by locating the number of electron domains in column 1. A central atom surrounded by three electron domains will have one of the three possible molecular geometries listed for that entry in column 5. Reading across the table, we find that the electrons are distributed about the central atom, in a trigonal planar fashion. When we focus on the location of the atoms bound to the central atom, we find that the molecule geometry of SO_2 is best described as bent.

Exercise 4.7

Molecular geometries based on an octahedral distribution of electron domains are easier to predict because all of the corners of an octahedron are identical. Use the Lewis structures of BrF_5 and XeF_4 to predict the shapes of these molecules.

Solution

The valence-shell electrons on the bromine atom in BrF_5 are distributed toward the corners of an octahedron to form a molecule with a **square pyramidal** geometry.

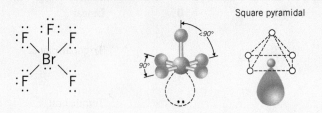

For XeF_4, two of the six domains in the valence shell of the central atom are nonbonding domains and four are bonding domains. Because they occupy considerably more space, the nonbonding domains are kept as far apart as possible. The ED model therefore predicts a **square planar** structure.

➤ **CHECKPOINT**

What is the shape of a molecule with three bonding domains and one nonbonding domain around the central atom?

The predictions of geometry around the central atom in a molecule that result from application of the ED model are consistent with a theoretical model that predicted what would happen when atomic orbitals on an atom were mixed or "hybridized." Two domains that are arranged in opposite directions through an angle of 180° produce the same geometry as the so-called sp hybrid orbitals. The three domains that give rise to the geometry of an equilateral triangle are equivalent to sp^2 hybrid orbitals. The four electron domains that give rise to a tetrahedron are equivalent to sp^3 hybrid orbitals. These designations are also given in Table 4.3. The SO_2 molecule is thus described as having three electron domains on the central atom, or as an sp^2 hybrid. A more detailed description of the use of these labels is given in the Special Topics section on hybridization at the end of this chapter.

4.16 Bond Angles

Just as bond lengths are important to the structure and hence the properties of a compound, so are bond angles. To define a bond angle, at least three atoms must be located. Consider, for example, the compound known as propyl alcohol or 1-propanol ($CH_3CH_2CH_2OH$).

$$H_d-\underset{\underset{H_e}{|}}{\overset{\overset{H_c}{|}}{C_3}}-\underset{\underset{H_f}{|}}{\overset{\overset{H_b}{|}}{C_2}}-\underset{\underset{H_g}{|}}{\overset{\overset{H_a}{|}}{C_1}}-\overset{..}{\underset{..}{O}}-H_h$$

It is difficult to imagine the three-dimensional structure of a molecule from the flat, two-dimensional representations on the printed page of a text. Consider the geometry around the carbon atom labeled C_3. The best way to arrange the four bonding domains on this atom is toward the corners of a tetrahedron. Each of the bond angles on this carbon is therefore 109.5°, regardless of whether we look at the H_c—C_3—H_d, H_d—C_3—H_e, or H_c—C_3—H_e angles. The bond angles on C_2 and C_1 are also 109.5° because there are four bonding domains on these atoms. The C_1—O—H_h angle, however, is slightly smaller than 109.5° because the oxygen atom is surrounded by two single-bond domains and two nonbonding domains. Because the nonbonding domains occupy more space than the bonding domains, the bonding pairs are somewhat squeezed to form an angle slightly less than 109.5°. The three-dimensional structure of propyl alcohol or 1-propanol is shown in Figure 4.21.

An ammonia molecule contains three bonding domains and one nonbonding domain, while a water molecule contains two bonding domains and two nonbonding domains. The ED model predicts that the four electron domains in water and ammonia will be arranged in a tetrahedral geometry around the central atom. We might therefore expect that the bond angles in ammonia and water would be the same as the 109.5° bond angle for the tetrahedral carbon atoms in methane, CH_4, and propyl alcohol, CH_3CH_2OH. The experimental values for the bond angles in ammonia and water are actually 107° and 104.5°, respectively. This can be explained by reminding ourselves that nonbonding domains occupy more space than bonding domains, thus pushing the bonding domains closer together and giving a smaller than expected bond angle around the central atom in these compounds.

Fig. 4.21 A so-called ball-and-stick model of the three-dimensional structure of propyl alcohol.

➤ **CHECKPOINT**

Predict the following bond angles for propyl alcohol:

$$C_1—C_2—C_3$$
$$C_2—C_1—H_g$$
$$H_g—C_1—O$$

Exercise 4.8

Assume that the atoms in acetic acid are given the following arbitrary labels.

$$
\begin{array}{ccc}
 & H_a & \ddot{O}_f: \\
 & | & \diagup\diagdown \\
H_b\!-\!\!\!\!& C_2\!-\!C_1 & \\
 & | & \diagdown\ddot{}\ddot{} \\
 & H_c & \ddot{O}_e\!-\!H_d \\
\end{array}
$$

Predict the following bond angles for acetic acid.

$$
\begin{array}{ll}
H_a—C_2—H_b & C_1—O_e—H_d \\
C_2—C_1—O_f & O_f—C_1—O_e \\
C_2—C_1—O_e & H_a—C_2—C_1 \\
\end{array}
$$

Solution

The total number of bonding and nonbonding domains around the central atom of the three atoms that form one of these bond angles determines the arrangement of the atoms in space. There are four bonding domains around C_2, three bonding domains around C_1, and a total of four domains (two bonding and two nonbonding) around O_e in the Lewis structure of acetic acid. The resulting bond angles are therefore:

$$H_a\text{—}C_2\text{—}H_b \approx 109.5° \qquad C_1\text{—}O_e\text{—}H_d < 109.5°$$
$$C_2\text{—}C_1\text{—}O_f \approx 120° \qquad O_f\text{—}C_1\text{—}O_e \approx 120°$$
$$C_2\text{—}C_1\text{—}O_e \approx 120° \qquad H_a\text{—}C_2\text{—}C_1 \approx 109.5°$$

- -

> ➤ CHECKPOINT
>
> The structure of benzene was discussed in the Checkpoint at the end of Section 4.9. Predict the bond angles around each carbon atom.

4.17 The Difference between Polar Bonds and Polar Molecules

The difference between the electronegativities (ΔEN) of chlorine ($EN = 2.87$) and hydrogen ($EN = 2.30$) is 0.57 and is sufficiently large that the bond in HCl is said to be *polar*. As we saw in Section 4.11, the partial charge on the hydrogen is positive and that on the chlorine is negative.

$$\overset{\delta+ \quad \delta-}{H\text{—}Cl}$$
$$\longmapsto$$

Because it contains only this one bond, the HCl molecule can be described as polar. The polarity of a bond in a structure is represented with an arrow having a plus sign at the positive end (the $\delta+$ and $\delta-$ shown in the above structure could be replaced by the $\longmapsto$ symbol). The arrow points in the direction of the more electronegative atom, and the plus sign is next to the least electronegative atom.

The polarity of a molecule can be predicted by considering the polarities of the individual bonds, the location of nonbonding pairs, and the three-dimensional shape of the molecule. The magnitude of the polarity of a molecule is measured using a quantity known as the **dipole moment,** μ. The dipole moment for a molecule depends on two factors: (1) the magnitude of the charge and (2) the distance between the negative and positive poles of the molecule. Dipole moments are reported in units of *debye* (D). The dipole moment for HCl is small, 1.08 D (Table 4.4). This can be understood by noting that the separation of charge in the HCl bond is relatively small because the H—Cl bond is relatively short.

C—Cl bonds ($\Delta EN = 0.33$) are not as polar as H—Cl bonds ($\Delta EN = 0.57$), but they are significantly longer. As a result, the dipole moment of CH_3Cl (1.89 D) shown in Figure 4.22 is greater than that for HCl.

At first glance, we might expect a similar dipole moment for carbon tetrachloride (CCl_4), which contains four polar C—Cl bonds. The experimental value of the dipole moment for CCl_4, however, is zero. This can be understood by considering the structure of CCl_4, shown in Figure 4.22. The individual C—Cl bonds in this molecule are polar. However, since the dipole moments of the four C—Cl bonds are equivalent and the bonds are symmetrically arranged at the corners of a tetrahedron, the four C—Cl dipoles cancel each other. Carbon tetrachloride therefore illustrates an important point: Not all molecules that contain polar bonds have a dipole moment.

Table 4.4
Dipole Moments of Common Compounds

Compound	Dipole Moment (D)
CH_4, methane	0
CH_3OH, methanol	1.70
CH_3Cl	1.89
CO, carbon monoxide	0.112
CO_2, carbon dioxide	0
HCl, hydrogen chloride	1.08
HI, hydrogen iodide	0.44
H_2O, water	1.85
H_2S, hydrogen sulfide	0.97

> ➤ CHECKPOINT
>
> Should chloroform have a dipole moment?
>
> $$\overset{\displaystyle :\overset{\cdot\cdot}{\underset{\cdot\cdot}{Cl}}:}{:\overset{\cdot\cdot}{\underset{\cdot\cdot}{Cl}}\text{—}\overset{\displaystyle |}{\underset{\displaystyle |}{C}}\text{—}\overset{\cdot\cdot}{\underset{\cdot\cdot}{Cl}}:}$$
> $$H$$

Exercise 4.9

The two-dimensional Lewis structure for dichloromethane is given below. On first inspection it would appear that the polarities of the two C—H bonds and the two C—Cl bonds should cancel one another, resulting in a nonpolar molecule with a dipole moment of zero. However, the measured dipole moment of dichloromethane is 1.60 D. Explain this discrepancy.

Fig. 4.22 Although the C—Cl bonds in both CH_3Cl and CCl_4 are polar, CCl_4 has no net dipole moment because the polarity of the four C—Cl bonds cancel each other.

$$
\overset{\displaystyle H}{\underset{\displaystyle H}{\overset{|}{\underset{|}{:\ddot{C}l—C—\ddot{C}l:}}}}
$$

Solution

Dichloromethane is not the flat two-dimensional molecule shown in the Lewis structure. It is tetrahedral, as shown below. The dipole moments of the C—H bonds point toward the carbon atom. The dipole moments of the C—Cl bonds point away from the carbon atom. Therefore the dipole moments of the bonds are additive to give a net dipole moment to the molecule, as shown in the following.

Net dipole

Exercise 4.10

Determine whether each of the following molecules is polar.

(a) NH_3 (b) CO_2

Solution

(a) The structure for ammonia is shown below. The dipoles of the N—H bonds are additive to give an overall dipole to the molecule. In addition, the nonbonding pair on the nitrogen adds to the overall dipole moment of ammonia, which is 1.47 D.

$\updownarrow$ *Net dipole*

$$
\overset{\displaystyle ..}{\underset{\displaystyle H}{H \overset{\displaystyle N}{\quad} H}}
$$

(b) The structure for carbon dioxide is shown below. Carbon dioxide is a linear molecule with an oxygen on each side of the carbon. Both C=O double bonds are polar, but their dipoles point in opposite directions, causing them to cancel one another and leading to a nonpolar molecule for which the dipole moment is 0 D.

$$
\ddot{O}=C=\ddot{O}
$$

Key Terms

Angular	Formal charge	Resonance hybrid
Bent	Isomers	Seesaw
Bond length	Lewis structures	Square planar
Bonding domains	Linear	Square pyramidal
Covalent bond	Nonbonding domains	T-shaped
Covalent molecules	Nonbonding electrons	Tetrahedron
Dipole moment	Octahedron	Trigonal bipyramid
Domains	Octet rule	Trigonal planar
Electron domain (ED) model	Partial charge	Trigonal pyramidal
Electronegativity	Polar covalent bond	Valence electrons

Problems

Valence Electrons

1. Define the term *valence electrons*.

2. Determine the number of valence electrons in neutral atoms of the following elements.

   (a) Li   (b) C   (c) Mg   (d) Ar

3. Determine the number of valence electrons in neutral atoms of the following elements.

   (a) Fe   (b) Zr   (c) Bi   (d) I

4. Determine the number of valence electrons in the following negative ions and describe any general trends.

   (a) C^{4-} (b) N^{3-} (c) S^{2-} (d) I^-

5. Determine the number of valence electrons in the following positive ions and describe any general trends.

   (a) Na^+ (b) Mg^{2+} (c) Al^{3+} (d) Sc^{3+}

6. Explain why filled *d* and *f* subshells are ignored when the valence electrons on an atom are counted.

7. What is meant by the term *octet rule*?

8. What is the difference between valence electrons and core electrons?

9. What is the relationship between the number of valence electrons and the group number for main-group elements.

The Covalent Bond

10. Write Lewis structures for the elements in the third row of the periodic table.

11. Describe the difference between the covalent bonds in O_2 and F_2. How many valence electrons surround an F atom in F_2? How many are around an O atom in O_2?

12. Draw the Lewis structure for N_2. Why is this a covalent molecule?

How Does the Sharing of Electrons Bond Atoms?

13. Why does the sharing of electrons between atoms offset the repulsive forces of the positively charged nuclei?

14. What must happen to the spin of two electrons if the electrons are to occupy a domain of space between two bonding nuclei?

15. What is a *bonding domain*?

16. If a stable bond is formed, what must be the relationship between attractive and repulsive forces between two nuclei?

17. Describe how the sharing of a pair of electrons by two chlorine atoms makes a Cl_2 molecule more stable than a pair of isolated Cl atoms.

Using Lewis Structures to Understand the Formation of Bonds

18. How many valence electrons does the hydrogen atom have? How many valence electrons does the chlorine atom have? Give a representation of these two neutral atoms using dots for electrons. Use dot representations to show how an HCl molecule is formed from H and Cl atoms.

19. Give the electron configurations for N and H. Use symbols with dots as electrons to represent the distribution of valence electrons around each atom. Combine the N and H atoms to form the NH_3 molecule. Identify the bonding and nonbonding domains in NH_3.

20. How many bonding electrons are there in an O_2 molecule? How many nonbonding electrons?

21. How is the number of valence electrons that must be assigned to each atom determined when writing Lewis structures?

22. How many bonding domains are there in the following molecules?

    (a) SO_2 (b) BeF_2 (c) CH_4 (d) Br_2

23. Which of the following molecules has two bonding domains?

    (a) BF_3 (b) NH_3 (c) O_3

Drawing Skeleton Structures

24. Draw skeleton structures of the following molecules.
    (a) CH_4 (b) CH_3Cl (c) H_2CO

25. Which of the following skeleton structures is best for SO_2? Explain.
    (a) O—S—O   (b) O—O—S

26. Which of the following skeleton structures is best for NO_2? Explain.
    (a) O—N—O   (b) N—O—O

27. Draw skeleton structures for each of the following molecules. (*Hint:* The central atom is underlined for each compound.)
    (a) $\underline{S}O_3$ (b) $\underline{S}O_2$ (c) $O\underline{C}S$
    (d) $\underline{N}H_3$ (e) $C\underline{C}l_3H$

A Step-by-Step Approach to Writing Lewis Structures

28. Determine the total number of valence electrons in the following molecules or ions.
    (a) BF_3 (b) CH_4 (c) NH_4^+ (d) H_2SO_4

29. Determine the total number of valence electrons in the following molecules or ions.
    (a) KrF_2 (b) SF_4 (c) SiF_6^{2-} (d) SO_4^{2-}

30. Which of the following molecules or ions contain the same number of valence electrons?
    (a) CO_2 (b) N_2O (c) CNO^-
    (d) NO_2^+ (e) SO_2 (f) O_3 (g) NO_2^-

31. Draw skeleton structures of the following ions.
    (a) NH_4^+ (b) NO_3^- (c) SO_4^{2-}

32. Write Lewis structures for the following ions or molecules.
    (a) NH_3 (b) CH_3^+ (c) H_3O^+ (d) BH_4^-

Molecules That Don't Seem to Satisfy the Octet Rule

33. Write Lewis structures for the following ions.
    (a) NO_3^- (b) SO_3^{2-} (c) CO_3^{2-} (d) NO_2^+

34. Write Lewis structures for the following ions or molecules.
    (a) C_2H_6 (b) C_2H_4 (c) C_2H_2 (d) C_2^{2-}

35. Write Lewis structures for the following nitrogen-containing molecules.
    (a) N_2O (b) N_2O_3 (ON—NO_2)

36. Write Lewis structures for the following nitrogen-containing molecules.
    (a) ClNO   (b) $ClNO_2$ (c) NO^+
    (d) NO_2^- (e) ONF_3

37. Explain why a satisfactory Lewis structure for N_2O_5 cannot be based on a skeleton structure that contains an N—N bond: O_2N—NO_3. Show how a satisfactory

Lewis structure can be written if we assume that the skeleton structure is O_2NONO_2.

38. Write Lewis structures for the following ions or molecules.
    (a) O_2 (b) O_3 (c) O_2^{2-} (d) O^{2-}

39. Write Lewis structures for the following ions or molecules.
    (a) SO_2 (b) SO_3 (c) SO_3^{2-} (d) SO_4^{2-}

40. Write Lewis structures for the following ions or molecules.
    (a) XeF_2 (b) XeF_4 (c) XeF_3^+ (d) $OXeF_4$

41. Which of the following ions or molecules have the same electron configurations as the N_2 molecule?
    (a) CO   (b) NO   (c) CN^- (d) NO^+ (e) NO^-

42. Which of the following are exceptions to the Lewis octet rule?
    (a) CO_2 (b) BeF_2 (c) SF_4 (d) SO_3

43. Which of the following are exceptions to the Lewis octet rule?
    (a) BF_3 (b) H_2CO (c) XeF_4 (d) IF_3

44. Use Lewis structures to explain why a pair of NO_2 molecules can combine to form N_2O_4.

$$2NO_2(g) \longrightarrow N_2O_4(g)$$

45. Use Lewis structures to decide whether any of the following molecules or ions obey the octet rule
    (a) PF_5 (b) SF_6 (c) IF_4^+ (d) ICl_4^-

Bond Lengths

46. Define the term *bond length*.

47. The nitrogen-to-nitrogen bond lengths in the following compounds are: N_2, 0.110 nm; HNNH, 0.125 nm; H_2NNH_2, 0.146 nm. In which compound is the nitrogen-to-nitrogen bond most likely to be a single bond? A double bond? A triple bond?

48. Use covalent radii from Figure 3.26 to estimate bond lengths for all the bonds in the following.
    (a) H_2S (b) OF_2 (c) NH_3 (d) BF_3

49. If the sulfur-to-oxygen bond lengths are about the same in SO_2 and SO_3, what can be said about the bonding between sulfur and oxygen in the two molecules?

Resonance Hybrids

50. Draw all the possible Lewis structures for the CO_3^{2-} ion.

51. Draw the three Lewis structures that may be used to describe the SCN^- ion.

52. Which of the following molecules or ions are best described as a resonance hybrid of two or more Lewis structures?
    (a) HCO_2^- (b) PH_3 (c) HCN   (d) C_2H_4

53. Which of the following molecules do not form a resonance hybrid?
    (a) CO_2 (b) C_2H_2 (c) $CHCl_3$ (d) SO_3

Electronegativity

54. Define *electronegativity* and *AVEE*. Why can electronegativity and AVEE be used more or less interchangeably?

55. Use the data in Figure 3.31 and trends in covalent and metallic radii to explain why electronegativity increases from left to right across a row in the periodic table.

56. Which of the following atoms is the most electronegative?
    (a) S   (b) As   (c) P   (d) Se   (e) Cl   (f) Br

57. Which of the following series of atoms are arranged in order of decreasing electronegativity?
    (a) $C > Si > P > As > Se$
    (b) $O > P > Al > Mg > K$
    (c) $Na > Li > B > N > F$
    (d) $K > Mg > Be > O > N$
    (e) $Li > Be > B > C > N$

58. What happens to the electronegativity of the element as we go down a column in the periodic table? Explain why.

Partial Charge

59. Calculate the partial charge on the fluorine atom in the following molecules.
    (a) F_2 (b) HF   (c) ClF

60. Calculate the partial charge on both atoms in the following molecules: IBr, ICl.

61. Calculate the partial charges on the two atoms in BF.

62. If ordinary purified table salt, NaCl, is heated to 801°C, the solid melts to produce Na^+ and Cl^-. Continued heating causes evaporation of the molten salt. Some of the species that are found in the gas phase are sodium chloride molecules, NaCl(*g*). Similar behavior is encountered for other solid salts when heated to sufficiently high temperatures. Calculate the partial charges on both atoms in the following gas-phase molecules: LiCl, NaCl, KCl, RbCl, CsCl. Is the trend in partial charge what you would have expected on the basis of electronegativities?

63. Assign relative partial charges (δ) to the oxygen and hydrogen atoms in the following molecule.

64. For the following molecules, assign relative partial charges (δ) to each atom.
    (a) SO_2 (b) ClO_2 (c) H_2O
    (d) OCS   (e) CH_3OH

65. Which atom in each of the following has the most positive partial charge?
    (a) CH_4 (b) CCl_4 (c) BH_3
    (d) NO_2 (e) ClNO

Formal Charge

66. Calculate the formal charge on the sulfur atom in the following molecules or ions.
    (a) SO_2 (b) SO_3 (c) SO_3^{2-}

67. Calculate the formal charge on the bromine atom in the following molecules.
    (a) HBr   (b) Br_2 (c) HOBr   (d) BrF_5

68. Calculate the formal charge on the nitrogen atom in the following ions, molecules, or compounds.
    (a) NH_3 (b) NH_4^+ (c) N_2H_4 (d) NH_2^-

69. Calculate the formal charge on the nitrogen atom for the best Lewis structure for each of the following molecules.
    (a) N_2O (b) N_2O_3 (c) N_2O_5

70. Calculate the formal charge on the boron and nitrogen atoms in BF_3, NH_3, and the $F_3B—NH_3$ molecule formed when the two gases combine.

71. Calculate the formal charge on the two different sulfur atoms in the thiosulfate ion, $S_2O_3^{2-}$. Assume a skeleton structure that could be described as $S—SO_3$.

72. Draw two correct Lewis structures for ClNO and use formal charge to decide which is the best representation. What experimental evidence could be used to support your choice?

73. Draw two correct Lewis structures for PO_4^{3-} and decide on the basis of formal charge which is the best. What experimental evidence would help support your choice of structure?

74. Which would you expect to be more stable: ozone (O_3) or oxygen (O_2)? Use formal charge to support your answer.

75. Below are two skeleton structures for nitrous acid, HNO_2. Which is the best arrangement of atoms, on the basis of formal charge?

$$H—O—N—O \qquad\qquad H—N—O$$
$$\qquad\qquad\qquad\qquad\qquad |$$
$$\qquad\qquad\qquad\qquad\qquad O$$

76. Draw all three Lewis structures for the SCN⁻ ion. Use formal charge to decide which is best. Explain why.

77. Draw all three Lewis structures for the NCO⁻ ion. Use formal charge to decide which is best. Explain why.

The Shapes of Molecules

78. If four pairs of electrons can be found in the valence shell of an atom, how will these four domains arrange themselves? How will three domains take position around a central atom? Two domains? Explain.

79. Predict the arrangement of geometry around the central atom in the following molecules or ions. Predict the angle the central atom makes with the atoms around it in each compound.
    (a) CO_3^{2-} (b) SO_4^{2-} (c) PF_6^- (d) AsF_5

80. Complete the following table for molecules for which the central atom has only bonding electrons.

Bonding Domains	Molecular Geometry	Example
—	linear	CO_2
3	—	—
—	tetrahedral	—
5	—	—
—	octahedral	—

81. How many bonding domains are on the underlined central atom in each of the following? What is the bond angle around the central atom in each case?
    (a) $\underline{N}H_4^+$ (b) $H\underline{C}Cl_3$ (c) $\underline{Be}H_2$
    (d) $O\underline{C}Cl_2$ (e) $O\underline{C}S$

The Role of Nonbonding Electrons in the ED Model

82. Determine the number of nonbonding pairs of electrons on the iodine atom in the following molecules or ions.
    (a) I_2 (b) I_3^- (c) IF_3 (d) ICl_4^-

83. Which of the following molecules or ions have one or more pairs of nonbonding electrons?
    (a) OH^- (b) O_2 (c) CO_3^{2-}
    (d) Br^- (e) NH_3

84. Which of the following molecules or ions have one or more pairs of nonbonding electrons?
    (a) H_2S (b) NH_4^+ (c) AlH_3 (d) CH_3^- (e) NH_2^-

85. Complete the following table.

Bonding Domains	Nonbonding Domains	Molecular Geometry
2	0	—
—	0	tetrahedral
2	—	bent
5	0	—
—	—	T-shaped
5	—	square pyramidal
—	2	square planar

Distinguish between the geometry associated with the arrangement of the electron domains around the central atom and the molecular geometry.

86. Predict the geometry around the central atom in the following molecules or ions.
    (a) PH_3 (b) GaH_3 (c) ICl_3 (d) XeF_3^+

87. Predict the geometry around the central atom in the following molecules or ions.
    (a) PO_4^{3-} (b) SO_4^{2-} (c) XeO_4 (d) MnO_4^-

88. The same elements often form compounds with very different shapes. Predict the geometry around the central atom in the following molecules or ions.
    (a) SnF_2 (b) SnF_3^- (c) SnF_4 (d) SnF_6^{2-}

89. Sulfur reacts with fluorine to form a pair of neutral molecules—SF_4 and SF_6—that in turn form positive and negative ions. Predict the geometry around the sulfur atom in each of the following.
    (a) SF_3^+ (b) SF_4 (c) SF_5^- (d) SF_6

90. Iodine and fluorine combine to form interhalogen compounds that can exist as either neutral molecules, positive ions, or negative ions. Predict the geometry around the iodine atom in each of the following.
    (a) IF_2^- (b) IF_3 (c) IF_4^+ (d) IF_4^-

91. Predict the geometry around the central atom in each of the following oxides of nitrogen.
    (a) N_2O (b) NO_2^- (c) NO_3^-

92. Predict the geometry around the central atom in the $Hg(CH_3)_2$ molecule.

93. Which of the following compounds is best described as T-shaped?
    (a) XeF_3^+ (b) NO_3^- (c) NH_3
    (d) ClO_3^- (e) SF_4

94. Which of the following molecules or ions have the same shape or geometry?
    (a) NH_2^- and H_2O (b) NH_2^- and BeH_2
    (c) H_2O and BeH_2 (d) NH_2^-, H_2O, and BeH_2

95. Which of the following molecules or ions have the same shape or geometry?
    (a) SF_4 and CH_4 (b) CO_2 and H_2O
    (c) CO_2 and BeH_2 (d) N_2O and NO_2
    (e) PCl_4^+ and PCl_4^-

96. Which of the following molecules are best described as bent, or angular?
    (a) H_2S (b) CO_2 (c) $ClNO$
    (d) NH_2^- (e) O_3

97. Which of the following molecules are planar?
    (a) SO_3 (b) SO_3^{2-} (c) NO_3^-
    (d) PF_3 (e) BH_3

98. Which of the following molecules are tetrahedral?
    (a) SiF_4 (b) CH_4 (c) NF_4^+
    (d) BF_4^- (e) TeF_4

99. Which of the following molecules are linear?
    (a) C_2H_2 (b) CO_2 (c) NO_2^-
    (d) NO_2^+ (e) H_2O

100. Which of the following elements would form a linear compound with the formula XO_2?

(a) Se (b) P (c) C (d) O (e) F

101. Explain why the nonbonding electrons occupy equatorial positions in ClF_3, not axial positions.

Bond Angles

102. How many atoms are needed to define a bond angle?

103. Complete the following table.

Bonding Domains	Nonbonding Domains	Bond Angle Central Atom
2	0	—
5	0	—
3	—	120°
2	1	—
—	3	180°
—	0	109°
3	—	109°
2	2	—
2	3	—
3	2	—
4	2	—

104. Give the bond angle around the central atom in the following.

(a) CO_3^{2-} (b) SO_2 (c) H_3O^+
(d) IF_4^+ (e) BrF_5 (f) ICl_4^-

105. Give the bond angles around each underlined atom in the following compounds.

(a) $H_3\underline{C}O\underline{C}H_3$ (b) $\underline{C}H_2Cl_2$ (c) $H\underline{O}NO$
(d) $H_3\underline{C}\underline{C}OH$ (e) $\underline{S}O_3$

106. The Cl—B—Cl bond angle in BCl_3 is 120° and the H—C—H bond angle in H_2CCH_2 is 121°. Explain the difference using the electron domain model. The H—O—H bond angle in H_2O is 104.5°. Explain this angle using the electron domain model.

107. Give an estimate of all bond angles in the following compounds.

(a) SF_4 (b) ClF_3 (c) I_3^- (d) XeF_4

108. Estimate the H—C—H bond angles in the following compounds.

(a) C_2H_2 (b) C_2H_4 (c) C_2H_6

The Difference between Polar Bonds and Polar Molecules

109. Explain why OCS is a polar molecule but CS_2 is not.

110. Use the Lewis structure of CO_2 to explain why the molecule has no dipole moment.

111. Which of the following molecules should be polar?

(a) CH_3OH (b) H_2O
(c) CH_3OCH_3 (d) CH_3CO_2H

112. Explain why formaldehyde (H_2CO) is a polar molecule.

113. Explain why both thionyl chloride ($SOCl_2$) and sulfuryl chloride (SO_2Cl_2) are polar molecules.

114. Draw Lewis structures and identify the polar species in the following:

(a) BF_3 (b) C_2H_6 (c) PH_3 (d) N_2O

Integrated Problems

115. Polymerization occurs when a relatively large number of identical units combine to produce a long chain. Phosphate detergent additives and many biologically important molecules consist of chains of individual PO_3^- units.

Draw the Lewis structure of PO_3^-. Use formal charges to explain why individual PO_3^- units come together to produce long chains. Would you expect SO_3 to polymerize?

116. Chemical formulas usually give a good idea of the skeleton structure of a compound, but some formulas are misleading. The formulas of common acids are often written as follows: HNO_3, H_2SO_4, H_3PO_4, and $HClO_4$. Write Lewis structures for those four acids, assuming that their skeleton structures are more appropriately described by the formulas $HONO_2$, $(HO)_2SO_2$, $(HO)_3PO$, and $HOClO_3$.

117. Write Lewis structures for the following compounds and then describe why these compounds are unusual.

(a) NO (b) NO_2 (c) ClO_2 (d) ClO_3

118. Which element forms a compound with the following Lewis structure?

$$\ddot{O}=\ddot{X}-\ddot{O}$$

(a) Al (b) Si (c) P (d) S (e) Cl

119. Which element forms a compound with the following Lewis structure?

$$\ddot{O}=\ddot{X}-\ddot{Cl}$$

(a) Be (b) B (c) C
(d) N (e) O (f) F

120. Which element forms a polyatomic ion with the formula XF_6^{2-} that has no nonbonding electrons in the valence shell of the central atom?

(a) N (b) C (c) Si (d) S (e) P

121. In which group of the periodic table does element X belong if there are two pairs of nonbonding electrons in the valence shell of the central atom in the XF_4^- ion?

122. In which group of the periodic table does element X belong if the distribution of electrons in the valence shell of the central atom in the XCl_4^- ion is octahedral?

123. Which of the following molecules does not contain a double bond?

(a) N_2 (b) CO_2 (c) C_2H_4
(d) NO_2 (e) SO_3

124. Which of the following contain only single bonds?

(a) CN^- (b) NO^+ (c) CO
(d) O_2^{2-} (e) Cl_2CO

125. Ingestion of oxalic acid ($H_2C_2O_4$), which can be found in a variety of vegetables and other plants, can produce nausea, vomiting, and diarrhea. When taken in excess, oxalic acid can be toxic. Draw the Lewis structure for molecule. (Assume that the skeleton structure can be described as HO_2CCO_2H.)

126. In which group does element X belong if the shape of the XF_2^- ion is linear?

127. The charged species OCN^- is well known in chemistry.

(a) Write all possible Lewis structures for the species.

(b) Which of the structures is best? Explain.

(c) Typical experimental carbon-to-oxygen bond lengths are as follows:

$$C{=}O \quad C{=}O \quad C{\equiv}O$$
$$150 \text{ pm} \quad 133 \text{ pm} \quad 120 \text{ pm}$$

What type of carbon–oxygen bond (single, double, triple, or somewhere in between) do you expect for OCN^-? Predict the carbon–oxygen bond length for OCN^-. Explain your answers. In addition to bond length, what other piece of experimental evidence would support your conclusion concerning the type of carbon–oxygen bond?

(d) What is the O—C—N bond angle? Explain your answer.

128. Draw the Lewis structures and select the one species for each of the following pairs that has the largest dipole moment. Show your work and explain your answer. The central atom is underlined.

(a) $C\underline{S}_2$, $\underline{S}O_2$ (b) $H\underline{C}N$, $N\underline{C}CN$ (C_2N_2)
(c) $\underline{Ge}F_4$, $\underline{Se}F_4$

129. The molecule thymine, one of four bases found in DNA, has the following structure:

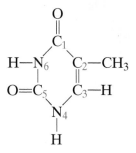

(a) Add the missing electrons to the structure.

(b) Give bond angles around each numbered atom.

130. In Section 4.12 we discussed two possible structures for SO_2 and stated that without experimental evidence we could not distinguish between the two. Structure I contains one single bond and one double bond, while structure II contains two double bonds. Explain why measurements of bond length cannot be used to distinguish between the two structures.

131. Each of the following Lewis structures is incorrect. Explain what is wrong with each one and give the correct Lewis structure.

(a) H—N—H   (b) H—C—O—H
            |
        :Cl: :O:

(c)
$$\left[\begin{array}{c} :Cl: \\ | \\ :Cl{-}I{-}Cl: \\ | \\ :Cl: \end{array}\right]^-$$

(d) :O—H—Cl:

(e) H=C=N:

132. Which of the following Lewis structures contributes the most to the description of the electron structure of HNO_3? Use formal charge to rationalize your choices

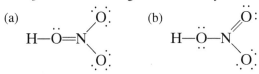

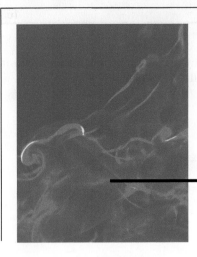

Chapter Four

SPECIAL TOPICS

4A.1 Valence Bond Theory

Our approach to covalent bonds in this chapter has been based on **valence bond theory,** which assumes that valence atomic orbitals on adjacent atoms can interact to form a bond that lies between the atoms. Covalent bonds are formed by a pair of electrons with opposite spins. Consider, for example, the formation of an H_2 molecule from a pair of hydrogen atoms, each of which has an unpaired electron in a $1s$ orbital.

A similar approach can be used to explain the covalent bond in HF. Each of the isolated atoms that come together to form the HF molecule has an orbital that contains a single, unpaired electron.

But now the bond is formed by the interaction between a $1s$ orbital on one atom and a $2p$ orbital on the other.

The valence bond model for the F_2 molecule is based on the head-to-head overlap between half-filled $2p$ orbitals on adjacent atoms.

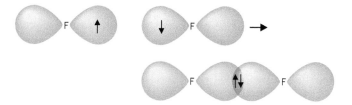

The covalent bonds in H_2, HF, and F_2 are called **sigma (σ) bonds** because they look like an s orbital when viewed along the bond. **Pi (π) bonds** are formed when orbitals interact to form a new orbital that looks like a p orbital when viewed along the bond.

One way to differentiate between sigma and pi bonds is to consider the bonding in the O_2 molecule. The electron configuration of the oxygen atom tells us that there are two unpaired electrons on each atom.

Let's assume that one of the unpaired electrons on each oxygen atom is in the $2p_z$ orbital that points toward the other oxygen atom. The interaction between the $2p_z$ orbitals on adjacent atoms leads to a sigma bond, just as it did in the F_2 molecule.

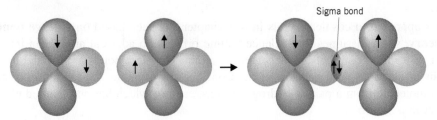

But this leaves us with an unpaired electron on each atom in one of the $2p$ orbitals perpendicular to the axis along which the sigma bond forms. The edge-on interaction between the half-filled orbitals leads to the formation of a bond that looks like a p orbital when viewed along the bond axis. In other words, it leads to a *pi bond*.

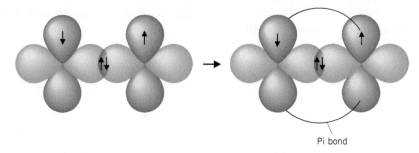

The combination of the sigma and pi bonds between the oxygen atoms suggests that the O_2 molecule is held together by a double bond, which is consistent with the Lewis structure for oxygen.

$$\ddot{\underset{..}{O}}=\ddot{\underset{..}{O}}$$

4A.2 Hybrid Atomic Orbitals

The simple version of the valence bond theory introduced in the previous section gives a satisfactory picture of the bonding in simple diatomic molecules such as H_2, HF, F_2, O_2, and N_2. But it gives less satisfactory results when applied to slightly more complex molecules, such as CH_4 and H_2O.

The electron configuration of carbon has only two unpaired electrons, and therefore carbon should form only two bonds.

$$C \quad \underset{1s^2}{\underline{\uparrow\downarrow}} \quad \underset{2s^2}{\underline{\uparrow\downarrow}} \quad \underset{2p^2}{\underline{\uparrow}\ \underline{\uparrow}\ \underline{}}$$

The fact that carbon atoms form four bonds can be explained by noting that it takes relatively little energy to excite one of the electrons from the filled $2s$ orbital into one of the empty $2p$ orbitals to give four orbitals that each contain an unpaired electron.

$$C^* \quad \underset{1s^2}{\underline{\uparrow\downarrow}} \quad \underset{2s^1}{\underline{\uparrow}} \quad \underset{2p^3}{\underline{\uparrow}\ \underline{\uparrow}\ \underline{\uparrow}}$$

The energy invested in moving the electron is more than repaid by the energy released when two more bonds are formed.

A more serious problem arises, however, when the valence bond theory introduced in the previous section is used to predict the geometry around an atom in a molecule. Consider water, for example. Combining a pair of hydrogen atoms with unpaired electrons in a $1s$ orbital with the two unpaired electrons on an oxygen atom ($2p$ orbitals) gives the correct number of bonds.

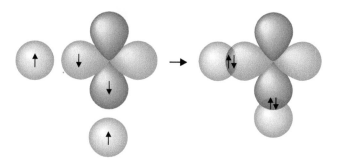

But this approach to explaining the formation of the bonds in an H_2O molecule predicts that the H—O—H bond angle would be 90° because that is the angle between the $2p$ orbitals on the oxygen atom. This is not consistent with experimental measurements, which give a bond angle of 105°.

This problem was solved by assuming that the valence atomic orbitals on an individual atom can be combined to form **hybrid atomic orbitals.** The geometry of a linear BeH_2 molecule can be explained, for example, by mixing the $2s$ orbital on the beryllium atom with one of the $2p$ orbitals to form a set of sp hybrid orbitals that point in opposite directions, as shown in Figure 4A.1. One of the valence electrons on the beryllium atom is then placed in each of these orbitals, and the orbitals are allowed to overlap with half-filled $1s$ orbitals on a pair of hydrogen atoms to form a linear BeH_2 molecule.

The geometry of trigonal planar molecules such as BF_3 can be explained by mixing a $2s$ orbital with a pair of $2p$ orbitals on the central atom to form three sp^2 hybrid orbitals that point toward the corners of an equilateral triangle. Molecules whose geometries are based on a tetrahedron, such as CH_4 and H_2O, can be understood by mixing a $2s$ orbital with all three $2p$ orbitals to obtain a set of four sp^3 orbitals that are oriented toward the corners of a tetrahedron.

The hybrid atomic orbital model can be extended to molecules whose shapes are based on trigonal bipyramidal or octahedral distributions of electrons by including valence shell d orbitals. When one of the $3d$ orbitals is mixed with the $3s$ and the three $3p$ orbitals on an atom, the resulting sp^3d hybrid orbitals point toward the corners of a trigonal bipyramid. When two of the $3d$ orbitals are mixed with the $3s$ and $3p$ orbitals, the result is a set of six sp^3d^2 hybrid orbitals that point toward the corners of an octahedron.

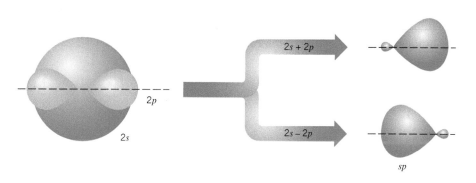

Fig. 4A.1 The sp hybrid orbitals used by the beryllium atom in the linear BeH_2 molecule are formed by combining the wave functions for the $2s$ and $2p$ orbitals on the beryllium atom. When the wave functions are added, we get an sp orbital that points in one direction. When one of the wave functions is subtracted from the other, we get an sp orbital that points in the opposite direction.

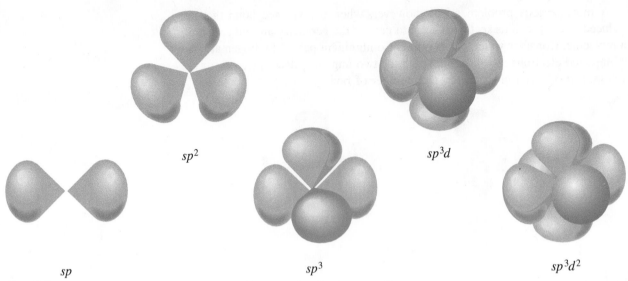

Fig. 4A.2 Shapes of the sp, sp^2, sp^3, sp^3d, and sp^3d^2 hybrid orbitals.

The number of hybrid atomic orbitals created by mixing atomic orbitals is always the same as the number of atomic orbitals combined. We get four hybrid orbitals, for example, by mixing a set of $2s$, $2p_x$, $2p_y$, and $2p_z$ atomic orbitals. Orbitals are therefore said to be "conserved" when they are hybridized.

The geometries of the five different sets of hybrid atomic orbitals (sp, sp^2, sp^3, sp^3d, and sp^3d^2) are shown in Figure 4A.2. The relationship between hybridization and the distribution of electrons in the valence shell of an atom is summarized in Table 4A.1

Table 4A.1

Relationship between the Distribution of Electron Domains on an Atom and the Hybridization of an Atom

Electron Domains	Distribution of Electron Domains	Hybridization	Examples
2	Linear	sp	BeF_2, CO_2, N_2
3	Trigonal planar	sp^2	BF_3, CO_3^{2-}, O_3
4	Tetrahedral	sp^3	CH_4, SO_4^{2-}, NH_3, H_2O
5	Trigonal bipyramidal	sp^3d	PF_5, ClF_3
6	Octahedral	sp^3d^2	SF_6, XeF_4

Exercise 4A.1

Determine the hybridization of the central atom in O_2, H_3O^+, $TeCl_4$, and ICl_4^- from the following Lewis structures

(a) O_2 $\overset{..}{O}{=}\overset{..}{O}$

(b) H_3O^+ $\left[\begin{array}{c} H{-}\overset{..}{O}{-}H \\ | \\ H \end{array} \right]^+$

(c) $TeCl_4$

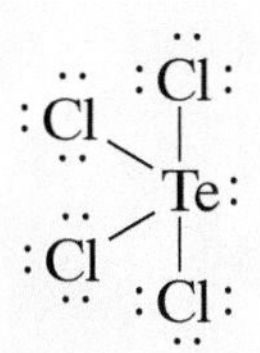

(d) ICl_4^-

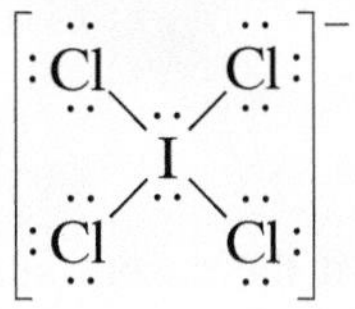

Solution

(a) The Lewis structure for the O_2 molecule suggests that the distribution of electron domains around each oxygen atom should be trigonal planar. The hybridization of the oxygen atoms in the O_2 molecule is therefore assumed to be sp^2.

(b) The Lewis structure suggests a tetrahedral distribution of electron domains around the oxygen atom in the ion. The oxygen atom in H_3O^+ is therefore sp^3 hybridized.

(c) The Lewis structure suggests a trigonal bipyramidal distribution of domains in the valence shell of the tellurium atom. The tellurium atom therefore forms an sp^3d hybrid.

(d) The Lewis structure suggests an octahedral distribution of domains in the valence shell of the iodine atom. The iodine atom is therefore sp^3d^2 hybridized.

• •

4A.3 Molecules with Double and Triple Bonds

The hybrid atomic orbital model can also be used to explain the formation of double and triple bonds. Let's consider the bonding in formaldehyde (H_2CO), for example, which has the following Lewis structure.

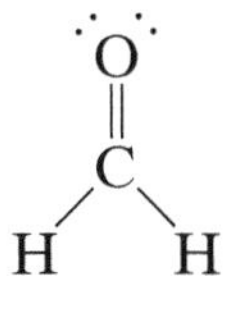

There are three places where electrons can be found in the valence shell of both the carbon and oxygen atoms in this molecule. As a result, the electron domain theory predicts that the valence electrons on the atom will be oriented toward the corners of an equilateral triangle. The carbon atom is therefore assumed to be sp^2 hybridized.

When we create a set of sp^2 hybrid orbitals, we combine the $2s$ and two of the $2p$ orbitals on the atom. One of the valence electrons on the carbon atom is placed in each of the three sp^2 hybrid orbitals. The fourth valence electron is placed in the $2p$ orbital that wasn't used during hybridization.

There are six valence electrons on a neutral oxygen atom. A pair of electrons is placed in each of two of the sp^2 hybrid orbitals. One electron is then

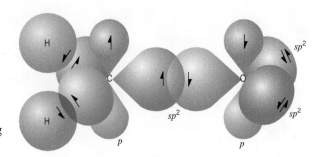

Fig. 4A.3 The hybrid atomic orbital picture of bonding in formaldehyde.

placed in the sp^2 hybrid orbital that points toward the carbon atom, and another is placed in the unhybridized $2p$ orbital.

The two C—H bonds are formed when the unpaired electron in one of the sp^2 hybrid orbitals on the carbon atom interacts with a $1s$ electron on a hydrogen atom, as shown in Figure 4A.3. A C—O bond is formed when the electron in the third sp^2 hybrid orbital on the carbon atom interacts with the unpaired electron in the sp^2 hybrid orbital on the oxygen atom. These bonds are called sigma (σ) bonds because they look like an s orbital when viewed along the bond.

The electron in the unhybridized $2p$ orbital on the carbon atom then interacts with the electron in the unhybridized $2p$ orbital on the oxygen atom to form a second covalent bond between these atoms. This is called a pi (π) bond because it looks like a p orbital when viewed along the bond.

When double bonds were first introduced, we noted that they occur most often in compounds that contain C, N, O, P, or S. There are two reasons for this. First, double bonds by their very nature are covalent bonds. They are therefore most likely to be found among the elements that form covalent compounds. Second, the interaction between p orbitals to form a π bond requires that the atoms come relatively close together, and π bonds therefore tend to be the strongest for atoms that are relatively small.

4A.4 Molecular Orbital Theory

The valence bond model for SO_2 can't adequately explain why the molecule contains two equivalent bonds with a bond order between that of an S—O single bond and an S=O double bond. The best it can do is suggest that SO_2 is a mixture, or hybrid, of the two Lewis structures that can be written for the molecule.

This problem, and many others, can be overcome by using a more sophisticated model of bonding based on **molecular orbitals.** Molecular orbital theory is more powerful than valence bond theory because the orbitals reflect the geometry of the molecule to which they are applied. But this power carries a significant cost in terms of the ease with which the model can be visualized. We can understand the basics of molecular orbital theory by constructing the molecular orbitals for the simplest possible molecules: homonuclear diatomic molecules such as H_2, O_2, and N_2.

Molecular orbitals are obtained by combining the atomic orbitals on the atoms to give orbitals that are characteristic of the molecule, not the individual atoms. Consider the H_2 molecule, for example. One of the molecular orbitals in the H_2 molecule is constructed by adding the mathematical functions for the two $1s$ atomic orbitals that come together to form the molecule. Another orbital is formed by subtracting one of these functions from the other, as shown in Figure 4A.4.

Note that hybrid orbitals are formed by mixing two or more orbitals on the same atom. Molecular orbitals are formed by mixing orbitals on more than one atom.

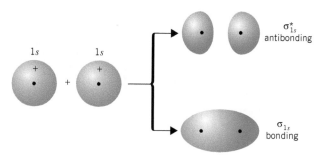

Fig. 4A.4 The interaction between a pair of $1s$ atomic orbitals on neighboring hydrogen atoms leads to the formation of bonding (σ) and antibonding (σ^*) molecular orbitals.

One of the orbitals is called a **bonding molecular orbital** because electrons that are placed in this orbital spend most of their time in the region directly between the two nuclei. It is called a *sigma* (σ) molecular orbital because it looks like an *s* orbital when viewed along the H—H bond. Electrons placed in the other orbital spend most of their time away from the region between the two nuclei. That orbital is therefore an **antibonding,** or *sigma star* (σ^*), molecular orbital.

The σ bonding molecular orbital concentrates electrons in the region directly between the two nuclei. Placing an electron in that orbital therefore stabilizes the H_2 molecule. Since the σ^* antibonding molecular orbital forces the electron to spend most of its time away from the area between the nuclei, placing an electron in that orbital makes the molecule less stable.

Note that when two atomic orbitals are combined, we get two molecular orbitals. Once again, we find that orbitals are "conserved" when they are combined. Energy is also "conserved" when orbitals are mixed. The energy of the σ^* antibonding orbital in Figure 4A.5 is raised above the energy of the $1s$ atomic orbitals on the isolated hydrogen atoms by exactly the same amount that the energy of the σ bonding orbital is lowered below the energy of the $1s$ atomic orbitals.

Electrons are added to molecular orbitals, one at a time, starting with the lowest-energy molecular orbital. The two electrons associated with a pair of hydrogen atoms are placed in the lowest-energy, or σ bonding, molecular orbital shown in Figure 4A.5. When this is done, the energy of an H_2 molecule is lower than that of a pair of isolated atoms. The H_2 molecule is therefore more stable than a pair of isolated atoms.

The molecular orbital model can be used to explain why He_2 molecules don't exist. Combining a pair of helium atoms with $1s^2$ electron configurations would produce a molecule with pairs of electrons in both the σ bonding and the σ^* antibonding molecular orbitals. The total energy of an He_2 molecule would be essentially the same as the energy of a pair of isolated helium atoms, and there would be nothing to hold the He_2 molecule together.

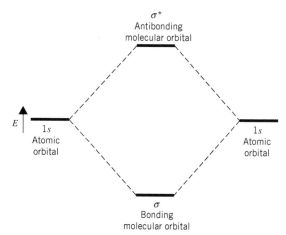

Fig. 4A.5 The relative energies of the $1s$ atomic orbitals on a pair of isolated hydrogen atoms and the σ and σ^* molecular orbitals they form when they interact.

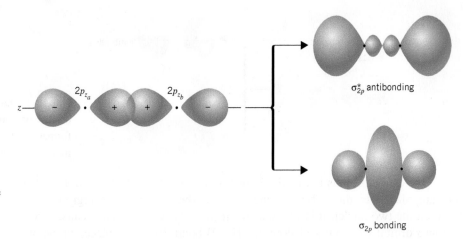

Fig. 4A.6 When a pair of $2p$ orbitals are combined so that they meet head-on, bonding (σ_{2p}) and antibonding ($\sigma_{2p}*$) molecular orbitals are formed. These are called σ orbitals because they concentrate the electrons along the axis between the two nuclei, just like the σ_{1s} and σ_{2s} orbitals.

The fact that a diatomic He_2 molecule is neither more nor less stable than a pair of isolated helium atoms illustrates an important principle: The core orbitals on an atom make no contribution to the stability of the molecules that contain the atom. The molecular orbitals that are important are those formed when valence-shell orbitals are combined. When writing the molecular orbital diagram for an O_2 molecule, we therefore ignore the $1s$ electrons on both oxygen atoms and concentrate on the interactions between the $2s$ and $2p$ valence orbitals.

The $2s$ orbitals on one oxygen atom combine with the $2s$ orbitals on another to form a σ_{2s} bonding and a $\sigma_{2s}*$ antibonding molecular orbital, just like the σ_{1s} and $\sigma_{1s}*$ orbitals formed from the $1s$ atomic orbitals. If we arbitrarily define the z axis of our coordinate system as the axis between the two atoms in an O_2 molecule, the $2p_z$ orbitals on the atoms in the molecule point directly to each other. When those orbitals interact, they meet head-on to form a σ_{2p} bonding and a $\sigma_{2p}*$ antibonding molecular orbital, as shown in Figure 4A.6.

The $2p_x$ and $2p_y$ orbitals on one oxygen atom interact with the $2p_x$ and $2p_y$ orbitals on the other atom to form molecular orbitals that have a different shape, as shown in Figure 4A.7. These molecular orbitals are called *pi* (π) orbitals. Whereas σ and $\sigma*$ orbitals concentrate the electrons along the axis on which the nuclei of the atoms lie, π and $\pi*$ orbitals concentrate the electrons above and below the axis. Because there are two $2p$ atomic orbitals on each atom that meet edge-on, we get two π orbitals and two $\pi*$ orbitals. These are called the π_x and π_y and the π_x* and π_y* molecular orbitals.

The interaction of four valence atomic orbitals on one atom with a set of four atomic orbitals on another atom therefore leads to the formation of a total of eight molecular orbitals: σ_{2s}, $\sigma_{2s}*$, σ_{2p}, $\sigma_{2p}*$, π_x, π_y, π_x*, and π_y*.

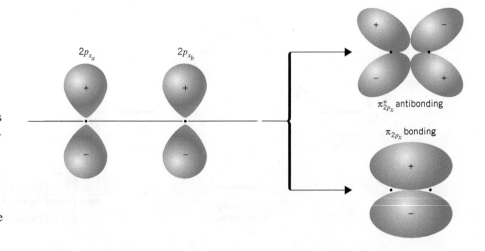

Fig. 4A.7 When a pair of $2p_x$ orbitals are combined so that they meet edge-on, bonding (π_x) and antibonding (π_x*) molecular orbitals are formed. A similar set of bonding (π_y) and antibonding (π_y*) orbitals is formed when a pair of $2p_y$ atomic orbitals are combined.

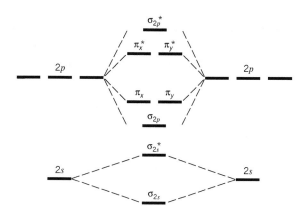

Fig. 4A.8 The interaction between $2p_z$ atomic orbitals that meet head-on is stronger than the interaction between $2p_x$ or $2p_y$ orbitals that meet edge-on. As a result, the difference between the energies of the σ_{2p} and $\sigma_{2p}*$ orbitals is larger than the difference between the π_x and π_x* or π_y and π_y* orbitals.

There is a small difference between the energies of the $2s$ and $2p$ orbitals on an oxygen atom, as shown in Table 3.5. As a result, the σ_{2s} and $\sigma_{2s}*$ orbitals both lie at lower energies than the σ_{2p}, $\sigma_{2p}*$, π_x, π_y, π_x*, and π_y*. To sort out the relative energies of the six molecular orbitals formed when the $2p$ atomic orbitals on a pair of atoms are combined, we need to understand the relationship between the strength of the interaction between a pair of orbitals and the relative energies of the molecular orbitals they form.

Because the $2p_z$ orbitals meet head-on, the interaction between these orbitals is stronger than the interactions between the $2p_x$ or $2p_y$ orbitals, which meet edge-on. As a result, we would expect the σ_{2p} orbital to lie at a lower energy than the π_x and π_y orbitals, and the $\sigma_{2p}*$ orbital to lie at a higher energy than the π_x* and π_y* orbitals, as shown in Figure 4A.8.

Unfortunately, an interaction is missing from the model we have constructed so far. It is possible for the $2s$ orbital on one atom to interact with the $2p_z$ orbital on the other. This interaction introduces a slight change in the relative energies of the molecular orbitals, to give the diagram shown in Figure 4A.9. Experiments have shown that O_2 and F_2 are best described by the model in Figure 4A.8, but B_2, C_2, and N_2 are best described by the model shown in Figure 4A.9.

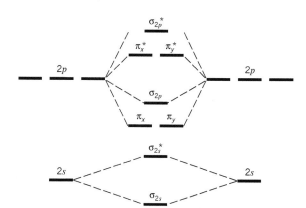

Fig. 4A.9 When hybridization is added to the molecular orbital theory, there is a slight change in the energies of the orbitals formed by overlap of the $2p$ atomic orbitals on neighboring atoms.

Exercise 4A.2

Construct a molecular orbital diagram for the O_2 molecule.

Solution

There are 6 valence electrons on a neutral oxygen atom and therefore 12 valence electrons in an O_2 molecule. These electrons are added to the diagram in Figure 4A.8, one at a time, starting with the lowest energy molecular orbital.

Because Hund's rules apply to the filling of molecular orbitals, molecular orbital theory predicts that there should be two unpaired electrons on this molecule—one electron in each of the π_x^* and π_y^* orbitals.

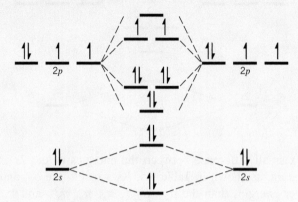

When writing the electron configuration of an atom, we usually list the orbitals in the order in which they fill.

$$\text{Pb} \qquad [\text{Xe}]\, 6s^2\, 4f^{14}\, 5d^{10}\, 6p^2$$

We can write the electron configuration of a molecule by doing the same thing. Concentrating only on the valence orbitals, we write the electron configuration of O_2 as follows.

$$O_2 \qquad \sigma_{2s}^{2}\, \sigma_{2s}^{*2}\, \sigma_{2p}^{2}\, \pi_x^{2}\, \pi_y^{2}\, \pi_x^{*1}\, \pi_y^{*1}$$

• •

The number of bonds between a pair of atoms is called the **bond order.** Bond orders can be calculated from Lewis structures, which are at the heart of the valence bond model. Oxygen, for example, has a bond order of two.

$$:\!\ddot{O}\!=\!\ddot{O}\!:$$

When there is more than one Lewis structure for a molecule, the bond order is an average of these structures. The bond order in sulfur dioxide, for example, is 1.5—the average of an S—O single bond in one Lewis structure and an S=O double bond in the other.

$$
\begin{array}{ccc}
\ddots\!\ddot{S}\diagdown & & \diagup\!\ddot{S}\ddots \\
:\!\ddot{O}\diagup \quad \ddot{O}: & \longleftrightarrow & :\!\ddot{O} \quad \diagdown\ddot{O}:
\end{array}
$$

In molecular orbital theory, we calculate bond orders by assuming that two electrons in a bonding molecular orbital contribute one net bond and that two electrons in an antibonding molecular orbital cancel the effect of one bond. We can calculate the bond order in the O_2 molecule by noting that there are eight valence electrons in bonding molecular orbitals and four valence electrons in antibonding molecular orbitals in the electron configuration of the molecule given in Exercise 4A.2. Thus, the bond order is 2.

$$\text{bond order} = \frac{\text{bonding electrons} - \text{antibonding electrons}}{2} = \frac{8-4}{2} = 2$$

Although the Lewis structure and molecular orbital models of oxygen give the same bond order, there is an important difference between the models. The electrons in the Lewis structure are all paired, but there are two unpaired electrons in the molecular orbital description of the molecule. As a result, we can test the predictions of the theories by studying the effect of a magnetic field on oxygen.

Atoms or molecules in which the electrons are paired are **diamagnetic;** that is, they are repelled by both poles of a magnet. Those that have one or more unpaired electrons are **paramagnetic,** or attracted to a magnetic field. Liquid oxygen is attracted to a magnetic field and can actually bridge the gap between the poles of a horseshoe magnet. The molecular orbital model of O_2 is therefore superior to the valence bond model because it suggests the presence of a strong, O=O double bond and yet also predicts the presence of unpaired electrons in the valence shell which explains why oxygen is paramagnetic.

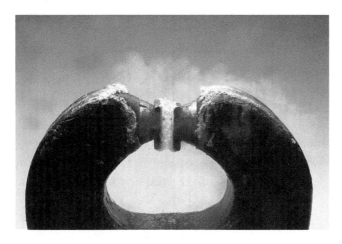

Liquid O_2 is so strongly attracted to a magnetic field that it will bridge the gap between the poles of a horseshoe magnet.

Exercise 4A.3

Use the molecular orbital diagram in Figure 4A.9 to calculate the bond order in nitrogen oxide (NO). Compare the results of this calculation with the bond order obtained from the Lewis structure.

Solution

There are 11 valence electrons in the NO molecule, and there are two possible Lewis structures, depending on whether we locate the unpaired electron on the nitrogen atom or the oxygen atom.

$$\ddot{N}{=}\ddot{O}: \longleftrightarrow :\ddot{N}{=}\ddot{O}:$$

Both Lewis structures contain an N-O double bond, however, so the bond order in the valence bond model for NO is 2.

To calculate the bond order in the molecular orbital model of NO, we have to predict the electron configuration of the molecule by adding electrons, one at a time, to the molecular orbitals in Figure 4A.9.

$$\text{NO} \qquad \sigma_{2s}^{2} \, \sigma_{2s}^{*2} \, \sigma_{2p}^{2} \, \pi_{x}^{2} \, \pi_{y}^{2} \, \pi_{x}^{*1}$$

There are eight valence electrons in bonding molecular orbitals and three valence electrons in antibonding molecular orbitals in the compound. The bond order in the molecular orbital model of NO is therefore 2.5.

$$\text{bond order} = \frac{\text{bonding electrons} - \text{antibonding electrons}}{2} = \frac{8 - 3}{2} = 2.5$$

The strength of the covalent bond in NO has been shown by experiment to be significantly stronger than a typical double bond, in agreement with the predictions of molecular orbital theory for the molecule. The strength of the bond in NO might explain why this compound is stable despite the presence of an unpaired electron.

Problems

Valence Bond Theory

A-1. What is meant by the term *orbital overlap*?

A-2. Which of the following orbitals could overlap?
(a) $2s + 2s$ (b) $2p_x + 2p_y$ (c) $2p_y + 2p_y$
(d) $2p_z + 2p_z$ (e) $2s + 2p_z$

A-3. Draw valence bond pictures showing the overlap of atomic orbitals in each of the following molecules.
(a) NH_3 (b) H_2S (c) NO_2 (d) C_2

Hybrid Atomic Orbitals

A-4. Determine the hybridization of the central atom in the following molecules or ions.
(a) NH_3 (b) NH_4^+ (c) NO
(d) NO_2 (e) NO_2^+

A-5. Determine the hybridization of the central atom in the following molecules or ions.
(a) CH_4 (b) H_2CO (c) HCO_2^-

A-6. Determine the hybridization of the central atom in the following molecules or ions.
(a) SF_4 (b) BrO_3^- (c) XeF_3^+ (d) Cl_2CO

Molecules with Double and Triple Bonds

A-7. Write the Lewis structures for molecules of ethylene, C_2H_4, and acetylene, C_2H_2. Use hybrid atomic orbitals to describe the bonding in these compounds.

A-8. Write the Lewis structures for carbon monoxide and carbon dioxide. Use hybrid atomic orbitals to describe the bonding in the compounds.

Molecular Orbital Theory

A-9. Describe the difference between σ and π molecular orbitals.

A-10. Describe the difference between bonding and antibonding molecular orbitals.

A-11. Explain why the difference between the energies of the σ_{2p} bonding and σ_{2p}^* antibonding orbitals is larger than the difference between the π_x or π_y bonding and π_x^* or π_y^* antibonding orbitals in Figure 4A.8.

A-12. Describe the molecular orbitals formed by the overlap of the following atomic orbitals. (Assume that the bond lies along the z axis of the coordinate system.)
(a) $2s + 2s$ (b) $2p_x + 2p_x$ (c) $2p_y + 2p_y$
(d) $2p_z + 2p_z$ (e) $2s + 2p_z$

A-13. Write the electron configuration for the following diatomic molecules and calculate the bond order in each molecule.
(a) H_2 (b) C_2 (c) N_2 (d) O_2 (e) F_2

A-14. Use molecular orbital theory to predict whether the H_2^+, H_2^-, and H_2^{2-} ions should be more stable or less stable than a neutral H_2 molecule.

A-15. Use molecular orbital theory to explain why the oxygen–oxygen bond is stronger in the O_2 molecule than in the O_2^{2-} (peroxide) ion.

A-16. Use molecular orbital theory to predict whether the bond order in the superoxide ion, O_2^-, should be larger or smaller than the bond order in a neutral O_2 molecule.

A-17. Use molecular orbital theory to predict whether the peroxide ion, O_2^{2-}, should be paramagnetic.

A-18. Write the electron configuration for the following diatomic molecules. Calculate the bond order in each molecule.
(a) HF (b) CO (c) CN^-
(d) ClO^- (e) NO^+

A-19. Classify the following molecules as paramagnetic or diamagnetic.
(a) HF (b) CO (c) CN^-
(d) NO (e) NO^+

Chapter Five

IONIC AND METALLIC BONDS

5.1 Metals, Nonmetals, and Semimetals

The tendency to divide elements into metals, nonmetals, and semimetals is based on differences between the chemical and physical properties of these three categories of elements. *Metals* have the following characteristic physical properties:

- They have a metallic shine or luster. (They look like metals!)
- They are usually solids at room temperature.
- They are *malleable.* They can be hammered, pounded, or pressed into different shapes without breaking.
- They are *ductile.* They can be drawn into thin sheets or wires without breaking.
- They conduct heat and electricity.

Nonmetals have the opposite physical properties:

- They seldom have a metallic luster. They tend to be colorless, like the oxygen and nitrogen in the atmosphere, or brilliantly colored, like bromine.
- They are often gases at room temperature.
- Nonmetallic elements that are solids at room temperature (such as carbon, phosphorus, sulfur, and iodine) are neither malleable nor ductile. They cannot be shaped with a hammer or drawn into sheets or wires.
- They are poor conductors of either heat or electricity. Nonmetals tend to be insulators, not conductors.

The *semimetals* have properties that lie between these extremes. They often look like metals but are brittle like nonmetals. They are neither conductors nor insulators but make excellent semiconductors.

5.2 The Active Metals

As a general rule, the elements toward the bottom-left corner of the periodic table are metals, and the elements toward the upper-right corner are nonmetals. There are no abrupt changes in the physical properties of the elements, however, as we go across a row of the periodic table or down a column. As a result, the change from metal to nonmetal must be gradual. Instead of arbitrarily dividing elements into metals and nonmetals, it might be more useful to describe some elements as being more metallic and others as more nonmetallic. Elements become less metallic and more nonmetallic as we go across a row of the periodic table from left to right.

⟵ **Metallic character increases**

Na Mg Al Si P S Cl Ar

Nonmetallic character increases ⟶

For the same reasons, elements toward the top of a column of the periodic table are the most nonmetallic, while elements toward the bottom of a column are the most metallic. This trend is evident in Group IVA, where a gradual transition is seen from a nonmetallic element (carbon) toward the well-known metals tin and lead.

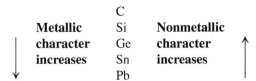

	C	
Metallic character increases ↓	Si Ge Sn Pb	**Nonmetallic character increases** ↑

➤ **CHECKPOINT**

Arrange the following metals in order of increasing metallic character: K, Ca, and Rb.

If elements become less metallic and more nonmetallic as we go from left to right across a row of the periodic table, we should encounter elements along each row that have properties that lie between the extremes of metals and nonmetals. These elements often look metallic, but they are brittle, like nonmetals. They are neither good conductors nor good insulators but serve as the basis for the semiconductor industry. These elements are often known as *metalloids* or *semimetals.*

The primary difference between the various metals in the periodic table is the ease with which they undergo chemical reactions. The elements toward the bottom-left corner of the periodic table, which have the most metallic character, are the metals that are the most **active** in the sense of being the most **reactive** (Figure 5.1). Lithium, sodium, and potassium all react with water, for example. But the reaction becomes more vigorous as we do down the column because they become more active as they become more metallic.

Metals are often divided into four classes on the basis of their activity, as shown in Table 5.1. The most active metals are so reactive that they readily combine with the O_2 and H_2O in the atmosphere and must therefore be stored under an inert liquid, such as mineral oil. Those metals are found in Groups IA and IIA of the periodic table.

Metals in the second class are slightly less active than Class I metals. They don't react with water at room temperature, but they react rapidly with acids. The less active metals can be used for a variety of purposes. Aluminum, for example, is used for aluminum foil and beverage cans. It is important to protect these metals from exposure to acids, however, so aluminum cans are lined with a plastic coating to prevent contact with the acidic soft drinks or fruit juices they contain.

The third class contains metals such as iron, tin, and lead that react only with strong acids. Some cereals that are described as "iron-enriched" contain tiny pieces of iron metal, not Fe^{3+} salts, as the source of the iron they provide because the iron metal dissolves in the strongly acidic stomach fluid to form the Fe^{3+} ions the body needs.

Metals in the fourth class are so unreactive they are essentially inert at room temperature. These metals are ideal for making jewelry and coins because they

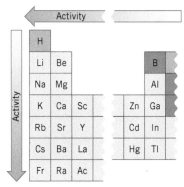

Fig. 5.1 The main-group metals in the bottom-left corner of the periodic table are known as the active metals because they are so reactive.

Aluminum metal is more "active" than iron. So much energy is given off when aluminum metal reacts with Fe_2O_3 to give iron metal and Al_2O_3 that this reaction is called the thermite reaction.

Table 5.1

Common Metals Divided into Classes on the Basis of Activity

Class I Metals: The Active Metals
Li, Na, K, Rb, Cs   (Group IA)
Ca, Sr, Ba   (Group IIA)

Class II Metals: The Less Active Metals
Mg, Al, Zn, Mn

Class III Metals: The Structural Metals
Cr, Fe, Sn, Pb, Cu

Class IV Metals: The Coinage Metals
Ag, Au, Pt

don't react with the majority of the substances with which they come into daily contact. As a result, they are often called the coinage metals.

5.3 Main-Group Metals and Their Ions

GROUP IA: THE ALKALI METALS

We begin our discussion of the formation of ions by examining three separate areas of the periodic table: main-group metals, main-group nonmetals, and transition metals. The **main-group elements** in the periodic table are made up of the elements in Groups IA through VIIIA. The periodic table and the electron configurations introduced in Chapter 3 can be used to explain the reactions by which metals form positive ions.

The metals in Group IA include lithium (Li), sodium (Na), potassium (K), rubidium (Rb), cesium (Cs), and francium (Fr). These elements are called the **alkali metals** because they form compounds, such as NaOH, that were once known as *alkalies.* Sodium and potassium are relatively common elements. In fact, they are among the eight most abundant elements in the earth's crust, as shown in Table 5.2. The other alkali metals are much less abundant. Discussions of the chemistry of alkali metals therefore tend to focus on sodium and potassium.

The electron configuration of a neutral sodium atom can be written as follows.

$$\text{Na} \qquad 1s^2 2s^2 2p^6 3s^1 \quad \text{or} \quad [\text{Ne}]3s^1$$

Sodium atoms lose their $3s$ electron to form positively charged Na^+ ions with the electron configuration $1s^2\ 2s^2\ 2p^6$. The electron configuration of the Na^+ ion is the same as that of a neutral Ne atom. Because they have the same number of electrons, a Ne atom and an Na^+ ion are said to be **isoelectronic.** Only one electron is lost from each sodium atom during the formation of an Na^+ ion because the electrons in the second shell are closer to the nucleus and are therefore more strongly attracted to the nucleus than the $3s$ electron.

The electron configurations of the alkali metals are all characterized by a single valence electron.

Li	$[\text{He}]2s^1$	Rb	$[\text{Kr}]5s^1$
Na	$[\text{Ne}]3s^1$	Cs	$[\text{Xe}]6s^1$
K	$[\text{Ar}]4s^1$	Fr	$[\text{Rn}]7s^1$

Group IA elements all have similar valence-shell electron configuration (ns^1) and are also characterized by unusually small AVEE values. This means that it is easier to remove an electron from these elements than from almost any other elements in the periodic table. Consequently, the chemistry of these elements is dominated by their tendency to lose an electron to form positively charged ions (Li^+, Na^+, K^+).

The periodic table can be used to predict the charge on many metal ions. As a general rule, the maximum charge on a metal ion can be found from the group number for that element. The alkali metals in Group IA, for example, form ions with a +1 charge.

GROUP IIA: THE ALKALINE EARTH METALS

The elements in Group IIA (Be, Mg, Ca, Sr, Ba, and Ra) are all metals, and all but Be and Mg are active metals. These elements are called the **alkaline earth metals.**

Table 5.2

Percent by Weight of the Most Common Elements in the Earth's Crust

Element	Percent by Weight
O	46.60
Si	27.72
Al	8.13
Fe	5.00
Ca	3.63
Na	2.83
K	2.59
Mg	2.09

The term *alkaline* reflects the fact that these metals form compounds such as MgO or CaO that can neutralize acids. The term *earth* was historically used to describe the fact that many of the compounds these elements form are insoluble in water.

Most of the chemistry of the alkaline earth metals (Group IIA) can be predicted from the behavior of the alkali metals (Group IA). Once again, these elements are characterized by relatively small average valence electron energies. As a result, the chemistry of these elements is dominated by their tendency to form positive ions. In this case, however, each element has two valence electrons, which means that they tend to form M^{2+} ions.

$$Mg \quad [Ne]\, 3s^2$$
$$Mg^{2+} \quad [Ne]$$

The alkaline earth metals in Group IIA are significantly less reactive than the alkali metals in Group IA. They are also harder and have higher melting points than the Group IA metals.

GROUP IIIA METALS

Group IIIA contains a semimetal (B) and four metals (Al, Ga, In, and Tl). Discussions of the chemistry of the Group IIIA metals often focus on aluminum because gallium, indium, and thallium are relatively scarce and therefore of limited interest. Aluminum, on the other hand, is the third most abundant element in the Earth's crust (see Table 5.2).

The average valence electron energy for aluminum is only slightly larger than that of the metals in Groups IA and IIA. Thus the chemistry of aluminum is dominated by the tendency of its atoms to lose electrons to form positive ions. Because the electron configuration of a neutral aluminum atom indicates the presence of three valence electrons, this metal forms the Al^{3+} ion in most of its compounds.

$$Al \quad [Ne]\, 3s^2 3p^1$$
$$Al^{3+} \quad [Ne]$$

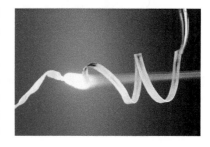

Magnesium metal is not as reactive as lithium, sodium, or potassium, but it will burst into flame when ignited.

➤ **CHECKPOINT**

Why do atoms of Group IIA elements tend to lose two electrons while those in Group IA lose only one?

The reaction between sodium metal and chlorine gas to form solid sodium chloride.

5.4 Main-Group Nonmetals and Their Ions

In Chapter 3, we noted that the ionization energy of the elements increases as we go from left to right across a row of the periodic table. We also noted that the second ionization energy for an element is larger than the first, and the third ionization energy is even larger.

As we go across a row of the periodic table, we therefore reach a point at which it takes too much energy to remove enough electrons from a neutral atom to form a positive ion with a noble gas electron configuration. The atoms of the main-group elements toward the right side of the periodic table therefore tend to gain electrons to form negatively charged ions with the electron configuration of one of the noble gases. **Ionic compounds,** or **salts,** that contain both positive and negative ions are often formed when main-group elements at opposite sides of the periodic table combine.

The alkali metals react with the nonmetals in Group VIIA (F_2, Cl_2, Br_2, I_2, and At_2) to form ionic compounds, or salts. Chlorine, for example, reacts with sodium metal to produce sodium chloride (table salt).

$$2\,Na(s) + Cl_2(g) \longrightarrow 2\,NaCl(s)$$

Because they form salts with so many metals, the elements in Group VIIA are known as the **halogens.** This name comes from the Greek word for "salt" (*hals*) and the Greek word meaning "to produce" (*gennan*). The halogens are therefore literally the "salt formers."

The salts formed by the halogens are called **halides.** These salts include **fluorides** (LiF), **chlorides** (NaCl), **bromides** (KBr), and **iodides** (NaI). The halide ions all carry a charge of -1 because the halogens add one more electron to achieve a filled-shell electron configuration.

$$Cl \qquad [Ne]3s^2 3p^5$$
$$Cl^- \qquad [Ne]3s^2 3p^6 = [Ar]$$

The halogens—F_2, Cl_2, Br_2, I_2, and At_2—are seldom, if ever, found in nature. These elements are found as part of ionic compounds in which they have a -1 charge. Fluoride ions are found in minerals such as fluorite (CaF_2) and cryolite (Na_3AlF_6). Sodium fluoride (NaF) has replaced stannous fluoride (SnF_2) as the most commonly used cavity-fighting agent in commercial toothpastes because it works well with the kinds of polishing agents used in modern translucent gel toothpastes. Chloride ions are found in rock salt (NaCl), in oceans (which are ~2% Cl^- by weight), and in lakes that have a high salt content, such as the Great Salt Lake in Utah. Both bromide and iodide ions are found at low concentrations in the oceans, as well as in brine wells.

Hydrogen is the only element in the periodic table that is listed in more than one group. In many periodic tables it can be found among both the metals in Group IA and the nonmetals in Group VIIA. This can be understood by looking at the position of hydrogen in the three-dimensional version of the periodic table shown in Figure 5.2. Hydrogen is almost exactly in the middle of the elements in terms of its AVEE and therefore its electronegativity. When it reacts with an element that is less electronegative, it tends to pick up one electron, like the halogens, to form a negative ion with the electron configuration of the noble gas helium.

$$H \qquad 1s^1$$
$$H^- \qquad 1s^2 = [He]$$

Compounds that contain the H^- ion are known as **hydrides.** Thus potassium reacts with hydrogen to form potassium hydride.

$$2\,K(s) + H_2(g) \longrightarrow 2\,KH(s)$$

The elements in Group VIA often react with metals to form compounds in which they contain a negatively charged ion with a -2 charge. Thus oxygen often forms **oxides** that contain the O^{2-} ion.

$$O \qquad [He]\,2s^2 2p^4$$
$$O^{2-} \qquad [He]\,2s^2 2p^6 = [Ne]$$

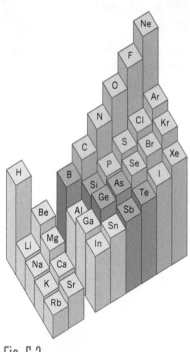

Fig. 5.2 Three-dimensional plot of the electronegativities of the main-group elements versus position in the periodic table. [Reprinted from L. C. Allen, *Journal of the American Chemical Society,* **111,** 9003 (1989).]

Oxygen is the most abundant element on the planet. The Earth's crust is 46.6% oxygen by weight, the oceans are 86% oxygen, and the atmosphere is 21% oxygen. In the Earth's crust oxygen is present primarily in ionic compound as an ion with a -2 charge. In the oceans oxygen is found in the covalently bonded H_2O molecule. The dominant form of oxygen in the atmosphere is the diatomic

O_2 molecule, but it also forms the triatomic ozone (O_3) molecule that plays an important role in protecting the planet from harmful ultraviolet radiation.

Like oxygen, sulfur tends to form a -2 ion. Compounds that contain the S^{2-} ion are called **sulfides.**

$$S \qquad [Ne]\, 3s^2 3p^4$$
$$S^{2-} \qquad [Ne]\, 3s^2 3p^6 = [Ar]$$

Sulfur also forms compounds with most of the other elements in the periodic table. Elemental sulfur usually consists of cyclic S_8 molecules in which each atom fills its valence shell by forming single bonds to two neighboring atoms. Because sulfur forms unusually strong S—S single bonds, it is capable of forming cyclic molecules that contain 6, 7, 8, 10, and 12 sulfur atoms.

Because of the strength of the N≡N triple bond, the N_2 in our atmosphere is virtually inert to chemical reactions at room temperature. The most active metals, however, react with nitrogen to form **nitrides,** such as lithium nitride, Li_3N. The nitride ion carries a charge of -3 because it adds three electrons to reach a filled-shell electron configuration.

$$N \qquad [He]\, 2s^2 2p^3$$
$$N^{3-} \qquad [He]\, 2s^2 2p^6 = [Ne]$$

Compounds that contain nitrogen are essential components of the proteins, nucleic acids, vitamins, and hormones that make life possible. Nature has therefore developed a mechanism to overcome the fact that N_2 is usually inert to chemical reactions at room temperature. Some plants contain bacteria that have an enzyme that catalyzes a reaction known as "nitrogen fixation" that allows them to pick up nitrogen from the atmosphere to form ammonia (NH_3). Most other plants require nitrogen in the form of a fertilizer. Animals then pick up the nitrogen they need from plants or other animals in their diet.

Exercise 5.1

Use the electron configuration of the following elements to predict the charge on the ions these elements form.

(a) Ca

(b) P

Solution

(a) Calcium forms Ca^{2+} ions by losing two electrons from its valence shell to achieve the electron configuration of argon.

$$Ca^{2+} \qquad 1s^2 2s^2 2p^6 3s^2 3p^6 = [Ar]$$

(b) Phosphorus forms the P^{3-} ion by gaining three electrons in order to achieve the electron configuration of argon.

$$P^{3-} \qquad 1s^2 2s^2 2p^6 3s^2 3p^6 = [Ar]$$

5.5 Transition Metals and Their Ions

The **transition metals** that lie between the main-group elements on either side of the periodic table form a variety of positively charged ions, such as the Ag^+, Cu^{2+}, and Fe^{3+} ions. They form these ions by losing electrons from either valence-shell s or d orbitals. Iron, for example, can form a $+2$ ion by losing its $4s$ electrons.

$$Fe \longrightarrow Fe^{2+} + 2e^-$$
$$[Ar]4s^2 3d^6 \quad\quad [Ar]3d^6$$

Iron can also form a $+3$ ion by losing one of the $3d$ electrons to form an electron configuration in which the d orbitals are half-filled, with one electron in each of the five orbitals in the subshell.

$$Fe \longrightarrow Fe^{3+} + 3e^-$$
$$[Ar]4s^2 3d^6 \quad\quad [Ar]3d^5$$

Transition metals are like main-group metals in many ways. They look like metals, they are malleable and ductile, they conduct heat and electricity, and they form positive ions.

Exercise 5.2

The most stable oxidation states of titanium correspond to the Ti^{2+} and Ti^{4+} ions. Predict the electron configuration of these ions.

Solution

Titanium has the following electron configuration.

$$Ti \quad\quad [Ar]4s^2 3d^2$$

This element forms Ti^{2+} ions by the loss of the two electrons in the valence $4s$ orbital.

$$Ti^{2+} \quad\quad [Ar]3d^2$$

Compounds in which the titanium can be thought to exist as Ti^{4+} ions are formed by the loss of the two valence electrons from the $4s$ subshell plus the loss of the two $3d$ electrons to give the noble gas electron configuration of argon.

$$Ti^{4+} \longrightarrow [Ar]$$

5.6 Chemistry and Color

PAINTS AND PIGMENTS

One of the joys of working with transition metal compounds is the extraordinary range of colors they exhibit. The characteristic colors of aqueous solutions of some common transition metals are shown in Table 5.3.

Table 5.3

Characteristic Colors of Common Transition Metal Ions

Cr^{2+}	Blue	Ni^{2+}	Green
Cr^{3+}	Blue violet	Zn^{2+}	Colorless
Co^{2+}	Pinkish red	CrO_4^{2-}	Yellow
Cu^{2+}	Light blue	$Cr_2O_7^{2-}$	Orange
Fe^{3+}	Yellow	MnO_4^{-}	Deep violet
Mn^{2+}	Very faint pink		

In the presence of various neutral molecules (such as NH_3) or negative ions (such as the SCN^- ion), the colors of these transition metal ions can change drastically. The characteristic light-blue color of the Cu^{2+} ion, for example, turns dark blue in the presence of ammonia due to the formation of the $Cu(NH_3)_4^{2+}$ ion. Adding the thiocyanate ion (SCN^-) to the faint-yellow solution of the Fe^{3+} ion gives rise to a blood-red solution of the $Fe(SCN)_2^+$ ion.

It is not surprising that early advances in chemistry were therefore accompanied by advances in paints and pigments available to artists. A handful of pigments have been known since antiquity. The first pigments, used in cave paintings as early as 17,000 years B.P. in the Lascaux Caves in France, were red and brown. Other pigments were gradually discovered.

The first synthetic pigment was Prussian blue $\{Fe_4[Fe(CN)_6]_3\}$, which was synthesized in 1704. Several new compounds, and therefore new pigments, became available toward the end of the eighteenth century. These included cobalt blue ($CoO \cdot Al_2O_3$) and Scheele's green [$Cu(AsO_2)_2$]. By the turn of the nineteenth century, new pigments were appearing, such as chrome yellow ($PbCrO_4$), emerald green [$Cu(OAc)_2 \cdot 3 Cu(AsO_2)_2$], and cerulean blue ($CoO \cdot nSnO_2$), which began to revolutionize the artist's palette.

Transition metals have also played an important role in the formation of both colored glass and the glaze that forms on pottery. An entire spectrum of colors can be produced by adding an oxide or carbonate salt of one of the transition metals, such as Ti, V, Cr, Mn, Fe, Co, Ni, Cu, and Au, to the components that form the glass or glaze. Firing a pot that contains copper in an open kiln, for example, produces the Cu^{2+} ion, which is blue or green; in a closed kiln, it produces the Cu^+ ion, which has a red color.

5.7 Predicting the Formulas of Ionic Compounds

A famous Sherlock Holmes story that revolves around a dog that didn't bark reminds us that we often forget to notice things that don't happen. As a chemical equivalent of the dog that doesn't bark, consider what happens when you pour a few crystals of table salt into the palm of your hand. Even though table salt is an ionic compound that is composed of electrically charged ions, this does not feel the same as making the mistake of touching a bare wire that had been plugged into an electric socket. In other words, ionic compounds carry no net electric charge. This means that they must contain just as many positive charges as negative charges.

The formulas of ionic compounds can be predicted by recognizing that the total charge on the positive ions must balance the total charge on the negative ions. The charges on the ions, therefore, dictate the formula of the ionic compound that is formed. Chemical formulas for ionic compounds are represented by the simplest whole-number ratio of positive to negative ions that will yield an electric charge of zero on the compound.

Exercise 5.3

Predict the formulas of the following ionic compounds.

(a) Calcium carbonate is the major component in chalk, limestone, and marble. It is composed of calcium ions, Ca^{2+}, and the polyatomic carbonate anion, CO_3^{2-}.

(b) Calcium fluoride contains Ca^{2+} and F^- ions, and exists in nature as the mineral known as fluorite.

(c) Strontium nitrate, which is composed of Sr^{2+} and NO_3^- ions, is added to fireworks to produce a red color when the fireworks burst into flame.

Solution

(a) Since the calcium ion has a charge of $+2$ and the carbonate ion has a charge of -2, the charge on one calcium ion cancels the charge on one carbonate ion. The formula of calcium carbonate is therefore $CaCO_3$.

(b) The charge on each calcium ion is $+2$, whereas the charge on each fluoride ion is -1. In order for the charges to cancel, there must be two fluoride ions for every calcium ion. The formula for this compound is therefore CaF_2.

(c) The $+2$ charge on the strontium requires two negatively charged nitrate ions in order to balance the charge. The chemical formula is written as $Sr(NO_3)_2$. Parentheses are used in this formula to indicate that the subscript "2" refers to the whole NO_3^- ion.

5.8 Predicting the Products of Reactions That Produce Ionic Compounds

The products of many reactions between main-group metals and other elements can be predicted from the electron configurations of the elements. Consider the reaction between sodium and chlorine to form sodium chloride, for example.

$$2\,Na(s)\ +\ Cl_2(g)\ \longrightarrow\ 2\,NaCl\,(s)$$

The net effect of the reaction is to transfer an electron from a neutral sodium atom to a neutral chlorine atom to form Na^+ and Cl^- ions that have filled-shell configurations.

$$Na\cdot\ +\ \cdot\ddot{\underset{..}{Cl}}:\ \longrightarrow\ [Na]^+[:\ddot{\underset{..}{Cl}}:]^-$$

We can often predict the product of reactions that form ionic compounds by using average valence electron energies (AVEE) or electronegativities and the electron configurations of the elements involved in the reaction. Consider the reaction between potassium metal and hydrogen gas, for example.

$$K(s)\ +\ H_2(g)\ \longrightarrow$$

Potassium and hydrogen have similar valence electron configurations.

$$K \qquad [Ar]\,4s^1$$
$$H \qquad 1s^1$$

To form an ionic compound, an electron has to be transferred from one element to the other. We can decide which element should lose an electron by comparing either the AVEE for potassium (0.42 MJ/mol) with that for hydrogen (1.31 MJ/mol) or the electronegativity of potassium (0.73) with that for hydrogen (2.30). The AVEE and electronegativity data both suggest that each potassium atom consumed in the reaction should lose an electron to a hydrogen atom to form K^+ and H^- ions.

$$K \cdot + \cdot H \longrightarrow [K]^+[:H]^-$$

The balanced equation for the reaction is written as follows.

$$2\,K(s) + H_2(g) \longrightarrow 2\,KH(s)$$

Exercise 5.4

Lithium metal is used in high-energy, lightweight batteries that have revolutionized lap-top computers and will soon revolutionize electric cars. These batteries are sealed to prevent the lithium metal from reacting with the oxygen in the air. Write a balanced equation for the reaction between lithium metal and oxygen.

$$Li(s) + O_2(g) \longrightarrow$$

Solution

We start by using the electron configurations of the elements to predict the ions that form in this reaction.

$$Li \qquad [He]\,2s^1$$
$$O \qquad [He]\,2s^2 2p^4$$

Everything we learned about the structure of the atom in Chapter 3 suggests that it should be easier to remove an electron from lithium than from oxygen. We can confirm this by comparing the average valence electron energies for lithium (0.52 MJ/mol) and oxygen (1.91 MJ/mol) or the electronegativities of lithium (0.91) and oxygen (3.61). We therefore expect that lithium will lose electrons and oxygen will gain electrons to form the Li^+ and O^{2-} ions.

$$Li^+ \qquad [He]$$
$$O^{2-} \qquad [He]\,2s^2 2p^6 = [Ne]$$

We then combine these ions to predict the product of the reaction—Li_2O—and write a balanced equation for the reaction, as follows.

$$4\,Li(s) + O_2(g) \longrightarrow 2\,Li_2O(s)$$

5.9 Oxides, Peroxides, and Superoxides

The method we have used to predict the products of reactions of the main-group metals is remarkably powerful. Exceptions to its predictions arise, however, when very active metals react with oxygen, which is one of the most reactive nonmetals.

Lithium is well behaved. It reacts with O_2 to form an oxide, as predicted in Exercise 5.4.

$$4\,Li(s) + O_2(g) \longrightarrow 2\,Li_2O(s)$$

Sodium, however, reacts with O_2 under normal conditions to form a compound that contains twice as much oxygen.

$$2\,Na(s) + O_2(g) \longrightarrow Na_2O_2(s)$$

Compounds that are unusually rich in oxygen are called **peroxides.** The prefixes *per-* or *hyper-* are used to indicate "above normal" or "excessive." Na_2O_2 is therefore a peroxide because it contains more than the expected amount of oxygen.

The difference between oxides and peroxides can be understood by remembering the basic assumption behind the chemistry of the alkali metals: These elements form compounds in which the metal is present as a +1 ion. Oxides, such as Li_2O, contain the O^{2-} ion.

$$[Li]^+\,[:\ddot{O}:]^{2-}\,[Li]^+$$

Peroxides, such as Na_2O_2, contain the polyatomic O_2^{2-} or peroxide ion.

$$[Na]^+\,[:\ddot{O}-\ddot{O}:]^{2-}\,[Na]^+$$

The formation of sodium peroxide can be explained by assuming that sodium is so reactive that the metal is consumed before each O_2 molecule can combine with enough sodium to form Na_2O. This explanation is supported by the fact that sodium reacts with O_2 in the presence of large excess of the metal, or in the presence of a limited amount of O_2, to form the oxide expected when the reaction goes to completion.

$$4\,Na(s) + O_2(g) \longrightarrow 2\,Na_2O(s)$$

It is also consistent with the fact that the very active alkali metals—potassium, rubidium, and cesium—react so rapidly with oxygen that they form **superoxides,** in which the alkali metal reacts with O_2 in a 1:1 mole ratio. Superoxides contain the polyatomic O_2^- ion.

$$K(s) + O_2(g) \longrightarrow KO_2(s)$$

Potassium superoxide forms on the surface of potassium metal, even when the metal is stored under an inert solvent. Thus old pieces of potassium metal are potentially dangerous. When someone tries to cut the metal, the pressure of the knife pushing down on the area where the superoxide touches the metal can induce the following reaction.

$$KO_2(s) + 3\,K(s) \longrightarrow 2\,K_2O(s)$$

Because potassium oxide is more stable than potassium superoxide, the reaction gives off enough energy to boil potassium metal off the surface, which reacts explosively with the oxygen and water vapor in the atmosphere.

The reactions between the alkali metals and oxygen raise an important point. We can predict the product of any reaction between a main-group metal and a main-group nonmetal from the AVEE values or electronegativities of the elements and their electron configurations. But we always have to check our predictions with an experiment because other factors, such as the rate at which one of the reactants is consumed, may affect the formula of the substance actually produced in the reaction.

The alkaline earth metals also react with oxygen to form oxides and peroxides. Because the activity of the alkaline earth metals increases as we go down the column of the periodic table, magnesium forms the oxide (MgO), whereas barium forms the peroxide (BaO_2).

> ➤ **CHECKPOINT**
>
> What is the formula of the peroxide of strontium, Sr?

5.10 The Ionic Bond

Ionic compounds are held together by the force of attraction between ions of opposite charge. The magnitude of the **electrostatic** or **coulombic** attraction between the particles depends on the product of the charges (q_1 and q_2) on the two ions and the square of the distance between the ions (r^2).

$$F = \frac{q_1 \times q_2}{r^2}$$

The bond that forms when the ions are brought together is called, logically enough, an **ionic bond.** In the formation of an ionic bond between a metal and nonmetal, there is a transfer of electron density from the metal to the nonmetal. This can be represented using Lewis structures, which show the transfer of valence-shell electrons.

$$\mathrm{K\cdot} + \cdot \ddot{\mathrm{I}}: \longrightarrow [\mathrm{K}]^+ [:\ddot{\mathrm{I}}:]^-$$

$$2\,\mathrm{Li} + \cdot \ddot{\mathrm{S}}: \longrightarrow [\mathrm{Li}]^+ [:\ddot{\mathrm{S}}:]^{2-} [\mathrm{Li}]^+$$

There is a fundamental difference between the covalent compounds discussed in the previous chapter and the ionic compounds in this chapter. Covalent compounds usually exist in the form of molecules that contain a limited number of atoms. Even compounds as complex as glucose ($C_6H_{12}O_6$) and vitamin B_{12} ($C_{63}H_{88}N_{14}O_{14}PCo$) consist of distinct molecules that contain a well-defined number of atoms.

The same can't be said of ionic compounds. They exist as huge, three-dimensional networks of ions of opposite charge that are held together by ionic bonds, as shown in Figure 5.3. Each Na^+ ion in this structure is surrounded by six Cl^- ions, and each Cl^- ion is surrounded by six Na^+ ions. The simplest whole-number ratio of sodium ions to chloride ions is therefore 1:1, so the formula of the compound is written as NaCl.

When ionic compounds dissolve in water, they break apart into aqueous solutions of the ions they contain. The resulting ions are free to move throughout the solution, and the aqueous solution can therefore conduct an electric current.

$$\mathrm{NaCl}(s) \xrightarrow{H_2O} \mathrm{Na}^+(aq) + \mathrm{Cl}^-(aq)$$

Similarily, when ionic compounds melt, they break apart into their ions; therefore, molten ionic compounds also conduct electricity.

$$\mathrm{NaCl}(s) \xrightarrow{heat} \mathrm{Na}^+(l) + \mathrm{Cl}^-(l)$$

NaCl(s)

Fig. 5.3 Three-dimensional network structure of NaCl. The smaller spheres represent Na^+ ions, and the larger spheres represent Cl^- ions.

5.11 Structures of Ionic Compounds

When we want to describe the structure of a covalent compound, we draw a picture of its molecules. When we want to envision the structure of an ionic compound, we draw a picture of the simplest repeating unit, or **unit cell,** in the three-dimensional network of its ions. By describing the size, shape, and contents of the unit cell—and the way the repeating units stack to form a three-dimensional solid—we can unambiguously describe the structure of the compound. Figure 5.4 shows the unit cells for several common ionic compounds. Because a unit cell is a part of a larger three-dimensional structure, ions at the corners, edges, and faces are shared by neighboring unit cells. Only when this is taken into consideration does the ionic formula match the ratio of ions in the unit cell, as will be shown in Chapter 9.

In 1850 Auguste Bravais showed that every crystal, no matter how complex its structure, could be classified as one of 14 unit cells that meet the following criteria:

- The unit cell is the simplest repeating unit in the crystal.
- Opposite faces of a unit cell are parallel.
- The edge of the unit cell connects equivalent points.

This chapter focuses on the three types of cubic unit cells—**simple cubic, body-centered cubic,** and **face-centered cubic**—shown in Figures 5.5 and 5.6.

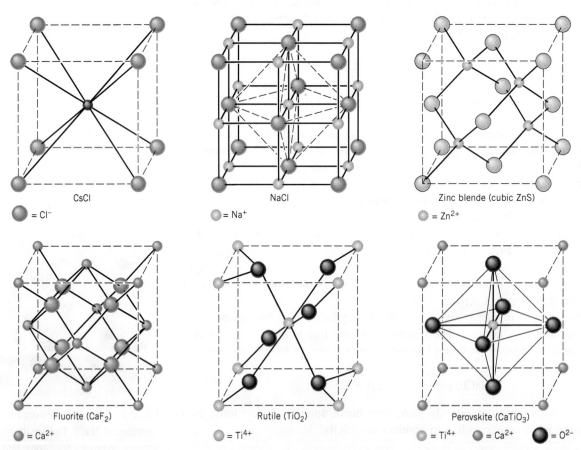

CsCl
⬤ = Cl⁻

NaCl
⬤ = Na⁺

Zinc blende (cubic ZnS)
⬤ = Zn²⁺

Fluorite (CaF₂)
⬤ = Ca²⁺

Rutile (TiO₂)
⬤ = Ti⁴⁺

Perovskite (CaTiO₃)
⬤ = Ti⁴⁺   ⬤ = Ca²⁺   ⬤ = O²⁻

Fig. 5.4 Unit cells of CsCl, NaCl (rock salt), ZnS (zinc blende), CaF₂ (fluorite), TiO₂ (rutile), and CaTiO₃ (perovskite).

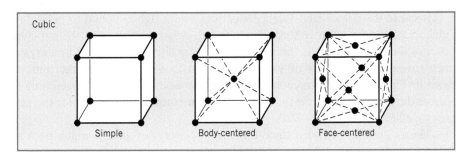

Cubic unit cells are important for two reasons. First, many metals, ionic solids, and intermetallic compounds crystallize in cubic unit cells. Second, these unit cells are relatively easy to visualize because the cell-edge lengths are all the same length and the cell angles are all 90°. Unit cells are defined in terms of **lattice points,** which are points at which a particle can be found in the lattice structure of the crystal. These particles can be atoms, ions, or molecules.

Unit cells are defined so that the cell edges always connect equivalent points. Therefore, an identical particle must be found at each of the eight corners of a cubic unit cell. The unit cell of NaCl shown in Figure 5.4, for example, contains an equivalent Cl^- ion at each of the eight corners of the unit cell.

Fig. 5.6 Models of simple cubic (left), body-centered cubic (middle), and face-centered cubic unit cells (right).

5.12 Metallic Bonds

The covalent bond that binds one chlorine atom to another to form a Cl_2 molecule has one thing in common with the ionic bond between the Na^+ and Cl^- ions in NaCl. In both cases, the electrons in the bond are **localized.** The electrons either reside on one of the atoms or ions that form the bond, or they are shared by a pair of atoms.

Metal atoms don't have enough valence electrons to reach a filled-shell configuration by sharing electrons with their neighbors. They are also relatively large and have relatively small AVEE values. This means that the valence electrons can be shared with many neighboring atoms, not just one. In effect, these valence electrons are **delocalized** over a number of metal atoms.

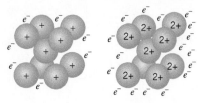

Fig. 5.7 The force of attraction between the positively charged metal ions and the surrounding sea of electrons in a metal is called a metallic bond. The electrons are not bound to individual atoms but may move throughout the metal.

► **CHECKPOINT**

Covalent and ionic compounds do not conduct electricity in the solid state. These compounds are called *insulators*. Metals do conduct electricity. Explain why substances that exhibit metallic bonding are conductors while covalent and ionic compounds are not.

Because the delocalized valence electrons aren't tightly bound to individual atoms, they are free to move through the metal. A useful picture of the structure of metals therefore envisions the metal atoms as positive ions locked in a crystal lattice surrounded by a sea of valence electrons that move among the ions, as shown in Figure 5.7. The force of attraction that would have to be overcome to separate the metal ions can be thought of as the **metallic bond** that holds the particles together.

Metals exist as extended three-dimensional arrays of atoms, which pack so that each atom can touch as many neighboring atoms as possible. When a metal is heated or beaten with a hammer, the planes of atoms that form the structure can slip past one another, which explains why metals are malleable and ductile (see Section 5.1) Each atom in the structure has a limited number of loosely held valence electrons that it can share with its neighbors. Due to the mobility of the electrons, metals can easily transfer kinetic energy from one atom to another. As a result, they are good conductors of heat.

Because the valence electrons in a metal are delocalized instead of being strictly associated with a single atom, they are free to move from one atom to another. The electrons that freely move through the metal are said to form **conduction bands.** If we draw the metal into a thin wire and connect the wire to a source of an electric current, electrons that enter the wire displace electrons that were already present on the atoms closest to the source of the current. Electrons flow through the conduction band until they eventually displace electrons from the other side of the wire. Metals are therefore good conductors of electricity.

Metals in their elemental form (Na, Cu, Fe) do not consist of individual atoms. They are composed of a three-dimensional network of positive metal ions surrounded by a sea of electrons. Consider a sample of pure iron. Because it would contain only iron atoms, we write the chemical formula as Fe.

5.13 The Relationship among Ionic, Covalent, and Metallic Bonds

There are significant differences in the physical properties of sodium chloride, sodium metal, and chlorine, as shown in Table 5.4. These differences result from significant differences in both the structure and bonding in these substances. Chemists therefore often classify compounds as ionic, metallic, or covalent on the basis of macroscopic, physical properties.

As we have seen, each Na^+ ion in NaCl is surrounded by six Cl^- ions, and vice versa. Removing one of these ions from the compound therefore involves breaking at least six bonds. As a result, ionic compounds such as NaCl tend to

Table 5.4
Some Physical Properties of NaCl, Na, and Cl$_2$

	NaCl	Na	Cl$_2$
Phase at room temperature	Solid	Solid	Gas
Density (g/cm^3)	2.2	0.97	0.0032
Melting point	801°C	97.81°C	−100.98°C
Boiling Point	1413°C	882.9°C	−34.6°C

have high melting points and boiling points. Ionic compounds are therefore almost always solids at room temperature.

Each Na atom in sodium metal is bound to as many neighboring atoms as possible. Because the bonds between metal atoms are relatively strong, metals are usually solids at room temperature. Mercury is the only metal that is a liquid at room temperature, although gallium metal has a low enough melting point that it can melt in the palm of one's hand. The melting point of sodium metal in Table 5.4 is less than 100°C. As a result, small pieces of sodium metal melt when they are first dropped onto the surface of water because of the heat given off in the vigorous reaction between this metal and water. As a rule, however, metals can be assumed to have relatively high boiling points because it takes a great deal of energy to disrupt all of the bonds necessary to remove individual atoms from the liquid to form a gas.

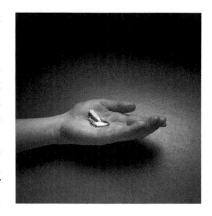

Elemental gallium has a melting point, which means it will melt on the palm of the hand.

$Cl_2(g)$ consists of molecules in which one chlorine atom is tightly bound to another chlorine atom to form a relatively strong Cl—Cl covalent bond. These aren't the bonds that are broken when chlorine melts or boils, however. It is the relatively weak forces between Cl_2 molecules that need to be broken. As a result, it is relatively easy to remove a Cl_2 molecule from its nearest neighbors, and chlorine is therefore a gas at room temperature.

The difference between the physical properties of NaCl and Cl_2 is so large that it is easy to believe that the bond between two atoms is either ionic or covalent. For many years, chemists have recognized that this isn't true. They have argued that words such as *ionic* and *covalent* referred to the extremes at either end of a continuous spectrum of bonding.

To see how they came to that conclusion, let's compare what happens when covalent and ionic bonds form. When two chlorine atoms come together to form a covalent bond, each atom contributes one electron to form a pair of electrons shared equally by the two chlorine atoms. The electrons are shared equally because both chlorine atoms have the same AVEE or electronegativity. When elements from the right-hand side of the periodic table that have about the same electronegativities combine, the substance is considered to be **covalent.** The formation of a covalent bond between chlorine atoms can be represented by Lewis structures as follows.

$$: \overset{..}{\underset{..}{Cl}} \cdot \; + \; \cdot \overset{..}{\underset{..}{Cl}} : \; \longrightarrow \; : \overset{..}{\underset{..}{Cl}} - \overset{..}{\underset{..}{Cl}} :$$

Methane (CH_4) and nitrogen dioxide (NO_2) are examples of compounds we would predict to be covalent on the basis of electronegativity data.

	CH_4		NO_2
C	$EN = 2.54$	O	$EN = 3.61$
H	$EN = 2.30$	N	$EN = 3.07$
	$\Delta EN = 0.24$		$\Delta EN = 0.54$

These compounds, which consist of discrete molecules, have relatively low melting points (MP) and boiling points (BP), and they are both gases at room temperature. They therefore have the characteristic properties of low-molecular-weight covalent compounds.

	CH_4	NO_2
MP	$-182.5°C$	$-163.6°C$
BP	$-161.5°C$	$-151.8°C$

If the electronegativities of the atoms in a substance are about the same and if the atoms come from the metallic side of the periodic table, the substance will be **metallic.** Typical metallic substances include pure metals, such as sodium, in which the atoms are held together by metallic bonds. They also include intermetallic compounds such as CdLi, in which the difference in electronegativity of the atoms is small ($\Delta EN = 0.6$) and both atoms in the compound can be found on the metallic side of the periodic table.

When a sodium atom combines with a chlorine atom to form an ionic bond, for example, each atom contributes one electron to form a pair of electrons, but that pair of electrons spends most of its time on the more electronegative chlorine. This can be represented using Lewis structures as follows.

$$\text{Na} \cdot + \cdot \overset{\cdot\cdot}{\underset{\cdot\cdot}{\text{Cl}}} : \longrightarrow [\text{Na}]^+ [: \overset{\cdot\cdot}{\underset{\cdot\cdot}{\text{Cl}}} :]^-$$

When the difference between the electronegativities of the elements in a compound is relatively large, the compound is best classified as **ionic.** Large differences in electronegativity are normally associated with bonding between metals from the left side of the periodic table and nonmetals from the right side. NaCl and SrF_2 are good examples of ionic compounds. In each case, the electronegativity of the nonmetal is at least two units larger than that of the metal.

NaCl		*SrF₂*	
Cl	$EN = 2.87$	F	$EN = 4.19$
Na	$EN = 0.87$	Sr	$EN = 0.96$
	$\Delta EN = 2.00$		$\Delta EN = 3.23$

We can therefore assume a net transfer of electrons from the metal to the nonmetal to form positive and negative ions that form a three-dimensional network. We therefore write the following Lewis structures for these compounds.

NaCl $\quad [\text{Na}]^+ [: \overset{\cdot\cdot}{\underset{\cdot\cdot}{\text{Cl}}} :]^-$

$SrF_2 \quad [: \overset{\cdot\cdot}{\underset{\cdot\cdot}{\text{F}}} :]^- [\text{Sr}]^{2+} [: \overset{\cdot\cdot}{\underset{\cdot\cdot}{\text{F}}} :]^-$

These compounds have high melting points and boiling points, which are characteristic properties of ionic compounds.

	NaCl	*SrF₂*
MP	801°C	1473°C
BP	1413°C	2489°C

Inevitably, there must be compounds that fall between the extremes of covalent and ionic compounds. For such compounds, the difference between the electronegativities of the elements is large enough to be significant but not large enough to classify the compound as ionic. Consider water, for example.

H₂O	
O	$EN = 3.61$
H	$EN = 2.30$
	$\Delta EN = 1.31$

Water is neither purely ionic nor purely covalent. It doesn't contain positive and negative ions, as indicated by the ionic Lewis structure in Figure 5.8. But the

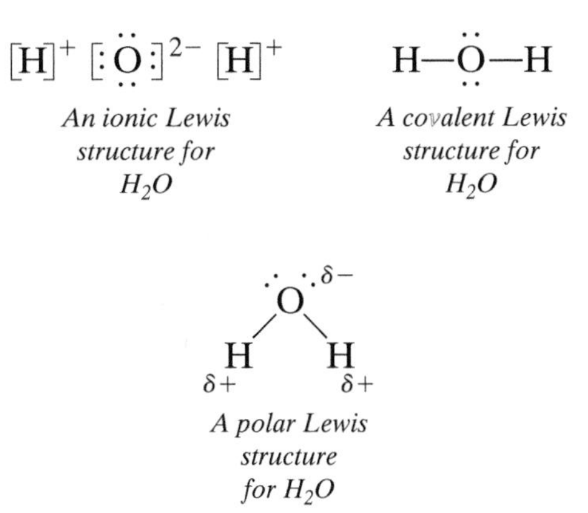

$$[H]^+ \; [:\!\ddot{O}\!:]^{2-} \; [H]^+ \qquad H-\ddot{O}-H$$

An ionic Lewis
structure for
H_2O

A covalent Lewis
structure for
H_2O

$$\overset{\cdot\cdot\;\cdot\cdot\;\delta-}{\ddot{O}}$$
$$\underset{\delta+}{H} \qquad \underset{\delta+}{H}$$

A polar Lewis
structure
for H_2O

Fig. 5.8   Water is neither purely ionic nor purely covalent; it is a polar covalent compound.

electrons are not shared equally, as indicated by the covalent Lewis structure in Figure 5.8. Water is best described as a **polar covalent compound.** One end, or pole, of the molecule has a partial positive charge ($\delta+$), and the other end has a partial negative charge ($\delta-$), as indicated in the polar Lewis structure at the bottom of Figure 5.8.

Ionic and covalent bonds differ in the extent to which a pair of electrons is shared by the atoms that form the bond. When one of the atoms is much better at drawing electrons toward itself, the bond is *ionic*. When the atoms are approximately equal in their ability to draw electrons toward themselves, the atoms share the pair of electrons more or less equally, and the bond is *covalent*. It is important to recognize that the terms *ionic* and *covalent* describe the extremes of a continuum of bonding. There is some covalent character in even the most ionic compounds. The chemistry of magnesium oxide, for example, can be understood if we assume that MgO contains Mg^{2+} and O^{2-} ions. But no compounds are 100% ionic. There is experimental evidence, for example, that the partial charge on the magnesium and oxygen atoms in MgO is +1.5 and −1.5, respectively.

Electronegativity is a powerful concept that summarizes the tendency of an atom of an element to gain, lose, or share electrons when it combines with another atom. The difference between the electronegativity of the atoms in a bond (ΔEN), however, isn't enough information to categorize the bond between atoms as primarily ionic, covalent, or metallic. Consider BF_3 ($\Delta EN = 2.14$) and SiF_4 ($\Delta EN = 2.27$), for example. The difference between the electronegativities of the atoms in these compounds leads us to expect these compounds to behave as if they were ionic, but both compounds are best classified as covalent. They are both gases at room temperature, and their boiling points are −99.9°C and −86°C, respectively.

Another example of the problem one encounters when trying to classify the bonding in a compound on the basis of the difference in electronegativities can be seen by comparing MgH_2 ($\Delta EN = 1.0$) and CO_2 ($\Delta EN = 1.1$). The value of ΔEN is nearly the same in these compounds. But MgH_2 is best described as an ionic compound that contains Mg^{2+} and H^- ions, whereas CO_2 is assumed to exist as covalent molecules held together by carbon—oxygen double bonds.

The problem of predicting the type of bond in a compound is complicated by the fact that the traditional approach to bonding—trying to differentiate between ionic and covalent bonds—is misleading. As we have seen, there is a third type of bond—the bond between atoms in a metal, such as sodium metal. Metallic bonds are found in more or less pure metals, such as sodium. They are also found in alloys, which are solutions of one metal dissolved in another. The alloy known as bronze, for example, is a solution of tin metal dissolved in copper. Finally, there is a family of so-called intermetallic compounds, such as CdLi, that are held together by metallic bonds.

> ➤ **CHECKPOINT**
>
> Zinc sulfide (ZnS) was used for many years for X-ray screens and in cathode ray tubes (CRT's) because it emits light when excited by X rays or beams of electrons. What is the difference between the electronegativities of zinc and sulfur? Using only the difference in electronegativity, would you expect ZnS to exhibit primarily ionic or covalent bonding?

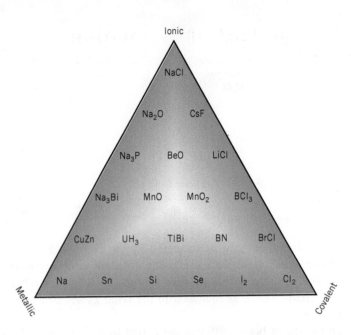

Fig. 5.9   The three different forms of chemical bonds—ionic, covalent, and metallic—form a two-dimensional plane, not a linear continuum.

One approach to the problem of bond types was taken in the 1970s by William L. Jolly, who constructed the diagram shown in Figure 5.9.[1] This figure assumes that the three different bond types—ionic, covalent, and metallic—form a two-dimensional triangle, not a linear continuum. Some elements combine to form ionic bonds, others form metallic bonds, and still others form covalent bonds.

In the 1990s, Leland Allen, William Jensen, and Gordon Sproul[2,3,4] used a triangular representation to describe the bonding in systems that don't fit the extremes of ionic, covalent, or metallic bonding. Whereas Jolly's triangle was descriptive, the more recent bond-type triangles have predictive power. Figure 5.10 shows a bond-type triangle constructed for the elements in the second row of the periodic table, those between lithium and fluorine. In this bond-type triangle, the difference between the electronegativity of the atoms in a bond (ΔEN) is plotted on the vertical axis, and the average electronegativity ($\overline{EN}$) of these atoms is plotted along the base of the triangle.

The vertex on the bottom-left corner of these bond-type triangles represents pure metallic bonding between identical metal atoms. The vertex on the bottom-right corner corresponds to a pure covalent bond between identical nonmetal atoms. Compounds near the top of the triangle represent ideal ionic bonds, such as LiF, in which the transfer of electrons is essentially complete.

Compounds near the three vertices are predominantly metallic, covalent, or ionic, respectively. Those that lie toward the middle of the diagram are the most difficult to describe in terms of any one of the three categories of chemical bonds. Compounds in the middle of the diagram have properties intermediate between the three types of bonding.

The elements to the left of boron on the baseline of Figure 5.10 are metals, whereas those on the right of boron are nonmetals. The two dark, shadowy lines that intersect at boron in Figure 5.10 divide the triangle into metallic, ionic, and covalent regions. The principal reason for categorizing substances by type of bonding is to more accurately predict the properties of the compound.

[1]W. L. Jolly, *The Principles of Inorganic Chemistry,* McGraw-Hill, New York, 1974, p. 187.

[2]L. C. Allen, *Journal of the American Chemical Society,* **114,** 1510 (1992).

[3]W. B. Jensen, *Bulletin of the History of Chemistry,* **13–14,** 47 (1992–1993).

[4]G. Sproul, *Journal of Physical Chemistry,* **98,** 13221 (1994).

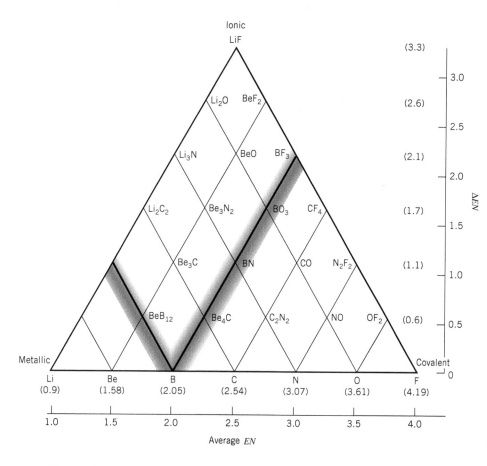

Fig. 5.10 A bond-type triangle for the second-row elements. ΔEN is plotted on the vertical axis, and the average electronegativity is plotted along the base. The bond type varies from ionic at the top of the triangle to metallic at the bottom left to covalent at the bottom right. A compound can be classified according to bond type by locating it within a bond-type triangle. The dark, shadowy lines separate the triangle into three regions: metallic, ionic, and covalent.

The position of a compound within a bond-type triangle can be used to visualize the ionic, covalent, or metallic character of the compound. As we proceed down the right side of this diagram, from LiF to F_2, we reach a point at which we encounter compounds that have some of the characteristics of both ionic and covalent compounds. As we go down the left side of the triangle, we encounter an interface between ionic and metallic compounds. Some compounds that may at first glance appear to be ionic may in fact be metallic. As we go up the triangle from the base to the vertex, there is a steady increase in ionic character.

5.14 Bond-Type Triangles

The **bond-type triangle** in Figure 5.10 was constructed by calculating both the electronegativity difference (ΔEN) and the average electronegativity ($\overline{EN}$) for a series of **binary compounds** formed from two elements in the second row of the periodic table. The vertical position of each compound in this diagram reflects the difference between the electronegativities of the elements; the horizontal position is determined by the average of these electronegativities. Because LiF has the largest electronegativity difference of any of the compounds in this figure, it is the most ionic compound in this triangle. It therefore appears at the top of the triangle.

Other compounds are placed in the triangle by calculating values of ΔEN and the average EN and plotting them at the appropriate positions. For example, the compound BN is located above the base of the triangle midway between boron and nitrogen. The electronegativity at this point on the base of the triangle (2.56) is the average of the electronegativity of boron (2.05) and the electronegativity of

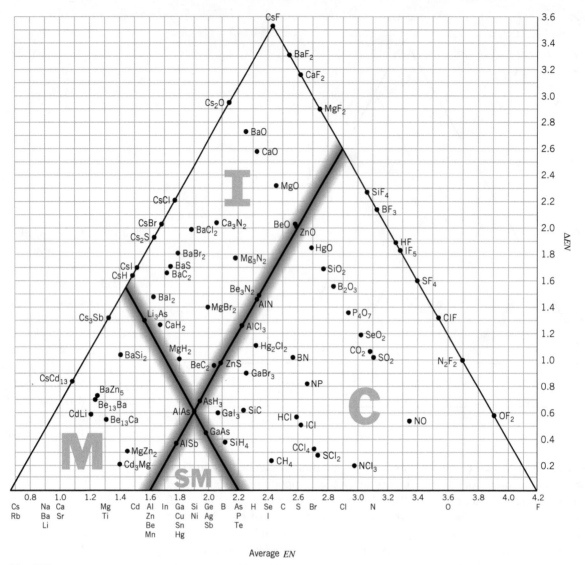

Fig. 5.11 Bond-type triangle for elements with electronegativities ranging from Cs to F. The lines separating the regions into ionic (I), metallic (M), covalent (C), and semimetallic compounds (SM) were determined empirically. These lines are parallel to the sides of the triangle and start at the positions occupied by the elements Al and Te on the base of the triangle.

nitrogen (3.07). ΔEN, which gives the vertical distance from the base, is 1.02. The location of compounds in the triangle gives information about the type of bonding. CO, for example, is best described as covalent, but perhaps not as covalent as NO, which is closer to the right corner of the triangle.

Figure 5.10 doesn't contain formulas for compounds that might be formed by combining lithium with either boron or beryllium because no binary compounds of these elements are known. However, the possible properties of such compounds could be predicted from their position on the triangle.

An enlarged and more detailed bond-type triangle is shown in Figure 5.11. This triangle has been expanded to include many of the most common elements and their binary compounds. By plotting more compounds on the triangle, we can make a better estimate of the position of the borders between the three types of bonding. The dark, shadowy lines separating the regions into ionic, metallic, covalent, and semimetallic are based on a classification of more than 300 binary compounds by Sproul.[5]

A diagonal stairstep line is used to separate the metals and nonmetals in most periodic tables. In Chapter 3 we noted that AVEE data can be used to explain the position of this line. Tellurium and arsenic, which have about the same AVEE values, define the right-hand limit of metallic behavior in the periodic table, whereas aluminum represents the left-hand limit of nonmetallic behavior. In Figure 5.11 the same elements are used to separate the triangle into different types of bonding. Dark, shadowy lines parallel to the sides of the triangle have been drawn in Figure 5.11 to separate elements below Al from those above Te.

This expanded representation of the bond-type triangle not only allows classification of a variety of binary compounds but also makes clear that changing atom combinations can dramatically alter the physical properties of a substance. A series of compounds of fluorine can be seen along the right side of the triangle. As we go down this line, we pass from ionic to covalent compounds. Compounds of Cs can be found along the left side of the triangle. As we go down the line, the compounds change from ionic to metallic bonding.

The area between tellurium and aluminum in Figure 5.11 is called the metalloid or semimetal region.[6] In this region of the triangle a change occurs from metallic to covalent bonding, and these compounds have properties that are intermediate between the two bonding types. Many compounds that fall in this region have electrical conductivities between conductors and insulators and are therefore known as **semiconductors.** Other compounds that fall outside this region but are close to the ionic–covalent border are also semiconductors.

The bond-type triangle shows why the value of ΔEN doesn't always provide enough information to categorize a compound. Both the average electronegativities of the elements in the compounds and the electronegativity differences must be taken into consideration. The bond-type triangle in Figure 5.11, for example, clearly shows both BF_3 and SiF_4 within the covalent region of the triangle, even though the difference between the electronegativities of the atoms, when used as the only criteria to predict the bond type, might lead one to expect these compounds to be ionic.

Predominantly covalent substances are formed when atoms that have relatively large electronegativities and small ΔEN values combine. Metallic substances are formed when we combine elements that have small electronegativities and a small ΔEN. Ionic compounds occur, as we might expect, when elements with very different electronegativities are combined.

The bond-type triangle enables us to think about bonds in terms of three contributing factors. Each compound on the triangle can be described in terms of the relative contributions of the three bonding types to the overall bond.

As we move from left to right on the triangle, the bonding electrons change from delocalization in metallic compounds to localized shared electrons in covalent compounds. The average electronegativity increases from left to right on the bond-type triangle and thus is a measure of the degree of covalent character of the bond. As we move from the bottom to the top of the triangle, the transfer of bonding electrons becomes more complete. The difference in electronegativity increases from the bottom to the top of the triangle and thus is a measure of the ionic character of a bond.

In the previous section, we encountered a pair of compounds that have more or less the same ΔEN but very different bonding: MgH_2 and CO_2. Because they have similar values of ΔEN, these compounds lie on the same horizontal line in

➤ **CHECKPOINT**

In the previous Checkpoint you were asked to use only ΔEN to describe the type of bonding in zinc sulfide. Now use the bond-type triangle to predict the bonding in ZnS.

---

[5]G. Sproul, *Journal of Physical Chemistry,* **98,** 6699 (1994).

[6]L. C. Allen, *Journal of the American Chemical Society,* **114,** 1510 (1992).

Figure 5.11. But MgH_2 lies in the ionic region, whereas CO_2 lies in the covalent region, as we would expect.

Another example of the power of Figure 5.11 can be seen by considering the compounds CdLi, AlAs, SiC, NO, and OF_2, which all have a value of ΔEN of about 0.6 and therefore lie along a narrow band parallel to the base of the triangle in Figure 5.11. The bond-type triangle suggests that these compounds have very different bonding characteristics and therefore very different properties. CdLi is a metallic compound, AlAs lies on the metallic–covalent border, SiC is just beyond the metallic–covalent border, and NO and OF_2 are predominantly covalent compounds. As you move to the right along this horizontal line, the covalent character of the bonding increases while the difference in electronegativity remains about the same.

Let's now consider a series of compounds that have more or less the same average electronegativities but different values of ΔEN: $GaBr_3$, Hg_2Cl_2, and CaO. The average electronegativity for these compounds is approximately 2.3. ΔEN increases, however, from about 0.9 for $GaBr_3$ to 1.1 for Hg_2Cl_2 and 2.6 for CaO. $GaBr_3$ lies toward the bottom of Figure 5.11, in the covalent region. Hg_2Cl_2 lies close to the border between ionic and covalent compounds, and so it is more ionic. CaO is toward the top of the triangle and is predominantly ionic.

Classification of bonding is most difficult for compounds near the borders between the three types of bonds. Compounds lying close to these borders have properties intermediate between the bonding classifications separated by the dividing line. They therefore exhibit interesting and exciting properties and are areas of current chemical research. These include semiconductors, high-temperature superconductors, and ceramics for automobile engines.

Exercise 5.5

Use electronegativities from Table B.7 in Appendix B and the bond-type triangle in Figure 5.11 to describe the bonding in the following compounds.

(a) $AlBr_3$

(b) $SnCl_4$

(c) CaS

(d) InNa

Solution

(a) The electronegativity of Al is 1.61, whereas the value for bromine is 2.69. This means that the average electronegativity is 2.15 and the difference between the electronegativities of the two elements is 1.08. If we start by finding the average electronegativity in this compound on the base of the triangle in Figure 5.11 and then move up the triangle to a value of ΔEN of 1.08, we reach a point that falls on the border between ionic and covalent bonding. Thus it isn't surprising to note that $AlBr_3$ has some of the characteristics of both ionic and covalent bonding. When it dissolves in aqueous solution, it forms the Al^{3+} ion, as would be expected for an ionic compound. But the structure of this compound in the solid state is best described by the empirical formula Al_2Br_6, which is consistent with some covalent character to the bonds.

(b) The electronegativity of tin is 1.82, whereas chlorine has an electronegativity of 2.87. The average electronegativity for these elements is 2.35, and the electronegativity difference is 1.05. This places $SnCl_4$ in the region of Figure 5.11 that suggests predominantly covalent bonding in this compound. Thus, it isn't surprising that $SnCl_4$ is a colorless liquid with a boiling point of 114°C.

(c) There is a significant difference between the electronegativities of calcium (1.03) and sulfur (2.59). The value of ΔEN is 1.56, and the value of $\overline{EN}$ for this compound is 1.81. These results suggest that the bonding in CaS is primarily ionic.

(d) The electronegativity of indium (1.66) is close to that of sodium (0.87). Because the average electronegativity is 1.27 and the difference between the electronegativity of these elements is only 0.79, Figure 5.11 suggests that the bond in this compound would be described as metallic.

● ●

5.15   Properties of Metallic, Covalent, and Ionic Compounds

The bond-type triangle allows the classification of compounds according to the way the compounds share electrons in a bond. The classification of bonding type is important because the properties of the compound depend on the degree of metallic, covalent, or ionic character in the bond between the atoms. A compound that is primarily ionic has a high melting point and will not conduct electricity as a solid but will conduct in the molten state or when dissolved in water. A covalent compound has a low melting point and, like a solid ionic compound, is an insulator. Metallic compounds have a range of melting points and conduct heat and electricity in the solid state.

By knowing the properties associated with a particular bond type, it is possible to select those atoms whose combination may produce a material having certain desired characteristics. Thus if we are interested in finding a substance that will conduct electricity as a solid, we would seek a material containing metallic bonds. If an insulator is desired, a compound located in the ionic region of the bond-type triangle would be appropriate. A low-melting insulator could be prepared from atoms whose ΔEN and $(\overline{EN})$ place it in the lower-right region of the triangle. Semiconductors are found in the lower-middle portion of the bond-type triangle in the area corresponding to the crossover from metallic to covalent bonding.

➤ **CHECKPOINT**

Classify the following compounds as ionic, covalent, or metallic. Which would be the best compound for an insulator that would stand up to high heat? $BaSi_2$, $BaBr_2$, $GaBr_3$.

5.16   Oxidation Numbers

The bond-type triangle helps us understand why it is so difficult to decide whether to talk about ZnS as if it contained Zn^{2+} and S^{2-} ions or polar covalent bonds in which zinc and sulfur atoms share a pair of electrons almost, but not quite, equally. Because it is sometimes difficult to distinguish between ionic and covalent compounds, chemists have developed the concept of **oxidation number** or **oxidation state.** Oxidation numbers treat all compounds as if they were ionic.

It doesn't matter whether the compound actually contains ions. The oxidation number is the charge an atom would have *if the compound were ionic.* Both the strontium atom in SrF_2 and the carbon atom in CO, for example, are assigned oxidation numbers of $+2$ or are described as being in the $+2$ oxidation state, even though one of the compounds is predominantly ionic and the other is predominantly covalent. The titanium atom in $TiCl_4$ is assigned an oxidation number of $+4$, even though the physical properties of this compound are consistent with what one would expect to be observed in a covalent compound. As we will see, oxidation numbers are nothing more than a bookkeeping method for keeping track of the flow of electrons in a chemical reaction.

For the active metals in Groups IA and IIA, oxidation numbers give a good description of the charge that the metal has in its compounds. The main-group metals in Groups IIIA and IVA, however, form compounds that have a significant amount of covalent character. Although we assign an oxidation number of $+3$ to aluminum and -1 to bromine, it is misleading to assume that aluminum bromide contains Al^{3+} and Br^- ions. It actually exists as Al_2Br_6 molecules.

The problem becomes even more severe when we turn to the chemistry of the transition metals. MnO, for example, is ionic enough to be considered a salt that contains Mn^{2+} and O^{2-} ions. Mn_2O_7, on the other hand, is a covalent compound that boils at room temperature. It is therefore more useful to think about Mn_2O_7 as if it contained manganese in a $+7$ oxidation state, not Mn^{7+} ions.

For many years, chemists have used a set of general rules to determine the oxidation number of an element:

- The oxidation number is 0 in any neutral substance that contains atoms of only one element. Aluminum foil, iron metal, and the H_2, O_2, O_3, P_4, and S_8 molecules all contain atoms that have an oxidation number of zero.

- The oxidation number is equal to the charge on the ion for ions that contain only a single atom. The oxidation number of the Na^+ ion, for example, is $+1$, whereas the oxidation number for the Cl^- ion is -1.

- The oxidation number of hydrogen is $+1$ when it is combined with a *more* electronegative element. Hydrogen is therefore in the $+1$ oxidation state in CH_4, NH_3, H_2O, and HCl.

- The oxidation number of hydrogen is -1 when it is combined with a *less* electronegative element. Hydrogen is therefore in the -1 oxidation state in LiH, NaH, CaH_2, and $LiAlH_4$.

- The elements in Groups IA and IIA form compounds in which the metal atoms have oxidation numbers of $+1$ and $+2$, respectively.

- Oxygen usually has an oxidation number of -2. Exceptions include molecules and polyatomic ions that contain O—O bonds, such as O_2, O_3, H_2O_2, and the O_2^{2-} ion.

- Elements in Group VIIA have an oxidation number of -1 when the atom is bonded to a *less* electronegative element.

- The sum of the oxidation numbers of the atoms in a neutral substance is zero.

$$H_2O \ (2 \text{ hydrogen} \times +1) + (1 \text{ oxygen} \times -2) = 0$$

- The sum of the oxidation numbers in a polyatomic ion is equal to the charge on the ion.

$$OH^- (1 \text{ oxygen} \times -2) + (1 \text{ hydrogen} \times +1) = -1$$

- The least electronegative element in a compound is assigned a positive oxidation state. Sulfur carries a positive oxidation state in SO_2, for example, because it is less electronegative than oxygen.

$$SO_2 \ (1 \text{ sulfur} \times +4) + (2 \text{ oxygen} \times -2) = 0$$

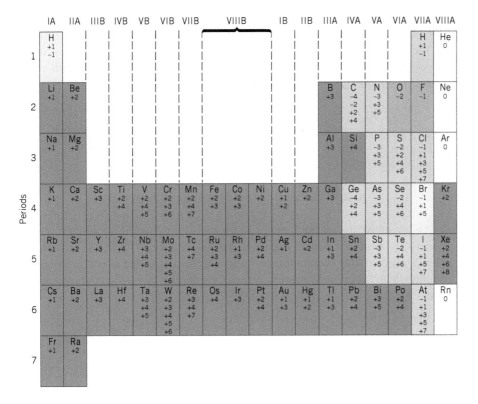

Fig. 5.12 Common oxidation numbers of the elements. Different shades are used to distinguish among elements that form only positive oxidation states, only negative oxidation states, and both positive and negative oxidation states.

Figure 5.12 shows the common oxidation numbers for many of the elements in the periodic table. There are several clear patterns in the data.

- Elements in the same group often have the same oxidation numbers.
- The largest, or most positive, oxidation number of an atom is often equal to the group number of the element. The largest oxidation number for phosphorus, in Group VA, for example, is +5.
- The smallest, or most negative, oxidation number of a nonmetal can often be found by subtracting 8 from the group number. The most negative oxidation number of phosphorus, for example, is $5 - 8 = -3$.

For compounds that contain transition metal ions, such as the Fe^{2+} and Fe^{3+} ions, these rules provide the easiest way of determining oxidation numbers. In the next section, an alternate method of determining oxidation numbers is introduced.

Exercise 5.6

Assign oxidation numbers to the atoms in the following compounds.

(a) Aluminum oxide or alumina (Al_2O_3), which is used in pigments, ceramics, and abrasives.

(b) Xenon tetrafluoride (XeF_4), one of the first rare gas compounds.

(c) Potassium dichromate ($K_2Cr_2O_7$), which is one of the principal components in the Breathalyzer test used to determine whether someone has been drinking alcohol.

Solution

(a) The sum of the oxidation numbers in Al_2O_3 must be zero because the compound is neutral. If we assume that oxygen is in the -2 oxidation state, the oxidation state of the aluminum must be $+3$.

$$Al_2O_3 \text{ (2 aluminum} \times +3) + (3 \text{ oxygen} \times -2) = 0$$

(b) Because the oxidation number of the fluorine atom is -1, the xenon atom must be present in the $+4$ oxidation state.

$$XeF_4 \text{ (1 xenon} \times +4) + (4 \text{ fluorine} \times -1) = 0$$

(c) Assigning oxidation numbers in $K_2Cr_2O_7$ is simplified if we recognize that it is an ionic compound that contains K^+ and $Cr_2O_7^{2-}$ ions. The oxidation state of the potassium is $+1$ in the K^+ ion. Because the oxidation number of oxygen is usually -2, the oxidation state of the chromium in the $Cr_2O_7^{2-}$ ion must be $+6$.

$$Cr_2O_7^{2-} \text{ (2 chromium} \times +6) + (7 \text{ oxygen} \times -2) = -2$$

Exercise 5.7

Arrange the following compounds in order of increasing oxidation state for the carbon atom.

(a) CO, carbon monoxide

(b) CO_2, carbon dioxide

(c) H_2CO, formaldehyde

(d) CH_3OH, methanol

(e) CH_4, methane

Solution

$$CH_4 < CH_3OH < H_2CO < CO < CO_2$$
$$-4 -2 0 +2 +4$$

5.17 Calculating Oxidation Numbers

Oxidation numbers for atoms in a compound can be calculated by starting with the following equation, which was used to calculate partial charge in Chapter 4.

$$\delta_a = V_a - N_a - B_a\left(\frac{EN_a}{EN_a + EN_b}\right)$$

The symbols in this equation represent the following quantities.

- δ_a is the partial charge on atom a.
- V_a is the number of valence electrons on a neutral atom of element a.
- N_a is the number of nonbonding electrons on atom a.
- B_a is the number of bonding electrons on the atom.
- EN is the electronegativity of the atom, and the symbol b is used to represent the atom to which a is bonded.

This equation can be transformed into the equation used to calculate formal charge in Chapter 4 by assuming that the electrons in a covalent bond are shared equally by the atoms that form the bond. In other words, the atoms were treated as if $EN_a = EN_b$. When calculating B_a within the context of a formal charge calculation, the bonding electrons are divided equally between the atoms that form covalent bonds. Since EN_a and EN_b are assumed to be equal, the value of $EN_a/(EN_a + EN_b)$ in a formal charge calculation is equal to ½. This gives us the following equation for formal charge.

$$FC_a = V_a - N_a - \frac{B_a}{2}$$

Let's now think about the relationship between partial charge, formal charge and oxidation state.

- When we calculate the *partial charge* on each atom in a bond, we try to estimate the fraction of the electrons in the bond that should be assigned to each atom.
- When we calculate *formal charge,* we divide the bonding electrons equally between the atoms that form covalent bonds.
- When we calculate *oxidation numbers,* we treat each bond as if it were an ionic bond in which the electrons in the bond are transferred to the more electronegative element.

In other words, when we calculate oxidation numbers we treat the atoms as if $EN_a \gg EN_b$. If the electronegativity of atom a is much larger than the electronegativity of the other element in the bond (atom b), the value of $EN_a/(EN_a + EN_b)$ becomes more or less equal to 1. After the electrons in a bond are transferred to the more electronegative atom, the only electrons left to be counted are nonbonding electrons. Thus the calculation of oxidation number is based on only two pieces of information from the partial charge equation: the number of valence electrons on a neutral atom of the element (V) and the number of electrons assigned to the atom in the Lewis structure of the molecule (N) after the electrons in a bond are assigned to the more electronegative atom. The oxidation number can therefore be calculated with the following simplified equation.

$$OX = V - N$$

The relationship among the three ways of describing the charge associated with atoms within a compound is summarized below. It should be remembered that partial charge is the best description of the actual distribution of electron density in a molecule. Formal charge and oxidation numbers are invented bookkeeping methods. Formal charge, which treats all bonds as if they were purely covalent, is particularly useful in assessing Lewis structures. Oxidation numbers, which treat all bonds as purely ionic, are useful in describing a class of reactions called oxidation–reduction reactions.

$$\delta_a = V_a - N_a - B_a \left(\frac{EN_a}{EN_a + EN_b} \right)$$

$$EN_a = EN_b \qquad\qquad EN_a \gg EN_b$$
$$FC_a = V_a - N_a - (B_a/2) \qquad OX_a = V_a - N_a$$

As we have seen, oxidation numbers can be assigned to relatively simple compounds on the basis of the rules outlined in the previous section. The equation we've derived for calculating oxidation numbers is particularly useful for molecules in which an element has more than one oxidation state, such as ethanol—the "alcohol" in alcoholic beverages. Ethanol has the following Lewis structure.

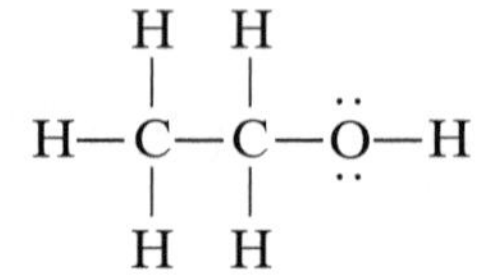

$$\begin{array}{c} \text{H} \quad \text{H} \\ | \quad | \quad \; .. \\ \text{H--C--C--O--H} \\ | \quad | \quad \; .. \\ \text{H} \quad \text{H} \end{array}$$

We start the calculation of oxidation states in this molecule by recognizing that O ($EN = 3.61$) is more electronegative than C ($EN = 2.54$), which, in turn, is more electronegative than H ($EN = 2.30$). We therefore assign the electrons in each covalent bond to the more electronegative atom. When a covalent bond exists between a pair of atoms of the same element, this calculation assumes that the electrons in the bond are divided equally between the atoms. In the case of the bond between the two carbon atoms, one electron is assigned to each carbon since both atoms have the same electronegativity

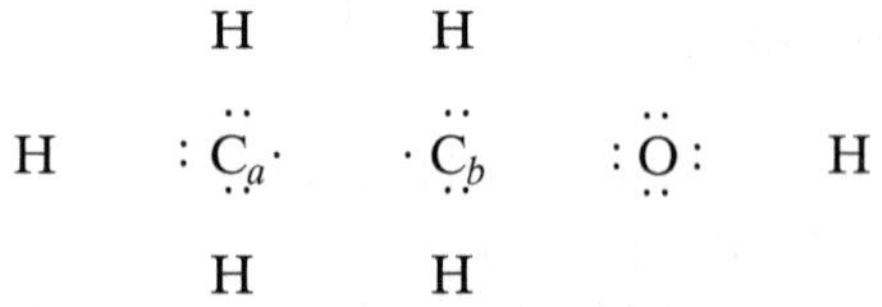

$$\begin{array}{ccccc} & \text{H} & & \text{H} & \\ & .. & & .. & .. \\ \text{H} & : C_a \cdot & \cdot C_b & : O : & \text{H} \\ & .. & & .. & .. \\ & \text{H} & & \text{H} & \end{array}$$

Oxygen is in Group VIA of the periodic table, which means that a neutral atom has six valence electrons. Oxygen is assigned eight electrons in the above structure. The oxidation number of the oxygen atom is therefore -2.

$$OX_{\text{oxygen}} = V - N = 6 - 8 = -2$$

Hydrogen, which is in Group IA, has no valence electrons once the electrons in the bonds are assigned to the more electronegative oxygen or carbon atoms. The oxidation number of the hydrogen atom is therefore $+1$.

$$OX_{\text{hydrogen}} = V - N = 1 - 0 = +1$$

The two carbon atoms in ethanol have different oxidation numbers. Carbon is in Group IVA and has four valence electrons. C_a is assigned seven electrons in the above structure, giving it an oxidation number of -3.

$$OX_{\text{carbon } a} = V - N = 4 - 7 = -3$$

C_b has only five electrons assigned to it in the above structure. It is therefore assigned an oxidation number of -1.

$$OX_{\text{carbon } b} = V - N = 4 - 5 = -1$$

Note that the sum of the oxidation numbers for all of the nine atoms in the CH_3CH_2OH molecule is zero, which is consistent with the fact that there is no charge on this molecule.

Exercise 5.8

Acetic acid, $C_2H_4O_2$, is the compound that gives vinegar its sour taste. Use the method described in this section to determine the oxidation numbers for all atoms in the Lewis structure of acetic acid shown here.

$$
\begin{array}{c}
\quad\quad H \quad\quad\quad \ddot{O}: \\
\quad\quad | \quad\quad\quad\; \nparallel \\
H-C-C \\
\quad\quad | \quad\quad\quad\; \diagdown \\
\quad\quad H \quad\quad\quad \ddot{O}-H
\end{array}
$$

Solution

The first step in the calculation involves assigning the electrons in each bond to the more electronegative element.

$$
\begin{array}{ccc}
& H & & & \ddot{O}\cdot \\
H & :\ddot{C_a}\cdot & \cdot C_b & \\
& H & & \ddot{O}: & H
\end{array}
$$

Note that all four electrons in the carbon–oxygen double bond are assigned to oxygen. Both oxygen atoms therefore end up with eight nonbonding electrons. In the carbon–carbon bond, one electron is assigned to each carbon since they have equal electronegativities.

Oxygen	$OX = 6-8 = -2$
Hydrogen	$OX = 1-0 = +1$
Carbon$_a$	$OX = 4-7 = -3$
Carbon$_b$	$OX = 4-1 = +3$

The sum of the oxidation numbers for all seven atoms is zero. If the molecule had a charge, the sum of the oxidation numbers would be equal to that charge.

> ➤ **CHECKPOINT**
>
> The partial charge gives the best description of the charge on atoms in a compound. What do formal charge and oxidation numbers represent?

5.18 Oxidation–Reduction Reactions

The principal use of oxidation numbers is to help us understand **oxidation–reduction reactions,** which are both common and important. Consider the reaction that occurs when magnesium metal burns in the presence of oxygen.

$$2\,Mg(s) + O_2(g) \longrightarrow 2\,MgO(s)$$

The term *oxidation* was originally used to describe reactions, such as this, in which an element combines with oxygen.

$$2\,Mg(s) + O_2(g) \longrightarrow 2\,MgO(s)$$

oxidation

The term **reduction** comes from the Latin stem meaning "to lead back." Anything that leads back to magnesium metal therefore involves reduction. The reaction between magnesium oxide and carbon at 2000°C to form magnesium metal and carbon monoxide is an example of the reduction of magnesium oxide to magnesium metal.

$$MgO(s) + C(s) \longrightarrow Mg(s) + CO(g)$$

reduction

The oxidation of magnesium metal to form magnesium oxide can be described in terms of the transfer of electrons from magnesium to oxygen. From this perspective, the reaction between magnesium and oxygen would be written as follows.

$$2\,Mg + O_2 \longrightarrow 2\,[Mg]^{2+}[O]^{2-}$$

In the course of the reaction, each magnesium atom loses two electrons to form an Mg^{2+} ion.

$$Mg \longrightarrow Mg^{2+} + 2\,e^-$$

Each O_2 molecule, on the other hand, gains four electrons to form a pair of O^{2-} ions.

$$O_2 + 4\,e^- \longrightarrow 2\,O^{2-}$$

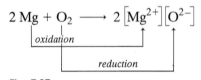

$$2\,Mg + O_2 \longrightarrow 2\left[Mg^{2+}\right]\left[O^{2-}\right]$$

oxidation

reduction

Fig. 5.13 Oxidation cannot occur in the absence of reduction.

Because electrons are neither created nor destroyed in a chemical reaction, oxidation and reduction are linked. The four electrons gained by each oxygen molecule require that four electrons be lost by magnesium, causing two Mg atoms to form two Mg^{2+} ions. The number of electrons gained during reduction must equal the number of electrons lost during oxidation. It is impossible to have oxidation without the reduction, as shown in Figure 5.13.

As their understanding of chemical reactions improved, chemists recognized that oxidation–reduction reactions don't always involve the transfer of electrons. They can also occur by the transfer of atoms. Consider the following reaction, for example, which serves as the basis for the multistep synthesis of gasoline from coal—a process known as coal liquefaction.

$$CO(g) + H_2O(g) \longrightarrow CO_2(g) + H_2(g)$$

The total number of electrons in the valence shell of each atom remains the same in the reaction.

$$:C\equiv O: + H-\overset{..}{\underset{..}{O}}-H \longrightarrow :\overset{..}{O}=C=\overset{..}{O}: + H-H$$

What changes in the reaction is the oxidation state of the atoms. The oxidation state of carbon increases from $+2$ to $+4$, while the oxidation state of the hydrogen decreases from $+1$ to 0.

$$CO + H_2O \longrightarrow CO_2 + H_2$$

$$+2+1+40$$

Oxidation and reduction are therefore best defined as follows.

> *Oxidation* **occurs when the oxidation number of an atom becomes more positive.**
> *Reduction* **occurs when the oxidation number of an atom becomes more negative.**

Thus, in the reaction between CO and H_2O the CO is said to be oxidized because the oxidation state of the carbon atom increases. The oxidation state of the hydrogen atoms in water decreases, however, and water is therefore said to be reduced.

➤ **CHECKPOINT**

Which of the following reactions are oxidation–reduction reactions?

$$H_3PO_4(aq) + NH_3(aq)$$
$$\longrightarrow (NH_4)H_2PO_4(aq)$$
$$4\,NH_3(g) + 5\,O_2(g)$$
$$\longrightarrow 4\,NO(g) + 6\,H_2O(g)$$
$$CH_3OH(l) + CO(g)$$
$$\longrightarrow \quad \underset{\underset{CH_3C-OH(l)}{\|}}{O}$$

5.19 Nomenclature

Long before chemists knew the formulas for chemical compounds, they developed a system of **nomenclature** (from the Latin words *nomen* meaning "name," and *calare* meaning "to call") that gave each compound a unique name. Today we often use chemical formulas, such as NaCl, $C_{12}H_{22}O_{11}$, and Co $(NO)_6(ClO_4)_3$, to describe chemical compounds. But we still need unique names that unambiguously identify each compound.

COMMON NAMES

Some compounds have been known for so long that a systematic nomenclature cannot compete with well-established common names. Examples of compounds for which common names are used include water (H_2O), ammonia (NH_3), and methane (CH_4).

NAMING IONIC COMPOUNDS, OR SALTS

The names of ionic compounds are written by listing the name of the positive ion followed by the name of the negative ion.

NaCl	sodium chloride
$(NH_4)_2SO_4$	ammonium sulfate
Fe_2O_3	iron(III) oxide
$NaHCO_3$	sodium hydrogen carbonate (sodium bicarbonate)
$Al(ClO_4)_3$	aluminum perchlorate

We therefore need a series of rules that allow us to unambiguously name the positive and negative ions before we can name ionic compounds.

NAMING POSITIVE IONS

Positive ions that consist of a single atom carry the name of the element from which they are formed.

Na^+	sodium	Zn^{2+}	zinc
Ca^{2+}	calcium	H^+	hydrogen
K^+	potassium	Sr^{2+}	strontium

As we have seen, some metals—particularly the transition metals—form positive ions in more than one oxidation state. One of the earliest methods of distinguishing between these ions used the suffixes *-ic* and *-ous* added to the Latin name of the element to represent the higher and lower oxidation states, respectively.

Fe^{2+} ferrous Fe^{3+} ferric
Sn^{2+} stannous Sn^{4+} stannic
Cu^{+} cuprous Cu^{2+} cupric

Chemists now use a simpler method, in which the charge on the ion is indicated by a Roman numeral in parentheses immediately after the name of the element.

Fe^{2+} iron(II) Fe^{3+} iron(III)
Sn^{2+} tin(II) Sn^{4+} tin(IV)
Cu^{+} copper(I) Cu^{2+} copper(II)

There are only a limited number of polyatomic positive ions. These ions often have common names ending with the suffix *-onium.*

H_3O^{+} hydronium
NH_4^{+} ammonium

NAMING NEGATIVE IONS

Negative ions that consist of a single atom are named by adding the suffix *-ide* to the stem of the name of the element.

F^{-} fluoride O^{2-} oxide
Cl^{-} chloride S^{2-} sulfide
H^{-} hydride P^{3-} phosphide

RULES FOR NAMING POLYATOMIC NEGATIVE IONS

The names of the common polyatomic negative ions are given in Table 1.6. At first glance, the nomenclature of these ions seems hopeless. There are several rules, however, that can bring some order to this apparent chaos:

- The name of the ion usually ends in *-ite* or *-ate.*
- The *-ite* ending indicates a low oxidation state. Thus the NO_2^{-} ion is the nitrite ion.
- The *-ate* ending indicates a high oxidation state. Thus the NO_3^{-} ion is the nitrate ion.
- When more than two polyatomic ions exist with the same central atom, the prefix *per-* (as in *hyper-*) and the prefix *hypo-* are used to indicate the very largest and very smallest number of oxygens.

Consider the ClO^{-}, ClO_2^{-}, ClO_3^{-}, and ClO_4^{-} ions, for example. The lowest oxidation state for the chlorine atom is in ClO^{-}, which is known as the hypochlorite ion. ClO_2^{-} and ClO_3^{-} are the chlorite and chlorate ions, respectively. The ion with the largest oxidation state for the chlorine atom is ClO_4^{-}, which is the perchlorate ion.

There are a handful of exceptions to these generalizations. The names of the hydroxide (OH^-), cyanide (CN^-), and peroxide (O_2^{2-}) ions, for example, have the -ide ending because they were once thought to be monoatomic ions.

Exercise 5.9

Name the following ionic compounds.

(a) $FePO_4$, which is used in the automobile industry as an anticorrosion film that improves the way paint adheres to metal

(b) $SrCO_3$, which is used as an X-ray absorber in the glass faceplate of color television tubes

(c) $Ca(ClO)_2$, which is used in bleaching and sanitizing applications and is the solid bleach in Clorox 2

Solution

(a) Iron compounds contain either Fe^{2+} or Fe^{3+} ions. Since the phosphate ion has a -3 charge, $FePO_4$ could only be neutral if it contained the Fe^{3+} ion. This compound is therefore known as iron(III) phosphate.

(b) Because strontium always has a $+2$ charge in its compounds, $SrCO_3$ is known as strontium carbonate.

(c) $Ca(ClO)_2$ contains the Ca^{2+} and ClO^- ions. Because calcium only forms Ca^{2+} ions, the compound is known as calcium hypochlorite.

NAMING SIMPLE COVALENT COMPOUNDS

One of the most difficult tasks faced in naming a compound is deciding whether to name it as an ionic or covalent compound. A bond-type triangle can be used to predict the predominant type of bonding in binary compounds. As a general rule, compounds formed by combining two or more nonmetals are classified as covalent compounds.

Chemists usually write formulas in which the least electronegative element is written first, followed by the more electronegative element(s). Covalent compounds are usually named by listing the name of the less electronegative element and then adding the suffix -ide to the stem of the name of the more electronegative atom. Consider HCl, for example.

   HCl   hydrogen chloride

The number of atoms of an element in simple covalent compounds is indicated by adding one of the following Greek prefixes to the name of the element.

1	mono-	6	hexa-
2	di-	7	hepta-
3	tri-	8	octa-
4	tetra-	9	nona-
5	penta-	10	deca-

The prefix mono- is seldom used because it is redundant. The principal exception to this rule is carbon monoxide (CO).

Exercise 5.10

Name the following compounds.

(a) NO_2, which is a component of a series of reactions responsible for Los Angeles smog

(b) SF_4, which is a highly reactive colorless gas

Solution

(a) Nitrogen dioxide

(b) Sulfur tetrafluoride

NAMING ACIDS

Some covalent compounds that contain hydrogen, such as HCl, HBr, and HCN, dissolve in water to produce acids. These solutions are named by replacing the hydrogen in the name of the compound with the prefix *hydro-* and then replacing the suffix *-ide* with *-ic*. For example, hydrogen chloride (HCl) dissolves in water to form hydrochloric acid, hydrogen bromide (HBr) forms hydrobromic acid, and hydrogen cyanide (HCN) forms hydrocyanic acid.

Many of the oxygen-rich polyatomic negative ions in Table 1.4 form acids that are named by replacing the suffix *-ate* with *-ic* and the suffix *-ite* with *-ous*.

$CH_3CO_2^-$	acetate	CH_3CO_2H	acetic acid
CO_3^{2-}	carbonate	H_2CO_3	carbonic acid
BO_3^{3-}	borate	H_3BO_3	boric acid
NO_3^-	nitrate	HNO_3	nitric acid
NO_2^-	nitrite	HNO_2	nitrous acid
SO_4^{2-}	sulfate	H_2SO_4	sulfuric acid
SO_3^{2-}	sulfite	H_2SO_3	sulfurous acid
ClO_4^-	perchlorate	$HClO_4$	perchloric acid
ClO_3^-	chlorate	$HClO_3$	chloric acid
ClO_2^-	chlorite	$HClO_2$	chlorous acid
ClO^-	hypochlorite	$HClO$	hypochlorous acid
PO_4^{3-}	phosphate	H_3PO_4	phosphoric acid
MnO_4^-	permanganate	$HMnO_4$	permanganic acid
CrO_4^{2-}	chromate	H_2CrO_4	chromic acid

Salts that contain anions with acidic hydrogen atoms can be named by indicating the presence of the acidic hydrogen as follows.

$NaHCO_3$	sodium hydrogen carbonate (also known as sodium bicarbonate)
$NaHSO_3$	sodium hydrogen sulfite (also known as sodium bisulfite)
KH_2PO_4	potassium dihydrogen phosphate

 Exercise 5.11

Name the following compounds.

(a) $NaClO_3$

(b) $Al_2(SO_4)_3$

(c) P_4S_3

(d) SCl_4

Solution

The key to naming these compounds is recognizing that the first two are primarily ionic compounds and the last two are predominantly covalent compounds.

(a) sodium chlorate

(b) aluminum sulfate

(c) tetraphosphorus trisulfide

(d) Sulfur tetrachloride

• •

Key Terms

Active metal	Halide	Oxidation–reduction reaction
Alkali metal	Halogen	Oxidation state
Alkaline earth metal	Hydride	Oxide
Binary compound	Iodide	Peroxide
Body-centered cubic	Ionic bond	Polar covalent compound
Bond-type triangle	Ionic compound	Reactive
Bromide	Isoelectronic	Reduction
Chloride	Lattice point	Salt
Conduction band	Localized	Semiconductor
Coulombic	Main-group element	Simple cubic
Covalent	Metallic bond	Sulfide
Delocalized	Nitride	Superoxide
Electrostatic	Nomenclature	Transition metal
Face-centered cubic	Oxidation number	Unit cell
Flouride		

Problems

The Active Metals

1. List the elements in the third row of the periodic table in order of decreasing metallic character. List the elements in order of increasing nonmetallic character. Explain the similarities between the trends. Identify each element as either a metal, nonmetal, or semimetal.

2. List the elements in Group VA of the periodic table in terms of increasing metallic character. Identify each element in the group as a metal, nonmetal, or semimetal.

3. Which of the following sets of elements is arranged in order of increasing nonmetallic character?

    (a) $Sr < Al < Ga < N$ (b) $K < Mg < Rb < Si$

    (c) $Ge < P < As < N$ (d) $Al < B < N < F$

4. What do we mean when we say that sodium is more metallic than lithium?

5. For each of the following pairs of metals, which is more active?

    (a) Mg or Ca   (b) Na or Mg

    (c) K or Mg   (d) Mg or Al

6. Which metal in each of the following pairs would you expect to react more rapidly with water?

    (a) Na or K   (b) Na or Mg

    (c) Mg or Ca   (d) Ca or Al

Main-Group Metals and Their Ions

Group IA: The Alkali Metals

7. Why is the chemistry of the alkali metals dominated by singly charged positive ions?

8. Give the electron configurations for each of the $+1$ ions of the alkali metals.

9. The Cs^+ ion is isoelectronic with what electrically neutral atom?

10. Which is more difficult to remove, an electron from K^+ or Ar? Explain.

11. How do the AVEE and the first ionization energy vary in Group IA from Li through Fr? Explain.

12. How does the size of the alkali metals change from Li to Fr? Explain.

13. How would you expect the second ionization energy of Li to compare to its first ionization energy? Explain.

Group IIA: The Alkaline Earth Metals

14. Why is the chemistry of the alkaline earth metals dominated by doubly charged positive ions?

15. Give the electron configurations for the +2 ions of the alkaline earth metals.

16. The Ca^{2+} ion is isoelectronic with what electrically neutral atom?

17. Which is more difficult to remove, an electron from Mg or Na?

18. How do the AVEE and second ionization energy vary in Group IIA from Be through Ra?

19. How does the size of the alkaline earth metals vary from Be to Ra? Explain.

20. How would you expect the third ionization energy of Ba to compare to the second ionization energy? Explain.

Group IIIA Metals

21. Which of the Group IIIA metals is the most abundant?

22. Explain why Al forms the +3 ion in most of its compounds.

23. Give the trend in AVEE and first ionization energy for B and Al in Group IIIA.

24. What is the trend in size of the atoms of the elements in Group IIIA from B through Tl? Explain.

25. Give the electron configuration for Ga, Ga^+, Ga^{2+}, and Ga^{3+}. How would you expect the fourth ionization energy for Ga to compare to the third? Explain.

26. What is the most likely charge on the positive ion formed by B?

Main-Group Nonmetals and Their Ions

27. Calculate the charge on the negative ions formed by the following elements.
    (a) As   (b) Te   (c) Se

28. Calculate the charge on the negative ions formed by the following elements.
    (a) C   (b) P   (c) S   (d) I

29. Describe the difference between the hydrogen atoms in metal hydrides such as LiH and nonmetal hydrides such as CH_4 and H_2O.

30. Write the formulas for the fluoride, hydride, sulfide, nitride, and oxide of barium.

31. Why do the halides tend to form ions with a -1 charge?

32. In what elemental form do the halogens occur in nature?

33. Why is hydrogen listed in two groups in the periodic table?

Transition Metals and Their Ions

34. Where are the transition metals located in the periodic table?

35. Give the electron configuration for Zn. What is the most likely charge on the zinc ion? Explain.

36. Give the electron configuration for Cr. What is the charge on the chromium ion that is most likely to form? Explain.

37. Give the electron configurations for Ni and Ni^{2+}.

38. Cobalt commonly forms two ions. What would you expect the charges on these ions to be? Explain.

Predicting the Formulas of Ionic Compounds

39. Fluoride toothpastes convert the mineral apatite in tooth enamel into fluorapatite, $Ca_5(PO_4)_3F$. If fluorapatite contains Ca^{2+} and PO_4^{3-} ions, what is the charge on the fluoride ion in this compound?

40. Verdigris is a green pigment used in paint. The simplest formula for the compound is $Cu_3(OH)_2(CH_3CO_2)_4$. What is the charge on the copper ions in this compound if the other polyatomic ions both carry a charge of -1?

41. Predict the formulas for neutral compounds containing the following pairs of ions.
    (a) Mg^{2+} and NO_3^- (b) Fe^{3+} and SO_4^{2-}
    (c) Na^+ and CO_3^{2-}

42. Predict the formulas for neutral compounds containing the following pairs of ions.
    (a) Na^+ and O_2^{2-} (b) Zn^{2+} and PO_4^{3-}
    (c) K^+ and $PtCl_6^{2-}$

43. Predict the formulas for sodium nitride and aluminum nitride if the formula for magnesium nitride is Mg_3N_2.

44. Compounds that contain the O^{2-} ion are called oxides. Those that contain the O_2^{2-} ion are called peroxides. If the formula for potassium oxide is K_2O, what is the formula for potassium peroxide?

45. What is the value of x in the $[Co(NO_2)_x]^{3-}$ ion if this complex ion contains Co^{3+} and NO_2^- ions?

46. Magnetic iron oxide has the formula Fe_3O_4. Explain this formula by assuming that Fe_3O_4 contains both Fe^{2+} and Fe^{3+} ions combined with O^{2-} ions.

Predicting the Products of Reactions That Produce Ionic Compounds

47. Which one of the following elements is most likely to form an oxide with the formula XO and also a hydride with the formula XH_2?
    (a) Na   (b) Mg   (c) Al   (d) Si   (e) P

48. A main-group metal reacts with hydrogen and oxygen to form compounds with the formulas XH_4 and XO_2. In which group of the periodic table does the element belong?

49. In which column of the periodic table do we find main-group metals that form sulfides with the formula M_2S_3 and react with acid to form M^{3+} ions and H_2 gas?

50. Explain why sodium reacts with chlorine to form $NaCl$, not $NaCl_2$ or $NaCl_3$.

51. Explain why magnesium reacts with chlorine to form $MgCl_2$, not $MgCl_3$.

52. Describe why lithium reacts with nitrogen to give Li_3N, not a compound with another formula.

53. Predict the ionic product of the reaction between aluminum and nitrogen.

54. Predict the ionic product of the reaction between strontium metal and phosphorus.

55. The light meters in automatic cameras are based on the sensitivity of gallium arsenide to light. Predict the formula of gallium arsenide.

56. Write balanced equations for the reactions of sodium metal with each of the following elements.

   (a) F_2 (b) O_2 (c) H_2
   (d) S_8 (e) P_4

57. Write balanced equations for the reactions of calcium metal with each of the following elements.

   (a) H_2 (b) O_2 (c) S_8
   (d) F_2 (e) N_2 (f) P_4

58. Predict the ionic product of the reactions of F_2 with each of the following metals.

   (a) Zn (b) Al (c) Sn
   (d) Mg (e) Bi

Oxides, Peroxides, and Superoxides

59. What is the difference between an oxide and a superoxide?

60. Write the formulas for the oxide and peroxide of Cs.

61. Give the Lewis dot structures for Rb, Rb^+, and O^{2-}. Use the dot structures to predict the formulas of the superoxide and oxide of rubidium.

62. What is the formula of the oxide of beryllium?

The Ionic Bond *Coulombic force*

63. NaCl is classified as an ionic compound. What forces act to hold the atoms of this compound together?

64. Use Lewis dot structures for Na and Cl to write the Lewis structure for NaCl.

65. Write Lewis structures for $BaCl_2$, Na_2S, and CaS.

66. What is one fundamental difference between a covalent and an ionic compound?

67. The attractive force, F, that exists between two oppositely charged ions can be expressed as

$$F = \frac{q_1 \times q_2}{r^2}$$

In an ionic compound the q refers to the charges on the ions and r is the distance between them.

(a) What factors does the attractive force depend on?

(b) For a fixed distance, is the force of attraction greater for two ions of charge $+1$ and -1 or for two ions of charge $+2$ and -2?

(c) If $q+$ is $+1$ and $q-$ is -1, is the force of attraction greater at a smaller distance or at a larger distance between the ions? Explain using the relationship given above.

(d) Which ionic compound would have the strongest attractive forces, NaCl or KCl? Refer to Tables 3.7 and 3.8. Explain.

(e) Refer to Appendix B, Table B.4. In which ionic compound, Li_2O or CaO, is the force of attraction between the positive and negative ions the largest? Explain.

Structures of Ionic Compounds

68. Describe the difference between simple cubic, body-centered cubic, and face-centered cubic unit cells.

69. Ca metal forms a face-centered cubic closest-packed unit cell. Roughly sketch this unit cell.

70. Sodium hydride crystallizes in a face-centered cubic unit cell of H^- ions with Na^+ ions at the center of the unit cell and in the center of each edge of the unit cell. How many Na^+ ions does each H^- ion touch? How many H^- ions does each Na^+ ion touch?

Metallic Bonds

71. What similarities do covalent bonds and ionic bonds share?

72. What is the difference between localized and delocalized electrons?

73. Define the term *metallic bond*.

74. What are the characteristics of atoms that tend to enter into metallic bonding?

75. Two characteristics of an atom are reflected in the AVEE that favor metallic bonding. What are they?

76. What are the differences between a metallic bond, an ionic bond, and a covalent bond?

The Relationship among Ionic, Covalent, and Metallic Bonds

77. Explain why chemists believe that ionic and covalent are the two extremes of a continuum of differences in the bonding between two atoms in which the atoms in the bond are localized between these atoms.

78. Which of the following elements forms bonds with fluorine that are the most covalent?

   (a) P (b) Ca (c) Al (d) O (e) Se

79. Which one of the following elements forms the most covalent bond with oxygen?

   (a) Sr (b) In (c) Sb (d) Te (e) Se

80. Which of the following compounds are best described as ionic?

    (a) CO   (b) H_2O (c) BeF_2 (d) $MgBr_2$
    (e) AlI_3 (f) ZnS   (g) CdLi

81. Which of the following compounds are best described as metallic?

    (a) CaH_2 (b) BrF_3 (c) NF_3
    (d) $SiCl_4$ (e) AsH_3 (f) $MgZn_2$

Answer problems 82–85—using the difference in electronegativity, ΔEN, only.

82. Which of the following compounds should be ionic?

    (a) ZnS   (b) $AlCl_3$ (c) SnF_2 (d) BH_3 (e) H_2S

83. Which of the following compounds should be covalent?

    (a) CH_4 (b) CO_2 (c) $SrCl_2$ (d) NaH   (e) SF_4

84. Which of the following pairs of elements should combine to give ionic compounds?

    (a) $Mg + O_2$ (b) $S_8 + F_2$
    (c) $Na + Hg$ (d) $K + I_2$

85. Which of the following pairs of elements should combine to give metallic compounds?

    (a) $N_2 + O_2$ (b) $Cl_2 + F_2$ (c) $Cl_2 + Cs$
    (d) $S_8 + Na$ (e) $Cu + Sn$

Bond-Type Triangles

86. Predict whether the following compounds are ionic or covalent by using both the general rule that metals combine with nonmetals to form ionic compounds and a bond-type triangle.

    (a) OF_2 (b) S_2 (c) MgO   (d) ZnS

87. Predict whether the following compounds are ionic or covalent by using both the general rule that metals combine with nonmetals to form ionic compounds and a bond-type triangle.

    (a) IF_3 (b) $SiCl_4$ (c) BF_3 (d) Na_2S

88. Use Figure 5.11 to determine whether there is an electronegativity difference, ΔEN, above which a compound can always be predicted to be ionic.

89. Use Figure 5.11 to determine whether there is an electronegativity difference, ΔEN, below which a binary compound can always be classified as covalent.

90. For each of the following characteristics, select three binary compounds from Figure 5.11 that exhibit the characteristic.

    (a) ΔEN is relatively the same for the three compounds, but the bonding type changes from metallic to semimetallic to covalent.

    (b) ΔEN ranges from about 3 to about 1, but all compounds are ionic.

    (c) The average EN of the atoms in the compounds stays the same, but the ionic character increases.

    (d) ΔEN of the atoms in the compounds stays the same, but the covalent character increases.

    (e) ΔEN is about the same, but the bonding type changes from metallic to ionic to covalent.

91. Classify the following compounds according to bonding type.

    (a) HgS   (b) GaSb   (c) Li_3N
    (d) NaBr   (e) $SnBr_4$ (f) Na_3P
    (g) InP   (h) InN   (i) TeO_2

Properties of Metallic, Covalent, and Ionic Compounds

92. For each of the following properties, list a binary solid compound that exhibits the property.

    (a) Conducts electricity in the solid state

    (b) Is an insulator

    (c) Is a semiconductor

    (d) Has a very high melting point

    (e) Has a low melting point

    (f) Electrons are shared unequally between the atoms composing the compound

93. The French have recently reported a new compound composed of Mn and Al. Is this compound a conductor, an insulator, or a semiconductor? Explain.

94. Select two compounds from Figure 5.11 that have about the same ΔEN, of which one is a conductor of electricity and one is an insulator.

95. (a) Describe the bonding in the following compounds as ionic, metallic, or covalent: BBr_3, Li_3P, SrF_2, $SrZn_5$, Cd_3N_2.

    (b) From the above compounds, pick the ones that satisfy the following criteria:

    (i) High melting and boiling points; dissolves in water to give solutions that conduct electricity

    (ii) Low melting and boiling points; localized electrons

    (iii) Low melting point and high boiling point; both solid and liquid conduct electricity

96. Consider the following three compounds: P_4O_6, Mg_3N_2, Mg_3Sb_2.

    (a) One of these compounds melts at 24°C, and the liquid phase does not conduct a electric current.

    (b) One melts at 961°C, and the solid phase does conduct an electric current.

    (c) One melts at 800°C and when melted conducts an electric current.

    Identify each compound and show how you arrived at your conclusions.

97. Which of the following compounds—Mg_3N_2, $BaSi_2$, B_2O_3—

    (a) are insulators?

(b) has a very high melting point but does not conduct electricity in the solid state?

(c) conducts electricity in the solid state?

Explain your answers and show all calculations.

Oxidation Numbers

98. An area of active research interest in recent years has involved ions such as $Re_2Cl_8^{2-}$, $Cr_2Cl_9^{3-}$, and $Mo_2Cl_8^{4-}$ that contain bonds between metal atoms. Calculate the oxidation number of the metal atom in each compound.

99. Determine the oxidation state of phosphorus in the following compounds or ions.
    (a) K_3P (b) Na_3PO_4 (c) PO_3^{3-} (d) P_2Cl_4

100. Calculate the oxidation number of the aluminum atom in the following compounds or ions.
    (a) $LiAlH_4$ (b) $Al(H_2O)_6^{3+}$ (c) $Al(OH)_4^-$

101. The active ingredient in Rolaids™ antacid tablets has the formula $NaAl(OH)_2CO_3$. Calculate the oxidation state of the aluminum atom in the compound.

102. Which of the following compounds contain hydrogen in a negative oxidation state?
    (a) H_2S (b) H_2O (c) NH_3
    (d) H_3PO_4 (e) $LiAlH_4$ (f) HF
    (g) CaH_2 (h) CH_4

103. Calculate the oxidation number of the chlorine atom in the following compounds or ions.
    (a) Cl_2 (b) Cl^- (c) ClO^-
    (d) ClO_2^- (e) ClO_3^- (f) ClO_4^-

104. Calculate the oxidation number of the iodine atom in the following compounds. Group together any compounds in which iodine has the same oxidation number.
    (a) HI (b) KI (c) I_2 (d) HOI
    (e) KIO_3 (f) I_2O_5 (g) KIO_4 (h) H_5IO_6

105. Xe forms a limited number of compounds. Calculate the oxidation number of the xenon atom in the following compounds or ions and describe any trends in the oxidation states.
    (a) XeF_2 (b) XeF_4 (c) $XeOF_2$
    (d) XeF_6 (e) $XeOF_4$ (f) XeO_3
    (g) XeO_4 (h) XeO_6^{4-}

106. Carbon can have any oxidation number between -4 and $+4$. Calculate the oxidation number of carbon in the following compounds. Group compounds with the same oxidation number and describe any trends.
    (a) CCl_4 (b) $COCl_2$ (c) CO (d) CO_2
    (e) CS_2 (f) CH_3Li (g) CH_4 (h) H_2CO
    (i) Na_2CO_3 (j) HCO_2H

107. Sulfur can have any oxidation number between $+6$ and -2. Calculate the oxidation number of sulfur in the following species. Group species with the same oxidation number and describe any trends.
    (a) S_8 (b) H_2S (c) ZnS
    (d) SF_4 (e) SF_6 (f) SO_2
    (g) SO_3 (h) SO_3^{2-} (i) SO_4^{2-}
    (j) H_2SO_3 (k) H_2SO_4

108. Calculate the oxidation number of manganese in the following compounds. Group compounds with the same oxidation number and describe any trends.
    (a) MnO (b) Mn_2O_3 (c) MnO_2
    (d) MnO_3 (e) Mn_2O_7 (f) $Mn(OH)_2$
    (g) $Mn(OH)_3$ (h) H_2MnO_4 (i) $HMnO_4$
    (j) $CaMnO_3$ (k) $MnSO_4$

109. Calculate the oxidation number of titanium in the following compounds. Group compounds with the same oxidation number and describe any trends.
    (a) TiO (b) TiO_2 (c) Ti_2O_3
    (d) Ti_2S_3 (e) $TiCl_3$ (f) $TiCl_4$
    (g) K_2TiO_3 (h) H_2TiCl_6 (i) $Ti(SO_4)_2$

110. Prussian blue is a pigment with the formula $Fe_4[Fe(CN)_6]_3$. If the compound contains the $Fe(CN)_6^{4-}$ ion, what is the oxidation state of the other four iron atoms? Turnbull's blue is a pigment with the formula $Fe_3[Fe(CN)_6]_2$. If the compound contains the $Fe(CN)_6^{3-}$ ion, what is the oxidation state of the other three iron atoms?

Calculating Oxidation Numbers

111. Determine the oxidation state of each atom in the following organic molecules.

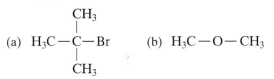

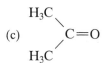 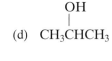

112. Determine the oxidation number of each atom in the following organic molecules.

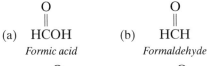

113. Determine the oxidation number of each atom in the following organic molecules.

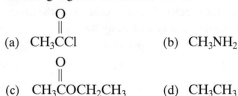

    (a) CH_3CCl (with O double bonded)   (b) CH_3NH_2

    (c) $CH_3COCH_2CH_3$ (with O double bonded)   (d) CH_3CH_3

    (e) $CH_2{=}CH{-}CH_3$ (f) $HC{\equiv}CH$

114. Describe the difference among partial charge, formal charge, and oxidation number.

115. Calculate the partial charge on the atoms in HCl using the methods of Section 4.11. Find the oxidation numbers of H and Cl in HCl. Give the formal charge on all the atoms in HCl. Compare and contrast these three results.

Oxidation–Reduction Reactions

116. Decide whether each of the following reactions involves oxidation–reduction. If it does, identify what is oxidized and what is reduced.
    (a) $CO_2(g) + H_2O(l) \longrightarrow H_2CO_3(aq)$
    (b) $Fe_2O_3(s) + 3\,CO(g) \longrightarrow 2\,Fe(s) + 3\,CO_2(g)$
    (c) $CO_2(g) + H_2(g) \longrightarrow CO(g) + H_2O(g)$
    (d) $CO(g) + 2\,H_2(g) \longrightarrow CH_3OH(l)$

117. Decide whether each of the following reactions involves oxidation–reduction. If it does, identify what is oxidized and what is reduced.
    (a) $Mg(s) + 2\,HCl(g) \longrightarrow MgCl_2(s) + H_2(g)$
    (b) $I_2(s) + 3\,Cl_2(g) \longrightarrow 2\,ICl_3(s)$
    (c) $NaOH(aq) + HCl(aq) \longrightarrow NaCl(aq) + H_2O(aq)$
    (d) $2\,Na(s) + 2\,H_2O(l) \longrightarrow 2\,NaOH(aq) + H_2(g)$

118. Decide whether each of the following reactions involves oxidation–reduction. If it does, identify what is oxidized and what is reduced.
    (a) $P_4O_{10}(g) + 10\,C\,(s) \longrightarrow 10\,CO(g) + P_4(g)$
    (b) $P_4(s) + 5\,O_2(g) \longrightarrow P_4O_{10}(s)$

119. What is the change in oxidation state of C and S in the following reaction? Which is oxidized and which is reduced?

$$4\,C(s) + S_8(l) \longrightarrow 4\,CS_2(l)$$

120. The Ag_2S that forms when silver tarnishes can be removed by polishing the silver with a source of cyanide ion or by wrapping it in aluminum foil and immersing it in salt water.

$$Ag_2S(s) + 4\,CN^-(aq)$$
$$\longrightarrow 2\,Ag(CN)_2^-(aq) + S^{2-}(aq)$$
$$3\,Ag_2S(s) + 2\,Al(s) \longrightarrow 6\,Ag(s) + Al_2S_3(s)$$

Which of the reactions involves oxidation–reduction?

Nomenclature

121. Describe what is wrong with the common names for the following compounds and write a better name for each compound.
    (a) phosphorus pentoxide (P_2O_5)
    (b) iron oxide (Fe_2O_3)
    (c) chlorine monoxide (Cl_2O)
    (d) copper bromide ($CuBr_2$)

122. Explain why calcium bromide is a satisfactory name for $CaBr_2$, but $FeBr_2$ must be called iron(II) bromide.

123. Write the formulas for the following compounds.
    (a) tetraphosphorus trisulfide
    (b) silicon dioxide
    (c) carbon disulfide
    (d) carbon tetrachloride
    (e) phosphorus pentafluoride

124. Write the formulas for the following compounds.
    (a) silicon tetrafluoride
    (b) sulfur hexafluoride
    (c) oxygen difluoride
    (d) dichlorine heptoxide
    (e) chlorine trifluoride

125. Write the formulas for the following compounds.
    (a) tin (II) chloride
    (b) mercury (II) nitrate
    (c) tin (IV) sulfide
    (d) chromium (III) oxide
    (e) iron (II) phosphide

126. Write the formulas for the following compounds.
    (a) beryllium fluoride
    (b) magnesium nitride
    (c) calcium carbide
    (d) barium peroxide
    (e) potassium carbonate

127. Write the formulas for the following compounds.
    (a) cobalt (III) nitrate
    (b) iron(III) sulfate
    (c) gold (III) chloride
    (d) manganese (IV) oxide
    (e) tungsten (VI) chloride

128. Name the following compounds.
    (a) KNO_3 (b) Li_2CO_3
    (c) $BaSO_4$ (d) PbI_2

129. Name the following compounds.
    (a) $AlCl_3$ (b) Na_3N (c) Ca_3P_2
    (d) Li_2S (e) MgO

130. Name the following compounds.
    (a) NH_4OH (b) H_2O_2 (c) $Mg(OH)_2$
    (d) $Ca(ClO)_2$ (e) $NaCN$

131. Name the following compounds.
    (a) Sb_2S_3 (b) $SnCl_2$ (c) SF_4
    (d) $SrBr_2$ (e) $SiCl_4$

132. Write the formulas of the following common acids.
    (a) acetic acid   (b) hydrochloric acid
    (c) sulfuric acid   (d) phosphoric acid
    (e) nitric acid

133. Write the formulas of the following less common acids.
    (a) carbonic acid   (b) hydrocyanic acid
    (c) boric acid   (d) phosphorous acid
    (e) nitrous acid

134. If sodium carbonate is Na_2CO_3 and sodium hydrogen carbonate is $NaHCO_3$, what are the formulas for sodium sulfite and sodium hydrogen sulfite?

135. The prefix *thio-* describes compounds in which sulfur replaces oxygen, for example, cyanate (OCN^-) and thiocyanate (SCN^-). If $SO_4{}^{2-}$ is the sulfate ion, what is the formula for the thiosulfate ion?

136. Name the compound in each of the following minerals.
    (a) fluorite (CaF_2)
    (b) galena (PbS)
    (c) quartz (SiO_2)
    (d) rutile (TiO_2)
    (e) hematite (Fe_2O_3)

137. Name the compound in each of the following minerals.
    (a) calcite ($CaCO_3$)
    (b) barite ($BaSO_4$)

Integrated Problems

138. The differences among the electronegativities of the atoms in the compounds SeO_2, CaH_2, and Cs_3Sb are all about the same. Yet only one of these compounds conducts electricity in the solid state, only one has a high melting point and dissolves in water to give a solution that conducts electricity, and only one has a relatively low melting point. Identify each of the compounds and explain your identification.

139. The common oxidation numbers for iron are $+2$ and $+3$. Iron reacts with O_2 (g) and Cl_2 (g) to form two different oxides and two different chlorides. What would be the formula of the four compounds formed? Classify the compounds according to bond type. Explain your answer.

140. Compare and contrast (what is the same and what is different) the distribution of electrons in covalent, ionic, and metallic bonds.

141. Partial charge, formal charge, and oxidation number are all used to describe the arrangement of electrons or charge associated with atoms within a compound. Determine the formal charge, partial charge, and oxidation number on both atoms in IBr. Give a short description for each of the three calculations describing what information the calculated value gives you. Which of the three calculations most accurately describes the distribution of charge between I and Br? What is the utility of the two calculations that are not the best description of the distribution of charge?

142. Describe what happens to the *distribution of electrons* in a bond between two elements as you move from left to right on a bond-type triangle. Describe what happens to the *distribution of electrons* in a bond between two elements as you move from the bottom to the top of a bond-type triangle.

143. Describe the types of bonding in $Ba(NO_3)_2$. What is different about the bonding in barium nitrate as compared to other compounds described in this chapter? Draw a Lewis structure for the cation and anion in this compound.

144. The following reaction is used in the Breathalyzer to determine the amount of ethyl alcohol on the breath of individuals suspected of driving while under the influence.

$$3 \, H{-}\overset{\overset{\displaystyle H}{|}}{\underset{\underset{\displaystyle H}{|}}{C}}{-}\overset{\overset{\displaystyle H}{|}}{\underset{\underset{\displaystyle H}{|}}{C}}{-}O{-}H(g) + 2 \, Cr_2O_7{}^{2-}(aq) + 16 \, H^+(aq)$$

$$\longrightarrow 3 \, H{-}\overset{\overset{\displaystyle H}{|}}{\underset{\underset{\displaystyle H}{|}}{C}}{-}C\overset{\displaystyle O}{\underset{\displaystyle O{-}H}{<}}(aq) + 4 \, Cr^{3+}(aq) + 11 \, H_2O(l)$$

    (a) If the reaction is an oxidation–reduction reaction, identify what is oxidized and what is reduced.
    (b) If the reaction is not an oxidation–reduction reaction, explain how you reached your conclusion.

145. The industrial manufacture of acrylonitrile, C_3H_3N, an important chemical used for the production of plastics, synthetic rubber, and fibers, is as follows.

$$4 \, C_3H_6(g) + 6 \, NO(g)$$
$$\longrightarrow 4 \, C_3H_3N(g) + 6 \, H_2O(g) + N_2(g)$$

The skeleton structures for C_3H_3N and C_3H_6 are

(a) Draw the Lewis structures for C_3H_6 and C_3H_3N and give the bond angle around each carbon atom.

(b) Which of the following statements concerning this reaction are true? Explain each.

   (i)   NO is oxidized.

   (ii) C_3H_6 is reduced.

   (iii) Both C_3H_6 and NO are reduced.

   (iv) None of the above.

146. Predict which of the following solid compounds would have the highest melting point on the basis of the information in Problem 67: HgO, ZrO, SrO, SeO$_2$. Explain fully and show all calculations.

147. (a) Lead (II) sulfide solid reacts with oxygen molecules to form lead (II) oxide solid and sulfur dioxide gas. Write a balanced reaction for this chemical change.

(b) Is the reaction in (a) an oxidation–reduction reaction? If so, what is oxidized and what is reduced?

148. Name all of the species in the following reaction.

$$4\,NaCl(s) + 2\,H_2SO_4(aq) + MnO_2(s)$$
$$\longrightarrow 2\,Na_2SO_4(aq) + MnCl_2(aq) + 2\,H_2O(l) + Cl_2(g)$$

Is the reaction an oxidation–reduction reaction? If so, what is oxidized and what is reduced?

149. Which of the following solid compounds would have a high melting point and not conduct electricity? HgO, MgO, HgCl.

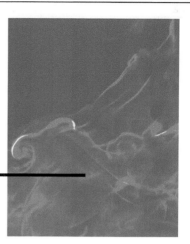

Chapter Six

GASES

The Nobel Prize–winning physicist Richard Feynman once asked, "If, in some cataclysm, all of scientific knowledge were to be destroyed, and only one sentence passed on to the next generations of creatures, what statement would contain the most information in the fewest words?" Feynman then answered the question, "I believe it is the atomic hypothesis that all things are made of atoms—little particles that move about in perpetual motion, attracting each other when they are a little distance apart, but repelling upon being squeezed into one another."[1] This is, as Feynman observed, a statement of enormous power. So far we have developed the idea that all things are made of atoms, but we haven't discussed the other parts of Feynman's statement—that atoms move about in perpetual motion and that particles attract and repel each other.

Atoms and molecules in a gas are never at rest. The N_2 and O_2 molecules in the atmosphere move through space at speeds of almost a thousand miles per hour until they collide with another particle in the gas or with the walls of the container. They tumble around their center of gravity, and the bonds that hold these molecules together vibrate at rates on the order of 7×10^{13} times per second. This chapter is devoted to building a model of the behavior of gases such as the atmosphere in which we live.

6.1 Temperature

We don't need to be told that the temperature is 95°F (35°C) on a summer afternoon to know it is hot. Nor do we need to be told that it is −15°F (−26°C) on a clear winter night to realize it is cold. For most purposes, we can rely on our senses to distinguish between hot and cold.

But there are times when our senses can be misled. Imagine that you get out of bed in the middle of a cold winter night, stepping onto a "cold" floor and then onto a small throw rug. Your senses tell you that the rug is warmer than the floor. But your senses have to be wrong. The floor can't be any colder than the rug. Both objects are just as warm (or just as cold) as the air in the room. Your senses can also mislead you when you reach into a freezer, in which case metal ice-cube trays feel colder than plastic trays. But this can't be true; all the trays in the freezer are equally cold. So why do they feel different?

Just as some materials conduct electricity better than others, some materials conduct heat better than others. Metals are good conductors of heat. The metal atoms are held in place by a sea of electrons that are free to move throughout the metal. As a metal is heated, its atoms vibrate more rapidly and the electrons transport heat from one part of the metal to another.

Materials whose electrons are not delocalized do not conduct heat as well as or in the same way as metals. Covalent and ionic compounds conduct heat by passing the heat along through vibrating strings of atoms and ions instead of through electrons. Different substances in contact with one another transfer heat only when the atoms in one substance acquire sufficient heat to cause them to vibrate so wildly that they collide with neighboring atoms in the other substance, causing its atoms to vibrate more. Good thermal conductors such as the floor of your room on a cold winter's night or a metal ice-cube tray feel cooler than poor thermal conductors such as a rug or a plastic ice-cube tray because

---

[1]From *The Feynman Lectures on Physics,* Volume 1, pp. 1–2 by R. Feynman and R Leighton Copyright © 1963 by the California Institute of Technology. Reprinted by permission of Addison-Wesley-Longman, Reading, Massachusetts.

good conductors transfer more heat away from our hands and feet, making us feel cool.

Because our senses can be misled, it is useful to have a reliable measure of the degree to which an object is either hot or cold. This quantity is called the **temperature** of the object.

6.2 Temperature as a Property of Matter

Temperature is nothing more than a quantitative measure of the degree to which an object is either "hot" or "cold." Temperature can be measured on either relative or absolute scales (Figure 6.1). The Celsius (°C) and Fahrenheit (°F) scales measure *relative* temperatures. These scales compare the temperature of a system with arbitrary standards such as boiling water (100°C or 212°F) and an ice-water bath (0°C or 32°F). The Kelvin (K) scale measures *absolute* temperatures in units of Kelvin above absolute zero on this scale. An object that has a temperature of 0°C has a temperature of 32°F and a temperature of 273 K.

At absolute zero on the Kelvin scale (-273.15°C), the atoms, molecules, or ions in a substance are packed as close together as they can get, and any motion of the units is at a minimum. What happens when a substance at absolute zero is warmed? The atoms, molecules, or ions begin to move. The warmer we make the substance, the more its components jostle about.

Consider what happens when ice melts, for example. Ice melts when it is heated because the average motion of the water molecules increases with temperature. Molecules can move in three ways: (1) vibration, (2) rotation, and (3) translation (Figure 6.2). Water molecules *vibrate* when H—O bonds are stretched or bent. *Rotation* involves the motion of a molecule around its center of gravity. *Translation* literally means to change from one place to another and describes the motion of molecules through space.

As the system becomes warmer, the motion of the water molecules eventually becomes too large to allow the molecules to be locked into the rigid structure of ice. At this point, the solid melts to form a liquid. The energy associated with the motion of molecules is called **kinetic energy.** Eventually, the kinetic

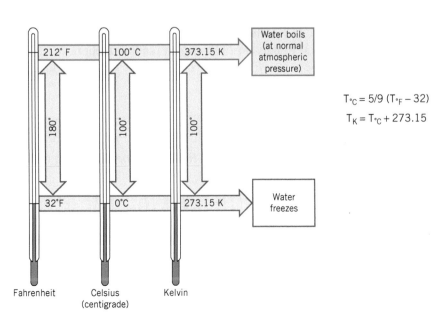

$$T_{°C} = 5/9 \, (T_{°F} - 32)$$
$$T_K = T_{°C} + 273.15$$

Fig. 6.1 The three common temperature scales. Note that water boils at 212 °F, 100°C, or 373.15 K and freezes at 32°F, 0°C, or 273.15 K.

Fig. 6.2 A water molecule can move in three different ways. (*a*) Vibration occurs when O—H bonds stretch or bend. (*b*) Rotation involves movement around the center of gravity of the molecule. (*c*) Translation occurs when a water molecule moves through space.

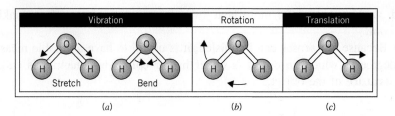

energy of the molecules becomes so large and their movement so rapid that the liquid boils to form a gas in which each particle moves more or less randomly through space.

The kinetic energy of the particles in a gas, liquid, or solid is a property that exists on the atomic scale (see Section 1.13). The temperature of a sample is a macroscopic property of matter (see Section 1.13). As we saw in Chapter 2, the concept of the mole is one of the links or bridges between matter on the atomic and macroscopic scale. Another link or bridge is based on the relationship between the average kinetic energy of the particles in a sample on the atomic scale and the temperature of the sample on the macroscopic scale. As we will see, the following is one of the basic assumptions of the kinetic theory of gases.

Temperature is a measure of the average kinetic energy of the atoms, molecules, or ions in a substance.

➤ **CHECKPOINT**

How does the average kinetic energy of the water molecules compare in samples of liquid water and gaseous water (or steam) at 100°C?

6.3 The States of Matter

There are three primary **states** of matter: gases, liquids, and solids. We'll begin our discussion of the states of matter with gases. There are two reasons for studying gases before liquids and solids. First, the behavior of gases is easier to describe because most of the properties of gases don't depend on the identity of the gas. We can therefore develop a model for a gas without worrying about whether the gas is O_2, N_2, H_2, or a mixture of gases. Second, a relatively simple yet powerful model known as the **kinetic molecular theory** is available that explains most of the behavior of gases.

The term *gas* comes from the Greek word for "chaos" because gases consist of a chaotic collection of particles in constant, random motion. In the course of our discussion of gases, we will provide the basis for answering the following questions.

- Why does popcorn "pop" when we heat it?
- Why does a hot-air balloon rise when the air in the balloon is heated?
- Why does a balloon filled with helium rise? Why does a balloon filled with CO_2 sink?
- Why does the pressure of the air in the tires on a car increase when the car is driven?
- Is the volume of the O_2 in the room in which you are sitting the same as the volume of the N_2? Is the pressure of the O_2 the same as the pressure of the N_2?
- At 25°C and 1 atmosphere of pressure, which weighs more: a liter of dry air or air that is saturated with water vapor?

6.4 Elements or Compounds That Are Gases at Room Temperature

Before examining the chemical and physical properties of gases, it may be useful to ask: What kinds of elements or compounds are gases at room temperature? To help answer this question, a list of some common compounds that are gases at room temperature is given in Table 6.1.

The data in Table 6.1 show several patterns:

- Common gases at room temperature include both elements (such as H_2 and O_2) and compounds (such as CO_2 and NH_3).

- The elements that are gases at room temperature are all *nonmetals* (such as He, Ar, N_2, and O_2).

- Compounds that are gases at room temperature are all *covalent compounds* (such as CO_2, SO_2, and NH_3) that contain two or more nonmetals.

- With rare exception, these gases have relatively small atomic or molecular weights.

As a general rule, elements and compounds that consist of relatively light, covalent molecules are most likely to be gases at room temperature.

Table 6.1
Common Gases at Room Temperature

Element or Compound	Molecular Weight (g/mol)
H_2 (hydrogen)	2.02
He (helium)	4.00
CH_4 (methane)	16.04
NH_3 (ammonia)	17.03
Ne (neon)	20.18
HCN (hydrogen cyanide)	27.03
CO (carbon monoxide)	28.01
N_2 (nitrogen)	28.01
NO (nitrogen oxide)	30.01
C_2H_6 (ethane)	30.07
O_2 (oxygen)	32.00
PH_3 (phosphine)	34.00
H_2S (hydrogen sulfide)	34.08
HCl (hydrogen chloride)	36.46
F_2 (fluorine)	38.00
Ar (argon)	39.95
CO_2 (carbon dioxide)	44.01
N_2O (dinitrogen oxide)	44.01
C_3H_8 (propane)	44.10
NO_2 (nitrogen dioxide)	46.01
O_3 (ozone)	48.00
C_4H_{10} (butane)	58.12
SO_2 (sulfur dioxide)	64.07
BF_3 (boron trifluoride)	67.81
Cl_2 (chlorine)	70.91
Kr (krypton)	83.80
CF_2Cl_2 (dichlorodifluoromethane)	120.92
Xe (xenon)	131.29
SF_6 (sulfur hexafluoride)	146.05

6.5 The Properties of Gases

Gases have three characteristic properties: (1) They are easy to compress, (2) they expand to fill their containers, and (3) they occupy much more space than the equivalent amount of a liquid or solid under normal conditions.

COMPRESSIBILITY

The internal combustion engine found in most cars provides a good example of the ease with which gases can be compressed. In a typical four-stroke engine, the piston is first pulled out of the cylinder to create a partial vacuum, which draws a mixture of gasoline vapor and air into the cylinder (Figure 6.3). The piston is then pushed into the cylinder, compressing the gasoline–air mixture to a fraction of the original volume.

The ratio of the volume of the gas in the cylinder after the first stroke to its volume after the second stroke is the *compression ratio* of the engine. Modern cars run at compression ratios of about 9:1, which means the gasoline–air mixture in the cylinder is compressed by a factor of 9 in the second stroke. After the gasoline–air mixture is compressed, the spark plug at the top of the cylinder fires, and the resulting explosion pushes the piston down the cylinder in the third stroke. Finally, the piston is pushed back up the cylinder in the fourth stroke, clearing out the exhaust gases.

Liquids are much harder to compress than gases. They are so difficult to compress that the hydraulic brake systems used in most cars operate on the principle that there is essentially no change in the volume of the brake fluid when pressure is applied to the liquid. Most solids are even harder to compress. The only exceptions belong to a unique class of compounds that includes natural and synthetic rubber. Most rubber balls that are easy to compress, such as a racquetball, are filled with air, which is compressed when the ball is squeezed.

EXPANDABILITY

Anyone who has walked into a kitchen where bread is baking has experienced the fact that gases expand to fill their containers, as the air in the kitchen becomes filled with wonderful aromas. Unfortunately, the same thing happens when someone breaks open a rotten egg and the characteristic odor of hydrogen sulfide (H_2S) rapidly diffuses through the room. Because gases expand to fill their containers, the volume of a gas is equal to the volume of its container. The volume of the O_2 in the atmosphere in a room, therefore, is indeed equal to the volume of the N_2 in the atmosphere in that room.

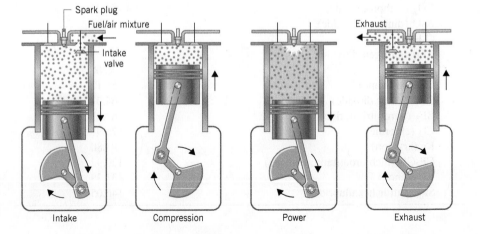

Fig. 6.3 The operation of a four-stroke engine can be divided into four cycles: intake, compression, power, and exhaust stages.

VOLUMES OF GASES VERSUS VOLUMES OF LIQUIDS OR SOLIDS

The difference between the volume of a gas and the volume of the liquid or solid from which it forms can be illustrated with the following examples. One gram of liquid oxygen at its boiling point ($-183°C$) has a volume of 0.894 mL. The same amount of O_2 gas at room temperature and atmospheric pressure has a volume of 753 mL—an increase of a factor of about 800. One gram of solid CO_2 at $-79°C$ has a volume of 0.641 mL. At room temperature and atmospheric pressure, the same amount of CO_2 gas has a volume of 548 mL—once again, an increase of a factor of about 800.

The consequences of the enormous change in volume that occurs when a liquid or solid is transformed into a gas are often used to do work. The steam engine, which brought about the Industrial Revolution, is based on the fact that water boils to form a gas (steam) that has a much larger volume. The gas therefore escapes from the container in which it has been generated, and the escaping steam can be made to do work. The same principle is at work when dynamite is used to blast rocks. In 1867, the Swedish chemist Alfred Nobel discovered that the highly dangerous liquid explosive known as nitroglycerin could be absorbed onto clay or sawdust to produce a solid that was much more stable and therefore safer to use. When dynamite is detonated, the nitroglycerin decomposes to produce a mixture of CO_2, H_2O, N_2, and O_2 gases.

$$4 \, C_3H_5N_3O_9(l) \longrightarrow 12 \, CO_2(g) + 10 \, H_2O(g) + 6 \, N_2(g) + O_2(g)$$

Because 29 mol of gas are produced for every 4 mol of liquid that decomposes, and each mole of gas occupies a volume hundreds of times larger than a mole of liquid, the reaction produces a shock wave that destroys anything in its vicinity.

The same phenomenon occurs on a much smaller scale when we "pop" popcorn. When kernels of popcorn are heated in oil, the liquids inside the kernel turn into gases. The pressure that builds up inside the kernel is enormous and eventually causes the kernel to explode.

Popcorn "pops" because of the enormous difference between the volume of the liquids inside the kernel and the volume of the gases these liquids produce when they boil. Ted Kinsman/Photo Researchers

6.6 Pressure versus Force

The volume of a gas is one of its characteristic properties. Another characteristic property is the **pressure** the gas exerts on its surroundings. Many of us got our first exposure to the pressure of a gas when we rode to the neighborhood gas station to check the pressure in our bicycle tires. Depending on the kind of bicycle we had, we added air to the tires until the pressure gauge read between 30 and 70 pounds per square inch (lb/in.^{2}), or psi. Two important properties of pressure can be gleaned from this example.

- The pressure of a gas increases as more gas is added to the container (as long as there is no change in the volume and temperature of the gas).
- Pressure is measured in units (such as lb/in.^{2}) that describe the **force** exerted by the gas divided by the **area** over which the force is distributed.[2]

The first conclusion can be summarized in the following relationship, where P is the pressure of the gas and n is the number of moles of gas in the container.

$$P \propto n \, (T \text{ and } V \text{ constant})$$

---

[2]The SI unit for pressure is the pascal (Pa), which is defined as a force of 1 newton averaged over an area of 1 m^2.

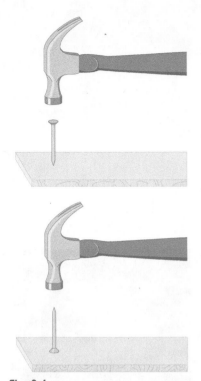

Fig. 6.4   The *force* exerted by a hammer hitting a nail is the same regardless of whether the hammer hits the nail on the head or on the point. But the *pressure* exerted on the wood is very different.

The symbol ∝ means "proportional to," and the relationship is read as "pressure is proportional to the number of moles of gas." Because the pressure increases as gas is added to a container with a fixed volume at a given temperature, P is directly proportional to n.

The second conclusion describes the relationship between pressure and force. Pressure is defined as the force exerted on an object divided by the area over which the force is distributed.

$$\text{pressure} = \frac{\text{force}}{\text{area}}$$

The difference between pressure and force can be illustrated with an analogy based on a 10-penny nail, a hammer, and a piece of wood (Figure 6.4). By resting the nail on its point and hitting the head with the hammer, we can drive the nail into the wood. But what happens if we turn the nail over and rest the head of the nail against the wood? If we hit the nail with the same force, we can't get the nail to penetrate into the wood.

When we hit the nail on the head, the force of the blow is applied to the very small area of the wood in contact with the point of the nail, and the nail slips easily into the wood. But when we turn the nail over and hit it on the point, the same force is distributed over a much larger area. The force is now distributed over the surface of the wood that touches any part of the nail head. As a result, the pressure applied to the wood is much smaller and the nail just bounces off the wood.

Exercise 6.1

(a) Calculate the pressure exerted by the shoes of a 200-lb man wearing size 10 shoes if each shoe makes contact with an area of the floor that is 20 in.^{2}.

(b) Calculate the pressure exerted by each heel of a 100-lb woman in high heels if the area beneath the heel of each shoe is 0.250 in.^{2}.

Solution

(a) The pressure is calculated by dividing the force by the area over which it is distributed. Because one-half of the weight of the man is applied to each shoe, the pressure in this case is 5.0 lb/in.^{2}.

$$\text{pressure} = \frac{\text{force}}{\text{area}} = \frac{100 \text{ lb}}{20 \text{ in.}^2} = 5.0 \text{ lb/in.}^2$$

(b) We can assume that about one-fourth of the weight of the woman is applied to each heel if her weight is evenly divided between the heel and sole of each shoe.

$$\text{pressure} = \frac{\text{force}}{\text{area}} = \frac{25.0 \text{ lb}}{0.250 \text{ in.}^2} = 100 \text{ lb/in.}^2$$

The pressure exerted by the heels of the 100-lb woman is 20 times greater than the pressure exerted by a man who weighs twice as much.

➤ CHECKPOINT

Is the pressure exerted on the floor by a woman wearing a spike heel greater than, less than, or equal to the pressure exerted when the same person wears a flat sandal? Why?

6.7 Atmospheric Pressure

What would happen if we bent a long piece of glass tubing into the shape of the letter U and then carefully filled one arm of the U-tube with water and the other arm with ethyl alcohol? Most people expect the height of the columns of liquid in the two arms of the tube would be the same. Experimentally, we find the results shown in Figure 6.5. A 100-cm column of water balances a 127-cm column of ethyl alcohol, regardless of the diameter of the glass tubing.

We can explain this by comparing the densities of water (1.00 g/cm³) and ethyl alcohol (0.789 g/cm³). The ratio of the mass of each liquid to the area on which the mass exerts a force for a column of water 100 cm tall is equal to the mass/area of a column of ethyl alcohol 127 cm tall.

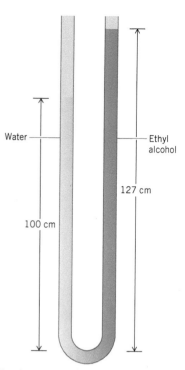

$$\text{water} \qquad 100 \text{ cm} \times \frac{1.00 \text{ g}}{\text{cm}^3} = 100 \text{ g/cm}^2$$

$$\text{ethyl alcohol} \qquad 127 \text{ cm} \times \frac{0.789 \text{ g}}{\text{cm}^3} = 100 \text{ g/cm}^2$$

As a result, the pressure of the water pushing down on one arm of the U-tube is equal to the pressure of the alcohol pushing down on the other arm of the tube. The system is therefore in balance. This demonstration provides the basis for understanding how a mercury barometer can be used to measure the pressure of the atmosphere.

Fig. 6.5 Because of the difference between the densities of water and ethyl alcohol, the weight of a column of water 100 cm long balances the weight of a column of ethyl alcohol 127 cm long in the other arm of a U-tube.

THE DISCOVERY OF THE BAROMETER

In the early 1600s, Galileo argued that suction pumps were able to draw water from a well because of the "force of vacuum" inside the pump. After Galileo's death, Evangelista Torricelli (1608–1647) proposed another explanation. He suggested that the air in the atmosphere has weight and that the force of the atmosphere pushing down on the surface of the water drives the water into the suction pump when it is evacuated.

In 1646 Torricelli described an experiment in which a glass tube about 1 m long was sealed at one end, filled with mercury, and then inverted into a dish filled with mercury, as shown in Figure 6.6. Some, but not all, of the mercury drained out of the glass tube into the dish to create a vacuum at the top of the tube. Torricelli explained this result by assuming that mercury drains from the glass tube until the pressure of the column of mercury pushing down on the *inside* of the tube exactly balances the pressure of the atmosphere pushing down on the surface of the liquid *outside* the tube.

Torricelli predicted that the height of the mercury column would change from day to day as the pressure of the atmosphere changed. Today, his apparatus is known as a *barometer,* from the Greek word *baros,* meaning, "weight," because it literally measures the weight of the atmosphere. Repeated experiments have shown that the average pressure of the atmosphere at sea level is equal to the pressure of a column of mercury 760 mm tall. Thus, a standard unit of pressure known as the *atmosphere* is defined as follows.

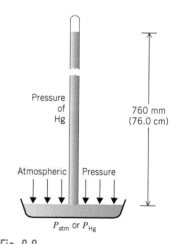

$$1 \text{ atm} \equiv 760 \text{ mmHg (exactly)}$$

To recognize Torricelli's contributions, some scientists describe pressure in units of torr, defined as follows.

$$1 \text{ torr} \equiv 1 \text{ mmHg}$$

Fig. 6.6 On a sunny day, at sea level the pressure of a 760-mm-tall column of mercury inside a glass tube balances the pressure of the atmosphere pushing down on the pool of mercury that surrounds the tube. The pressure of the atmosphere is therefore said to be equivalent to 760 mmHg.

Exercise 6.2

Calculate the pressure in units of atmospheres on a day when a barometer gives a measurement of 745.8 mmHg.

Solution

The conversion between mmHg and atmospheres is based on the following definition.

$$1 \text{ atm} \equiv 760 \text{ mmHg}$$

Using this relationship to generate an appropriate conversion factor gives the following result.

$$745.8 \text{ mmHg} \times \frac{1 \text{ atm}}{760 \text{ mmHg}} = 0.9813 \text{ atm}$$

Although chemists still work with pressures in units of atm or mmHg, neither unit is accepted in the SI system (see Appendix A). The SI unit of pressure is the pascal (Pa). The relationship between one standard atmosphere pressure and the pascal is given by the following:

$$1 \text{ atm} \equiv 101{,}325 \text{ Pa} \equiv 101.325 \text{ kPa} = 0.101325 \text{ MPa}$$

The pressure of the atmosphere can be demonstrated by connecting a 1-gallon can to a vacuum pump. Normally the pressure inside the can balances the pressure of the atmosphere pushing on the outside of the can. When the vacuum pump is turned on, however, the can rapidly collapses as it is evacuated. The surface

The pressure of the atmosphere can be demonstrated by connecting an empty paint-thinner can to a vacuum pump. Within seconds of turning on the vacuum pump, the can collapses.

area of a 1-gallon can is about 250 in.^{2}. At an atmospheric pressure of 14.7 lb/in.^{2} (1 atm), this corresponds to a total force over the surface of the can of about 3700 lb. For the sake of comparison, it might be noted that each of the 18 wheels of a 70,000-lb truck carries only about 3900 lb.

THE DIFFERENCE BETWEEN THE PRESSURE OF A GAS AND PRESSURE RESULTING FROM WEIGHT

There is an important difference between the pressure of a gas and the other examples of pressure discussed in this section. The pressure exerted by a 70,000-lb truck is directional. The truck exerts all of its pressure on the surface beneath its wheels. In contrast, gas pressure is the same in all directions. To demonstrate this, we can fill a glass cylinder with water and rest a glass plate on top of the cylinder. When we turn the cylinder over, the plate doesn't fall to the floor because the pressure of the air outside the cylinder pushing up on the bottom of the plate is larger than the pressure exerted by the water in the cylinder pushing down on the plate. It would take a column of water 33.9 ft tall to produce as much pressure as the gas in the atmosphere.

6.8 Boyle's Law

Torricelli's work with the mercury barometer caught the eye of the British scientist Robert Boyle. Boyle's most famous experiments were done in a J-tube apparatus similar to the one shown in Figure 6.7. By adding mercury to the open end of the tube, Boyle was able to trap a small volume of air in the sealed end.

Boyle was interested in a phenomenon he called the "spring of air." His experiments were based on the fact that gases are *elastic*. (They return to their original size and shape after being stretched or squeezed.) When he added more mercury to the open end of the J-tube, Boyle noticed that the air in the sealed end was compressed into a smaller volume.

Table 6.2 contains some of the experimental data Boyle reported in 1662. The first column in this table lists the volume of the gas in the sealed end of the J-tube, in arbitrary units. The second column is the difference in heights of the mercury in the sealed and open arms of the J-tube, to the nearest $\pm 1/16$ inch. The third column is the product of the volume of the gas (V) and the pressure (P) calculated from Boyle's data. Because of the way the experiment was done, the number of moles of gas and the temperature of the gas were held constant for these experiments.

The product of the pressure times the volume for any measurement in the table is equal to the product of the pressure times the volume for any other measurement, within experimental error.

$$P_1V_1 = P_2V_2$$

This expression, or its equivalent,

$$P \propto \frac{1}{V} \ (T \text{ and } n \text{ constant})$$

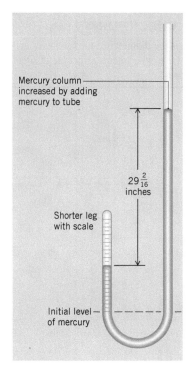

Fig. 6.7 Boyle's law is based on data obtained with a J-tube apparatus. The temperature and number of moles of gas were held constant. The volume is in arbitrary units, and the height is in inches.

Table 6.2

Boyle's Data on the Dependence of the Volume of a Gas on the Pressure of the Gas[a]

Volume	Pressure	$P \times V$
48	29²⁄₁₆	1398
46	30⁹⁄₁₆	1406
44	31¹⁵⁄₁₆	1405
42	33⁸⁄₁₆	1407
40	35⁵⁄₁₆	1413
38	37	1406
36	39⁵⁄₁₆	1413
34	41¹⁰⁄₁₆	1415
32	44³⁄₁₆	1414
30	47¹⁄₁₆	1412
⋮	⋮	⋮

[a]Because these are Boyle's original data from the seventeenth century, no attempt has been made to round off to the appropriate number of significant figures.

➤ **CHECKPOINT**
Does the diameter of the tube used in Boyle's experiments affect the results obtained?

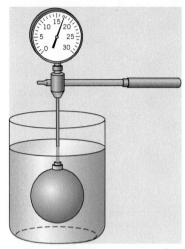

Fig. 6.8 The apparatus for demonstrating Amontons' law consists of a pressure gauge connected to a metal sphere of constant volume, which is immersed in water at different temperatures.

Table 6.3

The Dependence of the Pressure of a Gas on Its Temperature

Temperature (°C)	Pressure (lb/in.^{2})
100	18.1
74	16.7
24	14.5
0	13.2
−47	10.8

is now known as **Boyle's law.** Remember that the term *law* is used here to indicate a mathematical equation that fits the experimental data, more or less, under reasonable conditions, as noted in Section 1.1.

6.9 Amontons' Law

Toward the end of the 1600s, the French physicist Guillaume Amontons built a thermometer based on the fact that the pressure of a gas is directly proportional to its temperature. The relationship between the pressure and the temperature of a gas is therefore known as **Amontons' law.**

$$P \propto T \qquad (n \text{ and } V \text{ constant})$$

Amontons' law explains why car manufacturers recommend adjusting the pressure of your tires before you start on a trip. The flexing of the tire as you drive inevitably raises the temperature of the air in the tire. When this happens, the pressure of the gas inside the tires increases.

Amontons' law can be demonstrated with the apparatus shown in Figure 6.8. Data obtained with this apparatus at various temperatures are given in Table 6.3.

The relationship between the temperature and pressure of a gas provided the first definition for **absolute zero** on the temperature scale. In 1779 Joseph Lambert defined absolute zero as the temperature at which the pressure of a gas becomes zero. When the data in Table 6.3 are plotted, the pressure of a gas approaches zero when the temperature is about −270°C, as shown in Figure 6.9. When more accurate measurements are made, the pressure of a gas extrapolates to zero when the temperature is −273.15°C. Absolute zero on the Celsius scale is therefore −273.15°C.

The relationship between the temperature and pressure data in Table 6.3 can be simplified by converting the temperatures from the Celsius to the Kelvin scale.

$$T_K = T_{°C} + 273.15$$

When this is done, a plot of the temperature versus the pressure of a gas gives a straight line that passes through the origin. Any two points along the line therefore fit the following equation.

$$\frac{P_1}{P_2} = \frac{T_1}{T_2}$$

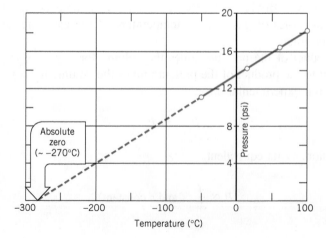

Fig. 6.9 When data obtained with the Amontons' law apparatus in Figure 6.8 are extrapolated, the pressure of a gas approaches zero when the temperature of the gas is approximately −270°C.

It is important to remember that this equation is valid only if the temperatures are converted from the Celsius to the Kelvin scale before calculations are done, and the volume of gas and number of moles of gas must be constant.

➤ CHECKPOINT

What relationship can be derived by combining Boyle's and Amontons' laws?

6.10 Charles' Law

On June 5, 1783, Joseph and Etienne Montgolfier used a fire to inflate a spherical balloon 30 feet in diameter that traveled about 1.5 miles before it came back to earth. News of this remarkable achievement spread throughout France, and Jacques-Alexandre-César Charles immediately tried to duplicate the performance. As a result of his work with hot-air balloons, Charles noticed that the volume of a gas is directly proportional to its temperature when the pressure and number of moles of gas are held constant.

$$V \propto T \qquad (n \text{ and } P \text{ constant})$$

This relationship between the temperature and volume of a gas, which became known as **Charles' law,** can be used to explain how a hot-air balloon works. Ever since the third century B.C., it has been known that an object floats when it weighs less than the fluid it displaces. Because a gas expands when heated, a given weight of hot air occupies a larger volume than the same weight of cold air. As a result, hot air is less dense than cold air. Once the air in a balloon gets hot enough, the net weight of the balloon plus the hot air is less than the weight of an equivalent volume of cold air, and the balloon starts to rise. When the gas in the balloon is allowed to cool, the balloon returns to the ground.

Charles' law can be demonstrated with the apparatus shown in Figure 6.10. A 30-mL syringe and a thermometer are inserted through a rubber stopper into a flask that has been cooled to 0°C. The ice bath is then removed, and the flask is immersed in a warm-water bath. The gas in the flask expands as it warms, slowly pushing the piston out of the syringe. The total volume of the gas in the system is equal to the volume of the flask plus the volume of the syringe. Table 6.4 contains a set of typical data obtained with this apparatus.

The data in Table 6.4 are plotted in Figure 6.11. This graph provides us with another way of defining absolute zero on the temperature scale: It is the temperature at which the volume of a gas becomes zero when a plot of volume versus temperature for a gas is extrapolated. As expected, the value of absolute zero obtained by extrapolating the data in Table 6.4 is essentially the same as the value obtained from the graph of pressure versus temperature in the preceding section. Absolute zero can therefore be more accurately defined as the temperature at which the pressure and the volume of a gas both extrapolate to zero.

When the temperatures in Table 6.4 are converted from the Celsius to the Kelvin scale, a plot of the volume versus the temperature of a gas becomes a straight line that passes through the origin. Any two points along the line can therefore be used to construct the following equation, which is known as Charles' law.

$$\frac{V_1}{V_2} = \frac{T_1}{T_2}$$

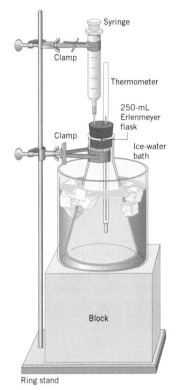

Fig. 6.10 Charles' law can be demonstrated with a simple apparatus. When the flask is removed from the ice bath and placed in a warm-water bath, the gas in the flask expands, slowly pushing up on the piston of the syringe.

Table 6.4

Dependence of the Volume of a Gas on Its Temperature

Temperature (°C)	Volume (mL)
0	107.9
5	109.7
10	111.7
15	113.6
20	115.5
25	117.5
30	119.4
35	121.3
40	123.2

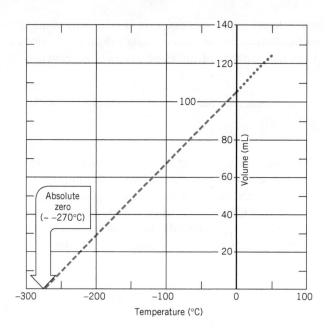

Fig. 6.11 When data obtained with the Charles' law apparatus in Figure 6.10 are extrapolated, the volume of the gas approaches zero when the temperature of the gas is approximately −270°C.

➤ **CHECKPOINT**

Could Charles' law have been predicted from Boyle's and Amontons' laws, or is it independent of those laws?

Before you use this equation, however, it is important to remember to convert temperatures from °C to K.

6.11   Gay-Lussac's Law

Joseph Louis Gay-Lussac (1778–1850) studied the volume of gases consumed or produced in a chemical reaction because he was interested in the reaction between hydrogen and oxygen to form water. Gay-Lussac found that 199.89 parts by volume of hydrogen were consumed for every 100 parts by volume of oxygen. Thus hydrogen and oxygen seemed to combine in a simple 2:1 ratio by volume.

hydrogen + oxygen $\longrightarrow$ water
2 volumes   1 volume

Gay-Lussac found similar whole-number ratios for the reactions between other pairs of gases. On December 31, 1808, Gay-Lussac announced his results at a meeting of the Societé Philomatique in Paris in terms of a **law of combining volumes.** Today, the law of combining volumes is stated as follows: The ratio of the volumes of gases consumed or produced in a chemical reaction is equal to the ratio of simple whole numbers when temperature and pressure are held constant.

6.12   Avogadro's Hypothesis

Gay-Lussac's law of combining volumes was announced in 1808 only a few years after John Dalton proposed his atomic theory. The link between the two ideas was first recognized by Amadeo Avogadro three years later, in 1811. Avogadro argued that Gay-Lussac's law of combining volumes could be explained by assuming that equal volumes of different gases collected under similar conditions contain the same number of particles.

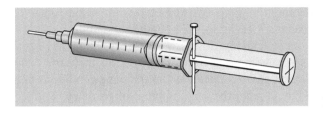

Fig. 6.12 Apparatus used to demonstrate Avogadro's hypothesis.

HCl and NH_3 combine in a 1:1 ratio by volume, for example, because one molecule of HCl is consumed for every molecule of NH_3 in the reaction and equal volumes of the gases contain the same number of molecules.

$$NH_3(g) + HCl(g) \longrightarrow NH_4Cl(s)$$

Anyone who has blown up a balloon can accept the notion that the volume of a gas is proportional to the number of particles in the gas.

$$V \propto n \qquad (T \text{ and } P \text{ constant})$$

The more air you add to a balloon, the bigger it gets. Unfortunately this example doesn't test **Avogadro's hypothesis**—that equal volumes of *different gases* contain the same number of particles. The best way to probe the validity of this hypothesis is to measure the number of molecules in a given volume of different gases at the same temperature and pressure. This can be done with the apparatus shown in Figure 6.12.

A small hole is drilled through the plunger of a 50-mL plastic syringe. The plunger is pushed into the syringe, and the syringe is sealed with a syringe cap. The plunger is then pulled out of the syringe until the volume reads 50 mL, and a nail is inserted through the hole in the plunger so that the plunger is not drawn back into the barrel of the syringe. The "empty" syringe is then weighed, the syringe is filled with 50 mL of a gas, the nail is inserted through the hole in the plunger, once again, and the syringe is reweighed. The difference between the measurements is the mass of 50 mL of the gas.

The results of experiments with six gases are given in Table 6.5. The number of molecules in a 50-mL sample of any one of the gases can be calculated from the mass of the sample, the molecular weight of the gas, and the number of molecules in a mole. Consider the following calculation of the number of H_2 molecules in 50 mL of hydrogen gas, for example.

$$0.005 \text{ g } H_2 \times \frac{1 \text{ mol } H_2}{2.02 \text{ g } H_2} \times \frac{6.02 \times 10^{23} \text{ molecules}}{1 \text{ mol}} = 1 \times 10^{21} \ H_2 \text{ molecules}$$

Table 6.5

Experimental Data for the Mass of 50-mL Samples of Different Gases (*T* and *P* are Constant)

Compound	Mass of 50 mL Gas (g)	Molecular Weight of Gas (g/mol)	Number of Molecules in 50 mL Gas
H_2	0.005	2.02	1×10^{21}
N_2	0.055	28.01	1.2×10^{21}
O_2	0.061	32.00	1.1×10^{21}
CO_2	0.088	44.01	1.2×10^{21}
C_4H_{10}	0.111	58.12	1.15×10^{21}
CCl_2F_2	0.228	120.91	1.14×10^{21}

The last column in Table 6.5 summarizes the results obtained when the calculation is repeated for each gas. The number of significant figures in the answer changes from one calculation to the next. But the number of molecules in each sample is the same, within experimental error. We therefore conclude that equal volumes of different gases collected under the same conditions of temperature and pressure do in fact contain the same number of particles.

6.13 The Ideal Gas Equation

So far, gases have been described in terms of four variables: pressure (P), volume (V), temperature (T), and the number of moles of gas (n). In the course of this chapter, five relationships between pairs of these variables have been discussed. In each case, two of the variables were allowed to change while the other two were held constant. The discussion of the bicycle tire, for example, showed that the pressure of a gas is directly proportional to the moles of gas when the temperature and volume of the gas are held constant.

$$P \propto n \qquad (T \text{ and } V \text{ constant})$$

Other relationships between pairs of variables include the following.

Boyle's law:	$P \propto 1/V$	$(T \text{ and } n \text{ constant})$
Amontons' law:	$P \propto T$	$(V \text{ and } n \text{ constant})$
Charles' law:	$V \propto T$	$(P \text{ and } n \text{ constant})$
Avogadro's hypothesis:	$V \propto n$	$(P \text{ and } T \text{ constant})$

Each of these relationships is a special case of a more general relationship known as the **ideal gas equation.**

$$PV = nRT$$

In this equation, R is a proportionality constant known as the **ideal gas constant** and T is the absolute temperature. The value of R depends on the units used to express the four variables P, V, n, and T. By convention, most chemists use the following set of units.

P: atmospheres T: kelvins
V: liters n: moles

According to the **ideal gas law,** the product of the pressure times the volume of an ideal gas divided by the product of the amount of gas times the absolute temperature is a constant.

$$\frac{PV}{nT} = R$$

We can calculate the value of R from data from experiments in which the number of moles of gas and the pressure of the gas are held constant. For exactly 1 mol of gas at 1.0000 atm and 0°C—a condition known as standard temperature and pressure, or **STP**—the volume is found to be 22.414 L. The value of R can therefore be calculated as follows.

$$\frac{(1.0000 \text{ atm})(22.414 \text{ L})}{(1.000 \text{ mol})(273.15 \text{ K})} = 0.08206 \text{ L} \cdot \text{atm/mol} \cdot \text{K}$$

For most ideal gas calculations, four significant figures are sufficient. The standard value of the ideal gas constant is therefore 0.08206 L · atm/mol · K.

6.14 Dalton's Law of Partial Pressures

The *CRC Handbook of Chemistry and Physics* describes the atmosphere as 78.084% N_2, 20.946% O_2, 0.934% Ar, and 0.033% CO_2 by volume, once the water vapor has been removed. What image does this description evoke in your mind? Do you believe that only 20.946% of the room in which you are sitting contains O_2? Or do you believe that the atmosphere in your room is a more or less homogeneous mixture of the gases?

In Section 6.5 we argued that gases expand to fill their containers. The volume of O_2 in your room is therefore the same as the volume of N_2. Both gases expand to fill the room.

What about the pressure of the different gases in your room? Is the pressure of the O_2 in the atmosphere the same as the pressure of the N_2? We can answer this question by rearranging the ideal gas equation as follows.

$$P = n \times \frac{RT}{V}$$

According to this equation, the pressure of a gas is proportional to the number of moles of gas if the temperature and volume are held constant. Because the temperature and volume of the O_2 and N_2 in the atmosphere are the same, the pressure of each gas is proportional to the number of the moles of the gas. Because there is more N_2 in the atmosphere than O_2, the contribution to the total pressure of the atmosphere from N_2 is larger than the contribution from O_2.

John Dalton was the first to recognize that the total pressure of a mixture of gases is the sum of the contributions of the individual components of the mixture. The part of the total pressure of a mixture that results from one component is called the **partial pressure** of that component. **Dalton's law of partial pressures** states that the total pressure (P_{tot}) of a mixture of gases is the sum of the partial pressures of the various components.

$$P_{tot} = P_1 + P_2 + P_3 + \cdots$$

$$P_{tot} = n_1\frac{RT}{V} + n_2\frac{RT}{V} + n_3\frac{RT}{V} + \cdots$$

$$P_{tot} = n_{tot}\frac{RT}{V}$$

where $\qquad\qquad n_{tot} = n_1 + n_2 + n_3 + \cdots$

Exercise 6.3

Calculate the total pressure of a mixture that contains 1.00 g of H_2 and 1.00 g of He in a 5.00-L container at 21°C.

Solution

We start, as always, by listing what we know about the problem.

$$V = 5.00\,L \qquad T = 21°C$$

We also know the masses of H_2 and He in the flask, however, so we can calculate the number of moles of each gas in the container.

$$1.00 \text{ g } H_2 \times \frac{1 \text{ mol } H_2}{2.016 \text{ g } H_2} = 0.496 \text{ mol } H_2$$

$$1.00 \text{ g He} \times \frac{1 \text{ mol He}}{4.003 \text{ g He}} = 0.250 \text{ mol He}$$

Because we know the volume, the temperature, and the number of moles of each component of the mixture, we can calculate the partial pressures of H_2 and He. We start by rearranging the ideal gas equation as follows.

$$P = \frac{nRT}{V}$$

We then use that equation to calculate the partial pressure of each gas.

$$P_{H_2} = \frac{(0.496 \text{ mol } H_2)(0.08206 \text{ L} \cdot \text{atm/mol} \cdot \text{K})(294 \text{ K})}{5.00 \text{ L}} = 2.39 \text{ atm}$$

$$P_{He} = \frac{(0.250 \text{ mol He})(0.08206 \text{ L} \cdot \text{atm/mol} \cdot \text{K})(294 \text{ K})}{5.00 \text{ L}} = 1.21 \text{ atm}$$

Dalton's law of partial pressures predicts that the total pressure in the mixture would be the sum of the partial pressures of the two components.

$$P_{tot} = P_{H_2} + P_{He} = 3.60 \text{ atm}$$

or

$$P_{tot} = n_{tot}\frac{RT}{V} = (0.496 \text{ mol}_{H_2} + 0.250 \text{ mol}_{He})\frac{\left(0.08206 \dfrac{\text{L} \cdot \text{atm}}{\text{mol} \cdot \text{K}}\right)(294 \text{ K})}{5.00 \text{ L}}$$

$$= 3.60 \text{ atm}$$

• •

Dalton derived the law of partial pressures from his work on the amount of water vapor that could be absorbed by air at different temperatures. It is therefore fitting that this law is most often used to correct for the amount of water vapor picked up when a gas is collected by displacing water. Suppose, for example, that we want to collect a sample of O_2 prepared by heating potassium chlorate until it decomposes.

$$2 \text{ KClO}_3(s) \longrightarrow 2 \text{ KCl}(s) + 3 \text{ O}_2(g)$$

The gas given off in this reaction can be collected by filling a flask with water, inverting the flask in a trough, and then letting the gas bubble into the flask, as shown in Figure 6.13.

Because some of the water in the flask will evaporate during the experiment, the gas that collects in the flask is going to be a mixture of O_2 and water vapor. The total pressure of the gas is the sum of the partial pressures of the two components.

$$P_{tot} = P_{O_2} + P_{water}$$

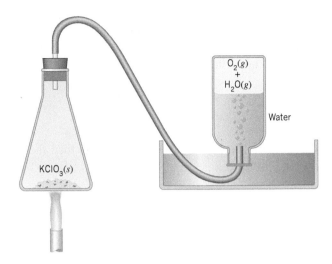

Fig. 6.13 Gases, such as O_2, that aren't very soluble in water can be collected by displacing water from a container.

The total pressure of the mixture must be equal to atmospheric pressure. (If it were any larger, the gas would push water out of the container. If it were any smaller, water would be forced into the container.) If we had some way to estimate the partial pressure of the water in the system, we could therefore calculate the partial pressure of the oxygen gas.

The partial pressure of the gas that collects in a closed container above a liquid is known as the **vapor pressure** of the liquid. Table B.3 in Appendix B gives the vapor pressure of water at various temperatures. If we know the temperature at which a gas is collected and we assume that the gas is saturated with water vapor at that temperature, we can calculate the partial pressure of the gas by subtracting the vapor pressure of water from the total pressure of the mixture of gases collected in the experiment.

Exercise 6.4

Calculate the number of grams of O_2 that can be collected by displacing water from a 250-mL bottle at 21°C and 746.2 mmHg.

Solution

It is often useful to work a problem backward. In this case, our goal is the mass of O_2 in the flask. To reach this goal we need to calculate the number of moles of O_2 in the sample. To do this, we need to know the pressure, volume, and temperature of the O_2. We already know some of this information.

We know the volume of the O_2 because a gas expands to fill its container. We also know the temperature of the O_2. But we don't know the pressure of the O_2.

$$V_{O_2} = 250 \, \text{mL} \qquad T_{O_2} = 21°\text{C} \qquad P_{O_2} = ?$$

All we know is the *total pressure* of the oxygen plus the water vapor that collects in the bottle.

$$P_{\text{tot}} = P_{O_2} + P_{\text{water}} = 746.2 \, \text{mmHg}$$

Table B.3 in Appendix B gives the vapor pressure of water at 21°C as 18.7 mmHg. If we assume that the gas collected in the experiment is saturated with water vapor, we can write the following.

$$P_{tot} = P_{O_2} + P_{water}$$
$$746.2 \text{ mmHg} = P_{O_2} + 18.7 \text{ mmHg}$$

In other words, the partial pressure of O_2 must be less than the total pressure in the bottle by 18.7 mmHg.

$$P_{O_2} = 746.2 \text{ mmHg} - 18.7 \text{ mmHg} = 727.5 \text{ mmHg}$$

We now know the pressure (727.5 mmHg), volume (250 mL), and temperature (21°C) of the O_2 collected in the experiment. Once we convert the measurements to appropriate units, we can use the ideal gas equation to calculate the number of moles of O_2 collected.

$$n = \frac{PV}{RT} = \frac{(0.9572 \text{ atm})(0.250 \text{ L})}{(0.08206 \text{ L} \cdot \text{atm/mol} \cdot \text{K})(294 \text{ K})} = 0.00992 \text{ mol } O_2$$

We now have enough information to calculate the number of grams of O_2 collected.

$$0.00992 \text{ mol } O_2 \times \frac{32.00 \text{ g } O_2}{1 \text{ mol}} = 0.317 \text{ g}$$

6.15 Ideal Gas Calculations: Part I

The ideal gas equation can be used to predict the value of any one of the four variables that describes a gas from known values of the other three.

Exercise 6.5

Divers wear tanks that have the nickname "scuba" from the term "Self-Contained Underwater Breathing Apparatus." A typical tank has a volume of 11.1 liters that can be filled with compressed air at pressures of up to about 200 atmospheres. Calculate the mass of the air that can be stored at 21°C and 170 atm in one of these cylinders. Assume that the air in the atmosphere is a mixture of roughly 78% N_2 (28.0 g/mol), 21% O_2 (32.0 g/mol), and 1% Ar (39.9 g/mol) and therefore has an average molecular weight of about 29.0 g/mol.

Solution

For this problem we know three of the four variables in the ideal gas equation. We know the pressure (170 atm), volume (11.1 L), and temperature (21°C) of the gas, as shown in Figure 6.14. This suggests that the ideal gas equation will play an important role in solving the problem.

$$PV = nRT$$

11.1 L
170 atm
21°C (294 K)

Fig. 6.14 Scuba tank of compressed air.

Before we can use the equation, we have to convert the temperature from °C to K.

$$T_K = T_{°C} + 273 = 294 \text{ K}$$

We can then substitute the known information into the ideal gas equation.

$$(170 \text{ atm})(11.1 \text{ L}) = (n)(0.08206 \text{ L} \cdot \text{atm/mol} \cdot \text{K})(294 \text{ K})$$

and then solve the equation for the number of moles of gas in the container.

$$n = \frac{(170 \text{ atm})(11.1 \text{ L})}{(0.08206 \text{ L} \cdot \text{atm/mol} \cdot \text{K})(294 \text{ K})} = 78.2 \text{ mol air}$$

We can then use the average mass of a mole of air to calculate the number of grams of air that can be stored in the cylinder.

$$78.2 \text{ mol air} \times \frac{29.0 \text{ g air}}{1 \text{ mol air}} = 2.27 \times 10^3 \text{ g air}$$

According to this calculation, 2.27 kg of air can be stored in a typical scuba tank. For a typical diver, working at reasonable depths, this amount of gas should last about 20 minutes.

The key to solving ideal gas problems often involves recognizing what is known and deciding how to use the information.

 Exercise 6.6

Calculate the mass of the air in a typical hot-air balloon that would have a volume of 5.00×10^5 L when the temperature of the gas is 30°C and the pressure is 748 mmHg. Assume that the average molecular weight of air is 29.0 g/mol.

Solution

We know the pressure (748 mmHg), volume (5.00×10^5 L), and temperature (30°C) of the gas in the hot-air balloon, as shown in Figure 6.15. Thus, once again, we know three of the four variables in the ideal gas equation. We might therefore consider using the ideal gas equation to calculate the number of moles of gas in the balloon.

$$PV = nRT$$

Before we can do that, however, we have to convert the pressure to units of atmospheres,

$$748 \text{ mmHg} \times \frac{1 \text{ atm}}{760 \text{ mmHg}} = 0.984 \text{ atm}$$

and the temperature to kelvins.

$$T_K = T_{°C} + 273 = 303 \text{ K}$$

5.00 × 10⁵ L
748 mmHg
(0.984 atm)
30°C
(303 K)

Fig. 6.15 Hot-air balloon.

We can then substitute this information into the ideal gas equation and solve for the variable we don't know.

$$n = \frac{PV}{RT} = \frac{(0.984 \text{ atm})(5.00 \times 10^5 \text{ L})}{(0.08206 \text{ L} \cdot \text{atm/mol} \cdot \text{K})(303 \text{ K})} = 1.98 \times 10^4 \text{ mol}$$

We can then go back and reread the problem to see whether we are getting any closer to its solution.

The problem asks for the mass of the air in the balloon, which can be calculated from the number of moles of gas and the average molecular weight of air.

$$1.98 \times 10^4 \text{ mol} \times \frac{29.0 \text{ g}}{1 \text{ mol}} = 5.73 \times 10^5 \text{ g}$$

The balloon therefore contains more than 1000 pounds of air.

The ideal gas equation can be applied to problems that don't seem to ask for one of the variables in this equation.

Exercise 6.7

Calculate the molecular weight of the butane found in a can of butane lighter fluid if 0.5813 g of the gas fills a 250.0-mL flask at a temperature of 24.4°C and a pressure of 742.6 mmHg.

Solution

We know something about the pressure (742.6 mmHg), volume (250.0 mL), and temperature (24.4°C) of the gas, as shown in Figure 6.16. We might therefore start by calculating the number of moles of gas in the sample. To do this, we need to convert the pressure, volume, and temperature into appropriate units.

$$742.6 \text{ mmHg} \times \frac{1 \text{ atm}}{760 \text{ mmHg}} = 0.9771 \text{ atm}$$

$$250.0 \text{ mL} \times \frac{1 \text{ L}}{1000 \text{ mL}} = 0.2500 \text{ L}$$

$$T_{\text{K}} = T_{\circ\text{C}} + 273.15 = 297.6 \text{ K}$$

Solving the ideal gas equation for the number of moles of gas and substituting the known values of the pressure, volume, and temperature into the equation give the following result.

$$n = \frac{PV}{RT} = \frac{(0.9771 \text{ atm})(0.2500 \text{ L})}{(0.08206 \text{ L} \cdot \text{atm/mol} \cdot \text{K})(297.6 \text{ K})} = 0.01000 \text{ mol}$$

The problem asks for the molecular weight of the gas, which would have the units of grams per mole. At this point, we know the mass of the gas in the sample and the number of moles of gas in the sample. We can therefore

250.0 mL
24.4°C
(297.6 K)

0.5813 g butane
742.6 mmHg
(0.9771 atm)

Fig. 6.16 Stoppered round-bottom flask containing butane.

calculate the molecular weight of butane by dividing the mass of the sample by the number of moles of gas in the sample.

$$\frac{0.5813 \text{ g}}{0.01000 \text{ mol}} = 58.13 \text{ g/mol}$$

The molecular weight of butane is therefore 58.13 g/mol. The chemical formula for the normal isomer of butane would be written as $CH_3CH_2CH_2CH_3$. "Butane lighter fluid" actually contains an isomer of butane, known as isobutane, with the following structure:

$$
\begin{array}{c}
CH_3 \\
\diagdown \\
\quad CH-CH_2 \\
\diagup \qquad\qquad \diagdown \\
CH_3 \qquad\qquad CH_3
\end{array}
$$

• •

The ideal gas equation can even be used to solve problems that don't seem to contain enough information.

Exercise 6.8

Dinitrogen oxide, N_2O, is a sweet-smelling, colorless gas that was originally given the name "nitrous oxide." Humphry Davy proposed calling it "laughing gas" after noticing the amusing effects it had on visitors to his institute to whom he administered the gas. Upon sampling this gas, the poet Robert Southey commented "I am sure the air in heaven must be this wonder-working gas of delight." It was not until the 1840's that a dentist named Horace Wells recognized that it could be used as a painkiller, or anesthetic. Nitrous oxide is a popular anesthetic still used by dentists today. Use the ideal gas law to predict the density of N_2O gas at 0°C and 1.00 atm.

Solution

This time we have information about only two of the variables in the ideal gas equation.

$$P = 1.00 \text{ atm} \qquad T = 0°\text{C}$$

What can we calculate from this information? One way to answer that question is to rearrange the ideal gas equation, putting the knowns (P and T) on one side of the equation and the unknowns (n and V) on the other.

$$\frac{n}{V} = \frac{P}{RT}$$

According to this equation, we have enough information to calculate the number of moles of gas per liter.

$$\frac{n}{V} = \frac{P}{RT} = \frac{(1.00 \text{ atm})}{(0.08206 \text{ L} \cdot \text{atm/mol} \cdot \text{K})(273 \text{ K})} = 0.0446 \text{ mol } O_2/L$$

The problem asks for the density of the gas in grams per liter. Because we know the number of moles of N_2O per liter, we can use the molar mass of N_2O to calculate the number of grams per liter.

$$\frac{0.0446 \text{ mol } N_2O}{1 \text{ L}} \times \frac{44.01 \text{ g } N_2O}{1 \text{ mol } N_2O} = 1.96 \text{ g } N_2O/L$$

The density of N_2O gas at 0°C and 1.00 atm is therefore 1.96 g/L.

Exercise 6.9

Explain why balloons filled with helium at room temperature (21°C) and 1.00 atm rise up toward the ceiling, whereas balloons filled with CO_2 sink toward the floor. Assume that the average molecular weight of dry air is 29.0 g/mol.

Solution

Avogadro's hypothesis suggests that equal volumes of different gases at the same temperature and pressure contain the same number of gas particles. But this hypothesis is nothing more than a special case of the ideal gas law, which predicts that the number of moles of gas per liter at room temperature (21°C) and 1 atm does not depend on the identity of the gas.

$$\frac{n}{V} = \frac{P}{RT} = \frac{(1.00 \text{ atm})}{(0.08206 \text{ L} \cdot \text{atm/mol} \cdot \text{K})(294 \text{ K})} = 0.0414 \text{ mol/L}$$

The number of grams per liter, however, depends on the mass of a mole of each gas. Helium is less dense than air, whereas CO_2 is more dense than air.

$$d_{He} = \frac{0.0414 \text{ mol}}{1 \text{ L}} \times \frac{4.00 \text{ g He}}{1 \text{ mol}} = 0.166 \text{ g He/L}$$

$$d_{air} = \frac{0.0414 \text{ mol}}{1 \text{ L}} \times \frac{29.00 \text{ g air}}{1 \text{ mol}} = 1.20 \text{ g air/L}$$

$$d_{CO_2} = \frac{0.0414 \text{ mol}}{1 \text{ L}} \times \frac{44.0 \text{ g } CO_2}{1 \text{ mol}} = 1.82 \text{ g } CO_2/L$$

A balloon filled with helium therefore weighs significantly less than the air it displaces, so it rises toward the ceiling. A balloon filled with CO_2, on the other hand, weighs more than the air it displaces, so it sinks toward the floor.

➤ **CHECKPOINT**

Use the difference between the average molecular weights of air (29.0 g/mol) and water (18.0 g/mol) to explain why 1 L of dry air weighs more than 1 L of air that has been saturated with water vapor.

6.16 Ideal Gas Calculations: Part II

Gas law problems often ask you to predict what happens when one or more changes are made in the variables that describe the gas. The most powerful approach to such problems is based on the fact that the ideal gas constant is, in fact, a constant.

We can start by solving the ideal gas equation for the ideal gas constant.

$$R = \frac{PV}{nT}$$

According to this equation, the ratio PV/nT under one set of conditions must be equal to this ratio under any other set of conditions.

$$\frac{P_1 V_1}{n_1 T_1} = \frac{P_2 V_2}{n_2 T_2}$$

We now substitute the known values of pressure, temperature, volume, and amount of gas into the equation for the two sets of conditions and solve for the appropriate unknown.

Exercise 6.10

Assume that a sample of NH_3 gas fills a 27.0-L container at $-15°C$ and 2.58 atm. Calculate the volume of the gas at 21°C and 751 mmHg.

Solution

We can start by listing what we know and what we don't know about the initial and final states of the system.

Initial Conditions	Final Conditions
$P_1 = 2.58$ atm	$P_2 = 751$ mmHg
$V_1 = 27.0$ L	$V_2 = ?$
$T_1 = -15°C$	$T_2 = 21°C$
$n_1 = ?$	$n_2 = ?$

We then convert the data to a consistent set of units for use in the ideal gas equation.

Initial Conditions	Final Conditions
$P_1 = 2.58$ atm	$P_2 = 0.988$ atm
$V_1 = 27.0$ L	$V_2 = ?$
$T_1 = 258$ K	$T_2 = 294$ K
$n_1 = ?$	$n_2 = ?$

We can then substitute this information into the following equation.

$$\frac{P_1 V_1}{n_1 T_1} = \frac{P_2 V_2}{n_2 T_2}$$

When this is done, we get one equation with three unknowns: n_1, n_2, and V_2.

$$\frac{(2.58 \text{ atm})(27.0 \text{ L})}{(n_1)(258 \text{ K})} = \frac{(0.988 \text{ atm})(V_2)}{(n_2)(294 \text{ K})}$$

It is impossible to solve one equation with three unknowns. But a careful reading of the problem suggests that the number of moles of NH_3 is the same before and after the temperature and pressure change.

$$n_1 = n_2$$

Substituting this equality into the unknown equation gives the following.

$$\frac{(2.58 \text{ atm})(27.0 \text{ L})}{(n_1)(258 \text{ K})} = \frac{(0.988 \text{ atm})(V_2)}{(n_1)(294 \text{ K})}$$

Multiplying both sides of the equation by n_1 gives an equation with one unknown.

$$\frac{(2.58 \text{ atm})(27.0 \text{ L})}{(258 \text{ K})} = \frac{(0.988 \text{ atm})(V_2)}{(294 \text{ K})}$$

We can therefore solve for the only unknown left in the equation.

$$V_2 = \frac{(2.58 \text{ atm})(27.0 \text{ L})(294 \text{ K})}{(258 \text{ K})(0.988 \text{ atm})} = 80.3 \text{ L}$$

According to the calculation, the volume of the gas should increase from 27.0 L at $-15°C$ and 2.58 atm to 80.3 L at 21°C and 751 mmHg.

6.17 The Kinetic Molecular Theory

The experimental observations about the behavior of gases discussed so far can be explained with a theoretical model known as the *kinetic molecular theory*. This theory is based on the following postulates, or assumptions.

1. Gases are composed of a large number of particles that behave like hard, spherical objects in a state of constant, random motion.
2. These particles move in a straight line until they collide with another particle or the walls of the container.
3. The particles in a gas are much smaller than the distance between these particles. Most of the volume of a gas is therefore empty space.
4. There is no force of attraction between gas particles or between the particles and the walls of the container.
5. Collisions between gas particles and collisions with the walls of the container are perfectly elastic. Energy can be transferred from one particle to another during a collision, but the total kinetic energy of the particles after the collision is the same as it was before the collision.
6. The average kinetic energy of a collection of gas particles depends on the temperature of the gas and nothing else.

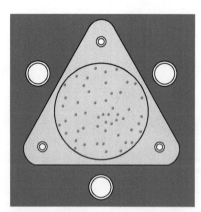

Fig. 6.17 The six postulates of the kinetic molecular theory can be demonstrated with a molecular dynamics simulator such as this.

The assumptions of the kinetic molecular theory can be illustrated with the apparatus shown in Figure 6.17, which consists of a glass plate surrounded by walls mounted on a vibrating motor. A handful of steel ball bearings are placed on top of the glass plate to represent the gas particles.

When the motor is turned on, the glass plate vibrates, which makes the ball bearings move in a constant, random fashion (postulate 1). Each ball moves in a straight line until it collides with another ball or with the walls of the container (postulate 2). Although collisions are frequent, the average distance between the ball bearings is much larger than the diameter of the balls (postulate 3). There is no force of attraction between the individual ball bearings or between the ball bearings and the walls of the container (postulate 4).

The collisions that occur in the apparatus are very different from those that occur when a rubber ball is dropped on the floor. Collisions between the rubber ball and the floor are *inelastic,* as shown in Figure 6.18. A portion of the energy of the ball is lost each time it hits the floor, until it eventually rolls to a stop. In our apparatus, however, the collisions are perfectly *elastic.* The balls have just as much kinetic energy—on average—after a collision as before (postulate 5).

Any object in motion has a **kinetic energy** that is defined as one-half of the product of its mass times its velocity squared.

$$KE = \frac{1}{2}mv^2$$

At any time, some of the ball bearings on the apparatus are moving faster than others, but the system can be described by an *average kinetic energy*. When we increase the "temperature" of the system by increasing the voltage to the motors, we find that the average kinetic energy of the ball bearings increases (postulate 6).

We have described gaseous systems in terms of an "average" kinetic energy. Use of the term *average* implies that not all particles are moving with the same kinetic energy. Some particles must be moving with less kinetic energy and some with more kinetic energy than the average. In other words, there is a distribution of kinetic energies among the particles.

Collisions of rapidly moving particles with slower ones result in a transfer of energy from the rapid particle to the slow particle. About 10^{27} collisions per second take place in 1 mL of gas at 1 atm pressure and room temperature. This ensures a continuous exchange of both kinetic energy and velocities among the particles of a gas. At any given instant a few particles are moving very slowly, a few are moving very rapidly, and the bulk of the particles have velocities and hence kinetic energies somewhere between the two extremes.

Figure 6.19 shows a plot of kinetic energy along the x axis versus the relative number of particles that possess a particular kinetic energy along the y axis for an ideal gas. When the temperature of the system increases, the number of particles having a relatively large kinetic energy also increases, as indicated in Figure 6.19 by the shift in the distribution curve to the right.

6.18 How the Kinetic Molecular Theory Explains the Gas Laws

The kinetic molecular theory can be used to explain each of the experimentally determined gas laws. All we have to do is assume that the pressure of a gas results from collisions between the gas particles and the walls of the container. Each time a gas particle hits the wall, it exerts a force on the wall. The frequency of collisions with the walls and the mass and velocity of the gas particles determine the magnitude of the force. Because the pressure of a gas is the force applied to the walls of the container per unit area, a change in either force or surface area results in a change in pressure.

> ► **CHECKPOINT**
> The kinetic molecular theory assumes that the temperature of a gas on the macroscopic scale is directly proportional to the average kinetic energy of the particles on the atomic scale. Use this assumption to explain what happens to the particles of a gas when the temperature of a gas increases.

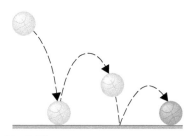

Fig. 6.18 Most collisions are inelastic. Some energy is transferred each time a ball collides with the floor, for example.

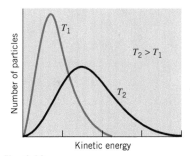

Fig. 6.19 The kinetic molecular theory states that the *average* kinetic energy of a gas is proportional to the temperature of the gas and nothing else. At any given temperature, however, some of the gas particles are moving faster than others.

THE LINK BETWEEN P AND n (T AND V CONSTANT)

Any increase in the number of gas particles in the container increases the frequency of collisions with the walls and therefore the pressure of the gas.

AMONTONS' LAW ($P \propto T$; n AND V CONSTANT)

The last postulate of the kinetic molecular theory states that the average kinetic energy of a gas particle depends only on the temperature of the gas. Thus, the average kinetic energy of the gas particles increases as the gas becomes warmer. Because the mass of the particles is constant, their kinetic energy can increase only if the average velocity of the particles increases. The faster the particles are moving when they hit the wall and the larger the frequency of collisions with the wall, the larger the force they exert on the wall. Because the volume is held constant, the surface area is constant. Because the impact per collision and the number of collisions increase as the temperature increases, the pressure of the gas must increase as well.

BOYLE'S LAW ($P \propto 1/V$; T AND n CONSTANT)

Gases can be compressed because most of the volume of a gas is empty space. If we compress a gas without changing its temperature, the average kinetic energy of the gas particles stays the same. There is no change in the speed with which the particles move, but the volume of the container is smaller. Thus the particles travel from one end of the container to the other in a shorter time. This means that they hit the walls more often. Any increase in the frequency of collisions with the walls must lead to an increase in the pressure of the gas. Thus the pressure of a gas increases as the volume of the gas decreases.

CHARLES' LAW ($V \propto T$; P AND n CONSTANT)

The average kinetic energy of the particles in a gas is proportional to the temperature of the gas. Because the mass of the particles is constant, the particles must move faster (on average) as the gas becomes warmer. If they move faster, the particles will have a larger impact on the container each time they hit the walls and they will strike the walls more frequently. These two factors lead to an increase in the pressure of the gas. If the walls of the container are flexible, the container will expand until the pressure of the gas once again balances the pressure of the atmosphere. The volume of the gas in a flexible container therefore increases as the temperature of the gas increases.

AVOGADRO'S HYPOTHESIS ($V \propto n$; T AND P CONSTANT)

As the number of gas particles increases, the frequency of collisions with the walls of the container must increase. This, in turn, leads to an increase in the pressure of the gas. Flexible containers, such as a balloon, will expand until the pressure of the gas inside the balloon once again balances the pressure of the gas outside. Thus the volume of the gas is proportional to the number of gas particles.

DALTON'S LAW OF PARTIAL PRESSURES ($P_{tot} = P_1 + P_2 + P_3 + \ldots$; V AND T CONSTANT)

Imagine what would happen if six ball bearings of a different size were added to the ball bearings already in the apparatus in Figure 6.17. The total pressure would increase because there would be more collisions with the walls of the container. But the pressure resulting from the collisions between the original ball bearings and the walls of the container would remain the same. There is so much empty space in the container that each type of ball bearing hits the walls of the container as often in the mixture as it would if there was only one kind of ball bearing on the glass plate. The total number of collisions with the wall in the mixture is therefore equal to the sum of the collisions that would occur when each size of ball bearing is present by itself. In other words, the total pressure of a mixture of gases is equal to the sum of the partial pressures of the individual gases.

RELATIVE VELOCITIES OF MOLECULES

The last postulate of the kinetic theory states that the temperature of a system is proportional to the average kinetic energy of its particles and nothing else. In other words, the temperature of a system increases if and only if there is an increase in the average kinetic energy of its particles.

Two gases at the same temperature, such as H_2 and O_2, therefore must have the same average kinetic energy. This can be represented by the following equation.

$$\frac{1}{2} m_{H_2} v_{H_2}^2 = \frac{1}{2} m_{O_2} v_{O_2}^2$$

This equation can be simplified by multiplying both sides by 2.

$$m_{H_2} v_{H_2}^2 = m_{O_2} v_{O_2}^2$$

It can then be rearranged to give the following.

$$\frac{v_{H_2}^2}{v_{O_2}^2} = \frac{m_{O_2}}{m_{H_2}}$$

Taking the square root of both sides of the equation gives a relationship between the ratio of the velocities at which the two gases move and the square root of the ratio of their masses.

$$\frac{v_{H_2}}{v_{O_2}} = \sqrt{\frac{m_{O_2}}{m_{H_2}}}$$

Because mass is proportional to molecular weight, this relationship can also be written in terms of molecular weight, as follows.

$$\frac{v_{H_2}}{v_{O_2}} = \sqrt{\frac{MW_{O_2}}{MW_{H_2}}}$$

Exercise 6.11

Calculate the average velocity of an H_2 molecule at 0°C if the average velocity of an O_2 molecule at that temperature is 425 m/s.

Solution

The relative velocities of the H_2 and O_2 molecules at a given temperature are described by the following equation.

$$\frac{v_{H_2}}{v_{O_2}} = \sqrt{\frac{MW_{O_2}}{MW_{H_2}}}$$

Table 6.6

The Average Speed (m/s) of Common Gases at 25ºC and 1 atm

Compound	Average Speed in m/s
H_2	1770
He	1260
H_2O	590
N_2	470
O_2	440
CO_2	380
Cl_2	300
HI	220
Hg	180

➤ **CHECKPOINT**

According to kinetic molecular theory, if two different gases are at the same temperature, they have the same kinetic energy. How can this be true if the particles composing the two gases are moving at different velocities?

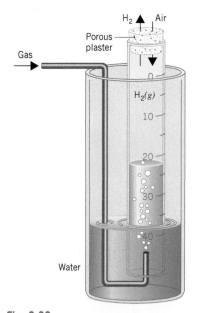

Fig. 6.20 The rate of diffusion, or mixing of a gas with air, can be studied with the apparatus shown. If the gas escapes from the tube faster than air enters the tube, the amount of water in the tube will increase. If air enters the tube faster than the gas escapes, water will be displaced from the tube.

Substituting the molecular weights of H_2 and O_2 and the average velocity of an O_2 molecule into the equation gives the following.

$$\frac{v_{H_2}}{425 \text{ m/s}} = \sqrt{\frac{32.0 \text{ g/mol}}{2.02 \text{ g/mol}}}$$

Solving this equation for the average velocity of an H_2 molecule gives a value of about 1690 m/s, or about 3780 mi/h.

Because it is easy to make mistakes when setting up a ratio problem such as this, it is important to check the answer to see whether it makes sense. This relationship suggests that light molecules move faster on the average than heavy molecules. In this case, the answer makes sense because H_2 molecules are much lighter than O_2 molecules and they should therefore travel much faster.

The average speed of various gases at 25° and 1 atmosphere pressure are given in Table 6.6. Note that a Boeing 747 traveling across the ocean at 600 miles per hour is moving through space at a rate of 267 meters per second. The average speed of an H_2 molecule at 25°C and 1 atm is therefore just under 4000 miles per hour.

6.19 Graham's Laws of Diffusion and Effusion

Most of the physical properties of a gas do not depend on the identity of the gas. But a few do. One of these physical properties can be seen when the movement of gases is studied.

In 1829 Thomas Graham used an apparatus similar to the one shown in Figure 6.20 to study the **diffusion** of gases—the rate at which two gases mix. The apparatus consists of a glass tube sealed at one end with plaster that has holes large enough to allow a gas to enter or leave the tube. When the tube is filled with H_2 gas, the level of water in the tube slowly rises because the H_2 molecules inside the tube escape through the holes in the plaster more rapidly than the molecules in the air outside the tube can enter the tube. By studying the rate at which the water level in the apparatus changed, Graham was able to obtain data on the rate at which different gases mixed with air.

Graham found that the rates at which gases diffuse are inversely proportional to the square root of their densities.

$$\text{rate}_{\text{diffusion}} \, \alpha \, \frac{1}{\sqrt{\text{density}}}$$

This relationship eventually became known as **Graham's law of diffusion.**

To understand the importance of this discovery, we have to remember that equal volumes of different gases at the same temperature and pressure contain the same number of particles. As a result, the number of moles of gas per liter at a given temperature and pressure is constant. This means that the density of a gas is directly proportional to its molecular weight (MW). Graham's law of diffusion can therefore also be written as follows.

$$\text{rate}_{\text{diffusion}} \; \alpha \; \frac{1}{\sqrt{MW}}$$

Similar results were obtained when Graham studied the rate of **effusion** of a gas, which is the rate at which the gas escapes through a pinhole into a vacuum. The rate of effusion of a gas is also inversely proportional to the square root of either the density or the molecular weight of the gas.

$$\text{rate}_{\text{effusion}} \; \alpha \; \frac{1}{\sqrt{\text{density}}}$$

$$\text{rate}_{\text{effusion}} \; \alpha \; \frac{1}{\sqrt{MW}}$$

Graham's law of effusion can be demonstrated with the apparatus in Figure 6.21. A thick-walled filter flask is evacuated with a vacuum pump. A syringe is filled with 25 mL of gas, and the time required for the gas to escape through the syringe needle into the evacuated filter flask is measured with a stopwatch.

The experimental data in Table 6.7 were obtained by using a special needle with a very small (0.015-cm) hole through which the gas could escape.

As we can see when the data are graphed in Figure 6.22, the *time* required for 25-mL samples of different gases to escape into a vacuum is proportional to the square root of the molecular weight of the gas. The *rate* at which the gases effuse is therefore inversely proportional to the square root of the molecular weight.

Graham's observations about the rate at which gases diffuse (mix) or effuse (escape through a pinhole) suggest that relatively light gas particles, such as H_2 molecules and He atoms, move faster than relatively heavy gas particles, such as CO_2 and SO_2 molecules.

Graham's experimental observations support the relation between velocities and molecular weight developed in Section 6.18 from the basic assumptions of the kinetic molecular theory. If the rate of diffusion or effusion of a gas is directly related to the average velocity with which the molecules move, then Graham's law can be obtained from the following mathematical relationship.

$$\frac{v_1}{v_2} = \sqrt{\frac{MW_2}{MW_1}} = \frac{\text{rate}_1}{\text{rate}_2}$$

Table 6.7

Time Required for 25-mL Samples of Gases to Escape through a 0.015-cm Hole into a Vacuum

Compound	Time (s)	Molar Mass (g/mol)
H_2	5.1	2.02
He	7.2	4.00
NH_3	14.2	17.0
Air	18.2	29.0
O_2	19.2	32.0
CO_2	22.5	44.0
SO_2	27.4	64.1

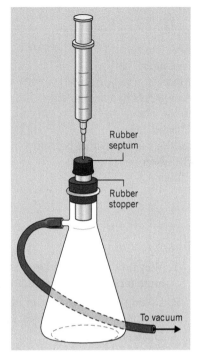

Fig. 6.21 The rate of effusion of a gas can be demonstrated with the apparatus shown. The time required for the gas in the syringe to escape into an evacuated flask is measured. The faster the gas molecules effuse, the less time it takes for a given volume of the gas to escape into the flask.

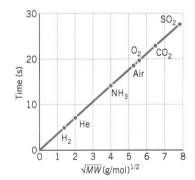

Fig. 6.22 Graph of the time required for 25-mL samples of different gases to escape into an evacuated flask versus the square root of the molecular weight of the gas. Relatively heavy molecules move more slowly, and it takes more time for the gas to escape.

Key Terms

Absolute zero
Amontons' law
Area
Avogadro's hypothesis
Boyle's law
Charles' law
Dalton's law of partial pressures
Diffusion
Effusion

Force
Gay-Lussac's law
Graham's law of diffusion
Graham's law of effusion
Ideal gas constant
Ideal gas equation
Ideal gas law
Kinetic energy

Kinetic molecular theory
Law of combining volumes
Partial pressure
Pressure
States
STP
Temperature
Vapor pressure

Problems

Temperature

1. Why do some objects known to be at the same temperature feel warmer or colder to the touch?

2. Could you use touch to arrange objects in order of increasing warmth or coldness? Explain why or why not.

3. Define *temperature*.

Temperature as a Property of Matter

4. What are the two most common temperature scales in everyday use? What temperature scale do scientists usually use?

5. What is the freezing point of water on three different temperature scales?

6. Describe what happens to the motion of water molecules as water at 25°C is heated to 373 K.

7. What is meant by *kinetic energy*?

8. If the average kinetic energies of the atoms composing two different substances are the same, what can be said about the temperatures of the two substances?

The States of Matter

9. Why are gases easier to study than other states of matter?

10. Liquid water and gaseous water coexist at 100°C and 1 atm pressure. How do the average kinetic energies of the molecules of water in these states compare under these conditions?

Elements or Compounds That Are Gases at Room Temperature

11. Which of the following elements and compounds are most likely to be gases room temperature?

    (a) Ar   (b) CO   (c) CH_4 (d) $C_{10}H_{22}$
    (e) Cl_2 (f) Fe_2O_3 (g) Na   (h) NaCl
    (i) Pt   (j) S_8

12. Is it true that all elements that have a molecular weight below 40 g/mol are gases at room temperature? Support your answer.

13. In general, common gases have two characteristics. What are they?

The Properties of Gases

14. Why is the volume of a gas the same as the volume of its container?

15. How does the volume of a mole of liquid water compare to that of a mole of gaseous water both at 25°C and 1 atm?

16. A helium atom is smaller than a xenon atom. Explain why the volumes of a mole of helium and xenon are the same at 25°C and 1 atm.

17. Predict what will happen to the mass of an evacuated cylinder when it is filled with helium gas. Will it increase, decrease, or remain the same?

Pressure versus Force

18. In what way is pressure related to force?

19. In a hurricane-strength wind, which has the greater pressure on it, a billboard or a stop sign? Which has the greater force on it?

20. If two marbles are pushed together, a great pressure can be produced with little applied force. Explain why.

Atmospheric Pressure

21. What is the pressure in units of atmospheres when a barometer reads 745.8 mmHg? What is the pressure in units of pascals?

22. One atmosphere pressure will support a column of mercury 760 mm tall in a barometer with a tube 1.00 cm in diameter. What would be the height of the column of mercury if the diameter of the tube were twice as large?

23. Atmospheric pressure is announced during weather reports in the United States in units of inches of mercury. How many inches of mercury would exert a pressure of 1.00 atm? In Canada, atmospheric pressure is reported in units of pascals. What is 1.00 atm in pascals?

24. If the directions that come with your car tell you to inflate the tires to 200 kPa pressure and you have a tire pressure gauge calibrated in pounds per square inch (psi), what pressure in psi should you use?

25. The vapor pressure of the mercury gas that collects at the top of a barometer is 2×10^{-3} mmHg. Calculate the vapor pressure of the gas in atmospheres.

26. Calculate the force exerted on the earth by the atmosphere if atmospheric pressure is 14.7 lb/in.^{2} and the surface area of the planet is 5.1×10^8 km.^{2}

Boyle's Law

27. A 425-mL sample of O_2 gas was collected at 742.3 mmHg. What would be the pressure in mmHg if the gas were allowed to expand to 975 mL at constant temperature?

28. What would happen to the volume of a balloon filled with 0.357 L of H_2 gas collected at 741.3 mmHg if the atmospheric pressure increased to 758.1 mmHg? T is constant.

29. What is the volume of a scuba tank if it takes 2000 L of air collected at 1 atm to fill the tank to a pressure of 150 atm? Assume that T is constant.

30. Calculate the volume of a balloon that could be filled at 1.00 atm with the helium in a 2.50-L compressed gas cylinder in which the pressure is 200 atm at 25°C.

Amontons' Law

31. A can is filled with 5.00 atm of a gas at 21°C. Calculate the pressure in the can when it is stored in a warehouse on a hot summer day when the temperature reaches 38°C.

32. An automobile tire was inflated to a pressure of 32 lb/in.^{2} at 21°C. At what temperature would the pressure reach 60 psi?

33. At 25°C, four-fifths of the pressure of the atmosphere is due to N_2 and one-fifth is due to O_2. At 100°C, what fraction of the pressure is due to N_2?

Charles' Law

34. Calculate the percent change in the volume of a toy balloon when the gas inside is heated from 22°C to 75°C in a hot-water bath.

35. A sample of O_2 gas with a volume of 0.357 liter was collected at 21°C. Calculate the volume of the gas when it is cooled to 0°C if the pressure remains constant.

36. Two balloons at room temperature are filled to the same pressure, one with 1 mol of He gas and one with 1 mol of Xe gas. If they are both cooled to the freezing point of water, which balloon will have the greater change in volume?

Gay-Lussac's Law

37. Calculate the volume of H_2 and N_2 gas formed when 1.38 L of NH_3 decomposes at a constant temperature and pressure.

$$2 NH_3(g) \longrightarrow N_2(g) + 3 H_2(g)$$

38. Ammonia burns in the presence of oxygen to form nitrogen oxide and water.

$$4 NH_3(g) + 5 O_2(g) \longrightarrow 4 NO(g) + 6 H_2O(g)$$

What volume of NO can be prepared when 15.0 L of ammonia reacts with excess oxygen if all measurements are made at the same temperature and pressure?

39. Acetylene burns in oxygen to form CO_2 and H_2O.

$$2 C_2H_2(g) + 5 O_2(g) \longrightarrow 4 CO_2(g) + 2 H_2O(g)$$

Calculate the total volume of the products formed when 15.0 L of C_2H_2 burns in the presence of 15.0 L of O_2 if all measurements are made at the same temperature and pressure.

40. Methane reacts with steam to form hydrogen and carbon monoxide.

$$CH_4(g) + H_2O(g) \longrightarrow CO(g) + 3 H_2(g)$$

It can also react with steam to form carbon dioxide.

$$CH_4(g) + 2 H_2O(g) \longrightarrow CO_2(g) + 4 H_2(g)$$

What are the products of a reaction if 1.50 L of methane is found by experiment to react with 1.50 L of water vapor? T and P are constant.

Avogadro's Hypothesis

41. Which weighs more, dry air at 25°C and 1 atm, or air at that temperature and pressure that is saturated with water vapor? (Assume that the average molecular weight of air is 29.0 g/mol.)

42. Which of the following samples would have the largest volume at 25°C and 750 mmHg?
    (a) 100 g CO_2 (b) 100 g CH_4
    (c) 100 g NO   (d) 100 g SO_2

43. Nitrous oxide decomposes to form nitrogen and oxygen. Use Avogadro's hypothesis to determine the formula for nitrous oxide if 2.36 L of the compound decomposes to form 2.36 L of N_2 and 1.18 L of O_2 at the same temperature and pressure.

44. Two equal-volume containers at the same temperature are filled with different gases to the same pressure. One contains N_2 and one contains H_2. Which has the greater number of molecules?

The Ideal Gas Equation

45. Predict the shape of the following graphs for an ideal gas (assume all other variables are held constant).
    (a) pressure versus volume
    (b) pressure versus temperature
    (c) volume versus temperature

(d) kinetic energy versus temperature

(e) pressure versus the number of moles of gas

46. Which of the following graphs could not give a straight line for an ideal gas? (Assume that all other variables are held constant.)

(a) V versus T (b) T versus P (c) P versus $1/V$

(d) n versus $1/T$ (e) n versus $1/P$

47. Which of the following statements is always true for an ideal gas?

(a) If the temperature and volume of a gas both increase at constant pressure, the number of moles of gas must also increase.

(b) If the pressure increases and the temperature decreases for a constant number of moles of gas, the volume must decrease.

(c) If the volume and the number of moles of gas both decrease at constant temperature, the pressure must decrease.

48. Calculate the value of the ideal gas constant in units of mL-psi/mol-K if 1.00 mol of an ideal gas at 0°C occupies a volume of 22,400 mL at 14.7 lb/in.^{2} pressure.

49. Chemical engineers are sometimes asked to solve problems when the volume is given in cubic centimeters cm^3 instead of liters. What is the value of R in the units of cm^3-atm/mol-K?

Dalton's Law of Partial Pressures

50. Calculate the partial pressure of propane in a mixture that contains equal weights of propane (C_3H_8) and butane (C_4H_{10}) at 20°C and 746 mmHg.

51. Calculate the partial pressure of helium in a 1.00-L flask that contains equal numbers of moles of N_2, O_2, and He at a total pressure of 7.5 atm.

52. Calculate the total pressure in a 10.0-L flask at 27°C of a sample of gas that contains 6.0 g of H_2, 15.2 g of N_2, and 16.8 g of He.

53. A 1.00-L flask is filled with carbon monoxide at 27°C until the pressure is 0.200 atm. Calculate the total pressure after 0.450 g of carbon dioxide has been added to the flask.

54. Calculate the volume of the hydrogen obtained when the water vapor is removed from 289 mL of H_2 gas collected by displacing water from a flask at 15°C and 0.988 atm.

Ideal Gas Calculations: Part I

55. Nitrogen gas sells for roughly $0.50 per 100 cubic feet at 0°C and 1 atm. What is the price per gram of nitrogen?

56. Calculate the pressure in atmospheres of 80 g of CO_2 in a 30-L container at 23°C.

57. Calculate the temperature at which 1.5 g of O_2 has a pressure of 740 mmHg in a 1.0-L container.

58. Calculate the number of kilograms of O_2 gas that can be stored in a compressed gas cylinder with a volume of 40 L when the cylinder is filled at 150 atm and 21°C.

59. Calculate the pressure in an evacuated 250-mL container at 0°C when the O_2 in 1.00 cm^3 of liquid oxygen evaporates. Liquid oxygen has a density of 1.118 g/cm^3.

60. A 1.00-L flask was evacuated, 5.00 g of liquid NH_3 was added to the flask, and the flask was sealed with a cork. If it takes 7.10 atm of pressure in the flask to blow out the cork, at what temperature will the cork be blown out?

61. Calculate the density of CH_2Cl_2 in the gas phase at 40°C and 1.00 atm. Compare this with the density of liquid CH_2Cl_2 (1.336 g/cm^3).

62. Calculate the density of methane gas, CH_4, in kilograms per cubic meter at 25°C and 956 mmHg.

63. Calculate the density of helium at 0°C and 1.00 atm and compare this with the density of air (1.29 g/L) at 0°C and 1.00 atm. Explain why 1.00 ft^3 of helium can lift a weight of 0.076 lb under these conditions.

64. Calculate the ratio of the densities of H_2 and O_2 at 0°C and 100°C at 1.00 atm.

65. Which of the noble gases in Group VIIIA of the periodic table has a density of 3.7493 g/L at 0°C and 1.00 atm?

66. Calculate the average molecular weight of air assuming a sample of air weighs 1.700 times as much as an equivalent volume of ammonia, NH_3.

Ideal Gas Calculations: Part II

67. What is the volume of the gas in a balloon at −195°C if the balloon has been filled to a volume of 5.0 L at 25°C?

68. CO_2 gas with a volume of 25.0 L was collected at 25°C and 0.982 atm. Calculate the pressure of the gas if it is compressed to a volume of 0.150 L and heated to 350°C.

69. Calculate the pressure of 4.80 g of ozone, O_3, in a 2.45-L flask at 25°C. Assume that the ozone completely decomposes to molecular oxygen.

$$2\,O_3(g) \longrightarrow 3\,O_2(g)$$

Calculate the pressure inside the flask once the reaction is complete.

70. 10.0 L of O_2 gas was collected at 120°C and 749.3 mmHg. Calculate the volume of the gas when it is cooled to 0°C and stored in a container at 1.00 atm.

71. 5.0 L of CO_2 gas was collected at 25°C and 2.5 atm. At what temperature would the gas have to be stored to fill a 10.0-L flask at 0.978 atm?

72. Two 10-L samples of O_2 collected at 120°C and 749.3 mmHg are combined and stored in a 1.25-L flask at 27°C. Calculate the pressure of the gas.

The Kinetic Molecular Theory

73. What would happen to a balloon if the gas molecules were in a state of constant motion but the motion was not random?

74. Explain why the pressure of a gas is evidence for the assumption that gas particles are in a state of constant random motion.

75. Use the kinetic molecular theory to explain why the pressure of a gas is proportional to the number of gas particles and the temperature of the gas but inversely proportional to the volume of the gas.

76. Which of the following are true according to the kinetic molecular theory of gases? Explain your answer in each case.

   (a) All of the molecules of a gas are moving with the same kinetic energy at a given temperature.

   (b) The kinetic energy of a gas can be increased by increasing the pressure while holding the temperature constant.

   (c) The average kinetic energy of the molecules increases as temperature increases.

   (d) Most of the volume of a gas is empty space.

How the Kinetic Molecular Theory Explains the Gas Laws

77. Two identical flasks are labeled A and B. Flask A contains NH_3 (g) at 50°C and flask B contains O_2 (g) at 50°C. The average kinetic energy of O_2 is 7×10^{-21} J/molecule.

   (a) What is the average kinetic energy of an NH_3 molecule?

   (b) Which molecule, NH_3 or O_2, is moving the most rapidly? Explain why.

   (c) With the information provided, can you tell in which flask the pressure is the largest? Explain why or why not.

   (d) How could the average kinetic energy of the O_2 molecules be doubled?

   (e) If the only change made were to double the volume of the flask containing O_2 molecules, explain what would happen to

      (i) the average kinetic energy and

      (ii) the pressure of the gas in the flask.

   (f) Give three ways the pressure in the flask containing NH_3 could be doubled.

78. You have three flasks with the following contents.

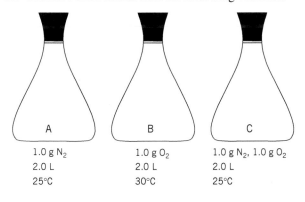

A	B	C
1.0 g N_2	1.0 g O_2	1.0 g N_2, 1.0 g O_2
2.0 L	2.0 L	2.0 L
25°C	30°C	25°C

   (a) $N_2(g)$ and $O_2(g)$ are placed into flasks A and B as indicated above. In which flask is the pressure the largest? Show all calculations.

   (b) In which flask, A or B, is the average kinetic energy the largest? Explain why.

   (c) If 1.0 g of N_2 and 1.0 g of O_2 are placed into the 2.0-L flask C at 25°C, what will be the total pressure? Explain your answer.

   (d) Which molecules are moving the fastest in flask C? Explain your answer.

   (e) Which molecules, N_2 or O_2, are colliding most frequently with the walls of the container in flask C? Explain your answer.

79. Three flasks of identical volume each contain argon gas.

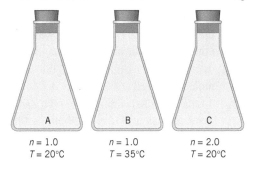

A	B	C
$n = 1.0$	$n = 1.0$	$n = 2.0$
$T = 20°C$	$T = 35°C$	$T = 20°C$

   (a) In which flask(s) is the pressure the largest? State your reasoning.

   (b) In which flask(s) is the average kinetic energy of the argon atoms the smallest? Explain your answer.

   (c) In which flask(s) are the argon atoms moving the most rapidly? Provide a clear explanation.

   (d) In which flask(s) are the argon atoms colliding the most frequently with the walls of the flask? Explain your answer.

80. Two identical flasks, one containing He and one containing Ne, are at the same temperature but different pressures.

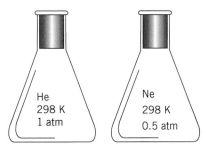

He
298 K
1 atm

Ne
298 K
0.5 atm

Indicate whether the following statements are true or false and explain your reasoning for each.

   (a) Both flasks contain the same number of atoms.

   (b) The atoms in the flask containing the gas at the highest pressure are moving more rapidly than the atoms in the low-pressure flask.

   (c) The kinetic energy of the atoms is greatest in the flask with the lowest pressure.

81. A 1.0-L flask contains CH_4 gas at 0°C and 1.0 atm.
    (a) Will the kinetic energy of the molecules increase, decrease, or remain the same if
        (i) the temperature is increased?
        (ii) the pressure is decreased at 0°C?
        (iii) the volume is decreased at 0°C?
        (iv) the number of moles of CH_4 is doubled and the temperature is decreased to $-10°C$?
    (b) Will the CH_4 molecules move more rapidly, less rapidly, or remain the same if the temperature is unchanged but the volume of the flask is doubled?
    (c) Will the pressure increase, decrease, or remain the same if
        (i) the temperature is increased while n and V are unchanged?
        (ii) the temperature is unchanged when the volume and number of moles are doubled?
        (iii) the volume is doubled and the number of moles is halved at 0°C?
    (d) A 1.0-L flask is filled with He gas at 0°C and 1.5 atm. An identical flask at 0°C and 1.0 atm is filled with CH_4. Which flask contains the most molecules?

Graham's Laws of Diffusion and Effusion

82. Define what is meant by the phrase *diffusion of a gas*.

83. Define what is meant by the phrase *effusion of a gas*.

84. What do Graham's observations suggest about the relative speeds of molecules that have different molecular weights?

85. List the following gases in order of increasing rate of diffusion.
    (a) Ar　　　(b) Cl_2　　　(c) CF_2Cl_2
    (d) SO_2　　　(e) SF_6

86. Bromine vapor is roughly five times as dense as oxygen gas. Calculate the relative rates at which Br_2 (g) and O_2 (g) diffuse.

87. Two flasks with the same volume are connected by a valve. One gram of hydrogen is added to one flask, and 1 g of oxygen is added to the other. What happens to the weight of the gas in the flask filled with hydrogen when the valve is opened?

88. What happens to the relative amounts of N_2, O_2, Ar, CO_2, and He in air as air diffuses from one flask to another through a pinhole?

89. N_2O and NO are often known by the common names *nitrous oxide* and *nitric oxide*. Associate the correct formula with the appropriate common name if nitric oxide diffuses through a pinhole 1.21 times as fast as nitrous oxide.

90. If it takes 6.5 s for 25.0 cm³ of helium gas to effuse through a pinhole into a vacuum, how long would it take for 25.0 cm³ of CH_4 to escape under the same conditions?

91. Calculate the molecular weight of an unknown gas if it takes 60.0 s for 250 cm³ of the gas to escape through a pinhole in a flask into a vacuum and if it takes 84.9 s for the same volume of oxygen to escape under identical conditions.

92. A lecture hall has 50 rows of seats. If laughing gas (N_2O) is released from the front of the room at the same time ammonia (NH_3) is released from the back of the room, in which row (counting from the front) will students first begin to laugh and smell the NH_3? (Assume Graham's law of diffusion is valid.)

93. The atomic weight of radon was first estimated by comparing its rate of diffusion with that of mercury vapor. What is the atomic weight of radon if mercury vapor diffuses 1.082 times as fast?

Integrated Problems

94. What are the molecular formulas for phosphine, PH_x, and diphosphine, P_2H_y, if the densities of the gases are 1.517 and 2.944 g/L, respectively, at 0°C and 1.00 atm?

95. Calculate the weight of magnesium that would be needed to generate 500 mL of hydrogen gas at 0°C and 1.00 atm.

$$Mg(s) + 2\,HCl(aq) \longrightarrow Mg^{2+}(aq) + 2\,Cl^-(aq) + H_2(g)$$

96. Calculate the formula of the oxide formed when 10.0 g of chromium metal reacts with 6.98 L of O_2 at 20°C and 0.994 atm.

97. Calculate the volume of CO_2 gas measured at 756 mmHg and 23°C given off when 150 kg of limestone is heated until it decomposes.

$$CaCO_3(s) \longrightarrow CaO(s) + CO_2(g)$$

98. Calculate the volume of O_2 that would have to be inhaled at 20°C and 1.00 atm to consume 1.00 kg of fat, $C_{57}H_{110}O_6$.

$$2\,C_{57}H_{110}O_6(s) + 163\,O_2(g) \longrightarrow 114\,CO_2(g) + 110\,H_2O(l)$$

99. Calculate the volume of CO_2 gas collected at 23°C and 0.991 atm that can be prepared by reacting 10.0 g of calcium carbonate with excess acid.

$$CaCO_3(aq) + 2\,H^+(aq) \longrightarrow Ca^{2+}(aq) + CO_2(g) + H_2O(l)$$

100. Determine the identity of an unknown metal if 1.00 g of the metal reacts with excess acid according to the

following equation to produce 374 mL of H_2 gas at 25°C and 1.00 atm.

$$M(s) + 2\,H^+(aq) \longrightarrow M^{2+}(aq) + H_2(g)$$

101. Imagine two identical flasks at the same temperature. One contains 2 g of H_2 and the other contains 28 g of N_2. Which of the following properties are the same for the two flasks?

(a) pressure   (b) average kinetic energy   (c) density

(d) number of molecules per container

(e) weight of the container

102. At 25°C, four-fifths of the pressure of the atmosphere is due to N_2 and one-fifth is due to O_2. What volume of a room will N_2 occupy?

103. Which of the noble gases in Group VIIIA of the periodic table has a density of 5.86 g/L at 0°C and 1.00 atm?

104. Two flasks with the same volume and temperature are connected by a valve. One gram of hydrogen is added to one flask, and 1 g of oxygen is added to the other.

(a) In which flask is the average kinetic energy of the gas molecules the largest?

(b) In which flask are the gas molecules moving the most rapidly?

105. When a meteorologist reports precipitation in the weather forecast, this is usually associated with a low-pressure system. Explain why precipitation would be associated with a low atmospheric pressure. (*Hint:* See Problem 41.)

106. If equal weights of O_2 and N_2 are placed in identical containers at the same temperature, which of the following statements is true?

(a) Both flasks contain the same number of molecules.

(b) The pressure in the flask that contains the N_2 will be greater than the pressure in the flask that contains the O_2.

(c) There will be more molecules in the flask that contains O_2 than the flask that contains N_2.

(d) This question cannot be answered unless we know the weights of O_2 and N_2 in the flask.

(e) None of the above are correct.

107. N_2H_4 decomposes at a fixed temperature in a closed container to form $N_2(g)$ and $H_2(g)$. If the reaction goes to completion, the final pressure will be

(a) the same as the initial pressure.

(b) twice the initial pressure.

(c) three times the initial pressure.

(d) one-half of the initial pressure.

(e) one-third of the initial pressure.

108. For the apparatus diagrammed in this problem, what will be the final partial pressures of O_2 and N_2 after the stopcock is opened? The temperature of both flasks is the same and does not change after opening the stopcock. What will be the final pressure in the apparatus?

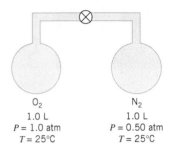

O_2
1.0 L
$P = 1.0$ atm
$T = 25°C$

N_2
1.0 L
$P = 0.50$ atm
$T = 25°C$

109. N_2 gas is in a 2.0-L container at 298 K and 1 atm. Explain what will happen to both the kinetic energy and the frequency of collisions of the nitrogen molecules for each of the following conditions.

(a) The number of moles of N_2 is halved.

(b) The temperature is changed to 15°C.

(c) The temperature is changed to 1500°C.

(d) The volume is decreased to 1 L.

110. Equal volumes of oxygen and an unknown gas at the same temperature and pressure weigh 3.00 g and 7.50 g, respectively. Which of the following is the unknown gas?

(a) CO   (b) CO_2 (c) NO

(d) NO_2 (e) SO_2 (f) SO_3

111. What is the molecular weight of acetone if 0.520 g of acetone occupies a volume of 275.5 mL at 100°C and 756 mmHg?

112. Boron forms a number of compounds with hydrogen, including B_2H_6, B_4H_{10}, B_5H_9, B_5H_{11}, and B_6H_{10}. For which compound would a 1.00-g sample occupy a volume of 390 cm^3 at 25°C and 0.993 atm?

113. Cyclopropane is an anesthetic that is 85.63% carbon and 14.37% hydrogen by mass. What is the molecular formula of the compound if 0.45 L of cyclopropane reacts with excess oxygen at 120°C and 0.72 atm to form 1.35 L of carbon dioxide and 1.35 L of water vapor?

114. Calculate the molecular formula of diazomethane, assuming the compound is 28.6% C, 4.8% H, and 66.6% N by mass and the density of the gas is 1.72 g/L at 25°C and 1.00 atm.

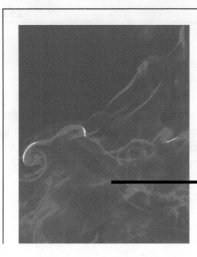

Chapter Six

SPECIAL TOPICS

6A.1 Deviations from Ideal Gas Law Behavior: The van der Waals Equation

The ideal gas equation is a mathematical model that fits experimental data, more or less, under normal conditions. The behavior of real gases usually agrees with the predictions of the ideal gas equation to within ±5% at normal temperatures and pressures. At low temperatures or high pressures, however, real gases deviate significantly from ideal gas behavior. In 1873, while searching for a way to link the behavior of liquids and gases, the Dutch physicist Johannes van der Waals developed an explanation of these deviations from ideal gas behavior and an equation that was able to fit the behavior of real gases over a much wider range of pressures.

van der Waals realized that two of the assumptions of the kinetic molecular theory were questionable. The ideal gas law is based on the kinetic theory of gases, which assumes that gas particles occupy a negligible fraction of the total volume of the gas. It also assumes that the force of attraction between gas molecules is zero.

The first assumption works at pressures close to 1 atm. But it becomes increasingly less valid as the gas is compressed. Imagine for the moment that the atoms or molecules in a gas were all clustered in one corner of a cylinder, as shown in Figure 6A.1. At normal pressures, the volume occupied by the particles is a negligibly small fraction of the total volume of the gas. Essentially all of the volume of the gas is empty space. But at high pressures, this is no longer true. For O_2, for example, the gas molecules occupy 0.13% of the total volume at 1.00 atm but 17% of the volume at 100 atm. As a result, real gases are not quite as compressible at high pressures as an ideal gas. The volume of a real gas at high pressures is therefore somewhat larger than expected from the ideal gas equation.

$$V_{real} \geq V_{ideal}$$

van der Waals recognized that we can correct for the fact that the volume of a real gas is larger than expected at high pressures by *subtracting* a term from the volume of the real gas before we substitute it into the ideal gas equation. He therefore introduced a constant (b) into the ideal gas equation that was related to the volume actually occupied by a mole of gas particles. Because the volume of the gas particles depends on the number of moles of gas in the container, the term that is subtracted from the real volume of the gas is equal to the number of moles of gas times b.

$$P(V_{real} - nb) = nRT$$

When the pressure is relatively low and the volume is reasonably large, the nb term is too small to make any difference in the calculation. But at high pressures, when the volume of the gas is small, the nb term corrects for the fact that the volume of a real gas is larger than expected from the ideal gas equation.

The assumption that there is no force of attraction between gas particles can't be true. If it were, gases would never condense to form liquids. In reality, there is a small force of attraction between gas molecules that tends to hold the molecules together. This force of attraction has two consequences: (1) Gases condense to form liquids at low temperatures, and (2) the pressure of a real gas is sometimes smaller than expected for an ideal gas.

$$P_{real} \leq P_{ideal}$$

To correct for the fact that the pressure of a real gas is smaller than expected from the ideal gas equation, van der Waals *added* a term to the pressure before

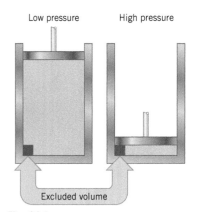

Fig. 6A.1 The volume occupied by the particles in a gas is relatively small at low pressures, but it can be a significant fraction of the total volume at high pressures.

Table 6A.1
van der Waals Constants for Various Gases

Compound	a (L$^2 \cdot$ atm/mol^2)	b (L/mol)
He	0.03412	0.02370
Ne	0.2107	0.01709
H$_2$	0.2444	0.02661
Ar	1.345	0.03219
O$_2$	1.360	0.03803
N$_2$	1.390	0.03913
CO	1.485	0.03985
CH$_4$	2.253	0.04278
CO$_2$	3.592	0.04267
NH$_3$	4.170	0.03707

it was substituted into the ideal gas equation. This term contained a second constant (a) and has the form an^2/V^2. The complete **van der Waals equation** is therefore written as follows.

$$\left(P_{\text{real}} + \frac{an^2}{V_{\text{real}}{}^2}\right)(V_{\text{real}} - nb) = nRT$$

At normal temperatures and pressures, we can use the ideal gas law to describe the behavior of most gases.

$$P_{\text{ideal}}V_{\text{ideal}} = nRT$$

Under other conditions, we can use the van der Waals equation.

$$\left(P_{\text{real}} + \frac{an^2}{V_{\text{real}}{}^2}\right)(V_{\text{real}} - nb) = nRT$$

The van der Waals equation is something of a mixed blessing. It provides a much better fit with the behavior of a real gas than the ideal gas equation. But it does this at the cost of a loss in generality. The ideal gas equation is equally valid for any gas, whereas the van der Waals equation contains a pair of constants (a and b) that change from gas to gas. Values of the van der Waals constants for certain gases are given in Table 6A.1.

The ideal gas equation predicts that a plot of PV/RT versus P for one mole of a gas at constant T would be a horizontal line because the product of pressure times volume should be constant for an ideal gas. Experimental data for PV/RT versus P for H$_2$ and N$_2$ gas at 0°C and CO$_2$ at 40°C are given in Figure 6A.2. As the

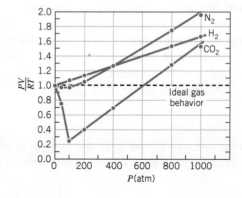

Fig. 6A.2 The ideal gas equation predicts that a plot of PV/RT versus P for one mole of a gas at constant T would be a horizontal line because PV should be a constant. Experimental data for PV/RT versus P for H$_2$ and N$_2$ at 0°C and CO$_2$ at 40°C.

pressure increases, the product of pressure times volume for N_2 and CO_2 first falls below the line expected from the ideal gas equation and then rises above the line.

This behavior can be understood by comparing the results of calculations using the ideal gas equation and the van der Waals equation for 1.00 mol of CO_2 at 0°C in containers of different volumes. Let's start with a 22.4-L container. According to the ideal gas equation, the pressure of the gas should be 1.00 atm.

$$P_{ideal} = \frac{nRT}{V_{ideal}} = \frac{(1.00 \text{ mol})(0.08206 \text{ L} \cdot \text{atm/mol} \cdot \text{K})(273 \text{ K})}{(22.4 \text{ L})} = 1.00 \text{ atm}$$

Substituting what we know about CO_2 into the van der Waals equation gives a much more complex equation.

$$\left(P_{real} + \frac{an^2}{V_{real}^2}\right)(V_{real} - nb) = nRT$$

$$\left[P_{real} + \frac{(3.592 \text{ L}^2 \cdot \text{atm/mol}^2)(1.00 \text{ mol})^2}{(22.4 \text{ L})^2}\right]\left[22.4 \text{ L} - (1.00 \text{ mol})(0.04267 \text{ L/mol})\right]$$
$$= (1.00 \text{ mol})(0.08206 \text{ L} \cdot \text{atm/mol} \cdot \text{K})(273 \text{ K})$$

The equation can be solved, however, for the pressure of the gas.

$$P_{real} = 0.995 \text{ atm}$$

At normal temperatures and pressures, the ideal gas and van der Waals equations give essentially the same results.

Let's now repeat the calculation, assuming that the gas is compressed so that it fills a container that has a volume of only 0.200 L. According to the ideal gas equation, the pressure would have to increase to 112 atm to compress 1.00 mol of CO_2 at 0°C to a volume of 0.200 L.

$$P_{ideal} = \frac{nRT}{V_{ideal}} = \frac{(1.00 \text{ mol})(0.08206 \text{ L} \cdot \text{atm/mol} \cdot \text{K})(273 \text{ K})}{(0.200 \text{ L})} = 112 \text{ atm}$$

The van der Waals equation, however, predicts that the pressure will only have to increase to 52.6 atm to achieve the same results.

$$\left(P_{real} + \frac{an^2}{V_{real}^2}\right)(V_{real} - nb) = nRT$$

$$\left[P_{real} + \frac{(3.592 \text{ L}^2 \cdot \text{atm/mol}^2)(1.00 \text{ mol})^2}{(0.200 \text{ L})^2}\right]\left[0.200 \text{ L} - (1.00 \text{ mol})(0.04267 \text{ L/mol})\right]$$
$$= (1.00 \text{ mol})(0.08206 \text{ L} \cdot \text{atm/mol} \cdot \text{K})(273 \text{ K})$$
$$P_{real} = 52.6 \text{ atm}$$

As the volume into which the CO_2 is compressed decreases, the van der Waals equation gives pressures that are *smaller* than the ideal gas equation, as shown in the left side of Figure 6A.2. This results from the force of attraction between CO_2 molecules that are now much closer together.

Let's now compress the gas even further, raising the pressure until the volume of the gas is only 0.0500 L. The ideal gas equation predicts that the pressure would have to increase to 448 atm to compress 1.00 mol of CO_2 at 0°C to a volume of 0.0500 L.

$$P = \frac{nRT}{V} = \frac{(1.00 \text{ mol})(0.08206 \text{ L} \cdot \text{atm/mol} \cdot \text{K})(273 \text{ K})}{(0.0500 \text{ L})} = 448 \text{ atm}$$

The van der Waals equation predicts that the pressure will have to reach 1.62×10^3 atm to achieve the same results.

$$\left[P_{real} + \frac{(3.592 \text{ L}^2 \cdot \text{atm/mol}^2)(1.00 \text{ mol})^2}{(0.0500 \text{ L})^2} \right] \left[0.0500 \text{ L} - (1.00 \text{ mol})(0.04267 \text{ L/mol}) \right]$$
$$= (1.00 \text{ mol})(0.08206 \text{ L} \cdot \text{atm/mol} \cdot \text{K})(273 \text{ K})$$
$$P_{real} = 1.62 \times 10^3 \text{ atm}$$

The van der Waals equation gives results that are *larger* than the ideal gas equation at very high pressures, as shown in Figure 6A.2, because of the volume occupied by the CO_2 molecules.

6A.2 Analysis of the van der Waals Constants

The van der Waals equation contains two constants, *a* and *b*, that are characteristic properties of a particular gas. The first of these constants, *a*, corrects for the force of attraction between gas particles. Compounds for which the force of attraction between particles is relatively strong have relatively large values for *a*. If you think about what happens when a liquid boils, you might expect compounds with large values of *a* to have higher boiling points. (As the force of attraction between gas particles becomes stronger, we have to go to higher temperatures before we can break the forces of attraction between the molecules in the liquid to form a gas.) It isn't surprising to find a correlation between the value of the *a* constant in the van der Waals equation and the boiling points of a number of simple compounds, as shown in Figure 6A.3. Gases with very small values of *a*, such as H_2 and He, must be cooled to almost absolute zero before they condense to form a liquid.

The other van der Waals constant, *b*, is a rough measure of the size of a gas particle. One of the first estimates of the size of an atom was extracted from this constant. The value of "b" for argon, for example, is 0.03219 L/mol, which means that the volume of a mole of argon atoms is 0.03219 liters. This number can be used to estimate the volume of an individual argon atom. We start by using Avogadro's number to calculate the volume of an individual atom.

$$\frac{0.03219 \text{ L}}{1 \text{ mol}} \times \frac{1 \text{ mol}}{6.022 \times 10^{23} \text{ atoms}} = 5.345 \times 10^{-26} \text{ L/atom}$$

The volume of an argon atom is then converted into cubic centimeters using the appropriate conversion factors.

$$5.345 \times 10^{-26} \text{ L} \times \frac{1000 \text{ mL}}{1 \text{ L}} \times \frac{1 \text{ cm}^3}{1 \text{ mL}} = 5.345 \times 10^{-23} \text{ cm}^3$$

If we assume that argon atoms are spherical, we can estimate the radius of these atoms. We start by noting that the volume of a sphere is related to its radius by the following formula.

$$V = \tfrac{4}{3}\pi r^3$$

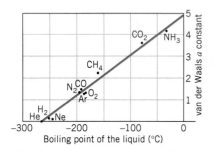

Fig. 6A.3 The boiling point of a liquid is an indirect measure of the force of attraction between its molecules. Thus it isn't surprising to find a correlation between the value of the van der Waals constant *a*, which measures the force of attraction between gas particles, and the boiling point of compounds that are gases at room temperature.

If we assume that the volume of an argon atom is 5.345×10^{-23} cm, we can estimate the radius of the atom.

$$r = 2.3 \times 10^{-8} \text{ cm}$$

According to this calculation, an argon atom should have a radius of about 2 Å.

Problems

Deviations from Ideal Gas Law Behavior: The van der Waals Equation

A-1. Predict whether the force of attraction between particles makes the volume of a real gas larger or smaller than that of an ideal gas.

A-2. Predict whether the fact that the volume of gas particles is not zero makes the volume of a real gas larger or smaller than that of an ideal gas.

A-3. Describe the conditions under which significant deviations from ideal gas behavior are observed.

A-4. Calculate the fraction of empty space in CO_2 gas, assuming 1 L of the gas at 0°C and 1.00 atm can be compressed until it changes to a liquid with a volume of 1.26 cm^3.

A-5. The following data were obtained in a study of the pressure and volume of a sample of acetylene.

P (atm):	1	45.8	84.2	110.5	176.0	282.2	398.7
V (L):	1	0.01705	0.00474	0.00411	0.00365	0.00333	0.00313

Calculate the product of pressure times volume for each measurement. Plot PV versus P and explain the shape of the curve.

A-6. Calculate the pressure of 1.00 mol of O_2 at 0°C in 1.0-L, 0.10-L, and 0.010-L containers using both the ideal gas equation and the van der Waals equation.

Analysis of the van der Waals Constants

A-7. Use the van der Waals constants for helium, neon, and argon to calculate the relative sizes of the atoms of these gases.

A-8. Identify the term in the van der Waals equation used to explain why gases become cooler when they are allowed to expand rapidly.

A-9. Refer to Figure 6A.2. At 100 atm pressure, which term in the van der Waals equation, an^2/V^2 or nb, is most important for CO_2? Which term is most important for H_2 at the same pressure? Explain.

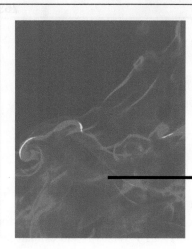

Chapter Seven

MAKING AND BREAKING OF BONDS

7.1 Energy

We need energy to run our automobiles, power our batteries, heat our homes, and fuel our bodies. But what do we mean when we say "energy," and what is the source of this energy?

The primary source of energy on which the United States has depended has changed several times during the course of history. For the first 100 years, the United States depended on wood as its primary source of energy. Inevitably, our economy switched from wood to coal. By the mid-1900s, petroleum products became the energy source of choice. Today, about 23% of our energy still comes from coal, most of which is used to generate electric power. About 24% of the energy we consume comes from natural gas, and 37% comes from petroleum products. Thus, almost 85% of the 100 quadrillion (100×10^{15}) BTUs of energy consumed in the United States each year results from the burning of a carbon-based fuel (coal, oil, natural gas, and petroleum).

The gasoline our automobiles consume is a complex mixture of more than 500 hydrocarbons, most of which contain between six and eight carbon atoms. Although consumers buy gasoline classified by "octane number," pure octane ($CH_3CH_2CH_2CH_2CH_2CH_2CH_3$) does not burn very well. Isooctane (or 2,2,4-trimethylpentane) is the standard to which different grades of gasoline are compared.

$$CH_3-\overset{\overset{\displaystyle CH_3}{|}}{\underset{\underset{\displaystyle CH_3}{|}}{C}}-CH_2-\overset{\overset{\displaystyle CH_3}{|}}{CH}-CH_3 \qquad \begin{array}{c} \text{2,2,4-Trimethylpentane} \\ \text{(isooctane)} \end{array}$$

The hydrocarbons in gasoline burst into flame when mixed with oxygen and ignited by a spark in the cylinder of an automobile. The products of the combustion of a hydrocarbon are carbon dioxide and water. The reaction between one of the C_8 hydrocarbons in gasoline and oxygen can be described by the following equation.

$$2\,C_8H_{18}(g) + 25\,O_2(g) \longrightarrow 16\,CO_2(g) + 18\,H_2O(g)$$

The hot gases produced in these combustion reactions expand, pushing the pistons out of the cylinder of the engine, and so the automobile's wheels turn.

Many people find the energy they need to heat their home by burning natural gas, which is a mixture of lightweight hydrocarbons that is about 95% methane (CH_4). Others use furnaces that burn fuel oil, which is a mixture of hydrocarbons that contain between 9 and 16 carbon atoms.

The energy we need to fuel our bodies comes from proteins, lipids, and carbohydrates. In a complex series of reactions, these materials are converted to simple sugars, such as glucose, that react with oxygen in the body in a sequence of reactions that can be summarized by the following overall equation.

$$C_6H_{12}O_6(s) + 6\,O_2(g) \longrightarrow 6\,CO_2(g) + 6\,H_2O(l)$$

As a result of this reaction, the body is provided with energy to do work.

Other common sources of energy include flashlight batteries, which produce light when electrons are driven through a wire in the lightbulb. The power to drive electrons through this wire results from a chemical reaction that takes place within the battery.

All of the processes discussed in this section occur because the energy needed to move pistons, drive electrons through wires, and heat our bodies and homes can be derived from energy-producing chemical reactions.

What is the ultimate source of the energy produced in those reactions? How much energy is available from such reactions? How can the reactions be used to obtain the maximum amount of energy from a given process? The answers to those questions form a major part of the remainder of the book.

Because this is a course in chemistry, all the examples that have been given relate to chemical processes. There are other kinds of energy, however, and some understanding of the different ways energy can be produced is essential to a more complete understanding of chemical energy.

Energy can be classified as either kinetic or potential energy. **Kinetic energy** is the energy of *motion*. Molecules that are moving through space (translational motion) or rotating around their center of gravity (rotational motion) possess kinetic energy. Physicists define the kinetic energy of an object in motion as one-half of the product of the mass of the object times the square of the velocity with which it is moving.

$$KE = 1/2 \; mv^2$$

Potential energy is the energy of *position*. A box lifted up a ladder is at a higher potential energy than a box left on the ground because the position of the box on the ladder is higher in the earth's gravitational field than a box on the ground.

A vibrating molecule has both kinetic and potential energy. The kinetic energy results from the fact that the atoms in the molecule are moving relative to one another. The potential energy results from changes in the distance between atoms.

Both the kinetic and potential energy of an object or system can change. Consider what happens when an object expands as it is heated, for example. The energy supplied by heat not only causes the atoms and molecules in the object to move more vigorously, it also causes the molecules to move away from each other. Thus both potential and kinetic energy change when the object is heated because adding energy increases the motion of the particles in the object and the increased motion leads to increased distance between the atoms and molecules that form the object.

Energy can also be transferred from one object to another. Consider what happens during a game of pool when a cue ball strikes another billiard ball. During the collision, some of the kinetic energy carried by the cue ball is transferred to the other billiard ball. Another example of the transfer of energy can be found in the technique known as photoelectron spectroscopy (PES) described in Chapter 3. In PES, the energy carried by a UV or X-ray photon is transferred to an electron, which is ejected from an atom. The kinetic energy of this electron is then measured and used to calculate the energy required to remove the electron from the atom.

The transfer of energy from one object to another in the course of a collision.

Energy can also be converted from one form to another. A car's battery, for example, converts chemical energy into electrical energy, which is then converted into mechanical energy. The spectrum of hydrogen studied in Chapter 3 provides another example of the conversion of energy from one form to another. In this case, the energy associated with electromagnetic radiation is used to move an electron from one energy level to another.

No matter what form energy takes or how it is transferred, the total energy before a process takes place and the total energy after the process is completed are the same. In other words, *energy is conserved*. Conservation of energy means that energy cannot be created or destroyed. In an elastic collision between a moving cue ball and a stationary billiard ball, the billiard ball moves away with increased kinetic energy. The cue ball, however, loses energy and therefore slows down. When a photon strikes an atom in a PES experiment, an electron is ejected from the atom. Because energy is conserved, the energy of the photon that is absorbed must be equal to the sum of the energy required to eject the electron (*IE*) and the kinetic energy of the ejected electron (*KE*).

$$KE + IE = h\nu$$

This chapter is concerned with the transfer of energy and the conversions between different forms of energy that are associated with chemical reactions. These processes are part of the area of study known as **thermodynamics.** Chemical energy is the energy associated with the force of attraction between the electrons and the nuclei in atoms, ions, and molecules. In other words, chemical energy is the energy that is due to chemical bonds.

As a general rule: *Changes in energy that occur during a chemical reaction are due to the making and breaking of chemical bonds.* The amount of energy associated with a chemical reaction is directly related to the strength of the chemical bonds that are broken and formed during that reaction. The relationship between energy and the strength of chemical bonds can be illustrated by the following hypothetical discussion of the process by which molecules of dimethyl ether and ethanol are constructed from isolated atoms in the gas phase.

Imagine a collection of isolated atoms in the gas phase consisting of 2 moles of carbon atoms, 6 moles of hydrogen atoms, and 1 mole of oxygen atoms. The atoms can be brought together to make two different molecules whose structures are consistent with the rules developed for Lewis structures described in Chapter 4. One of these molecules contains two C—O bonds and is known as dimethyl ether (CH_3OCH_3).

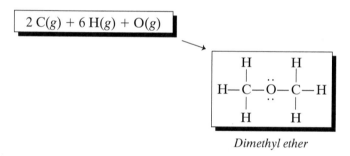

Dimethyl ether

If these atoms are brought together to make 1 mole of dimethyl ether in the gas phase, 3151 kJ of energy is released. What is the source of this energy? The energy is produced by making six C—H bonds and two C—O bonds per molecule of dimethyl ether, for a total of eight bonds for each molecule formed. This process gives off energy because *energy is always released during the formation of chemical bonds.*

The isolated atoms in the gas phase can also be brought together to form a different compound known as ethanol or ethyl alcohol (CH_3CH_2OH).

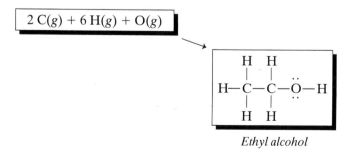

Ethyl alcohol

Each molecule of ethanol contains five C—H bonds, one C—C bond, one C—O bond, and one O—H bond for a total, once again, of eight bonds. The formation of 1 mole of ethanol in the gas phase from isolated atoms in the gas phase releases 3204 kJ of energy. Thus more energy is released in the formation of ethanol than in the formation of dimethyl ether. We can therefore conclude that the eight bonds in ethanol are stronger, on average, than the eight bonds in dimethyl ether.

The effect of bond strength on the energy released during a chemical reaction can be further illustrated with a reaction more practical than the formation of compounds from their gaseous atoms. Consider the reactions that occur when ethanol and dimethyl ether burn in the presence of oxygen. In each case, 1 mole of the alcohol or the ether reacts with 3 moles of oxygen to produce 2 moles of carbon dioxide and 3 moles of water.

$$CH_3CH_2OH(g) + 3\ O_2(g) \longrightarrow 2\ CO_2(g) + 3\ H_2O(g)$$
Ethyl alcohol

$$CH_3OCH_3(g) + 3\ O_2(g) \longrightarrow 2\ CO_2(g) + 3\ H_2O(g)$$
Dimethyl ether

The combustion of 1 mole of ethyl alcohol releases 1275 kJ of energy, whereas the combustion of 1 mole of dimethyl ether releases 1327 kJ of energy. The same products are formed in both reactions, but different bonds must be broken in the two reactant molecules.

Breaking chemical bonds always requires an input of energy. As we have seen, the eight bonds in ethanol are stronger than those in dimethyl ether. More energy must therefore be expended to break the bonds in ethanol. As a result, less energy is given off when ethanol is burned. In other words, dimethyl ether would be a better source of energy than ethanol.

> ➤ **CHECKPOINT**
>
> In the following reaction, are bonds broken or are they made?
>
> $$O(g) + O(g) \longrightarrow O_2(g)$$
>
> Will this reaction release energy or require an input of energy?

7.2 Heat

Heat can be defined as energy in transit. Heat is one way in which energy can be transferred from one object to another. The transfer of heat is usually associated with a change in temperature. Although heat and temperature are related to one another, they are not the same thing. **Temperature,** as discussed in Chapter 6, is a measure of the "hotness" or "coldness" of an object and may be measured using the Fahrenheit, Celsius, or Kelvin scales. Because heat is defined in terms of the transfer of energy, it must be measured in the same units as energy (joules). Because heat is associated with the transfer of energy, it is a mistake to consider a system or object as *containing* heat energy. A system that contains one of the forms of kinetic or potential energy discussed in the previous section can transfer some of that energy to another object as heat.

Consider what happens when a hot brick is placed in contact with a cold brick. Energy will be transferred as heat from the hot brick to the cold one. When this happens, the hot brick will get cooler and the cool brick will get warmer until the two bricks eventually come to the same temperature. When this happens, the hot brick will have lost energy in the form of heat and the cold brick will have gained energy in the form of heat.

7.3 Heat and the Kinetic Molecular Theory

Discussions of thermodynamics often involve dividing the universe into a system and its surroundings. The **system** is that small portion of the universe in which we are interested. It may consist of the water in a beaker or a gas trapped in a

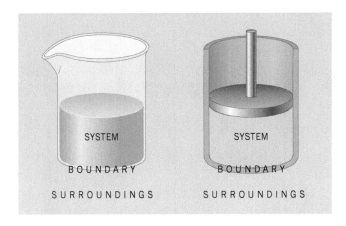

Fig. 7.1 In the kinetic theory, heat is transferred across the boundary between a system and its surroundings.

piston and cylinder, as shown in Figure 7.1. The **surroundings** are everything else—in other words, the rest of the universe.

The system and its surroundings are separated by a **boundary.** The boundary can be as real as the glass in a beaker or the walls of a balloon. It can also be imaginary, such as a line 200 nm from the surface of a metal that arbitrarily divides the air close to the metal surface from the rest of the atmosphere. The boundary can be rigid or it can be elastic. In the kinetic theory, heat is transferred across the boundary between a system and its surroundings.

As we have seen, when a hot brick is placed in contact with a cold brick, kinetic energy is transferred as heat from the hot brick to the cold one until they reach the same temperature. Initially, the particles in the hot brick are vibrating more rapidly than those in the cold brick. Eventually the particles in the two bricks come to the same average energy; in other words, the temperatures of the two bricks become the same.

The amount of energy needed to increase the temperature of a system by a given amount depends on the nature of the system. Energy supplied to a substance can increase the translational, rotational, and vibrational motion of the particles that form the substance. The increased motion can cause the particles in the substance to move a little farther apart from one another. Because there is always at least some force of attraction between these particles, moving them farther apart increases their potential energy—much like stretching a rubber band increases its potential energy. Thus the transfer of energy by means of heat can increase both the kinetic and potential energy of the system. The **kinetic theory of heat** can be summarized as follows: *Heat, when it enters a system, can increase the average motion (the average kinetic energy) with which the particles of the system move.*

The idea that atoms, molecules, and ions are in continuous random motion plays a very important role in much of the chemistry that is to follow. In addition to providing a means of understanding temperature and heat, the kinetic theory allows many properties of liquids, solids, and gases and the factors that influence how fast a chemical reaction occurs to be understood.

➤ **CHECKPOINT**

Use the kinetic theory to explain what happens to the gas particles in a balloon when heat enters the balloon from its surroundings.

7.4 Specific Heat

Experiments done by Joseph Black between 1759 and 1762 showed what happens when liquids at different temperatures are mixed. When equal volumes of water at 100°F and 150°F were mixed, Black found that the temperature of the mixture was the average of the two samples (125°F), as shown in Figure 7.2a. He

Fig. 7.2 (*a*) The temperature of a mixture of equal volumes of water is the average of the temperatures before the samples are combined. (*b*) When equal volumes of water and mercury are mixed, the temperature of the mixture is much closer to the temperature of the water before mixing.

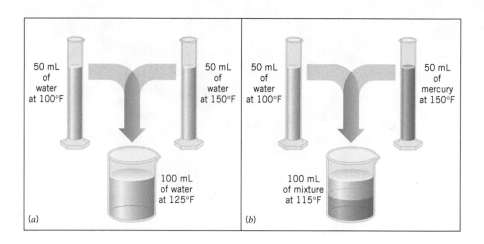

concluded that the amount of heat lost by the sample that was originally at 150°F was equal to the amount of heat absorbed by the sample that was at initially 100°F.

When water at 100°F was mixed with an equal volume of mercury at 150°F, however, the temperature of the liquids after mixing was only 115°F (Figure 7.2*b*). The temperature of the mercury fell by 35°F, but the temperature of the water increased by only 15°F. Black assumed that the heat lost by the mercury as it cooled down was equal to the heat gained by the water as it became warmer. He therefore concluded that the temperature of the water changed by a smaller amount because water has a larger "capacity for heat." In other words, it takes more heat to produce a given change in the temperature of water than it does to produce the same change in the temperature of an equivalent volume of mercury. Subsequent experiments have shown that it is not only the capacity for heat but also the amount of matter that determines the temperature change of a substance when it is heated.

Because water has a larger capacity for heat than mercury, it takes more heat to raise the temperature of a given mass of liquid water by one degree than it does to raise the temperature of the same mass of mercury by one degree. Any measurement of the capacity for heat therefore must take into account not only the mass of the sample being heated and the change in the temperature observed but also the identity of the substance being heated.

By convention, the heat needed to raise the temperature of *one gram* of substance by *one degree Celsius* is known as the **specific heat.** The units of specific heat were originally cal/g·°C. Because one degree on the Celsius scale is equal to one kelvin, specific heats can also be described in units of cal/g·K.

As might be expected, it is also possible to describe the heat required to raise the temperature of *one mole* of a substance by one degree. When this is done, we get the **molar heat capacity** in units of either cal/mol·°C or cal/mol·K. Molar heat capacities are generally used by chemists because they compare equal numbers of particles.

When the SI system (Appendix A) was introduced, the approved unit of heat became the *joule* and the calorie was defined as exactly 4.184 joules.

1 cal is defined as 4.184 J

The approved unit of temperature in the SI system is the kelvin. In the SI system, the units of specific heat are therefore J/g·K, and the units of molar heat capacity are J/mol·K.

When heat is absorbed by a substance, there is often (but not always) an increase in the average kinetic energy of the atoms, ions, and molecules that form

Table 7.1
Specific Heats and Molar Heat Capacities of Common Substances (25°C)

Substance	Specific Heat (J/g · K)	Molar Heat Capacity (J/mol · K)
Al(s)	0.901	24.3
C(s)	0.709	8.52
Cu(s)	0.3844	24.43
H_2O(l)	4.18	75.3
Fe(s)	0.449	25.1
Hg(l)	0.1395	27.98
O_2(g)	0.9172	29.35
N_2(g)	1.040	29.12
NaCl(s)	0.8641	50.50

the substance. If equal amounts of heat are absorbed by an equal number of moles of two substances, there will be a larger increase in the temperature of the substance with the smaller molar heat capacity. This occurs because the substance with the larger molar heat capacity will require more thermal energy to produce a given change in the temperature.

Table 7.1 lists the specific heats and molar heat capacities of a variety of substances. Note in particular the high specific heat and molar heat capacity of liquid water.

Exercise 7.1

Which metal will experience the largest increase in temperature when 10-mole samples of aluminum, copper, and iron absorb equivalent amounts of heat?

Solution

Since samples with an equal number of moles of atoms are being compared, molar heat capacity data with units based on the number of moles (J/mol · K) rather than specific heat with units based on mass (J/g · K) should be used. The molar heat capacities of aluminum, copper, and iron are given in Table 7.1. The molar heat capacity of aluminum is smaller than the molar heat capacities of either copper or iron. Therefore the aluminum sample will experience the largest increase in temperature.

➤ **CHECKPOINT**

Table 7.1 lists iron as having a smaller specific heat than aluminum but a larger molar heat capacity. Explain how this can happen.

We are now ready to derive a formula that can be used to calculate the amount of heat (q) given off or absorbed by a substance. Before we can do this, however, we need to introduce the symbol that will be used to indicate the change in temperature: ΔT. By definition, ΔT is the final value for the temperature minus the initial value for the temperature.

$$\Delta T = T_{\text{final}} - T_{\text{initial}}$$

We need to know three quantities to calculate the amount of heat given off or absorbed by a substance: the number of moles (n) or the mass of the substance in grams (m), the molar heat capacity of this substance (C) or the specific heat (s), and the change in the temperature (ΔT).

$$q = nC\Delta T$$
$$q = sm\Delta T$$

These equations can be understood by looking at the units with which the various quantities are expressed. Consider the first equation, for example. The amount of heat (in joules) is equal to the product of the amount of material in the sample (in units of moles) times the heat capacity (in units of joules per mole · Kelvin) times the change in the temperature of the sample (in units of Kelvin).

$$q = nC\Delta T$$
$$J = (\text{mol})\left(\frac{J}{\text{mol} \cdot K}\right)(K)$$

Exercise 7.2

(a) How much heat would have to be absorbed to raise the temperature of a 500-g iron bar from 25.0°C to 50.0°C?

(b) The 500-g iron bar at 50.0°C is then placed into 500 g of water at 25.0°C. If the iron bar loses 5.1×10^3 J of heat, what will be the final temperature of the water?

Solution

(a) The heat needed to raise the temperature of the iron bar can be found by applying the equation that describes the relationship among the amount of heat that would have to be absorbed, the number of grams of substance heated, and the specific heat of that substance.

$$q = 500 \text{ g} \times 0.449 \text{ J/g} \cdot K \times (323.2 \text{ K} - 298.2 \text{ K})$$
$$= 5.61 \times 10^3 \text{ J}$$

ΔT is the same whether °C or the Kelvin scale is used.
The same result is obtained if the calculation is done on the basis of the number of moles of iron in the bar and the molar heat capacity of iron.

$$q = 500 \text{ g} \times \frac{1 \text{ mole Fe}}{55.85 \text{ g}} \times 25.1 \text{ J/mol} \cdot K \times 25.0 \text{ K}$$
$$q = 5.61 \times 10^3 \text{ J}$$

(b) If the hot iron bar is placed into water at 25.0°C, the iron will lose heat and the water will gain heat. Thus the final temperature of the iron will be less than its initial temperature, and the final temperature of the water will be larger than its initial temperature. Both the iron and the water will eventually reach the same final temperature. All the heat lost by the iron is assumed to be gained by the water. Thus, we can assume that the water gains 5.1×10^3 J of energy in the form of heat. The change in the temperature of the water can therefore be calculated as follows:

$$q_{H_2O} = nC\Delta T$$

$$5.1 \times 10^3 \text{ J} = 500 \text{ g} \times \frac{1 \text{ mol } H_2O}{18.02 \text{ g}} \times 75.3 \text{ J/mol} \cdot \text{K} \times \Delta T$$

$$\Delta T = 2.4 \text{ K}$$

The final temperature of both the iron bar and the water will be 300.6 K or 27.4°C.

• •

7.5 State Functions

Every system can be described in terms of certain measurable properties. A gas, for example, can be described in terms of the number of moles of particles it contains, its temperature, its pressure, its volume, its mass, or its density. Those properties describe the **state** of the system at a particular moment in time. The properties of the system that depend on the size of the sample, such as mass and volume, are called **extensive properties.** The properties of the system that do not depend on the size of the sample, such as temperature and density, are known as **intensive properties.**

One way to determine whether a given property of the system is extensive or intensive involves asking: What happens to the magnitude of the property when we combine two samples? Mass and volume are both *extensive properties* that depend on the size of the sample. If we add two samples of water that both weigh 100 grams, we get a sample of water that weighs twice as much, that is, 200 grams, and occupies twice as much volume. Density and temperature, however, are both *intensive properties*. When we mix two samples of water at 0°C, the density is still 1 g/cm^3 and the temperature is still 0°C.

Properties can also be classified on the basis of whether or not they are **state functions.** By definition, one of the properties of a system is a state function if it depends only on the state of the system, not on the path used to get to that state.

Consider the temperature of a liquid. The fact that the temperature is 75.1°C at some moment doesn't tell us anything about the history of the system. It doesn't tell us how often the liquid has been heated or cooled before we take the measurement. Temperature is a state function because it reflects only the state of the system at the moment at which it is measured.

Because temperature is a state function, it is possible to calculate a unique value for the change in temperature when we compare the initial and final states of the system. As we saw in the previous section, ΔT is the difference between the final and initial values of the temperature of a system.

$$\Delta T = T_{final} - T_{initial}$$

Energy is also a state function: It depends only on the state of the system, not the path used to get to that state. We can therefore calculate the change in the energy of the system in much the same way.

$$\Delta E = E_{final} - E_{initial}$$

An analogy can be used to illustrate that energy is a state function. Suppose we have a crate on the first floor of a tall building and we wish to take the crate to the tenth floor. To do that, we will have to expend someone's or something's energy. The energy of the crate will change because the crate's potential energy is greater on the tenth floor than on the first by an amount that depends on the mass of the crate, the acceleration of gravity, and the difference in height between the tenth and first floors. The only variable factor influencing the energy of the crate is height. How the crate gets from one floor to the other doesn't change the energy of the crate. The energy change undergone by the crate depends only on the initial floor and the final floor.

It is important to remember that only the initial and final conditions matter for a state function. Suppose the crate were transported to the tenth floor from the first and then returned to the first floor. What would be the energy change of the crate? Because the initial and final conditions are the same, the energy of the crate wouldn't change. This provides us with a way of testing whether a given property of the system is a state function. X is a state function if and only if the magnitude of ΔX is zero when the system returns to its initial state.

➤ CHECKPOINT

Suppose a crate is dropped from the tenth floor of a building. When at rest on the tenth floor, what kind of energy does the crate have? During the fall, what kind of energy does the crate have? What happens to the energy acquired by the falling crate when it comes to rest on the ground?

7.6 The First Law of Thermodynamics

Scientists use the term *law* in two fundamentally different ways. When they talk about Boyle's law or Charles' law, they are describing a mathematical equation—or model (see Section 1.1)—that fits experimental data reasonably well. When they talk about the laws of thermodynamics, however, they are referring to statements for which there are no exceptions. The laws of thermodynamics are statements of universal validity.

The **first law of thermodynamics** can be summarized as follows: *Energy is conserved.* In other words, energy cannot be either created or destroyed. Thus the total energy before and after any process is carried out must be the same. By total energy, we mean the energy of everything that might conceivably be altered as a result of the process.

Energy is neither created nor destroyed. The energy absorbed by the system (the ice cubes, as they melt) in this example is exactly equal to the energy lost by its surroundings (the tea).

Consider the following experiment. A piston–cylinder apparatus is filled with an ideal gas, as shown in Figure 7.3. The piston is secured by stops, which prevent its movement. A brick is then brought into contact with the apparatus that is hotter than the gas in the piston–cylinder apparatus. As a result, the average kinetic energy of the particles in the brick is larger than that of the molecules of gas in the container. Some of the kinetic energy of the brick particles is transferred to the wall of the container and then, through collisions with the wall, to the molecules of the gas. As a result, the temperature of the gas gradually increases. As noted in Section 6.18, this leads to an increase in the average kinetic energy of the gas molecules. This, in turn, increases both the frequency and the force of collisions of the gas molecules with the walls of the container. As a result, the pressure of the gas in the container increases.

In this experiment, the gas is the system, and the brick and the piston–cylinder apparatus are the surroundings. Let's start the task of building a set of equations that represent what happens as this system comes to equilibrium by noting that the total energy in this experiment is the sum of the energy of the system and its surroundings.

$$E_{total} = E_{sys} + E_{surr}$$

The subscripts sys and surr in this equation stand for the system and its surroundings, respectively. If we assume that the system consists of the gas in the apparatus, this equation can be written as follows:

$$E_{total} = E_{gas} + E_{surr}$$

Fig. 7.3 A piston–cylinder apparatus containing an ideal gas in which the piston is held in place by stops.

As the brick in Figure 7.3 cools down, and the gas in the cylinder becomes warmer, the energy of both the system and its surroundings change. The first law of thermodynamics, however, assures us that there is no change in the total energy of the system and its surroundings. This can be represented by the following equation:

$$\Delta E_{total} = \Delta E_{sys} + \Delta E_{surr} = 0$$

In our piston–cylinder experiment, all the energy that entered the system went directly to increasing the energy of the molecules of gas. The piston didn't move despite the increased pressure in the container because the piston was held in place. We can summarize the changes in the energy of the system and its surrounding in the following way. Before the brick was brought into contact with the container, the total energy was the sum of the energy of the gas and its surroundings.

$$(E_{total})_{before} = (E_{gas})_{before} + (E_{surr})_{before}$$

After the brick exchanged energy with the gas and its container, the total energy was still the sum of the energy of the gas and its surroundings.

$$(E_{total})_{after} = (E_{gas})_{after} + (E_{surr})_{after}$$

The total change in energy can be related to the energy before and after the process as follows.

$$\Delta E_{total} = (E_{total})_{after} - (E_{total})_{before} = 0$$

Substituting what we know about the relationship between the total energy and the energy of the system and surroundings gives the following.

$$\Delta E_{total} = (E_{gas} + E_{surr})_{after} - (E_{gas} + E_{surr})_{before} = 0$$

➤ **CHECKPOINT**

Assume that a hot brick is thrown into cold water. According to the first law of thermodynamics, could the brick get hotter and the water colder?

This equation can be rearranged as follows.

$$\Delta E_{total} = \left[(E_{gas})_{after} - (E_{gas})_{before} \right] + \left[(E_{surr})_{after} - (E_{surr})_{before} \right] = 0$$

The term in the first set of brackets on the right side of this equation is equal to the change in the energy of the system, and the term in the second set of brackets is equal to the change in the energy of the surroundings.

$$\Delta E_{total} = \Delta E_{gas} + \Delta E_{surr} = 0$$

According to the first law, there is no change in the total energy of the system and its surroundings. Thus the change in the energy of the surroundings must be equal in magnitude and opposite in sign to the change in the energy of the system. In other words, the energy lost by the surroundings was gained by the gas in the system.

$$\Delta E_{gas} = -\Delta E_{surr}$$

ΔE_{gas} is positive when the system gains energy and negative when it loses energy.

The experiment shown in Figure 7.3 was done with a system that is said to be at *constant volume* because the piston was not allowed to move in the cylinder. Because there is no change in the volume of the system, no work of expansion can be done. Changes in the system and its surroundings that occur under conditions of constant volume therefore can only involve the transfer of heat. We can represent the relationship between the heat transferred from the surroundings into the system as follows.

$$\Delta E_{gas} = q_v \quad \text{(constant volume)}$$

The subscript v is used in this equation to indicate that the process occurred at constant volume.

7.7 Work

Now imagine a second experiment in which the process described in the previous section is the same, but the stops on the piston have been removed, as shown in Figure 7.4. Just as before, we must take account of all interactions between the system and its surroundings that can result in changes in the energy of the system. Because the stops have been removed, the piston can now move upward if the gas in the system expands. Or the piston can move further into the cylinder if the gas in the system contracts.

The work done by this system on its surroundings when the gas in the piston expands is known as **work of expansion.** The amount of **work** ($|w|$ absolute value of work) that is done when the gas expands is equal to the product of the force (F) that is applied to the piston times the distance the piston moves (Δx).

$$|w| = F \times \Delta x$$

This force is applied equally to every point on the surface area of the piston (A). The result is a pressure that is equal to the force per unit area.

$$P = \frac{F}{A}$$

Fig. 7.4 A piston–cylinder apparatus with an ideal gas with the stops removed, so that the piston can move.

Rearranging this equation to solve for the force and then substituting the result into the equation that defines the amount of work that is done gives the following result.

$$|w| = P \times A \times \Delta x$$

But the product of the surface area of the piston (A) times the distance the piston moves is equal to the change in the volume of the gas that occurs when the piston moves. If the pressure of the gas remains constant while it expands, the magnitude of the work done by the system while it is pushing up on the piston is equal to the product of the constant pressure times the change in the volume of the system.

$$|w| = P\Delta V$$

So far we have calculated the *magnitude* of the work that is done if the gas pushes the piston out of the cylinder at a constant pressure. Now we have to add the sign convention. As the piston moves up—out of the cylinder—the system will be doing work on its surroundings. Because this leads to a *decrease* in the energy of the system, the sign convention for work of expansion can be written as follows.

$$w = -P\Delta V$$

In order for the piston to move, some energy will have to be used to lift the weight of the piston. When the piston moves up because of the increased pressure in the container, it will be higher in the earth's gravitational field and thus have a larger potential energy. Let's use the symbol ΔE_{piston} to represent the change in the energy of the piston. Once again, the total change in energy is equal to the change in the energy of the system plus the change in the energy of the surroundings.

$$\Delta E_{total} = \Delta E_{sys} + \Delta E_{surr}$$

Here it is convenient to consider that the system consists of both the gas and the piston.

$$\Delta E_{total} = (\Delta E_{gas} + \Delta E_{piston}) + \Delta E_{surr} = 0$$

The increase in the potential energy of the piston is equal to $mg\Delta h$, where m is the mass of the piston, g is the acceleration due to gravity, and Δh is the difference between the final and initial height of the piston. The raising of a weight requires work. Because it takes energy to do work, we can write the change in the energy of the piston as follows:

$$\Delta E_{piston} = mg\Delta h = -\text{work} = -w$$

Thus the first law becomes

$$\Delta E_{total} = \Delta E_{gas} - w + \Delta E_{surr} = 0$$

This equation can also be written as follows:

$$\Delta E_{gas} - w = -\Delta E_{surr}$$

In the first experiment, all of the energy lost by the surroundings went into increasing the temperature of the gas in the system. This is not the case in this second experiment. In the second experiment, some of the energy from the surroundings goes into increasing the temperature of the gas, but some also goes into doing the work involved in lifting the piston. If the same amount of energy is supplied to the gas in both experiments, the gas in the first experiment will be warmer than in the second.

In the first experiment, the container was kept at a *constant volume*. In the second experiment, the volume of the gas changes, but the container is kept at *constant pressure*. In the second experiment, the piston moves to maintain a balance between the pressure of the gas in the container and the pressure pushing down on the gas from the piston.

Experiments done in the laboratory are usually done in open flasks at constant pressure, not in piston–cylinder types of containers at constant volume. Instead of pushing against a piston, an expanding system in an open flask pushes against the weight of the atmosphere. A system in an open flask therefore has to do work to expand and loses some energy when this happens. If a system in an open flask contracts, the system can actually gain energy as a result of this process. Regardless of whether the gas in the system expands or contract, the first can be written as follows for all processes that are carried out under conditions of constant pressure.

$$\Delta E_{total} = \Delta E_{sys} - w + \Delta E_{surr} = 0$$

And therefore,

$$\Delta E_{gas} - w = -\Delta E_{surr}$$

In the previous section, we noted that the experiment in Figure 7.3 was done at *constant volume*. As a result, no work of expansion was done in this experiment, and the relationship between the heat transferred from the surroundings into the system can be written as follows.

$$\Delta E_{gas} = q_v \quad \text{(constant volume)}$$

The experiment in Figure 7.4, however, was not done at constant volume. It was done under conditions that are said to be at *constant pressure*. In the second experiment, the interaction between the system and its surroundings involved both heat and work. We therefore need a new state function that is a measure of the effect of both heat and work. This state function is called **enthalpy** (*H*). The enthalpy of the system is defined as the sum of the energy of the system plus the product of the pressure times the volume of the gas in the system.

$$H = E + PV$$

For the piston–cylinder apparatus shown in Figure 7.4, the change in the enthalpy of the system is equal to the change in the energy of the system minus the work that is done when the system expands.

$$\Delta H_{sys} = (\Delta E_{sys} - w)$$

The concept of enthalpy is useful because it can be shown that the magnitude of the heat given off or absorbed by the system at constant pressure is equal to the change in the enthalpy of the system.

$$\Delta H_{sys} = q_p \quad \text{(constant pressure)}$$

➤ CHECKPOINT

If the same quantity of heat is transferred from the brick to the gas in a fixed–volume piston–cylinder apparatus (Figure 7.3) and in a movable piston–cylinder apparatus (Figure 7.4), in which container will the temperature of the gas increase the most?

Reactions that involve liquids and solids usually occur with little or no change in volume. As a result, little if any work of expansion occurs during the reaction. The difference between ΔH and ΔE for these reactions is usually small. The difference between ΔH and ΔE for the following reaction, for example, is only 0.16 J/mol$_{rxn}$.

$$H_2O(s) \rightleftharpoons H_2O(l)$$

Reactions that involve gases, however, can involve a significant change in the volume of the system. When that happens, the difference between ΔH and ΔE can be considerable. Consider the following reaction, for example, where the difference between ΔH and ΔE is 3100 J/mol$_{rxn}$.

$$H_2O(l) \rightleftharpoons H_2O(g)$$

Let's now consider a process that is of considerable interest to a chemist. The same apparatus will be used as in the first and second experiments except the ideal gas will be replaced with a mixture of ethane gas and oxygen gas (Figure 7.5). A spark will then be used to initiate the following combustion reaction.

$$2\ C_2H_6(g) + 7\ O_2(g) \longrightarrow 4\ CO_2(g) + 6\ H_2O(g)$$

The temperature in the container will increase as a result of the energy given off by the reaction. Heat is released by the chemical reaction because the overall bond strengths of the products are greater than those of the reactants. Part of this heat goes into heating the container and the gases, and part goes into raising the piston. Because this reaction is done under conditions of constant pressure, the heat given off is equal to the change in the enthalpy of the system that occurs during the reaction (ΔH_{rxn}).

$$\Delta H_{sys} = \Delta H_{rxn} = q_p \quad \text{(constant pressure)}$$

Table 7.2 presents the heat given off per mole of hydrocarbon when several hydrocarbons burn in the presence of oxygen at constant pressure. Because these reactions are run under conditions of constant pressure, the heat given off in each of these reactions is equal to the change in the enthalpy that occurs as a result of each reaction.

Fig. 7.5 The ideal gas in the piston–cylinder apparatus has been replaced with a mixture of ethane and oxygen. There are no stops on the piston, and the source of external heat—the brick—has been removed.

Exercise 7.3

Section 7.1 noted that it takes energy to break chemical bonds and that energy is released when these bonds are formed. What does this tell us about the relative strengths of the bonds in the products and reactants in the combustion of the hydrocarbons in Table 7.2?

Solution

In each case, these hydrocarbons burn to form $CO_2(g)$ and $H_2O(l)$.

$$C_5H_{12}(g) + 8\ O_2(g) \longrightarrow 5\ CO_2(g) + 6\ H_2O(l)$$

These combustion reactions all give off heat. Therefore, the sum of the bond strengths must be larger in the products than in the reactants. If this were not the case, more heat would be required to break the bonds in the reactants than was returned on formation of the products, and this would result in a net input of heat.

Table 7.2

Enthalpy of Reaction for the Combustion of Common Hydrocarbons at 298 K and 1 atm

Hydrocarbon	Name	Structure	Heat Released in kJ per Mole of Hydrocarbon that Reacts with Oxygen
$CH_4(g)$	Methane	$H\!-\!\underset{\underset{H}{\mid}}{\overset{\overset{H}{\mid}}{C}}\!-\!H$	890
$C_2H_6(g)$	Ethane	$H\!-\!\underset{\mid}{\overset{\mid}{C}}\!-\!\underset{\mid}{\overset{\mid}{C}}\!-\!H$	1560
$C_3H_8(g)$	Propane	$H\!-\!\underset{\mid}{\overset{\mid}{C}}\!-\!\underset{\mid}{\overset{\mid}{C}}\!-\!\underset{\mid}{\overset{\mid}{C}}\!-\!H$	2222
$C_4H_{10}(g)$	Butane	$H\!-\!C\!-\!C\!-\!C\!-\!C\!-\!H$	2877
$C_5H_{12}(g)$	Pentane	$H\!-\!C\!-\!C\!-\!C\!-\!C\!-\!C\!-\!H$	3540

7.8 The Enthalpy of a System

All chemical reactions, no matter how simple or complex, have one thing in common—they all involve the breaking and re-forming of bonds between atoms or ions. At some step, for example, the combustion of methane must involve breaking the C—H bonds in methane and the O=O bonds in oxygen. And the reaction must eventually involve the formation of the covalent C=O and O—H bonds that hold the CO_2 and H_2O molecules together.

$$CH_4(g) + 2\,O_2(g) \longrightarrow CO_2(g) + 2\,H_2O(g)$$

Because the making and breaking of bonds play such a central role in chemistry, a way to determine the energy consumed or produced in a chemical reaction is of extreme importance to chemists. The first law of thermodynamics provides us with a way to monitor changes in the energy of the system that accompany a chemical reaction.

$$\Delta E_{total} = \Delta E_{sys} + \Delta E_{surr} = 0$$

The heat given off or absorbed by a chemical reaction can be measured in a **calorimeter,** shown in Figure 7.6. The heat given off or absorbed by the reaction

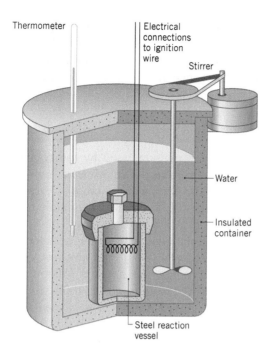

Fig. 7.6 A bomb calorimeter. Because the volume of the system is constant, no work of expansion can be done. As a result, $\Delta E_{sys} = q_V$.

that occurs in the steel reaction vessel (ΔE_{sys}) is measured by determining the change in the temperature of the water in the insulated container that surrounds the reaction vessel (ΔE_{surr}). Because the reaction is run in a sealed container at constant volume, no work can be done. As a result, the change in the energy of the system (ΔE_{sys}) must have the same magnitude but the opposite sign of the change in the energy of the surroundings (ΔE_{surr}).

$$\Delta E_{sys} = -\Delta E_{surr}$$

By monitoring the change in temperature of a water bath that surrounds the reaction container, we can calculate the change in the energy of the surroundings, ΔE_{surr}. This is equal in magnitude and opposite in sign to the change in the energy of the system that occurs during the chemical reaction.

As noted in Section 7.4, thermodynamicists use the symbol q to represent the heat transferred from the system to its surroundings or vice versa. Because no work is done when the reaction is run in a calorimeter at constant volume, the heat given off or absorbed in a chemical reaction is equal to the change in the energy of the system.

$$\Delta E_{sys} = q_v \quad \text{(constant volume)}$$

Chemists, however, seldom work in sealed containers at constant volume. They usually carry out reactions in containers such as beakers or flasks that are open to the atmosphere. These reactions occur under conditions of *constant pressure*. When the volume of a system can change, not all of the heat supplied to or taken from the system goes into producing a temperature change; some of the heat is used to do work. When describing a system at constant pressure, chemists therefore look for changes in the *enthalpy* of the system, ΔH_{sys}.

Enthalpy, like energy, is a state function. The change in enthalpy that occurs during a chemical reaction at constant pressure is exactly equal to the heat given off or absorbed by the reaction. A constant-pressure calorimeter can be used to collect data to calculate the value of ΔH for a reaction.

$$\Delta H_{sys} = q_p \quad \text{(constant pressure)}$$

Calorimeters are routinely used to determine the caloric content of food. The food in question may be a sugar, such as sucrose, which is table sugar. The heat released when sucrose reacts with oxygen in the calorimeter is measured by determining how much the temperature of the water surrounding the calorimeter goes up.

$$C_{12}H_{22}O_{11}(s) + 12\ O_2(g) \longrightarrow 12\ CO_2(g) + 11\ H_2O(l)$$

In this reaction, 5645 kJ of heat is released by the combustion of 1 mole of sucrose. The reaction that occurs in the calorimeter is the same as the overall reaction that takes place in the body when sucrose is eaten. Thus an equivalent amount of energy, 5645 kJ, would be supplied to or stored by the body when we digest sucrose.

Before the SI system was introduced, the common unit of measurement of heat was the calorie (1 cal is defined as 4.184 J). This unit was too small for convenient use in measurements of food energy, so the unit of Calorie (with a capital C) was introduced. By definition, one Cal is equivalent to 1000 cal. Thus one Cal is equivalent to 4184 J or 4.184 kJ. A soft drink that is labeled as providing 120 food calories (or 120 Cal) therefore provides more than 500,000 joules of energy.

The caloric value of 1 mole of sucrose is 5645 kJ, as noted above, which corresponds to 1349 food Cal. One mole of sucrose weighs 342 g; so if one teaspoon (about 5 g) of sucrose is ingested, it will provide about 20 Cal, or about 4 Cal per gram. If that energy is not expended by work or exercise, it remains in the body. Slow walking expends about 150 Cal per hour, so an 8-minute walk will use up the calories from a teaspoon of sugar.

> ➤ **CHECKPOINT**
>
> A tablespoon of sugar can provide several Calories of energy. What does this tell us about how the strengths of the bonds in the reactants compare to those of the products?

7.9 Enthalpies of Reaction

Chemical reactions are divided into two classes on the basis of whether they give off or absorb heat from their surroundings. **Exothermic** reactions give off heat to the surroundings. **Endothermic** reactions absorb heat from the surroundings. Use of the prefix *exo-* to indicate reactions that give off heat is consistent with its use in a variety of terms, from *exodus* (a mass departure of immigrants) to *exoskeleton* (the rigid external covering of insects). Use of the prefix *endo-* is also consistent with its use in other terms, from *endogenous* (to grow from within) to *endoskeleton* (the internal bone structure of vertebrates).

The heat given off or absorbed in a chemical reaction at constant pressure is known as the **enthalpy of reaction.** The enthalpy of the system decreases when a reaction gives off heat to its surroundings. This means that the final value of the enthalpy of the system is smaller than the initial value. Exothermic reactions are therefore characterized by negative values of ΔH.

Exothermic reactions: ΔH is negative   ($\Delta H < 0$)

Endothermic reactions, on the other hand, take in heat from their surroundings. As a result, the enthalpy of the system increases. Endothermic reactions are therefore characterized by positive values of ΔH.

Endothermic reactions: ΔH is positive   ($\Delta H > 0$)

An example of an exothermic reaction occurs when a balloon filled with a mixture of hydrogen and oxygen gas is ignited. The reaction is accompanied by

The reaction between NH_4SCN and $Ba(OH)_2$ is an example of a spontaneous endothermic reaction. This reaction absorbs so much heat from its surroundings that it can freeze a beaker to a wooden board if the outside of the beaker is moistened.

a large ball of fire and loud boom. The following reaction diagram can be used to describe the reaction.

$$2\,H_2(g) + O_2(g)$$

$\Delta H = -483.64\ kJ/mol_{rxn}$

$$2\,H_2O(g)$$

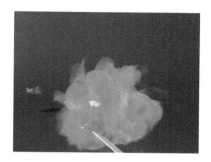

The reaction between hydrogen and oxygen to form water that occurs when a balloon filled with H_2 gas is ignited by a candle

When 2 moles of hydrogen react with 1 mole of oxygen to produce 2 moles of water, 483.64 kJ of heat is released. Changes in enthalpy are written in units of kilojoules per mole of reaction (kJ/mol$_{rxn}$), where "mol$_{rxn}$" represents the balanced chemical equation as a whole unit. To have meaning, a change in enthalpy must be associated with a specific chemical equation. Although the term *mol*$_{rxn}$ is used to describe the reaction, it does not necessarily mean that there is only 1 mole of reactant or product. In the above chemical equation there are 3 moles of reactants (2 moles of H_2 and 1 mole of O_2) and 2 moles of product, but the reaction as a whole is referred to as one unit or one mole of chemical reaction. The coefficients can be thought of as giving the number of moles of a substance per mole of reaction, for example, 2 mol H_2/mol$_{rxn}$.

If the chemical equation is changed, there must be an accompanying change in the enthalpy of reaction. If the coefficients in the above chemical equation are all doubled, for example, the change in enthalpy will also be doubled.

$$4\,H_2(g) + 2\,O_2(g)$$

$\Delta H = -967.28\ kJ/mol_{rxn}$

$$4\,H_2O(g)$$

Because twice as many moles of reactants undergo the reaction, twice as much energy is released. The term *mol*$_{rxn}$ in the enthalpy change of -967.28 kJ/mol$_{rxn}$ refers now to the following chemical equation.

$$4\,H_2(g) + 2\,O_2(g) \longrightarrow 4\,H_2O(g)$$

The reaction between hydrogen and oxygen to produce water is exothermic and therefore releases heat. If the reaction was written in the opposite direction, an input of heat would be required.

$$2\,H_2(g) + O_2(g)$$

$\Delta H = +483.64\ kJ/mol_{rxn}$

$$2\,H_2O(g)$$

The reaction in which water decomposes to form its elements in the gas phase has a positive ΔH and is therefore endothermic. Whenever a chemical equation is reversed, the magnitude of the change in enthalpy will remain the same, but the sign will change. In this text, when a process is exothermic, the reaction diagram will be drawn with the arrow pointing down. When a process is endothermic, the diagram will be drawn with an arrow pointing up.

► **CHECKPOINT**

What is the sign of the enthalpy change for the combustion reactions in Table 7.2?

The change in the enthalpy of the system that occurs for the following reaction at 298 K is -92.2 kJ/mol$_{rxn}$.

$$N_2(g) + 3 H_2(g) \longrightarrow 2 NH_3(g)$$

This value of ΔH_{rxn} corresponds to the energy given off by the reaction in which 1 mole of nitrogen reacts with 3 moles of hydrogen to produce 2 moles of ammonia.

What would be the change in the enthalpy of the system if 0.200 moles of nitrogen react with 0.600 moles of hydrogen to form 0.400 moles of ammonia? According to the equation that defines the reaction between nitrogen and hydrogen to form ammonia, 92.2 kJ of heat is produced when one mole of nitrogen is consumed. This is equivalent to stating that the change in the enthalpy of the system during the reaction is -92.2 kJ per mole of N_2 consumed. If only 0.200 moles of N_2 react, the change in enthalpy can be calculated as follows:

$$(0.200 \text{ mol } N_2) \times \left(\frac{-92.2 \text{ kJ}}{1 \text{ mol } N_2} \right) = -18.4 \text{ kJ}$$

The same result is obtained if we do the calculation based on the amount of hydrogen consumed in the reaction, assuming that 92.2 kJ of heat is given off for every 3 moles of hydrogen consumed.

$$(0.600 \text{ mol } H_2) \times \left(\frac{-92.2 \text{ kJ}}{3 \text{ mol } H_2} \right) = -18.4 \text{ kJ}$$

And the same result is obtained if we base the calculation on the number of moles of ammonia produced in the reaction.

$$(0.400 \text{ mol } NH_3) \times \left(\frac{-92.2 \text{ kJ}}{2 \text{ mol } NH_3} \right) = -18.4 \text{ kJ}$$

Exercise 7.4

Pentaborane, B_5H_9, was once considered as a potential rocket fuel. B_5H_9 burns in the presence of oxygen to form B_2O_3 and water vapor.

$$2 B_5H_9(g) + 12 O_2(g) \longrightarrow 5 B_2O_3(g) + 9 H_2O(g)$$

At 298 K the enthalpy change for the reaction is -8686.6 kJ/mol$_{rxn}$. Calculate the change in enthalpy when 0.600 moles of pentaborane are consumed.

Solution

As the reaction is written, 8686.6 kJ of heat is produced from the combustion of 2 moles of B_5H_9. Therefore

$$(0.600 \text{ mol } B_5H_9) \times \left(\frac{-8686.6 \text{ kJ}}{2 \text{ mol } B_5H_9} \right) = -2610 \text{ kJ} = -2.61 \times 10^3 \text{ kJ}$$

If 0.600 moles of B_5H_9 are consumed, 3.60 moles of O_2 must be consumed and 1.50 moles of B_2O_3 and 2.70 moles of H_2O are produced. It doesn't matter which chemical species is chosen for the calculation of ΔH because the value of ΔH_{rxn} is for "one mole of reaction." The same results would be obtained if the number of moles of water produced in the reaction was used to calculate the change in enthalpy.

$$(2.70 \text{ mol H}_2\text{O}) \times \left(\frac{-8686.6 \text{ kJ}}{9 \text{ mol H}_2\text{O}}\right) = -2610 \text{ kJ} = -2.61 \times 10^3 \text{ kJ}$$

Exercise 7.5

Water is far more abundant than petroleum. Why can't the decomposition of water into its elements be used to run an automobile?

$$2 \text{ H}_2\text{O}(g) \longrightarrow 2 \text{ H}_2(g) + \text{O}_2(g)$$

Cars are now being tested that use hydrogen as a fuel. These cars produce less pollution, and they don't consume petroleum. Explain why the following reaction can be used to power an automobile.

$$2 \text{ H}_2(g) + \text{O}_2(g) \longrightarrow 2 \text{ H}_2\text{O}(g)$$

Explain the origin of the heat released or absorbed in these reactions.

Solution

The reaction $2 \text{ H}_2\text{O}(g) \longrightarrow 2 \text{ H}_2(g) + \text{O}_2(g)$ is endothermic and couldn't be used to run an engine. The reverse reaction $2 \text{ H}_2(g) + \text{O}_2(g) \longrightarrow 2 \text{ H}_2\text{O}(g)$ is exothermic and could supply the energy needed to run a car. The origin of the heat is in the making and breaking of bonds. The Lewis structures of the three components of the reaction are written as follows.

$$2 \text{ H—H} + \ddot{\text{O}}{=}\ddot{\text{O}} \longrightarrow 2 \quad \begin{array}{c} \ddot{\text{O}} \\ \diagup \;\; \diagdown \\ \text{H} \quad\;\; \text{H} \end{array}$$

Two H—H bonds and one O=O double bond must be broken to form two $\text{H}_2\text{O}(g)$ molecules that each contain two H—O bonds. Because the reaction is exothermic, the sum total of the H—O bonds in the water molecules formed in this reaction must be greater than the sum total of the energy associated with the bonds that are broken in H_2 and O_2. The reverse reaction is endothermic because more enthalpy is required to break the bonds in water than is gained by the formation of the bonds in H_2 and O_2.

7.10 Enthalpy as a State Function

Both the energy and the enthalpy of a system are state functions. They depend only on the state of the system at any moment, not its history. To examine the consequences of this fact, let's consider the following reaction.

$$\text{H}_2(g) + \text{Cl}_2(g) \longrightarrow 2 \text{ HCl}(g)$$

Because enthalpy is a state function, we can visualize the reaction as if it occurred in two hypothetical steps. First, we break the bonds in the starting materials to form hydrogen and chlorine atoms in the gas phase. This step is endothermic because bonds are being broken.

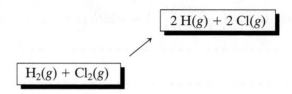

We can then assume that the atoms recombine to form the product of the reaction. This second step is exothermic because bonds are being made.

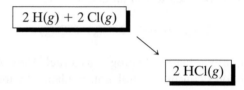

It doesn't matter whether or not the reaction actually occurs by these hypothetical steps. Because enthalpy is a state function, ΔH for this hypothetical reaction will be exactly the same as ΔH for the reaction whatever the pathway by which the starting materials are actually transformed into the products of the reaction.

The first step in predicting the amount of energy given off or absorbed in this reaction involves determining the energy associated with each of the individual steps. It takes 243.4 kJ to break the covalent bonds in a mole of Cl_2 molecules to form 2 moles of chlorine atoms.

$$Cl_2(g) \longrightarrow 2\,Cl(g) \qquad \Delta H = 243.4 \text{ kJ/mol}_{rxn}$$

It takes an additional 435.3 kJ to break apart a mole of H_2 molecules to form 2 moles of hydrogen atoms.

$$H_2(g) \longrightarrow 2\,H(g) \qquad \Delta H = 435.3 \text{ kJ/mol}_{rxn}$$

We therefore have to invest a total of 678.7 kJ of energy in the system to transform a mole of H_2 molecules and a mole of Cl_2 molecules into 2 moles of H atoms and 2 moles of Cl atoms.

Bond breaking:
$$
\begin{array}{ll}
Cl_2(g) \longrightarrow 2\,Cl(g) & \Delta H = 243.4 \text{ kJ/mol}_{rxn} \\
\underline{H_2(g) \longrightarrow 2\,H(g)} & \underline{\Delta H = 435.3 \text{ kJ/mol}_{rxn}} \\
H_2(g) + Cl_2(g) \longrightarrow 2\,H(g) + 2\,Cl(g) & \Delta H = 678.7 \text{ kJ/mol}_{rxn}
\end{array}
$$

ΔH is positive for the first step in our hypothetical pathway for the reaction *because any process that involves the breaking of bonds is endothermic.*

Let's now turn to the process by which the isolated atoms in the gas phase recombine to form HCl molecules. The bond between hydrogen and chlorine atoms is relatively strong. We get back 431.6 kJ for each mole of H—Cl bonds that are formed. Because we have 2 moles of hydrogen atoms and 2 moles of chlorine atoms, we get 2 moles of HCl molecules. Thus the bond-making process gives off a total of 863.2 kJ.

Bond making: $2\,H(g) + 2\,Cl(g) \longrightarrow 2\,HCl(g) \qquad \Delta H = -863.2 \text{ kJ/mol}_{rxn}$

ΔH for the second step in our hypothetical pathway is negative *because any process that involves the making of bonds is exothermic.*

The overall change in the enthalpy of the system that occurs during the reaction can be calculated by combining ΔH for the two hypothetical steps in the reaction.

$$
\begin{array}{ll}
678.7 \text{ kJ/mol}_{rxn} & \text{Bond breaking} \\
\underline{-863.2 \text{ kJ/mol}_{rxn}} & \text{Bond making} \\
-184.5 \text{ kJ/mol}_{rxn} &
\end{array}
$$

According to this calculation, the overall reaction is exothermic (ΔH is negative), releasing a total of 184.5 kJ of energy in the form of heat when 1 mole of H_2 reacts with 1 mole of Cl_2 to form 2 moles of HCl.

$$H_2(g) + Cl_2(g) \longrightarrow 2\, HCl(g) \qquad \Delta H = -184.5 \text{ kJ/mol}_{rxn}$$

Because enthalpy is a state function, it doesn't matter whether the reaction proceeds through these hypothetical steps. *The value of ΔH for a reaction only depends on the initial and final states of the system, not the path by which the reactants are transformed into the products of the reaction.*

7.11 Standard-State Enthalpies of Reaction

The amount of heat given off or absorbed by a chemical reaction depends on the conditions under which the reaction is carried out. Three factors influence the amount of heat associated with a given reaction: (1) the amounts of the starting materials and products involved in the reaction, (2) the temperature at which the reaction is run, and (3) the pressure of any gases involved in the reaction. The reaction in which methane is burned can be used to illustrate why the reaction conditions must be specified.

Assume that we start with a mixture of CH_4 and O_2 at 25°C in a container just large enough so that the pressure of each gas is 1 atm. In the presence of a spark, these gases react to form a mixture of CO_2 and H_2O.

$$CH_4(g) + 2\, O_2(g) \longrightarrow CO_2(g) + 2\, H_2O(g)$$

At this temperature and pressure, the reaction gives off 802.4 kJ/mol$_{rxn}$. If we start with the reactants at 1000°C and 1 atm pressure, however, and generate the products at 1000°C and 1 atm, the reaction gives off only 800.5 kJ/mol$_{rxn}$. It is therefore important to specify the conditions under which a reaction occurs when reporting thermodynamic data.

Thermodynamic data are often collected at 25°C (298 K). Measurements taken at other temperatures are identified by adding a subscript specifying the temperature in kelvins. The data collected for the combustion of methane at 1000°C, for example, would be reported as follows: $\Delta H_{1273} = -800.5$ kJ/mol$_{rxn}$.

The effect of pressure and the amount of materials on the heat given off or absorbed in a chemical reaction is controlled by defining a set of standard conditions for thermodynamic experiments. By definition, the **standard state** for thermodynamic measurements is defined in terms of the partial pressure of any gases and the concentrations of any solutions involved in the reaction.

Standard-state conditions:
   All solutes have concentrations of 1 _M_.
   All gases have a partial pressures of 1 bar, or 0.9869 atm. (This can be rounded to 1 atm for all but the most exact calculations.)

Enthalpy measurements done under the standard-state condition are indicated by adding a superscript $^\circ$ to the symbol for enthalpy. The _standard-state_ enthalpy of reaction for the combustion of methane at 25°C, for example, would be reported as follows.

$$\Delta H^\circ{}_{298} = -802.4 \text{ kJ/mol}_{rxn}$$

> **CHECKPOINT**

What does the symbol ΔH°_{373} mean?

7.12 Calculating Enthalpies of Reaction

The origin of the change in enthalpy that accompanies a chemical reaction can be seen more clearly by examining the bonds in the products and reactants in the following reaction.

$$CO(g) + H_2O(g) \longrightarrow CO_2(g) + H_2(g)$$

We can use Lewis structures to visualize the bonds that have to be broken and the bonds that have to be formed in this reaction.

$$:C\equiv O: + H-\overset{..}{\underset{..}{O}}-H \rightleftharpoons \overset{..}{\underset{..}{O}}=C=\overset{..}{\underset{..}{O}} + H-H$$

Once again, we can assume a hypothetical two-step process for converting reactants into products. The first step involves breaking all of the bonds in the starting materials to form isolated atoms in the gas phase. Note that a carbon–oxygen triple bond and two hydrogen–oxygen single bonds are broken in this step, which is endothermic because bonds are being broken.

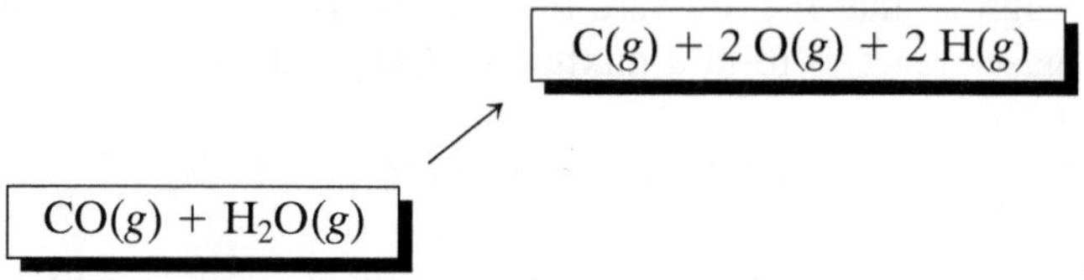

$$C(g) + 2\,O(g) + 2\,H(g)$$

$$CO(g) + H_2O(g)$$

The isolated atoms can then recombine to make the bonds necessary to form the products of the reaction. Because this step involves making bonds, it must be exothermic.

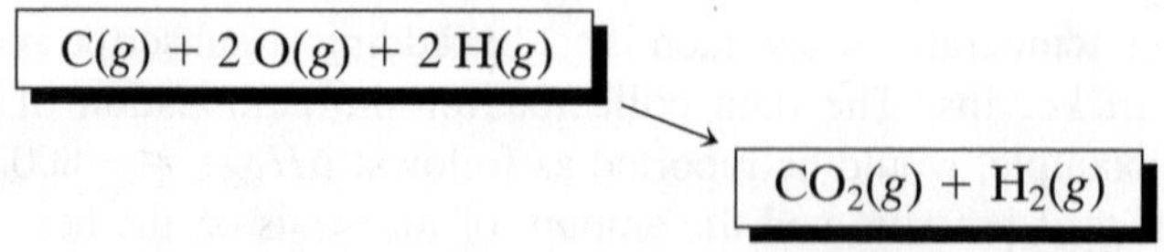

$$C(g) + 2\,O(g) + 2\,H(g)$$

$$CO_2(g) + H_2(g)$$

Experimentally we find that it takes 1076.4 kJ/mol$_{rxn}$ to break the $C\equiv O$ triple bonds in a mole of CO molecules to form isolated C and O atoms in the gas phase. It takes an additional 926.3 kJ/mol$_{rxn}$ to break the bonds in a mole of

H_2O molecules to form isolated H and O atoms. Thus the bond-breaking process takes a total of 2002.7 kJ per mole of reaction.

Bond breaking: $CO(g) \longrightarrow C(g) + O(g)$ $\Delta H° = 1076.4 \text{ kJ/mol}_{rxn}$
                  $H_2O(g) \longrightarrow 2 H(g) + O(g)$ $\Delta H° = 926.3 \text{ kJ/mol}_{rxn}$
                  _____
                  $CO(g) + H_2O(g) \longrightarrow C(g) + 2 H(g) + 2 O(g)$ $\Delta H° = 2002.7 \text{ kJ/mol}_{rxn}$

When the isolated atoms in the gas phase combine to form new bonds in the second step of the reaction, we see that two C=O double bonds are formed per CO_2 molecule, giving off a total of 1608.5 kJ per mole of CO_2. An H—H single bond is also created in this reaction for each H_2 molecule generated in the reaction. We have already found that we get 435.3 kJ per mole of H_2 molecules formed when hydrogen atoms combine. We therefore get a total of 2043.8 kJ back when the C, H, and O atoms combine to form one mole of CO_2 and H_2.

Bond making: $C(g) + 2 O(g) \longrightarrow CO_2(g)$ $\Delta H°_{ac} = -1608.5 \text{ kJ/mol}_{rxn}$
                $2 H(g) \longrightarrow H_2(g)$ $\Delta H°_{ac} = -435.3 \text{ kJ/mol}_{rxn}$
                _____
                $C(g) + 2 H(g) + 2 O(g) \longrightarrow CO_2(g) + H_2(g)$ $\Delta H° = -2043.8 \text{ kJ/mol}_{rxn}$

When we combine the change in the enthalpy of the system for the two hypothetical steps, we conclude that the overall enthalpy of reaction is relatively small compared with the enthalpy change associated with either bond breaking or bond making.

$$
\begin{array}{ll}
2002.7 \text{ kJ/mol}_{rxn} & \text{Bond breaking} \\
-2043.8 \text{ kJ/mol}_{rxn} & \text{Bond making} \\
\hline
-41.1 \text{ kJ/mol}_{rxn} &
\end{array}
$$

The reaction is exothermic, and it gives off heat because the sum of bond strengths in the products is larger than that in the reactants. But the amount of heat given off by the reaction is not large.

7.13 Enthalpies of Atom Combination

Chemists are often interested in knowing whether a reaction gives off or absorbs heat—and how much heat is given off or absorbed. The heat released or absorbed at a constant pressure is equal to the enthalpy change associated with the reaction and can be determined experimentally in the laboratory or calculated from experimentally measured values, as shown in the previous section.

To apply the technique used in the previous section to other reactions, we need a set of data that allows us to predict how much heat will be absorbed when we transform the starting materials into their isolated atoms in the gas phase and how much heat will be given off when the atoms recombine to give the products of the reaction. The data could be compiled in several ways. Consider ammonia (NH_3) for example. We could choose to compile the enthalpy change for the reaction in which the bonds in NH_3 molecules are broken to form isolated nitrogen and hydrogen atoms in the gas phase, which is 1171.76 kJ/mol_{rxn}.

$$NH_3(g) \longrightarrow N(g) + 3 H(g)$$

When the reaction is written in this way, the enthalpy change would be called the **enthalpy of atomization** because the reaction involves breaking apart the molecules in a sample into isolated atoms.

Another way to tabulate the data would be to start with the reverse of the reaction just shown.

$$N(g) + 3\,H(g) \longrightarrow NH_3(g)$$

In this case ammonia molecules are formed from isolated nitrogen and hydrogen atoms in the gas phase, and the enthalpy change for the reaction is -1171.76 kJ/mol$_{rxn}$. Such enthalpy changes are called **enthalpies of atom combination.**

It makes no difference how we tabulate the data as long as we specify the process, atomization or atom combination, to which the enthalpies refer. The authors of this text have chosen to use the atom combination reaction, and that is how data are compiled in Table B.13 in Appendix B. The enthalpy of atom combination (ac) for $NH_3(g)$ can be found in Table B.13 by looking for the data for compounds of the element nitrogen and then scanning these data until we find the entry for NH_3 as a gas. By definition, the symbol ΔH°_{ac} represents the enthalpy change under standard-state conditions at 25°C when 1 mole of the compound listed in the table is formed from its atoms in the gas phase.

$$N(g) + 3\,H(g) \longrightarrow NH_3(g)$$

The value of ΔH°_{ac} for $NH_3(g)$ at 1 atm pressure and 25°C is -1171.76 kJ/mol$_{rxn}$.

When we start with a substance that is a gas at 25°C and 1 atm, the enthalpy of atom combination only involves making the bonds within that substance, and nothing else. The value of ΔH°_{ac} for methane, for example, reflects the heat that is released when the 4 moles of C—H bonds in 1 mole of CH_4 molecules are formed from isolated carbon and hydrogen atoms in the gas phase.

$$C(g) + 4\,H(g) \longrightarrow CH_4(g) \qquad \Delta H^{\circ}_{ac} = -1662.09 \text{ kJ/mol}_{rxn}$$

Because enthalpy is a state function, the data in Table B.13 can be used to calculate ΔH for many processes without having to experimentally measure the heat given off or absorbed in the process. It is possible, for example, to use nothing more than the data in Table B.13 to calculate the economics of one fuel compared to another based on energy production.

Relatively few reactions involve only starting materials and products that exist as gases at room temperature and atmospheric pressure. The enthalpy of atom combination for liquid methanol (CH_3OH), for example, includes two terms in its measurement. The first is the heat released in the exothermic reaction that occurs when we convert the carbon, hydrogen, and oxygen atoms into CH_3OH molecules in the gas phase.

$$C(g) + 4\,H(g) + O(g) \longrightarrow CH_3OH(g) \qquad \Delta H^{\circ}_{ac} = -2037.11 \text{ kJ/mol}_{rxn}$$

The origin of the heat given off in this reaction is the formation of the *intramolecular* covalent bonds in methanol molecules.

The second part of the enthalpy of atom combination measurement is the heat given off when the gas condenses to form liquid methanol, which can be measured by calorimetry.

$$CH_3OH(g) \longrightarrow CH_3OH(l) \qquad \Delta H^{\circ} = -38.00 \text{ kJ/mol}_{rxn}$$

➤ **CHECKPOINT**

Write the chemical equations for the atom combination reaction of $N(g)$, $H(g)$, and $NH_3(g)$. What are the values of ΔH°_{ac} for $N(g)$ and $H(g)$? Explain your answers. What is the value of ΔH°_{ac} of $NH_3(g)$? Explain your answer.

The heat given off during a change of state isn't due to formation of new intramolecular bonds. (The covalent bonds in liquid and gaseous methanol are essentially the same.) What, then, is responsible for the enthalpy change when gaseous methanol is condensed to the liquid state?

Methanol molecules in the gas phase are relatively far apart and only touch for that brief moment in time when they collide. In the liquid state the methanol molecules are in constant contact with one another. As we will see in the next chapter, interactions between electrons on neighboring molecules in the liquid give rise to a force of attraction between these molecules. Such forces are called *intermolecular forces*[1] when they exist between neutral molecules and *interionic forces* when they occur between ions.

When a gas condenses to form a liquid, its particles come close enough that they are attracted to one another, and as a consequence heat is released. Thus the enthalpies of atom combination for liquids and solids are the sum of the energies associated with the formation of bonds between the atoms or ions and any additional interactions between molecules or pairs of ions held together by intermolecular or interionic forces.

The enthalpy of atom combination for the formation of gaseous methanol from its isolated atoms in the gas phase is -2037.11 kJ/mol$_{rxn}$, whereas the enthalpy of atom combination for liquid methanol from its atoms in the gas phase is -2075.11 kJ/mol$_{rxn}$.

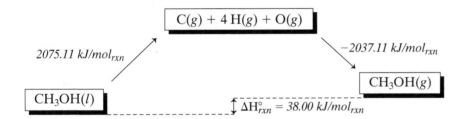

The difference between the two values (38.00 kJ/mol$_{rxn}$) represents the energy released when a mole of gaseous methanol condenses to form liquid methanol or the heat that must be absorbed to evaporate one mole of liquid methanol to form gaseous methanol. The change in enthalpy associated with the process by which liquid methanol forms from methanol in the gas phase is exothermic because of the formation of intermolecular forces. The change in enthalpy associated with the process by which liquid methanol is converted into a gas is endothermic because energy is required to overcome the force of attraction between the molecules in the liquid.

Enthalpies of atom combination data for more than 200 substances are given in Table B.13 in Appendix B. The first column of Table B.13 lists the chemical formulas and physical states of a variety of substances. The second column lists the enthalpies of atom combination associated with forming all of the bonds in 1 mole of the substance from isolated atoms in the gas phase. These data can be used to determine the change in enthalpy associated with any chemical reaction if data for all of the reactants and products are available.

Table 7.3 extends the number of compounds for which enthalpy of atom combination are known by providing data for various organic compounds in the gas phase.

---

[1]One way of remembering the difference between *intramolecular* and *intermolecular* forces is to remember that many college students participate in *intramural* sports, but the number who participate in *intercollegiate* sports is significantly smaller.

Table 7.3
Enthalpies of Atom Combination for Several Gaseous Organic Compounds

Name	Formula	ΔH_{ac}°, kJ/mol$_{rxn}$ at 298.15 K
Methane	$CH_4(g)$	−1662.09
Ethane	$C_2H_6(g)$	−2823.94
Propane	$C_3H_8(g)$	−3992.9
n-Butane	$C_4H_{10}(g)$	−5169.38
Isobutane	$C_4H_{10}(g)$	−5177.75
n-Pentane	$C_5H_{12}(g)$	−6337.9
n-Hexane	$C_6H_{14}(g)$	−7509.1
Methanol	$CH_3OH(g)$	−2037.11
Ethanol	$CH_3CH_2OH(g)$	−3223.53
n-Propanol	$CH_3CH_2CH_2OH(g)$	−4394.2
n-Butanol	$CH_3CH_2CH_2CH_2OH(g)$	−5564.5
n-Pentanol	$CH_3CH_2CH_2CH_2CH_2OH(g)$	−6735.9
Dimethyl ether	$CH_3OCH_3(g)$	−3171.3
Ethylmethyl ether	$CH_3OCH_2CH_3(g)$	−4354.6
Diethyl ether	$CH_3CH_2OCH_2CH_3(g)$	−5541.4
Dipropyl ether	$CH_3CH_2CH_2OCH_2CH_2CH_3(g)$	−7883.1
Ethylene	$CH_2{=}CH_2(g)$	−2251.70
Propene	$CH_3CH{=}CH_2(g)$	−3432.6
1-Butene	$CH_2{=}CHCH_2CH_3(g)$	−4604.9
1-Pentene	$CH_2{=}CHCH_2CH_2CH_3(g)$	−5777.4
1-Hexene	$CH_2{=}CHCH_2CH_2CH_2CH_3(g)$	−6947.7
1,3-Butadiene	$CH_2{=}CHCH{=}CH_2(g)$	−4058.9
Benzene	$C_6H_6(g)$	−5523.07
Toluene	$C_6H_5CH_3(g)$	−6690.0
Ethylbenzene	$C_6H_5CH_2CH_3(g)$	−7870.3
1,3,5-Trimethylbenzene	$C_6H_3(CH_3)_3(g)$	−9067.2

Source: J. D. Cox and G. Pilcher, *Thermochemistry of Organic and Organometallic Compounds*, Academic Press, New York, 1970.

To illustrate how atom combination enthalpy data are used, let's consider the synthesis of ammonia from nitrogen and hydrogen.

$$N_2(g) + 3\,H_2(g) \longrightarrow 2\,NH_3(g)$$

Because enthalpy is a state function, we can divide the reaction into a sequence of bond-breaking and bond-making steps. Heat would have to be absorbed to break the N≡N triple bonds in a mole of N_2 and the H—H single bonds in 3 moles of H_2 to form isolated nitrogen and hydrogen atoms in the gas phase.

$$\boxed{2\,N(g) + 6\,H(g)}$$

$\nearrow$ ΔH *is positive* (+)

$$\boxed{N_2(g) + 3\,H_2(g)}$$

Heat is then given off when the atoms come together to form NH_3 molecules in the gas phase.

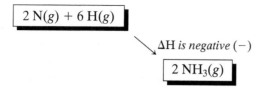

The primary reason for introducing the sign convention for ΔH is to remind us that endothermic reactions, which absorb heat from the surroundings, are represented by *positive* values of ΔH. Exothermic reactions, which give off heat, are described by *negative* values of ΔH. Note that if we want the enthalpy change for breaking the bonds in a compound, we must reverse the sign on the enthalpy given in Table B.13 in Appendix B.

It always takes energy to break the bonds in a molecule to form isolated atoms in the gas phase. When we calculate the heat that must be absorbed to break the bonds for this hypothetical reaction, the change in the enthalpy for each of the starting materials will have a positive value. Heat is always given off when atoms recombine to form molecules. The change in the enthalpy for each of the products of the reaction will therefore have a negative value.

The energetics of the bond breaking in the synthesis of ammonia can be calculated from the enthalpies tabulated in Table B.13 for H_2 ($\Delta H^{\circ}_{ac} = -435.30$ kJ/mol$_{rxn}$) and N_2 ($\Delta H^{\circ}_{ac} = -945.41$ kJ/mol$_{rxn}$). In the first hypothetical step in the reaction, 1 mole of N_2 and 3 moles of H_2 are transformed into hydrogen and nitrogen atoms in the gas phase. The *magnitude* of the enthalpy of reaction for breaking the N≡N triple bonds in a mole of N_2 is equal to the magnitude of ΔH°_{ac} in Table B.13 for the N_2 molecule. But the *sign* of the enthalpy of reaction for breaking the N≡N triple bonds is the opposite of the sign of ΔH°_{ac} in Table B.13 because we are breaking this bond, not combining atoms to form the bond.

The first step also shows that 3 moles of H_2 are converted into 6 moles of hydrogen atoms in the gas phase. The *magnitude* of the enthalpy of reaction for breaking the H—H single bonds in 3 moles of H_2 is equal to three times the value of ΔH°_{ac} in Table B.13 for the H_2 molecule. But we have to change the *sign* of the change in enthalpy because we are breaking the H—H bonds, not making them.

Bond breaking:
$$N_2(g) \longrightarrow 2\,N(g) \qquad \Delta H^{\circ} = 945.41 \text{ kJ/mol}_{rxn}$$
$$3\,H_2(g) \longrightarrow 6\,H(g) \qquad \Delta H^{\circ} = 1305.9 \text{ kJ/mol}_{rxn}$$

$$N_2(g) + 3\,H_2(g) \longrightarrow 2\,N(g) + 6\,H(g) \qquad \Delta H^{\circ} = 2251.3 \text{ kJ/mol}_{rxn}$$

The heat given off during bond formation can be calculated from the enthalpy of atom combination of NH_3 ($\Delta H^{\circ}_{ac} = -1171.76$ kJ/mol$_{rxn}$) found in Table B.13. When the isolated nitrogen and hydrogen atoms come together to form NH_3 molecules, 2 moles of NH_3 are formed. The enthalpy of reaction for the formation of 2 moles of NH_3 is therefore twice the enthalpy of atom combination of NH_3.

Bond making: $2\,N(g) + 6\,H(g) \longrightarrow 2\,NH_3(g) \qquad \Delta H^{\circ}_{ac} = -2343.52$ kJ/mol$_{rxn}$

Note that the step that involves bond breaking is endothermic (ΔH is positive), whereas the step that involves bond making is exothermic (ΔH is negative).

All we have to do to calculate the overall enthalpy of reaction for the synthesis of ammonia from nitrogen and hydrogen is to add the results of our calculations for the two hypothetical steps in the reaction.

$$2251.3 \text{ kJ/mol}_{rxn}$$
$$\underline{-2343.2 \text{ kJ/mol}_{rxn}}$$
$$-92.2 \text{ kJ/mol}_{rxn}$$

► CHECKPOINT

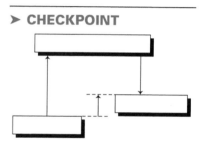

A chemical reaction can be diagrammed as shown here for the purpose of calculating enthalpy changes. Label the boxes in this reaction diagram as *reactants, products,* and *isolated atoms in the gas phase.* Also label the arrows that represent the *energy needed to break bonds,* the *energy given off when bonds form,* and the *enthalpy change* for the overall reaction. Use this diagram to predict whether the total strength of the bonds in the products is larger or smaller than the total strength of the bonds in the reactants.

The overall reaction is therefore exothermic.

   It is important to remember when doing enthalpy calculations using atom combination data that very few chemical reactions actually occur by first breaking all the bonds in the starting materials to form isolated atoms in the gas phase, followed by recombination of the atoms to form the products of the reaction. But, for our purposes, that doesn't matter. If all we want to know is whether the reaction gives off or absorbs heat, we can *assume* that it occurs by atomization and recombination steps. Because enthalpy is a state function, ΔH for a reaction doesn't depend on the path used to convert the starting materials into the products.

Exercise 7.6

Use enthalpy of atom combination data from Table B.13 in Appendix B to predict whether the reaction between graphite and oxygen to form carbon dioxide is endothermic or exothermic and to calculate the amount of heat given off or absorbed when the carbon in graphite reacts with oxygen to form carbon dioxide.

$$C(s) + O_2(g) \longrightarrow CO_2(g)$$

Solution

We can imagine a hypothetical pathway for this reaction by using the following diagram.

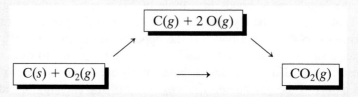

   The enthalpy of atom combination data for graphite suggests that a significant amount of heat would be given off when graphite is formed from its isolated carbon atoms in the gas phase.

$$C(g) \longrightarrow C(s) \qquad \Delta H^{\circ}_{ac} = -716.68 \text{ kJ/mol}_{rxn}$$

Likewise, atom combination data show that when the O=O double bond in an O_2 molecule is formed, 498.34 kJ/mol$_{rxn}$ of heat is given off.

$$2\,O(g) \longrightarrow O_2(g) \qquad \Delta H^{\circ}_{ac} = -498.34 \text{ kJ/mol}_{rxn}$$

Therefore the hypothetical step in which we break all the bonds in the starting materials would require a large input of energy in the form of heat.

Bond breaking: $\quad\quad\quad\quad C(s) \longrightarrow C(g) \quad\quad\quad\quad\quad \Delta H^{\circ} = 716.68 \text{ kJ/mol}_{rxn}$

$\quad\quad\quad\quad\quad\quad\quad\quad\quad\quad\quad O_2(g) \longrightarrow 2\,O(g) \quad\quad\quad\quad\quad \Delta H^{\circ} = 498.34 \text{ kJ/mol}_{rxn}$

$\quad\quad\quad\quad\quad\quad\quad\quad C(s) + O_2(g) \longrightarrow C(g) + 2\,O(g) \quad\quad \Delta H^{\circ} = 1215.02 \text{ kJ/mol}_{rxn}$

But the C=O double bonds in a CO_2 molecule are unusually strong ($\Delta H^{\circ}_{ac} = -1608.53$ kJ/mol$_{rxn}$). Thus a great deal of energy is given off when the isolated atoms come together to form CO_2 molecules.

Bond making: $\quad\quad C(g) + 2\,O(g) \longrightarrow CO_2(g) \quad\quad \Delta H^{\circ} = -1608.53 \text{ kJ/mol}_{rxn}$

The overall reaction is therefore strongly exothermic.

$$1215.02 \text{ kJ/mol}_{rxn}$$
$$\underline{-1608.53 \text{ kJ/mol}_{rxn}}$$
$$-393.51 \text{ kJ/mol}_{rxn}$$

 Exercise 7.7

Carbon in the form of graphite can also react with oxygen to produce carbon monoxide (CO). Is this reaction endothermic or exothermic?

Solution

This reaction is represented by the following equation.

$$2 \text{ C}(s) + \text{O}_2(g) \longrightarrow 2 \text{ CO}(g)$$

The enthalpy change can be understood in terms of the following diagram.

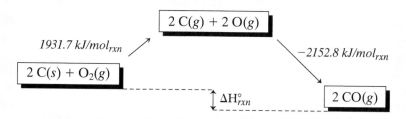

The overall enthalpy of reaction can be calculated by adding the enthalpy of reaction for the two steps. As always, the first step is highly endothermic ($\Delta H^\circ = 1931.7 \text{ kJ/mol}_{rxn}$) because of the carbon–carbon and oxygen–oxygen bonds that have to be broken. But the second step is highly exothermic ($\Delta H^\circ = -2152.8 \text{ kJ/mol}_{rxn}$) because of the carbon–oxygen triple bonds that are formed when the isolated atoms come together to form CO molecules.

$$\Delta H^\circ = 1931.7 \text{ kJ/mol}_{rxn} + (-2152.8 \text{ kJ/mol}_{rxn}) = -221.1 \text{ kJ/mol}_{rxn}$$

The overall reaction is therefore exothermic, but not as exothermic as the reaction in which graphite burns to form carbon dioxide.

$$\text{C}(s) + \text{O}_2(g) \longrightarrow \text{CO}_2(g) \qquad \Delta H^\circ = -393.51 \text{ kJ/mol}_{rxn}$$

The difference between the amounts of heat given off when carbon burns to form CO versus CO_2 can be understood by noting that the C≡O triple bond in CO isn't as strong as the two C=O double bonds in CO_2. The reaction of carbon with oxygen can and does produce carbon monoxide, particularly if inadequate supplies of oxygen are available for conversion to CO_2. Carbon monoxide can combine with the iron in human blood and is toxic. To cut down on the production of CO when automobiles burn gasoline, catalytic converters are used to convert any CO that is formed to CO_2 by catalyzing the following reaction.

$$2 \text{ CO}(g) + \text{O}_2(g) \longrightarrow 2 \text{ CO}_2(g)$$

➤ **CHECKPOINT**

Is the following reaction endothermic or exothermic?

$$2 \text{ CO}(g) + \text{O}_2(g) \longrightarrow 2 \text{ CO}_2(g)$$

7.14 Using Enthalpies of Atom Combination to Probe Chemical Reactions

Enthalpies of atom combination can be used in ways other than to calculate the enthalpy change for a reaction. These data can also be used as a measure of the total bond strengths for gaseous compounds. They therefore provide a direct means of comparing the strengths of the bonds holding a compound together.

In 1833 Jöns Jacob Berzelius suggested that compounds with the same formula but different structures should be called **isomers** (literally, "equal parts"). Butane and isobutene, for example, are isomers because they both contain the same number of carbon, hydrogen, and oxygen atoms.

Butane is known as a straight-chain hydrocarbon because it contains one continuous chain of C—C bonds. Isobutane, on the other hand, is an example of a branched hydrocarbon. The so-called butane lighter fluid in disposable lighters actually contains more isobutane than anything else. Although we tend to describe butane as a straight-chain hydrocarbon, it is important to remember that the electron domain model introduced in Section 4.14 would predict that the geometry around each carbon atom in the molecule would be tetrahedral. The shapes of the butane and isobutane molecules are shown in Figure 7.7.

Because they are isomers, these compounds must have the same molecular formula: C_4H_{10}. But there is an even greater similarity between the two compounds. Each compound contains 3 C—C bonds and 10 C—H bonds. Are the bonds in the two compounds of the same strength? Enthalpy of atom combination data from Table 7.3 show us that butane ($\Delta H^\circ_{ac} = -5169.38$ kJ/mol$_{rxn}$) has a slightly less negative enthalpy of atom combination than isobutane ($\Delta H^\circ_{ac} = -5177.75$ kJ/mol). If we form both compounds from their atoms, more heat would be released when the atoms combined to form isobutane. Thus isobutene molecules have slightly stronger bonds than butane molecules.

Let's consider the energetics of the reaction in which butane is converted into isobutane.

butane $\longrightarrow$ isobutane

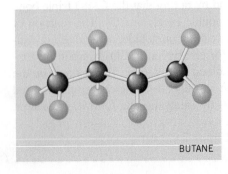
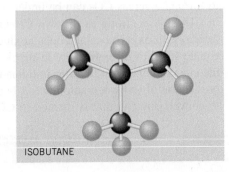

Fig. 7.7 Two isomers with the formula C_4H_{10} are possible: butane and isobutane. Butane is a straight-chain hydrocarbon, and isobutane is a branched hydrocarbon. All carbon atoms in both hydrocarbons have 109° bond angles with hydrogen.

We can calculate the enthalpy change for this reaction as follows.

Bond breaking in butane:	$\Delta H° = 5169.38 \text{ kJ/mol}_{rxn}$
Bond making in isobutane:	$\Delta H° = -5177.75 \text{ kJ/mol}_{rxn}$
	$\Delta H° = -8.37 \text{ kJ/mol}_{rxn}$

We see that the overall enthalpy change for the conversion of butane into isobutane is negative ($-8.37 \text{ kJ/mol}_{rxn}$), which means that the bonds formed in the product, isobutane, are stronger overall than those broken in the reactant, butane. This information tells us that it may be possible to convert straight-chain hydrocarbons to branched chains. Experimental data confirm this. At high temperatures (500°–600°C) and high pressures (25–50 atm), straight-chain hydrocarbons isomerize to form branched hydrocarbons. The reaction can be run under more moderate conditions in the presence of a catalyst, such as a mixture of silica (SiO_2) and alumina (Al_2O_3). Such reactions play a vital role in the refining of gasoline because branched alkanes burn more evenly than straight-chain alkanes and are therefore less likely to cause an engine to "knock."

Exercise 7.8

Fumaric acid and maleic acid are isomers that differ only in the orientation of substituents around the C=C double bond in the compounds. Organic chemists therefore describe these compounds as stereoisomers.

$$HO_2C \quad \quad CO_2H$$
$$C=C$$
$$H \quad \quad H$$
Maleic acid

$$HO_2C \quad \quad H$$
$$C=C$$
$$H \quad \quad CO_2H$$
Fumaric acid

In maleic acid, the two hydrogen atoms and the two —CO_2H groups are on the same side of a horizontal plane passing through the atoms in the C=C double bond. In fumaric acid, they are on opposite sides of the bond. Use the enthalpies of atom combination of fumaric acid ($\Delta H°_{ac} = -5545.03 \text{ kJ/mol}_{rxn}$) and maleic acid ($\Delta H°_{ac} = -5524.53 \text{ kJ/mol}_{rxn}$) to predict whether we should be able to convert fumaric acid to maleic acid, or maleic acid to fumaric acid.

Solution

More heat is released when fumaric acid is formed from its atoms than when maleic acid forms. This suggests that it may be possible to convert maleic acid to fumaric acid by some chemical process. Again, we can also calculate the enthalpy change for the conversion of maleic acid to fumaric acid.

Bond breaking in maleic acid:	$\Delta H° = 5524.53 \text{ kJ/mol}_{rxn}$
Bond making in fumaric acid:	$\Delta H°_{ac} = -5545.03 \text{ kJ/mol}_{rxn}$
	$\Delta H° = -20.50 \text{ kJ/mol}_{rxn}$

Experimentally we find that maleic acid can be converted to fumaric acid by heating maleic acid in the presence of a strong acid for about 30 minutes.

➤ CHECKPOINT

Compare the enthalpies of atom combination of $CH_3CH_2OH(g)$ and $CH_3CH_2OH(l)$. What can you conclude from the difference in the two enthalpies?

Exercise 7.9

The enthalpy change for the reaction of HF with SiO_2 is exothermic.

$$4\,HF(g) + SiO_2(s) \longrightarrow SiF_4(g) + 2\,H_2O(g) \quad \Delta H° = -103.4 \text{ kJ/mol}_{rxn}$$

The reaction of HCl with SiO_2, however, is endothermic.

$$4\,HCl(g) + SiO_2(s) \longrightarrow SiCl_4(g) + 2\,H_2O(g) \quad \Delta H° = +139.6 \text{ kJ/mol}_{rxn}$$

How can we account for the difference in the enthalpy change for these two similar reactions? What is the average bond enthalpy of Si—Cl and Si—F bonds?

Solution

The magnitude and sign of the enthalpy change are determined by the sum of the bond strengths in the products and reactants. These two reactions have two species in common, SiO_2 and H_2O, so the difference in the enthalpy change must be due to the difference in the bond strengths for HF and SiF_4 in one reaction and HCl and $SiCl_4$ in the other. Table B.13 in Appendix B provides the necessary data to determine the average bond enthalpies.

	$\Delta H°_{ac}$(kJ/mol$_{rxn}$)
HF(g)	−567.7
SiO_2(s)	−1864.9
SiF_4(g)	−2386.5
H_2O(g)	−926.29
HCl(g)	−431.64
$SiCl_4$(g)	−1599.3

Although the HF bond (567.7 kJ/mol$_{rxn}$) is stronger than the HCl bond (431.64 kJ/mol$_{rxn}$), the reaction between SiO_2 and HF liberates more heat (−103.4 kJ/mol$_{rxn}$) than the reaction between SiO_2 and HCl (139.6 kJ/mol$_{rxn}$). This implies that the SiF bond in SiF_4 must be stronger than the SiCl bond in $SiCl_4$.

The average bond strength for Si—F and Si—Cl bonds can be estimated from $\Delta H°_{ac}$ data. Because there are four Si—F bonds in SiF_4, the average Si—F bond strength is one-fourth of the enthalpy of atom combination for this compound, or 597 kJ/mol. A similar calculation for $SiCl_4$ gives an average value for the Si—Cl bond of 400 kJ/mol. Thus almost 200 kJ more heat is released when a mole of Si—F bonds are formed than when a mole of Si—Cl bonds are formed. The Si—F bonds in SiF_4 (597 kJ/mol$_{rx}$) are much stronger than the Si—Cl bonds (400 kJ/mol$_{rxn}$) in $SiCl_4$. Despite the weaker bond strength in HCl as compared to HF, the strength of the Si—F bonds in SiF_4 more than compensates for the enthalpy required to break the HF bond.

Chemists often need approximate data for compounds for which no thermodynamic data are available. One way of estimating the bond strengths in these compounds is to use average bond enthalpies such as those calculated for the Si—F and Si—Cl bonds in Exercise 7.9. We can illustrate how this is done by estimating the sum of the bond strengths in Cl_3SiOH.

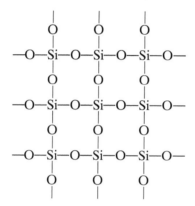

O—H
|
Si
Cl／ | ＼Cl
Cl

In Exercise 7.9 we saw that the average Si—Cl bond enthalpy was 400 kJ/mol. The O—H bond enthalpy can be estimated from the enthalpy of atom combination data for H_2O. There are two O—H bonds in each water molecule, so the average strength of an O—H should be half of the value of ΔH_{ac}° for water.

O
H／   ＼H

$$926.29/2 = 463 \text{ kJ per mole of O—H bonds}$$

The strength of the Si—O bond in SiO_2 (silica) is somewhat more difficult to determine. The empirical formula is SiO_2, but silica is actually a network of Si—O bonds. Each silicon atom is bonded to four oxygen atoms.

```
     |       |       |
     O       O       O
     |       |       |
 —O—Si—O—Si—O—Si—O—
     |       |       |
     O       O       O
     |       |       |
 —O—Si—O—Si—O—Si—O—
     |       |       |
     O       O       O
     |       |       |
 —O—Si—O—Si—O—Si—O—
     |       |       |
     O       O       O
     |       |       |
```

Thus, if the atom combination enthalpy for SiO_2 is -1864.9 kJ/mol$_{rxn}$, the strength of an average Si—O bond should be $\dfrac{1864.9}{4} = 466$ kJ/mol.

Now we can estimate the total bond strengths in Cl_3SiOH. There are three Si—Cl bonds, one Si—O bond, and one O—H bond for a total of 2129 kJ/mol.

$$3 \times (400 \text{ kJ/mol}) + (466 \text{ kJ/mol}) + (463 \text{ kJ/mol}) = 2129 \text{ kJ/mol}$$

➤ **CHECKPOINT**

Estimate the sum of the bond strengths in F_3SiOH.

7.15 Bond Length and the Enthalpy of Atom Combination

In Chapters 3 and 4 we noted that the distance between neighboring atoms, the bond length, could often be estimated from the relative size of atoms. We expect a C—O single bond to be longer than a C—H single bond, for example, because the covalent radius of oxygen is larger than that of hydrogen. We can now relate the strengths of certain bonds to the bond length. The bond lengths for the series of halogen acids given in Table 7.4 were estimated from the covalent radii of the atoms given in Figure 3.26. The enthalpies of atom combination in this table were taken from Table B.13 in the appendix.

The strongest hydrogen–halogen bond, H—F, has the shortest bond length. The other compounds in this table follow the trend of increasing bond length with decreasing bond strength.

➤ **CHECKPOINT**

Predict which molecule, ClF(g) or BrF(g), has the weakest bond.

Table 7.4
**Bond Lengths and Enthalpies of Atom Combination
for the Hydrogen Halides**

Molecule	Bond Length (nm)	ΔH°_{ac} (kJ/mol$_{rxn}$)
HF(g)	0.101	-567.7
HCl(g)	0.136	-431.64
HBr(g)	0.151	-365.93
HI(g)	0.170	-298.01

7.16 Hess's Law

As we have seen, enthalpy is a state function. As a result, the value of ΔH for a reaction doesn't depend on the path used to go from one state of the system to another.

In 1840, German Henri Hess, professor of chemistry at the University and Artillery School in St. Petersburg, Russia, came to the same conclusion on the basis of experimental data and proposed a general rule known as **Hess's law,** which states that ΔH_{rxn} is the same regardless of whether a reaction occurs in one step or in several steps, and regardless of the path by which the reactants are converted into the products of the reaction. Thus we can calculate the enthalpy of reaction by adding the enthalpies associated with a series of hypothetical steps into which the reaction can be broken regardless of whether or not the reaction actually occurs by these steps.

Exercise 7.10

The heat given off under standard-state conditions has been measured for the reactions in which water as both a liquid and a gas is formed from its elements.

$$H_2(g) + \tfrac{1}{2}O_2(g) \longrightarrow H_2O(l) \qquad \Delta H^\circ = -285.83 \text{ kJ/mol}_{rxn}$$
$$H_2(g) + \tfrac{1}{2}O_2(g) \longrightarrow H_2O(g) \qquad \Delta H^\circ = -241.82 \text{ kJ/mol}_{rxn}$$

Use these data and Hess's law to calculate ΔH° for the following reaction.

$$H_2O(l) \longrightarrow H_2O(g)$$

Solution

The key to solving this problem is finding a way to combine the two reactions for which experimental data are known to give the reaction for which ΔH° is unknown. We can do this by reversing the direction in which the first reaction is written and then adding it to the equation for the second reaction. In other words, we assume that a mole of water is decomposed into its elements in the first reaction and that a mole of water vapor is formed from its elements in the second reaction.

$$
\begin{array}{lll}
H_2O(l) \longrightarrow H_2(g) + \tfrac{1}{2}O_2(g) & \Delta H^\circ = & 285.83 \text{ kJ/mol}_{rxn} \\
\underline{H_2(g) + \tfrac{1}{2}O_2(g) \longrightarrow H_2O(g)} & \underline{\Delta H^\circ = -241.82 \text{ kJ/mol}_{rxn}} \\
H_2O(l) \longrightarrow H_2O(g) & \Delta H^\circ = & 44.01 \text{ kJ/mol}_{rxn}
\end{array}
$$

Because we reversed the direction in which the first reaction was written, we had to change the sign of $\Delta H°$ for this reaction. $H_2(g)$ and $\frac{1}{2} O_2(g)$ don't appear in the final chemical equation because they are on both sides of the equations that were added to each other. These terms therefore cancel when the equations are combined to obtain the overall equation for the reaction.

. . . .

There was just enough information in Exercise 7.10 to solve the problem. The next exercise forces us to choose the reactions that will be combined from a wealth of information.

 Exercise 7.11

Before pipelines were built to deliver natural gas, individual towns and cities contained plants that produced a fuel known as "town gas".

$$C(s) + H_2O(g) \longrightarrow CO(g) + H_2(g)$$

Calculate $\Delta H°$ for the reaction used to produce "town gas" from the following information.

$$C(s) + \frac{1}{2}O_2(g) \longrightarrow CO(g) \qquad \Delta H° = -110.53 \text{ kJ/mol}_{rxn}$$
$$C(s) + O_2(g) \longrightarrow CO_2(g) \qquad \Delta H° = -393.51 \text{ kJ/mol}_{rxn}$$
$$CO(g) + \frac{1}{2}O_2(g) \longrightarrow CO_2(g) \qquad \Delta H° = -282.98 \text{ kJ/mol}_{rxn}$$
$$H_2(g) + \frac{1}{2}O_2(g) \longrightarrow H_2O(g) \qquad \Delta H° = -214.82 \text{ kJ/mol}_{rxn}$$

Solution

In this case, the desired equation can be found by using the first reaction to generate $CO(g)$ from $C(s)$ and the reverse of the fourth reaction to generate $H_2(g)$ from $H_2O(g)$.

$$C(s) + \frac{1}{2}O_2(g) \longrightarrow CO(g) \qquad \Delta H° = -110.53 \text{ kJ/mol}_{rxn}$$
$$\underline{H_2O(g) \longrightarrow H_2(g) + \frac{1}{2}O_2(g) \qquad \Delta H° = 214.82 \text{ kJ/mol}_{rxn}}$$
$$C(s) + H_2O(g) \longrightarrow CO(g) + H_2(g) \qquad \Delta H° = 104.29 \text{ kJ/mol}_{rxn}$$

. . . .

7.17 Enthalpies of Formation

Hess's law suggests that we can save a lot of work measuring enthalpies of reaction by using a little imagination in choosing the reactions for which measurements are made. The question is: What is the best set of reactions to study so that we get the greatest benefit from the smallest number of experiments?

Exercise 7.11 predicted $\Delta H°$ for the following reaction,

$$C(s) + H_2O(g) \longrightarrow CO(g) + H_2(g)$$

by combining enthalpy of reaction measurements for these reactions.

$$C(s) + \tfrac{1}{2}O_2(g) \longrightarrow CO(g)$$
$$H_2(g) + \tfrac{1}{2}O_2(g) \longrightarrow H_2O(g)$$

The reactions that were combined in Exercise 7.11 have one thing in common. Both reactions lead to the formation of a compound from the elements in their most thermodynamically stable form at 25°C. The enthalpy of reaction for each of the reactions is therefore known as the **enthalpy of formation** of the compound, ΔH_f°. By definition, ΔH_f° is the enthalpy associated with the reaction that forms 1 mole of a compound from its elements in their most thermodynamically stable states at 25°C and 1 atm. Enthalpies of formation are similar to enthalpies of atom combination in that both represent the formation of a compound from its elements. However, the values differ because enthalpies of formation represent the formation of a compound from its *elements* in their most thermodynamically stable states at 25°C and 1 atm pressure as compared to enthalpies of atom combination in which the compounds are formed from *isolated atoms in the gas phase*. Because the enthalpy change for a reaction is a state function, enthalpies of formation can be used to calculate enthalpies of reaction in a manner similar to that used with enthalpies of atom combination.

Exercise 7.12

Which of the following equations describes a reaction for which ΔH° is equal to the enthalpy of formation ΔH_f° of a compound?

(a) $Mg(s) + \tfrac{1}{2}O_2(g) \longrightarrow MgO(s)$

(b) $MgO(s) + CO_2(g) \longrightarrow MgCO_3(s)$

(c) $Mg(s) + C(s) + \tfrac{3}{2}O_2(g) \longrightarrow MgCO_3(s)$

(d) $Mg(g) + O(g) \longrightarrow MgO(s)$

Solution

Equations (a) and (c) describe enthalpy of formation reactions. These reactions result in the formation of a compound from the most thermodynamically stable form of its elements. Equation (b) can't be an enthalpy of formation reaction because the product of the reaction isn't formed from its elements. Equation (d) is an enthalpy of atom combination reaction, not an enthalpy of formation reaction.

Exercise 7.13

Use Hess's law to calculate ΔH° for the reaction between magnesium oxide and carbon dioxide to form magnesium carbonate,

$$MgO(s) + CO_2(g) \longrightarrow MgCO_3(s)$$

from the following enthalpy of formation data.

$$Mg(s) + \tfrac{1}{2}O_2(g) \longrightarrow MgO(s) \qquad \Delta H_f^\circ = -601.70 \text{ kJ/mol}_{rxn}$$
$$C(s) + O_2(g) \longrightarrow CO_2(g) \qquad \Delta H_f^\circ = -393.51 \text{ kJ/mol}_{rxn}$$
$$Mg(s) + C(s) + \tfrac{3}{2}O_2(g) \longrightarrow MgCO_2(s) \qquad \Delta H_f^\circ = -1095.8 \text{ kJ/mol}_{rxn}$$

Solution

The reaction in which we are interested converts two reactants (MgO and CO_2) into a single product ($MgCO_3$). We might therefore start by reversing the direction in which we write the enthalpy of formation reactions for MgO and CO_2, thereby decomposing these substances into their elements in their most thermodynamically stable form. We can then add the enthalpy of formation reaction for $MgCO_3$, thereby forming the product from its elements in their most stable form.

$$
\begin{array}{lll}
MgO(s) \longrightarrow Mg(s) + \tfrac{1}{2}\,O_2(g) & \Delta H^\circ = & 601.70 \text{ kJ/mol}_{rxn} \\
CO_2(g) \longrightarrow C(s) + O_2(g) & \Delta H^\circ = & 393.51 \text{ kJ/mol}_{rxn} \\
\underline{Mg(s) + C(s) + \tfrac{3}{2}\,O_2 \longrightarrow MgCO_3(s)} & \underline{\Delta H_f^\circ = -1095.8 \text{ kJ/mol}_{rxn}} \\
MgO(s) + CO_2(g) \longrightarrow MgCO_3(s) & \Delta H^\circ = -100.6 \text{ kJ/mol}_{rxn}
\end{array}
$$

Adding the three equations gives the desired overall reaction. ΔH° for the overall reaction is therefore the sum of the enthalpies of the three hypothetical steps.

An important point was made during the discussion of calculations of the enthalpy of reaction from enthalpy of atom combination data in Section 7.13: It doesn't matter whether or not the reaction actually occurs by these hypothetical steps. Because enthalpy is a state function, ΔH for the hypothetical reaction will be exactly equal to ΔH for the reaction whatever the pathway by which the starting materials are actually transformed into the products of the reaction.

· ·

The procedure used in Exercise 7.13 works, no matter how complex the reaction. All we have to do as the reaction becomes more complex is add more intermediate steps. This approach works because enthalpy is a state function. Thus, ΔH° is the same regardless of the path used to get from the starting materials to the products of the reaction. Instead of considering the reaction as taking place in a single step,

$$MgO(s) + CO_2(g) \longrightarrow MgCO_3(s)$$

we can split it into two steps. In the first step, the starting materials are converted to the elements from which they form in their most thermodynamically stable states at 25°C and 1 atm pressure.

$$MgO(s) + CO_2(g) \longrightarrow Mg(g) + C(s) + \tfrac{3}{2}\,O_2(g)$$

In the second step, the elements combine to form the products of the reaction.

$$Mg(s) + C(s) + \tfrac{3}{2}\,O_2(g) \longrightarrow MgCO_3(s)$$

If we analyze the technique used to solve the problem in Exercise 7.13, we find that we calculated ΔH_{rxn}° for the reaction in which MgO and CO_2 combine to form $MgCO_3$ by *adding* the enthalpy of formation for the *products* of this reaction and *subtracting* the enthalpy of formation for each of the *reactants*. In other words, we can calculate ΔH_f° for any reaction,

$$aA + bB \longrightarrow cC + dD$$

(where A, B, C, and D represent the different components of the reaction and a, b, c, and d represent coefficients in the balanced chemical equation for the reaction) from enthalpy of formation (ΔH_f°) data using the following equation.

$$\Delta H_{rxn}^\circ = \left[c(\Delta H_f^\circ)_C + d(\Delta H_f^\circ)_D\right] - \left[a(\Delta H_f^\circ)_A + b(\Delta H_f^\circ)_B\right]$$

Standard-state enthalpy of formation data for a variety of elements and compounds can be found in Table B.16 in Appendix B. One point needs to be understood before this table can be used effectively. By definition, the enthalpy of formation of any element in its most thermodynamically stable form under standard-state conditions is zero because the initial and final states of the system are exactly the same.

Under standard-state conditions, the most thermodynamically stable form of oxygen, for example, is the diatomic molecule in the gas phase: $O_2(g)$. By definition, the enthalpy of formation of this substance is equal to the enthalpy associated with the reaction in which it is formed from its elements in their most thermodynamically stable form. For O_2 molecules in the gas phase, ΔH_f° is therefore equal to the heat given off or absorbed in the following reaction.

$$O_2(g) \longrightarrow O_2(g)$$

Because the initial and final states of the reaction are identical, no heat can be given off or absorbed, so ΔH_f° for $O_2(g)$ is zero.

We are now ready to use standard-state enthalpy of formation data to predict enthalpies of reaction.

 Exercise 7.14

In Exercise 7.4 we calculated the enthalpy change associated with the combustion of 0.600 moles of pentaborane, B_5H_9. Use enthalpy of formation data to calculate the heat given off when one mole of B_5H_9 reacts with excess oxygen according to the following equation.

$$2 B_5H_9(g) + 12 O_2(g) \longrightarrow 5 B_2O_3(s) + 9 H_2O(g)$$

Solution

We start by looking up the appropriate data in Table B.16 in Appendix B.

Compound	$\Delta H_f^\circ(\text{kJ/mol}_{rxn})$
$B_5H_9(g)$	73.2
$B_2O_3(s)$	-1272.77
$O_2(g)$	0
$H_2O(g)$	-241.82

We then substitute these data into the following equation

$$\Delta H_{rxn}^\circ = \left[c(\Delta H_f^\circ)_C + d(\Delta H_f^\circ)_D\right] - \left[a(\Delta H_f^\circ)_A + b(\Delta H_f^\circ)_B\right]$$
$$\Delta H_{rxn}^\circ = \left[5(\Delta H_f^\circ)_{B_2O_3} + 9(\Delta H_f^\circ)_{H_2O}\right] - \left[2(\Delta H_f^\circ)_{B_5H_9} + 12(\Delta H_f^\circ)_{O_2}\right]$$
$$\Delta H_{rxn}^\circ = \left[5(-1272.77 \text{ kJ/mol}_{rxn}) + 9(-241.82 \text{ kJ/mol}_{rxn})\right]$$
$$- \left[2(73.2 \text{ kJ/mol}_{rxn}) + 12(0 \text{ kJ/mol}_{rxn})\right]$$
$$\Delta H_{rxn}^\circ = -8686.6 \text{ kJ/mol}_{rxn}$$

According to the balanced equation, this is the enthalpy change when 2 moles of B_5H_9 are consumed. $\Delta H°$ for the reaction is therefore -4343.3 kJ per mole of B_5H_9. This is five times the molar enthalpy of reaction for the combustion of CH_4. On a per-gram basis, it is about 20% larger than the energy released when methane burns. It isn't surprising that B_5H_9 was once considered for use as a rocket fuel.

Key Terms

Boundary	First law of thermodynamics	Standard state
Calorimeter	Heat	State
Endothermic	Hess's law	State function
Enthalpy	Intensive property	Surroundings
Enthalpy of atom combination	Isomer	System
Enthalpy of atomization	Kinetic energy	Temperature
Enthalpy of formation	Kinetic theory of heat	Thermodynamics
Enthalpy of reaction	Molar heat capacity	Work
Exothermic	Potential energy	Work of expansion
Extensive property	Specific heat	

Problems

Energy

1. Define *kinetic* and *potential* energy.

2. A ball is attached to a spring and supported from the ceiling. When the ball is at rest, does it have kinetic or potential energy, or both? If the ball is pulled down and released, does it have kinetic or potential energy, or both?

3. What is one way energy can be transferred from one object to another?

4. Give an example of how energy can be converted from one form into another.

5. When certain chemical reactions occur, energy may be released. What is the source of this energy?

Heat

6. What is the difference between heat and temperature?

7. If a lead ball is dropped from the top of a building to the sidewalk, both the ball and the sidewalk will be warmer. What is the source of this heat?

Heat and the Kinetic Molecular Theory

8. Neon gas at 25°C is contained in a flask that is connected to an identical flask that contains neon gas at 50°C. If the connection between the flasks is opened, describe in molecular terms what will happen to the speed of the molecules. What will be the final temperature if equal quantities of Ne are in both flasks?

9. An iron bar at 25°C is placed in contact with an identical iron bar at 50°C. Describe on a microscopic level what will happen to the temperature of the bars.

10. Define *heat* according to the kinetic molecular theory.

11. What physical properties on both the atomic and macroscopic scales change when a balloon filled with helium is heated? Use the kinetic theory of heat to explain each of the changes.

12. It is often believed that things that are hot contain a lot of heat. Use the thermodynamic concepts of system, surroundings, and boundaries to explain why this notion is incorrect.

Specific Heat

13. What is the difference between specific heat and molar heat capacity?

14. The same quantity of heat is added to equal numbers of moles of CCl_4 and H_2O. The temperature increases more in the CCl_4 than in the H_2O. Which has the highest heat capacity? Explain.

15. If 1 mole of water at 20°C is placed in contact with 1 mole of Hg(l) at 50°C, will the final temperature be 35°C, greater than 35°C, or less than 35°C? Explain.

16. In a calorimeter experiment done at constant pressure in which all the heat from a chemical reaction was absorbed by the surrounding water bath, the temperature of the water went up by 2.31 K. If the size of the water bath was 200 g, what was the amount of heat transferred to the water? If the chemical reaction was

$$CH_4(g) + 2\,O_2(g) \longrightarrow CO_2(g) + 2\,H_2O(l)$$

and 2.17×10^{-3} moles of CH_4 were burned, what is ΔH for the reaction?

State Functions

17. Give examples of at least five physical properties that are state functions.

18. Which of the following descriptions of a trip are state functions?
    (a) work done
    (b) energy expended
    (c) cost
    (d) distance traveled
    (e) tire wear
    (f) gasoline consumed
    (g) change in location of a car
    (h) elevation change
    (i) latitude change
    (j) longitude change

19. Which of the following are state functions?
    (a) temperature
    (b) energy
    (c) pressure
    (d) volume
    (e) heat
    (f) work

Work and the First Law of Thermodynamics

20. An ideal gas in a fixed-volume container is heated. Describe what happens to the gas molecules and their energy.

21. The first law of thermodynamics is often described as saying that "energy is conserved." Describe why it is incorrect to assume that the first law suggests that the "energy of a *system* is conserved."

22. Describe what happens to the energy of a system when the system does work on its surroundings. What happens to the energy of the system when it loses heat to its surroundings?

23. Give examples of both a system doing work on its surroundings and a system losing heat to its surroundings. Describe what happens to the energy of the system in each case.

24. Give examples of both a system having work done on it by its surroundings and a system gaining heat from its surroundings. What happens to the energy of the system in each case?

25. Give a verbal definition of the first law of thermodynamics.

26. Describe what happens to the energy of the system when an exothermic reaction is run under conditions of constant volume.

The Enthalpy of a System

27. What one thing do all chemical reactions have in common?

28. The reaction between hydrogen and oxygen to produce water gives off heat. What is the source of this heat? How could this heat be measured?

29. What is meant by the change in enthalpy that accompanies a chemical reaction?

Enthalpies of Reaction

30. Oxyacetylene torches are fueled by the combustion of acetylene, C_2H_2.

    $$2\,C_2H_2(g) + 5\,O_2(g) \longrightarrow 4\,CO_2(g) + 2\,H_2O(g)$$

    If the enthalpy change for the reaction is -2511.14 kJ/mol_{rxn}, how much heat can be produced by the reaction of
    (a) 2 mol of C_2H_2?
    (b) 1 mol of C_2H_2?
    (c) 0.500 mol of C_2H_2?
    (d) 0.2000 mol of C_2H_2?
    (e) 10 g of C_2H_2?

31. If the enthalpy change for the following reaction

    $$C(s) + H_2O(g) \longrightarrow CO(g) + H_2(g)$$

    is 131.29 kJ/mol_{rxn}, how much heat will be absorbed by the reaction of
    (a) 1 mol of $H_2O(g)$?
    (b) 2 mol of $H_2O(g)$?
    (c) 0.0300 mol of $H_2O(g)$?
    (d) 0.0500 mol of $C(s)$?

32. How much heat is released when 1 mole of nitrogen reacts with 2 moles of O_2 to give 2 moles of $NO_2(g)$ if $\Delta H°$ for the reaction is 33.2 kJ/mol_{rxn}?

    $$N_2(g) + 2\,O_2(g) \longrightarrow 2\,NO_2(g)$$

33. Calculate the standard-state enthalpy change for the following reaction if 1.00 g of magnesium gives off 46.22 kJ of heat when it reacts with excess fluorine.

    $$Mg(s) + F_2(g) \longrightarrow MgF_2(s)$$

34. Calculate $\Delta H°$ for the following reaction, assuming that it gives off 4.65 kJ of heat when hydrogen reacts with 1.00 g of calcium.

    $$Ca(s) + H_2(g) \longrightarrow CaH_2(s)$$

Enthalpy as a State Function

35. Nitrogen and oxygen can react directly with one another to produce nitrogen dioxide according to

    $$N_2(g) + 2\,O_2(g) \longrightarrow 2\,NO_2(g)$$

    The reaction may also be imagined to take place by first producing nitrogen oxide

    $$N_2(g) + O_2(g) \longrightarrow 2\,NO(g)$$

    which then produces NO_2

    $$2\,NO(g) + O_2(g) \longrightarrow 2\,NO_2(g)$$

The overall reaction is found by summing reactions and gives

$$N_2(g) + 2\,O_2(g) \longrightarrow 2\,N_2O(g)$$

How does the enthalpy change for the first reaction compare to that for the fourth reaction? Does this illustrate that enthalpy is a state function? Explain.

36. Consider the following process. Argon gas is heated to 35°C and then cooled to 10°C. The gas is then brought back to its original state. Has the gas undergone an enthalpy change? Explain.

37. The enthalpy change for a chemical reaction does not depend on the way the reaction is carried out. What two things must be known in order to determine the enthalpy change for a reaction?

Standard-State Enthalpies of Reaction

38. What factors affect the quantity of heat given off or absorbed by a given chemical reaction?

39. At what temperature and pressure are thermodynamic data usually reported?

40. What does the superscript $°$ on the symbol $\Delta H°$ tell us about the conditions under which the enthalpy change is reported?

41. Explain why there is only one value of $\Delta H°$ for a reaction at 25°C but there are many values of ΔH.

Calculating Enthalpies of Reaction

42. For the reaction

$$2\,H_2(g) + O_2(g) \longrightarrow 2\,H_2O(g)$$

draw the Lewis structures for each species. What bonds are broken in the course of the reaction? What bonds are formed? The enthalpy change for this reaction is $-484\,kJ/mol_{rxn}$. Are the bonds in the reactants or products the strongest? Explain. What would be the enthalpy change for the following reaction? Explain.

$$2\,H_2O(g) \longrightarrow 2\,H_2(g) + O_2(g)$$

43. If a chemical reaction is exothermic, what can be said about the sums of the bond strengths in the products and reactants? What does an endothermic reaction tell us about the relative bond strengths in products and reactants?

Enthalpies of Atom Combination

44. Why is separating atoms in molecules an endothermic process?

45. What does the sign of ΔH tell you about the net flow of energy in a given chemical reaction?

46. Why is the energy released when a bond is formed precisely the same as the amount of energy needed to break the bond?

47. Predict whether each of the following reactions would be exothermic or endothermic.
    (a) $CO(g) \longrightarrow C(g) + O(g)$
    (b) $2\,H(g) + O(g) \longrightarrow H_2O(g)$
    (c) $Na^+(g) + Cl^-(g) \longrightarrow NaCl(s)$
    What is the sign of ΔH in each of the reactions?

48. What is the value of ΔH for the overall process of separating 1 mole of CH_4 into its constituent atoms and then re-forming 1 mole of CH_4?

49. Does the amount of enthalpy released when a molecule is formed from its gaseous atoms depend on the amount of substance formed? For example, how much enthalpy is released when 2 moles of $CH_4(g)$ are formed as opposed to 1 mole?

50. If the sum of the enthalpies of atom combination for all of the reactants is more negative than that for the products, will the value of ΔH be positive or negative?

51. Predict without using tables which of the following reactions would be endothermic.
    (a) $H_2(g) \longrightarrow 2\,H(g)$
    (b) $H_2O(g) \longrightarrow H_2O(l)$

52. Predict without using tables which of the following reactions would be endothermic.
    (a) $2\,C_8H_{18}(g) + 25\,O_2(g) \longrightarrow 16\,CO_2(g) + 18\,H_2O(g)$
    (b) $Na(g) + Cl(g) \longrightarrow NaCl(s)$
    (c) $Na^+(g) + e^- \longrightarrow Na(g)$

53. Use the enthalpy of atom combination data in Table B.13 in Appendix B to determine whether heat is given off or absorbed when limestone is converted to lime and carbon dioxide.

$$CaCO_3(s) \longrightarrow CaO(s) + CO_2(g)$$

54. Calculate $\Delta H°$ for the following reaction from the enthalpy of atom combination data in Table B.13 in Appendix B.

$$CO(g) + NH_3(g) \longrightarrow HCN(g) + H_2O(g)$$

55. Phosphine (PH_3) is a foul-smelling gas, which often burns on contact with air. Use the enthalpy of atom combination data in Table B.13 in Appendix B to calculate $\Delta H°$ for the reaction, to obtain an estimate of the amount of energy given off when the compound burns.

$$PH_3(g) + 2\,O_2(g) \longrightarrow H_3PO_4(s)$$

56. Carbon disulfide (CS_2) is a useful, but flammable, solvent. Calculate $\Delta H°$ for the following reaction from the

enthalpy of atom combination data in Table B.13 in Appendix B.

$$CS_2(l) + 3\,O_2(g) \longrightarrow CO_2(g) + 2\,SO_2(g)$$

57. The disposable lighters that so many smokers carry use butane as a fuel. Calculate $\Delta H°$ for the combustion of butane from the enthalpy of atom combination data in Table B.13 in Appendix B and Table 7.3.

$$2\,C_4H_{10}(g) + 13\,O_2(g) \longrightarrow 8\,CO_2(g) + 10\,H_2O(g)$$

58. The first step in the synthesis of nitric acid involves burning ammonia. Calculate $\Delta H°$ for the following reaction from the enthalpy of atom combination data in Table B.13 in Appendix B.

$$4\,NH_3(g) + 5\,O_2(g) \longrightarrow 4\,NO(g) + 6\,H_2O(g)$$

59. Lavoisier believed that all acids contained oxygen because so many compounds he studied that contained oxygen form acids when they dissolve in water. Calculate $\Delta H°$ for the reaction between tetraphosphorus decaoxide and water to form phosphoric acid from the enthalpy of atom combination data in Table B.13 in Appendix B.

$$P_4O_{10}(s) + 6\,H_2O(l) \longrightarrow 4\,H_3PO_4(aq)$$

60. Calculate $\Delta H°$ for the decomposition of hydrogen peroxide.

$$2\,H_2O_2(aq) \longrightarrow 2\,H_2O(l) + O_2(g)$$

61. Calculate $\Delta H°$ for the thermite reaction.

$$Fe_2O_3(s) + 2\,Al(s) \longrightarrow 2\,Fe(s) + Al_2O_3(s)$$

62. Calculate $\Delta H°$ for the reaction of Al with Cr_2O_3 and compare to Problem 61 to predict which reaction will liberate the most heat per mole of Al consumed.

$$Cr_2O_3(s) + 2\,Al(s) \longrightarrow 2\,Cr(s) + Al_2O_3(s)$$

63. The first step in extracting iron ore from pyrite, FeS_2, involves roasting the ore in the presence of oxygen to form iron(III) oxide and sulfur dioxide.

$$4\,FeS_2(s) + 11\,O_2(g) \longrightarrow 2\,Fe_2O_3(s) + 8\,SO_2(g)$$

Calculate $\Delta H°$ for the reaction.

64. In which of the following reactions is the sum of the bond strengths greater in the products than in the reactants?
    (a) $CH_3OH(l) \longrightarrow HCHO(g) + H_2(g)$
    (b) $2\,CH_3OH(l) \longrightarrow 2\,CH_4(g) + O_2(g)$
    (c) $CH_3OH(l) \longrightarrow CO(g) + 2\,H_2(g)$

65. Which compound, P_4 or P_2, has the strongest average P—P bonds?

Using Enthalpies of Atom Combination to Probe Chemical Reactions

66. Find the enthalpies of atom combination for the following species in Table B.13 in Appendix B.

$$H_2(g),\ H_2O(g),\ CO(g),\ CH_4(g),\ CO_2(g),\ O_2(g)$$

From these data, calculate the average bond strength for H—H, O—H, C—H, C≡O, O=O and C=O. List these bonds in order of increasing bond strength. Use your average bond strengths to calculate $\Delta H°$ for

$$CH_4(g) + 2\,O_2(g) \longrightarrow CO_2(g) + 2\,H_2O(g)$$

67. Would you expect the following reaction to be exothermic or endothermic? Explain. Use your data from Problem 66. The average C—O bond strength is 358 kJ/mol.

$$CO(g) + 2\,H_2(g) \longrightarrow CH_3OH(g)$$

68. Calculate the enthalpy change for the reaction

$$2\,ZnS(s) + 3\,O_2(g) \longrightarrow 2\,ZnO(s) + 2\,SO_2(g)$$

In terms of bond strengths, explain the sign of the enthalpy change.

69. Explain the sign of the enthalpy change for

$$2\,NO_2(g) \longrightarrow 2\,NO(g) + O_2(g)$$

by using bond strengths.

Bond Length and the Enthalpy of Atom Combination

70. (a) Predict the order of increasing bond length for
    (i) $H_2(g)$ (ii) $I_2(g)$ (iii) $F_2(g)$
    (b) Rank the same molecules in order of increasing bond strength.
    (c) Refer to Table B.13 in Appendix B to find the bond strengths for each molecule. Is your answer in (b) supported by these data?

71. Which of the carbon-to-carbon bonds in the following is shortest? Longest?
    (a) H_3CCH_3 (b) H_2CCH_2 (c) HCCH
    Which carbon–carbon bond is the strongest? Explain.

72. Is the following statement always true? Explain. *The shorter the bond, the stronger the bond.*

Hess's Law

73. Explain how Hess's law is a direct consequence of the fact that the enthalpy of a system is a state function.

74. Use the following data to calculate $\Delta H°$ for the conversion of graphite into diamond.

$$C(s, \text{graphite}) + O_2(g) \longrightarrow CO_2(g)$$
$$\Delta H° = -393.51 \text{ kJ/mol}_{\text{rxn}}$$

$$C(s, \text{diamond}) + O_2(g) \longrightarrow CO_2(g)$$
$$\Delta H° = -395.41 \text{ kJ/mol}_{\text{rxn}}$$

75. Use the following data

$$H_2(g) + \tfrac{1}{2} O_2(g) \longrightarrow H_2O(l)$$
$$\Delta H° = -285.83 \text{ kJ/mol}_{\text{rxn}}$$

$$H_2(g) + O_2(g) \longrightarrow H_2O_2(aq)$$
$$\Delta H° = -191.17 \text{ kJ/mol}_{\text{rxn}}$$

to calculate $\Delta H°$ for the decomposition of hydrogen peroxide.

$$2 H_2O_2(aq) \longrightarrow 2 H_2O(l) + O_2(g)$$

76. In the presence of a spark, nitrogen and oxygen react to form nitrogen oxide.

$$N_2(g) + O_2(g) \longrightarrow 2 NO(g)$$

Calculate $\Delta H°$ for the reaction from the following data.

$$\tfrac{1}{2} N_2(g) + O_2(g) \longrightarrow NO_2(g)$$
$$\Delta H° = 33.2 \text{ kJ/mol}_{\text{rxn}}$$

$$NO(g) + \tfrac{1}{2} O_2(g) \longrightarrow NO_2(aq)$$
$$\Delta H° = -57.1 \text{ kJ/mol}_{\text{rxn}}$$

77. Enthalpy of reaction data can be combined to determine $\Delta H°$ for reactions that are difficult, if not impossible, to study directly. Nitrogen and oxygen, for example, do not react directly to form dinitrogen pentoxide.

$$2 N_2(g) + 5 O_2(g) \longrightarrow 2 N_2O_5(g)$$

Use the following data to determine $\Delta H°$ for the hypothetical reaction in which nitrogen and oxygen combine to form N_2O_5.

$$N_2(g) + 3 O_2(g) + H_2(g) \longrightarrow 2 HNO_3(aq)$$
$$\Delta H° = -414.7 \text{ kJ/mol}_{\text{rxn}}$$

$$N_2O_5(g) + H_2O(l) \longrightarrow 2 HNO_3(aq)$$
$$\Delta H° = -140.24 \text{ kJ/mol}_{\text{rxn}}$$

$$2 H_2(g) + O_2(g) \longrightarrow 2 H_2O(l)$$
$$\Delta H° = -571.7 \text{ kJ/mol}_{\text{rxn}}$$

78. Use the following data

$$3 C(s) + 4 H_2(g) \longrightarrow C_3H_8(g)$$
$$\Delta H° = -103.85 \text{ kJ/mol}_{\text{rxn}}$$

$$C(s) + O_2(g) \longrightarrow CO_2(g)$$
$$\Delta H° = -393.51 \text{ kJ/mol}_{\text{rxn}}$$

$$H_2(g) + \tfrac{1}{2} O_2(g) \longrightarrow H_2O(g)$$
$$\Delta H° = -241.82 \text{ kJ/mol}_{\text{rxn}}$$

to calculate the heat of combustion of propane, C_3H_8.

$$C_3H_8(g) + 5 O_2(g) \longrightarrow 3 CO_2(g) + 4 H_2O(g)$$

79. Use the following data

$$C_4H_9OH(l) + 6 O_2(g) \longrightarrow 4 CO_2(g) + 5 H_2O(g)$$
$$\Delta H° = -2456.1 \text{ kJ/mol}_{\text{rxn}}$$

$$(C_2H_5)_2O(l) + 6 O_2(g) \longrightarrow 4 CO_2(g) + 5 H_2O(g)$$
$$\Delta H° = -2510.0 \text{ kJ/mol}_{\text{rxn}}$$

to calculate $\Delta H°$ for the following reaction.

$$(C_2H_5)_2O(l) \longrightarrow C_4H_9OH(l)$$

Enthalpies of Formation

80. For which of the following substances is $\Delta H_f°$ equal to zero?
    (a) $P_4(s)$ (b) $H_2O(g)$ (c) $H_2O(l)$ (d) $O_3(g)$
    (e) $Cl(g)$ (f) $F_2(g)$ (g) $Na(g)$

81. Use the enthalpy of formation data in Table B.16 in Appendix B to calculate the enthalpy change for the following reaction. Compare the results of the calculation to that of Problem 53.

$$CaCO_3(s) \longrightarrow CaO(s) + CO_2(g)$$

82. Calculate $\Delta H°$ for the following reaction from the enthalpy of formation data in Table B.16 in Appendix B. Compare to Problem 54.

$$CO(g) + NH_3(g) \longrightarrow HCN(g) + H_2O(g)$$

83. Use the enthalpy of formation data in Table B.16 in Appendix B to calculate $\Delta H°$ for the combustion of PH_3. Compare the answer to Problem 55.

$$PH_3(g) + 2 O_2(g) \longrightarrow H_3PO_4(s)$$

84. Carbon disulfide (CS_2) is a useful, but flammable, solvent. Calculate $\Delta H°$ for the following reaction from the enthalpy of formation data in Table B.16 in Appendix B.

$$CS_2(l) + 3 O_2(g) \longrightarrow CO_2(g) + 2 SO_2(g)$$

85. Calculate $\Delta H°$ for the combustion of butane from the enthalpy of formation data in Table B.16 in

Appendix B. The enthalpy of formation of butane is -126.2 kJ/mol$_{rxn}$.

$$2\ C_4H_{10}(g) + 13\ O_2(g) \longrightarrow 8\ CO_2(g) + 10\ H_2O(g)$$

86. The first step in the synthesis of nitric acid involves burning ammonia. Calculate $\Delta H°$ for the following reaction from the enthalpy of formation data in Table B.16 in Appendix B.

$$4\ NH_3(g) + 5\ O_2(g) \longrightarrow 4\ NO(g) + 6\ H_2O(g)$$

87. Calculate $\Delta H°$ for the following from the enthalpy of formation data in Table B.16 in Appendix B. Compare the answer to Problem 59.

$$P_4O_{10}(s) + 6\ H_2O(l) \longrightarrow 4\ H_3PO_4(aq)$$

88. Small quantities of oxygen can be prepared in the laboratory by heating potassium chlorate ($KClO_3$) until it decomposes. Calculate $\Delta H°$ for the following reaction from the enthalpy of formation data in Table B.16 in Appendix B. The enthalpy of formation of $KClO_3(s)$ is -391.2 kJ/mol$_{rxn}$.

$$2\ KClO_3(s) \longrightarrow 2\ KCl(s) + 3\ O_2(g)$$

89. Use enthalpies of formation to predict which of the following reactions gives off the most heat per mole of aluminum consumed.

$$Fe_2O_3(s) + 2\ Al(g) \longrightarrow 2\ Fe(s) + Al_2O_3(s)$$
$$Cr_2O_3(s) + 2\ Al(g) \longrightarrow 2\ Cr(s) + Al_2O_3(s)$$

Integrated Problems

90. Use the enthalpy of combustion for methane, given here, to estimate the energy released when 100 ft^3 of natural gas is burned.

$$CH_4(g) + 2\ O_2(g) \longrightarrow CO_2(g) + 2\ H_2O(g)$$
$$\Delta H° = -802.6\ \text{kJ/mol}_{rxn}$$

91. Which do you predict has the stronger bond, C—H or C—Cl? Calculate the average C—H bond enthalpy in CH_4 from $\Delta H°_{ac}$. Calculate the average C—Cl bond enthalpy in CCl_4 from $\Delta H°_{ac}$. Compare the two bond enthalpies. Is this the result you predicted?

92. If the enthalpy change for breaking all the bonds in the reactants is greater than the enthalpy change for making the bonds of the products, what is the sign of ΔH?

93. If the enthalpy change for breaking all the bonds in the reactants is less than the enthalpy change for making the bonds of the products, what is the sign for ΔH?

94. A measure of the forces that operate between molecules in a liquid can be obtained by comparing the enthalpy required to separate the molecules to the gaseous phase. In which of the following liquids are the intermolecular forces strongest?

(a) CH_3COOH (b) CH_3CH_2OH

(c) C_6H_6 (d) CCl_4

95. Determine the average C—H bond enthalpy for the following compounds (see Table 7.3).

(a) CH_4 (b) C_2H_6 (c) C_3H_8 (d) C_4H_{10}, n-butane

96. The enthalpy change per mole of hydrocarbon combusted with oxygen is given in Table 7.2. Calculate the amount of heat released per mole of covalent bonds broken in each hydrocarbon listed. Is there a relation between molecular structure and the heat released? If so, what is it?

97. Isomers are compounds that have the same number and kinds of atoms but have a different arrangement of the atoms. The enthalpies of atom combination for several pairs of gaseous isomers are given here. For each pair, decide which has the strongest bonds.

(a) $CH_3CH_2CH_2OH$ and $\underset{\underset{CH_3}{|}}{CH_3CHOH}$

-4394.2 kJ/mol$_{rxn}$ -4410.7 kJ/mol$_{rxn}$

(b) $CH_2{=}CHCH_2CH_3$ and $CH_3CH{=}CHCH_3$
-4604.9 kJ/mol$_{rxn}$ -4611.9 kJ/mol$_{rxn}$

(c) $H_2C{=}CHCH{=}CHCH_3$ and $H_2C{=}CHCH_2CH{=}CH_2$
-5243.3 kJ/mol$_{rxn}$ -5213.5 kJ/mol$_{rxn}$

98. When hydrocarbons are bonded only by single bonds, they are said to be saturated; pentane, $CH_3CH_2CH_2CH_2CH_3$, is an example. If carbon–carbon double or triple bonds are present, the compound is said to be *unsaturated*. Unsaturated hydrocarbons are generally better for human nutrition, hence the claims made by manufacturers to have reduced saturated fats in foods such as margarine. An unsaturated hydrocarbon can be saturated by adding hydrogen across the double bond.

$$CH_3CH{=}CHCH_3(g) + H_2(g) \longrightarrow CH_3CH_2CH_2CH_3(g)$$

Is the sum of the bond strengths greater in the products or the reactants in the above reaction? Draw Lewis structures for the reactants and products and give all bond angles. Which species are planar? The C=C bond is rigid and therefore the —CH$_3$ groups attached to those carbon atoms can appear either on the same side (in which case the compound is called *cis*-2-butene) or on opposite sides (in which case the compound is named *trans*-2-butene).

cis-2-Butene

$\Delta H^{\circ}_{ac} = -4611.86\ kJ/mol_{rxn}$

trans-2-Butene

$\Delta H^{\circ}_{ac} = -4616.58\ kJ/mol_{rxn}$

Given the enthalpies of atom combination above and data from Table B.13 in Appendix B, suggest a way that a chemical reaction could be used to differentiate between the two forms of 2-butene.

99. Both dimethyl ether, CH_3—O—CH_3, and ethyl alcohol, CH_3CH_2OH, have been suggested as possible fuels. When reacted with oxygen, O_2, both compounds yield $CO_2(g)$ and $H_2O(g)$. The reactions are called *combustion reactions*.

   (a) Write balanced chemical equations that describe the combustion reaction between dimethyl ether(g) and O_2. Write a second reaction for the combustion reaction between ethyl alcohol(g) and O_2.

   (b) Calculate ΔH for both reactions. (See Table 7.3.) Why is the heat different from that given in Section 7.1?

   (c) Which is the better fuel? In other words, which releases the most heat on combustion with O_2?

   (d) Which molecule, dimethyl ether or ethyl alcohol, has the stronger bonds? Explain.

100. For the following reaction

$$SiBr_4(g) + 2\ Cl_2(g) \longrightarrow SiCl_4(g) + 2\ Br_2(g)$$

   (a) Calculate ΔH° for the reaction. [ΔH°_{ac} for $SiBr_4(g)$ is $-1272\ kJ/mol_{rxn}$.]

   (b) Calculate the average Si—Br, Si—Cl, Cl—Cl, and Br—Br bond enthalpies.

   (c) Which do you expect to be the stronger bond, Si—Br or Si—Cl? Explain. Which do you expect to be the stronger bond, Br—Br or Cl—Cl? Explain. Do your predictions agree with the calculations in (b)?

   (d) Is the reaction endothermic or exothermic? Explain the sign of ΔH°.

101. When carbon is burned in air, the following reaction takes place and releases heat.

$$C(s) + O_2(g) \longrightarrow CO_2(g)$$

   Which of the following is responsible for the heat produced?

   (a) breaking oxygen–oxygen bonds

   (b) making carbon–oxygen bonds

   (c) breaking carbon–carbon bonds

   (d) both (a) and (c) are correct

   (e) (a), (b), and (c) are correct

   (f) none of the above are correct

102. In which molecule would you expect the nitrogen–nitrogen bond strengths to be the greatest? Explain.

   (a) H_2N—NH_2 (b) F_2N—NF_2

   (c) HN=NH (d) N≡N

103. Determine the average bond strengths in N_2, H_2, and NH_3. Would you expect the following reaction to be exothermic or endothermic? Use the average bond strengths to support your answer.

$$N_2(g) + 3\ H_2(g) \longrightarrow 2\ NH_3(g)$$

104. For the reaction

$$2\ H_2(g) + O_2(g) \longrightarrow 2\ H_2O(g)$$

   determine the average bond strengths for hydrogen–hydrogen, oxygen–oxygen, and oxygen–hydrogen bonds. Do you expect the reaction to be exothermic or endothermic? Explain.

105. In terms of the bonds made and the bonds broken, explain why the following reaction is endothermic.

$$Si(s) + 2\ H_2(g) \longrightarrow SiH_4(g)$$

   Use enthalpies of atom combination (Table B.13 in Appendix B) to support your answer.

106. Methane, CH_4, is commonly used in the laboratory as a fuel for Bunsen burners.

$$CH_4(g) + O_2(g) \longrightarrow CO_2(g) + H_2O(g)$$

   (a) Balance the equation.

   (b) Give the Lewis structures of all products and reactants.

   (c) Calculate the enthalpy change for the combustion of CH_4.

107. Iodine reacts with the halogens to form a wide variety of compounds. Two reactions are as follows.

$$I_2(g) + Cl_2(g) \longrightarrow 2\ ICl(g)$$
$$I_2(g) + Br_2(g) \longrightarrow 2\ IBr(g)$$

   (a) Based on bond lengths, which do you expect to have the strongest bond, Cl_2 or Br_2? Explain.

   (b) Based on bond lengths, which do you expect to have the weakest bonds, ICl or IBr. Explain.

   (c) Are your predictions consistent with ΔH°_{ac} data?

   (d) Which of the two reactions above is the most exothermic?

   (e) Explain why, in terms of bonds made and bonds broken, one of these reactions is more exothermic than the other.

108. Consider the reaction

$$SiCl_4(g) + 2 H_2O(g) \longrightarrow SiO_2(s) + 4 HCl(g)$$

(a) Find ΔH for this reaction. Show all work.

(b) How do the strengths of the bonds in the products compare to those of the reactants? Explain.

(c) How would you expect ΔH for the above reaction to compare to ΔH for the following? Explain.

$$SiF_4(g) + 2 H_2O(g) \longrightarrow SiO_2(s) + 4 HF(g)$$

(d) How would the bond strengths in $SiCl_4(g)$ compare to those in $SiF_4(g)$? Explain.

109. Magnesium reacts with chlorine according to the following equation.

$$Mg(s) + Cl_2(g) \longrightarrow MgCl_2(s)$$

(a) Classify the bonding type of each substance in this reaction according to whether it is ionic, covalent, metallic, or metalloid. Explain how you made your classification.

(b) Calculate ΔH for this reaction.

(c) Explain the sign and magnitude of ΔH for this reaction.

Chapter Eight

LIQUIDS AND SOLUTIONS

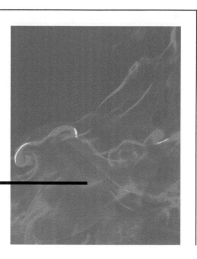

8.1 The Structure of Gases, Liquids, and Solids

Solids, liquids and gases are the three normal phases of matter represented by the diagrams in Figure 8.1. Pure substances that are liquids at room temperature and atmospheric pressure, such as water, are usually composed of covalent molecules such as H_2O molecules. Ionic and metallic compounds are generally solids at room temperature and atmospheric pressure. Although ionic and metallic substances can melt at high temperatures, we shall confine our discussion in this chapter to molecular liquids.

The kinetic molecular theory explains the characteristic properties of gases by assuming that gas particles are in a state of constant, random motion and that the diameter of the particles is very small compared with the distance between these particles (Figure 8.1c). Because most of the volume of a gas is empty space, the simplest analogy might be to compare the particles of a gas to a handful of fruit flies in a more or less empty jar.

Fig. 8.1 Particles in a solid (a) are packed tightly in a regular pattern. The particles in a liquid (b) do not pack as tightly as they do in a solid. The structure of liquids also contains small, particle-sized holes that enable the liquid to flow so that it can conform to the shape of its container. Gas particles (c) are in random motion and occupy only a small fraction of the volume of the gas.

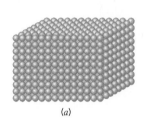
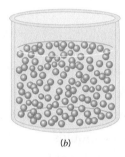

(a)   (b)   (c)

Many of the properties of solids have been captured in the way the term *solid* is used in English. It describes something that holds its shape, such as a solidly constructed house. It implies continuity; there is a definite position for each particle. It implies the absence of empty space, as in a solid chocolate Easter bunny. Finally, it describes things that occupy three dimensions, as in solid geometry. A solid might be compared to a brick wall, in which the individual bricks form a regular structure and the amount of empty space is kept to a minimum (Figure 8.1a).

Liquids have properties between the extremes of gases and solids. Like gases, they flow to conform to the shape of their containers (Figure 8.1b). Like solids, they can't expand to fill their containers, and they are very difficult to compress. The structure of a liquid might be compared to a bag full of marbles being shaken vigorously, back and forth. The model for liquids therefore assumes that there are small, particle-sized holes randomly distributed through the liquid. Particles that are close to one of these holes behave in much the same way as particles in a gas; those that are far from a hole act more like the particles in a solid. The difference in the structures of gases, liquids, and solids might best be understood by comparing the densities of substances in the three phases. As shown by the data in Table 8.1, typical solids are about 20% more dense than the corresponding liquid, whereas the liquid is about 800 times as dense as the gas.

Because water is the only substance that we routinely encounter as a solid, a liquid, and a gas, it may be useful to consider what happens to water as we change the temperature. At low temperatures, water is a solid in which the individual molecules are locked into a rigid structure. As we raise the temperature, the average kinetic energy of the molecules increases due to an increase in the motion with which these molecules move about their lattice positions in the solid.

Table 8.1
Densities of Solid, Liquid, and Gaseous Forms of Three Elements

	Solid (g/cm³)	Liquid (g/cm³)	Gas (g/cm³)
Ar	1.65	1.40	0.001784
N_2	1.026	0.8081	0.001251
O_2	1.426	1.149	0.001429

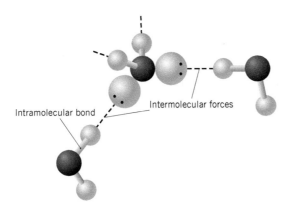

Intramolecular bond

Intermolecular forces

Fig. 8.2 The covalent bonds between the hydrogen and oxygen atoms in a water molecule are called *intramolecular bonds.* The attractive forces between water molecules are called *intermolecular forces.*

To understand the effect of molecular motion on the structure of water at different temperatures, we need to differentiate between intramolecular bonds and intermolecular forces, as shown in Figure 8.2. The covalent bonds between the hydrogen and oxygen atoms in a water molecule are called **intramolecular bonds.** The prefix *intra-* comes from a Latin stem meaning "within or inside." Intramural sports, for example, match teams from within the same institution. The attractive forces between the neighboring water molecules are called **intermolecular forces.** The prefix *inter-* comes from a Latin stem meaning "between." It is used in words such as *interact, intermediate, international,* and *intercollegiate.*

The *intramolecular* bonds that hold the atoms in H_2O molecules together are much stronger than the *intermolecular* forces between water molecules. It takes 463 kJ to break the H—O bonds in a mole of water molecules, but only about 45 kJ to break the intermolecular forces that hold a mole of water molecules to one another. As the temperature of a sample of water increases, so does the average kinetic energy of the water molecules. The increase in the motion of the water molecules that results from this increase in the average kinetic energy disrupts the intermolecular forces between water molecules.

As solid water (ice) becomes warmer, the kinetic energy of the water molecules eventually becomes too large to allow the molecules to be locked into the rigid structure of ice. At this point, the solid melts to form a liquid, in which the intermolecular forces between molecules are now sufficiently weak that the molecules may move through the liquid. As the temperature continues to increase, the kinetic energy of the water molecules becomes so large, and they move so rapidly, that most of the attractive intermolecular forces are overcome, and the liquid boils to form a gas in which each particle moves more or less randomly through space. At no point is the increased motion sufficient to overcome the strengths of the covalent bonds that hold the atoms in these molecules together. Water therefore exists as a molecular species in the gaseous state up to extremely high temperatures.

The increase in the average kinetic energy of the particles that form a liquid or solid as it is heated can provide enough energy to increase the separation between these particles. The distance between molecules is one factor that determines the strength of intermolecular forces. Thus how closely molecules may approach one another is an important factor in determining how strongly molecules attract each other. It follows then that, in addition to temperature, the size and shape of molecules become significant for the strength of intermolecular attractive forces.

We now have a means of explaining why a substance forms a solid, liquid, or gas at room temperature. The difference between the three phases of matter is based on a competition between the strength of intermolecular forces and the kinetic energy of the system. When the force of attraction between the particles is relatively weak, the substance is likely to be a gas at room temperature. When the force of attraction is strong, the substance is more likely to be a solid. As might be expected, a substance is a liquid at room temperature when the intermolecular forces are neither too strong nor too weak.

8.2 Intermolecular Forces

The kinetic theory of gases assumes that there is no force of attraction between the particles in a gas. If that assumption were correct, gases would never condense to form liquids and solids at low temperatures. In 1873 the Dutch physicist Johannes van der Waals derived an equation that not only included the force of attraction between gas particles but also corrected for the fact that the volume of the particles becomes a significant fraction of the total volume of the gas at high pressures.

The van der Waals equation, described in the Special Topics section of Chapter 6, gives a better description of the experimental data for real gases than can be obtained with the ideal gas equation. But that wasn't van der Waals's goal. He was trying to develop a model that would explain the behavior of liquids by including terms that reflected the size of the atoms or molecules in the liquid and the strength of the forces between the atoms or molecules.

The weak intermolecular forces in liquids and solids are therefore often called **van der Waals forces.** Intermolecular forces can be divided into four categories: (1) dipole–dipole, (2) dipole–induced dipole, (3) induced dipole–induced dipole, and (4) hydrogen bonding. Although we will discuss these van der Waals forces one at a time, it is important to recognize that a given substance may exhibit more than one of these intermolecular forces. Because the forces of attraction between particles vary greatly from one substance to another, it isn't possible to describe the properties of liquids or solids with just one equation ($PV = nRT$), as we did with gases in Chapter 6.

DIPOLE–DIPOLE FORCES

Many molecules, such as H_2O, are held together by intramolecular bonds that fall between the extremes of pure ionic and pure covalent bonds. The difference between the electronegativities of the atoms that form these molecules is large enough that the electrons aren't shared equally, and yet small enough that the electrons aren't drawn exclusively to one of the atoms to form positive and negative ions. The bonds in these molecules are said to be **polar** because they have positive and negative ends, or poles. As we saw in Section 4.17, differences in the way charge is distributed in a molecule can also give rise to negative and positive ends, or poles, for the molecule as a whole. The magnitude of this polarity is reflected in the **dipole moment** (μ) of the compound.

The acetone molecules in nail polish remover are polar ($\mu = 2.88$ D) because the carbon atom in the C=O double bond has a slight positive charge and the oxygen atom in this bond has a slight negative charge. The net result is a force of attraction between the positive end of one molecule and the negative end of another, as shown in Figure 8.3.

The dipole–dipole interaction in acetone is relatively weak; it takes only a small amount of energy to pull acetone molecules apart from one another. In con-

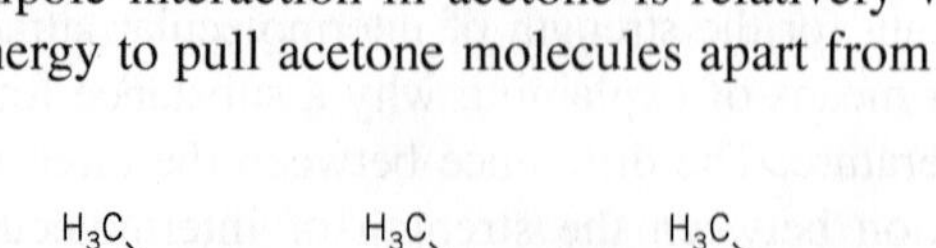

Fig. 8.3 A dipole–dipole force exists between adjacent acetone molecules. The dipole moment is generated by the polar bond. C=O

trast, the covalent bonds (C—H, C—C, and C=O) between the atoms in an acetone molecule are much stronger. The strength of the dipole–dipole attraction depends on the magnitude of the dipole moment and on how closely the molecules approach one another. The closer the molecules approach each other, the larger the intermolecular forces of attraction. If molecules approach too closely, however, a repulsive force occurs.

DIPOLE–INDUCED DIPOLE FORCES

What would happen if we mixed acetone with carbon tetrachloride, which has no dipole moment? The individual C—Cl bonds in carbon tetrachloride are polar due to the difference in electronegativity between carbon and chlorine atoms. Because the molecule has a symmetric tetrahedral molecular geometry, the polarities of the four bonds cancel one another to yield a uniform charge distribution with a dipole moment of zero. The electrons in a molecule aren't static, however, but are in constant motion. When a carbon tetrachloride molecule comes close to a polar acetone molecule, the electrons in carbon tetrachloride can shift to one side of the molecule to produce a very small dipole moment, as shown in Figure 8.4.

By distorting the distribution of electrons in carbon tetrachloride, the polar acetone molecule induces a small dipole moment in the CCl_4 molecule, which creates a dipole–induced dipole force of attraction between the acetone and carbon tetrachloride molecules. The strength of dipole–induced dipole interactions increases as the molecules approach one another and weaken rapidly as the molecules move apart. In general, a polar molecule can distort the electron cloud of any neighboring molecule, producing a dipole–induced dipole interaction.

INDUCED DIPOLE–INDUCED DIPOLE FORCES

Bromine has no dipole and therefore has neither dipole–dipole nor dipole–induced dipole forces, yet bromine is a liquid at room temperature. This means there must be a force of attraction between the Br_2 molecules to hold these molecules together. This force of attraction can be understood by noting that the electrons in bromine are in constant motion. Thus, there is some probability that for an instant in time there may be more electron density on one side of the molecular than on the other, producing a temporary dipole.

The temporary dipole can induce a temporary dipole in an adjacent bromine molecule, as shown in Figure 8.5. Such fluctuations in electron density occur constantly, creating temporary induced dipole–induced dipole forces of attraction—also known as **dispersion forces** or **London forces**—between pairs of molecules or atoms throughout a liquid. All molecules experience some degree of induced dipole–induced dipole interaction when the molecules get close enough together. This is the only type of intermolecular force present in substances, such as Br_2 or CCl_4 that are made up of nonpolar molecules.

Because intermolecular forces must be overcome to melt a molecular solid or to boil a liquid, melting points and boiling points can serve as a measure of the relative strengths of the intermolecular forces that hold the molecules together. Table 8.2 shows the melting points and boiling points of chlorine, bromine, and iodine. All three molecules are nonpolar and are only held together by dispersion forces, but they have significantly different melting and boiling points. At room temperature and pressure chlorine is a gas, bromine a liquid, and iodine a solid. The data in Table 8.2 indicate that the dispersion forces in iodine are much stronger than the dispersion forces in bromine, which are stronger than the dispersion forces in chlorine.

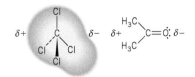

Fig. 8.4 When a CCl_4 molecule comes close to a polar acetone molecule, the distribution of electrons around the carbon tetrachloride molecule is distorted. The shaded area around the carbon tetrachloride molecule in this figure represents the electron density on the CCl_4 molecule. The electron density is attracted toward the partial positive charge of the acetone. A small dipole moment is induced in the carbon tetrachloride molecule, which allows a weak dipole–induced dipole force of attraction.

Fig. 8.5 Fluctuations in electron density occur around the nuclei of neighboring bromine molecules, creating induced dipole–induced dipole forces of attraction.

The dispersion forces between adjacent molecules are strong enough for iodine to be a solid at room temperature. Iodine vapor can be produced, however, by heating the sample.

Table 8.2

Melting Points and Boiling Points of Three Substances That Have Only Dispersion Forces

	Molecular Weight (g/mol)	Melting Point (°C)	Boiling Point (°C)
Cl_2	70.91	-101.0	-34.6
Br_2	159.81	-7.2	58.8
I_2	253.81	113.5	184.4

The magnitude of an induced dipole moment depends on the ease with which a molecule can be polarized—in other words, the ease with which electrons can be moved in the molecule to form a dipole moment. The polarizability of a molecule depends on the number of electrons in the molecule, how tightly these electrons are held, and the shape of the molecule. For similar substances, dispersion force interactions increase as the number of electrons in the molecule increases. As a result, dispersion forces increase with increasing molecular weight. There is an increase in molecular weight as we move from Cl_2 to Br_2 to I_2, for example, which indicates an increase in the strength of the dispersion forces between adjacent molecules. This increase is the result of an increase in the distance of the valence electrons from the nucleus as we move down the periodic table from Cl to Br to I. As the distance from the nucleus increases, the outermost electrons are held less tightly and can more easily be pulled to one side of the molecule. In general, polarizability and induced dipole forces increase as the number of electrons on an atom, ion, or molecule increases.

The shape of a molecule is also a factor in determining the magnitude of dispersion forces. This can be demonstrated using three compounds with the same molecular weights but different structures. As we have seen, molecules that contain the same atoms but have different structures are called *isomers*. The data in Figure 8.6 show how the shape of a molecule influences the boiling point of a compound.

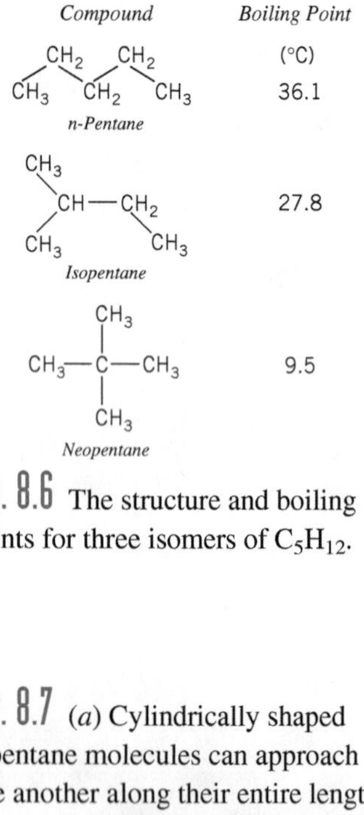

One of the compounds, neopentane, has very symmetrical molecules, with four identical CH_3 groups arranged in a tetrahedral geometry around the central carbon atom. As a result, the symmetrical neopentane molecule doesn't have as much surface area in contact with neighboring molecules as does the straight-chain *n*-pentane molecule, as shown in Figure 8.7. The dispersion forces are therefore weaker, and the boiling point is lower in neopentane than in *n*-pentane. Isopentane, being a branched molecule, falls between neopentane and *n*-pentane in how closely it can approach its neighbors, and therefore isopentane has a boiling point between that of neopentane and *n*-pentane.

The relationship between the molecular weight of a compound and its boiling point is shown in Figure 8.8. The compounds in Figure 8.8 all have the same

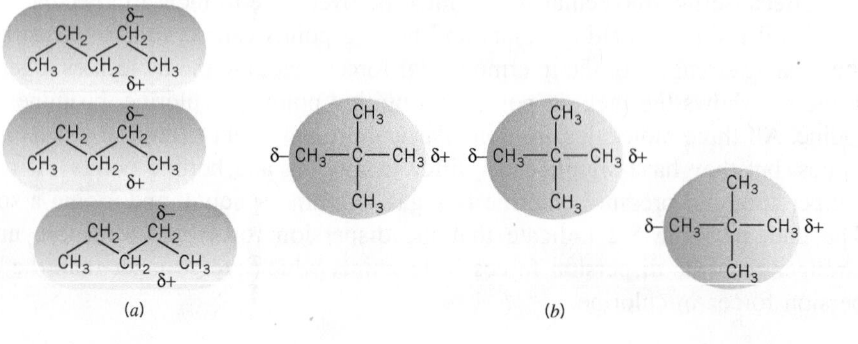

Fig. 8.6 The structure and boiling points for three isomers of C_5H_{12}.

Fig. 8.7 (*a*) Cylindrically shaped *n*-pentane molecules can approach one another along their entire length, allowing the formation of temporary dispersion force interactions. These interactions are larger than those found in neopentane due to the increased polarizability of their electron clouds. (*b*) Spherically shaped neopentane molecules do not have as much surface area in contact as *n*-pentane and undergo smaller dispersion force interactions.

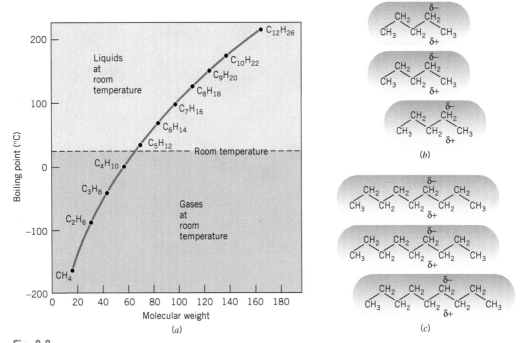

Fig. 8.8 (a) There is a gradual increase in the boiling points of compounds with the generic formula C_nH_{2n+2} as the molecules in the series increase in chain length. (b, c) Increasing chain length increases the area in contact.

generic formula, C_nH_{2n+2}, and are all straight-chain hydrocarbons. The only differences between the compounds are their sizes and their molecular weights. As shown in Figure 8.8(a), the relationship between the molecular weights of the compounds and their boiling points isn't a straight line, but rather a remarkably smooth curve.

The data in Figure 8.8 are for a series of molecules with increasing length from 1 to 12 carbons. Because these molecules are nonpolar, the only intermolecular interactions between them are due to dispersion interactions. As the length of the chain of carbon atoms increases, so does the number of dispersion interactions. The molecular weight of the compound also increases with the number of carbon atoms, which leads to additional polarizable electrons and hence to increased intermolecular interactions. The resulting increase in the forces of attraction between molecules leads to an increase in the boiling point of the compound, when similar compounds are compared.

➤ **CHECKPOINT**

Arrange the elements in Group VIIIA (He, Ne, Ar, Kr, Xe) in order of increasing dispersion interactions.

HYDROGEN BONDING

The intermolecular force known as **hydrogen bonding** is actually a type of dipole–dipole interaction. Hydrogen bonds are separated from other examples of van der Waals forces because they are unusually strong: 15–25 kJ/mol. However, it should be noted that even though the name of the interaction is hydrogen *bonding,* a hydrogen bond is not a covalent, ionic, or metallic bond. In water, it is an intermolecular force between adjacent molecules.[1]

Molecules that can form hydrogen bonds have relatively polar H—X bonds, such as NH_3, H_2O, and HF. The hydrogen bond is created when a hydrogen atom

---

[1]The hydrogen bonds between H_2O molecules in water or ice are examples of *intermolecular* interactions. It is important to recognize, however, that hydrogen bonds can also occur within molecules. The hydrogen bonds that hold the chains of DNA together, or that determine the secondary structure of a protein are examples of *intramolecular* hydrogen bonds.

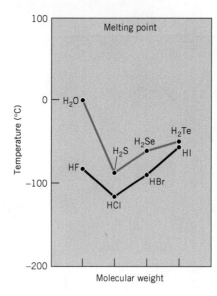
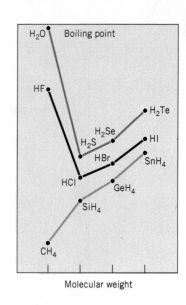

Fig. 8.9 Plots of the melting points and boiling points of hydrides of elements in Groups IVA, VIA, and VIIA. The melting points and boiling points of HF and H_2O are unusually large because of the strength of the hydrogen bonds between molecules in those compounds.

➤ **CHECKPOINT**

Explain why the hydrogen bonds between NH_3 molecules are weaker than those between H_2O molecules.

forms a bridge between two very electronegative atoms. The hydrogen is covalently bonded to one of the atoms and hydrogen-bonded to the other. The H—X bond must be polar to create the partial positive charge on the hydrogen atom that allows the bridging interactions to exist. As the X atom in the H—X bond becomes more electronegative, hydrogen bonding between molecules becomes more important. Hydrogen bonding is most important when the hydrogen atom is bonded to N, O, or F atoms. An illustration of hydrogen bonding is shown as the dashed lines between hydrogen and oxygen atoms on adjacent water molecules in Figure 8.2.

As previously discussed, melting and boiling points can be used to compare the strengths of intermolecular forces in similar compounds. Figure 8.9 shows the relationship between the melting points and boiling points of the hydrides of elements in Groups IVA, VIA, and VIIA. The boiling points of the hydrides of Group IVA (CH_4, SiH_4, GeH_4, and SnH_4) increase in a somewhat linear fashion with molecular weight, as would be predicted from our discussion of molecular weights and dispersion forces. However, the melting points of H_2O and HF don't follow the expected trends for the hydrides of elements in Groups VIA and VIIA. The unusually high melting and boiling points of H_2O and HF are due to the strong hydrogen bonds formed between water molecules and between HF molecules. The effect of hydrogen bonds in water is discussed in more detail in Section 8.9.

8.3 Relative Strengths of Intermolecular Forces

We can now classify the forces of attraction between particles in terms of four categories: dipole–dipole, dipole–induced dipole, induced dipole–induced dipole or dispersion forces, and hydrogen bonds. Consider the compounds in Table 8.3, for example.

Dispersion forces are present in all of these compounds. In fact, dispersion forces must always be present in molecular substances. All compounds contain electrons, and therefore all compounds will have dispersion forces. The other forces—dipole–dipole, dipole–induced dipole, and hydrogen bonding—depend on

Table 8.3
Intermolecular Forces

Molecule	Structure	Dipole–Dipole	Dipole–Induced Dipole	Dispersion Forces	Hydrogen Bonding
Propyl alcohol	$CH_3CH_2CH_2OH$	Yes	Yes	Yes	Yes
Diethyl ether	$CH_3CH_2OCH_2CH_3$	Yes	Yes	Yes	No
Ethyl fluoride	CH_3CH_2F	Yes	Yes	Yes	No
Tin tetrachloride	$SnCl_4$	No	No	Yes	No
Acetic acid	$CH_3-\overset{\overset{O}{\|\|}}{C}-OH$	Yes	Yes	Yes	Yes

the structure of the molecule, and their presence can be deduced only if a suitable structure such as the Lewis structure for the compound is known.

Table 8.4 compares three classes of organic compounds: alkanes, aldehydes, and carboxylic acids. There are two trends in these data. The first is the relationship between boiling point (BP) and molecular weight. For each class of compounds, the boiling point increases as the molecular weight of the compound increases ($BP_{butanal} < BP_{pentanal} < BP_{hexanal}$). Figure 8.8 shows the same trend graphically for the alkanes. When comparing similar compounds of significantly different molecular weights, we see that the compounds with the highest molecular weight will generally have the strongest intermolecular forces and, therefore, the highest boiling point. This relationship is due to the dispersion force interactions that increase with increasing molecular weight.

The second trend that can be observed in Table 8.4 is shown by comparing the boiling points of molecules from different categories that have similar molecular weights. Heptane, hexanal, and pentanoic acid have similar molecular weights (100.2, 100.2, and 102.1 g/mol, respectively), but their respective boiling points are 98.4, 128, and 186°C. To understand the difference in boiling points, we must examine the structure of the molecules and determine the types of intermolecular forces between molecules.

The structures of heptane, hexanal, and pentanoic acid are given in Figure 8.10. Alkanes such as heptane are composed of only carbon and hydrogen atoms.

Table 8.4
Boiling Points of Three Classes of Organic Compounds

Alkane	MW (g/mol)	BP (°C)	Aldehyde	MW (g/mol)	BP (°C)	Carboxylic Acid	MW (g/mol)	BP (°C)
Butane $CH_3(CH_2)_2CH_3$	58.1	−0.5	Butanal $CH_3(CH_2)_2CHO$	72.1	75.7	Butanoic acid $CH_3(CH_2)_2COOH$	88.1	164
Pentane $CH_3(CH_2)_3CH_3$	72.2	36.1	Pentanal $CH_3(CH_2)_3CHO$	86.1	103	Pentanoic acid $CH_3(CH_2)_3COOH$	102.1	186
Hexane $CH_3(CH_2)_4CH_3$	86.2	69.0	Hexanal $CH_3(CH_2)_4CHO$	100.2	128	Hexanoic acid $CH_3(CH_2)_4COOH$	116.2	205
Heptane $CH_3(CH_2)_5CH_3$	100.2	98.4	Heptanal $CH_3(CH_2)_5CHO$	114.2	153	Heptanoic acid $CH_3(CH_2)_5COOH$	130.2	223
Octane $CH_3(CH_2)_6CH_3$	114.2	126	Octanal $CH_3(CH_2)_6CHO$	128.2	171	Octanoic acid $CH_3(CH_2)_6COOH$	144.2	239

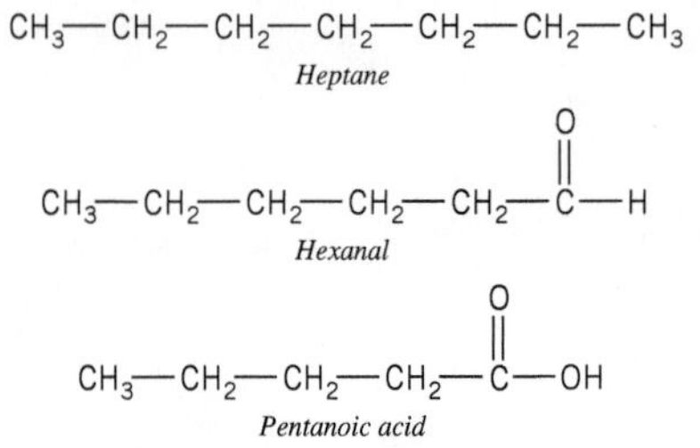

$CH_3-CH_2-CH_2-CH_2-CH_2-CH_2-CH_3$
Heptane

$CH_3-CH_2-CH_2-CH_2-CH_2-\overset{\overset{\displaystyle O}{\|}}{C}-H$
Hexanal

$CH_3-CH_2-CH_2-CH_2-\overset{\overset{\displaystyle O}{\|}}{C}-OH$
Pentanoic acid

Fig. 8.10 Structures of heptane, hexanal, and pentanoic acid.

Because carbon and hydrogen atoms have relatively similar electronegativities the covalent bonds aren't very polar. In addition, the symmetric structure of alkanes leads to a uniform distribution of charge within the molecule and results in a molecule with essentially no dipole moment. The addition of electronegative oxygen atoms to form aldehydes and carboxylic acids produces compounds such as hexanal and pentanoic acid that have a significant dipole moment. As a result, both aldehydes and carboxylic acids are polar and can have significant dipole–dipole interactions. Aldehydes and carboxylic acids therefore have higher boiling points than alkanes of similar molecular weights.

The boiling point of pentanoic acid, however, is considerably higher than that of hexanal. This indicates that the intermolecular forces attracting pentanoic acid molecules to one another are stronger than the forces found in hexanal. The structure of pentanoic acid in Figure 8.10 shows that a hydrogen atom is attached to a very electronegative oxygen atom. This allows for the formation of hydrogen bonds between this hydrogen atom on one molecule and an oxygen atom on an adjacent molecule. Therefore, pentanoic acid has dispersion, dipole–induced dipole, dipole–dipole, and hydrogen-bonding intermolecular forces, giving it a relatively high boiling point, when compared with other compounds in Table.8.4.

Exercise 8.1

The following compounds have similar molecular weights. Arrange them in order of increasing boiling point.

Formaldehyde: $H_2C=O$ Methanol: CH_3-OH Ethane: CH_3CH_3

Solution

Since these compounds all have similar molecular weights, they would be expected to have similar dispersion forces. The structures indicate that the following intermolecular forces would be present.

- Formaldehyde is polar and therefore would have dispersion, dipole–dipole, and dipole–induced dipole interactions.

- Methanol is polar and also has a hydrogen atom covalently bonded to an electronegative oxygen that is capable of forming hydrogen bonds with neighboring molecules. Therefore it would be expected to have dispersion, dipole–dipole, dipole–induced dipole, and hydrogen-bonding interactions.

- Ethane is nonpolar and would have only dispersion forces.

The order of increasing boiling points therefore would be ethane < formaldehyde < methanol.

➤ CHECKPOINT

Arrange the following compounds in order of increasing boiling point.
(a) carbon tetrachloride (CCl_4), acetone (C_3H_6O), and 1,2,3,4-tetrabromobutane ($C_4H_6Br_4$)

Arrange the following compounds in order of decreasing boiling point.
(b) octanoic acid $CH_3(CH_2)_6$ COOH, decane $CH_3(CH_2)_8CH_3$, and nonanal $CH_3(CH_2)_7CHO$

The relative strengths of intermolecular forces and bonds are summarized in Table 8.5. It is useful to compare the magnitude of these forces with a typical covalent bond, such as the bond in Cl_2 (243 kJ/mol), or a typical ionic bond, such as the bond that holds Na^+ and Cl^- ions together in NaCl (787 kJ/mol).

Table 8.5

Relative Strengths of Intermolecular Forces Compared with Ionic, Covalent, and Metallic Bonds

Force	Example	Energy (kJ/mol)
Dipole–dipole		≈ 5
Dipole–induced dipole		≈ 2
Induced dipole–induced dipole (dispersion)		≈ 5
Hydrogen bond		≈ 20
Ion–dipole force		≈ 342
Covalent bond	$:\ddot{C}l-\ddot{C}l:$	243
Ionic bond		787
Metallic bond		107

8.4 The Kinetic Theory of Liquids

Liquids are more complex than the gases discussed in Chapter 6, but the particles in a liquid are still in a state of constant, random motion. Because the *average kinetic energy* only depends on the temperature of a sample, the *average kinetic energy* of the particles in a liquid is the same as the *average kinetic energy* of the particles in a gas at the same temperature. We must include the term *average* in the statement because all molecules at the same temperature don't have the same kinetic energy. There is an enormous range of kinetic energies possessed by the molecules in a sample at a given temperature.

The force of attraction between the particles in a liquid, however, is large enough to hold the particles relatively close together. As a result, collisions between particles in a liquid occur much more frequently than in a gas. In a typical gas, a particle will collide with its neighbors about 10^9 times per second, whereas in a liquid the frequency of collisions is 10^{13} times per second.

The particles in a liquid undergo almost constant collisions with their neighbors to form clusters of particles that stick together for a moment, and then come apart. The force of attraction between the particles in a liquid can be estimated by measuring the amount of heat required to transform a given quantity of a liquid into the corresponding gas (or "vapor") at a given temperature. The result of this measurement is known as the **enthalpy of vaporization** of the liquid, ΔH°_{vap}. As the force of attraction between the particles in a liquid increases, so does the enthalpy of vaporization.

Chapter 7 described enthalpies of atom combination as the energy released due to the forces of attraction that result when atoms in the gas phase combine to form a compound. These forces of attraction consist of both intramolecular bonds and intermolecular or interionic forces. Therefore, enthalpies of atom combination can be used as a direct means of comparing the strength of the intermolecular forces between molecules.

The enthalpies of atom combination for liquid and gaseous carbon tetrachloride, for example, are -1338.84 and -1306.3 kJ/mol$_{rxn}$, respectively. The covalent intramolecular C—Cl bonds in CCl$_4$ in the liquid and the gaseous phases are the same. Any difference between the enthalpies of atom combination for the liquid and gas must therefore be due to intermolecular forces. The more negative enthalpy of atom combination for liquid CCl$_4$ shows that the intermolecular forces are stronger in the liquid than in the gaseous CCl$_4$.

As we saw in Chapter 7, the enthalpy change at 298 K for the reaction in which liquid CCl$_4$ is converted into gaseous CCl$_4$ can be calculated from the data in the following diagram.

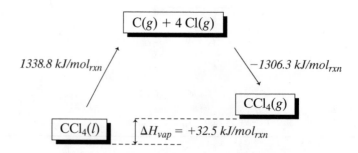

Remember that this diagram is only used to illustrate how the change in enthalpy can be calculated from the enthalpies of atom combination found in Table B.13

in Appendix B. It is *not* meant to imply that the CCl_4 phase change from liquid to gas involves the breaking of bonds to form atoms. No C—Cl bonds are broken during the phase change.

The enthalpy of vaporization of CCl_4 is 32.5 kJ/mol$_{rxn}$. The enthalpy change is positive, indicating that heat must be put into $CCl_4(l)$ to change it to the gaseous state; that is, intermolecular forces that attract the CCl_4 molecules to each other in the liquid state must be overcome to transform the liquid into a gas.

A similar approach can be taken to determining the relative magnitude of intermolecular forces in solids. In this case the reaction involves converting a solid to a liquid, and the enthalpy change that accompanies the process is called the **enthalpy of fusion,**[2] ΔH°_{fus}. The major difference is that for a phase change from liquid to gas essentially all of the intermolecular forces in the liquid must be broken when forming the gas because there is very little interaction of particles in the gas phase. This is not true when a solid melts; intermolecular interactions are still important in the liquid phase.

The kinetic theory of liquids can explain the effect of changes in the temperature of a liquid on many of its characteristic properties. Consider the density of a liquid, for example. The density of a substance is determined by the mass and shape of its particles and how close the particles are together. As the temperature of a liquid increases, the average kinetic energy of its particles increases. The increased thermal motion will cause the particles to move farther apart. As a result, liquids usually become less dense with increasing temperature.

> **➤ CHECKPOINT**
>
> The enthalpy of fusion for water is 6 kJ/mol$_{rxn}$ at 0°C.
>
> $$H_2O(s) \longrightarrow H_2O(l)$$
>
> Which phase change, liquid to gas or solid to liquid, involves breaking the stronger intermolecular forces? Which state—solid, liquid, or gas—has the strongest intermolecular forces?

8.5 The Vapor Pressure of a Liquid

In Latin, the term *vapor* meant "steam." In English, this term has been used to describe the gaseous state of matter produced when a liquid undergoes a phase change to form a gas. A liquid can form a gas or vapor by either *boiling* or *evaporation*. The phase change that occurs when a gas forms a liquid is known as *condensation*.

A liquid doesn't have to be heated to its boiling point before it can become a gas. Water, for example, evaporates from an open container at room temperature (20°C), even though the boiling point of water is 100°C. According to the kinetic molecular theory, the *average kinetic energy* of particles in the liquid, solid, and gas states depends on the temperature of the substance. However, not all molecules have the same energy at a given temperature. Instead, they possess a range of kinetic energies.

Figure 8.11 illustrates the fraction of the molecules in the liquid phase that have specific kinetic energies at two different temperatures. As the temperature increases from T_1 to T_2, the shape of the curve changes, and there is an overall increase in the average kinetic energy of the molecules. Even at temperatures well below the boiling point of the liquid, some of the particles are moving fast enough to escape from the liquid. The shaded portion of this graph begins at the minimum energy a molecule must possess in order to escape into the vapor state. Any kinetic energy larger than that value will be sufficient to allow the molecule to escape from the liquid state and move into the gaseous state. At temperature T_2 a larger number of molecules have sufficient energy to move into the vapor state than at temperature T_1.

---

[2]The word *fusion* is used because it literally means to melt something. We "blow a fuse" in an electric circuit, for example, by melting the thin piece of metal in the center of the fuse.

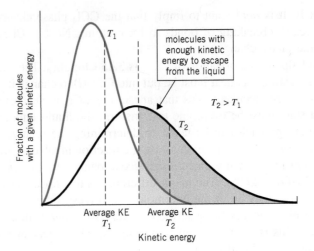

Fig. 8.11 At a given temperature, some of the particles in a liquid have enough energy to escape to form a gas. As the temperature increases, the fraction of the molecules moving fast enough to escape from the liquid (shaded area) increases. As a result, the vapor pressure of the liquid also increases.

When some of the molecules in a liquid evaporate, the average kinetic energy of the molecules that remain in the liquid decreases. As a result, the liquid becomes cooler. It therefore absorbs energy from its surroundings until it returns to thermal equilibrium. But as soon as that happens, some of the molecules in the liquid state acquire enough energy to escape from the liquid. In an open container, the process continues until all of the liquid evaporates.

Figure 8.12 illustrates what happens to a liquid placed in a closed container that is maintained at a constant temperature with respect to time. Water is used here as the example, but other liquids would behave in the same way. Suppose that the initial pressure in the closed container in Figure 8.12 at time t_1 is 1 atm. Furthermore, let's assume that the gas above the liquid is dry air, which is composed primarily of nitrogen and oxygen. As time passes (t_2), some of the molecules begin to escape from the surface of the liquid. The pressure exerted by the water vapor is called the partial pressure of the water vapor. At time t_2, enough water vapor has accumulated so that some of the molecules in the gas phase begin to condense to form the liquid state. But the rate at which the liquid is evaporating to form a vapor is still larger than the rate at which the vapor condenses to form a liquid.

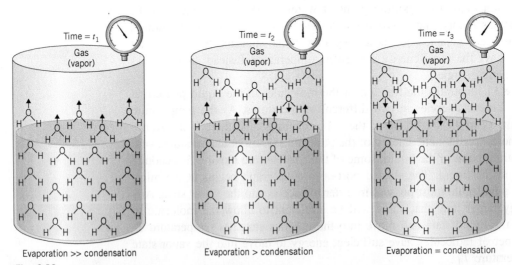

Fig. 8.12 Change in the pressure exerted by a vapor with respect to time. The vapor pressure (Time = t_3) of the liquid is the equilibrium partial pressure of the gas (or vapor) that collects above the liquid in a closed container at a given temperature.

By time t_3, the pressure exerted by the water vapor has continued to increase due to the evaporation of additional water. The rate at which the liquid evaporates to form a gas, however, has become equal to the rate at which the gas condenses to form the liquid (illustrated by an equal number of molecules entering the liquid state as enter the gas state). At this point, the system is said to be in **equilibrium** (from Latin, "a state of balance"). The space above the liquid is saturated with water vapor. The number of water molecules in the vapor phase remains constant, and hence the pressure in the container will no longer change as long as the temperature is constant. The pressure due to the water vapor in the closed container at equilibrium is called the **vapor pressure.**

The vapor pressure is the maximum pressure that can be exerted by a vapor at a given temperature. Individual molecules continue to move between the liquid and gas states, but at equilibrium the total number of molecules in the liquid and gas states remains constant and therefore the pressure remains constant. Note that the number of molecules in the liquid and gas states don't have to be equal. What is equal is the number of molecules in the gas phase entering the liquid state per second and the number of molecules in the liquid phase entering the gas state per second.

We have used two terms that are quite similar but are very important to differentiate. In our discussion we used the term *pressure exerted by the vapor* to describe the partial pressure exerted by the water vapor in the closed container. The *pressure exerted by the vapor* changes with respect to time. We then used the term *vapor pressure* to describe the pressure exerted by a vapor under the very special condition of equilibrium with its liquid. The *vapor pressure* is a constant for a given liquid at a given temperature.

The kinetic theory suggests that the vapor pressure of a liquid depends on its temperature because the fraction of the molecules that have enough energy to escape from a liquid increases with the temperature of the liquid, as shown in Figure 8.11. The vapor pressure of a liquid is determined by the strength of the intermolecular forces that hold the molecules of the liquid together. The stronger the force of attraction between molecules, the smaller the tendency for the molecules to escape from the liquid into the gas phase and therefore the lower the vapor pressure of the liquid. Any increase in the temperature of the liquid, however, will increase the kinetic energy of its molecules. The increased motion of the molecules offsets the intermolecular attractive forces. Thus the vapor pressure of the liquid will increase with increasing temperature.

The vapor pressure of water at temperatures from 0°C to 50°C is given in Table B.3 in Appendix B. Figure 8.13 shows that the relationship between vapor

> ➤ **CHECKPOINT**
>
> What will happen to the pressure exerted by the vapor if the volume of the container in Figure 8.12 is decreased without a change in temperature? Describe what must happen on the molecular level.

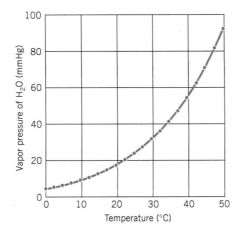

Fig. 8.13 Plot of the vapor pressure of water versus temperature.

Fig. 8.14 Molecules in a liquid feel a force of cohesion that pulls them into the body of the liquid. For molecules in the interior of the liquid, the forces are exerted from all directions, resulting in a net force of zero. However, molecules at the surface are attracted only by molecules to the side or below, resulting in a net force in the downward direction.

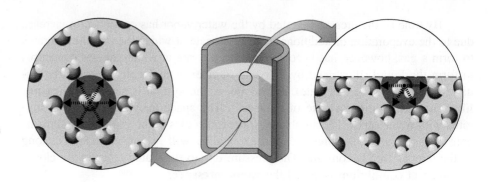

pressure and temperature is not linear. The vapor pressure of water increases more rapidly than the temperature of the system.

The "humidity" reported by meteorologists reflects the amount of water vapor in the atmosphere on a given day. The "humidity" should be known as the relative humidity, however, because it is equal to the ratio of the partial pressure of water vapor in the atmosphere to the vapor pressure of water at that temperature if the atmosphere was saturated with water. Thus, a relative humidity of 50% at 20°C would imply that the vapor pressure of the water in the atmosphere is 50% of the value in Figure 8.13 at this temperature.

Below the surface of a liquid, the force of **cohesion** (literally, "sticking together") between particles is the same in all directions, as shown in Figure 8.14. Particles on the surface of the liquid feel a net force attracting them back toward the body of the liquid. As a result, the liquid takes on the shape that has the smallest possible surface area. The force that controls the shape of the liquid is called the **surface tension.** The surface tension of the liquid becomes larger as the force of attraction between the particles in the liquid increases. As the temperature increases, the increased motion of the particles in a liquid partially overcomes the attractions between the particles. As a result, the surface tension of the liquid decreases with increasing temperature.

8.6 Melting Point and Freezing Point

Does water always become hotter when it is heated? Think about what happens when you heat a pot of water, for example. At first, the water gets hotter, but eventually it starts to boil. From that moment on, the temperature of the water remains the same (100°C at 1 atm), regardless of how much heat is added to the water, until all of the liquid has boiled away.

Imagine another experiment in which a thermometer is immersed in a snowbank on a day when the temperature finally gets above 0°C. The snow gradually melts as it gains heat from the air above it. But the temperature of the snow remains at 0°C until the last snow melts.

Apparently heat can sometimes enter a system without changing its temperature. This happens whenever there is a change in the state of matter. Heat can enter or leave a sample without any detectable change in its temperature when a solid melts, when a liquid freezes or boils, or when a gas condenses to form a liquid.

Figure 8.15 shows what would happen when ice initially at −100°C is heated in an expandable but closed container at 1 atm pressure. Initially, the heat that enters the system is used to increase the temperature of the ice from −100°C

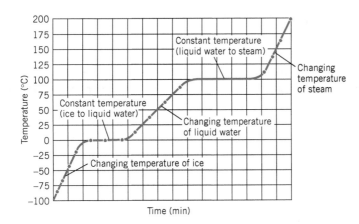

Fig. 8.15 Plot showing the change in temperature and changes in phase as ice is heated to form liquid water and then steam at 1 atm pressure.

to 0°C. At 0°C the heat entering the system is used to melt the ice, and there is no change in the temperature until all of the ice is gone. The amount of heat required to melt the ice is called the enthalpy of fusion.

Once the ice melts, the temperature of the water slowly increases from 0°C to 100°C. But once the water starts to boil, the heat that enters the sample is used to convert the liquid to a gas, and the temperature of the sample remains constant until the liquid has boiled away. The amount of heat required to vaporize a liquid is called the *enthalpy of vaporization*. As heat continues to enter the system, the temperature of the steam that collects in the closed container increases. This causes the container to expand.

At the boiling point of a liquid, the energy input from heating overcomes the intermolecular forces that attract the molecules to one another. Since the energy is used to overcome the intermolecular forces, there is no increase in the kinetic energy of the molecules and hence no change in the temperature of the system.

Pure, crystalline solids have a characteristic **melting point** (MP), which is the temperature at which the solid melts to become a liquid. The melting point of solid oxygen, for example, is −218.4°C at 1 atm pressure. Liquids have a characteristic temperature at which they turn into solids, known as the **freezing point** (FP). The freezing point and melting point of a given substance occur at the same temperature.

It is difficult, if not impossible, to heat a solid above its melting point because the heat that enters the solid at its melting point is used to convert the solid to a liquid. It is possible, however, to cool some liquids to temperatures below their freezing points without forming a solid. When this is done, the liquid is said to be *supercooled*.

Because it is difficult to heat solids to temperatures above their melting points, and because pure solids melt over a very small temperature range, melting points are often used to help identify compounds. We can distinguish between the three sugars known as *glucose* (MP = 150°C), *fructose* (MP = 103°– 105°C), and *sucrose* (MP = 185°– 186°C), for example, by determining the melting point of a small sample.

Measurements of the melting point of a solid can also provide information about the purity of the substance. Pure, crystalline solids melt over a very narrow range of temperatures, whereas mixtures melt over a broad temperature range. Mixtures also tend to melt at temperatures below the melting points of the pure solids. A common example is provided by adding salt to ice, which lowers the melting point of the ice.

8.7 Boiling Point

Boiling occurs when the intermolecular forces that hold molecules in the liquid phase to one another are broken and the molecules enter the gaseous state. Figure 8.16 shows a diagram illustrating the interactions that must be overcome to boil water. Note that no covalent bonds are broken during the boiling process. When a liquid is heated, it eventually reaches a temperature at which the vapor pressure is large enough that bubbles of vapor form inside the body of the liquid. This temperature is called the **boiling point.** Once the liquid starts to boil, the temperature remains constant until all of the liquid has been converted to a gas.

The *normal boiling point* of a liquid is defined as the temperature at which the liquid boils at 1 atm pressure. The normal boiling point of water is 100°C. However, water can boil at other temperatures depending on the pressure exerted on the liquid. If you try to cook an egg in boiling water while camping in the Rocky Mountains at an elevation of 10,000 feet, you will find that it takes longer for the egg to cook because water boils at only 90°C at that elevation.

Before microwave ovens became popular, pressure cookers were used to decrease the amount of time it took to cook food. In a typical pressure cooker, water can remain a liquid at temperatures as high as 120°C, and food cooks in as little as one-third the normal time.

To explain why water boils at 90°C in the mountains and at 120°C in a pressure cooker, even though the normal boiling point of water is 100°C, we have to understand why a liquid boils.

> **A liquid boils when the pressure of the vapor escaping from the liquid is equal to the pressure exerted on the liquid by its surroundings.**

We recognize that a liquid is boiling by the formation of bubbles in the liquid. The bubbles are balls of vapor formed by molecules that have acquired sufficient energy to enter the vapor phase. The vapor in the bubble is in equilibrium with the liquid. Therefore the pressure exerted by the vapor within the bubble is equal

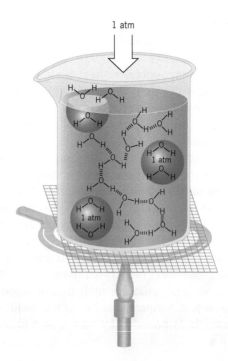

Fig. 8.16 Molecular scale diagram of water boiling. Bubbles of water vapor form inside the liquid when some of the hydrogen bonds between the water molecules are broken.

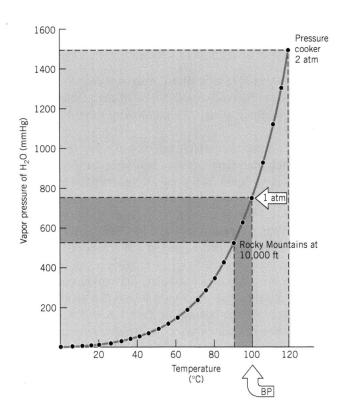

Fig. 8.17 A liquid boils when its vapor pressure is equal to the pressure exerted on the liquid by the surroundings. The normal boiling point of water is 100°C. In the mountains, atmospheric pressure is less than 1 atm, and water boils at a temperature below 100°C. In a pressure cooker at 2 atm, water doesn't boil until the temperature reaches 120°C.

to the vapor pressure of the liquid at the temperature of the liquid. If the external pressure is larger than the vapor pressure developed in the gaseous pockets, the pockets will be crushed and no visible evidence of boiling will be seen. If, however, the vapor pressure in the bubbles is equal to the external pressure, they won't collapse and will be seen to rise to the surface because they are less dense than the liquid from which they form.

This means that we can cause a liquid to boil in two ways: by increasing the temperature of the liquid or by decreasing the pressure exerted on the liquid. As the temperature of a liquid is increased, there is a corresponding increase in the vapor pressure, as shown for water in Figure 8.17. When the vapor pressure has increased to the point at which it is equal to the pressure exerted on the surface of the liquid, boiling will occur.

The normal boiling point of water is 100°C because this is the temperature at which the vapor pressure of water is 760 mmHg, or 1 atm. At 10,000 feet above sea level, the pressure of the atmosphere is only 526 mmHg. At that elevation, water boils when its vapor pressure is 526 mmHg. Figure 8.17 shows that water needs to be heated to only 90°C for its vapor pressure to reach 526 mmHg.

Pressure cookers are equipped with a valve that lets gas escape when the pressure inside the pot exceeds some fixed value. This valve is often set at 15 pounds/in.^{2} (psi), which, combined with the usual prevailing atmospheric pressure of approximately 15 psi, means that the water vapor inside the pot must reach a pressure of 2 atm before it can escape. Because water doesn't reach a vapor pressure of 2 atm until the temperature is 120°C, it boils in the pressurized container at 120°C.

A second way to make a liquid to boil is to reduce the pressure exerted on the surface of the liquid. The vapor pressure of water is roughly 20 mmHg at room temperature. We can therefore make water boil at room temperature by reducing the pressure in its container to less than 20 mmHg.

8.8 Phase Diagrams

Figure 8.18 shows an example of a **phase diagram,** which describes the state or phase of a substance at different combinations of temperature and pressure. This diagram is divided into three areas, which represent the solid, liquid, and gaseous states of the substance.

The best way to remember which area corresponds to each of these states is to remember the conditions of temperature and pressure that are most likely to be associated with a solid, a liquid, and a gas. Low temperatures and high pressures favor the formation of a solid. Gases, on the other hand, are most likely to be found at high temperatures and low pressures. Liquids lie between these extremes.

Phase diagrams can be used in several ways. We can focus on the regions separated by the lines in these diagrams and get some idea of the conditions of temperature and pressure that are most likely to produce a gas, a liquid, or a solid. Or we can focus on the lines that divide the diagram into states, which represent the combinations of temperature and pressure at which two states are in equilibrium.

The points along the line connecting points A and B in the phase diagram in Figure 8.18 represent the combinations of temperature and pressure at which the solid is in equilibrium with the gas. The solid line between points B and C is identical to the plot of the temperature dependence of the vapor pressure of the liquid shown in Figure 8.13. It contains all of the combinations of temperature and pressure at which the liquid boils. The solid line between points B and D contains the combinations of temperature and pressure at which the solid and liquid are in equilibrium.

The BD line is almost vertical because the melting point of a solid isn't particularly sensitive to changes in pressure. For most compounds, this line has a small positive slope, as shown in Figure 8.18. The slope of this line is slightly negative for water, however. As a result, water can melt at temperatures near but below its freezing point when subjected to increased pressure.

Figure 8.18 shows what happens when we draw a horizontal line across a phase diagram at a pressure of exactly 1 atm. This dashed line crosses the line between points B and D at the normal melting point of the substance because solids normally melt at the temperature at which the solid and liquid are in equilibrium at 1 atm pressure. The dashed line crosses the line between points B and C at the boiling point of the substance because the normal boiling point of a liquid is the temperature at which the liquid and gas are in equilibrium at 1 atm pressure and the vapor pressure of the liquid is therefore equal to 1 atm.

Fig. 8.18 A phase diagram, which describes the state of a substance at any possible combination of temperature and pressure. A horizontal line across a phase diagram at a pressure of 1 atm crosses the curve separating solids and liquids at the melting point of the solid, and it crosses the curve separating liquids and gases at the boiling point of the liquid.

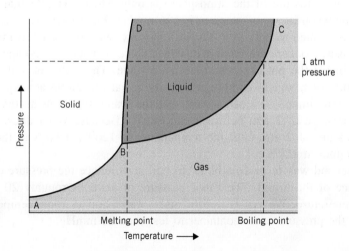

8.9 Hydrogen Bonding and the Anomalous Properties of Water

We are so familiar with the properties of water that it is difficult to appreciate the extent to which its behavior is unusual.

- Most solids expand when they melt. Water expands when it freezes.
- Most solids are more dense than the corresponding liquids. Ice (0.917 g/cm^3) is not as dense as water; therefore, ice floats on liquid water.
- Water has a melting point at least 100°C higher than expected on the basis of the melting points of H_2S, H_2Se, and H_2Te (see Figure 8.9).
- Water has a boiling point almost 200°C higher than expected from the boiling points of H_2S, H_2Se, and H_2Te (see Figure 8.9).
- Water has the largest surface tension of any common liquid except liquid mercury.
- Water is an excellent solvent. It can dissolve compounds, such as NaCl, that are insoluble or only slightly soluble in almost any other liquid.
- Liquid water has an unusually high specific heat. It takes more heat to raise the temperature of 1 g of water by 1°C than any other common liquid.

Ice floats because it is less dense than liquid water.

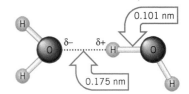

Fig. 8.19 *Hydrogen bonding* is the primary attraction between neighboring water molecules. The polarity of water molecules creates this unusually strong intermolecular force.

These anomalous properties all result from the strong intermolecular forces in water. In Section 5.13 we concluded that water is best described as a polar molecule in which there is a partial separation of charge to give positive and negative poles. The force of attraction between a positive partial charge on the hydrogen atom on one water molecule and the negative partial charge on the oxygen atom on another gives rise to the intermolecular force in water called hydrogen bonding, as shown in Figure 8.19. The hydrogen bonds in water are particularly important because of the dominant role that water plays in the chemistry of living systems.

The hydrogen bonds between water molecules in ice produce the open structure shown in Figure 8.20. When ice melts, some of the hydrogen bonds are broken, and the structure collapses to form a liquid that is about 10% denser. This unusual property of water has several important consequences. The expansion of water when it freezes is responsible for the cracking of concrete, which forms potholes in streets and highways. But it also means that ice floats on top of rivers and streams.

As discussed in Section 8.2, water has a much higher boiling point than would be predicted from its molecular weight. Figure 8.9 shows a steady increase in boiling point in the series CH_4, SiH_4, GeH_4, and SnH_4. However, in the other two series shown in the figure, the boiling points of H_2O and HF are anomalously large because of the strong hydrogen bonds between molecules in liquid H_2O and HF. If this doesn't seem important, try to imagine what life would be like if water boiled at −80°C.

The unusually large specific heat of water discussed in Section 7.4 is also related to the strength of the hydrogen bonds between water molecules. Anything that increases the motion of water molecules, and therefore the temperature of water, must interfere with the hydrogen bonds between the molecules. The fact that it takes so much energy to overcome hydrogen bonds means that water can store enormous amounts of thermal energy. Although the water in lakes and rivers gets warmer in the summer and cooler in the winter, the large specific heat of water limits the range of temperatures; otherwise, extremes of temperature would threaten the life that flourishes in those environments. The specific heat of water is also responsible for the ability of oceans to act as a thermal reservoir that moderates the swings in temperature that occur from winter to summer.

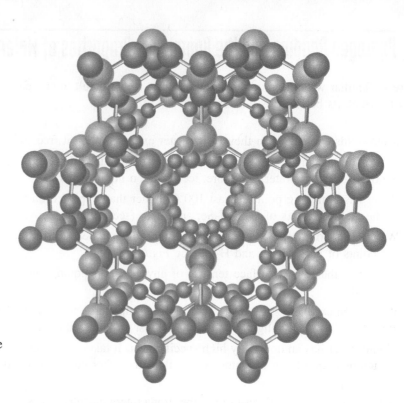

Fig. 8.20 Structure of ice. Note that in each of the hydrogen bonds the hydrogen atoms are closer to one of the oxygen atoms than the other.

8.10 Solutions: Like Dissolves Like

So far, this chapter has focused on the structure and properties of pure liquids. The rest of the chapter will deal with solutions prepared by dissolving a solute in a solvent. As we saw in Section 2.13, solutions can be prepared by mixing solvents and solutes from almost any combination of the states of matter. The discussion of solutions in this chapter, however, will focus on solutions formed by dissolving a solute in a solvent that is a liquid.

Figure 8.21 shows what happens when we add a pair of solutes to a pair of solvents.

Solutes: I_2 and $KMnO_4$

Solvents: H_2O and CCl_4

The solutes are both solids, and they both have an intense violet or purple color. The solvents are both colorless liquids that don't mix with one another.

The solutes are held together by different forces. Iodine consists of individual I_2 molecules held together by relatively weak intermolecular forces. Potassium permanganate consists of K^+ and MnO_4^- ions held together by the strong force of attraction between ions of opposite charge. The attraction between the K^+ and MnO_4^- ions in $KMnO_4$ is much stronger than the dispersion force that holds I_2 molecules together. It is therefore much easier to separate the I_2 molecules in solid iodine than it is to separate $KMnO_4$ into its constituent ions.

There is also a significant difference between the solvents, CCl_4 and H_2O. The difference between the electronegativities of the carbon and chlorine atoms in CCl_4 is so small ($\Delta EN = 0.33$) that there is relatively little ionic character in the C—Cl bonds. Even if there were some separation of charge in the bonds, the CCl_4 molecule wouldn't be polar because it has a symmetrical shape in which the

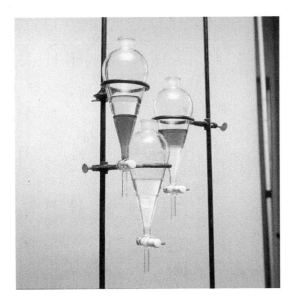

Fig. 8.21 Water and carbon tetrachloride form two separate liquid phases in a separatory funnel (center); $KMnO_4$ dissolves in the water layer on the top of the separatory funnel to form an intensely colored solution (right); I_2 dissolves in the CCl_4 layer on the bottom of the separatory funnel to form an intensely colored solution (left).

four chlorine atoms point toward the corners of a tetrahedron, as shown in Figure 8.22. CCl_4 is therefore best described as a **nonpolar solvent.**

The difference between the electronegativities of the hydrogen and oxygen atoms in water is much larger ($\Delta EN = 1.31$), and the H—O bonds in the molecule are therefore polar. If the H_2O molecule were linear, the polarity of the two O—H bonds would cancel, and the molecule would have no net dipole moment. Water molecules, however, have a bent, or angular, shape. As a result, they have distinct positive and negative poles, and water is a polar molecule, as shown in Figure 8.23. Water is therefore classified as a **polar solvent.**

Because water and carbon tetrachloride don't mix, two separate liquid phases are clearly visible when these solvents are added to each of the separatory funnels in Figure 8.21. We can use the densities at 25°C of CCl_4 (1.584 g/cm^3) and H_2O (1.0 g/cm^3) to decide which phase is water and which is carbon tetrachloride. The more dense CCl_4 settles to the bottom of the funnel.

When a few crystals of iodine are added and the contents of the funnel are shaken, the I_2 dissolves in the CCl_4 layer to form a violet solution, as shown in the separatory funnel on the left in Figure 8.21. The water layer stays essentially colorless, which suggests that little if any I_2 dissolves in water.

When the experiment is repeated with potassium permanganate, the water layer picks up the characteristic dark-purple color of the MnO_4^- ion, and the CCl_4 layer remains colorless, as shown in the separatory funnel on the right in Figure 8.21. This suggests that $KMnO_4$ dissolves in water but not in carbon tetrachloride. The results of the experiment are summarized in Table 8.6. Two important questions are raised. Why does $KMnO_4$ dissolve in water, but not in carbon tetrachloride? Why does I_2 dissolve in carbon tetrachloride but to only a very small extent in water?

It takes a lot of energy to separate K^+ and MnO_4^- ions in potassium permanganate. The ions can form strong attractive interactions with neighboring water molecules, however, as shown in Figure 8.24. The energy released from the formation of the **ion–dipole** interactions compensates for the energy that has to be invested to separate individual ions in the $KMnO_4$ crystal, which takes 342 kJ/mol$_{rxn}$. No such forces are possible between the K^+ or MnO_4^- ions and the nonpolar CCl_4 molecules. As a result, $KMnO_4$ does not dissolve in CCl_4.

The I_2 molecules in iodine and the CCl_4 molecules in carbon tetrachloride are both held together by weak intermolecular forces. These intermolecular forces must be broken in order to separate solute molecules from one another and solvent

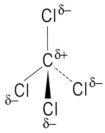

Fig. 8.22 Although there is some separation of charge within the individual bonds in CCl_4, the symmetrical shape of the CCl_4 molecule ensures that there is no net dipole moment. CCl_4 is therefore *nonpolar.*

Fig. 8.23 Because water molecules are bent, or angular, they have distinct negative and positive poles. H_2O is therefore an example of a *polar molecule.*

Table 8.6
Solubilities of I_2 and $KMnO_4$ in CCl_4 and Water

	H_2O	CCl_4
I_2	Very slightly soluble	Very soluble
$KMnO_4$	Very soluble	Insoluble

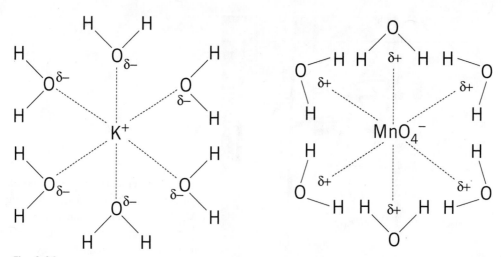

Fig. 8.24 $KMnO_4$ dissolves in water because the energy released when ion–dipole forces form between the K^+ ion and the negative end of neighboring water molecules and between the MnO_4^- ion and the positive end of the solvent molecules compensates for the energy it takes to separate the K^+ and MnO_4^- ions.

molecules from one another. Similar intermolecular forces are formed, however, between I_2 and CCl_4 molecules when the I_2 is dissolved in CCl_4. I_2 therefore readily dissolves in CCl_4 because the intermolecular forces that are broken in the solute and solvent are very similar to the intermolecular forces that are formed between the solute and solvent molecules. The molecules in water are held together by hydrogen bonds that are stronger than most intermolecular forces. No interaction between I_2 and H_2O molecules is strong enough to compensate for the hydrogen bonds between water molecules that have to be broken to dissolve iodine in water, so relatively little I_2 dissolves in H_2O.

We can summarize the results of the experiment by noting that *nonpolar solutes* (such as I_2) dissolve in *nonpolar solvents* (such as CCl_4), whereas many *polar* or *ionic solutes* (such as $KMnO_4$) dissolve in *polar solvents* (such as H_2O). As a general rule, *like dissolves like.*

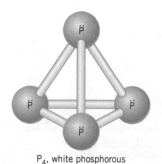

P_4, white phosphorous

Fig. 8.25 Pure elemental phosphorus is a white, waxy solid that consists of tetrahedral P_4 molecules in which the P—P—P bond angle is only 60°.

Exercise 8.2

Elemental phosphorus is often stored under water because it doesn't dissolve in water. Elemental phosphorus is very soluble in carbon disulfide, however. Use the structure of the P_4 molecule shown in Figure 8.25 to explain why P_4 is soluble in CS_2 but not in water.

Solution

The P_4 molecule is a perfect example of a nonpolar solute. It is therefore more likely to be soluble in nonpolar solvents than in polar solvents such as water.

The Lewis structure of CS_2 suggests that the molecule is linear.

Thus, even if there is some separation of charge in the C=S double bond, the molecule would have no net dipole moment because of its symmetry. The electronegativities of carbon ($EN = 2.54$) and sulfur ($EN = 2.59$), however, suggest that the C=S double bonds are almost perfectly covalent. CS_2 is therefore a nonpolar solvent, which readily dissolves nonpolar P_4.

Exercise 8.3

The iodide ion reacts with iodine in aqueous solution to form the I_3^-, or triiodide, ion.

$$I^-(aq) + I_2(aq) \longrightarrow I_3^-(aq)$$

What would happen if CCl_4 were added to an aqueous solution that contained a mixture of KI, I_2, and KI_3?

Solution

Two layers would form, with CCl_4 on the bottom and the aqueous layer on top. KI and KI_3 are both ionic compounds. One contains K^+ and I^- ions; the other contains K^+ and I_3^- ions. The ionic compounds are more soluble in polar solvents, such as water, than in nonpolar solvents, such as CCl_4. KI and KI_3 would therefore remain in the aqueous solution. I_2 is a nonpolar molecule, which is more soluble in a nonpolar solvent, such as carbon tetrachloride. Most of the I_2 would therefore leave the aqueous layer and enter the CCl_4 layer, where it would exhibit the characteristic violet color of solutions of molecular iodine.

A frequently encountered example of the rule that like dissolves like can be seen in salad dressings that are made by mixing vegetable oils and vinegar. Vegetable oils, such as olive oil, are complex mixtures of nonpolar molecules that are insoluble in polar solvents such as water. Vinegar, on the other hand, is a solution of acetic acid (CH_3CO_2H) dissolved in water. Because one of these liquids is polar and the other is not, they cannot dissolve in each other. Because vegetable oils are less dense than water, the oil floats on top of the vinegar solution when the two liquids separate. The bottle of salad dressing therefore has to be shaken before it is used, to get a momentary mixture of small droplets of oil dispersed in the vinegar solution.

At one time, bottles of milk provided another example of the principle that like dissolves like. Milk is a mixture of fats and liquids that used to separate, leaving a layer of cream containing the nonpolar fats floating on the water soluble components. To prevent this, milk is "homogenized" by forcing the liquid through a narrow opening at high pressures. This breaks the globules of fat into very small droplets that are spread out more evenly through the liquid.

8.11 Hydrophilic and Hydrophobic Molecules

Hydrocarbons, such as the alkanes discussed in Section 8.3, are compounds that contain only carbon and hydrogen. Because of the small difference between the electronegativities of carbon and hydrogen ($\Delta EN = 0.24$), hydrocarbons are nonpolar. As a result, they don't dissolve in polar solvents such as water. Hydrocarbons are therefore described as insoluble in water.

Table 8.7
Solubilities of Alcohols in Water

Formula	Name	Solubility in Water (g/100 g)
CH_3OH	Methanol	Infinitely soluble
CH_3CH_2OH	Ethanol	Infinitely soluble
$CH_3(CH_2)_2OH$	Propanol	Infinitely soluble
$CH_3(CH_2)_3OH$	Butanol	9
$CH_3(CH_2)_4OH$	Pentanol	2.7
$CH_3(CH_2)_5OH$	Hexanol	0.6
$CH_3(CH_2)_6OH$	Heptanol	0.18
$CH_3(CH_2)_7OH$	Octanol	0.054
$CH_3(CH_2)_9OH$	Decanol	Insoluble in water

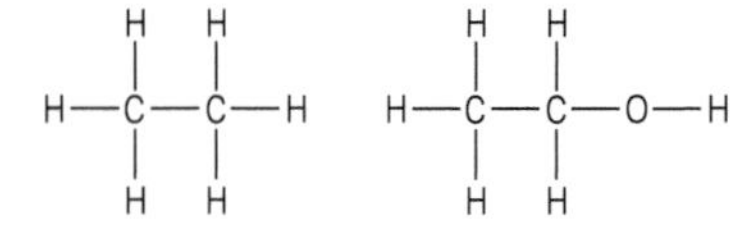

Fig. 8.26 When a hydrogen atom in a hydrocarbon such as ethane is replaced by an —OH group, the resulting compound is called an alcohol.

➤ CHECKPOINT

Amino acids are classified as either hydrophilic or hydrophobic on the basis of their side chains. Use the structure of the side chains of the following amino acids to justify the classifications shown below.

	Amino Acid	Side Chain
Hydrophobic	Alanine	—CH_3
	Cysteine	—$CH_2CH_2SCH_3$
Hydrophilic	Lysine	—$CH_2CH_2CH_2CH_2NH_3^+$
	Serine	—CH_2OH

When one of the hydrogen atoms in a hydrocarbon is replaced with an —OH group, the compound is known as an **alcohol,** as shown in Figure 8.26. Because alcohols contain the same —OH group as water, alcohols have properties between the extremes of hydrocarbons and water. When the hydrocarbon chain is short, the alcohol is soluble in water. Methanol (CH_3OH) and ethanol (CH_3CH_2OH) are infinitely soluble in water, for example. There is no limit to the amount of these alcohols that can dissolve in a given quantity of water. The alcohol in beer, wine, and hard liquors is ethanol, and mixtures of ethanol and water can have any concentration between the extremes of pure alcohol (200 proof) and pure water (0 proof).

As the hydrocarbon chain becomes longer, the alcohol becomes less soluble in water, as shown in Table 8.7. One end of the longer alcohol molecules has so much nonpolar character it is called **hydrophobic** (literally, "water hating"), as shown in Figure 8.27. The other end contains an —OH group that can form hydrogen bonds to neighboring water molecules and is therefore said to be **hydrophilic** (literally, "water loving"). As the hydrocarbon chain becomes longer, the hydrophobic character of the molecule increases and the solubility of the alcohol in water gradually decreases until it becomes essentially insoluble in water.

People encountering the terms *hydrophilic* and *hydrophobic* for the first time sometimes have difficulty remembering which stands for "water hating" and which stands for "water loving." If you remember that Hamlet's girlfriend was named Ophelia (not Ophobia), you can remember that the prefix *philo-* is commonly used to describe love—for example, in *philanthropist, philharmonic, philosopher*—and that *phobia* means "dislike."

Table 8.8 shows that the ionic compound NaCl is relatively soluble in water. Water molecules, being polar, are able to cluster around the positively and negatively charged ions formed when NaCl dissolves. As the solvent becomes more like a hydrocarbon, the solubility of NaCl decreases because the longer-chain hydrocarbon solvents do not interact as strongly with Na^+ and Cl^-.

Hydrophilic head

$$CH_3CH_2CH_2CH_2CH_2CH_2CH_2CH_2CH_2CH_2OH$$

Hydrophobic tail

Fig. 8.27 One end of the decanol molecule is nonpolar and therefore *hydrophobic;* the other end is polar and therefore *hydrophilic.*

Table 8.8
Solubility of Sodium Chloride in Water and in Alcohols

Formula of Solvent	Solvent Name	Solubility of NaCl (g/100 g solvent)
H_2O	Water	35.92
CH_3OH	Methanol	1.40
CH_3CH_2OH	Ethanol	0.065
$CH_3(CH_2)_2OH$	Propanol	0.012
$CH_3(CH_2)_3OH$	Butanol	0.005
$CH_3(CH_2)_4OH$	Pentanol	0.0018

8.12 Soaps, Detergents, and Dry-Cleaning Agents

The chemistry behind the manufacture of soap hasn't changed since it was made from animal fat and the ash from wood fires almost 5000 years ago. Solid animal fats (such as the tallow obtained during the butchering of sheep and cattle) and liquid plant oils (such as palm oil and coconut oil) are still heated in the presence of a strong base to form a soft, waxy material that enhances the ability of water to wash away the grease and oil that form on our bodies and our clothes.

Animal fats and plant oils contain compounds known as *fatty acids.* Fatty acids, such as stearic acid (Figure 8.28), have small, polar, hydrophilic heads attached to long, nonpolar, hydrophobic tails. Fatty acids are seldom found by themselves in nature. They are usually bound to molecules of glycerol ($HOCH_2CHOHCH_2OH$) to form triglycerides, such as the one shown in Figure 8.29. The triglycerides break down in the presence of a strong base to form the Na^+ or K^+ salt of the fatty acid, as shown in Figure 8.30. This reaction is called *saponification,* which literally means "the making of soap."

The cleaning action of soap results from the fact that soaps are *surfactants—* they tend to concentrate on the surface of water. They cling to the surface because they try to orient their polar CO_2^- heads toward water molecules and their nonpolar $CH_3CH_2CH_2...$ tails away from neighboring water molecules.

Water by itself can't wash the soil out of clothes because the soil particles that cling to textile fibers are covered by a layer of nonpolar grease or oil molecules, which repel water. The nonpolar tails of the soap dissolve in the grease or oil that surrounds a soil particle, as shown in Figure 8.31. The soap therefore disperses or emulsifies the soil particles, which makes it possible to wash the particles out of the clothes.

Most soaps are more dense than water. They can be made to float, however, by incorporating air into the soap during its manufacture. Most soaps are also opaque; they absorb rather than transmit light. Translucent soaps can be made by adding alcohol, sugar, and glycerol, which slow down the growth of

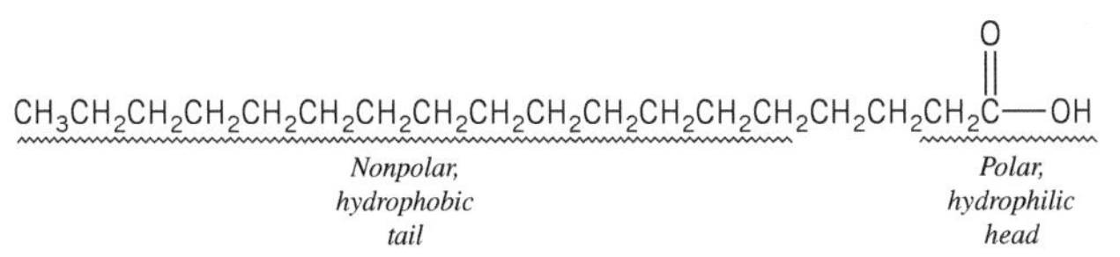

Soap and detergents are surfactants that concentrate on the surface of water.

$$CH_3CH_2CH_2CH_2CH_2CH_2CH_2CH_2CH_2CH_2CH_2CH_2CH_2CH_2CH_2CH_2C\overset{\displaystyle O}{\overset{\displaystyle \|}{{}}}\text{—OH}$$

Nonpolar, hydrophobic tail   *Polar, hydrophilic head*

Fig. 8.28 The hydrocarbon chain on one end of a fatty acid molecule is nonpolar and hydrophobic, whereas the —CO_2H group on the other end of the molecule is polar and hydrophilic.

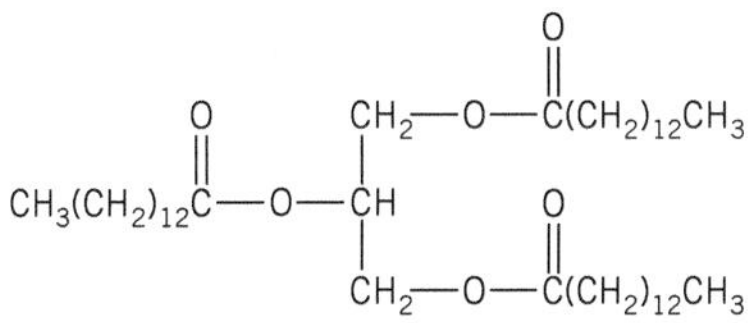

$$CH_3(CH_2)_{12}\overset{\displaystyle O}{\overset{\displaystyle \|}{C}}\text{—O—CH}\begin{matrix}CH_2\text{—O—}\overset{\displaystyle O}{\overset{\displaystyle \|}{C}}(CH_2)_{12}CH_3 \\ \\ CH_2\text{—O—}\overset{\displaystyle O}{\overset{\displaystyle \|}{C}}(CH_2)_{12}CH_3\end{matrix}$$

Fig. 8.29 Structure of the triglyceride known as *trimyristin,* which can be isolated in high yield from nutmeg.

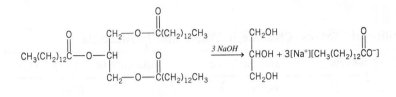

Fig. 8.30 Saponification of the trimyristin extracted from nutmeg.

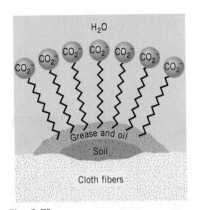

Fig. 8.31 Soap molecules disperse, or emulsify, soil particles coated with a layer of nonpolar grease or oil molecules.

soap crystals while the soap solidifies. Liquid soaps are made by replacing the sodium salts of the fatty acids with the more soluble K^+ or NH_4^+ salts.

In the 1950s, more than 90% of the cleaning agents sold in the United States were soaps. Today soap represents less than 20% of the market for cleaning agents. The primary reason for the decline in the popularity of soap is the reaction between soap and "hard" water. The most abundant positive ions in tap water are Na^+, Ca^{2+}, and Mg^{2+}. Water that is particularly rich in Ca^{2+}, Mg^{2+}, or Fe^{3+} ions is said to be "hard." Hard water interferes with the action of soap because the Ca^{2+}, Mg^{2+}, and Fe^{3+} ions combine with soap to form insoluble precipitates that have no cleaning power. The precipitates not only decrease the concentration of the soap in solution, but actually bind soil particles to clothing, leaving a dull, gray film.

One way around the problem is to "soften" the water by replacing the Ca^{2+} and Mg^{2+} ions with Na^+ ions. Many water softeners are filled with a resin that contains $—SO_3^-$ ions attached to a polymer, as shown in Figure 8.32. The resin is treated with NaCl until each $—SO_3^-$ ion picks up an Na^+ ion. When hard water flows over the resin, Ca^{2+} and Mg^{2+} ions bind to the $—SO_3^-$ ions on the polymer chain and Na^+ ions are released into the solution. Periodically, the resin becomes saturated with Ca^{2+} and Mg^{2+} ions. When that happens, the resin has to be regenerated by being washed with a concentrated solution of NaCl.

There is another way around the problem of hard water. Instead of removing Ca^{2+} and Mg^{2+} ions from water, we can find a cleaning agent that doesn't form insoluble salts with those ions. Synthetic detergents are an example of such cleaning agents. Detergents consist of long, hydrophobic hydrocarbon tails attached to polar, hydrophilic $—SO_3^-$ or $—OSO_3^-$ heads. By themselves, detergents don't have the cleaning power of soap. "Builders" were therefore added to synthetic detergents to increase their strength. The builders were often salts of highly charged ions, such as the triphosphate ion ($P_3O_{10}^{5-}$).

Cloth fibers swell when they are washed in water. This leads to changes in the dimensions of the cloth that can cause wrinkles—which are local distortions in the structure of the fiber—or even more serious damage, such as shrinking. These problems can be avoided by "dry cleaning," which uses a nonpolar solvent that doesn't adhere to, or wet, the cloth fibers. The nonpolar solvents used in dry cleaning dissolve the nonpolar grease or oil layer that coats soil particles, freeing the soil particles to be removed by detergents added to the solvent or by the tum-

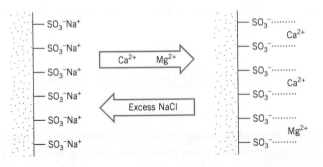

Fig. 8.32 When a water softener is "charged," it is washed with a concentrated NaCl solution until all of the $—SO_3^-$ ions pick up an Na^+ ion. The softener then picks up Ca^{2+} and Mg^{2-} from hard water, replacing those ions with Na^+ ions.

bling action inside the machine. Dry cleaning has the added advantage that it can remove oily soil at lower temperatures than soap or detergent dissolved in water, so it is safer for delicate fabrics.

When dry cleaning was first introduced in the United States between 1910 and 1920, the solvent was a mixture of hydrocarbons isolated from petroleum when gasoline was refined. Over the years, those flammable hydrocarbon solvents have been replaced by halogenated hydrocarbons, such as trichloroethane (Cl_3C—CH_3), trichloroethylene (Cl_2C═$CHCl$), and perchloroethylene (Cl_2C═CCl_2).

8.13 Why Do Some Solids Dissolve in Water?

When asked to describe what happens when a teaspoon of sugar is stirred into a cup of coffee, people often answer, "The sugar initially settles to the bottom of the cup. When the coffee is stirred, it dissolves to produce a sweeter cup of coffee." When asked to extend the description to the molecular level, they hesitate. If they are chemistry students, like yourself, they often ask to be reminded of the formula for sugar. When told that the sugar used in cooking is *sucrose,* $C_{12}H_{22}O_{11}$, they write equations such as the following.

$$C_{12}H_{22}O_{11}(s) \xrightarrow{H_2O} C_{12}H_{22}O_{11}(aq)$$

Although there is nothing wrong with this equation, it doesn't explain what happens at the molecular level when sugar dissolves. Nor does it provide any hints about why sugar dissolves but other solids do not. To understand what happens at the molecular level when a solid dissolves, we have to refer back to the discussion of molecular and ionic solids in Chapter 5.

The sugar we use to sweeten coffee or tea is a *molecular solid,* in which the individual molecules are held together by intermolecular forces. When sugar dissolves in water, the intermolecular forces between the individual sucrose molecules are broken, and the individual $C_{12}H_{22}O_{11}$ molecules are released into solution, as shown in Figure 8.33.

It takes energy to overcome the intermolecular forces between the $C_{12}H_{22}O_{11}$ molecules in sucrose. It also takes energy to break the hydrogen bonds in water that must be disrupted to insert one of the sucrose molecules into the solution. Sugar dissolves in water because the slightly polar sucrose molecules form hydrogen bonds with the polar water molecules. The intermolecular forces that form between the solute and the solvent help compensate for the energy needed to disrupt the structure of both the pure solute and the solvent. In the case of sugar and water, the process works so well that up to 1800 g of sucrose can dissolve in a liter of water.

Ionic solids (or salts) contain positive and negative ions that are held together by the strong force of attraction between particles with opposite charges. In 1887, Svante Arrhenius suggested that salts, such as NaCl, dissociate into their ions when they dissolve in water.

$$NaCl(s) \xrightarrow{H_2O} Na^+(aq) + Cl^-(aq)$$

When an ionic solid dissolves in water, the ions that form the solid are released into solution, where they are attracted to the polar solvent molecules by an

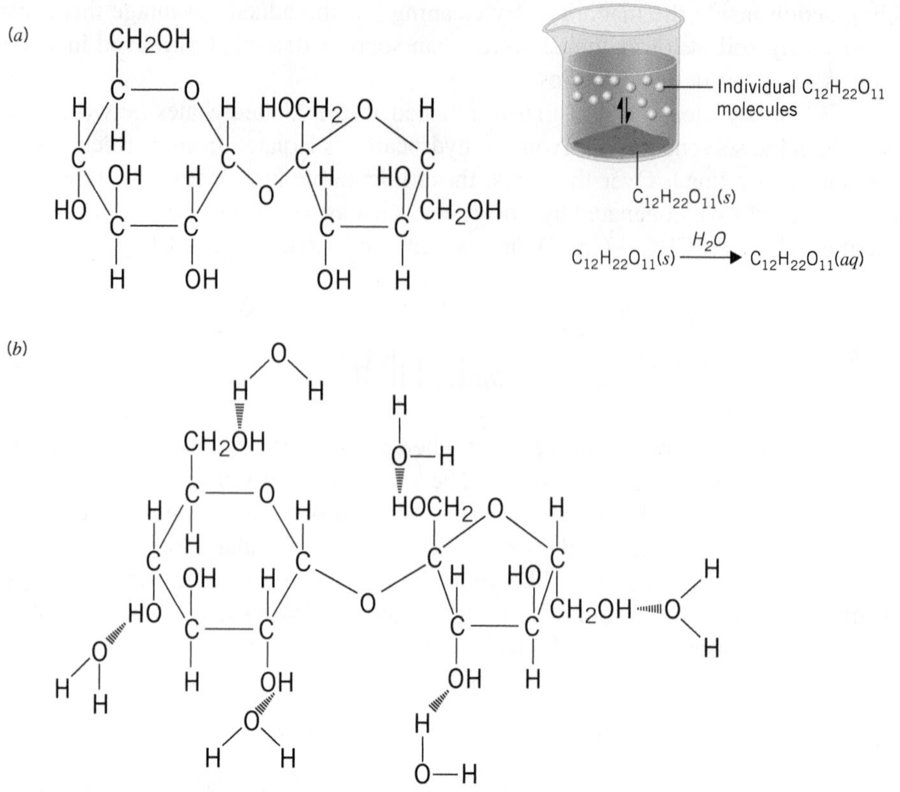

(a)

(b)

Fig. 8.33 Sucrose (a) dissociates into (b) individual $C_{12}H_{22}O_{11}$ molecules hydrogen bonded to water when it dissolves in water.

ion–dipole force, as shown in Figure 8.34. There is a force of attraction between the neutral water molecules in the solution and the Na^+ and the Cl^- ions that form when the salt dissociates. The positively charged Na^+ ion is attracted to the negative end of the polar water molecule. The negatively charged Cl^- ion is attracted to the positive end of the water molecule. As a result, we can generally assume that salts dissociate into their ions when they dissolve in water. Ionic compounds dissolve in water if the energy given off when the ions form ion–dipole forces of attraction with water molecules compensates for the energy needed to break the ionic bonds in the solid and for the energy required to separate the water molecules so that the ions can be inserted into the solution.

It takes an enormous amount of energy to break apart an ionic crystal. It takes 861 kJ/mol, for example, to transform a mole of solid LiCl into Li^+ and Cl^- ions in the gas phase.

$$LiCl(s) \longrightarrow Li^+(g) + Cl^-(g) \qquad \Delta H° = 861 \, kJ/mol_{rxn}$$

It takes so much energy to separate the Li^+ and Cl^- ions in LiCl that we might not expect the compound to dissolve in water. The force of attraction between Li^+ and Cl^- ions with water molecules is so large, however, that $898 \, kJ/mol_{rxn}$ of energy is released when the gaseous Li^+ and Cl^- ions interact with water molecules.

$$Li^+(g) + Cl^-(g) \xrightarrow{H_2O} Li^+(aq) + Cl^-(aq) \qquad \Delta H° = -898 \, kJ/mol_{rxn}$$

The overall enthalpy of reaction for the process in which solid LiCl dissolves in water is therefore negative. The interaction of the ions with water compensates for the energy needed to break apart the ionic structure.

$$LiCl(s) \xrightarrow{H_2O} Li^+(aq) + Cl^-(aq) \qquad \Delta H° = -37 \, kJ/mol_{rxn}$$

➤ **CHECKPOINT**

Use the atom combination data in Table B.13 in Appendix B to calculate $\Delta H°$ for the reaction in which $BaCl_2(s)$ dissolves in water to form $Ba^{2+}(aq)$ and $2 \, Cl^-(aq)$.

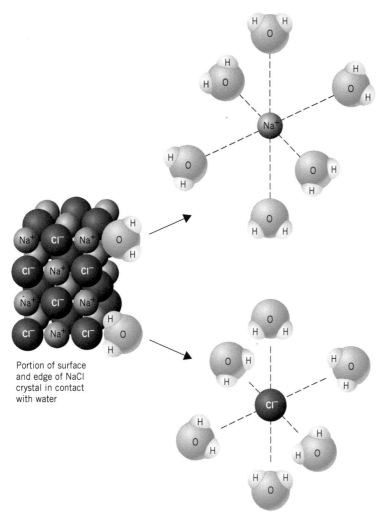

Fig. 8.34 Interaction of water and ionic compounds. The electrostatic interaction between the polar water molecules and the ions is an ion-dipole force. [Reprinted with permission from L. J. Malone, *Basic Concepts of Chemistry,* John Wiley & Sons, New York, 1994, p. 351.]

As a result, LiCl is very soluble in water.

Silver chloride is very insoluble in water. As was previously seen for LiCl, a lot of heat is required to separate AgCl(s) into its gas-phase ions.

$$\text{AgCl}(s) \longrightarrow \text{Ag}^+(g) + \text{Cl}^-(g) \qquad \Delta H° = 915.7 \text{ kJ/mol}_{rxn}$$

The heat released when the gaseous Ag^+ and Cl^- ions interact with water is also large.

$$\text{Ag}^+(g) + \text{Cl}^-(g) \xrightarrow{H_2O} \text{Ag}^+(aq) + \text{Cl}^-(aq) \qquad \Delta H° = -850.2 \text{ kJ/mol}_{rxn}$$

But the heat released when the ions interact with water isn't large enough to compensate for the heat needed to separate the ions in the crystal. As a result, the overall enthalpy of reaction is very unfavorable.

$$\text{AgCl}(s) \xrightarrow{H_2O} \text{Ag}^+(aq) + \text{Cl}^-(aq) \qquad \Delta H° = 65.5 \text{ kJ/mol}_{rxn}$$

With a positive $\Delta H°$ of this magnitude, we would expect relatively little AgCl to dissolve in water. In fact, less than 0.002 g of AgCl dissolves in a liter of water at room temperature. The solubility of silver chloride in water is so small that AgCl is often said to be "insoluble" in water, even though that term is misleading.

➤ **CHECKPOINT**

Calculate $\Delta H°$ for the following reaction using the data from Table B.13 in Appendix B.

$$\text{AgCl}(s) \xrightarrow{H_2O} \text{Ag}^+(aq) + \text{Cl}^-(aq)$$

8.14 Solubility Equilibria

Discussions of solubility equilibria are based on the following assumption: *When they dissolve, solids break apart to give solutions of the molecules or ions from which they are formed.* Most molecular compounds dissolve to give individual covalent molecules,

$$C_{12}H_{22}O_{11}(s) \xrightarrow{H_2O} C_{12}H_{22}O_{11}(aq)$$

and ionic solids dissociate to give solutions of the positive and negative ions they contain.

$$NaCl(s) \xrightarrow{H_2O} Na^+(aq) + Cl^-(aq)$$

Solutes such as NaCl that break up almost completely into ions when they dissolve are called **strong electrolytes.** Solutes such as sucrose that don't break up into ions when they dissolve are called **nonelectrolytes.** There are some solutes that partially break up into ions and partially exist as soluble molecules in solution; they are called **weak electrolytes.**

We can detect the presence of Na^+ and Cl^- ions in an aqueous solution with the conductivity apparatus shown in Figure 8.35. This apparatus consists of a lightbulb connected to a pair of metal wires that can be immersed in a beaker of water. The circuit in the conductivity apparatus isn't complete. In order for the lightbulb to glow when the apparatus is plugged into an electrical outlet, there must be a way for electrical charge to flow through the solution from one of the metal wires to the other.

When an ionic solid dissolves in the beaker of water, the resulting positive and negative ions are free to move through the aqueous solution. The net result is a flow of electric charge through the solution that completes the circuit and lets the lightbulb glow. As might be expected, the brightness of the bulb is proportional to the concentration of the ions in the solution. Slightly soluble ionic compounds such as calcium sulfate make the lightbulb glow dimly. When the wires of the conductivity apparatus are immersed in a solution of a very soluble ionic compound (such as NaCl), the lightbulb glows brightly.

The conductivity apparatus in Figure 8.35 gives us only qualitative information about the relative concentrations of the ions in different solutions. It is possible to build a more sophisticated instrument that gives quantitative measurements of the conductivity of a solution, which is directly proportional to the concentration of the ions in the solution. Figure 8.36 shows what happens to the conductivity of water as we gradually add infinitesimally small amounts of AgCl to water and wait for the solid to dissolve before taking measurements.

The system conducts a very small electric current even before any AgCl is added because of small quantities of H_3O^+ and OH^- ions in water. The solution becomes a slightly better conductor when AgCl is added because some of this ionic compound dissolves to give Ag^+ and Cl^- ions, which can carry an electric current through the solution. The conductivity continues to increase as more AgCl is added, until about 0.002 g of the ionic compound has dissolved per liter of solution.

The fact that the conductivity doesn't increase after the solution has reached a concentration of 0.002 g AgCl/liter tells us that there is a limit on the solubility of the ionic compound in water. Once the solution reaches that limit, no more AgCl dissolves, regardless of how much solid we add to the system. This is

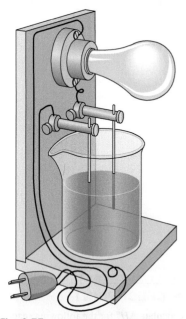

Fig. 8.35 A conductivity apparatus can be used to demonstrate the difference between aqueous solutions of ionic and covalent solids.

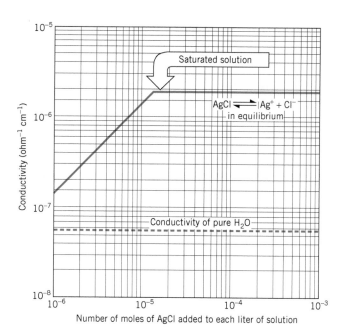

Fig. 8.36 The conductivity of a solution of AgCl in water increases at first as the AgCl dissolves and dissociates into Ag^+ ions and Cl^- ions. Once the solution has become saturated with AgCl, however, the conductivity remains the same no matter how much solid is added.

exactly what we would expect if the solubility of AgCl were controlled by an equilibrium. Once the solution reaches equilibrium, the rate at which AgCl dissolves to form Ag^+ and Cl^- ions is equal to the rate at which the ions recombine to form AgCl.

Figure 8.37 shows what happens to the concentrations of Ag^+ and Cl^- when a large excess of solid silver chloride is added to the water. When the ionic compound is first added, it dissolves and dissociates rapidly. The Ag^+ and Cl^- concentrations increase rapidly at first.

$$AgCl(s) \xrightarrow[\textit{dissociate}]{\textit{dissolve}} Ag^+(aq) + Cl^-(aq)$$

The concentrations of the ions soon become large enough that the reverse reaction starts to compete with the forward reaction, which leads to a decrease in the rate at which Ag^+ and Cl^- ions enter the solution. The reverse reaction is the

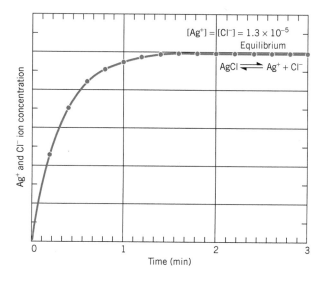

Fig. 8.37 Plot of concentration of Ag^+ and Cl^- ions versus time as solid AgCl dissolves in water.

A precipitation reaction that occurs when a colorless solution of KI is added to a colorless solution of $Pb(NO_3)_2$.

formation of insoluble silver chloride. Reactions in which soluble species form an insoluble product are called **precipitation reactions.**

$$Ag^+(aq) + Cl^-(aq) \xrightarrow[\text{precipitate}]{\text{associate}} AgCl(s)$$

Eventually, the Ag^+ and Cl^- ion concentrations become large enough that the rate at which precipitation occurs exactly balances the rate at which AgCl dissolves. Once that happens, there is no change in the concentration of the ions with time, and the reaction is at equilibrium. When the system reaches equilibrium, it is called a **saturated solution** because it contains the maximum concentration of ions that can exist in equilibrium with the solid ionic compound at a given temperature. The amount of substance that must be added to a given volume of solvent to form a saturated solution is called the **solubility** of the substance. Chemists use double arrows to indicate a reaction at equilibrium.

$$AgCl(s) \xrightarrow{H_2O} Ag^+(aq) + Cl^-(aq)$$

8.15 Solubility Rules

It is not a simple task to predict whether an ionic compound will be soluble in water. There are a number of patterns in the data obtained from measuring the solubilities of different ionic compounds, however. These patterns form the basis for the rules outlined in Table 8.9, which can guide predictions of whether a given ionic compound will dissolve in water. The rules are based on the following

Table 8.9
Solubility Rules for Ionic Compounds in Water

Soluble Ionic Compounds

The Na^+, K^+, and NH_4^+ ions form *soluble ionic compounds*. Thus, NaCl, KNO_3, and $(NH_4)_2CO_3$ are *soluble ionic compounds*.

The nitrate ion (NO_3^-) forms *soluble ionic compounds*. Thus, $Cu(NO_3)_2$ and $Fe(NO_3)_3$ are soluble.

The chloride (Cl^-), bromide (Br^-), and iodide (I^-) ions usually form *soluble ionic compounds*. Exceptions include ionic compounds of the Pb^{2+}, Hg_2^{2+}, Ag^+, and Cu^+ ions. $CuBr_2$ is soluble, but CuBr is not.

The sulfate ion (SO_4^{2-}) usually forms *soluble ionic compounds*. Exceptions include $BaSO_4$, $SrSO_4$, and $PbSO_4$, which are insoluble, and Ag_2SO_4, $CaSO_4$, and Hg_2SO_4, which are slightly soluble.

Insoluble Ionic Compounds

Sulfides (S^{2-}) are usually *insoluble*. Exceptions include Na_2S, K_2S, $(NH_4)_2S$, MgS, CaS, SrS, and BaS.

Oxides (O^{2-}) are usually *insoluble*. Exceptions include Na_2O, K_2O, SrO, and BaO, which are soluble, and CaO, which is slightly soluble.

Hydroxides (OH^-) are usually *insoluble*. Exceptions include NaOH, KOH, $Sr(OH)_2$, and $Ba(OH)_2$, which are soluble, and $Ca(OH)_2$, which is slightly soluble.

Chromates (CrO_4^{2-}), phosphates (PO_4^{3-}), and carbonates (CO_3^{2-}) are usually *insoluble*. Exceptions include ionic compounds of the Na^+, K^+, and NH_4^+ ions, such as Na_2CrO_4, K_3PO_4, and $(NH_4)_2CO_3$.

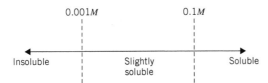

Fig. 8.38 Solubilities of compounds in water cover a wide range that is divided into the categories insoluble, slightly soluble, and soluble.

definitions of the terms *soluble, insoluble,* and *slightly soluble,* as indicated in Figure 8.38.

- A compound is defined as *soluble* if it dissolves in water to give a solution with a concentration of at least 0.1 mol/L at room temperature.
- A compound is defined as *insoluble* if the concentration of a saturated aqueous solution is less than 0.001 M at room temperature.
- Slightly soluble compounds give solutions that fall between the extremes.

It is important to remember that there is a limit to the solubility of even the most soluble ionic compounds and that ionic compounds that are labeled "insoluble" will dissolve to a very small extent.

8.16 Net Ionic Equations

The concept of a balanced chemical equation was introduced in Section 1.17. These equations can be written in three different forms: condensed, ionic, and net ionic equations. Most of the equations that you have encountered thus far have been **condensed equations** in which all reactants and products have been written as electrically neutral molecules and compounds. The reaction of aqueous barium chloride with aqueous sodium sulfate, for example, can be written as the following condensed equation.

$$BaCl_2(aq) + Na_2SO_4(aq) \longrightarrow BaSO_4(s) + 2\,NaCl(aq)$$

Three of the four components of this reaction, however, are strong electrolytes (see Section 8.14). Therefore, a better description of this reaction on the atomic or molecular level might be provided by an **ionic equation.** Ionic equations are written by assuming that strong electrolytes dissociate in aqueous solution into the corresponding ions. The ionic equation for the reaction between $BaCl_2$ and Na_2SO_4 would be written as follows.

$$Ba^{2+}(aq) + 2\,Cl^-(aq) + 2\,Na^+(aq) + SO_4^{2-}(aq)$$
$$\longrightarrow BaSO_4(s) + 2\,Na^+(aq) + 2\,Cl^-(aq)$$

The three soluble ionic compounds—$BaCl_2$, Na_2SO_4, and NaCl—are all shown as having dissociated into their ions. $BaSO_4$, however, is insoluble in water according to Table 8.9 and is therefore listed in this equation as a solid. Pure liquids and gases are also treated as electrically neutral molecules or compounds.

Certain ions appear on both sides of the ionic equation shown above. Because these ions remain unchanged during the chemical reaction, they are referred to as **spectator ions.** Spectator ions can be canceled from both sides of the equation on a one-to-one basis—for every ion eliminated from the product side of the equation, the corresponding ion must be eliminated from the reactant side. This yields the following **net ionic equation** for the reaction between aqueous solutions of $BaCl_2$ and Na_2SO_4.

$$Ba^{2+}(aq) + SO_4^{2-}(aq) \longrightarrow BaSO_4(s)$$

Net ionic equations contain all the information necessary to understand what is happening on the atomic scale during a chemical reaction. Regardless of whether chemical reactions are represented by condensed equations, ionic equations, or net ionic equations, the number of atoms or ions of each element and the net charge on both sides of the equation must be balanced.

The most difficult task in writing net ionic equations is determining which components of the reaction are strong electrolytes and should be written as ions. Strong electrolytes include soluble ionic compounds (refer to Table 8.9 to determine whether an ionic compound is soluble), strong acids (HCl, HBr, HI, HNO_3, $HClO_4$, and H_2SO_4), and strong bases [$LiOH$, $NaOH$, KOH, $RbOH$, $CsOH$, $Ca(OH)_2$, $Sr(OH)_2$, and $Ba(OH)_2$]. The strong acids and bases will be discussed in Chapter 11.

Exercise 8.4

Write the condensed, ionic, and net ionic equations that describe the following chemical reactions.

(a) Aqueous solutions of sodium chromate and lead(II) nitrate react to form a yellow precipitate of lead(II) chromate and an aqueous solution of sodium nitrate.

(b) An aqueous solution of the weak acid acetic acid (CH_3CO_2H) reacts with an aqueous solution of the strong base strontium hydroxide to form the ionic compound strontium acetate and liquid water.

Solution

(a) Condensed equation:

$$Na_2CrO_4(aq) + Pb(NO_3)_2(aq) \longrightarrow PbCrO_4(s) + 2\,NaNO_3(aq)$$

Ionic equation:

$$2\,Na^+(aq) + CrO_4^{2-}(aq) + Pb^{2+}(aq) + 2\,NO_3^-(aq)$$
$$\longrightarrow PbCrO_4(s) + 2\,Na^+(aq) + 2\,NO_3^-(aq)$$

Net ionic equation:

$$CrO_4^{2-}(aq) + Pb^{2+}(aq) \longrightarrow PbCrO_4(s)$$

(b) Condensed equation:

$$2\,CH_3CO_2H(aq) + Sr(OH)_2(aq) \longrightarrow Sr(CH_3CO_2)_2(aq) + 2\,H_2O(l)$$

Ionic equation:

$$2\,CH_3CO_2H(aq) + Sr^{2+}(aq) + 2\,OH^-(aq)$$
$$\longrightarrow Sr^{2+}(aq) + 2\,CH_3CO_2^-(aq) + 2\,H_2O(l)$$

Acetic acid is not broken into its ions because it is not a strong electrolyte. Water is not broken into ions because it is a liquid.

Only the strontium ion is found on both sides of the equation in exactly the same form, and therefore only the strontium ion can be eliminated from the ionic equation. Eliminating the spectator ion yields

$$2\,CH_3CO_2H(aq) + 2\,OH^-(aq) \longrightarrow 2\,CH_3CO_2^-(aq) + 2\,H_2O(l)$$

This can be simplified to give the following net ionic equation.

$$CH_3CO_2H(aq) + OH^-(aq) \longrightarrow CH_3CO_2^-(aq) + H_2O(l)$$

Key Terms

Alcohol	Hydrophilic	Polar
Boiling point	Hydrophobic	Polar solvent
Cohesion	Intermolecular forces	Precipitation reaction
Condensed equation	Intramolecular bonds	Saturated solution
Dipole moment	Ion–dipole	Solubility
Dispersion forces (London forces)	Ionic equation	Spectator ion
Enthalpy of fusion	Melting point	Strong electrolyte
Enthalpy of vaporization	Net ionic equation	Surface tension
Equilibrium	Nonelectrolyte	van der Waals forces
Freezing point	Nonpolar solvent	Vapor pressure
Hydrocarbons	Phase diagram	Weak electrolyte
Hydrogen bonding		

Problems

The Structure of Gases, Liquids, and Solids

1. Describe the differences in the properties of gases, liquids, and solids on the atomic scale. Explain how the differences give rise to the observed differences in the macroscopic properties of the three states of matter.

2. Describe the difference between *intermolecular* forces and *intramolecular* bonds, giving examples of each. Which are stronger?

3. Solids consist of particles locked into a rigid structure. Describe what happens on a microscopic scale when a solid melts. Describe what happens when a liquid is heated to boiling.

4. Why are most substances solids at very low temperatures? Why are most substances gases at very high temperatures?

5. What determines whether a substance is a solid, liquid, or gas at room temperature?

6. Why are liquids usually less dense than their corresponding solids?

Intermolecular Forces

7. There are four categories of intermolecular forces. What are they, and how do they differ? Give an example of each.

8. Why do induced dipole–induced dipole forces increase as the number of electrons in a molecule increases?

9. What structural features are necessary for a molecule to exhibit hydrogen bonding?

10. List all of the types of intermolecular forces that each of the following molecules would have.

    (a) SO_2 (b) CH_3OH (c) ICl_3 (d) SF_4

    Note that the Lewis structure of each molecule must be determined.

11. If a molecule is known to have a dipole moment, what types of intermolecular forces must be present?

Relative Strengths of Intermolecular Forces

12. Predict the order in which the boiling points of the following compounds should increase. Explain your reasoning.

    (a) NH_3 (b) PH_3 (c) AsH_3

13. Which compound would you expect to have the highest boiling point? Explain your reasoning.

    (a) methane, CH_4

    (b) chloromethane, CH_3Cl

    (c) dichloromethane, CH_2Cl_2

    (d) chloroform, $CHCl_3$

    (e) carbon tetrachloride, CCl_4

14. Explain why the boiling points of hydrocarbons that have the generic formula C_nH_{2n+2} increase with molecular weight.

15. Explain why propane (C_3H_8) is a gas but pentane (C_5H_{12}) is a liquid at room temperature.

16. Explain why methane (CH_4) is a liquid only over a very narrow range of temperatures.

17. Why is the boiling point of n-pentane larger than that of isopentane?

18. Explain the trends in the following data that show the melting point and dipole moment for the hydrogen halides.

	MP,°C	Dipole Moment (D)
HCl	−114	1.08
HBr	−87	0.81
HI	−51	0.45

The Kinetic Theory of Liquids

19. The average velocity of a molecule in a gas is about 1000 miles per hour. How fast are the same molecules in the liquid phase moving when they are at the same temperature?

20. If a liquid has a high enthalpy of vaporization, what does this mean about the forces holding the liquid together?

21. What is the difference between an enthalpy of fusion and an enthalpy of vaporization?

22. What effect does decreasing the temperature have on the density of most liquids? Explain.

The Vapor Pressure of a Liquid

23. Explain why it is important to specify the temperature at which the vapor pressure of a liquid is measured.

24. Explain why water eventually evaporates from an open container at room temperature ($\sim20°C$), even though it normally boils at 100°C.

25. One postulate of the kinetic theory of gases suggests that the temperature of a gas is directly proportional to the *average kinetic energy* of the particles in the gas. Why is the term *average* used?

26. Use the kinetic molecular theory to explain why the vapor pressure of water becomes larger as the temperature of the water increases.

27. Explain why a cloth soaked in water feels cool when placed on your forehead.

28. What would happen to the vapor pressure of liquid bromine, Br_2, at 20°C if the liquid was transferred from a narrow 10-mL graduated cylinder into a wide petri dish or crystallizing dish? Would it increase, decrease, or remain the same? What would happen to the vapor pressure in a closed container if more liquid were added to the container? Would it increase, decrease, or remain the same?

29. Explain what it means to say that the liquid and vapor in a closed container are in *equilibrium*.

30. Explain why each of the following increases the rate at which water evaporates from an open container.

    (a) increasing the temperature of the water

    (b) increasing the surface area of the water

    (c) blowing air over the surface of the water

    (d) decreasing the atmospheric pressure on the water

31. On a very dry day, snow changes directly to a gas through a process known as *sublimation*. Use the kinetic molecular theory to explain how this can happen.

32. Why does a pressure cooker cook food more rapidly than an open cooker?

33. The force of cohesion between mercury atoms is much larger than the force of cohesion between water molecules. Conversely, the attractive force between water molecules and glass is much larger than the attractive force between mercury atoms and glass. Use these observations to explain why mercury forms small drops when it spills on glass rather than forming a single large puddle such as water does.

34. Describe how the surface tension of water can be used to explain the fact that a steel sewing needle floats on the surface of water.

35. Explain why a drop of water seems to bead up on the surface of a freshly waxed car.

36. Explain the advantages and disadvantages in a plant of having leaves that have a large surface area. Explain why plants that grow in arid climates seldom have leaves as broad as those found on maple trees.

37. Which of each pair of the following compounds would have the lower boiling point? The lower vapor pressure?

    (a) $CH_3CH_2CH_3$ and CH_3CH_2OH

    (b) CH_3CH_3 and CH_3Br

    (c) $HOCH_2CH_2OH$ and CH_3CH_2OH

    (d) $CH_3CH_2CH_2CH_2CH_3$ and $CH_3CH_2OCH_2CH_3$

38. Predict which of the following liquids should have the lowest boiling point from their vapor pressures (VP) at 0°C.

    (a) acetone, VP = 67 mmHg

    (b) benzene, VP = 24.5 mmHg

(c) ether, VP = 183 mmHg

(d) methyl alcohol, VP = 30 mmHg

(e) water, VP = 4.6 mmHg

Melting Point and Freezing Point

39. Does a solid always get hotter when heat is added? Explain.

40. The enthalpy of fusion of water is 6.0 kJ/mol$_{rxn}$, and that of methanol (CH_3OH) is 3.2 kJ/mol$_{rxn}$. Which solid, water or methanol, has the largest intermolecular forces? Predict whether the enthalpy of fusion of CH_3CH_3 will be larger than, less than, or the same as that of methanol. Explain.

41. Why do different substances have different melting points?

42. Explain why a "3-minute egg" cooked while camping in the Rocky Mountains does not taste as good as it does when cooked while camping near the Great Lakes.

43. Explain why it takes more time to boil food in Denver than in Miami.

44. Explain why water boils when the pressure on the system is reduced.

45. At what temperature does water boil when the pressure is 50 mmHg? Use the data in Figure 8.13 or Table B.3 in Appendix B.

46. What pressure has to be achieved before water can boil at 20°C? Use the data in Figure 8.13 or Table B.3 in Appendix B.

47. Increasing the temperature of a liquid will do which of the following?

    (a) increase the boiling point

    (b) increase the melting point

    (c) increase the vapor pressure

    (d) increase the amount of heat required to boil a mole of the liquid

    (e) all of the above

48. Liquid air is composed primarily of liquid oxygen (BP = −183°C) and liquid nitrogen (BP = −196°C). It can be separated into its component gases by increasing the temperature until one of the gases boils off. Which gas boils off first?

49. According to the data in Table 8.4, butane (C_4H_{10}) should be a gas at room temperature (BP = −0.5°C). Use this to explain why you can hear gas escape when a can of butane lighter fluid is opened with the nozzle pointed up. If you shake the can, however, you can hear a liquid bounce against either end of the can. Furthermore, when you open the can with the nozzle pointed down, you can see a liquid escape. Explain how butane is stored as a liquid in cans at room temperature.

50. (a) Which of the following molecules has the most polar bonds? Explain.

$$CH_4, \; CCl_4, \; CBr_4$$

    (b) What type of bonding is present in each of these compounds?

    (c) Which of the liquids of these molecules has the highest vapor pressure? Which has the lowest? Explain why.

    (d) Arrange these compounds in order of increasing boiling point. State which is the highest. Explain why.

Phase Diagrams

51. At what conditions of temperature and pressure is the formation of a solid favored?

52. At a low pressure and a high temperature, what state of a substance is most likely to exist?

53. A horizontal dotted line is drawn across the phase diagram of Figure 8.18 at a pressure less than 1 atm. Describe what information can be obtained from the intersection of the dotted line with a solid line on the diagram. How will the freezing point and the boiling point be changed from what they were at 1 atm?

54. According to Figure 8.18, what effect does decreasing the pressure have on the freezing point of most compounds?

Hydrogen Bonding and the Anomalous Properties of Water

55. Why are the boiling point and melting point of water much higher than you would expect from the boiling points and melting points of H_2S, H_2Se, and H_2Te?

56. Why is hydrogen bonding very strong in HF and H_2O? Explain why hydrogen bonding is much weaker in HCl and H_2S.

57. Why is ice less dense than liquid water?

58. Explain why the strength of hydrogen bonds decreases in the following order: $HF > H_2O > NH_3$.

59. Why does water have an unusually large specific heat?

Solutions: Like Dissolves Like

60. Use a drawing to describe what happens when I_2 molecules dissolve in CCl_4 and when $KMnO_4$ dissolves in water.

61. One way of screening potential anesthetics involves testing whether the compound dissolves in olive oil because all common anesthetics, including dinitrogen oxide (N_2O), cyclopropane (C_3H_6), and halothane ($CF_3CHBrCl$), are soluble in olive oil. What property do the compounds have in common?

62. Carboxylic acids with the general formula $CH_3(CH_2)_nCO_2H$ have a nonpolar $CH_3CH_2\ldots$ tail and a polar $\ldots CO_2H$ head. What effect does increasing the value of n have on the solubility of carboxylic acids in polar solvents, such as water? What is the effect on their solubility in nonpolar solvents, such as CCl_4?

63. Which of the following compounds would be the most soluble in a nonpolar solvent, such as CCl_4?

    (a) H_2O

    (b) CH_3OH

    (c) $CH_3CH_2CH_2OH$

    (d) $CH_3CH_2CH_2CH_2CH_2OH$

    (e) $CH_3CH_2CH_2CH_2CH_2CH_2CH_2OH$

64. Potassium iodide reacts with iodine in aqueous solution to form aqueous potassium triiodide, KI_3

$$KI(aq) + I_2(aq) \longrightarrow KI_3(aq)$$

    What would happen if we added CCl_4 to the reaction mixture?

    (a) The KI and KI_3 would dissolve in the CCl_4 layer.

    (b) The I_2 would dissolve in the CCl_4 layer.

    (c) Both KI and I_2, but not KI_3, would dissolve in the CCl_4 layer.

    (d) Neither KI, KI_3, nor I_2 would dissolve in the CCl_4 layer.

    (e) No distinct CCl_4 layer would form because CCl_4 is soluble in water.

65. Phosphorus pentachloride can react with itself in an equilibrium reaction to form an ionic compound that contains the PCl_4^+ and PCl_6^- ions.

$$2\,PCl_5 \rightleftharpoons PCl_4^+ + PCl_6^-$$

    The extent to which the reaction occurs depends on the solvent in which it is run. Predict whether a nonpolar solvent, such as CCl_4, favors the products or the reactants of the reaction. Predict what would happen to the reaction if we used a polar solvent, such as acetonitrile (CH_3CN).

Hydrophilic and Hydrophobic Molecules

66. Why are hydrocarbons not soluble in water?

67. Why are some alcohols soluble in water while others are not?

68. Arrange the following compounds in order of increasing solubility in water. Explain your order.

    (a) $NaCl$

    (b) $CH_3CH_2CH_2CH_3$

    (c) CH_3CH_2OH

    (d) CH_3COOH

Soaps, Detergents, and Dry-Cleaning Agents

69. Why are soaps composed of hydrophilic and hydrophobic groups?

70. What is hard water? How can it be softened?

71. What is meant by "dry" cleaning? What advantages does dry cleaning have over soaps or detergents?

Why Do Some Solids Dissolve in Water?

72. Explain why some molecular solids are soluble in water but others are not.

73. Explain why $BaCl_2$ is soluble in water but $AgCl$ is not.

74. On the microscopic level, explain what happens when an ionic compound dissolves in water.

Solubility Equilibria

75. Explain why the lightbulb in the conductivity apparatus in Figure 8.35 glows more brightly when the wires are immersed in a solution of $NaCl$ than when the wires are immersed in tap water.

76. Explain why the addition of a few small crystals of silver chloride makes water a slightly better conductor of electricity. Explain why the conductivity gradually increases as more $AgCl$ is added, until it eventually reaches a maximum. Describe what is happening in the solution when its conductivity reaches the maximum.

Solubility Rules

77. Which of the following ionic compounds are *insoluble* in water?

    (a) $Ba(NO_3)_2$

    (b) $BaCl_2$

    (c) $BaCO_3$

    (d) BaS

    (e) $Ba(C_2H_3O_2)_2$

78. Which of the following ionic compounds are *insoluble* in water?

    (a) $(NH_4)_2SO_4$

    (b) K_2CrO_4

    (c) Na_2S

    (d) $Pb(NO_3)_2$

    (e) $Cr(OH)_3$

79. Which of the following ionic compounds are *soluble* in water?

    (a) PbS

    (b) PbO

    (c) $PbCrO_4$

    (d) $PbCO_3$

    (e) $Pb(NO_3)_2$

Net Ionic Equations

80. Write net ionic equations for each of the following condensed equations.

    (a) $Zn(s) + Cu(NO_3)_2(aq)$
    $$\longrightarrow Zn(NO_3)_2(aq) + Cu(s)$$

    (b) $MnCl_2(aq) + (NH_4)_2S(aq)$
    $$\longrightarrow MnS(s) + 2\,NH_4Cl(aq)$$

81. Write condensed and net ionic equations for each of the following chemical reactions.

    (a) Magnesium metal reacts with the strong acid HCl to produce an aqueous solution of magnesium chloride and hydrogen gas.

    (b) Aqueous solutions of sodium carbonate and calcium nitrate react to form a precipitate of calcium carbonate and an aqueous solution of sodium nitrate.

82. Use the following net ionic equations to write condensed equations for these reactions. Assume that the nitrate anion and potassium cation are present in solution.

    (a) $Ag^+(aq) + Cl^-(aq) \longrightarrow AgCl(s)$
    (b) $Ni^{2+}(aq) + CrO_4{}^{2-}(aq) \longrightarrow NiCrO_4(s)$

Integrated Problems

83. How would you explain the difference between the bonds that are broken when NaCl boils and the forces that are broken when water boils? How would you explain the difference between the boiling points of NaCl (BP = 1465°C) and water (BP = 100°C)?

84. Because they are isomers, ethanol (CH_3CH_2OH) and dimethyl ether (CH_3OCH_3) have the same molecular weight. Explain why ethanol (BP = 78.5°C) has a much higher boiling point than dimethyl ether (BP = −23.6°C).

85. The following compounds have the same molecular weight. Explain why one of the compounds has a higher boiling point than the other.

$$\underset{\substack{|\\ CH_3}}{CH_3NCH_3} \qquad \underset{\substack{|\\ CH_3}}{CH_3CH_2NH}$$

    BP = 3°C   *BP = 35°C*

    Trimethyl amine   *Ethylmethyl amine*

86. Assume you like a really hot cup of tea. Would you prefer to live in Denver or in Miami if hot tea is your only consideration? Why?

87. A thermometer is taped to the outside of an open flask filled with water. The water is heated to boiling and allowed to continue to boil until it is gone. Sketch a rough plot of how the temperature would change with respect to time for water initially at 25°C. Suppose the same setup is used except that no heat is supplied to the open flask, which is initially at 25°C. If you come back some time later, all of the water will have evaporated. How do you think the temperature recorded by the thermometer would have changed with respect to time? What is the difference between the phase change of water at 100°C and at 25°C?

88. Which illustration most correctly describes the boiling of water at 100°C?

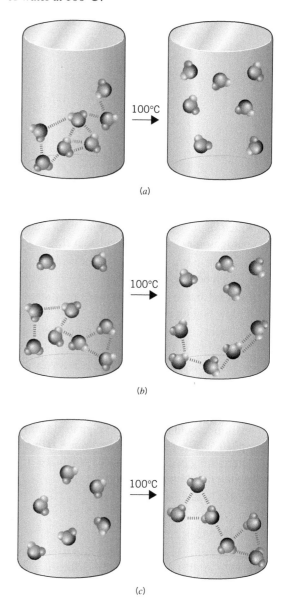

89. Compare the enthalpy change associated with boiling 1 mol of water (converting 1 mol of water into steam at 100°C) to evaporating 1 mol of water (converting 1 mol of water to water vapor at room temperature).

90. The diagram in this problem shows liquid methanol, CH_3OH, being prepared to be poured into a beaker of water. Draw a figure that illustrates the solution that will result. Clearly show the intermolecular forces. Describe the intermolecular forces that are broken in

the solute and solvent and formed in the solution. Do you expect methanol to be very soluble in water?

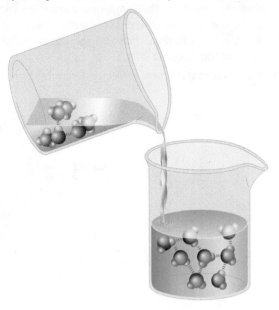

91. Which of the following statements is true? Explain your reasoning on the molecular level.
    (a) A substance with a high vapor pressure will have a high boiling point.
    (b) A substance with a high vapor pressure will have a low boiling point.
    (c) There can be no relationship between vapor pressure and boiling point.

92. An insoluble solid that forms when two or more soluble species are mixed together is known as a precipitate. Use the solubility rules for ionic compounds in water to predict whether each of the following aqueous mixtures would produce a precipitate. Identify the precipitate in cases where one is formed. (*Hint*: Begin by determining what compounds might be formed by mixing the solutions together.)
    (a) $FeCl_3$ and KOH
    (b) Na_2CO_3 and $(NH_4)_2SO_4$
    (c) $Pb(NO_3)_2$ and $NaCl$

93. Liquid benzene, C_6H_6, has a vapor pressure of 325 mmHg at 80°C. If 1.00 g of benzene is placed in an evacuated 1.00-L flask at 80°C, determine the mass of benzene that will evaporate and the final pressure in the container. Determine the same variables if the experiment is repeated using a 500-mL evacuated container. Explain what is different about the two experiments.

94. Use data from Table 8.4 to answer the following questions.
    (a) 1-Pentanol is an alcohol with the molecular formula $CH_3CH_2CH_2CH_2CH_2OH$ and a molecular mass of 88.2 g/mol. How would you expect the boiling point of pentanol to be related to the

boiling points of butanoic acid, pentanal, and hexane? Explain your reasoning.
    (b) Predict the boiling point for nonanal, $CH_3(CH_2)_7CHO$. Explain your reasoning.

95. A sealed container equipped with a movable piston contains a liquid in equilibrium with its vapor. A pressure gauge is attached to the container to measure the pressure of the vapor. Explain what is happening on the molecular level for each of the following experimental observations.
    (a) A valve on the container is opened, and some of the gas is allowed to escape. The pressure is observed to immediately drop when the valve is opened. When the valve is closed, however, the pressure is observed to increase and then stabilize. (The volume of the container and the temperature remain constant.)
    (b) The movable piston on the container is depressed, causing the volume to decrease. The pressure increases as the piston moves in but then stabilizes at a constant pressure even as the piston continues to be depressed. (The temperature of the container remains constant.)

96. Octane, dibutyl ether, and 1-octanol have the following structures.

$$CH_3CH_2CH_2CH_2CH_2CH_2CH_2CH_3$$
Octane

$$CH_3CH_2CH_2CH_2{-}O{-}CH_2CH_2CH_2CH_3$$
Dibutyl ether

$$CH_3CH_2CH_2CH_2CH_2CH_2CH_2CH_2OH$$
1-Octanol

(a) Arrange the three compounds in order of increasing vapor pressure.
    (b) Arrange the three compounds in order of decreasing boiling point.

97. Which of the following properties would you expect to generally increase as temperature increases: vapor pressure, surface tension, and heat of vaporization? Explain.

98. The specific heat of liquid water at room temperature is about 4.2 J/g · K. Would you expect the specific heat of gaseous water to be larger than that of the liquid at room temperature? Explain.

99. The surface tension of liquid water is larger than that of liquid methanol (CH_3OH) at room temperature. How can you explain the difference?

100. Two molecular compounds have the same molecular weight, but one boils at 195°C and the other at 142°C. What factors can account for the difference in boiling point? Which compound has the lowest vapor pressure? Do either or both of the compounds have a dipole moment?

101. Consider the following three compounds:

$CH_3CH_2CH_2CH_2CH_2CH_3$
n-Hexane

$$CH_3\overset{\overset{\textstyle O}{\|}}{C}CH_2CH_2CH_2CH_3$$
2-Hexanone

$$CH_3\overset{\overset{\textstyle OH}{|}}{C}HCH_2CH_2CH_2CH_3$$
2-Hexanol

(a) What types of intermolecular forces exist between like molecules of each compound?

(b) Which compound will have the highest boiling point? Explain.

(c) Which compound will have the lowest vapor pressure? Explain.

(d) Which compound will be the most soluble in water? Explain.

102. The following data for the hydrides of Group V are given:

	Boiling Point, °C
NH_3	−33
PH_3	−88
AsH_3	−57
SbH_3	−17

(a) List these compounds in order of increasing vapor pressure. Which has the highest vapor pressure? Which has the lowest?

(b) Which of the following would be the best solvent for NH_3? Explain why.

$CH_3CH_2CH_2CH_2CH_2CH_3$, CCl_4, CH_3OH, $BrCH_2CH_2CH_3$

103. The molecular weight and dipole moment of three compounds are as follows.

	MW (g/mol)	$\mu(D)$

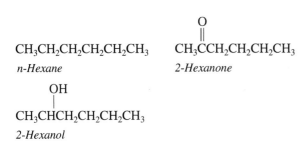

	MW (g/mol)	$\mu(D)$
(A)	72	2.5
(B)	74	1.7
(C)	74	1.6

(a) Arrange these compounds in order of increasing boiling point. Specify which has the highest boiling point and which has the lowest. Explain your reasoning in detail.

(b) Which of the above would have the lowest vapor pressure? The highest vapor pressure? Explain.

(c) Which should be the most soluble in water? Explain.

104. Figure 8.36 shows a plot of conductivity versus the moles of AgCl (a very slightly soluble ionic compound) added to the solution. Make a rough sketch of this plot. The *x* axis will be moles of added compound. Add a second plot that describes what would happen to conductivity as NaCl, a soluble salt, is added to a beaker of pure water. Add a third plot that describes what would happen to conductivity as table sugar ($C_{12}H_{22}O_{11}$), a molecular solid, is added to a beaker of pure water.

105. When a molecular solid is heated and melts to form a liquid, must all of the intermolecular forces between the solid molecules be broken? Explain your reasoning.

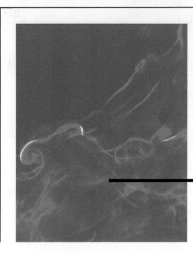

Chapter Eight

SPECIAL TOPICS

8A.1 Colligative Properties

Dissolving a solute in a solvent results in a solution with physical properties that are different from those of either the pure solute or the pure solvent. **Colligative properties** are physical properties of solutions that depend on the number of solute particles in a solution but not on the identity of the solute particles. This means that two solutions that contain different solutes dissolved in the same solvent that have the same concentrations of solute particles would exhibit the same colligative properties.

To begin our discussion of colligative properties, we must introduce a new unit for measuring concentration, **mole fraction.** By definition, the mole fraction of any component of a solution is the number of moles of that component divided by the total number of moles of solute and solvent. The symbol for mole fraction is the Greek letter chi, χ. The mole fraction of the *solute,* for example, is equal to the number of moles of solute divided by the total number of moles of solute and solvent.

$$\chi_{solute} = \frac{\text{moles of solute}}{\text{moles of solute + moles of solvent}}$$

Conversely, the mole fraction of the *solvent* is the number of moles of solvent divided by the total number of moles of solute and solvent.

$$\chi_{solvent} = \frac{\text{moles of solvent}}{\text{moles of solute + moles of solvent}}$$

In a solution that contains a single solute dissolved in a solvent, the sum of the mole fractions of solute and solvent must be equal to 1.

$$\chi_{solute} + \chi_{solvent} = 1$$

Exercise 8A.1

A solution of hydrogen sulfide in water can be prepared by bubbling H_2S gas into water until no more gas dissolves. Calculate the mole fraction of both H_2S and H_2O in this solution if 0.385 g of H_2S gas dissolves in 100 g of water at 20°C and 1 atm.

Solution

The number of moles of solute in the solution can be calculated from the mass of H_2S that dissolves.

$$0.385 \text{ g } H_2S \times \frac{1 \text{ mol } H_2S}{34.08 \text{ g } H_2S} = 0.0113 \text{ mol } H_2S$$

To determine the mole fraction of the solute and solvent, we also need to know the number of moles of water in the solution.

$$100 \text{ g } H_2O \times \frac{1 \text{ mol } H_2O}{18.02 \text{ g } H_2O} = 5.55 \text{ mol } H_2O$$

The mole fraction of the solute is the number of moles of H_2S divided by the total number of moles of both H_2S and H_2O.

$$\chi_{\text{solute}} = \frac{0.0113 \text{ mol } H_2S}{0.0113 \text{ mol } H_2S + 5.55 \text{ mol } H_2O} = 0.00203$$

The mole fraction of the solvent is the number of moles of H_2O divided by the moles of both H_2S and H_2O.

$$\chi_{\text{solvent}} = \frac{5.55 \text{ mol } H_2O}{0.0113 \text{ mol } H_2S + 5.55 \text{ mol } H_2O} = 0.998$$

Note that the sum of the mole fractions of the two components of the solution is 1.

$$\chi_{\text{solute}} + \chi_{\text{solvent}} = 0.00203 + 0.998 = 1.000$$

● ●

In Chapter 6, we saw that the ideal gas law was valid only for ideal gases. In much the same way, the equations we'll use to describe the colligative properties of solutions are valid only for **ideal solutions.** An ideal solution is one in which the forces that hold the solute particles together are similar to those that hold the solvent particles together. When this is true, the forces of attraction that must be broken to separate the particles in the solute and in the solvent are similar to the forces formed when the solute and solvent particles interact. This is not the case for most real solutions. As a result, the equations we develop to describe colligative properties will give good approximations for real-world solutions, but not exact answers.

8A.2 Depression of the Partial Pressure of a Solvent

The vapor pressure of a liquid was defined in Section 8.5 as the pressure of the gas or vapor in equilibrium with the corresponding liquid. When a solute is added to a liquid solvent, there is a decrease in the pressure exerted by the vapor of the solvent above the solution. We'll define $P°$ as the vapor pressure of the pure liquid—the solvent—and P as the pressure exerted by the vapor of the solvent over a solution.

$$P < P°$$

Partial pressure	*Vapor pressure*
of the solvent	*above the*
above a solution	*pure solvent*

Between 1887 and 1888, François-Marie Raoult showed that the pressure of the solvent escaping from a solution is equal to the mole fraction of the solvent times the vapor pressure of the pure solvent. This equation is known as **Raoult's law.**

$$P = \chi_{\text{solvent}}P°$$

Partial pressure	*Vapor pressure*
of the solvent	*above the*
above a solution	*pure solvent*

When the solvent is pure and the mole fraction of the solvent is equal to 1, P is equal to $P°$. As the mole fraction of the solvent becomes smaller, the partial pressure of the solvent escaping from the solution also becomes smaller.

Let's assume, for the moment, that the solvent is the only component of the solution volatile enough to have a measurable vapor pressure. This would be true, for example, for a nonvolatile solute, such as NaCl, dissolved in water. The partial pressure of the solution would be equal to the pressure produced by the solvent escaping from the solution. Raoult's law states that the difference between the vapor pressure of the pure solvent and the partial pressure over the solution increases as the mole fraction of the solvent decreases.

When a solute is added to a pure solvent, the change in the partial pressure of the solvent above the solution is the difference between the vapor pressure of the pure solvent and the partial pressure exerted by the solvent above the solution.

$$\Delta P = P° - P$$

Substituting Raoult's law into the above equation gives the following result

$$\Delta P = P° - \chi_{\text{solvent}}P° = (1 - \chi_{\text{solvent}})P°$$

This equation can be simplified by remembering the relationship between the mole fraction of the solute and the mole fraction of the solvent.

$$\chi_{\text{solute}} + \chi_{\text{solvent}} = 1$$

Substituting this relationship into the equation that defines ΔP gives another form of Raoult's law.

$$\Delta P_{\text{solvent}} = \chi_{\text{solute}}P°_{\text{solvent}}$$

This equation reminds us that, for an ideal solution, as more solute is dissolved in the solvent, the change in pressure, ΔP, increases.

One consequence of vapor pressure depression of solvents by solutes can be seen in Figure 8A.1. This figure shows a beaker containing pure solvent and a second beaker containing a solution with a nonvolatile solute. Both beakers are inside a sealed container. The pure solvent and the solvent in the solution begin to evaporate, attempting to establish equilibrium between the liquid and the vapor phase. However, the vapor pressure associated with the pure solvent will be larger

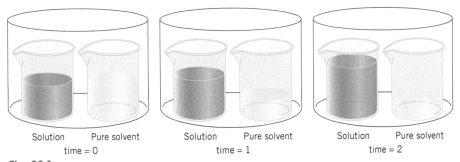

Solution   Pure solvent   Solution   Pure solvent   Solution   Pure solvent
    time = 0   time = 1   time = 2

Fig. 8A.1 As evaporation takes place, the sealed container fills with the vapor of the solvent. Due to the difference in the partial pressures of the solvent over the pure solvent and over the solution that contains a nonvolatile solute, the solvent gradually evaporates from the beaker containing the pure solvent, while vapor condenses into the beaker containing the solution.

than the pressure exerted by the solvent in the solution. As time passes, the solvent will gradually evaporate from the beaker that contains pure solvent and will condense in the beaker that contains the solution.

The vapor that accumulates above a solution that contains a *volatile* solute will have two components: the solvent and the solute. The total pressure of this vapor will be the sum of the partial pressure of the solvent and the partial pressure of the solute.

Exercise 8A.2

Assume that a solution is prepared by mixing 500 mL of ethanol (C_2H_6O) and 500 mL of water at 25°C. The vapor pressures of pure water and pure ethanol at that temperature are 23.76 and 59.76 mmHg, respectively. The densities of water and ethanol at 25°C are 0.9971 and 0.786 g/mL, respectively. Determine the partial pressure of each component in the solution and the total pressure. Assume that the solution is ideal.

Solution

We begin by determining the number of moles of each of the components of the solution.

$$500 \text{ mL H}_2\text{O} \times \frac{0.9971 \text{ g H}_2\text{O}}{1 \text{ mL}} \times \frac{1 \text{ mol H}_2\text{O}}{18.02 \text{ g H}_2\text{O}} = 27.7 \text{ mol H}_2\text{O}$$

$$500 \text{ mL C}_2\text{H}_6\text{O} \times \frac{0.786 \text{ g C}_2\text{H}_6\text{O}}{1 \text{ mL}} \times \frac{1 \text{ mol C}_2\text{H}_6\text{O}}{46.07 \text{ g C}_2\text{H}_6\text{O}} = 8.53 \text{ mol C}_2\text{H}_6\text{O}$$

We then calculate the mole fractions of the two components of the solution.

$$\chi_{\text{water}} = \frac{27.7 \text{ mol water}}{8.53 \text{ mol ethanol} + 27.7 \text{ mol water}} = 0.765$$

$$\chi_{\text{ethanol}} = \frac{8.53 \text{ mol ethanol}}{8.53 \text{ mol ethanol} + 27.7 \text{ mol water}} = 0.235$$

According to Raoult's law, the partial pressure of the water escaping from the solution is equal to the product of the mole fraction of water and the vapor pressure of pure water.

$$P_{\text{water}} = \chi_{\text{water}} P°_{\text{water}} = (0.765)(23.76 \text{ mmHg}) = 18.2 \text{ mmHg}$$

The partial pressure of ethanol is found in a similar fashion.

$$P_{\text{ethanol}} = \chi_{\text{ethanol}} P°_{\text{ethanol}} = (0.235)(59.76 \text{ mmHg}) = 14.0 \text{ mmHg}$$

The total pressure of the gases escaping from the solution is the sum of the partial pressures of the two gases.

$$P_{\text{total}} = P_{\text{water}} + P_{\text{ethanol}} = 32.2 \text{ mmHg}$$

Although the solution is a 50:50 mixture by volume, slightly more than three-quarters of the particles in the solution are water molecules. As a result, the total pressure of the solution more closely resembles the vapor pressure of pure water than it does pure ethanol. The magnitude of the change in the partial pressure of ethanol is also much larger than the change in the partial pressure of water.

8A.3 Boiling Point Elevation

Section 8.7 described how a pure liquid could be boiled by heating the liquid until the vapor pressure became equal to the pressure pushing down on the surface of the liquid. If a nonvolatile solute is added to the pure liquid, the vapor pressure of the solvent escaping from the solution will be reduced. This means that it will be necessary to increase the temperature even more in order to increase the pressure of the solvent's vapor until it becomes equal to the pressure pushing down on the surface of the solution. Therefore the boiling point of a solution will be larger than the boiling point of the pure solvent. This is known as boiling point elevation.

Because changes in the boiling point of the solvent (ΔT_{BP}) that occur when a solute is added result from changes in the partial pressure of the solvent, the magnitude of the change in the boiling point is also proportional to the mole fraction of the solute.

In very dilute solutions, the mole fraction of the solute is proportional to the molality of the solution. Molality, m, is defined as the number of moles of solute per kilogram of solvent.

$$m = \frac{\text{moles of solute}}{\text{kilograms of solvent}}$$

Molality is similar to molarity except the denominator is *kilograms of solvent* instead of *liters of solution*. Molality has an important advantage over molarity. The molarity of an aqueous solution changes with temperature because the density of water is sensitive to temperature. The molality of a solution doesn't change with temperature because it is defined in terms of the mass of the solvent, not its volume.

The equation that describes the magnitude of the boiling point elevation that occurs when a solute is added to a solvent can be written as follows.

$$\Delta T_{BP} = k_b m$$

Here, ΔT_{BP} is the **boiling point elevation**—that is, the change in boiling point that occurs when a solute dissolves in the solvent— and k_b is a proportionality constant known as the *molal boiling point elevation constant* for the solvent. Molal boiling point elevation constants for selected compounds are given in Table 8A.1.

Because colligative properties depend on the number of particles in solution, but not their identity, colligative properties can be used to determine molecular weights. Consider elemental sulfur, for example. The chemical formula is written as S_8. How do we know that sulfur forms S_8 molecules?

Table 8A.1

Freezing Point Depression and Boiling Point Elevation Constants

Compound	Freezing Point (°C)	k_f (°C/m)
Water	0	1.853
Acetic acid	16.66	3.90
Benzene	5.53	5.12
p-Xylene	13.26	4.3
Naphthalene	80.29	6.94
Cyclohexane	6.54	20.0
Carbon tetrachloride	−22.95	29.8
Camphor	179.8	40

Compound	Boiling Point (°C)	k_b (°C/m)
Water	100	0.515
Ethyl ether	34.55	2.02
Carbon disulfide	46.23	2.35
Benzene	80.10	2.53
Carbon tetrachloride	76.75	5.03
Camphor	207.42	5.95

Exercise 8A.3

Calculate the molecular weight of sulfur if 35.5 g of sulfur dissolves in 100 g of CS_2 to produce a solution that has a boiling point of 49.48°C.

Solution

The relationship between the boiling point of the solution and the molecular weight of sulfur isn't immediately obvious. We therefore start by asking: What do we know about the problem? We might start by drawing a figure, such as Figure 8A.2, that helps us organize the information in the problem.

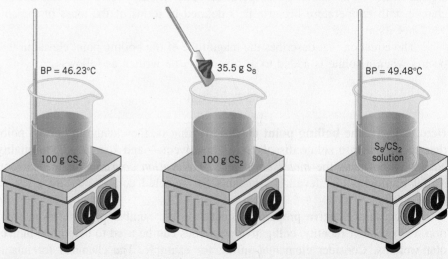

Fig. 8A.2 Diagram for Exercise 8A.3, where 35.5 g of S_8 is dissolved in 100 g of CS_2.

We know the boiling point of the solution, so we might start by looking up the boiling point of the pure solvent (Table 8A.1) in order to calculate the change in the boiling point that occurs when the sulfur is dissolved in CS_2.

$$\Delta T_{BP} = 49.48°C - 46.23°C = 3.25°C$$

We also know that the change in the boiling point is proportional to the molality of the solution.

$$\Delta T_{BP} = k_b m$$

Since we know the change in the boiling point (ΔT_{BP}) and we can look up the boiling point elevation constant for the solvent (k_b) in Table 8A.1, we might decide to calculate the molality of the solution at this point.

$$m = \frac{\Delta T_{BP}}{k_b} = \frac{3.25°C}{2.35°C/m} = 1.38 \ m$$

In the search for the solution to a problem, it is periodically useful to consider what we have achieved so far. At this point, we know the molality of the solution and the mass of the solvent used to prepare the solution. We can therefore calculate the number of moles of sulfur present in the carbon disulfide solution.

$$\frac{1.38 \text{ mol sulfur}}{1000 \text{ g } CS_2} \times 100 \text{ g } CS_2 = 0.138 \text{ mol sulfur}$$

We now know the number of moles of sulfur in the solution and the mass of the sulfur. We can therefore calculate the number of grams per mole of sulfur.

$$\frac{35.5 \text{ g}}{0.138 \text{ mol}} = 257 \text{ g/mol}$$

From the periodic table, we know that the atomic mass of sulfur is 32.07 g/mol of sulfur atoms. Dividing this into the mass of a mole of sulfur molecules tells us that a molecule of sulfur must contain eight sulfur atoms.

⋅ ⋅

8A.4 Freezing Point Depression

As we have seen, adding a solute to a solvent lowers the vapor pressure and raises the boiling point of the solution. The addition of a solute will also lower the freezing point or melting point of the solution. Examples of the use of freezing point depression include adding salt to ice to decrease the temperature of melting ice when making homemade ice cream or salting highways to prevent the formation of ice during the winter.

An equation, similar to the boiling point elevation equation, can be written to describe what happens to the freezing point (or melting point) of a solvent when a solute is added to the solvent.

$$\Delta T_{FP} = -k_f m$$

In this equation, ΔT_{FP} is the **freezing point depression,** that is, the change in freezing point that occurs when the solute dissolves in the solvent, and k_f is the *molal freezing point depression constant* for the solvent. A negative sign is used in the equation to indicate that the freezing point of the solvent decreases when a solute is added. Molal freezing point depression constants for selected compounds are given in Table 8A.1.

Exercise 8A.4

Determine the molecular weight of acetic acid if a solution containing 30.0 g of acetic acid in 1000 g of water freezes at $-0.93°C$. Do the results agree with the assumption that acetic acid has the formula CH_3CO_2H?

Solution

The freezing point depression for the solution is equal to the difference between the freezing point of the solution ($-0.93°C$) and the freezing point of pure water ($0°C$).

$$\Delta T_{FP} = -0.93°C - 0.0°C = -0.93°C$$

We now turn to the equation that defines the relationship between freezing point depression and the molality of the solution.

$$\Delta T_{FP} = -k_f m$$

Because we know the change in the freezing point and we can find the freezing point depression constant for water in Table 8A.1, we have enough information to calculate the molality of the solution.

$$m = \frac{\Delta T_{FP}}{k_f} = \frac{0.93°C}{1.853°C/m} = 0.50\,m$$

At this point, we might return to the statement of the problem to see if we are making any progress toward an answer. According to this calculation, there is 0.50 mol of acetic acid per kilogram of solvent. The problem stated that there was 30.0 g of acetic acid per 1000 g of solvent. Because we know the number of grams and the number of moles of acetic acid in the sample, we can calculate the molecular weight of acetic acid.

$$\frac{30.0\,g}{0.50\,mol} = 60\,g/mol$$

The results of the experiment are in good agreement with the molecular weight (60.05 g/mol) expected if the formula for acetic acid is CH_3CO_2H.

Exercise 8A.5

Explain why a 0.100 m solution of HCl dissolved in benzene has a freezing point depression of 0.512°C, whereas a 0.100 m solution of HCl in water has a freezing point depression of 0.369°C.

Solution

We can predict the change in the freezing point that should occur in the solutions from the freezing point depression constant for the solvent and the molality of the solution. For the 0.100 m solution of HCl in benzene, the results of the calculation agree with the experimental value.

$$\Delta T_{FP} = -k_f m = -(5.12°C/m)(0.100\ m) = -0.512°C$$

For water, however, the calculation gives a predicted value for the freezing point depression that is half of the observed value.

$$\Delta T_{FP} = -k_f m = -(1.853°C/m)(0.100\ m) = -0.185°C$$

To explain the results, it is important to remember that colligative properties depend on the relative number of solute particles in a solution, not their identity. If the acid dissociates (breaks up into its ions) to an appreciable extent, the solution will contain more solute particles than we might expect from its molality.

If HCl dissociates more or less completely in water, the total concentration of solute particles (H^+ and Cl^- ions) in the solution will be twice as large as the molality of the solution. The freezing point depression for the solution therefore will be twice as large as the change that would be observed if HCl did not dissociate.

$$HCl(g) \xrightarrow{H_2O} H^+(aq) + Cl^-(aq)$$

If we assume that 0.100 m HCl dissociates to form H^+ and Cl^- ions in water, the freezing point depression for the solution should be $-0.371°C$, which is slightly larger than what is observed experimentally.

$$\Delta T_{FP} = -k_f m = -(1.853°C/m)(2 \times 0.100\ m) = -0.371°C$$

This exercise suggests that HCl doesn't dissociate into ions when it dissolves in benzene, but dilute solutions of HCl dissociate more or less quantitatively in water.

• •

In 1884 Jacobus Henricus van't Hoff introduced another term into the freezing point depression and boiling point elevation expressions to explain the colligative properties of solutions of compounds that dissociate when they dissolve in water.

$$\Delta T_{FP} = -k_f(i)m$$

Substituting the experimental value for the freezing point depression of a 0.100 m HCl solution into the equation gives a value for the i term of 1.99. If HCl did not dissociate in water, i would be 1. If it dissociates completely, i would be 2. The experimental value of 1.99 suggests that about 99% of the HCl molecules dissociate in this solution.

Problems

Colligative Properties

A-1 Define *colligative properties*.

A-2 5.62 g of methanol, CH_3OH, is dissolved in 50.0 g of water. Calculate the mole fraction of water and methanol in the solution. How can you check your answer?

A-3 Which pairs of the following substances would be expected to mix to form ideal solutions?
  (a) hexane and heptanes
  (b) hexane and hexanoic acid
  (c) NaCl and water
  (d) methanol and water

Depression of the Partial Pressure of a Solvent

A-4. Predict what will happen to the rate at which water evaporates from an open flask when salt is dissolved in the water, and explain why the rate of evaporation changes. If you place a beaker of pure water (I) and a beaker of a saturated solution of sugar in water (II) in a sealed bell jar, the level of water in beaker I will slowly decrease, and the level of the sugar solution in beaker II will slowly increase. Explain why.

A-5. Explain why the vapor pressure of a liquid at a particular temperature is not a colligative property, but the change in the partial pressure of the liquid when a solute is added is a colligative property.

A-6. 2.56 g of the nonvolatile solute sucrose $C_{12}H_{22}O_{11}$ is added to 500 g of water at 25°C. What will be the partial pressure of the water over this solution?

A-7. If 500 g of pentane is mixed with 500 g of heptane at 20°C, what will be the total pressure above the solution? The vapor pressure of pentane is 420 mmHg, and that of heptane is 36.0 mmHg at 20°C.

A-8. The partial pressure of water above a solution of water and a nonvolatile solute at 25°C is 19.5 mmHg. What is the mole fraction of the solute?

Boiling Point Elevation

A-9. Explain how the decrease in the vapor pressure of a solvent that occurs when a solute is added to the solvent leads to an increase in the solvent's boiling point.

A-10. What changes in the vapor pressure of a solvent occur when a solute is added?

A-11. What is the boiling point of a solution of 10.0 g of P_4 in 25.0 g of carbon disulfide?

A-12. The boiling point elevation of a solution consisting of 2.50 g of a nonvolatile solute in 100.0 g of benzene is 0.686°C. What is the molecular weight of the unknown solute?

A-13. What is the boiling point of a solution containing 3.41 g of the nonvolatile solute I_2 in 50.0 g of carbon tetrachloride?

A-14. The boiling point elevation of a solution of 0.120 mol of a sugar in 50.0 g of water is 1.23°C. Calculate k_b from these data.

Freezing Point Depression

A.15 Predict the shape of a plot of the freezing point of a solution versus the molality of the solution.

A.16. Explain why salt is added to the ice that surrounds the container in which ice cream is made.

A.17. What is the approximate freezing point of a saturated solution of caffeine ($C_8H_{10}O_2N_4 \cdot H_2O$) in water if it takes 45.6 g of water to dissolve 1.00 g of caffeine?

A.18. A 0.100 m solution of sulfuric acid in water freezes at −0.371°C. Which of the following statements is consistent with this observation?

(a) H_2SO_4 does not dissociate in water.

(b) H_2SO_4 dissociates into H^+ and HSO_4^- ions in water.

(c) H_2SO_4 dissociates in water to form two H^+ ions and one SO_4^{2-} ion.

(d) H_2SO_4 dissociates in water to form $(H_2SO_4)_2$ molecules.

A.19. The "Tip of the Week" in a local newspaper suggested using a fertilizer such as ammonium nitrate or ammonium sulfate instead of salt to melt snow and ice on sidewalks, because salt can damage lawns. Which of the following compounds would give the largest freezing point depression when 100 g is dissolved in 1 kg of water?

(a) NaCl

(b) NH_4NO_3

(c) $(NH_4)_2 SO_4$

A.20. p-Dichlorobenzene (PDCB) is replacing naphthalene as the active ingredient in mothballs. Calculate the value of k_f for camphor if a 0.260 m solution of PDCB in camphor decreases the freezing point of camphor by 9.8°C.

A.21. Explain why many cities and states spread salt on icy highways.

A.22. We usually assume that salts such as KCl dissociate completely when they dissolve in water

$$KCl \xrightarrow{H_2O} K^+(aq) + Cl^-(aq)$$

Estimate the percentage of the KCl that actually dissociates in water if the freezing point of a 0.100 m solution of the salt in water is −0.345°C.

A.23. Calculate the freezing point of a 0.100 m solution of acetic acid in water if the CH_3CO_2H molecules are 1.33% ionized in the solution.

A.24. Compare the values of k_f and k_b for water, benzene, carbon tetrachloride, and camphor. Explain why measurements of molecular weight based on freezing point depression might be more accurate than those based on boiling point elevation.

A.25. If an aqueous solution containing a nonvolatile solute boils at 100.50°C, at what temperature does it freeze?

Chapter Nine

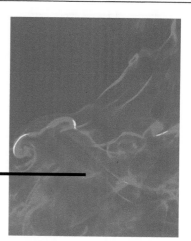

SOLIDS

9.1 Types of Solids

Solids can be divided into three categories on the basis of the way the particles pack together. **Crystalline solids** can be thought of as three-dimensional analogs of a brick wall. They have a regular structure in which particles pack in a repeating pattern, row upon row and layer upon layer, from one edge of the solid to the other. **Amorphous solids** (literally, "solids without form"), such as glass or the carbon black used in ink, have a random structure with little if any long-range order. Many solids, such as aluminum and steel, have a structure that falls between the two extremes of crystalline and amorphous solids. Such **polycrystalline solids** are aggregates of large numbers of small crystals or grains within which the structure is regular, but the crystals or grains are arranged in a random fashion.

Solids can also be classified on the basis of the forces that hold the particles together. This approach categorizes solids as either molecular, network covalent, ionic, or metallic. As shown in Chapter 5, the bonding between atoms in a substance can be predicted using a bond-type triangle based on the average electronegativies and the difference between the electronegativities of the atoms in the bond.

Ionic and covalent bonds are often imagined as if they were opposite ends of a two-dimensional model of bonding in which compounds that contain polar bonds such as H_2O fall somewhere between the two extremes.

ionic . . . polar . . . covalent

In reality, there are three kinds of bonds between adjacent atoms: ionic, covalent, and metallic, as shown in the bond-type triangle in Figure 9.1. Nonmetals combine to form elements and compounds that contain primarily covalent bonds, such as F_2, HCl, and CH_4. Metals combine with nonmetals to form salts, such as CsF and CaO, which are held together by predominantly ionic bonds. The primary force of attraction between atoms in metals (such as copper), alloys (such as brass

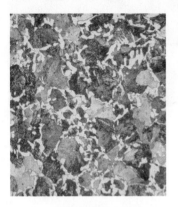

A polarized light micrograph of a tin section of brass showing the random distribution of microcrystals that gives rise to the polycrystallization structure of this alloy

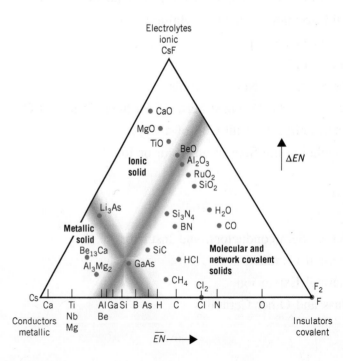

Fig. 9.1 Bond-type triangle for selected compounds. Such triangles may be used to classify solids as molecular, network covalent, ionic, or metallic.

and bronze), or intermolecular compounds (such as Li_3As) are metallic bonds. Distinguishing among solids that are primarily held together by covalent, ionic, or metal bonds is useful because it allows us to predict many of the physical properties of the solid.

9.2 Molecular and Network Covalent Solids

MOLECULAR SOLIDS

The iodine (I_2) that dissolves in alcohol to make the antiseptic known as tincture of iodine, the cane sugar ($C_{12}H_{22}O_{11}$) found in a sugar bowl, and the polyethylene used to make garbage bags all have one thing in common. They are all examples of compounds that are **molecular solids** at room temperature. Water and bromine are liquids that form molecular solids when cooled slightly; H_2O freezes at 0°C and Br_2 freezes at −7°C. Many substances that are gases at room temperature will form molecular solids when cooled far enough; F_2, at the extreme right of the bond-type triangle in Figure 9.1, freezes to form a molecular solid at −220°C.

Molecular solids contain both *intramolecular* bonds and *intermolecular* forces, as described in Chapter 8. The atoms within the individual molecules are held together by relatively strong intramolecular covalent bonds. Molecular solids are therefore found in the covalent region of a bond-type triangle. The molecules in these solids are held together by much weaker intermolecular forces. Because intermolecular forces are relatively weak, molecular solids are often soft substances with low melting points.

Dry ice, or solid carbon dioxide, is a perfect example of a molecular solid. The van der Waals forces holding the CO_2 molecules together are weak enough that at dry ice **sublimes** at a temperature of −78°C—it goes directly from the solid to the gas phase.

$$Strong\ intramolecular\ bond \searrow \quad {\overset{\delta+}{O}}{=}{C}{=}{\overset{\delta-}{O}}$$

$$\vdots \leftarrow Weak\ intermolecular\ forces$$

$$O{=}\underset{\delta+}{C}{=}\underset{\delta-}{O}$$

Changes in the strength of the van der Waals forces that hold molecular solids together can have important consequences for the properties of the solid. Polyethylene ($-CH_2-CH_2-$)$_n$ is a soft plastic used in sandwich bags and garbage bags that melts at relatively low temperatures. Replacing one of the hydrogens on every other carbon atom with a chlorine atom produces a plastic known as poly(vinyl chloride), or PVC, which is hard enough to be used to make the plastic pipes that are slowly replacing metal pipes for plumbing.

$$\begin{array}{cccccc} H & H & H & Cl & H & H \\ | & | & | & | & | & | \\ -C- & C- & C- & C- & C- & C- \\ | & | & | & | & | & | \\ H & Cl & H & H & H & Cl \end{array}$$

Poly(vinyl chloride), or PVC

Much of the strength of PVC can be attributed to the van der Waals force of attraction between the chlorine atoms on the chains of ($-CH_2-CHCl-$)$_n$ molecules that form the solid. A copolymer of poly(vinyl chloride) ($-CH_2-CHCl-$)$_n$ and poly(vinylidene chloride) ($-CH_2-CCl_2-$)$_n$ is sold under the trade name *Saran*. The same increase in the force of attraction between polymer chains that makes PVC harder than polyethylene gives a thin film of *Saran* a tendency to be attracted to itself. Saran wrap therefore clings to itself, whereas the polyethylene in sandwich bags does not.

The *halogens* (F_2, Cl_2, Br_2, and I_2) and *hydrogen halides* (e.g., HCl and HBr) can provide a basis for understanding the effect of differences in the strengths of intermolecular forces on the properties of a molecular solid. Consider chlorine, for example, which exists as diatomic Cl_2 molecules in the gas phase at room temperature. When the gas is cooled, the average kinetic energy of the Cl_2 molecules becomes smaller. As the motion of the molecules decreases, the force of attraction between the molecules becomes large enough to hold the molecules together, and the gas condenses to form a liquid. Further cooling transforms the liquid into a molecular solid, as predicted by the position of Cl_2 in the bond-type triangle in Figure 9.1.

The unit on which this molecular solid is built is the diatomic Cl_2 molecule. The covalent bond holding one chlorine atom to another in Cl_2 is relatively strong (243 kJ/mol). The intermolecular forces that hold one Cl_2 molecule to another are much smaller (18 kJ/mol).

Because the Cl_2 molecule has no dipole moment, the weak intermolecular forces that hold Cl_2 molecules together result solely from induced dipole–induced dipole or dispersion forces. These dispersion forces are nondirectional, and the molecules pack in the solid in the geometry that allows them to come as close together as possible.

Covalent molecules with a dipole moment, such as HCl and HBr, also form molecular solids, when cooled, in which the molecules pack as tightly as possible. Polar molecules, however, also have a directional component to the intermolecular forces, namely, dipole–dipole interactions. This force controls the orientation of the HCl and HBr molecules as they pack, so that the negative end of one dipole is oriented toward the positive end of the other.

The primary difference between the solid and liquid phases for covalent molecules is the regular pattern of packing in the solid versus the random structure in the liquid. Consider ice, for example. The individual H_2O molecules in the molecular solid are held together by a combination of dipole, dispersion, and hydrogen-bond forces.

Two parameters can be used to estimate the relative strength of intermolecular forces—the melting point of the compound and the enthalpy of fusion, ΔH_{fus}. The melting point, as we saw in Section 8.6, is the temperature at which the solid melts at atmospheric pressure. The enthalpy of fusion is the heat required to melt the substance [e.g., $H_2O(s) \longrightarrow H_2O(l)$] in units of kilojoules per mole.

The enthalpy of fusion of H_2O is relatively small, only 6.00 kJ/mol$_{rxn}$. This is a small fraction of the strength of the hydrogen bonds between water molecules because melting the solid breaks only some, not all, of the hydrogen bonds between the water molecules. To break all of the hydrogen bonds we have to boil water [$H_2O(l) \longrightarrow H_2O(g)$]; the enthalpy of vaporization of H_2O is 40.88 kJ/mol$_{rxn}$ at the boiling point.

Melting points and enthalpies of fusion are convenient measures of the relative strengths of the intermolecular interactions that hold molecular solids together. Table 9.1 gives the melting points and the enthalpies of fusion of the halogens. The only forces that hold the crystals together are dispersion forces. Because

Table 9.1
Melting Points and Enthalpies of Fusion of the Halogens

Halogen	Molecular Weight (g/mol)	MP (°C)	ΔH_{fus} (kJ/mol$_{rxn}$)
F_2	38	−219.6	0.51
Cl_2	71	−101	6.41
Br_2	160	−7.2	10.8
I_2	254	113.5	15.3

dispersion forces depend on the number of electrons in the atoms or molecules, as the size of the halogen atoms increases, the dispersion force interactions should become stronger. This is reflected in the increase in both the melting point and the enthalpy of fusion with increasing molecular weight of the halogens.

The effect of adding dipole–dipole and hydrogen-bond interactions to the intermolecular forces that hold molecules together can be seen in the data for dimethyl ether (CH_3OCH_3), methanol (CH_3OH), and water in Table 9.2. Dispersion and dipole forces exist in all three compounds. Two of the compounds, H_2O and CH_3OH, also form hydrogen bonds. As the number of hydrogen atoms that can form hydrogen bonds increases from zero in CH_3OCH_3 to one per molecule in CH_3OH and then two per molecule in H_2O, there is a significant increase in the melting point. The enthalpy of fusion is determined by the intermolecular attractive forces. The number of electrons, and hence the polarizability, of the compounds decreases from CH_3OCH_3 to CH_3OH to H_2O, and it might be expected that water would have the smallest enthalpy of fusion. The fact that water has the highest enthalpy of fusion shows the relative importance of hydrogen bonding.

NETWORK COVALENT SOLIDS

Network covalent solids include substances such as diamond and quartz whose crystals can be viewed as a single giant molecule made up of an almost endless number of covalent bonds. Network covalent solids are often very hard, and they are notoriously difficult to melt. Both molecular solids and network covalent solids are located in the covalent region of a bond-type triangle. A bond-type triangle, such as Figure 9.1, therefore can't be used to distinguish between these two types of solids.

Each carbon atom in diamond is covalently bound to four other carbon atoms oriented toward the corners of a tetrahedron, as shown in Figure 9.2. Because all of the bonds in the structure are equally strong, diamond is the hardest natural substance, and it melts at 3550°C. Quartz is a network covalent solid composed of SiO_2. This structure is consistent with predictions that would be made with the bond-type triangle in Figure 9.1, which indicates that SiO_2 is located in the covalent region of the triangle.

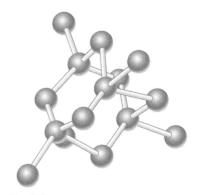

Fig. 9.2 The simplest repeating unit in the structure of a perfect diamond is a single molecule in which each carbon atom is tightly bound to four neighboring carbon atoms arranged toward the corners of a tetrahedron.

➤ **CHECKPOINT**

Describe the differences between molecular solids and network covalent solids on the atomic and macroscopic scales.

Table 9.2
Melting Points and Enthalpies of Fusion

Compound	Molecular Weight (g/mol)	MP (°C)	ΔH_{fus} (kJ/mol$_{rxn}$)
CH_3OCH_3	46	−141.5	4.94
CH_3OH	32	−97.9	3.18
H_2O	18	0	6.00

9.3 The Physical Properties of Molecular and Network Covalent Solids

The relationship between the physical properties of molecular and network covalent solids can be illustrated by the physical properties of the different elemental forms of carbon. Carbon occurs as a variety of allotropes. There are two common crystalline forms—diamond and graphite—and a number of amorphous (noncrystalline) forms, such as charcoal, coke, and carbon black.

The characteristic properties of diamond (MP = 3550°C, BP = 4827°C, and hardness) are a direct result of its structure, shown in Figure 9.2. A perfect diamond is a single giant molecule. The strength of the individual C—C bonds and their arrangement in space give rise to the remarkable properties of diamond.

In some ways, the properties of graphite are like those of diamond. But graphite also has properties that are very different from diamond. Graphite is much less dense than diamond. Whereas diamond is the hardest substance known, graphite is one of the softest. Diamond is an excellent **insulator;** graphite is such a good conductor of electricity that graphite electrodes are used in electrical cells.

The physical properties of graphite can be understood from the structure of the solid shown in Figure 9.3. Graphite consists of extended planes of carbon atoms in which each carbon forms strong covalent bonds to three other carbon atoms. (The strong bonds between carbon atoms *within each plane* explain the exceptionally high melting point and boiling point of graphite.) These planes of atoms, however, are held together by relatively weak van der Waals forces. Because the bonds *between planes of atoms* are weak, it is easy to deform the solid by allowing one plane of atoms to move relative to another. Graphite is therefore soft enough to be used in pencils and as a lubricant in motor oil.

The characteristic properties of graphite and diamond might lead us to expect that diamond would be more stable than graphite. This isn't what is observed experimentally. At 25°C and 1 atm pressure, graphite is slightly more stable than diamond. (The enthalpy of atom combination of graphite is -716.7 kJ/mol$_{rxn}$, and that of diamond is -714.8 kJ/mol$_{rxn}$). At very high temperatures and pressures, however, diamond becomes more stable than graphite. In 1955 General Electric developed a process to make industrial-grade diamonds by treating graphite with a metal catalyst at temperatures of 2000 to 3000 K and pressures above 125,000 atm. Roughly 40% of industrial-quality diamonds are now synthetic. Although gem-quality diamonds can be synthesized, until recently the costs were prohibitive.

Both diamond and graphite occur as regularly packed crystals. Other forms of carbon are *amorphous*—they lack a regular structure. *Charcoal* results from heating wood in the absence of oxygen. To make *carbon black,* natural gas is burned

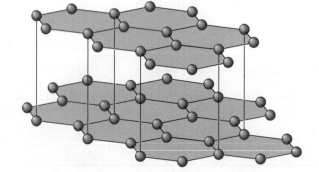

Fig. 9.3 Graphite consists of planes of atoms that contain very strong carbon–carbon bonds. These planes of atoms are held together by relatively weak van der Waals forces.

in a limited amount of air to give a thick, black smoke that contains extremely small particles of carbon that can be collected when the gas is cooled and passed through an electrostatic precipitator. *Coke* is a more regularly structured material, closer in structure to graphite than either charcoal or carbon black, which is made from coal.

In 1985 a third stable, crystalline form of carbon was made by vaporizing graphite with a laser. The product of this reaction is a molecule with the formula C_{60} that has a structure with the symmetry of a soccer ball. Because this structure resembles the geodesic dome invented by R. Buckminster Fuller, C_{60} was named *buckminsterfullerene,* or "buckyball" for short.

Some of the fascination of C_{60} can be understood by contrasting this form of elemental carbon with diamond and graphite. C_{60} is unique because it exists as distinct molecules, not extended arrays of atoms. Equally important, C_{60} can be obtained as a pure substance, whereas the surfaces of diamond and graphite are inevitably contaminated by hydrogen atoms that bind to the carbon atoms on the surface.

C_{60} is now known to be a member of a family of compounds known as the *fullerenes.* C_{60} may be the most important of the fullerenes because it is the most perfectly symmetric molecule possible, spinning in the solid at a rate of more than 100 million times per second. Because of their symmetry, C_{60} molecules pack as regularly as Ping-Pong balls. The resulting solid has unusual properties. Initially, it is as soft as graphite, but when compressed by 30%, it becomes harder than diamond. When this pressure is released, the solid springs back to its original volume. C_{60} therefore has the remarkable property that it bounces back when shot at a metal surface at high speeds.

C_{60} also has the remarkable ability to form compounds in which it is an insulator, a conductor, a semiconductor, or a superconductor. By itself, C_{60} is a semiconductor. When mixed with just enough potassium to give a compound with the empirical formula K_3C_{60}, it conducts electricity like a metal. When excess potassium is added, this solid becomes an insulator. When K_3C_{60} is cooled to 18 K, the result is a superconductor. The potential of fullerene chemistry for both practical materials and laboratory curiosities is large enough to explain why this molecule has been described as "exocharmic"—it exudes charm.

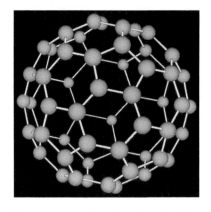

Computer graphic of a molecule of buckminsterfullerene (C_{60}) first synthesized in 1985.

9.4 Metallic Solids

Molecular, ionic, and network covalent solids all have one thing in common. With only rare exceptions, the electrons in the solids are *localized.* They either reside on one of the atoms or ions, or they are shared by a pair of atoms in a covalent bond.

As we saw in Section 5.12, metal atoms can't acquire enough electrons to fill their valence shells by sharing electrons with their immediate neighbors. Electrons in the valence shell are therefore shared among many atoms, instead of just two. In effect, the valence electrons are *delocalized* over many metal atoms (Figure 9.4). Because the electrons aren't tightly bound to individual atoms, they are free to move through the metal. As a result, metals are good conductors of heat and electricity. Electrons that enter the metal at one edge can displace other electrons in the metal to give rise to a net flow of electrons through the metal. As a result, metals are good conductors of both heat and electricity.

The bonds that hold metals together are very different from ionic and covalent bonds and are therefore placed in a category of their own: **metallic bonds.** Metallic bonding occurs when both the difference between the electronegativity of the atoms in a bond (ΔEN) and the average EN of these atoms are relatively small (i.e., the substance is placed in the lower-left corner of Figure 9.1). Because

Fig. 9.4 The force of attraction between the positively charged metal ions and the surrounding sea of electrons in a metal is called a metallic bond. The electrons are not bound to individual atoms but may move throughout the metal.

the atoms in a metal form bonds with many neighboring atoms, metals are usually solids in which each atom is surrounded by as many neighboring atoms as possible. Lithium, for example, crystallizes in a structure in which each atom touches eight nearest neighbors. The distance between the nuclei of adjacent atoms in lithium metal is 0.304 nm.

Lithium has three electrons: $1s^2\,2s^1$. There is a significant difference, however, between the ease with which an electron can be removed from the $1s$ and $2s$ orbitals on a lithium atom. According to the data in Table 3.4, it takes 0.52 MJ/mol to remove an electron from the $2s$ orbital on lithium but more than 10 times as much energy (6.26 MJ/mol) to remove one of the electrons from the $1s$ orbital. The core electrons in the $1s$ orbitals on a lithium atom are bound so tightly to the nucleus of the atom that they are unaffected by neighboring atoms. Thus there is only one valence electron per lithium atom that can be shared with neighboring atoms, the $2s^1$ electron.

In the gas phase, lithium can form a diatomic Li_2 molecule that is held together by the sharing of a pair of electrons by the two lithium nuclei.[1] The distance between the lithium atoms in the Li_2 molecule is 0.267 nm, which is considerably smaller than the distance between lithium atoms in the metal. This suggests that the covalent bond in the Li_2 molecule is significantly stronger than the metallic bonds in lithium metal. However, there are more bonds per lithium atom in the metal. As a result, the enthalpy of atomization [$Li(s) \longrightarrow Li(g)$], which can be calculated from the data in Table B.13 in the Appendix, for lithium metal is 159 kJ/mol$_{rxn}$, whereas the bond holding the two atoms together in an Li_2 molecule is only 57 kJ/mol$_{rxn}$.

Figure 4.2 and the AVEE values in Figure 3.31 show that the difference between the energies of the subshells within a given shell becomes smaller as one moves down a column of the periodic table. In other words, the s, p, and d subshells become closer in energy. This difference becomes even smaller when an atom forms a bond with another atom or group of atoms. The energies of the valence subshells of neighboring atoms in a metal are therefore very similar. This allows electrons to move easily between all available subshells from atom to atom. The electrons are therefore said to be *delocalized*. The electrons are no longer confined to the space between the nuclei of neighboring atoms. When this happens, bonding becomes nondirectional, and the atoms pack together as tightly as possible. Thus a substance held together by metallic bonds can be considered to be made up of metallic cations immersed in a sea of delocalized electrons.

As one goes down a column of the periodic table, the size of the atoms increases. This makes it easier to remove outer-shell electrons, and the energy gaps between subshells become smaller. These two factors explain why the elements become more metallic as we move toward the bottom-left corner of the periodic table.

> **CHECKPOINT**
>
> The AVEE value or electronegativity of an atom is made up of two important contributions. What are they, and why are they important for understanding metallic behavior?

9.5　Physical Properties That Result from the Structure of Metals

Metals have certain characteristic physical properties.

- They have a metallic shine, or luster.
- They are usually solids at room temperature.

---

[1]Although dilithium molecules can exist in the gas phase, the famous "dilithium crystals" that fueled the Starship *Enterprise* existed only in the imagination of Gene Roddenberry.

- They are *malleable* (from the Latin word for "hammer"): They can be hammered, pounded, or pressed into different shapes.
- They are *ductile:* They can be drawn into thin sheets or wires without breaking.
- They conduct heat and electricity.

The structures of metals can be used to explain their characteristic properties.

A chrome-plated surface has a characteristic metallic luster because the metal reflects (literally, "throws back") a significant fraction of the light that hits its surface. Silver is better than any other metal at reflecting light: Roughly 88% of the light that hits the surface of a silver mirror is reflected.

Why are metals solid? Some nonmetals, such as hydrogen and oxygen, are gases at room temperature because these elements form molecules that are held together by weak intermolecular forces between adjacent molecules. Metal atoms are held closely together by strong metallic bonds in a three-dimensional network, and most metals (with the exception of mercury) are solids at room temperature.

Metals are malleable and ductile because they pack in structures that contain planes of atoms. In theory, changing the shape of the metal is simply a matter of applying a force that makes the atoms in one of these planes slide past the atoms in an adjacent plane, as shown in Figure 9.5. In practice, it is easier to do this when the metal is hot. The layers of atomic cores in metals can slip easily over one another because there are no directional forces tending to keep them in locked positions, and thus metals are both ductile and malleable.

Why are metals good conductors of heat and electricity? As we have already seen, the delocalization of valence electrons in a metal allows the solid to conduct an electric current. Metals conduct heat by the movement of electrons through the metal. Because the electrons are relatively free to move from atom to atom in the metal, they can quickly transport heat throughout the metal.

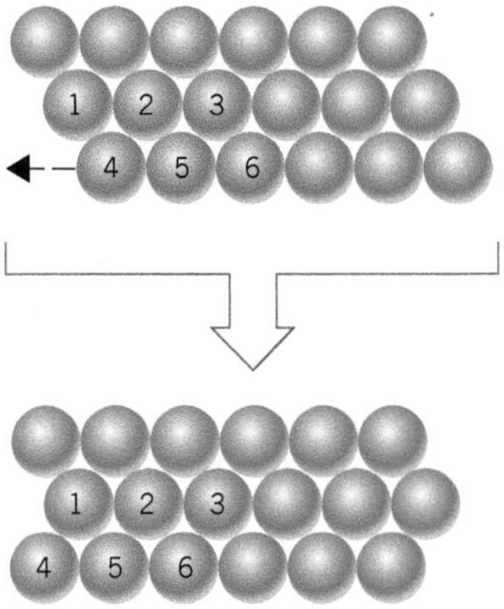

Fig. 9.5 Metals are malleable and ductile because planes of atoms can slip past one another to reach equivalent positions.

9.6 The Structure of Metals

We can describe the structure of pure metals by assuming that the atoms of the metals are identical perfect spheres that pack in regular patterns. The same model can be used to describe the structure of the solid noble gases (He, Ne, Ar, Kr, Xe) at low temperatures. These substances all crystallize in one of four basic structures, known as simple cubic (SC), body-centered cubic (BCC), hexagonal closest packed (HCP), and cubic closest packed (CCP).

Solids are very difficult to compress because the amount of space between particles in a solid is at a minimum. As a rule, we can conclude that the most probable structure for a solid is the structure that makes the most effective use of space. To illustrate the principle that atoms in a solid pack as tightly as possible, let's try to imagine the best way to pack identical spheres, such as Ping-Pong balls, into an empty box.

One approach involves carefully packing the Ping-Pong balls to form a square-packed plane of spheres, as shown in Figure 9.6.

A second plane of spheres can be stacked directly on top of the first. The result is a regular structure in which the simplest repeating unit is a cube of eight spheres, as shown in Figure 9.7. This structure is called **simple cubic packing.** Each sphere in the structure touches four identical spheres in the same plane. It also touches one sphere in the plane above and one in the plane below. Each sphere is therefore said to have a **coordination number** of 6. If the spheres represent

Fig. 9.6 A square-packed plane of spheres.

Fig. 9.7 A simple cubic packing of spheres.

atoms, each atom in the structure can form bonds to its six nearest neighbors arranged toward the corners of an octahedron.

One way to decide whether the simple cubic structure is an efficient way of packing spheres is to ask: What happens when we shake the box? Do the Ping-Pong balls stay in the same positions, or do they settle into a different structure? It is fairly easy to show that a simple cubic structure isn't an efficient way of using space. Only 52% of the available space is actually occupied by the spheres in a simple cubic structure. The rest is empty space. Because the structure is inefficient, only one element—polonium—crystallizes in a simple cubic structure.

This raises an interesting question: How can we use space more efficiently? Another approach starts by doing something that seems irrational, at first: separating the spheres to form a square-packed plane in which the spheres do not quite touch each other, as shown in Figure 9.8.

The spheres in a second plane are now packed above the holes in the first plane, as shown in Figure 9.9. Spheres in the third plane pack above holes in the second plane. Spheres in the fourth plane pack above holes in the third plane, and so on. The result is a structure in which the odd-numbered planes of atoms are identical and the even-numbered planes are identical. The *ABABABAB...* repeating structure of square-packed planes is known as **body-centered cubic packing.**

This structure is called *body-centered cubic* because each sphere touches four spheres in the plane above and four more in the plane below, arranged toward the corners of a cube. Thus the repeating unit in the structure is a cube of eight spheres with a ninth identical sphere in the center of the body—in other words, a body-centered cube, as shown in Figure 9.10. The coordination number in this structure is 8.

Although we generated this structure by first separating the spheres in each plane so that they were not quite touching, it can be shown that body-centered cubic packing is a more efficient way of using space than simple cubic packing

Fig. 9.8 A square-packed plane in which the spheres do not quite touch.

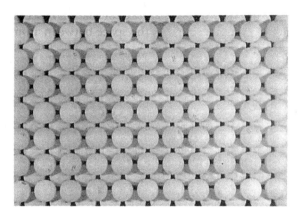

Fig. 9.9 The spheres in the second plane of a body-centered cubic structure pack above the holes in the plane shown in Figure 9.8.

because 68% of the space in the structure is now filled. Body-centered cubic packing is an important structure for metals. All of the metals in Group IA (Li, Na, K, Rb, Cs, and Fr), barium in Group IIA, and a number of the early transition metals (such as V, Cr, Mo, W, and Fe) pack in a body-centered cubic structure at room temperature.

Two structures pack spheres so efficiently they are called **closest-packed structures.** Both start by packing the spheres in planes in which each sphere touches six others oriented toward the corners of a hexagon, as shown in Figure 9.11. A second plane is then formed by packing spheres above the triangular holes in the first plane, as shown in Figure 9.12.

What about the next plane of spheres? The spheres in the third plane could pack directly *above the spheres* in the first plane to form an *ABABABAB . . .* repeating structure. Because such a structure is composed of alternating planes of hexagonal-closest-packed spheres, it is called a **hexagonal-closest-packed** structure. Each sphere touches six spheres in the same plane, three spheres in the plane above, and three spheres in the plane below, as shown in Figure 9.13. Thus the coordination number in a hexagonal-closest-packed structure is 12, and 74% of the space in a hexagonal-closest-packed structure is filled. No more efficient way of packing spheres is known, and the hexagonal-closest-packed structure is important for such metals as Be, Co, Mg, and Zn, as well as the rare gas He at low temperatures.

There is another way of stacking hexagonal-closest-packed planes of spheres. The atoms in the third plane can be packed *above the holes* in the first plane that weren't used to form the second plane. The fourth hexagonal-closest-packed plane of atoms then packs directly above the first. The net result is an *ABCABCABC. . .* structure, which is called **cubic closest packed.** Each sphere in the structure touches six others in the same plane, three in the plane above, and three in the plane below, as shown in Figure 9.14. Thus the coordination number is still 12.

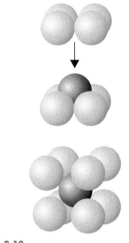

Fig. 9.10 Body-centered cubic structure. All spheres represent identical atoms.

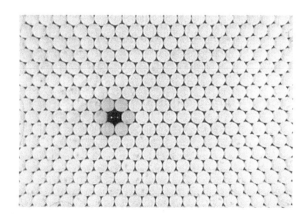

Fig. 9.11 A closest-packed plane in which each sphere touches six others oriented toward the corners of a hexagon.

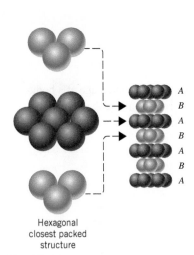

Hexagonal
closest packed
structure

Fig. 9.13 Each atom in a hexagonal-closest-packed structure touches six atoms in the same plane, three in the plane above, and three in the plane below. The result is an *ABABAB...* repeating pattern of closest-packed planes. All spheres represent identical atoms.

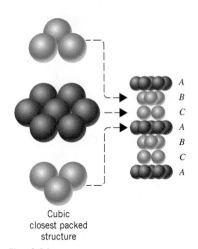

Cubic
closest packed
structure

Fig. 9.14 Each atom in a cubic-closest-packed structure also touches six atoms in the same plane, three in the plane above, and three in the plane below. But the atoms in the top plane are rotated by 180° relative to the bottom plane. The planes of atoms therefore form an *ABCABCABC...* repeating pattern. All spheres represent identical atoms.

Fig. 9.12 Atoms in the second plane of closest-packed structures pack above the triangular holes in the first plane shown in Figure 9.11.

The difference between hexagonal- and cubic-closest-packed structures can be understood by comparing Figures 9.13 and 9.14. In the hexagonal-closest-packed structure, the atoms in the first and third planes lie directly above each other. In the cubic-closest-packed structure, the atoms in those planes are oriented in different directions.

The cubic-closest-packed structure is just as efficient as the hexagonal-closest-packed structure. (Both use 74% of the available space.) Many metals, including Ag, Al, Au, Ca, Cu, Ni, Pb, and Pt, crystallize in a cubic-closest-packed structure. All the rare gases except helium behave in the same manner when cooled to temperatures low enough to allow solidification.

Hexagonal closest-packed planes of spheres can be seen in other situations where the goal is to hold as many identical spheres as close together as possible. If one takes a warm bottle of diet soda and very slowly cracks open the cap so that the bubbles that form on the surface are all of essentially the same size, the bubbles often form a structure that is remarkably similar to a closest-packed plane of atoms. There are often imperfections in this structure that resemble the imperfections that occur in samples of metals that are not pure.

9.7 Coordination Numbers and the Structures of Metals

The coordination numbers of the four structures described in the preceding section are summarized in Table 9.3.

Most metals pack in hexagonal- or cubic-closest-packed structures. Not only do those structures use space as efficiently as possible, they also have the largest

Table 9.3
Coordination Numbers for Common Crystal Structures

Structure	Coordination Number	Stacking Pattern
Simple cubic	6	*AAAAAAAA...*
Body-centered cubic	8	*ABABABAB...*
Hexagonal closest packed	12	*ABABABAB...*
Cubic closest packed	12	*ABCABCABC...*

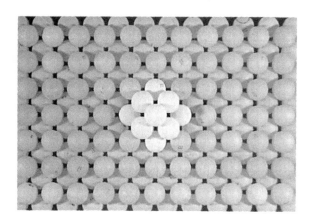

Fig. 9.15 Each atom in a body-centered cubic structure touches four atoms in the plane above and four in the plane below. In addition, each atom *almost* touches six more atoms.

possible coordination numbers, which allows each metal atom to form bonds to the largest number of neighboring metal atoms.

It is less obvious why one-third of the metals pack in a body-centered cubic structure, in which the coordination number is only 8. The reason metals sometimes pack in a body-centered cubic structure can be understood by referring to Figure 9.15. The coordination number for body-centered cubic structures in Table 9.3 only counts the neighboring atoms that actually touch a given atom in the structure. Figure 9.15 shows that each atom *almost touches* four more atoms in the same plane, a fifth atom two planes above, and a sixth atom two planes below. The distance from each atom to the nuclei of these "nearby" atoms is only 15% larger than the distance to the nuclei of the atoms that it actually touches. Each atom in a body-centered cubic structure therefore interacts with 14 other atoms—eight strong interactions to the atoms that it touches and six weaker interactions to the atoms it almost touches.

This makes it easier to understand why a metal might prefer the body-centered cubic structure to the hexagonal- or cubic-closest-packed structure. Each metal atom in the closest-packed structures interacts with 12 neighboring atoms. In the body-centered cubic structure, each atom interacts with a total of 14 neighboring atoms, although six of these interactions are somewhat weaker.

9.8 Unit Cells: The Simplest Repeating Unit in a Crystal

So far, our description of solids has focused on the way the particles pack to fill space. Another way of describing the structures of solids assumes that crystals are three-dimensional analogs of a piece of wallpaper (see Section 5.11). Wallpaper has a regular repeating design that extends from one edge to the other. Crystals have a similar repeating design, but in this case the design extends in three dimensions from one edge of the solid to the other.

We can unambiguously describe a piece of wallpaper by specifying the size, shape, and contents of the simplest repeating unit in the design. In a similar manner, we can describe a three-dimensional crystal by specifying the size, shape, and contents of the simplest repeating unit and the way the repeating units stack to form the crystal. The simplest repeating unit in a crystal is called a **unit cell,** which is defined in terms of **lattice points**—the points in space about which the particles are free to vibrate in a crystal.

This section focuses on the three unit cells shown in Figure 9.16: simple cubic, body-centered cubic, and face-centered cubic. These unit cells are important for two reasons. First, many metals, ionic solids, and intermetallic compounds

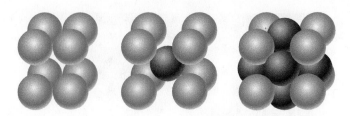

Fig. 9.16 Models of simple cubic (left), body-centered cubic (center), and face-centered cubic unit cells (right).

crystallize in cubic unit cells. Second, because these unit cells have identical edge lengths for a given cubic cell and the cell angles are all 90°, calculations based on these structures are somewhat easier to do than with more complex unit cells.

As might be expected, the **simple cubic unit cell** is the simplest repeating unit in a simple cubic structure. Each corner of the unit cell is defined by a lattice point at which an identical particle can be found. By convention, the edge of a unit cell always connects equivalent points. Each of the eight corners of the unit cell therefore must contain an identical particle. Other particles can be present on the edges or faces of the unit cell or within the body of the unit cell. But the minimum that must be present for the unit cell to be classified as simple cubic is eight equivalent particles on the eight corners.

The **body-centered cubic unit cell** is the simplest repeating unit in a body-centered cubic structure. Once again, there are eight identical particles on the eight corners of the unit cell. In this case, however, there is a ninth identical particle in the center of the body of the unit cell. It is important to remember that the particle at the center of the body in this unit cell must be the same as the particles that define the eight corners of the cube.

The **face-centered cubic unit cell** also starts with identical particles on the eight corners of the cube. But the structure also contains the same particles in the centers of the six faces of the unit cell, for a total of 14 identical lattice points. The face-centered cubic unit cell is the simplest repeating unit in a cubic-closest-packed structure. In fact, the presence of face-centered cubic unit cells in the structure explains why the structure is known as *cubic* closest packed.

The lattice points in a unit cell can be described in terms of a three-dimensional graph. For the sake of argument, let's define the *a* axis as the vertical axis of our coordinate system, as shown in Figure 9.17. The *b* axis will describe movement across the front of the unit cell, and the *c* axis will represent movement toward the back of the unit cell.

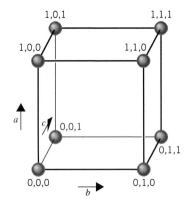

Fig. 9.17 The coordinates of the eight lattice points that define a cubic unit cell.

The bottom-left corner of the unit cell is now the origin (0,0,0) of our coordinate system. The coordinates 1,0,0 indicate a lattice point one cell-edge length away from the origin along the *a* axis. Similarly, 0,1,0 and 0,0,1 represent lattice points that are displaced by one cell-edge length from the origin along the *b* and *c* axes, respectively.

9.9 Solid Solutions and Intermetallic Compounds

When asked to give an example of a solution, chemists usually think in terms of solutions of a gas such as HCl or a solid such as NaCl dissolved in a liquid such as water. It is also possible, however, to prepare solutions in which one solid dissolves in another. The most important examples of these solid solutions are copper dissolved in aluminum and carbon dissolved in iron.

The solubility of one solid in another depends on temperature. At room temperature, copper doesn't dissolve in aluminum. But at 550°C, aluminum can form solutions that contain up to 5.6% copper by weight. Aluminum metal that has been saturated with copper at 550°C will try to reject the copper atoms as it cools to room temperature. When this happens, the copper atoms combine with aluminum atoms as the solution cools to form an **intermetallic compound** with the formula $CuAl_2$.

Intermetallic compounds such as $CuAl_2$ are the key to a process known as **precipitation hardening.** Aluminum metal packs in a cubic-closest-packed structure in which one plane of atoms can slip past another. As a result, pure aluminum metal is too weak to be used as a structural metal in cars or airplanes. Precipitation hardening produces alloys that are five to six times as strong as aluminum and make excellent structural metals.

The first step in precipitation hardening of aluminum involves heating the metal to 550°C. Copper is then added to form a solution that is quenched with cold water. The solution cools so fast that the copper atoms can't come together to form microcrystals of copper metal. Over a period of time, copper atoms move through the quenched solution to form microcrystals of the $CuAl_2$ intermetallic compound that are so small they are hard to see with a microscope.

These $CuAl_2$ particles are both hard and strong—so hard they inhibit the flow of the aluminum metal that surrounds them. These microcrystals of $CuAl_2$ strengthen aluminum metal by interfering with the way planes of atoms slip past each other. The result is a metal that is both harder and stronger than pure aluminum.

Copper dissolved in aluminum at high temperature is an example of a **substitution solution,** in which copper atoms pack in the positions normally occupied by aluminum atoms. There is another way in which a solid solution can be made. Atoms of one element can pack in the holes, or *interstices,* between atoms of the host element because even the most efficient crystal structures use only 74% of the available space in the crystal. The result is an **interstitial solution.**

Steel at high temperatures is a good example of an interstitial solution. It is formed by dissolving carbon in iron. At very high temperatures, iron packs in a cubic-closest-packed structure that leaves just enough space to allow carbon atoms to fit in the holes between the iron atoms. Below 910°C, however, iron metal packs in a body-centered cubic structure, in which the holes are too small to hold carbon atoms.

This has important consequences for the properties of steel. At temperatures above 910°C, carbon readily dissolves in iron to form a solid solution that contains as much as 1% carbon by weight. This material is both malleable and ductile, and it can be rolled into thin sheets or hammered into various shapes. When this solution cools below 910°C, the iron changes to a body-centered cubic structure and the carbon atoms are rejected from the metal. If the solution is allowed to cool gradually, the carbon atoms migrate through the metal to form a compound with the formula Fe_3C, which precipitates from the solution. These Fe_3C crystals serve the same role in steel that the $CuAl_2$ crystals play in aluminum—they inhibit the flow of the planes of metal atoms and thereby make steel significantly stronger than iron metal.

➤ CHECKPOINT

It is often difficult to replace the U-trap beneath a sink if it was installed by an amateur instead of a professional plumber. Use the fact that atoms can slowly diffuse through a metal to explain what happens when one of these U-traps "freezes."

9.10 Semimetals

The labels *electrolytes, conductors,* and *insulators* that appear in the three corners of the bond-type triangle in Figure 9.1 are observable properties that are closely related to the primary bond type in an element or a compound. Metals are conductors because they conduct electricity in both the solid and liquid states. Ionic substances are electrolytes because the ions they contain that are released into solution when they dissolve in water can conduct an electric current. Covalent compounds are poor conductors of electricity and are often insulators because the electrons are localized. They are held tightly between the nuclei that form the covalent bond. Compounds or materials that lie relatively far from one of the vertices of the bond-type triangle can exhibit properties that seem to be a mixture of these categories.

In general, elements that have small AVEE values have small energy gaps in their valence subshells and tend to form delocalized (metallic) bonds. Elements that have large AVEE values have a large energy separation of their valence subshells, and these elements tend to form covalent bonds. Semimetals fall between the extremes of delocalized (metallic) and localized (covalent or ionic) bonding.

Most periodic tables contain a line that separates the elements that are more likely to be metals from those that are more likely to be nonmetals. The elements along the dividing line are called *semimetals* or *metalloids* and have properties between those of metals and nonmetals. The semimetals can be found on the bond-type triangle in Figure 9.1 in the region between Al and As. Because these elements lie between those that conduct electricity, on one hand, and those that are insulators, on the other, they often form **semiconductors.**

9.11 Ionic Solids

Ionic solids are salts, such as NaCl, that form an extended three-dimensional network of ions held together by the strong force of attraction between ions of opposite charge, as shown in Figure 9.18.

The structures of six ionic solids are shown in Figure 9.19. Other compounds are often described in terms of one of these structures. The magnetic oxide tapes that were once used to record music, for example, contain CrO_2, which has a structure similar to that of TiO_2 shown in Figure 9.19.

Because the force of attraction depends inversely on the square of the distance between the positive and negative charges, the strength of an ionic bond depends inversely on the size of the ions that form the solid.

$$F = \frac{q_1 \times q_2}{r^2}$$

When the ions are large, the bond is relatively weak. But the ionic bond is still strong enough to ensure that salts have relatively high melting points and boiling points. Sodium chloride, for example, melts at 801°C and boils at 1413°C. These solids are often brittle, however, easily breaking into smaller parts when hit with a hammer.

Solids retain their shape, are difficult to compress, and are denser than liquids and gases. These characteristic properties suggest that solids contain particles that are packed as tightly as possible. Ionic compounds form solids in which the force of attraction between the ions of opposite charge is maximized by keeping the ions as close together as possible. Ionic solids are located in the ionic region of a bond-type triangle as shown in Figure 9.1.

Some understanding of the strength of the bonds in an ionic compound can be obtained by considering the enthalpy of the process in which the struc-

Fig. 9.18 Ionic compounds are made up of a three-dimensional network of positive and negative ions.

➤ **CHECKPOINT**

Describe the differences on the atomic and macroscopic scales between molecular solids and ionic solids.

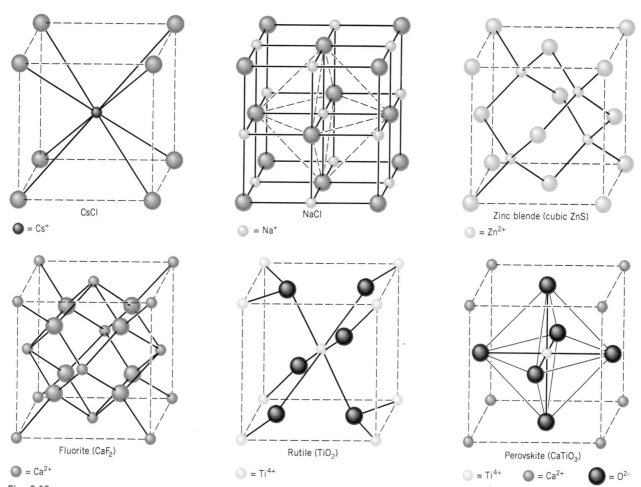

Fig. 9.19 The unit cells of CsCl, NaCl (rock salt), ZnS (zinc blende), CaF$_2$ (fluorite), TiO$_2$ (rutile), and CaTiO$_3$ (perovskite).

ture of an ionic solid is completely disrupted to form isolated ions in the gas phase.

$$NaCl(s) \longrightarrow Na^+(g) + Cl^-(g)$$

The energy required to break an ionic compound into isolated ions in the gas phase is known as the **lattice energy.** The lattice energy for NaCl is 787 kJ/mol$_{rxn}$.

The lattice energies of compounds formed by combining one of the alkali metals with a halogen are given in Table 9.4. The lattice energies of the compounds decrease as the size of the ions increases because the distance between the centers of the positive and negative charges on the ions increases. It therefore takes less energy to break one of the solids apart as the ions become larger, or less energy is given off when one of the compounds is formed from the corresponding positive and negative ions in the gas phase.

The lattice energies for ionic compounds formed when one of the alkaline earth metals combines with oxygen to form an oxide (MgO, CaO) show a trend similar to the halides given in Table 9.4. The lattice energy for MgO ($\Delta H^\circ_{LE} = 3791$ kJ/mol$_{rxn}$), however, is about five times as large as the lattice energy for NaCl. Part of the difference can be explained by noting that MgO contains ions with charges of +2 and −2. Thus the product of the charge on the positive and negative ions is four times larger in MgO than it is in NaCl,

Table 9.4

Lattice Energies of Alkali Metal Halides (kJ/mol$_{rxn}$)

	F$^-$	Cl$^-$	B$^-$	I$^-$
Li$^+$	1046	861	818	762
Na$^+$	923	787	747	704
K$^+$	821	718	682	649
Rb$^+$	785	689	660	630
Cs$^+$	740	659	631	604

➤ **CHECKPOINT**

Which has the larger lattice energy, $MgCl_2$ or MgF_2?

➤ **CHECKPOINT**

Describe the differences between metallic solids and ionic solids on the atomic and macroscopic scales.

➤ **CHECKPOINT**

Iron metal and cesium chloride have similar structures. The simplest repeating unit in iron is a cube of eight iron atoms with a ninth iron atom in the center of the body of the cube. The simplest repeating unit in CsCl is a cube of Cl^- ions with a Cs^+ ion in the center of the body. Explain why one of the structures is classified as a body-centered cubic unit cell and the other as a simple cubic unit cell.

which contains +1 and −1 ions. The remainder of the difference results from the fact that the Mg^{2+} ion is smaller than the Na^+ ion and the O^{2-} ion is smaller than the Cl^- ion.

Exercise 9.1

Arrange the following ionic compounds in order of increasing lattice energy: KF, CaF_2, $CaCl_2$, CaO.

Solution

The lattice energy for ionic compounds is directly related to the product of the charges on the ions and inversely related to the distance between the ions. Often the distance between ions can be estimated by using the principles established in Section 3.22. In some cases it may be necessary to refer to data in Table B.4 in Appendix B.

In order to answer this question, we need to examine these compounds one pair at a time.

- KF has a smaller product of charges on its ions than either CaF_2 or $CaCl_2$, and the distance between its ions is similar to CaF_2 and $CaCl_2$. KF therefore has a smaller lattice energy than either CaF_2 or $CaCl_2$.

- CaF_2 and $CaCl_2$ have the same charges on their ions, but the Cl^- ion is larger than the F^- ion, which means the distance between the ions will be larger in $CaCl_2$ than CaF_2. As a result, the lattice energy in $CaCl_2$ will be smaller than in CaF_2.

- CaO has ions of charge +2 and −2 and about the same distance between ions as CaF_2. CaO therefore has a much larger lattice energy than either CaF_2 or $CaCl_2$.

The compounds arranged in order of increasing lattice energy are: KF < $CaCl_2$ < CaF_2 < CaO.

Thinking about the unit cell as a three-dimensional graph allows us to describe the structure of a crystal with a remarkably small amount of information. We can specify the structure of cesium chloride, for example, with only four pieces of information.

- CsCl crystallizes in a cubic unit cell.
- The length of the unit cell edge is 0.4123 nm.
- There is a Cl^- ion at the coordinates 0,0,0.
- There is a Cs^+ ion at the coordinates ½, ½, ½.

Because the cell edge must connect equivalent lattice points, the presence of a Cl^- ion at one corner of the unit cell (0,0,0) implies the presence of a Cl^- ion at every corner of the cell. The coordinates ½,½,½ describe a lattice point at the center of the cell. CsCl is therefore a simple cubic unit cell of Cl^- ions with a Cl^+ in the center of the body of the cell, as shown in Figure 9.19.

The structure and physical properties of the solids discussed in this chapter are summarized in Table 9.5.

Table 9.5

Classifications and Properties of Solids

Classification of the Solid	Primary Type of Bonding	Force Holding Solid Together	Physical Properties	Examples
Molecular	Covalent	Intermolecular forces	Low melting points, electrical insulators	H_2O, Cl_2, HCl, CO_2
Network covalent	Covalent	Covalent bonds	Very high melting points, electrical insulators, very hard	Diamond (C), quartz (SiO_2)
Ionic	Ionic	Ionic bonds	High melting points, electrical conductor in the molten and aqueous state	NaCl, CaO, LiF, $BaCl_2$
Metallic	Metallic	Metallic bonds	Range of melting points, conductors of heat and electricity, lustrous, malleable, ductile	Na, Fe, Al, $CuAl_2$, $BaZn_5$

9.12 The Search for New Materials

Materials science is an interdisciplinary field that involves chemists, physics, and engineers in the study of the properties of new materials. Materials chemistry is concerned with the relationship between structure, properties and performance of materials. Examples of these materials include piezoelectric crystals that deform when an electric field is applied, which are used in loudspeakers, pressure gauges, and buzzers. Floppy disks and hard drives for computers use other innovative materials that have come from materials science research. Catalytic converters, sunglasses, and superconductivity all have resulted from an understanding of the structure of solids on the atomic scale.

The properties of solids depend on several factors. One of the first considerations, however, is whether the material is likely to be an ionic, metallic, or covalent solid. There are also intermediate possibilities, such as semimetals, semiconductors, and semielectrolytes.

The bond-type triangle shown in Figure 9.1 can help us understand the remarkable differences in the properties of solids that might seem similar from the positions of their elements in the periodic table. Consider BeO and CO, for example. BeO is also known as beryllia; it melts at 2250°C, is very hard, and is a ceramic. Carbon monoxide falls in the covalent region of the triangle and forms a molecular solid at temperatures below -200°C. The electronegativity difference between Be and O is larger than that for C and O, and the average electronegativity is smaller for BeO than for CO. These two conditions place BeO in a very different region of the triangle.

Compounds such as $CuAl_2$ are located in the metallic region. As we move away from the metallic region, atom combinations produce compounds that become increasingly insulating. Ceramic materials such as SiC, BN, and BeO are found near or along the interface between ionic and covalent areas. Semiconductors are located along the semimetal–covalent boundary, reflecting the gradual change from conducting toward insulating materials as one moves to the right across the bond-type triangle.

When an electric current is passed through any material at room temperature, some of the energy of the electrons is dissipated in the form of heat. In metals, resistance to an electric current becomes smaller as the metal is cooled. In 1911, Heike Kamerlingh Onnes found that when mercury is cooled to

> ➤ **CHECKPOINT**
>
> What types of solids are B_4C and MoC? Suggest an application for each of the compounds

temperatures below 4.1 K, its resistance falls to zero. Above that temperature, mercury conducts electricity. Below this temperature, it becomes a **superconductor.** By 1913, he had found that tin and lead also become superconductors at temperatures below 4 K.

Kamerlingh Onnes recognized the potential of superconductivity for constructing magnets with unusually strong magnetic fields. Standard electromagnets are made by winding a coil of insulated copper wire around an iron alloy core. As the current passes through the copper wire, a magnetic field is created. The field induces an alignment of electrons in the iron alloy core, which in turn produces a magnetic field in the core that is up to 1000 times larger than the field produced by the copper wire. There is an upper limit to the strength of the field that iron alloy magnets can produce, however. The magnets "saturate" at a magnetic field above 2 tesla, which is 40,000 times larger than the Earth's magnetic field.

Kamerlingh Onnes believed that superconducting magnets could be produced that would achieve much higher fields. Unfortunately, none of the superconducting metals he studied were able to carry enough electric current. It took 50 years before alloys of niobium and tantalum were discovered that could carry the current needed to produce high-field magnets. Commercial superconducting magnets made from niobium–tantalum alloys became available toward the end of the 1960s. The primary disadvantage of these magnets was the fact that the alloy has to be cooled to the temperature of liquid helium (4.2 K) before it becomes a superconductor. Chemists routinely use superconducting magnets in instruments known as nuclear magnetic resonance spectrometers. They are used by physicists in high-energy particle accelerators such as the Large Hadron Collider recently completed on the French-Swiss border.

The cost of maintaining a superconducting magnet could be decreased by as much as a factor of 1000 if it could operate at liquid nitrogen temperatures (77 K). The search for "high-temperature" superconductors is an important object lesson in the proper role of theory and experiment. At first glance, we might expect ReO_3 and RuO_2 to be insulators, like other metal oxides. In practice, those oxides conduct electricity the way a metal would. In 1964, it was found that other metal oxides, such as NbO and TiO, conduct electricity so well that they become superconductors when cooled to extremely low temperatures (1 K). This is also unexpected, based on their position in a bond-type triangle.

A major step in the evolution of high-temperature superconductors occurred in 1986, when Alex Müller and Georg Bednorz at the IBM Research Laboratory in Zurich discovered that certain ceramic materials that contained lanthanum, barium, copper, and oxygen became superconductors when cooled to temperatures below 35 K. Their results contained two surprises. First, ceramics—such as the plates on which we eat dinner—are usually insulators, not conductors. Second, the transition temperature for superconductivity in these new materials was higher than that for any known metal or metal alloy.

Within a few years, a family of superconducting ceramics had been discovered that were all based on compounds of copper and oxygen. Müller and Bednorz worked with ceramics that were derivatives of a compound with the formula La_2CuO_4. If forced to assign oxidation states to the compound, most chemists would write it as $[La^{3+}]_2[Cu^{2+}][O^{2-}]_4$. The parent compound is an insulator. When some of the lanthanum atoms are replaced with barium atoms, however, a nonstoichiometric superconductor with the formula $La_{2-x}Ba_xCuO_4$ is obtained.

Applying the concept of oxidation states to the compound, we are formally replacing an La^{3+} ion with a Ba^{2+} ion each time a barium atom is incorporated into the structure. If the net charge on the compound is going to stay the same, the oxidation state of the copper atom must increase. Each time an La^{3+} ion is replaced with a Ba^{2+} ion, a Cu^{2+} ion therefore has to become a Cu^{3+} ion. Müller and Bednorz found that when enough barium had been incorporated to raise the average oxidation state of the copper to +2.2, the compound became a super-conductor at low temperatures. The electron that is formally removed from the copper atom is apparently delocalized and therefore capable of moving through the solid when it is cooled to low temperatures. Similar results can be obtained by incorporating either strontium or calcium into La_2CuO_4. $La_{1.8}Sr_{0.2}CuO_4$ has the highest transition temperature of any member of the family: 40 K.

Superconductivity has also been observed with a family of compounds known as 1–2–3 superconductors. The first member of the family was discovered in 1987, when $YBa_2Cu_3O_7$ was found to become a superconductor when cooled to 95 K—above the temperature of liquid nitrogen. (The common name of these superconductors is based on the fact that there are three metals in a 1:2:3 ratio.) This compound also contains copper in a fractional oxidation state. The yttrium atom can be assumed to exist in the +3 oxidation state. Thus one Y^{3+} and two Ba^{2+} ions contribute a charge of +7 toward balancing the charge of −14 on the seven oxygens. The remaining charge of +7 has to be distributed over the three copper atoms, for an average oxidation state of +2.33.

Clay pots are one form of the class of materials known as ceramics.

Exercise 9.2

Describe how you would determine what elements might be used to prepare the following materials.

(a) A heat-resistant, insulating ceramic material, such as glass or a piece of pottery

(b) A new semiconductor

(c) A crystalline material that would stand up to high temperatures and not conduct electricity or heat

Solution

(a) The best place to search for such ceramics would be toward the bottom and center of a bond-type triangle. Compounds that lie too close to the bottom-right corner would tend to form molecules that lack the long-range order needed to form a ceramic. Compounds that lie close to the bottom-left corner would be more likely to form metallic bonds that would make the material a conductor. Compounds that lie close to the top of the triangle would have the long-range order and insulating properties that are desired, but such solids are often too brittle to form useful materials.

Both silicon carbide (SiC) and boron nitride (BN) make ceramic materials. Characteristics of the two ceramics are that they are strong but brittle. Both are poor conductors of electricity, and both are very hard materials. BN is, in fact, comparable in hardness to diamond.

(b) You would look toward the bottom of the triangle. This time you might shift the focus slightly to the left of center. GaAs, for example, lies on the border of the semimetal region and toward the covalent (insulator) region and is a semiconductor (see Figure 9.1). The element silicon is a semiconductor used in the manufacture of silicon chips for integrated circuits. GaAs is a new semiconducting material that has certain advantages over silicon. GaP is also promising for use as a semiconductor.

(c) These characteristics are most likely to be met by materials whose atoms have electronegativities that place them in the upper-middle region of a bond-type triangle. Such combinations might be aluminum and oxygen or magnesium and oxygen.

9.13   Measuring the Distance between Particles in a Unit Cell

Nickel was identified in Section 9.6 as one of the metals that crystallizes in a cubic-closest-packed structure. When we consider that a nickel atom has a mass of only 9.75×10^{-23} g, it is a remarkable achievement for us to be able to describe the structure of this metal on an atomic scale. The obvious question is: How do we know that nickel packs in a cubic-closest-packed structure? The only way to answer this question is to find a quantity that can be measured for a sample on both the atomic and macroscopic scale.

The only way to determine the structure of matter on an atomic scale is to use a probe that is even smaller. As we have seen, one of the most useful probes for studying matter on the atomic scale is electromagnetic radiation. In 1912, Max von Laue found that X rays that strike the surface of a crystal are diffracted into patterns that resemble the patterns produced when light passes through a very narrow slit. Shortly thereafter, William Lawrence Bragg, who was just completing an undergraduate degree in physics at Cambridge University, explained von Laue's results. Bragg argued that X rays were reflected from planes of atoms near the surface of the crystal, as shown in Figure 9.20. He then concluded that the only way the X rays could stay in phase was if some integer (n) times the wavelength of the radiation (l) was twice the distance (d) between adjacent planes of atoms times the sine of the angle θ.

$$n\lambda = 2d \sin \theta$$

This relationship, which became known as the **Bragg equation,** allows us to calculate the distance between planes of atoms in a crystal from the pattern of diffraction of X rays of known wavelength.

The pattern by which X rays are diffracted by nickel metal suggests that the metal packs in a cubic unit cell with a distance between planes of atoms of 0.3524 nm. Thus the length of the edge of a unit cell in the crystal must be 0.3524 nm. Knowing that nickel crystallizes in a cubic unit cell isn't enough. We still have to decide whether it is a simple cubic, body-centered cubic or face-centered cubic unit cell. As we'll see in the next section, we can do this by measuring the density of the metal.

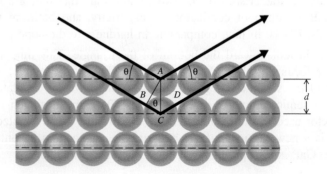

Fig. 9.20 Diffraction of X rays by the first and second planes in a crystal.

9.14 Determining the Unit Cell of a Crystal

Atoms on the corners, edges, and faces of a unit cell are shared by more than one unit cell, as shown in Figure 9.21. An atom on one of the faces is shared by two unit cells, so one-half of that atom belongs to each of the cells. An atom on an edge is shared by four unit cells, and an atom on a corner is shared by eight unit cells. Thus, only one-quarter of an atom on an edge and one-eighth of an atom on a corner can be assigned to each of the unit cells that share the atoms.

If nickel crystallized in a simple cubic unit cell, there would be a nickel atom on each of the eight corners of the cell. Because only one-eighth of these atoms can be included in a given unit cell, each unit cell in a simple cubic structure would have one net nickel atom.

Simple cubic structure:

$$8 \text{ corners} \times \frac{1}{8} = 1 \text{ } net \text{ atom/unit cell}$$

If nickel formed a body-centered cubic structure, there would be two atoms per unit cell because the nickel atom in the center of the unit cell isn't shared with any other unit cells.

Body-centered cubic structure:

$$\left(8 \text{ corners} \times \frac{1}{8}\right) + 1 \text{ body} = 2 \text{ net atoms/unit cell}$$

If nickel crystallized in a face-centered cubic structure, the atoms on the six faces of the unit cell would contribute three net nickel atoms, for a total of four atoms per unit cell.

Face-centered cubic structure:

$$\left(8 \text{ corners} \times \frac{1}{8}\right) + \left(6 \text{ faces} \times \frac{1}{2}\right) = 4 \text{ net atoms/unit cell}$$

Because they have different numbers of atoms in a unit cell, each of these structures would have a significantly different density. Let's therefore compare the experimentally determined density of nickel metal (8.90 g/cm^3) with the results of predictions of the density based on the three possible structures for this metal. To do this, we need to know the volume of the unit cell in cubic centimeters and the weight of a nickel atom.

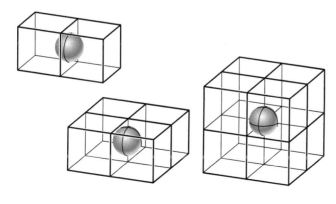

Fig. 9.21 Because an atom on the face of a unit cell is shared by two unit cells, only one-half of the atom belongs to each of the cells. For similar reasons, one-quarter of an atom on the edge of a unit cell and one-eighth of an atom on the corner of a unit cell belong to the unit cell.

We can start by noting that the unit cell edge length for nickel was given in the previous section. According to the results of an X-ray diffraction study of nickel metal, the unit cell edge is 0.3524 nm. The volume (V) of the unit cell is therefore equal to the cell-edge length (a) cubed.

$$V = a^3 = (0.3524\,\text{nm})^3 = 0.04376\,\text{nm}^3$$

Because the results of experimental measurements of the density of nickel metal are given in grams per cubic centimeter, it would be useful to convert the results of this calculation into units of cm^3. This can be done by noting that there are 10^9 nm in a meter and 100 cm in a meter. Thus, there must be 10^7 nm in a cm.

$$\frac{10^9\,\text{nm}}{1\,\text{m}} \times \frac{1\,\text{m}}{100\,\text{cm}} = 10^7\,\text{nm/cm}$$

Converting the volume of the unit cell to cubic centimeters gives the following result.

$$4.376 \times 10^{-2}\,\text{nm}^3 \times \frac{(1\,\text{cm})^3}{(10^7\,\text{nm})^3} = 4.376 \times 10^{-23}\,\text{cm}^3$$

We now know the volume of the unit cell. If we are going to base our decision about the structure of nickel metal on measurements of density, we need to know the weight of a single nickel atom. This can be calculated from the atomic weight of the metal and Avogadro's constant.

$$\frac{58.69\,\text{g Ni}}{1\,\text{mol}} \times \frac{1\,\text{mol}}{6.022 \times 10^{23}\,\text{atoms}} = 9.746 \times 10^{-23}\,\text{g/atom}$$

Now that the volume of the unit cell and the weight of a single nickel atom are known, theoretical values for the density of nickel metal can be predicted for each of the three possible structures of the metal. There would be only one nickel atom per unit cell if nickel crystallized in a simple cubic unit cell. The density of nickel, if it crystallized in a simple cubic structure, would therefore be 2.227 g/cm^3.

Simple cubic structure:
$$\frac{9.746 \times 10^{-23}\,\text{g/unit cell}}{4.376 \times 10^{-23}\,\text{cm}^3/\text{unit cell}} = 2.227\,\text{g/cm}^3$$

Because there would be twice as many nickel atoms per unit cell if nickel crystallized in a body-centered cubic structure, the density of nickel if it crystallized in this structure would be twice as large as predicted for the simple cubic structure.

Body-centered cubic structure:
$$\frac{2(9.746 \times 10^{-23}\,\text{g/unit cell})}{4.376 \times 10^{-23}\,\text{cm}^3/\text{unit cell}} = 4.454\,\text{g/cm}^3$$

There would be four nickel atoms per unit cell in a face-centered cubic structure, and the density of nickel in this structure therefore would be four times as large as the value predicted for the simple cubic structure.

Face-centered cubic structure:
$$\frac{4(9.746 \times 10^{-23}\,\text{g/unit cell})}{4.376 \times 10^{-23}\,\text{cm}^3/\text{unit cell}} = 8.909\,\text{g/cm}^3$$

The experimental value for the density of nickel is 8.90 g/cm³. The only possible conclusion is that nickel crystallizes in a face-centered cubic unit cell and therefore has a cubic-closest-packed structure.

9.15 Calculating the Size of an Atom or Ion

Estimates of the radii of most metal atoms can be found in Table B.4 in Appendix B. Where do these data come from? How do we know, for example, that the metallic radius of a nickel atom is 0.1246 nm? The starting point for calculating the metallic radius of an atom uses the results of the previous two sections. We now know that the nickel crystallizes in a cubic unit cell with a cell-edge length of 0.3524 nm, and we know that the unit cell for the crystal is face-centered cubic.

One of the faces of a face-centered cubic unit cell is shown in Figure 9.22 According to Figure 9.22, the diagonal across the face of the unit cell is equal to four times the metallic radius of a nickel atom.

$$d_{face} = 4\, r_{Ni}$$

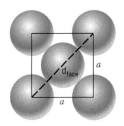

Fig. 9.22 The diagonal across the face of a face-centered cubic unit cell is equal to four times the radius of the atoms that form the cell.

The Pythagorean Theorem states that the square of the length of the long edge of a right triangle is equal to the sum of the squares of the other sides. The diagonal across the face of the unit cell is therefore related to the unit cell-edge length by the following equation.

$$d_{face}^2 = a^2 + a^2 = 2\, a^2$$

Taking the square root of both sides gives the following result.

$$d_{face} = \sqrt{2}\, a$$

Because the diagonal across the face is four times the metallic radius of a nickel atom, the following substitution can be made.

$$4\, r_{Ni} = \sqrt{2}\, a$$

Thus the metallic radius of a nickel atom is 0.1246 nm.

$$r_{Ni} = \frac{\sqrt{2}\, a}{4} = \frac{\sqrt{2}\ \times 0.3524\ \text{nm}}{4} = 0.1246\ \text{nm}$$

A similar approach can be taken to estimate the size of an ion. Consider cesium chloride, for example, which crystallizes in a simple cubic unit cell of Cl⁻ ions with a Cs⁺ ion in the center of the body of the cell, as shown in Figure 9.23.

Figure 9.23 is based on the assumption that the positively charged Cs⁺ ions and the negatively charged Cl⁻ in CsCl can be thought of as perfect spheres that pack so that ions of opposite charge touch each other.

The Cs⁺ ion in the center of the CsCl unit cell must touch the Cl⁻ ions at the eight corners. The diagonal across the body of the CsCl unit cell is therefore equal to the sum of the radii of two Cl⁻ ions and two times the radius of a Cs⁺ ion.

$$d_{body} = 2\, r_{Cs^+} + 2\, r_{Cl^-}$$

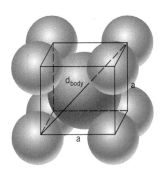

Fig. 9.23 The diagonal across the body of the CsCl unit cell is equal to twice the sum of the radii of the Cs⁺ and Cl⁻ ions.

The three-dimensional equivalent of the Pythagorean Theorem suggests that the square of the diagonal across the body of a cube is the sum of the squares of the three sides.

$$d_{body}^2 = a^2 + a^2 + a^2 = 3\,a^2$$

Taking the square root of both sides of the equation gives the following result.

$$d_{body} = \sqrt{3}\,a$$

If the cell-edge length in CsCl is experimentally determined to be 0.4123 nm, the diagonal across the body in the unit cell is 0.7141 nm.

$$d_{body} = \sqrt{3}\,a = \sqrt{3} \times 0.4123 \text{ nm} = 0.7141 \text{ nm}$$

The sum of the ionic radii of Cs^+ and Cl^- ions is half that distance, or 0.3571 nm.

$$r_{Cs^+} + r_{Cl^-} = \frac{d_{body}}{2} = \frac{0.7141 \text{ nm}}{2} = 0.3571 \text{ nm}$$

If we had an estimate of the size of either the Cs^+ or Cl^- ion, we could use the results of our calculation to estimate the size of the other ion. By combining the analysis of many ionic compounds, it is possible to create a set of consistent data for the size of the ions that form the crystals. Some of the data were reported in Section 3.22; a more complete set can be found in Table B.4 in Appendix B. The small discrepancy between the sum of the ionic radii of the Cs^+ ion (0.169 nm) and the Cl^- ion (0.181 nm) reported in tables of ionic radii and the results of the calculation for CsCl reflect the fact that ionic radii seem to vary slightly from one crystal to another.

► **CHECKPOINT**

For the three cubic structures studied, which has the simplest relationship between the radii of the ions and the length of the edge of the unit cell?

Key Terms

Amorphous solid	Insulators	Polycrystalline solid
Body-centered cubic packing	Intermetallic compounds	Precipitation hardening
Body-centered cubic unit cell	Interstitial solution	Semiconductor
Bragg equation	Ionic solid	Simple cubic packing
Closest-packed structure	Lattice energy	Simple cubic unit cell
Coordination number	Lattice points	Sublimes
Crystalline solid	Metallic bond	Substitution solution
Cubic closest packing	Metallic solid	Superconductor
Face-centered cubic unit cell	Molecular solid	Unit cell
Hexagonal closest packing	Network covalent solid	

Problems

Types of Solids

1. What are the three categories of solids that are based on the forces that hold the particles together?

2. What three types of bonds can hold particles together?

3. What category of compounds is formed by nonmetals? By metals? By metals combining with nonmetals?

Molecular and Network Covalent Solids

4. Which of the following solids are held together by an extended network of covalent bonds?
   (a) sodium chloride   (b) $CuAl_2$
   (c) gold   (d) calcium carbonate
   (e) diamond   (f) dry ice (solid CO_2)

5. Which force must be overcome to sublime dry ice, solid CO_2?

    (a) metallic bonding   (b) ionic bonding

    (c) covalent bonding   (d) dispersion forces

6. Which force must be overcome to melt solid pentane, C_5H_{12}?

    (a) metallic bonding   (b) ionic bonding

    (c) covalent bonding   (d) dispersion forces

Physical Properties of Molecular and Network Covalent Solids

7. Compare and contrast the bonding in molecular solids and network covalent solids.

8. Which would you expect to have a higher melting point, molecular solids or network covalent solids?

9. What are the three stable crystalline forms of carbon? How do their structures differ?

Metallic Solids

10. What are delocalized electrons?

11. What is a metallic bond? What types of atoms are most likely to form metallic bonds?

12. Describe the bonding in a substance that is held together by metallic bonds.

13. Molecular, ionic, and network covalent solids all have one characteristic in common that makes them different from metallic solids. What is this characteristic?

Physical Properties That Result from the Structure of Metals

14. Explain why metals are solids at room temperature.

15. Explain why metals are malleable and ductile.

16. Explain why metals conduct heat and electricity.

17. Which of the following categories is most likely to contain a compound that is a poor conductor of electricity when solid but a very good conductor when molten?

    (a) molecular solids

    (b) covalent solids

    (c) ionic solids

    (d) metallic solids

The Structure of Metals

18. List three common structures for metals.

19. Describe the difference in the way planes of atoms stack to form *hexagonal-closest-packed*, *cubic-closest-packed*, *body-centered cubic*, and *simple cubic structures*.

20. Explain why the structure of polonium is called *simple cubic*, why the structure of iron is called *body-centered cubic*, and why the structure of cobalt is called *hexagonal closest packed*.

Coordination Numbers and the Structures of Metals

21. Define *coordination number*.

22. Roughly sketch a simple cubic packing of spheres. Show how to find the coordination number of a representative sphere.

23. Roughly sketch a body-centered cubic structure. Show how to find the coordination number of a representative sphere.

24. Determine the coordination numbers of the metal atoms in each of the following structures.

    (a) cubic-closest-packed aluminum

    (b) hexagonal-closest-packed magnesium

    (c) body-centered cubic chromium

    (d) simple cubic polonium

25. Sodium crystallizes in a structure in which the coordination number is 8. Which structure best describes the crystal?

    (a) simple cubic   (b) body-centered cubic

    (c) cubic closest packed   (d) hexagonal closest packed

Unit Cells: The Simplest Repeating Unit in a Crystal

26. What is a lattice point?

27. Define *unit cell*.

28. There are three common types of unit cells. What are they? What are the coordination numbers in each?

Solid Solutions and Intermetallic Compounds

29. Describe the difference between an intermetallic compound, such as $CuAl_2$, and an alloy, such as brass or bronze.

30. Describe how the formation of the intermetallic compounds $CuAl_2$ and Fe_3C hardens aluminum and steel, respectively.

31. Describe the difference between solid solutions and interstitial solutions. Give an example of each. Predict the effect of changes in the relative size of a pair of atoms on their ability to form either of these solutions.

Semimetals

32. Why do ionic compounds conduct electricity better in the liquid state than the solid state?

33. Which of the following compounds should conduct an electric current when dissolved in water?

    (a) $MgCl_2$ (b) CO_2 (c) CH_3OH

    (d) KNO_3 (e) Ca_3P_2

34. Which of the following would you expect to conduct an electric current?

    (a) solid Na metal   (b) liquid Na metal

    (c) solid NaCl   (d) liquid NaCl

    (e) NaCl dissolved in water

35. Use Figures 5.11 and 9.1 to describe the characteristics of a semimetal.

Ionic Solids

36. Define the term *lattice energy*.

37. The lattice energy of NaCl refers to which of the following reactions?
    (a) $2\,Na(s) + Cl_2(s) \longrightarrow 2\,NaCl(s)$
    (b) $NaCl(s) \longrightarrow Na(g) + Cl(g)$
    (c) $Na(g) + Cl(g) \longrightarrow NaCl(g)$
    (d) $NaCl(s) \longrightarrow Na^+(g) + Cl^-(g)$
    (e) $Na^+(g) + Cl^-(g) \longrightarrow NaCl(s)$

38. Which of the following salts has the largest lattice energy?
    (a) LiF　　(b) LiCl　　(c) LiBr　　(d) LiI

39. Which of the following salts has the largest lattice energy?
    (a) NaCl　　(b) NaI　　(c) KI
    (d) MgO　　(e) MgS

40. Explain the following trends in lattice energies.
    (a) $MgF_2 > MgCl_2 > MgBr_2 > MgI_2$
    (b) $BeF_2 > MgF_2 > CaF_2 > SrF_2 > BaF_2$

41. Use the *CRC Handbook of Chemistry and Physics*[2] to determine the solubility in water of NaF, NaCl, NaBr, and NaI. Describe the relationship between the solubilities of the salts and their lattice energies.

42. Use lattice energies to explain why MgO is much less soluble in water than is CaO.

43. One of the simplest ways of distinguishing between two covalent compounds is to measure their melting points or boiling points. Naphthalene melts at 80.5°C and camphor melts at 179.8°C, for example. Would you expect the melting points and boiling points of ionic compounds to be higher, lower, or about the same as those of covalent compounds? Explain.

44. Draw the unit cell for CsCl and for Cs. Classify each according to bond type. What are the differences between the two unit cells? Specify the identity of the particles that occupy the lattice points for each.

The Search for New Materials

45. Where in the periodic table are the atoms that are most likely to be involved in the formation of ceramics located?

46. Where in the periodic table are the atoms that are most likely to be involved in the formation of semiconductors located?

47. Use a bond-type triangle to classify the following compounds and describe the characteristics of each.
    (a) CrO_2　　(b) SiC　　(c) GaP　　(d) BeO

48. Suggest two elements that might be combined to produce the following.
    (a) a material with a high melting point that is not an electrical conductor
    (b) an insulating material that has a low melting point
    (c) a conductor of electricity in the solid state

Measuring the Distance between Particles in a Unit Cell

49. In addition to knowing the length of an edge of a unit cell, what else must be known to calculate the size of an atom in the cell?

50. Why is the density of a metal related to how close the particles are to one another?

Determining the Unit Cell of a Crystal

51. Silver crystallizes in a face-centered cubic unit cell with an edge length of 0.40862 nm. Calculate the density of Ag metal in grams per cubic centimeter.

52. Potassium crystallizes in a cubic unit cell with an edge length of 0.5247 nm. The density of potassium is 0.856 g/cm^3. Determine whether the element crystallizes in a simple cubic, a body-centered cubic, or a face-centered cubic unit cell.

53. Determine whether calcium crystallizes in a simple cubic, a body-centered cubic, or a face-centered cubic unit cell, assuming that the cell-edge length is 0.5582 nm and the density of the metal is 1.55 g/cm^3.

54. Determine whether molybdenum crystallizes in a simple cubic, a body-centered cubic, or a face-centered cubic unit cell, assuming that the cell-edge length is 0.3147 nm and the density of the metal is 10.2 g/cm^3.

55. Which of the following metals crystallizes in a face-centered cubic unit cell with an edge length of 0.3608 nm if the density of the metal is 8.95 g/cm^3?
    (a) Na　　(b) Ca　　(c) Tl　　(d) Cu　　(e) Au

56. CdO crystallizes in a cubic unit cell with a cell-edge length of 0.4695 nm. Calculate the number of Cd^{2+} and O^{2-} ions per unit cell, assuming that the density of the crystal is 8.15 g/cm^3.

57. LiF crystallizes in a cubic unit cell with a cell-edge length of 0.4017 nm. Calculate the number of Li^+ and F^- ions per unit cell, assuming that the density of the salt is 2.640 g/cm^3.

58. The metallic radius of a vanadium atom is 0.1321 nm. What is the density of vanadium if the metal crystallizes in a body-centered cubic unit cell?

Calculating the Size of an Atom or Ion

59. Chromium metal ($d = 7.20$ g/cm^3) crystallizes in a body-centered cubic unit cell. Calculate the volume of the unit cell and the radius of a chromium atom.

[2]CRC Press, Boca Raton, Florida.

60. Calculate the atomic radius of an Ar atom, assuming that argon crystallizes at low temperature in a face-centered cubic unit cell with a density of 1.623 g/cm^3.

61. Barium crystallizes in a body-centered cubic structure in which the cell-edge length is 0.5025 nm. Calculate the shortest distance between neighboring barium atoms in the crystal.

62. NaH crystallizes in a structure similar to that of NaCl. If the cell-edge length in the crystal is 0.4880 nm, what is the average length of the Na—H bond?

63. TlI crystallizes in a structure similar to that of CsCl with a cell-edge length of 0.4198 nm. Calculate the average Tl—I bond length in the crystal. If the ionic radius of an I^- ion is 0.216 nm, what is the ionic radius of the Tl^+ ion?

64. Calculate the ionic radius of the Cs^+ ion, assuming that the cell-edge length for CsCl is 0.4123 nm and the ionic radius of a Cl^- ion is 0.181 nm.

Integrated Problems

65. From the enthalpy data in Table B.13 in Appendix B, calculate the enthalpy change required to break the bond in the following: $F_2(g)$, $Cl_2(g)$, $Br_2(g)$, and $I_2(g)$. Compare the enthalpies to the enthalpy required to *melt* each of the halogens given in Table 9.1. What conclusions can you reach concerning the forces that hold the atoms together and those that hold the molecules to one another?

66. The enthalpies of fusion of the alkali metals are given below.

Metal	ΔH_{fus} (kJ/mol$_{rxn}$)
Li	2.9
Na	2.6
K	2.4
Rb	2.2

Identify the type of solid formed and explain the trend seen in the enthalpy required to melt the solids. Arrange the solids in order of increasing melting point.

67. For each of the following properties (1 through 11), choose the appropriate electronegativity characteristic [(a) through (f) below] for a binary compound.
    1. a good conductor of electricity
    2. a hard material that conducts

    3. electricity in the melted state
    4. an insulator
    5. a material that conducts electricity when dissolved in water
    6. a semiconductor
    7. a ceramic
    8. a molecular crystal
    9. a metallic compound
    10. an ionic compound
    11. a hard material that is an insulator

    (a) large ΔEN, low EN for both atoms
    (b) small ΔEN, high EN for both atoms
    (c) moderate ΔEN, moderate EN for both atoms
    (d) moderate ΔEN, high EN for both atoms
    (e) small ΔEN, low EN for both atoms
    (f) large ΔEN

68. Classify the following binary compounds as primarily metallic, covalent, or ionic. Figures 9.1 and 5.11 may be helpful.
    (a) B_2H_6 (b) B_4C (c) InAs   (d) HgI_2
    (e) Hg_2Na_3 (f) K_2S (g) Cd_3Mg (h) KBr
    (i) MgH_2 (j) GaS   (k) LiH   (l) Be_3P_2

69. (a) Two of the following compounds have very high melting points. Which two? Explain your reasoning.

    BaO, MgO, HgO

    (b) Of the two compounds that have very high melting points, which one would have the higher melting point? Explain.

70. Why are the bottoms of stainless-steel cooking pans generally clad with copper?

71. Tiles used to insulate the space shuttles must stand up to high heat. Use the bond-type triangle to suggest what combinations of atoms might be used to make such materials.

72. Gallium arsenide (GaAs) has replaced silicon in many uses. What are the expected properties of this material as compared to silicon? Comment on thermal conductivity, electrical conductivity, structure, and melting point.

73. Nickel aluminide (Ni_3Al) is a new material developed to compete with the heat-resistant ceramics. Describe some of the properties of Ni_3Al.

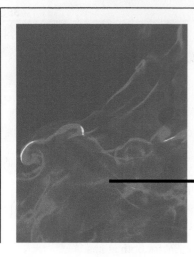

Chapter Nine

SPECIAL TOPICS

9A.1 Defects

As we have seen, it is useful to think about solids in terms of a regular repeating pattern of planes of particles. But it is important to recognize that solids are seldom perfectly ordered. There are four basic mechanisms for introducing a **point defect** into the structure of a solid, as shown in Figure 9A.1. When a particle is missing at one or more lattice sites, we get a **vacancy.** When a particle forces its way into a hole between lattice sites, we get an **interstitial impurity. Substitutional impurities** result from replacing the particle that should occupy a lattice site with a different particle, such as substituting a K^+ ion for a Na^+ ion in NaCl. (If an ion with a different charge is substituted, the electrical neutrality of the crystal must be maintained. If a Ca^{2+} ion is substituted for a Na^+ ion, for example, a second Na^+ ion must leave the crystal so that it doesn't pick up an electric charge.) **Dislocations** are one-dimensional defects caused by holes that are not large enough to be a vacancy.

When a significant fraction of the original particles is replaced by impurities, it is possible to get a **solid solution. Alloys,** such as bronze and brass, are examples of solid solutions. Bronze is a solution of tin dissolved in copper. Brass is a mixture of copper and zinc that can contain as little as 10% or as much as 45% zinc.

Distortions of the crystal lattice often occur when impurities are added to a solid. As a result, point defects often determine the properties of a material. They can change the ease with which a material conducts electricity, its mechanical strength, and its ability to be shaped by hammering (malleability) or to be drawn into wires (ductility). Dissolving small amounts of carbon in iron, for example, produces steel, which is significantly stronger than iron. But higher percentages of carbon make steel so brittle that it can shatter when dropped.

Point defects distort the lattice and provide a way for atoms to move about the solid. Atoms can move from a lattice site into a vacancy, for example, creating a new vacancy, as shown in Figure 9A.2.

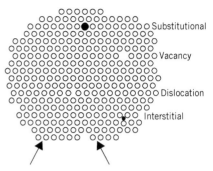

Fig. 9A.1 Defects in a solid include vacancies, interstitial impurities, substitutional impurities, and extended imperfections such as dislocations (sight along arrows). [Reprinted with permission from A. B. Ellis et al., *Teaching General Chemistry: A Materials Science Companion,* p. 161. Copyright 1993, American Chemical Society.]

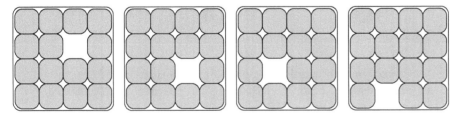

Fig. 9A.2 Vacancies allow the rearrangement of atoms in the solid state as the empty space moves from one lattice point to another. [Reprinted with permission from P. A. Thrower, *Materials in Today's World,* rev. ed., McGraw-Hill, New York, 1992, p. 69.]

Theoretical calculations of the ease with which one plane of atoms should slip over another suggest that metals should be much more resistant to stress than they are. In other words, metals are softer than one would expect. Metallurgists have explained this by assuming that metals contain defects that allow planes of atoms to slip past each other more readily than expected. This hypothesis has been confirmed by microscopic analysis, which shows dislocations that run through the crystal. There are two types of dislocations: edge or screw dislocations. An **edge dislocation** is an extra half-plane of atoms that goes part way through a solid structure, as shown in Figure 9A.3.

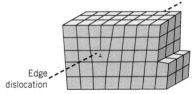

Fig. 9A.3 A schematic representation of a three-dimensional crystal containing an edge dislocation. Reprinted with permission from A. B. Ellis et al., *Teaching General Chemistry: A Materials Science Companion,* p. 161. Copyright 1993, American Chemical Society.]

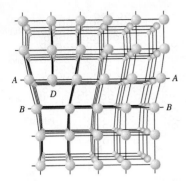

Fig. 9A.4 Plane *A* slides over plane *B* by using the dislocation defect to move gradually, a little at a time. [W. F. Smith, *Principles of Materials Science and Engineering,* McGraw-Hill, New York, 1986, Fig. 4.18. Reproduced with permission of the publisher.]

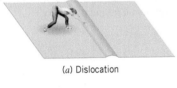

(*a*) Dislocation

(*b*) Work hardening

Fig. 9A.5 (*a*) A rug can be moved across a floor by pushing on a wrinkle because this doesn't require movement of the entire rug at the same time. (*b*) It is more difficult to move the rug when two or more wrinkles meet. The presence of one wrinkle "pins down" the other. [Reprinted with permission from A. B. Ellis et al., *Teaching General Chemistry: A Materials Science Companion,* p. 168. Copyright 1993 American Chemical Society.]

Imagine, for example, a single playing card inserted halfway into a deck of cards. The line formed by the inserted card would be a dislocation line. The presence of a dislocation defect allows one plane of atoms to slip more easily over its neighboring plane of atoms, as shown in Figure 9A.4. Not all the atoms in the two planes move past each other simultaneously; they move one row at a time.

An often-quoted analogy is that of moving a carpet. Dragging the carpet across the floor is difficult because of the friction developed from the contact of the surface of the carpet with the floor. Imagine what would happen, however, if a wrinkle is put into the carpet, as shown in Figure 9A.5*a*. The carpet can now be moved by pushing the wrinkle across the floor, because only the friction between a small section of carpet and the floor has to be overcome. A similar phenomenon occurs when one plane of atoms moves past another by means of a dislocation defect.

Because they allow planes of atoms in a solid to move one row at a time, dislocations can weaken a metal. Paradoxically, they can also strengthen a metal when the dislocations intersect to produce knots similar to the intersecting wrinkles in Figure 9A.5*b*. This phenomenon is encountered with metals that have been work hardened. Consider what happens, for example, when a piece of iron is heated, hammered, cooled, reheated, and reworked to form wrought iron. In the course of work hardening the metal, intersecting dislocations are generated that hinder the movement of planes of atoms.

Screw dislocations are more difficult to visualize than edge dislocations. Figure 9A.6 shows how a screw dislocation is produced when one side of a crystal is displaced relative to the other side. For either edge or screw dislocations, a distortion is produced around the dislocation with a corresponding stress produced within the material.

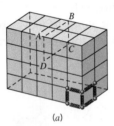
(*a*)
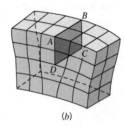
(*b*)
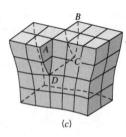
(*c*)

Fig. 9A.6 The formation of a screw dislocation. (*a*) A cubic lattice; (*b*) edge dislocation formed by inserting an extra half-plane of atoms in *ABCD*; (*c*) a left-handed screw dislocation formed by displacing the faces *ABCD* relative to each other in the *AB* direction. [Reprinted with permission Derek Hull, *Introduction to Dislocations,* 2nd ed., Pergamon Press Inc., New York, p. 20. 1975.

9A.2 Metals, Semiconductors, and Insulators

A significant fraction of the gross national product (GNP) of the United States, and all of the contribution to the GNP from high-technology industries, can be traced to efforts to harness differences in the way metals, semiconductors, and insulators conduct electricity. This difference can be expressed in terms of **electrical conductivity,** which measures the ease with which materials conduct an electric current. It can also be expressed in terms of **electrical resistivity,** the inverse of conductivity, which measures the resistance of a material to carrying an electric charge.

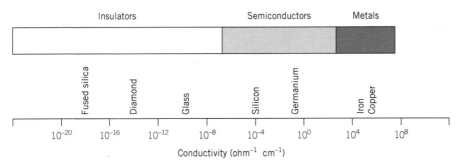

Fig. 9A.7 The range of conductivities of solids span roughly 24 orders of magnitude. [Reprinted with permission from A. B. Ellis et al., *Teaching General Chemistry: A Materials Science Companion*, p. 197. Copyright 1993 American Chemical Society.]

Silver and copper metal are among the best conductors of electricity, with a conductivity of 10^6 ohm^{-1}-cm^{-1}. (This is why copper is the metal most often used in electric wires.) The conductivity of semiconductors such as silicon and germanium is 10^8 to 10^{10} times smaller. (When pure, these semimetals have a conductivity of 10^{-2} to 10^{-4} ohm^{-1}-cm^{-1}.) Insulators include glass (10^{-10} ohm^{-1}-cm^{-1}), diamond (10^{-14} ohm^{-1}-cm^{-1}), and quartz (10^{-18} ohm^{-1}-cm^{-1}), which all have an extremely small tendency to carry an electric current.

The 10^{24}-fold range of conductivity is not the only difference among metals, semiconductors, and insulators. Metals become better conductors when they are cooled to lower temperatures. Some metals are such good conductors at very low temperatures that they no longer have a measurable resistance and therefore become **superconductors.** Semiconductors show the opposite behavior—they become much better conductors as the temperature increases. The difference between the temperature dependence of metals and semiconductors is so significant that it is often the best criterion for distinguishing between these materials. The large range of conductivities of solids is shown in Figure 9A.7.

Semiconductors are very sensitive to impurities. The conductivity of silicon or germanium can be increased by a factor of up to 10^6 by adding as little as 0.01% of an impurity. Metals, on the other hand, are fairly insensitive to impurities. It takes a lot of impurity to change the conductivity of a metal by as much as a factor of 10; and unlike semiconductors, metals become poorer conductors when impure.

To explain the behavior of metals, semiconductors, and insulators, we need to understand the bonding in solids in more detail. Because it is the lightest element in the periodic table that is a solid at room temperature, let's start by building a model of what happens when lithium atoms interact. As a first step, we can consider what happens when a pair of lithium atoms with a $1s^2\,2s^1$ configuration interact to form a gas-phase Li_2 molecule. The Li_2 molecule is formed by placing two electrons in the bonding domain between the two Li nuclei.

Now let's imagine what happens when enough lithium atoms come together to form a piece of lithium metal. The valence electrons are no longer confined to the region between pairs of lithium nuclei, as was the case for an isolated Li_2 molecule in the gas phase. In the metal, each lithium atom is perturbed by its neighbors, and the energy states of each atom are slightly altered. The $1s$ orbitals on the various metal atoms interact to form a band of orbitals whose energy falls within a range from slightly below the energy of the isolated $1s$ orbital to slightly above this energy, as shown in Figure 9A.8. The same thing happens to the $2s$ orbitals.

Each of the orbitals in these bands can hold two electrons of opposite spin. Because there were two electrons in each of the $1s$ orbitals that formed the lower-energy band, the "$1s$" band is filled. But there was only one electron in each of the $2s$ orbitals that formed the higher-energy band, which means that the "$2s$" band is only half-filled. It takes little, if any, energy to excite one of the electrons in the $2s$ band from one orbital to another in the band. (The energy gap between

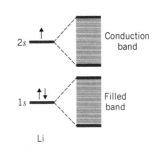

Fig. 9A.8 The overlap of atomic orbitals on a large number of Li atoms to form bands of orbitals.

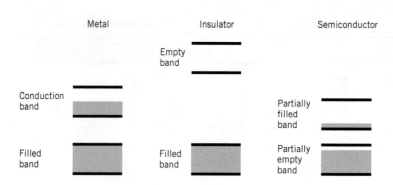

Fig. 9A.9 Band-theory diagrams for a conductor, an insulator, and a semiconductor.

orbitals in the 2s band in lithium is only about 10^{-45} kJ.) By moving from orbital to orbital within the 2s band, electrons can move from one end of the crystal to the other. This band of orbitals is therefore called a *conduction band* because it enables lithium metal to conduct electricity.

Let's now turn to magnesium, which has a [Ne] $3s^2$ configuration. The 3s orbitals on the neighboring magnesium atoms would overlap to form a band of 3s orbitals. Because there are two electrons in each 3s orbital, this band is totally filled. The empty 3p orbitals on magnesium, however, also interact to form a band of orbitals. This empty 3p overlaps the 3s band in magnesium, so that the combined band is only partially filled, allowing magnesium to conduct electricity.

The differences in the way metals, semiconductors, and insulators conduct electricity can be explained with the diagram in Figure 9A.9. Metals have filled bands of core electrons, such as the 1s band in lithium or the 1s and 2s bands in magnesium. But they also have partially filled bands of orbitals that allow electrons to move from one end of the crystal to the other. They therefore conduct an electric current. All of the bands in an insulator are either filled or empty. Furthermore, the gap between the highest-energy filled band and the lowest-energy empty band in an insulator is so large that it is difficult to excite electrons from one of these bands to the other. As a result, it is difficult to move electrons through an insulator.

Semiconductors also have a band structure that consists of filled and empty bands. The gap between the highest-energy filled band and the lowest-energy empty band is small enough, however, that electrons can be excited into the empty band by the thermal energy the electrons carry at room temperature. Semiconductors therefore fall between the extremes of metals and insulators in their ability to conduct an electric current.

To understand why metals become better conductors at low temperature, it is important to remember that temperature is a macroscopic reflection of the kinetic energy of the individual particles. Much of the resistance of a metal to an electric current at room temperature is the result of scattering of the electrons by the thermal motion of the metal atoms as they vibrate back and forth around their lattice points. As the metal is cooled, and this thermal motion slows down, there is less scattering and the metal becomes a better conductor.

Semiconductors become better conductors at high temperatures because the number of electrons with enough thermal energy to be excited from the filled band to the empty band increases.

To understand why semiconductors are sensitive to impurities, let's look at what happens when we add a small amount of a Group VA element, such as arsenic, to one of the Group IVA semiconductors. Arsenic atoms have one more valence electron than germanium and silicon atoms. Arsenic atoms can therefore lose an electron to form As^+ ions that can occupy some of the lattice points in the crystal where silicon or germanium atoms are normally found.

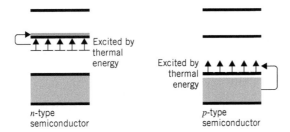

Fig. 9A.10 Band-theory diagrams for *n*-type and *p*-type doped semiconductors.

If the amount of arsenic is kept very small, the distance between arsenic atoms is so large that they don't interact. As a result, the extra electrons from the arsenic atoms occupy orbitals in a very narrow band of energies that lie between the filled and empty bands of the semiconductor, as shown in Figure 9A.10. This decreases the amount of energy required to excite an electron into the lowest-energy empty band in the semiconductor and therefore increases the number of electrons that have enough energy to cross this gap. As a result, this "doped" semiconductor becomes a very much better conductor of electricity than the pure semiconductor. Because the electric charge is carried by a flow of *negative* particles, these semiconductors are called *n-type*.

It is also possible to dope a Group IVA semiconductor with one of the elements in Group IIIA, such as indium. These atoms have one less valence electron than silicon or germanium atoms, and they can capture electrons from the highest-energy filled band to form holes in this band. The presence of holes in a filled band has the same effect as the presence of electrons in an empty band—it allows the solid to carry an electric current. The electric charge is now carried by a flow of positive particles, or holes, so these semiconductors are called *p-type*.

Bringing *n*-type and *p*-type semiconductors together produces a device that has a natural one-directional flow of electrons, which can be turned off by applying a small voltage in the opposite direction. This junction between *n*-type and *p*-type semiconductors was the basis of the revolution in industrial technology that followed the discovery of the transistor by William Shockley, John Bardeen, and Walter Brattain at Bell Laboratories in 1948.

9A.3 Thermal Conductivity

You may have noticed that metal ice-cube trays feel significantly colder then plastic ice-cube trays when you remove them from the freezer. Your senses are obviously misleading you because the trays are at the same temperature—the temperature of the freezer. The metal trays feel colder because metals are much better conductors of heat than plastic.

The ease with which metals conduct heat is related to their ability to conduct an electric current. Most of the energy absorbed by a metal when it is heated is used to increase the rate at which the atoms vibrate around their lattice sites. But some of this energy is absorbed by electrons in the metal, which move from orbital to orbital through the conduction band. The net result is a transport of kinetic energy from one portion of the metal surface to another. Metals feel cold to the touch because the electrons in the conduction band carry heat away from our bodies and distribute this energy through the metal object.

Plastics, on the other hand, are thermal insulators. They are poor conductors of heat because orbitals in which electrons are held tend to be localized on an individual atom or between pairs of atoms. The only way for electrons to carry

Table 9A.1
Thermal Conductivities of Various Substances

Material	Thermal Conductivity $(J/s \cdot cm \cdot K)^a$	Material	Thermal Conductivity $(J/s \cdot cm \cdot K)^a$
Air	0.00026	Carbon tetrachloride	0.0010
Glass wool	0.00042	White pine	0.0011
Cotton	0.00057	Oak	0.0015
Styrofoam	0.00079	He	0.001520
Cardboard	0.0021	Fe	0.804
Nylon	0.0025	Li	0.848
Water	0.0061	K	1.025
Brick	0.0063	C (graphite)^b	1.1–2.2
Glass	0.0072–0.0088	Zn	1.16
Concrete	0.0086–0.013	Brass	1.2
Hg	0.083	Na	1.42
SiC	0.090 (100°F)	Mg	1.56
NaCl	0.092 (0°C)	Be	2.01
ZnS (zinc blende)	0.264 (0°C)	BeO	2.20 (100°F)
Al_2O_3	0.303 (100°C)	Al	2.37
Pb	0.353	Au	3.18
Cs	0.359	Cu	4.01
MgO	0.360 (100°F)	Ag	4.29
Rb	0.582	C (diamond)^c	9.9–23.2

^{a}All values are at room temperature unless otherwise noted.
^{b}Value is dependent on the impurities in graphite and on the orientation of graphite, being larger in the direction parallel to the layers of carbon atoms.
^{c}Value is highly dependent on impurities and defects.

energy through a plastic is to use this energy to excite an electron from a filled orbital to an empty orbital. But the difference between the energies of the filled and empty orbitals is so large that this rarely happens.

The difference between thermal conductors and thermal insulators can be quantified by defining the **thermal conductivity** of a substance as the quantity of heat transmitted per second through a plate of the material 1 centimeter thick and 1 square centimeter in area when the temperature differential between the two sides of the plate is 1°C or 1 K. The copper used for pots and pans has a thermal conductivity that is more than 5000 times the value for the Styrofoam used for coffee cups, as shown by the data in Table 9A.1. This table is consistent with experience, which suggests that the air that gets trapped in the fibers of a down-filled jacket is a better insulator than cotton, which is a much better insulator than nylon.

9A.4 Thermal Expansion

It is tempting to think about solids as if the particles were locked into position, the way bricks are used to build a wall. This would be a mistake, however, because the particles in a solid are in more or less constant motion—rocking back and forth and rotating about their positions in the crystal. This motion depends on two factors: the temperature of the system and the strength of the interactions that hold the particles together. The higher the temperature, the faster the particles are

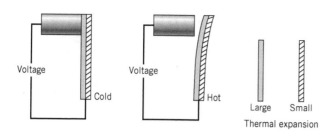

Fig. 9A.11 A voltage supply can be used to heat a bimetallic strip, which bends in the direction of the metal with the smallest expansion. [Taken from P. A. Thrower, *Materials in Today's World,* rev. ed., McGraw-Hill, New York, 1992, p. 125.]

moving. The stronger the force of attraction between the particles, the smaller the distances the particles move apart. Because the van der Waals forces that hold molecules together are much weaker than the bonds between atoms in a metal or between positive and negative ions in an ionic compound, molecular crystals expand more when heated than metals or ionic compounds.

The difference between the coefficients of **thermal expansion** of iron and copper was the source of a major problem for the Statue of Liberty, which consists of copper plates supported by an iron skeleton. The insulating material used to keep these two metals from coming into contact was inevitably rubbed away because of differences in the rate at which these two metals expand when heated and contract when cooled. (For each degree change in the temperature of the statue, the volume of the copper metal changes by 40% more than the iron metal.) When this happened, the two metals came into contact, forming an electric cell that greatly increased the rate at which the iron skeleton corroded.

The same phenomenon, however, is used to form the thermostats that turn electrical appliances on and off. When two metals with very different coefficients of thermal expansion are joined to form a bimetallic strip, the metal with the greater expansion when heated forces the adjoined metal strip to bend toward the metal with the smaller thermal expansion. This bimetallic strip can be used to make a device that will turn a heater on or off as contact is made or broken with an electrical contact, as shown in Figure 9A.11.

Thermal expansion and thermal conductivity can work together to weaken a material. If heat isn't transported quickly through an object that is heated, one part expands more rapidly than another. If any cracks or flaws are present, the hotter part of the substance will pull on the colder part and widen the crack, causing breakage.

9A.5 Glass and Other Ceramics

One of the characteristic properties of a substance is its **viscosity**, which is a measure of its resistance to flow. Motor oils are more viscous than gasoline, for example, and the maple syrup used on pancakes is more viscous than the vegetable oils used in salad dressings. Viscosity depends on any factor that can influence the ease with which molecules slip past each other. Liquids tend to become more viscous as the molecules become larger or as the intermolecular forces become stronger. They also become more viscous when cooled.

Imagine what would happen if you cooled a liquid until it became so viscous that it was rigid and yet it lacked any of the long-range order that characterizes the solids discussed in this chapter. You would have something known as a **glass.** Glasses have three characteristics that make them more closely resemble "frozen liquids" than crystalline solids. First and foremost, there is no long-range

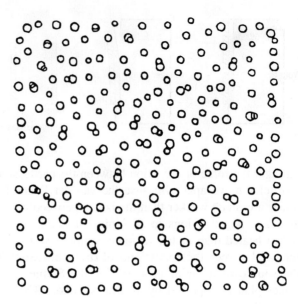

Fig. 9A.12 The structure of a glassy metal on the atomic scale.

order. Second, there are numerous empty sites or vacancies. Finally, glasses don't contain planes of atoms.

The simplest way to understand the difference between a glass and a crystalline solid is to look at the structure of glassy metals at the atomic scale. By rapidly condensing metal atoms from the gas phase, or by rapidly quenching a molten metal, it is possible to produce glassy metals that have the structure shown in Figure 9A.12.

The amorphous structure of glass makes it brittle. Because glass doesn't contain planes of atoms that can slip past each other, there is no way to relieve stress. Excessive stress therefore forms a crack that starts at a point where there is a surface flaw. Particles on the surface of the crack become separated. The stress that formed the crack is now borne by particles that have fewer neighbors over which the stress can be distributed. As the crack grows, the intensity of the stress at its tip increases. This allows more bonds to break, and the crack widens until the glass breaks. Thus, if you want to cut a piece of glass, start by scoring the glass with a file to produce a scratch along which it will break when stressed.

Glass has been made for at least 6000 years, since the Egyptians coated figurines made from sand (SiO_2) with sediment from the Nile River, heated these objects until the coating was molten, and then let them cool. Calcium oxide or "lime" (CaO) and sodium oxide or "soda" (Na_2O) from the sediment flowed into the sand to form a glass on the surface of the figurines. Trace amounts of copper oxide (CuO) in the sediment gave rise to a random distribution of Cu^{2+} ions in the glass that produced a characteristic blue color.

Sand is still the most common ingredient from which glass is made. (More than 90% of the sand consumed each year is used by the glass industry.) Sand consists of an irregular network of silicon atoms held together by Si—O—Si bonds. If the network was perfectly regular, each silicon atom would be surrounded by four oxygen atoms arranged toward the corner of a tetrahedron. (See Figure 9A.13.) Because each oxygen atom in this network is shared by two silicon atoms, the empirical formula of this solid would be SiO_2 and the material would have the structure of quartz. In sand, however, some of the Si—O—Si bridges are broken, in a random fashion.

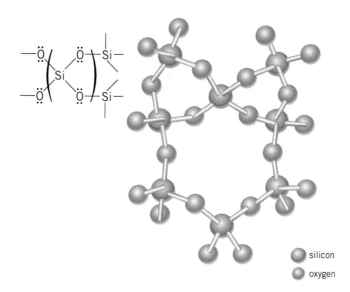

silicon

oxygen

Fig. 9A.13 Sand consists of an irregular network of silicon atoms held together by Si—O—Si bonds.

Modifiers (or fluxes) such as Na_2O and CaO are added to the sand to alter the network structure by replacing Si—O—Si bonds with Si—O⁻Na⁺ or Si—O⁻Ca²⁺ bonds. This separates the SiO_2 tetrahedra from each other, which makes the mixture more fluid and therefore more likely to form a glass after it has been melted and then cooled. These so-called soda-lime glasses account for 90% of the glass produced.

Al_2O_3 is added to some glasses to increase their durability; MgO is added to slow down the rate at which the glass crystallizes. Replacing Na_2O with B_2O_3 produces a borosilicate glass that expands less on heating. Adding PbO produces lead glasses that are ideally suited for high-quality optical glass.

The most common way of preparing a glass is to heat the mixture of sand and modifiers until it melts, and then cool it quickly so that it solidifies to produce a glass. If the cooling is rapid enough, the particles in the liquid state can't return to the original crystalline arrangement of the starting materials. Instead, they occupy randomly arranged lattice sites in which no planes of atoms can be identified. The result is an amorphous (literally "without shape") material.

GLASS-CERAMICS

An accidental overheating of a glass furnace led to the discovery of materials known as glass-ceramics. When the glass was overheated, small crystals formed in the amorphous material that prevented cracks from propagating through the glass.

The first step toward glass-ceramics involves conventional techniques for preparing a glass. The product is then heated to 750°–1150°C, until a portion of the structure is transformed into a fine-grained crystalline material. Glass-ceramics are at least 50% crystalline after they have been heated. In some cases, the final product is more than 95% crystalline.

Because glass-ceramics are more resistant to thermal shock, cookware made of this material can be transferred directly from a hot stove burner to the refrigerator without breaking. Because they are more crystalline, glass-ceramics are also slightly better at conducting heat than conventional glasses. Glass-ceramics are also stronger at high temperatures than glasses. Thus the glass-ceramic MgO-Al_2O_3-SiO_2 is used to make electrical insulators that have to operate at high temperatures, such as spark plug insulators. The properties and uses of some glasses and glass-ceramics are given in Table 9A.2.

Table 9A.2
Properties and Uses of Some Glasses and Glass-Ceramics

Composition	Property	Use
Glasses		
Al_2O_3, MgO, CaO, SiO_2	Translucent, chemically resistant	Window glass, bottles
PbO, SiO_2	High refractive index	Lead crystal
B_2O_3, SiO_2, Na_2O	Acid resistant, low expansion on heating	Pyrex
Glass-Ceramics		
MgO, Al_2O_3, SiO_2	Insulator with high mechanical strength at high temperatures	Spark plug insulators
$CaSiO_3$, $CaMgSi_2O_6$, $CaAl_2Si_2O_8$	Wear resistant	Building materials
$Li_2Si_2O_5$	Resistant to thermal shock	Nose cones on rockets, cookware

CERAMICS

The term *ceramic* comes from the Greek word for "pottery." It is used to describe a broad range of materials that include glass, enamel, concrete, cement, pottery, brick, porcelain, and chinaware. This class of materials is so broad that it is often easier to define ceramics as all solid materials except metals and their alloys that are made by the high-temperature processing of inorganic raw materials.

Ceramics can be either crystalline or glasslike. They can be either pure, single-phase materials or mixtures of two or more discrete substances. Most ceramics are polycrystalline materials, with abrupt changes in crystal orientation or composition across each grain in the structure. Ceramics can have electrical conductivities that resemble metals, such as ReO_3 and CrO_2. Ceramics can also make excellent insulators, such as the glass-ceramics used in spark plugs.

One of the most distinctive features of ceramics is their resistance to being worked or shaped after they are fired. With certain exceptions, such as glass tubing or plate glass, they can't be sold by the foot or cut to fit on the job. Their size and shape must be decided on before they are fired, and they must be replaced, rather than repaired, when they break.

The primary difference between ceramics and other materials is the chemical bonds that hold these materials together. Although they can contain covalent bonds, such as the Si—O—Si linkages in glass, they are often characterized by ionic bonds between positive and negative ions. When they form crystals, the strong force of attraction between ions of opposite charge in the planes of ions makes it difficult for one plane to slip past another. Ceramics are therefore brittle. They resist compression, but they are much weaker to stress applied in the form of bending.

The use of ceramics can be traced back to Neolithic times, when clay was first used to make bowls that were baked in campfires. Clay is formed by the weathering of rock to form shingle-like particles of alumina and silica that cling together when wet to form clay minerals, such as kaolinite, which has the formula $Al_4Si_4O_{10}(OH)_8$.

Today, ceramics play an important role in the search for materials that can resist thermal shock, act as abrasives, or have a better strength/weight ratio. Alumina ceramics are used for missile and rocket nose cones; silicon carbide (SiC) and molybdenum disilicide ($MoSi_2$) are used in rocket nozzles; and ceramic tiles are used for thermal insulation to protect the space shuttle on reentry through the Earth's atmosphere.

Ceramics made from uranium dioxide (UO_2) are being used as the fuel elements for nuclear power plants. Ceramics are also used as laser materials, from the chromium-doped crystals that emit a coherent monochromatic pulse of light to the optics through which the light passes. $BaTiO_3$ is used to make ceramic capacitors that have a very high capacitance. It is also used to make piezoelectric materials that develop an electric charge when subjected to a mechanical stress, which are the active elements of phonograph cartridges, sonar, and ultrasonic devices. Magnetic ceramics formed by mixing ZnO, FeO, MnO, NiO, BaO, or SrO with Fe_2O_3 are used in permanent magnets, computer memory devices, and telecommunications.

Key Terms

Alloy
Dislocation
Edge dislocation
Electrical conductivity
Electrical resistivity
Glass

Interstitial impurity
Point defect
Screw dislocation
Solid solution
Substitutional impurity

Superconductor
Thermal conductivity
Thermal expansion
Vacancy
Viscosity

Problems

Defects

A-1. Distinguish among the following point defects: vacancy, interstitial impurity, and dislocations.

A-2. What is a solid solution?

A-3. How can dislocations weaken a metal?

A-4. Why are metals softer than would be expected?

A-5. Metals are able to deform when a stress is applied. Explain why.

Metals, Semiconductors, and Insulators

A-6. Describe the differences among metals, semiconductors, and insulators.

A-7. Explain how metals conduct an electric current.

A-8. Explain why metals become better conductors of electricity as the temperature decreases but semiconductors become better conductors of electricity as the temperature increases.

A-9. Explain why adding small quantities of arsenic or gallium increases the conductivity of silicon.

A-10. Describe the difference between n-type and p-type semiconductors.

Thermal Conductivity

A-11. Why do metals feel cold even though they are at room temperature?

A-12. Why is air a good insulator?

A-13. Which would feel colder to the touch, concrete or cardboard? Both are at room temperature.

Thermal Expansion

A-14. When a garage door opener is activated, a light comes on to light the interior of the garage. Explain how you might make a device that would turn the light on and then would turn the light off after a few minutes.

A-15. If copper and iron metals are the components of a bimetallic strip, in which direction will the strip bend when heated?

Glass and Other Ceramics

A-16. How does a substance characterized as a glass differ from a ceramic?

A-17. In what region on the bond-type triangle, Figure 5.10, would ceramics be found?

A-18. Why do ceramics tend to be brittle?

A-19. Which of the compounds in Problem 68 of Chapter 9 could be classified as ceramic?

A-20. Ceramics or glasses may break if heated quickly. Explain why. What can be done to strengthen these materials?

A-21. What would be the advantages of a ceramic automobile engine over the conventional engine?

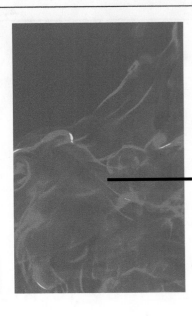

Chapter Ten

THE CONNECTION BETWEEN KINETICS AND EQUILIBRIUM

10.1 Reactions That Don't Go to Completion

Suppose that you were asked to describe the steps involved in calculating the mass of the finely divided white solid produced when a 2.00-g strip of magnesium metal is burned. You might organize your work as follows.

- Start by assuming that the magnesium reacts with oxygen in the atmosphere when it burns.
- Predict that the formula of the product is MgO.
- Use this formula to generate the following balanced equation.

$$2\,Mg(s) + O_2(g) \longrightarrow 2\,MgO(s)$$

- Use the mass of a mole of magnesium to convert grams of magnesium into moles of magnesium.

$$2.00\,\text{g Mg} \times \frac{1\,\text{mol Mg}}{24.31\,\text{g Mg}} = 0.0823\,\text{mol Mg}$$

- Use the balanced equation to convert moles of magnesium into moles of magnesium oxide.

$$0.0823\,\text{mol Mg} \times \frac{2\,\text{mol MgO}}{2\,\text{mol Mg}} = 0.0823\,\text{mol MgO}$$

- Finally, use the mass of a mole of magnesium oxide to convert moles of MgO into grams of MgO.

$$0.0823\,\text{mol MgO} \times \frac{40.30\,\text{g MgO}}{1\,\text{mol MgO}} = 3.32\,\text{g MgO}$$

Before you read any further, ask yourself, "How confident am I in the answer?"

Before we can trust the answer, we have to consider whether there are any hidden assumptions behind the calculation and then check the validity of these assumptions. Three assumptions were made in the calculation.

- We assumed that the metal strip was pure magnesium.
- We assumed that the magnesium reacts only with the oxygen in the atmosphere to form MgO, ignoring the possibility that some of the magnesium might react with the nitrogen in the atmosphere to form Mg_3N_2.
- We assumed that the reaction didn't stop until all of the magnesium metal had been consumed.

It is relatively easy to correct for the fact that the starting material may not be pure magnesium by weight. We can also correct for the fact that as much as 5% of the product of the reaction is Mg_3N_2 instead of MgO. But it is the third assumption that is of particular importance in this chapter.

The assumption that chemical reactions proceed until the last of the limiting reagent is consumed is fostered by calculations such as predicting the amount of MgO produced by burning a known amount of magnesium. It is reinforced by demonstrations such as the reaction in which a copper penny dissolves in concentrated nitric acid, which seems to continue until the copper penny disappears.

A copper penny reacting with nitric acid to form NO_2 gas.

Chemical reactions, however, don't always go to completion. The following equation provides an example of a chemical reaction that seems to stop prematurely.

$$2\,NO_2(g) \rightleftharpoons N_2O_4(g)$$

At 25°C, when 1 mol of NO_2 is added to a 1.00-L flask, the reaction seems to stop when 95% of the NO_2 has been converted to N_2O_4. Once it has reached that point, the reaction doesn't go any further. As long as the reaction is left at 25°C, about 5% of the NO_2 that was present initially will remain in the flask. Reactions that seem to stop before the limiting reagent is consumed are said to reach **equilibrium.**

It is useful to recognize the difference between reactions that come to equilibrium and those that stop when they run out of the limiting reagent. The reaction between a copper penny and nitric acid is an example of a reaction that continues until it has essentially run out of the limiting reagent. We indicate this by writing the equation for the reaction with a single arrow pointing from the reactants to the products.

$$Cu(s) + 4\,HNO_3(aq) \longrightarrow Cu(NO_3)_2(aq) + 2\,NO_2(g) + 2\,H_2O(l)$$

We indicate that a reaction comes to equilibrium by writing a pair of arrows pointing in opposite directions between the two sides of the equation.

$$2\,NO_2(g) \rightleftharpoons N_2O_4(g)$$

In order to work with reactions that come to equilibrium, we need a way to specify the amount of each reactant or product that is present in the system at any moment in time. By convention, this information is specified in terms of the concentration of each component of the system in units of moles per liter. This quantity is indicated by a symbol that consists of the formula for the reactant or product written in parentheses. For example,

(NO_2) = concentration of NO_2 in moles per liter at some moment in time

We then need a way to describe the system when it is at equilibrium. This is done by writing the symbols for each component of the system in square brackets.

$[NO_2]$ = concentration of NO_2 in moles per liter if, and only if, the reaction is at equilibrium

The fact that some reactions come to equilibrium raises a number of interesting questions.

- Why do these reactions seem to stop before all of the starting materials are converted into the products of the reaction?
- What is the difference between reactions that seem to go to completion and reactions that reach equilibrium?
- Is there any way to predict whether a reaction will go to completion or reach equilibrium?
- How does a change in the conditions of the reaction influence the amount of product formed at equilibrium?

Before we can understand how and why a chemical reaction comes to equilibrium, we have to build a model of the factors that influence the rate of a chemical reaction. We'll then apply this model to the simple chemical reactions that occur in the gas phase. The next chapter will include interactions between the components of the reaction and the solvent in which the reaction is run.

10.2 Gas-Phase Reactions

The simplest chemical reactions are those that occur in the gas phase in a single step, such as the following reaction in which a compound known as *cis*-2-butene is converted into its isomer, *trans*-2-butene.

$$CH_3 \quad CH_3 \qquad\qquad CH_3 \quad H$$
$$\underset{H}{\overset{\quad}{\diagdown}}C=C\underset{\quad H}{\overset{\quad}{\diagup}} \quad \rightleftharpoons \quad \underset{H}{\overset{\quad}{\diagdown}}C=C\underset{\quad CH_3}{\overset{\quad}{\diagup}}$$
$$\textit{cis-}2\text{-Butene} \qquad\qquad \textit{trans-}2\text{-Butene}$$

The two isomers of 2-butene have different physical properties. *cis*-2-Butene melts at $-139°C$, whereas the *trans* isomer melts at $-106°C$. The *cis* isomer boils at $4°C$, whereas the *trans* form boils at $1°C$. One of these isomers can be converted into the other by rotating one end of the $C=C$ double bond relative to the other. Rotation around $C=C$ double bonds, however, doesn't occur at room temperature. We can therefore obtain a pure sample of the starting material—*cis*-2-butene—and wait as it is transformed into the product—*trans*-2-butene—when we heat the sample.

Table 10.1 shows the amount of both *cis*- and *trans*-2-butene present in the system at various moments in time when a 1.000-mol sample of the *cis* isomer is heated to 400°C in a 10.0-L flask. Initially, there is no *trans*-2-butene in the system. With time, however, the amount of *cis*-2-butene gradually decreases as this compound is converted into the *trans* isomer.

Because one molecule of *trans*-2-butene is produced each time a molecule of the *cis* isomer is consumed in this reaction, the total number of moles of the two isomers must always be the same as the number of moles of the *cis* isomer that were present at the start of the experiment. The amount of the *trans* isomer present at any moment in time can therefore be calculated from the amount of the *cis* isomer that remains in the system, and vice versa.

➤ **CHECKPOINT**

Assume that $N_2O_4(g)$ is added to an evacuated container at 25°C. What compounds would be present in the container at equilibrium?

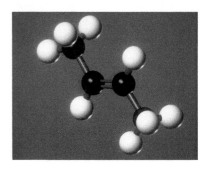

Molecular model of *trans*-2-butene

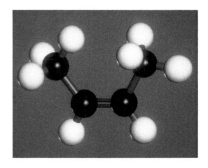

Molecular model of *cis*-2-butene

Table 10.1

Experimental Data for the Isomerization of *cis*-2-Butene to *trans*-2-Butene at 400°C in a 10.0-L Flask

Time	Moles of *cis*-2-Butene	Moles of *trans*-2-Butene
0	1.000	0
5.00 days	0.919	0.081
10.00 days	0.848	0.152
15.00 days	0.791	0.209
20.00 days	0.741	0.259
40.00 days	0.560	0.440
60.00 days	0.528	0.472
120.00 days	0.454	0.546
1 year	0.441	0.559
2 years	0.441	0.559
3 years	0.441	0.559

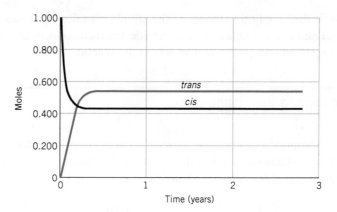

Fig. 10.1 A plot of the number of moles of the *cis* and *trans* isomers of 2-butene versus time at 400°C. Initially there is no *trans*-2-butene, but as time passes the concentration of the *cis* isomer decreases and the concentration of the *trans* isomer increases.

In this case, we started with 1.000 mol of the *cis* isomer. At any moment in time, the number of moles of the *cis* isomer that remain in the system is $1.000 - x$,

$$n_{cis\text{-}2\text{-butene}} = 1.000 - x$$

and the number of moles of the *trans* isomer is equal to x.

$$n_{trans\text{-}2\text{-butene}} = x$$

Figure 10.1 shows a plot of the number of moles of the *cis* and *trans* isomers of 2-butene versus time. This figure shows that there is no change in the number of moles of either component once the reaction reaches the point at which the system contains 0.441 mol of *cis*-2-butene and 0.559 mol of *trans*-2-butene. No matter how long we wait, no more *cis*-2-butene is converted into *trans*-2-butene. This indicates that the reaction has come to equilibrium.

Chemical reactions at equilibrium are described in terms of the number of moles per liter of each component of the system. For example,

[*cis*-2-butene] = concentration of *cis*-2-butene at equilibrium in units of
moles per liter

[*trans*-2-butene] = concentration of *trans*-2-butene at equilibrium in units of
moles per liter

The concentrations of *cis*- and *trans*-2-butene at equilibrium depend on the initial conditions of the experiment. But, at a given temperature, the ratio of the equilibrium concentrations of the two components of the reaction is always the same. It doesn't matter whether we start with a great deal of *cis*-2-butene or a relatively small amount, or whether we start with a pure sample of *cis*-2-butene or one that already contains some of the *trans* isomer. When the reaction reaches equilibrium at 400°C, the concentration of the *trans* isomer divided by that of the *cis* isomer is always 1.27.

The equation that describes the relationship between the concentrations of the two components of the reaction at equilibrium is known as the **equilibrium constant expression**, where K_c is the **equilibrium constant** for the reaction.

$$K_c = \frac{[trans\text{-}2\text{-butene}]}{[cis\text{-}2\text{-butene}]} = 1.27$$

The subscript c in the equilibrium constant indicates that the constant has been calculated from the concentrations of the reactants and products in units of moles per liter.

➤ **CHECKPOINT**

Assume that 1.0 mol of *trans*-2-butene is placed in an empty flask at 400°C. What will be the ratio of the concentration of *trans*-2-butene to *cis*-2-butene when this system reaches equilibrium?

Exercise 10.1

Calculate the equilibrium constant for the conversion of *cis*-2-butene to *trans*-2-butene for the following sets of experimental data.

(a) A 5.00-mol sample of the *cis* isomer was added to a 10.0-L flask and heated to 400°C until the reaction came to equilibrium. At equilibrium, the system contained 2.80 mol of the *trans* isomer.

(b) A 0.100-mol sample of the *cis* isomer was added to a 25.0-L flask and heated to 400°C until the reaction came to equilibrium. At equilibrium, the system contained 0.0559 mol of the *trans* isomer.

Solution

(a) If a sample that started with 5.00 mol of *cis*-2-butene forms 2.80 mol of *trans*-2-butene at equilibrium, then 2.20 mol of the *cis* isomer must remain after the reaction reaches equilibrium. Because the experiment was done in a 10.0-L flask, the equilibrium concentrations of the two components of the reaction have the following values.

$$[\textit{trans}\text{-2-butene}] = \frac{2.80 \text{ mol}}{10.0 \text{ L}} = 0.280 \, M$$

$$[\textit{cis}\text{-2-butene}] = \frac{2.20 \text{ mol}}{10.0 \text{ L}} = 0.220 \, M$$

The equilibrium constant, K_c, for the reaction is therefore 1.27.

$$K_c = \frac{[\textit{trans}\text{-2-butene}]}{[\textit{cis}\text{-2-butene}]} = \frac{0.280 \, M}{0.220 \, M} = 1.27$$

(b) In this case, the system comes to equilibrium at the following concentrations of the *cis* and *trans* isomers.

$$[\textit{trans}\text{-2-butene}] = \frac{0.0559 \text{ mol}}{25.0 \text{ L}} = 0.00224 \, M$$

$$[\textit{cis}\text{-2-butene}] = \frac{0.0441 \text{ mol}}{25.0 \text{ L}} = 0.00176 \, M$$

Even though the concentrations of the two isomers at equilibrium are very different from the values obtained in the previous experiment, the ratio of the concentrations at equilibrium is exactly the same.

$$K_c = \frac{[\textit{trans}\text{-2-butene}]}{[\textit{cis}\text{-2-butene}]} = \frac{0.00224 \, M}{0.00176 \, M} = 1.27$$

10.3 The Rate of a Chemical Reaction

Experiments such as the one that gave rise to the data in Table 10.1 are classified as measurements of **chemical kinetics** (from the Greek stem meaning "to move"). The goal of these experiments is to describe the **rate of reaction,** that is, the rate at which the reactants are transformed into the products of the reaction.

The term *rate* is often used to describe the change in a quantity that occurs per unit of time. The rate of inflation, for example, is the change in the average cost of a collection of standard items per year. The rate at which an object travels through space is the distance traveled per unit of time in units of miles per hour or kilometers per second.

In chemical kinetics, the distance traveled is the change in the concentration of one of the components of the reaction. The rate of a reaction is therefore the change in the concentration of one of the components, $\Delta(X)$, that occurs during a given time, Δt. The concentration of X is written in parentheses here because the system isn't at equilibrium.

$$\text{rate} = \frac{\Delta(X)}{\Delta t} \qquad \text{(where } X \text{ is one of the products)}$$

By convention, the symbol Δ represents a change calculated by subtracting the initial conditions from the final conditions. Thus, the $\Delta(X)$ represents the final concentration minus the initial concentration.

$$\Delta(X) = (X)_{\text{final}} - (X)_{\text{initial}}$$

If X is one of the products of the reaction, then $\Delta(X)$ is positive. If X is a reactant, $\Delta(X)$ is negative and a minus sign is added to the rate equation to turn the rate into a positive number.

$$\text{rate} = -\frac{\Delta(X)}{\Delta t} \qquad \text{(if } X \text{ is one of the reactants)}$$

Let's use the data in Table 10.1 to calculate the rate at which *cis*-2-butene is transformed into *trans*-2-butene during each of the following periods.

- During the first time interval, when the number of moles of *cis*-2-butene in the 10.0-L flask falls from 1.000 to 0.919
- During the second interval, when the amount of *cis*-2-butene falls from 0.919 to 0.848 mol
- During the third interval, when the amount of *cis*-2-butene falls from 0.848 to 0.791 mol

Before we can calculate the rate of the reaction during each of these time intervals, we have to remember that the rate of reaction is defined in terms of changes in the number of moles per liter (M) of one of the components of the reaction, not the number of moles of that reactant. Thus we have to recognize that 1.000 mol of *cis*-2-butene in a 10.0-L flask corresponds to a concentration of 0.1000 M.

During the first time period, the rate of the reaction is 1.6×10^{-3} M/day.

$$\text{rate} = -\frac{\Delta(X)}{\Delta t} = -\frac{(0.0919\,M - 0.1000\,M)}{(5.00\,\text{days} - 0\,\text{days})} = 1.6 \times 10^{-3}\,M/\text{day}$$

During the second time period, the rate of reaction is slightly smaller.

$$\text{rate} = -\frac{\Delta(X)}{\Delta t} = -\frac{(0.0848\,M - 0.0919\,M)}{(10.00\,\text{days} - 5.00\,\text{days})} = 1.4 \times 10^{-3}\,M/\text{day}$$

During the third time period, the rate of reaction is even smaller.

$$\text{rate} = -\frac{\Delta(X)}{\Delta t} = -\frac{(0.0791\,M - 0.0848\,M)}{(15.00\,\text{days} - 10.00\,\text{days})} = 1.1 \times 10^{-3}\,M/\text{day}$$

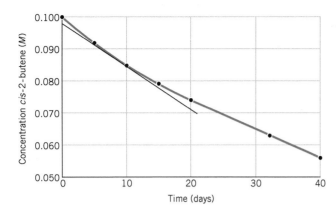

Fig. 10.2 The rate of reaction at a given time for the isomerization of *cis*-2-butene is the negative of the slope of a tangent drawn to the concentration curve at that particular point in time.

These calculations illustrate an important point: The rate of the reaction isn't constant; it changes with time. The rate of the reaction gradually decreases as the starting materials are consumed, which means that the rate of reaction changes while it is being measured.

We can minimize the error this introduces into our measurements by measuring the rate of reaction over periods of time that are short compared with the time it takes for the reaction to occur. We might try, for example, to measure the infinitesimally small change in reactant concentration, $d(X)$, that occurs over an infinitesimally short period of time, dt. This ratio is known as the *instantaneous rate of reaction.*

$$\text{rate} = -\frac{d(X)}{dt}$$

The instantaneous rate of reaction at any moment in time can be calculated from a graph of the concentration of the reactant (or product) versus time. Figure 10.2 shows how the rate of reaction for the isomerization of *cis*-2-butene can be calculated from such a graph. The rate of reaction at any moment is equal to the negative of the slope of a tangent drawn to the curve at that moment.

An interesting result is obtained when the instantaneous rate of reaction is calculated at various points along the curve in Figure 10.2. The rate of reaction at every point on the curve is directly proportional to the concentration of *cis*-2-butene at that moment in time.

$$\text{rate} = k(\textit{cis-}2\text{-butene})$$

This equation, which is determined from experimental data, describes the rate of the reaction. It is therefore called the **rate law** for the reaction. The proportionality constant k is known as the **rate constant.**

> ➤ **CHECKPOINT**
>
> The following data were obtained for the rate constant for the decomposition of one of the metabolites that supplies energy in living systems.
>
> | Temperature (°C) | Rate Constant (s^{-1}) |
> | --- | --- |
> | 15 | 2.5×10^{-2} |
> | 20 | 4.5×10^{-2} |
> | 25 | 8.1×10^{-2} |
> | 30 | 1.6×10^{-1} |
>
> What do these data suggest happens to the rate of the reaction as the temperature at which the reaction is run increases?

10.4 The Collision Theory Model of Gas-Phase Reactions

One way to understand why some reactions come to equilibrium is to consider a simple gas-phase reaction that occurs in a single step, such as the transfer of a chlorine atom from $ClNO_2$ to NO to form NO_2 and ClNO.

$$ClNO_2(g) + NO(g) \rightleftharpoons NO_2(g) + ClNO(g)$$

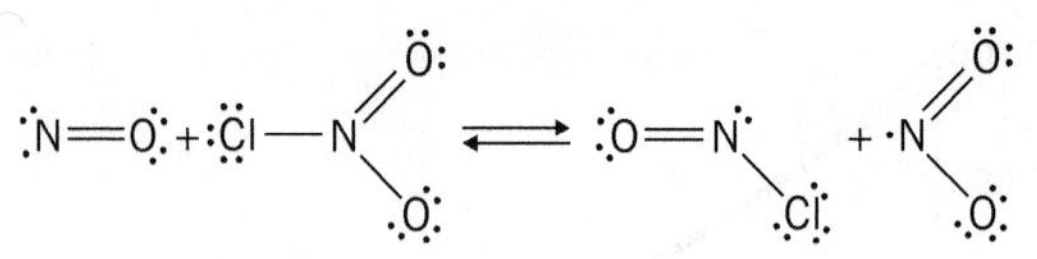

Fig. 10.3 The reaction between $ClNO_2$ and NO to form NO_2 and ClNO is a simple, one-step reaction that involves the transfer of a chlorine atom.

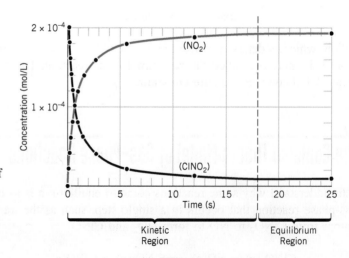

This reaction can be understood by writing the Lewis structures for the four components of the reaction. Because they contain an odd number of electrons, both NO and NO_2 can combine with a neutral chlorine atom to form a molecule in which all of the electrons are paired. The reaction therefore involves the transfer of a chlorine atom from one molecule to another, as shown in Figure 10.3.

Figure 10.4 combines a plot of the concentration of $ClNO_2$ versus time as this reactant is consumed with a plot of the concentration of NO_2 versus time as this product of the reaction is formed. The data in Figure 10.4 are consistent with the following rate law for the reaction.

$$\text{rate} = k(ClNO_2)(NO)$$

According to this rate law, the rate at which $ClNO_2$ and NO are converted into NO_2 and ClNO is proportional to the product of the concentrations of the two reactants. Initially, the rate of reaction is relatively fast. As the reactants are converted to products, however, the $ClNO_2$ and NO concentrations become smaller, and the reaction slows down.

We might expect the reaction to stop when it runs out of either $ClNO_2$ or NO. In practice, the reaction seems to stop before this happens. This is a very fast reaction—the concentration of $ClNO_2$ drops by a factor of 2 in about 1 second. And yet, no matter how long we wait, some residual $ClNO_2$ and NO remain in the reaction flask.

Figure 10.4 divides the plot of the change in the concentrations of NO_2 and $ClNO_2$ into a **kinetic region** and an **equilibrium region.** The kinetic region marks the period during which the concentrations of the components of the reaction are constantly changing. The equilibrium region is the period after which the reaction seems to stop, when there is no further significant change in the concentrations of the components of the reaction.

The fact that this reaction seems to stop before all of the reactants are consumed can be explained with the **collision theory** of chemical reactions. The collision theory assumes that $ClNO_2$ and NO molecules must collide before a chlorine atom can be transferred from one molecule to the other. This assumption

Fig. 10.4 Plot of the change in the concentration of $ClNO_2$ superimposed on a plot of the change in concentration of NO_2 as $ClNO_2$ reacts with NO to produce NO_2 and ClNO. The graph can be divided into a kinetic and an equilibrium region.

explains why the rate of the reaction is proportional to the concentration of both $ClNO_2$ and NO.

$$\text{rate} = k(ClNO_2)(NO)$$

The number of collisions per second between $ClNO_2$ and NO molecules depends on their concentrations. As $ClNO_2$ and NO are consumed in the reaction, the number of collisions per second between molecules of these reactants becomes smaller, and the reaction slows down.

Suppose that we start with a mixture of $ClNO_2$ and NO, but no NO_2 or ClNO. The only reaction that can occur at first is the transfer of a chlorine atom from $ClNO_2$ to NO.

$$ClNO_2(g) + NO(g) \longrightarrow NO_2(g) + ClNO(g)$$

Eventually, NO_2 and ClNO build up in the reaction flask, and collisions between these molecules can result in the transfer of a chlorine atom in the opposite direction.

$$ClNO_2(g) + NO(g) \longleftarrow NO_2(g) + ClNO(g)$$

The collision theory model of chemical reactions assumes that the rate of a simple, one-step reaction is proportional to the product of the concentrations of the substances consumed in that reaction. The rate of the forward reaction is therefore proportional to the product of the concentrations of the two starting materials.

$$\text{rate}_{\text{forward}} = k_f(ClNO_2)(NO)$$

The rate of the reverse reaction, on the other hand, is proportional to the product of concentrations of the compounds formed in the reaction.

$$\text{rate}_{\text{reverse}} = k_r(NO_2)(ClNO)$$

Initially, the rate of the forward reaction is much larger than the rate of the reverse reaction because the system contains $ClNO_2$ and NO but virtually no NO_2 or ClNO.

$$\text{Initially:} \quad \text{rate}_{\text{forward}} \gg \text{rate}_{\text{reverse}}$$

As $ClNO_2$ and NO are consumed, the rate of the forward reaction slows down. At the same time, NO_2 and ClNO accumulate, and the reverse reaction speeds up.

The system eventually reaches a point at which the rates of the forward and reverse reactions are the same.

$$\text{Eventually:} \quad \text{rate}_{\text{forward}} = \text{rate}_{\text{reverse}}$$

At this point, the concentrations of reactants and products remain the same, no matter how long we wait. On the molecular scale, $ClNO_2$ and NO are consumed in the forward reaction at the same rate at which they are produced in the reverse reaction. The same thing happens to NO_2 and ClNO. When the rates of the forward and reverse reactions are the same, there is no longer any change in the concentrations of the reactants or products of the reaction on the macroscopic scale. In other words, the reaction is at equilibrium.

We can now see that there are two ways to define equilibrium.

- A system in which there is no change in the concentrations of the reactants and products of a reaction with time.
- A system in which the rates of the forward and reverse reactions are the same.

➤ **CHECKPOINT**

If the rate of the reverse reaction is greater than the rate of the forward re-action at a given moment in time, does this mean that the reverse rate constant is larger than the forward rate constant?

The first definition is based on the results of experiments that tell us that some reactions seem to stop prematurely—they reach a point at which no more reactants are converted to products before the limiting reagent is consumed. The other definition is based on a theoretical model of chemical reactions that explains why reactions reach equilibrium.

10.5 Equilibrium Constant Expressions

Reactions don't stop when they come to equilibrium. But the rate of the forward and reverse reactions at equilibrium are the same, so there is no net change in the concentrations of the reactants or products, and the reactions appear to stop on the macroscopic scale. Chemical equilibrium is an example of a *dynamic* balance between the forward and reverse reactions, not a static balance.

Let's look at the logical consequences of the assumption that the reaction between $ClNO_2$ and NO eventually reaches equilibrium at a given temperature.

$$ClNO_2(g) + NO(g) \rightleftharpoons NO_2(g) + ClNO(g)$$

The rates of the forward and reverse reactions are the same when the system is at equilibrium.

At equilibrium: $rate_{forward} = rate_{reverse}$

Substituting the rate laws for the forward and reverse reactions into the equality gives the following result.

At equilibrium: $k_f(ClNO_2)(NO) = k_r(NO_2)(ClNO)$

But this equation is valid only when the system is at equilibrium, so we should replace the $(ClNO_2)$, (NO), (NO_2), and $(ClNO)$ terms with symbols indicating that the reaction is at equilibrium. By convention, we use square brackets for this purpose. The equation describing the balance between the forward and reverse reactions when the system is at equilibrium should therefore be written as follows.

At equilibrium: $k_f[ClNO_2][NO] = k_r[NO_2][ClNO]$

Rearranging the equation gives the following result.

$$\frac{k_f}{k_r} = \frac{[NO_2][ClNO]}{[ClNO_2][NO]}$$

Because k_f and k_r are constants, the ratio of k_f divided by k_r must also be a constant. This ratio is the **equilibrium constant** for the reaction, K_c. As we have seen, the ratio of the concentrations of the reactants and products is known as the **equilibrium constant expression.**

Equilibrium constant expression

$$K_c = \frac{k_f}{k_r} = \frac{[NO_2][ClNO]}{[ClNO_2][NO]}$$

Equilibrium constant

No matter what combination of concentrations of reactants and products we start with, the reaction reaches equilibrium when the ratio of the concentrations defined by the equilibrium constant expression is equal to the equilibrium constant for the reaction at a given temperature. We can start with a lot of $ClNO_2$ and very little NO, or a lot of NO and very little $ClNO_2$. It doesn't matter. When the reaction reaches equilibrium, the relationship between the concentrations of the reactants and products described by the equilibrium constant expression will always be the same. At 25°C, this reaction always reaches equilibrium when the ratio of the concentrations is 1.3×10^4.

$$K_c = \frac{[NO_2][ClNO]}{[ClNO_2][NO]} = 1.3 \times 10^4 \quad \text{(at 25°C)}$$

K_c is always reported without units. However, any calculations using K_c require the concentration of the products and reactants to be in units of molarity (moles/liter).

What happens if we approach equilibrium from the other direction? In this case, we start with a system that contains the products of the reaction—NO_2 and ClNO—and then let the reaction come to equilibrium. The rate laws for the forward and reverse reactions will still be the same.

$$\text{rate}_{forward} = k_f(ClNO_2)(NO)$$
$$\text{rate}_{reverse} = k_r(NO_2)(ClNO)$$

Now, however, the rate of the forward reaction initially will be much smaller than the rate of the reverse reaction.

$$\text{Initially:} \quad \text{rate}_{forward} \ll \text{rate}_{reverse}$$

As time passes, however, the rate of the reverse reaction will slow down and the rate of the forward reaction will speed up until they become equal. At that point, the reaction will have reached equilibrium.

$$\text{At equilibrium:} \quad k_f[ClNO_2][NO] = k_r[NO_2][ClNO]$$

Rearranging the equation gives us the same equilibrium constant expression.

$$K_c = \frac{k_f}{k_r} = \frac{[NO_2][ClNO]}{[ClNO_2][NO]} = \frac{[\text{products}]}{[\text{reactants}]}$$

We get the same equilibrium constant expression and the same equilibrium constant no matter whether we start with only reactants, only products, or a mixture of reactants and products.

Exercise 10.2

The rate constants for the forward and reverse reactions for the following reaction have been measured. At 25°C, k_f is 7.3×10^3 liters per mole-second and k_r is 0.55 liter per mole-second. Calculate the equilibrium constant for the reaction at this temperature.

$$ClNO_2(g) + NO(g) \rightleftharpoons NO_2(g) + ClNO(g)$$

Solution

We start by recognizing that the rates of the forward and reverse reactions at equilibrium are the same.

$$\text{At equilibrium:} \qquad \text{rate}_{\text{forward}} = \text{rate}_{\text{reverse}}$$

We then substitute the rate laws for the reaction into the equality.

$$\text{At equilibrium:} \qquad k_f[ClNO_2][NO] = k_r[NO_2][ClNO]$$

We then rearrange the equation to get the equilibrium constant expression for the reaction.

$$K_c = \frac{k_f}{k_r} = \frac{[NO_2][ClNO]}{[ClNO_2][NO]}$$

The equilibrium constant for the reaction is therefore equal to the rate constant for the forward reaction divided by the rate constant for the reverse reaction.

$$K_c = \frac{k_f}{k_r} = \frac{7.3 \times 10^3 \text{ L/mol-s}}{0.55 \text{ L/mol-s}} = 1.3 \times 10^4$$

➤ **CHECKPOINT**

If the forward rate constant for a given reaction is twice the reverse rate constant, what is the equilibrium constant for the reaction?

Any reaction that reaches equilibrium, no matter how simple or complex, has an equilibrium constant expression that satisfies the following rules.

Rules for Writing Equilibrium Constant Expressions

- Even though chemical reactions that reach equilibrium occur in both directions, the substances on the right side of the equation are assumed to be the "products" of the reaction and the substances on the left side of the equation are assumed to be the "reactants."
- The products of the reaction are always written above the line, in the numerator.
- The reactants are always written below the line, in the denominator.
- For systems in which all species are either gases or aqueous solutions, the equilibrium constant expression contains a term for every reactant and every product of the reaction.
- The numerator of the equilibrium constant expression is the product of the concentrations of each species on the right side of the equation raised to a

power equal to the coefficient for that component in the balanced equation for the reaction.

- The denominator of the equilibrium constant expression is the product of the concentrations of each species on the left side of the equation raised to a power equal to the coefficient for that component in the balanced equation for the reaction.

Exercise 10.3

Write equilibrium constant expressions for the following reactions.

(a) $2 NO_2(g) \rightleftharpoons N_2O_4(g)$

(b) $2 SO_3(g) \rightleftharpoons 2 SO_2(g) + O_2(g)$

(c) $N_2(g) + 3 H_2(g) \rightleftharpoons 2 NH_3(g)$

Solution

In each case, the equilibrium constant expression is the product of the concentrations of the species on the right side of the equation divided by the product of the concentrations of those on the left side of the equation. All concentrations are raised to the power equal to the coefficient for the species in the balanced equation.

(a) $K_C = \dfrac{[N_2O_4]}{[NO_2]^2}$

(b) $K_C = \dfrac{[SO_2]^2[O_2]}{[SO_3]^2}$

(c) $K_C = \dfrac{[NH_3]^2}{[N_2][H_2]^3}$

Exercise 10.4

The first step in the series of reactions by which the sugar known as glucose is metabolized involves the formation of a compound known as glucose-6-phosphate. In theory, this could occur by the direct reaction between glucose and inorganic phosphate (PO_4^{3-}), which is labeled P_i by biochemists.

$$\text{glucose}(aq) + P_i(aq) \rightleftharpoons \text{glucose-6-phosphate}(aq)$$

Write the equilibrium constant expression for this reaction.

Solution

No matter whether a reaction occurs in the gas phase or, as in this reaction, in a cell in the body, the same principles apply. All the species in this reaction exist in solution, and hence all can be expressed as concentrations in moles per liter. The equilibrium constant expression for this reaction would therefore be written as follows.

$$K_c = \frac{[\text{glucose-6-phosphate}]}{[\text{glucose}][P_i]}$$

What happens to the magnitude of the equilibrium constant for a reaction when we turn the equation around? Consider the following reaction, for example.

$$ClNO_2(g) + NO(g) \rightleftharpoons NO_2(g) + ClNO(g)$$

The equilibrium constant expression for the equation is written as follows.

$$K_c = \frac{[NO_2][ClNO]}{[ClNO_2][NO]} = 1.3 \times 10^4 \quad \text{(at 25°C)}$$

This reaction, however, can also be represented by an equation written in the opposite direction.

$$NO_2(g) + ClNO(g) \rightleftharpoons ClNO_2(g) + NO(g)$$

The equilibrium constant expression is now written as follows.

$$K_c' = \frac{[ClNO_2][NO]}{[NO_2][ClNO]}$$

Each of these equilibrium constant expressions is the inverse of the other. We can therefore calculate K_c' by dividing K_c into 1.

$$K_c' = \frac{1}{K_c} = \frac{1}{1.3 \times 10^4} = 7.7 \times 10^{-5}$$

We can also calculate equilibrium constants by combining two or more reactions for which the values of K_c are known. For example, we know the equilibrium constants for the following gas-phase reactions at 200°C.

$$N_2(g) + O_2(g) \rightleftharpoons 2\,NO(g) \qquad K_{c1} = \frac{[NO]^2}{[N_2][O_2]} = 2.3 \times 10^{-19}$$

$$2\,NO(g) + O_2(g) \rightleftharpoons 2\,NO_2(g) \qquad K_{c2} = \frac{[NO_2]^2}{[NO]^2[O_2]} = 3 \times 10^6$$

We can combine these reactions to obtain an equation for the reaction between N_2 and O_2 to form NO_2.

$$
\begin{array}{c}
N_2(g) + O_2(g) \rightleftharpoons 2\,NO(g) \\
+\ 2\,NO(g) + O_2(g) \rightleftharpoons 2\,NO_2(g) \\
\hline
N_2(g) + 2\,O_2(g) \rightleftharpoons 2\,NO_2(g) \qquad K_c = ?
\end{array}
$$

The equilibrium constant expression for the overall reaction is equal to the product of the equilibrium constant expressions for the two steps in the reaction.

$$K_c = K_{c1} \times K_{c2} = \frac{[NO]^2}{[N_2][O_2]} \times \frac{[NO_2]^2}{[NO]^2[O_2]} = \frac{[NO_2]^2}{[N_2][O_2]^2}$$

The equilibrium constant for the overall reaction is therefore equal to the product of the equilibrium constants for the individual reactions.

$$K_c = K_{c1} \times K_{c2} = (2.3 \times 10^{-19})(3 \times 10^6) = 7 \times 10^{-13}$$

Exercise 10.5

Given

$$N_2(g) + \tfrac{1}{2} O_2(g) \rightleftharpoons N_2O(g) \qquad K_{c1} = 2.7 \times 10^{-18}$$
$$N_2(g) + O_2(g) \rightleftharpoons 2\,NO(g) \qquad K_{c2} = 2.3 \times 10^{-19}$$

Determine K_c for the reaction

$$N_2(g) + 2\,NO(g) \rightleftharpoons 2\,N_2O(g) \qquad K_c = ?$$

Solution

We need to manipulate the chemical equations so that they sum together to give the equation for which we wish to determine K_c. We need $2\,N_2O$ as a product so the first reaction must be doubled resulting in squaring its equilibrium constant.

$$2 \times (N_2(g) + \tfrac{1}{2} O_2(g) \rightleftharpoons N_2O(g)) \qquad K'_{c1} = (2.7 \times 10^{-18})^2$$
$$2\,N_2(g) + O_2(g) \rightleftharpoons 2\,N_2O(g) \qquad K'_{c1} = 7.3 \times 10^{-36}$$

In the second reaction we need NO as a reactant so the reaction must be reversed, resulting in taking the inverse of its equilibrium constant.

$$2\,NO(g) \rightleftharpoons N_2(g) + O_2(g) \qquad K'_{c2} = 1/(2.3 \times 10^{-19})$$
$$2\,NO(g) \rightleftharpoons N_2(g) + O_2(g) \qquad K'_{c2} = 4.3 \times 10^{18}$$

The two new equations may now be added together.

$$2\,N_2(g) + O_2(g) \rightleftharpoons 2\,N_2O(g) \qquad\qquad K'_{c1} = 7.3 \times 10^{-36}$$
$$\underline{\quad 2\,NO(g) \rightleftharpoons N_2(g) + O_2(g) \qquad\qquad K'_{c2} = 4.3 \times 10^{18}}$$
$$2\,N_2(g) + \cancel{O_2(g)} + 2\,NO(g) \rightleftharpoons 2\,N_2O(g) + \cancel{N_2(g)} + \cancel{O_2(g)}$$

$$N_2(g) + 2\,NO(g) \rightleftharpoons 2\,N_2O(g) \quad K_c = K'_{c1} \times K'_{c2} = (7.3 \times 10^{-36})(4.3 \times 10^{18})$$
$$K_c = 3.1 \times 10^{-17}$$

10.6   Reaction Quotients: A Way to Decide Whether a Reaction Is at Equilibrium

We now have a model that describes what happens when a reaction reaches equilibrium. At the molecular level, the rate of the forward reaction is equal to the rate of the reverse reaction. Because the reaction proceeds in both directions at the same rate, there is no apparent change in the concentrations of the reactants or the products on the macroscopic scale (i.e., the level of objects visible to the naked eye).

This model can also be used to predict the direction in which a reaction has to shift to reach equilibrium. If the concentrations of the reactants are too large for the reaction to be at equilibrium, the rate of the forward reaction will be faster than that of the reverse reaction, and some of the reactants will be converted to products until equilibrium is achieved. Conversely, if the concentrations of the reactants are too small, the rate of the reverse reaction will exceed that of the

The kinetics of reactions that involve molecular iodine can be monitored by watching the formation or disappearance of the intense violet color of elemental iodine.

forward reaction, and the reaction will convert some of the excess products back into reactants until the system reaches equilibrium.

We can determine the direction in which a reaction has to shift to reach equilibrium by comparing the **reaction quotient (Q_c)** for the reaction at some moment in time with the equilibrium constant (K_c) for the reaction. The reaction quotient expression is written in much the same way as the equilibrium constant expression. But the concentrations used to calculate Q_c describe the system at any moment in time, whereas the concentrations used to calculate K_c describe the system only when it is at equilibrium.

To illustrate how the reaction quotient is used, consider the following gas-phase reaction.

$$H_2 + I_2(g) \rightleftharpoons 2\,HI(g)$$

The equilibrium constant expression for the reaction is written as follows.

$$K_c = \frac{[HI]^2}{[H_2][I_2]} = 60 \qquad \text{(at 350°C)}$$

By analogy, we can write the expression for the reaction quotient as follows.

$$Q_c = \frac{(HI)^2}{(H_2)(I_2)}$$

There are three important differences between the equilibrium constant expression and the reaction quotient expression. First, we use brackets, such as [HI], in the equilibrium constant expression to indicate that the reaction is at equilibrium. We then use parentheses, such as (HI), in the reaction quotient expression to indicate that the reaction quotient can be calculated at any moment in time. The most important difference between these expressions revolves around the results of the calculation. There is only one possible value for K_c because the equilibrium constant expression is valid only when the reaction is at equilibrium. Q_c, on the other hand, can take on any value from zero to infinity.

If the system contains a large amount of HI and very little H_2 and I_2, the reaction quotient is very large. If the system contains relatively little HI and a large amount of H_2 and/or I_2, the reaction quotient is very small.

At any moment in time, there are three possibilities.

- **Q_c is smaller than K_c.** The system contains too much reactant and not enough product to be at equilibrium. The value of Q_c must increase in order for the reaction to reach equilibrium. Thus the reaction has to convert some of the reactants into products to come to equilibrium.

- **Q_c is equal to K_c.** If this is true, then the reaction is at equilibrium.

- **Q_c is larger than K_c.** The system contains too much product and not enough reactant to be at equilibrium. The value of Q_c must decrease before the reaction can come to equilibrium. Thus the reaction must convert some of the products into reactants to reach equilibrium.

Exercise 10.6

Assume that the concentrations of H_2, I_2, and HI can be measured for the following reaction at any moment in time.

$$H_2 + I_2(g) \rightleftharpoons 2\,HI(g) \qquad K_c = 60 \text{ (at 350°C)}$$

For each of the following sets of concentrations, determine whether the reaction is at equilibrium. If it isn't, decide in which direction it must go to reach equilibrium.

(a) $(H_2) = (I_2) = (HI) = 0.010\ M$

(b) $(HI) = 0.30\ M$, $(H_2) = 0.010\ M$, $(I_2) = 0.15\ M$

(c) $(H_2) = (HI) = 0.10\ M$, $(I_2) = 0.0010\ M$

Solution

(a) The best way to decide whether the reaction is at equilibrium is to compare the reaction quotient with the equilibrium constant for the reaction.

$$Q_c = \frac{(HI)^2}{(H_2)(I_2)} = \frac{(0.010)^2}{(0.010)(0.010)} = 1.0 < K_c$$

The reaction quotient in this case is *smaller* than the equilibrium constant. The only way to get the system to equilibrium is to increase the magnitude of the reaction quotient. This can be done by converting some of the H_2 and I_2 into HI. The reaction therefore has to shift to the right to reach equilibrium.

(b) The reaction quotient for this set of concentrations is equal to the equilibrium constant for the reaction.

$$Q_c = \frac{(HI)^2}{(H_2)(I_2)} = \frac{(0.30)^2}{(0.010)(0.15)} = 60 = K_c$$

The reaction is therefore at equilibrium.

(c) The reaction quotient for this set of concentrations is larger than the equilibrium constant for the reaction.

$$Q_c = \frac{(HI)^2}{(H_2)(I_2)} = \frac{(0.10)^2}{(0.10)(0.0010)} = 1.0 \times 10^2 > K_c$$

To reach equilibrium, the concentrations of the reactants and products must change until the reaction quotient is equal to the equilibrium constant. This involves converting some of the HI back into H_2 and I_2. Thus the reaction has to shift to the left to reach equilibrium.

• •

10.7 Changes in Concentration That Occur as a Reaction Comes to Equilibrium

The values of Q_c and K_c for a reaction tell us whether a reaction is at equilibrium at any moment in time. If it isn't, the relative sizes of Q_c and K_c tell us the direction in which the reaction must shift to reach equilibrium. Now we need a way to predict how far the reaction has to go to reach equilibrium. To illustrate how this is done, let's look at the following reaction, which is known as the water-gas shift reaction.

$$CO(g) + H_2O(g) \rightleftharpoons CO_2(g) + H_2(g)$$

"Water gas" is a mixture of CO and H_2 prepared by blowing alternating blasts of steam and either air or oxygen through a bed of white-hot coal. The exothermic reactions between coal and oxygen to produce CO and CO_2 provide enough energy to drive the reaction between steam and coal to form "water gas." Water gas is also known as "syngas" because it can be used as the starting materials to produce synthetic natural gas or liquid fuels such as gasoline. The water-gas shift reaction can be used to produce high-purity hydrogen for the synthesis of ammonia.

Suppose that you are faced with the following problem.

The reaction between CO and H_2O in the presence of a catalyst at a temperature of 400°C produces a mixture of CO_2 and H_2 via the water-gas shift reaction.

$$CO(g) + H_2O(g) \rightleftharpoons CO_2(g) + H_2(g)$$

The equilibrium constant for this reaction is 0.080 at an elevated temperature. Assume that the initial concentration of both CO and H_2O is 0.100 mole per liter and that there is no CO_2 or H_2 in the system when we start. Calculate the concentrations of CO, H_2O, CO_2, and H_2 when the reaction reaches equilibrium.

The first step toward solving this problem involves organizing the information so that it provides clues as to how to proceed. The problem contains the following information: (1) a balanced equation, (2) an equilibrium constant for the reaction, (3) a description of the initial conditions, and (4) an indication of the goal of the calculation, namely, to figure out the equilibrium concentrations of the four components of the reaction.

The following format offers a useful way to summarize this information.

$$CO(g) + H_2O(g) \rightleftharpoons CO_2(g) + H_2(g) \qquad K_c = \frac{[CO_2][H_2]}{[CO][H_2O]} = 0.080$$

Initial: 0.100 *M*   0.100 *M*   0   0
Equilibrium: ?   ?   ?   ?

We start with the balanced equation and the equilibrium constant for the reaction and then add what we know about the initial and equilibrium concentrations of the various components of the reaction. Initially, the flask contains 0.100 mol/L of CO and H_2O in the gas phase and no CO_2 or H_2. Our goal is to calculate the equilibrium concentrations of the four substances.

Before we do anything else, we have to decide whether the reaction is at equilibrium. We can do this by comparing the reaction quotient for the initial conditions with the equilibrium constant for the reaction.

$$Q_c = \frac{(CO_2)(H_2)}{(CO)(H_2O)} = \frac{(0)(0)}{(0.100)(0.100)} = 0 < K_c$$

Although the equilibrium constant is small ($K_c = 8.0 \times 10^{-2}$), the reaction quotient is even smaller ($Q_c = 0$). The only way for the reaction to get to equilibrium is for some of the CO and H_2O to react to form CO_2 and H_2.

Because the reaction isn't at equilibrium, one thing is certain: The concentrations of CO, H_2O, CO_2, and H_2 will all change as the reaction comes to equilibrium. Because the reaction has to shift to the right to reach equilibrium, the concentrations of CO and H_2O will become smaller, while the concentrations of CO_2 and H_2 will become larger.

At first glance, the problem appears to have four unknowns—the equilibrium concentrations of CO, H_2O, CO_2, and H_2—and we only have one equation, the equilibrium constant expression. Because it is impossible to solve one equation for four unknowns, we need to look for relationships between the unknowns that can simplify the problem. One way of achieving this goal is to look at the relationship between the changes that occur in the concentrations of CO, H_2O, CO_2, and H_2 as the reaction approaches equilibrium.

Exercise 10.7

Calculate the increase in the CO_2 and H_2 concentrations that occurs when the concentrations of CO and H_2O decrease by 0.022 mol/L as the following reaction comes to equilibrium from a set of initial conditions in which the concentration of both CO and H_2O were 0.100 M.

$$CO(g) + H_2O(g) \rightleftharpoons CO_2(g) + H_2(g)$$

Solution

This reaction has a 1:1:1:1 stoichiometry, as shown in Figure 10.5. For every mole of CO that reacts with a mole of H_2O, we get 1 mole of CO_2 and 1 mole of H_2. Thus the magnitude of the change in the concentration of CO and H_2O that occurs as the reaction comes to equilibrium is equal to the magnitude of the change in the CO_2 and H_2 concentrations. The only difference is the sign of this change. The concentrations of CO and H_2O become smaller, whereas the concentrations of CO_2 and H_2 become larger as the reaction comes to equilibrium. If 0.022 mol/L of CO and H_2O are consumed as the reaction comes to equilibrium, 0.022 mol/L of CO_2 and H_2 must be formed at the same time.

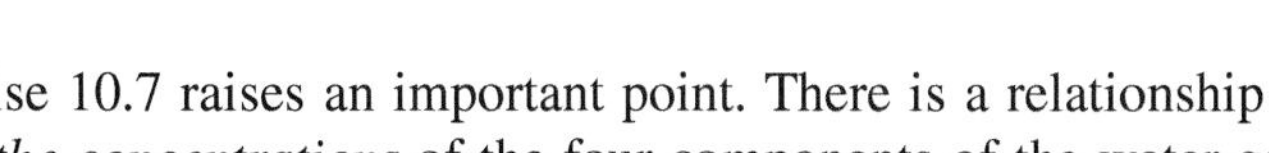

$$:C{\equiv}O: + H{-}\overset{..}{O}{-}H \rightleftharpoons \overset{..}{:}O{=}C{=}O\overset{..}{:} + H{-}H$$

Fig. 10.5 The reaction between CO and H_2O in the gas phase to form CO_2 and H_2 is a reversible reaction with a 1:1:1:1 stoichiometry.

> **CHECKPOINT**

Assume that 1.00 mol of PCl_3 and 1.00 mol of Cl_2 are added to an empty 1.00-L container. Furthermore, assume that the concentration of PCl_3 decreases by 0.96 mol/L as the reaction comes to equilibrium. What are the concentrations of PCl_5 and Cl_2 at equilibrium?

$$PCl_5 \rightleftharpoons PCl_3 + Cl_2$$

Exercise 10.7 raises an important point. There is a relationship between the *changes in the concentrations* of the four components of the water-gas shift reaction as it comes to equilibrium because of the stoichiometry of the reaction. Let's now continue to examine the problem posed at the beginning of this section.

It would be useful to have a symbol that represents the change that occurs in the concentration of one of the components of a reaction as it goes from the initial conditions to equilibrium. Let's define $\Delta(X)$ as the magnitude of the change in the concentration of X as the reaction comes to equilibrium. Thus, $\Delta(CO)$ is the magnitude of the change in the concentration of CO that occurs when this compound reacts with water in the gas phase to form CO_2 and H_2.

The important quantities for reactions that come to equilibrium are the concentrations of each reactant and/or product at equilibrium. Let's look, once again, at the water-gas shift reaction. By definition, the concentration of CO at equilibrium is equal to the initial concentration of CO minus the amount of CO consumed as the reaction comes to equilibrium.

$$[CO] \qquad = \qquad (CO)_i \qquad - \qquad \Delta(CO)$$

Concentration of	*Initial concentration*	*CO consumed as*
CO in moles per liter	*of CO in*	*the reaction comes*
at equilibrium	*moles per liter*	*to equilibrium*

In a similar fashion, we can define the concentration of H_2O at equilibrium as the initial concentration of water minus the amount of water consumed as the reaction comes to equilibrium.

$$[H_2O] \qquad = \qquad (H_2O)_i \qquad - \qquad \Delta(H_2O)$$

Concentration of	*Initial concentration*	*H_2O consumed as*
H_2O in moles per liter	*of H_2O in*	*the reaction comes*
at equilibrium	*moles per liter*	*to equilibrium*

We can then define $\Delta(CO_2)$ and $\Delta(H_2)$ as the changes that occur in the concentrations of CO_2 and H_2 as the reaction comes to equilibrium. Because CO_2 and H_2 are formed as this reaction comes to equilibrium, the concentrations of both substances at equilibrium will be larger than their initial concentrations.

$$[CO_2] = (CO_2)_i + \Delta(CO_2)$$
$$[H_2] = (H_2)_i + \Delta(H_2)$$

Because of the 1:1:1:1 stoichiometry of the reaction, the magnitude of the change in the concentrations of CO and H_2O as the reaction comes to equilibrium is equal to the changes in the concentrations of CO_2 and H_2, as we saw in Exercise 10.7.

$$\Delta(CO) = \Delta(H_2O) = \Delta(CO_2) = \Delta(H_2)$$

We can therefore rewrite the equations that define the equilibrium concentrations of CO, H_2O, CO_2, and H_2 in terms of a single unknown: ΔC.

$$[CO] = (CO)_i - \Delta C$$
$$[H_2O] = (H_2O)_i - \Delta C$$
$$[CO_2] = (CO_2)_i + \Delta C$$
$$[H_2] = (H_2)_i + \Delta C$$

Substituting what we know about the initial concentrations of CO, H_2O, CO_2, and H_2 into the equations that define the concentration of each component of the reaction at equilibrium gives the following result.

$$[CO] = [H_2O] = 0.100 - \Delta C$$
$$[CO_2] = [H_2] = 0 + \Delta C$$

We can now summarize what we know about the water-gas shift reaction as follows.

	$CO(g)$	$+$	$H_2O(g)$	$\rightleftharpoons$	$CO_2(g)$	$+$	$H_2(g)$
Initial:	0.100 *M*		0.100 *M*		0		0
Change:	$-\Delta C$		$-\Delta C$		$+\Delta C$		$+\Delta C$
Equilibrium:	$0.100 - \Delta C$		$0.100 - \Delta C$		ΔC		ΔC

We now have only one unknown, ΔC, and we need only one equation to solve for one unknown. The obvious equation to turn to is the equilibrium constant expression for the reaction.

$$K_c = \frac{[CO_2][H_2]}{[CO][H_2O]} = 0.080$$

Substituting what we know about the equilibrium concentrations of CO, H_2O, CO_2, and H_2 into the equation gives the following result.

$$\frac{[\Delta C][\Delta C]}{[0.100 - \Delta C][0.100 - \Delta C]} = 0.080$$

This equation can be expanded and then rearranged to give an equation

$$0.92[\Delta C]^2 + 0.016[\Delta C] - 0.00080 = 0$$

that can be solved with the quadratic formula.

$$\Delta C = \frac{-b \pm \sqrt{b^2 - 4ac}}{2a} = \frac{-(0.016) \pm \sqrt{(0.016)^2 - 4(0.92)(-0.00080)}}{2(0.92)}$$

$$\Delta C = 0.022 \text{ or } -0.039$$

Although two answers come out of the calculation, only the positive root makes any physical sense because we can't have a negative concentration. Thus the magnitude of the change in the concentrations of CO, H_2O, CO_2, and H_2 as the reaction comes to equilibrium is 0.022 mol/L.

$$\Delta C = 0.022$$

Substituting the value of ΔC back into the equations that define the equilibrium concentrations of CO, H_2O, CO_2, and H_2 gives the following results for the question posed at the beginning of this section.

$$[CO] = [H_2O] = 0.100\ M - 0.022\ M = 0.078\ M$$
$$[CO_2] = [H_2] = 0\ M + 0.022\ M = 0.022\ M$$

In other words, slightly less than one-quarter of the CO and H_2O react to form CO_2 and H_2 when the reaction comes to equilibrium.

To check whether the results of the calculation represent legitimate values for the equilibrium concentrations of the four components of this reaction, we can substitute these values into the equilibrium constant expression.

$$\frac{[CO_2][H_2]}{[CO][H_2O]} = \frac{[0.022][0.022]}{[0.078][0.078]} = 0.080$$

The results of our calculation must be legitimate because the equilibrium constant calculated from these concentrations is equal to the value of K_c given in the problem, within experimental error.

Exercise 10.8

Suppose that 1.00 mol of *cis*-2-butene was placed in a 1.00-L flask that contained no *trans*-2-butene at 400°C. What would be the concentrations of the *cis*-2-butene and *trans*-2-butene at equilibrium if K_c = 1.27 for the reaction in which the *cis*-2-butene isomer is converted into *trans*-2-butene?

$$\underset{\text{cis-2-butene}}{\overset{\text{H}\qquad\text{H}}{\underset{\text{CH}_3\qquad\text{CH}_3}{C=C}}} \rightleftharpoons \underset{\text{trans-2-butene}}{\overset{\text{CH}_3\qquad\text{H}}{\underset{\text{H}\qquad\text{CH}_3}{C=C}}}$$

Solution

We start, as always, by representing the information in the problem in the following format.

	cis-2-butene	$\rightleftharpoons$	*trans*-2-butene
Initial:	1.00 *M*		0 *M*
Equilibrium:	1.00 − ΔC		ΔC

We then write the equilibrium constant expression for the reaction.

$$K = \frac{[\textit{trans-2-butene}]}{[\textit{cis-2-butene}]} = 1.27$$

Substituting the expressions for the equilibrium concentrations of *cis*-2-butene and *trans*-2-butene into this equation gives

$$\frac{[\Delta C]}{[1.00 - \Delta C]} = 1.27$$

We then solve for ΔC and use the results of this calculation to determine the concentrations of butane and isobutane at equilibrium.

$$[\textit{trans-2-butene}] = \Delta C = 0.559\ M$$
$$[\textit{cis-2-butene}] = 1.00 - \Delta C = 0.441\ M$$

➤ **CHECKPOINT**

Do the equilibrium concentrations found in Exercise 10.8 correctly reproduce the K_c?

10.8 Hidden Assumptions That Make Equilibrium Calculations Easier

The water-gas shift reaction has a stoichiometry that can be described as 1:1:1:1. For each mole of carbon monoxide consumed in the reaction, one mole of water is consumed and one mole of carbon dioxide and one mole of hydrogen are produced. It is easy to imagine gas-phase reactions with a more complex stoichiometry. Consider the equilibrium between sulfur trioxide and a mixture of sulfur dioxide and oxygen, for example.

Sulfur trioxide decomposes to give sulfur dioxide and oxygen with an equilibrium constant of 1.6×10^{-10} at 300°C.

$$2\,SO_3(g) \rightleftharpoons 2\,SO_2(g) + O_2(g)$$

Calculate the equilibrium concentrations of the three components of the system if the initial concentration of SO_3 is 0.100 M.

Once again, the first step in the problem involves building a representation of the information in the problem.

$$2\,SO_3(g) \rightleftharpoons 2\,SO_2(g) + O_2(g) \qquad K_c = 1.6 \times 10^{-10}$$

Initial:	0.100 *M*	0	0
Equilibrium:	?	?	?

We then compare the reaction quotient for the initial conditions with the equilibrium constant for the reaction.

$$Q_C = \frac{(SO_2)^2(O_2)}{(SO_3)^2} = \frac{(0)^2(0)}{(0.100)^2} = 0 < K_C$$

Because the initial concentrations of SO_2 and O_2 are zero, the reaction has to shift to the right to reach equilibrium. As might be expected, some of the SO_3 has to decompose to SO_2 and O_2.

The stoichiometry of the reaction is more complex than the reaction in the previous section, but the changes in the concentrations of the three components of the reaction are still related, as shown in Figure 10.6. For every 2 moles of SO_3 that decompose, we get 2 moles of SO_2 and 1 mole of O_2.

The signs of the ΔC terms in the problem are determined by the fact that the reaction has to shift from left to right to reach equilibrium. The coefficients in the ΔC terms mirror the coefficients in the balanced equation for the reaction. Because twice as many moles of SO_2 are produced as moles of O_2, the change in the concentration of SO_2 as the reaction comes to equilibrium must be twice as large as the change in the concentration of O_2. Because two moles of SO_3 are consumed for every mole of O_2 produced, the change in the SO_3 concentration must be twice as large as the change in the concentration of O_2.

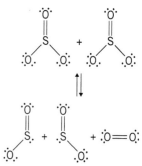

Fig. 10.6 The stoichiometry of this reaction requires that the change in concentrations of both SO_3 and SO_2 must be twice as large as the change in the concentration of O_2 that occurs as the reaction comes to equilibrium.

$$2\,SO_3(g) \rightleftharpoons 2\,SO_2(g) + O_2(g) \qquad K_c = 1.6 \times 10^{-10}$$

Initial:	0.100 *M*	0	0
Change:	$-2\Delta C$	$+2\Delta C$	$+\Delta C$
Equilibrium:	$0.100 - 2\Delta C$	$2\Delta C$	ΔC

Substituting what we know about the problem into the equilibrium constant expression for the reaction gives the following equation.

$$K_c = \frac{[SO_2]^2[O_2]}{[SO_3]^2} = \frac{[2\Delta C]^2[\Delta C]}{[0.100 - 2\Delta C]^2} = 1.6 \times 10^{-10}$$

This equation is a bit more of a challenge to expand, but it can be rearranged to give the following cubic equation.

$$4[\Delta C]^3 - (6.4 \times 10^{-10})[\Delta C]^2 + (6.4 \times 10^{-11})[\Delta C] - (1.6 \times 10^{-12}) = 0$$

Solving cubic equations is difficult, however. This problem is therefore an example of a family of problems that are difficult, if not impossible, to solve exactly. Such problems are solved with a general strategy that consists of making an assumption or approximation that turns them into simpler problems.

➤ **CHECKPOINT**

The three lines of information given under the balanced chemical equation help to organize the concentrations of reactants and products when solving an equilibrium problem. Which relationship between concentrations (initial, change, equilibrium) must follow the stoichiometry of the balanced chemical equation?

What assumption can be made to simplify the problem? Let's go back to the first thing we did after building a representation for the problem. We started our calculation by comparing the reaction quotient for the initial concentrations with the equilibrium constant for the reaction.

$$Q_C = \frac{(SO_2)^2(O_2)}{(SO_3)^2} = \frac{(0)^2(0)}{(0.100)^2} = 0 < K_C$$

We then concluded that the reaction quotient ($Q_c = 0$) was smaller than the equilibrium constant ($K_c = 1.6 \times 10^{-10}$) and decided that some of the SO_3 would have to decompose in order for the reaction to come to equilibrium.

But what are the relative sizes of the reaction quotient and the equilibrium constant for this reaction? The initial values of Q_c and K_c are both relatively small, which means that the initial conditions are reasonably close to equilibrium. As a result, the reaction doesn't have far to go to reach equilibrium. It is therefore reasonable to assume that ΔC is relatively small in this problem.

It is essential to understand the nature of the assumption being made. We aren't assuming that ΔC is zero. If we did that, some of the unknowns would disappear from the equation! We are only assuming that ΔC is so small compared with the initial concentration of SO_3 that it doesn't make a significant difference when $2\Delta C$ is subtracted from that number. We can write the assumption as follows.

$$0.100\,M - 2\Delta C \approx 0.100\,M$$

Let's now go back to the equation we are trying to solve.

$$\frac{[2\Delta C]^2[\Delta C]}{[0.100 - 2\Delta C]^2} = 1.6 \times 10^{-10}$$

By assuming that $2\Delta C$ is very much smaller than 0.100, we can replace this equation with the following approximate equation.

$$\frac{[2\Delta C]^2[\Delta C]}{[0.100]^2} \approx 1.6 \times 10^{-10}$$

We do not assume that ΔC is zero, in which case the $2\Delta C$ and ΔC terms in the numerator would disappear. We are assuming that ΔC is much smaller than 0.100. Expanding this equation gives an equation that is much easier to solve for ΔC.

$$4\Delta C^3 \approx 1.6 \times 10^{-12}$$
$$\Delta C \approx 7.4 \times 10^{-5}M$$

Before we can go any further, we have to check our assumption that $2\Delta C$ is so small compared with 0.100 that it doesn't make a significant difference when it is subtracted from that number. Is the assumption valid? Is $2\Delta C$ small enough compared with 0.100 to be ignored?

$$0.100M - 2(0.000074M) \approx 0.100M$$

ΔC is so small that $2\Delta C$ is smaller than the experimental error in the measurement of the initial concentration of SO_3 and can therefore be legitimately ignored. As a general rule, the change in concentration is small enough to be ignored if the change in concentration is less than 5% of the initial concentration. In this example, the change in the concentration of SO_3 as the reaction comes to equilibrium is only 0.15%, which is much smaller than 5%.

$$\frac{2\Delta C}{0.100} \times 100\% = \frac{2(0.000074)}{0.100} \times 100\% = 0.15\% < 5\%$$

We can now use the approximate value of ΔC to calculate the equilibrium concentrations of SO_3, SO_2, and O_2.

$$[SO_3] = 0.100\,M - 2\Delta C \approx 0.100M$$
$$[SO_3] = 2\Delta C \approx 1.5 \times 10^{-4}M$$
$$[O_2] = \Delta C \approx 7.4 \times 10^{-5}M$$

The equilibrium between SO_3 and mixtures of SO_2 and O_2 therefore strongly favors SO_3, not SO_2.

We can check the results of our calculation by substituting these results into the equilibrium constant expression for the reaction.

$$K_c = \frac{[SO_2]^2[O_2]}{[SO_3]^2} = \frac{[1.5 \times 10^{-4}]^2[7.4 \times 10^{-5}]}{[0.100]^2} = 1.7 \times 10^{-10}$$

The value of the equilibrium constant that comes out of the calculation agrees closely with the value given in the problem. Our assumption that $2\Delta C$ is negligibly small compared with the initial concentration of SO_3 is therefore valid, and we can feel confident in the answers it provides.

We can also use the equilibrium expression to solve for the concentration of products and reactants at equilibrium when a mixture of both products and reactants is present initially. Consider the same reaction, in which SO_3 decomposes to form SO_2 and O_2. But this time let's assume that the initial concentrations of both SO_3 and O_2 are 0.100 M. We start, as always, by arranging the relevant information in the problem in the following format.

$$2\,SO_3(g) \rightleftharpoons 2\,SO_2(g) + O_2(g) \qquad K_c = 1.6 \times 10^{-10}$$

Initial:	0.100M	0	0.100M
Change:	$-2\Delta C$	$+2\Delta C$	$+\Delta C$
Equilibrium:	0.100–2ΔC	$+2\Delta C$	0.100 + ΔC

We then compare the reaction quotient for the initial conditions with the equilibrium constant for the reaction.

$$Q_C = \frac{(SO_2)^2(O_2)}{(SO_3)^2} = \frac{(0)^2(0.100)}{(0.100)^2} = 0 < K_C$$

The reaction must proceed to the right to reach equilibrium because there is no SO_2 present initially. Substituting what we know about the concentrations of the three components of the reaction at equilibrium into the equilibrium constant expression gives the following result.

$$K_c = \frac{[SO_2]^2[O_2]}{[SO_3]^2} = \frac{[2\Delta C]^2[0.100 + \Delta C]}{[0.100 - 2\Delta C]^2} = 1.6 \times 10^{-10}$$

Because both Q_c and K_c are relatively small, we can assume that ΔC and $2\Delta C$ will be small compared with the initial concentrations of SO_3 and O_2, which gives us the following approximate equation.

$$\frac{[2\Delta C]^2[0.100]}{[0.100]^2} \approx 1.6 \times 10^{-10}$$

Rearranging this equation gives the following result,

$$4[\Delta C]^2 = 1.6 \times 10^{-11}$$

which can be solved for the approximate value of ΔC.

$$\Delta C \approx 2.0 \times 10^{-6} M$$

ΔC is much smaller than the initial concentration of either O_2 or SO_3, confirming the validity of our approximation. We can therefore use this approximate value of ΔC to calculate the equilibrium concentrations of the three components of the reaction.

$$[SO_3] = 0.100 M - 2\Delta C \approx 0.100 M$$
$$[SO_2] = 2\Delta C \approx 4.0 \times 10^{-6} M$$
$$[O_2] = 0.100 M + 2\Delta C \approx 0.100 M$$

The concentration of SO_2 (4.0×10^{-6} M) produced in this calculation when O_2 was initially present is considerably smaller than the concentration of SO_2 (1.5×10^{-4} M) produced in the previous calculations when there was no O_2 or SO_2 initially present.

> **CHECKPOINT**

For which of the following equilibrium constants could it be safely assumed that ΔC is small compared with the initial concentration of A when A decomposes to form B and C?

$$A(g) \rightleftharpoons B(g) + C(g)$$

(a) $K = 1.0 \times 10^5$
(b) $K = 1.0 \times 10^{-5}$
(c) $K = 1.0 \times 10^{-1}$
(d) $K = 1.0 \times 10^{-10}$

10.9   What Do We Do When the Assumption Fails?

What do we do when we encounter a problem for which the assumption that ΔC is small compared with the initial concentrations cannot possibly be valid? Consider the following problem, for example, which plays an important role in the chemistry of the atmosphere.

The equilibrium constant for the reaction between nitrogen oxide and oxygen to form nitrogen dioxide is 3.0×10^6 at 200°C

$$2\,NO(g) + O_2(g) \rightleftharpoons 2\,NO_2(g)$$

Assume initial concentrations of 0.100 M for NO and 0.050 M for O_2. Calculate the concentrations of the three components of the reaction at equilibrium

We start, as always, by representing the information in the problem as follows.

	$2\,NO(g) + O_2(g) \rightleftharpoons 2\,NO_2(g)$		$K_c = 3.0 \times 10^6$
Initial:	0.100 M 0.050 M	0	
Equilibrium:	? ?	?	

The first step is always the same: Compare the initial value of the reaction quotient with the equilibrium constant.

$$Q_c = \frac{(NO_2)^2}{(NO)^2(O_2)} = \frac{(0)^2}{(0.100)^2(0.050)} = 0 \ll K_c$$

The relationship between the initial reaction quotient ($Q_c = 0$) and the equilibrium constant ($K_c = 3.0 \times 10^6$) tells us something we may already have suspected: The reaction must shift to the right to reach equilibrium.

Some might ask: Why calculate the initial value of the reaction quotient for the reaction? Isn't it obvious that the reaction has to shift to the right to produce at least some NO_2? Yes, it is. But calculating the value of Q_c does more than tell us in which direction the reaction has to shift to reach equilibrium. It also gives us an indication of how far the reaction has to go to reach equilibrium.

In this case, Q_c is so very much smaller than K_c for the reaction that we have to conclude that the initial conditions are very far from equilibrium. It would therefore be a mistake to expect that ΔC is small when compared with the initial concentrations of the reactants.

We can't assume that ΔC is negligibly small in this problem, but we can redefine the problem so that the assumption becomes valid. The key to achieving this goal is to remember the conditions under which we can assume that ΔC is small enough to be ignored. This assumption is valid only when Q_c is of the same relative order of magnitude as K_c (i.e., when Q_c and K_c are both much larger than 1 or much smaller than 1). We can solve problems for which Q_c isn't close to K_c by redefining the initial conditions so that Q_c becomes close to K_c (Figure 10.7). To show how this can be done, let's return to the problem given in this section.

The equilibrium constant for the reaction between NO and O_2 to form NO_2 is much larger ($K_c = 3.0 \times 10^6$) than Q_c. This means that the equilibrium favors the products of the reaction. The best way to handle the problem is to drive the reaction as far as possible to the right, and then let it come back to equilibrium. Let's therefore define an intermediate set of conditions that correspond to what would happen if we push the reaction as far as possible to the right.

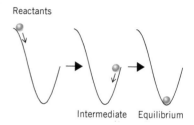

Reactants

Intermediate   Equilibrium

Redefine problem so that Δ is small

Fig. 10.7 When the initial conditions are very far from equilibrium, it is often useful to redefine the problem. This involves driving the reaction as far as possible in the direction favored by the equilibrium constant. When the reaction returns to equilibrium from the intermediate conditions, changes in the concentrations of the components of the reaction are often small enough compared with the initial concentration to be ignored.

	$2\,NO(g) + O_2(g) \rightleftharpoons 2\,NO_2(g)$		$K_c = 3.0 \times 10^6$
Initial:	$0.100\,M$ $0.050\,M$	0	
Change:	$-0.100\,M$ $-0.050\,M$	$+0.100\,M$	
Intermediate:	0 0	$0.100\,M$	

We can see where this gets us by calculating the reaction quotient for the intermediate conditions.

$$Q_c = \frac{(NO_2)^2}{(NO)^2(O_2)} = \frac{(0.100)^2}{(0)^2(0)} = \infty$$

The reaction quotient is now larger than the equilibrium constant, and the reaction has to shift back to the left to reach equilibrium. Some of the NO_2 must now decompose to form NO and O_2.

The relationship between the changes in the concentrations of the three components of this reaction is determined by the stoichiometry of the reaction, as shown in Figure 10.8. We therefore set up the problem as follows.

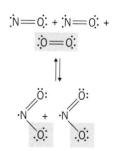

Fig. 10.8 Once again, the stoichiometry of the reaction determines the relationship among the magnitudes of the changes in the concentrations of the three components of the reaction as it comes to equilibrium.

	$2\,NO(g) + O_2(g) \rightleftharpoons 2\,NO_2(g)$		$K_c = 3.0 \times 10^6$
Intermediate:	0 0	$0.100\,M$	
Change:	$+2\Delta C$ $+\Delta C$	$-2\Delta C$	
Equilibrium:	$2\Delta C$ ΔC	$0.100 - 2\Delta C$	

We then substitute what we know about the reaction into the equilibrium constant expression.

$$K_c = \frac{[NO_2]^2}{[NO]^2[O_2]} = \frac{[0.100 - 2\Delta C]^2}{[2\Delta C]^2[\Delta C]} = 3.0 \times 10^6$$

Because the reaction quotient for the intermediate conditions and the equilibrium constant are both relatively large, we can assume that the reaction doesn't have very far to go to reach equilibrium. In other words, we assume that $2\Delta C$ is small compared with the intermediate concentration of NO_2, and we derive the following approximate equation.

$$\frac{[0.100]^2}{[2\Delta C]^2[\Delta C]} \approx 3.0 \times 10^6$$

We then solve the equation for an approximate value of ΔC.

$$\Delta C \approx 9.4 \times 10^{-4} M$$

We now check our assumption that $2\Delta C$ is small enough compared with the intermediate concentration of NO_2 to be ignored.

$$\frac{2(0.00094)}{0.100} \times 100\% = 1.9\%$$

The value of $2\Delta C$ is less than 2% of the intermediate concentration of NO_2, which means that it can be legitimately ignored in the calculation.

Since the approximation is valid, we can use the value of ΔC to calculate the equilibrium concentrations of NO_2, NO, and O_2.

$$[NO_2] = 0.100 - 2\Delta C \approx 0.098 \; M$$
$$[NO] = 2\Delta C \approx 0.0019 \; M$$
$$[O_2] = \Delta C \approx 0.00094 \; M$$

The assumption that ΔC is small compared with the initial concentrations of the reactants or products works best under the following conditions.

- When both Q_c and K_c are both much smaller than one.
- When both Q_c and K_c are both much larger than one.

10.10 The Effect of Temperature on an Equilibrium Constant

The temperature at which the reaction was run has been reported each time an equilibrium constant has been given in this chapter. If the equilibrium constant is really constant, why do we have to worry about the temperature of the reaction?

Although the value of K_c for a reaction is constant at a given temperature, it can change with temperature. Consider the equilibrium between NO_2 and its dimer, N_2O_4, for example.

$$2\,NO_2(g) \rightleftharpoons N_2O_4(g)$$

The equilibrium constant for this reaction decreases significantly with increasing temperature, as shown in Table 10.2.

As noted in Exercise 10.3, the equilibrium constant for this reaction is equal to the concentration of the product of the reaction divided by the square of the concentration of the reactant.

$$K_c = \frac{[N_2O_4]}{[NO_2]^2}$$

As a result, an increase in the equilibrium constant implies a shift toward the product of the reaction at equilibrium. In other words, at low temperatures, the equilibrium favors the dimer, N_2O_4. At high temperatures, the equilibrium favors NO_2. The fact that equilibrium constants are temperature dependent explains why you may find different values for the equilibrium constant for the same chemical reaction.

Table 10.2

Temperature Dependence of the Equilibrium Constant for the Formation of N_2O_4

Temperature ($°C$)	K_c
−78	4.0×10^8
0	1.4×10^3
25	1.7×10^2
100	2.1

10.11 Le Châtelier's Principle

In 1884, the French chemist and engineer Henry-Louis Le Châtelier proposed one of the central concepts of chemical equilibria. Le Châtelier's principle suggests that a change in one of the variables that describe a system at equilibrium produces a shift in the position of the equilibrium. The following rules summarize **Le Châtelier's principle** for closed systems at either constant pressure or constant temperature.

A closed system is one in which no matter can pass into or out of the system.

An increase in temperature for a closed system at constant pressure shifts the equilibrium in the direction in which the system absorbs heat from its surroundings.

For example, in an endothermic reaction, the equilibrium shifts to the right as temperature increases. In an exothermic reaction, the equilibrium shifts to the left as temperature increases.

An increase in pressure for a closed system at constant temperature shifts the equilibrium in the direction in which the volume of the system decreases.[1]

Our attention so far has been devoted to describing what happens as a system comes to equilibrium. Le Châtelier's principle describes what happens to a system at equilibrium when something momentarily takes it away from equilibrium. This section focuses on three ways in which we can change the conditions of a chemical reaction at equilibrium: (1) changing the concentration of one of the components of the reaction, (2) changing the pressure or volume of the system, and (3) changing the temperature.

CHANGES IN CONCENTRATION

To illustrate what happens when we change the concentration of one of the reactants or products of a reaction at equilibrium, let's consider a system that consists of 0.500 mol of *cis*-2-butene in a 1.0 L flask at 400°C.

---

[1]Ira N. Levine, *Physical Chemistry* 6th Edition, McGraw-Hill, New York, pp. 195–196 (2009).

	cis-2-butene	trans-2-butene
Initial:	0.500 *M*	0
Equilibrium:	0.500 − ΔC	ΔC

Substituting the expressions for the equilibrium concentration of *cis*-2-butene and *trans*-2-butene into the equilibrium constant expression gives:

$$K_c = \frac{[\text{trans-2-butene}]}{[\text{cis-2-butene}]} = \frac{[\Delta C]}{[0.500 - \Delta C]} = 1.27$$

This equation can then be solved for the equilibrium concentrations of *cis*-2-butene and *trans*-2-butene.

$$[\text{trans-2-butene}] = \Delta C = 0.280\ M$$
$$[\text{cis-2-butene}] = 0.500 - \Delta C = 0.220\ M$$

Now suppose that 0.200 mol of *trans*-2-butene is added to the reaction while it is at equilibrium. The system is no longer at equilibrium because the concentration of *trans*-2-butene is now 0.480 *M*. The reaction quotient at the instant the *trans*-2-butene is added is larger than the equilibrium constant for the reaction.

$$Q_c = \frac{[\text{trans-2-butene}]}{[\text{cis-2-butene}]} = \frac{0.480}{0.220} = 2.18 > K_c$$

The system must shift to reestablish the equilibrium ratio of 1.27, and so some of the *trans*-2-butene will be converted into *cis*-2-butene. When equilibrium is reestablished,

	cis-2-butene (g) ⇌	trans-2-butene
Initial:	0.220 *M*	0.480
Equilibrium:	0.220 + ΔC	0.480 − ΔC

Substituting the new conditions for equilibrium into the equilibrium constant expression gives the following equation:

$$K_c = \frac{[\text{trans-2-butene}]}{[\text{cis-2-butene}]} = \frac{[0.480 - \Delta C]}{[0.220 + \Delta C]} = 1.27$$

from which ΔC can be found to be 0.0885 *M*.

$$[\text{trans-2-butene}] = 0.480 - 0.0885 = 0.392\ M$$
$$[\text{cis-2-butene}] = 0.220 + 0.0885 = 0.309\ M$$

The calculation can be checked to determine if the ratio [*trans*-2-butene]/ [*cis*-2-butene] correctly gives K_c.

$$K_c = \frac{[0.392]}{[0.309]} = 1.27$$

By comparing the new equilibrium concentrations with those obtained before adding *trans*-2-butene, we can see the effect of increasing the concentration of the *trans*-2-butene on the equilibrium mixture.

Before	*After*
[*trans*-2-butene] = 0.480	[*trans*-2-butene] = 0.392
[*cis*-2-butene] = 0.220	[*cis*-2-butene] = 0.309

The addition of *trans*-2-butene shifted the equilibrium in such a way as to produce more *cis*-2-butene and remove some of the added *trans*-2-butene. Addition of more *cis*-2-butene would have the opposite effect, causing the equilibrium to shift toward the products.

The addition of a reactant or product will shift the equilibrium to reduce the amount of added reactant or product. The removal of a reactant or product will cause the equilibrium to shift to produce that reactant or product. Reaction quotients (Q_c) introduced in Section 10.6 provide a way to understand the shift in equilibrium that occurs when one component of the reaction is either added or removed from the reaction mixture.

The equilibrium between *cis*-2-butene and *trans*-2-butene can be used to show how to predict the effect of changes in either the concentration or volume of a system at constant temperature. Consider a syringe containing an equilibrium mixture of the two gases. What is the effect on the equilibrium of pulling the plunger partway out of the syringe, thereby increasing the volume of the system?

The first step toward answering this question involves considering the definition of the terms in the equilibrium constant expression for the reaction. By definition, the concentration of either component of the reaction is equal to the number of moles of that substance divided by the volume of the sample.

$$K_c = \frac{[\textit{trans}\text{-2-butene}]}{[\textit{cis}\text{-2-butene}]} = \frac{[n_{\textit{trans}}/V]}{[n_{\textit{cis}}/V]} = \frac{n_{\textit{trans}}}{n_{\textit{cis}}}$$

Because the volumes cancel, this equilibrium is not affected by any change in volume.

What would happen to this system if more *trans*-2-butene was added to the mixture? The reaction quotient would now be larger than the equilibrium constant for the reaction because the value of Q_c for the reaction would be equal to the ratio of moles of *trans*-2-butene to moles of *cis*-2-butene, and the equilibrium has been perturbed by adding more *trans*-2-butene.

$$Q_c = \frac{n_{\textit{trans}}}{n_{\textit{cis}}} > K_c$$

So in order to restore equilibrium, $n_{\textit{trans}}$ must decrease and $n_{\textit{cis}}$ must increase.

For reactions in which the sum of the coefficients of the gaseous products is equal to the sum of the coefficients for the gaseous reactants, no effect will be observed due to volume changes alone. If the sums of the coefficients are not the same, any change in volume will produce a shift in the reaction direction.

Exercise 10.9

The following reaction is at equilibrium in a syringe. The plunger is pushed into the syringe, thereby decreasing the volume of the system at constant temperature.

$$3\ O_2(g) \rightleftharpoons 2\ O_3(g)$$

In which direction would the reaction have to shift to get back to equilibrium?

Solution

The equilibrium constant expression for this reaction can be analyzed as follows.

$$K_c = \frac{[O_3]^2}{[O_2]^3} = \frac{[n_{O_3}/V]^2}{[n_{O_2}/V]^3} = \frac{[n_{O_3}]^2}{[n_{O_2}]^3} \times V$$

When the volume is changed, the system will no longer be at equilibrium. To determine the direction that the reaction must shift in order to reestablish equilibrium, we need the Q_c expression for this reaction.

$$Q_c = \frac{(O_3)^2}{(O_2)^3} = \frac{[n_{O_3}/V]^2}{[n_{O_2}/V]^3} = \frac{[n_{O_3}]^2}{[n_{O_2}]^3} \times V$$

We can define Q_n as the ratio of the moles of product raised to their appropriate powers divided by the moles of reactants raised to their appropriate powers. For this reaction,

$$Q_c = Q_n \times V$$

Pushing the plunger into the syringe decreases the volume but does not immediately change the moles of O_2 or O_3. When the volume, V, is decreased at constant temperature, $Q_c < K_c$ because the volume term is smaller but the moles of O_2 and O_3 are initially the same. The reaction must therefore shift to the right in order to establish a new equilibrium. This means that some of the O_2 will be consumed and some O_3 will be produced in order to increase the value of Q_c until it is equal to K_c and the system is at a new equilibrium.

A decrease in volume at constant temperature for a closed system also corresponds to an increase in pressure. This is consistent with Le Chatelier's principle that an increase in pressure at constant temperature shifts the equilibrium in the direction in which the volume of the system decreases. The volume of the system decreases because the volume of two moles of O_3 is less than the volume of three moles of O_2.

● ●

CHANGES IN PRESSURE

Sometimes it is convenient to discuss gas-phase chemical reactions in terms of the partial pressures of individual species rather than their concentrations. However, this makes no difference to the general conclusions about equilibrium that have been discussed. The reason is that partial pressures are related directly to concentrations through the ideal gas law equation.

The effect of changing the pressure on a gas-phase reaction depends on the stoichiometry of the reaction. We can demonstrate this by looking at the result of increasing the total pressure on the following reaction at equilibrium.

$$N_2(g) + 3\,H_2(g) \rightleftharpoons 2\,NH_3(g)$$

Let's start with a system that initially contains 2.5 atm of N_2 and 7.5 atm of H_2 at 500°C, and allow the reaction to come to equilibrium. Let's then compress the system by increasing the pressure by a factor of 10 and allow the system to return

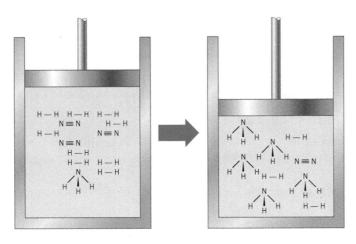

Fig. 10.9 The total number of molecules in the system decreases when N_2 reacts with H_2 to form NH_3. Shifting the equilibrium toward NH_3 decreases the total pressure of the gaseous mixture.

to equilibrium. The partial pressures at equilibrium of all three components of the reaction change when the system is compressed.

Before Compression	After Compression
$P_{NH_3} = 0.12$ atm	$P_{NH_3} = 8.4$ atm
$P_{N_2} = 2.4$ atm	$P_{N_2} = 21$ atm
$P_{H_2} = 7.3$ atm	$P_{H_2} = 62$ atm

Before the system was compressed, the partial pressure of NH_3 was only about 1% of the total pressure. After the system is compressed, the partial pressure of NH_3 is almost 10% of the total.

These data provide another example of Le Châtelier's principle. A reaction at equilibrium was subjected to an increase in the total pressure on the system. The reaction then shifted toward the products because this reduced the total number of molecules in the gaseous mixture, as shown in Figure 10.9. This in turn decreased the total pressure exerted by the gases.

Whenever the pressure exerted on a system, at constant temperature, containing gaseous reactants or products is changed, the equilibrium will shift. If the pressure is increased, the equilibrium will shift in the direction of fewer moles of gas. If the pressure is decreased, it will shift to produce more moles of gas.

CHANGES IN TEMPERATURE

Changes in the concentrations of the reactants or products of a reaction shift the position of the equilibrium, but they don't change the equilibrium constant for the reaction. Similarly, a change in the pressure on a reaction shifts the position of the equilibrium without changing the magnitude of the equilibrium constant. Changes in the temperature of the system, however, affect the position of the equilibrium by changing the magnitude of the equilibrium constant for the reaction, as shown in Section 10.10.

The reaction in which NO_2 dimerizes to form N_2O_4 provides an example of the effect of changes in temperature on the equilibrium constant for a reaction and the resulting shift in equilibrium. The reaction is exothermic.

$$2\,NO_2(g) \rightleftharpoons N_2O_4(g) \quad \Delta H° = -57.2\,kJ/mol_{rxn}$$

The equilibrium constant therefore decreases with increasing temperature, as shown in Table 10.2. This results in a shift in the equilibrium toward the left

➤ **CHECKPOINT**

Which way will the equilibrium for the following reaction shift if more P_2 is added to the system at equilibrium? Which way will the equilibrium shift if the pressure is increased? The temperature remains constant.

$$2\,P_2(g) \rightleftharpoons P_4(g)$$

to increase the concentration of NO_2 at equilibrium. For endothermic reactions, an increase in temperature will cause an increase in the equilibrium constant and therefore a shift in equilibrium toward the products.

Exercise 10.10

Predict the effect of the following changes on the equilibrium for the decomposition of SO_3 to form SO_2 plus O_2 for a sample contained in a syringe.

$$2\, SO_3(g) \rightleftharpoons 2\, SO_2(g) + O_2(g) \qquad \Delta H° = 197.84\ \text{kJ/mol}_{\text{rxn}}$$

(a) Increasing the pressure by decreasing the volume of the syringe at constant temperature.

(b) Decreasing the pressure by increasing the volume of the syringe at constant temperature.

(c) Adding an inert gas at constant volume and constant temperature.

(d) Adding an inert gas at constant temperature and constant pressure.

(e) Adding O_2 at constant volume and constant temperature.

(f) Adding SO_3 at constant pressure and constant temperature.

(g) Increasing the temperature at constant pressure.

Solution

We start by analyzing the equilibrium constant and reaction quotient expressions for the reaction.

$$K_c = \frac{[SO_2]^2[O_2]}{[SO_3]^2} = \frac{[n_{SO_2}/V]^2[n_{O_2}/V]}{[n_{SO_3}/V]^2} = \frac{[n_{SO_2}]^2[n_{O_2}]}{[n_{SO_3}]^2} \times \frac{1}{V}$$

$$Q_c = \frac{(SO_2)^2(O_2)}{(SO_3)^2} = \frac{(n_{SO_2}/V)^2(n_{O_2}/V)}{(n_{SO_3}/V)^2} = \frac{(n_{SO_2})^2(n_{O_2})}{(n_{SO_3})^2} \times \frac{1}{V}$$

$$Q_c = Q_n \times \frac{1}{V}$$

(a) Because the volume is decreased, $Q_c > K_c$. The reaction therefore shifts to the left.

(b) Because the volume is increased, $Q_c < K_c$. The reaction therefore shifts to the right.

(c) Because the volume is constant, $Q_c = K_c$ and there is no change in the equilibrium. Although the addition of an inert gas at constant volume and temperature does increase the pressure, the total moles also increase. As shown in Chapter 6

$$P_{tot} \times V = n_{tot} \times RT$$

from which we see:

$$\frac{p_{tot}}{n_{tot}} = \frac{RT}{V}$$

Thus the ratio P_{tot}/n_{tot} is unchanged because T and V are constant.

(d) Because the pressure remains constant, the volume must increase. If the volume increases, $Q_c < K_c$ and the reaction shifts to the right.

(e) Adding O_2 increases the numerator in Q_c so that $Q_c > K_c$, which means the reaction shifts to the left.

(f) If SO_3 is added at constant pressure, the volume must increase. This change combined with the effect of the increase in the number of moles of SO_3 makes Q_c smaller, so that $Q_c < K_c$. The reaction therefore shifts to the right.

(g) The reaction is endothermic. Therefore, according to the first rule introduced in Section 10.11, the equilibrium will shift toward the products of the reaction.

• •

For a general gas phase reaction

$$aA + bB \rightleftharpoons cC + dD$$

the relationship between Q_c and Q_n can be generalized as

$$Q_C = \frac{n_C^c\, n_D^d}{n_A^a\, n_B^b} \times \frac{1}{V^{\Delta n}} = Q_n \times \frac{1}{V^{\Delta n}}$$

where $\Delta n = (c + d) - (a + b)$.

10.12 Le Châtelier's Principle and the Haber Process

Ammonia has been produced commercially from N_2 and H_2 since 1913, when Badische Anilin und Soda Fabrik (BASF) built a plant that used the Haber process to make 30 metric tons of synthetic ammonia per day.

$$N_2(g) + 3\,H_2(g) \rightleftharpoons 2\,NH_3(g) \quad \Delta H° = -92.2 \text{ kJ/mol}_{rxn}$$

Until that time, the principal source of nitrogen for use in farming had been animal and vegetable waste. Today, almost 20 million tons of ammonia worth $2.5 billion is produced in the United States each year, about 80% of which is used for fertilizers. Ammonia is usually applied directly to the fields as a liquid at or near its boiling point of $-33.35°C$. By using this so-called anhydrous ammonia, farmers can apply a fertilizer that contains 82% nitrogen by weight.

The **Haber process** was the first example of the use of Le Châtelier's principle to optimize the yield of an industrial chemical. An increase in the pressure at which the reaction is run favors the products of the reaction because there is a net reduction in the number of molecules in the system as N_2 and H_2 combine to form NH_3. Because the reaction is exothermic, the equilibrium constant increases as the temperature of the reaction decreases.

Table 10.3 shows the mole percent of NH_3 at equilibrium when the reaction is run at different combinations of temperature and pressure. The mole percent of NH_3 under a particular set of conditions is equal to the number of moles of NH_3 at equilibrium divided by the total number of moles of all three components of the reaction times 100. As the data in Table 10.3 demonstrate, the best yields of ammonia are obtained at low temperatures and high pressures.

A photograph of the first high-pressure reactor for the synthesis of ammonia.

Unfortunately, low temperatures slow down the rate of the reaction, and the cost of building plants rapidly escalates as the pressure at which the reaction is run is increased. When commercial plants are designed, a temperature is chosen that allows the reaction to proceed at a reasonable rate without decreasing the equilibrium concentration of the product by too much. The pressure is also adjusted so that it favors the production of ammonia without excessively increasing the cost of building and operating the plant. The optimum conditions for running the reaction at present are a pressure between 140 atm and 340 atm and a temperature between 400°C and 600°C.

Despite all efforts to optimize reaction conditions, the percentages of hydrogen and nitrogen converted to ammonia are still relatively small. Another form of Le Châtelier's principle is therefore used to drive the reaction to completion. Periodically, the reaction mixture is cycled through a cooling chamber. The boiling point of ammonia (BP = −33°C) is much higher than that of either hydrogen (BP = −252.8°C) or nitrogen (BP = −195.8°C). Ammonia can be removed from the reaction mixture, forcing the equilibrium to the right. The remaining hydrogen and nitrogen gases are then recycled through the reaction chamber, where they react to produce more ammonia.

Table 10.3
Mole Percentage of NH_3 at Equilibrium

Temperature (°C)	Pressure (atm)			
	200	300	400	500
400	38.74	47.85	58.87	60.61
450	27.44	35.93	42.91	48.84
500	18.86	26.00	32.25	37.79
550	12.82	18.40	23.55	28.31
600	8.77	12.97	16.94	20.76

10.13 What Happens When a Solid Dissolves in Water?

Silver chloride is categorized as an insoluble ionic compound because the maximum amount of silver chloride that will dissolve in water is less than $0.1M$. However, if we add small amounts of silver chloride solid to water, the salt dissolves to form Ag^+ and Cl^- ions. The addition of solid silver chloride to a liter of water is shown in Table 10.4.

$$AgCl(s) \xrightleftharpoons{H_2O} Ag^+(aq) + Cl^-(aq)$$

If we added very small amounts of silver chloride (1.0×10^{-6} moles), we find that no solid will remain. All of the silver chloride has dissolved to form Ag^+ and Cl^-. In the same manner when we add 5.0×10^{-6} or 1.0×10^{-5} moles of solid AgCl, we observe that all of the solid dissolves. However, as the concentrations of the ions become larger, the reverse reaction starts to compete with the forward reaction, which leads to a decrease in the rate at which Ag^+ and Cl^- ions enter the solution.

As we continue to add solid AgCl, the Ag^+ and Cl^- ion concentrations become large enough that the rate at which **precipitation** (formation of solid AgCl) occurs exactly balances the rate at which AgCl dissolves. At this point the solid is in equilibrium with its ions in solution, and no additional solid will dissolve. As we add more solid (5.0×10^{-5} to 5.0×10^{-4} moles in Table 10.4), we observe that the concentrations of Ag^+ and Cl^- do not change and that solid AgCl that did not dissolve is in the bottom of the beaker.

When the system reaches equilibrium, it is called a **saturated solution** because it contains the maximum concentration of ions that can exist in equilibrium with the solid salt. The amount of salt that must be added to a given volume of solvent to form a saturated solution is called the **solubility** of the salt. A set of solubility rules for ionic compounds in water can be found in Table 8.9.

When an ionic compound dissolves in water it breaks up into its ions (cations and anions).

$$\text{ionic compound} \xrightleftharpoons{H_2O} \text{cations}(aq) + \text{anions}(aq)$$

Table 10.4
Solubility of AgCl(s) in 1.0 L of Water

Moles of AgCl(s) Added to 1.0 L H_2O	Moles of AgCl(s) Remaining in the Beaker	Ag^+ Concentration in the Solution	Cl^- Concentration in the Solution
1.0×10^{-6}	0	1.0×10^{-6}	1.0×10^{-6}
5.0×10^{-6}	0	5.0×10^{-6}	5.0×10^{-6}
1.0×10^{-5}	0	1.0×10^{-5}	1.0×10^{-5}
5.0×10^{-5}	3.7×10^{-5}	1.3×10^{-5}	1.3×10^{-5}
1.0×10^{-4}	8.7×10^{-5}	1.3×10^{-5}	1.3×10^{-5}
5.0×10^{-4}	4.9×10^{-4}	1.3×10^{-5}	1.3×10^{-5}

Estimates based on equilibrium calculations.

Exercise 10.11

Write chemical equations that describe the process in which the following ionic compounds dissolve in water.

(a) $Cu_2S(s)$

(b) $SrF_2(s)$

(c) $PbCO_3(s)$

(d) $Ag_2SO_4(s)$

(e) $Cr(OH)_3(s)$

Solution

(a) $Cu_2S(s) \overset{H_2O}{\rightleftharpoons} 2\,Cu^+(aq) + S^{2-}(aq)$

(b) $SrF_2(s) \overset{H_2O}{\rightleftharpoons} Sr^{2+}(aq) + 2\,F^-(aq)$

(c) $PbCO_3(s) \overset{H_2O}{\rightleftharpoons} Pb^{2+}(aq) + CO_3^{2-}(aq)$

(d) $Ag_2SO_4(s) \overset{H_2O}{\rightleftharpoons} 2\,Ag^+(aq) + SO_4^{2-}(aq)$

(e) $Cr(OH)_3(s) \overset{H_2O}{\rightleftharpoons} Cr^{3+}(aq) + 3\,OH^-(aq)$

10.14 The Solubility Product Expression

Silver chloride is so insoluble in water (≈ 0.002 g/L) that a saturated solution contains only about 1.3×10^{-5} moles of AgCl per liter of water.

$$AgCl(s) \overset{H_2O}{\rightleftharpoons} Ag^+(aq) + Cl^-(aq)$$

The rules for writing equilibrium constant expressions given in Section 10.5 do not address the inclusion of either pure liquids or pure solids in an equilibrium constant expression.

- **The concentrations of solids are never included in an equilibrium constant expression because the concentration of a solid is constant (it does not change).**

- **The concentrations of liquids are included in an equilibrium constant expression only when the concentration of the liquid changes during the chemical reaction.**

The dissolution of AgCl in water will serve as an example of why the concentration of a solid is not included in an equilibrium expression. Strict adherence to the rules discussed in Section 10.5 gives the following expression.

$$K_c = \frac{[Ag^+][Cl^-]}{[AgCl]}$$

(Water isn't included in the equilibrium constant expression because it is neither consumed nor produced in this reaction, even though it is a vital component of the system.)

Two of the terms in this expression are easy to interpret. The $[Ag^+]$ and $[Cl^-]$ terms represent the concentrations of the Ag^+ and Cl^- ions in units of moles per liter when the solution is at equilibrium. The third term—$[AgCl]$—is more ambiguous. It doesn't represent the concentration of AgCl dissolved in water because we assume that AgCl dissociates into Ag^+ ions and Cl^- ions when it dissolves.

The $[AgCl]$ term has to be translated quite literally as the number of moles of AgCl in a liter of the solid AgCl that lies at the bottom of the beaker. This quantity is a constant, however. The number of moles per liter in *solid* AgCl is the same at the start of the reaction as it is when the reaction reaches equilibrium. Since the $[AgCl]$ term is a constant, which has no effect on the equilibrium, it is built into the equilibrium constant for the reaction.

The concentration of solid silver chloride in units of moles per liter can be calculated from the density of the solid and its molar mass. The density of solid silver chloride is 5.56 g/cm^3, and the molar mass of silver chloride is 143.32 g/mol.

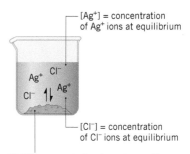

[Ag$^+$] = concentration of Ag$^+$ ions at equilibrium

[Cl$^-$] = concentration of Cl$^-$ ions at equilibrium

[AgCl] = concentration in moles per liter of AgCl in the solid at the bottom of the container

$$M = \left(\frac{5.56\text{ g}}{\text{cm}^3}\right)\left(\frac{1\text{ cm}^3}{1\text{ mL}}\right)\left(\frac{1000\text{ mL}}{\text{L}}\right)\left(\frac{1\text{ mol}}{143.32\text{ g}}\right) = 38.8\,M$$

The concentration of the solid does not change, even though the amount of the solid may become smaller as some of the solid dissolves. Although both the number of moles of solid and the volume of the solid decrease when it dissolves, the concentration doesn't change because there is no change in the ratio of moles of solid to liters of solid.

Because the $[AgCl]$ term is a constant, it has no effect on the equilibrium and is built into the equilibrium constant for the reaction.

$$[Ag^+][Cl^-] = K_c \times [AgCl]$$
$$[Ag^+][Cl^-] = K_c \times 38.8\,M = K_{sp}$$

This equation suggests that the product of the equilibrium concentrations of the Ag^+ and Cl^- ions in this solution is equal to a constant. Since this constant is proportional to the solubility of the salt, it is called the **solubility product equilibrium constant** for the reaction, or **K$_{sp}$**.

$$K_{sp} = [Ag^+][Cl^-] = [1.34 \times 10^{-5}][1.34 \times 10^{-5}] = 1.8 \times 10^{-10}$$

The K_{sp} expression for a salt is the product of the concentrations of the ions, with each concentration raised to a power equal to the coefficient of that ion in the balanced equation for the solubility equilibrium. Solubility product constants for a number of sparingly soluble salts are given in Table B.10 in Appendix B.

Exercise 10.12

Calcium fluoride (CaF$_2$) was considered as a possible source of the fluoride ion when toothpaste was first fluoridated. Write the K_{sp} expression for a saturated solution of CaF$_2$ in water.

Solution

The K_{sp} expression for a salt is the product of the concentrations of the ions formed when this salt dissolves in water, with each concentration raised to a power equal to the coefficient of that ion in the balanced equation for the solubility equilibrium.

We start with a balanced equation for the equilibrium we want to describe.

$$CaF_2(s) \underset{}{\overset{H_2O}{\rightleftharpoons}} Ca^{2+}(aq) + 2\,F^-(aq)$$

Because two F^- ions are produced for each Ca^{2+} ion when this salt dissolves in water, the K_{sp} expression for CaF_2 is

$$K_{sp} = [Ca^{2+}][F^-]^2$$

Calcium fluoride is a naturally occurring mineral known as *fluorite*.

10.15 The Relationship between K_{sp} and the Solubility of a Salt

K_{sp} is called the solubility product because it is literally the product of the concentrations of the ions in moles per liter raised to their appropriate powers. The solubility product of a salt can therefore be estimated from its solubility, or vice versa. It is important to remember that equilibrium calculations are models of what is happening during a chemical process. Particularly in the case of solubility calculations, these models are not exact and give only estimates. However, solubility calculations can still help us better understand the equilibrium associated with an ionic solid and its ions in solution, even though we realize the results of the calculations are not exact.

Photographic films are based on the sensitivity of AgBr to light. When light hits a crystal of AgBr, a small fraction of the Ag^+ ions are reduced to silver metal. The rest of the Ag^+ ions in these crystals are reduced to silver metal when the film is developed. AgBr crystals that don't absorb light are then removed from the film to "fix" the image. Let's calculate the solubility of AgBr in water in grams per liter, to see whether AgBr can be removed by simply washing the film.

We start with the balanced equation for the equilibrium.

$$AgBr(s) \underset{}{\overset{H_2O}{\rightleftharpoons}} Ag^+(aq) + Br^-(aq)$$

We then write the solubility product expression for this reaction and find the value of K_{sp} for this salt in Table B.10 in Appendix B.

$$K_{sp} = [\text{Ag}^+][\text{Br}^-] = 5.0 \times 10^{-13}$$

We can't solve one equation for two unknowns—the Ag^+ and Br^- ion concentrations. We are given the equilibrium constant and the initial concentrations of the products. It is not necessary to know the concentration of the solid because its concentration does not change and is incorporated as part of the equilibrium constant, K_{sp}. We can therefore set up a table like those used in previous equilibrium calculations.

$$\text{AgBr}(s) \rightleftharpoons \text{Ag}^+(aq) + \text{Br}^-(aq)$$

Initial:	____	0	0
Change:	ΔC	ΔC	ΔC
Equilibrium:	____	ΔC	ΔC

Substituting this equality into the K_{sp} expression gives the following result.

$$K_{sp} = [\text{Ag}^+][\text{Br}^-] = 5.0 \times 10^{-13} = \Delta C^2$$
$$\text{Because } [\text{Ag}^+] = [\text{Br}^-]$$
$$[\text{Ag}^+]^2 = 5.0 \times 10^{-13}$$

Taking the square root of both sides of this equation gives the equilibrium concentrations of the Ag^+ and Br^- ions.

$$[\text{Ag}^+] = [\text{Br}^-] = 7.1 \times 10^{-7} M$$

Once we know how many moles of AgBr dissolve in a liter of water, we can calculate the solubility in grams per liter.

$$\frac{7.1 \times 10^{-7} \text{ mol AgBr}}{1 \text{ L}} \times \frac{187.8 \text{ g AgBr}}{1 \text{ mol}} = 1.3 \times 10^{-4} \frac{\text{g AgBr}}{\text{L}}$$

The solubility of AgBr in water is only 0.00013 gram per liter. It therefore isn't practical to try to wash the unexposed AgBr off photographic film with water.

Solubility product calculations with 1:1 salts such as AgBr are relatively easy to perform. In order to extend such calculations to compounds with more complex formulas, we have to understand the relationship between the solubility of a salt and the concentrations of its ions at equilibrium. We will use the symbol ΔC to represent the solubility of a salt in a saturated solution at equilibrium in units of moles per liter.

Exercise 10.13

Write equations that describe the relationship between the solubility of CaF_2 and the equilibrium concentrations of the Ca^{2+} and F^- ions in a saturated solution as a first step toward evaluating its use as a fluoridating agent.

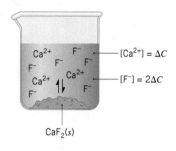

$CaF_2(s)$

Solution

As always, we start with the balanced equation for the reaction.

$$CaF_2(s) \overset{H_2O}{\rightleftharpoons} Ca^{2+}(aq) + 2\ F^-(aq)$$

Initial:	0	0
Change:	ΔC	$2\Delta C$
Equilibrium:	ΔC	$2\Delta C$

No values are given for the initial and/or equilibrium concentrations of CaF_2 because it is a solid and the concentration of a solid does not change.

Salts dissociate into their ions when they dissolve in water. For every mole of CaF_2 that dissolves, we get a mole of Ca^{2+} ions. The equilibrium concentration of the Ca^{2+} ion is therefore equal to the solubility of this compound in moles per liter.

$$[Ca^{2+}] = \Delta C$$

For every mole of CaF_2 that dissolves, we get twice as many moles of F^- ions. The F^- ion concentration at equilibrium is therefore equal to twice the solubility of the compound in moles per liter.

$$[F^-] = 2\Delta C$$

Exercise 10.14

Use the K_{sp} for calcium fluoride to calculate its solubility in grams per liter. Use the results of this calculation to explain why calcium fluoride wasn't used as a source of F^- ions in toothpaste, in spite of the fact that Ca^{2+} ions are good for bones (CaF_2: $K_{sp} = 4.0 \times 10^{-11}$).

Solution

According to Exercise 10.12, the solubility product expression for CaF_2 is written as follows.

$$K_{sp} = [Ca^{2+}][F^-]^2$$

Exercise 10.13 gave us the following equations for the relationship between the solubility of this salt and the concentrations of the Ca^{2+} and F^- ions.

$$[Ca^{2+}] = \Delta C$$
$$[F^-] = 2\Delta C$$

Substituting this information into the K_{sp} expression gives the following result.

$$[Ca^{2+}][F^-]^2 = 4.0 \times 10^{-11}$$
$$[\Delta C][2\Delta C]^2 = 4.0 \times 10^{-11}$$
$$4\Delta C^3 = 4.0 \times 10^{-11}$$

This equation can be solved for the solubility of CaF_2 in units of moles per liter.

$$\Delta C = 2.2 \times 10^{-4} M$$

Once we know how many moles of CaF_2 dissolve in a liter, we can calculate the solubility in units of grams per liter.

$$\frac{2.2 \times 10^{-4} \text{ mol } CaF_2}{1 \text{ L}} \times \frac{78.1 \text{ g } CaF_2}{1 \text{ mol}} = 0.017 \frac{\text{g } CaF_2}{\text{L}}$$

The solubility of calcium fluoride is fairly small: 0.017 gram per liter. Stannous fluoride, or tin (II) fluoride, is over 10,000 times as soluble, so SnF_2 was chosen as the first fluoridating agent used in fluoride toothpastes.

● ●

10.16 The Role of the Ion Product [Q_{sp}] in Solubility Calculations

Consider a saturated solution of AgCl in water.

$$AgCl(s) \underset{}{\overset{H_2O}{\rightleftharpoons}} Ag^+(aq) + Cl^-(aq)$$

Because AgCl is a 1:1 salt, the concentrations of the Ag^+ and Cl^- ions in this solution are equal.

Saturated solution of AgCl in water: $[Ag^+] = [Cl^-]$

Imagine what happens when a few crystals of solid $AgNO_3$ are added to a saturated solution of AgCl in water. According to the solubility rules in Table 8.9, silver nitrate is a soluble salt. It therefore dissolves and dissociates into Ag^+ and NO_3^- ions. As a result, there are two sources of the Ag^+ ion in this solution.

$$AgNO_3(s) \longrightarrow Ag^+(aq) + NO_3^-(aq)$$
$$AgCl(s) \overset{H_2O}{\rightleftharpoons} Ag^+(aq) + Cl^-(aq)$$

Adding $AgNO_3$ to a saturated AgCl solution therefore increases the Ag^+ ion concentration. When this happens, the solution is no longer at equilibrium because the product of the concentrations of the Ag^+ and Cl^- ions is too large. In other words, the **ion product (Q_{sp})** for the solution is larger than the solubility product (K_{sp}) for AgCl.

$$Q_{sp} = (Ag^+)(Cl^-) > K_{sp}$$

When Q_{sp} is larger than K_{sp}, the reaction has to shift toward the solid AgCl to come to equilibrium. Thus, AgCl will precipitate from the solution as shown by the following equation.

$$Ag^+(aq) + Cl^-(aq) \longrightarrow AgCl(s)$$

The ion product is literally the product of the concentrations of the ions raised to their appropriate powers. When the ion product is equal to the solubility product for the salt, the system is at equilibrium. Silver chloride will precipitate from solution until the concentrations of the Ag^+ and Cl^- ions decrease to the point at which the ion product is equal to K_{sp} and a new equilibrium has been established.

After the excess ions precipitate from solution as solid AgCl, the reaction comes back to equilibrium. When equilibrium is reestablished, however, the concentrations of the Ag^+ and Cl^- ions are no longer the same. Because there are two sources of the Ag^+ ion in this solution, there will be more Ag^+ ion at equilibrium than Cl^- ion.

Saturated solution of AgCl to which $AgNO_3$ has been added: $[Ag^+] > [Cl^-]$

Now imagine what happens when a few crystals of NaCl are added to a saturated solution of AgCl in water. There are two sources of the chloride ion in this solution.

$$NaCl(s) \xrightarrow{H_2O} Na^+(aq) + Cl^-(aq)$$

$$AgCl(s) \xrightleftharpoons{H_2O} Ag^+(aq) + Cl^-(aq)$$

Once again, the ion product is larger than the solubility product.

$$Q_{sp} = (Ag^+)(Cl^-) > K_{sp}$$

Therefore, AgCl will precipitate from solution until the product of the concentrations of the Ag^+ and Cl^- ions is equal to the K_{sp} and a new equilibrium has been established. This time, when the reaction comes back to equilibrium, there will be more Cl^- ion in the solution than Ag^+ ion.

Saturated solution of AgCl to which NaCl has been added: $[Ag^+] < [Cl^-]$

Figure 10.10 shows a small portion of the possible combinations of the Ag^+ and Cl^- ion concentrations in an aqueous solution. The solid line in this graph is called the saturation curve for AgCl. Any point along this curve corresponds to a system at equilibrium because the product of the Ag^+ and Cl^- ion concentrations for these solutions is equal to K_{sp} for AgCl.

Point A in Figure 10.10 represents a saturated solution at equilibrium that could be produced by dissolving two sources of the Ag^+ ion—such as $AgNO_3$ and AgCl—in water. Point B represents a saturated solution of AgCl in pure water, in which the $[Ag^+]$ and $[Cl^-]$ terms are equal. Point C describes a solution at

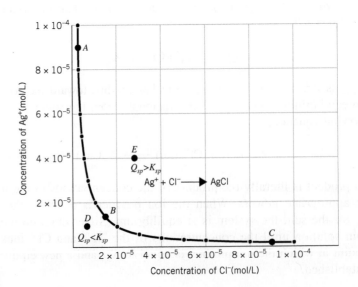

Fig. 10.10   Possible combinations of Ag^+ and Cl^- ion concentrations in an aqueous solution.

Table 10.5
The Addition of solid AgNO$_3$ to a 1.0 L Solution of 1.0 $\times$ 10^{-6} M Cl$^-$

Moles of AgNO$_3$ Added	Moles of AgCl Formed	Concentration of Ag$^+$ in Solution	Concentration of Cl$^-$ in Solution
1.0×10^{-5}	0	1.0×10^{-5}	1.0×10^{-6}
5.0×10^{-5}	0	5.0×10^{-5}	1.0×10^{-6}
1.0×10^{-4}	0	1.0×10^{-4}	1.0×10^{-6}
5.0×10^{-4}	6.4×10^{-7}	5.0×10^{-4}	3.6×10^{-7}
1.0×10^{-3}	8.2×10^{-7}	1.0×10^{-3}	1.8×10^{-7}
5.0×10^{-3}	1.0×10^{-6}	5.0×10^{-3}	3.6×10^{-8}

Estimates based on equilibrium calculations.

equilibrium that was prepared by dissolving two sources of the Cl$^-$ ion in water, such as NaCl and AgCl.

Points that don't lie on the solid line in Figure 10.10 represent solutions that aren't at equilibrium. Any point *below* the solid line (such as point *D*) represents a solution for which the ion product is smaller than the solubility product.

$$\text{point } D: \qquad Q_{sp} < K_{sp}$$

If more AgCl were added to the solution at point *D*, it would dissolve.

$$\text{if } Q_{sp} < K_{sp}: \qquad AgCl(s) \longrightarrow Ag^+(aq) + Cl^-(aq)$$

Points *above* the solid line (such as point *E*) represent solutions for which the ion product is larger than the solubility product.

$$\text{point } E: \qquad Q_{sp} > K_{sp}$$

The solution described by point *E* will eventually come to equilibrium after enough solid AgCl has precipitated.

$$\text{If } Q_{sp} > K_{sp}: \qquad Ag^+(aq) + Cl^-(aq) \longrightarrow AgCl(s)$$

Table 10.5 shows the addition of small amounts of soluble AgNO$_3$ to a solution that is not saturated in Ag$^+$ and Cl$^-$, but instead contains only dissolved NaCl, Na$^+$(aq) and Cl$^-$(aq).

Initially, when small amounts (1.0×10^{-5} to 1.0×10^{-4} moles) of AgNO$_3$ are added to the NaCl solution, the resulting concentrations of Ag$^+$ and Cl$^-$ yield Q_{sp} values that are less than the K_{sp} of AgCl. Therefore, no solid precipitant forms. However, as additional AgNO$_3$ is added (5.0×10^{-4} to 5.0×10^{-3}), the solution becomes saturated with Ag$^+$ and Cl$^-$ and solid AgCl begins to form. The Q_{sp} values have become greater than the K_{sp} of AgCl.

10.17 The Common-Ion Effect

When AgNO$_3$ is added to a saturated solution of AgCl, it is often described as a source of a common ion, the Ag$^+$ ion. By definition, a *common ion* is an ion that enters the solution from two different sources. Solutions to which both NaCl and

AgCl have been added also contain a common ion—in this case, the Cl^- ion. This section focuses on a phenomenon known as the **common-ion effect**—the effect of common ions on solubility product equilibria.

 Exercise 10.15

Calculate the solubility of AgCl in pure water (AgCl: $K_{sp} = 1.8 \times 10^{-10}$).

Solution

The solubility product expression for AgCl is written as follows.

$$K_{sp} = [Ag^+][Cl^-] = 1.8 \times 10^{-10}$$

Because there is only one source of the Ag^+ and Cl^- ions in this solution, the concentrations of these ions at equilibrium must be the same. Furthermore, the concentrations of both ions are equal to the solubility of AgCl in units of moles per liter: ΔC.

$$[Ag^+] = [Cl^-] = \Delta C$$

Substituting this information into the solubility product expressions leads to the conclusion that the solubility of AgCl is equal to the square root of K_{sp} for this salt.

$$[\Delta C][\Delta C] = 1.8 \times 10^{-10}$$
$$\Delta C^2 = 1.8 \times 10^{-10}$$
$$\Delta C = 1.3 \times 10^{-5} M$$

The common-ion effect can be understood by considering the following question: What happens to the solubility of AgCl when we dissolve this salt in a solution that is already $0.10 M$ NaCl? As a rule, we can assume that salts dissociate into their ions when they dissolve. A $0.10 M$ NaCl solution therefore contains 0.10 mole of the Cl^- ion per liter of solution. Because the Cl^- ion is one of the products of the solubility equilibrium, Le Châtelier's principle leads us to expect that AgCl will be even less soluble in a $0.10 M$ Cl^- solution than it is in pure water.

 Exercise 10.16

Calculate the solubility of AgCl in $0.10 M$ NaCl (AgCl: $K_{sp} = 1.8 \times 10^{-10}$).

Solution

The Ag^+ and Cl^- ion concentrations at equilibrium will no longer be the same because there are now two sources of the Cl^- ion in this solution: AgCl and NaCl.

$$[Ag^+] \neq [Cl^-]$$

Initially, there is no Ag^+ ion in the solution, but the Cl^- ion concentration is $0.10\,M$. As the reaction comes to equilibrium, some of the AgCl will dissolve and the concentrations of both the Ag^+ and Cl^- ions will increase. Both concentrations will increase by an amount equal to the solubility of AgCl in this solution: ΔC.

$$AgCl(s) \rightleftharpoons Ag^+(aq) + Cl^-(aq) \qquad K_{sp} = 1.8 \times 10^{-10}$$

Initial:	0	$0.10\,M$
Equilibrium:	ΔC	$0.10 + \Delta C$

We now write the solubility product expression for this reaction.

$$K_{sp} = [Ag^+][Cl^-] = 1.8 \times 10^{-10}$$

We then substitute what we know about the equilibrium concentrations of the Ag^+ and Cl^- ions into this equation.

$$[\Delta C][0.10 + \Delta C] = 1.8 \times 10^{-10}$$

We could expand the equation and solve it with the quadratic formula, but that would involve a lot of work. Let's see if we can find an assumption that makes the calculation easier.

What do we know about ΔC? In pure water, the solubility of AgCl is $0.000013\,M$. In this solution, we expect it to be even smaller. It therefore seems reasonable to expect that ΔC should be small compared with the initial concentration of the Cl^- ion.

$$[0.10 + \Delta C] \approx [0.10]$$

Substituting this approximation into the solubility product expression gives the following approximate equation.

$$[\Delta C][0.10] \approx 1.8 \times 10^{-10}$$

Solving this approximate equation gives the following result.

$$\Delta C \approx 1.8 \times 10^{-9} M$$

The assumption used to generate the approximate equation is valid. (The Cl^- ion concentration from the dissociation of AgCl is about 50 million times smaller than the initial Cl^- ion concentration.) This assumption works very well with common-ion problems involving insoluble salts because the K_{sp} values for these salts are so small.

Let's compare the results of Exercises 10.15 and 10.16.

AgCl in pure water:	$\Delta C = 1.3 \times 10^{-5} M$
AgCl in $0.10\,M$ NaCl:	$\Delta C = 1.8 \times 10^{-9} M$

These calculations show how the common-ion effect can be used to make an "insoluble" salt even less soluble in water.

Exercise 10.17

Which salt—$CaCO_3$ or Ag_2CO_3—is more soluble in water in units of moles per liter? ($CaCO_3$: $K_{sp} = 2.8 \times 10^{-9}$, Ag_2CO_3: $K_{sp} = 8.1 \times 10^{-12}$)

SOLUTION

We might expect $CaCO_3$ to be more soluble than Ag_2CO_3 because it has a larger K_{sp}. The only way to test this prediction is to calculate the solubilities of both compounds.

The solubility product expression for $CaCO_3$ has the following form.

$$K_{sp} = [Ca^{2+}][CO_3^{2-}]$$

Because $CaCO_3$ is a 1:1 salt, the concentrations of the Ca^{2+} and CO_3^{2-} would be the same at equilibrium.

$$[Ca^{2+}] = [CO_3^{2-}] = \Delta C$$

Substituting this information into the solubility product expressions leads to the conclusion that the solubility of $CaCO_3$ is equal to the square root of K_{sp} for this salt.

$$[\Delta C][\Delta C] = 2.8 \times 10^{-9}$$
$$\Delta C^2 = 2.8 \times 10^{-9}$$
$$\Delta C = 5.3 \times 10^{-5} M$$

Ag_2CO_3 is a 2:1 salt, for which the following solubility product expression is written.

$$K_{sp} = [Ag^+]^2[CO_3^{2-}]$$

The CO_3^{2-} ion concentration is equal to the solubility of the salt, but the Ag^+ ion concentration is twice as large. If we define ΔC as the solubility of the salt, then:

$$[Ag^+] = 2\Delta C$$
$$[CO_3^{2-}] = \Delta C$$

Substituting this information into the K_{sp} expression gives the following results.

$$[2\Delta C]^2[\Delta C] = 8.1 \times 10^{-12}$$
$$4[\Delta C]^3 = 8.1 \times 10^{-12}$$

This equation can be solved for the solubility of Ag_2CO_3.

$$\Delta C = 1.3 \times 10^{-4} M$$

In spite of the fact that K_{sp} for $CaCO_3$ is larger than K_{sp} for Ag_2CO_3, $CaCO_3$ is less soluble than Ag_2CO_3.

Ag_2CO_3:   Solubility $= 1.3 \times 10^{-4} M$
$CaCO_3$:   Solubility $= 5.3 \times 10^{-5} M$

Exercise 10.18

A solution contains Cd^{2+} and Cr^{3+} ions both at a concentration of 1.0×10^{-2} mol/L. If a solution of NaOH is slowly added to the mixture, the insoluble hydroxide of each ion will be formed. Which will precipitate first?

Solution

K_{sp} values are given in Table B.10 as 2.5×10^{-14} for cadmium hydroxide and 6.3×10^{-31} for the hydroxide of chromium.

We begin by writing the K_{sp} relations for both hydroxides:

$$K_{sp} = [Cd^{2+}][OH^-]^2 = 2.5 \times 10^{-14}$$
$$K_{sp} = [Cr^{3+}][OH^-]^3 = 6.3 \times 10^{-31}$$

Substituting the Cd^{2+} and Cr^{3+} concentrations into the K_{sp} expressions gives

$$[1.0 \times 10^{-2}][OH^-]^2 = 2.5 \times 10^{-14}$$
$$[1.0 \times 10^{-2}][OH^-]^3 = 6.3 \times 10^{-31}$$

The concentrations of hydroxide needed to cause precipitation can then be calculated to be

$$Cd(OH)_2 \quad [OH^-] = 1.6 \times 10^{-6}\,M$$
$$Cr(OH)_3 \quad [OH^-] = 4.0 \times 10^{-10}\,M$$

A lower concentration of hydroxide ion will cause a precipitate to first form with Cr^{+3}.

10.18 Selective Precipitation

Solutions that contain a mixture of many ions can be qualitatively analyzed by taking advantage of the different solubilities of the ions. For example, suppose a solution contains Ag^+, Cd^{2+}, and Ba^{2+} ions. Table B.10 provides solubility product constants for these and many other insoluble compounds. From this table it can be determined that of the three ions in this solution, only Ag^+ can be precipitated as the chloride, AgCl. Thus if a solution of NaCl is added, AgCl will precipitate and can be filtered away and separated from the remaining solution. Similarly, Ba^{2+} forms a precipitate with the sulfate ion $SO_4{}^{2-}$, and if a Na_2SO_4 solution is carefully added, the Ba^{2+} ion will precipitate as $BaSO_4$ and can be separated by filtration. Only the Cd^{2+} ion would then remain in solution, and if desired, this ion could be precipitated as the carbonate $CdCO_3$ by adding a Na_2CO_3 solution.

Exercise 10.19

A solution contains Pb^{2+}, Ca^{2+}, and Sn^{2+} ions. Devise a scheme to qualitatively separate each species from the solution.

Solution

From the K_{sp} values listed in Table B.10, it can be determined that only the Pb^{2+} ion forms a precipitate with Cl^- ion. Thus if a solution of NaCl is added to the mixture, $PbCl_2$ will precipitate and can be removed by filtration. The Sn^{2+} ion does not form a precipitate when mixed with the $CO_3{}^{2-}$ ion, but the Ca^{2+} ion does. If a solution of Na_2CO_3 is added to the solution after removal of Pb^{2+}, only $CaCO_3$ would precipitate, and this precipitate could be removed by filtration. If it is desired, Sn^{2+} can be precipitated as the hydroxide, $Sn(OH)_2$, by addition of a NaOH solution.

Key Terms

Chemical kinetics
Collision theory
Common-ion effect
Equilibrium
Equilibrium constant (K_c)
Equilibrium constant expression
Equilibrium region

Haber process
Ion product (Q_{sp})
Kinetic region
Le Châtelier's principle
Precipitation
Rate constant
Rate law

Rate of reaction
Reaction quotient (Q_c)
Saturated solution
Solubility
Solubility product equilibrium
   constant (K_{sp})

Problems

Reactions That Don't Go to Completion

1. Describe the difference between reactions that go to completion and reactions that come to equilibrium.

2. Describe the meaning of the symbols [NO] and (NO).

Gas-Phase Reactions

3. Define the terms *equilibrium constant* and *equilibrium constant expression*.

4. If 10.0 mol of *trans*-2-butene is placed into an empty flask at 400°C, what will be the equilibrium ratio of *trans*-2-butene to *cis*-2-butene? What if 15.0 mol is placed into an empty flask at the same temperature?

5. If K_c is greater than 1 for the reaction A $\rightleftharpoons$ B, will the equilibrium concentrations of the products be smaller or larger than the equilibrium concentrations of the reactants? What if K_c is less than 1?

6. Is the following statement true or false? The equilibrium concentrations depend on the initial concentrations, but the ratio of the equilibrium concentrations specified by the equilibrium constant expression is independent of the initial concentrations.

The Rate of a Chemical Reaction

7. Translate the following equation into an English sentence that carries the same meaning.

$$\text{rate of reaction} = -\frac{\Delta(X)}{\Delta t}$$

8. What does the rate law tell us about a chemical reaction? Does the rate law tell us anything about the ratio of reactants to products at equilibrium?

9. What is a rate constant? How does it differ from the rate law?

The Collision Theory Model of Gas-Phase Reactions

10. Use the collision theory to explain why the rate of the reaction of $ClNO_2$ with NO to form ClNO and NO_2 depends on the concentrations of the reactants $ClNO_2$ and NO.

11. Explain how the rates of the forward and reverse reactions change as the reaction between $ClNO_2$ and NO proceeds to equilibrium. Assume that no ClNO or NO_2 are present initially.

12. Sketch a graph of what happens to the concentrations of N_2, H_2, and NH_3 versus time as the following reaction comes to equilibrium.

$$N_2(g) + 3\,H_2(g) \rightleftharpoons 2\,NH_3(g)$$

Assume that the initial concentrations of N_2 and H_2 are both 1.00 mol/L and that no NH_3 is present initially. Label the kinetic and the equilibrium regions of the graph.

13. Give two ways to define *equilibrium*.

14. On the molecular level, do chemical reactions stop at equilibrium? Explain why or why not.

Equilibrium Constant Expressions

15. Which of the following is the correct equilibrium constant expression for the following reaction?

$$Cl_2(g) + 3 F_2(g) \rightleftharpoons 2 ClF_3(g)$$

(a) $K_c = \dfrac{2[ClF_3]}{[Cl_2] + 3[F_2]}$ (b) $K_c = \dfrac{[Cl_2] + 3[F_2]}{2[ClF_3]}$

(c) $K_c = \dfrac{[ClF_3]}{[Cl_2][F_2]}$ (d) $K_c = \dfrac{[ClF_3]^2}{[Cl_2][F_2]^3}$

(e) $K_c = \dfrac{[Cl_2][F_2]^3}{[ClF_3]^2}$

16. Which of the following is the correct equilibrium constant expression for the following reaction?

$$2 NO_2(g) \rightleftharpoons 2 NO(g) + O_2(g)$$

(a) $K_c = \dfrac{[NO_2]}{[NO][O_2]}$ (b) $K_c = \dfrac{[NO][O_2]}{[NO_2]}$

(c) $K_c = \dfrac{[NO_2]^2}{[NO]^2[O_2]}$ (d) $K_c = \dfrac{[NO]^2[O_2]}{[NO_2]^2}$

(e) $K_c = \dfrac{[2NO]^2[O_2]}{[2NO_2]^2}$

17. Write equilibrium constant expressions for the following reactions.
    (a) $O_2(g) + 2 F_2(g) \rightleftharpoons 2 OF_2(g)$
    (b) $2 SO_2(g) + O_2(g) \rightleftharpoons 2 SO_3(g)$
    (c) $2 SO_3(g) + 2 Cl_2(g) \rightleftharpoons 2 SO_2Cl_2(g) + O_2(g)$

18. Write equilibrium constant expressions for the following reactions.
    (a) $2 NO(g) + 2 H_2(g) \rightleftharpoons N_2(g) + 2 H_2O(g)$
    (b) $2 NOCl(g) \rightleftharpoons 2 NO(g) + Cl_2(g)$
    (c) $2 NO(g) + O_2(g) \rightleftharpoons 2 NO_2(g)$

19. Write equilibrium constant expressions for the following reactions.
    (a) $2 NO_2(g) \rightleftharpoons 2 NO(g) + O_2(g)$
    (b) $2 NO(g) + O_2(g) \rightleftharpoons 2 NO_2(g)$

    Calculate the value of K_c at 500 K for reaction (a) if the value of K_c for reaction (b) is 6.2×10^5 at 500 K.

20. Use the equilibrium constants for reaction (a) and (b) at 200°C to calculate the equilibrium constant for reaction (c) at that temperature.
    (a) $2 NO(g) \rightleftharpoons N_2(g) + O_2(g)$
    $$K_c = 4.3 \times 10^{18}$$
    (b) $2 NO_2(g) \rightleftharpoons 2 NO(g) + O_2(g)$
    $$K_c = 3.4 \times 10^{-7}$$
    (c) $2 NO_2(g) \rightleftharpoons N_2(g) + 2 O_2(g)$ $K_c = ?$

21. Use the equilibrium constants for reactions (a) and (b) at 1000 K to calculate the equilibrium constant for reaction (c), the water-gas shift reaction, at that temperature.
    (a) $CO(g) + \frac{1}{2} O_2(g) \rightleftharpoons CO_2(g)$
    $$K_c = 1.1 \times 10^{18}$$
    (b) $H_2O(g) \rightleftharpoons H_2(g) + \frac{1}{2} O_2(g)$
    $$K_c = 7.1 \times 10^{-12}$$
    (c) $CO(g) + H_2O(g) \rightleftharpoons CO_2(g) + H_2(g)$
    $$K_c = ?$$

22. Calculate K_c for the following reaction at 400 K if 1.000 mol/L of NOCl decomposes at that temperature to give equilibrium concentrations of 0.0222 M NO, 0.0111 M Cl_2, and 0.978 M NOCl.

$$2 NOCl(g) \rightleftharpoons 2 NO(g) + Cl_2(g)$$

23. Taylor and Crist [*Journal of the American Chemical Society*, **63**, 1381 (1941)] studied the reaction between hydrogen and iodine to form hydrogen iodide.

$$H_2(g) + I_2(g) \rightleftharpoons 2 HI(g)$$

They obtained the following data for the concentrations of H_2, I_2, and HI at equilibrium in units of moles per liter.

Trial	$[H_2]$	$[I_2]$	[HI]
I	0.0032583	0.0012949	0.015869
II	0.0046981	0.0007014	0.013997
III	0.0007106	0.0007106	0.005468

Calculate the value of K_c for each of the trials. Realizing that there will be deviation due to experimental error, is K_c constant for the reaction?

Reaction Quotients: A Way to Decide Whether a Reaction Is at Equilibrium

24. Suppose that the reaction quotient (Q_c) for the following reaction at some moment in time is 1.0×10^{-8} and the equilibrium constant for the reaction (K_c) at the same temperature is 3×10^{-7}.

$$2 NO_2(g) \rightleftharpoons 2 NO(g) + O_2(g)$$

Which of the following is a valid conclusion?
    (a) The reaction is at equilibrium.
    (b) The reaction must shift toward the products to reach equilibrium.
    (c) The reaction must shift toward the reactants to reach equilibrium.

25. Which of the following statements correctly describes a system for which Q_c is larger than K_c?
    (a) The reaction is at equilibrium.
    (b) The reaction must shift to the right to reach equilibrium

(c) The reaction must shift to the left to reach equilibrium.

(d) The reaction can never reach equilibrium.

26. Under which set of conditions will the following reaction shift to the right to reach equilibrium?

$$2\ SO_2(g) + O_2(g) \rightleftharpoons 2\ SO_3(g)$$

(a) $K_c < 1$ (b) $K_c > 1$ (c) $Q_c < K_c$
(d) $Q_c = K_c$ (e) $Q_c > K_c$

27. Carbon monoxide reacts with chlorine to form phosgene.

$$CO(g) + Cl_2(g) \rightleftharpoons COCl_2(g)$$

The equilibrium constant, K_c, for the reaction is 1.5×10^4 at 300°C. Is the system at equilibrium at the following concentrations: 0.0040 M $COCl_2$, 0.00021 M CO, and 0.00040 M Cl_2? If not, in which direction does the reaction have to shift to reach equilibrium?

Changes in Concentration That Occur as a Reaction Comes to Equilibrium

28. Explain why the change in the N_2 concentration that occurs when the following reaction comes to equilibrium is related to the change in the H_2 concentration.

$$N_2(g) + 3\ H_2(g) \rightleftharpoons 2\ NH_3(g)$$

Derive an equation that describes the relationship between the changes in the concentrations of the two reagents.

29. When confronted with the task in the previous problem, the following incorrect answer is often given.

$$\Delta(N_2) = 3\Delta(H_2)$$

Explain why this equation is wrong. Write the correct form of the relationship.

30. Calculate the changes in the CO and Cl_2 concentrations that occur if the concentration of $COCl_2$ decreases by 0.250 mol/L as the following reaction comes to equilibrium.

$$COCl_2(g) \rightleftharpoons CO(g) + Cl_2(g)$$

31. Calculate the changes in the N_2 and H_2 concentrations that occur if the concentration of NH_3 decreases by 0.234 mol/L as the following reaction comes to equilibrium.

$$2\ NH_3(g) \rightleftharpoons N_2(g) + 3\ H_2(g)$$

32. Which of the following equations describes the relationship between the magnitude of the changes in the

NO_2 and O_2 concentrations as the following reaction comes to equilibrium?

$$2\ NO(g) + O_2(g) \rightleftharpoons 2\ NO_2(g)$$

(a) $\Delta(NO_2) = \Delta(O_2)$ (b) $\Delta(NO_2) = 2\Delta(O_2)$
(c) $\Delta(O_2) = 2\Delta(NO_2)$

33. Which of the following equations correctly describes the relationship between the changes in the Cl_2 and F_2 concentrations as the following reaction comes to equilibrium?

$$Cl_2(g) + 3\ F_2(g) \rightleftharpoons 2\ ClF_3(g)$$

(a) $\Delta(Cl_2) = \Delta(F_2)$ (b) $\Delta(Cl_2) = 2\Delta(F_2)$
(c) $\Delta(Cl_2) = 3\Delta(F_2)$ (d) $\Delta(F_2) = 2\Delta(Cl_2)$
(e) $\Delta(F_2) = 3\Delta(Cl_2)$

34. Which of the following describes the change that occurs in the concentration of H_2O when ammonia reacts with oxygen to form nitrogen oxide and water according to the following equation if the change in the NH_3 concentration is ΔC?

$$4\ NH_3(g) + 5\ O_2(g) \rightleftharpoons 4\ NO(g) + 6\ H_2O(g)$$

(a) ΔC (b) $1.5\Delta C$ (c) $2\Delta C$
(d) $4\Delta C$ (e) $6\Delta C$

35. Calculate the concentrations of H_2 and NH_3 at equilibrium if a reaction that initially contained 1.000 M concentrations of both N_2 and H_2 is found to have an N_2 concentration of 0.922 M at equilibrium.

$$N_2(g) + 3\ H_2(g) \rightleftharpoons 2\ NH_3(g)$$

Initial:	1.000 M 1.000 M	0 M
Equilibrium: 0.922	?	?

36. Calculate the equilibrium constant for the reaction in the previous problem.

Hidden Assumptions That Make Equilibrium Calculations Easier

37. Calculate the equilibrium concentrations of N_2O_4 and NO_2 when 0.100 M N_2O_4 decomposes to form NO_2 at 25°C.

$$N_2O_4(g) \rightleftharpoons 2\ NO_2(g) \qquad K_c = 5.8 \times 10^{-5}$$

38. Without doing detailed equilibrium calculations, estimate the equilibrium concentration of N_2O_4 present when 1.00 M NO_2 reacts to form N_2O_4 at 25°C.

$$N_2O_4(g) \rightleftharpoons 2\ NO_2(g) \qquad K_c = 5.8 \times 10^{-5}$$

39. Calculate the equilibrium concentrations of N_2, H_2, and NH_3 present when a mixture that was initially

0.10 M N_2, 0.10 M H_2, and 0.10 M NH_3 comes to equilibrium at 500°C.

$$N_2(g) + 3 H_2(g) \rightleftharpoons 2 NH_3(g)$$
$$K_c = 0.040 \text{ (at } 500°C)$$

40. Calculate the equilibrium concentrations of CO, H_2O, CO_2, and H_2 present in the water-gas shift reaction at 800°C if the initial concentrations of CO and H_2O are 1.00 M.

$$CO(g) + H_2O(g) \rightleftharpoons CO_2(g) + H_2(g)$$
$$K_c = 0.72 \text{ (at } 800°C)$$

41. Calculate the equilibrium concentrations of N_2, O_2, and NO present when a mixture that was initially 0.100 M in N_2 and 0.090 M in O_2 comes to equilibrium at 600°C.

$$N_2(g) + O_2(g) \rightleftharpoons 2 NO(g) \qquad K_c = 3.3 \times 10^{-10}$$

42. Sulfuryl chloride decomposes to sulfur dioxide and chlorine. Calculate the concentrations of the three components of the system at equilibrium if 6.75 g of SO_2Cl_2 in a 1.00-L flask decomposes at 25°C.

$$SO_2Cl_2(g) \rightleftharpoons SO_2(g) + Cl_2(g)$$
$$K_c = 1.4 \times 10^{-5}$$

43. Without detailed equilibrium calculations, estimate the concentrations of NO and NOCl at equilibrium if a mixture that was initially 0.50 M in NO and 0.10 M in Cl_2 combined to form nitrosyl chloride, NOCl.

$$2 NO(g) + Cl_2(g) \rightleftharpoons 2 NOCl(g)$$
$$K_c = 2.1 \times 10^3 \text{ (at } 500 \text{ K)}$$

44. Calculate the concentrations of PCl_5, PCl_3, and Cl_2 that are present when the following gas-phase reaction comes to equilibrium. Calculate the percent of the PCl_5 that decomposes when the reaction comes to equilibrium. $K_c = 0.0013$ at 450 K.

$$PCl_5(g) \rightleftharpoons PCl_3(g) + Cl_2(g)$$

Initial: 1.0 M 0   0

45. Calculate the concentrations of PCl_5, PCl_3, and Cl_2 present when the following gas-phase reaction comes to equilibrium. Calculate the percent decomposition in the reaction and explain any difference between the results of this calculation and the results obtained in Section 10.7. Assume that $K_c = 0.0013$ for this reaction at 450 K.

$$PCl_5(g) \rightleftharpoons PCl_3(g) + Cl_2(g)$$

Initial: 1.00 M 0   0.20 M

46. Calculate the concentrations of NO, NO_2, and O_2 present when the following gas-phase reaction reaches equilibrium. Assume that $K_c = 3.4 \times 10^{-7}$ for this reaction at 200°C.

$$2 NO_2(g) \rightleftharpoons 2 NO(g) + O_2(g)$$

Initial: 0.100 M 0   0

47. Calculate the concentrations of NO, NO_2, and O_2 present when the following gas-phase reaction reaches equilibrium. Assume that $K_c = 3.4 \times 10^{-7}$ for this reaction at 200°C.

$$2 NO_2(g) \rightleftharpoons 2 NO(g) + O_2(g)$$

Initial: 0.100 M 0   0.050 M

48. Calculate the equilibrium concentrations of SO_3, SO_2, and O_2 present when 0.100 mol of SO_3 in a 250-mL flask at 300°C decomposes to form SO_2 and O_2. Assume that $K_c = 1.6 \times 10^{-10}$ for this reaction at 300°C.

$$2 SO_3(g) \rightleftharpoons 2 SO_2(g) + O_2(g)$$

49. Without detailed equilibrium calculations, estimate the equilibrium concentration of SO_3 when a mixture of 0.100 mol of SO_2 and 0.050 mol of O_2 in a 250-mL flask at 300°C combine to form SO_3. Assume that $K_c = 6.3 \times 10^9$ for this reaction at 300°C.

$$2 SO_2(g) + O_2(g) \rightleftharpoons 2 SO_3(g)$$

50. Sometimes the technique used in this chapter to simplify equilibrium problems is incorrectly stated as follows: "Assume that ΔC is zero." Explain why this is wrong. What is the correct way of describing the assumption?

51. What is the advantage of setting up equilibrium problems so that ΔC is small compared with the initial concentrations?

52. Describe how to test whether ΔC is small enough compared with the initial concentrations to be legitimately ignored.

53. At 600°C the equilibrium constant for the following reaction is 3.3×10^{-10}.

$$N_2(g) + O_2(g) \rightleftharpoons 2 NO(g)$$

(a) Is ΔC likely to be small or large for this reaction? Explain your answer.

(b) Find K_c for the following reaction and decide whether ΔC is likely to be large or small for the decomposition of NO.

$$2 NO(g) \rightleftharpoons N_2(g) + O_2(g)$$

What Do We Do When the Assumption Fails?

54. Describe what happens if you make the assumption that ΔC is zero in the following equation.

$$\frac{[0.125 - \Delta C][2.40 - 2\Delta C]^2}{[0.200 + 3\Delta C]^3} = 1.3 \times 10^{-8}$$

Explain how to get around this problem.

55. Explain why ΔC is relatively small when the reaction quotient (Q_c) is reasonably close to the equilibrium constant for the reaction (K_c).

56. Explain why the assumption that ΔC is small compared with the initial concentrations of the reactants and products is doomed to fail when the reaction quotient (Q_c) is very different from the equilibrium constant for the reaction (K_c).

57. Describe the technique used to solve problems for which the reaction quotient is very different from the equilibrium constant.

58. Before we can solve the following problem, we have to define a set of intermediate conditions under which the concentration of one of the reactants or products is zero.

$$2\,NO_2(g) \rightleftharpoons 2\,NO(g) + O_2(g)$$

Initial: 0.10 M   0.10 M   0.005 M

$$K_c = 5.3 \times 10^{-6}(\text{at } 250°C)$$

Which of the following goals determines whether we push the reaction as far as possible to the right or as far as possible to the left?

(a) To make both Q_C and K_c large
(b) To make both Q_C and K_c small
(c) To bring Q_C as close as possible to K_c
(d) To make the difference between Q_C and K_c *as* large as possible

The Effect of Temperature on an Equilibrium Constant

59. Why is it important to specify the temperature at which an equilibrium constant is reported?

60. If an equilibrium constant gets smaller as temperature increases, will increasing the temperature favor the products or the reactants?

61. If K_c decreases with decreasing temperature, will increasing the temperature favor the reactants or products?

Le Châtelier's Principle

62. Le Châtelier's principle has been applied to many fields, ranging from economics to psychology to political science. Give an example of Le Châtelier's principle in a field outside the physical sciences.

63. Predict the effect of increasing the pressure at constant temperature on the following reactions at equilibrium.

(a) $2\,SO_3(g) + 2\,Cl_2(g) \rightleftharpoons 2\,SO_2Cl_2(g) + O_2(g)$
(b) $O_2(g) + 2\,F_2(g) \rightleftharpoons 2\,OF_2(g)$
(c) $2\,NO(g) + O_2(g) \rightleftharpoons 2\,NO_2(g)$

64. Predict the effect of decreasing the pressure at constant temperature on the following reactions at equilibrium.

(a) $N_2O_4(g) \rightleftharpoons 2\,NO_2(g)$
(b) $N_2(g) + O_2(g) \rightleftharpoons 2\,NO(g)$
(c) $NO(g) + NO_2(g) \rightleftharpoons N_2O_3(g)$

65. Predict the effect of increasing the concentration of the reagent indicated in boldface on each of the following reactions at equilibrium. Assume that temperature and pressure are both constant.

(a) $\mathbf{2\,NO_2}(g) \rightleftharpoons N_2O_4(g)$
(b) $2\,SO_3(g) \rightleftharpoons 2\,SO_2(g) + \mathbf{O_2}(g)$
(c) $\mathbf{PF_5}(g) \rightleftharpoons PF_3(g) + F_2(g)$

66. Use Le Châtelier's principle to predict the effect of an increase in pressure on the solubility of a gas in water at a constant temperature.

Le Châtelier's Principle and the Haber Process

67. List as many ways as possible of increasing the yield of ammonia in the Haber process.

$$N_2(g) + 3\,H_2(g) \rightleftharpoons 2\,NH_3(g)$$

68. Explain why an increase in pressure favors the formation of ammonia in the Haber process.

69. Predict how an increase in the volume of the container by a factor of 2 would affect the concentrations of ammonia and oxygen in the following reaction. T is constant.

$$4\,NH_3(g) + 5\,O_2(g) \rightleftharpoons 4\,NO(g) + 6\,H_2O(g)$$

70. How are the data in Table 10.3 consistent with Le Châtelier's principle? Consider a fixed temperature with a changing pressure and a fixed pressure with a changing temperature.

What Happens When a Solid Dissolves in Water?

71. Write a chemical equation that describes the relationship between the concentrations of the Ag^+ and CrO_4^{2-} ions in a saturated solution of Ag_2CrO_4.

72. Write an equation that describes the relationship between the concentrations of the Bi^{3+} and S^{2-} ions in a saturated solution of Bi_2S_3.

The Solubility Product Expression

73. Explain why the $[Ag^+]$ and $[Cl^-]$ terms are variables but the $[AgCl]$ term is a constant no matter how much AgCl is added to a saturated solution of silver chloride in water.

74. What is the correct solubility product expression for the following reaction?

$$Ca_3(PO_4)_2(s) \rightleftharpoons 3\,Ca^{2+}(aq) + 2\,PO_4^{3-}(aq)$$

(a) $K_{sp} = \dfrac{[Ca^{2+}][PO_4^{3-}]}{[Ca_3(PO_4)_2]}$ (b) $K_{sp} = \dfrac{[Ca^{2+}]^3[PO_4^{3-}]^2}{[Ca_3(PO_4)_2]}$

(c) $K_{sp} = [Ca^{2+}][PO_4^{3-}]$ (d) $K_{sp} = [Ca^{2+}]^3[PO_4^{3-}]^2$

(e) $K_{sp} = [Ca^{2+}]^2[PO_4^{3-}]^3$

75. Which of the following is the correct solubility product expression for $Al_2(SO_4)_3$?

    (a) $K_{sp} = [Al^{3+}][SO_4^{2-}]$

    (b) $K_{sp} = [2\,Al^{3+}][3\,SO_4^{2-}]$

    (c) $K_{sp} = [Al^{3+}]^2[SO_4^{2-}]^3$

    (d) $K_{sp} = [2\,Al^{3+}]^2[3\,SO_4^{2-}]^3$

76. Write the solubility product expression for each of the following salts.

    (a) $BaCrO_4$ (b) $CaCO_3$

    (c) PbF_2 (d) Ag_2S

The Relationship between K_{sp} and the Solubility of a Salt

77. Write a chemical equation that describes the dissolution of Ag_2CO_3. Write a mathematical equation that describes the relationship between the concentrations of the Ag^+ and CO_3^{2-} ions in a saturated solution of Ag_2CO_3.

78. Write a chemical equation that describes the dissolution of Cu_2S. Write a mathematical equation that describes the relationship between the concentrations of the Cu^+ and S^{2-} ions in a saturated solution of Cu_2S.

79. Write a chemical equation that describes the dissolution of SrF_2. Calculate the K_{sp} constant for the dissolution of strontium fluoride if the solubility of SrF_2 in water is 8.5×10^{-4} mol/L.

80. Silver acetate, $Ag(CH_3CO_2)$, is marginally soluble in water. What is the K_{sp} for silver acetate if 6.6×10^{-2} moles of $Ag(CH_3CO_2)$ dissolve in 1000 mL of water?

81. For the dissolution of magnesium hydroxide in water:

    (a) Write a chemical equation for the dissolution process.

    (b) Write the K_{sp} equilibrium expression for this reaction.

    (c) If K_{sp} is 1.8×10^{-11}, calculate the solubility of magnesium hydroxide in moles per liter.

    (d) Calculate the solubility of magnesium hydroxide in grams per 100 mL.

82. What is the solubility of BaF_2 in water in grams per 100 mL if the K_{sp} is 1.0×10^{-6}?

83. What is the solubility in water for each of the following salts in grams per 100 mL?

    (a) Cu_2S ($K_{sp} = 2.5 \times 10^{-48}$)

    (b) CuS ($K_{sp} = 6.3 \times 10^{-36}$)

84. What is the solubility of Hg_2S in mol/L in a solution that contains an S^{2-} concentration of 0.10 M? What is the solubility of HgS in a solution that contains an S^{2-} concentration of 0.10 M? Hg_2S ($K_{sp} = 1.0 \times 10^{-47}$) HgS ($K_{sp} = 4 \times 10^{-53}$).

85. How many grams of AgBr will dissolve in 1.0 L of water containing a Br^- concentration of 0.050 M? Table B.10 in Appendix B contains solubility product constant values.

86. What is the solubility of Ag_2CO_3 in water in mol/L? What is the solubility in mol/L of Ag_2CO_3 in a solution containing an Ag^+ concentration of 0.15 M? See Table B.10 in Appendix B.

87. Which of the following equations describes the relationship between the solubility product for MgF_2 and the solubility of this compound?

    (a) $K_{sp} = 2\Delta C$ (b) $K_{sp} = \Delta C^2$ (c) $K_{sp} = 2\Delta C^2$

    (d) $K_{sp} = \Delta C^3$ (e) $K_{sp} = 4\Delta C^3$

88. Hg_2Cl_2 contains the Hg_2^{2+} and Cl^- ions. Which of the following equations describes the relationship between the solubility product and the solubility of this compound?

    (a) $K_{sp} = \Delta C$ (b) $K_{sp} = \Delta C^3$ (c) $K_{sp} = 4\Delta C^3$

    (d) $K_{sp} = \Delta C^4$ (e) $K_{sp} = 16\Delta C^4$

89. Which is more soluble, Ag_2S or HgS? (Ag_2S: $K_{sp} = 6.3 \times 10^{-50}$; HgS: $K_{sp} = 4 \times 10^{-53}$)

90. Which is more soluble, $PbSO_4$ or PbI_2? ($PbSO_4$: $K_{sp} = 1.6 \times 10^{-8}$; PbI_2: $K_{sp} = 7.1 \times 10^{-9}$)

91. Mercury forms salts that contain either the Hg^{2+} ion or the Hg_2^{2+} ion. Which is more soluble, HgS or Hg_2S? (HgS: $K_{sp} = 4 \times 10^{-53}$; Hg_2S: $K_{sp} = 1.0 \times 10^{-47}$)

92. What is the concentration of the CN^- ion in a saturated solution of zinc cyanide dissolved in water if the Zn^{2+} ion concentration is 4.0×10^{-5} M?

93. What is the concentration of the CrO_4^{2-} ion in a saturated solution of silver chromate dissolved in water if the Ag^+ ion concentration is 1.3×10^{-4} M?

94. What is the solubility product for strontium fluoride if the solubility of SrF_2 in water is 0.107 gram per liter?

95. Silver acetate, $Ag(CH_3CO_2)$, is marginally soluble in water. What is the K_{sp} for silver acetate if 1.190 grams of $Ag(CH_3CO_2)$ dissolve in 99.40 mL of water?

96. Lithium salts, such as lithium carbonate, are used to treat manic-depressives. What is the solubility product for lithium carbonate if 1.36 grams of Li_2CO_3 dissolve in 100 mL of water?

97. People who have the misfortune of going through a series of X rays of the gastrointestinal tract are often given a suspension of solid barium sulfate in water to drink. $BaSO_4$ is used instead of other Ba^{2+} salts, which also reflect X rays, because it is relatively insoluble in water. (Thus the patient is exposed to the minimum amount of toxic Ba^{2+} ion.) What is the solubility product for barium sulfate if 1 gram of $BaSO_4$ dissolves in 400,000 grams of water?

98. What is the solubility of silver sulfide in water in grams per 100 mL if the solubility product for Ag_2S is 6.3×10^{-50}?

99. What is the solubility in water for each of the following salts in grams per 100 mL?
    (a) Hg_2S ($K_{sp} = 1.0 \times 10^{-47}$)
    (b) HgS ($K_{sp} = 4 \times 10^{-53}$)

100. What is the solubility in water for each of the following salts in grams per 100 mL?
    (a) $Ca_3(PO_4)_2$ ($K_{sp} = 2.0 \times 10^{-29}$)
    (b) $Pb_3(PO_4)_2$ ($K_{sp} = 8.0 \times 10^{-43}$)
    (c) Ag_3PO_4 ($K_{sp} = 1.4 \times 10^{-16}$)

101. List the following salts in order of increasing solubility in water.
    (a) Ag_2S ($K_{sp} = 6.3 \times 10^{-50}$)
    (b) Bi_2S_3 ($K_{sp} = 1 \times 10^{-97}$)
    (c) CuS ($K_{sp} = 6.3 \times 10^{-36}$)
    (d) HgS ($K_{sp} = 4 \times 10^{-53}$)

The Role of the Ion Product (Q_{sp}) in Solubility Calculations

102. If a solution contains an Ag^+ concentration of 1.0×10^{-8} M and an I^- concentration of 1.0×10^{-8} M, will a precipitate form? Explain. (AgI: $K_{sp} = 8.3 \times 10^{-17}$)

103. If a solution contains a Pb^{2+} concentration of 1.9×10^{-4} M and an F^- concentration of 1.9×10^{-4} M, will a precipitate form? Explain. (PbF_2: $K_{sp} = 2.7 \times 10^{-8}$)

104. A 50.0-mL solution of 0.10 M $Ca(NO_3)_2$ is added to 50.0 mL of a 0.25 M solution of NaOH. Will a precipitate form? [K_{sp} $Ca(OH)_2 = 5.5 \times 10^{-6}$]

105. A 100-mL solution of 0.0015 M $AgNO_3$ is added to 50.0 mL of a 0.0030 M solution of Na_2CO_3. Will a precipitate form? (Ag_2CO_3: $K_{sp} = 8.1 \times 10^{-12}$)

The Common-Ion Effect

106. Define the term *common-ion effect*. Describe how Le Châtelier's principle can be used to explain the common-ion effect.

107. Describe what happens to the equilibrium concentrations of the Ag^+ and Cl^- ions when 10 grams of NaCl are added to a liter of a saturated solution of silver chloride in water.

108. Which of the following statements is true? (a) MgF_2 is more soluble in 0.100 M NaF than in pure water. (b) MgF_2 is less soluble in 0.100 M NaF than in pure water. (c) MgF_2 is just as soluble in 0.100 M NaF as in pure water.

109. Calculate the equilibrium concentration of the Ag^+ ion in a solution prepared by dissolving 3.21 grams of potassium iodide in 350 mL of water and then adding silver iodide until the solution is saturated with AgI. (AgI: $K_{sp} = 8.3 \times 10^{-17}$)

110. How many grams of silver sulfide will dissolve in 500 mL of a 0.050 M S^{2-} solution? (Ag_2S: $K_{sp} = 6.3 \times 10^{-50}$)

111. In which of the following solutions is $Pb(OH)_2$ most soluble? [$Pb(OH)_2$: $K_{sp} = 1.2 \times 10^{-15}$]
    (a) pure water
    (b) 0.010 M NaOH

Integrated Problems

112. Which of the following diagrams best represents the concentrations of the reactants and products for the following reaction at equilibrium? Explain what is wrong with each incorrect diagram. (⬤ represents isobutane, and ⬤ represents *n*-butane.)

$$\underset{\text{Isobutane}}{\overset{\displaystyle CH_3 \atop \displaystyle |}{CH_3CHCH_3(g)}} \rightleftharpoons \underset{\text{n-Butane}}{CH_3CH_2CH_2CH_3(g)} \qquad K_c = 0.4$$

113. A sparingly soluble hypothetical ionic compound, MX_2, is placed into a beaker of distilled water. Which of the following diagrams best describes what happens in solution? Explain what is wrong with each incorrect diagram.

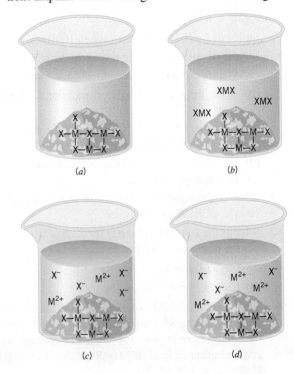

114. Describe the relationship between k_f and k_r for the following one-step reaction at equilibrium.

$$Z(g) + X(g) \underset{k_r}{\overset{k_f}{\rightleftharpoons}} Y(g) \qquad K_c = 1 \times 10^{-3}$$

Which is true: $k_f = k_r$, $k_f < k_r$, or $k_f > k_r$? Explain your reasoning.

115. For the reaction $A \rightleftharpoons B$, match the graphs of concentration versus time to the appropriate set of rate constants.

$$\text{rate}_{\text{forward}} = k_A(A) \qquad \text{rate}_{\text{reverse}} = k_B(B)$$

(a) $k_A = k_B$

(b) $k_A = 1.0/s$, $k_B = 0.5/s$

(c) $k_A = 0.5/s$, $k_B = 1.0/s$

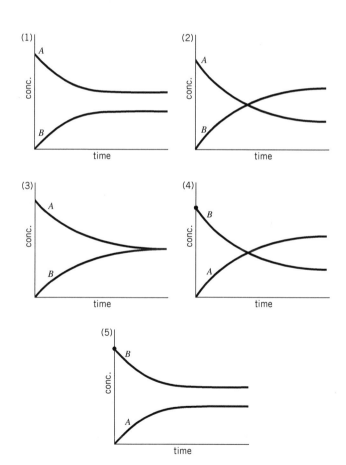

116. Write the equilibrium constant expression for the dissolving of strontium fluoride in water.

$$SrF_2(s) \rightleftharpoons Sr^{2+}(aq) + 2\,F^-(aq)$$

(a) When $SrF_2(s)$ is placed in water, the compound dissolves to produce equilibrium Sr^{2+} concentration of 5.8×10^{-4} mol/L. What is K_{sp} for the reaction?

(b) If 50.00 mL of 0.100 M Sr $(NO_3)_2$ is mixed with 50.00 mL of 0.100 M NaF, will a precipitate form? Explain your answer.

117. Several plots of concentration versus time for the reaction $A \rightleftharpoons B$ are given below. $K_c = 2$. Only one of the plots can be correct. Which one is it? Explain what is wrong with each of the incorrect plots.

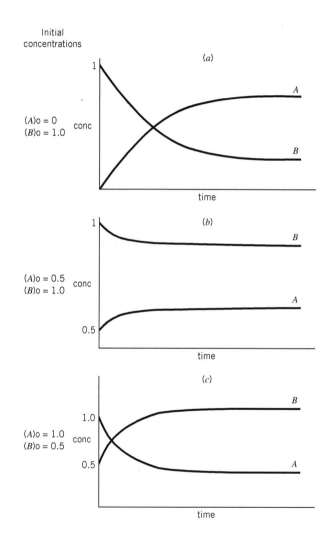

118. Molecular iodine dissociates into iodine atoms at 1000 K.

$$I_2(g) \rightleftharpoons 2\,I(g) \qquad K_c = 3.8 \times 10^{-5}$$

(a) If 0.456 mol of I_2 is placed into a 2.30-L flask at 1000 K, what will be the equilibrium concentrations of I_2 and I?

(b) If 0.912 mol of I is placed into a 2.30-L flask containing no I_2 at 1000 K, estimate the concentration of I_2 at equilibrium. Do no detailed equilibrium calculations, but clearly explain your answer.

119. When equilibrium is reached for the dissolving of solid calcium sulfate in water at 25°C, it is found that $[Ca^{2+}] = [SO_4^{2-}] = 4.9 \times 10^{-3} M$.

$$CaSO_4(s) \rightleftharpoons Ca^{2+}(aq) + SO_4^{2-}(aq)$$

(a) Calculate the equilibrium constant (K_{sp}) for the above reaction.

(b) A solution contains $Ca^{2+}(aq)$ at a concentration of $3.6 \times 10^{-3} M$ and $SO_4^{2-}(aq)$ at a concentration of $8.0 \times 10^{-3} M$. Will solid $CaSO_4$ be formed? Show your calculations.

(c) $CaSO_4$ is allowed to dissolve in 1.0 L of water at 25°C until equilibrium is reached. Then the water is allowed to evaporate to half its original volume. What are the equilibrium concentrations of Ca^{2+} and SO_4^{2-} in this solution? Explain.

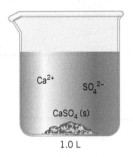

1.0 L

0.50 L

120. The equilibrium constant for the following reaction is 1.6×10^{-5} at 35°C.

$$2 NOCl(g) \rightleftharpoons 2 NO(g) + Cl_2(g)$$

If 1.0 mol of NOCl is placed into an empty 1.0-L flask, what will be the equilibrium concentrations of all species? State all assumptions and show all work. Provide a justification for any assumptions.

121. The rate constant for the following reaction in the forward direction, k_f, is $2.10 \times 10^{-7} s^{-1}$, and that in the reverse direction, k_r, is $1.65 \times 10^{-7} s^{-1}$.

$$cis\text{-}2\text{-butene}(g) \rightleftharpoons trans\text{-}2\text{-butene}(g)$$

Which of the following graphs could represent the change in concentrations with time? More than one graph could be correct. Explain your reasoning.

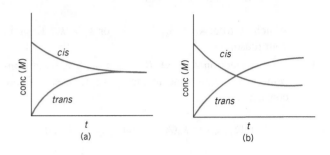

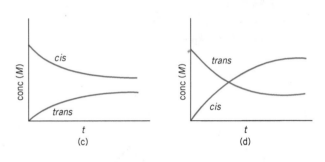

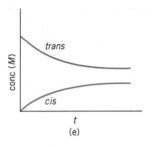

122. At 25°C, 1.5×10^{-2} mol of Ag_2SO_4 dissolves in 1.0 L of water.

$$Ag_2SO_4(s) \rightleftharpoons 2 Ag^+(aq) + SO_4^{2-}(aq)$$

(a) How many moles of Ag^+ are present?

(b) How many moles of SO_4^{2-} are present?

(c) Calculate the equilibrium constant for the reaction.

123. The following equilibrium concentrations were found for the reaction between NO and O_2 to form NO_2 at 230°C: $[NO] = 0.0542 M$, $[O_2] = 0.127 M$, $[NO_2] = 15.5 M$.

$$2 NO(g) + O_2(g) \rightleftharpoons 2 NO_2(g)$$

(a) What does it mean to say that a reaction has come to equilibrium? Must all reactions eventually come to equilibrium?

(b) What is the equilibrium constant, K_c, for the reaction?

(c) If sufficient O_2 and NO_2 are added to increase $[O_2]$ and $[NO_2]$ to 1.127 and 16.5 M, respectively, while keeping $[NO]$ at 0.0542 M, in which direction will the reaction proceed?

124. At a certain temperature, the following reaction has an equilibrium constant of 5.0×10^{-9}.

$$N_2F_4(g) \rightleftharpoons 2\,NF_2(g)$$

(a) If 1.0 mol of N_2F_4 is placed in a 1.0-L flask with no NF_2 present, what will be the equilibrium concentrations of NF_2 and N_2F_4?

(b) If 1.0 mol of NF_2 is placed in a 1.0-L flask with no N_2F_4 present, what will be approximately the equilibrium concentration of N_2F_4? No detailed equilibrium calculations are necessary.

125. The following equation represents a system at equilibrium.

$$Cl_2(g) \rightleftharpoons 2\,Cl(g)$$

Which of the following could be a valid representation of this system? ($\bullet\!\bullet$ = Cl_2 and $\bullet$ = Cl).

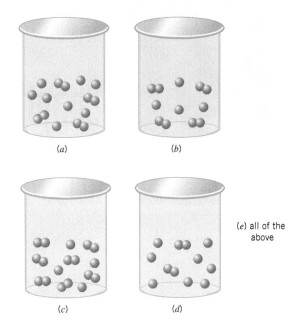

(a) (b)

(c) (d)

(e) all of the above

126. A student devises the following qualitative analysis scheme for a solution containing the ions Ni^{2+}, Mg^{2+}, and Ba^{2+}. First add a solution of NaF, filter, then add a solution of Na_2CO_3, filter, and finally add a solution of Na_2SO_4. Will this scheme succeed? If not, explain how the scheme could be modified to selectively separate the ions.

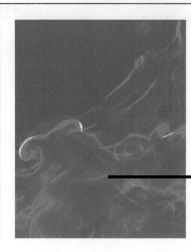

Chapter Eleven

ACIDS AND BASES

11.1 Properties of Acids and Bases

For more than 300 years, chemists have classified substances that behave like vinegar as **acids** and substances that have properties like wood ash as **bases** (or **alkalies**). The word *acid* comes from the Latin *acidus,* which means "sour," and refers to the sharp odor and sour taste of many acids. Vinegar, for example, tastes sour because it is a dilute solution of acetic acid in water. Lemon juice tastes sour because it contains citric acid. Milk turns sour when it spoils because lactic acid is formed, and the unpleasant, sour odor of rotten meat or butter can be attributed to compounds such as butyric acid that form when fat spoils.

One of the characteristic properties of an acid is its ability to dissolve most metals. Zinc metal, for example, rapidly reacts with hydrochloric acid to form an aqueous solution of $ZnCl_2$ and hydrogen gas.

$$Zn(s) + 2\,HCl(aq) \longrightarrow ZnCl_2(aq) + H_2(g)$$

Another characteristic property of acids is their ability to change the color of vegetable dyes such as litmus. Litmus is a mixture of blue dyes that turns red in the presence of acid. Litmus has been used to test for acids for more than 300 years.

Bases also have characteristic properties. They taste bitter and often feel slippery. They change the color of litmus from red to blue, thereby reversing the change in color that occurs when litmus comes in contact with an acid. Bases become less alkaline when they react with acids, and acids lose their characteristic sour taste and ability to dissolve metals when they are mixed with bases or alkalies.

11.2 The Arrhenius Definition of Acids and Bases

In 1887, Svante Arrhenius took a major step toward answering the important question, "What factors determine whether a compound is an acid or a base?" Arrhenius suggested that acids *dissociate* or *ionize* when they dissolve in water to give H^+ ions and a corresponding negative ion. According to this model, hydrogen chloride is an acid because it dissociates, or ionizes, when it dissolves in water to give H^+ and Cl^- ions (Figure 11.1). This aqueous solution is known as hydrochloric acid and is often written as HCl(aq). It is important to recognize, however, that HCl is assumed to dissociate almost completely to form the H^+ and Cl^- ions when it dissolves in water.

$$HCl(g) \xrightarrow{H_2O} H^+(aq) + Cl^-(aq)$$

Arrhenius argued that bases are compounds that dissociate in water to give OH^- ions and a positive ion. NaOH is an Arrhenius base because it dissociates in water to give the hydroxide (OH^-) and sodium (Na^+) ions.

$$NaOH(s) \xrightarrow{H_2O} Na^+(aq) + OH^-(aq)$$

An **Arrhenius acid** therefore can be defined as any substance that ionizes when it dissolves in water to give the hydrogen ion, H^+. An **Arrhenius base** is any substance that gives the hydroxide ion, OH^-, when it dissolves in water. Arrhe-

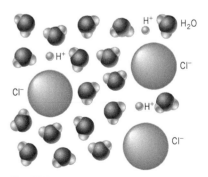

Fig. 11.1 The Arrhenius model assumes that HCl dissociates into H^+ and Cl^- ions when it dissolves in water.

Classify the following compounds as Arrhenius acids or bases: HNO_3, $Mg(OH)_2$, CH_3CO_2H.

$$HNO_3(aq) \xrightarrow{H_2O}$$
$$H^+(aq) + NO_3^-(aq)$$

$$Mg(OH)_2(s) \xrightleftharpoons{H_2O}$$
$$Mg^{2+}(aq) + 2\,OH^-(aq)$$

$$CH_3CO_2H(aq) \xrightleftharpoons{H_2O}$$
$$H^+(aq) + CH_3CO_2^-(aq)$$

nius acids include compounds such as HCl, HCN, and H_2SO_4 that ionize in water to give the H^+ ion. Arrhenius bases include ionic compounds that contain the OH^- ion, such as NaOH, KOH, and $Ca(OH)_2$.

11.3 The Brønsted–Lowry Definition of Acids and Bases

In 1923, Johannes Brønsted and Thomas Lowry independently proposed a more powerful set of definitions of acids and bases. The Brønsted, or Brønsted–Lowry, model is based on the assumption that acids donate H^+ ions to another ion or molecule, which acts as a base. According to this model, HCl doesn't dissociate in water to form H^+ and Cl^- ions. Instead, an H^+ ion is transferred from HCl to a water molecule to form an H_3O^+ ion and a Cl^- ion.

$$HCl(aq) + H_2O(l) \longrightarrow H_3O^+(aq) + Cl^-(aq)$$

The H_3O^+ ion is known as the **hydronium ion.** The Brønsted model of the reaction between HCl and water is shown in Figure 11.2.

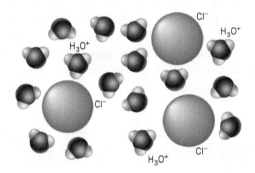

Fig. 11.2 The Brønsted model assumes that HCl molecules donate an H^+ ion to water molecules to form H_3O^+ and Cl^- ions when HCl dissolves in water.

Because it is a proton, an H^+ ion is several orders of magnitude smaller than the smallest atom. As a result, the charge on an isolated H^+ ion is distributed over such a small surface area that the H^+ ion is attracted toward any source of negative charge that exists in the solution. Thus the instant that an H^+ ion is created in an aqueous solution, it bonds to the electronegative oxygen atom of a water molecule. The Brønsted model, in which H^+ ions are transferred from one ion or molecule to another, therefore seems more reasonable than the Arrhenius model, which assumes that H^+ ions exist in aqueous solution.

Even the Brønsted model is naive, however. Each H^+ ion that an acid donates to water is actually bound in a complex of four neighboring water molecules, as shown in Figure 11.3. A more realistic formula for the substance produced when an acid loses an H^+ ion in water is therefore $H(H_2O)_4^+$, or $H_9O_4^+$. For practical purposes, however, this substance can be represented as the H_3O^+ ion.

The reaction between HCl and water provides the basis for understanding the definitions of a Brønsted acid and a Brønsted base. According to the Brønsted model, when HCl dissociates in water, HCl acts as an H^+ ion donor and H_2O acts as an H^+ ion acceptor.

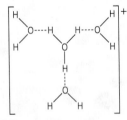

Fig. 11.3 Structure of the $H(H_2O)_4^+$ ion formed when an acid reacts with water. For practical purposes, the ion can be thought of as an H_3O^+ ion.

$$HCl(aq) + H_2O(l) \longrightarrow H_3O^+(aq) + Cl^-(aq)$$
H⁺ ion   *H⁺ ion*
donor   *acceptor*

A **Brønsted acid** is therefore any substance (such as HCl) that can donate an H^+ ion to a base. A **Brønsted base** is any substance (such as H_2O) that can accept an H^+ ion from an acid.

There are two ways of naming the H^+ ion. Some chemists call it a hydrogen ion; others call it a proton. As a result, Brønsted acids are known as either **hydrogen-ion donors** or **proton donors.** Brønsted bases are **hydrogen-ion acceptors** or **proton acceptors.** The main body of this chapter will deal primarily with **monoprotic acids** (and the corresponding monoprotic bases) that have a single H^+ ion that they can donate. The Special Topics section at the end of the chapter will discuss more complex acids and bases.

From the perspective of the Brønsted model, reactions between acids and bases always involve the transfer of an H^+ ion from a proton donor to a proton acceptor. Acids can be uncharged molecules.

$$HCl(aq) + NH_3(aq) \longrightarrow Cl^-(aq) + NH_4^+(aq)$$
$$\quad\;\; Acid \qquad\quad Base$$

They can also be positive ions

$$NH_4^+(aq) + OH^-(aq) \rightleftharpoons NH_3(aq) + H_2O(l)$$
$$\quad\;\; Acid \qquad\quad Base$$

or negative ions.

$$H_2PO_4^-(aq) + H_2O(l) \rightleftharpoons H_3O^+(aq) + HPO_4^{2-}(aq)$$
$$\quad\;\; Acid \qquad\quad Base$$

Brønsted bases can be identified from their Lewis structures. According to the Brønsted model, a base is any ion or molecule that can accept a proton. To understand the implications of this definition, look at how the prototypical base, the OH^- ion, accepts a proton.

$$H^+ + :\ddot{O}-H^- \longrightarrow H-\ddot{O}-H$$

The only way to accept an H^+ ion is to form a covalent bond to it. In order to form a covalent bond to an H^+ ion that has no valence electrons, the base must provide both of the electrons needed to form the covalent bond. Thus only compounds that have pairs of nonbonding valence electrons can act as H^+ ion acceptors, or Brønsted bases. The following compounds, for example, can all act as Brønsted bases because they all contain nonbonding pairs of electrons.

The Brønsted model therefore includes within the category of bases any ion or molecule that contains one or more pairs of nonbonding valence electrons that can accept a proton. Many molecules and ions satisfy the definition of a Brønsted base. However, the following substances are not Brønsted bases because they have no nonbonding valence electrons.

$$CH_4 \qquad H-\underset{\underset{H}{|}}{\overset{\overset{H}{|}}{C}}-H$$

$$H_2 \qquad H-H$$

$$NH_4^+ \qquad \left[H-\underset{\underset{H}{|}}{\overset{\overset{H}{|}}{N}}-H \right]^+$$

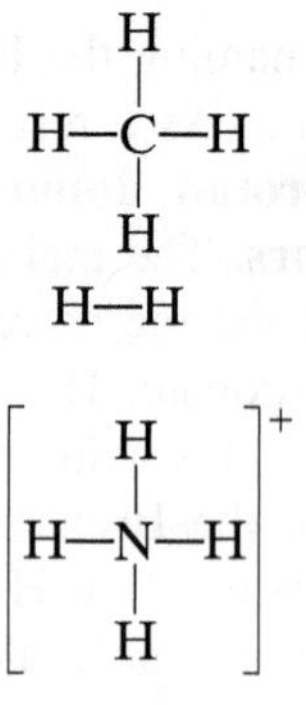

Exercise 11.1

Identify the reactant that behaves as a Brønsted acid and the reactant that behaves as a Brønsted base in each of the following reactions.

(a) $HF(aq) + OH^-(aq) \rightleftharpoons H_2O(l) + F^-(aq)$

(b) $CH_3CO_2H(aq) + H_2O(l) \rightleftharpoons CH_3CO_2^-(aq) + H_3O^+(aq)$

(c) $C_6H_5NH_2(aq) + HNO_3(aq) \rightleftharpoons C_6H_5NH_3^+(aq) + NO_3^-(aq)$

Solution

(a) acid: HF   base: OH^-

(b) acid: CH_3CO_2H base: H_2O

(c) acid: HNO_3 base: $C_6H_5NH_2$

11.4 Conjugate Acid–Base Pairs

An important consequence of the Brønsted theory is the recognition that acids and bases are linked or coupled to form **conjugate acid–base pairs.** The term *conjugate* comes from the Latin stem meaning "joined together" and refers to things that are joined, particularly in pairs. It is therefore the perfect term to describe the relationship between Brønsted acids and bases.

We can write the formula for a generic acid as HA. When the acid donates an H^+ ion to water, one product of the reaction is A^-, which could act as an H^+ ion acceptor, or Brønsted base. In other words, every time a Brønsted acid acts as a proton donor, it forms a conjugate base.

$$HA(aq) + H_2O(l) \rightleftharpoons H_3O^+(aq) + A^-(aq)$$
$$\textit{Acid} \qquad\qquad\qquad\qquad\qquad \textit{Conjugate base}$$

Conversely, if A^- accepts an H^+ ion from water, it forms HA, which could act as a proton donor or Brønsted acid. Thus, every time a base gains an H^+ ion, it forms the conjugate acid.

$$A^-(aq) + H_2O(l) \rightleftharpoons HA(aq) + OH^-(aq)$$
$$\textit{Base} \qquad\qquad\qquad\quad \textit{Conjugate acid}$$

Acids and bases in the Brønsted model therefore exist as **conjugate acid–base pairs** whose formulas are related by the gain or loss of a hydrogen ion.

In order to determine whether a substance behaves as a Brønsted acid or base, the substance must be examined in the context of the chemical reaction in which it participates. In the chemical equation shown above in which HA reacts with water, water behaves as a base. In the second equation in which A^- reacts with water, water behaves as an acid.

Our use of the symbols HA and A^- for a conjugate acid–base pair doesn't mean that all acids are electrically neutral molecules or that all bases are negative ions. It signifies only that the acid contains an H^+ ion that isn't present in the conjugate base. As noted earlier, Brønsted acids and bases can be electrically neutral molecules, positive ions, or negative ions. Various Brønsted acids and their conjugate bases are given in Table 11.1.

It is important to recognize that some compounds can be both a Brønsted acid and a Brønsted base. H_2O and HSO_4^-, for example, can be found in both columns in Table 11.1. Water is the perfect example of this behavior because it simultaneously acts as an acid and a base when it reacts with itself to form the H_3O^+ and OH^- ions.

$$H_2O(l) + H_2O(l) \rightleftharpoons H_3O^+(aq) + OH^-(aq)$$

The concept of conjugate acid–base pairs plays a vital role in explaining reactions between acids and bases. According to the Brønsted model, an acid always reacts with a base to form the conjugate base and conjugate acid. Consider the following reaction, for example.

$$HNO_3(aq) + NH_3(aq) \longrightarrow NH_4^+(aq) + NO_3^-(aq)$$

| *Acid* | *Base* | *Conjugate acid* | *Conjugate base* |

In the course of this reaction, nitric acid donates an H^+ ion to form its conjugate base, the nitrate ion (NO_3^-). At the same time, ammonia acts as a base, accepting an H^+ ion to form its conjugate acid, the ammonium ion (NH_4^+).

The products of the reaction between nitric acid and ammonia are often combined and written as an aqueous solution of an ionic compound, or salt.

$$HNO_3(aq) + NH_3(aq) \longrightarrow NH_4NO_3(aq)$$

Because the products of the reaction are neither as acidic as nitric acid nor as basic as ammonia, the reaction is often called a **neutralization reaction.** This doesn't imply that the products have no acid or base properties. It only suggests that the products are less acidic and less basic than the starting materials.

Water is often one of the products of a neutralization reaction. Consider the reaction between formic acid and sodium hydroxide, for example.

$$HCO_2H(aq) + NaOH(aq) \longrightarrow H_2O(l) + Na^+(aq) + HCO_2^-(aq)$$

| *Acid* | *Base* | *Conjugate acid* | *Conjugate base* |

The salt produced in this reaction is sodium formate, $NaHCO_2$. The formate ion, HCO_2^-, is the conjugate base of formic acid, and water is the conjugate acid of the hydroxide ion in sodium hydroxide. The Na^+ is a spectator ion.

Table 11.1

Typical Brønsted Acids and Their Conjugate Bases

Acid	Base
H_3O^+	H_2O
H_2O	OH^-
HCl	Cl^-
H_2SO_4	HSO_4^-
HSO_4^-	SO_4^{2-}
NH_4^+	NH_3

► CHECKPOINT

Phosphoric acid, H_3PO_4, is an important component of many carbonated beverages. Write the chemical equation for the dissociation of phosphoric acid in water and predict the chemical formula of its conjugate base. Aniline, $C_6H_5NH_2$, is a base used in the manufacture of dyes. Write the chemical equation for the reaction of aniline with water and predict the chemical formula of its conjugate acid.

Exercise 11.2

Write the chemical equation for the reaction of chlorous acid, $HClO_2$, with potassium hydroxide. Identify the acid and base among the reactants and the conjugate acid and conjugate base formed in this reaction.

Solution

$$HClO_2(aq) + KOH(aq) \longrightarrow H_2O(l) + K^+(aq) + ClO_2^-(aq)$$

Reactants		Products	
acid:	$HClO_2$	conjugate base:	ClO_2^-
base:	KOH	conjugate acid:	H_2O

11.5 The Role of Water in the Brønsted Model

The Lewis structure of water can help us understand why H_3O^+ and OH^- ions play such an important role in the chemistry of aqueous solutions. The Lewis structure suggests that the hydrogen and oxygen atoms in a water molecule are bound together by sharing a pair of electrons. But oxygen ($EN = 3.61$) is much more electronegative than hydrogen ($EN = 2.30$), so the electrons in the covalent bonds are not shared equally by the hydrogen and oxygen atoms. The electrons are drawn toward the oxygen atom in the center of the molecule and away from the hydrogen atoms on either end. As a result, the water molecule is *polar*. The oxygen atom carries a partial negative charge (-0.8), and each hydrogen atom carries a partial positive charge ($+0.4$).

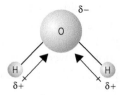

When neutral water molecules dissociate to form ions, positively charged H_3O^+ ions and negatively charged OH^- ions are formed.

$$2\,H_2O(l) \longrightarrow H_3O^+(aq) + OH^-(aq)$$

The opposite reaction can also occur: H_3O^+ ions can combine with OH^- ions to form neutral water molecules.

$$H_3O^+(aq) + OH^-(aq) \longrightarrow 2\,H_2O(l)$$

The fact that water molecules dissociate to form H_3O^+ and OH^- ions, which can then recombine to form water molecules, is indicated by the following equation.

$$2\,H_2O(l) \rightleftharpoons H_3O^+(aq) + OH^-(aq)$$

The pair of arrows that separate the "reactants" and the "products" of this reaction indicate that the reaction occurs in both directions.

The Brønsted model explains water's role in acid–base reactions.

- Water dissociates to form ions by transferring an H^+ ion from one molecule acting as an acid to another molecule acting as a base.

$$H_2O(l) + H_2O(l) \rightleftharpoons H_3O^+(aq) + OH^-(aq)$$
$$\qquad\;\; Acid \qquad\;\;\; Base$$

- Acids react with water by donating an H^+ ion to a neutral water molecule acting as a base to form the H_3O^+ ion.

$$HF(aq) + H_2O(l) \rightleftharpoons H_3O^+(aq) + F^-(aq)$$
$$\quad\; Acid \qquad\; Base$$

- Bases react with water by accepting an H^+ ion from a water molecule acting as an acid to form the OH^- ion.

$$NH_3(aq) + H_2O(l) \rightleftharpoons NH_4^+(aq) + OH^-(aq)$$
$$\quad\; Base \qquad\;\; Acid$$

- Water molecules can act as intermediates in acid–base reactions by gaining H^+ ions from an acid.

$$CH_3CO_2H(aq) + H_2O(l) \rightleftharpoons CH_3CO_2^-(aq) + H_3O^+(aq)$$
$$\qquad Acid \qquad\qquad Base$$

and then losing the H^+ ions to a base.

$$NH_3(aq) + H_3O^+(aq) \rightleftharpoons NH_4^+(aq) + H_2O(l)$$
$$\quad\; Base \qquad\;\; Acid$$

Adding these reactions so that species that are on both sides of the equation cancel gives the overall equation for the acid–base reaction.

$$CH_3CO_2H(aq) + H_2O(l) \rightleftharpoons CH_3CO_2^-(aq) + H_3O^+(aq)$$
$$\underline{NH_3(aq) + H_3O^+(aq) \rightleftharpoons NH_4^+(aq) + H_2O(l)}$$
$$CH_3CO_2H(aq) + NH_3(aq) \rightleftharpoons NH_4^+(aq) + CH_3CO_2^-(aq)$$

It is important to recognize that acid–base reactions don't have to occur in water; they can also occur in the gas phase. Many students encounter this phenomenon in their first chemistry courses. When bottles of concentrated hydrochloric acid and concentrated aqueous ammonia are both open at the same time, a white cloud of ammonium chloride often forms in the air above these bottles.

$$HCl(g) + NH_3(g) \rightleftharpoons NH_4Cl(s)$$

HCl gas and NH_3 gas that collect above open bottles of concentrated solutions of these compounds react in the gas phase to form a cloud of solid NH_4Cl.

11.6 To What Extent Does Water Dissociate to Form Ions?

At 25°C, the density of water is 0.9971 g/cm³, or 0.9971 g/mL. The concentration of pure water is therefore 55.35 *M*.

$$\frac{0.9971 \text{ g } H_2O}{1 \text{ mL}} \times \frac{1000 \text{ mL}}{1 \text{ L}} \times \frac{1 \text{ mol } H_2O}{18.015 \text{ g } H_2O} = 55.35 \text{ mol } H_2O/L$$

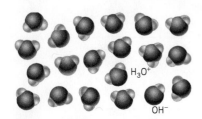

Fig. 11.4 If this figure is thought of as a high-resolution snapshot of 20 H_2O molecules, one of which has dissociated to form H_3O^+ and OH^- ions, we would have to look at an average of 25 million such snapshots before we encountered another pair of these ions.

The concentration of the H_3O^+ and OH^- ions formed by the dissociation of pure neutral H_2O at that temperature is only 1.0×10^{-7} mol/L. The ratio of the concentration of the H_3O^+ (or OH^-) ions to the concentration of the neutral H_2O molecules is therefore 1.8×10^{-9}.

$$\frac{1.0 \times 10^{-7} M\ H_3O^+}{55.35\ M\ H_2O} = 1.8 \times 10^{-9}$$

In other words, only about 2 parts per billion (ppb) of the water molecules dissociate into ions at room temperature.

It is difficult to imagine what 2 ppb means. One way to visualize this number is to assume that 2500 letters appear on a typical page of this book. If typographical errors occurred with a frequency of 2 ppb, the book would have to be 400,000 pages long in order to contain two errors. Figure 11.4 shows a model of 20 water molecules, one of which has dissociated to form H_3O^+ and OH^- ions. If this illustration were a high-resolution photograph of the structure of water, we would encounter a pair of H_3O^+ and OH^- ions on the average of only once for every 25 million such photographs.

The rate at which water molecules react to form the H_3O^+ and OH^- ions in pure water is equal to the rate at which these ions combine to form a pair of neutral water molecules.

$$2\ H_2O(l) \rightleftharpoons H_3O^+(aq) + OH^-(aq)$$

This reaction is therefore said to be at equilibrium. The extent to which water dissociates to form ions can be described in terms of the following equilibrium constant expression.

$$K_c = \frac{[H_3O^+][OH^-]}{[H_2O]^2}$$

As we have seen, the concentration of water at 25°C is 55.35 mol/L. The concentration of the H_3O^+ and OH^- ions, on the other hand, is only about $1.0 \times 10^{-7} M$. The concentration of H_2O molecules is therefore so much larger than the concentration of the H_3O^+ and OH^- ions that it remains effectively constant when water dissociates to form these ions. Chemists therefore rearrange the equilibrium constant expression for the dissociation of water to give the following equation.

$$K_c \times [H_2O]^2 = [H_3O^+][OH^-]$$

This effectively places two variables—the H_3O^+ and OH^- ion concentrations—on one side of the equation and two constants—K_c and $[H_2O]^2$—on the other side of the equation. We then replace the terms on the left-hand side of the equation with a constant known as the **water dissociation equilibrium constant, K_w.**

$$K_w = [H_3O^+][OH^-]$$

Because the H_3O^+ and OH^- concentrations in pure water at 25°C are each $1.0 \times 10^{-7} M$, the value of K_w at that temperature is 1.0×10^{-14}.

$$K_w = [1.0 \times 10^{-7}][1.0 \times 10^{-7}] = 1.0 \times 10^{-14} \quad \text{(at 25°C)}$$

Equilibrium constants are determined experimentally and are subject to some uncertainty. Therefore, equilibrium constants in this text are generally reported to only one or at most two significant figures. The equilibrium constant expression for the dissociation of water is still useful, however, for obtaining approximate concentrations of the H_3O^+ and OH^- ions in water. Regardless of the source of the H_3O^+ and OH^- ions in water, the product of the concentrations of these ions at equilibrium at 25°C is approximately 1.0×10^{-14}.

What happens to the H_3O^+ and OH^- ion concentrations that come from the dissociation of water when we add a strong acid to water? Suppose, for example, that we add enough acid to a beaker of water to raise the H_3O^+ concentration to 0.010 M. Note that there are now two sources of H_3O^+. One source is the acid and the other is the dissociation of water.

$$HA(aq) + H_2O(l) \rightleftharpoons H_3O^+(aq) + A^-(aq)$$
$$2\,H_2O(l) \rightleftharpoons H_3O^+(aq) + OH^-(aq)$$

According to Le Châtelier's principle, the additional H_3O^+ from the acid should drive the equilibrium between water and its ions to the left, reducing the number of H_3O^+ and OH^- ions from the dissociation of water. Adding an acid to water therefore decreases the extent to which the water dissociates into ions.

This is another example of the "common-ion effect" encountered in the previous chapter during the discussion of solubility product equilibria. The addition of a second source of an ion that is shared in common with an ion formed in an equilibrium that involves the dissociation of a substance decreases the amount of dissociation. In Section 11.15 we will see that the "common-ion effect" also occurs in so-called buffer solutions.

Because the dissociation of water shifts to the left in the presence of an acid, the concentration of H_3O^+ and OH^- ions from the dissociation of water will be smaller than the $1.0 \times 10^{-7} M$ concentration in pure water. The total concentration of the H_3O^+ ion will be equal to the H_3O^+ ion from dissociation of the acid (0.010 M) plus the H_3O^+ ion from the dissociation of water ($< 1.0 \times 10^{-7} M$). The H_3O^+ ion concentration from the dissociation of water, however, is so small that it is negligible when compared to the amount from the acid. Therefore, when the system returns to equilibrium, the H_3O^+ ion concentration is still about 0.010 M. Furthermore, when the reaction returns to equilibrium, the product of the H_3O^+ and OH^- ion concentrations is once again approximated by K_w.

$$[H_3O^+][OH^-] = 1.0 \times 10^{-14}$$

If the concentration of the H_3O^+ ion is 0.010 M, the concentration of the OH^- ion when the system returns to equilibrium is only $1.0 \times 10^{-12} M$.

$$[OH^-] = \frac{K_w}{[H_3O^+]} = \frac{1.0 \times 10^{-14}}{0.010} = 1.0 \times 10^{-12}\,M$$

Some of the H_3O^+ ions in the solution come from the dissociation of water; others come from the acid that has been added to the water. All of the OH^- ions, on the other hand, come from the dissociation of water. In pure water, dissociation of H_2O molecules gives us an OH^- ion concentration of $1.0 \times 10^{-7} M$. In the acid solution, the OH^- ion concentration is 5 orders of magnitude smaller. This means that adding enough acid to water to increase the concentration of the H_3O^+ ion to 0.010 M decreases the dissociation of water by a factor of about 100,000.

Adding an acid to water therefore has an effect on the concentration of both the H_3O^+ and OH^- ions. Some H_3O^+ is already present as a result of the dissociation of water. Adding an acid to water increases the concentration of that ion.

When an acid is added to water, the equilibrium between neutral water molecules and their ions shifts to the left, decreasing the amount of H_3O^+ from the dissociation of water.

$$2\,H_2O(l) \rightleftharpoons H_3O^+(aq) + OH^-(aq)$$

Does K_w still equal 1.0×10^{-14}?

However, adding an acid to water decreases the extent to which water dissociates and therefore leads to a significant decrease in the concentration of the OH^- ion.

As might be expected, the opposite effect is observed when a base is added to water. Because we are adding a base, the OH^- ion concentration increases. Once the system returns to equilibrium, the product of the H_3O^+ and OH^- ion concentrations is once again equal to K_w. The only way this can be achieved, of course, is by decreasing the concentration of the H_3O^+ ion from the dissociation of water.

11.7 pH as a Measure of the Concentration of the H₃O⁺ Ion

The range of concentrations of the H_3O^+ and OH^- ions in aqueous solutions is enormous. Typical solutions in the laboratory have concentrations of the H_3O^+ or OH^- ion as large as $1.0\ M$ (or larger) and as small as $1 \times 10^{-14}\ M$. Perhaps the best way to bring a range of 14 orders of magnitude into perspective is to note that it is comparable to the difference between 10¢ and the national debt of more than \$10 trillion, or the difference between the radius of a gold atom and a distance of 8 miles.

In 1909, the Danish biochemist S. P. L. Sørenson proposed a way of conveniently describing a large range of concentrations. Sørenson worked at a laboratory set up by the Carlsberg Brewery to apply scientific methods to the study of the fermentation reactions in brewing beer. Faced with the task of constructing graphs of the activity of the malt versus the H_3O^+ ion concentration, Sørenson suggested using logarithmic mathematics to condense the range of H_3O^+ and OH^- concentrations to a more convenient scale. By definition, the logarithm of a number is the power to which a base must be raised to obtain that number. The logarithm to the base 10 of 10^{-7}, for example, is -7.

$$\log(10^{-7}) = -7$$

Because the concentrations of the H_3O^+ and OH^- ions in aqueous solutions are usually smaller than $1\ M$, the logarithms of these concentrations are negative numbers. Because he considered positive numbers more convenient, Sørenson suggested that the sign of the logarithm should be changed after it had been calculated. He therefore introduced the symbol p to indicate the negative of the logarithm of a number. Thus, **pH** is the negative of the logarithm of the H_3O^+ ion concentration.

$$pH = -\log[H_3O^+]$$

Similarly, **pOH** is the negative of the logarithm of the OH^- ion concentration.

$$pOH = -\log[OH^-]$$

It should be noted that the equations for pH and pOH are valid only for dilute solutions of acid or base in pure water. Most solutions used in the laboratory do not meet this criterion, so the above equations can only be used to approximate the pH and pOH of real solutions. Since values from these calculations are approximations, they are normally reported to no more than two digits after the decimal. A pH meter, however, can accurately determine the pH of a solution to two digits after the decimal. If the pH is given, the H_3O^+ ion concentration can be calculated with the following equation.

$$[H_3O^+] = 10^{-pH}$$

Exercise 11.3

What is the pH of Pepsi Cola if the concentration of the H_3O^+ ion in the solution is 0.0035 M?

Solution

The pH of a solution is the negative of the logarithm of the H_3O^+ ion concentration. The pH of this solution is therefore 2.5.

$$pH = -\log[H_3O^+]$$
$$pH = -\log(3.5 \times 10^{-3})$$
$$pH = -(-2.5) = 2.5$$

The advantage of the pH scale is illustrated in Figure 11.5, which shows the possible combinations of H_3O^+ ion and OH^- concentrations in an aqueous solution. In Figure 11.5a, the concentration of the OH^- ion is plotted on the vertical axis, and the concentration of the H_3O^+ ion is plotted on the horizontal axis. For the plot to be a reasonable size, it has to be restricted to only 1 order of magnitude of concentration.

The concept of pH compresses the range of H_3O^+ concentrations onto a scale that is much easier to handle. As the H_3O^+ concentration decreases from roughly 1 M to $10^{-14}M$, the pH of the solution increases from 0 to 14. The relationship between the H_3O^+ concentration, OH^- concentration, and the pH of a solution is shown in Table 11.2. The advantage of the pH scale is illustrated in Figure 11.5. The only way to plot the concentration of the OH^- ion versus the concentration of the H_3O^+ ion is to restrict the data to a narrow range of concentrations, as shown in Figure 11.5a. The entire range of data in Table 11.2, however, can be plotted in a graph of pH versus pOH, as shown in Figure 11.5b. Note that not only

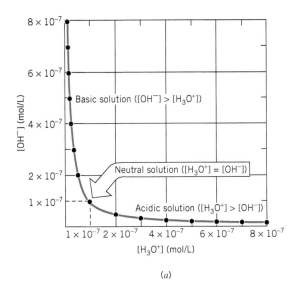

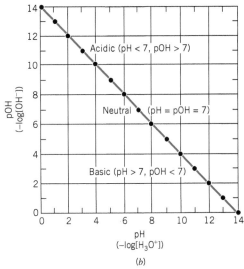

(a) $\qquad\qquad\qquad\qquad$ (b)

Fig. 11.5 (a) A small fraction of data showing the relationship between the H_3O^+ and OH^- ion concentrations in aqueous solutions. Every point on the solid line represents a pair of H_3O^+ and OH^- concentrations when the solution is at equilibrium. (b) We can fit the entire range of H_3O^+ and OH^- concentrations from Table 11.2 on a single graph by plotting pH versus pOH. Any point on the solid line corresponds to a solution at equilibrium.

Table 11.2
Pairs of Equilibrium Concentrations of the H_3O^+ and OH^- Ions with the Corresponding pH Values in Water at 25°C

$[H_3O^+]$ (mol/L)	$[OH^-]$ (mol/L)	pH	
1	1×10^{-14}	0	
1×10^{-1}	1×10^{-13}	1	
1×10^{-2}	1×10^{-12}	2	
1×10^{-3}	1×10^{-11}	3	Acidic solution
1×10^{-4}	1×10^{-10}	4	
1×10^{-5}	1×10^{-9}	5	
1×10^{-6}	1×10^{-8}	6	
1×10^{-7}	1×10^{-7}	7	Neutral solution
1×10^{-8}	1×10^{-6}	8	
1×10^{-9}	1×10^{-5}	9	
1×10^{-10}	1×10^{-4}	10	
1×10^{-11}	1×10^{-3}	11	Basic solution
1×10^{-12}	1×10^{-2}	12	
1×10^{-13}	1×10^{-1}	13	
1×10^{-14}	1	14	

is the scale compressed by using pH and pOH data, but the shape of the graph has changed from a sharp curve in Figure 11.5*a* to a straight line in Figure 11.5*b*.

The concentration of the H_3O^+ ion in pure water at 25°C is $1.0 \times 10^{-7} M$. Thus the pH of pure water is 7.

$$pH = -\log[H_3O^+] = -\log(1.0 \times 10^{-7}M) = 7.0$$

Solutions for which the concentrations of the H_3O^+ and OH^- ions are equal are said to be *neutral*. Solutions in which the concentration of the H_3O^+ ion is larger than $1 \times 10^{-7}M$ at 25°C are described as *acidic*. Those in which the concentration of the H_3O^+ ion is smaller than $1 \times 10^{-7}M$ are *basic*. Thus, at 25°C when the pH of a solution is less than 7, the solution is acidic. When the pH is more than 7, the solution is basic.

$$\text{Acidic:} \quad [H_3O^+] > 1 \times 10^{-7}M \quad pH < 7$$
$$\text{Basic:} \quad [H_3O^+] < 1 \times 10^{-7}M \quad pH > 7 \quad \text{(at 25°C)}$$

Measurements of pH in the laboratory were historically done with **acid–base indicators,** which are weak acids or weak bases that change color when they gain or lose an H^+ ion. Acid–base indicators are still used in the laboratory for rough pH measurements. An example of an acid–base indicator is litmus, which turns pink in solutions whose pH is below 5 and turns blue when the pH is above 8. To a large extent, these indicators have been replaced by pH meters, which are more accurate. The actual measuring device in a pH meter is an electrode that consists of a resin-filled tube with a thin glass bulb at one end. When the electrode is immersed in a solution to be measured, it produces an electric potential that is related to the H_3O^+ ion concentration in the solution.

As we have seen, adding an acid to water increases the H_3O^+ concentration and decreases the OH^- concentration. Adding a base does the opposite. Regardless of what is added to water, however, the product of the concentrations of the ions at equilibrium is 1.0×10^{-14} at 25°C.

$$[H_3O^+][OH^-] = 1.0 \times 10^{-14}$$

The relationship between the pH and pOH of an aqueous solution can be derived by taking the logarithm of both sides of the K_w expression.

$$\log([H_3O^+][OH^-]) = \log(1.0 \times 10^{-14})$$

The log of the product of two numbers is equal to the sum of their logs. Thus the sum of the logs of the H_3O^+ and OH^- ion concentrations is equal to the log of 10^{-14}.

$$\log[H_3O^+] + \log[OH^-] = -14.0$$

Both sides of the equation can now be multiplied by -1.

$$-\log[H_3O^+] - \log[OH^-] = 14.0$$

Substituting the definitions of pH and pOH into the equation gives the following result.

$$pH + pOH = 14.0$$

This equation can be used to convert from pH to pOH, and vice versa, for any aqueous solution at 25°C, regardless of how much acid or base has been added to the solution.

> ➤ **CHECKPOINT**
>
> Describe what happens to the pH of a solution as the concentration of the H_3O^+ ion in the solution increases.

Exercise 11.4

The pH values of samples of lemon juice (pH 2.2) and vinegar (pH 2.5) are quite similar. Calculate the H_3O^+ and OH^- concentrations for both solutions and compare them to one another.

Solution

Lemon juice: $pH = -\log[H_3O^+] = 2.2$ $pOH = -\log[OH^-] = 11.8$
        $[H_3O^+] = 10^{-2.2} = 6 \times 10^{-3}\,M$ $[OH^-] = 10^{-11.8} = 2 \times 10^{-12}\,M$

Vinegar: $pH = -\log[H_3O^+] = 2.5$ $pOH = -\log[OH^-] = 11.5$
        $[H_3O^+] = 10^{-2.5} = 3 \times 10^{-3}\,M$ $[OH^-] = 10^{-11.5} = 3 \times 10^{-12}\,M$

Note that the concentration of the H_3O^+ ion in lemon juice is twice as large as that in vinegar even though the two solutions have similar pH values.

11.8 Relative Strengths of Acids and Bases

Many hardware stores sell muriatic acid—a 6 M solution of hydrochloric acid, $HCl(aq)$—to clean bricks and concrete and control the pH of swimming pools. Grocery stores sell vinegar, which is a 1 M solution of acetic acid, CH_3CO_2H. Although both substances are acids, you wouldn't use muriatic acid in salad dressing, and vinegar is ineffective in cleaning bricks or concrete.

The difference between the two acids can be explored with the apparatus shown in Figure 11.6. Two metal electrodes connected to a source of electricity are dipped into a beaker that contains a solution of one of these acids. If the solution conducts an electric current because of the presence of positive and negative ions, the flow of these ions between the electrodes completes the electric circuit

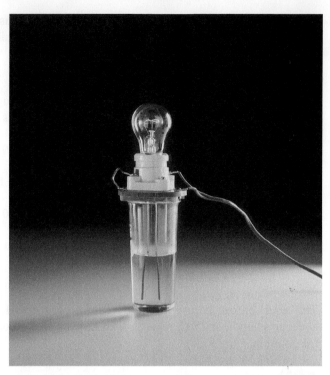

Fig. 11.6 The conductivity apparatus shows that even though the solution of HCl on the left and the solution of acetic acid on the right have the same concentrations, the acid in the solution on the left with the brightly glowing bulb has dissociated to produce more ions.

and the lightbulb starts to glow. The intensity with which the bulb glows depends on the ability of the solution to conduct a current. This, in turn, depends on the concentration of the positive and negative ions in the solution that conducts the electric current. The larger the number of positive and negative ions in the solution, the brighter the lightbulb glows.

When the electrodes of this apparatus are immersed in pure water, the lightbulb doesn't glow. This isn't surprising because the concentrations of H_3O^+ and OH^- ions in pure water are very small, only about $10^{-7} M$ at room temperature. When the electrodes are immersed in a 1 M solution of acetic acid, the bulb glows, but only dimly. When the electrodes are immersed in a 1 M solution of hydrochloric acid, the lightbulb glows very brightly. Although the concentration of the acid in both solutions is the same, 1 M, hydrochloric acid contains far more ions than the equivalent acetic acid solution.

The difference between these acids is a result of differences in the abilities of the two acids to donate a proton to water. The hydrochloric acid in muriatic acid is a **strong acid** because HCl it is very good at transferring an H^+ ion to a water molecule. Virtually all of the HCl molecules in this solution react with water to form H_3O^+ and Cl^- ions.

$$HCl(aq) + H_2O(l) \longrightarrow H_3O^+(aq) + Cl^-(aq)$$

In a 0.1 M solution of hydrochloric acid, very few HCl molecules remain in solution. Hydrochloric acid is therefore a strong acid indeed.

Note the difference between the ideas of a strong acid and a concentrated acid. A strong acid dissociates more or less completely when dissolved in water. A concentrated acid is one that has a relatively large value of the molarity. It is possible to have a concentrated solution of a weak acid. Consider concentrated orange juice, for example. A dilute solution of a strong acid will dissociate essentially 100% in

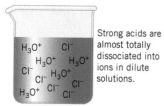

Strong acids are almost totally dissociated into ions in dilute solutions.

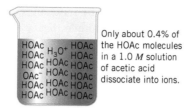

Only about 0.4% of the HOAc molecules in a 1.0 M solution of acetic acid dissociate into ions.

Fig. 11.7 (a) A dilute HCl solution. Essentially all of the HCl that dissolves in water dissociates to form H_3O^+ and Cl^- ions. (b) A concentrated solution of acetic acid. Only about 0.4% of the acetic acid molecules dissociate to form H_3O^+ and OAc^- ions when HOAc dissolves in water.

water and is therefore a strong acid even though it has a relatively small concentration, as shown on the left in Figure 11.7.

Acetic acid (often abbreviated as HOAc) is a **weak acid** because it is not very good at transferring H^+ ions to water. In a 1 M solution, less than 0.4% of the CH_3CO_2H molecules react with water to form H_3O^+ and $CH_3CO_2^-$ ions.

$$CH_3CO_2H(aq) + H_2O(l) \rightleftharpoons H_3O^+(aq) + CH_3CO_2^-(aq)$$

More than 99.6% of the acetic acid molecules remain undissociated, as shown on the right in Figure 11.7, even though this is a more concentrated solution than the strong acid shown on the left.

It would be useful to have a quantitative measure of the relative strengths of acids to replace the labels *strong* and *weak*. The extent to which an acid dissociates in water to produce the H_3O^+ ion is described in terms of an acid-dissociation equilibrium constant, K_a. To understand the nature of this equilibrium constant, let's assume that the reaction between an acid and water can be represented by the following generic equation.

$$HA(aq) + H_2O(l) \rightleftharpoons H_3O^+(aq) + A^-(aq)$$

In other words, we will assume that some of the HA molecules react to form H_3O^+ and A^- ions, as shown in Figure 11.8. By convention, the equilibrium concentrations of the ions in units of moles per liter are represented by the symbols $[H_3O^+]$ and $[A^-]$. The concentration of the undissociated HA molecules that remain in solution is represented by the symbol $[HA]$.

The equilibrium constant expression for the reaction between HA and water would be written as follows.

$$K_c = \frac{[H_3O^+][A^-]}{[HA][H_2O]}$$

Most acid solutions are so dilute that the concentration of H_2O at equilibrium is effectively the same as before the acid was added. The equilibrium constant for the reaction is therefore written as follows.

$$\frac{[H_3O^+][A^-]}{[HA]} = K_c \times [H_2O]$$

The result is an equilibrium constant for the equation known as the **acid dissociation equilibrium constant, K_a.**

$$K_a = \frac{[H_3O^+][A^-]}{[HA]} = K_c \times [H_2O]$$

When a strong acid dissolves in water, it reacts extensively with water to form H_3O^+ and A^- ions. (Essentially 100% of the strong acid dissociates, leaving only a very small residual concentration of HA molecules in solution.) The product of the concentrations of the H_3O^+ and A^- ions is therefore much larger than

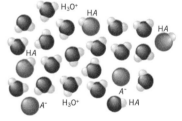

Fig. 11.8 Some, but not all, of the HA molecules in a typical acid react with water to form H_3O^+ and A^- ions when the acid dissolves in water. In a strong acid, there are very few undissociated HA molecules. In a weak acid, most of the HA molecules remain.

the concentration of the H*A* molecules, so K_a for a strong acid is greater than 1. Hydrochloric acid, for example, has a K_a of roughly 1×10^6.

$$\frac{[H_3O^+][Cl^-]}{[HCl]} = 1 \times 10^6$$

Weak acids, on the other hand, react only slightly with water. The product of the concentrations of the H_3O^+ and A^- ions is therefore smaller than the concentration of the residual H*A* molecules. As a result, K_a for a weak acid is less than 1. Acetic acid, for example, has a K_a of only 1.8×10^{-5}.

$$\frac{[H_3O^+][CH_3CO_2^-]}{[CH_3CO_2H]} = 1.8 \times 10^{-5}$$

K_a can therefore be used to distinguish between strong acids and weak acids. As a general rule,

Strong acids: $K_a > 1$

Weak acids: $K_a < 1$

A list of common acids and their acid dissociation constants, K_a, is given in Table 11.3. A more complete list can be found in Table B.8 in Appendix B. Also included in Table B.8 are pK_a values ($pK_a = -\log K_a$). Commonly accepted K_a values are reported for the strong acids. Experimental determination of K_a values for strong acids is difficult, resulting in considerable variation in the values reported in the literature.

The ionization of bases in water can be described in a similar manner. A generic base, *B*, reacting with water can be described by the following chemical equation.

$$B(aq) + H_2O(l) \rightleftharpoons BH^+(aq) + OH^-(aq)$$

Strict adherence to the rules for writing equilibrium constant expressions gives the following result.

$$K_c = \frac{[BH^+][OH^-]}{[B][H_2O]}$$

As we saw in our discussion of acids, the concentration of water at equilibrium is essentially constant, allowing the equilibrium expression to be written as follows.

$$K_b = \frac{[BH^+][OH^-]}{[B]} = K_c \times [H_2O]$$

The new equilibrium constant is known as the **base ionization equilibrium constant, K_b.** Just as K_a values describe the relative strengths of acids, K_b values describe the relative strengths of bases. The values of K_b for a limited number of bases are given in Table B.9 in Appendix B. This table also lists the pK_b values of these bases ($pK_b = -\log K_b$).

Strong bases are defined as ions or molecules that ionize more or less completely in water to produce the OH^- ion. Examples of strong bases include the following:

- Group IA metal hydroxides: LiOH, NaOH, KOH, RbOH, and CsOH

$$NaOH(s) \longrightarrow Na^+(aq) + OH^-(aq)$$

► CHECKPOINT

Write the equilibrium expressions that describe the K_a and the K_w of water. Show how the two expressions are different.

Table 11.3

Common Acids and Their Acid Dissociation Equilibrium Constant for the Loss of One Proton

	K_a
Strong Acids	
HI	3×10^9
HBr	1×10^9
$HClO_4$	1×10^8
HCl	1×10^6
H_2SO_4	1×10^3
H_3O^+	55
HNO_3	28
H_2CrO_4	9.6
Weak Acids	
H_3PO_4	7.1×10^{-3}
HF	7.2×10^{-4}
Citric acid	7.5×10^{-4}
CH_3CO_2H	1.8×10^{-5}
H_2S	1.0×10^{-7}
H_2CO_3	4.5×10^{-7}
H_3BO_3	7.3×10^{-10}
H_2O	1.8×10^{-16}

- Group IIA metal hydroxides that are soluble or slightly soluble in water: $Ca(OH)_2$ (slightly soluble), $Sr(OH)_2$, and $Ba(OH)_2$

$$Ca(OH)_2(aq) \longrightarrow Ca^{2+}(aq) + 2\,OH^-(aq)$$

- Soluble metal oxides: Li_2O, Na_2O, K_2O, and CaO

$$Li_2O(s) + H_2O(l) \longrightarrow 2\,Li^+(aq) + 2\,OH^-(aq)$$

In Section 11.7, **pH** was defined as the negative of the logarithm of the H_3O^+ ion concentration.

$$pH = -\log[H_3O^+]$$

Similarly, **pOH** was defined as the negative of the logarithm of the OH^- ion concentration.

$$pOH = -\log[OH^-]$$

Mathematicians would describe "p" as an operator. It implies the same operation whenever it is invoked. In each case, we take the negative of the logarithm of the variable to which it is applied. Thus we can also calculate pK_a's from the value of the acid-dissociation equilibrium constant for a reaction

$$pK_a = -\log(K_a)$$

and pK_b's from the value of the base-ionization equilibrium constant.

$$pK_b = -\log(K_b)$$

11.9 Relative Strengths of Conjugate Acid–Base Pairs

The relationship between the strength of an acid and its conjugate base can be understood by considering the implications of the fact that HCl is a strong acid. If it is a strong acid, HCl must be a good proton donor. HCl can only be a good proton donor, however, if the Cl^- ion is a poor proton acceptor. Thus, the Cl^- ion must be a **weak base.**

$$\underset{\text{Strong acid}}{HCl(aq)} + H_2O(l) \longrightarrow H_3O^+(aq) + \underset{\text{very very weak base}}{Cl^-(aq)}$$

Because HCl is a strong acid, it dissociates essentially completely, and there is little tendency for the reaction to go in the reverse direction. Thus the chloride anion is such a weak conjugate base that it has essentially no base properties.

Let's now consider the relationship between the strength of the ammonium ion (NH_4^+) and its conjugate base, ammonia (NH_3). The NH_4^+ ion is a weak acid, and its conjugate base ammonia is a reasonably good base.

$$\underset{\text{Weak acid}}{NH_4^+(aq)} + H_2O(l) \rightleftharpoons H_3O^+(aq) + \underset{\text{Good base}}{NH_3(aq)}$$

These examples can be summarized in the form of the following general rules:

The stronger the acid, the weaker the conjugate base.

The stronger the base, the weaker the conjugate acid.

> **CHECKPOINT**
>
> Which solution has the smallest pH, a solution with a low concentration of an acid with a small K_a or a solution with a low concentration of an acid with a high K_a?

11.10 Relative Strengths of Different Acids and Bases

We can summarize the relative strengths of Brønsted acids and bases by organizing the reactions in which a given acid is converted into its conjugate base, as shown in Table 11.4. The strongest acids are in the upper-left corner of this table, and the strongest bases are in the bottom-right corner.

This table can be used to compare the relative strengths of a pair of acids, to decide which is stronger. Consider HCl and the H_3O^+ ion, for example.

$$HCl: \quad K_a = 1 \times 10^6$$
$$H_3O^+: \quad K_a = 55$$

The K_a values suggest that both are strong acids, but HCl is a stronger acid than the H_3O^+ ion.

We can now understand why such a large proportion of the HCl molecules in an aqueous solution react with water to form H_3O^+ and Cl^- ions. The Brønsted

Table 11.4
Relative Strengths of Typical Brønsted Acids and Bases

Acid	Conjugate Base	K_a	
Best	$HI \rightleftharpoons H^+ + I^-$	3×10^9	
Brønsted	$HClO_4 \rightleftharpoons H^+ + ClO_4^-$	1×10^8	
acids	$HCl \rightleftharpoons H^+ + Cl^-$	1×10^6	
	$H_2SO_4 \rightleftharpoons H^+ + HSO_4^-$	1×10^3	
	$HClO_3 \rightleftharpoons H^+ + ClO_3^-$	5×10^2	
	$H_3O^+ \rightleftharpoons H^+ + H_2O$	55	
	$HNO_3 \rightleftharpoons H^+ + NO_3^-$	28	
	$H_2CrO_4 \rightleftharpoons H^+ + HCrO_4^{2-}$	9.6	
	$HSO_4^- \rightleftharpoons H^+ + SO_4^{2-}$	1.2×10^{-2}	
	$HClO_2 \rightleftharpoons H^+ + ClO_2^-$	1.1×10^{-2}	
	$H_3PO_4 \rightleftharpoons H^+ + H_2PO_4^-$	7.1×10^{-3}	
	$HF \rightleftharpoons H^+ + F^-$	7.2×10^{-4}	
	$CH_3CO_2H \rightleftharpoons H^+ + CH_3CO_2^-$	1.8×10^{-5}	
	$H_2CO_3 \rightleftharpoons H^+ + HCO_3^-$	4.5×10^{-7}	
	$H_2S \rightleftharpoons H^+ + HS^-$	1.0×10^{-7}	
	$HClO \rightleftharpoons H^+ + ClO^-$	2.9×10^{-8}	
	$H_2PO_4^- \rightleftharpoons H^+ + HPO_4^{2-}$	6.3×10^{-8}	
	$H_3BO_3 \rightleftharpoons H^+ + H_2BO_3^-$	7.3×10^{-10}	
	$NH_4^+ \rightleftharpoons H^+ + NH_3$	5.6×10^{-10}	
	$HCO_3^- \rightleftharpoons H^+ + CO_3^{2-}$	4.7×10^{-11}	
	$HPO_4^{2-} \rightleftharpoons H^+ + PO_4^{3-}$	4.2×10^{-13}	
	$HS^- \rightleftharpoons H^+ + S^{2-}$	1.3×10^{-13}	
	$H_2O \rightleftharpoons H^+ + OH^-$	1.8×10^{-16}	
	$CH_3OH \rightleftharpoons H^+ + CH_3O^-$	1×10^{-18}	
	$HC{\equiv}CH \rightleftharpoons H^+ + HC{\equiv}C^-$	1×10^{-25}	
	$NH_3 \rightleftharpoons H^+ + NH_2^-$	1×10^{-33}	
	$H_2 \rightleftharpoons H^+ + H^-$	1×10^{-35}	Best
	$CH_2{=}CH_2 \rightleftharpoons H^+ + CH_2{=}CH^-$	1×10^{-44}	Brønsted
	$CH_4 \rightleftharpoons H^+ + CH_3^-$	1×10^{-49}	Bases

model suggests that every acid–base reaction converts an acid into its conjugate base and a base into its conjugate acid.

$$HCl(aq) + H_2O(l) \longrightarrow H_3O^+(aq) + Cl^-(aq)$$

Acid   Base   Acid   Base

There are two acids and two bases in the reaction. The stronger acid, however, is on the left side of the equation, which suggests that the reaction should proceed to the right. HCl should react with water to form the H_3O^+ and Cl^- ions.

$$HCl(aq) + H_2O(l) \longrightarrow H_3O^+(aq) + Cl^-(aq)$$

Stronger acid   Weaker acid

What about the two bases: H_2O and the Cl^- ion? The general rules given in the previous section suggest that the stronger of a pair of acids must form the weaker of a pair of conjugate bases. The fact that HCl is a stronger acid than the H_3O^+ ion implies that the Cl^- ion is a weaker base than water.

Acid strength: $HCl > H_3O^+$

Base strength: $Cl^- < H_2O$

Thus the equation for the reaction between HCl and water can be written as follows.

$$HCl(aq) + H_2O(l) \longrightarrow H_3O^+(aq) + Cl^-(aq)$$

Stronger   Stronger   Weaker   Weaker
acid   base   acid   base

It isn't surprising that almost all of the HCl molecules in an aqueous solution react with water to give H_3O^+ ions and Cl^- ions. The stronger of a pair of acids should react with the stronger of a pair of bases to form a weaker acid and a weaker base.

As a general rule, acids with K_a values greater than the H_3O^+ ion ($K_a = 55$) can be assumed to dissociate completely (100%) in water.

Let's now look at the relative strengths of acetic acid and the H_3O^+ ion.

CH_3CO_2H: $K_a = 1.8 \times 10^{-5}$

H_3O^+: $K_a = 55$

These K_a values suggest that acetic acid is a much weaker acid than the H_3O^+ ion, which explains why acetic acid is a weak acid in water. Once again, the reaction between the acid and water converts the acid into its conjugate base and the base into its conjugate acid.

$$CH_3CO_2H(aq) + H_2O(l) \rightleftharpoons H_3O^+(aq) + CH_3CO_2^-(aq)$$

Acid   Base   Acid   Base

But in this case, the stronger acid and the stronger base are on the right side of the equation.

$$CH_3CO_2H(aq) + H_2O(l) \rightleftharpoons H_3O^+(aq) + CH_3CO_2^-(aq)$$

Weaker   Weaker   Stronger   Stronger
acid   base   acid   base

As a result, only a small percentage of the CH_3CO_2H molecules actually donate an H^+ ion to a water molecule to form the H_3O^+ and $CH_3CO_2^-$ ions. At equilibrium an acid–base reaction should lie predominantly on the side of the chemical equation that contains the weaker acid and weaker base.

Thus, Table 11.4 can be used to predict whether certain acid–base reactions should occur. Each base in this table is strong enough to deprotonate the acid in any line above it. According to this table, the acetate ($CH_3CO_2^-$) ion, for example, should be a strong enough base to remove a proton from nitric acid.

$$HNO_3(aq) + CH_3CO_2^-(aq) \rightleftharpoons CH_3CO_2H(aq) + NO_3^-(aq)$$

| Stronger acid | Stronger base | Weaker acid | Weaker base |

This table also predicts that ammonia (NH_3) should be a strong enough base to remove both protons from sulfuric acid (H_2SO_4) to form ammonium sulfate.

$$H_2SO_4(aq) + 2\,NH_3(aq) \rightleftharpoons 2\,NH_4^+(aq) + SO_4^{2-}(aq)$$

| Stronger acid | Stronger base | Weaker acid | Weaker base |

The usefulness of Table 11.4 might best be illustrated by asking whether the following reaction should occur as written.

$$H_2PO_4^-(aq) + HSO_4^-(aq) \rightleftharpoons H_3PO_4(aq) + SO_4^{2-}(aq)$$

The data in Table 11.4 suggest that $H_2PO_4^-$ ion is a stronger base than the SO_4^{2-} ion. These data also suggest that the HSO_4^- ion is a stronger acid than H_3PO_4. Because the stronger acid and the stronger base are on the left side of the reaction, we expect the reaction to occur as written.

$$H_2PO_4^-(aq) + HSO_4^-(aq) \rightleftharpoons H_3PO_4(aq) + SO_4^{2-}(aq)$$

| Stronger base | Stronger acid | Weaker acid | Weaker base |

The data in the table do not indicate that $H_2PO_4^-$ is a strong acid, in the absolute sense. These data merely indicate that $H_2PO_4^-$ is a stronger acid than HSO_4^-.

Exercise 11.5

The data in Table 11.4 are particularly useful for predicting some of the acid–base reactions that occur among organic compounds. Which of the following acid–base reactions would you expect to occur?

(a) The reaction between methanol (CH_3OH) and sodium hydride (NaH)

(b) The reaction between acetylene ($HC{\equiv}CH$) and sodium amide ($NaNH_2$)

Solution

(a) Sodium hydride is a source of the H^- or hydride ion. According to Table 11.4, the H^- ion should be a strong enough base to remove the acidic proton from methanol.

$$CH_3OH + H^- \rightleftharpoons CH_3O^- + H_2$$

| Stronger acid | Stronger base | Weaker base | Weaker acid |

(b) Sodium amide is a source of the NH_2^- or amide ion, which is a strong enough base to remove a proton from acetylene to form the acetylide ion.

$$HC\equiv CH + NH_2^- \rightleftharpoons HC\equiv C^- + NH_3$$

| *Stronger* | *Stronger* | *Weaker* | *Weaker* |
| *acid* | *base* | *base* | *acid* |

● ●

The magnitude of K_a can also be used to explain why some compounds that are potentially Brønsted acids or bases don't act like acids or bases when they dissolve in water. As long as K_a for the acid is significantly larger than the value of K_a for water (1.8×10^{-16}), the acid will ionize to some extent when it dissolves in water. As the K_a value for the acid approaches the K_a for water, the compound becomes more like water in its acidity. Although it is still potentially a Brønsted acid, it is so weak that we may be unable to detect the acidity in aqueous solution.

Measurements of the pH of dilute solutions are good indicators of the relative strengths of acids and bases. Experimental values of the pH of 0.10 *M* solutions of a number of common acids and bases are given in Table 11.5. Note that the pH of an 0.1 *M* solution of HCl is 1.1, not 1.0 as would be expected from the concentration of the solution. That can be understood by noting that HCl does not dissociate as completely as our model, so far, would predict.

➤ **CHECKPOINT**

HOBr is a weaker acid than HOCl. Which conjugate base, OBr^- or OCl^-, is stronger?

THE LEVELING EFFECT OF WATER

Because all strong acids dissociate in water to produce the same H_3O^+ ion, they all seem to have the same strength when dissolved in water, regardless of the

Table 11.5
pH of 0.10 *M* Solutions of Common Acids and Bases

Compound	pH
HCl (hydrochloric acid)	1.1
H_2SO_4 (sulfuric acid)	1.2
$NaHSO_4$ (sodium hydrogen sulfate)	1.4
H_2SO_3 (sulfurous acid)	1.5
H_3PO_4 (phosphoric acid)	1.5
HF (hydrofluoric acid)	2.1
CH_3CO_2H (acetic acid)	2.9
H_2CO_3 (carbonic acid)	3.8 (saturated solution)
H_2S (hydrogen sulfide)	4.1
NaH_2PO_4 (sodium dihydrogen phosphate)	4.4
NH_4Cl (ammonium chloride)	4.6
HCN (hydrocyanic acid)	5.1
NaCl (sodium chloride)	6.4
H_2O (freshly boiled distilled water)	7.0
$NaCH_3CO_2$ (sodium acetate)	8.4
$NaHCO_3$ (sodium hydrogen carbonate)	8.4
Na_2HPO_4 (sodium hydrogen phosphate)	9.3
Na_2SO_3 (sodium sulfite)	9.8
NaCN (sodium cyanide)	11.0
NH_3 (aqueous ammonia)	11.1
Na_2CO_3 (sodium carbonate)	11.6
Na_3PO_4 (sodium phosphate)	12.0
NaOH (sodium hydroxide, lye)	13.0

value of K_a. This phenomenon is known as the **leveling effect** of water—the tendency of water to limit the strengths of strong acids and bases. Close to 100% of the HCl molecules in hydrochloric acid react with water to form H_3O^+ and Cl^- ions, for example. Thus a 0.10 M solution of HCl produces approximately 0.10 M H_3O^+. In the same manner, a 0.10 M solution of the strong acid $HClO_4$ will also produce 0.10 M H_3O^+.

$$HCl(aq) + H_2O(l) \longrightarrow H_3O^+(aq) + Cl^-(aq)$$
$$HClO_4(aq) + H_2O(l) \longrightarrow H_3O^+(aq) + ClO_4^-(aq)$$

As a result, there is no difference between the acidity of these solutions even though $HClO_4$ ($K_a = 1 \times 10^8$) is supposedly a stronger acid than HCl ($K_a = 1 \times 10^6$). The acidity of both solutions is limited by the strength of the acid (H_3O^+) formed when water molecules accept an H^+ ion.

A similar phenomenon occurs in solutions of strong bases. Once the base reacts with water to form the OH^- ion, the solution cannot become any more basic. The strength of a strong base is limited by the strength of the base (OH^-) formed when water molecules lose an H^+ ion.

11.11 Relationship of Structure to Relative Strengths of Acids and Bases

Several factors influence the probability of having a heart attack. Heart attacks occur more often among those who smoke than among those who don't, among the elderly more often than the young, among those who are overweight more often than among those who aren't, and among those who never exercise more often than among those who exercise regularly. It is possible to sort out the relative importance of these factors, however, by trying to keep the other factors as constant as possible.

The same approach can be used to identify the factors that control the relative strengths of acids and bases. Three factors affect the acidity of the X—H bond in an acid: (1) the polarity of the bond, (2) the size of the atom X, and (3) the charge on the ion or molecule. A fourth factor has to be considered to understand the acidity of the X—OH group in a family of compounds known as "oxyacids."

THE POLARITY OF THE X—H BOND

When all other factors are kept constant, acids become stronger as the X—H bond becomes *more polar*. Consider the following electrically neutral compounds, for example, which become more acidic as the difference between the electronegativities of the X and H atoms (ΔEN) increases and therefore the partial charge on the hydrogen atom in these compounds (δ_H) increases. HF is the strongest of the four acids, and CH_4 is the weakest.

	K_a	ΔEN	δ_H
HF	7.2×10^{-4}	1.9	+0.29
H_2O	1.8×10^{-16}	1.3	+0.22
NH_3	1×10^{-33}	0.8	+0.14
CH_4	1×10^{-49}	0.2	+0.05

In all four of the compounds, the size of the central atom, X, to which the H is bonded is about the same. However, there are large differences in ΔEN among the four compounds. The molecule containing the central atom with the largest electronegativity, F, also has the H with the most positive partial charge, 0.29. As the central atom becomes more electronegative, electron density is pulled away from the hydrogen, resulting in a more positive partial charge on the hydrogen and a more polar H—X bond.

When these compounds act as acids, the H—X bond is broken to form H^+ and X^- ions. From the preceding data, it is apparent that the more polar this bond, the easier it is to form ions, and therefore the larger the value of K_a. Thus, all else being equal, acids become stronger as the H—X bond becomes more polar.

The data in Table 11.5 illustrate the magnitude of this effect. A 0.10 M HF solution is moderately acidic. Water is much less acidic, and the acidity of ammonia is so low that the chemistry of aqueous solutions of NH_3 is dominated by the ability of NH_3 to act as a base.

0.10 M HF	pH = 2.1
H_2O	pH = 7
0.10 M NH_3	pH = 11.1

THE SIZE OF THE X ATOM

We might expect HF, HCl, HBr, and HI to become weaker acids as we go down the column of the periodic table because the X—H bond becomes less polar. Experimentally, we find the opposite trend. The acids actually become stronger as we go down the column.

This can be explained by recognizing that the size of the X atom also influences the acidity of the X—H bond. Acids become stronger as the X—H bond becomes weaker, and bonds generally become weaker as the atoms get larger (Figure 11.9). The K_a data for HF, HCl, HBr, and HI reflect the fact that the magnitude of the enthalpy of atom combination (ΔH_{ac}°) of these compounds decreases as the X atom becomes larger. This indicates a weaker X—H bond, which means it will be easier to break this bond to lose an H^+ ion.

➤ **CHECKPOINT**

Which acid is stronger, H_2O or H_2S? Why?

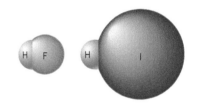

	K_a	ΔH_{ac}° (kJ/mol$_{rxn}$)	δ_H	X—H Bond Length (nm)
HF	7.2×10^{-4}	−567.7	+0.29	0.101
HCl	1×10^6	−431.6	+0.11	0.136
HBr	1×10^9	−365.9	+0.08	0.151
HI	3×10^9	−298.0	+0.01	0.170

Fig. 11.9 HI is a much stronger acid than HF because of the relative sizes of the fluorine and iodine atoms. The covalent radius of iodine is more than twice as large as that of fluorine, which means that the HI bond strength is less than that of HF.

THE CHARGE ON THE ACID OR BASE

The charge on a molecule or an ion can influence its ability to act as an acid or a base. This is clearly shown when the pH of 0.1 M solutions of H_3PO_4 and the $H_2PO_4^-$, HPO_4^{2-}, and PO_4^{3-} ions are compared.

H_3PO_4	pH = 1.5
$H_2PO_4^-$	pH = 4.4
HPO_4^{2-}	pH = 9.3
PO_4^{3-}	pH = 12.0

Compounds become less acidic and more basic as the negative charge increases.

Acidity: $\quad H_3PO_4 > H_2PO_4^- > HPO_4^{2-}$
Basicity: $\quad H_2PO_4^- < HPO_4^{2-} < PO_4^{3-}$

The decrease in acidity with increasing negative charge occurs because it is much easier to remove a positive H^+ ion from a neutral H_3PO_4 molecule than it is to remove an H^+ ion from a negatively charged $H_2PO_4^-$ ion. It is even harder to remove an H^+ ion from a negatively charged HPO_4^{2-} ion.

The increase in basicity with increasing negative charge can be similarly explained. There is a strong force of attraction between the negative charge on a PO_4^{3-} ion and the positive charge on an H^+ ion. As a result, PO_4^{3-} is a reasonably good base. The force of attraction between the HPO_4^{2-} ion and an H^+ ion is smaller because HPO_4^{2-} carries a smaller charge. The HPO_4^{2-} ion is therefore a weaker base than PO_4^{3-}. The charge on the $H_2PO_4^-$ ion is even smaller than that of HPO_4^{2-}, so this ion is an even weaker base than HPO_4^{2-}.

RELATIVE STRENGTHS OF OXYACIDS

There is no difference in either the size of the atom bonded to hydrogen or the charge on the acid when we compare the O—H bonds in oxyacids of the same element, such as H_2SO_4 and H_2SO_3 or HNO_3 and HNO_2 (Figure 11.10), yet there is a significant difference in the strengths of the acids. Consider the following K_a data, for example.

H_2SO_4	$K_a = 1 \times 10^3$	HNO_3	$K_a = 28$
H_2SO_3	$K_a = 1.7 \times 10^{-2}$	HNO_2	$K_a = 5.1 \times 10^{-4}$

The acidity of the oxyacids increases significantly as the number of oxygen atoms attached to the central atom increases. H_2SO_4 is a much stronger acid than H_2SO_3, and HNO_3 is a much stronger acid than HNO_2.

This trend is easiest to see in the four oxyacids of chlorine. At first glance, we might expect these acids to have more or less the same acidity. The same O—H bond is broken in each case, and the O atom in this bond is bound to a Cl atom in all four compounds. Oxygen, however, is second only to fluorine in electronegativity, and it tends to draw electron density toward itself. As more oxygen atoms are added to the central atom, more electron density is drawn toward the oxygens. This draws electrons away from the O—H bond in the acids, as shown in Figure 11.11. Pulling electron density away from the hydrogen results in the partial charge on the hydrogen, δ_H, becoming more positive, as can be seen in the accompanying table.

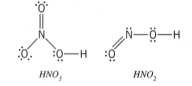

Fig. 11.10 Lewis structures of HNO_3 and HNO_2.

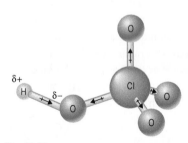

Fig. 11.11 The large difference between the K_a values for HOCl and $HOClO_3$ is the result of the addition of oxygen atoms to the chlorine. The electronegative oxygen atoms draw electron density toward themselves, resulting in electron density being pulled away from the hydrogen and toward the oxygen in the O—H bond. The net result is an increase in the polarity of the O—H bond, which leads to an increase in the acidity of the compound.

	K_a	Number of Oxygen Atoms	δ_H	Name of Oxyacid
HOCl	2.9×10^{-8}	1	+0.24	Hypochlorous acid
HOClO	1.1×10^{-2}	2	+0.29	Chlorous acid
$HOClO_2$	5.0×10^2	3	+0.31	Chloric acid
$HOClO_3$	1×10^8	4	+0.33	Perchloric acid

Exercise 11.6

For each of the following pairs, predict which compound is the stronger acid and explain why.

(a) H_2O or NH_3

(b) NH_4^+ or NH_3

(c) NH_3 or PH_3

(d) H_2SeO_4 or H_2SeO_3

Solution

(a) Oxygen and nitrogen atoms are about the same size, and H_2O and NH_3 are both neutral molecules. The difference between these compounds lies in the polarity of the X—H bond. Because oxygen is more electronegative, the O—H bond is more polar, and H_2O ($K_a = 1.8 \times 10^{-16}$) is a much stronger acid than NH_3 ($K_a = 1 \times 10^{-33}$).

(b) The difference between these compounds is the charge on the ion. The NH_4^+ ion is a stronger acid than NH_3 because it is easier to remove an H^+ ion from a positively charged NH_4^+ ion than from a neutral NH_3 molecule.

(c) PH_3 is a stronger acid than NH_3 because the phosphorous atom is larger than the nitrogen atom. As a result, the P—H bond is longer than the N—H bond. Longer bonds are often weaker bonds and therefore more easily broken. As a result, PH_3 is more acidic than NH_3.

(d) These compounds are both oxyacids that differ in the number of oxygen atoms attached to the selenium atom. More electron density is drawn away from the O—H bond in H_2SeO_4 because of the extra oxygen atom on the Se atom in the center of the molecule. Thus H_2SeO_4 is the stronger acid.

➤ **CHECKPOINT**

Which of the following acids is stronger, HOCl or HOI? Why?

The relative strengths of Brønsted bases can be predicted from the relative strengths of their conjugate acids, combined with the general rule that the stronger of a pair of acids always has the weaker conjugate base.

Exercise 11.7

For each of the following pairs of compounds, predict which compound is the weaker base.

(a) OH^- or NH_2^-

(b) NH_3 or NH_2^-

(c) NH_2^- or PH_2^-

(d) NO_3^- or NO_2^-

Solution

(a) The OH^- ion is the conjugate base of water, and the NH_2^- ion is the conjugate base of ammonia. Because H_2O is a stronger acid than NH_3, the OH^- ion must be a weaker base than the NH_2^- ion.

(b) Because it carries a negative charge, the NH_2^- ion is a stronger base than a neutral NH_3 molecule.

(c) PH_3 is a stronger acid than NH_3, which means the PH_2^- ion must be a weaker base than the NH_2^- ion.

(d) HNO_3 is a stronger acid than HNO_2, which means that the NO_3^- ion is the weaker base.

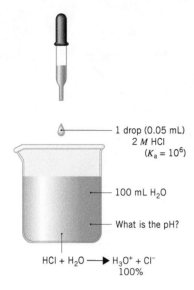

1 drop (0.05 mL)
2 M HCl
($K_a = 10^6$)

100 mL H_2O

What is the pH?

HCl + H_2O $\longrightarrow$ H_3O^+ + Cl^-
100%

11.12 Strong Acid pH Calculations

The simplest acid–base equilibria are those in which a strong acid (or base) is dissolved in water. Consider the calculation of the pH of a solution formed by adding a single drop of 2 M hydrochloric acid to 100 mL of water, for example. The key to this calculation is remembering that HCl is a strong acid ($K_a = 10^6$) and that acids as strong as HCl can be assumed to dissociate more or less completely in water.

$$HCl(aq) + H_2O(l) \longrightarrow H_3O^+(aq) + Cl^-(aq)$$

The H_3O^+ ion concentration at equilibrium can therefore be assumed to be essentially the same as the initial concentration of the acid. The details of this calculation are illustrated in the drawing in the margin.

A useful rule of thumb states that there are about 20 drops in each milliliter. One drop of 2 M HCl therefore has a volume of 0.05 mL, or 5×10^{-5} L. Now that we know the volume of the acid added to the beaker of water and the concentration of the acid, we can calculate the number of moles of HCl that were added to the water.

$$\frac{2 \text{ mol HCl}}{1 \text{ L}} \times 5 \times 10^{-5} \text{ L} = 1 \times 10^{-4} \text{ mol HCl}$$

Once we know the number of moles of HCl that have been added to the beaker and the volume of water in the beaker, we can calculate the concentration of the solution prepared by adding one drop of 2 M HCl to 100 mL of water.

$$\frac{1 \times 10^{-4} \text{ mol HCl}}{0.100 \text{ L}} = 1 \times 10^{-3} M \text{ HCl}$$

According to this calculation, the initial concentration of HCl before it dissociates into ions is 1×10^{-3} M.

If we assume that the acid dissociates completely, the H_3O^+ concentration at equilibrium is 1×10^{-3} M. We therefore expect the pH of the solution prepared by adding one drop of 2 M HCl to 100 mL of water to be 3.0.

$$pH = -\log[H_3O^+]$$
$$pH = -\log(1 \times 10^{-3}) = -(-3.0) = 3.0$$

Calculations such as this give results that are close to experimental values for solutions that are reasonably dilute.

➤ **CHECKPOINT**

When a strong acid is added to water, why can the pH usually be determined directly from the quantity of strong acid added?

11.13 Weak Acid pH Calculations

Unlike strong acids, which dissociate almost completely, weak acids dissociate only slightly in water. An aqueous solution of a weak acid therefore not only contains the dissociated ($H_3O^+ + A^-$) form of the acid but also significant amounts of the undissociated (HA) form of the acid. Equilibrium problems involving weak acids such as acetic acid (CH_3CO_2H) can be solved by applying the techniques developed in Chapter 10.

As noted in Section 11.8, the concentration of the H_3O^+ ion in an aqueous solution of an acid depends on two factors:

- The strength of the acid, as represented by the value of K_a for the acid.
- The concentration of the acid, in moles per liter.

These two factors were illustrated in Figure 11.7. The beaker on the left shows a strong acid in which all of the acid molecules have reacted with water to produce H_3O^+ ions. However, the concentration of this solution is much smaller than the concentration of the weak acid shown in the beaker on the right. The beaker on the right shows a solution that is much more concentrated because it contains more moles of the acid per liter of solution. However, because this solution is a weak acid, only a small fraction of the acid molecules have combined with water to form H_3O^+ ions. The importance of considering both the strength of an acid and its concentration can be understood by comparing an 0.0010 M HCl solution with a solution that is 1.0 M in acetic acid. HCl is a much stronger acid ($K_a = 10^6$), but the concentration of the acetic acid solution is so much larger that there are actually more H_3O^+ ions at equilibrium in the acetic acid solution.

Consider, for example, the dissociation of acetic acid, CH_3CO_2H, in water. The formula of acetic acid is often abbreviated as HOAc, and that of its conjugate base, the acetate ion ($CH_3CO_2^-$), as OAc^-. We can start the calculation of the H_3O^+, OAc^-, and HOAc concentrations at equilibrium in a 0.10 M solution of acetic acid in water by building a representation of what we know about the reaction.

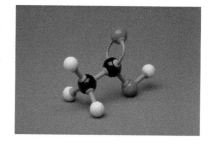

A ball-and-stick model of the structure of acetic acid, CH_3CO_2H.

$$HOAc(aq) + H_2O(l) \rightleftharpoons H_3O^+(aq) + OAc^-(aq) \quad K_a = 1.8 \times 10^{-5}$$

Initial: 0.10 M ≈ 0 0

Equilibrium: ?   ?   ?

The relevant information for this calculation is summarized in the drawing shown in the margin.

We now compare the initial reaction quotient (Q_a) with the equilibrium constant (K_a) for the reaction. In this case, the initial reaction quotient is smaller than the equilibrium constant.

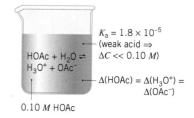

$K_a = 1.8 \times 10^{-5}$
(weak acid $\Rightarrow$
$\Delta C \ll 0.10$ M)

Δ(HOAc) = $\Delta(H_3O^+)$ = $\Delta(OAc^-)$

0.10 M HOAc

$$Q_a = \frac{(H_3O^+)(OAc^-)}{(HOAc)} = \frac{(0)(0)}{0.10} = 0 < K_a = 1.8 \times 10^{-5}$$

We therefore conclude that the reaction must shift to the right to reach equilibrium.

The balanced equation for the reaction states that we get one H_3O^+ ion and one OAc^- ion each time an HOAc molecule dissociates. This allows us to write equations for the equilibrium concentrations of the three components of the reaction. Let's start by writing an expression that describes the concentration of acetic acid when the reaction comes to equilibrium.

The concentration of acetic acid at equilibrium will be equal to the initial concentration of the acid, 0.10 M, minus the change in the concentration of this acid, ΔC.

$$[HOAc] = 0.10\,M - \Delta C$$

Because there is very little H_3O^+ ion in pure water, we can assume that the initial concentration of this acid is effectively zero (≈ 0). Because we get one H_3O^+ ion and one OAc^- ion every time an acetic acid molecule dissociates, the concentrations of these ions at equilibrium are equal to the change in the concentration of acetic acid as this acid dissociates.

$$[H_3O^+] = \Delta C$$

$$[OAc^-] = \Delta C$$

We can therefore represent the problem that needs to be solved as follows.

$$HOAc(aq) + H_2O(l) \rightleftharpoons H_3O^+(aq) + OAc^-(aq) \quad K_a = 1.8 \times 10^{-5}$$

Initial: 0.10 M 0   0
Equilibrium: 0.10 $-\Delta C$ ΔC ΔC

Substituting what we know about the system at equilibrium into the K_a expression gives the following equation.

$$K_a = \frac{[H_3O^+][OAc^-]}{[HOAc]} = \frac{[\Delta C][\Delta C]}{[0.10 - \Delta C]} = 1.8 \times 10^{-5}$$

Although we could rearrange the equation and solve it with the quadratic formula, let's test the assumption that acetic acid is a relatively weak acid. In other words, let's assume that ΔC is very small compared with the initial concentration of acetic acid. If this assumption is true, then subtracting ΔC from the initial concentration of the acid (0.10 M) will not significantly reduce that value. This assumption gives us the following approximate equation:

$$\frac{[\Delta C][\Delta C]}{[0.10]} \approx 1.8 \times 10^{-5}$$

Or

$$\frac{[\Delta C]^2}{[0.10]} \approx 1.8 \times 10^{-5}$$

And therefore

$$[\Delta C]^2 \approx 1.8 \times 10^{-6}$$

We then solve the approximate equation for the value of ΔC.

$$\Delta C \approx 0.0013 \, M$$

We can now test our assumption. Is ΔC small enough to be ignored in this problem? Yes, because it is less than 5% of the initial concentration of acetic acid.

$$\frac{0.0013}{0.10} \times 100\% = 1.3\%$$

Our assumption is valid because subtracting 0.0013 from 0.10 would give us 0.10, to two significant figures. We can therefore use the value of ΔC obtained from this calculation to estimate the equilibrium concentrations of H_3O^+, OAc^-, and HOAc.

$$[HOAc] = 0.10 \, M - \Delta C \cong 0.10 \, M$$

$$[H_3O^+] = [OAc^-] = \Delta C \cong 0.0013 \, M$$

The pH of a 0.10 M HOAc is therefore 2.9.

$$pH = -\log[H_3O^+] = 2.9$$

Two assumptions were made in the above calculation.

- We assumed that the amount of acid that dissociates is small compared with the initial concentration of the acid.
- We assumed that enough acid dissociates to allow us to ignore the contribution to the total H_3O^+ ion concentration from the dissociation of water.

In this case, both assumptions are valid. Only about 1.3% of the acetic acid molecules actually dissociate at equilibrium. But this gives us an H_3O^+ concentration that is four orders of magnitude larger than the concentration of this ion in pure water. Thus we can legitimately ignore the dissociation of water.

The assumptions made in this calculation work for acids that are neither too strong nor too weak. Fortunately, many acids fall into that category. If the acid is too strong to assume that ΔC is small compared with the initial concentration, the problem can be solved with the quadratic equation. If the acid is too weak or too dilute to ignore the dissociation of water, variations in the pH of the water from one sample to another because of impurities dissolved in the water are probably as significant as the effect on the pH of adding the very weak acid.

To illustrate the importance of paying attention to the assumptions that are made in pH calculations, consider the task of calculating the pH of a solution prepared by dissolving 1.0×10^{-8} mol of a strong acid, such as hydrochloric acid, in a liter of water. It is tempting to assume that the acid dissociates completely in water to give a solution with the following H_3O^+ ion concentration.

$$[H_3O^+] = 1.0 \times 10^{-8}M$$

If this concentration is used with the definition of pH, we get the following result.

$$pH = -\log[H_3O^+] = -\log(1.0 \times 10^{-8}M) = 8.0$$

The result of this calculation, however, does not make sense. It is impossible to create a basic solution (pH > 7) by adding an acid to water!

The mistake, of course, was ignoring the concentration of the H_3O^+ ion from the dissociation of water. In pure water, we start with 1.0×10^{-7} M H_3O^+ ion. If we add 10^{-8} mol/L from the strong acid, the total concentration of H_3O^+ from both sources will be about $1.1 \times 10^{-7}M$. Thus the pH of the solution is actually slightly less than 7.

$$pH = -\log[H_3O^+] = -\log(1.1 \times 10^{-7}M) = 6.96$$

Two factors control the concentration of the H_3O^+ ion in a solution of a weak acid: (1) the acid dissociation equilibrium constant, K_a, of the acid and (2) the concentration of the solution. When solutions of equivalent concentrations are compared, the amount of the H_3O^+ ion at equilibrium increases as the value of K_a of the acid increases, as shown in Exercise 11.8.

> **CHECKPOINT**

Assume that you have two solutions. Solution 1 has a low concentration of an acid with a large K_a. Solution 2 has a high concentration of an acid with a small K_a. Do you have enough information to determine which solution has the largest pH? Explain your answer.

Exercise 11.8

Determine the approximate pH of 0.10 *M* solutions of the following acids.

(a) Hypochlorous acid, HOCl, $K_a = 2.9 \times 10^{-8}$

(b) Hypobromous acid, HOBr, $K_a = 2.4 \times 10^{-9}$

(c) Hypoiodous acid, HOI, $K_a = 2.3 \times 10^{-11}$

Solution

The first calculation can be set up as follows.

$$HOCl(aq) + H_2O(l) \rightleftharpoons H_3O^+(aq) + OCl^-(aq)$$

Initial:	0.10 *M*	≈0	0
Equilibrium:	0.10 − ΔC	ΔC	ΔC

Substituting this information into the K_a expression gives the following equation.

$$K_a = \frac{[H_3O^+][OCl^-]}{[HOCl]} = \frac{[\Delta C][\Delta C]}{[0.10 - \Delta C]} = 2.9 \times 10^{-8}$$

Because the equilibrium constant for the reaction is relatively small, we can try the assumption that ΔC is small compared with the initial concentration of the acid.

$$\frac{[\Delta C][\Delta C]}{[0.10]} \approx 2.9 \times 10^{-8}$$

Solving the equation for ΔC gives the following result.

$$\Delta C = 5.4 \times 10^{-5} M$$

Both assumptions made in the calculation are legitimate. ΔC is small compared with the initial concentration of HOCl, but it is several orders of magnitude larger than the H_3O^+ ion concentration from the dissociation of water. We can therefore use this value of ΔC to calculate the pH of the solution.

$$pH = -\log[H_3O^+] = -\log(5.4 \times 10^{-5} M) = 4.3$$

Repeating the calculation with the same assumptions for hypobromous and hypoiodous acid gives the following results.

HOCl	$[H_3O^+] \approx 5.4 \times 10^{-5} M$	pH ≈ 4.3
HOBr	$[H_3O^+] \approx 1.5 \times 10^{-5} M$	pH ≈ 4.8
HOI	$[H_3O^+] \approx 1.5 \times 10^{-6} M$	pH ≈ 5.8

As expected, the H_3O^+ ion concentration at equilibrium—and therefore the pH of the solution—depends on the value of K_a for the acid. The H_3O^+ ion concentration increases, and the pH of the solution decreases as the value of K_a becomes larger.

The next exercise shows how the H_3O^+ ion concentration at equilibrium depends on the initial concentration of the acid.

Exercise 11.9

Determine the H_3O^+ ion concentration and pH of acetic acid solutions with the following concentrations: 1.0 M, 0.10 M, and 0.010 M.

The vinegar that is often mixed with vegetable oils as the basis of salad dressings is an acetic acid solution that has a concentration of approximately 1 M.

Solution

The first calculation can be set up as follows.

$$HOAc(aq) + H_2O(l) \rightleftharpoons H_3O^+(aq) + OAc^- aq \quad K_a = 1.8 \times 10^{-5}$$

Initial: 1.0 M ≈ 0 0
Equilibrium: $1.0 - \Delta C$ ΔC ΔC

Substituting this information into the K_a expression gives the following result.

$$K_a = \frac{[H_3O^+][OAc^-]}{[HOAc]} = \frac{[\Delta C][\Delta C]}{[1.0 - \Delta C]} = 1.8 \times 10^{-5}$$

We now assume that ΔC is small compared with the initial concentration of acetic acid,

$$\frac{[\Delta C][\Delta C]}{[1.0]} \approx 1.8 \times 10^{-5}$$

and then solve the equation for an approximate value of ΔC.

$$\Delta C \approx 0.0042 \ M$$

Once again, both of the characteristic assumptions of weak acid equilibrium calculations are legitimate in this case. ΔC is small compared with the initial concentration of acid but large compared with the concentration of H_3O^+ ion from the dissociation of water. We can therefore use this value of ΔC to calculate the pH of the solution.

$$pH = -\log[H_3O^+] = -\log(4.2 \times 10^{-3} M) = 2.4$$

Repeating the calculation for the different initial concentrations gives the following results.

1.0 M HOAc	$[H_3O^+] \approx 4.2 \times 10^{-3} M$	pH ≈ 2.4
0.10 M HOAc	$[H_3O^+] \approx 1.3 \times 10^{-3} M$	pH ≈ 2.9
0.010 M HOAc	$[H_3O^+] \approx 4.2 \times 10^{-4} M$	pH ≈ 3.4

The concentration of the H_3O^+ ion in an aqueous solution of a weak acid gradually increases and the pH of the solution decreases as the solution becomes more concentrated.

► **CHECKPOINT**

Which of the acetic acid solutions in Exercise 11.9 is the most acidic?

Common window cleaners are often dilute solutions of ammonia in water.

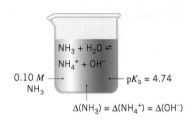

11.14 Base pH Calculations

With minor modifications, the techniques applied to equilibrium calculations for acids are valid for solutions of bases in water. Consider, for example, the calculation of the pH of a 0.10 M NH_3 solution. The details of this calculation are illustrated in the drawing in the margin. We can start by writing an equation for the reaction between ammonia and water.

$$NH_3(aq) + H_2O(l) \rightleftharpoons NH_4^+(aq) + OH^-(aq)$$

The ionization of NH_3 in water can be described by the following **base ionization equilibrium constant expression.**

$$K_b = \frac{[NH_4^+][OH^-]}{[NH_3]}$$

The values of K_b for a limited number of bases can be found in Table B.9 in Appendix B. For NH_3, K_b is 1.8×10^{-5}.

We can organize what we know about the equilibrium with the format we used for equilibria involving acids.

$$NH_3(aq) + H_2O(l) \rightleftharpoons NH_4^+(aq) + OH^-(aq) \qquad K_b = 1.8 \times 10^{-5}$$

Initial: 0.10 M	0	≈ 0
Equilibrium: 0.10 $- \Delta C$	ΔC	ΔC

Substituting what we know about the concentrations of the components of this equilibrium into the K_b expression gives the following equation.

$$K_b = \frac{[NH_4^+][OH^-]}{[NH_3]} = \frac{[\Delta C][\Delta C]}{[0.10 - \Delta C]} = 1.8 \times 10^{-5}$$

K_b for ammonia is small enough to consider the assumption that ΔC is small compared with the initial concentration of the base.

$$\frac{[\Delta C][\Delta C]}{[0.10]} \approx 1.8 \times 10^{-5}$$

Solving the approximate equation gives the following result.

$$\Delta C \approx 1.3 \times 10^{-3} M$$

The value of ΔC is small enough compared with the initial concentration of NH_3 to be ignored and yet large enough compared with the OH^- ion concentration in water to ignore the dissociation of water. ΔC can therefore be used to calculate the pOH of the solution.

$$pOH = -\log(1.3 \times 10^{-3}) = 2.9$$

This, in turn, can be used to calculate the pH of the solution.

$$pH = 14.0 - pOH = 11.1$$

➤ **CHECKPOINT**

Which is more basic, a 1 M solution of a base that has a small K_b or a 1 M solution of a base that has a large K_b?

THE RELATIONSHIP BETWEEN K_a AND K_b

Equilibrium problems involving bases can be solved if the value of K_b for the base is known. Values of K_b are listed in Table B.9 in Appendix B, however, for only a limited number of compounds. The first step in many base equilibrium calculations therefore involves determining the value of K_b for the reaction from the value of K_a for the conjugate acid.

The relationship between K_a and K_b for a conjugate acid–base pair can be understood by using a specific example of this relationship. Consider a solution containing the acetate ion (OAc⁻), for example. We already know the value of K_a for the conjugate acid, acetic acid (HOAc)

$$HOAc(aq) + H_2O(l) \rightleftharpoons H_3O^+(aq) + OAc^-(aq) \qquad K_a = 1.8 \times 10^{-5}$$

What is the value of K_b for the conjugate base, OAc⁻?

$$OAc^-(aq) + H_2O(l) \rightleftharpoons HOAc(aq) + OH^-(aq) \qquad K_b = ?$$

The K_a and K_b expressions for acetic acid and its conjugate base both contain the ratio of the equilibrium concentrations of the acid and its conjugate base. K_a is proportional to [OAc⁻] divided by [HOAc], whereas K_b is proportional to [HOAc] divided by [OAc⁻].

$$K_a = \frac{[H_3O^+][OAc^-]}{[HOAc]} \qquad K_b = \frac{[HOAc][OH^-]}{[OAc^-]}$$

Two changes need to be made to derive the K_b expression from the K_a expression: We need to remove the [H_3O^+] term and introduce an [OH⁻] term. We can do this by multiplying the top and bottom of the K_a expression by the OH⁻ ion concentration.

$$K_a = \frac{[H_3O^+][OAc^-]}{[HOAc]} \times \frac{[OH^-]}{[OH^-]}$$

Rearranging this equation gives the following result.

$$K_a = \frac{[OAc^-]}{[HOAc][OH^-]} \times [H_3O^+][OH^-]$$

The two terms on the right side of the equation should look familiar. The first is the inverse of the K_b expression, and the second is the expression for K_w.

$$K_a = \frac{1}{K_b} \times K_w$$

Rearranging this equation gives the following result.

$$K_a K_b = K_w = 1.0 \times 10^{-14}$$

According to this equation, the value of K_b for the reaction between the acetate ion and water can be calculated from the K_a for acetic acid.

$$K_b = \frac{K_w}{K_a} = \frac{1.0 \times 10^{-14}}{1.8 \times 10^{-5}} = 5.6 \times 10^{-10}$$

The relationship between K_a and K_b for a conjugate acid–base pair can be used to rationalize why strong acids tend to have weak conjugate bases and strong bases tend to have weak conjugate acids. The product of K_a for any Brønsted acid times K_b for its conjugate base is equal to K_w for water.

$$K_a K_b = K_w = 1.0 \times 10^{-14}$$

But K_w is a relatively small number (1.0×10^{-14}). As a result, the value of K_a and K_b for a conjugate acid–base pair can't both be large. When K_a is relatively large—indicating a strong acid—K_b for the conjugate base must be relatively small—indicating a weak conjugate base. Conversely, if K_b is large—indicating a strong base—K_a for the conjugate acid must be small—indicating a weak acid. As noted in Section 11.9, the relative strengths of a conjugate acid–base pair can be summarized as follows.

The stronger the acid, the weaker its conjugate base.

The stronger the base, the weaker its conjugate acid.

K_a and K_b for a conjugate acid–base pair can't both be large, but they can both be relatively small. Consider ammonia (NH_3) and its conjugate acid, the ammonium ion (NH_4^+), for example. NH_3 is a moderately weak base ($K_b = 1.8 \times 10^{-5}$), and the NH_4^+ ion is a relatively weak acid ($K_a = 5.6 \times 10^{-10}$).

ACID–BASE PROPERTIES OF A SALT

The reaction of an acid with a base produces an ionic compound referred to as a *salt*. A salt is an ionic compound with a cation other than H^+ and an anion other than OH^- or O^{2-}. Dilute solutions of soluble salts are assumed to ionize completely in water. The acid–base properties of these compounds are therefore determined by the acid–base properties of the aqueous ions formed when the salt dissolves in water. As a general rule, Group I and II cations (Li^+, Na^+, K^+, Rb^+, Cs^+, Ca^{2+}, Sr^{2+}, Ba^{2+}) have no significant acid or base properties. The conjugate bases of strong monoprotic acids (Cl^-, Br^-, I^-, ClO_4^-, NO_3^-) are so extremely weak that they have essentially no basic properties in water.

Other salts, however, can exhibit the properties of an acid or a base. Consider potassium fluoride, for example.

$$KF(aq) \xrightarrow{H_2O} K^+(aq) + F^-(aq)$$

The potassium ion is neither an acid nor a base. However, the F^- ion is the conjugate base of a weak acid, HF. The fluoride ion will therefore behave as a base.

$$F^-(aq) + H_2O(l) \rightleftharpoons HF(aq) + OH^-(aq)$$

A solution of KF in water is therefore mildly basic ($0.10\ M$ NaF, pH = 8.0), and potassium fluoride is referred to as a basic salt.

Ammonium nitrate, NH_4NO_3, is an example of an acidic salt. A dilute solution will ionize completely when it dissolves in water.

$$NH_4NO_3(aq) \longrightarrow NH_4^+(aq) + NO_3^-(aq)$$

The nitrate anion, NO_3^-, is the conjugate base of the strong acid HNO_3 and therefore has essentially no base properties. The ammonium cation, NH_4^+, however, is the conjugate acid of the weak base ammonia (NH_3) and should therefore behave as an acid in water. Thus an aqueous solution of NH_4NO_3 is slightly acidic (0.10 M NH_4NO_3, pH = 5.4).

$$NH_4^+(aq) + H_2O(l) \rightleftharpoons H_3O^+(aq) + NH_3(aq)$$

Sodium chloride, NaCl, is an example of a salt that produces essentially a neutral solution (0.10 M NaCl, pH = 6.4 $\approx$ 7). In dilute aqueous solutions, sodium chloride ionizes completely.

$$NaCl(aq) \longrightarrow Na^+(aq) + Cl^-(aq)$$

The chloride ion is the conjugate base of a strong acid and is therefore much too weak a base to be noticed. The sodium ion is neither an acid nor a base in the Brønsted model of acids and bases. Aqueous solutions of NaCl are therefore essentially neutral.

Exercise 11.10

Calculate the HOAc, OAc^-, and OH^- concentrations and pH at equilibrium in a 0.10 M NaOAc solution. (For HOAc, $K_a = 1.8 \times 10^{-5}$.)

Solution

We start with a solution of an ionic salt—NaOAc—dissolved in water and assume that the salt ionizes completely.

$$NaOAc(aq) \longrightarrow Na^+(aq) + OAc^-(aq)$$

Because the Na^+ ion is neither an acid nor a base, only the acetate ion will significantly affect the pH. We therefore start, as always, by building a representation of the information in the statement of the problem. The relevant information for this calculation is illustrated in the drawing in the margin.

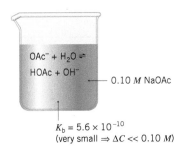

$OAc^- + H_2O \rightleftharpoons$
$HOAc + OH^-$ — 0.10 M NaOAc

$K_b = 5.6 \times 10^{-10}$
(very small $\Rightarrow \Delta C \ll 0.10$ M)

$$OAc^-(aq) + H_2O(l) \rightleftharpoons HOAc(aq) + OH^-(aq) \qquad K_b = ?$$

	OAc^-	HOAc	OH^-
Initial:	0.10 M	0	≈ 0
Equilibrium:	$0.10 - \Delta C$	ΔC	ΔC

The only other information we need to solve the problem is the value of K_b for the reaction between the acetate ion and water, which can be calculated from the value of K_a for acetic acid.

$$K_b = \frac{K_w}{K_a} = \frac{1.0 \times 10^{-14}}{1.8 \times 10^{-5}} = 5.6 \times 10^{-10}$$

Substituting what we know into the equilibrium constant expression gives the following result.

$$K_b = \frac{[HOAc][OH^-]}{[OAc^-]} = \frac{[\Delta C][\Delta C]}{[0.10 - \Delta C]} = 5.6 \times 10^{-10}$$

Because the value of K_b for the OAc$^-$ ion is relatively small, we can feel reasonably confident that ΔC is small compared with the initial concentration of base.

$$\frac{[\Delta C][\Delta C]}{[0.10]} = 5.6 \times 10^{-10}$$

We can then solve the approximate equation the assumption generates.

$$\Delta C \approx 7.5 \times 10^{-6} M$$

ΔC is very much smaller than the initial concentration of the base, which means that one of the key assumptions of acid–base equilibrium calculations is legitimate. But [OH$^-$] is still at least 75 times larger than the concentration of the OH$^-$ ion in pure water, which means that we can also legitimately ignore the dissociation of water in the calculation. We can therefore use the value of ΔC to calculate the equilibrium concentrations of HOAc, OAc$^-$, and OH$^-$.

$$[OAc^-] = 0.10M - \Delta C \approx 0.10M$$
$$[HOAc] = [OH^-] = \Delta C \cong 7.5 \times 10^{-6} M$$

The pH of this solution is therefore 8.9.

$$[H_3O^+] = \frac{K_w}{[OH^-]} = \frac{1.0 \times 10^{-14}}{[7.5 \times 10^{-6}]} = 1.3 \times 10^{-9}$$
$$pH = -\log[H_3O^+] = 8.9$$

► **CHECKPOINT**

If K_a for HNO$_2$ is 5.1×10^{-4}, what is K_b for the NO$_2^-$ ion?

11.15 Mixtures of Acids and Bases: Buffers

Mixtures of a weak acid and its conjugate base are called **buffers.** So far, we have examined 0.10 M solutions of both acetic acid and sodium acetate.

0.10 M HOAc	[H$_3$O$^+$] $\approx 1.3 \times 10^{-3}M$	pH ≈ 2.9
0.10 M NaOAc	[H$_3$O$^+$] $\approx 7.5 \times 10^{-6}M$	pH ≈ 8.9

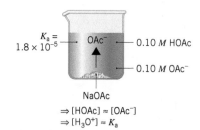

$K_a = 1.8 \times 10^{-5}$ OAc$^-$ 0.10 M HOAc

0.10 M OAc$^-$

NaOAc

$\Rightarrow$ [HOAc] $\approx$ [OAc$^-$]
$\Rightarrow$ [H$_3$O$^+$] $\approx K_a$

What would happen if we prepared a buffer from enough sodium acetate and acetic acid solution so that the solution is simultaneously 0.10 M in both HOAc and NaOAc? The relevant information for this calculation is illustrated in the drawing in the margin.

The first step toward answering this question is recognizing that there are two sources of the OAc$^-$ ion in the solution. Acetic acid, of course, dissociates to give the H$_3$O$^+$ and OAc$^-$ ions.

$$HOAc(aq) + H_2O(l) \rightleftharpoons H_3O^+(aq) + OAc^-(aq)$$

The salt, sodium acetate, however, also dissociates in water to give the OAc$^-$ ion.

$$NaOAc(s) \xrightarrow{H_2O} Na^+(aq) + OAc^-(aq)$$

These reactions share a common ion: the OAc⁻ ion. Le Châtelier's principle predicts that adding sodium acetate to acetic acid should shift the equilibrium between HOAc and the H_3O^+ and OAc⁻ ions to the left, another example of the common-ion effect. Thus adding NaOAc reduces the extent to which HOAc dissociates. The mixture is therefore less acidic than 0.10 M HOAc by itself.

In Exercise 11.10 we saw that the OAc⁻ ion behaves as a base. The K_b for OAc⁻ is several orders of magnitude smaller than K_a for HOAc. Therefore, the reaction of OAc⁻ with water is small compared to the reaction of HOAc with water.

We can quantify the discussion by setting up the problem as follows.

$$\text{HOAc}(aq) + \text{H}_2\text{O}(l) \rightleftharpoons \text{H}_3\text{O}^+(aq) + \text{OAc}^-(aq) \qquad K_a = 1.8 \times 10^{-5}$$

Initial: 0.10M   ≈0   0.10M
Equilibrium: ?   ?   ?

We don't have any basis for predicting whether the dissociation of water can be ignored in this calculation, so let's make that assumption and check its validity later.

If most of the H_3O^+ ion concentration at equilibrium comes from dissociation of the acetic acid, the reaction has to shift to the right to reach equilibrium.

$$\text{HOAc}(aq) + \text{H}_2\text{O}(l) \rightleftharpoons \text{H}_3\text{O}^+(aq) + \text{OAc}^-(aq) \qquad K_a = 1.8 \times 10^{-5}$$

Initial: 0.10M   ≈0   0.10M
Equilibrium: 0.10 − ΔC   ΔC   0.10 + ΔC

Substituting this information into the K_a expression gives the following result.

$$K_a = \frac{[\text{H}_3\text{O}^+][\text{OAc}^-]}{[\text{HOAc}]} = \frac{[\Delta C][0.10 + \Delta C]}{[0.10 - \Delta C]} = 1.8 \times 10^{-5}$$

Let's assume that ΔC is small compared with the initial concentrations of HOAc and OAc⁻.

$$\frac{[\Delta C][0.10]}{[0.10]} \approx 1.8 \times 10^{-5}$$

We then solve the approximate equation for ΔC.

$$\Delta C \approx 1.8 \times 10^{-5}$$

Are our two assumptions legitimate? Is ΔC small enough compared with 0.10 to be ignored? Is ΔC large enough so that the dissociation of water can be ignored? The answer to both questions is yes. We can therefore use the value of ΔC to calculate the H_3O^+ ion concentration at equilibrium and the pH of the solution.

$$[\text{H}_3\text{O}^+] \approx \Delta C \approx 1.8 \times 10^{-5}M$$
$$\text{pH} = -\log(1.8 \times 10^{-5}) = 4.7$$

We therefore conclude that the buffer solution is acidic, but less acidic than HOAc by itself. As we would expect from Le Châtelier's principle, adding a source of the acetate ion shifts the equilibrium for the dissociation of acetic acid in water toward the reactants. This results in a decrease in the amount of H_3O^+. In other words, the HOAc solution is less acidic due to the addition of a base, OAc⁻.

11.16 Buffers and Buffer Capacity

The term *buffer* means "to lessen or absorb shock." Mixtures of weak acids and their conjugate bases are buffers because they lessen or absorb the drastic change in pH that occurs when small amounts of acids or bases are added to water.

In Section 11.12 we concluded that adding a single drop of 2 *M* HCl to 100 mL of pure water would change the pH by 4 units, from pH 7 to pH 3. If we add the same amount of HCl to a buffer solution that is 0.10 *M* in both acetic acid (HOAc) and a salt of the acetate ion (OAc$^-$), there is no observable change in the pH of the solution. In fact, we can add 10 times as much 2 *M* HCl to the buffer solution without noticing a significant change in the pH of the solution. Thus the pH of the buffer solution is truly "buffered" against the effect of small amounts of acid or base.

To understand how the buffer works, it is important to recognize that a buffer is made by mixing a weak acid and its conjugate base so that significant amounts of both a weak acid and weak base are present in solution.

$$HOAc(aq) + H_2O(l) \rightleftharpoons H_3O^+(aq) + OAc^-(aq)$$

Weak acid   *Conjugate base*

At equilibrium, the concentrations of acetic acid and the H$_3$O$^+$ and OAc$^-$ ions in the solution are related by the following equation.

$$K_a = \frac{[H_3O^+][OAc^-]}{[HOAc]}$$

If we rearrange the equation, by solving for the H$_3$O$^+$ ion concentration at equilibrium, we get the following result.

$$[H_3O^+] = K_a \times \frac{[HOAc]}{[OAc^-]}$$

This equation suggests that the H$_3$O$^+$ ion concentration in the solution will stay more or less constant as long as there is no significant change in the ratio of the concentrations of the undissociated HOAc molecules and the OAc$^-$ ions. The buffer solution discussed in Section 11.15 was prepared by adding enough sodium acetate (NaOAc) to acetic acid (HOAc) to make the concentration of both the acid (HOAc) and its conjugate base (OAc$^-$) the same. (The solution is simultaneously 0.10 *M* HOAc and 0.10 *M* OAc$^-$.) The pH of this solution as determined by a pH meter is 4.74.

If a small amount of a strong acid, H$_3$O$^+$, is added to the solution, it will react with the OAc$^-$ ion to form a bit more acetic acid. This is illustrated by the drawing in the margin.

$$OAc^-(aq) + H_3O^+ \longrightarrow HOAc(aq) + H_2O(l)$$

The equilibrium constant for this reaction is the inverse of the value of K_a for acetic acid because the reaction is written in the opposite direction from the equation that describes the dissociation of this acid.

$$K' = \frac{1}{K_a} = \frac{1}{1.8 \times 10^{-5}} = 5.6 \times 10^4$$

The equilibrium constant is large enough that the reaction can be assumed to proceed essentially to completion, forming acetic acid.

The H_3O^+ concentration after the acid is added to the buffer solution can be calculated as follows.

$$[H_3O^+] = 1.8 \times 10^{-5} \frac{[HOAc]}{[OAc^-]} = 1.8 \times 10^{-5} \frac{[0.100 + \Delta C]}{[0.100 - \Delta C]}$$

Because the extent of the reaction is so large, adding one drop of 2 M HCl to 100 mL of this HOAc/OAc⁻ buffer gives a change in the concentration, ΔC, of only 0.001 M and the new H_3O^+ concentration is

$$[H_3O^+] = 1.8 \times 10^{-5} \frac{0.101}{0.099} = 1.8 \times 10^{-5}$$

The pH of the solution is therefore still 4.74

The net result of adding one drop of 2 M HCl to the buffer solution is a slight increase in the concentration of the undissociated HOAc molecules and a slight decrease in the concentration of the OAc⁻ ion. But if only a reasonably small amount of acid is added to the solution, the ratio of the concentrations would be roughly the same. As a result, the H_3O^+ ion concentration in the solution remains essentially the same, which means there is no change in the pH of the solution. The addition of a strong acid, H_3O^+, to the buffer results in the H_3O^+ being converted into an equivalent amount of undissociated weaker acid, HOAc.

If a small amount of a strong base, OH⁻, is added to the solution, it will react with some of the undissociated HOAc molecules to form a bit more of the OAc⁻ ion.

$$HOAc(aq) + OH^-(aq) \longrightarrow OAc^-(aq) + H_2O(l)$$

But, once again, if we add only a reasonably small amount of the base, the ratio of the concentrations of HOAc and OAc⁻ in the solution will be more or less the same. This means that the pH of the solution is more or less the same. The addition of a strong base, OH⁻, to the buffer converts the OH⁻ into an equivalent amount of weaker base, OAC⁻.

Because it contains significant amounts of both HOAc molecules and the OAc⁻ ion, the solution can resist a change in pH when reasonably small amounts of either a strong acid or a strong base are added to the solution. The concentrations of acetic acid and the acetate ion will change slightly when the acid or base is added to the buffer solution, but the ratio of the concentrations of these two components of the equilibrium will not change by a large enough amount to produce a significant change in the pH of the solution.

As long as the concentrations of HOAc and the OAc⁻ ion in the buffer are larger than the amount of acid or base added to the solution, the pH remains constant. Table 11.6 compares the effects of adding different amounts of hydrochloric acid to water and two buffer solutions of different concentrations. The first column gives the number of moles of HCl added per liter. The second column shows the effect of this much acid on the pH of water. The third column shows the effect on a buffer solution that contains 0.10 M HOAc and 0.10 M NaOAc. The fourth column shows the result of adding the acid to a more concentrated buffer that contains twice as much acetic acid and twice as much sodium acetate. Even when as much as 0.01 mol/L of acid is added to the first buffer, there is very little change in the pH of the buffer solution, as shown in Figure 11.12.

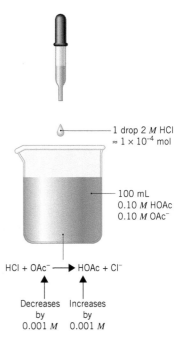

— 1 drop 2 M HCl
≈ 1 × 10⁻⁴ mol

— 100 mL
0.10 M HOAc
0.10 M OAc⁻

HCl + OAc⁻ ⟶ HOAc + Cl⁻

Decreases   Increases
by   by
0.001 M 0.001 M

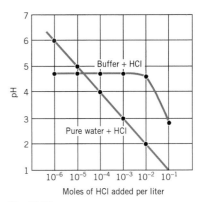

Moles of HCl added per liter

Fig. 11.12 The pH of water is very sensitive to the addition of hydrochloric acid. Adding as little as 10⁻³ mol of HCl to a liter of water can drop the pH by 4 units. The pH of a buffer, however, remains constant until relatively large amounts of hydrochloric acid have been added. The buffer contains 0.10 M OAc⁻ and HOAc.

Table 11.6
Effect of Adding Hydrochloric Acid to Water and Two Buffer Solutions

Mol HCl/L Added to Each Solution	pH of Water	pH of Buffer 1 (0.10 M HOAc/ 0.10 M OAc$^-$)	pH of Buffer 2 (0.20 M HOAc/ 0.20 M OAc$^-$)
0	7	4.74	4.74
0.000001	6	4.74	4.74
0.00001	5	4.74	4.74
0.0001	4	4.74	4.74
0.001	3	4.74	4.74
0.01	2	4.66	4.70
0.1	1	2.72	4.27

By definition, the **buffer capacity** of a buffer is the amount of strong acid or base that can be added before the pH of the solution changes significantly. The capacity of the second buffer solution in Table 11.6 is even larger than that of the first. Buffer 2 protects against quantities of acid or base almost as large as 0.1 mol/L.

The acetic acid/acetate buffer was described using the K_a expression for acetic acid.

$$K_a = \frac{[H_3O^+][OAc^-]}{[HOAc]}$$

This expression can be rearranged to give a more useful form of the equation for describing buffers.

$$[H_3O^+] = K_a \times \frac{[HOAc]}{[OAc^-]}$$

Taking the negative log of both sides of the previous equation yields the following.

$$-\log[H_3O^+] = -\log K_a - \log\left(\frac{[HOAc]}{[OAc^-]}\right)$$

The term on the left side of this equation is the expression for pH. The first term on the right side of the equation is the expression for pK_a. Substituting pH and pK_a into this equation and then inverting the logarithm term gives the following result.

$$pH = pK_a + \log\left(\frac{[OAc^-]}{[HOAc]}\right)$$

This equation can be generalized, as follows, to obtain an expression known as the Henderson–Hasselbalch equation.

$$pH = pK_a + \log\left(\frac{[\text{conjugate base}]}{[\text{conjugate acid}]}\right)$$

If the concentrations of the conjugate acid and the conjugate base used to prepare a buffer are the same, the logarithm term in the Henderson–Hasselbalch equation is equal to zero. The pH of this solution is therefore equal to the pK_a of the acid. By adjusting the ratio of the concentrations of the conjugate acid and its conjugate base it is possible to prepare a buffer with a pH slightly above or slightly below the pK_a.

Suppose a buffer with a pH of 4.74 is desired; this would suggest that an acetic acid/acetate conjugate pair with a pK_a of 4.76 would be appropriate.

If a buffer with a pH of 10.0 is needed, the $NaHCO_3/Na_2CO_3$ conjugate acid–base pair, which has a pK_a of 9.25 could be used. When the conjugate acid–base pair has a pK_a close to the desired pH, the ratio of the concentrations of acid to base can be calculated from the previous equation.

Buffers can also be made from a weak base and its conjugate acid, such as ammonia and a salt of the ammonium ion.

$$NH_4^+(aq) + H_2O(l) \rightleftharpoons H_3O^+(aq) + NH_3(aq)$$

Conjugate acid   *Weak base*

There is an important difference, however, between this buffer and one made by mixing sodium acetate and acetic acid. Mixtures of HOAc and OAc^- ion form an **acidic buffer,** with a pH below 7. Mixtures of NH_3 and the NH_4^+ ion form a **basic buffer,** with a pH above 7.

We can predict whether a buffer will be acidic or basic by comparing the values of K_a and K_b for the conjugate acid–base pair. K_a for acetic acid is significantly larger than K_b for the acetate ion. We therefore expect mixtures of the conjugate acid–base pair to be acidic.

$$HOAc \qquad K_a = 1.8 \times 10^{-5}$$
$$OAc^- \qquad K_b = 5.6 \times 10^{-10}$$

K_b for ammonia, on the other hand, is much larger than K_a for the ammonium ion. Mixtures of this acid–base pair are therefore basic.

$$NH_3 \qquad K_b = 1.8 \times 10^{-5}$$
$$NH_4^+ \qquad K_a = 5.6 \times 10^{-10}$$

In general, if K_a for the acid is larger than 1×10^{-7}, the buffer will be acidic. If K_b is larger than 1×10^{-7}, the buffer is basic.

The choice of the conjugate acid/base pair used to prepare a buffer solution can be made by comparing the desired pH with a table of values of K_a for weak acids.

Exercise 11.11

Describe how to prepare a pH 9.35 buffer solution.

Solution

The data in either Table 11.4 or Table B.8 in the Appendix suggest that the NH_4^+/NH_3 acid–base pair could be used to prepare this buffer because the pK_a for the ammonium ion is 9.25.

$$NH_4^+(aq) + H_2O(l) \rightleftharpoons NH_3(aq) + H_3O^+(aq)$$

We can start by substituting the values of the pH of the buffer solution and the pK_a value for the conjugate acid into the Henderson–Hasselbalch equation.

$$pH = pK_a + \log\left(\frac{[\text{conjugate base}]}{[\text{conjugate acid}]}\right)$$

$$9.35 = 9.25 + \log\left(\frac{[\text{conjugate base}]}{[\text{conjugate acid}]}\right)$$

Solving for the ratio of the concentrations of the conjugate acid–base pair gives the following result.

$$\log\left(\frac{[\text{conjugate base}]}{[\text{conjugate acid}]}\right) = 9.35 - 9.25 = 0.10$$

$$\frac{[\text{conjugate base}]}{[\text{conjugate acid}]} = 1.3$$

Thus if the concentration of the base is 1.3 times that of the concentration of the acid, the solution would be an effective buffer at a pH of 9.35.

Suppose that 1.0 L of 0.10 M NH_3 was available. The equilibrium constant for the reaction between NH_3 and water is so small that the concentration of NH_3 at equilibrium is essentially 0.10 M.

$$NH_3(aq) + H_2O(l) \rightleftharpoons NH_4^+(aq) + OH^-(aq) \qquad K_b = 1.8 \times 10^{-5}$$

We can calculate the concentration of the NH_4^+ needed to prepare a pH 9.35 buffer as follows.

$$\frac{[NH_3]}{[NH_4^+]} = 1.3$$

$$\frac{[0.10\ M]}{[NH_4^+]} = 1.3$$

$$[NH_4^+] = 0.077\ M$$

Because adding 0.077 mole of solid NH_4Cl to the 0.10 M NH_3 solution would produce a negligible change in the volume of the solution, the result would be 1.0 L of a buffer with the desired pH.

pH meters are often calibrated with buffer solutions because the pH of these solutions is stable. [Copyright of Thermo Fisher Scientific Inc., 2010.]

➤ CHECKPOINT

The pH of a buffered solution is 7.4. What happens to the pH if an acid is added to the buffer? What happens to the pH if a base is added to the buffer? Does the resulting pH depend on how much acid or base is added?

Buffers are used extensively in the chemistry laboratory. They can be purchased from chemical suppliers as solutions with known concentrations and pH values. They can also be purchased as packets of a mixture of a solid conjugate acid–base pair. Dissolving the packet in water yields a buffer with a pH equal to the value stated on the package. In addition, buffers can be prepared by mixing measured amounts of an appropriate conjugate acid–base pair and then adding strong acid or base to adjust the pH to the desired value. The pH of the buffer is normally monitored with a pH meter as the strong acid or strong base is added.

11.17 Buffers in the Body

Buffers are very important in living organisms for maintaining the pH of biological fluids within the very narrow ranges necessary for the biochemical reactions of life processes. Three primary buffer systems maintain the pH of blood in the human body within the limits 7.36 to 7.42. One of these buffers is composed of carbonic acid, H_2CO_3, and its conjugate base, HCO_3^-, the hydrogen carbonate ion. Some of the CO_2 that enters a red blood cell from respiring tissue reacts with water and is converted to carbonic acid by an enzyme known as carbonic anhydrase.

$$CO_2(g) + H_2O(l) \rightleftharpoons H_2CO_3(aq)$$

The carbonic acid then partially dissociates to form an equilibrium mixture with the hydrogen carbonate (or bicarbonate) ion.

$$H_2CO_3(aq) + H_2O(l) \rightleftharpoons HCO_3^-(aq) + H_3O^+(aq)$$

When the concentration of the H_3O^+ ion in the blood is too high (too low a pH), a condition known as *acidosis* results. This condition can be caused by diseases such as diabetes mellitus, in which large amounts of acids are produced. As the concentration of H_3O^+ in blood increases, the H_2CO_3/HCO_3^- equilibrium shifts to the left to maintain the proper pH.

A low concentration of H_3O^+ in the blood (too high a pH) is known as *alkalosis*. This condition can result during hyperventilation (rapid breathing) when excess amounts of CO_2 are lost from the body through the lungs. The loss of CO_2 shifts the CO_2/H_2CO_3 equilibrium to the left, resulting in a decrease in the concentration of carbonic acid in the blood. This disrupts the H_2CO_3/HCO_3^- buffer by shifting the equilibrium to the left and therefore reducing the amount of H_3O^+ in the blood. The resulting increase in pH is alkalosis.

> ➤ CHECKPOINT
>
> Lactic acid, $HC_3H_5O_3$, is produced in the body during strenuous exercise. Write a chemical reaction that can occur in blood to maintain the proper pH when lactic acid is produced.

11.18 Acid–Base Reactions

STRONG ACID–STRONG BASE

The reaction between an acid and a base can produce a salt and water. Consider, for example, the reaction of the strong acid hydrochloric acid with the strong base sodium hydroxide.

$$HCl(aq) + NaOH(aq) \longrightarrow H_2O(l) + NaCl(aq)$$

The ionic equation for this reaction is

$$H^+(aq) + Cl^-(aq) + Na^+(aq) + OH^-(aq) \longrightarrow H_2O(l) + Na^+(aq) + Cl^-(aq)$$

For acid and base reactions, $H^+(aq)$ is best represented by $H_3O^+(aq)$, and the net ionic equation for this reaction is usually written as follows.

$$H_3O^+(aq) + OH^-(aq) \longrightarrow H_2O(l) + H_2O(l)$$
Stronger acid   Stronger base   Weaker acid   Weaker base

H_3O^+ is a stronger acid than H_2O, and OH^- is a stronger base than H_2O. As shown in Section 11.10, the equilibrium in the above reaction lies almost entirely to the right. The equilibrium constant for this reaction is the reciprocal of K_w, or 1.0×10^{14}. No matter which strong acid or strong base is reacted, the net ionic reaction is always that given by the preceding equation, with $K = 1.0 \times 10^{14}$.

When just enough strong acid has been added to completely react with a strong base, the resultant solution will have equal concentrations of H_3O^+ and OH^- and thus give a pH of 7.0.

WEAK ACID–STRONG BASE

If a weak acid is reacted with a strong base, the reaction will go almost to completion.

$$HA(aq) + OH^-(aq) \rightleftharpoons H_2O(l) + A^-(aq)$$

Weak acid   Strong base

In contrast to the reaction of a strong acid with a strong base, the pH of a solution produced from the complete reaction of a weak acid and strong base will not be 7.0 but will depend on the interaction of the A^- ion with water.

The weak acid HNO_2 reacts with a sodium hydroxide solution, for example, to form water and the NO_2^- ion.

$$HNO_2(aq) + OH^-(aq) \longrightarrow H_2O(l) + NO_2^-(aq)$$

Weak acid   Strong base

When this reaction is essentially complete, the pH of the resulting solution is due to the interaction of the weak base NO_2^- with H_2O.

The equilibrium constant for this reaction can be calculated from tabulated values of the equilibrium constants for HNO_2 (see Table B.8 in Appendix B) and K_w,

$$HNO_2(aq) + H_2O(l) \rightleftharpoons H_3O^+(aq) + NO_2^-(aq) \qquad K_a = 5.1 \times 10^{-4}$$
$$H_3O^+(aq) + OH^-(aq) \rightleftharpoons 2\,H_2O(l) \qquad K = 1.0 \times 10^{14}$$

where K for the second reaction is the inverse of K_w. Adding the two reactions gives

$$HNO_2(aq) + OH^-(aq) \rightleftharpoons H_2O(l) + NO_2^-(aq)$$
$$K_{rxn} = K_a \times K = 5.1 \times 10^{-4} \times 1.0 \times 10^{14} = 5.1 \times 10^{10}$$

The equilibrium constant is so large that the reaction is essentially complete. In general, the reaction of a weak acid with a strong base proceeds far to the right.

STRONG ACID–WEAK BASE

The weak base methyl amine, CH_3NH_2, reacts with a strong acid such as HI according to the following equation.

$$CH_3NH_2(aq) + HI(aq) \longrightarrow CH_3NH_3^+(aq) + I^-(aq)$$

Assuming that HI dissociates more or less completely in water, we can write the equation for this reaction as follows.

$$CH_3NH_2(aq) + H_3O^+(aq) \longrightarrow CH_3NH_3^+(aq) + H_2O(l)$$

The equilibrium constant for this reaction can be found by combining appropriate reactions to give the desired reaction for which the equilibrium constants are known. Tables B.8 and B.9 in Appendix B contain a tabulation of many of these useful equilibrium constants.

$$CH_3NH_2(aq) + H_2O(l) \rightleftharpoons CH_3NH_3^+(aq) + OH^-(aq) \qquad K_b = 4.8 \times 10^{-4}$$
$$\underline{H_3O^+(aq) + OH^-(aq) \rightleftharpoons 2\,H_2O(l) \qquad\qquad\qquad K = 1.0 \times 10^{14}}$$
$$CH_3NH_2(aq) + H_3O^+(aq) \rightleftharpoons CH_3NH_3^+(aq) + H_2O(l) \qquad K_{rxn} = K_b \times K = 4.8 \times 10^{10}$$

The equilibrium constant for the reaction is so large that the reaction proceeds nearly to completion. In general, a strong acid will react completely with a weak base. The pH of the resulting mixture will depend on the nature of the acid, in this case $CH_3NH_3^+$, produced in the reaction.

WEAK BASE–WEAK ACID

The reaction between the weak base methyl amine and the weak acid nitrous acid can be described by the following equation.

$$CH_3NH_2(aq) + HNO_2(aq) \longrightarrow CH_3NH_3^+(aq) + NO_2^-(aq)$$

$\qquad$ *Weak base* $\qquad$ *Weak acid*

This reaction will proceed to an extent that depends on the magnitude of the equilibrium constant. For such reactions in general, K_{rxn} can be found by the following procedure.

$$
\begin{array}{ll}
HA(aq) + H_2O(l) \rightleftharpoons H_3O^+(aq) + A^-(aq) & K_a = ? \\
B(aq) + H_2O(l) \rightleftharpoons BH^+(aq) + OH^-(aq) & K_b = ? \\
\underline{H_3O^+(aq) + OH^-(aq) \rightleftharpoons 2\,H_2O(l)} & \underline{K = 1.0 \times 10^{14}} \\
HA(aq) + B(aq) \rightleftharpoons BH^+(aq) + A^-(aq) & K_{rxn} = K_a \times K_b \times 1.0 \times 10^{14}
\end{array}
$$

For the CH_3NH_2 and HNO_2 reaction, Tables B.8 and B.9 in Appendix B give the necessary K_a and K_b values and

$$K_{rxn} = 5.1 \times 10^{-4} \times 4.8 \times 10^{-4} \times 1.0 \times 10^{14} = 2.5 \times 10^7$$

In this case K_{rxn} is quite large, and the reaction proceeds nearly to completion. However, this is not always the case. For the reaction of HCN with hydrazine, H_2NNH_2, the equilibrium constant is only 0.072; thus this reaction hardly proceeds at all. Weak acid–weak base reactions are difficult to generalize, and in each case an examination of the equilibrium constants is necessary.

11.19 pH Titration Curves

We can measure the concentration of a solution by a technique known as **titration.** A solution of known concentration is slowly added to a known quantity of a reagent with which it reacts until we observe something that tells us that exactly equivalent numbers of moles of the reagents are present. Titrations therefore depend on the existence of a class of compounds known as *indicators.* A common type of titration is an acid–base titration in which an acid of unknown concentration is titrated with a base that has an accurately known concentration. Titrations can therefore be used to determine the concentration of an acid or base solution.

Indicators are weak acids or weak bases whose conjugate acid–base pairs have different colors in aqueous solution. A commonly used indicator is phenolphthalein, which is colorless in its acid form and pink in its base form. The two forms of an indicator can be represented by HIn for the acid form and In$^-$ for the base form. The ionization of the indicator can therefore be represented by the following equation,

$$HIn(aq) + H_2O(aq) \rightleftharpoons H_3O^+(aq) + In^-(aq)$$

Table 11.7
Common Acid–Base Indicators

Indicator	pH-interval	Color change (Acid$_{form}$ to Base$_{form}$)
Thymol blue	1.2–2.8	red-yellow
Methyl orange	3.1–4.4	orange-yellow
Methyl red	4.2–6.2	red-yellow
Bromthymol blue	6.0–7.6	yellow-blue
Phenol red	6.8–8.2	yellow-red
Cresol red	7.2–8.8	yellow-red
Phenolphthalein	8.3–10.0	colorless-pink
Alizarin yellow	10.1–12.0	yellow-red

for which K_a could be written as follows:

$$K_a = [H_3O^+]\frac{[In^-]}{[HIn]}$$

This equation can be rearranged to give:

$$\frac{[In^-]}{[HIn]} = \frac{K_a}{[H_3O^+]}$$

The color of the indicator solution depends on the concentration of the H_3O^+ ion. When the concentration of the H_3O^+ ion is large, the value of [HIn] is larger than [In$^-$], and the solution has the characteristic color of the acid form of the indicator. If phenolphthalein is the indicator, the solution is colorless in acid. When the concentration of the H_3O^+ ion is small, the value of [In$^-$] is larger than [HIn], and the color of the base form of the indicator predominates. For phenolphthalein, the color of the base form is pink. For most indicators color changes occur when the ratio [In$^-$]/[HIn] changes from 1/10 to about 10. Some common indicators are listed in Table 11.7 along with the pH interval over which the indicator changes color.

 Exercise 11.12

Calculate the pH of a solution of methyl red ($K_a = 5.0 \times 10^{-6}$) for which the ratio [HIn]/[In$^-$] is 0.10, 1.0, and 10. Predict the color of the solution for each pH if the acid form of the indicator is red and the base form is yellow.

Solution

The H_3O^+ ion concentration can be predicted from the [HIn]/[In$^-$] ratio using the following equation.

$$[H_3O^+] = K_a\frac{[HIn]}{[In^-]} = 5.0 \times 10^{-6}\frac{[HIn]}{[In^-]}$$

When the [HIn]/[In$^-$] ratio is 0.10, the pH of the solution is 6.3, and the indicator is yellow.

$$[H_3O^+] = 5.0 \times 10^{-6} \times \frac{[1]}{[10]} = 5.0 \times 10^{-7}; pH = 6.3$$

When the ratio is 1.0, the pH is 5.3 and the color of the solution should be a mixture of red and yellow, and therefore orange.

$$[H_3O^+] = 5.0 \times 10^{-6} \times \frac{[1]}{[1]} = 5.0 \times 10^{-6}; pH = 5.3$$

When the ratio is 10, the pH is 4.3 and the indicator is red.

$$[H_3O^+] = 5.0 \times 10^{-6} \times \frac{[10]}{[1]} = 5.0 \times 10^{-5}; pH = 4.3$$

The **end point** of an acid–base titration is the point at which the indicator turns color. The **equivalence point** is the point at which exactly enough base has been added to neutralize the acid. Chemists try to use indicators for which the end point is as close as possible to the point at which equivalent amounts of acid and base are present.

Phenolphthalein provides an example of how chemists compensate for the fact that the indicator doesn't always turn color exactly at the equivalence point of the titration. Acid–base titrations that use phenolphthalein as the indicator are stopped at the point at which a pink color appears and remains for 10 seconds as the solution is stirred. Because this is one drop before the indicator turns a permanent color, it is as close as we can get to the equivalence point of the titration.

The selection of an indicator for the end-point determination depends on the nature of the acid and base. Consider a titration in which a strong acid such as HCl is titrated with a strong base such as NaOH.

$$HCl(aq) + NaOH(aq) \rightleftharpoons Na^+(aq) + Cl^-(aq) + H_2O(l).$$

Because neither the Na^+ ion nor the Cl^- ion affects the pH of the water solvent, the pH will be 7.0 at the equivalence point.

If a weak acid such as acetic acid is titrated by a strong base, the solution will be basic at the equivalence point because of the reaction between the OAc^- ion and water.

$$NaOH(aq) + HOAc(aq) \rightleftharpoons H_2O(l) + Na^+(aq) + OAc^-(aq)$$
$$OAc^-(aq) + H_2O(l) \rightleftharpoons HOAc(aq) + OH^-(aq)$$

Exercise 11.13

Select an indicator from Table 11.7 that would change color at the equivalence point for each of the following titrations.

(a) HNO_3 and NaOH

(b) HOBr and KOH

(c) HCl and NH_3

Solution

(a) This titration involves the reaction between a strong acid and a strong base. The equivalence point should therefore occur at a pH of about 7. Bromthymol blue would therefore be an appropriate indicator.

(b) This titration involves a weak acid and a strong base, which will suggest a pH larger than 7 at the equivalence point. Cresol red would be an appropriate indicator.

(c) This titration involves a strong acid and a weak base, which suggests a pH smaller than 7 at the equivalence point. Methyl red is therefore an appropriate indicator.

• •

Let's consider what would happen if we followed the reaction between sulfuric acid (H_2SO_4) and potassium hydroxide (KOH) using an appropriate titration indicator. The reaction can be described by the following equation. Because this is a reaction between a strong acid and a strong base, the reaction can be assumed to go to completion.

$$H_2SO_4(aq) + 2\,KOH(aq) \longrightarrow 2\,K^+(aq) + SO_4^{2-}(aq) + 2\,H_2O(l)$$

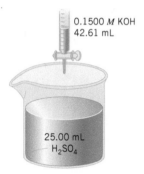

0.1500 *M* KOH
42.61 mL

25.00 mL
H₂SO₄

A 25.00-mL sample of sulfuric acid solution requires 42.61 mL of a 0.1500 *M* KOH solution to titrate to the phenolphthalein end point. What does this information tell us about the concentration of the sulfuric acid solution?

We only know the volume of the sulfuric acid solution, but we know both the volume and the concentration of the KOH solution. It therefore seems reasonable to start by calculating the number of moles of KOH in this solution.

$$\frac{0.1500 \text{ mol KOH}}{1 \text{ L}} \times 0.04261 \text{ L} = 6.392 \times 10^{-3} \text{ mol KOH}$$

We now know the number of moles of KOH consumed in the reaction, and we have a balanced chemical equation for the reaction that occurs when the two solutions are mixed. We can therefore calculate the number of moles of H_2SO_4 needed to consume that much KOH.

$$6.392 \times 10^{-3} \text{ mol KOH} \times \frac{1 \text{ mol } H_2SO_4}{2 \text{ mol KOH}} = 3.196 \times 10^{-3} \text{ mol } H_2SO_4$$

We now know the number of moles of H_2SO_4 in the sulfuric acid solution and the volume of the solution. We can therefore calculate the number of moles of sulfuric acid per liter, or the molarity of the solution.

$$\frac{3.196 \times 10^{-3} \text{ mol } H_2SO_4}{0.02500 \text{ L}} = 0.1278 M\ H_2SO_4$$

The sulfuric acid solution therefore has a concentration of 0.1278 mole per liter.

Acid–base titrations can be described by using a titration curve, a plot of solution pH versus amount of titrant added. Buffers and buffering capacity play a major role in determining the shape of the titration curve in Figure 11.13, which shows the pH of a solution of 25.00 mL of 0.10 *M* acetic acid as it is titrated with a strong base, 0.10 *M* sodium hydroxide. Four points (*A*, *B*, *C*, and *D*) on the curve will be discussed in some detail.

Point *A* in Figure 11.13 represents the pH at the start of the titration, that is the pH of a 0.10 *M* HOAc solution.

$$HOAc(aq) + H_2O(l) \rightleftharpoons OAc^-(aq) + H_3O^+(aq)$$

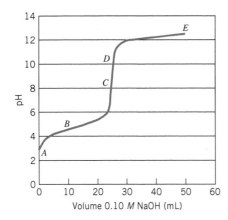

Fig. 11.13 Titration curve for the titration of 25.00 mL of a 0.10 M acetic acid solution by a 0.10 M NaOH solution.

The pH can be calculated from

$$K_a = 1.8 \times 10^{-5} = \frac{[OAc^-][H_3O^+]}{[HOAc]} \approx \frac{\Delta C^2}{0.10\ M}$$

$$\Delta C = 1.3 \times 10^{-3} = [H_3O^+],\ pH \approx 2.9$$

As NaOH solution is added to the HOAc solution, some of the acid is converted to its conjugate base, the OAc^- ion. The equilibrium constant for the reaction between a weak acid and a strong base is large, and therefore the equilibrium lies almost entirely to the right.

$$HOAc(aq) + OH^-(aq) \longrightarrow OAc^-(aq) + H_2O(l)$$

As the titration proceeds, a buffer of HOAc and OAc^- forms, and the pH changes slowly as the buffer capacity is exceeded with each acid addition.

Point B, corresponding to the addition of 12.50 mL of NaOH solution, is the point at which exactly half of the HOAc molecules have been converted to OAc^- ions. The concentration of the H_3O^+ ions in the solution is controlled by the equilibrium between HOAc and OAc^-, which is described by the following equilibrium constant expression.

$$K_a = \frac{[H_3O^+][OAc^-]}{[HOAc]}$$

At point B, the concentration of HOAc molecules is equal to the concentration of the OAc^- ions. Because the ratio of the HOAc and OAc^- concentrations is 1:1, the concentration of the H_3O^+ ion at point B is equal to the K_a of the acetic acid.

$$K_a = [H_3O^+],\ \text{and the pH is 4.7 at point B.}$$

This provides us with a way of measuring K_a for an acid. We can titrate a sample of the acid with a strong base, plot the titration curve, and then find the point along the curve at which exactly half of the acid has been consumed. The H_3O^+ ion concentration at that point will be equal to the value of K_a for the acid.

Point C is the equivalence point for the titration, the point at which enough base has been added to the solution to consume the acid present at the start of the titration. The goal of any titration is to find the equivalence point. What is actually observed is the end point of the titration, the point at which the indicator changes color.

At point C the acetic acid has been completely reacted, and what remains in solution are Na^+ ions and OAc^- ions. The Na^+ ions have no effect on the pH of the solution, but the OAc^- ions react with water to produce a basic solution.

$$OAc^-(aq) + H_2O(l) \rightleftharpoons HOAc(aq) + OH^-(aq)$$

The moles of OAc^- can be found from the initial moles of HOAc.

$$0.025 \text{ L} \times 0.10 \text{ mol/L HOAc} = 2.5 \times 10^{-3} \text{ mol HOAc}$$

At the equivalence point, 2.5×10^{-3} mol of OAc^- have been formed, and there is essentially no HOAc remaining. The total volume in the titration flask is equal to the initial 25.00 mL of HOAc plus the 25.00 ml of NaOH required to reach the equivalence point. The concentration of the OAc^- is then given by

$$[OAc^-] = \frac{2.5 \times 10^{-3} \text{ mol } OAc^-}{0.05000 \text{ L}} = 5.0 \times 10^{-2} M \, OAc^-$$

The pH of the solution at this point can be calculated from the equilibrium constant, K_b, for the above reaction.

$$K_b = 5.6 \times 10^{-10} = \frac{[HOAc][OH^-]}{[OAc^-]} \approx \frac{\Delta C^2}{5.0 \times 10^{-2} M}$$

$\Delta C = [OH^-] = 5.3 \times 10^{-6} M$, pOH ≈ 5.3, and the pH is 8.7 at point C.

Every effort is made to bring the end point as close as possible to the equivalence point of a titration. Thus in this titration an indicator whose color change occurs around a pH of 8.7 should be chosen. Table 11.7 shows that phenolphthalein indicator turns from colorless to pink over a pH range of 8.3 to 10.0. Figure 11.13 shows the range from points C to D over which the color change occurs. Therefore phenolphthalein would be a good indicator choice.

At point E, 50.00 mL of NaOH solution has been added. This is 25.00 mL in excess of the amount required to react will the acetic acid. Thus the solution now has a volume of 75.00 mL, and the excess moles of NaOH can be calculated from

$$0.10 \text{ mol/L} \times 0.02500 \text{ L} = 0.0025 \text{ mol NaOH}$$

The OH^- concentration is then

$$[OH^-] = 0.0025 \text{ mol}/0.07500 \text{ L} = 0.033 M$$

The pOH is 1.5 and the pH at point E is 12.5.

Note the shape of the pH titration curve in Figure 11.13. The pH rises rapidly at first because we are adding a strong base to a weak acid and the base neutralizes some of the acid. The curve then levels off, and the pH remains more or less constant as we add base because some of the HOAc present initially is converted into OAc^- ions to form a buffer solution. The pH of the buffer solution stays relatively constant until most of the acid has been converted to its conjugate base. At that point the pH rises rapidly because essentially all of the HOAc in the system has been converted into OAc^- ions and the buffer is exhausted. The pH then gradually levels off as the solution begins to look like a 0.10 M NaOH solution.

The titration curve for a weak base, such as ammonia, titrated with a strong acid, such as hydrochloric acid, would be analogous to the curve in Figure 11.13.

The principal difference is that the pH value for this titration curve would be high at first and decrease as acid is added.

Exercise 11.14

Figure 11.14 is the titration curve for the strong acid–weak base titration of 25.00 mL of a 0.1000 M NH$_3$ solution titrated by a 0.1000 M solution of HCl. Calculate the pH at points A, B, C, and D.

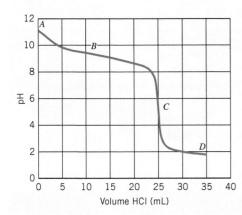

Figure 11.14 Titration curve for the titration of 25.00 mL of a 0.1000 M NH$_3$ solution by a 0.1000 M HCl solution.

Solution

Point A: At start of the titration the NH$_3$ concentration is 0.1000 M. The pH is calculated from:

$$NH_3(aq) + H_2O(l) \rightleftharpoons NH_4^+(aq) + OH^-(aq)$$

$$K_b = 1.8 \times 10^{-5} = \frac{[NH_4^+][OH^-]}{[NH_3]}$$

$\Delta C^2/0.1000\ M = 1.8 \times 10^{-5}$, $\Delta C = 1.3 \times 10^{-3} M = [OH^-]$, pOH = 2.9 and pH = 11.1

Point B: Halfway to the equivalence point $[NH_4^+] = [NH_3]$ and

$$1.8 \times 10^{-5} = [OH^-], \text{ pOH} = 4.7, \text{ pH} = 9.3$$

Point C. At the equivalence point all of NH$_3$ had been titrated, and only NH$_4^+$ remains in solution. The concentration of the NH$_4^+$ is given by

$$[NH_4^+] = \frac{(0.1000\ M)(0.02500\ L)}{0.05000\ L} = 0.05000\ M$$

$$NH_4^+(aq) + H_2O(l) \rightleftharpoons H_3O^+(aq) + NH_3(aq)$$

$$K_a = 5.6 \times 10^{-10} = \frac{[H_3O^+][NH_3]}{[NH_4^+]} = \frac{\Delta C^2}{0.0500 M}$$

$$\Delta C = [H_3O^+] = 5.3 \times 10^{-6} M, \text{ pH} = 5.3$$

> **CHECKPOINT**

Describe what happens to the pH of an HCl solution as NaOH is slowly added.

Point *D*. 35.00 mL of HCl solution have been added; this is 10.00 mL in excess of the end point:

$$\frac{0.01000\,L \times 0.1000\,M}{0.0600\,L} = [H_3O^+] = 1.70 \times 10^{-2}M \qquad pH = 1.8$$

Key Terms

Acid–base indicator
Acid dissociation equilibrium
 constant (K_a)
Acidic buffer
Acids
Alkalies
Arrhenius acid/base
Base ionization equilibrium
 constant (K_b)
Base ionization equilibrium constant
 expression
Bases

Basic buffer
Brønsted acid/base
Buffer
Buffer capacity
Conjugate acid–base pair
End point
Equivalence point
Hydrogen-ion acceptor/donor
Hydronium ion
Leveling effect
Monoprotic acid
Neutralization reaction

pH
pOH
Proton acceptor/donor
Strong acid
Strong base
Titration
Water dissociation equilibrium
 constant (K_w)
Weak acid
Weak base

Problems

Properties of Acids and Bases

1. Describe how acids and bases differ. Give examples of substances from daily life that fit each category.

2. Many metals dissolve in acids. Write an equation for the reaction between magnesium metal in hydrochloric acid.

3. How can litmus be used to distinguish between an acid and a base?

4. What happens when acids and bases are combined?

The Arrhenius Definition of Acids and Bases

5. Assume that the reaction between water and either HBr, H_2SO_4, $FeCl_3$, or $Fe(OH)_3$ can be described by the following equations.

$$HBr(aq) \xrightarrow{H_2O} H^+(aq) + Br^-(aq)$$

$$H_2SO_4(aq) \xrightarrow{H_2O} HSO_4^-(aq) + H^+(aq)$$

$$FeCl_3(s) \xrightarrow{H_2O} Fe^{3+}(aq) + 3\,Cl^-(aq)$$

$$Fe(OH)_3(s) \rightleftharpoons Fe^{3+}(aq) + 3\,OH^-(aq)$$

Which of the following compounds are Arrhenius acids?

(a) HBr (b) H_2SO_4 (c) $FeCl_3$ (d) $Fe(OH)_3$

6. Assume that the reaction between water and either LiOH, H_2S, Na_2SO_4, or NH_4Cl can be described by the following equations.

$$LiOH(s) \xrightarrow{H_2O} Li^+(aq) + OH^-(aq)$$

$$H_2S(aq) \xrightarrow{H_2O} H^+(aq) + HS^-(aq)$$

$$Na_2SO_4(s) \xrightarrow{H_2O} 2\,Na^+(aq) + SO_4^{2-}(aq)$$

$$NH_4Cl(s) \xrightarrow{H_2O} NH_4^+(aq) + Cl^-(aq)$$

Which of the following compounds are Arrhenius bases?

(a) LiOH (b) H_2S
(c) Na_2SO_4 (d) NH_4Cl

The Brønsted–Lowry Definition of Acids and Bases

7. Write balanced equations to show what happens when hydrogen bromide dissolves in water to form an acidic solution and when ammonia dissolves in water to form a basic solution.

8. Give an example of an acid whose formula carries a positive charge, an acid that is electrically neutral, and an acid that carries a negative charge.

9. The following compounds are all oxyacids. In each case, the acidic hydrogen atoms are bound to oxygen atoms. Write the skeleton structures for the acids.

(a) H_3PO_4 (b) HNO_3 (c) $HClO_4$
(d) H_3BO_3 (e) H_2CrO_4

10. Describe how the Brønsted definition of an acid can be used to explain why compounds that contain hydrogen with an oxidation number of $+1$ are often acids.

11. Which of the following compounds cannot be Brønsted bases?
    (a) H_3O^+ (b) MnO_4^- (c) BH_4^-
    (d) CN^- (e) S^{2-}

12. Which of the following compounds cannot be Brønsted bases?
    (a) O_2^{2-} (b) CH_4 (c) PH_3
    (d) SF_4 (e) CH_3^+

13. Label the Brønsted acids and bases in the following reactions.
    (a) $HSO_4^-(aq) + H_2O(l)$
       $\rightleftharpoons H_3O^+(aq) + SO_4^{2-}(aq)$
    (b) $CH_3CO_2H(aq) + OH^-(aq)$
       $\rightleftharpoons CH_3CO_2^-(aq) + H_2O(l)$
    (c) $CaF_2(s) + H_2SO_4(aq)$
       $\rightleftharpoons CaSO_4(aq) + 2\,HF(aq)$
    (d) $HNO_3(aq) + NH_3(aq) \rightleftharpoons NH_4NO_3(aq)$
    (e) $LiCH_3(l) + NH_3(l)$
       $\rightleftharpoons CH_4(g) + LiNH_2(s)$

14. Write the chemical equation for the Brønsted bases, reacting with water.
    (a) NH_3 (b) CH_3NH_2 (c) IO^-

Conjugate Acid–Base Pairs

15. Identify the conjugate acid–base pairs in the following reactions.
    (a) $HCl(aq) + H_2O(l) \rightleftharpoons H_3O^+(aq) + Cl^-(aq)$
    (b) $HCO_3^-(aq) + H_2O(l)$
       $\rightleftharpoons H_2CO_3(aq) + OH^-(aq)$
    (c) $NH_3(aq) + H_2O(l) \rightleftharpoons NH_4^+(aq) + OH^-(aq)$
    (d) $CaCO_3(s) + HCl(aq)$
       $\rightleftharpoons Ca^{2+}(aq) + 2\,Cl^-(aq) + H_2CO_3(aq)$

16. Which of the following is not an example of a conjugate acid–base pair?
    (a) NH_4^+/NH_3 (b) H_2O/OH^-
    (c) H_3O^+/OH^- (d) CH_4/CH_3^-

17. Write Lewis structures for the following Brønsted acids and their conjugate bases.
    (a) formic acid, HCO_2H (b) methanol, CH_3OH

18. Write Lewis structures for the following Brønsted bases and their conjugate acids.
    (a) methylamine, CH_3NH_2
    (b) acetate ion, $CH_3CO_2^-$

19. Identify the conjugate base of each of the following Brønsted acids.
    (a) H_3O^+ (b) H_2O (c) OH^- (d) NH_4^+

20. Identify the conjugate base of each of the following Brønsted acids.
    (a) HPO_4^{2-} (b) $H_2PO_4^-$
    (c) HCO_3^- (d) HS^-

21. Identify the conjugate acid of each of the following Brønsted bases.
    (a) O^{2-} (b) OH^-
    (c) H_2O (d) NH_2^-

22. Acids (such as hydrochloric acid and sulfuric acid) react with bases (such as sodium hydroxide and ammonia) to form salts (such as sodium chloride, sodium sulfate, ammonium chloride, and ammonium sulfate). Write balanced equations for each acid reacting with each base.

The Role of Water in the Brønsted Model

23. True or false? Water is both an acid and a base. Provide support for your answer.

24. Write an equation that shows how a base reacts with water. Do the same for an acid.

25. Give the Lewis structures for water and acetic acid, CH_3CO_2H. Write an equation that describes the interaction between water and acetic acid using Lewis structures.

To What Extent Does Water Dissociate to Form Ions?

26. Explain why the concentrations of the H_3O^+ ion and the OH^- ion in pure water are the same.

27. Explain why it is impossible for water to be at equilibrium when it contains large quantities of both the H_3O^+ and OH^- ions.

28. The dissociation of water is an endothermic reaction.

    $$2\,H_2O(l) \rightleftharpoons H_3O^+(aq) + OH^-(aq)$$
    $$\Delta H° = 55.84 \text{ kJ/mol}_{rxn}$$

    Use Le Châtelier's principle to predict what should happen to the fraction of water molecules that dissociate into ions as the temperature of water increases.

29. Use Le Châtelier's principle to explain why adding either an acid or a base to water suppresses the dissociation of water.

30. What is the difference between K_c and K_w for water? Calculate a value of K_c for the dissociation of H_2O to H_3O^+ and OH^-.

31. When an acid is added to pure water, there are two sources for H_3O^+. What are they? Similarly, when a base is added to pure water, there are two sources for OH^-. What are they?

32. Does the product of $[H_3O^+]$ and $[OH^-]$ always have to be 10^{-14} at 25°C? Explain.

33. If the $[H_3O^+]$ concentration in an aqueous solution is 1.0×10^{-12}, what is the $[OH^-]$?

pH as a Measure of the Concentration of H_3O^+ Ion

34. Calculate the number of H_3O^+ and OH^+ ions in 1.00 mL of pure water.

35. Calculate the H_3O^+ and OH^- ion concentrations in a solution that has a pH of 3.72.

36. Explain why the pH of a solution decreases when the H_3O^+ ion concentration increases.

37. What happens to the concentration of the H_3O^+ ion when a strong acid is added to water? What happens to the concentration of the OH^- ion? What happens to the pH of the solution?

38. What happens to the OH^- ion concentration when a strong base is added to water? What happens to the concentration of the H_3O^+ ion? What happens to the pH of the solution?

39. Which of the following solutions is the most acidic?
    (a) $0.10\ M$ acetic acid, pH = 2.9
    (b) $0.10\ M$ hydrogen sulfide, pH = 4.1
    (c) $0.10\ M$ sodium acetate, pH = 8.4
    (d) $0.10\ M$ ammonia, pH = 11.1

40. Which of the following solutions is the most acidic?
    (a) $0.10\ M\ H_3PO_4$, pH = 1.5
    (b) $0.10\ M\ H_2PO_4^-$, pH = 4.4
    (c) $0.10\ M\ HPO_4^{2-}$, pH = 9.3
    (d) $0.10\ M\ PO_4^{3-}$, pH = 12.0

41. Calculate the pH and pOH of a solution for which $[H_3O^+]$ is $1.5 \times 10^{-6}\ M$.

42. The pH of a $0.10\ M$ solution of Na_2CO_3 is 11.6. What are $[H_3O^+]$ and $[OH^-]$?

43. The pH of a saturated solution of H_2CO_3 is 3.8. What are $[H_3O^+]$ and $[OH^-]$?

44. Calculate the pH of the following solutions
    (a) $1.0\ M$ HOAc $[H_3O^+] = 4.2 \times 10^{-3}\ M$
    (b) $0.10\ M$ HOAc $[H_3O^+] = 1.3 \times 10^{-3}\ M$
    (c) $0.010\ M$ HOAc $[H_3O^+] = 4.2 \times 10^{-4}\ M$

45. The pH of a $0.10\ M$ solution of ammonia is 11.1. Find the hydroxide concentration.

Relative Strengths of Acids and Bases

46. Define the following terms: *weak acid, strong acid, weak base,* and *strong base.*

47. Describe the difference between strong acids (such as hydrochloric acid) and weak acids (such as acetic acid) and the difference between strong bases (such as sodium hydroxide) and weak bases (such as ammonia).

48. Describe the relationship between the acid dissociation equilibrium constant for an acid, K_a, and the strength of the acid.

49. Use the table of acid dissociation equilibrium constants in Table B.8 in Appendix B to classify the following acids as either strong or weak.
    (a) acetic acid, CH_3CO_2H
    (b) boric acid, H_3BO_3
    (c) chromic acid, H_2CrO_4
    (d) formic acid, HCO_2H
    (e) hydrobromic acid, HBr

50. Which of the following is the weakest Brønsted acid?
    (a) $H_2S_2O_3$, $K_a = 0.3$
    (b) H_2CrO_4, $K_a = 9.6$
    (c) H_3BO_3, $K_a = 7.3 \times 10^{-10}$
    (d) C_6H_5OH, $K_a = 1.0 \times 10^{-10}$

51. Which of the following solutions is the most acidic?
    (a) $0.10\ M\ CH_3CO_2H$, $K_a = 1.8 \times 10^{-5}$
    (b) $0.10\ M\ HCO_2H$, $K_a = 1.8 \times 10^{-4}$
    (c) $0.10\ M\ ClCH_2CO_2H$, $K_a = 1.4 \times 10^{-3}$
    (d) $0.10\ M\ Cl_2CHCO_2H$, $K_a = 5.1 \times 10^{-2}$

52. Which of the following compounds is the strongest base?
    (a) $CH_3CO_2^-$ (for CH_3CO_2H, $K_a = 1.8 \times 10^{-5}$)
    (b) HCO_2^- (for HCO_2H, $K_a = 1.8 \times 10^{-4}$)
    (c) $ClCH_2CO_2^-$ (for $ClCH_2CO_2H$, $K_a = 1.4 \times 10^{-3}$)
    (d) $Cl_2CHCO_2^-$ (for Cl_2CHCO_2H, $K_a = 5.1 \times 10^{-2}$)

53. List acetic acid, chlorous acid, hydrofluoric acid, and nitrous acid in order of increasing strength if $0.10\ M$ solutions of the acids contain the following equilibrium concentrations.

Acid	$[HA]$	$[A^-]$	$[H_3O^+]$
HOAc	$0.099\ M$	$0.0013\ M$	$0.0013\ M$
HOClO	$0.072\ M$	$0.028\ M$	$0.028\ M$
HF	$0.092\ M$	$0.0081\ M$	$0.0081\ M$
HNO_2	$0.093\ M$	$0.0069\ M$	$0.0069\ M$

Relative Strengths of Conjugate Acid–Base Pairs

54. What is the relationship between the strength of a base and its conjugate acid?

55. Which of the following conjugate bases have essentially no base properties in water solvent? I^-, ClO_4^-, F^-, NO_2^-, NO_3^-. Explain why not.

56. Is the conjugate base of a weak acid a relatively strong or weak base? Explain.

57. In the Brønsted model of acid–base reactions, what does the statement that HCl is a stronger acid than H_2O mean in terms of the following reaction?

$$HCl(aq) + H_2O(l) \longrightarrow H_3O^+(aq) + Cl^-(aq)$$

58. In the Brønsted model of acid–base reactions, what does it mean to say that NH_3 is a weaker acid than H_2O?

59. What can you conclude from experiments suggesting that the following reaction proceeds far to the right, as written?

$$HBr(aq) + H_2O(l) \longrightarrow H_3O^+(aq) + Br^-(aq)$$

60. What can you conclude from experiments suggesting that the following reaction proceeds far to the right, as written?

$$HCl(aq) + NH_3(aq) \longrightarrow NH_4^+(aq) + Cl^-(aq)$$

61. What can you conclude from experiments suggesting that the following reaction does not occur to an appreciable extent?

$$HCO_2^-(aq) + H_2O(l) \longrightarrow HCO_2H(aq) + OH^-(aq)$$

Relative Strengths of Different Acids and Bases

62. Refer to Table 11.3 and Tables B.8 and B.9 in Appendix B. For each pair of reactants, state whether the reaction proceeds nearly to completion, makes only a small amount of product, or doesn't proceed to a detectable amount.
    (a) $H_2SO_4(aq) + NaOH(aq)$
    (b) $H_2O(aq) + HNO_3(aq)$
    (c) $HF(aq) + NaOH(aq)$
    (d) $HCl(aq) + NH_3(aq)$
    (e) $C_5H_5NH(aq) + HNO_2(aq)$

63. Describe what is meant by the leveling effect of water.

64. Why does a 0.10 M solution of HI contain the same $[H_3O^+]$ as a 0.10 M solution of HBr? Would a 0.10 M solution of HI have the same $[H_3O^+]$ as a 0.10 M solution of HF? Explain.

65. Explain why the H_3O^+ ion concentration in a strong acid solution depends on the concentration of the solution but not on the value of K_a for the acid.

Relationship of Structure to Relative Strengths of Acids and Bases

66. Use the structures of the acids to explain the following observations.
    (a) H_2SO_4 is a stronger acid than HSO_4^-.
    (b) HNO_3 is a stronger acid than HNO_2.
    (c) H_2S is a stronger acid than H_2O.
    (d) H_2S is a stronger acid than PH_3.

67. Arrange the following in order of increasing acid strength. Explain.
    (a) H_3AsO_4 (b) $H_2AsO_4^-$
    (c) $HAsO_4^{2-}$ (d) AsO_4^{3-}

68. Arrange the following in order of increasing acid strength. Explain.
    (a) SiH_4 (b) HCl   (c) H_2S (d) PH_3

69. Which of the following is the strongest Brønsted acid? Explain.
    (a) H_2Se (b) H_2O (c) H_2S (d) H_2Te

70. Which of the following is the weakest Brønsted acid? Explain.
    (a) $HBrO_2$ (b) HBrO
    (c) $HBrO_4$ (d) $HBrO_3$

71. Arrange the following acids in order of increasing acidity. Explain. HOCl, HOI, HOBr. Look up the K_a values in Appendix B8. Does your answer agree? Explain why.

72. Which acid, NH_4^+ or PH_4^+, has the largest K_a? Explain.

73. Arrange the following acids in order of increasing K_a. Explain your order.

$$CH_3COOH, CH_2ClCOOH, CCl_3COOH$$

Strong Acid pH Calculations

74. Calculate the pH and pOH of a 0.035 M HCl solution.

75. Calculate the pH and pOH of a solution that contains 0.568 g of HCl per 250 mL of solution.

76. What would be the pH of a 0.010 M solution of any strong acid? Explain.

77. What would be the pH of the following solutions?
    (a) 1.0 M $HClO_4$
    (b) 0.568 g of $HClO_4$ per 250 mL of solution
    (c) 0.568 g of HNO_3 per 250 mL of solution
    (d) 1.14 g of HBr per 500 mL of solution

78. Calculate the pH of a 0.056 M solution of hydroiodic acid, HI.

79. Nitric acid is often grouped with sulfuric acid and hydrochloric acid as one of the strong acids. Calculate the pH of 0.10 M nitric acid, assuming that it is a strong acid that dissociates completely. Calculate the pH of the solution using the value of K_a for the acid (for HNO_3, $K_a = 28$).

Weak Acid pH Calculations

80. Describe the two assumptions that are commonly made in weak acid equilibrium problems. Describe how you can test whether the assumptions are valid for a particular calculation.

81. Explain why the techniques used to calculate the equilibrium concentrations of the components of a weak acid solution can't be used for either strong acids or very weak acids.

82. Explain why the H_3O^+ ion concentration in a weak acid solution depends on both the value of K_a for the acid and the concentration of the acid.

83. Which of the following solutions has the largest H_3O^+ ion concentration?
    (a) 0.10 M HOAc   (b) 0.010 M HOAc
    (c) 0.0010 M HOAc

84. Calculate the percentage of HOAc molecules that dissociate in 0.10 M, 0.010 M, and 0.0010 M solutions of acetic acid. What happens to the percent ionization as the solution becomes more dilute? (For HOAc, $K_a = 1.8 \times 10^{-5}$.)

85. Formic acid (HCO_2H) was first isolated by the destructive distillation of ants. In fact, the name comes from the Latin word for "ants," *formi*. Calculate the HCO_2H, HCO_2^-, and H_3O^+ concentrations in a 0.100 M solution of formic acid in water. (For HCO_2H, $K_a = 1.8 \times 10^{-4}$.)

86. Hydrogen cyanide (HCN) is a gas that dissolves in water to form hydrocyanic acid. Calculate the H_3O^+, HCN, and CN^- concentrations in a 0.174 M solution of hydrocyanic acid. (For HCN, $K_a = 6 \times 10^{-10}$.)

87. The first disinfectant used by Joseph Lister was called *carbolic acid*. The substance is now known as *phenol*. Calculate the H_3O^+ ion concentration in a 0.0167 M solution of phenol (C_6H_5OH). ($K_a = 1.0 \times 10^{-10}$.)

88. Calculate the concentration of acetic acid that would give an H_3O^+ ion concentration of $2.0 \times 10^{-3} M$. (For HOAc, $K_a = 1.8 \times 10^{-5}$.)

89. Calculate the value of K_a for ascorbic acid (vitamin C) if 2.8% of the ascorbic acid molecules in a 0.100 M solution dissociate.

90. Calculate the value of K_a for nitrous acid (HNO_2) if a 0.100 M solution is 7.1% dissociated at equilibrium.

91. Why is it wrong to assume that the pH of a $10^{-8} M$ HCl solution is 8?

92. The pH of a 0.10 M solution of HOAc is 2.9. Calculate K_a for acetic acid.

93. The $[H_3O^+]$ of a 0.10 M solution of HOCl is $5.4 \times 10^{-5} M$. Calculate the pH of this solution and the K_a for HOCl. How would the pH of this solution compare to that of a 0.10 M solution of HNO_3?

94. The pOH of a 0.10 M solution of HF is 11.9. Calculate K_a for HF.

Base pH Calculations

95. Which of the following equations correctly describes the relationship between K_b for the formate ion (HCO_2^-) and K_a for formic acid (HCO_2H)?
    (a) $K_a = K_w \times K_b$ (b) $K_b = K_a/K_w$
    (c) $K_b = K_w/K_a$ (d) $K_b = K_w + K_a$
    (e) $K_b = K_w - K_a$

96. Use the relationship between K_a for an acid and K_b for its conjugate base to explain why strong acids have weak conjugate bases and weak acids have strong conjugate bases.

97. Calculate the HCO_2H, OH^-, and HCO_2^- ion concentrations in a solution that contains 0.020 mol of sodium formate ($NaHCO_2$) in 250 mL of solution. (For HCO_2H, $K_a = 1.8 \times 10^{-4}$.)

98. Calculate the OH^-, HOBr, and OBr^- ion concentrations in a solution that contains 0.050 mol of sodium hypobromite (NaOBr) in 500 mL of solution. (For HOBr, $K_a = 2.4 \times 10^{-9}$.)

99. Calculate the pH of a 0.756 M solution of NaOAc. (For HOAc, $K_a = 1.8 \times 10^{-5}$.)

100. A solution of NH_3 dissolved in water is known as both aqueous ammonia and ammonium hydroxide. Use the value of K_b for the following reaction to explain why aqueous ammonia is the better name.

$$NH_3(aq) + H_2O(l) \rightleftharpoons NH_4^+(aq) + OH^-(aq)$$
$$K_b = 1.8 \times 10^{-5}$$

101. At 25°C, a 0.10 M aqueous solution of methylamine (CH_3NH_2) is 6.8% ionized.

$$CH_3NH_2(aq) + H_2O(l)$$
$$\rightleftharpoons CH_3NH_3^+(aq) + OH^-(aq)$$

Calculate K_b for methylamine. Is methylamine a stronger base or a weaker base than ammonia?

102. What is the molarity of an aqueous ammonia solution that has an OH^- ion concentration of $1.0 \times 10^{-3} M$?

103. What is the pH of a 0.10 M solution of methylamine? See Problem 101.

104. Calculate K_b for hydrazine (H_2NNH_2) if the pH of a 0.10 M aqueous solution of the rocket fuel is 10,54.

105. Calculate the pH of a 0.016 M aqueous solution of calcium acetate, $Ca(OAc)_2$. (For HOAc, $K_a = 1.8 \times 10^{-5}$.)

106. Arrange the following 0.10 M solutions in order of increasing acidity.
    (a) KCl   (b) KF   (c) KOAc

107. Which of the following 0.10 M solutions will be acidic? Explain.
    (a) NH_4Cl (b) NaF   (c) $RbNO_3$

108. Which of the following 0.10 M solutions will be basic? Explain.
    (a) Li_2SO_4 (b) $KClO_4$ (c) $NaNO_2$

109. Identify the conjugate acid produced by the dissociation of the anion of each of the following salts.
    (a) Na_2HPO_4 (b) $NaHCO_3$
    (c) Na_2SO_4 (d) $NaNO_2$

Mixtures of Acids and Bases: Buffers

110. What is meant by the term *common ion*? If NaOAc is added to a solution of HOAc, which way does the acid equilibrium shift? What happens to the pH?

111. What is the pH of a 0.50 M solution of HF? What is the pH of a solution containing 0.50 M HF and 0.50 M NaF? See Appendix B8 for the equilibrium constant.

112. Determine the pH of a mixture containing 0.10 M formic acid and 0.10 M sodium formate. See Appendix B8 for the equilibrium constant.

113. Determine the pH of a mixture containing 0.10 M ammonia and 0.10 M ammonium chloride. See Appendix B8 for the equilibrium constant.

114. Determine the pH of a solution formed by adding 15 g of benzoic acid and 10 g of sodium benzoate to water to form 1.0 liter of solution. See Appendix B8 for the equilibrium constant.

Buffers and Buffer Capacity

115. Explain how buffers resist changes in pH.

116. Explain why a mixture of HOAc and NaOAc is an acidic buffer, but a mixture of NH_3 and NH_4Cl is a basic buffer. (For HOAc, $K_a = 1.8 \times 10^{-5}$; for NH_4^+, $K_a = 5.6 \times 10^{-10}$.)

117. Which of the following mixtures in equal concentrations would make the best buffer?
    (a) HCl and NaCl
    (b) NaOAc and NH_3
    (c) HOAc and NH_4Cl
    (d) NaOAc and NH_4Cl
    (e) NH_3 and NH_4Cl

118. Which of the following solutions is an acidic buffer?
    (a) 0.10 M HCl and 0.10 M NaOH
    (b) 0.10 M HCl and 0.10 M NaCl
    (c) 0.10 M HCO_2H and 0.10 M $NaHCO_2$
    (d) 0.10 M NH_3 and 0.10 M NH_4Cl

119. Which of the following solutions is a basic buffer?
    (a) 0.10 M HCl and 0.10 M NaOH
    (b) 0.10 M HCl and 0.10 M NaCl
    (c) 0.10 M HCO_2H and 0.10 M $NaHCO_2$
    (d) 0.10 M NH_3 and 0.10 M NH_4Cl

120. What is the best way to increase the capacity of a buffer made by dissolving $NaHCO_2$ in an aqueous solution of HCO_2H? Explain.
    (a) Increase the concentration of HCO_2H.
    (b) Increase the concentration of $NaHCO_2$.
    (c) Increase the concentrations of both HCO_2H and $NaHCO_2$.
    (d) Increase the ratio of the concentration of HCO_2H to the concentration of $NaHCO_2$.
    (e) Increase the ratio of the concentration of $NaHCO_2$ to the concentration of HCO_2H.

121. In order to maintain a pH of 5.60 what must be the ratio $[HA]/[A^-]$ for the following weak acids?

acetic acid	$K_a = 1.75 \times 10^{-5}$
chlorous acid	$K_a = 1.1 \times 10^{-2}$
formic acid	$K_a = 1.8 \times 10^{-4}$

122. What must be the ratio of the conjugate base to that of the conjugate acid if the pH of a solution based on the following reaction is to maintained at 9.00? Note that $K_b = 1.7 \times 10^{-9}$ for this base.

$$C_5H_5N(aq) + H_2O(l) \rightleftharpoons C_5H_5NH^+(aq) + OH^-(aq)$$

123. What would be the pH of a buffer prepared from 0.50 M Na_2CO_3 and 0.50 M $NaHCO_3$ solutions?

124. Calculate the pH of a solution containing 0.50 M acetic acid and 0.25 M sodium acetate. $K_a = 1.75 \times 10^{-5}$

Acid–Base Reactions

125. Which of the following reactions would be expected to go nearly to completion?
    (a) strong acid–strong base
    (b) weak acid–weak base
    (c) weak base–strong acid
    (d) weak acid–strong base

126. It is possible to predict the pH of the solution formed in the following reaction by knowing only the concentrations and volumes of the reactants. Why can this be done?

$$HClO_4(aq) + KOH(aq) \longrightarrow H_2O(l) + KClO_4(aq)$$

127. Suppose 100 mL each of 0.10 M $HClO_4$ and KOH are mixed. What will be the pH of the solution?

128. If 50 mL of 0.10 M $HClO_4$ is mixed with 100 mL of 0.10 M KOH, what will be the resulting pH?

129. Write a net ionic equation for the reaction of the following.
    (a) HOBr and NaOH
    (b) HOAc and NH_3
    (c) HBr and KOH
    (d) HCO_2H and $Ba(OH)_2$
    Which of these reactions will go to completion?

130. What will be the equilibrium constant for the reaction between $HOI(aq)$ and $C_5H_5N(aq)$ (pyridine)? What does the magnitude of the equilibrium constant tell us about this reaction?

pH Titration Curves

131. Define the following: *titration, indicator, end point,* and *equivalence point*.

132. Describe in detail the experiment you would use to measure the value of K_a for formic acid, HCO_2H.

133. Sketch a titration curve of 0.10 M aniline ($K_b = 4.0 \times 10^{-10}$) being titrated with a 0.10 M HCl solution. Label the equivalence point and the point at which the OH^- ion concentration is equal to K_b for the base.

134. An indicator has a $K_a = 1.0 \times 10^{-7}$. At what pH will the indicator change color?

135. For a solution with a pH of 10.0, what will be the color of the solution in the presence of the following indicators?

    (a) methyl orange

    (b) cresol red

    (c) phenolphthalein

    (d) alizarin yellow

Integrated Problems

136. Explain why pure (nonpolluted) rainwater has a pH of 5.6. Explain why boiling water to drive off the CO_2 raises the pH.

137. On April 10, 1974, at Pitlochry, Scotland, the rain was found to have a pH of 2.4. Calculate the H_3O^+ ion concentration in the rain and compare it with the H_3O^+ concentration in 0.10 M acetic acid, which has a pH of 2.9.

138. Which of the following statements are true for a 0.10 M solution of the weak acid CH_3CO_2H in water? Explain what is wrong with any incorrect answers.

    (a) pH = 1.00

    (b) $[H_3O^+] \gg [CH_3COO^-]$

    (c) $[H_3O^+] = [CH_3COO^-]$

    (d) pH < 1.00

139. Which of the following statements are true for a 1.0 M solution of the strong acid HCl in water? Explain what is wrong with any incorrect answers.

    (a) $[Cl^-] > [H_3O^+]$ (b) pH = 0.00

    (c) $[H_3O^+] = 1.0\ M$ (d) [HCl] = 1.0 M

140. Which of the following statements are true for a 1.0 M solution of the weak base NH_3 in water? Explain what is wrong with any incorrect answers.

    (a) $[OH^-] = [H_3O^+]$ (b) $[NH_4^+] = [OH^-]$

    (c) pH < 7.00   (d) $[NH_3] = 1.0\ M$

141. The pH of a 0.10 M solution of formic acid is 2.37.

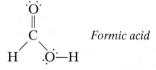

Formic acid

    (a) Which of the two hydrogens in the formic acid structure is the "acidic" hydrogen? Explain.

    (b) Write a chemical equation that describes what happens when formic acid is placed in water.

    (c) Calculate the K_a of formic acid.

    (d) Estimate the pH of a 0.10 M solution of the following acid in water. Explain your estimate.

142. Methylamine, CH_3NH_2, is a weak base with $K_b = 3.6 \times 10^{-4}$.

    (a) Write a chemical equation that describes what happens when methylamine is placed into water.

    (b) What is the pOH of a 0.10 M solution of methylamine?

    (c) What is the pH of a 0.10 M solution of methylamine?

143. Rank the following solutions in order of increasing pH. Explain your reasoning.

    (a) 0.10 M $HOBrO_2$ (b) 0.10 M HOBrO

    (c) 0.10 M HBr   (d) 0.10 M NaBr

    (e) 0.10 M $NaBrO_2$ (f) 0.10 M $NaBrO_3$

144. Describe a neutral (in terms of acid–base characteristics) solution. The K_w for the self-hydrolysis of H_2O at room temperature (25°C) is 1×10^{-14}, and that at body temperature (37°C) is 2.5×10^{-14}.

    (a) Describe the quantitative difference in the self-hydrolysis of pure water at 25°C and 37°C.

    (b) Determine the approximate pH of pure water at 25°C and 37°C.

    (c) Again, describe a neutral solution. Is this the same description that you gave at the beginning of the problem?

145. Match each of the following to its corresponding particulate representation. Water molecules are not shown. Explain your reasoning.

    (a) a dilute solution of a strong acid

    (b) a concentrated solution of a weak acid

    (c) a good buffer

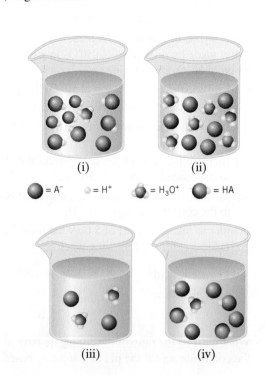

146. When NH_4Cl is dissolved in water, the following species will be present: NH_4^+, Cl^-, H_2O, OH^-, H_3O^+, and NH_3. Using the symbols $\ll$, $<$, and $\approx$, arrange the species in order of increasing concentration in the solution.

147. The weak acid HB can be represented as $\circ\bullet$, where $\bullet$ represents B^- and $\circ$ represents H^+. Match the conjugate base of HB and the conjugate acid of HB with its appropriate particulate representation.

    (a) $\circ\bullet\circ$ (b) $\bullet\circ\bullet$ (c) $\bullet$ (d) $\bullet\circ\bullet\circ$ (e) $\circ$

148. Which of the following compounds could dissolve in water to give a 0.10 M solution with a pH of about 5?

    (a) NH_3 (b) NaCl   (c) HCl

    (d) KOH   (e) NH_4Cl

149. Write balanced chemical equations for the following acid–base reactions.

    (a) $Al_2O_3(s) + HCl(aq) \longrightarrow$

    (b) $CaO(s) + H_2SO_4(aq) \longrightarrow$

    (c) $Na_2O(s) + H_2O(l) \longrightarrow$

    (d) $MgCO_3(s) + HCl(aq) \longrightarrow$

    (e) $NaOH(s) + H_3PO_4(aq) \longrightarrow$

150. Which of the following factors influences the value of K_a for the dissociation of formic acid?

$$HCO_2H(aq) + H_2O(l)$$
$$\rightleftharpoons HCO_2^-(aq) + H_3O^+(aq)$$

    (a) temperature

    (b) pressure

    (c) pH

    (d) the initial concentration of HCO_2H

    (e) the initial concentration of the HCO_2^- ion

151. A. 1.0 mol of the acid HBr is placed in sufficient water to make 1.0 L of solution. $K_a = 1 \times 10^9$.

    (a) Write an equation that describes the reaction that takes place.

    (b) What will be $[H_3O^+]$ in the solution?

    (c) What is the pH?

    (d) What will be $[OH^-]$ in the solution?

    (e) What is the conjugate base of HBr?

  B. 1.0 mol of the base ammonia, NH_3, is placed in sufficient water to make 1.0 L of solution. $K_a = 1.8 \times 10^{-5}$.

    (a) Write an equation that describes the reaction that takes place.

    (b) What is $[OH^-]$?

    (c) What is $[NH_3]$?

    (d) What is the concentration of the conjugate acid of ammonia?

    (e) What is the pH?

  C. 1.0 mol of NaBr is placed in sufficient water to make 1.0 L of solution.

$$NaBr(s) \rightleftharpoons Na^+(aq) + Br^-(aq) \qquad K_c \gg 1$$

    (a) What is $[Br^-]$ in this solution?

    (b) Write an equation describing the possible reaction of Br^- with water. What do you think the equilibrium constant for this reaction might be? Explain your reasoning. Part A of this question might be useful.

    (c) What is the pH of this solution?

152. The following K_a values have been experimentally determined for formic and acetic acids.

                      Formic acid
                      $K_a = 1.8 \times 10^{-4}$

                      Acetic acid
                      $K_a = 1.8 \times 10^{-5}$

    (a) Which is the stronger acid? State your reasoning.

    (b) Write a chemical equation that describes what happens when formic acid is placed into water. Use Lewis structures for products and reactants similar to those above.

    (c) What is the pH of a 0.10 M solution of formic acid? Show all calculations. State and justify any assumptions made.

    (d) Which is the stronger base, HCO_2^- or $CH_3CO_2^-$? Explain your answer.

    (e) Will the equilibrium constant for the reaction

$$CH_3CO_2H(aq) + HCO_2^-(aq)$$
$$\rightleftharpoons CH_3CO_2^-(aq) + HCO_2H$$

    be greater than 1 or less than 1? Explain.

    (f) Which of the following acids is the strongest? Explain.

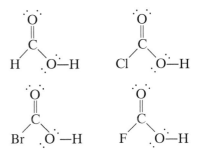

153. K_a for HNO_2 is 5.1×10^{-4}.

(a) Write an equation that describes the reaction of HNO_2 with H_2O.

(b) What is the pH of a 1.0 M solution of HNO_2?

(c) What is the pH of a 1.0 M solution of $NaNO_2$?

(d) What is the pH of a 1.0 M solution of HNO_3?

(e) Which of the following diagrams best describes a 1.0 M solution of HNO_3? Explain.

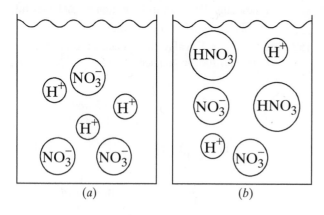

(a)   (b)

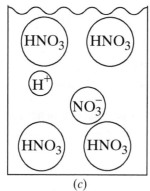

(c)

154. Arrange the following 0.10 M solutions in order of increasing pH. Show clearly which has the highest and which has the lowest pH. Explain.

CsCl

HOCl

HOClO

NaOCl

NaOClO

155. Estimate the pH of each of the following solutions. Explain in each case.

(a) 0.10 M HCl   (b) 0.10 M NaCl

(c) 0.10 M HOCl   (d) 0.10 M NaF

(e) 0.10 M NaOH

156. Calculate the pH of the following solutions.

(a) a 2.0 M solution of HOCl; $K_a = 2.9 \times 10^{-8}$

(b) a 1.0 M solution of HF; $K_a = 7.2 \times 10^{-4}$

(c) a 0.02 M solution of HCl

(d) a 1.0 M solution of NaF

157. Arrange the following 0.10 M solutions in order of increasing acidity. Explain.

KI

$KClO_3$

$HOClO_2$ ($HClO_3$)

KOH

NH_4ClO_4

$HOClO_3$ ($HClO_4$)

158. 50.0 mL of 0.100 M HCl is titrated with 0.100 M NaOH. What is the pH at the start of the titration? What will be the pH after the addition of 20.0, 50.0, and 60.0 mL of NaOH? What is the pH at the equivalence point?

159. 50.0 mL of 0.100 M HOAc is titrated with 0.100 M NaOH. Calculate the pH at the beginning of the titration and after the addition of 20.0, 50.0, and 60.0 mL of NaOH.

160. Roughly sketch the titration curve for the titration of 50.0 mL of 0.250 M NH_3 with 0.500 M HCl. What will be the initial pH? What volume of HCl is required to reach the equivalence point? What will be the pH at the equivalence point?

161. Which of the following ions is the conjugate base of a strong acid?

(a) OH^- (b) HSO_4^- (c) NH_2^-

(d) S^{2-} (e) H_3O^+

162. Many insects leave small quantities of formic acid, HCO_2H, behind when they bite. Explain why the itching sensation can be relieved by treating the bite with an aqueous solution of ammonia (NH_3) or baking soda ($NaHCO_3$).

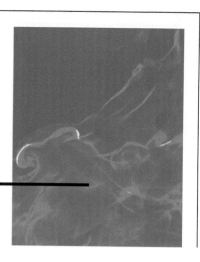

Chapter Eleven

SPECIAL TOPICS

11A.1 Diprotic Acids

The problems discussed so far that have involved acids have focused on **monoprotic acids** that each have a single H^+ ion, or proton, that can be donated when they act as Brønsted acids. Hydrochloric acid (HCl), acetic acid (CH_3CO_2H), and nitric acid (HNO_3) are all monoprotic acids.

There are also important acids that can be classified as **polyprotic acids** because they can lose more than one H^+ ion when they act as Brønsted acids. **Diprotic acids,** such as sulfuric acid (H_2SO_4), carbonic acid (H_2CO_3), hydrogen sulfide (H_2S), chromic acid (H_2CrO_4), and oxalic acid ($H_2C_2O_4$), have two acidic hydrogen atoms. **Triprotic acids,** such as phosphoric acid (H_3PO_4) and citric acid ($C_6H_8O_7$), have three.

Naturally occurring diprotic and triprotic acids such as citric acid, fumaric acid, maleic acid, oxalic acid, and succinic acid play an important role in biological systems. Furthermore, all of the amino acids used to form proteins are polyprotic acids. Glycine, for example, is a diprotic acid, whereas other amino acids such as aspartic acid are triprotic acids. Other diprotic acids such as adipic acid and terephthalic acid are used as starting materials from which synthetic fibers such as nylon and polyester are made. Each year more than 4 billion pounds of terephthalic acid and more than 5 billion pounds of adipic acid are used in the production of synthetic fibers.

Table 11A.1 gives values of K_a for some common polyprotic acids. There is usually a large difference in the ease with which polyprotic acids lose the first and second (or second and third) protons. Because sulfuric acid is classified as a strong acid, it is often assumed that it loses both of its protons when it reacts with water. That isn't a legitimate assumption. Sulfuric acid is a strong acid because K_a for the loss of the first proton is much larger than 1. We therefore assume that essentially all the H_2SO_4 molecules in an aqueous solution lose the first proton to form the HSO_4^- (hydrogen sulfate) ion.

$$H_2SO_4(aq) + H_2O(l) \longrightarrow H_3O^+(aq) + HSO_4^-(aq) \qquad K_{a1} = 1 \times 10^3$$

Table 11A.1
Acid Dissociation Equilibrium Constants for Common Polyprotic Acids

Acid	K_{a1}	K_{a2}	K_{a3}
Sulfuric acid (H_2SO_4)	1.0×10^3	1.2×10^{-2}	
Chromic acid (H_2CrO_4)	9.6	3.2×10^{-7}	
Oxalic acid (HO_2CCO_2H)	5.4×10^{-2}	5.4×10^{-5}	
Sulfurous acid (H_2SO_3)	1.7×10^{-2}	6.4×10^{-8}	
Maleic acid (*cis*- $HO_2CCH{=}CHCO_2H$)	1.2×10^{-2}	5.4×10^{-7}	
Phosphoric acid (H_3PO_4)	7.1×10^{-3}	6.3×10^{-8}	4.2×10^{-13}
Glycine ($H_2NCH_2CO_2H$)	4.6×10^{-3}	2.5×10^{-10}	
Fumaric acid (*trans*-$HO_2CCH{=}CHCO_2H$)	9.3×10^{-4}	3.6×10^{-5}	
Citric acid ($C_6H_8O_7$)	7.5×10^{-4}	1.7×10^{-5}	4.0×10^{-7}
Terephthalic acid ($HO_2CC_6H_4CO_2H$)	2.9×10^{-4}	3.5×10^{-5}	
Adipic acid ($HO_2C(CH_2)_4CO_2H$)	3.7×10^{-5}	3.9×10^{-6}	
Carbonic acid (H_2CO_3)	4.5×10^{-7}	4.7×10^{-11}	
Hydrogen sulfide (H_2S)	1.0×10^{-7}	1.3×10^{-13}	

But K_a for the loss of the second proton is only 1.2×10^{-2}, which means that only 10% of the H_2SO_4 molecules in a 1 M solution lose a second proton.

$$HSO_4^-(aq) + H_2O(l) \rightleftharpoons H_3O^+(aq) + SO_4^{2-}(aq) \qquad K_{a2} = 1.2 \times 10^{-2}$$

H_2SO_4 loses both H^+ ions only when it reacts with a base stronger than water, such as ammonia.

Table 11A.1 gives values of K_a for some common polyprotic acids. The large difference between the values of K_a for the sequential loss of protons by a polyprotic acid is important because it means we can assume that the acids dissociate one step at a time, this assumption is known as **stepwise dissociation.**

Let's look at the consequence of the assumption that polyprotic acids lose protons one step at a time by examining the chemistry of a saturated solution (<0.10 M) of H_2S in water. Hydrogen sulfide is the foul-smelling gas that gives rotten eggs their unpleasant odor. It is an excellent source of the S^{2-} ion, however, and is therefore commonly used in introductory chemistry laboratories. H_2S is a weak acid that dissociates in steps. Some of the H_2S molecules lose a proton in the first step to form the HS^- (or hydrogen sulfide) ion.

First step: $\qquad H_2S(aq) + H_2O(l) \rightleftharpoons H_3O^+(aq) + HS^-(aq)$

A small fraction of the HS^- ions formed in the reaction then go on to lose additional H^+ ions in a second step.

Second step: $\qquad HS^-(aq) + H_2O(l) \rightleftharpoons H_3O^+(aq) + S^{2-}(aq)$

Since there are two steps in the reaction, we can write two equilibrium constant expressions.

$$K_{a1} = \frac{[H_3O^+][HS^-]}{[H_2S]} = 1.0 \times 10^{-7}$$

$$K_{a2} = \frac{[H_3O^+][S^{2-}]}{[HS^-]} = 1.3 \times 10^{-13}$$

Although each of these equations contains three terms, there are only four unknowns—$[H_3O^+]$, $[H_2S]$, $[HS^-]$, and $[S^{2-}]$—because the $[H_3O^+]$ and $[HS^-]$ terms appear in both equations. The $[H_3O^+]$ term represents the total H_3O^+ ion concentration from both steps and therefore must have the same value in both equations. Similarly, the $[HS^-]$ term, which represents the balance between the HS^- ions formed in the first step and the HS^- ions consumed in the second step, must have the same value for both equations.

It takes four equations to solve for four unknowns. We already have two equations: the K_{a1} and K_{a2} expressions. We are going to have to find either two more equations or a pair of assumptions that can generate two equations. We can base one assumption on the fact that the value of K_{a1} for H_2S is almost a million times larger than the value of K_{a2}.

$$K_{a1} \gg K_{a2}$$

This means that only a small fraction of the HS^- ions formed in the first step go on to dissociate in the second step. If this is true, most of the H_3O^+ ions at equilibrium come from the dissociation of H_2S, and most of the HS^- ions formed in this reaction remain in solution. As a result, we can assume that the H_3O^+ and HS^- ion concentrations at equilibrium are more or less equal.

First assumption: $[H_3O^+] \approx [HS^-]$

We need one more equation, and therefore one more assumption. Note that H_2S is a weak acid ($K_{a1} = 1.0 \times 10^{-7}$; $K_{a2} = 1.3 \times 10^{-13}$). Thus we can assume that most of the H_2S that dissolves in water will still be present when the solution reaches equilibrium. In other words, we can assume that the equilibrium concentration of H_2S is approximately equal to the initial concentration. Because a saturated solution of H_2S in water has an initial concentration of about 0.10 M, we can summarize the second assumption as follows.

Second assumption: $[H_2S] \approx 0.10\ M$

We now have four equations in four unknowns.

$$K_{a1} = \frac{[H_3O^+][HS^-]}{[H_2S]} = 1.0 \times 10^{-7}$$

$$K_{a2} = \frac{[H_3O^+][S^{2-}]}{[HS^-]} = 1.3 \times 10^{-13}$$

$$[H_3O^+] \approx [HS^-]$$

$$[H_2S] \approx 0.10\ M$$

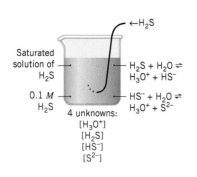

Saturated solution of H_2S

0.1 M H_2S

$\leftarrow H_2S$

$H_2S + H_2O \rightleftharpoons H_3O^+ + HS^-$

$HS^- + H_2O \rightleftharpoons H_3O^+ + S^{2-}$

4 unknowns:
$[H_3O^+]$
$[H_2S]$
$[HS^-]$
$[S^{2-}]$

Because there is always a unique solution to four equations in four unknowns, we are now ready to calculate the H_3O^+, H_2S, HS^-, and S^{2-} concentrations at equilibrium in a saturated solution of H_2S in water. The relevant information for this calculation is illustrated in the drawing in the margin.

K_{a1} is so much larger than K_{a2} for H_2S that we can assume stepwise dissociation. We start by assuming that we can work with the equilibrium expression for the first step without worrying about the second step for the moment. We therefore start with the expression for K_{a1}.

$$K_{a1} = \frac{[H_3O^+][HS^-]}{[H_2S]} = 1.0 \times 10^{-7}$$

We then invoke one of our assumptions.

$$[H_2S] \approx 0.10\ M$$

Substituting this approximation into the K_{a1} expression gives the following equation.

$$\frac{[H_3O^+][HS^-]}{[0.10]} \approx 1.0 \times 10^{-7}$$

We then invoke the other assumption.

$$[H_3O^+] \approx [HS^-] \approx \Delta C$$

Substituting this approximation into the K_{a1} expression gives the following result.

$$\frac{[\Delta C][\Delta C]}{[0.10]} \approx 1.0 \times 10^{-7}$$

We now solve the approximate equation for ΔC.

$$\Delta C \approx 1.0 \times 10^{-4}$$

If our assumptions are valid, we are three-fourths of the way to our goal. If our assumptions are legitimate, we now know the H_2S, H_3O^+, and HS^- concentrations.

$$[H_2S] \approx 0.10 \, M$$
$$[H_3S^+] \approx [HS^-] \approx 1.0 \times 10^{-4} \, M$$

Having extracted the values of three unknowns from the first equilibrium expression, we turn to the second equilibrium expression.

$$K_{a2} = \frac{[H_3O^+][S^{2-}]}{[HS^-]} = 1.3 \times 10^{-13}$$

Substituting the known values of the H_3O^+ and HS^- ion concentrations into this expression gives the following equation.

$$\frac{[1.0 \times 10^{-4}][S^{2-}]}{[1.0 \times 10^{-4}]} = 1.3 \times 10^{-13}$$

Because the equilibrium concentrations of the H_3O^+ and HS^- ions are about the same, the S^{2-} ion concentration at equilibrium is approximately equal to the value of K_{a2} for this acid.

$$[S^{2-}] \approx 1.3 \times 10^{-13} \, M$$

It is now time to check our assumptions. Is the dissociation of H_2S small compared with the initial concentration? Yes. The HS^- and H_3O^+ ion concentrations obtained from the calculation are $1.0 \times 10^{-4} \, M$, which is only 0.1% of the initial concentration of H_2S. The following assumption is therefore valid.

$$[H_2S] \approx 0.10 \, M$$

Is the difference between the S^{2-} and HS^- ion concentrations large enough to allow us to assume that essentially all of the H_3O^+ ions at equilibrium are formed in the first step and that essentially all of the HS^- ions formed in this step remain in solution? Yes. The S^{2-} ion concentration obtained from these calculations is 10^9 times smaller than the HS^- ion concentration. Thus our other assumption is also valid.

$$[H_3O^+] \approx [HS^-]$$

We can therefore summarize the concentrations of the various components of the equilibrium as follows.

$$[H_2S] \approx 0.10 \, M$$
$$[H_3O^+] \approx [HS^-] \approx 1.0 \times 10^{-4} \, M$$
$$[S^{2-}] \approx 1.3 \times 10^{-13} \, M$$

11A.2 Diprotic Bases

The techniques we have used with diprotic acids can be extended to diprotic bases. The only challenge is calculating the values of K_b for the base. Suppose we are given the following problem:

Calculate the H_2CO_3, HCO_3^-, CO_3^{2-}, and OH^- concentrations at equilibrium in a solution that is initially 0.10 M in Na_2CO_3. (For H_2CO_3, $K_{a1} = 4.5 \times 10^{-7}$ and $K_{a2} = 4.7 \times 10^{-11}$.) The relevant information for this calculation are illustrated in the drawing in the margin.

In the margin:
$0.10\ M$ CO_3^{2-} ; $CO_3^{2-} + H_2O \rightleftharpoons HCO_3^- + OH^-$; $HCO_3^- + H_2O \rightleftharpoons H_2CO_3 + OH^-$; Na_2CO_3

Sodium carbonate dissociates into its ions when it dissolves in water.

$$Na_2CO_3(s) \xrightarrow{H_2O} 2\,Na^+(aq) + CO_3^{2-}(aq)$$

The carbonate ion then acts as a base toward water, picking up a pair of protons—one at a time—to form the bicarbonate ion, HCO_3^-, and then eventually carbonic acid, H_2CO_3.

$$CO_3^{2-}(aq) + H_2O(l) \rightleftharpoons HCO_3^-(aq) + OH^-(aq) \qquad K_{b1} = ?$$
$$HCO_3^-(aq) + H_2O(l) \rightleftharpoons H_2CO_3(aq) + OH^-(aq) \qquad K_{b2} = ?$$

The first step in calculating the concentrations of the various components of this solution at equilibrium involves determining the values of K_{b1} and K_{b2} for the carbonate ion. We start by writing the K_b expressions for the carbonate ion and comparing them to the K_a expressions for carbonic acid.

$$K_{b1} = \frac{[HCO_3^-][OH^-]}{[CO_3^{2-}]} \qquad K_{a2} = \frac{[H_3O^+][CO_3^{2-}]}{[HCO_3^-]}$$
$$K_{b2} = \frac{[H_2CO_3][OH^-]}{[HCO_3^-]} \qquad K_{a1} = \frac{[H_3O^+][HCO_3^-]}{[H_2CO_3]}$$

The expressions for K_{b1} and K_{a2} have something in common: They both depend on the concentrations of the HCO_3^- and CO_3^{2-} ions. The expressions for K_{b2} and K_{a1} also have something in common: They both depend on the HCO_3^- and H_2CO_3 concentrations. We can therefore calculate K_{b1} from K_{a2} and K_{b2} from K_{a1}.

We start by multiplying the top and bottom of the K_{a1} expression by the OH^- ion concentration to introduce the $[OH^-]$ term.

$$K_{a1} = \frac{[H_3O^+][HCO_3^-]}{[H_2CO_3]} \times \frac{[OH^-]}{[OH^-]}$$

We then group the terms in the equation as follows.

$$K_{a1} = \frac{[HCO_3^-]}{[H_2CO_3][OH^-]} \times [H_3O^+][OH^-]$$

The first term in the equation is the inverse of the K_{b2} expression, and the second term is the K_w expression.

$$K_{a1} = \frac{1}{K_{b2}} \times K_w$$

Rearranging this equation gives the following result.

$$K_{a1}K_{b2} = K_w$$

Similarly, we can multiply the top and bottom of the K_{a2} expression by the OH^- ion concentration.

$$K_{a2} = \frac{[H_3O^+][CO_3^{2-}]}{[HCO_3^-]} \times \frac{[OH^-]}{[OH^-]}$$

Collecting terms gives the following equation.

$$K_{a2} = \frac{[CO_3^{2-}]}{[HCO_3^-][OH^-]} \times [H_3O^+][OH^-]$$

The first term in the equation is the inverse of K_{b1}, and the second term is K_w.

$$K_{a2} = \frac{1}{K_{b1}} \times K_w$$

This equation can therefore be rearranged as follows.

$$K_{a2}K_{b1} = K_w$$

We can now calculate the values of K_{b1} and K_{b2} for the carbonate ion from the corresponding values of K_{a1} and K_{a2} for carbonic acid.

$$K_{b1} = \frac{K_w}{K_{a2}} = \frac{1.0 \times 10^{-14}}{4.7 \times 10^{-11}} = 2.1 \times 10^{-4}$$

$$K_{b2} = \frac{K_w}{K_{a1}} = \frac{1.0 \times 10^{-14}}{4.5 \times 10^{-7}} = 2.2 \times 10^{-8}$$

We are now ready to do the calculations needed to determine the H_2CO_3, HCO_3^-, CO_3^{2-}, and OH^- concentrations at equilibrium. We start with the K_{b1} expression because the CO_3^{2-} ion is the strongest base in the solution and is therefore the best source of the OH^- ion.

$$K_{b1} = \frac{[HCO_3^-][OH^-]}{[CO_3^{2-}]}$$

The difference between K_{b1} and K_{b2} for the carbonate ion is large enough to suggest that most of the OH^- ions at equilibrium come from the first step and most of the HCO_3^- ions formed in the first step remain in solution. We can therefore make the following assumption.

$$[OH^-] \approx [HCO_3^-] \approx \Delta C$$

The value of K_{b1} is small enough to assume that ΔC should be small compared with the initial concentration of the carbonate ion. If this is true, the concentration of the CO_3^{2-} ion at equilibrium will be roughly equal to the initial concentration of Na_2CO_3, which was described in the statement of the problem as 0.10 M.

$$[CO_3^{2-}] \approx 0.10\ M$$

Substituting this information into the K_{b1} expression gives the following result.

$$\frac{[\Delta C][\Delta C]}{[0.10]} \approx 2.1 \times 10^{-4}$$

This approximate equation can now be solved for ΔC.

$$\Delta C \approx 0.0046 \ M$$

We then use the value of ΔC to calculate the equilibrium concentrations of the OH^-, HCO_3^-, and CO_3^{2-} ions.

$$[CO_3^{2-}] \approx 0.10 \ M$$
$$[OH^-] \approx [HCO_3^-] \approx 0.0046 \ M$$

We now turn to the K_{b2} expression.

$$K_{b2} = \frac{[H_2CO_3][OH^-]}{[HCO_3^-]}$$

Substituting what we know about the OH^- and HCO_3^- ion concentrations into this equation gives the following result.

$$K_{b2} = \frac{[H_2CO_3][0.0046]}{[0.0046]} \approx 2.2 \times 10^{-8}$$

According to this equation, the H_2CO_3 concentration at equilibrium is approximately equal to K_{b2} for the carbonate ion.

$$[H_2CO_3] \approx 2.2 \times 10^{-8} \ M$$

Summarizing the results of our calculations allows us to test the assumptions made generating the following results.

$$[CO_3^{2-}] \approx 0.10 \ M$$
$$[OH^-] \approx [HCO_3^-] \approx 0.0046 \ M$$
$$[H_2CO_3] \approx 2.2 \times 10^{-8} \ M$$

All of our assumptions are valid. The extent of the reaction between the CO_3^{2-} ion and water to give the HCO_3^- ion is less than 5% of the initial concentration of Na_2CO_3. Furthermore, most of the OH^- ion comes from the first step, and most of the HCO_3^- ion formed in the first step remains in solution.

11A.3 Compounds That Could Be Either Acids or Bases

Sometimes the hardest part of a calculation is deciding whether the compound is an acid or a base. Consider sodium bicarbonate, for example, which dissolves in water to give the bicarbonate ion. Relevant information for this calculation is illustrated in the drawing in the margin.

$$NaHCO_3(s) \xrightarrow{H_2O} Na^+(aq) + HCO_3^-(aq)$$

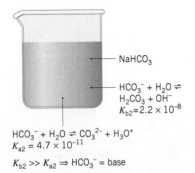

NaHCO$_3$

$HCO_3^- + H_2O \rightleftharpoons$
$H_2CO_3 + OH^-$
$K_{b2} = 2.2 \times 10^{-8}$

$HCO_3^- + H_2O \rightleftharpoons CO_3^{2-} + H_3O^+$
$K_{a2} = 4.7 \times 10^{-11}$

$K_{b2} \gg K_{a2} \Rightarrow HCO_3^- = base$

In theory, the bicarbonate ion can act as either a Brønsted acid or a Brønsted base toward water.

$$HCO_3^-(aq) + H_2O(l) \rightleftharpoons H_3O^+(aq) + CO_3^{2-}(aq)$$
$$HCO_3^-(aq) + H_2O(l) \rightleftharpoons H_2CO_3(aq) + OH^-(aq)$$

Which reaction predominates? Is the HCO_3^- ion more likely to act as an acid or as a base when it dissolves in water?

We can answer those questions by comparing the equilibrium constants for the reactions. The equilibrium in which the HCO_3^- ion acts as a Brønsted acid is described by K_{a2} for carbonic acid.

$$K_{a2} = \frac{[H_3O^+][CO_3^{2-}]}{[HCO_3^-]} = 4.7 \times 10^{-11}$$

The equilibrium in which the HCO_3^- acts as a Brønsted base is described by K_{b2} for the carbonate ion.

$$K_{b2} = \frac{[H_2CO_3][OH^-]}{[HCO_3^-]} = 2.2 \times 10^{-8}$$

Since K_{b2} is significantly larger than K_{a2}, we conclude that the HCO_3^- ion is a stronger base than it is an acid. An aqueous solution of $NaHCO_3$ should therefore be basic.

Exercise 11A.1

Phosphoric acid, H_3PO_4, is obviously an acid. The phosphate ion, PO_4^{3-}, is a base. Predict whether solutions of the $H_2PO_4^-$ and HPO_4^{2-} ions are more likely to be acidic or basic.

Solution

Let's start by looking at the stepwise dissociation of phosphoric acid.

$$H_3PO_4(aq) + H_2O(l) \rightleftharpoons H_3O^+(aq) + H_2PO_4^-(aq) \qquad K_{a1} = 7.1 \times 10^{-3}$$
$$H_2PO_4^-(aq) + H_2O(l) \rightleftharpoons H_3O^+(aq) + HPO_4^{2-}(aq) \qquad K_{a2} = 6.3 \times 10^{-8}$$
$$HPO_4^{2-}(aq) + H_2O(l) \rightleftharpoons H_3O^+(aq) + PO_4^{3-}(aq) \qquad K_{a3} = 4.2 \times 10^{-13}$$

We can then look at the steps by which the PO_4^{3-} ion picks up protons from water to form phosphoric acid.

$$PO_4^{3-}(aq) + H_2O(l) \rightleftharpoons HPO_4^{2-}(aq) + OH^-(aq) \qquad K_{b1} = ?$$
$$HPO_4^{2-}(aq) + H_2O(l) \rightleftharpoons H_2PO_4^-(aq) + OH^-(aq) \qquad K_{b2} = ?$$
$$H_2PO_4^-(aq) + H_2O(l) \rightleftharpoons H_3PO_4(aq) + OH^-(aq) \qquad K_{b3} = ?$$

Applying the procedure used for sodium carbonate in Section 11A.2 gives the following relationships among the values of the six equilibrium constants.

$$K_{a1}K_{b3} = K_w$$
$$K_{a2}K_{b2} = K_w$$
$$K_{a3}K_{b1} = K_w$$

We can predict whether the $H_2PO_4^-$ ion should be an acid or a base by looking at the values of the K_a and K_b constants that characterize its reactions with water.

$$H_2PO_4^-(aq) + H_2O(l) \rightleftharpoons H_3O^+(aq) + HPO_4^{2-}(aq) \qquad K_{a2} = 6.3 \times 10^{-8}$$
$$H_2PO_4^-(aq) + H_2O(l) \rightleftharpoons H_3PO_4(aq) + OH^-(aq) \qquad K_{b3} = 1.4 \times 10^{-12}$$

Because K_{a2} is significantly larger than K_{b3}, solutions of the $H_2PO_4^-$ ion in water should be acidic.

We can use the same method to decide whether HPO_4^{2-} is an acid or a base when dissolved in water. We start by looking at the values of K_a and K_b that characterize its reactions with water.

$$HPO_4^{2-}(aq) + H_2O(l) \rightleftharpoons H_3O^+(aq) + PO_4^{3-}(aq) \qquad K_{a3} = 4.2 \times 10^{-13}$$
$$HPO_4^{2-}(aq) + H_2O(l) \rightleftharpoons H_2PO_4^-(aq) + OH^-(aq) \qquad K_{b2} = 1.6 \times 10^{-7}$$

Because K_{b2} is significantly larger than K_{a3}, solutions of the HPO_4^{2-} ion in water should be basic.

Measurements of the pH of 0.10 M solutions of the species yield the following results, which are consistent with the predictions of this exercise.

H_3PO_4	$H_2PO_4^-$	HPO_4^{2-}	PO_4^{3-}
pH = 1.5	pH = 4.4	pH = 9.3	pH = 12.0

In the 1880s, a variety of syrups started to appear that could be added to carbonated water at the local drugstore. They included a root beer abstract marketed by Charles Hires, James Vernor's Ginger Ale, R. S. Lazenby's formula for Dr. Pepper, John Pemberton's formula for Coca-Cola, and Brad's Drink which was later renamed Pepsi-Cola. These beverages were designed to quench thirst and provide energy during hot summer days. The energizing feeling came from the fact that Coca-Cola originally contained cocaine extracted from the coca plant, caffeine extracted from the kola bean, and a considerable amount of sugar. It has also been noted that the gentle release of CO_2 gas that occurs when the liquid is swallowed has a physiologically soothing effect.

The per capita consumption of bottled carbonated beverages in the United States is almost 50 gallons per year. In order to create an acidic medium that enhances the absorption of CO_2 when these soft drinks are bottled, both phosphoric acid ($K_{a1} = 7.1 \times 10^{-3}$) and citric acid ($K_{a1} = 7.5 \times 10^{-4}$) are added to soft drinks to produce a beverage with a pH of 2.8. The high level of phosphate in soft drinks is associated with the loss of calcium from the blood, which eventually has to be replenished by calcium from bone. The National Academy of Sciences recently increased the daily recommended intake of calcium from 800 mg to 1200 mg to compensate for the effect of soft drinks on the level of calcium in the blood.

Conventional wisdom once assumed that the sugar in soft drinks could lead to tooth decay. Today we recognize that the real danger is the acid in soft drinks, which can leach enamel from the teeth, leaving a softened matrix for bacteria to enter the teeth.

Problems

Diprotic Acids

A-1. Calculate the H_3O^+, CO_3^{2-}, HCO_3^-, and H_2CO_3 concentrations at equilibrium in a solution that initially contained 0.100 mol of carbonic acid per liter. (For H_2CO_3, $K_{a1} = 4.5 \times 10^{-7}$ and $K_{a2} = 4.7 \times 10^{-11}$.)

A-2. Calculate the equilibrium concentrations of the important components of a 0.250 M malonic acid ($HO_2CCH_2CO_2H$) solution. Use the symbol H_2M as an abbreviation for malonic acid and assume stepwise dissociation of the acid. (For H_2M, $K_{a1} = 1.4 \times 10^{-5}$ and $K_{a2} = 2.1 \times 10^{-8}$.)

A-3. Check the validity of the assumption in the previous problem that malonic acid dissociates in a stepwise fashion by comparing the concentrations of the HM^- and M^{2-} ions obtained in the problem. Is this assumption valid?

A-4. Glycine, the simplest of the amino acids found in proteins, is a diprotic acid with the formula $HO_2CCH_2NH_3^+$. If we symbolize glycine as H_2G^+, we can write the following equations for its stepwise dissociation.

$$H_2G^+(aq) + H_2O(l) \rightleftharpoons HG(aq) + H_3O^+(aq)$$
$$K_{a1} = 4.6 \times 10^{-3}$$
$$HG(aq) + H_2O(l) \rightleftharpoons G^-(aq) + H_3O^+(aq)$$
$$K_{a2} = 2.5 \times 10^{-10}$$

Calculate the concentrations of H_3O^+, G^-, HG, and H_2G^+ in a 2.00 M solution of glycine in water.

A-5. Which of the following equations accurately describes a 2.0 M solution of glycine? See A-4.

(a) $[H_3O^+] \approx [H_2G^+]$ (b) $[H_3O^+] < [H_2G^+]$
(c) $[H_3O^+] > [HG]$ (d) $[H_3O^+] \approx [HG]$
(e) $[H_3O^+] < [HG]$

A-6. Which of the following equations results from the fact that glycine is a weak diprotic acid? See A-4.

(a) $[G^-] \approx [H_2G^+]$ (b) $[G^-] \approx K_{a2}$
(c) $[G^-] \approx [HG]$ (d) $[G^-] > [HG]$
(e) $[G^-] \approx [H_3O^+]$

A-7. Oxalic acid ($H_2C_2O_4$) has been implicated in diseases such as gout and kidney stones. Calculate the H_3O^+, $H_2C_2O_4$, $HC_2O_4^-$, and $C_2O_4^{2-}$ concentrations in a 1.25 M solution of oxalic acid.

$$H_2C_2O_4(aq) + H_2O(l)$$
$$\rightleftharpoons H_3O^+(aq) + HC_2O_4^-(aq)$$
$$K_{a1} = 5.4 \times 10^{-2}$$
$$HC_2O_4^-(aq) + H_2O(l)$$
$$\rightleftharpoons H_3O^+(aq) + C_2O_4^{2-}(aq)$$
$$K_{a2} = 5.4 \times 10^{-5}$$

Diprotic Bases

A-8. Which of the following sets of equations can be used to calculate K_{b1} and K_{b2} for sodium oxalate ($Na_2C_2O_4$) from K_{a1} and K_{a2} for oxalic acid ($H_2C_2O_4$)?

(a) $K_{b1} = K_w \times K_{a1}$ and $K_{b2} = K_w \times K_{a2}$
(b) $K_{b1} = K_w \times K_{a2}$ and $K_{b2} = K_w \times K_{a1}$
(c) $K_{b1} = K_w/K_{a1}$ and $K_{b2} = K_w/K_{a2}$
(d) $K_{b1} = K_w/K_{a2}$ and $K_{b2} = K_w/K_{a1}$
(e) $K_{b1} = K_{a1}/K_w$ and $K_{b2} = K_{a2}/K_w$

A-9. Calculate the pH of a 0.028 M solution of sodium oxalate ($Na_2C_2O_4$) in water.

$$H_2C_2O_4(aq) + H_2O(l)$$
$$\rightleftharpoons H_3O^+(aq) + HC_2O_4^-(aq)$$
$$K_{a1} = 5.4 \times 10^{-2}$$
$$HC_2O_4^-(aq) + H_2O(l)$$
$$\rightleftharpoons H_3O^+(aq) + C_2O_4^{2-}(aq)$$
$$K_{a2} = 5.4 \times 10^{-5}$$

A-10. Calculate the H_3O^+, OH^-, H_2CO_3, HCO_3^-, and CO_3^{2-} concentrations in a 0.150 M solution of sodium carbonate (Na_2CO_3) in water. (For H_2CO_3, $K_{a1} = 4.5 \times 10^{-7}$ and $K_{a2} = 4.7 \times 10^{-11}$.)

Compounds That Could Be Either Acids or Bases

A-11. According to the data in Table 11.5, the pH of a 0.10 M $NaHCO_3$ solution is 8.4. Explain why the HCO_3^- ion forms solutions that are basic, not acidic.

A-12. Which of the following statements concerning sodium hydrogen sulfide (NaSH) is correct? (For H_2S, $K_{a1} = 1.0 \times 10^{-7}$ and $K_{a2} = 1.3 \times 10^{-13}$.)

(a) NaSH is an acid because K_{a1} for H_2S is much larger than K_{a2}.
(b) NaSH is an acid because K_{a1} for H_2S is smaller than K_{b1} for Na_2S.
(c) NaSH is a base because K_{b1} for Na_2S is much larger than K_{b2}.
(d) NaSH is a base because K_{b2} for Na_2S is larger than K_{a2} for H_2S.

A-13. Calculate the pH of 0.10 M solutions of H_3PO_4, NaH_2PO_4, Na_2HPO_4, and Na_3PO_4. (For H_3PO_4, $K_{a1} = 7.1 \times 10^{-3}$, $K_{a2} = 6.3 \times 10^{-8}$, and $K_{a3} = 4.2 \times 10^{-13}$.)

A-14. Predict whether an aqueous solution of sodium hydrogen sulfate ($NaHSO_4$) should be acidic, basic, or neutral. (For H_2SO_4, $K_{a1} = 1 \times 10^3$ and $K_{a2} = 1.2 \times 10^{-2}$.)

A-15. Predict whether an aqueous solution of sodium hydrogen sulfite ($NaHSO_3$) should be acidic, basic, or neutral. (For H_2SO_3, $K_{a1} = 1.7 \times 10^{-2}$ and $K_{a2} = 6.4 \times 10^{-8}$.)

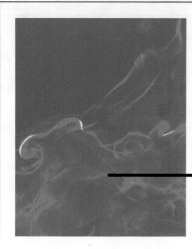

Chapter Twelve

OXIDATION—REDUCTION REACTIONS

12.1 Common Oxidation–Reduction Reactions

We find examples of oxidation–reduction or **redox reactions** almost every time we analyze the reactions used as sources of either heat or work. When natural gas (methane, CH_4) burns, for example, an oxidation–reduction reaction occurs that releases more than 800 kJ/mol_{rxn} of energy.

$$CH_4(g) + 2\,O_2(g) \rightleftharpoons CO_2(g) + 2\,H_2O(g)$$

Within our bodies, a sequence of oxidation–reduction reactions is used to burn sugars such as glucose ($C_6H_{12}O_6$) and the fatty acids such as stearic acid ($CH_3(CH_2)_{16}CO_2H$) in the foods we eat.

$$C_6H_{12}O_6(aq) + 6\,O_2(g) \rightleftharpoons 6\,CO_2(g) + 6\,H_2O(l)$$
$$CH_3(CH_2)_{16}CO_2H(aq) + 26\,O_2(g) \rightleftharpoons 18\,CO_2(g) + 18\,H_2O(l)$$

We don't have to restrict ourselves to reactions that can be used as a source of energy, however, to find examples of oxidation–reduction reactions. Silver metal is oxidized when it comes in contact with either trace quantities of H_2S or SO_2 in the atmosphere or foods, such as eggs, that are rich in sulfur compounds.

$$4\,Ag(s) + 2\,H_2S(g) + O_2(g) \rightleftharpoons 2\,Ag_2S(s) + 2\,H_2O(l)$$

Fortunately, the film of Ag_2S that collects on the metal surface forms a protective coating that slows down further oxidation of the silver metal.

The tarnishing of silver is just one example of a broad class of oxidation–reduction reactions that fall under the general heading of **corrosion.** Another example is the series of reactions that occur when iron or steel rusts. Iron does not rust at room temperature when exposed to either oxygen or water, by themselves. It is only in the presence of both oxygen and water that rust forms.

Rust is the generic term for a series of iron(III) oxides. The first step in this process involves the oxidation of iron metal to give a hydrated form of iron(II) oxide in which water is trapped in holes in the iron(II) oxide crystal lattice.

$$2\,Fe(s) + O_2(aq) + 2\,H_2O(l) \longrightarrow 2\,FeO \cdot H_2O(s)$$

Because this compound has the same empirical formula as $Fe(OH)_2$, it is often mistakenly called iron(II), or ferrous, hydroxide. The $FeO \cdot H_2O$ formed in this reaction is further oxidized by O_2 dissolved in water to give a hydrated form of iron(III), or ferric, oxide, known as rust.

$$4\,FeO \cdot H_2O(s) + O_2(aq) + 2\,H_2O(l) \longrightarrow 2\,Fe_2O_3 \cdot 3\,H_2O(s)$$

These reactions only occur in the presence of both water and oxygen; therefore, cars tend to rust where water collects. The simplest way to prevent iron from rusting is to coat the metal so that it doesn't come in contact with water. Cars, therefore, were originally painted for only one reason—to slow down the formation of rust.

Oxidation–reduction reactions or redox reactions can involve the transfer of electrons. The element or compound that gains electrons is said to undergo **reduction.** The element or compound that loses electrons undergoes **oxidation.** The following reaction, for example, takes place when sodium metal reacts with chlorine gas. Sodium loses electrons to form Na^+ ions, while chlorine gains electrons to form Cl^- ions.

$$2\,Na + Cl_2 \rightleftharpoons 2\,[Na^+][Cl^-]$$

Redox reactions are often analyzed by considering the processes of oxidation and reduction separately. When this is done, the oxidation or reduction reactions are written as **half-reactions.** It is important to remember, however, that oxidation cannot occur unless reduction also takes place. The electrons that are gained by the substance being reduced are lost by the substance being oxidized.

When oxidation–reduction reactions involve the transfer of electrons, the number of electrons gained and lost in the reaction must be the same because electrons cannot be created or destroyed during a reaction. In the reaction between sodium and chlorine, for example, the electrons lost by the sodium atoms are gained by chlorine molecules to form chloride ions. The reduction and oxidation half-reactions for the reaction between sodium metal and chlorine gas are as follows. Combining these two half-reactions gives the overall reaction.

Oxidation half-reaction: $2\,Na \rightleftharpoons 2\,Na^+ + 2e^-$

Reduction half-reaction: $Cl_2 + 2\,e^- \rightleftharpoons 2\,Cl^-$

Oxidation–reduction reactions do not have to involve the transfer of electrons. They can also occur by the transfer of atoms. Consider the following atom-transfer reactions, for example.

$$ClNO_2(g) + NO(g) \rightleftharpoons ClNO(g) + NO_2(g)$$
$$CO_2(g) + H_2(g) \rightleftharpoons CO(g) + H_2O(g)$$

Organic chemists often think about oxidation–reduction in terms of the gain or loss of oxygen atoms,

$$\underset{\displaystyle CH_3CH(l)}{\overset{\displaystyle O}{\|}} + H_2O_2(l) \rightleftharpoons \underset{\displaystyle CH_3COH(l)}{\overset{\displaystyle O}{\|}} + H_2O(l)$$

or the gain or loss of hydrogen atoms.

$$C_2H_4(g) + H_2(g) \rightleftharpoons C_2H_6(g)$$

➤ **CHECKPOINT**

Is it possible to have oxidation without reduction?

The concept of **oxidation numbers** or **oxidation states** gives us a way to identify redox reactions. Any reaction that results in a change in oxidation number of one or more atoms is classified as a redox reaction, regardless of whether it involves the transfer of electrons or the transfer of atoms.

12.2 Determining Oxidation Numbers

Oxidation numbers are assigned to the atoms in a compound by considering the compound as if it were ionic, assigning the shared electrons in each bond to the more electronegative element. Chemists often refer to an atom that has been assigned an oxidation number as having a particular *oxidation state.*

Two methods for assigning oxidation numbers of atoms were introduced in Chapter 5. In Section 5.17, oxidation numbers were calculated by first assigning all of the electrons in covalent bonds between nonequivalent atoms in the Lewis structure to the more electronegative atom. The oxidation number for each atom in the structure was then calculated using the following equation.

$$OX_a = V_a - N_a$$

In this equation, OX_a is the oxidation number of atom a, V_a is the number of valence electrons on a neutral atom of that element, and N_a is the number of electrons on the atom after the bonding electrons in the Lewis structure have been assigned to the more electronegative atom.

This method of assigning oxidation numbers is particularly useful for organic compounds in which carbon atoms have several different oxidation numbers. Consider 1-propanol ($CH_3CH_2CH_2OH$), for example, which contains three carbon atoms we will label a, b, and c.

$$
\begin{array}{ccc}
\text{H} & \text{H} & \text{H} \\
| & | & | \\
\text{H}-\text{C}_a-\text{C}_b-\text{C}_c-\text{O}-\text{H} \\
| & | & | \\
\text{H} & \text{H} & \text{H}
\end{array}
$$

There are two kinds of covalent bonds in this compound. First, there are covalent C—C bonds between equivalent atoms that form the skeleton structure of the molecule. Second, there are C—H and C—O bonds between atoms that are not equivalent. We can assume that the electrons in the covalent bonds between equivalent atoms are split evenly between the atoms that form the bond. If we then formally assign the electrons in the C—H and C—O bond to the more electronegative atom, we obtain the following result.

$$
\begin{array}{ccccc}
\text{H} & \text{H} & \text{H} \\
\ddot{} & \ddot{} & \ddot{} \\
\text{H} \ :\!\ddot{\text{C}}_a\!\cdot & \cdot\ddot{\text{C}}_b\!\cdot & \cdot\ddot{\text{C}}_c & :\!\ddot{\text{O}}\!: & \text{H} \\
\text{H} & \text{H} & \text{H}
\end{array}
$$

The oxidation state of each of the carbon atoms can now be calculated as follows.

$$OX_{C_a} = 4 - 7 = -3$$
$$OX_{C_b} = 4 - 6 = -2$$
$$OX_{C_c} = 4 - 5 = -1$$

The oxidation number of the carbon atoms becomes more negative as the number of hydrogen atoms on the carbon increases. Thus, C_a, which has three hydrogen atoms, has an oxidation number of -3, while C_b with two hydrogen atoms has an oxidation number of -2. Furthermore, an increase in the number of oxygen atoms to which a carbon atom is bound leads to a more positive oxidation number. Thus, C_b with two hydrogen atoms and no oxygen atoms has an oxidation state of -2, whereas C_c with one oxygen atom and two hydrogen atoms has an oxidation state of -1.

Another method for assigning oxidation numbers is based on a series of rules introduced in Section 5.16. Consider how these rules can be applied to determining the oxidation number of the nitrogen atom in the NO_3^- or nitrate anion, for example. We start by noting that the sum of the oxidation numbers of

➤ **CHECKPOINT**

What are the oxidation numbers for all the atoms in the following compounds?

$$FeCl_3, CH_4, O_2, MnO_4^-, HCNH_2$$
$$\overset{\|}{O}$$

the individual atoms must be equal to the overall charge on the molecule or ion. The sum of the oxidation numbers of the four atoms in the NO_3^- ion therefore must be equal to -1. According to the rules in Section 5.16, the oxidation number of the oxygen atoms is -2. The oxidation number of the nitrogen atom therefore has to be $+5$.

$$OX_N + 3(OX_O) = -1$$
$$OX_N + 3(-2) = -1$$
$$OX_N = +5$$

12.3 Recognizing Oxidation–Reduction Reactions

The most general model of oxidation–reduction reactions is based on the following definitions:

> **Oxidation occurs when the oxidation number of an atom becomes more positive. Changes in oxidation number from -3 to -1 or from $+2$ to $+4$ would be labeled as oxidation.**

> **Reduction occurs when the oxidation number of an atom becomes more negative. Changes in oxidation number from -1 to -3 or from $+4$ to $+2$ would be labeled as reduction.**

Oxidation–reduction reactions can therefore be recognized by looking for a change in the oxidation number of one or more atoms.

We can use this model of oxidation–reduction reactions to describe reactions such as the following in which electrons are formally transferred from one atom to another to form an ionic compound. The oxidation number for each atom is given directly beneath the symbol for that atom.

$$2\,Na + Cl_2 \rightleftharpoons 2\,[Na^+][Cl^-]$$

0   0   +1   −1

The oxidation number of sodium increases from zero to $+1$ in this reaction, indicating that sodium is oxidized. The oxidation number of chlorine decreases from zero to -1, indicating that it is reduced.

This model for redox reactions is particularly useful for describing reactions such as the following, where the oxidation numbers are given beneath the symbol for each atom.

$$CH_4(g) + 2\,O_2(g) \rightleftharpoons CO_2(g) + 2\,H_2O(g)$$

−4 +1   0   +4 −2   +1 −2

In this reaction the oxidation number of carbon increases from -4 to $+4$, indicating that carbon is oxidized. The oxygen is reduced because its oxidation number decreases from zero to -2.

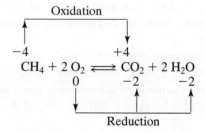

Exercise 12.1

Which of the following are oxidation–reduction reactions? For each oxidation–reduction reaction, indicate which substance is oxidized and which substance is reduced.

(a) $Cu(s) + 2\,Ag^+(aq) \rightleftharpoons Cu^{2+}(aq) + 2\,Ag(s)$

(b) $CH_3CO_2H(aq) + OH^-(aq) \rightleftharpoons CH_3CO_2^-(aq) + H_2O(l)$

(c) $SF_4(g) + F_2(g) \rightleftharpoons SF_6(g)$

(d) $CuSO_4(aq) + BaCl_2(aq) \rightleftharpoons BaSO_4(s) + CuCl_2(aq)$

Solution

(a) This is an oxidation–reduction reaction. The copper is oxidized from copper 0 to copper +2. The silver is reduced from silver +1 to silver 0.

(b) This is not an oxidation–reduction reaction; it is a Brønsted acid–base reaction. There is no change in the oxidation numbers of any of the atoms in this reaction

(c) This is an oxidation–reduction reaction. The SF_4 is oxidized as sulfur changes from the +4 oxidation number in SF_4 to the +6 oxidation number in SF_6. The fluorine molecule, F_2, is reduced as the oxidation number decreases from zero in F_2 to −1 in SF_6.

(d) This is not an oxidation–reduction reaction because there is no change in oxidation number of any of the elements in this reaction.

Oxidation–reduction reactions of organic compounds can often be recognized by looking for a change in the number of hydrogen and/or oxygen atoms. Organic compounds can be oxidized by either removing hydrogen atoms or adding oxygen atoms. When a bottle of wine is opened and allowed to sit in contact with air, for example, the wine eventually turns into vinegar. The reaction involves the oxidation of the alcohol in wine, ethanol, to form acetic acid.

$$H-\underset{\underset{H}{|}}{\overset{\overset{H}{|}}{C_a}}-\underset{\underset{H}{|}}{\overset{\overset{H}{|}}{C_b}}-O-H + O_2 \longrightarrow H-\underset{\underset{H}{|}}{\overset{\overset{H}{|}}{C_a}}-\overset{\displaystyle O}{\underset{\displaystyle O-H}{C_b}} + H_2O$$

Ethanol   *Acetic acid*

In the course of this reaction, the number of hydrogen atoms bound to C_b decreases and the number of oxygen atoms increases. C_b has therefore been oxidized in this reaction.

We can examine the reaction more quantitatively by assigning oxidation numbers to the atoms that are oxidized or reduced. We start by assigning the electrons in each C—H, C—O, and O—H bond in both ethanol and acetic acid to the more electronegative atom. The electrons in bonds between equivalent carbon atoms are then divided evenly between the two atoms.

$$H-\underset{\underset{H}{|}}{\overset{\overset{H}{|}}{C_a}}\cdot C_b \quad :\ddot{O}: \ H \longrightarrow H-\underset{\underset{H}{|}}{\overset{\overset{H}{|}}{C_a}}\cdot C_b \quad \overset{:\ddot{O}:}{\underset{:\ddot{O}:}{}} H$$

Ethanol   *Acetic acid*

Once this is done, C_b is assigned five electrons in ethanol and only one electron in acetic acid.

$$OX_{C_b \text{ in ethanol}} = 4 - 5 = -1$$
$$OX_{C_b \text{ in acetic acid}} = 4 - 1 = +3$$

The oxidation number of C_b has therefore increased from -1 to $+3$. At the same time, the oxidation number of oxygen has decreased from zero in elemental oxygen to -2 for the oxygen atom in acetic acid and the oxygen in water.

$$OX_{\text{elemental oxygen}} = 6 - 6 = 0$$
$$OX_{\text{oxygen in acetic acid}} = 6 - 8 = -2$$
$$OX_{\text{oxygen in water}} = 6 - 8 = -2$$

Oxidation–reduction reactions that involve organic compounds are not always written as balanced equations, as shown above, but instead show only the initial organic reactant and the final organic product. Acetic acid can be prepared from ethanol, for example, by oxidation with potassium dichromate, $K_2Cr_2O_7$. The reaction is often written as follows:

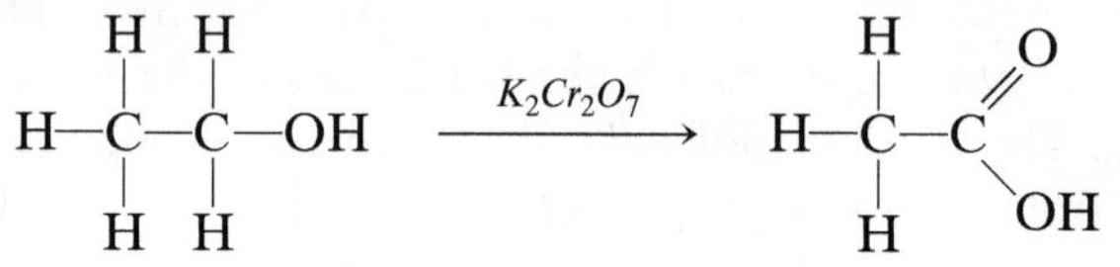

The organic reactant (ethanol) and the organic product (acetic acid) are shown. The reagent that produces the reaction plus any additional reaction conditions are written above the arrow that connects the two sides of the equation. When this is done, the numbers of hydrogen and oxygen atoms on both sides of the equation are not balanced.

Exercise 12.2

For each of the following reactions, determine whether it is an oxidation–reduction reaction. For those that are oxidation–reduction reactions, determine whether the organic compound is oxidized or reduced. (Note that the equations are not balanced.)

(a) $H_2C{=}CH_2(g) \xrightarrow{H_2} CH_3CH_3(g)$
    Ethene *Ethane*

(b) $HCOH(aq) + NaOH(aq) \longrightarrow [HCO^-]Na^+(aq) + H_2O(l)$
    Formic acid   *Sodium formate*

(c) $CH_3CH_2CH(aq) \xrightarrow{KMnO_4} CH_3CH_2COH(aq)$
    Propanal   *Propanoic acid*

Solution

(a) For this reaction, the oxidation number of both carbon atoms changes from -2 to -3. The reaction therefore involves the reduction of ethene to ethane. (The H_2 is oxidized in this reaction.)

(b) There is no change in oxidation number for any of the atoms in the reaction. This is an acid–base reaction, not an oxidation–reduction reaction.

(c) In this reaction, the oxidation number of the carbon atom bonded to the oxygen changes from +1 in propanal to +3 in propanoic acid. Propanal is therefore oxidized to propanoic acid. The $KMnO_4$ is reduced in this reaction.

• •

Biochemical pathways such as the citric acid (or Krebs) cycle shown in Figure 12.1 often focus on the starting material and the product of each step in the reaction pathway. The citric acid cycle consists of a series of reactions that serve as the principal source of metabolic energy for an organism. Nutrients from carbohydrates, proteins, and lipids are converted into intermediates that can be

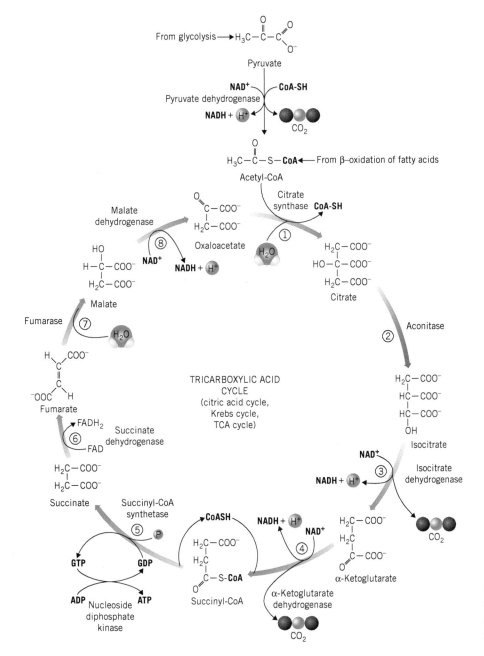

Fig. 12.1 The citric acid cycle. [Reprinted from R. H. Garrett and C. M. Grisham, *Biochemistry*-HSIE, ©1994 Brooks/Cole, a part of Cengage Learning, Inc. Reproduced by permission.]

oxidized in the citric acid cycle to carbon dioxide and water. The carbon in carbon dioxide is in the +4 oxidation state, carbon's highest oxidation state. Carbon in this state therefore has been completely oxidized.

We will examine just one of the reactions from the cycle in Figure 12.1. The first oxidation–reduction reaction in this cycle is reaction 3, in which the isocitrate ion is oxidized to the α-ketoglutarate ion. The oxidation numbers of the carbon atoms in the starting material and product of this reaction are indicated with circled numbers in the following diagram.

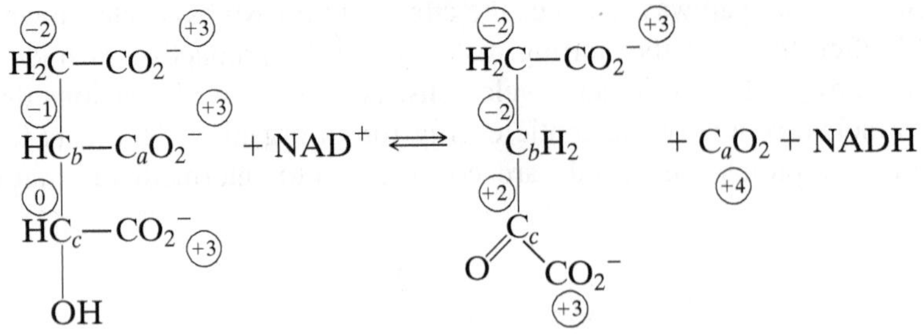

Isocitrate   *α-Ketoglutarate*

Three of the carbon atoms of the isocitrate molecule undergo a change in oxidation number during the reaction (labeled C_a, C_b, C_c). Carbon atom C_c is oxidized from 0 to +2, carbon atom C_b is reduced from −1 to −2, and carbon atom C_a is oxidized from +3 to +4. As a result, there is a net increase in oxidation of +3 for two carbon atoms and a decrease of −1 for the third. The net change in oxidation number for the molecule is therefore +2, indicating that the isocitrate ion has been oxidized. The enzyme cofactor known as NAD^+ (nicotinamide adenine dinucleotide) is reduced in this reaction to produce NADH. NAD^+ undergoes a −2 change in oxidation number to balance the +2 change for isocitrate.

12.4 Voltaic Cells

Section 5.2 noted that the common metals are divided into four classes on the basis of their activity (or reactivity). The "active metals" such as Na and K burst into flame in the presence of water. Metals such as Mg, Al, and Zn are described as "less active metals." They don't react with water at room temperature, but they react rapidly with acids. Zinc metal, for example, reacts with acid to form an aqueous solution of the Zn^{2+} ions and H_2 gas.

$$Zn(s) + 2\,H^+(aq) \rightleftharpoons Zn^{2+}(aq) + H_2(g)$$

This reaction has some of the characteristic features of oxidation–reduction reactions.

- It is exothermic, $\Delta H° = -153.89$ kJ/mol$_{rxn}$.
- The equilibrium constant for the reaction is very large ($K_c = 6 \times 10^{25}$), and chemists therefore often write the equation for the reaction as if essentially all of the reactants were converted to products.
- It can be formally divided into separate oxidation and reduction half-reactions.

Oxidation: $Zn \rightleftharpoons Zn^{2+} + 2\,e^-$

Reduction: $2\,H^+ + 2\,e^- \rightleftharpoons H_2$

- By separating the two half-reactions, so that the electrons have to pass through an external circuit, the energy given off by this reaction can be used to do work.

Electrochemical cells in which an oxidation–reduction reaction produces an electric current that can be used to do work are known as **voltaic cells** or **galvanic cells.** As we will see, these cells serve as the basis for the construction of a battery. Figure 12.2 shows a voltaic cell formed by immersing a strip of zinc metal into a 1 M $Zn(NO_3)_2$ solution and gently bubbling H_2 gas over the surface of a platinum wire that has been immersed in a 1 M H^+ ion solution. The zinc metal and platinum wire are connected with an electrical conductor, such as a piece of copper wire. The circuit is then completed with a **salt bridge,** a U-tube filled with a saturated solution of a soluble salt such as KNO_3.

Oxidation takes place at the zinc electrode in the beaker shown on the left in Figure 12.2.

$$Zn(s) \rightleftharpoons Zn^{2+}(aq) + 2\,e^-$$

As the reaction proceeds, the zinc electrode gradually dissolves as the zinc metal is oxidized to aqueous Zn^{2+} ions. The electrons given off in this reaction collect on the zinc electrode. Because negative charge builds up on this metal, this electrode becomes negatively charged.

The half-reaction in which zinc metal is oxidized to Zn^{2+} ions cannot occur without an accompanying reduction half-reaction. Reduction takes place at the platinum electrode, and the reduction half-reaction is written as follows.

$$2\,H^+(aq) + 2\,e^- \rightleftharpoons H_2(g)$$

H^+ ions in the solution that surrounds the platinum electrode migrate toward the surface of this electrode where they are reduced to form H_2 gas, which bubbles out of solution. Because the platinum electrode supplies the electrons for this reaction, the electrode becomes positively charged.

The excess electrons on the zinc electrode move through the wire connecting the two electrodes toward the positively charged platinum electrode. This produces an electric current that can be used to light a lightbulb or run an electric motor. If a voltmeter is connected between the two electrodes, a **cell potential** for the voltaic cell can be measured.

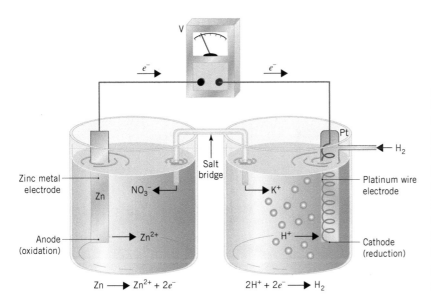

Fig. 12.2 The reaction between zinc metal and hydrogen ions produces an electric potential resulting in the flow of electrons from the zinc electrode to the inert platinum electrode immersed in the acid solution. K^+ ions move from the salt bridge into the solution to replace H^+ ions as these ions are converted to H_2 gas at the cathode. NO_3^- ions move from the salt bridge into solution to balance the charge of the Zn^{2+} produced from the zinc metal at the anode.

The cell potential is the potential of the cell to do work on its surroundings by driving an electric current through a wire. By definition, a potential of 1 volt is produced when 1 joule of energy is used to transport 1 coulomb, C, of electrical charge across the potential.

$$1 \text{ V} = \frac{1 \text{ J}}{1 \text{ C}}$$

The cell potential for the voltaic cell in Figure 12.2 is 0.7628 V. The magnitude of the cell potential is a measure of the driving force behind an electrochemical reaction. The larger the cell potential, the larger the driving force pushing the reaction toward the products.

The function of the salt bridge in Figure 12.2 is to maintain electric neutrality in the cell. Negatively charged nitrate ions diffuse from the salt bridge into the zinc solution to balance the positive charge of the Zn^{2+} ions produced at this electrode when the zinc metal is oxidized. Positively charged potassium ions diffuse from the salt bridge into the acid solution to balance the negative charge created when H^+ ions are reduced to form H_2 gas.

The electrode at which oxidation occurs in an electrochemical cell is called the **anode.** The electrode at which reduction takes place is called the **cathode.** The identity of the cathode and anode can be remembered by recognizing that positive ions, or **cations,** flow toward the cathode, and negative ions, or **anions,** flow toward the anode. In the voltaic cell shown in Figure 12.2, the zinc electrode is the anode and the platinum electrode is the cathode.

➤ **CHECKPOINT**

What is the charge on the anode in Figure 12.2? What is the charge on the cathode?

BALANCING OXIDATION–REDUCTION EQUATIONS

Figure 12.3 shows a voltaic cell based on the following half-reactions.

$$Cu(s) \rightleftharpoons Cu^{2+}(aq) + 2\,e^-$$
$$Ag^+(aq) + e^- \rightleftharpoons Ag(s)$$

Copper metal is oxidized to form Cu^{2+} ions in the beaker on the left; Ag^+ ions are reduced to silver metal in the beaker on the right. Using the law of conservation of

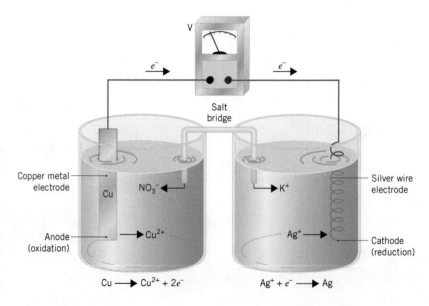

Fig. 12.3 The reaction between copper metal and silver ions produces an electric potential resulting in the flow of electrons from the copper electrode to the silver electrode. K^+ ions move from the salt bridge into the solution to replace Ag^+ ions as the silver ions are converted to silver metal at the cathode. NO_3^- ions move from the salt bridge into solution to balance the charge of the Cu^{2+} ions produced from the copper metal at the anode.

mass, it would be tempting to write the following complete oxidation–reduction reaction.

$$Cu(s) + Ag^+(aq) \rightleftharpoons Cu^{2+}(aq) + Ag(s)$$

There are an equal number of silver and copper atoms on both sides of the equation, so atoms are conserved. However, the electric charge on the two sides of the equation is not balanced.

The overall reaction in the voltaic cell shown in Figure 12.3 involves the oxidation of copper and the reduction of silver. For every copper atom that is oxidized, two electrons will be released. However, only one electron is required to reduce a silver ion to silver metal. Because neither matter nor charge can be created or destroyed in a chemical reaction, more electrons cannot be produced at the anode than are consumed at the cathode. In a balanced chemical reaction, not only atoms, but also electrons, must be conserved. The balanced oxidation–reduction equation for this reaction requires that the number of electrons gained during reduction must equal the number of electrons lost during oxidation. Therefore, for every copper atom oxidized at the anode, two silver ions must be reduced at the cathode.

$$Cu(s) \rightleftharpoons Cu^{2+}(aq) + 2\,e^-$$
$$\underline{2 \times [Ag^+(aq) + e^- \rightleftharpoons Ag(s)]}$$
$$Cu(s) + 2\,Ag^+(aq) \rightleftharpoons Cu^{2+}(aq) + 2\,Ag(s)$$

➤ **CHECKPOINT**

Is the following oxidation–reduction reaction balanced? Explain your answer.

$$Zn^{2+}(aq) + Cr(s)$$
$$\rightleftharpoons Cr^{3+}(aq) + Zn(s)$$

12.5 Standard Cell Potentials

The cell potential for a voltaic cell depends on the concentrations of any species present in solution, the partial pressure of any gases involved in the reaction, and the temperature at which the reaction is run. To provide a basis for comparing the results of one experiment with another, the following set of standard conditions for electrochemical measurements has been defined

Standard-state conditions:

Each solute has a concentration of 1 M.

All gases are present at a partial pressure of 1 bar (0.1 MPa or 0.9869 atm).

Although standard-state measurements can be made at any temperature, they are most often made at 25°C.

When the voltaic cell in Figure 12.2 was introduced in the previous section, we assumed that the zinc metal was immersed in a 1 M solution of the Zn^{2+} ion and that the platinum wire had been immersed in a 1 M H^+ ion solution. Furthermore, we assumed that H_2 gas was gently being passed over the platinum electrode, which suggests that the partial pressure of this gas is only slightly greater than 1 atm, and therefore close to 1 bar. The potential for this cell is therefore equal to the **standard cell potential, $E°$.**

The experimental value for the standard cell potential measures the *relative* reducing power of zinc metal compared with hydrogen gas. But it doesn't tell us anything about the *absolute* value of the reducing power for either zinc metal or H_2 gas. We can solve this problem by arbitrarily defining the standard cell potential for the reduction of H^+ ions to H_2 gas as exactly zero volts.

$$2\,H^+(aq) + 2\,e^- \rightleftharpoons H_2(g) \qquad E° = 0.000\ldots V$$

The sum of the half-cell potentials for the oxidation and reduction half-cells is equal to the overall cell potential for the voltaic cell.

$$E^\circ_{\text{cell}} = E^\circ_{\text{ox}} + E^\circ_{\text{red}}$$

The overall potential for the reaction between zinc and acid is 0.7628 V, and the half-cell potential for the reduction of H^+ ions is 0.000 . . . V. Therefore, the half-cell potential for the oxidation of zinc metal must be 0.7628 V.

$$
\begin{array}{ll}
\text{Zn}(s) \rightleftharpoons \text{Zn}^{2+}(aq) + 2\,e^- & E^\circ_{\text{ox}} = 0.7628 \text{ V} \\
\underline{+2\,\text{H}^+(aq) + 2\,e^- \rightleftharpoons \text{H}_2(g)} & \underline{E^\circ_{\text{red}} = 0.000 \ldots \text{ V}} \\
\text{Zn}(s) + 2\,\text{H}^+(aq) \rightleftharpoons \text{Zn}^{2+}(aq) + \text{H}_2(g) & E^\circ_{\text{cell}} = 0.7628 \text{ V}
\end{array}
$$

It is important to recognize that the units of half-cell potentials are volts, not volts per mole or volts per electron. Thus, when combining half-reactions, it is only necessary to add the two half-cell potentials. We do not have to multiply the half-cell potentials by the integers used to balance the number of electrons transferred in the reaction.

The *magnitude* of the cell potential is a measure of the driving force behind a reaction. The larger the value of the cell potential, the farther the reaction is from equilibrium. The *sign* of the cell potential tells us the direction in which the reaction must shift to reach equilibrium. The fact that E° is positive for the zinc–acid cell tells us that when the system is at standard conditions, it has to shift to the right to reach equilibrium. If the reaction is reversed, the sign of the cell potential changes. Reactions for which E° is positive therefore have equilibrium constants that favor the formation of the products of the reaction. A reaction with a positive E° should occur naturally and is referred to as **spontaneous** or **favorable.**

Exercise 12.3

Use the overall cell potentials to predict which of the following reactions are favorable.

(a) $\text{Cu}(s) + 2\,\text{Ag}^+(aq) \rightleftharpoons \text{Cu}^{2+}(aq) + 2\,\text{Ag}(s)$ $E^\circ \doteq 0.46 \text{ V}$

(b) $2\,\text{Fe}^{3+}(aq) + 2\,\text{Cl}^-(aq) \rightleftharpoons 2\,\text{Fe}^{2+}(aq) + \text{Cl}_2(g)$ $E^\circ = -0.59 \text{ V}$

(c) $2\,\text{Fe}^{3+}(aq) + 2\,\text{I}^-(aq) \rightleftharpoons 2\,\text{Fe}^{2+}(aq) + \text{I}_2(aq)$ $E^\circ = 0.24 \text{ V}$

(d) $2\,\text{H}_2\text{O}_2(aq) \rightleftharpoons 2\,\text{H}_2\text{O}(l) + \text{O}_2(aq)$ $E^\circ = 1.09 \text{ V}$

(e) $\text{Cu}(s) + 2\,\text{H}^+(aq) \rightleftharpoons \text{Cu}^{2+}(aq) + \text{H}_2(g)$ $E^\circ = -0.34 \text{ V}$

Solution

Any reaction for which the overall cell potential is positive is favorable. Reactions (a), (c), and (d) are therefore favorable.

Exercise 12.4

Use the standard cell potential for the following reaction

$$\text{Cu}(s) + 2\,\text{H}^+(aq) \rightleftharpoons \text{Cu}^{2+}(aq) + \text{H}_2(g) \qquad E^\circ = -0.34 \text{ V}$$

Whiskers of silver metal form when a piece of copper metal is immersed in a solution of Ag^+ ions.

to predict the standard cell potential for the opposite reaction

$$Cu^{2+}(aq) + H_2(g) \rightleftharpoons Cu(s) + 2\,H^+(aq) \qquad E° = ?$$

Solution

Turning the reaction around doesn't change the *magnitude* of the cell potential. But turning the equation around changes the *sign* of the cell potential and can therefore turn an unfavorable reaction into one that is spontaneous, or vice versa. The standard cell potential for the reduction of Cu^{2+} ions by H_2 gas is therefore $+0.34$ V.

$$Cu^{2+}(aq) + H_2(g) \rightleftharpoons Cu(s) + 2\,H^+(aq) \qquad E° = -(-0.34\ V) = +0.34\ V$$

➤ **CHECKPOINT**

What happens to the cell potential when the direction of the reaction is reversed?

12.6 Oxidizing and Reducing Agents

So far we have focused on what happens when a particular substance undergoes a change in oxidation state in an oxidation–reduction reaction. Let's now consider the role that each substance plays in the reaction.

When a piece of zinc metal is immersed in an acidic solution, the zinc metal donates electrons to the H^+ ions and therefore reduces the H^+ ions. Zinc therefore acts as a **reducing agent** in the reaction.

$$Zn(s) + 2\,H^+(aq) \rightleftharpoons Zn^{2+}(aq) + H_2(g)$$
Reducing agent

The H^+ ions, on the other hand, gain electrons from the zinc metal and thereby oxidize the zinc. The H^+ ion is therefore an **oxidizing agent.**

$$Zn(s) + 2\,H^+(aq) \rightleftharpoons Zn^{2+}(aq) + H_2(s)$$
Oxidizing agent

In general, oxidizing and reducing agents can be defined as follows.

> An *oxidizing agent* undergoes a decrease in oxidation number and causes the oxidation of another species. A change in oxidation number from 0 to -1 or from $+2$ to 0 would be characteristic of an oxidizing agent.

> A *reducing agent* undergoes an increase in oxidation number and causes the reduction of another species. A change in oxidation number from -1 to 0 or from 0 to $+2$ would be characteristic of a reducing agent.

Zinc metal is the reducing agent when it reacts with acid because it reduces H^+ ions to H_2 gas. The H^+ ion is the oxidizing agent in this reaction because it oxidizes the zinc metal to form Zn^{2+} ions.

When zinc metal loses electrons, it forms a conjugate oxidizing agent that could gain electrons if the reaction were reversed.

$$Zn \rightleftharpoons Zn^{2+} + 2\,e^-$$
*Reducing   Oxidizing
agent   agent*

Conversely, when the H^+ ions gain electrons, they form a conjugate reducing agent that could lose electrons if the reaction went in the opposite direction.

$$2\,H^+ + 2\,e^- \rightleftharpoons H_2$$

Oxidizing　　　*Reducing*
agent　　　*agent*

Oxidizing agents and reducing agents are therefore linked or coupled in much the same way that Brønsted acids and Brønsted bases are linked or coupled (see Section 11.4). Every oxidizing agent has a conjugate reducing agent, and vice versa.

The relationship between the strength of a reducing agent and its conjugate oxidizing agent can be understood by extending the argument used in Section 11.9 to link the strengths of Brønsted acids and bases. If zinc metal is a relatively good reducing agent, we must conclude that the Zn^{2+} ion is a relatively weak oxidizing agent.

$$Zn(s) \rightleftharpoons Zn^{2+}(aq) + 2\,e^-$$

Relatively　　　*Relatively*
Strong　　　*Weak*
reducing　　　*oxidizing*
agent　　　*agent*

If, on the other hand, the Cu^{2+} ion is a relatively good oxidizing agent, then copper metal must be a relatively weak reducing agent.

$$Cu^{2+}(aq) + 2\,e^- \rightleftharpoons Cu(s)$$

Relatively　　　*Relatively*
Strong　　　*Weak*
oxidizing　　　*reducing*
agent　　　*agent*

As was seen in Figure 5.12, many atoms have more than one oxidation number. The ability of an element to behave as an oxidizing or reducing agent is related to its oxidation state.

When an element is in its highest oxidation state, it cannot be oxidized further. As a result, it can only act as an oxidizing agent.

When an element is in its lowest oxidation state, it cannot be reduced further. Thus it can only act as a reducing agent.

When an element is in an intermediate oxidation state, it can act as either an oxidizing or reducing agent.

► **CHECKPOINT**

Identify the oxidizing and reducing agents in the following oxidation–reduction reaction.

$Sn(s) + 4\,HNO_3(aq)$
$\rightleftharpoons SnO_2(s) + 4\,NO_2(g) + 2\,H_2O(l)$

Carbon can have any oxidation number between -4 and $+4$. In carbon dioxide, CO_2, the carbon atom has an oxidation number of $+4$ and therefore cannot be oxidized further; the carbon in carbon dioxide can only be reduced. As a result, the carbon atom in carbon dioxide can act only as an oxidizing agent. Conversely, in methane, CH_4, carbon has an oxidation number of -4 and can undergo only an oxidation reaction. Methane therefore can act only as a reducing agent in a redox reaction. Elemental carbon in the form of graphite, which has an oxidation number of 0, can be either oxidized or reduced depending on the substance with which it reacts.

12.7　Relative Strengths of Oxidizing and Reducing Agents

The following rule can be used to predict whether a redox reaction should occur and which substance should be oxidized and which reduced if a reaction does occur.

Oxidation–reduction reactions should occur when they convert the stronger of a pair of oxidizing agents and the stronger of a pair of reducing agents into a weaker oxidizing agent and a weaker reducing agent.

Given that the following reaction occurs as written, zinc must be a stronger reducing agent than H_2 gas, and the H^+ ion must be a stronger oxidizing agent than the Zn^{2+} ion.

$$Zn(s) + 2\,H^+(aq) \rightleftharpoons Zn^{2+}(aq) + H_2(g)$$

| Stronger reducing agent | Stronger oxidizing agent | Weaker oxidizing agent | Weaker reducing agent |

It is important to recognize that the fact that this reaction occurs as written does not imply that zinc is a *strong* reducing agent, in an absolute sense. It only implies that zinc is a *stronger* reducing agent than H_2 gas.

On the basis of many such experiments, the common oxidation–reduction half-reactions have been organized into a table in which the strongest reducing agents are at one end and the strongest oxidizing agents are at the other, as shown in Table 12.1. By convention, all of the half-reactions are written as reduction half-reactions. The relationship between the half-reactions is quantified by listing the *reduction potential* for each half-reaction. The *reduction potential* is a measure of the driving force behind a reduction half-reaction and is measured in units of volts, V. **Standard reduction potentials** for other substances can be found in Table B.12 in Appendix B.

The data in Table 12.1 provide an explanation for the arbitrary assignment of exactly zero volts to the reduction half-cell potential for the H^+ ion. This arbitrary standard provides a set of data in which the magnitude of the half-cell potential for the strongest reducing agent ($E^\circ = -2.924$ V) is approximately the same as the magnitude of the half-cell potential for the strongest oxidizing agent ($E^\circ = 3.03$ V).

The E°_{red} values in Table 12.1 can be used to predict the direction in which an oxidation–reduction reaction should be spontaneous. Consider the standard half-cell potentials for the following reactions, for example.

$$Zn^{2+} + 2\,e^- \rightleftharpoons Zn \qquad E^\circ = -0.7628\text{ V}$$
$$Ag^+ + e^- \rightleftharpoons Ag \qquad E^\circ = 0.7996\text{ V}$$

The magnitudes of the two half-cell potentials are similar, but the two half-cell potentials have different signs. Because the silver reduction half-reaction has a much more positive potential ($+0.7996$ V) than the zinc half-reaction (-0.7628 V), the silver ion is much more likely to undergo the reduction half-reaction. On the basis of these half-cell potentials, we can conclude that zinc metal is a stronger reducing agent than silver metal and that the Ag^+ ion is a stronger oxidizing agent than the Zn^{2+} ion.

$$Zn(s) + 2\,Ag^+(aq) \rightleftharpoons Zn^{2+}(aq) + 2\,Ag(s)$$

| Stronger reducing agent | Stronger oxidizing agent | Weaker oxidizing agent | Weaker reducing agent |

The zinc half-reaction in Table 12.1 is written as a reduction half-reaction and therefore must be reversed to become an oxidation half-reaction. *When a reaction is reversed, the magnitude of its potential remains the same but its sign*

Table 12.1
Standard Reduction Potentials, E°_{red}

	Half-Reaction[a]	E°_{red} (Volts)	
	$K^+ + e^- \rightleftharpoons K$	−2.924	Best
	$Ca^{2+} + 2\,e^- \rightleftharpoons Ca$	−2.76	reducing
	$Na^+ + e^- \rightleftharpoons Na$	−2.7109	agents
	$Mg^{2+} + 2\,e^- \rightleftharpoons Mg$	−2.375	
	$Al^{3+} + 3\,e^- \rightleftharpoons Al$	−1.706	
	$Mn^{2+} + 2\,e^- \rightleftharpoons Mn$	−1.18	
	$Zn^{2+} + 2\,e^- \rightleftharpoons Zn$	−0.7628	
	$Cr^{3+} + 3\,e^- \rightleftharpoons Cr$	−0.74	
	$S + 2\,e^- \rightleftharpoons S^{2-}$	−0.508	
	$Cr^{3+} + e^- \rightleftharpoons Cr^{2+}$	−0.41	
	$Fe^{2+} + 2\,e^- \rightleftharpoons Fe$	−0.409	
	$Co^{2+} + 2\,e^- \rightleftharpoons Co$	−0.28	
	$Ni^{2+} + 2\,e^- \rightleftharpoons Ni$	−0.23	
	$Sn^{2+} + 2\,e^- \rightleftharpoons Sn$	−0.1364	
	$Pb^{2+} + 2\,e^- \rightleftharpoons Pb$	−0.1263	
	$Fe^{3+} + 3\,e^- \rightleftharpoons Fe$	−0.036	
	$2\,H^+ + 2\,e^- \rightleftharpoons H_2$	0.0000 . . .	
Oxidizing	$Sn^{4+} + 2\,e^- \rightleftharpoons Sn^{2+}$	0.15	↑
power	$Cu^{2+} + e- \rightleftharpoons Cu^+$	0.158	Reducing
increases	$Cu^{2+} + 2\,e^- \rightleftharpoons Cu$	0.3402	Power
↓	$O_2 + 2\,H_2O + 4\,e^- \rightleftharpoons 4\,OH^-$	0.401	Increases
	$Cu^+ + e- \rightleftharpoons Cu$	0.522	
	$MnO_4^- + 2\,H_2O + 3\,e^- \rightleftharpoons MnO_2 + 4\,OH^-$	0.588	
	$O_2 + 2\,H^+ + 2\,e^- \rightleftharpoons H_2O_2$	0.682	
	$Fe^{3+} + e^- \rightleftharpoons Fe^{2+}$	0.770	
	$Hg_2^{2+} + 2\,e^- \rightleftharpoons 2\,Hg$	0.7961	
	$Ag^+ + e^- \rightleftharpoons Ag$	0.7996	
	$Hg^{2+} + 2\,e^- \rightleftharpoons Hg$	0.851	
	$HNO_3 + 3\,H^+ + 3\,e^- \rightleftharpoons NO + 2\,H_2O$	0.96	
	$Br_2(aq) + 2\,e^- \rightleftharpoons 2\,Br^-$	1.087	
	$CrO_4^{2-} + 8\,H^+ + 3\,e^- \rightleftharpoons Cr^{3+} + 4\,H_2O$	1.195	
	$O_2 + 4\,H^+ + 4\,e^- \rightleftharpoons 2\,H_2O$	1.229	
	$Cr_2O_7^{2-} + 14\,H^+ + 6\,e^- \rightleftharpoons 2\,Cr^{3+} + 7\,H_2O$	1.33	
	$Cl_2(g) + 2\,e^- \rightleftharpoons 2\,Cl^-$	1.3583	
	$PbO_2 + 4\,H^+ + 2\,e^- \rightleftharpoons Pb^{2+} + 2\,H_2O$	1.467	
	$MnO_4^- + 8\,H^+ + 5\,e^- \rightleftharpoons Mn^{2+} + 4\,H_2O$	1.491	
	$Au^+ + e^- \rightleftharpoons Au$	1.68	
Best	$Co^{3+} + e^- \rightleftharpoons Co^{2+}$	1.842	
oxidizing	$O_3(g) + 2\,H^+ + 2\,e^- \rightleftharpoons O_2(g) + H_2O$	2.07	
agents	$F_2(g) + 2\,H^+ + 2\,e^- \rightleftharpoons 2\,HF(aq)$	3.03	

[a]In tables of standard reduction potential, the symbol H^+ is used instead of H_3O^+.

changes. The cell potential for the oxidation half-reaction for zinc metal is therefore +0.7628 V.

$$Zn \rightleftharpoons Zn^{2+} + 2e^- \qquad E^\circ = +0.7628\ V$$

The oxidation half-reaction now has a positive potential, indicating that it is much more likely to occur than the reduction half-reaction of Zn^{2+} ion. In this voltaic cell, zinc metal therefore undergoes oxidation. Using the relationship

$$E^\circ_{cell} = E^\circ_{ox} + E^\circ_{red}$$

we can now determine the standard potential for the cell.

$$Zn(s) \rightleftharpoons Zn^{2+}(aq) + 2\,e^- \qquad E^{\circ}_{ox} = 0.7628\ V$$
$$\underline{+2\,Ag^+(aq) + 2\,e^- \rightleftharpoons 2\,Ag(s) \qquad E^{\circ}_{red} = 0.7996\ V}$$
$$Zn(s) + 2\,Ag^+(aq) \rightleftharpoons Zn^{2+}(aq) + 2\,Ag(s) \qquad E^{\circ}_{cell} = 1.5624\ V$$

The overall standard cell potential for this reaction is positive, which implies that this reaction should be spontaneous as written.

There is no need to remember that reducing agents become stronger toward the upper-right corner of this table or that the strength of the oxidizing agents increases toward the bottom-left corner. All you have to do is remember some of the chemistry of the elements at the top and bottom of this table.

Consider the half-reaction at the top of the table.

$$K^+ + e^- \rightleftharpoons K \qquad E^{\circ}_{red} = -2.924\ V$$

What do we know about potassium metal? Potassium is one of the most reactive metals—it bursts into flame when added to water, for example. We can therefore conclude that potassium metal is listed among the strongest reducing agents in this table.

The strongest reducing agents can therefore be found toward the top of Table 12.1. But it is important to remember that the strongest reducing agent in this table is potassium metal, not the K^+ ion. When the time comes to use this half-reaction in a voltaic cell, we have to reverse the direction in which it is written so that it becomes an oxidation half-reaction.

$$K \rightleftharpoons K^+ + e^- \qquad E^{\circ}_{ox} = +2.924\ V$$

Now consider the last reaction in the table.

$$F_2(g) + 2\,H^+ + 2\,e^- \rightleftharpoons 2\,HF(aq) \qquad E^{\circ}_{red} = 3.03\ V$$

Fluorine is one of the most electronegative elements in the periodic table. It shouldn't be surprising to find that F_2 is the strongest oxidizing agent in Table 12.1.

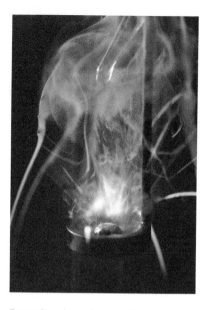

Potassium is such a good reducing agent that it bursts into flame when added to water.

Exercise 12.5

Arrange the following oxidizing and reducing agents in order of increasing strength.

>Reducing agents: Cl^-, Cu, H_2, HF, Pb, and Zn
>Oxidizing agents: Cr^{3+}, $Cr_2O_7^{2-}$, Cu^{2+}, H^+, O_2, O_3, and Na^+

Solution

According to Table 12.1, these reducing agents become stronger in the following order.

$$HF < Cl^- < Cu < H_2 < Pb < Zn$$

These oxidizing agents become stronger in the following order.

$$Na^+ < Cr^{3+} < H^+ < Cu^{2+} < O_2 < Cr_2O_7^{2-} < O_3$$

Exercise 12.6

Use Table 12.1 to predict whether the following oxidation–reduction reactions should occur as written.

(a) $2 Ag(s) + S(s) \rightleftharpoons Ag_2S(s)$

(b) $2 Ag(s) + Cu^{2+}(aq) \rightleftharpoons 2 Ag^+(aq) + Cu(s)$

(c) $MnO_4^-(aq) + 3 Fe^{2+}(aq) + 2 H_2O(l)$
    $\rightleftharpoons MnO_2(s) + 3 Fe^{3+}(aq) + 4 OH^-(aq)$

(d) $MnO_4^-(aq) + 5 Fe^{2+}(aq) + 8 H^+(aq)$
    $\rightleftharpoons Mn^{2+}(aq) + 5 Fe^{3+}(aq) + 4 H_2O(l)$

Solution

(a) No. The S^{2-} ion in Ag_2S is a better reducing agent than Ag metal, and the Ag^+ ion in Ag_2S is a better oxidizing agent than S.

(b) No. Cu is a better reducing agent than Ag, and the Ag^+ ion is a better oxidizing agent than the Cu^{2+} ion.

(c) No. MnO_4^- in basic solution is not a strong enough oxidizing agent to oxidize Fe^{2+} to Fe^{3+}.

(d) Yes. MnO_4^- in acidic solution is a strong enough oxidizing agent to oxidize Fe^{2+} to Fe^{3+}.

Exercise 12.7

Use cell potential data to explain why copper metal does not dissolve in a 1 M solution of a typical strong acid, such as hydrochloric acid,

$$Cu(s) + 2 H^+(aq) \not\rightleftharpoons$$

but will dissolve in 1 M nitric acid.

$$3 Cu(s) + 2 HNO_3(aq) + 6 H^+(aq) \rightleftharpoons 3 Cu^{2+}(aq) + 2 NO(g) + 4 H_2O(l)$$

Solution

Copper does not dissolve in a typical strong acid because the overall cell potential for the oxidation of copper metal to Cu^{2+} ions coupled with the reduction of H^+ ions to H_2 is negative.

$$
\begin{array}{ll}
Cu \rightleftharpoons Cu^{2+} + 2 e^- & E_{ox}^\circ = -(0.34 \text{ V}) \\
2 H^+ + 2 e^- \rightleftharpoons H_2 & E_{red}^\circ = 0.000\ldots \text{V} \\
\hline
Cu(s) + 2H^+(s) \rightleftharpoons Cu^{2+}(aq) + H_2(g) & E^\circ = E_{ox}^\circ + E_{red}^\circ = -0.34 \text{ V}
\end{array}
$$

Copper dissolves in I M nitric acid, however, because the reaction at the cathode now involves the reduction of nitric acid to NO gas, and the potential for that half-reaction is strong enough to overcome the half-cell potential for oxidation of copper metal to Cu^{2+} ions.

$$3\,(Cu \rightleftharpoons Cu^{2+} + 2\,e^-) \qquad E^{\circ}_{ox} = -(0.34\ V)$$
$$2\,(HNO_3 + 3\,H^+ + 3\,e^- \rightleftharpoons NO + 2\,H_2O) \qquad E^{\circ}_{red} = 0.96\ V$$
$$\overline{3\,Cu(s) + 2\,HNO_3(aq) + 6\,H^+(aq) \rightleftharpoons 3\,Cu^{2+}(aq) + 2\,NO(g) + 4\,H_2O(l)}$$
$$E^{\circ} = E^{\circ}_{ox} + E^{\circ}_{red} = 0.62\ V$$

A copper penny reacts with concentrated nitric acid to form the brown NO_2 gas.

12.8 Batteries

The term *battery* is used to describe a set of similar or connected items, such as an artillery battery or a battery of telephones. How, then, did it also come to mean a device that converts chemical energy into electrical energy? The first voltaic cell stable enough to be used in a battery was based on the following reaction between zinc metal and Cu^{2+} ions, and it produced a standard-state cell potential of 1.10 V.

$$Zn(s) + Cu^{2+}(aq) \rightleftharpoons Zn^{2+} + Cu(s)$$

When a number of these cells are connected in series, we get a "battery" of cells with a voltage equal to 1.10 V times the number of cells.

Until recently, the market for disposable batteries was dominated by batteries based on the *Leclanché cell,* which was introduced in 1860. Reduction in these batteries occurs at the cathode, which is a mixture of solid MnO_2 and carbon. Oxidation occurs at the anode, which is a sheet of zinc metal immersed in a solution of NH_4Cl and $ZnCl_2$. Because the electrolyte is trapped inside the battery case, these batteries are often referred to as "dry cell" batteries. The initial cell potential of a single cell in these batteries is 1.54 V. The popularity of manganese dioxide–zinc batteries was based on the fact that they were relatively inexpensive, available in many voltages and sizes, and suitable for intermittent use.

In 1949, the first "alkaline" dry cell battery was produced. This cell uses a mixture of MnO_2 and carbon as the cathode and amalgamated zinc as the anode. However, KOH is used as the electrolyte. The cell reactions that occur during the discharge of a manganese dioxide–zinc primary battery can be written as follows.

Cathode (+): $2\,MnO_2 + 2\,H_2O + 2\,e^- \longrightarrow 2\,MnO(OH) + 2\,OH^-$

Anode (−): $Zn + 2\,OH^- \longrightarrow ZnO + H_2O + 2\,e^-$

Overall reaction: $Zn + 2\,MnO_2 + H_2O \longrightarrow ZnO + 2\,MnO(OH)$

Alkaline dry cells (Figure 12.4) have several advantages. They can be used over a wider range of temperatures because the electrolyte is more stable. They also require very little electrolyte and can therefore be very compact. More importantly, the batteries maintain a constant voltage for a longer period of time and therefore last longer. The market for alkaline storage batteries has been estimated to be $5 billion per year.

LEAD–ACID BATTERIES

The lead–acid storage battery used in cars was first demonstrated to the French Academy of Sciences by Gaston Planté in 1860. It contained nine cells that were

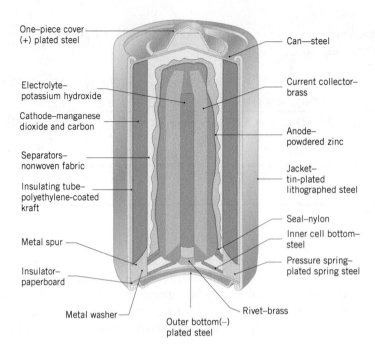

Fig. 12.4 Cutaway drawing of a typical alkaline battery.

constructed from lead plates separated by layers of flannel immersed in a 10% H_2SO_4 solution.

When the lead–acid battery in Figure 12.5 discharges, PbO_2 reacts with sulfuric acid at the cathode to form $PbSO_4$ and water,

$$PbO_2(s) + 3\,H^+(aq) + HSO_4^-(aq) + 2\,e^- \underset{charge}{\overset{discharge}{\rightleftharpoons}} PbSO_4(aq) + 2\,H_2O(l)$$

and lead metal reacts with the HSO_4^- ion at the anode to form $PbSO_4$.

$$Pb(s) + HSO_4^-(aq) \underset{charge}{\overset{discharge}{\rightleftharpoons}} PbSO_4(aq) + H^+ + 2\,e^-$$

The total cell reaction can therefore be written as follows.

$$PbO_2(s) + Pb(s) + 2\,H^+(aq) + 2\,HSO_4^{2-}(aq) \underset{charge}{\overset{discharge}{\rightleftharpoons}} 2\,PbSO_4(aq) + 2\,H_2O(l)$$

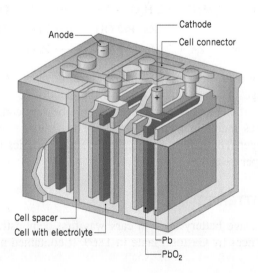

Fig. 12.5 Cutaway drawing of a typical lead–acid battery.

When the battery is charged, these reactions are reversed. The magnitude of the standard potential for a single cell in lead–acid batteries is slightly larger than 2 V. The typical 12-V lead storage battery therefore contains six 2-V cells.

During the charging of a lead–acid storage battery, water in the battery cells can decomposed into its elements.

$$2 \, H_2O(l) \rightleftharpoons 2 \, H_2(g) + O_2(g)$$

Lead storage batteries therefore represent a potential threat of hydrogen explosions. Because such explosions can spray the 10% sulfuric acid electrolyte onto the individual working on the battery, safety goggles should always be worn when working with these batteries.

RECHARGEABLE (NICD AND NIMH) BATTERIES

The first rechargeable nickel–cadmium or *Nicad* batteries were produced in the United States in 1946. They have become so popular that about 1.5 billion NiCd batteries are produced each year. When these batteries discharge, cadmium metal is oxidized at the anode to form $Cd(OH)_2$. The reaction at the cathode involves reduction of nickel from the +3 oxidation state [$NiO(OH)$] to the +2 oxidation state [$Ni(OH)_2$]. The net equation for the discharge–charge reactions can be written as follows.

$$2 \, NiO(OH) + Cd + 2 \, H_2O \underset{charge}{\overset{discharge}{\rightleftharpoons}} 2 \, Ni(OH)_2 + Cd(OH)_2$$

The net potential for this cell is 1.29 V.

An alternative approach to rechargeable batteries is based on nickel–metal hydride (NiMH) cells. NiMH batteries were once more expensive than NiCd batteries, but they have become significantly less expensive with time. NiMH batteries are available in the common AA and AAA sizes, but they can also be found in hybrid electric cars. When NiMH cells discharge, a metal hydride is oxidized at the anode. The metal is abbreviated as "M" in the equation for the reaction at this electrode because it is usually an intermetallic compound formed by combining one of the rare earth elements and either nickel, cobalt or manganese.

$$MH + OH^- \underset{charge}{\overset{discharge}{\rightleftharpoons}} M + H_2O + e^-$$

When NiMH batteries discharge, $NiO(OH)$ is reduced at the cathode to $Ni(OH)_2$.

$$NiO(OH) + H_2O + e^- \underset{charge}{\overset{discharge}{\rightleftharpoons}} Ni(OH)_2 + OH^-$$

The typical cell voltage for NiMH batteries is 1.25 V.

NiMH batteries are becoming increasingly important because of their use in hybrid electric cars. NiMH batteries have been shown to be capable of lasting for the life of the vehicle. Hybrid electric cars often come with warranties for the batteries of 8–10 years and 100,000 or more miles.

LITHIUM–ION BATTERIES

Lithium–ion batteries are commonly used in consumer electronics such as laptop computers. At one time, lithium–ion batteries used lithium metal for the anode.

These batteries had serious safety problems, including the possibility of either bursting into flame or exploding. Today, the anode is a substance such as graphite into which lithium ions can be embedded or materials such as $LiFePO_4$ that contain lithium ions. Lithium–ion batteries differ from the batteries discussed so far because they involve the flow of Li^+ ions, not electrons, between the anode and cathode of the cell.

FUEL CELLS

Ever since the start of the space program, enthusiasm has run high for batteries that are *fuel cells*. By definition, a fuel cell is an electrochemical cell that converts the energy associated with burning a fuel into electrical and thermal energy. Fuel cells therefore require a continuous feed of a fuel and either oxygen or air.

In a typical fuel cell, hydrogen gas is fed into the anode, and oxygen (or air) is fed into the cathode to produce a cell potential of 1.10 V, as shown in Figure 12.6.

$$2\,H_2(g) + O_2(g) \rightleftharpoons 2\,H_2O(g) \qquad E° = 1.10\ V$$

The major attraction of hydrogen-based fuel cells is their efficiency. Fuel cells have an efficiency of 60%, compared with 22% for gasoline or 45% for diesel internal combustion engines. Coupling fuel cells to electric motors, which are 90% efficient, would convert the chemical energy of hydrogen to mechanical work with a minimal amount of energy lost in the form of heat.

The primary disadvantage of fuel cells is the lack of a "hydrogen infrastructure" that would allow the user to obtain and safely store reasonable amounts of hydrogen gas. Other problems with fuel cells involve the tendency of the catalysts used to promote the chemical reaction from degrading and being poisoned by impurities in the reactant gas.

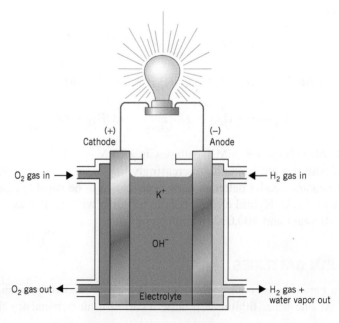

Fig. 12.6 A hydrogen–oxygen fuel cell.

12.9 Electrochemical Cells at Nonstandard Conditions: The Nernst Equation

In 1836, a chemistry professor from London named John Frederic Daniell developed the first voltaic cell stable enough to be used in a battery. The Daniell cell was based on the following reaction. A line drawing of the original Daniell cell is shown in Figure 12.7.

$$Zn(s) + Cu^{2+}(aq) \rightleftharpoons Zn^{2+}(aq) + Cu(s)$$

What happens when this oxidation–reduction reaction is used to do work?

- The zinc electrode becomes lighter as zinc atoms are oxidized to Zn^{2+} ions, which go into solution.
- There is a net flow of electrons through the external electric circuit from the negatively charged anode to the positively charged cathode.
- The copper electrode becomes heavier as Cu^{2+} ions in the solution are reduced to copper metal that plates out on the copper electrode.
- The concentration of Zn^{2+} ions at the anode increases, and the concentration of the Cu^{2+} ions at the cathode decreases.
- Negative ions flow from the salt bridge toward the anode to balance the charge on the Zn^{2+} ions produced at that electrode.
- Positive ions flow from the salt bridge toward the cathode to compensate for the Cu^{2+} ions consumed in the reaction.

An important property of the cell is missing from this list. Over a period of time, the cell runs down and eventually has to be replaced.

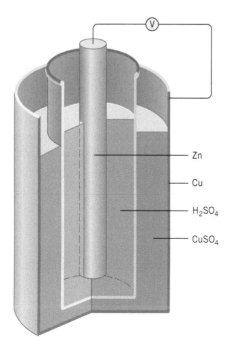

Fig. 12.7 Daniell cell.

Let's assume that our cell is initially a standard cell in which the concentrations of the Zn^{2+} and Cu^{2+} ions are both 1 M. As the reaction goes forward—as copper ions are consumed and zinc ions are produced—the driving force behind the reaction must become weaker. Therefore, the cell potential must become smaller.

In Section 12.5, we argued that the magnitude of the cell potential measures the driving force behind a reaction. The cell potential is therefore zero if and only if the reaction is at equilibrium. When the reaction is at equilibrium, there is no net change in the amount of zinc ions or copper ions in the system, so no electrons flow through the external circuit from the anode to the cathode. If there is no longer a net flow of electrons, the cell can no longer do electrical work. Its potential for doing work must therefore be zero.

In 1889 Hermann Walther Nernst showed that the potential for an electrochemical reaction is described by the following equation.

$$E = E° - \frac{RT}{nF} \ln Q_c$$

In the **Nernst equation,** E is the cell potential at some moment in time, $E°$ is the cell potential when the reaction is at standard conditions, R is the ideal gas constant in units of joules per mol $\cdot$ K, T is the temperature in kelvins, n is the number of electrons transferred in the balanced equation for the reaction, F is the charge on a mole of electrons, and Q_c is the reaction quotient at that moment in time. The symbol ln in this equation indicates a natural logarithm to the base e, where e is an irrational number equal to 2.71828. . . .

Two terms in the Nernst equation are constants: R and F. The ideal gas constant expressed in units of joules is 8.314 J/mol $\cdot$ K, and the charge on a mole of electrons can be calculated from Avogadro's number and the charge on a single electron.

$$F = \frac{6.022137 \times 10^{23} \, e^-}{1 \text{ mol } e^-} \times \frac{1.60217733 \times 10^{-19} \, C}{1 \, e^-} = \frac{96,485.31 \, C}{1 \text{ mol } e^-}$$

By convention, the temperature at which the Nernst equation is used is assumed to be 25°C. Substituting this information into the Nernst equation gives the following.

$$E = E° - \frac{0.02569}{n} \ln Q_c$$

A Daniell cell is made by immersing a piece of zinc metal in a solution of $ZnSO_4$ in a porous cup that is then immersed in a solution of $CuSO_4$ into which a piece of copper metal has been inserted.

Three of the remaining terms in this equation are characteristics of a particular reaction: n, $E°$, and Q_c. The standard potential for the Daniell cell is 1.10 V. Two moles of electrons are transferred from zinc metal to Cu^{2+} ions in the balanced equation for the reaction, so n is 2 for the cell. Because we never include the concentrations of solids in either reaction quotient or equilibrium constant expressions, Q_c for the reaction is equal to the concentration of the Zn^{2+} ion produced in the reaction divided by the concentration of the Cu^{2+} ion consumed in the reaction.

$$Q_c = \frac{(Zn^{2+})}{(Cu^{2+})}$$

Substituting what we know about the Daniell cell into the Nernst equation gives the following result, which represents the cell potential for the Daniell cell at 25°C at any moment in time.

$$E = 1.10 \, V - \frac{0.02569}{2} \ln \frac{(Zn^{2+})}{(Cu^{2+})}$$

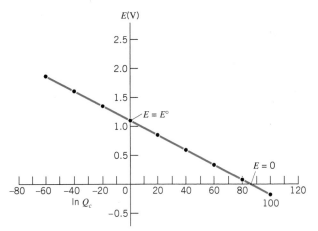

Fig. 12.8 Plot of $\ln Q_c$ versus cell potential for the Daniell cell. The cell potential is equal to $E°$ when the cell is at standard conditions. The cell potential is equal to zero if and only if the reaction is at equilibrium.

Figure 12.8 shows a plot of the potential for the Daniell cell as a function of the natural logarithm of the reaction quotient. When the reaction quotient is small, and the value of $\ln Q_c$ is negative, the cell potential is positive and relatively large. This isn't surprising because the reaction is far from equilibrium and the driving force behind the reaction should be relatively large. As the reaction quotient becomes larger, and $\ln Q_c$ gradually becomes positive, the cell potential becomes smaller. Eventually, Q_c becomes so large that the cell potential becomes negative, which means that the reaction would have to shift back toward the reactants to reach equilibrium.

The Nernst equation can be used to calculate the potential of a cell that operates at nonstandard conditions. But it can also be used to determine the equilibrium constant for a reaction. To understand how this is done, we have to recognize what happens to the cell potential when an oxidation–reduction reaction comes to equilibrium. Because the reaction is at equilibrium, there is no driving force pushing the reaction either toward the reactants or toward the products. Thus, the overall cell potential for the reaction becomes zero ($E = 0$) when the reaction quotient is equal to the equilibrium constant ($Q_c = K_c$).

At equilibrium, the Nernst equation therefore takes the following form.

$$0 = E° - \frac{RT}{nF} \ln K_c$$

Rearranging this equation gives the following result.

At equilibrium: $nFE° = RT \ln K_c$

According to this equation, we can calculate the equilibrium constant for any oxidation–reduction reaction from its standard cell potential. We start by solving this equation for the natural logarithm of the equilibrium constant.

$$\ln K_c = \frac{nFE°}{RT}$$

We then take advantage of the mathematics of logarithms to conclude that the equilibrium constant for the reaction can be calculated from the following equation.

$$K_c = e^{nFE°/RT}$$

Exercise 12.8

Calculate the potential at 25°C for the following reaction when the concentration of the Ag^+ ion is $4.8 \times 10^{-3}M$ and the concentration of the Cu^{2+} ion is $2.4 \times 10^{-2}M$.

$$Cu(s) + 2\,Ag^+(aq) \rightleftharpoons Cu^{2+}(aq) + 2\,Ag(s)$$

Solution

We start by calculating the standard cell potential for the overall reaction from the standard potentials for the two half-reactions.

Oxidation: $Cu \rightleftharpoons Cu^{2+} + 2\,e^-$ $E^\circ_{ox} = -(0.3402 \text{ V})$
Reduction: $Ag^+ + e^- \rightleftharpoons Ag$ $\underline{E^\circ_{red} = 0.7996 \text{ V}}$
$$E^\circ_{cell} = E^\circ_{red} + E^\circ_{ox} = 0.4594 \text{ V}$$

We now set up the Nernst equation for the cell, noting that n is 2 because two electrons are transferred in the balanced equation for the reaction.

$$E = E^\circ - \frac{0.02569}{2} \ln \left[\frac{(Cu^{2+})}{(Ag^+)^2} \right]$$

Substituting the concentrations of the Cu^{2+} and Ag^+ ions into this equation gives the value for the cell potential at these particular concentrations of the Cu^{2+} and Ag^+ ions.

$$E = 0.4594 \text{ } V - \frac{0.02569}{2} \ln \left[\frac{(0.024)}{(0.0048)^2} \right] = 0.37 \text{ } V$$

Exercise 12.9

Calculate the equilibrium constant at 25°C for the reaction between zinc metal and acid.

$$Zn(s) + 2\,H^+(aq) \rightleftharpoons Zn^{2+}(aq) + H_2(g)$$

Solution

We start by calculating the overall standard cell potential for the reaction from the half-cell potentials for the half-reactions.

$Zn \rightleftharpoons Zn^{2+} + 2\,e^-$ $E^\circ_{ox} = -(-0.7628 \text{ V})$
$\underline{2\,H^+ + 2\,e^- \rightleftharpoons H_2}$ $\underline{E^\circ_{red} = 0.0000 \text{ V}}$
$Zn(s) + 2\,H^+(aq) \rightleftharpoons Zn^{2+}(aq) + H_2(g)$ $E^\circ_{cell} = 0.7628 \text{ V}$

At equilibrium, $Q_c = K_c$ and $E = 0$.

$$0 = E^\circ - \frac{RT}{nF} \ln K_c$$

We then rearrange the equation

$$nFE° = RT \ln K_c$$

and solve for the natural logarithm of the equilibrium constant.

$$\ln K_c = \frac{nFE°}{RT}$$

Substituting what we know about the reaction into the equation gives the following result because the product of coulombs time volts is equal to joules (C × V = J).

$$\ln K_c = \frac{(2)(96485 \text{ C/mol})(0.7628 \text{ V})}{(8.314 \text{ J/mol-K})(298 \text{ K})} = 59.4$$

The equilibrium constant for the reaction can therefore be calculated by raising e to the power of 59.4.

$$K_c = e^{59.4} = 6 \times 10^{25}$$

This is a very large equilibrium constant, which means that equilibrium lies heavily on the side of the products. The equilibrium constant is so large that the equation for the reaction is often written as if it proceeds to completion, as noted in Section 12.4.

$$Zn(s) + 2 H^+(aq) \longrightarrow Zn^{2+}(aq) + H_2(g)$$

➤ **CHECKPOINT**

Would you expect the following reaction run under nonstandard conditions to have a potential greater than, equal to, or less than its standard potential, $E°$? Explain your reasoning.

$$Ni(s) + Cu^{2+}(aq)$$
$$\rightleftharpoons Ni^{2+}(aq) + Cu(s)$$

Initial concentrations of Cu^{2+} and Ni^{2+} are 1.5 and 0.010 M, respectively.

12.10 Electrolysis and Faraday's Law

The cells discussed so far have one thing in common: They all use a spontaneous chemical reaction to produce an electric current in an external circuit. As we have seen, these voltaic cells are important because they are the basis for the batteries that place such a major role in modern society. But they are not the only kind of electrochemical cell. It is also possible to construct a cell that does work on a chemical system by applying a potential to the cell that causes an electric current to flow through the cell. These cells are called **electrolytic cells.** Electrolysis is used to drive an oxidation–reduction reaction in the direction in which it doesn't occur spontaneously.

One application of an electrolytic process is the electroplating of one metal onto another. This is done both to protect against corrosion and to improve appearance. For example, a thin layer of silver can be plated onto jewelry or tableware made from a less expensive metal. Figure 12.9 is a schematic diagram of the plating of silver onto a fork. Oxidation occurs at the silver anode to produce Ag^+ ions in solution. The Ag^+ ions react with CN^- ions in the solution to form the $[Ag(CN)_2]^-$ complex ion. The $[Ag(CN)_2]^-$ complex ion is then reduced at the cathode to form a thin layer of silver on the iron fork. The half-reaction at the cathode is:

$$[Ag(CN)_2]^-(aq) + e^- \rightleftharpoons Ag(s) + 2 CN^-(aq)$$

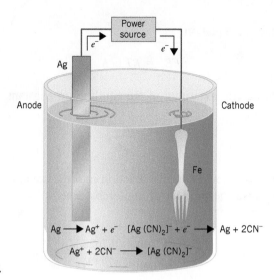

Fig. 12.9 The plating of silver onto a fork.

The $[Ag(CN)_2]^-$ complex ion is used because it produces a more uniform plating than the direct reduction of a metal cation.

The relationship between the amount of current passed through a solution and the mass of the substance consumed or produced by the current was developed by Michael Faraday, a famous nineteenth-century English chemist. **Faraday's law** of electrolysis can be stated as follows.

> **The amount of a substance either consumed or produced at one of the electrodes in an electrolytic cell is directly proportional to the amount of electricity that passes through the cell.**

In order to use Faraday's law, we need to recognize the relationship among current, time, and the amount of electric charge that flows through a circuit. By definition, 1 coulomb of charge is transferred when a 1-ampere (A) current is passed through the cell for 1 second.

$$1\text{ C} = 1\text{ A} \times 1\text{ s}$$

To illustrate how Faraday's law can be used, let's calculate the length of time that a 10-amp current has to be driven through a molten solution of the Al^{3+} ion to produce just enough aluminum metal to make a single soft-drink can.

Roughly 30 soft-drink cans can be made from a pound of aluminum metal. Aluminum is produced commercially by the electrolysis of a mixture of an ore known as bauxite ($Al_2O_3 \cdot 3\text{ H}_2O$) and a mineral known as cryolite (Na_3AlF_6). Both of these starting materials contain aluminum as Al^{3+} ions.

Let's assume that we are trying to make enough aluminum for one can, in other words, 15.0 grams.

$$15.0\text{ g Al} \times \frac{1\text{ mol Al}}{27.0\text{ g}} = 0.556\text{ mol Al}$$

But reduction of Al^{3+} ions at the cathode of an electrolytic cell requires three moles of electrons per mole of aluminum metal produced.

$$\text{Cathode:} \quad Al^{3+} + 3\,e^- \longrightarrow Al$$

Thus, we need three times as many moles of electrons.

$$0.556 \text{ mol Al} \times \frac{3 \text{ mol } e^-}{1 \text{ mol Al}} = 1.67 \text{ mol } e^-$$

We can now take advantage of the definition of Faraday's constant, F, which was introduced in the previous section. The value of Faraday's constant used in most calculations in this text is 96,485 coulombs per mole of electrons. We can therefore calculate the charge on 1.67 moles of electrons as follows.

$$1.67 \text{ mol } e^- \times \frac{96,485 \text{ C}}{1 \text{ mol } e^-} = 1.61 \times 10^5 \text{ C}$$

By definition, a coulomb of electric charge is carried by a current of 1 amp that flows for one sec. Thus, 1 C = 1 amp·sec. It would therefore take 1.61×10^4 seconds for a 10.0-amp current to deliver 1.61×10^5 C.

$$\frac{1.61 \times 10^5 \text{ amp·s}}{10.0 \text{ amp}} = 1.61 \times 10^4 \text{ s}$$

This means that it would take 4.5 hours using a 10-amp current to produce enough aluminum to make just one soft-drink can. Thus, it isn't surprising that the cost of the aluminum in a soft-drink can is roughly twice the cost of the ingredients that go into making the soft drink. Nor is it surprising to find that the cost of recycling the aluminum needed to make a soft-drink can is only 5% of the cost of making aluminum by electrolysis.

 Exercise 12.10

Calculate the volume of H_2 gas at 25°C and 1.00 atm that will collect at the cathode when water is electrolyzed for 2.00 hours with a 10.0-amp current.

$$2 \text{ H}_2\text{O}(l) \xrightarrow{\text{electrolysis}} 2 \text{ H}_2(g) + \text{O}_2(g)$$

Solution

We can start by calculating the amount of electrical charge that passes through the solution.

$$10.0 \text{ A} \times 2.00 \text{ h} \times \frac{60 \text{ min}}{1 \text{ h}} \times \frac{60 \text{ s}}{1 \text{ min}} \times \frac{1 \text{ C}}{1 \text{ A·}s} = 7.20 \times 10^4 \text{ C}$$

We then calculate the number of moles of electrons that carry this charge.

$$7.20 \times 10^4 \text{ C} \times \frac{1 \text{ mole } e^-}{96,485 \text{ C}} = 0.746 \text{ mol } e^-$$

We can now write a balanced equation for the half-reaction that produces H_2 gas at the cathode.

$$\text{Cathode}(-): \quad 2 \text{ H}_2\text{O} + 2 \, e^- \rightleftharpoons \text{H}_2 + 2 \text{ OH}^-$$

The equation for this half-reaction predicts that we will get one mole of H_2 gas at the cathode for every two moles of electrons that flow through the cell.

$$0.746 \text{ mole } e^- \times \frac{1 \text{ mol } H_2}{2 \text{ mole}} = 0.373 \text{ mol } H_2$$

We now have all of the information we need to calculate the volume of the gas produced in this reaction at 25°C and 1 atm pressure.

$$V = \frac{nRT}{P} = \frac{(0.373 \text{ mol})(0.08206 \text{ L} \cdot \text{atm/mol-K})(298 \text{ K})}{(1.00 \text{ atm})} = 9.12 \text{ L}$$

● ●

We can extend the general pattern outlined in this section to answer questions that might seem impossible at first glance.

 Exercise 12.11

Determine the oxidation number of the chromium in an unknown salt if electrolysis of a molten sample of the salt for 1.50 hours with a 10.0-amp current deposits 9.71 grams of chromium metal at the cathode.

Solution

We start, as before, by calculating the number of moles of electrons that passed through the cell during electrolysis.

$$10.0 \text{ A} \times 1.50 \text{ h} \times \frac{60 \text{ min}}{1 \text{ h}} \times \frac{60 \text{ min}}{1 \text{ min}} \times \frac{1 \text{ C}}{1 \text{ A} \cdot s} = 5.40 \times 10^4 \text{ C}$$

$$5.40 \times 10^4 \text{ C} \times \frac{1 \text{ mol } e^-}{96,485 \text{ C}} = 0.560 \text{ mol } e^-$$

Since we don't know the balanced equation for the reaction at the cathode in this cell, it doesn't seem obvious how we are going to use this information. We therefore write down this result in a conspicuous location and go back to the original statement of the question to see what else can be done.

The problem tells us the mass of chromium deposited at the cathode. We can therefore use this information to calculate the number of moles of chromium metal generated.

$$9.71 \text{ g Cr} \times \frac{1 \text{ mol Cr}}{52.00 \text{ g Cr}} = 0.187 \text{ mol Cr}$$

We now know the number of moles of chromium metal produced and the number of moles of electrons it took to produce the metal. We might therefore look at the relationship between the moles of electrons consumed in the reaction and the moles of chromium produced.

$$\frac{0.560 \text{ mol } e^-}{0.187 \text{ mol Cr}} \approx 3$$

Three moles of electrons are consumed for every mole of chromium metal produced. The only way to explain this is to assume that the net reaction at the cathode involves reduction of Cr^{3+} ions to chromium metal.

$$\text{Cathode}(-): \quad Cr^{3+} + 3\,e^- \rightleftharpoons Cr$$

Thus the oxidation number of chromium in the unknown salt must be $+3$.

12.11 Electrolysis of Molten NaCl

An idealized cell for the electrolysis of sodium chloride is shown in Figure 12.10. In this cell, a source of direct current is connected to a pair of inert electrodes immersed in molten sodium chloride. The Na^+ ions flow toward the negative electrode, and the Cl^- ions flow toward the positive electrode.

When Na^+ ions collide with the negative electrode, the battery provides a large enough potential to force these ions to pick up electrons to form sodium metal. This electrode is called the *cathode* because reduction always occurs at the cathode, regardless of whether the cell is a voltaic cell or an electrolytic cell.

$$\text{Negative electrode (cathode):} \quad Na^+ + e^- \rightleftharpoons Na$$

Cl^- ions that collide with the positive electrode are oxidized to Cl_2 gas, which bubbles off at this electrode. This electrode is called the *anode* because oxidation always occurs at the anode of an electrochemical cell.

$$\text{Positive electrode (anode):} \quad 2\,Cl^- \rightleftharpoons Cl_2 + 2\,e^-$$

The positively charged Na^+ ions, or *cations,* flow toward the *cathode,* and the negatively charged Cl^- ions, or *anions,* flow toward the anode.

The net effect of passing an electric current through the molten salt in the electrolytic cell is to decompose sodium chloride into its elements, sodium metal and chlorine gas.

Electrolysis of NaCl:

$$\text{Cathode}(-): \quad Na^+ + e^- \rightleftharpoons Na$$
$$\text{Anode}(+): \quad 2\,Cl^- \rightleftharpoons Cl_2 + 2\,e^-$$

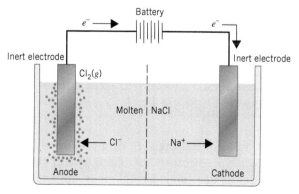

Fig. 12.10 Electrolysis of molten sodium chloride involves using an electric current to reduce Na^+ ions to sodium metal at the cathode and to oxidize Cl^- ions to Cl_2 gas at the anode.

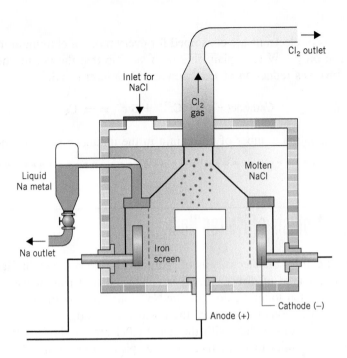

Fig. 12.11 Cross section of the Downs cell used for the electrolysis of a molten mixture of calcium chloride and sodium chloride.

This example explains why the process is called **electrolysis.** The suffix *-lysis* comes from a Greek stem meaning "to loosen or split up." Electrolysis literally uses an electric current to split a compound into its elements.

$$2\,NaCl(l) \xrightarrow{\;electrolysis\;} 2\,Na(l)\,+\,Cl_2(g)$$

This example also illustrates the difference between voltaic cells and electrolytic cells. Voltaic cells use the energy given off in a spontaneous reaction to do electrical work. Electrolytic cells use electrical work as a source of energy to drive the reaction in the opposite direction.

The dotted vertical line in the center of Figure 12.10 represents a diaphragm that keeps the Cl_2 gas produced at the anode from coming into contact with the sodium metal generated at the cathode. The function of the diaphragm can be understood by turning to a more realistic drawing of the commercial Downs cell used to electrolyze sodium chloride, shown in Figure 12.11.

Chlorine gas that forms on the graphite anode inserted into the bottom of the Downs cell bubbles through the molten sodium chloride into a funnel at the top of the cell. Sodium metal that forms at the cathode floats up through the molten sodium chloride into a sodium-collecting ring, from which it is periodically drained. The diaphragm that separates the two electrodes is a screen of iron gauze, which prevents the explosive reaction between sodium metal and chlorine from forming sodium chloride that would occur if the products of the electrolysis reaction came in contact.

The feedstock for the Downs cell is a 3:2 mixture by mass of $CaCl_2$ and NaCl. This mixture is used because it has a melting point of 580°C, whereas pure sodium chloride has to be heated to more than 800°C before it melts.

12.12 Electrolysis of Aqueous NaCl

Figure 12.12 shows an idealized drawing of a cell in which an aqueous solution of sodium chloride is electrolyzed. Once again, the Na^+ ions migrate toward the negative electrode, and the Cl^- ions migrate toward the positive

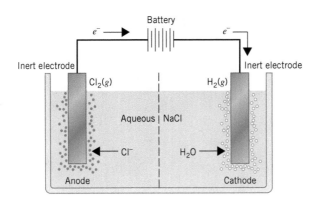

Fig. 12.12 Electrolysis of aqueous sodium chloride results in reduction of water to form H_2 gas at the cathode and oxidation of Cl^- ions at the anode.

electrode. But now two substances can be reduced at the cathode: Na^+ ions and water molecules.

Cathode(−):

$$Na^+ + e^- \rightleftharpoons Na \qquad E^\circ_{red} = -2.71 \text{ V}$$

$$2\,H_2O + 2\,e^- \rightleftharpoons H_2 + 2\,OH^- \qquad E^\circ_{red} = -0.83 \text{ V}$$

Because it is much easier to reduce water than Na^+ ions, the products formed at the cathode are hydrogen gas and OH^-.

Cathode(−): $2\,H_2O(l) + 2\,e^- \rightleftharpoons H_2(g) + 2\,OH^-(aq)$

There are also two substances that can be oxidized at the anode: Cl^- ions and water molecules.

Anode(+):

$$2\,Cl^- \rightleftharpoons Cl_2 + 2\,e^- \qquad E^\circ_{ox} = -1.36 \text{ V}$$

$$2\,H_2O \rightleftharpoons O_2 + 4\,H^+ + 4\,e^- \qquad E^\circ_{ox} = -1.23 \text{ V}$$

The magnitude of the standard potentials for the two half-reactions are so close to each other that we might expect to see a mixture of Cl_2 and O_2 gas collect at the anode. In practice, the only product is Cl_2.

Anode(+): $2\,Cl^- \rightleftharpoons Cl_2 + 2\,e^-$

At first glance, it would seem easier to oxidize water ($E^\circ_{ox} = -1.23$ volts) than Cl^- ions ($E^\circ_{ox} = -1.36$ volts). It is worth noting, however, that these are standard half-cell potentials, and the cell is never allowed to reach standard conditions. The solution is typically 25% NaCl by mass, which significantly decreases the potential required to oxidize the Cl^- ion. The deciding factor is *overvoltage*, which is the extra voltage that must be applied to a reaction to get it to occur at the rate at which it would occur in an ideal system. Under ideal conditions, a potential of 1.23 volts is large enough to oxidize water to O_2 gas. Under real conditions, however, it can take a much larger voltage to initiate this reaction. (The overvoltage for the oxidation of water can be as large as 1 volt.) By carefully choosing the electrode to maximize the overvoltage for the oxidation of water and then carefully controlling the potential at which the cell operates, we can ensure that only chlorine is produced in the reaction.

In summary, electrolysis of aqueous solutions of sodium chloride doesn't give the same products as electrolysis of molten sodium chloride. Electrolysis of molten NaCl decomposes the compound into its elements.

$$2\,NaCl(l) \xrightarrow{\ \ electrolysis\ \ } 2\,Na(l) + Cl_2(g)$$

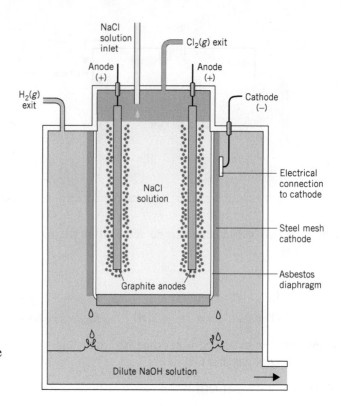

Fig. 12.13 About half of the commercial cells for the electrolysis of aqueous sodium chloride use the diaphragm cell shown here.

Electrolysis of aqueous NaCl solutions gives a mixture of hydrogen and chlorine gas and an aqueous sodium hydroxide solution.

$$2\,NaCl(aq) \;+\; 2\,H_2O(l) \xrightarrow{\;\text{electrolysis}\;} 2\,Na^+(aq) \;+\; 2\,OH^-(aq) \;+\; H_2(g) \;+\; Cl_2(g)$$

Because the demand for chlorine is much larger than the demand for sodium, electrolysis of aqueous sodium chloride is a more important process commercially.

Electrolysis of an aqueous NaCl solution has two other advantages. It produces H_2 gas at the cathode, which can be collected and sold. It also produces NaOH, which can be drained from the bottom of the electrolytic cell and sold, as shown in Figure 12.13.

The dotted vertical line in the idealized cell shown in Figure 12.12 represents a diaphragm that prevents the Cl_2 produced at the anode in the cell from coming into contact with the NaOH that accumulates at the cathode. When the diaphragm is removed from the cell, the products of the electrolysis of aqueous sodium chloride react to form sodium hypochlorite, which is the first step in the preparation of hypochlorite bleaches, such as Clorox.

$$Cl_2(g) \;+\; 2\,OH^-(aq) \xrightarrow{\;\text{electrolysis}\;} Cl^-(aq) \;+\; OCl^- \;+\; H_2O(l)$$

12.13 Electrolysis of Water

Electrolysis can be used to decompose water into its elements.

$$2\,H_2O(l) \xrightarrow{\;\text{electrolysis}\;} 2\,H_2(g) \;+\; O_2(g)$$

A pair of inert electrodes are sealed in opposite ends of a container designed to collect the H_2 and O_2 gas given off in the reaction. The electrodes are then connected to a battery or another source of electric current.

By itself, water is a very poor conductor of electricity. We therefore add an electrolyte to water to provide ions that can flow through the solution, thereby completing the electric circuit. The electrolyte must be soluble in water. It should also be relatively inexpensive. Most importantly, it must contain ions that are harder to oxidize or reduce than water.

$$2 H_2O + 2 e^- \rightleftharpoons H_2 + 2 OH^- \qquad E^\circ_{red} = -0.83 \text{ V}$$
$$2 H_2O \rightleftharpoons O_2 + 4 H^+ + 4 e^- \qquad E^\circ_{red} = -1.23 \text{ V}$$

According to the reduction half-cell potentials in Table B.12 in Appendix B, the following cations are harder to reduce than water: Li^+, Rb^+, K^+, Cs^+, Ba^{2+}, Sr^{2+}, Ca^{2+}, Na^+, and Mg^{2+}. Two of these cations are more likely candidates than the others because they form inexpensive, soluble salts: Na^+ and K^+.

Table B.12 in Appendix B suggests that the SO_4^{2-} ion might be the best anion to use because it is one of the most difficult anions to oxidize. The potential for oxidation of the SO_4^{2-} ion to $S_2O_8^{2-}$ ion is -2.05 volts.

$$2 SO_4^{2-} \rightleftharpoons S_2O_8^{2-} + 2 e^- \qquad E^\circ_{red} = -2.05 \text{ V}$$

When an aqueous solution of either Na_2SO_4 or K_2SO_4 is electrolyzed in the apparatus shown in Figure 12.14, H_2 gas collects at one electrode and O_2 gas collects at the other.

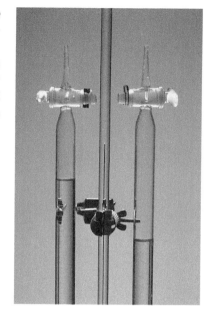

Electrolysis of an aqueous Na_2SO_4 solution results in the reduction of water to form H_2 gas at the cathode and the oxidation of water to form O_2 gas at the anode.

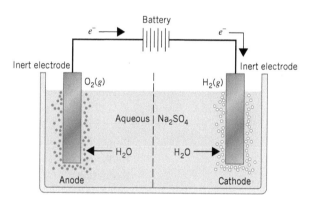

Fig. 12.14 Electrolysis of an aqueous Na_2SO_4 solution results in the reduction of water to form H_2 gas at the cathode and the oxidation of water to form O_2 gas at the anode.

12.14 The Hydrogen Economy

It has been estimated that the United States consumes about 100 quadrillion (100×10^{15}) BTU or about 1×10^{20} joules of energy each year. Almost 85% of this energy comes from fossil fuels: 37% from oil, 24% from natural gas, and 23% from coal. As the cost of petroleum-based fuels rises because of a combination of both increased demand and concern about the greenhouse effect associated with increasing the amount of CO_2 in the atmosphere, an increased focus has been placed on the search for alternative sources of energy.

One of the alternatives to fossil fuels that has been proposed is a **hydrogen economy.** Hydrogen is an example of an *energy carrier* that can be used to move energy in a usable form from one place to another. The most well-known energy carrier is electricity, which is called a secondary energy source because it has to be produced by the consumption of primary sources of energy. Hydrogen has the highest energy content of any common fuel when measured by weight—three times more than gasoline. But it has the lowest energy content when measured by volume—about four times less than gasoline.

The use of hydrogen as an energy carrier has potential benefits. First, and perhaps foremost, it reduces the amount of CO_2 that directly enters the atmosphere because it forms water vapor when used as a fuel. When allowed to burn in an internal combustion engine, it rapidly releases energy in the form of heat. When used in fuel cells, it produces energy more gently with water as the only by-product. Unlike other energy carriers such as electricity, hydrogen can be stored for later use.

Technological advances need to occur in three areas to achieve a hydrogen economy: production, storage, and use. The most promising route for production is by electrolysis of water. To eliminate fossil fuels from this energy cycle and to avoid releasing CO_2 into the atmosphere, the energy needed to split water must come from noncarbon sources. These sources could include photovoltaic energy captured by solar collectors, the heat from nuclear reactors, or electricity generated by renewable sources such as hydropower or wind power.

One disadvantage of using hydrogen as a routine energy carrier concerns its safe production and storage. As shown in Exercise 12.10, it takes a significant amount of energy in the form of an electric current to produce reasonable amounts of hydrogen gas. This gas can be stored in pressured gas cylinders or as a liquid at cryogenic temperatures, but not at a sufficient density to allow practical applications such as driving 300 miles on a single tank of fuel. As we have seen, hydrogen can be used in fuel cells, but the cost of producing these cells remains high. It has been estimated that the cost of prototype fuel cells is $3000 per kilowatt of power, which might be reduced by a factor of 10 if and when these fuel cells were mass produced. But this does not compete with the cost of operating gasoline engines, which is estimated to be $30 per kilowatt of power.

Key Terms

Anion
Anode
Cathode
Cation
Cell potential
Corrosion
Electrolysis
Electrolytic cells
Faraday's law

Favorable
Galvanic cell
Half-reaction
Hydrogen economy
Nernst equation
Oxidation
Oxidation number
Oxidation state
Oxidizing agent

Redox reactions
Reducing agent
Reduction
Salt bridge
Spontaneous
Standard cell potentials
Standard state conditions
Standard reduction potential (E_{red}°)
Voltaic cell

Problems

Common Oxidation–Reduction Reactions

1. Define *oxidation* and *reduction*.

2. What is a half-reaction?

3. Which of the following are oxidation half-reactions, and which are reduction half-reactions?

(a) $Zn^{2+}(aq) + 2\,e^- \rightleftharpoons Zn(s)$

(b) $Ag(s) \rightleftharpoons Ag^+(aq) + e^-$

(c) $Cu^{2+}(aq) + 2\,e^- \rightleftharpoons Cu(s)$

(d) $Al(s) \rightleftharpoons Al^{3+}(aq) + 3\,e^-$

Determining Oxidation Numbers

4. Calculate the oxidation number of chromium in the following compounds and ions.

   (a) CrO_4^{2-} (b) $Cr_2O_7^{2-}$ (c) CrO_2

5. Calculate the oxidation number of the aluminum atom in the following compounds and ions.

   (a) $Al(OH)_3$ (b) $AlCl_3$

   (c) $AlCl_4^-$ (d) $Al_2(SO_4)_3$

6. Which of the following compounds or ions contain hydrogen with a negative oxidation number?

   (a) HCl (b) NH_4^+ (c) LiH

   (d) C_2H_6 (e) H_2O

7. Arrange the following compounds in order of increasing oxidation number of the carbon atom.

   (a) C (b) CO (c) CO_2

   (d) H_2CO (e) CH_3OH (f) CH_4

8. Carbon can have any oxidation number between -4 and $+4$. Calculate the oxidation number of carbon in the following compounds.

   (a) $CH_3{-}CH_2{-}\overset{\displaystyle O}{\overset{\|}{C}}{-}H$

   (b) $CH_3{-}\overset{\displaystyle CH_3}{\underset{\displaystyle CH_3}{\overset{\displaystyle |}{\underset{\displaystyle |}{C}}}}{-}OH$

   (c) (C_6H_6)

9. Sulfur can have any oxidation number between $+6$ and -2. Calculate the oxidation number of sulfur in the following compounds.

   (a) Na_2S (b) MgS (c) CS_2

   (d) H_2SO_4 (e) SF_6 (f) $BaSO_3$

   (g) $NaHSO_4$ (h) SCl_2

10. Calculate the oxidation number of each atom in the following compounds.

    (a) $CH_3{-}\overset{\displaystyle O}{\overset{\|}{C}}{-}NH_2$

    (b) $CH_2{=}CH{-}CH_2{-}OH$

    (c) $CH_3{-}CH_2{-}\overset{\displaystyle O}{\overset{\|}{C}}{-}O{-}CH_3$

11. What is the relationship between the oxidation number of chlorine in each of the following acids and the relative strength of the acids? $HClO_4$, $HClO_3$, $HClO_2$, $HClO$.

Recognizing Oxidation–Reduction Reactions

12. Which statement(s) correctly describe(s) the following reaction?

$$3\,Sn^{2+}(aq) + Cr_2O_7^{2-}(aq) + 14\,H^+(aq)$$
$$\rightleftharpoons 3\,Sn^{4+}(aq) + 2\,Cr^{3+}(aq) + 7\,H_2O(l)$$

   (a) Both the Sn^{2+} and H^+ ions are reduced.

   (b) The $Cr_2O_7^{2-}$ ion is reduced.

   (c) The Sn^{2+} ion is reduced.

   (d) None of the above are true.

13. The H_3O^+ or hydronium ion participates in reactions other than acid–base reactions. Which of the following involves oxidation or reduction of H_3O^+?

$$Mg(s) + 2\,H_3O^+(aq)$$
$$\rightleftharpoons Mg^{2+}(aq) + H_2(g) + 2\,H_2O(l)$$
$$H_3O^+(aq) + NH_3(aq) \rightleftharpoons NH_4^+(aq) + H_2O(l)$$

14. Decide whether each of the following reactions involves oxidation–reduction. If it does, identify what is oxidized and what is reduced.

    (a) $4\,CH_3\overset{\displaystyle O}{\overset{\|}{C}}CH_3 + LiAlH_4 + 4\,H_2O$

    $$\rightleftharpoons 4\,CH_3\overset{\displaystyle OH}{\overset{|}{C}}HCH_3 + LiOH + Al(OH)_3$$

    (b) $CH_3CH_2OH \xrightarrow{\;H_2SO_4\;} CH_2{=}CH_2 + H_2O$

    (c) $CH_3\overset{\displaystyle O}{\overset{\|}{C}}OH + CH_3NH_2$

    $$\rightleftharpoons CH_3\overset{\displaystyle O}{\overset{\|}{C}}O^- + CH_3NH_3^+$$

15. Use Lewis structures to explain what happens when CO_2 reacts with water to form carbonic acid, H_2CO_3. Is this an oxidation–reduction reaction?

16. Use Lewis structures to assign oxidation numbers to all the atoms in the following reaction and show that this reaction is a reduction half-reaction.

$$NO_3^-(aq) + 4\,H^+(aq) + 3\,e^-$$
$$\rightleftharpoons NO(g) + 2\,H_2O(l)$$

17. Use Lewis structures to assign oxidation numbers to all the atoms in the following reaction using the methods introduced in Section 5.15.

$$2\,SO_3^{2-}(aq) + O_2(g) \rightleftharpoons 2\,SO_4^{2-}(aq)$$

18. Organic chemists often think about oxidation in terms of the loss of a pair of hydrogen atoms. Ethyl alcohol,

CH_3CH_2OH, is oxidized to acetaldehyde, CH_3CHO, for example, by removing a pair of hydrogen atoms. Show that this reaction obeys the general rule that oxidation is any process in which the oxidation number of an atom increases.

19. Organic chemists often think about reduction in terms of the gain of a pair of hydrogen atoms. Ethylene, C_2H_4, is reduced when it reacts with H_2 to form ethane, C_2H_6, for example. Show that this reaction obeys the general rule that reduction is any process in which the oxidation number of an atom decreases.

20. Step 7 in the citric acid cycle in Figure 12.1 is the conversion of fumarate to malate. Is this an oxidation–reduction reaction? Why or why not?

Voltaic Cells

21. Describe the function of a salt bridge in an electrochemical cell. Explain what happens when the salt bridge is removed from the system and why.

22. Describe the relationships between pairs of the following terms: *cathode, anode, cation,* and *anion.*

23. Explain why cations move from the salt bridge into the solution containing the cathode of a voltaic cell.

24. Balance the following reactions:
    (a) $Al(s) + Cr^{3+}(aq) \rightleftharpoons Al^{3+}(aq) + Cr(s)$
    (b) $Fe^{2+}(aq) + I_2(aq) \rightleftharpoons Fe^{3+}(aq) + I^-(aq)$
    (c) $Cr(s) + Fe^{3+}(aq) \rightleftharpoons Cr^{3+}(aq) + Fe^{2+}(aq)$
    (d) $Zn(s) + H^+(aq) \rightleftharpoons Zn^{2+}(aq) + H_2(g)$

25. For each of the reactions in Problem 24, write the corresponding half-reactions.

Standard Cell Potentials

26. Describe the difference between E and $E°$ for a cell.

27. Describe what the sign and magnitude of $E°$ for an oxidation–reduction reaction tells us about the reaction.

28. Explain why cell potentials measure the relative strengths of a pair of oxidizing agents and the relative strengths of a pair of reducing agents but not their absolute strengths.

29. Describe what happens to the sign and magnitude of $E°$ for an oxidation–reduction reaction when the direction in which the reaction is written is reversed.

30. Which of the following oxidation–reduction reactions should occur as written when run under standard conditions?
    (a) $Al(s)+Cr^{3+}(aq) \rightleftharpoons Al^{3+}(aq)+Cr(s)$
    $$E° = 0.966 \text{ V}$$
    (b) $3 Cr^{2+}(aq) \rightleftharpoons Cr(s)+2 Cr^{3+}(aq)$
    $$E° = 0.33 \text{ V}$$
    (c) $Fe(s)+Cr^{3+}(aq) \rightleftharpoons Cr(s)+Fe^{3+}(aq)$
    $$E° = -0.70 \text{ V}$$
    (d) $3 H_2(g)+2 Cr^{3+}(aq) \rightleftharpoons 2 Cr(s)+6 H^+(aq)$
    $$E° = -0.74 \text{ V}$$

Oxidizing and Reducing Agents

31. In each of the following reactions, state which species is oxidized and which species is reduced. Identify the oxidizing and reducing agents for each reaction.
    (a) $Al(s) + Cr^{3+}(aq) \rightleftharpoons Al^{3+}(aq) + Cr(s)$
    (b) $3 Cr^{2+}(aq) \rightleftharpoons Cr(s) + 2 Cr^{3+}(aq)$
    (c) $Fe(s) + Cr^{3+}(aq) \rightleftharpoons Cr(s) + Fe^{3+}(aq)$
    (d) $3 H_2(g) + 2 Cr^{3+}(aq) \rightleftharpoons 2 Cr(s) + 6 H^+(aq)$

32. Identify the oxidizing and reducing agents in each of the following reactions.
    (a) $Co(s) + 2 Cr^{3+}(aq) \rightleftharpoons Co^{2+}(aq) + 2 Cr^{2+}(aq)$
    (b) $H_2(g) + Zn^{2+}(aq) \rightleftharpoons Zn(s) + 2 H^+(aq)$
    (c) $Sn(s) + 2 H^+(aq) \rightleftharpoons Sn^{2+}(aq) + H_2(g)$
    (d) $3 Na(l) + AlCl_3(l) \rightleftharpoons 3 NaCl(l) + Al(l)$

33. Can carbon in the oxidation state +4 act as an oxidizing agent? Explain.

34. Some of the following can serve as oxidizing agents, some as reducing agents, and some as both. Identify each.

Element	Oxidation State
C	−4
N	+5
S	+2
Ge	+4
As	+3
Sb	+5
N	−3
I	+7

35. Which of the following can't be an oxidizing agent?
    (a) Cl^- (b) Br_2 (c) Fe^{3+}
    (d) Zn (e) CaH_2

36. Which of the following can't be a reducing agent?
    (a) H_2 (b) Cl_2 (c) Fe^{3+}
    (d) Al (e) LiH

37. Which of the following can be both an oxidizing agent and a reducing agent?
    (a) H_2 (b) I_2 (c) H_2O_2 (d) P_4 (e) S_8

38. Identify the oxidizing agents on both sides of the following unbalanced equation.

$$S_2O_3^{2-}(aq) + MnO_4^-(aq)$$
$$\rightleftharpoons SO_4^{2-}(aq) + Mn^{2+}(aq)$$

39. Why can an atom in its lowest oxidation state act only as a reducing agent?

Relative Strengths of Oxidizing and Reducing Agents

40. Which of the following transition metals is the strongest reducing agent?
    (a) Cr   (b) Mn   (c) Fe   (d) Co   (e) Ni

41. Which of the following oxidation–reduction reactions should occur as written?

(a) $Co(s) + 2 Cr^{3+}(aq) \rightleftharpoons Co^{2+}(aq) + 2 Cr^{2+}(aq)$

(b) $H_2(g) + Zn^{2+}(aq) \rightleftharpoons Zn(s) + 2 H^+(aq)$

(c) $Sn(s) + 2 H^+(aq) \rightleftharpoons Sn^{2+}(aq) + H_2(g)$

42. Which of the following oxidation–reduction reactions should occur as written? See Table B.12 in Appendix B.

(a) $Pb(s) + Mn^{2+}(aq) \rightleftharpoons Pb^{2+}(aq) + Mn(s)$

(b) $6 Fe^{2+}(aq) + 14 H^+(aq) + Cr_2O_7^{2-}(aq)$
   $\rightleftharpoons 6 Fe^{3+}(aq) + 2 Cr^{3+}(aq) + 7 H_2O(l)$

(c) $2 H_2O_2(aq) \rightleftharpoons 2 H_2O(l) + O_2(g)$

43. Use Table 12.1 to predict the products of the following oxidation–reduction reactions.

(a) $CrO_4^-(aq) + H^+(aq) + Fe(s) \rightleftharpoons$

(b) $Mg(s) + Cr^{3+}(aq) \rightleftharpoons$

(c) $MnO_4^-(aq) + H^+(aq) + Au^+(aq) \rightleftharpoons$

44. Use the data in Table B.12 in Appendix B to determine which of the following pairs of ions will react with each other in aqueous solution.

(a) Cr^{2+} and MnO_4^- (b) Fe^{3+} and $Cr_2O_7^{2-}$

(c) Ag^+ and Cu^{2+} (d) Fe^{2+} and Cr^{3+}

45. Which of the following solutions contains the strongest oxidizing agent?

(a) $CrO_4^{2-}(aq) + H^+(aq)$

(b) $Cr_2O_7^{2-}(aq) + H^+(aq)$

(c) $MnO_4^-(aq) + H^+(aq)$

(d) $F_2(g) + H^+(aq)$

(e) $O_2(g) + H^+(aq)$

46. What do the following half-cell reduction potentials tell us about the relative strengths of zinc and copper metal as reducing agents? What do they tell us about the relative strengths of Zn^{2+} and Cu^{2+} ions as oxidizing agents?

$$Zn^{2+} + 2 e^- \rightleftharpoons Zn \quad E° = -0.7628 \text{ V}$$
$$Cu^{2+} + 2 e^- \rightleftharpoons Cu \quad E° = 0.3402 \text{ V}$$

47. Describe an experiment that would allow you to determine the relative strengths of copper and iron metal as reducing agents.

48. Which of the following is the strongest reducing agent?

(a) Zn   (b) Fe   (c) H_2

(d) Cu   (e) Ag

49. Which of the following is the strongest oxidizing agent? See Table B.12 in Appendix B.

(a) Zn^{2+} (b) Zn   (c) Sn^{4+}

(d) Sn^{2+} (e) Cl_2

50. Which of the following is the strongest reducing agent? See Table B.12 in Appendix B.

(a) H^+ (b) H_2 (c) H^-

(d) H_2O (e) O_2

51. Which is the better reducing agent? See Table B.12 in Appendix B.

(a) H^- or K   (b) Sn or Fe^{2+}

(c) Ag^+ or Au^+

52. Which of the following oxidation–reduction reactions should occur as written when run under standard conditions?

(a) $2 Fe^{2+}(aq) + H_2O_2(aq)$
   $\rightleftharpoons 2 Fe^{3+}(aq) + 2 OH^-(aq) \quad E° = 0.11 \text{ V}$

(b) $2 Fe^{2+}(aq) + Cl_2(aq)$
   $\rightleftharpoons 2 Fe^{3+}(aq) + 2 Cl^-(aq) \quad E° = 0.588 \text{ V}$

(c) $2 Fe^{2+}(aq) + Br_2(aq)$
   $\rightleftharpoons 2 Fe^{3+}(aq) + 2 Br^-(aq) \quad E° = 0.317 \text{ V}$

(d) $2 Fe^{2+}(aq) + I_2(aq)$
   $\rightleftharpoons 2 Fe^{3+}(aq) + 2 I^-(aq) \quad E° = -0.235 \text{ V}$

53. Use the results of the previous problem to determine the relative strengths of H_2O_2, Cl_2, Br_2, I_2, and the Fe^{3+} ion as oxidizing agents.

54. Predict which of the following reactions should occur spontaneously as written when run under standard conditions.

(a) $Zn(s) + 2 H^+(aq) \rightleftharpoons Zn^{2+}(aq) + H_2(g)$

(b) $Cr(s) + 3 Fe^{3+}(aq) \rightleftharpoons Cr^{3+}(aq) + 3 Fe^{2+}(aq)$

(c) $Mn(s) + Mg^{2+}(aq) \rightleftharpoons Mn^{2+}(aq) + Mg(s)$

55. Predict which of the following reactions should occur spontaneously as written when run under standard conditions.

(a) $5 Fe^{3+}(aq) + Mn^{2+}(aq) + 4 H_2O(l)$
   $\rightleftharpoons 5 Fe^{2+}(aq) + MnO_4^-(aq) + 8 H^+(aq)$

(b) $3 Cu(s) + 2 HNO_3(aq) + 6 H^+(aq)$
   $\rightleftharpoons 3 Cu^{2+}(aq) + 2 NO(g) + 4 H_2O(l)$

56. Calculate $E°$ for the following reaction and predict whether the reaction should occur spontaneously as written when run under standard conditions.

$$6 Fe^{2+}(aq) + Cr_2O_7^{2-}(aq) + 14 H^+(aq)$$
$$\rightleftharpoons 6 Fe^{3+}(aq) + 2 Cr^{3+}(aq) + 7 H_2O(l)$$

57. Calculate $E°$ for the following reaction and predict whether the reaction should occur spontaneously as written when run under standard conditions.

$$Al^{3+}(aq) + Fe(s) \rightleftharpoons Fe^{3+}(aq) + Al(s)$$

58. Some metals react with water at room temperature, and others do not. Calculate $E°$ for the following reaction and predict whether the reaction should occur spontaneously as written when run under standard conditions.

$$Ca(s) + 2 H_2O(l)$$
$$\rightleftharpoons Ca^{2+}(aq) + 2 OH^-(aq) + H_2(g)$$

59. Calculate $E°$ for the following reaction to predict whether the Cu^+ ion should spontaneously undergo disproportionation under standard conditions.

$$2\,Cu^+(aq) \rightleftharpoons Cu(s) + Cu^{2+}(aq)$$

60. Use standard half-cell reduction potentials to predict whether an $Fe^{2+}(aq)$ solution should react with an $H^+(aq)$ solution to produce $Fe^{3+}(aq)$ and $H_2(g)$ under standard conditions.

61. Exercise 12.7 explained why Cu metal dissolves in nitric acid but not in hydrochloric acid. Use half-cell potentials to explain why Zn metal dissolves in both acids.

62. Explain why copper, gold, mercury, platinum, and silver can be found in the metallic state in nature. Explain why Li is not found in the metallic state in nature.

Batteries

63. What happens when a lead–acid battery is discharged? How can this battery be charged?

64. Why were Leclanché cell batteries popular?

65. What is a fuel cell?

66. What is the danger associated with charging a lead–acid battery?

Electrochemical Cells at Nonstandard Conditions: The Nernst Equation

67. Explain why there is only one value of $E°$ for a cell at a given temperature but many different values of E.

68. Explain why the magnitude of the cell potential becomes smaller as the cell comes closer to equilibrium.

69. What does it mean when E for a cell is equal to zero? What does it mean when $E°$ is equal to zero?

70. Which of the following describes an oxidation–reduction reaction at equilibrium?
    (a) $E = 0$ (b) $E° = 0$ (c) $E = E°$
    (d) $Q_c = 0$ (e) $\ln K_c = 0$

71. Write the Nernst equation for the following reaction and calculate the cell potential, assuming that the Al^{3+} ion concentration is 1.2 M and the Fe^{3+} ion concentration is $1.0 \times 10^{-4}M$ at 298 K.

$$Al^{3+}(aq) + Fe(s) \rightleftharpoons Al(s) + Fe^{3+}(aq)$$

Calculate the equilibrium constant for this reaction. Is Fe(s) oxidized by Al^{3+}?

72. Calculate $E°$ for the following reaction.

$$2\,Fe^{2+}(aq) + H_2O_2(aq)$$
$$\rightleftharpoons 2\,Fe^{3+}(aq) + 2\,OH^-(aq)$$

Write the Nernst equation for the reaction and calculate the cell potential for a system in a pH 10 buffer for which all other components are present at standard conditions.

73. Assume that we start with a Daniell cell at standard conditions.

$$Zn(s) + Cu^{2+}(aq) \rightleftharpoons Zn^{2+}(aq) + Cu(s)$$

Calculate the cell potential under the following sets of conditions.
    (a) In all, 99% of the Cu^{2+} ions have been consumed.
    (b) The reaction has reached 99.99% completion.
    (c) The reaction has reached 99.9999% completion.
    (d) The Cu^{2+} ion concentration is only $1 \times 10^{-8}M$.
    (e) The reaction has reached equilibrium.

74. Calculate the standard cell potential for the voltaic cell built around the following oxidation–reduction reaction.

$$2\,MnO_4^-(aq) + 5\,H_2O_2(aq) + 6\,H^+(aq)$$
$$\rightleftharpoons 2\,Mn^{2+}(aq) + 8\,H_2O(l) + 5\,O_2(g)$$

Use the Nernst equation to predict the effect on the cell potential of an increase in the pH of the solution. Predict the effect of an increase in the H_2O_2 concentration.

75. What would be the effect of an increase in temperature on the potential of the Daniell cell? Would it increase, decrease, or remain the same?

76. Predict the standard cell potential for the following reaction at 25°C.

$$Zn(s) + Fe^{2+}(aq) \rightleftharpoons Zn^{2+}(aq) + Fe(s)$$

Calculate the equilibrium constant for this reaction.

77. Calculate the equilibrium constant at 25°C for the Daniell cell. Use the results of the calculation to explain why a single arrow rather than a double-headed arrow could be used in the following equation.

$$Zn(s) + Cu^{2+}(aq) \rightleftharpoons Zn^{2+}(aq) + Cu(s)$$

Electrolysis and Faraday's Law

78. What is the difference between an electrolytic cell and a voltaic cell?

79. Describe an experiment that uses Faraday's law to determine the number of moles of electrons used in an electrolysis reaction.

80. Calculate the mass of silver metal that can be prepared from Ag^+ ions in solution with 1.0000 C of electric charge.

81. At a current of 1 amp, how many seconds will it take to prepare 1 mole of Na metal from molten NaCl?

82. If a 5.0-A current were used, how long would it take to prepare 1.0 mole of sodium metal? Of magnesium metal? Of aluminum metal?

83. Predict the products of the electrolysis of molten LiBr and calculate the number of grams of each product formed by electrolysis for 1.0 h with a 2.5-A current.

84. Calculate the number of coulombs necessary to produce a metric tonne (1000 kg) of Cl_2 gas from Cl^-.

85. Calculate the mass of copper metal deposited after a 2.0 M solution of $CuSO_4$ has been electrolyzed for 2.5 h with a 4.5-A current.

86. Calculate the ratio of the mass of O_2 to the mass of H_2 produced when water is electrolyzed for 2.56 h with a 1.34-A current.

87. Under a particular set of conditions, electrolysis of an aqueous $AgNO_3$ solution generated 1.00 g of silver metal. How many grams of I_2 would be produced by the electrolysis of an aqueous solution of NaI under the same conditions?

88. Calculate the ratio of the mass of Br_2 that collects at the anode to the mass of aluminum metal plated out at the cathode when molten $AlBr_3$ is electrolyzed for 10.5 h with a 20.0-A current.

89. Aqueous solutions of the following salts are electrolyzed for 20.0 min with a 10.0-A current. Which solution will deposit the most grams of metal at the cathode?

    (a) $ZnCl_2$ (b) $ZnBr_2$ (c) WCl_6

    (d) $ScBr_3$ (e) $HfCl_4$

90. An electric current is passed through a series of three cells filled with aqueous solutions of $AgNO_3$, $Cu(NO_3)_2$, and $Fe(NO_3)_3$. How many grams of each metal will be deposited by a 1.58-A current for 3.54 h?

91. What are the values of x and y in the formula Ce_xCl_y if electrolysis of the molten salt for 16.5 h with a 1.00-A current deposits 21.6 g of cerium metal?

92. What is the oxidation number of the manganese in an unknown salt if electrolysis of an aqueous solution of the salt for 30 min with a 10.0-A current generates 2.56 g of manganese metal?

93. Gold forms compounds in the +1 and +3 oxidation states. What is the oxidation number of gold in a compound that deposits 1.53 g of gold metal when electrolyzed for 15 min with a 2.50-A current?

Electrolysis of Molten NaCl

94. Molten NaCl contains Na^+ and Cl^- ions. Why does Na^+ migrate to the cathode and Cl^- to the anode when a battery supplies a current to the molten salt?

95. What would be produced at the anode and cathode if molten CaF_2 were electrolyzed? Show the half-reactions.

96. How does the electrolysis of molten NaCl illustrate the difference between voltaic and electrolytic cells?

Electrolysis of Aqueous NaCl

97. Explain why electrolysis of aqueous NaCl produces H_2 gas, not sodium metal, at the cathode, and Cl_2 gas, not O_2 gas, at the anode.

98. Calculate the standard cell potential necessary to electrolyze an aqueous solution of NaCl.

Electrolysis of Water

99. Explain why an electrolyte, such as Na_2SO_4, has to be added to water before the water can be electrolyzed to H_2 and O_2.

100. Describe what would happen if $NiSO_4$ were used instead of Na_2SO_4 when water was electrolyzed.

101. Why does the solution around the cathode become basic when an aqueous solution of Na_2SO_4 is electrolyzed?

102. What are the most likely products of the electrolysis of an aqueous solution of KBr?

    (a) $K^+(aq) + Br^-(aq)$

    (b) $K(s) + Br_2(l)$

    (c) $OH^-(aq) + H_2(g) + Br_2(l)$

    (d) $K^+(aq) + OH^-(aq) + O_2(g) + Br_2(l)$

103. Why can't aluminum metal be prepared by the electrolysis of an aqueous solution of the Al^{3+} ion?

104 Which of the following reactions occurs at the anode during electrolysis of an aqueous Na_2SO_4 solution?

    (a) $SO_4^{2-}(aq) \rightleftharpoons SO_2(g) + O_2(g) + 2\ e^-$

    (b) $2\ H_2O(l) \rightleftharpoons O_2(g) + 4\ H^+(aq) + 4\ e^-$

    (c) $2\ H_2O(l) + O_2(g) + 4\ e^- \rightleftharpoons 4\ OH^-(aq)$

    (d) $SO_4^{2-}(aq) + 4\ H^+(aq) + 2\ e^-$
    $\rightleftharpoons SO_2(g) + 2\ H_2O(l)$

Integrated Problems

105. An electrochemical cell is constructed from a Zn/Zn^{2+} electrode and a Cr/Cr^{3+} electrode as shown here. The voltage measured for the cell is 0.02 V. When the switch is closed, the chromium electrode increases in mass.

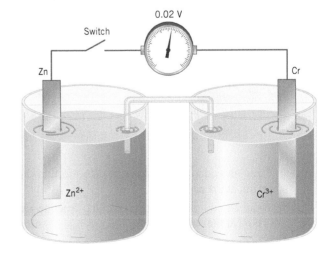

(a) Which electrode is the anode, and which is the cathode?

(b) What is the direction of electron flow?

(c) Which electrode is positive? Which electrode is negative?

(d) What are the chemical reactions occurring at each electrode?

(e) What is the overall chemical reaction of the cell?

(f) If the Zn/Zn^{2+} couple is replaced by a standard hydrogen electrode [$2\,H^+(aq) + 2\,e^- \rightleftharpoons H_2(g)$], which is then connected to the Cr/Cr^{3+} couple, what will be the direction of the electron flow and the spontaneous overall chemical reaction?

106. An iron/tin galvanic cell is constructed. An iron strip is placed into a 1.0 *M* solution of $Fe(NO_3)_3$, and this electrode is connected by a salt bridge to an electrode consisting of a strip of tin in a 1.0 *M* $Sn(NO_3)_2$ solution. The tin strip is observed to dissolve.

(a) Roughly sketch the galvanic cell.

(b) Identify the anode and cathode.

(c) Which electrode is positive and which is negative?

(d) What is the overall spontaneous chemical reaction?

(e) What is the standard potential of the cell?

107. The heroine of a movie has been captured by the villains and thrown into a storage room of an old mansion. There are steel bars on the windows, a silver door knob on the door, brass hinges on the skylight, and exposed copper plumbing in the room. In the corner of the room she finds a container of muriatic acid (6 *M* HCl), used for cleaning bricks. Suggest some options for her escape. Explain your reasoning.

108. The following electrochemical cell is set up. When the circuit is closed, the tin electrode gains weight.

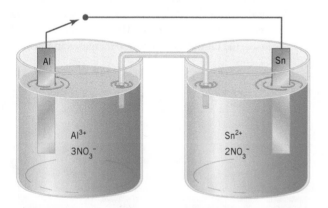

(a) Write an equation that describes the reaction that occurs at the cathode.

(b) Write an equation that describes the reaction that occurs at the anode.

(c) What is the overall cell reaction?

(d) What is the direction of electron flow in the external circuit?

(e) If the aluminum reduction potential is –1.66 V and the cell voltage is 1.52 V, what is the reduction potential for the tin electrode?

109. Given the following half-cell reduction potentials:

$$Ni^{2+}(aq) + 2\,e^- \rightleftharpoons Ni(s) \qquad E°(V) = -0.23$$
$$Pt^{2+}(aq) + 2\,e^- \rightleftharpoons Pt(s) \qquad E°(V) = +1.2$$
$$Pd^{2+}(aq) + 2\,e^- \rightleftharpoons Pd(s) \qquad E°(V) = 0.99$$

(a) Roughly sketch the cell for which the overall cell potential is the greatest.

(b) Identify the cathode and anode, and then show the direction of electron flow for the cell in (a).

(c) Will $Pt(s)$ reduce $Pd^{2+}(aq)$? Explain.

(d) What will happen if a strip of $Ni(s)$ is placed into a solution containing $Pt^{2+}(aq)$? Explain.

110. Which of the following statements are true for the reaction

$$5\,CrO_4^{2-}(aq) + 3\,Mn^{2+}(aq) + 16\,H^+(aq)$$
$$\rightleftharpoons 5\,Cr^{3+}(aq) + 3\,MnO_4^-(aq) + 8\,H_2O(l)$$

(a) Both Mn^{2+} and H^+ are oxidizing agents.

(b) CrO_4^{2-} is the oxidizing agent.

(c) Mn^{2+} is reduced.

(d) Cr^{3+} is a reducing agent.

(e) MnO_4^- is an oxidizing agent.

111. The following electrochemical cell is set up.

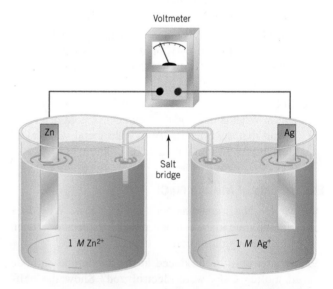

$$Zn^{2+}(aq) + 2\,e^- \rightleftharpoons Zn(s) \qquad E°(V) = -0.76$$
$$Ag^+(aq) + e^- \rightleftharpoons Ag(s) \qquad E°(V) = +0.80$$

(a) Which bar, Ag or Zn, will gain weight when the circuit is closed?

(b) What is the voltage of the cell?

(c) Identify the anode and the cathode.

(d) What is the direction of electron flow in the circuit?

(e) What is the overall chemical reaction?

(f) Calculate the equilibrium constant for the overall reaction.

112. The stoichiometry of the reaction between hydrogen peroxide and the MnO_4^- ion depends on the pH of the solution in which the reaction is run.

$$2\ MnO_4^-(aq) + 5\ H_2O_2(aq) + 6\ H^+(aq)$$
$$\rightleftharpoons 2\ Mn^{2+}(aq) + 5\ O_2(g) + 8\ H_2O(l)$$
$$2\ MnO_4^-(aq) + 3\ H_2O_2(aq) \rightleftharpoons 2\ MnO_2(s)$$
$$+\ 3\ O_2(g) + 2\ OH^-(aq) + 2\ H_2O(l)$$

Assume that it took 32.45 mL of 0.145 M $KMnO_4$ to titrate 28.46 mL of 0.248 M H_2O_2. Was the reaction run in acidic or basic solution?

113. Oxalic acid reacts with the chromate ion in acidic solution to give CO_2 and Cr^{3+} ions.

$$3\ H_2C_2O_4(aq) + 2\ CrO_4^{2-}(aq) + 10\ H^+(aq)$$
$$\xrightarrow{H^+} 6\ CO_2(g) + 2\ Cr^{3+}(aq) + 8\ H_2O(l)$$

If 10.0 mL of oxalic acid consumes 40.0 mL of 0.0250 M CrO_4^{2-} solution, state the molarity of the oxalic acid solution.

114. A 20.00-mL sample of a $K_2C_2O_4$ solution was acidified and then titrated with an acidic 0.256 M $KMnO_4$ solution. What is the molarity of the oxalate solution if it took 14.6 mL of the MnO^{4-} solution to reach the end point of the titration?

$$5\ C_2H_2O_4(aq) + 2\ MnO_4^-(aq) + 6\ H^+(aq)$$
$$\rightleftharpoons 10\ CO_2(g) + 2\ Mn^{2+}(aq) + 8\ H_2O(l)$$

115. Calculate the number of grams of iron(II) chloride ($FeCl_2$) that can be oxidized by 3.2 mL of 3.0 M $KMnO_4$.

$$5\ Fe^{2+}(aq) + MnO_4^-(aq) + 8\ H^+(aq)$$
$$\rightleftharpoons Mn^{2+}(aq) + 4\ H_2O(l) + 5\ Fe^{3+}(aq)$$

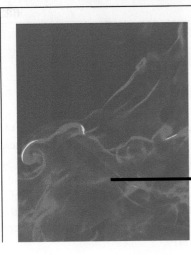

Chapter Twelve

SPECIAL TOPICS

12A.1 Balancing Oxidation–Reduction Equations

A trial-and-error approach to balancing chemical equations was introduced in Chapter 1. This approach involves playing with the equation—adjusting the ratio of the reactants and products—until the following goals have been achieved.

Goals for Balancing Chemical Equations

- The same number of atoms of each element can be found on both sides of the equation, and therefore mass is conserved. (Mass is conserved because atoms are neither created nor destroyed in a chemical reaction.)
- The sum of the positive and negative charges is the same on both sides of the equation, and therefore charge is conserved. (Charge is conserved because electrons are neither created nor destroyed in the reaction.)

At times, an equation is too complex to be solved by trial and error. Consider the following reaction, for example.

$$3\,Cu(s) + 8\,HNO_3(aq) \rightleftharpoons 3\,Cu^{2+}(aq) + 2\,NO(g) + 6\,NO_3^-(aq) + 4\,H_2O(l)$$

At other times, more than one balanced equation can be written. The following are just a few of the balanced equations that can be written for the reaction between the permanganate ion and hydrogen peroxide, for example.

$$2\,MnO_4^-(aq) + H_2O_2(aq) + 6\,H^+(aq) \rightleftharpoons 2\,Mn^{2+}(aq) + 3\,O_2(g) + 4\,H_2O(l)$$
$$2\,MnO_4^-(aq) + 3\,H_2O_2(aq) + 6\,H^+(aq) \rightleftharpoons 2\,Mn^{2+}(aq) + 4\,O_2(g) + 6\,H_2O(l)$$
$$2\,MnO_4^-(aq) + 5\,H_2O_2(aq) + 6\,H^+(aq) \rightleftharpoons 2\,Mn^{2+}(aq) + 5\,O_2(g) + 8\,H_2O(l)$$
$$2\,MnO_4^-(aq) + 7\,H_2O_2(aq) + 6\,H^+(aq) \rightleftharpoons 2\,Mn^{2+}(aq) + 6\,O_2(g) + 10\,H_2O(l)$$

Equations such as these have to be balanced by a more systematic approach than trial and error.

12A.2 Redox Reactions in Acidic Solutions

A powerful technique for balancing oxidation–reduction equations involves dividing the reactions into separate oxidation and reduction half-reactions. We then balance the half-reactions, one at a time, and combine them so that electrons are neither created nor destroyed in the reaction.

Consider the oxidation of sulfur dioxide by the dichromate ion in acidic solution, for example.

$$SO_2(aq) + Cr_2O_7^{2-}(aq) \rightleftharpoons SO_4^{2-}(aq) + Cr^{3+}(aq)$$

The reason why this equation is inherently difficult to balance has nothing to do with the ratio of moles of SO_2 to moles of $Cr_2O_7^{2-}$. It results from the fact that the solvent takes an active role in both half-reactions. Let's therefore use this reaction to illustrate the steps in the half-reaction method for balancing equations.

We start by writing a skeleton equation for the reaction and assigning oxidation numbers to atoms on both sides of the equation. When the rules for assigning oxidation numbers described in Section 5.16 are applied to the skeleton equation for this reaction, we get the following results.

$$SO_2 + Cr_2O_7^{2-} \rightleftharpoons SO_4^{2-} + Cr^{3+}$$
$$+4 \; -2 \quad +6 \; -2 \qquad +6 \; -2 \qquad +3$$

The next step involves determining which atoms are oxidized and which are reduced. In the course of the reaction, there is an increase in the oxidation number of the sulfur atom and a decrease in the oxidation number of the chromium. Thus sulfur is oxidized and chromium is reduced.

$$SO_2 + Cr_2O_7^{2-} \rightleftharpoons SO_4^{2-} + Cr^{3+}$$
$$+4 \qquad +6 \qquad\qquad +6 \qquad +3$$

Oxidation

Reduction

We now divide the reaction into oxidation and reduction half-reactions and balance the half-reactions, one at a time. The reaction between SO_2 and the $Cr_2O_7^{2-}$ ion can be divided into the following half-reactions.

Oxidation: $SO_2 \rightleftharpoons SO_4^{2-}$
                     $+4 \qquad\quad +6$

Reduction: $Cr_2O_7^{2-} \rightleftharpoons Cr^{3+}$
                     $+6 \qquad\quad +3$

It doesn't matter which half-reaction we balance first, so let's start with the reduction half-reaction. Because the $Cr_2O_7^{2-}$ ion contains two chromium atoms that must be reduced from the +6 to the +3 oxidation state, six electrons are consumed in this half-reaction.

Reduction: $Cr_2O_7^{2-} + 6\,e^- \rightleftharpoons 2\,Cr^{3+}$

What happens to the oxygen atoms when the chromium atoms are reduced? The seven oxygen atoms in the $Cr_2O_7^{2-}$ ion are in the -2 oxidation state. If these atoms were released into solution in the -2 oxidation state when the chromium was reduced, they would be present as O^{2-} ions. But the O^{2-} ion is a very strong base, and the reaction is being run in an acidic solution. Any O^{2-} ions liberated when the dichromate ion is reduced would instantly react with the H^+ ions in the acidic solution to form water. Because there are seven oxygen atoms in a -2 oxidation state in the dichromate ion, 14 H^+ ions are consumed and seven H_2O molecules are produced in this half-reaction.

Reduction: $Cr_2O_7^{2-} + 14\,H^+ + 6\,e^- \rightleftharpoons 2\,Cr^{3+} + 7\,H_2O$

We can now turn to the oxidation half-reaction, and start by noting that two electrons are given off when sulfur is oxidized from the +4 to the +6 oxidation state.

Oxidation: $SO_2 \rightleftharpoons SO_4^{2-} + 2\,e^-$

The charge on both sides of this equation is balanced by remembering that the reaction is run in acid, which contains H^+ ions and H_2O molecules. We can therefore add H^+ ions or H_2O molecules to either side of the equation, as needed.

The only way to balance the charge on both sides of this equation is to add four H^+ ions to the products of the reaction.

$$\text{Oxidation:} \quad SO_2 \rightleftharpoons SO_4^{2-} + 2\,e^- + 4\,H^+$$

We then balance the number of hydrogen and oxygen atoms on both sides of the equation by adding a pair of H_2O molecules to the reactants.

$$\text{Oxidation:} \quad SO_2 + 2\,H_2O \rightleftharpoons SO_4^{2-} + 2\,e^- + 4\,H^+$$

We are now ready to combine the two half-reactions so that electrons are neither created nor destroyed. Six electrons are consumed in the reduction half-reaction, and two electrons are given off in the oxidation half-reaction. We can combine the half-reactions so that electrons are conserved by multiplying the oxidation half-reaction by 3.

$$Cr_2O_7^{2-} + 14\,H^+ + 6\,e^- \rightleftharpoons 2\,Cr^{3+} + 7\,H_2O$$
$$3\,(SO_2 + 2\,H_2O \rightleftharpoons SO_4^{2-} + 2\,e^- + 4\,H^+)$$
$$\overline{Cr_2O_7^{2-} + 3\,SO_2 + 14\,H^+ + 6\,H_2O \rightleftharpoons 2\,Cr^{3+} + 3\,SO_4^{2-} + 12\,H^+ + 7\,H_2O}$$

Although the equation appears balanced, we aren't quite finished with it. We can simplify the equation by subtracting 12 H^+ ions and 6 H_2O molecules from each side to generate the following balanced equation. We indicate that the reaction is reversible by writing a double-headed arrow between the left and right sides of the equation.

$$Cr_2O_7^{2-}(aq) + 3\,SO_2(aq) + 2\,H^+(aq) \rightleftharpoons 2\,Cr^{3+}(aq) + 3\,SO_4^{2-}(aq) + H_2O(l)$$

Exercise 12A.1

As we have seen, an endless number of balanced equations can be written for the reaction between the permanganate ion and hydrogen peroxide in acidic solution to form the Mn^{2+} ion and oxygen gas.

$$MnO_4^-(aq) + H_2O_2(aq) \xrightarrow{H^+} Mn^{2+}(aq) + O_2(g)$$

Use the half-reaction method to determine the correct stoichiometry for the reaction.

Solution

Once again, we start by writing a skeleton equation and assigning oxidation numbers to atoms on both sides of the equation,

$$MnO_4^- + H_2O_2 \rightleftharpoons Mn^{2+} + O_2$$
$$_{+7\,-2}\quad _{+1\,-1}\qquad _{+2}\qquad _{0}$$

and then determining which atoms are oxidized and which are reduced.

$$MnO_4^- + H_2O_2 \rightleftharpoons Mn^{2+} + O_2$$
$$_{+7}\qquad _{-1}\qquad _{+2}\quad _{0}$$

$$\underset{\text{Reduction}}{\underline{\qquad\qquad}}$$
$$\underset{\text{Oxidation}}{\underline{\qquad\qquad\qquad}}$$

> CHECKPOINT

On an exam, students were asked to use half-reactions to write a balanced equation for the reaction between SO_2 and the $Cr_2O_7^{2-}$ ion in the presence of a strong acid.

$$SO_2(g) + Cr_2O_7^{2-}(aq)$$
$$\xrightarrow{H^+} SO_4^{2-}(aq) + Cr^{3+}(aq)$$

One student wrote the following half-reactions.

Oxidation: $\quad SO_2 + 4\,OH^-$
$$\rightleftharpoons SO_4^{2-} + 2\,H_2O + 2\,e^-$$

Reduction: $\quad Cr_2O_7^{2-} + 7\,H_2O + 6\,e^-$
$$\rightleftharpoons 2\,Cr^{3+} + 14\,OH^-$$

The student then wrote the following overall equation.

$$3\,SO_2(g) + Cr_2O_7^{2-}(aq) + H_2O(l)$$
$$\rightleftharpoons 3\,SO_4^{2-}(aq) + 2\,Cr^{3+}(aq)$$
$$+ 2\,OH^-(aq)$$

Explain why the instructor gave no credit for this answer, even though the overall equation is balanced.

We then divide the reaction into oxidation and reduction half-reactions and balance the half-reactions.

This reaction can be divided into the following half-reactions.

$$\text{Oxidation:} \quad \underset{-1}{H_2O_2} \quad \rightleftharpoons \quad \underset{0}{O_2}$$

$$\text{Reduction:} \quad \underset{+7}{MnO_4^-} \rightleftharpoons \underset{+2}{Mn^{2+}}$$

Let's balance the reduction half-reaction first. It takes five electrons to reduce manganese from +7 to +2.

$$\text{Reduction:} \quad MnO_4^- + 5\,e^- \rightleftharpoons Mn^{2+}$$

Because the reaction is run in acid, we can add H^+ ions or H_2O molecules to either side of the equation, as needed. Two hints tell us which side of the equation gets H^+ ions and which side gets H_2O molecules. The only way to balance the charge on both sides of the equation is to add eight H^+ ions to the left side of the equation.

$$\text{Reduction:} \quad MnO_4^- + 8\,H^+ + 5\,e^- \rightleftharpoons Mn^{2+}$$

The only way to balance the number of oxygen atoms is to add four H_2O molecules to the right side of the equation.

$$\text{Reduction:} \quad MnO_4^- + 8\,H^+ + 5\,e^- \rightleftharpoons Mn^{2+} + 4\,H_2O$$

To balance the oxidation half-reaction, we have to remove two electrons from a pair of oxygen atoms in the -1 oxidation state to form a neutral O_2 molecule.

$$\text{Oxidation:} \quad H_2O_2 \rightleftharpoons O_2 + 2\,e^-$$

We can then add a pair of H^+ ions to the products to balance both charge and mass in the half-reaction.

$$\text{Oxidation:} \quad H_2O_2 \rightleftharpoons O_2 + 2\,H^+ + 2\,e^-$$

Two electrons are given off during oxidation, and five electrons are consumed during reduction. We can combine the half-reactions so that electrons are conserved by using the lowest common multiple of 5 and 2.

$$\begin{array}{l} 2\,(MnO_4^- + 8\,H^+ + 5\,e^- \rightleftharpoons Mn^{2+} + 4\,H_2O) \\ \underline{5\,(H_2O_2 \rightleftharpoons O_2 + 2\,H^+ + 2\,e^-)} \\ 2\,MnO_4^- + 5\,H_2O_2 + 16\,H^+ \rightleftharpoons 2\,Mn^{2+} + 5\,O_2 + 10\,H^+ + 8\,H_2O \end{array}$$

The simplest balanced equation for the reaction is obtained when 10 H^+ ions are subtracted from each side of the equation derived in the previous step.

$$2\,MnO_4^-(aq) + 5\,H_2O_2(aq) + 6\,H^+(aq) \rightleftharpoons 2\,Mn^{2+}(aq) + 5\,O_2(g) + 8\,H_2O(l)$$

12A.3 Redox Reactions in Basic Solutions

The key to balancing half-reactions in basic solutions is to recognize that basic solutions contain H_2O molecules and OH^- ions. We can therefore add water molecules or hydroxide ions to either side of the equation, as needed.

In the previous section, we obtained the following equation for the reaction between the permanganate ion and hydrogen peroxide in an acidic solution.

$$2\,MnO_4^-(aq) + 5\,H_2O_2(aq) + 6\,H^+(aq) \rightleftharpoons 2\,Mn^{2+}(aq) + 5\,O_2(g) + 8\,H_2O(l)$$

It might be interesting to see what happens when the reaction occurs in a basic solution.

Exercise 12A.2

Write a balanced equation for the reaction between the permanganate ion and hydrogen peroxide in a basic solution to form manganese dioxide and oxygen.

$$MnO_4^-(aq) + H_2O_2(aq) \overset{OH^-}{\rightleftharpoons} MnO_2(s) + O_2(g)$$

Solution

We start, as always, with a skeleton equation to which oxidation numbers have been assigned.

$$\underset{+7\,-2}{MnO_4^-} + \underset{+1\,-1}{H_2O_2} \rightleftharpoons \underset{+4\,-2}{MnO_2} + \underset{0}{O_2}$$

We then determine which atoms are oxidized and which are reduced,

$$\underset{+7}{MnO_4^-} + \underset{-1}{H_2O_2} \rightleftharpoons \underset{+4}{MnO_2} + \underset{0}{O_2}$$

Reduction

Oxidation

divide the reaction into oxidation and reduction half-reactions, and then balance the half-reactions. This reaction can be divided into the following half-reactions.

$$\text{Reduction:}\quad \underset{+7}{MnO_4^-} \rightleftharpoons \underset{+4}{MnO_2}$$

$$\text{Oxidation:}\quad \underset{-1}{H_2O_2} \rightleftharpoons \underset{0}{O_2}$$

Let's start by balancing the reduction half-reaction. It takes three electrons to reduce manganese from $+7$ to the $+4$ oxidation state.

$$\text{Reduction:}\quad MnO_4^- + 3\,e^- \rightleftharpoons MnO_2$$

We now try to balance the number of atoms and the charge on both sides of the equation. Because the reaction is run in basic solution, we can add OH^- ions or H_2O molecules to either side of the equation, as needed. The only way to balance the net charge of -4 on the left side of the equation is to add four OH^- ions to the products.

$$\text{Reduction:} \quad MnO_4^- + 3\,e^- \rightleftharpoons MnO_2 + 4\,OH^-$$

We can then balance the number of hydrogen and oxygen atoms by adding two H_2O molecules to the reactants.

$$\text{Reduction:} \quad MnO_4^- + 3\,e^- + 2\,H_2O \rightleftharpoons MnO_2 + 4\,OH^-$$

We now turn to the oxidation half-reaction. Two electrons are lost when hydrogen peroxide is oxidized to form O_2 molecules.

$$\text{Oxidation:} \quad H_2O_2 \rightleftharpoons O_2 + 2\,e^-$$

We can balance the charge on the half-reaction by adding a pair of OH^- ions to the reactants.

$$\text{Oxidation:} \quad H_2O_2 + 2\,OH^- \rightleftharpoons O_2 + 2\,e^-$$

The only way to balance the number of hydrogen and oxygen atoms is to add two H_2O molecules to the products.

$$\text{Oxidation:} \quad H_2O_2 + 2\,OH^- \rightleftharpoons O_2 + 2\,H_2O + 2\,e^-$$

Two electrons are given off during the oxidation half-reaction, and three electrons are consumed in the reduction half-reaction. We can therefore combine the half-reactions so that electrons are neither created nor destroyed by using the lowest common multiple of 2 and 3.

$$2\,(MnO_4^- + 3\,e^- + 2\,H_2O \rightleftharpoons MnO_2 + 4\,OH^-)$$
$$3\,(H_2O_2 + 2\,OH^- \rightleftharpoons O_2 + 2\,H_2O + 2\,e^-)$$
$$\overline{2\,MnO_4^- + 3\,H_2O_2 + 6\,OH^- + 4\,H_2O \rightleftharpoons 2\,MnO_2 + 3\,O_2 + 8\,OH^- + 6\,H_2O}$$

The simplest balanced equation is obtained by subtracting four H_2O molecules and six OH^- ions from each side of the equation derived in the previous step.

$$2\,MnO_4^-(aq) + 3\,H_2O_2(aq) \rightleftharpoons 2\,MnO_2(s) + 3\,O_2(g) + 2\,OH^-(aq) + 2\,H_2O(l)$$

Note that the ratio of moles of MnO_4^- to moles of H_2O_2 consumed is different in acidic and basic solutions. This difference results from the fact that MnO_4^- is reduced all the way to Mn^{2+} in acid, but the reaction stops at MnO_2 in base.

· ·

12A.4 Molecular Redox Reactions

Lewis structures can play a vital role in understanding the oxidation–reduction reactions of covalent molecules. Consider the following reaction, for example, which is used in the Breathalyzer to determine the amount of ethyl alcohol or

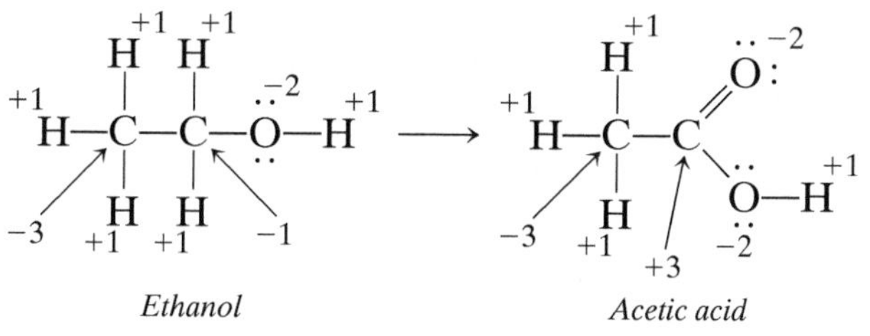

Fig. 12A.1 Oxidation of ethanol to acetic acid.

ethanol on the breath of individuals who are suspected of driving while under the influence.

$$3\,CH_3CH_2OH(g) + 2\,Cr_2O_7^{2-}(aq) + 16\,H^+(aq)$$
$$\Longleftrightarrow 3\,CH_3CO_2H(aq) + 4\,Cr^{3+}(aq) + 11\,H_2O(l)$$

We start by assigning oxidation numbers to each of the atoms in the Lewis structures of ethanol and acetic acid, as shown in Figure 12A.1.

The carbon in the CH_3 group has the same oxidation state in both ethanol and acetic acid. There is a change in the oxidation number of the other carbon atom, however, from -1 to $+3$. The oxidation half-reaction therefore corresponds to the loss of four electrons by one of the carbon atoms.

Oxidation: $CH_3CH_2OH \Longleftrightarrow CH_3CO_2H + 4\,e^-$

Because the reaction is run in acidic solution, we can add H^+ and H_2O as needed to balance the equation.

Oxidation: $CH_3CH_2OH + H_2O \Longleftrightarrow CH_3CO_2H + 4\,e^- + 4\,H^+$

The other half of the reaction involves a six-electron reduction of the $Cr_2O_7^{2-}$ ion in acidic solution to form a pair of Cr^{3+} ions.

Reduction: $Cr_2O_7^{2-} + 6\,e^- \Longleftrightarrow 2\,Cr^{3+}$

Adding H^+ ions and H_2O molecules as needed gives the following balanced equation for the reduction half-reaction.

Reduction: $Cr_2O_7^{2-} + 14\,H^+ + 6\,e^- \Longleftrightarrow 2\,Cr^{3+} + 7\,H_2O$

We are now ready to combine the two half-reactions by recognizing that electrons are neither created nor destroyed in the reaction.

$$3\,(CH_3CH_2OH + H_2O \Longleftrightarrow CH_3CO_2H + 4\,e^- + 4\,H^+)$$
$$\underline{2\,(Cr_2O_7^{2-} + 14\,H^+ + 6\,e^- \Longleftrightarrow 2\,Cr^{3+} + 7\,H_2O)}$$
$$3\,CH_3CH_2OH + 2\,Cr_2O_7^{2-} + 28\,H^+ + 3\,H_2O \Longleftrightarrow 3\,CH_3CO_2H + 4\,Cr^{3+} + 12\,H^+ + 14\,H_2O$$

Simplifying the equation by removing 3 H_2O molecules and 12 H^+ ions from both sides of the equation gives the balanced equation for the reaction.

$$3\,CH_3CH_2OH(g) + 2\,Cr_2O_7^{2-}(aq) + 16\,H^+(aq)$$
$$\Longleftrightarrow 3\,CH_3CO_2H(aq) + 4\,Cr^{3+}(aq) + 11\,H_2O(l)$$

Problems

Balancing Oxidation–Reduction Equations

A-1. Some of the earliest matches consisted of white phosphorus and potassium chlorate glued to wooden sticks. Write a balanced equation for the following reaction, which occurred when the match was rubbed against a hard surface.

$$P_4(s) + KClO_3(s) \Longrightarrow P_4O_{10}(s) + KCl(s)$$

Is this an oxidation–reduction reaction? Explain your answer.

Redox Reactions in Acidic Solutions

A-2. Write balanced equations for the following reactions, in which an acidic solution of the CrO_4^{2-} or $Cr_2O_7^{2-}$ ion is used as an oxidizing agent.
(a) $I^-(aq) + CrO_4^{2-}(aq) \Longrightarrow I_2(aq) + Cr^{3+}(aq)$
(b) $Fe^{2+}(aq) + Cr_2O_7^{2-}(aq)$
$$\Longrightarrow Fe^{3+}(aq) + Cr^{3+}(aq)$$
(c) $H_2S(g) + CrO_4^{2-}(aq) \Longrightarrow S_8(s) + Cr^{3+}(aq)$

A-3. Write balanced equations for the following oxidation–reduction reactions, which involve an acidic solution of the MnO_4^- ion.
(a) $Fe^{2+}(aq) + MnO_4^-(aq)$
$$\Longrightarrow Fe^{3+}(aq) + Mn^{2+}(aq)$$
(b) $S_2O_3^{2-}(aq) + MnO_4^-(aq)$
$$\Longrightarrow SO_4^{2-}(aq) + Mn^{2+}(aq)$$
(c) $Mn^{2+}(aq) + PbO_2(s)$
$$\Longrightarrow MnO_4^-(aq) + Pb^{2+}(aq)$$
(d) $SO_2(g) + MnO_4^-(aq)$
$$\Longrightarrow H_2SO_4(aq) + Mn^{2+}(aq)$$

A-4. The two most common oxidation states of chromium are Cr(III) and Cr(VI). Balance the following oxidation–reduction equations that involve one of these oxidation states of chromium. Assume that the reactions occur in acidic solution.
(a) $Cr(s) + O_2(g) + H^+(aq) \Longrightarrow Cr^{3+}(aq)$
(b) $Fe^{2+}(aq) + Cr_2O_7^{2-}(aq)$
$$\Longrightarrow Fe^{3+}(aq) + Cr^{3+}(aq)$$
(c) $BrO_3^-(aq) + Cr^{3+}(aq)$
$$\Longrightarrow Br_2(aq) + HCrO_4^-(aq)$$

A-5. Reactions in which a reagent undergoes both oxidation and reduction are known as *disproportionation* reactions. Write balanced equations for the following disproportionation reactions in acidic solution.
(a) $H_2O_2(aq) \Longrightarrow H_2O(l) + O_2(g)$
(b) $NO_2(g) + H_2O(l) \Longrightarrow HNO_3(aq) + NO(g)$

A-6. The term *conproportionation* has been proposed to describe reactions that are the opposite of dispropor-

tionation reactions. Write balanced equations for the following conproportionation reactions in acidic solution.
(a) $HIO_3(aq) + HI(aq) \Longrightarrow I_2(aq)$
(b) $SO_2(g) + H_2S(g) \Longrightarrow S_8(s) + H_2O(l)$

A-7. Acids that are also oxidizing agents play an important role in the preparation of small quantities of the halogens in the laboratory. Write balanced equations for the following oxidation–reduction equations in acidic solution.
(a) $HCl(aq) + HNO_3(aq) \Longrightarrow NO(g) + Cl_2(g)$
(b) $HBr(aq) + H_2SO_4(aq) \Longrightarrow SO_2(g) + Br_2(g)$
(c) $HCl(aq) + MnO_4^-(aq) \Longrightarrow Cl_2(g) + Mn^{2+}(aq)$

A-8. The product of the reaction between copper metal and nitric acid depends on the concentration of the acid. In dilute nitric acid, NO is formed. Concentrated nitric acid produces NO_2. Write balanced equations for the following oxidation–reduction reactions between copper metal and nitric acid.
(a) $Cu(s) + HNO_3(aq) \Longrightarrow Cu^{2+}(aq) + NO(g)$
(b) $Cu(s) + HNO_3(aq) \Longrightarrow Cu^{2+}(aq) + NO_2(g)$

A-9. Use half-reactions to balance the following oxidation–reduction equation, which seems to give only a single product in acid solution.

$$H_2S(g) + H_2O_2(aq) \Longrightarrow S_8(s)$$

Redox Reactions in Basic Solutions

A-10. Balance the following oxidation–reduction equations that occur in basic solution.
(a) $NO(g) + MnO_4^-(aq)$
$$\Longrightarrow NO_3^-(aq) + MnO_2(s)$$
(b) $NH_3(g) + MnO_4^-(aq) \Longrightarrow N_2(aq) + MnO_2(s)$

A-11. Most organic compounds react with the MnO_4^- ion in the presence of acid to form CO_2 and water. MnO_4^- in the presence of base, however, is a weaker and therefore gentler oxidizing agent. Balance the following redox reactions in which the MnO_4^- ion is used to oxidize an organic compound.
(a) $CH_3OH(aq) + MnO_4^-(aq)$
$$\xrightarrow{H^+} CO_2(g) + Mn^{2+}(aq)$$
(b) $CH_3OH(aq) + MnO_4^-(aq)$
$$\xrightarrow{OH^-} HCO_2H(aq) + MnO_2(s)$$

A-12. The Cr^{3+} ion is not soluble in basic solutions because it precipitates as $Cr(OH)_3$. Chromium metal will dissolve in excess base, however, because it can form a soluble $Cr(OH)_4^-$ complex ion. $Cr(OH)_3$ can also be dissolved by oxidizing the chromium to the

+6 oxidation state. Write balanced equations for the following reactions that occur in basic solution.

(a) $Cr(s) + OH^-(aq) \rightleftharpoons Cr(OH)_4^-(aq) + H_2(g)$

(b) $Cr(OH)_3(s) + H_2O_2(aq) \rightleftharpoons CrO_4^{2-}(aq)$

A-13. Write balanced equations for the following reactions to see whether there is any difference between the stoichiometry of the reaction in which the sulfite ion is oxidized to the sulfate ion when the reaction is run in acidic versus basic solution.

(a) $SO_3^{2-}(aq) + O_2(g) \xrightarrow{H^+} SO_4^{2-}(aq)$

(b) $SO_3^{2-}(aq) + O_2(g) \xrightarrow{OH^-} SO_4^{2-}(aq)$

A-14. Write a balanced equation for the oxidation–reduction reaction used to extract gold from its ores.

$$Au(s) + CN^-(aq) + O_2(g)$$
$$\rightleftharpoons Au(CN)_2^-(aq) + OH^-(aq)$$

A-15. Write a balanced equation for the Raschig process, in which ammonia reacts with the hypochlorite ion in a basic solution to form hydrazine.

$$NH_3(aq) + OCl^-(aq) \rightleftharpoons N_2H_4(aq) + Cl^-(aq)$$

A-16. Use half-reactions to balance the following disproportionation reactions that occur in basic solution.

(a) $Cl_2(g) + OH^-(aq) \rightleftharpoons Cl^-(aq) + OCl^-(aq)$

(b) $Cl_2(g) + OH^-(aq) \rightleftharpoons Cl^-(aq) + ClO_3^-(aq)$

(c) $P_4(s) + OH^-(aq) \rightleftharpoons PH_3(g) + H_2PO_2^-(aq)$

(d) $H_2O_2(aq) \rightleftharpoons O_2(g) + H_2O(l)$

A-17. The chemistry of nitrogen is unusual because so many of its compounds undergo reactions that give nitrogen in its elemental form. Balance the following oxidation–reduction reaction in which ammonia is converted into molecular nitrogen in basic solution.

$$CuO(s) + NH_3(g) \rightleftharpoons Cu(s) + N_2(g) + H_2O(l)$$

Molecular Redox Reactions

A-18. What is oxidized and what is reduced in the following reactions?

(a) $CH_4(g) + 2\,O_2(g) \rightleftharpoons CO_2(g) + 2\,H_2O(g)$

(b) $CH_3CH{=}CH_2(g) + H_2(g)$
$$\rightleftharpoons CH_3CH_2CH_3(g)$$

A-19. Which, if any, of the following in each pair of species represents the oxidized species?

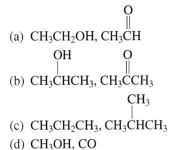

(a) $CH_3CH_2OH, CH_3\overset{\displaystyle O}{\overset{\|}{C}}H$

(b) $CH_3\overset{OH}{\underset{|}{C}}HCH_3, CH_3\overset{\displaystyle O}{\overset{\|}{C}}CH_3$

(c) $CH_3CH_2CH_3, CH_3\overset{CH_3}{\underset{|}{C}}HCH_3$

(d) CH_3OH, CO

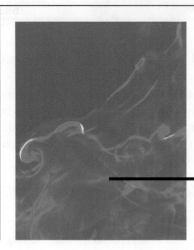

Chapter Thirteen

CHEMICAL THERMODYNAMICS

13.1 Spontaneous Chemical and Physical Processes

No one is surprised when a cup of hot coffee gradually loses heat to its surroundings as it cools, or when the ice in a glass of lemonade melts as it absorbs heat from its surroundings. But we would be surprised if a cup of coffee suddenly grew hotter until it boiled, or if the water in a glass of lemonade suddenly froze on a hot summer's day.

In a similar fashion, no one should be surprised when a piece of zinc metal dissolves in a strong acid to give bubbles of hydrogen gas that rise through the solution to escape from the top surface of the solution.

$$Zn(s) + 2 H^+(aq) \longrightarrow Zn^{2+}(aq) + H_2(g)$$

But if we saw a video in which H_2 bubbles formed on the surface of a solution and then sank through the solution until they disappeared, while a strip of zinc metal formed in the middle of the solution, we would conclude that the video was being run backward. Philosophers have argued that conventional wisdom gives us some idea of **time's arrow;** the direction in which time flows. Time's arrow is the direction in which a process is said to be spontaneous.

Many chemical and physical processes proceed in a direction in which they are said to be **spontaneous.** This raises the question: What factor or factors determine the direction in which a process is spontaneous? What drives a chemical reaction in one direction and not the other?

Section 7.9 noted that chemical reactions that give off heat to their surroundings are said to be **exothermic.** So many spontaneous processes are exothermic that it seems reasonable to assume that one of the driving forces that determine whether a reaction is spontaneous is a tendency to give off heat. The oxidation of ethanol by the body is an example of a spontaneous chemical reaction that is exothermic.

$$CH_3CH_2OH(l) + 3 O_2(g) \longrightarrow 2 CO_2(g) + 3 H_2O(l) \qquad \Delta H° = -1367 \text{ kJ/mol}_{rxn}$$

So is the reaction in which iron is oxidized to iron (III) oxide.

$$4 Fe(s) + 3 O_2(g) \longrightarrow 2 Fe_2O_3(s) \qquad \Delta H° = -1648.4 \text{ kJ/mol}_{rxn}$$

There are spontaneous processes, however, that absorb energy from their surroundings. At 100°C, for example, water boils spontaneously, even though the process is **endothermic.**

$$H_2O(l) \rightleftharpoons H_2O(g) \qquad \Delta H°_{373} = 40.88 \text{ kJ/mol}_{rxn}$$

Another example of a spontaneous reaction that absorbs heat from its surroundings is the process by which ammonium nitrate dissolves in water. This reaction is spontaneous even though heat is absorbed when the reaction takes place.

$$NH_4NO_3(s) \overset{H_2O}{\rightleftharpoons} NH_4^+(aq) + NO_3^-(aq) \qquad \Delta H° = 28.1 \text{ kJ/mol}_{rxn}$$

The temperature of the system can play a role in determining the direction in which a process is spontaneous. At temperatures below 0°C, water spontaneously freezes.

$$T < 0°C \qquad H_2O(l) \longrightarrow H_2O(s)$$

At temperatures above 0°C, however, the process is spontaneous in the opposite direction.

$$T > 0°C \qquad H_2O(s) \longrightarrow H_2O(l)$$

Thus the tendency of a process to give off heat can't be the only driving force that determines the direction in which the process is spontaneous. There must be another factor that helps determine whether a process is spontaneous. This factor, known as **entropy,** is a measure of the randomness or disorder of the system.

➤ **CHECKPOINT**

When a teaspoon of table salt is poured into a glass of water, the salt dissolves, even though heat is absorbed during the process. Is this a spontaneous process?

13.2   Entropy and Disorder

Entropy is commonly considered to be a measure of disorder or randomness. To understand what we mean by the terms **disorder** and *randomness,* try to imagine the process by which a sample of ice first melts and then eventually boils as the sample is heated. The water molecules in ice are locked into a rigid structure, as shown in Figure 8.20. Any translational, rotational, or vibrational motion of the individual molecules is severely restricted by the hydrogen bonds between the water molecules locked into this rigid structure. We can therefore conclude that the water molecules in ice can be described as having a high degree of order or a low degree of randomness.

Now imagine what happens when the ice melts. The water molecules are no longer locked into a rigid structure; they are freer to move through space. Liquid water can therefore be assumed to be more disordered than ice, and the process by which ice melts can be said to produce an increase in the entropy of the system. Another significant increase in entropy should occur when the liquid boils, to form a gas in which the motion of the water molecules can best be described as random or chaotic.

In 1877 Ludwig Boltzmann proposed the following relationship between entropy and microscopic disorder of a system.

$$S = k \ln W$$

In this equation, S is a state function corresponding to the entropy of the system, k is a proportionality constant, ln represents a logarithm to the base e, and W is a measure of disorder on the microscopic or atomic scale.

According to this equation, the entropy of the system increases as the value of W increases. W can be thought of as the number of energetically equivalent ways one can put together a given macroscopic state. Imagine a perfect crystal at 0 K. The value of W will be small, if not quite equal to one, because there is only one way in which a perfect crystal can be formed. As the crystal is warmed, the translational, rotational, and vibrational motion of the particles will increase. This leads to an increase in the number of possible ways in which the particles that form the crystal might exist and therefore an increase in the value of W. When the crystal melts, to form a liquid, the number of possible ways of arranging the particles that form the system increases, as does the value of W. The number of energetically equivalent microscopic states for a gas would be significantly larger, and therefore so would be the entropy of the system.

The particles of an ideal gas discussed in Chapter 6 have a distribution of kinetic energies as shown in Figure 13.1. At the temperature T_1, some of the particles have more kinetic energy than average and some have less. If the gas is heated without allowing the volume to change, the kinetic energy of the gas particles will increase. At the higher temperature, T_2, the average temperature is higher, and the energy is shared differently among the particles than at the lower temperature. In Figure 13.1 the energy distribution curve at the higher temperature is broadened, showing that the distribution of energies has changed so that more particles are moving with a higher kinetic energy. The hotter gas has the higher entropy because there is now more energy in the system and there is more randomness in the way the energy is distributed. A similar situation occurs as a result of a chemical reaction. Because of the energy change produced by the making and breaking of bonds, at equilibrium, the energy is distributed among the species present in the most probable way.

Microscopic disorder can be realized in ways that are not always obvious. In some cases, such as the melting of a solid or the breaking apart of a larger molecule into smaller molecules, an increase in disorder can be readily inferred.

Because entropy is a state function, the change in entropy that occurs as a system is transformed from some initial state to some final state can be calculated as follows.

$$\Delta S = S_{final} - S_{initial}$$

When a process produces an increase in the disorder of the universe, the entropy change, ΔS_{univ}, is *positive*.

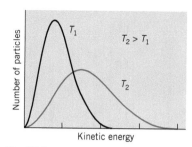

Fig. 13.1 For an ideal gas at a given temperature, not all particles have the same kinetic energy. As temperature increases, the average kinetic energy of the particles increases and the distribution of kinetic energies of the particles broadens.

13.3 Entropy and the Second Law of Thermodynamics

As noted in Section 7.1, *thermodynamics* is the study of energy and its transformations. The **first law of thermodynamics,** discussed in Chapter 7, provides us with ways of understanding energy changes in chemical reactions. The **second law of thermodynamics** helps us determine the direction in which natural processes such as chemical reactions should occur.

The role that disorder plays in determining the direction in which a natural process occurs is summarized by the second law, which states that for a process to occur, the entropy of the universe must increase. The term *universe* means everything that might conceivably have its entropy altered as a result of the process. The second law of thermodynamics can therefore be written as follows.

Second law: $\Delta S_{univ} \geq 0$ (for natural processes)

This statement means that processes for which $\Delta S_{univ} > 0$ will occur naturally, but that processes for which $\Delta S_{univ} < 0$ will probably never occur. If $\Delta S_{univ} = 0$, the process may occur in either direction.

It is important to recognize that the second law of thermodynamics describes what happens to the entropy of the *universe,* not the system. The universe is divided into two components, the system and its surroundings. The change in the entropy of the universe is the sum of the change in the entropy of the system and the surroundings.

$$\Delta S_{univ} = \Delta S_{sys} + \Delta S_{surr}$$

Chemists generally choose the chemical reaction in which they are interested as the system. The term *surroundings* is then used to describe the environment that is altered as a consequence of the reaction. The second law does not mean that if $\Delta S_{sys} > 0$ then all processes will occur naturally. Consider, for example, the equilibrium between solid water and liquid water at 0°C and 1 atm. The following data shows how ΔS_{sys}, ΔS_{surr}, and ΔS_{univ} vary with temperature.

$$H_2O(s) \rightleftharpoons H_2O(l)$$

	+10°C	0°C	−10°C
ΔS_{surr} (J/Kmol$_{rxn}$)	−22.6	−22.1	−21.5
ΔS_{sys} (J/Kmol$_{rxn}$)	+23.5	+22.1	+20.7
ΔS_{univ} (J/Kmol$_{rxn}$)	+0.9	0	−0.8

At 0°C water (s) and water (l) are in equilibrium and ΔS_{univ} is zero, which means that both solid and liquid water can coexist at this temperature. But if the temperature of the surroundings is increased to +10°C, ΔS_{univ} is +0.9 J/Kmol$_{rxn}$ and the process occurs naturally, that is, ice melts. However, at a surrounding temperature of −10°C, ΔS_{univ} is −0.8 J/Kmol$_{rxn}$ and ice will not melt. In all cases $\Delta S_{sys} > 0$, but the conditions in the surroundings influence the direction of the reaction.

The regular lattice structure of the solid and the closeness of the water molecules in this structure make it energetically more stable than the liquid at temperatures below 0°C. When the temperature is sufficiently high, the randomness of the liquid can dominate the energy stability of the solid liquid structure and the solid melts.

Unless otherwise specified, the symbol ΔS will be used from now on in this text to refer to the system. Entropy, like **enthalpy,** is a state function (see Section 7.5). Thus the value of ΔS for a process depends only on the initial and final conditions of the system.

> $\Delta S > 0$ implies that the system becomes *more disordered* during the reaction.
>
> $\Delta S < 0$ implies that the system becomes *less disordered* during the reaction.

The following generalizations can help us decide when a chemical reaction leads to an increase in the disorder of the system.

- Solids have a much more regular structure than liquids. Liquids therefore have a larger entropy than the corresponding solids.
- The particles in a gas are distributed over a larger volume than those in a liquid. Gases therefore have a larger entropy than the corresponding liquids.
- A process that increases the number of particles increases the entropy of the system.

Exercise 13.1

Which of the following would lead to an increase in the entropy of the system?

(a) The production of ammonia, NH_3, from its elements:

$$N_2(g) + 3\,H_2(g) \rightleftharpoons 2\,NH_3(g)$$

(b) The evaporation of water:

$$H_2O(l) \rightleftharpoons H_2O(g)$$

(c) The decomposition of limestone ($CaCO_3$) to produce lime (CaO):

$$CaCO_3(s) \rightleftharpoons CaO(s) + CO_2(g)$$

Solution

(a) The system becomes more ordered in this gas-phase reaction because the total number of particles in the system decreases. The entropy of the system therefore decreases.

(b) Gases are more disordered than the corresponding liquids, so the entropy of the system increases.

(c) Reactions in which a compound decomposes into two products generally lead to an increase in entropy because the system becomes more disordered. The increase in entropy in this reaction is even larger because the starting material is a solid and one of the products is a gas.

• •

Section 13.1 suggested that the sign of ΔH for a chemical reaction affects the direction in which the reaction occurs.

Exothermic reactions ($\Delta H < 0$) are often, but not always, spontaneous.

This section introduced the notion that ΔS for a reaction can also play a role in the direction of the reaction.

Reactions that lead to an increase in the disorder of the system ($\Delta S > 0$) are often, but not always, spontaneous.

To decide in which direction a chemical reaction will occur, it is therefore important to consider the effect of changes in both the enthalpy and entropy of the system that occur during the reaction.

• •

Exercise 13.2

Use the Lewis structures of NO_2 and N_2O_4 and the stoichiometry of the following reaction to decide whether ΔH and ΔS favor the reactants or the products of the reaction.

$$2\,NO_2(g) \rightleftharpoons N_2O_4(g)$$

Solution

NO_2 has an unpaired electron in its Lewis structure.

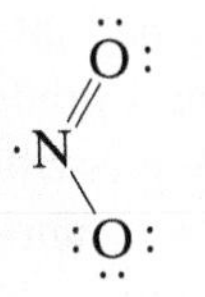

When a pair of NO_2 molecules collides in the proper orientation, a covalent bond can form between the nitrogen atoms to produce the N_2O_4 dimer.

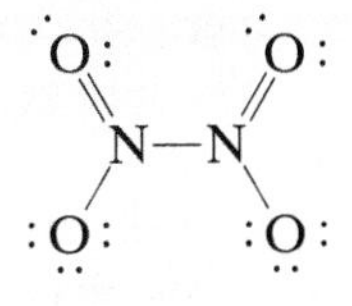

► **CHECKPOINT**

Is it possible for a process to proceed if the entropy of the system decreases? Explain why or why not.

Because a new bond is formed, the reaction should be exothermic ($\Delta H < 0$). ΔH for the reaction therefore favors the products.

The reaction leads to a decrease in the entropy of the system because the system becomes more ordered when two molecules come together to form a larger molecule ($\Delta S < 0$). ΔS for the reaction therefore favors the reactants.

13.4 Standard-State Entropies of Reaction

When the change in the entropy of a system that accompanies a chemical reaction is measured under **standard-state conditions,** the result is the **standard-state entropy of reaction, $\Delta S°$.** By convention, the standard state for thermodynamic measurements is characterized by the following conditions.

> **Standard-state conditions:**
>   **All solutes have a concentration of 1 *M*.**
>
>   **All gases have a partial pressure of 1 bar, or 0.9869 atm. (This can be rounded to 1 atm for all but the most exact calculations.)**

Although standard-state entropies can be measured at any temperature, they are often measured at 25°C. Standard-state **entropies of atom combination** at 25°C for a variety of compounds are given in Table B.13 in Appendix B.

13.5 The Third Law of Thermodynamics

The **third law of thermodynamics** defines absolute zero on the entropy scale.

> **The entropy of a perfect crystal is zero when the temperature of the crystal is equal to absolute zero (0 K).**

The crystal must be perfect, or else there will be some inherent disorder. It also must be at 0 K.

As the crystal warms to temperatures above 0 K, the particles in the crystal begin to move more and more vigorously, generating some disorder. The entropy of the crystal gradually increases with temperature as the average kinetic energy of the particles increases. At the melting point, the entropy of the system increases abruptly as the compound is transformed into a liquid, which isn't as well ordered as the solid. The entropy of the liquid gradually increases as the liquid becomes warmer because of the increase in the vibrational, rotational, and translational motion of the particles. There is another abrupt increase in the entropy of the substance at the boiling point, as the liquid is transformed into a random, chaotic gas.

Matter is never at rest. Even at the coldest temperature (absolute zero) there is a residual motion of the atoms that make up all matter. On warming, the motion and hence the disorder of the atoms increase. Thermal entropies are a measure of the increase in disorder over that at absolute zero that are produced by warming the substance to a higher temperature.

Table 13.1 provides an example of the difference between the entropy of a substance in the solid, liquid, and gaseous phases. The units of entropy are joules

Table 13.1

Entropies of Solid, Liquid, and Gaseous Forms of Sulfur Trioxide

Compound	$S°(\text{J/mol} \cdot \text{K})$
$SO_3(s)$	70.7
$SO_3(l)$	113.8
$SO_3(g)$	256.76

per mole kelvin (J/mol · K). A typical plot of the entropy of a system versus temperature is shown in Figure 13.2, which describes the entropy, $S°$, of a single substance at different temperatures.

13.6 Calculating Entropy Changes for Chemical Reactions

In Exercise 13.2 we determined that ΔH favors formation of the products while ΔS favors formation of the reactants for the reaction in which two molecules of NO_2 come together to produce N_2O_4.

$$2\,NO_2(g) \rightleftharpoons N_2O_4(g)$$

In order to gain additional insight into the meaning of the change in entropy that accompanies this reaction, we will calculate a quantitative value for ΔS for this reaction in units of $J/mol_{rxn} \cdot K$. We will approach the calculation of ΔS using the method that was developed in Chapter 7 for calculating ΔH.

Because entropy is a state function, the change in entropy that occurs during a chemical reaction does not depend on the path used to transform the starting reactants into the products of the reaction. This means that we can envision a reaction occurring by the same path that we have used for calculating ΔH, even though this is not the path by which the chemical reaction occurs.

We begin by building a reaction diagram that represents the process by which all of the bonds in the starting materials are broken to form isolated atoms in the gas phase.

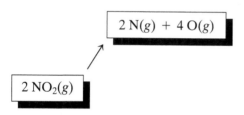

The arrow is drawn pointing up in this reaction diagram because there is an increase in the entropy of the system when NO_2 molecules are transformed into isolated atoms in the gas phase. The entropy change produced by this process can be calculated from the entropy of atom combination data in Table B.13 in Appendix B.

The tabulated values of $\Delta S°_{ac}$ in the Appendix B refer to the process whereby one mole of the substance listed in the table is *formed from its atoms in the gas phase*. Thus, if we want the entropy change for *breaking a compound apart* into its atoms, we must reverse the sign of the entropy data in the appendix. Because the overall reaction consumes 2 moles of NO_2 for every mole of N_2O_4 produced, the change in the entropy of the system that accompanies this step in the reaction must be multiplied by 2.

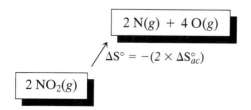

➤ **CHECKPOINT**

Why is the entropy of a substance the smallest at absolute zero?

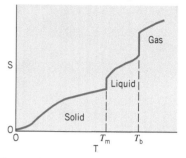

Fig. 13.2 There is a gradual increase in the entropy of a solid when it is heated until the temperature of the system reaches the melting point of the solid. At this point, the entropy of the system increases abruptly as the solid melts to form a liquid. The entropy of the liquid gradually increases until the system reaches the boiling point of the liquid, at which point it increases abruptly once more. The entropy of the gas then increases gradually with temperature.

According to the data in Table B.13, the value of ΔS°_{ac} for 1 mole of NO_2 is -235 J/mol$_{rxn} \cdot$ K. This value represents the change in the entropy for the formation of NO_2 molecules from isolated nitrogen and oxygen atoms. The dissociation of NO_2 into nitrogen and oxygen atoms in the gas phase, on the other hand, must lead to an increase in the disorder of the system. The value of ΔS° for two moles of NO_2 in this hypothetical bond-breaking process is therefore $+470$ J/mol$_{rxn} \cdot$ K.

Now consider what happens when the isolated atoms in the gas phase recombine to form N_2O_4.

$$2\,N(g)\ +\ 4\,O(g) \qquad \Delta S^\circ = \Delta S^\circ_{ac}$$
$$N_2O_4(g)$$

Six isolated atoms in the gas phase have to come together to form each N_2O_4 molecule. As a result, there is a significant decrease in the disorder in the system. The entropy of atom combination for N_2O_4 is relatively large, -647 J/mol$_{rxn} \cdot$ K, and ΔS° for this bond-making process is therefore both large and negative: $\Delta S^\circ = -647$ J/mol$_{rxn} \cdot$ K.

The overall entropy change for the dimerization of NO_2 to form N_2O_4 can be calculated by adding the values of ΔS° for the breaking and making of bonds in this hypothetical mechanism for the reaction.

Bond breaking:	$\Delta S^\circ = 470$ J/mol$_{rxn} \cdot$ K
Bond making:	$\Delta S^\circ = -647$ J/mol$_{rxn} \cdot$ K
Overall:	$\Delta S^\circ = -177$ J/mol$_{rxn} \cdot$ K

The overall entropy change in the reaction is negative, which isn't surprising because the reaction brings two isolated molecules together to form a more organized molecule.

The entropy of atom combination of a compound can tell us something about the way the atoms are held together to form the compound. Consider ΔS°_{ac} for cyclopentane and 1-pentene in the gas phase, for example.

$$5\,C(g) + 10\,H(g) \longrightarrow \underset{\text{\textit{Cyclopentane}}}{\begin{array}{c} CH_2 \\ CH_2 \quad CH_2 \\ | \qquad | \\ CH_2 \!\!-\!\! CH_2 \end{array}} (g)$$

$$\Delta S^\circ_{ac} = -1645 \text{ J/mol}_{rxn} \cdot \text{K}$$

$$5\,C(g) + 10\,H(g) \longrightarrow \underset{\text{\textit{1-Pentene}}}{CH_2\!\!=\!\!CHCH_2CH_2CH_3(g)}$$

$$\Delta S^\circ_{ac} = -1590 \text{ J/mol}_{rxn} \cdot \text{K}$$

Both compounds have the same number of carbon and hydrogen atoms, but they differ in the way the atoms are arranged, as shown in Figure 13.3. The entropy of atom combination of cyclopentane is -1645 J/mol$_{rxn} \cdot$ K, and that of 1-pentene is -1590 J/mol$_{rxn} \cdot$ K. What do these entropies tell us about the two compounds?

The change in entropy associated with a chemical reaction is essentially a measure of changes in the constraints to the motion of atoms, ions, or molecules that occur in the course of the reaction. Entropies of atom combination

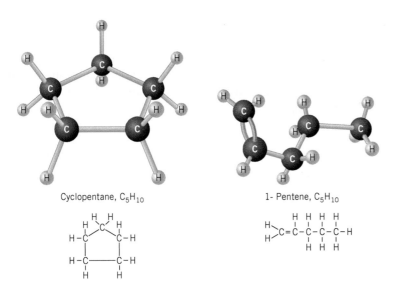

Fig. 13.3 Structures of cyclopentane and 1-pentene.

are therefore a measure of the freedom of motion lost when ions or molecules are formed from their gaseous atoms.

There are several types of motion associated with a molecule, as shown in Figure 13.4. First, there is the movement of the molecule as a whole from one position in space to another, which is known as *translational* motion. The larger the mass of the molecule, the larger the contribution to entropy from translational motion will be.

Molecules also tumble and twist as they move from place to place. This motion is called *rotation,* and the amount of rotational motion depends on both the molecular weight and the shape of the molecule. A turning or twisting motion occurs when atoms rotate around a bond. Molecules that are symmetrical have less rotational freedom than molecules whose atoms are arranged in a less symmetric fashion.

Vibrational motion occurs within the molecule because the bonds holding atoms together behave like a spring. Vibration can occur by either stretching and then compressing a bond, as shown in Figure 13.4, or by changing the bond angles.

When atoms form molecules, the magnitude of the decrease in the entropy of the system will depend on the number of atoms in the molecule, the symmetry

➤ **CHECKPOINT**

What would be the sign of ΔS for the conversion of glucose from its cyclic into its open-chain form?

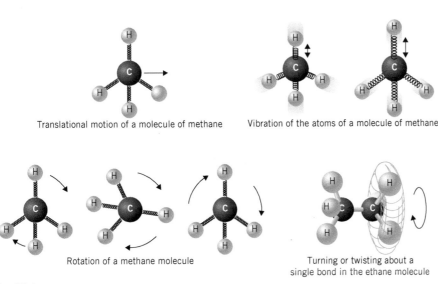

Fig. 13.4 Molecules have three types of motion: translational, rotational, and vibrational.

of the molecule, and the nature of the bonds between atoms. Let's consider how these general principles apply to the particular case of cyclopentane and 1-pentene.

First note that the number and kind of atoms in cyclopentane and 1-pentene are the same. Thus any difference in the entropy of atom combination must result from the way the atoms are held together in the molecule. The more negative value of ΔS°_{ac} for cyclopentane tells us that the atoms in cyclopentane have lost more freedom of motion than those of 1-pentene during the process that leads to the formation of these molecules. This, in turn, means that the atoms of cyclopentane have a more constrained motion than those of 1-pentene.

The carbon atoms in 1-pentene are free to vibrate relative to one another, and the various segments of the molecule twist around the single bonds between carbon atoms. Because cyclopentane is a cyclic structure, the twisting of segments of the molecule would require covalent bonds to break. Thus the carbon atoms in cyclopentane are restricted from undergoing that type of motion.

Let's apply entropy of atom combination data to the following reaction in which cyclopentane opens up to form 1-pentene.

$$\text{cyclopentane}(g) \rightleftharpoons \text{1-pentene}(g)$$

The entropy change for this reaction can be calculated by combining the entropy associated with decomposing cyclopentane into its atoms in the gas phase and the entropy associated with bringing these atoms together to form 1-pentene.

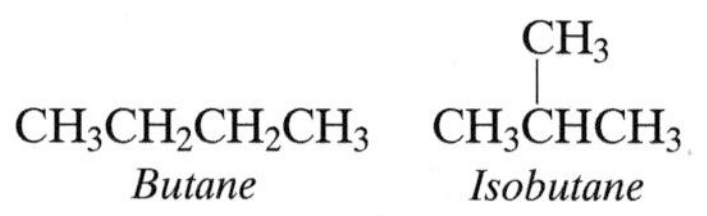

Bond breaking:	$\Delta S^\circ = -(-1645 \text{ J/mol}_{rxn} \cdot \text{K})$
Bond making:	$\Delta S^\circ = -1590 \text{ J/mol}_{rxn} \cdot \text{K}$
Overall:	$\Delta S^\circ = +55 \text{ J/mol}_{rxn} \cdot \text{K}$

The conversion of cyclopentane to 1-pentene is accompanied by an increase in entropy, which means that there is greater freedom of motion (greater disorder) for the atoms in the 1-pentene molecule than for those in cyclopentane.

ΔS°_{ac} data can provide a useful tool for studying a chemical reaction. There is always a positive entropy change when the bonds in the starting material are broken to produce isolated atoms in the gas phase. The gaseous atoms can then combine to re-form the starting materials, or they can rearrange to form the products of the reaction. In either case, the system becomes less disordered. As a result, the change in the entropy of the system when the atoms recombine is negative.

This raises an interesting question: Which process does entropy favor—recombination of the atoms in the gas phase to re-form the starting materials or rearrangement of the atoms to form the products of the reaction? Let's consider the two nitrogen atoms and four oxygen atoms in the example discussed at the beginning of this section. These atoms can combine to form either two molecules of NO_2 or one molecule of N_2O_4.

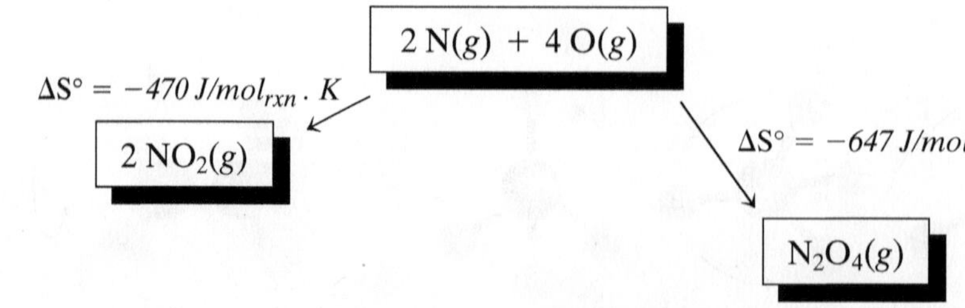

The second law of thermodynamics suggests that natural processes tend toward maximum disorder. Entropy therefore favors the process in which the

> **CHECKPOINT**

The entropies of atom combination of butane and isobutane are -1469 and $-1485 \text{ J/mol}_{rxn} \cdot \text{K}$, respectively. How can you account for this difference?

$$\text{CH}_3\text{CH}_2\text{CH}_2\text{CH}_3 \quad \overset{\overset{\displaystyle \text{CH}_3}{|}}{\text{CH}_3\text{CHCH}_3}$$
$$\textit{Butane} \qquad\quad \textit{Isobutane}$$

isolated gas-phase atoms combine to form the materials that involve the smallest decrease in the entropy of the system. We can therefore conclude that entropy favors the starting materials (NO_2), not the products (N_2O_4), of this reaction.

Entropy of atom combination data can be used by thinking about chemical reactions as if all bonds in the reactants are broken to form isolated atoms in the gas phase, which then combine to form the products of the reaction. Relatively few reactions occur through this mechanism, but entropy is a state function. This means that the value of ΔS obtained this way is the same as the value of ΔS that would be measured experimentally.

Exercise 13.3

Use the entropy of atom combination data given in this exercise to calculate the value of $\Delta S°$ at 298 K for the following reactions. Explain the sign of $\Delta S°$ for each reaction.

(a) $CH_3OH(l) \rightleftharpoons CH_3OH(g)$

Compound	$\Delta S°_{ac}(J/mol_{rxn} \cdot K)$
$CH_3OH(l)$	−651.2
$CH_3OH(g)$	−538.19

(b) $N_2(g) + O_2(g) \rightleftharpoons 2\,NO(g)$

Compound	$\Delta S°_{ac}(J/mol_{rxn} \cdot K)$
$N_2\,(g)$	−114.99
$O_2\,(g)$	−116.97
$NO(g)$	−103.59

Solution

(a) For liquid CH_3OH we can imagine a process that involves overcoming the intermolecular forces that hold the molecules together in the liquid and then breaking the intramolecular bonds within the CH_3OH molecules. This leads to an increase in the entropy of the system by 651.2 $J/mol_{rxn} \cdot K$. If the atoms then combine to form a mole of CH_3OH gas, the entropy of the system would decrease by 538.19 $J/mol_{rxn} \cdot K$.

$$\boxed{C(g) + 4\,H(g) + O(g)}$$

$\Delta S° = 651.2\ J/mol_{rxn} \cdot K$ $\Delta S° = 113.0\ J/mol_{rxn} \cdot K$ $\Delta S° = -538.19\ J/mol_{rxn} \cdot K$

$$\boxed{CH_3OH(g)}$$

$$\boxed{CH_3OH(l)}$$

When we combine the change in entropy for the steps in which bonds are broken and bonds are made, we find that the overall entropy of reaction is positive.

Bond breaking:	651.2 $J/mol_{rxn} \cdot K$
Bond making:	−538.19 $J/mol_{rxn} \cdot K$
	113.0 $J/mol_{rxn} \cdot K$

The entropy of the system increases in the course of this reaction because we are transforming a relatively ordered liquid into a more disordered gas.

(b) The change in entropy of the system that accompanies the process by which a mole of N_2 and a mole of O_2 are transformed into isolated nitrogen and oxygen atoms in the gas phase is equal to the sum of the entropy changes required to break the bonds in the two starting materials.

Bond breaking:

$$N_2(g) \longrightarrow 2\,N(g) \qquad\qquad \Delta S° = 114.99\ \text{J/mol}_{rxn} \cdot \text{K}$$
$$O_2(g) \longrightarrow 2\,O(g) \qquad\qquad \Delta S° = 116.97\ \text{J/mol}_{rxn} \cdot \text{K}$$
$$\overline{N_2(g) + O_2(g) \longrightarrow 2\,N(g) + 2\,O(g) \qquad \Delta S° = 231.96\ \text{J/mol}_{rxn} \cdot \text{K}}$$

The magnitude of $\Delta S°$ for the step in which the atoms in the gas phase combine to form two moles of NO is equal to twice the entropy of atom combination for the product of the reaction. The sign of this step is negative because the system becomes more ordered.

Bond making:

$$2\,N(g) + 2\,O(g) \longrightarrow 2\,NO(g) \qquad \Delta S°_{ac} = -(2 \times 103.59\ \text{J/mol}_{rxn} \cdot \text{K})$$

The overall change in entropy that accompanies the reaction can be calculated by combining the two steps in the hypothetical reaction.

$$
\begin{array}{ll}
\text{Bond breaking:} & 231.96\ \text{J/mol}_{rxn} \cdot \text{K} \\
\text{Bond making:} & \underline{-207.18\ \text{J/mol}_{rxn} \cdot \text{K}} \\
& 24.78\ \text{J/mol}_{rxn} \cdot \text{K}
\end{array}
$$

$\Delta S°$ for the reaction is small, but positive, because the freedom for rotational motion is larger in NO because it is less symmetric than N_2 or O_2.

● ●

13.7   Gibbs Free Energy

Section 13.3 noted that some reactions are spontaneous because they give off heat ($\Delta H < 0$) and others are spontaneous because they lead to an increase in the disorder of the system ($\Delta S > 0$). The following exercise shows how calculations of ΔH and ΔS can be used to determine the driving force behind a particular reaction.

● ●

Exercise 13.4

Use the enthalpy and entropy of atom combination data given in this exercise to calculate $\Delta H°$ and $\Delta S°$ for the following reaction. Determine the direction in which each of these factors would drive the reaction.

$$N_2(g) + 3\,H_2(g) \rightleftharpoons 2\,NH_3(g)$$

Compound	ΔH_{ac}° (kJ/mol$_{rxn}$)	ΔS_{ac}° (J/mol$_{rxn}\cdot$K)
$N_2(g)$	-945.41	-114.99
$H_2(g)$	-435.30	-98.74
$NH_3(g)$	-1171.76	-304.99

Solution

Both calculations can be based on the following reaction diagram.

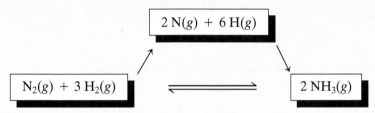

The enthalpy associated with breaking apart a mole of N_2 and 3 moles of H_2 molecules to form isolated nitrogen and hydrogen atoms in the gas phase is equal to the negative of the enthalpy of atom combination of N_2 ($\Delta H_{ac}^{\circ} = -945.41$ kJ/mol$_{rxn}$) plus three times the negative of the enthalpy of atom combination of H_2 ($\Delta H_{ac}^{\circ} = -435.30$ kJ/mol$_{rxn}$)

Bond breaking:

$N_2(g) \longrightarrow 2\,N(g)$	$\Delta H^{\circ} = -(-945.41$ kJ/mol$_{rxn})$
$3\,H_2(g) \longrightarrow 6\,H(g)$	$\Delta H^{\circ} = -(3 \times -435.30$ kJ/mol$_{rxn})$
$N_2(g) + 3\,H_2(g) \longrightarrow 2\,N(g) + 6\,H(g)$	$\Delta H^{\circ} = 2251.3$ kJ/mol$_{rxn}$

The magnitude of the enthalpy given off when the atoms come together to form 2 moles of NH_3 is equal to twice the enthalpy of atom combination of ammonia ($\Delta H_{ac}^{\circ} = -1171.76$ kJ/mol$_{rxn}$).

Bond making:

$$2\,N(g) + 6\,H(g) \longrightarrow 2\,NH_3 \qquad \Delta H_{ac}^{\circ} = (2 \times -1171.76 \text{ kJ/mol}_{rxn})$$

When the bond-breaking and bond-making steps are combined, we find that the overall reaction is exothermic ($\Delta H^{\circ} < 0$).

$$
\begin{array}{r}
2251.3 \ \text{kJ/mol}_{rxn} \\
-2343.52 \ \text{kJ/mol}_{rxn} \\
\hline
-92.2 \ \text{kJ/mol}_{rxn}
\end{array}
$$

The entropy change associated with transforming 1 mole of N_2 and 3 moles of H_2 to isolated nitrogen and hydrogen atoms in the gas phase can be found by a similar procedure.

Bond breaking:

$N_2(g) \longrightarrow 2\,N(g)$	$\Delta S^{\circ} = -(-114.99$ J/mol$_{rxn}\cdot$K$)$
$3\,H_2(g) \longrightarrow 6\,H(g)$	$\Delta S^{\circ} = -(3 \times -98.74$ J/mol$_{rxn}\cdot$K$)$
$N_2(g) + 3\,H_2(g) \longrightarrow 2\,N(g) + 6\,H(g)$	$\Delta S^{\circ} = 411.2$ J/mol$_{rxn}\cdot$K

The entropy change associated with the formation of 2 moles of NH_3 can be calculated from the entropy of atom combination of ammonia.

Bond making:

$$2\,N(g) + 6\,H(g) \longrightarrow 2\,NH_3 \qquad \Delta S_{ac}^{\circ} = (2 \times -304.99 \text{ J/mol}_{rxn}\cdot\text{K})$$

The overall entropy of reaction is negative because the reaction transforms 4 moles of reactants into 2 moles of products.

$$411.2 \quad J/mol_{rxn} \cdot K$$
$$\underline{-609.98 \ J/mol_{rxn} \cdot K}$$
$$-198.8 \quad J/mol_{rxn} \cdot K$$

According to these calculations, $\Delta H°$ for the reaction is -92.2 kJ/mol$_{rxn}$ and $\Delta S°$ for the reaction is -198.8 J/mol$_{rxn} \cdot$ K. Enthalpy ($\Delta H° < 0$) therefore drives the reaction toward the products, but entropy ($\Delta S° < 0$) drives the reaction toward the reactants.

●●●

What happens when one of the two thermodynamic quantities that determine the direction in which a chemical reaction occurs is favorable and the other is not? We can answer this question by defining a new quantity known as the **Gibbs free energy (G)** of the system, which reflects the balance between the forces. The name of this quantity recognizes the contributions of J. Willard Gibbs, a professor of mathematical physics at Yale from 1871 until the early 1900s, who some consider to be the greatest scientist produced by the United States.

The Gibbs free energy of a system at any moment in time is defined as the enthalpy of the system minus the product of the temperature times the entropy of the system.

$$G = H - TS$$

The Gibbs free energy is therefore a state function because it is defined in terms of thermodynamic properties (enthalpy, entropy, and temperature) that are state functions. At a given temperature, the change in the Gibbs free energy of the system that occurs during a reaction is therefore equal to the change in the enthalpy of the system minus the product of the temperature times the change in entropy of the system.

$$\Delta G = \Delta H - T\Delta S$$

The beauty of the equation defining changes in the free energy of a system is its ability to determine the relative importance of the enthalpy and entropy terms for a particular reaction at a given temperature. The change in the free energy of the system that occurs during a reaction measures the balance between the two driving forces that determine whether a reaction is spontaneous.

As we have seen, the enthalpy and entropy terms have different sign conventions. The enthalpy term is favorable when it is negative, whereas the entropy term is favorable when positive.

Favorable	Unfavorable
$\Delta H < 0$	$\Delta H > 0$
$\Delta S > 0$	$\Delta S < 0$

The entropy term is therefore *subtracted* from the enthalpy term when calculating ΔG for a reaction.

Because of the way the free energy of the system is defined, ΔG is negative for any reaction for which ΔH is negative and ΔS is positive. Thus, ΔG is negative for any reaction that is favored by both the enthalpy and entropy terms. We can therefore conclude that any reaction for which ΔG is negative should be favorable, or spontaneous.

For favorable, or spontaneous, reactions, $\Delta G < 0$.

Conversely, ΔG is positive for any reaction for which ΔH is positive and ΔS is negative. Any reaction for which ΔG is positive is therefore unfavorable in the direction in which it is written.

For unfavorable, or nonspontaneous, reactions, $\Delta G > 0$.

Reactions are classified as either **exothermic** ($\Delta H < 0$) or **endothermic** ($\Delta H > 0$) on the basis of whether they give off or absorb heat. Reactions can also be classified as either **exergonic** ($\Delta G < 0$) or **endergonic** ($\Delta G > 0$) on the basis of whether the free energy of the system decreases or increases during the reaction.

When a reaction is favored by both enthalpy ($\Delta H < 0$) and entropy ($\Delta S > 0$), there is no need to calculate the value of ΔG to decide whether the reaction should proceed as written. The same can be said for reactions favored by neither enthalpy ($\Delta H > 0$) nor entropy ($\Delta S < 0$). Free energy calculations are important, however, for reactions favored by only one of these factors.

The change in the free energy of a system that occurs during a reaction can be measured under any set of conditions. If the data are collected under standard-state conditions, the result is the **standard-state free energy of reaction ($\Delta G°$).**

$$\Delta G° = \Delta H° - T\Delta S°$$

As always, the standard state for thermodynamic measurements assumes that all gases are present at a pressure of 1 bar (0.987 atm) and that the concentrations of all solutes are 1 M. As noted previously, we can assume that a pressure of 1 atm is effectively equal to 1 bar.

> **CHECKPOINT**
>
> The Checkpoint at the end of Section 13.1 noted that table salt dissolves in water even though heat is absorbed during the process ($\Delta H > 0$). What can we conclude about ΔS for the reaction?

Exercise 13.5

Use the atom combination data given here to calculate $\Delta H°$ and $\Delta S°$ for the following reaction at 25°C.

$$2\,NO_2(g) \rightleftharpoons N_2O_4(g)$$

Compound	$\Delta H°_{ac}(kJ/mol_{rxn})$	$\Delta S°_{ac}(J/mol_{rxn} \cdot K)$
$NO_2(g)$	−937.86	−235.35
$N_2O_4(g)$	−1932.93	−646.53

Use the results of these calculations to determine the value of $\Delta G°$ for the reaction and to predict whether the reaction will occur as written.

Solution

The following reaction diagram can be used to calculate ΔH° and ΔS° for this reaction from enthalpy and entropy of atom combination data.

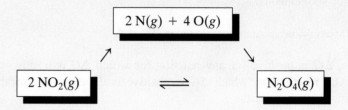

The enthalpy of reaction calculation can be set up as follows.

Bond breaking: $2\,NO_2(g) \longrightarrow 2\,N(g) + 4\,O(g)$ $\Delta H^\circ = -(2 \times -937.86\ \text{kJ/mol}_{rxn})$
Bond making: $2\,N(g) + 4\,O(g) \longrightarrow N_2O_4(g)$ $\underline{\Delta H^\circ = -1932.93\ \text{kJ/mol}_{rxn}}$

$\Delta H^\circ = -57.20\ \text{kJ/mol}_{rxn}$

The overall reaction is exothermic, and the enthalpy of reaction is therefore favorable.

The entropy of reaction calculation is done in much the same way.

Bond breaking: $2\,NO_2(g) \longrightarrow 2\,N(g) + 4\,O(g)$ $\Delta S^\circ = -(2 \times -235.35\ \text{J/mol}_{rxn} \cdot \text{K})$
Bond making: $2\,N(g) + 4\,O(g) \longrightarrow N_2O_4(g)$ $\underline{\Delta S^\circ = -646.53\ \text{J/mol}_{rxn} \cdot \text{K}}$

$\Delta S^\circ = -175.83\ \text{J/mol}_{rxn} \cdot \text{K}$

The reaction leads to a significant decrease in the disorder of the system and therefore isn't favored by the entropy of reaction.

To decide whether the reaction should proceed at 25°C, we have to compare the ΔH° and $T\Delta S^\circ$ terms to see which is larger. Before we can do this, we have to convert the temperature to kelvins.

$$T_K = 25°C + 273 = 298\ \text{K}$$

We also have to recognize that the units of ΔH° for the reaction are *kilojoules* and the units of ΔS° are *joules per kelvin*. At some point in the calculation, we have to convert these quantities to a consistent set of units. One way of doing this is to convert ΔS° to kilojoules. We then multiply the entropy term by the absolute temperature and subtract that quantity from the enthalpy term.

$$\Delta G^\circ = \Delta H^\circ - T\Delta S^\circ$$
$$= (-57.2\ \text{kJ/mol}_{rxn}) - (298\ \text{K})(-0.1758\ \text{kJ/mol}_{rxn} \cdot \text{K})$$
$$= (-57.2\ \text{kJ/mol}_{rxn}) + (52.4\ \text{kJ/mol}_{rxn})$$
$$= -4.8\ \text{kJ/mol}_{rxn}$$

At 25°C, the standard-state free energy for the reaction is negative because the ΔH term at that temperature is larger in magnitude than the $T\Delta S^\circ$ term.

It is important to note that standard-state conditions were used in this calculation. As a result, our conclusion that the reaction is spontaneous at 25°C only applies when all gases are present at roughly 1 atm pressure.

Exercise 13.6

Glucose is broken down into lactic acid in muscle cells undergoing vigorous exercise.

Glucose   *Lactic acid*

Use the following thermodynamic data collected at 298.15 K to determine the driving forces (if any) pushing the reaction forward and whether the reaction is favorable at this temperature.

Compound	ΔH°_{ac} (kJ/mol$_{rxn}$)	ΔS°_{ac} (J/mol$_{rxn}$ · K)
$C_6H_{12}O_6(aq)$	-9670.0	-3013.9
$CH_3CHOHCO_2H(aq)$	-4889.7	-1417

Solution

This exercise provides an example of how tables of thermochemical data can be used to determine whether a reaction should occur.

	ΔH° (kJ/mol$_{rxn}$)	ΔS° (J/mol$_{rxn}$ · K)
Bond breaking:	$-(-9670.0)$	$-(-3013.9)$
Bond making:	(2×-4889.7)	(2×-1417)
	$\Delta H^\circ = -109.4$ kJ/mol$_{rxn}$	$\Delta S^\circ = 180$ J/mol$_{rxn}$ · K

ΔH° for the reaction is negative, which means that the enthalpy of reaction for the conversion of glucose to lactic acid is favorable. ΔS° for the reaction is positive because there is more disorder in the products of the reaction. Lactic acid is therefore favored by the entropy of reaction. Both the enthalpy and entropy changes indicate that this reaction should proceed under standard-state conditions.

We could obtain the same conclusion by calculating ΔG° for the reaction.

$$\begin{aligned} \Delta G^\circ &= \Delta H^\circ - T\Delta S^\circ \\ &= (-109.4 \text{ kJ/mol}_{rxn}) - (298.15 \text{ K} \times 0.180 \text{ kJ/mol}_{rxn} \cdot \text{K}) \\ &= -163 \text{ kJ/mol}_{rxn} \end{aligned}$$

The sign of ΔG° is negative, which suggests that the reaction is favorable, and the magnitude of ΔG° is relatively large.

> ➤ **CHECKPOINT**
>
> Think about what happens when cement hardens. (Does it become more ordered or less ordered?) Use the results of predictions of the sign of ΔS for this process to predict whether it is possible to make cement that will harden without releasing heat. Explain your reasoning.

13.8 The Effect of Temperature on the Free Energy of a Reaction

The balance between the contributions of the enthalpy and entropy terms to the free energy of a reaction depends on the temperature at which the reaction is run.

 ## Exercise 13.7

Use the values of $\Delta H°$ and $\Delta S°$ calculated in Exercise 13.4 to predict whether the following reaction should occur at 25°C and 1 atm partial pressure for each gas.

$$N_2(g) + 3\,H_2(g) \rightleftharpoons 2\,NH_3(g)$$

Solution

According to the values obtained in Exercise 13.4, the reaction is favored by enthalpy but not by entropy.

$$\Delta H° = -92.2 \text{ kJ/mol}_{rxn} \qquad \text{(favorable)}$$
$$\Delta S° = -198.8 \text{ J/mol}_{rxn}\cdot K \qquad \text{(unfavorable)}$$

Before we can determine which of these terms dominates the free energy of the reaction, we have to incorporate into our calculation the temperature at which the reaction is run.

$$T_K = 25°C + 273 = 298 \text{ K}$$

We then multiply the entropy of reaction by the absolute temperature and subtract the $T\Delta S°$ term from the $\Delta H°$ term.

$$
\begin{aligned}
\Delta G° &= \Delta H° - T\Delta S° \\
&= (-92.2 \text{ kJ/mol}_{rxn}) - (298 \text{ K})(-0.1988 \text{ kJ/mol}_{rxn}\cdot K) \\
&= (-92.2 \text{ kJ/mol}_{rxn}) + (59.2 \text{ kJ/mol}_{rxn}) \\
&= -33.0 \text{ kJ/mol}_{rxn}
\end{aligned}
$$

The sign of $\Delta G°$ for this reaction is negative, which means the reaction should be spontaneous under standard-state conditions at 25°.

What happens as we raise the temperature of the reaction? The equation used to define free energy suggests that the entropy term will become more important as the temperature increases and may dominate the enthalpy term at high temperatures.

$$\Delta G° = \Delta H° - T\Delta S°$$

 ## Exercise 13.8

Predict whether the following reaction should occur at 500°C and 1 atm partial pressure of each gas.

$$N_2(g) + 3\,H_2(g) \rightleftharpoons 2\,NH_3(g)$$

Assume that the values of $\Delta H°$ and $\Delta S°$ used in the previous exercise are still valid at that temperature.

Solution

We need to calculate the temperature on the Kelvin scale.

$$T_K = 500°C + 273 = 773 \text{ K}$$

We can then multiply the entropy term by the absolute temperature and subtract the result from the value of $\Delta H°$ for the reaction.

$$\begin{aligned}
\Delta G°_{773} &= \Delta H°_{298} - T\Delta S°_{298} \\
&= (-92.2 \text{ kJ/mol}_{rxn}) - (773 \text{ K})(-0.1988 \text{ kJ/mol}_{rxn} \cdot \text{K}) \\
&= (-92.2 \text{ kJ/mol}_{rxn}) - (-154 \text{ kJ/mol}_{rxn}) \\
&= 62 \text{ kJ/mol}_{rxn}
\end{aligned}$$

Because the entropy term becomes more important as the temperature increases, the reaction changes from one that is spontaneous at low temperatures to one that isn't spontaneous at high temperatures.

13.9 Beware of Oversimplifications

The calculation of $\Delta G°$ for the following reaction at 25°C in Exercise 13.7 is perfectly legitimate.

$$N_2(g) + 3 H_2(g) \rightleftharpoons 2 NH_3(g)$$

The enthalpy and entropy data used in the calculation were standard-state measurements made at 25°C, which can legitimately be used to predict the standard-state free energy of reaction at that temperature. The calculation of $\Delta G°_{773}$ for the reaction at 500°C in Exercise 13.8, however, was based on the assumption that the enthalpy and entropy data measured at 25°C are still valid at 500°C.

This is a useful assumption, but it isn't always valid. Changes in $\Delta H°$ and $\Delta S°$ are often small over moderate temperature ranges. However, when the temperature range over which the data are extrapolated is large, as it is in Exercise 13.8, the results of the calculation represent an estimate, not a prediction, of the magnitude of the effect of the change in temperature. In this case, the assumption that $\Delta H°$ and $\Delta S°$ are constant underestimates the magnitude of $\Delta G°$. $\Delta G°_{773}$ for the reaction is 73 kJ/mol$_{rxn}$, not 62 kJ/mol$_{rxn}$.

13.10 Standard-State Free Energies of Reaction

$\Delta G°$ for a reaction can be calculated from tabulated standard-state Gibbs free energy data, such as the **standard-state free energy of atom combination,** $\Delta G°_{ac}$, data in Table B.13 in Appendix B. These values correspond to the formation of one mole of the substance listed in the table from its isolated atoms in the gas phase.

Consider the reaction involved in the combustion of methane in the presence of oxygen to form carbon dioxide and water vapor at 298 K and 1 atm partial pressure for each gas.

$$CH_4(g) + 2\,O_2(g) \longrightarrow CO_2(g) + 2\,H_2O(g)$$

The following data for this reaction can be obtained from Table B.13 in Appendix B.

Compound	ΔG_{ac}° (kJ/mol$_{rxn}$)
$CH_4(g)$	-1535.00
$O_2(g)$	-463.46
$CO_2(g)$	-1529.08
$H_2O(g)$	-866.80

The following reaction diagram can be used as the basis for calculating the overall change in the free energy of the system that occurs in this reaction from the ΔG_{ac}° data.

The bond-breaking step in the reaction involves atomization of 2 moles of O_2 for each mole of CH_4 consumed in the reaction.

Bond breaking:

$$CH_4(g) \longrightarrow C(g) + 4\,H(g) \qquad \Delta G^{\circ} = -(-1535.00 \text{ kJ/mol}_{rxn})$$
$$2\,O_2(g) \longrightarrow 4\,O(g) \qquad \Delta G^{\circ} = -(2 \times -463.46 \text{ kJ/mol}_{rxn})$$
$$\overline{\qquad\qquad\qquad\qquad \Delta G^{\circ} = 2461.92 \text{ kJ/mol}_{rxn}}$$

The magnitude of the change in the free energy of the system as the isolated atoms in the gas phase combine to form the products of the reaction is equal to the sum of the free energy of atom combination of CO_2 and twice that of H_2O.

Bond making:

$$C(g) + 2\,O(g) \longrightarrow CO_2(g) \qquad \Delta G^{\circ} = -1529.08 \text{ kJ/mol}_{rxn}$$
$$4\,H(g) + 2\,O(g) \longrightarrow 2\,H_2O(g) \qquad \Delta G^{\circ} = (2 \times -866.80 \text{ kJ/mol}_{rxn})$$
$$\overline{\qquad\qquad\qquad\qquad \Delta G^{\circ} = -3262.7 \text{ kJ/mol}_{rxn}}$$

When the results of these calculations are combined, we find that the overall free energy of reaction is negative. The reaction is therefore spontaneous at 298 K and 1 atm.

$$
\begin{aligned}
\text{Bond breaking:} \quad & \Delta G^{\circ} = 2461.92 \text{ kJ/mol}_{rxn} \\
\text{Bond making:} \quad & \Delta G^{\circ} = -3262.7 \text{ kJ/mol}_{rxn} \\
\hline
& \Delta G^{\circ} = -800.8 \text{ kJ/mol}_{rxn}
\end{aligned}
$$

If all reactant molecules are atomized to gaseous atoms and if the gaseous atoms may recombine, what factors determine whether the recombination is to form the original reactants or the products? Those molecules having the largest total bond strengths are enthalpically preferred, whereas those molecules having the most freedom associated with their molecular structure are entropically preferred. The balance between stability (enthalpy) and freedom (entropy) determines the position of the equilibrium.

13.11 Equilibria Expressed in Partial Pressures

Equilibrium constants can be calculated from standard Gibbs free energies or from a combination of enthalpy and entropy data. Equilibria are often described by equilibrium constant expressions that are written in terms of the concentrations of the reactants and products in units of moles per liter. When this is done, a subscript c is added to the symbol for the equilibrium constant to show that it was calculated from concentrations of the components of the reaction. Consider the following reaction, for example.

$$N_2(g) + 3\,H_2(g) \rightleftharpoons 2\,NH_3(g)$$

The expression for K_c would be written as follows.

$$K_c = \frac{[NH_3]^2}{[N_2][H_2]^3}$$

Equilibrium constant expressions for gas-phase reactions can also be written in terms of the partial pressures of the components of the reaction. We can understand why this is possible by rearranging the ideal gas equation,

$$PV = nRT$$

to give the following relationship between the pressure of a gas and its concentration in moles per liter (n/V)

$$P = \frac{n}{V} \times RT$$

Because the pressure of a gas involved in a chemical reaction is proportional to its concentration, in units of moles per liter, we can characterize the reaction between N_2 and H_2 to form NH_3 as follows.

$$K_P = \frac{P_{NH_3}^{\,2}}{P_{N_2}P_{H_2}^{\,3}}$$

K_p, like K_c, is always reported without units. However, any calculation involving K_p requires the partial pressures of the products and reactants to be in units of atmospheres.

What is the relationship between K_p and K_c for a gas-phase reaction? According to the rearranged version of the ideal gas equation shown above, the pressure of a gas is equal to the concentration of the gas times the product of the ideal gas constant and the temperature in units of kelvins.

$$P = \left[\frac{n}{V}\right] \times RT$$

We can therefore calculate the value of K_p for a reaction by multiplying each of the terms in the K_c expression by RT.

$$K_P = \frac{P_{NH_3}^2}{P_{N_2}P_{H_2}^3} = \frac{([NH_3] \times RT)^2}{([N_2] \times RT)([H_2] \times RT)^3}$$

Collecting terms in this example gives the following result.

$$K_p = K_c \times (RT)^{-2}$$

In general, the value of K_p for a reaction can be calculated from K_c with the following equation.

$$K_p = K_c \times (RT)^{\Delta n}$$

In this equation, Δn is the difference between the number of moles of gaseous products and the number of moles of gaseous reactants in the balanced equation ($\Delta n = n_{prod} - n_{react}$).

Exercise 13.9

Calculate the value of K_p for the following reaction at 650°C if the value of K_c for the reaction at that temperature is 0.040.

$$N_2(g) + 3\,H_2(g) \rightleftharpoons 2\,NH_3(g)$$

Solution

The value of K_p for a reaction can be calculated from K_c with the following equation.

$$K_p = K_c \times (RT)^{\Delta n}$$

To use this equation, we need to know the value of Δn for the reaction. The balanced equation for this gas-phase reaction generates 2 moles of products for every 4 moles of reactants consumed. Δn for the reaction is therefore $2 - 4$, or -2.

$$K_p = K_c \times (RT)^{-2}$$

We can now use the known value of the ideal gas constant and the temperature in kelvins to calculate the value of K_p for the reaction at the temperature given.

$$K_p = (0.040) \times [(0.08206 \text{ L-atm/mol} \cdot \text{K})(923 \text{ K})]^{-2} = 7.0 \times 10^{-6}$$

The techniques for working problems using K_p expressions are the same as those described for K_c problems, as can be seen from Exercise 13.10.

Exercise 13.10

K_p for the following reaction is 60 at 350°C.

$$H_2(g) + I_2(g) \rightleftharpoons 2\,HI(g)$$

If the initial partial pressures of H_2 and I_2 are 1.00 atm for each of the three gases involved in this reaction, what will be the equilibrium pressures of HI, H_2, and I_2?

Solution

No HI is present initially, thus, the reaction will proceed to the right to form HI. The partial pressure of H_2 and I_2 will decrease by an amount that can be represented as ΔP, and the partial pressure of HI will increase by an amount that is twice as large.

	$H_2(g)$	$+$	$I_2(g)$	$\rightleftharpoons$	$2\,HI(g)$
Initial:	1.00 atm		1.00 atm		0 atm
Change:	$-\Delta P$		$-\Delta P$		$2\Delta P$
Equilibrium:	$1.00 - \Delta P$		$1.00 - \Delta P$		$2\Delta P$

Substituting this information into the equilibrium constant expression gives the following result.

$$K_P = \frac{P_{HI}^2}{P_{H_2}P_{I_2}} = \frac{(2\Delta P)^2}{(1.00 - \Delta P)(1.00 - \Delta P)} = 60$$

The equilibrium constant is reasonably large, and the reaction quotient for the initial conditions is equal to zero. This means that ΔP isn't likely to be small compared with the initial partial pressures of H_2 and I_2. We are therefore going to have to expand this equation as follows.

$$\frac{4\Delta P^2}{\Delta P^2 - 2\Delta P + 1.00} = 60$$

Rearranging this equation gives the following quadratic equation.

$$56\Delta P^2 - 120\Delta P + 60 = 0$$

Using the quadratic formula to solve this equation gives the following result.

$$\Delta P = \frac{120 \pm \sqrt{(120)^2 - 4(56)(60)}}{2(56)} = \frac{120 \pm 31}{112}$$

Two possible values for ΔP are obtained in this calculation: 0.79 and 1.35. If we choose the larger of these values ($\Delta P = 1.35$), we would predict that more H_2 and I_2 are consumed as the reaction comes to equilibrium than were present initially, which is impossible. Thus the smaller of these values ($\Delta P = 0.79$) is used. The equilibrium pressure of HI is $2\Delta P = 1.6$ atm, and those of H_2 and I_2 are $1.00 - \Delta P = 0.21$ atm.

• •

As shown in Sections 10.6 and 10.11, the reaction quotient expressed in terms of the concentrations of the various components of a reaction (Q_C) can be used to predict the shift in the position of equilibrium that occurs in response to changes in either temperature, pressure, or the concentrations of different components of the reaction. Similarly, a comparison of K_P and Q_p can describe how a gas-phase reaction at equilibrium will respond to a change in temperature, pressure, or concentration.

In Section 10.11, it was shown that $Q_c = Q_n/V^{\Delta n}$.

As noted earlier, the relationship between K_p and K_c for a gas-phase reaction can be written as follows.

$$K_p = K_c \times (RT)^{\Delta n}$$

A similar equation can be used to describe the relationship between Q_p and Q_c for the reaction.

$$Q_p = Q_c \times (RT)^{\Delta n}$$

$$Q_p = \frac{Q_n}{V^{\Delta n}} \times (RT)^{\Delta n} = Q_n \times \left(\frac{RT}{V}\right)^{\Delta n}$$

Further, as was shown in Chapter 6, the total pressure, P_{tot}, is related to the total number of moles, n_{tot}, by

$$P_{tot}V = n_{tot}RT$$

$$Q_p = Q_n\left(\frac{P_{tot}}{n_{tot}}\right)^{\Delta n}$$

These relationships are useful for examining the effects of changes in pressure, volume, temperature, or concentration for gas-phase reactions.

Exercise 13.11

Consider the reaction between N_2 and H_2 to form NH_3 at equilibrium.

$$N_2(g) + 3 H_2(g) \rightleftharpoons 2 NH_3(g)$$

Predict the direction in which the equilibrium will shift for each of the following changes.

(a) The total pressure is increased at constant temperature by reducing the volume of the container.

(b) An inert gas is added at constant temperature and volume.

(c) An inert gas is added at constant temperature and pressure.

(d) $H_2(g)$ is added to the equilibrium mixture at constant temperature and pressure.

Solution

The stoichiometry of this reaction indicates that we get 2 moles of product from 4 moles of starting material. Thus, the value of Δn for this reaction is -2. We can therefore write the following equation that describes the relationship between Q_n and Q_p.

$$Q_p = Q_n\left(\frac{P_{tot}}{n_{tot}}\right)^{-2} = Q_n\left(\frac{n_{tot}}{P_{tot}}\right)^2$$

$$Q_p = \frac{(n_{NH_3})^2}{(n_{N_2})(n_{H_2})^3} \times \frac{(n_{tot})^2}{(P_{tot})^2}$$

(a) If the total pressure is increased, the $(n_{tot})^2/(P_{tot})^2$ ratio will decrease. When this happens, the reaction quotient will no longer be equal to the equilibrium constant. Q_P will now be smaller than the equilibrium constant, so the reaction will have to shift to the right to get back to equilibrium. This is consistent with Le Châtelier's principle for a closed system.

(b) In this case both n_{tot} and P_{tot} will increase, but the n_{tot}/P_{tot} ratio remains the same because at constant volume and temperature the n_{tot}/P_{tot} ratio is equal to V/RT, which does not change. Thus, no shift in the position of the equilibrium will occur. Another way of analyzing this effect due to the addition of an inert gas is to consider that if the temperature and volume are constant, the species participating in the reaction do not experience any increase in their kinetic energy or frequency of collision with the walls of the container. When an inert gas is added, the total pressure increases, but the partial pressure of each reacting species remains the same.

(c) When an inert gas is added to the system at constant temperature and pressure, the increase in the value of n_{tot} results in an increase in the n_{tot}/P_{tot} ratio. When this happens, $Q_p > K_p$. The reaction therefore has to shift to the left to get back to equilibrium.

(d) If H_2 is added at constant pressure, both n_{H_2} and n_{tot} increase, but the direction of the reaction depends on the ratio $(n_{tot})^2/(n_{H_2})^3$. Because n_{H_2} must be smaller than n_{tot}, the addition of more H_2 produces a larger percent change for n_{H_2} than for n_{tot}. In addition n_{H_2} is also raised to a higher power than n_{tot}. Thus the ratio $(n_{tot})^2/(n_{H_2})^3$ decreases and $Q_P < K_P$. The reaction therefore has to shift to the right to get back to equilibrium.

• •

13.12 Interpreting Standard-State Free Energy of Reaction Data

We are now ready to ask the question: What does the value of $\Delta G°$ tell us about a reaction? Consider the following reaction, for example.

$$N_2(g) + 3\,H_2(g) \rightleftharpoons 2\,NH_3(g) \qquad \Delta G° = -33.0 \text{ kJ/mol}_{rxn}$$

By definition, the value of $\Delta G°$ for a reaction measures the difference between the free energies of the reactants and products *when all components of the reaction are present at standard-state conditions.*

$\Delta G°$ therefore describes the reaction only when all three components are present at roughly 1 atm pressure. Note that the appropriate **reaction quotient** for this reaction would be Q_p because the tabulated values of the thermodynamic parameters for gases are given in units of pressure.

$$\text{Standard state:} \qquad Q_p = \frac{P_{NH_3}{}^2}{P_{N_2}P_{H_2}{}^3} = \frac{(1)^2}{(1)(1)^3} = 1$$

The *sign* of $\Delta G°$ tells us the direction in which the reaction has to shift from standard-state conditions to come to equilibrium. The fact that $\Delta G°$ is negative for this reaction at 25°C means that the system under standard-state conditions at that temperature would have to shift to the right, converting some of the reactants into products, before it can reach equilibrium.

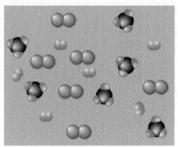

Standard state
$P_{NH_3} = P_{N_2} = P_{H_2} = 1$ atm

Fig. 13.5 At standard-state conditions, the partial pressures of N_2, H_2, and NH_3 in this system are 1 atm. Thus Q_p for the system is equal to 1.

The *magnitude* of $\Delta G°$ for a reaction tells us how far the standard state is from equilibrium. The larger the value of $\Delta G°$, the farther the reaction has to go to get from the standard-state conditions to equilibrium.

Assume, for example, that we start with the following reaction under standard-state conditions, as shown in Figure 13.5.

$$N_2(g) + 3\,H_2(g) \rightleftharpoons 2\,NH_3(g)$$

The value of ΔG at that moment in time is equal to the standard-state free energy for the reaction, $\Delta G°$.

When $Q_p = 1$, $\Delta G = \Delta G°$

This raises an interesting question: Is there any way to predict the value of ΔG for this reaction at any moment in time as the reaction proceeds to equilibrium?

13.13 The Relationship between Free Energy and Equilibrium Constants

When a reaction leaves the standard state because of a change in the concentrations or partial pressures of the reactants and products, we have to describe the system in terms of a nonstandard-state free energy of reaction (ΔG). The difference between $\Delta G°$ and ΔG for a reaction is important. There is only one value of $\Delta G°$ for a reaction at a given temperature, but there are an infinite number of possible values of ΔG.

In general, the relationship between the free energy of reaction at any moment in time (ΔG) and the standard-state free energy of reaction ($\Delta G°$) is described by the following equation.

$$\Delta G = \Delta G° + RT \ln Q$$

In this equation, R is the ideal gas constant measured in joules (8.314 J/mol$_{rxn}$ · K), T is the temperature in kelvins, ln represents a logarithm to the base e, and Q is the reaction quotient measured in either pressure or concentration units.

This equation contains two variables (ΔG and $\ln Q$) and two constants ($\Delta G°$ and RT) when it is applied to a given reaction at a given temperature. $\Delta G°$ for the reaction is constant because there is only one standard state for any reaction and RT is a constant because the equation is being applied to the reaction at a given temperature. ΔG is a variable because it changes the moment the reaction leaves standard-state conditions, and $\ln Q$ is a variable because it reflects the concentrations or partial pressures of the components of the reaction at that moment in time. This equation can therefore be treated as an example of an equation for a straight line: $y = mx + b$.

Figure 13.6 shows the relationship between ΔG and $\ln Q_p$ for the following reaction.

$$N_2(g) + 3\,H_2(g) \rightleftharpoons 2\,NH_3(g)$$

Data on the far left side of Figure 13.6 correspond to relatively small values of Q_p. They therefore describe systems in which there is far more reactant than product. The sign of ΔG for these systems is negative, and the magnitude of ΔG is

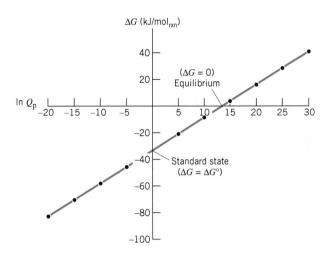

Fig. 13.6 Plot of $\ln Q_P$ versus ΔG for the reaction in which N_2 and H_2 combine to form NH_3. $\Delta G = \Delta G^\circ$ when the reaction quotient is 1 ($\ln Q_p = 0$). $\Delta G = 0$ when the reaction is at equilibrium.

large. The system is therefore relatively far from equilibrium, and the reaction must shift to the right to reach equilibrium.

Data on the far right side of Figure 13.6 describe systems in which there is more product than reactant. The sign of ΔG is now positive, and the magnitude of ΔG is moderately large.

The sign of ΔG for any point in Figure 13.6 tells us the direction in which the reaction would have to shift to reach equilibrium. Any point for which ΔG is negative describes a system that would have to shift toward to the right—toward the products of the reaction—to reach equilibrium. Any point for which ΔG is positive describes a system that would have to shift toward the reactants to reach equilibrium. The magnitude of ΔG at any point in this figure tells us how far we have to go to reach equilibrium.

The points at which the straight line in Figure 13.6 crosses the horizontal and vertical axes of the diagram are particularly important. The straight line crosses the vertical axis when the reaction quotient for the system is equal to 1 ($\ln Q_p = 0$). That point therefore describes the standard-state conditions, and the value of ΔG at that point is equal to the standard-state free energy of reaction, ΔG°.

$$\text{When } Q_p = 1, \Delta G = \Delta G^\circ$$

The point at which the straight line crosses the horizontal axis describes a system for which ΔG is equal to zero. Because there is no driving force behind the reaction, the system must be at equilibrium.

$$\text{When } Q_p = K_p, \Delta G = 0$$

As we have seen, the driving force behind a chemical reaction is zero ($\Delta G = 0$) when the reaction is at equilibrium ($Q = K$). Substituting this information into the equation that defines the relationship between ΔG and ΔG° gives the following result.

$$0 = \Delta G^\circ + RT \ln K$$

This equation can be solved for the relationship between ΔG° and K.

$$\Delta G^\circ = -RT \ln K$$

This equation allows us to calculate the equilibrium constant for any reaction from the standard-state free energy of reaction, or vice versa.

➤ CHECKPOINT

Which way does a reaction proceed when $Q_p > K_p$?

We can also get the equilibrium constant for a reaction from electrode potentials, using the Nernst equation.

$$E = E° - \frac{RT}{nF} \ln Q_c$$

In Section 12.9, we noted that the Nernst equation takes the following form at equilibrium.

$$0 = E° - \frac{RT}{nF} \ln K_c$$

Or

$$nFE° = RT \ln K_c$$

Thus, we can calculate the equilibrium constant for any oxidation–reduction from its standard cell potential using the following equation:

$$\ln K_c = \frac{nFE°}{RT}$$

where F is Faraday's constant (96,485 C/mol e^-) and $E°$ is the standard electrode potential.

Combining the relationship between the standard-state Gibbs free energy and the equilibrium constant for a reaction with the relationship between the equilibrium

$$\frac{-\Delta G°}{RT} = \ln K_c = \frac{nFE°}{RT}$$

Rearranging this equation gives the relationship between the Gibbs standard free energy and the standard electrode potential for a reaction.

$$\Delta G° = -nFE°$$

It is important to recognize that the Gibbs free energy of a reaction and the cell potential have opposite sign conventions. The half-reactions given in Table 12.1 whose standard reduction potentials have a positive value for $E°$ will have a negative value for $\Delta G°$. This means that the half-reactions with the most positive values of $E°$ will have the greatest tendency to occur. The half-reaction $Fe^{3+}(aq) + e^- \rightleftharpoons Fe^{2+}(aq)$ that has an $E°$ of 0.770 V will proceed as written compared to the half-reaction $Sn^{2+}(aq) + 2\ e^- \rightleftharpoons Sn(s)$ for which $E°$ is -0.1364 V. In a cell consisting of these two electrodes, Fe^{3+} will gain electrons and be reduced, while the Sn^{2+}/Sn electrode will supply electrons through the external circuit.

$$2\ Fe^{3+}(aq) + Sn(s) \rightleftharpoons 2\ Fe^{2+}(aq) + Sn^{2+}(aq)$$

Exercise 13.12

Calculate the standard-state Gibbs free energy and the equilibrium constant for the following cell reaction at 298.15 K from the half-cell reduction potential data in Table 12.1.

$$2\ Ag^+(aq) + Cu(s) \rightleftharpoons 2\ Ag(s) + Cu^{2+}(aq)$$

Solution

The reduction potential for the Ag^+/Ag half cell is more positive than that for the Cu^{2+}/Cu half cell.

$$Ag^+(aq) + e^- \rightleftharpoons Ag(s) \qquad E° = 0.7996 \text{ V}$$
$$Cu^{2+}(aq) + 2e^- \rightleftharpoons Cu(s) \qquad E° = 0.3402 \text{ V}$$

The Ag^+/Ag half cell will therefore be the cathode of the cell.

The overall standard-state cell potential for the reaction that involves these half cells can be calculated as follows.

$$2\,Ag^+(aq) + 2e^- \rightleftharpoons 2\,Ag(s) \qquad\qquad E°_{red} = 0.7996 \text{ V}$$
$$\underline{\quad Cu(s) \rightleftharpoons Cu^{2+}(aq) + 2e^- \qquad\qquad E°_{ox} = -(0.3402 \text{ V})\quad}$$
$$2\,Ag^+(aq) + Cu(s) \rightleftharpoons 2\,Ag(s) + Cu^{2+}(aq) \qquad E° = 0.4594 \text{ V}$$

The positive value of $E°$ indicates that the reaction proceeds to the right, as written.

The standard-state Gibbs free energy for this reaction can now be calculated as follows.

$$\Delta G° = -nFE°$$
$$\Delta G° = -(2)(96{,}485 \text{ C/mol } e^-)(0.4594 \text{ V}) = -8.865 \times 10^4 \text{ J/mol}_{rxn}$$

The equilibrium constant for the reaction can then be calculated as follows.

$$\ln K_c = \frac{-\Delta G°}{RT}$$
$$\ln K_c = \frac{8.865 \times 10^4 \text{ J/mol}_{rxn}}{(8.314 \text{ J/mol}_{rxn}\text{K})(298.15\text{K})} = 35.76$$
$$K_c = 3.4 \times 10^{15}$$

• •

To understand the relationship between $\Delta G°$ and K, we must recognize that the magnitude of $\Delta G°$ tells us how far the standard state is from equilibrium. The smaller the absolute value of $\Delta G°$, the closer the standard state is to equilibrium. The larger the absolute value of $\Delta G°$, the farther the reaction has to go to reach equilibrium. The sign of ΔG tells us whether the equilibrium constant is smaller than 1 or larger than 1. When ΔG is positive, the equilibrium constant is smaller than 1; when it is negative, the equilibrium constant is larger than 1.

The relationship between $\Delta G°$ and the equilibrium constant for a chemical reaction is illustrated by the data in Table 13.2.

The data in Table 13.2 raise an interesting point: How do we know whether the equilibrium constant that comes out of calculations based on free energy data is K_p or K_c? The equilibrium constant calculated from Gibbs free energies that involve only gases or only gases and pure solids and liquids is always K_p. When the reaction involves only aqueous solutions and pure liquids or solids, the equilibrium constant calculated from free energy data is always K_c.

Table 13.2
Values of $\Delta G°$ and K for Common Reactions at 25°C

Reaction	$\Delta G°$(kJ/mol$_{rxn}$)	K
$2\ SO_3(g) \rightleftharpoons 2\ SO_2(g) + O_2(g)$	141.7	1.5×10^{-25}
$H_2O(l) \rightleftharpoons H^+(aq) + OH^-(aq)$	79.9	1.0×10^{-14}
$AgCl(s) \overset{H_2O}{\rightleftharpoons} Ag^+(aq) + Cl^-(aq)$	55.7	1.7×10^{-10}
$HOAc(aq) \overset{H_2O}{\rightleftharpoons} H^+(aq) + OAc^-(aq)$	27.1	1.8×10^{-5}
$N_2(g) + 3\ H_2(g) \rightleftharpoons 2\ NH_3(g)$	-33.0	6×10^5
$HCl(g) \overset{H_2O}{\rightleftharpoons} H^+(aq) + Cl^-(aq)$	-35.9	2×10^6
$Cu^{2+}(aq) + 4\ NH_3(aq) \longrightarrow Cu(NH_3)_4{}^{2+}(aq)$	-70.6	2.3×10^{12}
$Zn(s) + Cu^{2+}(aq) \longrightarrow Zn^{2+}(aq) + Cu(s)$	-212.6	1.8×10^{37}

Exercise 13.13

Calculate the equilibrium constant for the following reaction at 25°C from the value of $\Delta G°$ for this reaction calculated in Exercise 13.7.

$$N_2(g) + 3\ H_2(g) \rightleftharpoons 2\ NH_3(g)$$

Solution

Exercise 13.7 gave the following value for $\Delta G°$ for this reaction at 25°C.

$$\Delta G° = -33.0\ \text{kJ/mol}_{rxn}$$

We now turn to the relationship between $\Delta G°$ and the equilibrium constant for the reaction.

$$\Delta G° = -RT \ln K$$

Solving for the natural log of the equilibrium constant gives the following equation.

$$\ln K = -\frac{\Delta G°}{RT}$$

Substituting the known values of $\Delta G°$, R, and T into this equation gives the following result.

$$\ln K_p = -\frac{(-33.0 \times 10^3\ \text{J/mol}_{rxn})}{(8.314\ \text{J/mol}_{rxn} \cdot \text{K})(298\ \text{K})} = 13.3$$

The equilibrium constant for the reaction at 25°C is therefore 6×10^5.

$$K_p = e^{13.3} = 6 \times 10^5$$

Because this is a gas-phase reaction, the equilibrium constant that results from this calculation is K_p.

It is easy to make mistakes when handling the sign of the relationship between $\Delta G°$ and K. It is therefore a good idea to check the final answer to see whether it makes sense. $\Delta G°$ for the reaction in this exercise is negative, and the equilibrium should lie on the side of the products. The equilibrium constant should therefore be much larger than 1, which it is.

Exercise 13.14

(a) Calculate the acid dissociation equilibrium constant (K_a) for formic acid at 25°C from the value of $\Delta G°$ given below.

$$HCO_2H(aq) \rightleftharpoons H^+(aq) + HCO_2^-(aq) \qquad \Delta G° = 21.3 \text{ kJ/mol}_{rxn}$$

(b) What concentration of H^+ is required for the reaction to favor the products?

Solution

(a) We start with the relationship between $\Delta G°$ and the equilibrium constant for the reaction

$$\Delta G° = -RT \ln K$$

and solve for the natural logarithm of the equilibrium constant.

$$\ln K = -\frac{\Delta G°}{RT}$$

Substituting the known values of $\Delta G°$, R, and T into this equation gives the following result.

$$\ln K_c = -\frac{(21.3 \times 10^3 \text{ J/mol}_{rxn})}{(8.314 \text{ J/mol}_{rxn} \cdot \text{K})(298 \text{ K})} = -8.60$$

We can now calculate the value of the equilibrium constant.

$$K_c = e^{-8.60} = 1.8 \times 10^{-4}$$

In this case $\Delta G°$ is positive, and the equilibrium will favor the reactants; in other words, the equilibrium constant will be less than 1. The equilibrium constant that comes out of this calculation is the same as the value of K_a for formic acid reported in Table B.8 in Appendix B.

(b) $\Delta G°$ is used to calculate the equilibrium constant and refers in this reaction to all reactants and products present at a 1 M concentration. If the concentration of formic acid and formate ion remain at 1 M, we can calculate the concentration of H^+ necessary for the reaction to favor the products. We can begin by solving for the concentration of H^+ at equilibrium when ΔG is zero.

$$\Delta G = \Delta G° + RT \ln Q$$
$$0 = 21.3 \text{ kJ/mol}_{rxn} + 8.314 \times 10^{-3} \text{ kJ/mol}_{rxn} \cdot \text{K} (298 \text{ K}) \ln Q_c$$

$$-21.3 \text{ kJ/mol}_{rxn} = 8.314 \times 10^{-3} \text{ kJ/mol}_{rxn} \cdot \text{K} (298 \text{ K}) \ln \left(\frac{(1 \text{ M})(H^+)}{(1 \text{ M})} \right)$$

$$(H^+) = 1.8 \times 10^{-4} M$$

If the H^+ concentration is just slightly less than $1.8 \times 10^{-4} M$, ΔG will be negative and the reaction will proceed to the right.

Exercise 13.15

At body temperatures, 37°C, glucose reacts with adenosine 6-diphosphate ion, (ADP^{4-}) to form glucose 6-phosphate^{2-}, ADP^{3-}, and H_3O^+.

$$\text{Glucose}(aq) + ATP^{4-}(aq) \rightleftharpoons \text{glucose 6-phosphate}^{-2}(aq) + ADP^{3-}(aq) + H_3O^+(aq)$$

The standard free energy change at 37°C is 24.8 kJ/mol$_{rxn}$.

(a) If all concentrations are 1 M what is the value of Q_c?

(b) Will K_c be larger than or smaller than Q_c?

(c) Use K_c and Q_c to determine the direction in which the reaction will proceed.

Solution

(a) $Q_c = \dfrac{[\text{glu 6-phosphate}^{-2}][ADP^{3-}][H_3O^+]}{[\text{glu}][ATP^{4-}]}$ If all concentrations are 1 M the value of Q_c is 1.0.

(b) K_c for this reaction is found from $\Delta G° = -RT \ln K_c = 24.8 \times 10^3 \text{ J/mol}_{rxn}$. This gives a K_c of 6.6×10^{-5}. Because $\Delta G°$ is positive K_c will be less than one.

(c) Q_c is larger than K_c and the reaction will proceed toward the reactants.

13.14 The Temperature Dependence of Equilibrium Constants

When equilibrium constants were introduced in Chapter 10, we noted that they aren't strictly constant because they change with temperature. We are now ready to understand why.

The standard-state free energy of reaction is a measure of how far the standard state is from equilibrium.

$$\Delta G° = -RT \ln K$$

But the magnitude of $\Delta G°$ depends on the temperature of the reaction.

$$\Delta G° = \Delta H° - T\Delta S°$$

As a result, the equilibrium constant must depend on the temperature of the reaction.

An example of this phenomenon is the reaction in which NO_2 dimerizes to form N_2O_4.

$$2\,NO_2(g) \rightleftharpoons N_2O_4(g)$$
Brown   *Colorless*

NO_2 is a brown gas and N_2O_4 is colorless. We can therefore monitor the extent to which NO_2 dimerizes to form N_2O_4 by examining the intensity of the brown color in a sealed tube of the gas. What should happen to the equilibrium between NO_2 and N_2O_4 as the temperature is lowered?

Figure 13.7 shows what happens to the intensity of the brown color when a sealed tube containing NO_2 gas in equilibrium with N_2O_4 at 25°C is immersed in liquid nitrogen. There is a drastic decrease in the amount of NO_2 in the tube as it is cooled to −196°C.

The dimerization of NO_2 is exothermic ($\Delta H° = -57.2$ kJ/mol$_{rxn}$). In Chapter 10 we described how decreasing the temperature of an exothermic reaction increases the equilibrium constant. The equilibrium constant for this reaction increases as we decrease the temperature of the system, as shown by the data in Table 13.3. The reaction therefore shifts toward N_2O_4 as we lower the temperature.

We can now develop a quantitative relationship between temperature and the equilibrium constant. We have seen that

$$\Delta G° = -RT \ln K = \Delta H° - T\Delta S°$$

Solving for ln K, we obtain

$$\ln K = -\frac{\Delta H°}{RT} + \frac{\Delta S°}{R}$$

Because both $\Delta H°$ and $\Delta S°$ are essentially constant for moderate changes in temperature, this equation describes the change in K that results from changes in temperature. Thus the natural logarithm of the equilibrium constant depends on the magnitude and sign of $\Delta H°$ and $\Delta S°$. However, only the $-\Delta H°/RT$ term depends on temperature. As temperature increases, for example, this term becomes smaller. If $\Delta H°$ is < 0, $-\Delta H°/RT$ is positive and contributes to making ln K larger no matter what the sign of $\Delta S°$. If $\Delta H° > 0$, $-\Delta H°/RT$ is negative and contributes to making ln K smaller. At higher temperatures this negative influence becomes less significant and ln K is larger than at low temperatures.

It is tempting to predict the effect of temperature on the equilibrium constant by considering only the expression

$$\Delta G° = \Delta H° - T\Delta S°$$

Suppose that in a given situation $\Delta H°$ is positive and $\Delta S°$ is negative. Then $\Delta G°$ would become more positive with increasing temperature, and it might be expected that the equilibrium constant would decrease. However, a positive $\Delta H°$ means that the equilibrium constant should increase with increasing temperature. This dilemma is resolved by noting that it is not $\Delta G°$ but $\Delta G°/T$ that determines the change of the equilibrium constant with temperature:

$$\Delta G° = -RT \ln K \quad \text{therefore} \quad \ln K = -\frac{\Delta G°}{RT}$$

The above equations *cannot* be used to describe the change in K as temperature is changed because $\Delta G°$ is *not* constant with respect to temperature.

Fig. 13.7 The equilibrium between NO_2 and N_2O_4 changes with temperature. The characteristic brown color of NO_2 gas disappears when a tube containing NO_2 is lowered into a liquid nitrogen bath (−196°C).

Table 13.3

Temperature Dependence of the Equilibrium Constant for the Dimerization of NO_2

Temperature (°C)	K_c	K_p
100	2.1	0.069
25	170	6.95
0	1.4×10^3	62.5
−78	4.0×10^8	2.5×10^7

Table 13.4

Temperature Dependence of the Equilibrium Constant for the Reaction of Methane to Produce Ethane

Temperature (K)	$\Delta G°$ (kJ/mol$_{rxn}$)	$\Delta G°/T$ (J/mol$_{rxn}$·K)	ln K	K
298	68.58	230	−27.7	9×10^{-13}
400	69.84	175	−21.0	8×10^{-10}
600	72.28	120	−14.4	6×10^{-7}

Consider the following reaction in which methane produces ethane and hydrogen gas.

$$2\,CH_4(g) \rightleftharpoons C_2H_6(g) + H_2(g)$$

Thermodynamic calculations for this reaction give the following results: $\Delta H° = 64.94$ kJ/mol$_{rxn}$ and $\Delta S° = -12.24$ J/mol$_{rxn}$·K. If $\Delta H°$ and $\Delta S°$ are assumed to be more or less constant as the temperature changes, $\Delta G°$ will become more positive with increasing temperature, but K will nevertheless increase.

$$\Delta G° = 64.94 \text{ kJ/mol}_{rxn} - T(-0.01224 \text{ J/mol}_{rxn} \cdot K)$$

The data in Table 13.4 illustrate the temperature dependence of the equilibrium constant for the above reaction, for which the free energy of reaction becomes more positive with increasing temperature.

Exercise 13.16

Most enzymes are proteins that consist of long polymeric chains of amino acids (compounds with both a carboxylic acid, —COOH, and an amine,—NH$_2$, functional group). Trypsin, for example, is a protein that acts as an enzyme to break the bonds between amino acids in other proteins. It is found in the small intestines and aids in digestion. The active form of most proteins is one in which the amino acid chain folds to form structures such as the structure of trypsin shown in Figure 13.8.

When either heated or exposed to a change in pH, many proteins lose at least part of their folded structure. When this happens, the protein can't carry out its normal biochemical function and is said to have been denatured. The equilibrium constant for the denaturation of trypsin is 7.20 at 50°C.

$$\text{trypsin} \rightleftharpoons \text{denatured trypsin} \qquad K_c = 7.20$$

(a) Calculate $\Delta G°$ for the denaturation of trypsin.

(b) $\Delta H°$ for the denaturation of trypsin at 50°C is +278 kJ/mol$_{rxn}$. What is $\Delta S°$ at this temperature?

(c) What do the sign and magnitude of $\Delta S°$ imply about the difference between denatured trypsin and trypsin?

(d) Under standard-state conditions, should trypsin denature at 50°C?

(e) Use $\Delta H°$ and $\Delta S°$ to describe what happens when a protein such as trypsin denatures.

Fig. 13.8 The structure of trypsin is typical of the folding and twisting found in protein chains.

(f) As temperature increases, does the equilibrium constant for the denaturation of trypsin increase, decrease, or remain the same? Will the equilibrium shift and, if so, in which direction?

Solution

(a) $\Delta G°$ can be calculated from the equilibrium constant for the denaturation of trypsin.

$$\Delta G° = -RT \ln K$$
$$= -(8.314 \text{ J/mol}_{rxn} \cdot \text{K})(323 \text{ K})(\ln 7.20)$$
$$= -5.30 \text{ kJ/mol}_{rxn}$$

(b) We can start with the equation that defines the relationship among free energy, enthalpy, and entropy.

$$\Delta G° = \Delta H° - T\Delta S°$$

Substituting the known values of $\Delta G°$, $\Delta H°$, and T gives the following result,

$$-5.30 \text{ kJ/mol}_{rxn} = 278 \text{ kJ/mol}_{rxn} - (323 \text{ K})(\Delta S°)$$

which can be solved for the value of $\Delta S°$ for the reaction.

$$\Delta S° = 877 \text{ J/mol}_{rxn} \cdot \text{K}$$

(c) $\Delta S°$ is relatively large and positive. This means that a significant amount of freedom of motion has been gained by the denaturing of trypsin.

(d) $\Delta G°$ is negative for the denaturation of trypsin at 50°C, so heating trypsin to that temperature should cause it to unfold.

(e) The positive sign of $\Delta H°$ shows that the protein is more tightly bonded in its folded structure than in its unfolded state. The large positive sign of $\Delta S°$ means that the unfolding process has produced rotational and vibrational freedom of motion not possible in the folded position. Thus denaturation requires the disruption of strong forces of attraction to produce weaker ones. For denaturation to occur, there must be a gain in entropy that offsets the disruption of the forces of attraction.

(f) This is an endothermic reaction ($\Delta H° = 278 \text{ kJ/mol}_{rxn}$). As temperature increases, the equilibrium constant will increase. As the equilibrium constant becomes larger, the reaction will shift to the right to produce more product.

> **CHECKPOINT**
> If $\Delta H° > 0$ and $\Delta S° > 0$ for a reaction, which way will the equilibrium shift if the temperature is decreased? If $\Delta H° > 0$ and $\Delta S° < 0$ for a reaction, which way will the equilibrium shift if the temperature is decreased?

If $\Delta G°$ for the denaturation of trypsin is favorable at 50°C, what is the highest temperature under standard-state conditions at which we would expect trypsin to be stable?

Trypsin should denature at any temperature at which $\Delta G°$ is negative. It would therefore be useful to find the temperature at which $\Delta G°$ for this reaction is zero. We start with the equation that defines the relationship among $\Delta G°$, $\Delta H°$, and $\Delta S°$.

$$\Delta G° = \Delta H° - T\Delta S°$$

We then set $\Delta G° = 0$ and solve for T, assuming that $\Delta H°$ and $\Delta S°$ don't change much with temperature.

$$0 = \Delta H° - T\Delta S°$$

$$T = \frac{\Delta H°}{\Delta S°} = \frac{(278 \text{ kJ/mol}_{rxn})}{(0.877 \text{ kJ/mol}_{rxn} \cdot K)} = 317 \text{ K} = 44°C$$

According to this calculation, the maximum temperature at which trypsin is stable should be 44°C. At any temperature *above* 44°C, $\Delta G°$ for the denaturation reaction is negative and the protein should denature.

13.15 Gibbs Free Energies of Formation and Absolute Entropies

We saw in Chapter 7 that enthalpy changes can be calculated from either enthalpies of atom combination or enthalpies of formation. Enthalpies of atom combination use the free gaseous atoms as a point of reference, while enthalpies of formation use the element in its most stable state of aggregation at 1 bar and usually at 298 K. Because the bar and the atmosphere differ by so little, we will, as in Chapter 7, assume that the standard-state pressure is effectively 1 atm.

It is also possible to use enthalpy and entropy of formation data to calculate $\Delta G°$ for a given process. Free energies of formation, just as enthalpies of formation, are based on differences in free energy between the compound of interest and the elements that compose the compound, each element being in its stable state. These elements are assigned a free energy of formation of zero. For oxygen this stable state is $O_2(g)$, for hydrogen it is $H_2(g)$, for bromine it is a liquid $Br_2(l)$, and for iron the stable state is the solid, $Fe(s)$.

$\Delta G°$ for a chemical reaction may be found from **standard-state free energies of formation** ($\Delta G_f^°$) by calculating the difference between the free energies of reactants and products. For the general reaction

$$aA + bB \rightleftharpoons cC + dD$$

$$\Delta G° = [c(\Delta G_f^°)_C + d(\Delta G_f^°)_D] - [a(\Delta G_f^°)_A + b(\Delta G_f^°)_B]$$

the superscript zero sign refers to a standard state of roughly 1 atm for gases, solids, and liquids and a 1 M concentration of all solutes. A table of free energies of formation is given in Table B.16 in Appendix B.

Exercise 13.17

Calculate the free energy change for the following reaction at 298 K and 1 atm from the data tabulated in Table B.16 in Appendix B.

$$CH_4(g) + 2 O_2(g) \longrightarrow CO_2(g) + 2 H_2O(g)$$

Solution

The following Gibbs free energies of formation for the four components of this reaction can be obtained from Table B.16 in Appendix B.

	$\Delta G_f^\circ (kJ/mol_{rxn})$
$CH_4(g)$	-50.752
$O_2(g)$	0
$CO_2(g)$	-394.359
$H_2O(g)$	-228.572

ΔG° for this reaction can be calculated as follows.

$$\Delta G^\circ = [2(\Delta G_f^\circ)_{H_2O(g)} + (\Delta G_f^\circ)_{CO_2(g)}] - [2(\Delta G_f^\circ)_{O_2(g)} + (\Delta G_f^\circ)_{CH_4(g)}]$$
$$\Delta G^\circ = [2(-228.527) - (394.359)] - [0 + (-50.752)]$$
$$= -800.751 \; kJ/mol_{rxn}$$

Tabulations of entropies calculated from experimental data are also available. These tables are called **absolute entropies** or third law entropies. The standard state is 1 atm and usually 298 K. Entropies, however, are reported as absolute values, S°, rather than as comparisons with stable states of the elements. The entropy change for a process can be calculated directly from S° values. For the general chemical reaction

$$aA + bB \rightleftharpoons cC + dD$$
$$\Delta S = [cS_C^\circ + dS_D^\circ] - [aS_A^\circ + bS_B^\circ]$$

Entropies are tabulated along with free energies and enthalpies of formation in Table B.16 in Appendix B.

ΔG°, ΔH°, and ΔS° for processes and chemical reactions can be calculated from either atom combination data or from free energy of formation data, enthalpy of formation data, or absolute entropy data.

Exercise 13.18

Use the data in Table B.16 in Appendix B to calculate the entropy and enthalpy change of reaction for the following reaction at 298 K and 1 atm.

$$CH_4(g) + 2\,O_2(g) \longrightarrow CO_2(g) + 2\,H_2O(g)$$

Use the enthalpy and entropy change to calculate ΔG° for the above reaction and compare ΔG° to that found in Exercise 13.17.

Solution

Enthalpy of formation data and absolute entropy data from Table B.16 are as follows.

	$\Delta H_f^\circ (kJ/mol_{rxn})$	$S^\circ (J/mol_{rxn} \cdot K)$
$CH_4(g)$	-74.81	186.264
$O_2(g)$	0	205.138
$CO_2(g)$	-393.509	213.74
$H_2O(g)$	-241.818	188.25

$$\Delta H° = [2(\Delta H_f°)_{H_2O(g)} + (\Delta H_f°)_{CO_2(g)}] - [2(\Delta H_f°)_{O_2(g)} + (\Delta H_f°)_{CH_4(g)}]$$

$$\Delta H° = [2(-241.818) + (-393.509)] - [0 + (-74.81)]$$

$$= -802.34 \text{ kJ/mol}_{rxn}$$

$$\Delta S° = [2 S°_{H_2O(g)} + S°_{CO_2(g)}] - [2 S°_{O_2(g)} + S°_{CH_4(g)}]$$

$$= [2(188.25) + (213.74)] - [2(205.138) + (186.264)]$$

$$= -6.30 \text{ J/mol}_{rxn} \cdot \text{K}$$

$$\Delta G° = \Delta H° - T\Delta S°$$

$$= -802.34 - (298)(-0.00630)$$

$$= -800.46 \text{ kJ/mol}_{rxn}$$

There is a very small difference (0.04%) in the value of $\Delta G°$ calculated by the two methods used in Exercises 13.17 and 13.18. This difference is typical for thermodynamic calculations of this type and can be considered insignificant.

Key Terms

Absolute entropy
Disorder
Endergonic ($\Delta G > 0$)
Endothermic
Enthalpy
Entropy (S)
Entropy of atom combination
Exergonic ($\Delta G < 0$)
Exothermic

First law of thermodynamics
Gibbs free energy (G)
Reaction quotient (Q_P)
Second law of thermodynamics
Spontaneous
Standard-state conditions
Standard-state entropy of reaction
$\Delta S°$(J/mol$_{rxn}$ · K)

Standard-state free energy of atom
combination ($\Delta G_{ac}°$)
Standard-state free energy of
formation ($\Delta G_f°$)
Standard-state free energy of reaction
($\Delta G°$)
Third law of thermodynamics

Problems

Spontaneous Chemical and Physical Processes

1. What is meant by the term *spontaneous*?

2. Are exothermic processes always spontaneous? Give an example of a spontaneous endothermic process.

3. Are there processes that are spontaneous at some temperatures but not at other temperatures? Give an example to support your answer.

Entropy and Disorder

4. What is the relationship between the entropy of a system and microscopic disorder?

5. Which has a greater entropy, liquid water or gaseous water?

6. A drop of red food-coloring solution is added to a beaker of water. As time passes, the entire beaker of water becomes a very pale red. Has there been an increase or a decrease in disorder? Has there been an increase or a decrease in entropy?

Entropy and the Second Law of Thermodynamics

7. Describe what happens on a microscopic scale when an ice-cold penny is dropped into water at room temperature.

8. Which of the following reactions leads to an increase in the entropy of the system?
    (a) $H_2(g) + Cl_2(g) \rightleftharpoons 2HCl(g)$
    (b) $2 NO_2(g) \rightleftharpoons N_2O_4(g)$
    (c) $CaO(s) + CO_2(g) \rightleftharpoons CaCO_3(s)$
    (d) $2 H_2(g) + O_2(g) \rightleftharpoons 2 H_2O(g)$
    (e) $2 NH_3(g) \rightleftharpoons N_2(g) + 3 H_2(g)$

9. In which of the following processes is ΔS negative?
    (a) $2 H_2O_2(aq) \rightleftharpoons 2 H_2O(l) + O_2(g)$
    (b) $CO_2(s) \rightleftharpoons CO_2(g)$
    (c) $H_2O(l) \rightleftharpoons H_2O(g)$
    (d) $4 Al(s) + 3 O_2(g) \rightleftharpoons 2 Al_2O_3(s)$

10. In which of the following processes is ΔS positive?
    (a) $2\,NO(g) + Cl_2(g) \rightleftharpoons 2\,NOCl(g)$
    (b) $NaCl(s) \rightleftharpoons NaCl(l)$
    (c) $3\,O_2(g) \rightleftharpoons 2\,O_3(g)$
    (d) $C_2H_4(g) + H_2(g) \rightleftharpoons C_2H_6(g)$

11. Which of the following processes should have the most positive value of ΔS?
    (a) $N_2(g) + O_2(g) \rightleftharpoons 2\,NO(g)$
    (b) $3\,C_2H_2(g) \rightleftharpoons C_6H_6(l)$
    (c) $H_2O(l) \rightleftharpoons H_2O(g)$
    (d) $4\,Al(s) + 3\,O_2(g) \rightleftharpoons 2\,Al_2O_3(s)$

12. Which of the following processes should have the most positive value of ΔS?
    (a) $H_2O(l) \rightleftharpoons H_2O(s)$
    (b) $NaNO_3(s) \xrightleftharpoons{H_2O} Na^+(aq) + NO_3^-(aq)$
    (c) $2\,HCl(g) \rightleftharpoons H_2(g) + Cl_2(g)$
    (d) $2\,H_2(g) + O_2(g) \rightleftharpoons 2\,H_2O(g)$

13. For some processes, even though the entropy change is favorable, the process doesn't occur. Explain why.

14. For some processes, even though the enthalpy change is favorable, the process doesn't occur. Explain why.

Standard-State Entropies of Reaction

15. What are the standard-state conditions for entropy for a chemical reaction?

16. Explain the difference between the symbols ΔS and $\Delta S°$.

The Third Law of Thermodynamics

17. What is the entropy of a perfect crystal at absolute zero? What happens to the entropy of the crystal as the temperature increases?

18. What happens to the motion of the particles composing a crystal as the temperature is raised from absolute zero to room temperature?

19. Which state of matter for a given substance—solid, liquid, or gas—has the highest entropy? Explain.

Calculating Entropy Changes for Chemical Reactions

20. Which favors the formation of products in a reaction, an increase or a decrease in entropy? Explain.

21. If $\Delta S°$ for a reaction is positive, does this mean that the reactants or the products have a larger entropy? What if $\Delta S°$ is negative?

22. One of the key steps toward transforming coal into a liquid fuel involves reducing carbon monoxide with H_2 gas to form methanol. Calculate $\Delta S°$ at 298 K for the following reaction to determine whether there is a favorable change in the entropy of the system.

$$CO(g) + 2\,H_2(g) \rightleftharpoons CH_3OH(l)$$

23. Tetraphosphorus decaoxide is often used as a dehydrating agent because of its tendency to pick up water. Calculate $\Delta S°$ at 25°C for the following reaction to determine whether entropy might be a measure of the driving force behind this reaction.

$$P_4O_{10}(s) + 6\,H_2O(l) \rightleftharpoons 4\,H_3PO_4(aq)$$

24. Calculate $\Delta S°$ at 25°C for the following reaction. Comment on both the sign and the magnitude of $\Delta S°$. The entropy of atom combination of $NH_4NO_2(s)$ is -949.5 J/mol$_{rxn}$ · K.

$$NH_4NO_2(s) \rightleftharpoons N_2(g) + 2\,H_2O(g)$$

25. Calculate $\Delta S°$ at 298 K for the following reaction in which oxygen in the atmosphere reacts to form ozone. Is there more disorder in the products or reactants?

$$3\,O_2(g) \rightleftharpoons 2\,O_3(g)$$

26. Compare the entropies of atom combination for the various forms of elemental phosphorus. Explain why $\Delta S°_{ac}$ is more negative for $P(s)$ than $P_2(g)$. Why is $\Delta S°_{ac}$ for $P_4(g)$ more negative than for $P_2(g)$?

Compound	$\Delta S°_{ac}$(J/mol$_{rxn}$ · K)
P(s)	-122.10
P(g)	0
$P_2(g)$	-108.257
$P_4(g)$	-372.79

Gibbs Free Energy

27. Explain the difference between ΔG and $\Delta G°$ for a chemical reaction.

28. What does it mean when ΔG for a reaction is zero?

29. Which of the following combinations of $\Delta H°$ and $\Delta S°$ always indicates a spontaneous reaction?
    (a) $\Delta H° > 0, \Delta S° < 0$ (b) $\Delta H° < 0, \Delta S° > 0$
    (c) $\Delta H° > 0, \Delta S° > 0$ (d) $\Delta H° < 0, \Delta S° < 0$
    (e) $\Delta H° = 0, \Delta S° = 0$

30. Predict the signs of $\Delta H°$ and $\Delta S°$ for the following reactions without referring to a table of thermodynamic data, and explain your predictions.
    (a) $2\,H_2(g) + O_2(g) \rightleftharpoons 2\,H_2O(g)$
       (*Hint:* The reaction is explosive under suitable conditions.)
    (b) $2\,Na(s) + Cl_2(g) \rightleftharpoons 2\,NaCl(s)$
       (*Hint:* Is table salt stable?)
    (c) $N_2(g) + 3\,H_2(g) \rightleftharpoons 2\,NH_3(g)$
       (*Hint:* The reaction occurs in the presence of a suitable catalyst.)

(d) $2 \, Cu(NO_3)_2(s)$
$$\rightleftharpoons 2 \, CuO(s) + 4 \, NO_2(g) + O_2(g)$$

(*Hint:* The reaction occurs spontaneously to produce CuO when the starting material is heated.)

31. Predict the signs of $\Delta H°$ and $\Delta S°$ for the following reaction without referring to a table of thermodynamic data, and explain your predictions.

$$NH_3(g) \overset{H_2O}{\rightleftharpoons} NH_3(aq)$$

Explain why the odor of NH_3 gas that collects above an NH_3 solution becomes more intense as the temperature increases.

32. Limestone decomposes to form lime and carbon dioxide when it is heated. Calculate $\Delta H°$ and $\Delta S°$ at 25°C for the decomposition of limestone and decide whether enthalpy or entropy is favorable for the reaction as written.

$$CaCO_3(s) \rightleftharpoons CaO(s) + CO_2(g)$$

33. Calculate $\Delta H°$ and $\Delta S°$ at 298 K for the thermite reaction and determine which is the most important measure of the driving force behind the reaction.

$$Fe_2O_3(s) + 2 \, Al(s) \rightleftharpoons Al_2O_3(s) + 2 \, Fe(s)$$

34. The first step in extracting iron ore from pyrite, FeS_2, involves roasting the ore in the presence of oxygen to form iron(III) oxide and sulfur dioxide.

$$4 \, FeS_2(s) + 11 \, O_2(g) \rightleftharpoons 2 \, Fe_2O_3(s) + 8 \, SO_2(g)$$

Calculate $\Delta H°$ and $\Delta S°$ at 25°C for the reaction and identify the most important measure of the driving force for the reaction.

35. Calculate $\Delta H°$ and $\Delta S°$ at 298 K for the following reaction. Are products or reactants favored?

$$2 \, KMnO_4(s) + 5 \, H_2O_2(aq) + 6 \, H^+(aq) \longrightarrow$$
$$2 \, K^+(aq) + 2 \, Mn^{2+}(aq) + 5 \, O_2(g) + 8 \, H_2O(l)$$

36. What determines whether a reaction is spontaneous?

37. It is possible to make hydrogen chloride by reacting phosphorus pentachloride with water and boiling the HCl out of the solution. Calculate $\Delta H°$ and $\Delta S°$ for this reaction at 25°C. What is the driving force behind the reaction: the enthalpy of reaction, the entropy of reaction, or Le Châtelier's principle?

$$PCl_5(g) + 4 \, H_2O(l) \rightleftharpoons H_3PO_4(aq) + 5 \, HCl(g)$$

38. Which of the following processes is spontaneous when all species are in their standard states at 298 K?

(a) $CH_3OH(l) \rightleftharpoons CH_3OH(g)$
(b) $CH_3OH(l) \rightleftharpoons HCHO(g) + H_2(g)$
(c) $2 \, CH_3OH(l) \rightleftharpoons 2 \, CH_4(g) + O_2(g)$
(d) $CH_3OH(l) \rightleftharpoons CO(g) + 2 \, H_2(g)$

39. Calculate $\Delta H°$ and $\Delta S°$ for the following reaction at 298 K. Explain why it is a mistake to use water to put out a fire that contains white-hot iron metal.

$$3 \, Fe(s) + 4 \, H_2O(l) \longrightarrow Fe_3O_4(s) + 4 \, H_2(g)$$

40. Use thermodynamic data to predict whether sodium or silver will react with water at 298 K.

$$2 \, Na(s) + 2 \, H_2O(l)$$
$$\longrightarrow 2 \, Na^+(aq) + 2 \, OH^-(aq) + H_2(g)$$
$$2 \, Ag(s) + 2 \, H_2O(l)$$
$$\longrightarrow 2 \, Ag^+(aq) + 2 \, OH^-(aq) + H_2(g)$$

41. Use thermodynamic data at 25°C to explain why zinc reacts with 1 M acid but not with water.

$$Zn(s) + 2 \, H^+(aq) \longrightarrow Zn^{2+}(aq) + H_2(g)$$
$$Zn(s) + 2 \, H_2O(l)$$
$$\longrightarrow Zn^{2+}(aq) + 2 \, OH^-(aq) + H_2(g)$$

42. Calculate $\Delta H°$ and $\Delta S°$ at 25°C for the following reaction. Why is ammonium nitrate a potential explosive?

$$2 \, NH_4NO_3(s) \longrightarrow 2 \, N_2(g) + O_2(g) + 4 \, H_2O(g)$$

43. Silane, SiH_4, decomposes to form elemental silicon and hydrogen. Calculate $\Delta H°$ and $\Delta S°$ for this reaction at 25°C and predict the effect on $\Delta G°$ of an increase in the temperature at which it is run.

$$SiH_4(g) \rightleftharpoons Si(s) + 2 \, H_2(g)$$

44. For which of the following reactions would you expect $\Delta H°$ and $\Delta G°$ to be about the same?

(a) $4 \, Fe(s) + 3 \, O_2(g) \longrightarrow 2 \, Fe_2O_3(s)$
(b) $2 \, Na(s) + 2 \, H_2O(l)$
$\longrightarrow 2 \, Na^+(aq) + 2 \, OH^-(aq) + H_2(g)$
(c) $Fe_2O_3(s) + 2 \, Al(s) \longrightarrow Al_2O_3(s) + 2 \, Fe(s)$
(d) $N_2O_4(g) \rightleftharpoons 2 \, NO_2(g)$
(e) $CaC_2(s) + 2 \, H_2O(l)$
$\rightleftharpoons Ca^{2+}(aq) + 2 \, OH^-(aq) + C_2H_2(g)$

The Effect of Temperature on the Free Energy of a Reaction

45. For each of the following reactions, how will the $\Delta G°$ change as the temperature increases?

(a) $2 \, CO(g) + O_2(g) \rightleftharpoons 2 \, CO_2(g)$
$$\Delta H° = -565.97 \text{ kJ/mol}_{rxn}$$
$$\Delta S° = -173.00 \text{ J/mol}_{rxn} \cdot K$$

(b) $2\,H_2O(g) \rightleftharpoons 2\,H_2(g) + O_2(g)$
$$\Delta H° = 483.64\,kJ/mol_{rxn}$$
$$\Delta S° = 90.01\,J/mol_{rxn}\cdot K$$

(c) $2\,N_2O(g) \rightleftharpoons 2\,N_2(g) + O_2(g)$
$$\Delta H° = -164.1\,kJ/mol_{rxn}$$
$$\Delta S° = 148.66\,J/mol_{rxn}\cdot K$$

(d) $PbCl_2(s) \overset{H_2O}{\rightleftharpoons} Pb^{2+}(aq) + 2\,Cl^-(aq)$
$$\Delta H° = 23.39\,kJ/mol_{rxn}$$
$$\Delta S° = -12.5\,J/mol_{rxn}\cdot K$$

46. Explain why the $\Delta G°$ for the following reaction becomes more positive as the temperature increases.

$$N_2(g) + 3\,H_2(g) \rightleftharpoons 2\,NH_3(g)$$

47. Which of the following diagrams best describes the relationship between $\Delta G°$ and temperature for the following reaction?

$$2\,H_2(g) + O_2(g) \rightleftharpoons 2\,H_2O(g)$$

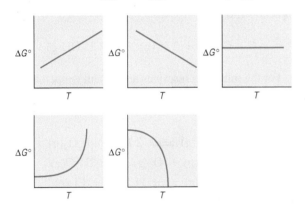

48. Calculate $\Delta G°$ for the following reaction from $\Delta H°$ and $\Delta S°$ data for the reaction at 298 K.

$$CS_2(l) + 3\,O_2(g) \rightleftharpoons CO_2(g) + 2\,SO_2(g)$$

49. What is the sign of $\Delta G°$ for the following process at $-10°C$, $0°C$, and $10°C$ at 1 atm pressure?

$$H_2O(s) \rightleftharpoons H_2O(l)$$

50. What is the sign of $\Delta G°$ for the following process at $90°C$, $100°C$, and $110°C$ at 1 atm pressure of $H_2O(g)$?

$$H_2O(l) \rightleftharpoons H_2O(g)$$

51. For the process $H_2O(l) \rightleftharpoons H_2O(g)$, what are the signs of ΔH and ΔS?

Beware of Oversimplifications

52. Is it usually correct to assume that $\Delta H°$ and $\Delta S°$ do not change very much with temperature? When might this assumption be in serious error?

53. For the reaction

$$2\,H_2(g) + O_2(g) \longrightarrow 2\,H_2O(g)$$

$\Delta G°$ is $-457\,kJ/mol_{rxn}$, $\Delta H°$ is $-484\,kJ/mol_{rxn}$, $\Delta S°$ is $-89\,J/mol_{rxn}\cdot K$, all at 298 K. Assume that both $\Delta H°$ and $\Delta S°$ are independent of temperature and calculate $\Delta G°$ at 500 K and 1000 K. The experimentally determined values for $\Delta G°$ are $-438\,kJ/mol_{rxn}$ at 500 K and $-385\,kJ/mol_{rxn}$ at 1000 K. Compare your calculations with the experimental values numbers. Why are these values different?

Standard-State Free Energies of Reaction

54. What are standard-state conditions for free energies of atom combination?

55. Using free energies of atom combination, calculate $\Delta G°$ for the following reactions.

(a) $CO(g) + 2\,H_2(g) \rightleftharpoons CH_3OH(l)$

(b) $3\,O_2(g) \rightleftharpoons 2\,O_3(g)$

(c) $CaCO_3(s) \rightleftharpoons CaO(s) + CO_2(g)$

56. Calculate the free energy change for the following reaction using free energies of atom combination at 298 K and 1 atm.

$$H_2O(l) \rightleftharpoons H_2O(g)$$

In which direction should this reaction proceed? What would be the sign of the free energy change for the reverse of this reaction? Explain what this means.

Equilibria Expressed in Partial Pressures

57. Explain why pressure can be used instead of concentration to describe equilibrium constant expressions for gas-phase reactions.

58. Which equation correctly describes the relationship between K_p and K_c for the following reaction?

$$Cl_2(g) + 3\,F_2(g) \rightleftharpoons 2\,ClF_3(g)$$

(a) $K_p = K_c$ (b) $K_p = K_c \times (RT)^{-1}$
(c) $K_p = K_c \times (RT)^{-2}$ (d) $K_p = K_c \times (RT)$
(e) $K_p = K_c \times (RT)^2$

59. Which equation correctly describes the relationship between K_p and K_c for the following reaction?

$$2\,NO_2(g) \rightleftharpoons 2\,NO(g) + O_2(g)$$

(a) $K_p = K_c$ (b) $K_p = 1/K_c$
(c) $K_p = K_c(RT)$ (d) $K_p = K_c(RT)^{-1}$

60. Calculate K_p for the decomposition of NOCl at 500 K if 27.3% of a 1.00-atm sample of NOCl decomposes to NO and Cl_2 at equilibrium.

$$2\,NOCl(g) \rightleftharpoons 2\,NO(g) + Cl_2(g)$$

61. Calculate the partial pressures of phosgene, carbon monoxide, and chlorine at equilibrium when a system that was initially 0.124 atm in $COCl_2$ decomposes at 300°C according to the following equation.

$$COCl_2(g) \rightleftharpoons CO(g) + Cl_2(g) \qquad K_p = 3.2 \times 10^{-3}$$

62. Without detailed equilibrium calculations, estimate the partial pressure of SO_3 that would be present at equilibrium when a mixture that was initially 0.490 atm in SO_2 and 0.245 atm in O_2 comes to equilibrium at 700 K.

$$2 SO_2(g) + O_2(g) \rightleftharpoons 2 SO_3(g)$$
$$K_p = 6.7 \times 10^4$$

63. Calculate the partial pressures of SO_3, SO_2, and O_2 that would be present at equilibrium when a mixture that was initially 0.490 atm in SO_3 reacts at 700 K.

$$2 SO_3(g) \rightleftharpoons 2 SO_2(g) + O_2(g)$$
$$K_p = 1.5 \times 10^{-5}$$

64. Estimate the partial pressures of N_2 and H_2 present at equilibrium when a mixture that was initially 0.50 atm in N_2, 0.60 atm in H_2, and 0.20 atm in NH_3 comes to equilibrium at a temperature at which K_p for the following reaction is 1.0×10^{-5}.

$$N_2(g) + 3 H_2(g) \rightleftharpoons 2 NH_3(g)$$

65. A mixture of NH_3, O_2, NO_2, and H_2O is initially 0.50 atm in each of the four gases. In which direction will the reaction proceed?

$$4 NO_2(g) + 6 H_2O(g) \rightleftharpoons 4 NH_3(g) + 7 O_2(g)$$
$$K_p = 1.8 \times 10^{-28}$$

If the pressure is increased at constant temperature on the equilibrium mixture of these gases, in which direction will the reaction proceed?

If the volume is reduced on the equilibrium mixture at constant temperature in which direction will the reaction proceed?

66. The partial pressures of N_2, H_2, and NH_3 present at equilibrium at 1000 K are 2.0, 6.0, and 0.018 atm, respectively. Calculate $\Delta G°$ for the following reaction.

$$2 NH_3(g) \rightleftharpoons N_2(g) + 3 H_2(g)$$

If an inert gas is added at constant temperature and volume, what will be the effect on the equilibrium?

67. Calculate the concentrations of N_2, O_2, and NO present when a mixture that was initially 0.40 M N_2 and 0.60 M O_2 reacts to form NO at 700°C. (*Hint:* Look

carefully at the symbol for the equilibrium constant for the reaction.)

$$N_2(g) + O_2(g) \rightleftharpoons 2 NO(g)$$
$$K_p = 4.3 \times 10^{-9} (\text{at } 700°C)$$

If the volume of the equilibrium mixture is decreased at constant temperature, in which direction will the reaction proceed?

68. Industrial chemicals can be made from coal by a process that starts with the reaction between red-hot coal and steam to form a mixture of CO and H_2. This mixture, which is known as synthesis gas, can be converted to methanol (CH_3OH) in the presence of a ruthenium/cobalt catalyst. The methanol produced in the reaction can then be converted into a host of other products, ranging from acetic acid to gasoline. The partial pressure of methanol when a mixture of 1.00 atm CO and 2.00 atm H_2 comes to equilibrium at 65°C is 0.98 atm. Calculate the standard free energy of the reaction at 65°C.

$$CO(g) + 2 H_2(g) \rightleftharpoons CH_3OH(g)$$

69. For the following reactions at constant temperature and volume, predict the effect of increasing the concentration of the reagent in boldface.

$$2 NO_2(g) \rightleftharpoons 2 NO(g) + \mathbf{O_2(g)}$$
$$2 H_2(g) + \mathbf{O_2(g)} \rightleftharpoons 2H_2O(g)$$
$$\mathbf{PCl_5(g)} \rightleftharpoons PCl_3(g) + Cl_2(g)$$

For which of these reactions would increasing the temperature at a constant pressure cause the reaction to proceed to the right? Show all necessary calculations.

Predict the effect of the addition of an inert gas at constant temperature and volume to each of the above reactions.

Interpreting Standard-State Free Energy of Reaction Data

70. What does $\Delta G°$ measure for a chemical reaction?
71. If $\Delta G°$ is large, what does this tell us about how far from equilibrium the standard-state conditions are?
72. Under what conditions does $\Delta G = \Delta G°$?

The Relationship between Free Energy and Equilibrium Constants

73. Which of the following correctly describes a reaction at equilibrium?
    (a) $\Delta G = 0$ (b) $\Delta G° = 1$ (c) $\Delta G = \Delta G°$
    (d) $Q = 0$ (e) $\ln K = 0$

74. Calculate the equilibrium constant at 25°C for the following reaction from the standard-state free energies of atom combination of the reactants and products.

$$CO(g) + 2 H_2(g) \rightleftharpoons CH_3OH(g)$$

75. Calculate K_p at 1000 K for the following reaction. Describe the assumptions you had to make to do the calculation.

$$2 CH_4(g) \rightleftharpoons C_2H_6(g) + H_2(g)$$

76. Calculate the equilibrium constant at 25°C for the following reaction.

$$2 HI(g) + Cl_2(g) \rightleftharpoons 2 HCl(g) + I_2(s)$$

77. Calculate $\Delta H°$, $\Delta S°$, and $\Delta G°$ for the following reactions at 298 K. Use the data to calculate the values of K_a for hydrochloric and acetic acid. Compare the results of the calculations with the K_a data in Table 11.3 in Chapter 11.

$$HCl(g) \overset{H_2O}{\rightleftharpoons} H^+(aq) + Cl^-(aq)$$
$$CH_3CO_2H(aq) \rightleftharpoons H^+(aq) + CH_3CO_2^-(aq)$$

78. Calculate $\Delta H°$, $\Delta S°$, and $\Delta G°$ for the following reaction at 298 K. Use the data to calculate the value of K_b for ammonia. Compare the result of the calculation with the value of K_b in Table B.9 in Appendix B.

$$NH_3(aq) + H_2O(l) \rightleftharpoons NH_4^+(aq) + OH^-(aq)$$

79. For the reaction $Zn(s) + 2 H^+(aq) \rightleftharpoons Zn^{2+}(aq) + H_2(g)$ at 298 K $E° = 0.76$ V. What are $\Delta G°$ and the equilibrium constant for the reaction?

80. The standard free energy change for $3H_2(g) + 2 Cr^{3+}(aq) \rightleftharpoons 2Cr(s) + 6 H^+(aq)$ is 4.3×10^2 kJ/mol$_{rxn}$. Calculate the standard electrode potential for this reaction.

81. At 298 K $I_2(s) + 2 Br^-(aq) \rightleftharpoons Br_2(l) + 2I^-(aq)$ $K_c = 1.8 \times 10^{-19}$. Find the standard cell potential for this reaction.

The Temperature Dependence of Equilibrium Constants

82. Synthesis gas can be made by reacting red-hot coal with steam. Estimate the temperature at which the equilibrium constant for the reaction is equal to 1.

$$C(s) + H_2O(g) \rightleftharpoons CO(g) + H_2(g)$$

83. Calculate $\Delta H°$, $\Delta S°$, $\Delta G°$, and the equilibrium constant at 25°C for the following reaction. Predict the effect of an increase in the temperature of the system on the equilibrium constant for the reaction.

$$CO_2(g) + H_2(g) \rightleftharpoons CO(g) + H_2O(g)$$

84. What happens to the equilibrium constant for the following reaction as the temperature increases?

$$PCl_5(g) \rightleftharpoons PCl_3(g) + Cl_2(g)$$

85. At approximately what temperature does the equilibrium constant for the following reaction become larger than 1?

$$PCl_5(g) \rightleftharpoons PCl_3(g) + Cl_2(g)$$

86. For which of the following reactions would the equilibrium constant increase with increasing temperature?
    (a) $2 H_2(g) + O_2(g) \rightleftharpoons 2 H_2O(g)$
    (b) $2 HCl(g) \rightleftharpoons H_2(g) + Cl_2(g)$
    (c) $2 NH_3(g) \rightleftharpoons N_2(g) + 3 H_2(g)$

87. Calculate the equilibrium constant for the following reaction at 25°C, 200°C, 400°C, and 600°C. Explain any trend you observe. Assume $\Delta H°$ and $\Delta S°$ do not change with temperature.

$$2 SO_3(g) \rightleftharpoons 2 SO_2(g) + O_2(g)$$

88. Calculate $\Delta H°$ and $\Delta S°$ for the following reaction. Predict what will happen to the equilibrium constant of the reaction as the temperature increases.

$$Cu(s) + 2 H^+ \rightleftharpoons Cu^{2+}(aq) + H_2(g)$$

Gibbs Free Energies of Formation and Absolute Entropies

89. Calculate $\Delta S°$ for the following reactions using absolute entropies.
    (a) $2 H_2(g) + O_2(g) \longrightarrow 2 H_2O(g)$
    (b) $2 Na(s) + Cl_2(g) \longrightarrow 2 NaCl(s)$
    (c) $N_2(g) + 3 H_2(g) \longrightarrow 2 NH_3(g)$

    Calculate $\Delta H°$ for the above reactions from enthalpies of formation. Calculate $\Delta G°$ from $\Delta H°$ and $\Delta S°$ obtained from enthalpy of formation data and absolute entropy data. Compare this $\Delta G°$ with $\Delta G°$ calculated from free energy of formation data. Compare the results to the predictions made in Problem 30.

90. For Problems 48, 55, 56, and 74, calculate $\Delta G°$ from free energies of formation.

91. Calculate $\Delta H°$ and $\Delta S°$ from enthalpies of formation and absolute entropies for the following process.

$$NH_3(g) \rightleftharpoons NH_3(aq)$$

Compare your results with the answer to Problem 31.

Integrated Problems

92. $\Delta S°$ for the following reaction is favorable even though $\Delta H°$ is not.

$$CH_3OH(l) \rightleftharpoons CH_3OH(g)$$

Assume that methanol boils at the temperature at which ΔG for this reaction is equal to zero. Use the values of $\Delta H°$ and $\Delta S°$ at 25°C for the reaction to estimate the boiling point of methanol.

93. Use the equation that describes the relationship between ΔG and $\Delta G°$ for a reaction to construct a graph of ΔG versus $\ln Q_c$ at 25°C for the following reactions over a range of values of Q_c between 10^{-10} and 10^{10}.

$$HOAc(aq) \overset{H_2O}{\rightleftharpoons} H^+(aq) + OAc^-(aq)$$
$$\Delta G° = 27.1 \text{ kJ/mol}_{rxn}$$

$$HCl(aq) \overset{H_2O}{\rightleftharpoons} H^+(aq) + Cl^-(aq)$$
$$\Delta G° = -35.9 \text{ kJ/mol}_{rxn}|$$

Explain the significance of the difference between the points at which the straight lines for the data cross the horizontal and vertical axes.

94. From the following $\Delta H°$ and $\Delta S°$ values, predict whether each of the reactions would lead to $K > 1$ or $K < 1$ at 25°C. If a reaction has $K < 1$, at what temperature might K become greater than 1? Explain your reasoning.
   (a) $\Delta H° = 10 \text{ kJ/mol}_{rxn}$; $\Delta S° = 30 \text{ J/mol}_{rxn} \cdot K$
   (b) $\Delta H° = 2 \text{ kJ/mol}_{rxn}$; $\Delta S° = 100 \text{ J/mol}_{rxn} \cdot K$
   (c) $\Delta H° = -125 \text{ kJ/mol}_{rxn}$; $\Delta S° = 80 \text{ J/mol}_{rxn} \cdot K$
   (d) $\Delta H° = -10 \text{ kJ/mol}_{rxn}$; $\Delta S° = -30 \text{ J/mol}_{rxn} \cdot K$

95. For the reaction

$$2 NO(g) + O_2(g) \rightleftharpoons 2 NO_2(g)$$

   (a) Find $\Delta H°$.
   (b) Calculate $\Delta S°$.
   (c) What is $\Delta G°$?
   (d) At 298 K, are the products or reactants favored? Explain.
   (e) Calculate the equilibrium constant for the reaction at 25°C.
   (f) If the partial pressures of NO, O_2, and NO_2 are 1.0 atm, in which direction will the reaction proceed at 298 K?
   (g) If the temperature is decreased, will the equilibrium constant increase, decrease, or stay the same? Explain.

96. The thermodynamic parameters at 25°C are given for the following two reactions.

   (I) $2 CO(g) + O_2(g) \rightleftharpoons 2 CO_2(g)$
   $$\Delta H° = -565 \text{ kJ/mol}_{rxn}$$
   $$\Delta S° = -173 \text{ J/mol}_{rxn} \cdot K$$

   (II) $2 H_2O(g) \rightleftharpoons 2 H_2(g) + O_2(g)$
   $$\Delta H° = +484 \text{ kJ/mol}_{rxn}$$
   $$\Delta S° = +90 \text{ J/mol}_{rxn} \cdot K$$

   (a) For which of these reactions is $\Delta H°$ favorable? Explain.
   (b) For which of these reactions is $\Delta S°$ favorable? Explain.
   (c) For which reaction will the equilibrium constant be greater than 1 at 25°C?
   (d) Explain the sign of $\Delta S°$ for each of these reactions.
   (e) If the temperature is increased at constant pressure, which way will the equilibrium shift for each reaction?

97. The following reaction is involved in the production of acid rain.

$$2 NO(g) + O_2(g) \rightleftharpoons 2 NO_2(g)$$

This reaction occurs in lightning flashes in the upper atmosphere. If the temperature in the lightning flash is much higher than 25°C, will $\Delta G°$ be greater, lesser, or the same as $\Delta G°$ at 25°C? Explain.

98. The following enthalpy and entropy changes are known for the reactions shown at 298 K.

$$Cu(s) + 2 H^+(aq) \rightleftharpoons Cu^{2+}(aq) + H_2(g)$$
$$\Delta H° = 65 \text{ kJ/mol}_{rxn}$$
$$\Delta S° = -2.1 \text{ J/mol}_{rxn} \cdot K$$

$$Zn(s) + 2H^+(aq) \rightleftharpoons Zn^{2+}(aq) + H_2(g)$$
$$\Delta H° = -153 \text{ kJ/mol}_{rxn}$$
$$\Delta S° = -23.1 \text{ J/mol}_{rxn} \cdot K$$

   (a) Which of these two metals, Zn or Cu, should dissolve in a 1 M acid solution? Explain.
   (b) Which of these two reactions has an equilibrium constant that increases with temperature?
   (c) Would you have expected both reactions to have had a negative entropy change? How can you explain $\Delta S° < 0$ for these reactions?

99. Use the following thermodynamic data to predict which should be the strongest acid at 25°C, H_2O or CH_3COOH.

$$H_2O(aq) \rightleftharpoons H^+(aq) + OH^-(aq)$$
$$CH_3CO_2H(aq) \rightleftharpoons H^+(aq) + CH_3CO_2^-(aq)$$

	ΔH°_{ac}(kJ/mol$_{rxn}$)	ΔS°_{ac}(J/mol$_{rxn}$ · K)
$H_2O(aq)$	-970	-321
$H^+(aq)$	-218	-115
$OH^-(aq)$	-696	-286
$CH_3COOH(aq)$	-3288	-919
$CH_3COO^-(aq)$	-3071	-896

Show all calculations and explain your reasoning.

100. For the gas-phase reaction

$$2\ NO_2(g) \rightleftharpoons N_2O_4(g)$$

In which direction will the equilibrium reaction proceed if

(a) The volume of the container is increased at constant temperature.

(b) The volume is decreased at constant temperature.

(c) Under conditions of constant volume and temperature some NO_2 is introduced into the equilibrium mixture.

(d) An inert gas is added under conditions of constant temperature and pressure.

101. The following reaction is at equilibrium under conditions of constant temperature and pressure.

$$N_2(g) + 3\ H_2(g) \rightleftharpoons 2\ NH_3(g)$$

In some cases when N_2 is added to the equilibrium reaction mixture, the reaction shifts to the left. Explain how this could happen.

102. (a) For the reaction

$$Zn(s) + Cu^{2+}(aq) \rightleftharpoons Zn^{2+}(aq) + Cu(s)$$
$$K_c = 2.0 \times 10^{37}$$

Which half-reaction has the most positive E°?

(b) For the reaction

$$2\ Br^-\ (aq) + Cu^{2+}(aq) \rightleftharpoons Br_2(aq) + Cu(s)$$
$$K_c = 5.5 \times 10^{-26}$$

Identify the two half-cells in this reaction. Which of the two half-cells for this reaction has the higher standard reduction potential?

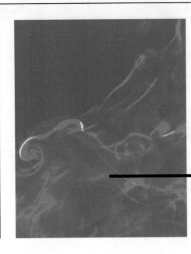

Chapter Fourteen

KINETICS

14.1 The Forces That Control a Chemical Reaction

The previous chapter showed how thermodynamics can explain why some reactions occur but not others. In that chapter, we saw that thermodynamics is a powerful tool for predicting what should (or should not) happen in a chemical reaction. It is ideally suited for answering questions that begin, "What if . . . ?" Thermodynamics predicts, for example, that hydrogen should react with oxygen to form water.

$$2 H_2(g) + O_2(g) \longrightarrow 2 H_2O(g)$$

It also predicts that iron (III) oxide should react with aluminum metal to form aluminum oxide and iron metal.

$$Fe_2O_3(s) + 2 Al(s) \longrightarrow Al_2O_3(s) + 2 Fe(s)$$

Both of these reactions can, in fact, occur. We can experience the first by touching a balloon filled with H_2 gas with a candle tied to the end of a meter stick. We can demonstrate the other reaction by igniting a mixture of Fe_2O_3 and powdered aluminum.

This doesn't mean that H_2 and O_2 burst spontaneously into flame the instant the gases are mixed. Nor does it mean that scrupulous attention must be paid to prevent Fe_2O_3 from coming into contact with aluminum metal, lest some violent reaction take place. In the absence of a spark, mixtures of H_2 and O_2 can be stored for years with no detectable change. In the absence of a source of heat, mixtures of Fe_2O_3 and powdered aluminum are so stable they are sold in 50-pound lots under the trade name Thermite.

It is a mistake to lose sight of the fact that thermodynamics can never tell us what *will* happen. It can only tell us what *should* or *might* happen. Reactions that behave as thermodynamics predicts they should are said to be under **thermodynamic control.** Potassium metal, for example reacts violently when it comes in contact with water, as one would predict from the sign and magnitude of $\Delta G°$ for this reaction.

$$2 K(s) + 2 H_2O(l) \longrightarrow 2 K^+(aq) + 2 OH^-(aq) + H_2(g)$$
$$\Delta G° = -406.77 \text{ kJ/mol}_{rxn}$$

Other reactions are said to be under **kinetic control.** Thermodynamics predicts that these reactions should occur, but the rate of these reactions is negligibly slow. Examples of reactions under kinetic control include the reactions in which elemental sulfur and its compounds burn.

$$S_8(s) + 8 O_2(g) \longrightarrow 8 SO_2(g)$$
$$CS_2(g) + 3 O_2(g) \longrightarrow CO_2(g) + 2 SO_2(g)$$
$$2 H_2S(g) + 3 O_2(g) \longrightarrow 2 H_2O(g) + 2 SO_2(g)$$

Thermodynamics predicts that these reactions should form SO_3, not SO_2, because SO_3 is more stable than SO_2. Under normal conditions, however, these reactions invariably stop at SO_2 because the reaction between SO_2 and O_2 to form SO_3 is extremely slow.

To fully understand chemical reactions, we need to combine the predictions of thermodynamics with studies of the factors that influence the rates of chemical reactions. These factors fall under the general heading of *chemical kinetics* (from a Greek stem meaning "to move"), which is the subject of this chapter.

The reaction between Fe_2O_3 and powdered aluminum gives off so much energy it is called the thermite reaction.

➤ **CHECKPOINT**

Thermodynamics predicts that under normal circumstances carbon in the form of graphite is more stable than carbon in the form of diamond. Why don't you have to worry that a diamond ring will turn to graphite?

14.2 Chemical Kinetics

Chemical kinetics can be defined as the search for answers to the following questions.

- What is the rate at which the reactants are converted into the products of a reaction?
- What factors influence the rate of the reaction?
- What is the sequence of steps, or the mechanism, by which the reactants are converted into products?

As we saw in Section 10.3, the term *rate* is used to describe the change in a quantity that occurs per unit of time. The rate at which cars are sold, for example, is equal to the number of cars that change hands divided by the period of time during which this number is counted. The rate at which an object travels through space is the distance traveled per unit of time, such as miles per hour or kilometers per second. In chemical kinetics, the distance traveled can be thought of as the change in the concentration of one of the components of the reaction. The rate of a reaction is therefore the change in the concentration of one of the reactants or products, $\Delta(X)$, that occurs during a given period of time, Δt

$$\text{rate} = \frac{(X)_{\text{final}} - (X)_{\text{intial}}}{t_{\text{final}} - t_{\text{intial}}} = \frac{\Delta(X)}{\Delta t}$$

One of the first kinetic experiments was done in 1850 by Ludwig Wilhelmy, a lecturer in physics at the University of Heidelberg. Wilhelmy studied the rate at which sucrose, or cane sugar, reacts with acid to form a mixture of fructose and glucose known as invert sugar.

$$\text{sucrose}(aq) + H_3O^+(aq) \longrightarrow \underbrace{\text{glucose}(aq) + \text{fructose}(aq)}_{\textit{Invert sugar}}$$

Wilhelmy found that the rate at which sucrose is consumed in this reaction is directly proportional to the concentration of the sucrose in the solution when the pH and temperature of the solution are held constant. The rate of the reaction at any moment in time can therefore be described by the following equation, in which k is a proportionality constant and (sucrose) is the concentration of sucrose in units of moles per liter at that moment.

$$\text{rate} = k(\text{sucrose})$$

Because the sucrose concentration decreases as sucrose is consumed in the reaction, the rate of the reaction gradually decreases with time.

14.3 Is the Rate of Reaction Constant?

Acid–base titrations often use phenolphthalein as the end-point indicator, as shown in Figure 14.1. Phenolphthalein is colorless in the presence of acid but turns pink in the presence of excess base. An interesting thing happens, however, if you let

Fig. 14.1 Acid–base titration using phenolphthalein as the indicator.

the solution stand once the end point has been reached: The solution gradually turns colorless once again as the phenolphthalein reacts with the excess base.

Table 14.1 contains experimental data that show what happens to the concentration of phenolphthalein in a solution at 25°C that was initially 0.00500 *M* in phenolphthalein and 0.61 *M* in OH⁻ ion. As you can see when the data are plotted in Figure 14.2, the phenolphthalein concentration decreases by a factor of 10 over a period of about 230 seconds.

To find the rate of reaction over any period of time, we divide the change in the concentration of phenolphthalein by the time interval over which the change occurred. Because phenolphthalein is consumed in the reaction, the change in the phenolphthalein concentration is a negative number. (The concentration of phenolphthalein is smaller at the end of the time interval than at the beginning of the time interval.) By convention, rates are reported as positive numbers. When the rate of a reaction is defined in terms of the rate at which one of the reactants disappears, we therefore add a negative sign to the equation that defines the rate of the reaction.

$$\text{rate} = -\frac{\Delta(\text{phenolphthalein})}{\Delta t}$$

Table 14.1
Experimental Data for the Reaction between Phenolphthalein and Excess Base

Concentration of Phenolphthalein (M)	Time (s)
0.00500	0.0
0.00450	10.5
0.00400	22.3
0.00350	35.7
0.00300	51.1
0.00250	69.3
0.00200	91.6
0.00150	120.4
0.00100	160.9
0.000500	230.2
0.000250	299.5
0.000150	350.7
0.000100	391.2

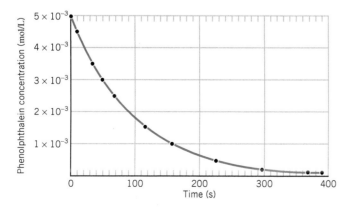

Fig. 14.2 Plot of the concentration of phenolphthalein versus time for the reaction between this acid–base indicator and excess OH⁻ ion.

Exercise 14.1

Use the data in Table 14.1 to calculate the rate at which phenolphthalein reacts with the OH$^-$ ion during each of the following periods.

(a) During the first time interval, when the phenolphthalein concentration falls from 0.00500 M to 0.00450 M

(b) During the second interval, when the concentration falls from 0.00450 M to 0.00400 M

(c) During the third interval, when the concentration falls from 0.00400 M to 0.00350 M

Solution

The rate of reaction is equal to the change in the phenolphthalein concentration divided by the length of time over which the change occurs.

(a) During the first time period, the rate of the reaction is 4.8 × 10^{-5} moles per liter per second

$$\text{rate} = -\frac{\Delta(X)}{\Delta t} = -\frac{(0.00450\ M) - (0.00500\ M)}{(10.5\ \text{s} - 0\ \text{s})} = 4.8 \times 10^{-5}\ M/s$$

(b) During the second time period, the rate of reaction is slightly slower.

$$\text{rate} = -\frac{\Delta(X)}{\Delta t} = -\frac{(0.00400\ M) - (0.00450\ M)}{(22.3\ \text{s} - 10.5\ \text{s})} = 4.2 \times 10^{-5}\ M/s$$

(c) During the third time period, the rate of reaction is even slower.

$$\text{rate} = -\frac{\Delta(X)}{\Delta t} = -\frac{(0.00350\ M) - (0.00400\ M)}{(35.7\ \text{s} - 22.3\ \text{s})} = 3.7 \times 10^{-5}\ M/s$$

14.4 Instantaneous Rates of Reaction

Exercise 14.1 illustrates an important point: The rate of the reaction between phenolphthalein and the OH$^-$ ion isn't constant; it changes with time. Like most reactions, the rate of the reaction gradually decreases as the reactants are consumed.

The rate of reaction not only changes from one time period to another, it changes while it is being measured. To minimize the error this introduces into our measurements, it seems advisable to measure the rate of reaction over periods of time that are short compared with the time it takes for the reaction to occur. We might try, for example, to measure the infinitesimally small change in concentration, $d(X)$, that occurs over an infinitesimally short period of time, dt. The ratio of these quantities is known as the **instantaneous rate of reaction.**

$$\text{rate} = -\frac{d(X)}{dt}$$

The instantaneous rate of reaction can be calculated from a graph of the concentration of one of the components of the reaction versus time. Figure 14.3

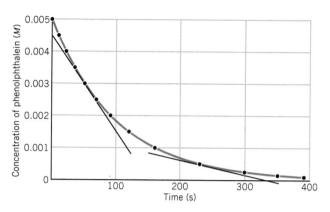

Fig. 14.3 The instantaneous rate of reaction at any moment in time can be calculated from the slope of a tangent to the graph of concentration versus time. At a point 50 s after the beginning of the reaction, for example, the instantaneous rate of this reaction is 2.7×10^{-5} mole per liter per second. At a point 250 s after the beginning of the reaction, the instantaneous rate of reaction is 4.1×10^{-6} mole per liter per second.

shows how the instantaneous rate can be calculated for the reaction between phenolphthalein and excess base. The rate of reaction at any moment in time is equal to the negative of the slope of a tangent drawn to the curve at that moment.

The instantaneous rate of reaction can be measured at any time between the moment at which the reactants are mixed and the time when the reaction reaches equilibrium. The **initial instantaneous rate of reaction** is the rate at which the reaction occurs at the moment the reactants are mixed.

➤ **CHECKPOINT**

What is the problem with using a large value for Δt when determining the rate of a reaction using the following equation?

$$\text{rate} = -\frac{\Delta(X)}{\Delta t}$$

14.5 Rate Laws and Rate Constants

An interesting result is obtained when the instantaneous rate of reaction is calculated at various points along the curve in Figure 14.3. The rate of reaction at every point on the curve is directly proportional to the concentration of phenolphthalein at that moment in time.

$$\text{Rate} = k(\text{phenolphthalein})$$

Because this equation is an experimental law that describes the rate of the reaction, it is called the **rate law** for the reaction. The proportionality constant, k, is known as the **rate constant.**

➤ **CHECKPOINT**

Use the rate law for the reaction between phenolphthalein and excess base to explain why the rate of the reaction gradually slows down.

Exercise 14.2

Calculate the rate constant for the reaction between phenolphthalein and the OH^- ion if the instantaneous rate of reaction is 2.5×10^{-5} moles per liter per second when the concentration of phenolphthalein is $0.00250 \ M$.

Solution

We start with the rate law for the reaction.

$$\text{Rate} = k(\text{phenolphthalein})$$

We then substitute the known rate of reaction and the known concentration of phenolphthalein into the equation.

$$2.5 \times 10^{-5} \frac{\text{mol/L}}{\text{s}} = k(0.00250 \text{ mol/L})$$

➤ **CHECKPOINT**

Describe the difference between the *rate* of a chemical reaction and the *rate constant*. How do the rate and rate constant vary with respect to concentration and time?

➤ **CHECKPOINT**

What units will the rate constant have in a reaction for which the rate law is: rate = $k(X)^2$?

Solving for the rate constant gives the following result.

$$k = 0.010 \text{ s}^{-1}$$

The rate of reaction is the change in the concentration of phenolphthalein divided by the time over which the change occurs, it is reported in units of moles per liter per second. Because the number of moles of phenolphthalein per liter is the molarity of the solution, the rate can also be reported in terms of the change in molarity per second: M/s. The units of the rate constant will depend on the form of the rate law. When substituted into the rate law, the rate constant must have units that yield a rate in units of M/s.

• •

 Exercise 14.3

Use the rate constant for the reaction between phenolphthalein and the OH⁻ ion calculated in the previous exercise to calculate the initial instantaneous rate of reaction for the data in Table 14.1.

Solution

We start, once again, with the rate law for the reaction.

$$\text{rate} = k(\text{phenolphthalein})$$

We then substitute into this equation the rate constant for the reaction and the initial concentration of phenolphthalein.

$$\text{rate} = (0.010 \, s^{-1})(0.00500 \, \text{mol/L}) = 5.0 \times 10^{-5} \frac{\text{mol/L}}{\text{s}}$$

• •

14.6 The Rate Law versus the Stoichiometry of a Reaction

In the 1930s, Sir Christopher Ingold and co-workers at the University of London studied the kinetics of substitution reactions such as the following.

$$\underset{\substack{|\\ \text{H}}}{\overset{\substack{\text{H}\\|}}{\text{H}-\text{C}-\text{Br}}}(aq) + \text{OH}^-(aq) \longrightarrow \underset{\substack{|\\ \text{H}}}{\overset{\substack{\text{H}\\|}}{\text{H}-\text{C}-\text{OH}}}(aq) + \text{Br}^-(aq)$$

They found that the rate of this reaction is proportional to the concentrations of both reactants.

$$\text{rate} = k(\text{CH}_3\text{Br})(\text{OH}^-)$$

When they ran a similar reaction on a slightly different starting material, they got similar products.

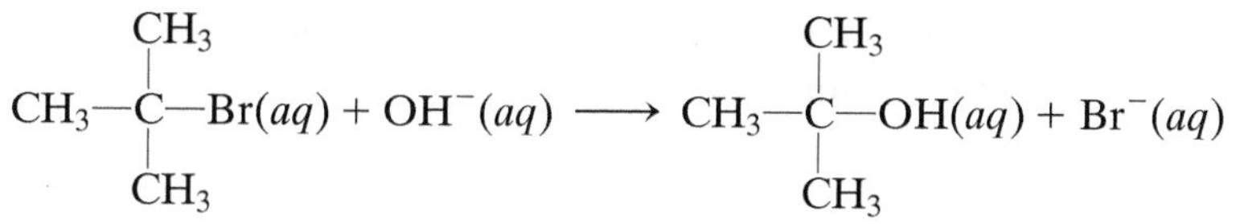

$$CH_3-\underset{\underset{CH_3}{|}}{\overset{\overset{CH_3}{|}}{C}}-Br(aq) + OH^-(aq) \longrightarrow CH_3-\underset{\underset{CH_3}{|}}{\overset{\overset{CH_3}{|}}{C}}-OH(aq) + Br^-(aq)$$

But now the rate of reaction was proportional to the concentration of only one of the reactants.

$$\text{rate} = k((CH_3)_3CBr)$$

These results illustrate an important point: *The rate law for a reaction can't be predicted from the stoichiometry of the reaction; it must be determined experimentally.* Sometimes the rate law is consistent with what we expect from the stoichiometry of the reaction. The balanced equation for the following reaction, for example, involves two molecules of HI, and the rate law for the reaction depends on the square of the concentration of HI.

$$2\,HI(g) \longrightarrow H_2(g) + I_2(g) \qquad \text{rate} = k(HI)^2$$

Often, however, the rate law doesn't follow the stoichiometry. The observed rate law for the decomposition of N_2O_5, for example, is proportional to the concentration of N_2O_5, not to the square of the concentration of N_2O_5, as we might expect from the stoichiometry of the reaction.

$$2\,N_2O_5(g) \longrightarrow 4\,NO_2(g) + O_2(g) \qquad \text{rate} = k(N_2O_5)$$

The decomposition of N_2O_5 raises another interesting point. We could study the kinetics of this reaction by monitoring the rate at which N_2O_5 is consumed. Or we could study the rate at which NO_2 or O_2 is formed. We'll get different values for the rate of reaction, however, depending on which reagent we choose. According to the balanced equation for this reaction, we get 4 moles of NO_2 for every 2 moles of N_2O_5 consumed. NO_2 is therefore formed at a rate that is twice as fast as the rate at which N_2O_5 is consumed.

$$\frac{d(NO_2)}{dt} = -2\frac{d(N_2O_5)}{dt}$$

We get only 1 mole of O_2, however, for every 2 moles of N_2O_5 consumed in this reaction. Thus O_2 is formed at a rate that is only one-half the rate at which N_2O_5 is consumed.

$$\frac{d(O_2)}{dt} = -\frac{1}{2}\frac{d(N_2O_5)}{dt}$$

> ➤ **CHECKPOINT**
>
> Predict the relationship between the rate of disappearance of HI and the rate of formation of H_2 in the following reaction.
>
> $$2\,HI(g) \longrightarrow H_2(g) + I_2(g)$$
>
> Is the rate of disappearance of HI faster, slower, or equal to the rate of formation of H_2?

14.7 Order and Molecularity

Some reactions occur in a single step. The reaction in which a chlorine atom is transferred from $ClNO_2$ to NO to form NO_2 and ClNO is a good example of a one-step reaction.

$$ClNO_2(g) + NO(g) \longrightarrow NO_2 + ClNO(g)$$

Other reactions occur by a series of individual steps. N_2O_5, for example, decomposes to NO_2 and O_2 by a three-step mechanism.

$$\text{Step 1:} \qquad N_2O_5 \longrightarrow NO_2 + NO_3$$
$$\text{Step 2:} \qquad NO_2 + NO_3 \longrightarrow NO_2 + NO + O_2$$
$$\text{Step 3:} \qquad NO + NO_3 \longrightarrow 2\,NO_2$$

The steps in a reaction are classified in terms of their **molecularity,** which describes the number of molecules consumed in that step. When a single molecule is consumed, the step is called **unimolecular.** When two molecules are consumed, it is **bimolecular.**

Exercise 14.4

Determine the molecularity of each step in the reaction by which N_2O_5 decomposes to NO_2 and O_2.

Solution

All we have to do is count the number of molecules consumed in each step in the reaction to decide that the first step is unimolecular and the other two steps are bimolecular.

$$\text{Step 1:} \qquad N_2O_5 \longrightarrow NO_2 + NO_3 \qquad\qquad \text{(unimolecular)}$$
$$\text{Step 2:} \qquad NO_2 + NO_3 \longrightarrow NO_2 + NO + O_2 \quad \text{(bimolecular)}$$
$$\text{Step 3:} \qquad NO + NO_3 \longrightarrow 2\,NO_2 \qquad\qquad \text{(bimolecular)}$$

Reactions can also be classified in terms of their **order.** Let's start our discussion by looking at the decomposition of dinitrogen oxide, N_2O, in the presence of platinum metal.

$$2\,N_2O(g) \xrightarrow{\;Pt\;} 2\,N_2(g) + O_2(g)$$

This is called a **zero-order reaction** because the rate of reaction is constant. In other words, it depends on the concentration of N_2O to the zero-order.

$$\text{rate} = k(N_2O)^0 = k$$

The decomposition of N_2O_5 is a **first-order reaction** because the rate of reaction depends on the concentration of N_2O_5 raised to the first power.

$$\text{rate} = k(N_2O_5)$$

The decomposition of HI is a **second-order reaction** because the rate of reaction depends on the concentration of HI raised to the second power.

$$\text{rate} = k(HI)^2$$

When the rate of a reaction depends on more than one reagent, we classify the reaction in terms of the order of each reagent. The reaction between CH_3Br

and the OH$^-$ ion, for example, would be described as first-order in both CH$_3$Br and the OH$^-$ ion, or second-order overall.

$$\text{rate} = k(\text{CH}_3\text{Br})(\text{OH}^-)$$

In general, for a hypothetical reaction such as

$$A + B \longrightarrow C + D$$

The rate law is written in terms of the concentrations of the reactants, each raised to an exponent. The exponent gives the order of the reaction with respect to a particular reactant. Thus for the general reaction above

$$\text{rate} = k(A)^m(B)^n$$

where k is the rate constant for the reaction, m is the order with respect to A and n is the order with respect to B. the overall order of the reaction is given by $m + n$.

Exercise 14.5

When there are only trace quantities of NO in the atmosphere, it doesn't react with O$_2$ to form NO$_2$. As the concentration of NO in the atmosphere builds up, however, it reaches a point at which this reaction can and does occur.

$$2\,\text{NO}(g) + \text{O}_2(g) \longrightarrow 2\,\text{NO}_2(g)$$

The experimentally determined rate law for the reaction is

$$\text{rate} = k(\text{NO})^2(\text{O}_2)$$

Classify the order of this reaction.

Solution

The reaction is first-order in O$_2$ and second-order in NO. The overall reaction is third-order.

When the rate of a reaction doesn't depend on the concentration of one or more of the substances consumed in that reaction, it is said to be *zero-order* in that substance. In a zero-order reaction, the rate of reaction literally depends on the concentration of that reagent to the zeroth power. Since any quantity raised to the zeroth power is equal to 1, the concentration of the reagent has no effect on the rate of reaction.

We have already seen an example of a reaction that is zero-order in one of the reactants.

$$\text{CH}_3\text{—}\underset{\underset{\text{CH}_3}{|}}{\overset{\overset{\text{CH}_3}{|}}{\text{C}}}\text{—Br}(aq) + \text{OH}^-(aq) \longrightarrow \text{CH}_3\text{—}\underset{\underset{\text{CH}_3}{|}}{\overset{\overset{\text{CH}_3}{|}}{\text{C}}}\text{—OH}(aq) + \text{Br}^-(aq)$$

The rate law for this reaction is first-order in $(CH_3)_3CBr$, but it is zero-order in the OH^- ion.

$$\text{rate} = k((CH_3)_3CBr)^1(OH^-)^0 = k((CH_3)_3CBr)$$

It is important to recognize the difference between the molecularity and the order of a reaction. The molecularity of a reaction, or a step within a reaction, describes what happens on the molecular level. The order of a reaction describes what happens on the macroscopic scale. We determine the order of a reaction experimentally by watching the products of a reaction appear or the starting materials disappear. The molecularity of the reaction is something we deduce to explain the experimental results.

14.8 A Collision Theory Model of Chemical Reactions

The **collision theory** model of chemical reactions introduced in Section 10.4 can be used to explain the observed rate laws for both one-step and multistep reactions. This model of chemical reactions assumes that the rate of any step in a reaction depends on the frequency of collisions between the particles involved in that step. Consider the following simple, one-step reaction, for example.

$$ClNO_2(g) + NO(g) \longrightarrow NO_2(g) + ClNO(g)$$

Figure 14.4 provides a basis for understanding the implications of the collision theory model for this reaction. The kinetic molecular theory assumes that the number of collisions per second in a gas depends on the number of particles per unit volume. Thus the rate at which NO_2 and $ClNO$ are formed in the reaction should be directly proportional to the concentrations of both $ClNO_2$ and NO.

$$\text{rate} = k(ClNO_2)(NO)$$

The collision theory model suggests that the rate of any step in a reaction is proportional to the concentrations of the reagents consumed in that step. The rate law for a one-step reaction should therefore agree with the stoichiometry of the reaction. The following reaction, for example, occurs in a single step.

$$CH_3Br(aq) + OH^-(aq) \longrightarrow CH_3OH(aq) + Br^-(aq)$$

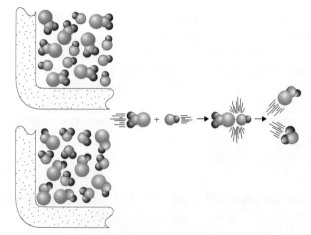

Fig. 14.4 "Snapshot" of a small portion of a container in which $ClNO_2$ reacts with NO to form NO_2 and ClNO. The collision theory of reactions assumes that molecules must collide in order to react. Anything that increases the frequency of the collisions increases the rate of reaction, so the rate of reaction must be proportional to the concentrations of both of the reactants consumed in this reaction.

When the reactant molecules collide in the proper orientation, a pair of non-bonding electrons on the OH⁻ ion can be donated to the carbon atom at the center of the CH₃Br molecule, as shown in Figure 14.5. During the collision, a carbon–oxygen bond forms at the same time that the carbon–bromine bond is broken. The net result of the reaction is the substitution of an OH⁻ ion for a Br⁻ ion. Because the reaction occurs in a single step, which involves collisions between the two reactants, the rate of the reaction is proportional to the concentration of both reactants.

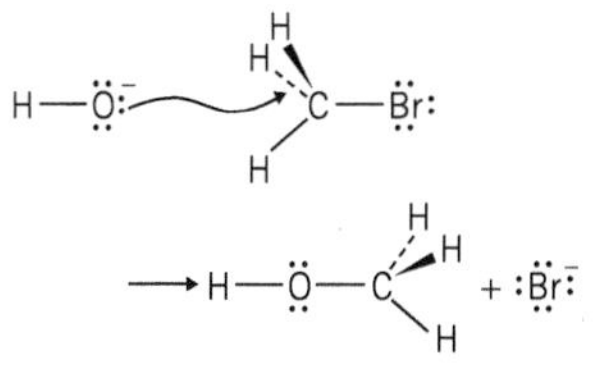

$$Rate = k(CH_3Br)(OH^-)$$

Not all reactions, however, occur in a single step. The following reaction occurs in three steps.

$$(CH_3)_3CBr(aq) + OH^-(aq) \longrightarrow (CH_3)_3COH(aq) + Br^-(aq)$$

In the first step, the $(CH_3)_3CBr$ molecule dissociates into a pair of ions.

Fig. 14.5 The reaction between CH₃Br and the OH⁻ ion occurs in a single step, which involves attack by the OH⁻ ion on the carbon atom. In the course of the reaction, a carbon–oxygen bond forms at the same time that the carbon–bromine bond is broken.

$$\underset{\underset{CH_3}{|}}{\overset{\overset{CH_3}{|}}{CH_3-C-Br}} \longrightarrow \underset{\underset{CH_3}{|}}{\overset{\overset{CH_3}{|}}{CH_3-C^+}} + Br^- \qquad \text{(first step)}$$

The positively charged $(CH_3)_3C^+$ ion then reacts with water in a second step.

$$\underset{\underset{CH_3}{|}}{\overset{\overset{CH_3}{|}}{CH_3-C^+}} + H_2O \longrightarrow \underset{\underset{CH_3}{|}}{\overset{\overset{CH_3}{|}}{CH_3-C-OH_2^+}} \qquad \text{(second step)}$$

The product of the reaction then loses a proton to either the OH⁻ ion or water in the final step.

$$\underset{\underset{CH_3}{|}}{\overset{\overset{CH_3}{|}}{CH_3-C-OH_2^+}} + OH^- \longrightarrow \underset{\underset{CH_3}{|}}{\overset{\overset{CH_3}{|}}{CH_3-C-OH}} + H_2O \qquad \text{(third step)}$$

The rate constant for the first step, k_1, is much smaller than the rate constants, k_2 and k_3, for the other two steps.

$$(CH_3)_3CBr \longrightarrow (CH_3)_3C^+ + Br^- \qquad \text{(rate-limiting step)}$$
$$(CH_3)_3C^+ + H_2O \longrightarrow (CH_3)_3COH_2^+$$
$$(CH_3)_3COH_2^+ + OH^- \longrightarrow (CH_3)_3COH + H_2O$$

The overall rate of this reaction is therefore very close to the rate of the first step. The first step in this reaction is called the **rate-limiting step (RLS)** because it literally limits the rate at which the products of the reaction can be formed. Because only one reagent is involved in the rate-limiting step, the overall rate of reaction is proportional to the concentration of only that reagent.

$$rate = k((CH_3)_3CBr)$$

A rate-limiting step can only be identified if there is adequate information as to the relative magnitudes of the rate constants and reactant concentrations. The RLS is not necessarily the slowest step in a reaction sequence.

The rate law for this reaction therefore differs from what we would predict from the stoichiometry of the reaction. Although the reaction consumes both $(CH_3)_3CBr$ and OH^-, the rate of the reaction is proportional only to the concentration of $(CH_3)_3CBr$.

The rate laws for chemical reactions can be summarized by the following general rules.

- The rate of any step in a reaction is directly proportional to the concentrations of the reagents consumed in that step.
- The overall rate law for a reaction is determined by the sequence of steps, or the **mechanism,** by which the reactants are converted into the products of the reaction.
- The overall rate law for a reaction is dominated by the rate-limiting step in the reaction.

14.9 The Mechanisms of Chemical Reactions

What happens when the first step in a multistep reaction is not the rate-limiting step? Consider the reaction between NO and O_2 to form NO_2, for example.

$$2\,NO(g) + O_2(g) \longrightarrow 2\,NO_2(g)$$

This reaction occurs by a two-step mechanism. In the first step, two NO molecules combine to form a dimer, N_2O_2. The product of this step then reacts with an O_2 molecule to form a pair of NO_2 molecules.

Step 1: $2\,NO \rightleftharpoons N_2O_2$

Step 2: $N_2O_2 + O_2 \longrightarrow 2\,NO_2$ (rate-limiting step)

The net effect of these reactions is the transformation of two NO molecules and one O_2 molecule into a pair of NO_2 molecules.

$$\frac{(2\,NO \rightleftharpoons N_2O_2)}{+(N_2O_2 + O_2 \longrightarrow 2\,NO_2)}$$
$$2\,NO + O_2 \longrightarrow 2\,NO_2$$

In this reaction, the second step is the rate-limiting step. As we have seen, the rate of any step in a reaction is directly proportional to the concentrations of the reactants consumed in that step. The rate law for the second step in this reaction is therefore proportional to the concentrations of both N_2O_2 and O_2.

Step 2: $rate_{2nd} = k_2(N_2O_2)(O_2)$

Because the second step in the reaction is the rate-limiting step, the overall rate of reaction is more or less equal to the rate of the second step.

$$rate_{overall} \approx k_2(N_2O_2)(O_2)$$

NO_2 is a brown-colored gas formed by the reaction of NO with O_2.

This rate law isn't very useful because it is difficult to measure the concentrations of **intermediates,** such as N_2O_2, that are simultaneously formed and consumed in the reaction. It would therefore be better to have an equation that related the overall rate of reaction to the concentrations of the original reactants.

Let's take advantage of the fact that the first step in the reaction occurs rapidly and can therefore reach equilibrium.

$$\text{Step 1:} \quad 2\,NO \rightleftharpoons N_2O_2$$

The collision theory predicts that the rate of the forward reaction in this step will depend on the concentration of NO raised to the second power.

$$\text{Step 1:} \quad \text{rate}_{\text{forward}} = k_f(NO)^2$$

The rate of the reverse reaction, however, will depend only on the concentration of N_2O_2.

$$\text{Step 1:} \quad \text{rate}_{\text{reverse}} = k_r(N_2O_2)$$

Because the first step in the reaction occurs fast enough to come to equilibrium, we can write the following equality.

$$k_f(NO)^2 = k_r(N_2O_2)$$

Let's rearrange this equation to solve for one of the terms that appears in the rate law for the second step in the reaction

$$(N_2O_2) = \frac{k_f}{k_r}(NO)^2$$

Substituting this equation for (N_2O_2) into the rate law for the second step gives the following result.

$$\text{rate}_{2\text{nd}} = k_2\left(\frac{k_f}{k_r}\right)(NO)^2(O_2)$$

Note that k_f/k_r is equal to the equilibrium constant for the reaction as shown in Section 10.5. Since k_2, k_f, and k_r are all constants, they can be replaced by a single constant, k', to give the experimental rate law for this reaction, described in Exercise 14.5.

$$\text{rate}_{\text{overall}} \approx \text{rate}_{2\text{nd}} = k'(NO)^2(O_2)$$

The two-step mechanism for this reaction explains some of the observed properties of the oxides of nitrogen. In the presence of a spark, the nitrogen and oxygen in the atmosphere can react to form nitrogen oxide, NO.

$$N_2(g) + O_2(g) \xrightarrow{spark} 2\,NO(g)$$

NO is a colorless gas that can be collected by displacing water from a container because it is virtually insoluble in water. When a flask filled with NO gas is opened, an almost instantaneous reaction occurs between the NO and oxygen in the air to form nitrogen dioxide.

$$2\,NO(g) + O_2(g) \longrightarrow 2\,NO_2(g)$$

NO is a colorless gas that does not dissolve in water. When exposed to air, NO instantly reacts with O_2 to form NO_2, a dark-brown gas.

The product of the reaction is a brown gas, and the progress of the reaction can be followed by noting the appearance of this gas.

This raises an interesting question. When people talk about sources of pollution, they often refer to the oxides of nitrogen as NO_x to suggest that a mixture of NO and NO_2 is formed. If the reaction between NO and O_2 to form NO_2 is virtually instantaneous, why don't we assume that the product formed when a stroke of lightning passes through the atmosphere or when the spark plug in an internal combustion engine ignites the reaction between gasoline and oxygen is NO_2? Why do people assume that the pollution that is formed is a mixture of NO and NO_2, rather than just pure NO_2?

The answer to this question can be found in the mechanism of the reaction between NO and O_2. As we have seen, the first step in this reaction is a fast, reversible reaction in which NO dimerizes to form N_2O_2.

$$\text{Step 1:} \quad 2\,NO \rightleftharpoons N_2O_2$$

This step can be understood by noting that the Lewis structure of NO suggests the presence of an unpaired electron. One of these electrons on each of the NO molecules can combine to form a new covalent bond, as shown in Figure 14.6.

The second step in the reaction occurs when molecules of N_2O_2 and O_2 collide.

$$\text{Step 2:} \quad N_2O_2 + O_2 \longrightarrow 2\,NO_2$$

In an open flask filled with NO, there is enough N_2O_2 to allow the second step to occur quite rapidly, and we see the rapid formation of the brown NO_2 gas. However, the concentration of NO emitted from the tailpipe of a car or truck is relatively small. Therefore, relatively little formation of N_2O_2 occurs when the NO is formed as a pollutant. The second step is the rate-limiting step. Or, in other words, there is time for the NO to circulate through the atmosphere before it is converted to NO_2. As a result, it is correct to think about the pollutants formed when a spark passes through a mixture of N_2 and O_2 as NO_x—a mixture of NO and NO_2.

Fig. 14.6 The dimerization of NO to form N_2O_2.

14.10 Zero-Order Reactions

We can now understand why some reactions are zero-order in one of the components of the reaction. The rate law for the following reaction, for example, is zero-order in the OH^- ion because the OH^- ion isn't involved in the rate-limiting step.

One question remains unanswered: How do we explain the fact that some reactions are zero-order in *all* of the starting materials? Consider the reaction in which nitrogen oxide decomposes to its elements on the surface of a piece of platinum metal, for example.

$$2\ NO(g) \xrightarrow{\ Pt\ } N_2(g) + O_2(g)$$

In the presence of a large excess of NO, the rate of this reaction is zero-order in NO.

$$\text{rate} = k(NO)^0 = k$$

A simple analogy can help us understand why the decomposition of NO to its elements on a solid surface is zero-order in NO, as long as a large excess of NO is present. Assume that you are waiting in line to eat at a restaurant. No one else can be seated until there is an empty table. The rate at which people are seated doesn't depend on the number of people standing in line. It depends only on the rate at which tables are vacated.

The same thing is true for the decomposition of NO on the platinum metal surface. For NO to react, it must find a site on the metal at which it can bind to the metal surface. Reaction can occur at only a limited number of sites, no matter how much NO is present, and only those NO molecules that occupy one of these sites can undergo a reaction. As a result, the rate of the reaction does not increase as we add more NO to the system and the reaction is zero-order in NO.

Any chemical reaction that occurs by a mechanism by which the reactant has to bind to a catalyst that has a limited number of active sites will exhibit similar behavior. Consider the following reaction, for example.

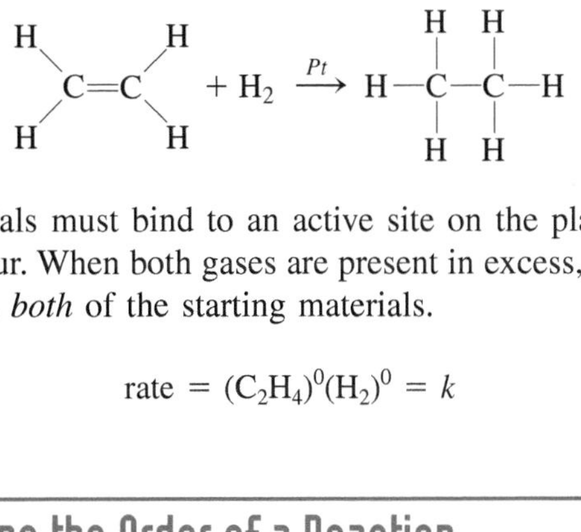

Both starting materials must bind to an active site on the platinum metal before the reaction can occur. When both gases are present in excess, the rate of the reaction is zero-order in *both* of the starting materials.

$$\text{rate} = (C_2H_4)^0(H_2)^0 = k$$

> ➤ **CHECKPOINT**
>
> How does the dependence of rate on concentration differ between a zero-order and first-order reaction? Describe what happens to the rate of a reaction with respect to time for both a zero- and a first-order reaction.

14.11 Determining the Order of a Reaction from Rates of Reaction

The rate law for a reaction can be determined by studying what happens to the instantaneous rate of reaction when we start with different initial concentrations of the reactants. To show how this is done, let's determine the rate law for the reaction in which hydrogen iodide decomposes to give a mixture of hydrogen and iodine in the gas phase.

$$2\ HI(g) \longrightarrow H_2(g) + I_2(g)$$

Data on initial rates of reaction for three experiments run at different initial concentrations of HI are given in Table 14.2.

Table 14.2
Rate of Reaction Data for the Decomposition of HI

Trial	Initial Concentration of HI (M)	Initial Instantaneous Rate of Reaction (M/s)
Trial 1	1.0×10^{-2}	4.0×10^{-6}
Trial 2	2.0×10^{-2}	1.6×10^{-5}
Trial 3	3.0×10^{-2}	3.6×10^{-5}

The only difference in the three experiments is the initial concentration of HI. Let's compare the experiments two at a time. The difference between trial 1 and trial 2 is a twofold increase in the initial HI concentration, which leads to a fourfold increase in the initial rate of reaction.

$$\frac{\text{rate for trial 2}}{\text{rate for trial 1}} = \frac{1.6 \times 10^{-5} \, M/s}{4.0 \times 10^{-6} \, M/s} = 4.0$$

The only difference between trials 1 and 3 is a threefold increase in the initial concentration of HI. The initial rate of reaction, however, increases by a factor of 9.

$$\frac{\text{rate for trial 3}}{\text{rate for trial 1}} = \frac{3.6 \times 10^{-5} \, M/s}{4.0 \times 10^{-6} \, M/s} = 9.0$$

The data suggest that the rate of the reaction is proportional to the *square of the HI concentration*. The reaction is therefore second-order in HI, as noted in Section 14.6.

$$\text{rate} = k(\text{HI})^2$$

Exercise 14.6

Use the experimental data shown below to determine the rate law for the following reaction.

$$\text{NH}_4^+(aq) + \text{NO}_2^-(aq) \longrightarrow \text{N}_2(g) + 2 \, \text{H}_2\text{O}(l)$$

	Initial Concentration of NH_4^+ (M)	Initial Concentration of NO_2^- (M)	Initial Instantaneous Rate of Reaction (M/s)
Trial 1	5.0×10^{-2}	2.0×10^{-2}	2.7×10^{-7}
Trial 2	5.0×10^{-2}	4.0×10^{-2}	5.4×10^{-7}
Trial 3	1.0×10^{-1}	2.0×10^{-2}	5.4×10^{-7}

Solution

When the experimental data for several trials of a reaction differ only in the initial concentrations of just one reactant, the rate law can often be determined from a simple analysis of the data. The only difference between the initial conditions for trials 1 and 2 is the initial concentration of the NO_2^- ion. When the initial concentration of the NO_2^- ion is doubled from trial 1 to trial 2, the rate of the reaction increases by the same amount—a factor of two. The only way to explain this observation is to assume that the rate law for the reaction is first-order in the NO_2^- ion.

The only difference between the initial conditions of trial 1 and trial 3 is a doubling of the initial concentration of the NH_4^+ ion. This also leads to a doubling of the initial instantaneous rate of reaction, which suggests that the rate law is also first-order in the NH_4^+ ion. We therefore conclude that the rate law is first-order in both the NH_4^+ and NO_2^- ions.

$$\text{rate} = k(NH_4^+)(NO_2^-)$$

However, if the data are not as straightforward, it is often necessary to mathematically determine the order with respect to each reactant. As shown in Section 14.7, the rate law for trial 1 can be written as

$$\text{rate}_1 = k(NH_4^+)_1{}^m(NO_2^-)_1{}^n$$

and the rate law for trial 2 is

$$\text{rate}_2 = k(NH_4^+)_2{}^m(NO_2^-)_2{}^n$$

The order with respect to NH_4^+ and NO_2^- can be found from the experimental data if we know how the initial rates of reaction change when the concentrations of only one of the reactants is varied. In trials 1 and 2 the (NH_4^+) remains constant while the (NO_2^-) is doubled. Thus a ratio of trial 1 to trial 2 can be used to determine how the changing (NO_2^-) affects the initial rate. This allows the determination of the order, n, with respect to (NO_2^-).

$$\frac{\text{rate}_1}{\text{rate}_2} = \frac{k(NH_4^+)_1{}^m(NO_2^-)_1{}^n}{k(NH_4^+)_2{}^m(NO_2^-)_2{}^n}$$

$$\frac{(2.7 \times 10^{-7})}{(5.4 \times 10^{-7})} = \frac{k(5.0 \times 10^{-2})^m(2.0 \times 10^{-2})^n}{k(5.0 \times 10^{-2})^m(4.0 \times 10^{-2})^n}$$

The rate constant and the (NH_4^+) terms are the same in the numerator and the denominator and therefore cancel out of the equation.

$$\frac{(2.7 \times 10^{-7})}{(5.4 \times 10^{-7})} = \frac{(2.0 \times 10^{-2})^n}{(4.0 \times 10^{-2})^n} = \left(\frac{1}{2}\right)^n$$

$$\frac{1.0}{2.0} = \left(\frac{1.0}{2.0}\right)^n$$

$$n = 1.0 = \text{first-order}$$

Similarly, we can use the ratio of trials 1 and 3 to find the order with respect to NH_4^+.

$$\frac{\text{rate}_1}{\text{rate}_3} = \frac{k(5.0 \times 10^{-2})^m(2.0 \times 10^{-2})^n}{k(1.0 \times 10^{-1})^m(2.0 \times 10^{-2})^n}$$

$$\frac{(2.7 \times 10^{-7})}{(5.4 \times 10^{-7})} = \frac{(5.0 \times 10^{-2})^m}{(1.0 \times 10^{-1})^m}$$

$$m = 1.0 = \text{first-order}$$

We therefore conclude that the rate law is first-order in both the NH_4^+ and NO_2^- ions.

$$\text{rate} = k(NH_4^+)(NO_2^-)$$

The overall order of the reaction is therefore, second order.

➤ **CHECKPOINT**

Why are trials 2 and 3 not used to determine the order of the reaction in terms of either the NH_4^+ or NO_2^- ion? Can trials 1 and 2 be used to determine the order with respect to NH_4^+?

➤ **CHECKPOINT**

Determine the value of the rate constant for the reaction in Exercise 14.6.

14.12 The Integrated Form of Zero-, First-, and Second-Order Rate Laws

As we have seen, the rate law for a reaction provides a basis for studying the mechanism of the reaction. It is also useful for predicting how much reactant will remain or how much product will be formed in a given amount of time. For these calculations, we use the **integrated form of the rate laws.** A derivation of the integrated forms of the rate laws can be found in the Special Topic section at the end of this chapter.

Let's start with the simplest of all rate laws, a reaction that is zero-order.

$$\frac{d(X)}{dt} = k(X)^0 = k$$

When this equation is rearranged and both sides are integrated, we get the following result.

Integrated form of the zero-order rate law:

$$(X) - (X)_0 = -kt$$

In this equation, (X) is the concentration of X at any moment in time, $(X)_0$ is the initial concentration of this reagent, k is the rate constant for the reaction, and t is the time since the reaction started.

Exercise 14.7

How long will it take for the concentration of a solution that was initially 2.0 M to decrease to 1.0 M if the reaction is zero-order and the rate constant, k, for the reaction is 1.5×10^{-2} mol/L·s?

Solution

For a zero-order reaction, the concentration at any time (X) after the reaction has begun is related to the initial concentration $(X)_0$ by the following equation.

$$(X) - (X)_0 = -kt$$

Substituting the known values for the initial and final concentration of X and the rate constant for the reaction into this equation gives the amount of time it takes for the change in the concentration of X to occur.

$$(1.0 \text{ mol/L}) - (2.0 \text{ mol/L}) = -(1.5 \times 10^{-2} \text{ mol/L·s}) \times t$$

Solving for t gives the following result.

$$t = \frac{1.0 \text{ mol/L}}{1.5 \times 10^{-2} \text{ mol/L·s}} = 67 \text{ s}$$

Let's now turn to the rate law for a reaction that is first-order in the disappearance of a single reactant, X.

$$-\frac{d(X)}{dt} = k(X)$$

When the equation is rearranged and both sides are integrated, we get the following result.

Integrated form of the first-order rate law:

$$\ln\left[\frac{(X)}{(X)_0}\right] = -kt$$

Once again, (X) is the concentration of X at any moment in time, $(X)_0$ is the initial concentration of the reagent, k is the rate constant for the reaction, and t is the time since the reaction started.

Exercise 14.8

The decomposition of N_2O_5 is a first-order reaction with a rate constant of 0.420 min^{-1}. If the initial concentration of N_2O_5 is 1.0 M, what will be the concentration after 5.00 minutes?

Solution

The integrated form of the rate law for reactions that follow first-order kinetics is written as follows.

$$\ln\left[\frac{(X)}{(X)_0}\right] = -kt$$

Taking advantage of the mathematics of logarithms, we can write this equation as follows.

$$\ln(X) - \ln(X)_0 = -kt$$

Substituting what is known about the decomposition of N_2O_5, we obtain

$$\ln(N_2O_5) - \ln(1.0) = -(0.420 \text{ min}^{-1})(5.00 \text{ min})$$
$$\ln(N_2O_5) = -(0.420 \text{ min}^{-1})(5.00 \text{ min})$$
$$\ln(N_2O_5) = -2.10$$
$$(N_2O_5) = 0.12 \, M$$

The concentration of N_2O_5 therefore decreases by a factor of about 8 over the course of 5 minutes.

Let's now turn to the rate law for a reaction that is second-order in a single reactant.

$$-\frac{d(X)}{dt} = k(X)^2$$

The integrated form of the rate law for this reaction is written as follows.

Integrated form of the second-order rate law:

$$\frac{1}{(X)} - \frac{1}{(X)_0} = kt$$

Once again, (X) is the concentration of X at any moment in time, $(X)_0$ is the initial concentration of X, k is the rate constant for the reaction, and t is the time since the reaction started.

Exercise 14.9

NO_2 decomposes to form NO and O_2 by second-order kinetics with a rate constant of 32.6 L/mol · min.

$$2\,NO_2(g) \longrightarrow 2\,NO(g) + O_2(g)$$

If the initial NO_2 concentration is 0.15 M, what will be the concentration after 1.0 min?

Solution

We can start with the rate law for this reaction

$$\text{rate} = k(NO_2)^2$$

and then write the integrated form of the rate law as follows.

$$\frac{1}{(NO_2)} - \frac{1}{(NO_2)_0} = kt$$

Substituting what is known for this reaction into this equation gives

$$\frac{1}{(NO_2)} - \frac{1}{(0.15)_0} = (32.6 \text{ L/mol} \cdot \text{min})(1.0 \text{ min})$$

Solving for the concentration of NO_2 after 1 minute gives the following result.

$$(NO_2) = 0.026 \, M$$

It is important to recognize the difference between a rate law that is second-order in a single reactant and a rate law that is first-order in two reactants and therefore second-order overall. A reaction that is first-order in two different reactants would obey the following rate law.

$$\text{rate} = k(X)(Y)$$

The integrated form of the rate law for this reaction would be written as follows.

$$\frac{1}{(X)_0 - (Y)_0} \ln \frac{(Y)_0(X)}{(X)_0(Y)} = kt$$

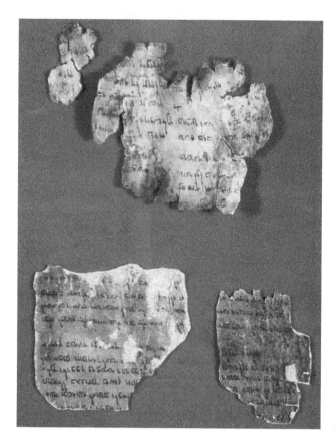

Fragments of the Dead Sea Scrolls, which were authenticated by ^{14}C dating.

In this equation, (X) and (Y) are the concentrations of X and Y at any moment in time, $(X)_0$ and $(Y)_0$ are the initial concentrations of X and Y, respectively, k is the rate constant for the reaction, and t is the time since the reaction started.

To further illustrate the power of the integrated form of the rate law for a reaction, let's use the integrated form of the equation for a first-order reaction to calculate how long it would take for the ^{14}C in a piece of charcoal to decay to half of its original concentration.

All radioactive decay processes are first-order. Thus, ^{14}C decays by first-order kinetics.

$$\text{rate} = k(^{14}\text{C})$$

The experimental value for the rate constant for this reaction is $1.210 \times 10^{-4} \text{ y}^{-1}$.

The integrated form of the rate law for this reaction would be written as follows.

$$\ln\left[\frac{(^{14}\text{C})}{(^{14}\text{C})_0}\right] = -kt$$

But we are interested in the moment when the concentration of ^{14}C in the charcoal is half of its initial value.

$$(^{14}\text{C}) = \frac{1}{2}(^{14}\text{C})_0$$

Substituting this relationship into the integrated form of the rate law gives the following equation.

$$\ln\left[\frac{\frac{1}{2}(^{14}C)_0}{(^{14}C)_0}\right] = -kt$$

We now simplify the equation as follows.

$$\ln\left[\frac{1}{2}\right] = -kt$$

Or, taking advantage of the mathematics of logarithms,

$$\ln(2) = kt$$

and then solving for t gives the following result.

$$t = \frac{\ln(2)}{k} = \frac{0.6931}{1.210 \times 10^{-4}\,\text{yr}^{-1}} = 5730\;\text{yr}$$

➤ **CHECKPOINT**

Which radioactive isotope would have the longer half-life, ^{15}O or ^{19}O?

^{15}O	$k = 5.63 \times 10^{-3}\,\text{s}^{-1}$
^{19}O	$k = 2.38 \times 10^{-2}\,\text{s}^{-1}$

It therefore takes 5730 years for half of the ^{14}C in the sample to decay. This is called the **half-life** ($t_{1/2}$) of ^{14}C. In general, the half-life for a first-order kinetic process can be calculated from the rate constant as follows.

$$t_{1/2} = \frac{\ln(2)}{k} = \frac{0.6931}{k}$$

Similar equations can be derived for the half-life of zero-order and second-order reactions, as shown in Table 14.3. There is an important difference between the equations for calculating the half-lives of zero-order, first-order, and second-order reactions, however. The half-lives for zero-order and second-order reactions are not constant; they depend on the initial concentration of X. Discussions of reaction half-lives are therefore confined to first-order processes because this is the only reaction order for which the half-life is independent of reactant concentration.

Table 14.3

Equations That Describe Zero-, First-, and Second-Order Reactions

➤ **CHECKPOINT**

Suppose a reaction was carried out twice, once with an initial concentration of 1 M and then with an initial concentration of 2 M. How will the half-life for the second trial compare to the half-life of the first trial for a zero-, first-, and second-order reaction?

Order	Rate Law	Integrated Form of the Rate Law	Half-Life
Zero	Rate = k	$(X) - (X)_0 = -kt$	$t_{1/2} = \dfrac{(X)_0}{2k}$
First	Rate = $k(X)$	$\ln\left[\dfrac{(X)}{(X)_0}\right] = -kt$	$t_{1/2} = \dfrac{0.6931}{k}$
Second	Rate = $k(X)^2$	$\dfrac{1}{(X)} - \dfrac{1}{(X)_0} = kt$	$t_{1/2} = \dfrac{1}{k(X)_0}$

14.13 Determining the Order of a Reaction with the Integrated Form of Rate Laws

The integrated forms of the rate laws for zero-, first-, and second-order reactions provide another way of determining the order of a reaction. Let's start by considering a reaction that is zero-order in X.

$$\text{rate} = k(X)^0 = k$$

If the reaction is zero-order, a plot of experimental data for the concentration of x versus time should fit the integrated form of the zero-order rate law.

$$(X) - (X)_0 = -kt$$

This equation contains two variables—(X) and t—and two constants—$(X)_0$ and k. It can therefore be set up in terms of the equation for a straight line.

$$y = mx + b$$
$$(X) = -kt + (X)_0$$

If the reaction is zero-order in X, a plot of the concentration of X versus time will be a straight line with a slope equal to $-k$, as shown in Figure 14.7.

If the plot of (X) versus time isn't a straight line, the reaction can't be zero-order in X. We could therefore examine the data to see if the reaction is first-order in X.

$$\text{rate} = k(X)$$

If the reaction is first-order in X, a plot of the experimental data should fit the integrated form of the first-order rate law.

$$\ln\left[\frac{(X)}{(X)_0}\right] = -kt$$

To see if a particular set of experimental data fits this equation, we can rearrange the integrated form of the first-order rate law as follows.

$$\ln(X) - \ln(X)_0 = -kt$$

Once again, we have an equation that contains two variables—$\ln(X)$ and t—and two constants—$\ln(X)_0$ and k. It can therefore be set up in the form of an equation for a straight line.

$$y = mx + b$$
$$\ln(X) = -kt + \ln(X)_0$$

If the reaction is first-order in X, a plot of the natural logarithm of the concentration of X versus time will be a straight line with a slope equal to $-k$, as shown in Figure 14.8.

If the plot of $\ln(X)$ versus time isn't a straight line, the reaction can't be first-order in X. We could therefore examine the data to see if the reaction is second-order in X.

$$\text{rate} = k(X)^2$$

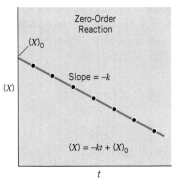

Fig. 14.7 A plot of the concentration of X versus the elapsed time since the reaction started for a zero-order reaction will be a straight line with a slope equal to the negative of the rate constant for the reaction.

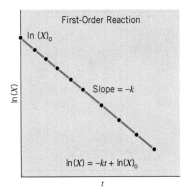

Fig. 14.8 For a reaction that is first-order in a single reactant, a plot of the natural log of the concentration of X versus the elapsed time since the reaction started will be a straight line with a slope equal to the negative of the rate constant.

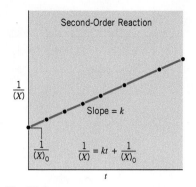

Fig. 14.9 For a reaction that is second-order in a single reactant, a plot of the inverse of the concentration of X versus the elapsed time since the reaction started will be a straight line with a slope equal to the rate constant.

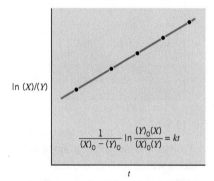

Fig. 14.10 For a reaction that is first-order in two different reactants, a plot of the natural logarithm of the ratio of the concentration of the two reactants versus the elapsed time since the reaction started would be a straight line.

If the reaction is second-order, the experimental data should fit the integrated form of the second-order rate law.

$$\frac{1}{(X)} - \frac{1}{(X)_0} = kt$$

This equation also contains two variables, (X) and t, and two constants, $(X)_0$ and k. Thus it also can be set up in terms of the equation for a straight line.

$$y = mx + b$$
$$\frac{1}{(X)} = kt + \frac{1}{(X)_0}$$

If the reaction is second-order in X, a plot of the reciprocal of the concentration of X versus time will be a straight line with a slope equal to k, as shown in Figure 14.9. If the plot of $1/(X)$ versus time is not linear, the reaction cannot be second-order.

What about reactions that are first-order in two different reactants, X and Y? In this case, a plot of the natural logarithm of the ratio of the concentrations of the two reactants versus time would give a straight line, as shown in Figure 14.10.

Exercise 14.10

Use the data in Table 14.1 to determine whether the reaction between phenolphthalein and the OH^- ion is zero-order, first-order, or second-order with respect to phenolphthalein when the reaction is run in the presence of excess OH^- ion.

Solution

The first step in solving this problem involves calculating the natural log of the phenolphthalein concentration, $\ln(PHTH)$, and the reciprocal of the concentration, $1/(PHTH)$, for each point at which a measurement was taken.

(PHTH) (mol/L)	ln(PHTH)	1/(PHTH)	Time (s)
0.00500	−5.298	200	0
0.00450	−5.404	222	10.5
0.00400	−5.521	250	22.3
0.00350	−5.655	286	35.7
0.00300	−5.809	333	51.1
0.00250	−5.991	400	69.3
0.00200	−6.215	500	91.6
0.00150	−6.502	667	120.4
0.00100	−6.908	1.00×10^3	160.9
0.000500	−7.601	2.00×10^3	230.2
0.000250	−8.294	4.00×10^3	299.5
0.000150	−8.805	6.67×10^3	350.7
0.000100	−9.210	1.00×10^4	391.2

We then construct graphs of (PHTH) versus t (Figure 14.11), of $\ln(PHTH)$ versus t (Figure 14.12), and of $1/(PHTH)$ versus t (Figure 14.13).

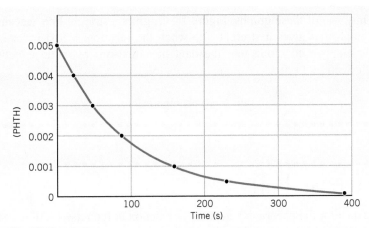

Fig. 14.11 A plot of the concentration of phenolphthalein versus time for the reaction between phenolphthalein and OH⁻ ion is *not* a straight line, which shows that the reaction is not a zero-order reaction.

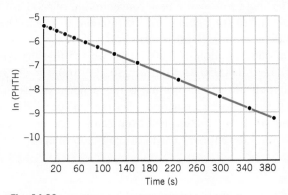

Fig. 14.12 A plot of the natural log of the concentration of phenolphthalein versus time for the reaction between phenolphthalein and OH⁻ ion is a straight line, which shows that the reaction is first-order in phenolphthalein.

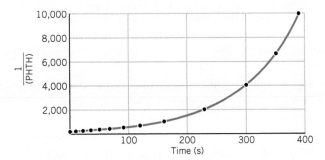

Fig. 14.13 A plot of the reciprocal of the concentration of phenolphthalein versus time for the reaction between phenolphthalein and OH⁻ ion *isn't* a straight line, which shows that the reaction isn't second-order in phenolphthalein.

Only one of these graphs, Figure 14.12, gives a straight line. Because this is the graph that gives a straight line when the reaction is first-order in phenolphthalein, we conclude that the data fit a first-order kinetic equation, as noted in Section 14.5.

$$\text{rate} = k(\text{phenolphthalein})$$

Exercise 14.11

Use the data for the phenolphthalein–OH$^-$ reaction in Exercise 14.10 to calculate the half-life and show that this reaction is first-order.

Solution

The data given in Exercise 14.10 can be used to find the half-life for several different initial concentrations. The time taken for the concentration to drop from 0.00400 M to 0.00200 M is 91.6 − 22.3 = 69.3 s. We can now imagine that we are starting again with a new concentration of 0.00200 M, and we can see how long it takes for one-half of that amount to react. The concentration drop from 0.00200 M to 0.00100 M takes 160.9 − 91.6 = 69.3 s. Starting again from an initial concentration of 0.00100 M, we see that the concentration fall from 0.00100 M to 0.00050 M requires 230.2 − 160.9 = 69.3 s.

Thus the half-life of the reaction is independent of the initial concentration. It is only for first-order reactions that the half-life is independent of initial concentration. If the reaction had been zero- or second-order, the half-life would have been different over each concentration interval.

14.14 Reactions That Are First-Order in Two Reactants

What about reactions that are first-order in two reactants, X and Y, and therefore second-order overall?

$$\text{rate} = k(X)(Y)$$

A plot of $1/(X)$ versus time won't give a straight line because the reaction isn't second-order in X. Unfortunately, neither will a plot of $\ln(X)$ versus time because the reaction isn't strictly first-order in X. It is simultaneously first-order in both X and Y.

One way around the problem is to turn the reaction into a **pseudo-first-order reaction** by making the concentration of one of the reactants so large that it is effectively constant. The rate law for the reaction is still first-order in both reactants, but the initial concentration of one reactant is so much larger than the other that the rate of reaction seems to be sensitive only to changes in the concentration of the reagent present in limited quantities.

Assume, for the moment, that the reaction between reagents X and Y is studied under conditions for which there is a large excess of Y. If this is true, the concentration of Y will remain essentially constant during the reaction. As a result, the rate of the reaction will be zero-order for the excess reagent. Instead, it will appear to be first-order in the other reactant, X. A plot of ln(X) versus time will therefore give a straight line with a slope of $-k'$.

$$\text{rate} = k'(X)$$

If we now run the reaction in the presence of a large excess of X, the reaction will appear to be first-order in Y. Under these conditions, a plot of ln(Y) versus time will be linear with a slope of $-k''$.

$$\text{rate} = k''(Y)$$

The value of the rate constant obtained from either of these equations won't be the actual rate constant for the reaction. It will be the product of the rate constant for the reaction times the concentration of the reagent that is present in excess.

The reaction between phenolphthalein and excess OH^- ion described in Section 14.3 is an example of a peudo-first-order reaction because the initial concentration of the OH^- ion was roughly 120 times the initial concentration of phenolphthalein. There was so much OH^- ion in the solution that the concentration of this ion remained virtually constant throughout the course of the data collection. The reaction therefore appeared to be first-order in phenolphthalein when the data were analyzed.

14.15 The Activation Energy of Chemical Reactions

Only a small fraction of the collisions between reactant molecules convert the reactants into the products of the reaction. This can be understood by turning, once again, to the reaction between $ClNO_2$ and NO.

$$ClNO_2(g) + NO(g) \longrightarrow NO_2(g) + ClNO(g)$$

In the course of this reaction, a chlorine atom is transferred from one nitrogen atom to another. For the reaction to occur, the nitrogen atom in NO must collide with the chlorine atom in $ClNO_2$.

The reaction won't occur if the oxygen from the NO molecule collides with the chlorine atom on $ClNO_2$.

No reaction

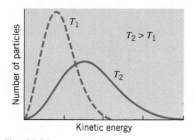

Fig. 14.14 The kinetic molecular theory states that the average kinetic energy of a gas is proportional to the temperature of the gas, and nothing else. At any given temperature, however, some of the gas particles are moving faster than others.

Nor will it occur if one of the oxygen atoms on $ClNO_2$ collides with the nitrogen atom on NO.

No reaction

Another factor that influences whether a reaction will occur is the energy the molecules have when they collide. Not all of the molecules have the same kinetic energy, as shown in Figure 14.14. This is important because the kinetic energy that molecules have when they collide is the principal source of the energy that must be invested in a reaction to get it started.

The overall standard free energy change for the reaction between $ClNO_2$ and NO is favorable.

$$ClNO_2(g) + NO(g) \longrightarrow NO_2(g) + ClNO(g) \qquad \Delta G° = -23.6 \text{ kJ/mol}_{rxn}$$

But before the reactants can be converted into products, the energy of the system must overcome the **activation energy** ($E_{a, \text{forward}}$) for the forward reaction, as shown in Figure 14.15. The vertical axis in Figure 14.15 represents the energy of a pair of molecules undergoing collision as a chlorine atom is transferred from $ClNO_2$ to NO. The horizontal axis represents the sequence of infinitesimally small changes that must occur to convert the reactants into the products of the reaction and is called the *reaction coordinate*.

The energy of activation for the reverse reaction, $E_{a, \text{reverse}}$, is the difference between the energy of the products and the top of the energy of activation curve as shown in Figure 14.15. The difference between the energy of the reactants and the products is the enthalpy of reaction, $\Delta H°$, which is also shown in this figure.

To understand why reactions have an activation energy, consider what has to happen for $ClNO_2$ to react with NO. In order for the reaction to occur, the two molecules have to collide with sufficient energy and they have to collide in exactly the right orientation relative to one another. Some energy also must be invested to begin breaking the $Cl—NO_2$ bond so that the $Cl—NO$ bond can form.

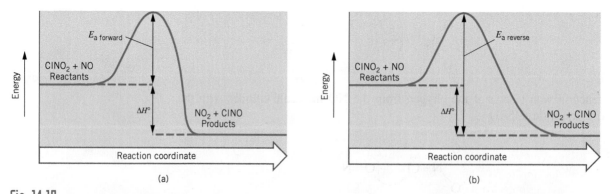

Fig. 14.15 (a) The activation energy for a reaction, $E_{a, \text{forward}}$, is the change in the potential energy of the reactant molecules that must be overcome before the forward reaction can occur. (b) $E_{a, \text{reverse}}$ is the activation energy for the conversion of products back to reactants.

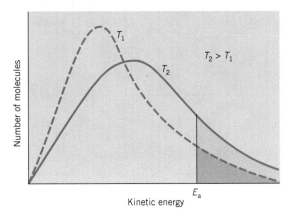

Fig. 14.16 As temperature changes, the activation energy remains essentially constant. As the temperature of the system increases, the number of molecules that have enough energy to react when they collide also increases, as shown by the shaded area.

NO and ClNO$_2$ molecules that collide in the correct orientation, with enough kinetic energy to overcome the activation energy barrier, can react to form NO$_2$ and ClNO. As the temperature of the system increases, the number of molecules that have enough energy to react when they collide also increases, as shown in Figure 14.16. The rate of reaction generally increases with temperature. As a rule, the rate of a reaction doubles for every 10°C increase in the temperature of the system.

> **CHECKPOINT**
>
> For an exothermic reaction, the activation energy for the forward reaction is smaller than the activation energy for the reverse reaction. For example, $E_{a \text{ forward}} < E_{a \text{ reverse}}$. Is this also true for an endothermic reaction? Draw a reaction coordinate diagram like the diagram in Figure 14.15 to explain your reasoning.

14.16 Catalysts and the Rates of Chemical Reactions

Aqueous solutions of hydrogen peroxide are stable until we add a small quantity of the I$^-$ ion, a piece of platinum metal, a few drops of blood, or a freshly cut slice of turnip, at which point the hydrogen peroxide rapidly decomposes.

$$2 H_2O_2(aq) \longrightarrow 2 H_2O(l) + O_2(g)$$

This reaction therefore provides the basis for understanding the effect of a catalyst on the rate of a chemical reaction. Five criteria must be satisfied in order for a substance to be classified as a **catalyst.**

- Catalysts increase the rate of reaction.
- Catalysts aren't consumed by the reaction.
- A small quantity of catalyst should be able to affect the rate of reaction for a large amount of reactant.
- Catalysts don't change the equilibrium constant for the reaction. A catalyst alters the path of the reaction but does not affect the reactants or products.
- Catalysts do not change ΔH or ΔS for a reaction.

The first criterion provides the basis for defining a catalyst as something that increases the rate of a reaction. The second reflects the fact that anything consumed in the reaction is a reactant, not a catalyst. The third criterion is a consequence of the second; because catalysts aren't consumed in the reaction, they can catalyze the reaction over and over again. The fourth criterion results from the fact that catalysts speed up the rates of the forward and reverse reactions equally, so the equilibrium constant for the reaction remains the same. The fifth criterion results from ΔH and ΔS being state functions that are not dependent on the path of the reaction.

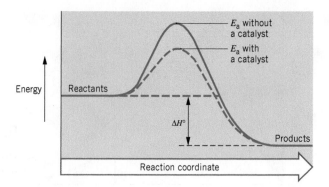

Fig. 14.17 A catalyst increases the rate of a reaction by providing an alternative mechanism that has smaller activation energy.

Catalysts increase the rate of reaction by providing a new mechanism that has a smaller activation energy, E_a, as shown in Figure 14.17. A larger proportion of the collisions that occur between reactants now have enough energy to overcome the activation energy for the reaction as shown in Figure 14.18. As a result, the rate of reaction increases.

The effect of several different catalysts on the activation energy for the decomposition of hydrogen peroxide and the relative rate of the reaction are summarized in Table 14.4. Adding a source of I^- ion to the solution decreases the activation energy by 25%, which increases the rate of the reaction by a factor of 2000. A piece of platinum metal decreases the activation energy even further, increasing the rate of reaction by a factor of 40,000. The *catalase* enzyme in blood and other tissue decreases the activation energy by almost a factor of 10, which leads to a 600-billion-fold increase in the rate of reaction.

The bombardier beetle uses the enzyme-catalyzed decomposition of hydrogen peroxide as a defense mechanism. When attacked, it mixes the contents of a sac that is about 25% H_2O_2 with a suspension of a catalase enzyme in a turret-like mixing tube. The reaction is both so fast and so exothermic that the mixture is vaporized and the beetle can direct a hot spray at its attacker.

To illustrate how a catalyst can decrease the activation energy for a reaction by providing another pathway for the reaction, let's look at the mechanism for the decomposition of hydrogen peroxide catalyzed by the I^- ion. In the presence of the I^- ion, the decomposition of H_2O_2 can occur in two steps, both of which are easier and therefore faster than the uncatalyzed reaction. In the first step, the I^- ion is oxidized by H_2O_2 to form the hypoiodite ion, OI^-.

$$H_2O_2(aq) + I^-(aq) \longrightarrow H_2O(l) + OI^-(aq)$$

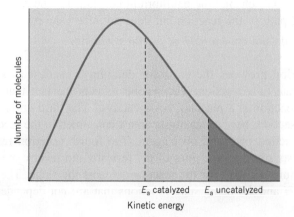

Fig. 14.18 Adding a catalyst decreases the activation energy for a reaction. When the activation energy is lowered, a larger proportion of the molecules will have the necessary energy to react.

Table 14.4
Effect of Catalysts on the Activation Energy for the Decomposition of Hydrogen Peroxide

Catalyst	$E_a(kJ/mol_{rxn})$	Relative Rate of Reaction
None	75.3	1
I^-	56.5	2.0×10^3
Pt	49.0	4.1×10^4
Catalase	8	6.3×10^{11}

In the second step, the OI^- ion is reduced to I^- by H_2O_2.

$$OI^-(aq) + H_2O_2(aq) \longrightarrow H_2O(l) + O_2(g) + I^-(aq)$$

Because there is no net change in the concentration of the I^- ion as a result of these reactions, the I^- ion satisfies the criteria for a catalyst. Because H_2O_2 and I^- are both involved in the first step in the reaction, and because the first step in the reaction is the rate-limiting step, the overall rate of reaction is first-order in both reagents. Because the OI^- ion produced in the first step is consumed in the second step, it doesn't appear in the rate law for the reaction or as one of the products of the reaction.

> ➤ **CHECKPOINT**
>
> The reaction coordinate diagrams in Figures 14.15 and 14.17 refer to the same reaction. Locate ΔH on both diagrams. Is ΔH the same for both? Explain.

14.17 Determining the Activation Energy of a Reaction

The rate of a reaction depends on the temperature at which it is run. As the temperature increases, the average kinetic energy of the particles that form the system increase. As a result, the molecules move faster and therefore collide more frequently. Thus the proportion of collisions that can overcome the activation energy for the reaction increases with temperature.

The relationship between temperature and the rate of a reaction can be explained by assuming that the rate constant depends on the temperature at which the reaction is run. In 1889, Svante Arrhenius suggested that the relationship between temperature and the rate constant for a reaction obeyed the following equation.

$$k = Ze^{-E_a/RT}$$

In this equation, k is the rate constant for the reaction, Z is a proportionality constant that varies from one reaction to another, e is the base of natural logarithms, E_a is the activation energy for the reaction, R is the ideal gas constant in joules per mole kelvin, and T is the temperature in kelvins. For a given Z and T, the rate constant, k, is therefore determined by E_a.

The **Arrhenius equation** can be used to determine the activation energy for a reaction. We start by taking the natural logarithm of both sides of the equation.

$$\ln k = \ln Z - \frac{E_a}{RT}$$

We then rearrange the equation to fit the equation for a straight line.

$$y = mx + b$$

$$\ln k = -\frac{E_a}{R}\left(\frac{1}{T}\right) + \ln Z$$

According to this equation, a plot of the natural logarithm of the rate constant for the reaction versus $1/T$ should give a straight line with a slope of $-E_a/R$, as shown in Figure 14.19.

By dividing the Arrhenius equation for the reaction at one temperature (k_1 and T_1) by the Arrhenius equation for the reaction at a second temperature (k_2 and T_2), it is possible to derive another form of the Arrhenius equation that can be used to predict the effect of a change in temperature on the rate constant for a reaction.

$$\ln\left(\frac{k_1}{k_2}\right) = \frac{E_a}{R}\left(\frac{1}{T_2} - \frac{1}{T_1}\right)$$

> **CHECKPOINT**

According to the Arrhenius relationship between E_a and k, is k large or small when E_a is large? What does a small k mean concerning how fast the reaction proceeds? Use this relationship to suggest how catalysts speed up reactions.

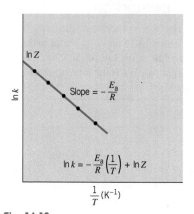

Fig. 14.19 A plot of the natural log of the rate constant for a reaction at different temperatures versus the inverse of the temperature in kelvins is a straight line with a slope equal to $-E_a/R$.

Exercise 14.12

Use the following data to determine the activation energy for the zero-order decomposition of HI on a platinum surface.

Temperature (K)	Rate Constant (M/s)
573	2.91×10^{-6}
673	8.38×10^{-4}
773	7.65×10^{-2}

Solution

We can determine the activation energy for a reaction from a plot of the natural log of the rate constants versus the reciprocal of the absolute temperature. We therefore start by calculating $1/T$ and the natural logarithm of the rate constants.

$\ln k$	$1/T$ (K^{-1})
-12.75	0.00175
-7.08	0.00149
-2.57	0.00129

When we construct a graph of the data, we get a straight line with a slope of -2.2×10^4 K. According to the Arrhenius equation, the slope of the line is equal to $-E_a/R$.

$$-2.2 \times 10^4 \text{ K} = -\frac{E_a}{8.314 \text{ J/mol}_{rxn} \cdot \text{K}}$$

When the equation is solved, we get the following value for the activation energy for the reaction.

$$E_a = 1.8 \times 10^2 \text{ kJ/mol}_{rxn}$$

Exercise 14.13

Calculate the rate constant for the decomposition of HI at 600°C.

Solution

We start with the following form of the Arrhenius equation.

$$\ln\left(\frac{k_1}{k_2}\right) = \frac{E_a}{R}\left(\frac{1}{T_2} - \frac{1}{T_1}\right)$$

We then pick any one of the three data points used in the preceding exercise as T_1 and allow the value of T_2 to be 873 K.

$$T_1 = 573 \text{ K} \qquad k_1 = 2.91 \times 10^{-6} \text{ M/s}$$
$$T_2 = 873 \text{ K} \qquad k_2 = ?$$

Substituting what we know about the system into the equation given above gives the following result.

$$\ln\left(\frac{2.91 \times 10^{-6} \text{ M/s}}{k_2}\right) = \frac{1.8 \times 10^5 \text{ J/mol}_{rxn}}{8.314 \text{ J/mol}_{rxn} \cdot \text{K}}\left(\frac{1}{873 \text{ K}} - \frac{1}{573 \text{ K}}\right)$$

We can simplify the right-hand side of this equation as follows.

$$\ln\left(\frac{2.91 \times 10^{-6} \text{ M/s}}{k_2}\right) = -13$$

We then take the antilog of both sides of the equation.

$$\left(\frac{2.91 \times 10^{-6} \text{ M/s}}{k_2}\right) = 2 \times 10^{-6}$$

Solving for k_2 gives the rate constant for the reaction at 600°C.

$$k_2 = 1 \text{ M/s}$$

Increasing the temperature of the reaction from 573 K to 873 K therefore increases the rate constant for the reaction by a factor of almost a million.

> ➤ **CHECKPOINT**
>
> Chapter 8 noted that the boiling point of water depends on atmospheric pressure. As a result, water boils at a lower temperature at 10,000 feet in the Rocky Mountains than at sea level. Use the Arrhenius equation to describe why it takes longer to cook a hard-boiled egg at the top of a mountain than it does at sea level.

14.18 The Kinetics of Enzyme-Catalyzed Reactions

Enzymes are proteins that catalyze reactions in living systems. In 1913, Leonor Michaelis and his student Maud Menten studied the rate at which an enzyme isolated from yeast catalyzed the hydrolysis of sucrose into fructose and glucose.

$$\text{sucrose}(aq) + \text{H}_2\text{O}(l) \xrightarrow{\text{enzyme}} \text{fructose}(aq) + \text{glucose}(aq)$$

At first glance, this reaction seems similar to the reaction between sucrose and acid, which had been studied by Ludwig Wilhelmy 60 years earlier.

$$\text{sucrose}(aq) + \text{H}_3\text{O}^+(aq) \longrightarrow \text{fructose}(aq) + \text{glucose}(aq)$$

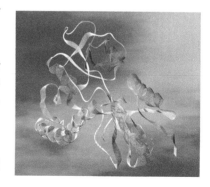

A schematic of the structure of the porphobilinogen deaminase enzyme.

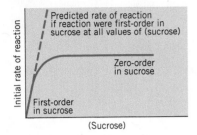

Fig. 14.20 If the reaction in which an enzyme hydrolyzes sucrose to fructose and glucose were first-order in the concentration of sucrose, the initial rate of reaction should be directly proportional to the initial concentration of sucrose. Michaelis and Menten, however, showed that the reaction is first-order in sucrose at low concentrations of sucrose but zero-order in sucrose at high concentrations.

Wilhelmy found that the reaction between sucrose and acid was first-order in sucrose at constant pH.

$$\text{rate} = k(\text{sucrose})$$

Michaelis and Menten, however, found that the initial rate of the enzyme-catalyzed reaction was first-order in sucrose only at low concentrations of sucrose. At high concentrations, the initial rate of reaction didn't increase when more sucrose was added to the system.

As we have seen, when the rate of reaction doesn't depend on the concentration of one of the reactants, the reaction is said to be *zero-order* in that reactant. The enzyme-catalyzed hydrolysis of sucrose apparently changes from a first-order reaction to a zero-order reaction as the amount of sucrose in the system increases, as shown in Figure 14.20. There is a maximum initial rate of reaction for the enzyme-catalyzed reaction that can't be exceeded no matter how much sucrose we add to the solution. Once the reaction reaches this maximum rate, the only way to increase the rate of reaction is to add more enzyme to the solution.

Michaelis and Menton explained this behavior by assuming that the reaction proceeds through a two-step mechanism. In the first step, the enzyme combines with sucrose to form a complex between the enzyme and sucrose that we can symbolize as *ES*. This reaction is reversible. The rate constant for the forward reaction is often labeled k_1, whereas the rate constant for the reverse reaction is k_{-1}

$$E + S \underset{k_{-1}}{\overset{k_1}{\rightleftharpoons}} ES$$

The enzyme–sucrose complex can decompose to form the product of the reaction, *P*, and regenerate the enzyme.

$$ES \underset{k_{-2}}{\overset{k_2}{\rightleftharpoons}} E + P$$

The rate-limiting step in the enzyme-catalyzed reaction is the decomposition of *ES* into enzyme and product. The rate of the overall reaction is therefore determined by the concentration of the enzyme–substrate complex:

$$\text{rate}_{\text{overall}} \approx \text{rate}_{\text{2nd}} = k_2(ES)$$

This means that the maximum rate of reaction will be seen when essentially all the enzyme is tied up as the *ES* complex.

There is a limit to the rate at which the enzyme can consume sucrose. When there is a large excess of sucrose in the solution, every time the enzyme transforms a molecule of sucrose into the products of the reaction, the enzyme will immediately pick up another sucrose molecule. No matter how much sucrose is added to the solution, the reaction can't occur any faster. The reaction no longer depends on the sucrose concentration and is therefore zero-order in sucrose.

$$\text{rate} = k(\text{sucrose})^0 = k$$

Enzyme-catalyzed reactions are therefore similar to the metal-catalyzed reactions described in Section 14.10. When there is enough reactant in the system so that essentially all of the catalytic sites are occupied at any moment in time, the reaction will exhibit zero-order kinetics. No matter how much reactant is added to the system, the reaction can't go any faster.

Key Terms

Activation energy (E_a)

Arrhenius equation

Bimolecular

Catalyst

Chemical kinetics

Collision theory

Enzyme

First-order reaction

Half-life

Initial instantaneous rate of reaction

Instantaneous rate of reaction

Integrated form of the rate laws

Intermediates

Kinetic control

Mechanism

Molecularity

Order

Pseudo-first-order reaction

Rate constant

Rate law

Rate-limiting step

Second-order reaction

Thermodynamic control

Unimolecular

Zero-order reaction

Problems

The Forces That Control a Chemical Reaction

1. What is meant by *thermodynamic control of a reaction*? What is meant by *kinetic control of a reaction*? Give an example of each.

2. The reaction of H_2 and O_2 to produce water is thermodynamically favorable. What does this mean? Why don't we observe this reaction taking place at room temperature and 1 atm?

Chemical Kinetics

3. How is the rate of a chemical reaction expressed?

4. What does the rate law for the reaction of sucrose with acid suggest happens to the rate of the reaction as the reaction transforms sucrose into invert sugar.

Is the Rate of Reaction Constant?

5. Why does the expression for the rate of the phenolphthalein reaction contain a negative sign?

6. Describe what happens to the rate at which a reactant is consumed during the course of a reaction. (Does it increase, decrease, or remain the same?)

Instantaneous Rates of Reaction

7. How can the instantaneous rate of reaction be calculated?

8. Define the instantaneous rate of reaction.

9. Describe the difference between the rate of reaction measured over a finite period of time and the instantaneous rate of reaction. Explain the advantage of measuring the instantaneous rate of reaction.

Rate Laws and Rate Constants

10. Describe the difference between the terms *rate of reaction, rate law,* and *rate constant*.

11. What are the units of the rate constant for the following reaction?

$$N_2O_4(g) \rightleftharpoons 2\ NO_2(g) \qquad rate = k(N_2O_4)$$

12. What are the units of the rate constant for the following reaction?

$$2\ NO_2(g) \rightleftharpoons N_2O_4(g) \qquad rate = k(NO_2)^2$$

13. What are the units of the rate constant for the following reaction?

$$2\ Br^-(aq) + H_2O_2(aq) + 2\ H^+(aq)$$
$$\rightleftharpoons Br_2(aq) + 2H_2O(l)$$
$$rate = k(Br^-)(\ H_2O_2)(\ H^+)$$

14. A reaction is said to be diffusion controlled when it occurs as fast as the reactants diffuse through the solution. The following is an example of a diffusion-controlled reaction.

$$H_3O^+(aq) + OH^-(aq) \rightleftharpoons 2\ H_2O(l)$$

The rate constant for this reaction is $1.4 \times 10^{11} M^{-1}$ s^{-1} at 25°C, and the reaction obeys the following rate law. Calculate the rate of reaction in a neutral solution (pH = 7.00).

$$rate = k(H_3O^+)(OH^-)$$

The Rate Law versus the Stoichiometry of a Reaction

15. Which equation describes the relationship between the rates at which Cl_2 and F_2 are consumed in the following reaction?

$$Cl_2(g) + 3\ F_2(g) \rightleftharpoons 2\ ClF_3(g)$$

(a) $-d(Cl_2)/dt = -d(F_2)/dt$
(b) $-d(Cl_2)/dt = 2[-d(F_2)/dt]$
(c) $2[-d(Cl_2)/dt] = -d(F_2)/dt$
(d) $-d(Cl_2)/dt = 3\ [-d(F_2)/dt]$
(e) $3\ [-d(Cl_2)/dt] = -d(F_2)/dt$

16. Which equation describes the relationship between the rates at which Cl_2 is consumed and ClF_3 is produced in the following reaction?

$$Cl_2(g) + 3 F_2(g) \rightleftharpoons 2 ClF_3(g)$$

(a) $-d(Cl_2)/dt = d(ClF_3)/dt$

(b) $-d(Cl_2)/dt = 2[d(ClF_3)/dt]$

(c) $2[-d(Cl_2)/dt] = d(ClF_3)/dt$

(d) $-d(Cl_2)/dt = 3[d(ClF_3)/dt]$

(e) $3[-d(Cl_2)/dt] = d(ClF_3)/dt$

17. Calculate the rate at which N_2O_4 is formed in the following reaction at the moment in time when NO_2 is being consumed at a rate of 0.0592 M/s.

$$2 NO_2(g) \rightleftharpoons N_2O_4(g)$$

18. Calculate the rate for the formation of I_2 in the following reaction if the rate for the disappearance of HI is 0.039 $M\,s^{-1}$.

$$2 HI(g) \rightleftharpoons H_2(g) + I_2(g) \qquad rate = k(HI)^2$$

19. Ammonia reacts in the gas phase to form nitrogen oxide and water.

$$4 NH_3(g) + 5 O_2(g) \longrightarrow 4 NO(g) + 6 H_2O(g)$$

Derive the relationship between the rates at which NH_3 and O_2 are consumed in the reaction. Derive the relationship between the rates at which O_2 is consumed and H_2O is produced.

20. Calculate the rate of formation of NO and the rate of disappearance of NH_3 for the following reaction at the moment in time when the rate of formation of water is 0.040 M/s.

$$4 NH_3(g) + 5 O_2(g) \longrightarrow 4 NO(g) + 6 H_2O(g)$$

21. Assume that the instantaneous rate of disappearance of the MnO_4^- ion in the following reaction is 4.56×10^{-3} M/s at some moment in time.

$$10 I^-(aq) + 2 MnO_4^-(aq) + 16 H^+(aq)$$
$$\rightleftharpoons 2 Mn^{2+}(aq) + 5 I_2(aq) + 8 H_2O(l)$$

What is the rate of appearance of I_2 at the same moment?

Order and Molecularity

22. Describe the difference between the molecularity and the order of a reaction.

23. Describe the conditions under which the rate law for a reaction is most likely to reflect the stoichiometry of the reaction.

24. List one or more factors that can make the rate law for a reaction differ from what the stoichiometry of the reaction leads us to expect.

25. Describe the difference between unimolecular and bimolecular reactions. Give an example of each.

A Collision Theory of Chemical Reactions

26. What would be the rate law for a simple one-step reaction such as the following?

$$CH_3Br + SH^- \longrightarrow CH_3SH + Br^-$$

27. What is a rate-limiting step?

28. Define the term *mechanism*.

29. If a reaction does not occur in a single step, what must be known in order to write a rate law for the reaction?

The Mechanisms of Chemical Reactions

30. Each of the following reactions was found experimentally to be second-order in a single reactant. Which of them is most likely to be an elementary reaction that occurs in a single step?

(a) $2 NO_2(g) + Cl_2(g) \longrightarrow 2 NO_2Cl(g)$

(b) $N_2O_3(g) \longrightarrow NO(g) + NO_2(g)$

(c) $3 O_2(g) \longrightarrow 2 O_3(g)$

(d) $2 NO(g) \longrightarrow N_2(g) + O_2(g)$

31. NO reacts with H_2 according to the following equation.

$$2 NO(g) + 2 H_2(g) \rightleftharpoons N_2(g) + 2 H_2O(g)$$

The mechanism for the reaction involves two steps.

$$2 NO + H_2 \longrightarrow N_2 + H_2O_2$$
$$H_2O_2 + H_2 \longrightarrow 2 H_2O$$

What is the predicted rate law for the reaction if the first step is the rate-limiting step?

32. The following reaction is first-order in both NO_2 and F_2.

$$2 NO_2(g) + F_2(g) \longrightarrow 2 NO_2F(g)$$
$$rate = k(NO_2)(F_2)$$

The rate law is consistent with which of the following mechanisms?

(a) $NO_2 + F_2 \rightleftharpoons NO_2F$

(b) $NO_2 + F_2 \rightleftharpoons NO_2F + F$

    $NO_2 + F \longrightarrow NO_2F$ (rate-limiting step)

(c) $NO_2 + F_2 \longrightarrow NO_2F + F$ (rate-limiting step)

    $NO_2 + F \longrightarrow NO_2F$

(d) $F_2 \longrightarrow 2 F$ (rate-limiting step)

    $2 NO_2 + 2 F \longrightarrow 2 NO_2F$

33. The disproportionation of NO to N_2O and NO_2 is third-order in NO.

$$3\ NO(g) \rightleftharpoons N_2O(g) + NO_2(g)$$
$$rate = k(NO)^3$$

The rate law is consistent with which of the following mechanisms?

(a) $NO + NO + NO \longrightarrow N_2O + NO_2$

(one-step reaction)

(b) $2\ NO \longrightarrow N_2O_2$ (rate-limiting step)

$N_2O_2 + NO \longrightarrow N_2O + NO_2$

(c) $2\ NO \rightleftharpoons N_2O_2$

$N_2O_2 + NO \longrightarrow N_2O + NO_2$

(rate-limiting step)

34. Is a rate law for the following reaction that is first-order in I^- and second-order in OCl^- consistent with the mechanism shown below?

$$I^-(aq) + OCl^-(aq) \longrightarrow Cl^-(aq) + OI^-(aq)$$

Mechanism:

$OCl^- + H_2O \rightleftharpoons HOCl + OH^-$

$I^- + HOCl \longrightarrow HOI + Cl^-$ (rate-limiting step)

$HOI + OH^- \longrightarrow OI^- + H_2O$

35. N_2O_5 decomposes to form NO_2 and O_2.

$$2\ N_2O_5(g) \longrightarrow 4\ NO_2(g) + O_2(g)$$

The rate law for the reaction is first-order in N_2O_5.

$$rate = k(N_2O_5)$$

Show how the rate law is consistent with the following three-step mechanism for the reaction.

Step 1:
$N_2O_5 \rightleftharpoons NO_2 + NO_3$

Step 2:
$NO_2 + NO_3 \longrightarrow NO_2 + NO + O_2$

(rate-limiting step)

Step 3:
$NO + NO_3 \longrightarrow 2\ NO_2$ plus additional fast steps

36. Hydrogen and bromine react together to form hydrogen bromide, HBr.

$$rate = k(H_2)(Br_2)^{1/2}$$

Show how the following mechanism, in which the second step is the rate-limiting step, is consistent with the rate law.

$$Br_2 \rightleftharpoons 2\ Br$$
$$Br + H_2 \longrightarrow HBr + H$$
$$H + Br_2 \longrightarrow HBr + Br$$

37. Describe the relationship between the forward and reverse rate constants and the equilibrium constant for a one-step reaction.

38. The following is a one-step reaction.

$$CH_3Cl(aq) + I^-(aq) \rightleftharpoons CH_3I(aq) + Cl^-(aq)$$

What is the equilibrium constant for the reaction if the rate constant for the forward reaction is $5.2 \times 10^{-7} M^{-1}\ s^{-1}$ and the rate constant for the reverse reaction is $1.5 \times 10^{-11} M^{-1}\ s^{-1}$?

Zero-Order Reactions

39. What effect does decreasing the temperature have on the rate of a zero-order reaction?

40. How can the collision theory account for a zero-order reaction?

41. What effect does increasing the concentrations have on the rate of a zero-order reaction that has the following rate law?

$$rate = k(C_2H_4)^0(H_2)^0$$

Determining the Order of a Reaction from Rates of Reaction

42. Use the following data to determine the rate law for the reaction between nitrogen oxide and chlorine to form nitrosyl chloride.

$$2\ NO(g) + Cl_2(g) \longrightarrow 2\ NOCl(g)$$

Initial(NO)(M)	Initial (Cl_2) (M)	Initial Rate of Reaction (M/s)
0.10	0.10	0.117
0.20	0.10	0.468
0.30	0.10	1.054
0.30	0.20	2.107
0.30	0.30	3.161

43. Use the results of the preceding problem to determine the rate constant for the reaction. Predict the initial instantaneous rate of reaction when the initial NO and Cl_2 concentrations are both 0.50 M.

44. Use the following data to determine the rate law for the reaction between nitrogen oxide and oxygen to form nitrogen dioxide. (*Hint:* How are pressures related to concentrations?)

$$2\ NO(g) + O_2(g) \rightleftharpoons 2\ NO_2(g)$$

Initial Pressure NO (mmHg)	Initial Pressure O_2 (mmHg)	Initial Rate of Reaction (mmHg/s)
100	100	0.355
150	100	0.800
250	100	2.22
150	130	1.04
150	180	1.44

45. Use the results of the preceding problem to determine the rate constant for the reaction. Predict the initial rate of reaction when the initial pressures for both NO and O_2 are 250 mmHg.

46. Use the following data to determine the rate law for the reaction between methyl iodide and the OH^- ion in aqueous solution to form methanol and the iodide ion.

$$CH_3I(aq) + OH^-(aq) \rightleftharpoons CH_3OH(aq) + I^-(aq)$$

Initial (CH_3I) (M)	Initial (OH^-) (M)	Initial Rate of Reaction (M/s)
1.35	0.10	8.78×10^{-6}
1.00	0.10	6.50×10^{-6}
0.85	0.10	5.53×10^{-6}
0.85	0.15	8.29×10^{-6}
0.85	0.25	1.38×10^{-5}

47. Use the results of the preceding problem to determine the rate constant for the reaction. Predict the initial instantaneous rate of reaction when the initial CH_3I concentration is 0.10 M and the initial OH^- concentration is 0.050 M.

The Integrated Form of Zero-, First-, and Second-Order Rate Laws

48. Describe the kinds of problems that are best solved by using the rate law for a reaction, such as the following.

$$rate = k(N_2O_5)$$

49. Describe the kinds of problems that are best solved with the integrated form of the rate law.

$$\ln\left[\frac{(N_2O_5)}{(N_2O_5)_0}\right] = -kt$$

50. The decomposition of NH_3 gas on tungsten metal follows zero-order kinetics with a rate constant of 3.4×10^{-6} mol/L·s. If the initial concentration of $NH_3(g)$ is 0.0068 M, what will be the concentration after 1000 s?

51. The following reaction, in which NO_2 forms a dimer, is second-order in NO_2.

$$2\,NO_2(g) \rightleftharpoons N_2O_4(g) \qquad rate = k(NO_2)^2$$

(a) Calculate the rate constant for the reaction if it takes 0.0050 s for the initial concentration of NO_2 to decrease from 0.50 M to 0.25 M.

(b) How long will it take for the concentration to decrease from 0.50 M to 0.15 M?

52. The decomposition of hydrogen peroxide is first-order in H_2O_2.

$$2\,H_2O_2(aq) \longrightarrow 2\,H_2O(l) + O_2(g)$$
$$rate = k(H_2O_2)$$

(a) How long will it take for half of the H_2O_2 in a 10-gallon sample to be consumed if the rate constant for the reaction is $5.6 \times 10^{-2}\,s^{-1}$?

(b) What will be the concentration of a 1.5 M H_2O_2 solution after 20 s?

53. Which graph best describes the rate of the following reaction if the reaction is first-order in N_2O_4?

$$N_2O_4(g) \rightleftharpoons 2\,NO_2(g)$$

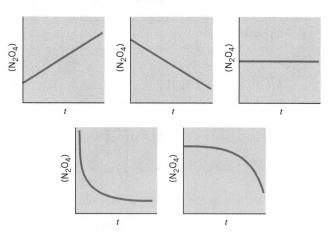

54. Calculate the rate constant for the following acid–base reaction if the half-life for the reaction is 0.0282 s at 25°C and the reaction is first-order in the NH_4^+ ion.

$$NH_4^+(aq) + H_2O(l) \rightleftharpoons NH_3(aq) + H_3O^+(aq)$$

55. The following reaction is second-order in NO_2.

$$2\,NO_2(g) \rightleftharpoons N_2O_4(g)$$

What effect would doubling the initial concentration of NO_2 have on the half-life for the reaction?

56. The reaction of HI(g) to produce $H_2(g)$ and $I_2(g)$ is second-order with a rate constant of 9.6×10^{-2} L/mol·min. If the initial concentration of HI is 0.55 M, what will be the concentration of HI after 10 minutes?

57. The age of a rock can be estimated by measuring the amount of ^{40}Ar trapped inside. The calculation is based on the fact that ^{40}K decays to ^{40}Ar by a first-

order process. It also assumes that none of the ^{40}Ar produced by the reaction has escaped from the rock since the rock was formed.

$$^{40}K + e^- \longrightarrow {}^{40}Ar \qquad k = 5.81 \times 10^{-11} \text{ year}^{-1}$$

Calculate the half-life of the radioactive decay.

58. Another way of determining the age of a rock involves measuring the extent to which the ^{87}Rb in the rock has decayed to ^{87}Sr.

$$^{87}Rb \longrightarrow {}^{87}Sr + e^- \qquad k = 1.42 \times 10^{-11} \text{ year}^{-1}$$

What fraction of the ^{87}Rb would still remain in a rock after 1.19×10^{10} years?

59. ^{14}C measurements on the linen wrappings from the book of Isaiah in the Dead Sea Scrolls suggest that the scrolls contain about 79.5% of the ^{14}C expected in living tissue. How old are the scrolls if the half-life for the decay of ^{14}C is 5.730×10^3 years?

60. The Lascaux cave near Montignac in France contains a series of cave paintings. Radiocarbon dating of charcoal taken from the site suggests an age of 15,520 years. What fraction of the ^{14}C present in living tissue is still present in the sample? (For ^{14}C, $t_{1/2} = 5.730 \times 10^3$ years.)

61. A skull fragment found in 1936 at Baldwin Hills, California, was dated by ^{14}C analysis. Approximately 100 g of bone was cleaned and treated with 1 M HCl(aq) to destroy the mineral content of the bone. The bone protein was then collected, dried, and pyrolyzed. The CO_2 produced was collected and purified, and the ratio of ^{14}C to ^{12}C was measured. If the sample contained roughly 5.7% of the ^{14}C present in living tissue, how old was the skeleton? (For ^{14}C, $t_{1/2} = 5.730 \times 10^3$ years.)

62. Charcoal samples from Stonehenge in England emit 62.3% of the disintegrations per gram of carbon per minute expected for living tissue. What is the age of the samples? (For ^{14}C, $t_{1/2} = 5.730 \times 10^3$ years.)

63. A lump of beeswax was excavated in England near a collection of Bronze Age objects roughly 2500 to 3000 years of age. Radiocarbon analysis of the beeswax suggests an activity equal to roughly 90.3% of the activity observed for living tissue. Determine whether the beeswax was part of the hoard of Bronze Age objects. (For ^{14}C, $t_{1/2} = 5.730 \times 10^3$ years.)

64. The activity of the ^{14}C in living tissue is 15.3 disintegrations per minute per gram of carbon. The limit for reliable determination of ^{14}C ages is 0.10 disintegration per minute per gram of carbon. Calculate the maximum age of a sample that can be dated accurately by radiocarbon dating. Assume the half-life of ^{14}C is 5.730×10^3 years.

Determining the Order of a Reaction with the Integrated Form of Rate Laws

65. Plot the following data to determine the rate law for the decomposition of N_2O.

$$2 N_2O(g) \longrightarrow 2 N_2(g) + O_2(g)$$

(N_2O) (M)	0.100	0.086	0.079	0.075	0.066	0.059	0.050	0.025
Time (s)	0	80	120	160	240	320	480	960

Use the relation between half-life and order shown in Table 14.3 to support your answer.

66. Use the results of the preceding problem to calculate the rate constant for the reaction. Predict the concentration of N_2O after 900 s.

67. Plot the following data to determine the rate law for the hydrolysis of the BH_4^- ion.

$$BH_4^-(aq) + 4 H_2O(l) \longrightarrow B(OH)_4^-(aq) + 4 H_2(g)$$

(BH_4^-) (M)	0.100	0.088	0.077	0.068	0.060	0.052	0.046
Time (h)	0	24	48	72	96	120	144

68. Use the results of the preceding problem to calculate the half-life for the reaction.

69. Triphenylphosphine, PPh_3, reacts with nickel tetracarbonyl, $Ni(CO)_4$, to displace a molecule of carbon monoxide.

$$Ni(CO)_4 + PPh_3 \rightleftharpoons Ph_3PNi(CO)_3 + CO$$

The following data were obtained when the reaction was run at 25°C in the presence of a large excess of triphenylphosphine.

($Ni(CO)_4$) (M)	10.0	7.6	5.8	4.4	3.3	2.5
Time (s)	0	40	80	120	160	200

Use the data to determine the reaction order.

70. The rate of the reaction in the preceding problem does not depend on the concentration of PPh_3. Combine this fact with the results of the preceding problem to determine whether the rate law for the reaction is consistent with the following mechanism.

$$Ni(CO)_4 \longrightarrow Ni(CO)_3 + CO \qquad \text{(rate limiting step)}$$
$$Ni(CO)_3 + PPh_3 \rightleftharpoons Ph_3PNi(CO)_3$$

71. Butadiene, C_4H_6, in the gas phase undergoes a reaction to produce C_8H_{12}. The following data were obtained for the reaction.

Time (min)	0	20	100	500	1000	2000	3000
(C_4H_6) (M)	1	0.98	0.91	0.66	0.49	0.32	0.25

Determine the order of the reaction. Calculate the rate constant for the reaction. How long does it take for the initial concentration to decrease by one-half? How long does it take for this value to decrease to half its value? What does this tell you about the order of the reaction?

72. For the gas-phase reaction

$$2\,NO_3(g) \longrightarrow 2\,NO_2(g) + O_2(g)$$

the following data were determined.

Time (s)	(NO_3) (M)
0	0.50
500	0.32
1000	0.24
2000	0.16
3000	0.12

What is the order of the reaction with respect to NO_3? Would it be a mistake to try to calculate the half-life of NO_3 for this reaction?

73. For the gas-phase reaction in the presence of platinum metal

$$A(g) \longrightarrow B(g) + C(g)$$

the following data were obtained.

Time (s)	(A) (M)
0	0.50
2	0.47
5	0.43
10	0.35
15	0.28
20	0.20
25	0.13
30	0.05

Use these data to determine the order and the rate constant for the reaction.

74. Dimethyl ether, CH_3OCH_3, decomposes at high temperatures as shown in the following equation.

$$CH_3OCH_3(g) \longrightarrow CH_4(g) + H_2(g) + CO(g)$$

The following data were obtained when the partial pressure of CH_3OCH_3 was studied as the compound decomposed at 500°C. Use the data to determine the order of the reaction. (*Hint:* How is pressure related to concentration?)

$P_{CH_3OCH_3}$ (mmHg)	312	278	251	227	156	78
Time (s)	0	390	777	1195	3155	6310

Use a plot of the data to determine the order of the reaction. Support your answer by application of the half-life relationships given in Table 14.3.

Reactions That Are First-Order in Two Reactants

75. The following reaction is first-order in both CH_3I and OH^-.

$$CH_3I(aq) + OH^-(aq) \rightleftharpoons CH_3OH(aq) + I^-(aq)$$

Describe how to turn the reaction into one that is pseudo-first-order in CH_3I.

76. $Cr(NH_3)_5Cl^{2+}$ reacts with the OH^- ion in aqueous solution to displace Cl^- from the complex ion.

$$Cr(NH_3)_5Cl^{2+}(aq) + OH^-(aq)$$
$$\rightleftharpoons Cr(NH_3)_5(OH)^{2+}(aq) + Cl^-(aq)$$

The following data were obtained when the reaction was run at 25°C in a buffer solution at constant pH.

$(Cr(NH_3)_5Cl^{2+})$ (M)	Time (min)
1.00	0
0.81	3
0.66	6
0.53	9
0.50	10
0.43	12
0.35	15
0.25	20

Use the data to determine whether the reaction is first-order or second-order in $Cr(NH_3)_5Cl^{2+}$.

77. The rate of the reaction in the preceding problem is proportional to the pH of the buffer solution in which the reaction is run. Each time the buffer is changed so that the OH^- ion concentration is doubled, the rate of reaction increases by a factor of 2. Combine this observation with the results of the preceding problem to determine the rate law for the reaction.

78. Show that the rate law derived in the preceding problem is consistent with the following mechanism.

$$Cr(NH_3)_5Cl^{2+} + OH^-$$
$$\longrightarrow Cr(NH_3)_4(NH_2)(Cl)^+ + H_2O$$
$$\text{(rate-limiting step)}$$
$$Cr(NH_3)_4(NH_2)(Cl)^+$$
$$\rightleftharpoons Cr(NH_3)_4(NH_2)^{2+} + Cl^- \quad \text{(fast step)}$$
$$Cr(NH_3)_4(NH_2)^{2+} + H_2O$$
$$\rightleftharpoons Cr(NH_3)_5(OH)^{2+} \quad \text{(fast step)}$$

79. The following reaction is first-order in both reactants and therefore second-order overall.

$$CH_3I(aq) + OH^-(aq) \rightleftharpoons CH_3OH(aq) + I^-(aq)$$

When the reaction is run in a buffer solution, however, it is pseudo-first-order in CH_3I.

$$rate = k(CH_3I)$$

What is the half-life of the reaction in a pH 10.00 buffer if the rate constant for this pseudo-first-order reaction is $6.5 \times 10^{-9}\ s^{-1}$?

80. In an experiment such as that described in Problem 79, if the concentration of CH_3I is 0.0010 M, which of the following OH^- concentrations would be best for turning the reaction into a pseudo-first-order reaction in CH_3I? Explain.

  (a) 0.0010 M (b) 0.010 M

  (c) 0.0001 M (d) 1.0 M

81. Why is it useful to arrange a kinetics experiment so that a reaction exhibits pseudo-order kinetics?

The Activation Energy of Chemical Reactions

82. Use Figure 14.14 to explain why the rate of a reaction generally increases with increasing temperature.

83. Explain why the larger the activation energy for a reaction, the slower the rate of the reaction.

84. Describe the factors that determine whether a collision between two molecules will lead to a reaction.

Catalysts and the Rates of Chemical Reactions

85. Describe the five properties of a catalyst. Give an example of a catalyzed reaction and show how the catalyst meets the criteria.

86. If the activation energy for a reaction is lowered, will the rate of the reaction speed up, slow down, or remain the same? Explain.

87. Why is ΔH for a reaction unaffected by a catalyst?

88. Draw a reaction coordinate diagram for which ΔH is endothermic. Add a dotted line showing how a catalyst changes the diagram.

89. Assume that the activation energy was measured for both the forward ($E_a = 120\ kJ/mol_{rxn}$) and reverse ($E_a = 185\ kJ/mol_{rxn}$) directions of a reversible reaction. What would be the activation energy for the reverse reaction in the presence of a catalyst that decreased the activation energy for the forward reaction to 90 kJ/mol_{rxn}?

Determining the Activation Energy of a Reaction

90. The rate constant for the decomposition of N_2O_5 increases from $1.52 \times 10^{-5}\ s^{-1}$ at 25°C to $3.83 \times 10^{-3}\ s^{-1}$ at 45°C. Calculate the activation energy for the reaction.

91. Calculate the activation energy for the following reaction if the rate constant for the reaction increases from $87.1\ M^{-1}\ s^{-1}$ at 500 K to $1.53 \times 10^3\ M^{-1}\ s^{-1}$ at 650 K.

$$2\ NO_2(g) \rightleftharpoons 2\ NO(g) + O_2(g)$$

92. Calculate the activation energy for the decomposition of NO_2 from the temperature dependence of the rate constant for the reaction.

$$2\ NO_2(g) \longrightarrow N_2(g) + 2\ O_2(g)$$

Temperature (K)	319	329	352	381	389
$k\ (M^{-1}\ s^{-1})$	0.522	0.755	1.70	4.02	5.03

93. Calculate the rate constant at 780 K for the following reaction if the rate constant for the reaction is $3.5 \times 10^{-7}\ M^{-1}\ s^{-1}$ at 550 K and the activation energy is 183 kJ/mol_{rxn}.

$$2\ HI(g) \rightleftharpoons H_2(g) + I_2(g)$$

94. Calculate the rate constant at 75°C for the following reaction if the rate constant for the reaction is $6.5 \times 10^{-5}\ M^{-1}\ s^{-1}$ at 25°C and the activation energy is 92.9 kJ/mol_{rxn}.

$$CH_3I(aq) + OH^-(aq) \rightleftharpoons CH_3OH(aq) + I^-(aq)$$

95. According to the Arrhenius equation, what happens to the rate constant if the activation energy for a particular reaction is lowered, as in the case of the introduction of a catalyst? Explain why the rate of a reaction changes when a catalyst is added.

The Kinetics of Enzyme-Catalyzed Reactions

96. Explain why the rate of the enzyme-catalyzed hydrolysis of sucrose is first-order in sucrose at low concentrations of the substance.

97. Explain why the rate of enzyme-catalyzed reactions becomes zero-order at very high concentrations of the substrate.

Integrated Problems

98. The activation energy for the following reaction in the forward direction is 200 kJ/mol_{rxn}.

$$2\ N_2O_5(g) \rightleftharpoons 4\ NO_2(g) + O_2(g)$$

Use enthalpy of atom combination data from Table B.13 in Appendix to answer the following questions.

  (a) What is the activation energy for the reverse reaction?

  (b) If a catalyst is added and the activation energy for the forward reaction is reduced to 150 kJ/mol_{rxn}, what will be the activation energy for the reverse reaction?

99. I. From the reaction coordinate diagrams below, and assuming constant temperature and Z for all the

diagrams, select the diagram for the conversion of reactants to products that has the following:

(a) the smallest rate constant for an endothermic reaction

(b) the largest rate constant for an exothermic reaction

(c) the largest rate constant for a reverse reaction

(d) the most rapid establishment of equilibrium

II. Of diagrams II and IV, which is most likely to have an equilibrium constant greater than 1? Can you decide which of the two reactions proceeds the most rapidly?

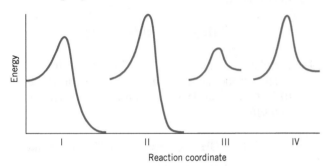

100. Enzymes act as catalysts in biochemical reactions. They are not consumed in the reaction and therefore do not appear as a reactant in the chemical equation. Their function is to provide a site where reactants can be brought together in the proper orientation to lead to a reaction. Predict the order of a biochemical reaction to which a very small amount of the appropriate enzyme has been added. Explain your reasoning.

101. Which of the following graphs best describes the relationship between the rate of a reaction and the temperature of the reaction?

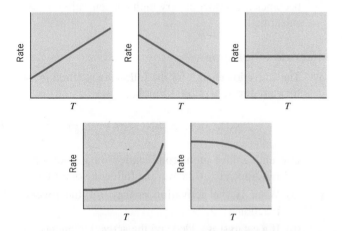

102. Explain why mixtures of H_2 and O_2 gas do not react when stored at room temperature for several years, whereas the reaction is complete within a few days at 300°C, within a few hours at 500°C, and almost instantaneously at 700°C.

103. Which of the following will be affected by the addition of a catalyst to a chemical reaction?

(a) the magnitude of the equilibrium constant

(b) the value of k_f, the forward rate constant

(c) the value of k_r, the reverse rate constant

(d) the value of the forward rate

(e) the value of the reverse rate

(f) the value of the ratio k_f/k_r

(g) $\Delta H°$

(h) $\Delta S°$

(i) $\Delta G°$

104. The mechanism shown below has been suggested for the decomposition of hydrogen peroxide, H_2O_2.

$$2\ H_2O_2(aq) \longrightarrow 2\ H_2O(l) + O_2(g)$$
$$H_2O_2 \longrightarrow 2\ OH \qquad \text{(rate-limiting step)}$$
$$H_2O_2 + OH \longrightarrow H_2O + HO_2$$
$$HO_2 + OH \longrightarrow H_2O + O_2$$

(a) What rate law would be expected from this mechanism? Explain why.

(b) The concentration of H_2O_2 was determined at various times as shown below.

Time (s)	0	300	835	1200	1670	2460	3000	3835
Conc (M)	1.0	0.78	0.50	0.37	0.25	0.13	0.082	0.041

Do these data agree with the rate law suggested by the mechanism? Explain.

(c) What is the half-life of the reaction? Show how you obtain your answer.

(d) If $\Delta H°$ for the reaction is < 0, which is larger, the activation energy in the forward direction or the activation energy in the reverse direction? Explain.

(e) If a piece of a turnip is added to $H_2O_2(aq)$, the rate of the decomposition is increased by about 600 billion times. Roughly sketch the reaction coordinate diagrams for the decomposition of H_2O_2 in the presence and absence of turnips. Label the diagram clearly.

105. The following data were obtained for the decomposition of benzoyl peroxide [$(C_6H_5COO)_2$].

$$\underset{\substack{\| \\ C_6H_5COOCC_6H_5}}{\overset{\substack{O \quad\ O}}{}} \longrightarrow 2\ \underset{\substack{\| \\ C_6H_5CO\cdot}}{\overset{O}{}}$$

Time (min)	0	7.83	19.8	27.6	39.6
Conc (M)	0.100	0.076	0.050	0.038	0.025

(a) What is the half-life of the reaction in minutes?

(b) Various plots of the concentration of the reactant versus time are given below. What is the order of the reaction? Explain.

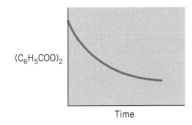

$(C_6H_5COO)_2$ vs. Time

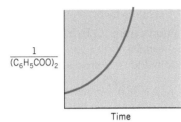

$\dfrac{1}{(C_6H_5COO)_2}$ vs. Time

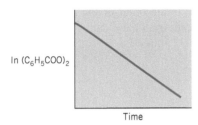

$\ln (C_6H_5COO)_2$ vs. Time

(c) What is the value of the rate constant at 100°C?

(d) Would you expect the half-life of the reaction to be longer, shorter, or the same at 70°C? Explain.

(e) The activation energy in the forward direction is 100 kJ/mol, and enthalpy change for the reaction is 25 kJ/mol. What is the activation energy in the reverse direction? Is the equilibrium constant for the reaction greater than 1 or less than 1? Explain.

106. The rate law for the reaction

$$2\,H_2(g) + 2\,NO(g) \longrightarrow N_2(g) + 2\,H_2O(g)$$

is: rate $= k(H_2)(NO)^2$. Are any of the following mechanisms consistent with the observed rate law? Explain.

(a) $H_2 + NO \longrightarrow H_2O + N$ (rate-limiting step)
    $N + NO \longrightarrow N_2 + O$
    $O + H_2 \longrightarrow H_2O$

(b) $H_2 + 2\,NO \longrightarrow N_2O + H_2O$ (rate-limiting step)
    $N_2O + H_2 \longrightarrow N_2 + H_2O$

(c) $2\,NO \rightleftharpoons N_2O_2$ (fast equilibrium)
    $N_2O_2 + H_2 \longrightarrow N_2O + H_2O$ (rate-limiting step)
    $N_2O + H_2 \longrightarrow N_2 + H_2O$

107. For the reaction

$$2\,A + B \longrightarrow \text{products}$$

the experimentally determined rate law is

$$\text{rate} = k\,(A)(B)$$

which of the following mechanisms are possible? Explain fully.

(a) $2\,A + B \longrightarrow$ products

(b) $A + B \longrightarrow C$ (rate-limiting step)
    $C + A \longrightarrow$ products

(c) $A \rightleftharpoons C$
    $C + A + B \longrightarrow$ products (rate-limiting step)

(d) $A \longrightarrow C$ (rate-limiting step)
    $C + A + B \longrightarrow$ products

108. The results shown in this problem were obtained when the following reaction was studied at $-10°C$.

$$2\,NO(g) + Cl_2(g) \rightleftharpoons 2\,NOCl(g)$$

Initial Concentration (M)

$(NO)_0$	$(Cl_2)_0$	Initial Rate (M/s)
0.10	0.10	0.18
0.10	0.20	0.36
0.20	0.20	1.44

(a) What is the rate law for this reaction? Show all work.

(b) What is the value (with units) of the rate constant?

(c) Is the following mechanism consistent with the rate law? Explain why or why not.

$NO + Cl_2 \rightleftharpoons NOCl_2$ (fast equilibrium)
$NOCl_2 + NO \longrightarrow 2\,NOCl$ (rate-limiting step)

109. The activation energy in the forward direction for the following reaction is 134 kJ/mol$_{rxn}$.

$$CO(g) + NO_2(g) \rightleftharpoons CO_2(g) + NO(g)$$

$\Delta H°$ for the reaction is -226 kJ/mol$_{rxn}$. The reaction occurs in a single step.

(a) What is the activation energy in the reverse direction?

(b) Give two reasons why all collisions between molecules are not effective in causing a reaction.

(c) Draw the best-activated complex for the above reaction and explain why you think this activated complex is best.

(d) What effect does increasing the temperature usually have on the rate of the reaction? Explain.

110. Assume that the following hypothetical reaction gives rise to the set of data shown below. Assume that the reaction is known to be either first-order or second-order.

$$2\,A \longrightarrow 2\,B + C$$

Time (s)	0	10	20	30	40	50	120
$(A)(M)$	1.000	0.800	0.667	0.571	0.500	0.444	0.250

(a) What is the order of the reaction with respect to A?

(b) What is the rate law for the reaction?

(c) Find the rate constant with units.

(d) What effect, if any, would a catalyst have on the rate constant? Explain.

111. Assume that the reaction between CH_4 and Cl_2 could proceed by the mechanism shown in the following.

$$CH_4(g) + Cl_2(g) \longrightarrow CH_3Cl(g) + HCl(g)$$

$Cl_2 \rightleftharpoons 2\,Cl$ \qquad (fast equilibrium)

$CH_4 + Cl \longrightarrow CH_3 + HCl$ \qquad (rate-limiting step)

$CH_3 + Cl_2 \longrightarrow CH_3Cl + Cl$

(a) Some reactions have orders that are not whole-number integers. What rate law would be expected from this mechanism?

(b) The following data were obtained for this reaction.

Expt. #	$(CH_4)_0$	$(Cl_2)_0$	Rate (M/s)
1	0.100	0.100	1.00×10^{-5}
2	0.200	0.100	2.00×10^{-5}
3	0.200	0.200	2.83×10^{-5}

Is the mechanism given above consistent with these data? Explain why or why not.

112. The fluorocarbon C_2F_4 reacts with itself to form a cyclic species in the gas phase. The following data were found for this process.

Time (min)	0	10	15	20	40	60
(C_2F_4) (M)	1.0	0.56	0.45	0.38	0.23	0.17

What is the order of the reaction? Calculate the rate constant from these data.

Chapter Fourteen

SPECIAL TOPIC

14A.1 Deriving the Integrated Rate Laws

To derive the integrated form of the zero-order rate law, we start with the equation that describes the rate law for the reaction.

$$-\frac{d(X)}{dt} = k(X)^0 = k$$

We then rearrange the equation as follows.

$$d(X) = -k\,dt$$

Our goal is to integrate both sides of the equation. Mathematically, this is equivalent to finding the area under the curve that would be produced if the function were graphed. This process is indicated with integral signs, as follows.

$$\int d(X) = \int -k\,dt = -k\int dt$$

We are interested in the area under the curve between the time when the reaction starts ($t = 0$) and some later time (t).

$$\int_{(X)_0}^{(X)} d(X) = -k\int_0^t dt$$

When these integrals are evaluated, we get the following equation.

Integrated form of the zero-order rate law:

$$(X) - (X)_0 = -kt$$

When using this equation, remember that (X) is the concentration of the reactant at any moment in time, $(X)_0$ is the initial concentration of the reactant, k is the rate constant for the reaction, and t is the time since the reaction started.

    The derivation of the integrated form of the first-order rate law also starts with the equation that defines the rate law of the reaction.

$$-\frac{d(X)}{dt} = k(X)$$

We then rearrange the equation as follows,

$$\frac{1}{(X)}d(X) = -k\,dt$$

and then integrate both sides of the equation.

$$\int_{(X)_0}^{(X)} \frac{1}{(X)}d(X) = -k\int_0^t dt$$

The integral of $1/(X)\,d(X)$ is equal to the natural logarithm of (X). The integrated form of the first-order rate law can therefore be written as follows.

Integrated form of the first-order rate law:

$$\ln\left[\frac{(X)}{(X)_0}\right] = -kt$$

Once again, (X) is the concentration of the reactant at any moment in time, $(X)_0$ is the initial concentration of the reactant, k is the rate constant for the reaction, and t is the time since the reaction started.

The derivation of the integrated form of the second-order rate law starts with the equation that defines the rate law of the reaction.

$$-\frac{d(X)}{dt} = k(X)^2$$

We start by rearranging the equation as follows,

$$-\frac{1}{(X)^2}d(X) = k\,dt$$

And then integrate both sides of the equation.

$$-\int_{(X)_0}^{(X)}\frac{1}{(X)^2}d(X) = k\int_0^t dt$$

The integral of $1/(X)^2 d(X)$ is $-1/(X)$. The integrated form of the second-order rate law is therefore written as follows.

Integrated form of the second-order rate law:

$$\frac{1}{(X)} - \frac{1}{(X)_0} = kt$$

Once again, the (X) term is the concentration of X at any moment in time, $(X)_0$ is the initial concentration of X, k is the rate constant for the reaction, and t is the time since the reaction started.

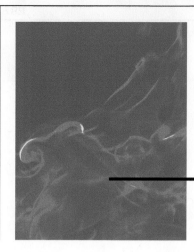

Chapter Fifteen

NUCLEAR CHEMISTRY

15.1 Radioactivity

The discovery of X rays by William Conrad Roentgen in November of 1895 excited the imagination of a generation of scientists who rushed to study the phenomenon of radiation that was able to pass through solid matter. Within a few months, Henri Becquerel found that both uranium metal and salts of this element gave off a different form of radiation, which could also pass through solids. By 1898, Marie Curie found that compounds of thorium were also "radioactive." After painstaking effort, she eventually isolated small quantities of two more radioactive elements—polonium and radium—from ores that contained uranium.

In 1899 Ernest Rutherford discovered at least two different forms of radioactivity when he studied the absorption of radioactivity by thin sheets of metal foil. One, which he called *alpha* (α) *particles,* was absorbed by metal foil that was a few hundredths of a centimeter thick. The other, *beta* (β) *particles,* could pass through 100 times as much metal foil before they became absorbed. Shortly thereafter, a third form of radiation, *gamma* (γ) *rays,* was discovered that could penetrate as much as several centimeters of lead.

The results of early experiments on these three forms of radiation are summarized in Figure 15.1. The direction in which a particles were deflected by an electric field suggested that they were positively charged. The magnitude of this deflection suggested that they had the same charge-to-mass ratio as an He^{2+} ion. To test the equivalence between α particles and He^{2+} ions, Rutherford built an apparatus that allowed α particles to pass through a very thin glass wall into an evacuated flask that contained a pair of metal electrodes. After a few days, he connected these electrodes to a battery and noted that the gas in the flask did indeed give off the characteristic emission spectrum of helium.

Experiments with electric and magnetic fields demonstrated that β particles were negatively charged. Furthermore, they had the same charge-to-mass ratio as an electron. To date, no detectable difference has been found between β particles and electrons. The only reason to retain the name β *particle* is to emphasize the fact that these particles are ejected from the nucleus of an atom when it undergoes radioactive decay.

The fact that γ rays are not deflected by either electric or magnetic fields suggests that these rays don't carry an electric charge. Since they travel at the speed of light, they are classified as a form of electromagnetic radiation that carries even more energy than X rays.

At the turn of the twentieth century, when radioactivity was discovered, atoms were assumed to be indestructible. Ernest Rutherford and Frederick Soddy,

Marie Curie in her laboratory ca. 1905.

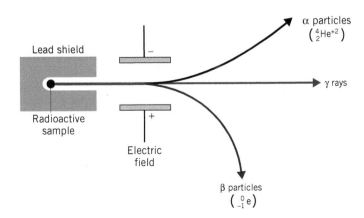

Fig. 15.1 The effect of an electric field on α, β, and γ radiation.

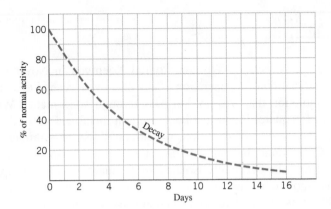

Fig. 15.2 Decay curve reported by Rutherford and Soddy for "uranium X," produced when uranium undergoes radioactive decay.

however, found that radioactive substances became less active with time (Figure 15.2). More importantly, they noticed that radioactivity was always accompanied by the formation of atoms of a different element. By 1903, they concluded that radioactivity was accompanied by a change in the structure of the atom. They therefore assumed that radiation was emitted when an element decayed into a different kind of atom.

By 1910, 40 radioactive elements had been isolated that were associated with the process by which uranium metal decays to lead. This created a problem, however, because there was space for only 11 elements between lead and uranium. In 1913, Kasimir Fajans and Frederick Soddy proposed an explanation for these results: They assumed that radioactive elements that fall in the same place in the periodic table are different forms of the same element. The radioactive thorium produced by the α-particle decay of uranium, for example, is a different form of the element than the radioactive thorium obtained by the β-particle decay of actinium.

Soddy proposed the name *isotope* to describe different radioactive atoms that occupy the same position in the periodic table. J. J. Thomson and Francis Aston then used a mass spectrometer to show that isotopes are atoms of the same element that have different atomic masses.

15.2 The Structure of the Atom

J. J. Thomson's discovery of the electron in 1897 suggested that there is an internal structure to the "indivisible" building blocks of matter known as atoms. This raised an obvious question: How many electrons does an atom contain? By studying the scattering of light, X rays, and α particles, Thomson concluded that the number of electrons in an atom is between 0.2 and 2 times the weight of the atom.

In 1911, Rutherford concluded that the scattering of α particles by extremely thin pieces of metal foil could be explained by assuming that all of the positive charge and most of the mass of the atom are concentrated in an infinitesimally small fraction of the total volume of the atom, for which he proposed the name *nucleus*. Rutherford's data also suggested that the nucleus of a gold atom has a positive charge that is about 80 times the charge on an electron. This is essentially equal to the atomic number of gold, which is 79.

The discovery of the neutron in 1932 explained the discrepancy between the charge on the nucleus and the mass of an atom. A neutral gold atom that has a mass of 197 amu consists of a nucleus that contains 79 protons and 118 neutrons surrounded by 79 electrons. By convention, this information is specified by the following symbol, which describes the only naturally occurring isotope of gold.

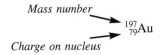

Mass number

$^{197}_{79}Au$

Charge on nucleus

This convention can also be applied to elementary particles. Lowercase letters are used to identify the electron, proton, and neutron.

$$^{4}_{2}He \qquad ^{0}_{-1}e \qquad ^{1}_{1}p \qquad ^{1}_{0}n$$

α particle   *electron*   *proton*   *neutron*
        (*β⁻ particle*)

Because anyone with access to a periodic table can find the atomic number of an element, and therefore the charge on the nucleus, a shorthand notation is often used that reports only the mass number of the atom and the symbol of the element. The shorthand notation for the naturally occurring isotope of gold is ^{197}Au.

Exercise 15.1

Determine the number of protons, neutrons, and electrons in a $^{210}Pb^{2+}$ ion.

Solution

The atomic number of lead is 82, which means that this ion contains 82 protons in the nucleus of the atom. Since it has a charge of $+2$, this ion must contain 80 electrons. Because neutrons and protons both have a mass of about 1 amu, the difference between the mass number (210) and the atomic number (82) is equal to the number of neutrons in the nucleus of the atom. This ion therefore contains 128 neutrons.

A particular combination of protons and neutrons is called a **nuclide.** Nuclides with the same number of protons are called **isotopes.** Nuclides with the same mass number are **isobars.** Nuclides with the same number of neutrons are **isotones.**

Exercise 15.2

Classify the following sets of nuclides as examples of isotopes, isobars, or isotones.

(a) ^{12}C, ^{13}C, and ^{14}C (b) ^{40}Ar, ^{40}K, and ^{40}Ca (c) ^{14}C, ^{15}N, and ^{16}O

Solution

(a) ^{12}C, ^{13}C, and ^{14}C are *isotopes* because they all contain six protons.

(b) ^{40}Ar, ^{40}K, and ^{40}Ca are *isobars* because they have the same mass number.

(c) ^{14}C, ^{15}N, and ^{16}O can't be isotopes because they contain different numbers of protons. They can't be isobars because they have different mass numbers. They all contain eight neutrons, however, so they are examples of *isotones*.

15.3 Modes of Radioactive Decay

Early studies of radioactivity indicated that three different kinds of radiation were emitted, symbolized by the first three letters of the Greek alphabet α, β, and γ. With time, it became apparent that this classification scheme was much too simple. The emission of a negatively charged β^- particle, for example, is only one example of a family of radioactive transformations known as β decay. Other categories such as spontaneous fission and induced fission also had to be added to describe the process by which certain radioactive nuclides decompose into fragments of different weight.

ALPHA DECAY

Alpha decay is usually restricted to the heavier elements in the periodic table. (Only a handful of nuclides with atomic numbers less than 83 emit an α particle.) The product of α decay can be predicted by recognizing that both mass and charge are conserved in nuclear reactions. Alpha decay of the ^{238}U "parent" nuclide, for example, produces ^{234}Th as the "daughter" nuclide.

$$^{238}_{92}U \longrightarrow {}^{234}_{90}Th + {}^4_2He \qquad (\alpha \text{ decay})$$

The sum of the mass numbers of the products (234 + 4) is equal to the mass number of the parent nuclide (238), and the sum of the charges on the nuclei of the products (90 + 2) is equal to the charge on the nucleus of the parent nuclide.

BETA DECAY

Three different modes of **beta decay** can be observed in nature: (1) electron (β^-) emission, (2) electron capture, and (3) positron (β^+) emission

Electron (β^-) emission is literally the process in which an electron is ejected or emitted from the nucleus. When this happens, the charge on the nucleus increases by one. Electron (β^-) emitters are found throughout the periodic table, from the lightest elements (^3H) to the heaviest (^{255}Es). The product of β^- emission can be predicted because both mass number and charge are conserved in nuclear reactions. If ^{40}K is a β^- emitter, for example, the product of this reaction must be ^{40}Ca.

$$^{40}_{19}K \longrightarrow {}^{40}_{20}Ca + {}^{\ 0}_{-1}e \qquad [\text{electron } (\beta^-) \text{ emission}]$$

Once again the sum of the mass numbers of the products is equal to the mass number of the parent nuclide, and the sum of the charge on the products is equal to the charge on the parent nuclide.

Nuclei can also decay by capturing one of the electrons that surround the nucleus. **Electron capture** leads to a decrease of one in the charge on the nucleus. The energy given off in this reaction is carried by an X-ray photon, which is represented by the symbol $h\nu$, where h is Planck's constant and ν is the frequency of the X ray. The product of this reaction can be predicted, once again, because mass and charge are conserved.

$$^{40}_{19}K + {}^{\ 0}_{-1}e \longrightarrow {}^{40}_{18}Ar + h\nu \qquad (\text{electron capture})$$

The electron captured by the nucleus in this reaction is usually a $1s$ electron because electrons in this orbital are the closest to the nucleus.

A third form of beta decay is called **positron (β^+) emission.** The positron is the antimatter equivalent of an electron. It has the same mass as an electron but the opposite charge. Positron (β^+) decay produces a daughter nuclide with one fewer positive charge on the nucleus than on the parent.

$$^{40}_{19}\text{K} \longrightarrow {}^{40}_{18}\text{Ar} + {}^{0}_{+1}e \qquad [\text{positron } (\beta^+) \text{ emission}]$$

Positrons have a very short lifetime. They rapidly lose their kinetic energy as they pass through matter. As soon as they come to rest, they combine with an electron to form two γ-ray photons in a matter–antimatter annihilation reaction.

$$^{0}_{+1}e + {}^{0}_{-1}e \longrightarrow 2\,\gamma$$

Thus, although it is theoretically possible to observe a fourth mode of beta decay corresponding to the capture of a positron, this reaction does not occur in nature.

The three forms of β decay for the ^{40}K nuclide are summarized in Figure 15.3. Note that there is no change in the mass number of the parent and daughter nuclides for electron emission, electron capture, and position emission. All three forms of β decay therefore interconvert isobars.

$$^{40}_{19}\text{K} \longrightarrow {}^{40}_{20}\text{Ca} + {}^{0}_{-1}e$$

$$^{40}_{19}\text{K} + {}^{0}_{-1}e \longrightarrow {}^{40}_{18}\text{Ar} + h\nu$$

$$^{40}_{19}\text{K} \longrightarrow {}^{40}_{18}\text{Ar} + {}^{0}_{+1}e$$

Fig. 15.3 The three forms of β decay for ^{40}K.

> ➤ **CHECKPOINT**
>
> Explain why electron capture, electron emission, and positron emission all interconvert nuclides that have the same mass but different numbers of protons.

GAMMA EMISSION

The daughter nuclides produced by α decay or β decay are often obtained in an excited state. The excess energy associated with this excited state is released when the nucleus emits a photon in the γ-ray portion of the electromagnetic spectrum. Most of the time, the γ ray is emitted within 10^{-12} seconds after the α or β particle. In some cases, **gamma emission** is delayed, and a short-lived, or **metastable,** nuclide is formed, which is identified by a lowercase letter m written after the mass number. Perhaps the most important metastable nuclide for students from the life sciences would be ^{99m}Tc. It is produced by the electron emission from the ^{99}Mo parent nuclide.

$$^{99}_{42}\text{Mo} \longrightarrow {}^{99m}_{43}\text{Tc} + {}^{0}_{-1}e$$

The metastable ^{99m}Tc nuclide has a half-life for γ-ray emission of 6.04 hours. This means that about 94% of this metastable isotope undergoes decay within 24 hours. Since electromagnetic radiation carries neither charge nor mass, the product of γ-ray emission by ^{99m}Tc is ^{99}Tc.

$$^{99m}_{43}\text{Tc} \longrightarrow {}^{99}_{43}\text{Tc} + \gamma \qquad (\gamma\text{-ray emission})$$

The metastable ^{99m}Tc isotope is commonly used as a radioactive tracer in diagnostic tests because of its short half-life, which keeps the patient's exposure to radiation relatively low. The primary use of ^{99m}Tc is for diagnostic medical imaging of blood flow during cardiac stress tests used to detect myocardial infarctions. Only a handful of reactors in the world can produce the ^{99}Mo parent nuclide, however, because it is made from heavily enriched uranium, which is strictly regulated to avoid the proliferation of nuclear weapons. The demand for ^{99}Mo, at times, therefore exceeds the available supply.

SPONTANEOUS FISSION

Nuclides with atomic numbers of 90 or more undergo a form of radioactive decay known as **spontaneous fission** in which the parent nucleus splits into a pair of smaller nuclei. The reaction is usually accompanied by the ejection of one or more neutrons, and one of the daughter nuclides is usually significantly larger than the other.

$$\ce{^{252}_{98}Cf} \longrightarrow \ce{^{140}_{54}Xe} + \ce{^{108}_{44}Ru} + 4\,\ce{^{1}_{0}\mathit{n}}$$

For all but the very heaviest isotopes, spontaneous fission is a very slow reaction. Spontaneous fission of ^{238}U, for example, is almost 2 million times slower than the rate at which this nuclide undergoes α decay.

Exercise 15.3

Predict the products of the following nuclear reactions.

(a) Electron emission by ^{14}C

(b) Positron emission by ^{8}B

(c) Electron capture by ^{125}I

(d) Alpha emission by ^{210}Rn

(e) Gamma-ray emission by ^{56m}Ni

Solution

We can predict the product of each reaction by writing an equation in which both mass number and charge are conserved.

(a) $\ce{^{14}_{6}C} \longrightarrow \ce{^{14}_{7}N} + \ce{^{0}_{-1}\mathit{e}}$

(b) $\ce{^{8}_{5}B} \longrightarrow \ce{^{8}_{4}Be} + \ce{^{0}_{+1}\mathit{e}}$

(c) $\ce{^{125}_{53}I} + \ce{^{0}_{-1}\mathit{e}} \longrightarrow \ce{^{125}_{52}Te} + h\nu$

(d) $\ce{^{210}_{86}Rn} \longrightarrow \ce{^{206}_{84}Po} + \ce{^{4}_{2}He}$

(e) $\ce{^{56m}_{28}Ni} \longrightarrow \ce{^{56}_{28}Ni} + \gamma$

15.4 Neutron-Rich versus Neutron-Poor Nuclides

In 1934 Enrico Fermi proposed a theory that explained the three forms of beta decay observed in nature. He argued that a neutron could decay to form a proton by emitting an electron.

$$\ce{^{1}_{0}\mathit{n}} \longrightarrow \ce{^{1}_{1}\mathit{p}} + \ce{^{0}_{-1}\mathit{e}} \qquad \text{[electron (β^{-}) emission]}$$

A proton, on the other hand, could be transformed into a neutron by two pathways. First, it can capture an electron.

$$\ce{^{1}_{1}\mathit{p}} + \ce{^{0}_{-1}\mathit{e}} \longrightarrow \ce{^{1}_{0}\mathit{n}} \qquad \text{(electron capture)}$$

Or, Second, it can emit a positron.

$$\ce{^{1}_{1}\mathit{p}} \longrightarrow \ce{^{1}_{0}\mathit{n}} + \ce{^{0}_{+1}\mathit{e}} \qquad \text{[positron (β^{+}) emission]}$$

Electron emission therefore leads to an *increase* in the atomic number of the nucleus.

$$\ce{^{14}_{6}C} \longrightarrow \ce{^{14}_{7}N} + \ce{^{0}_{-1}\mathit{e}}$$

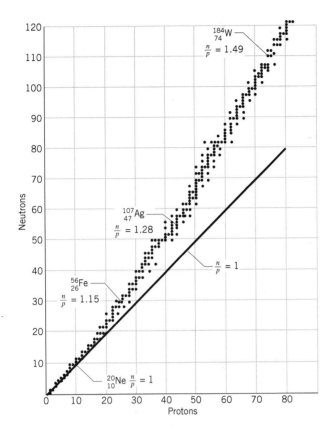

Fig. 15.4 A graph of the number of neutrons versus the number of protons for all stable naturally occurring nuclei. Nuclei that lie to the right of this band of stability are *neutron-poor;* nuclei to the left of the band are *neutron-rich.* The solid line represents a neutron-to-proton ratio of 1:1.

Electron capture and positron emission, on the other hand, both result in a *decrease* in the atomic number of the nucleus.

$$^{7}_{4}\text{Be} + {}^{0}_{-1}e \longrightarrow {}^{7}_{3}\text{Li} + h\nu$$
$$^{11}_{6}\text{C} \longrightarrow {}^{11}_{5}\text{B} + {}^{0}_{+1}e$$

A plot of the number of neutrons versus the number of protons for all of the stable naturally occurring isotopes is shown in Figure 15.4. Several conclusions can be drawn from this plot.

- The stable nuclides lie in a very narrow band of neutron-to-proton ratios.
- The ratio of neutrons to protons in stable nuclides gradually increases as the number of protons in the nucleus increases.
- Light nuclides, such as ^{12}C, contain about the same number of neutrons and protons. Heavy nuclides, such as ^{238}U, contain up to 1.6 times as many neutrons as protons.
- There are no stable nuclides with atomic numbers larger than 83.
- Nuclei that lie above the narrow band of stable nuclides have too many neutrons and are therefore **neutron-rich.**
- Nuclei that lie below the narrow band of stable nuclides don't have enough neutrons and are therefore **neutron-poor.**

➤ **CHECKPOINT**

Use the fact that the average mass of a sulfur atom is 32.06 amu to explain why the ^{35}S isotope can be described as *neutron-rich.*

The most likely mode of decay for a neutron-rich nucleus is one that converts a neutron into a proton. *Neutron-rich* radioactive isotopes with an atomic number smaller than 83 all decay by electron (β^-) emission. ^{14}C, ^{32}P, and ^{35}S, for example, are all neutron-rich nuclei that decay by the emission of an electron.

$$^{35}_{16}\text{S} \longrightarrow {}^{35}_{17}\text{Cl} + {}^{0}_{-1}e \qquad (\beta^- \text{ emission})$$

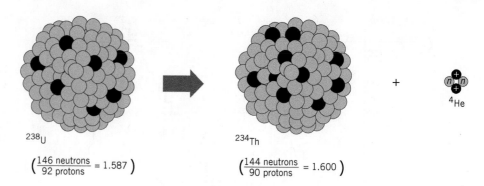

Fig. 15.5 Because the loss of an α particle leads to a small increase in the neutron-to-proton ratio, α-particle decay occurs in relatively heavy, neutron-poor nuclides such as ^{238}U.

Neutron-poor nuclides decay by modes that convert a proton into a neutron. Neutron-poor nuclides with atomic numbers less than 83 tend to decay by either electron capture or positron emission. Many of these nuclides decay by both routes, but positron emission is more often observed in the lighter nuclides, such as ^{22}Na.

$$^{22}_{11}\text{Na} \longrightarrow \, ^{22}_{10}\text{Ne} + \, ^{0}_{+1}e \qquad (\beta^+ \text{ emission})$$

Electron capture is more common among heavier nuclides, such as ^{125}I, because the 1s electrons are held closer to the nucleus of an atom as the charge on the nucleus increases.

$$^{125}_{53}\text{I} + \, ^{0}_{-1}e \longrightarrow \, ^{125}_{52}\text{Te} + h\nu \qquad (\text{electron capture})$$

A third mode of decay is observed in *neutron-poor* nuclides that have atomic numbers larger than 83. Although it is not obvious at first, α decay increases the ratio of neutrons to protons. Consider what happens during the α decay of ^{238}U, for example.

$$^{238}_{92}\text{U} \longrightarrow \, ^{234}_{90}\text{Th} + \, ^{4}_{2}\text{He} \qquad (\alpha \text{ decay})$$

The parent nuclide (^{238}U) in this reaction has 92 protons and 146 neutrons, which means that the neutron-to-proton ratio is 1.587. The daughter nuclide (^{234}Th) has 90 protons and 144 neutrons, so its neutron-to-proton ratio is 1.600. The daughter nuclide is therefore slightly less likely to be neutron-poor, as shown in Figure 15.5.

Exercise 15.4

Predict the most likely modes of decay and the products of decay of the following nuclides.

(a) ^{17}F   (b) ^{105}Ag   (c) ^{185}Ta

Solution

The first step in predicting the mode of decay of a nuclide is to decide whether the nuclide is neutron-rich or neutron-poor. This can be done by comparing the mass number of the nuclide with the atomic weight of the element.

(a) The atomic weight of fluorine is 18.998 amu. Because the mass of an ^{17}F nuclide is *smaller* than that of the average fluorine atom, the ^{17}F nuclide must contain fewer neutrons. It is therefore likely to be *neutron-poor*.

Because it is a relatively light nuclide, ^{17}F might be expected to decay by positron emission.

$$^{17}_{9}F \longrightarrow \, ^{17}_{8}O + \, ^{0}_{+1}e \qquad (\beta^+ \text{ emission})$$

(b) The atomic weight of silver is 107.868 amu. Because the mass of the ^{105}Ag nuclide is *smaller* than that of the average silver atom, it contains fewer neutrons than the stable isotopes of silver. ^{105}Ag is therefore likely to behave as expected for a *neutron-poor* nuclide. Since ^{105}Ag is a relatively heavy nuclide, we might expect it to decay by electron capture.

$$^{105}_{47}Ag + \, ^{0}_{-1}e \longrightarrow \, ^{105}_{46}Pd + h\nu \qquad (\text{electron capture})$$

(c) The atomic weight of tantalum is 180.948 amu. The mass of the ^{185}Ta isotope is therefore significantly *larger* than that of the average tantalum atom, which means it contains an unusually large number of neutrons. Because this isotope is *neutron-rich*, it decays by electron emission.

$$^{185}_{73}Ta \longrightarrow \, ^{185}_{74}W + \, ^{0}_{-1}e \qquad (\beta^- \text{ emission})$$

● ●

15.5 Binding Energy Calculations

We should be able to predict the mass of an atom from the masses of the subatomic particles it contains. A helium atom, for example, contains two protons, two neutrons, and two electrons. We therefore predict that the mass of a helium atom should be 4.0329802 amu.

$$2(1.0072765) \text{ amu} = 2.0145530 \text{ amu}$$
$$2(1.0086650) \text{ amu} = 2.0173300 \text{ amu}$$
$$\underline{2(0.00054858) \text{ amu} = 0.0010972 \text{ amu}}$$
$$\text{total mass} = 4.0329802 \text{ amu}$$

When the mass of a helium atom is measured, we find that the experimental value is smaller than the predicted mass by 0.0303769 amu.

$$\text{predicted mass} = 4.0329802 \text{ amu}$$
$$\underline{-\text{observed mass} = 4.0026033 \text{ amu}}$$
$$\text{mass defect} = 0.0303769 \text{ amu}$$

The difference between the mass of an atom and the sum of the masses of its protons, neutrons, and electrons is called the **mass defect.** The mass defect of an atom is equal to the energy released when the nucleus is formed from its protons and neutrons. The mass defect is therefore also known as the **binding energy** of the nucleus.

The binding energy serves the same function for nuclear reactions as $\Delta H°$ for a chemical reaction. It measures the difference between the stability of the products of the reaction and the starting materials. As the binding energy of a series of nuclides increases, the nuclei become more stable. The binding energy can also be viewed as the amount of energy it would take to rip the nucleus apart to form isolated neutrons and protons. It is therefore literally the energy that binds together the neutrons and protons in the nucleus.

The binding energy of a nuclide can be calculated from its mass defect with Einstein's equation that relates mass and energy.

$$E = mc^2$$

To obtain the binding energy in units of joules, we must convert the mass defect from atomic mass units to kilograms.

$$0.0303769 \text{ amu} \times \frac{1.6605655 \times 10^{-24} \text{g}}{1 \text{ amu}} \times \frac{1 \text{ kg}}{1000 \text{ g}} = 5.04428 \times 10^{-29} \text{kg}$$

Multiplying the mass defect in kilograms by the square of the speed of light in units of meters per second gives a binding energy for a single helium atom of 4.53357×10^{-12} joules.

$$E = (5.04428 \times 10^{-29} \text{kg})(2.9979246 \times 10^8 \text{m/s})^2$$
$$= 4.53357 \times 10^{-12} \text{J}$$

Multiplying the result of this calculation by the number of atoms in a mole gives a binding energy for helium of 2.730×10^{12} joules per mole, or 2.730 billion kilojoules per mole.

$$\frac{4.53357 \times 10^{-12} \text{J}}{1 \text{ atom}} \times \frac{6.022 \times 10^{23} \text{atoms}}{1 \text{ mol}} = 2.730 \times 10^{12} \text{J/mol}$$

This calculation helps us understand the fascination of nuclear reactions. The energy released when natural gas is burned is about 800 kJ/mol. The synthesis of a mole of helium releases 3.4 million times as much energy.

Since most nuclear reactions are carried out on very small samples of material, the mole is not a reasonable basis of measurement. Binding energies are usually expressed in units of electron volts (eV) or million electron volts (MeV) *per atom*. The binding energy of helium is 28.3×10^6 eV/atom, or 28.3 MeV/atom.

$$\frac{4.53357 \times 10^{-12} \text{J}}{1 \text{ atom}} \times \frac{1 \text{ eV}}{1.602 \times 10^{-19} \text{J}} = 28.30 \times 10^6 \text{eV/atom}$$

Calculations of the binding energy can be simplified by using the following conversion factor between the mass defect in atomic mass units and the binding energy in million electron volts.

$$1 \text{ amu} = 931.5016 \text{ MeV}$$

Exercise 15.5

Calculate the binding energy of ^{235}U if the mass of this nuclide is 235.0439 amu.

Solution

A neutral ^{235}U atom contains 92 protons, 92 electrons, and 143 neutrons. The predicted mass of a ^{235}U atom is therefore 236.960 amu, to four decimal places.

$$92(1.00728) \text{ amu} = 92.6698 \text{ amu}$$
$$92(0.0005486) \text{ amu} = 0.0505 \text{ amu}$$
$$\underline{143(1.00867) \text{ amu} = 144.240 \text{ amu}}$$
$$\text{total mass} = 236.960 \text{ amu}$$

To calculate the mass defect for this nucleus, we subtract the observed mass from the predicted mass.

$$\begin{array}{r} \text{predicted mass} = 236.960\,\text{amu} \\ -\text{observed mass} = 235.0439\,\text{amu} \\ \hline \text{mass defect} = 1.916\,\text{amu} \end{array}$$

Using the conversion factor that relates the binding energy to the mass defect, we obtain a binding energy for ^{235}U of 1785 MeV per atom.

$$\frac{1.916\,\text{amu}}{1\,\text{atom}} \times \frac{931.50\,\text{MeV}}{1\,\text{amu}} = 1785\,\text{MeV/atom}$$

•••

Binding energies gradually increase with atomic number, although they tend to level off near the end of the periodic table. A more useful quantity is obtained by dividing the binding energy for a nuclide by the total number of protons and neutrons it contains. This quantity is known as the **binding energy per nucleon.**

The binding energy per nucleon ranges from about 7.5 to 8.8 MeV for most nuclei, as shown in Figure 15.6. It reaches a maximum, however, at an atomic mass of about 60 amu. The largest binding energy per nucleon is observed for ^{56}Fe, which is the most stable nuclide in the periodic table.

The graph of binding energy per nucleon versus atomic mass explains why energy is released when relatively small nuclei combine to form larger nuclei in **fusion reactions.**

$$^{12}_{6}C + {}^{12}_{6}C \longrightarrow {}^{24}_{12}Mg \quad \text{(fusion)}$$

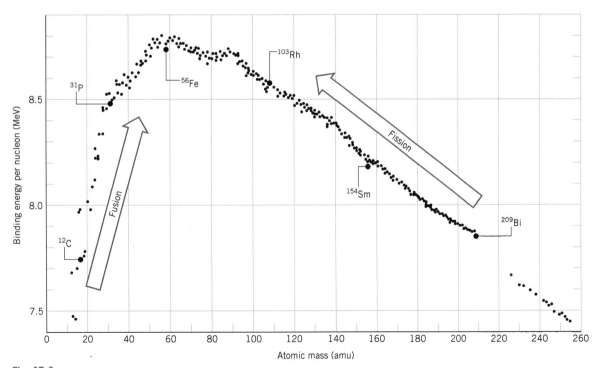

Fig. 15.6 The binding energy of the nucleus increases with atomic number. The binding energy *per nucleon,* however, reaches a maximum at ^{56}Fe. Nuclei lighter than ^{56}Fe can become more stable by fusing together; nuclei significantly heavier than ^{56}Fe can become more stable by splitting apart.

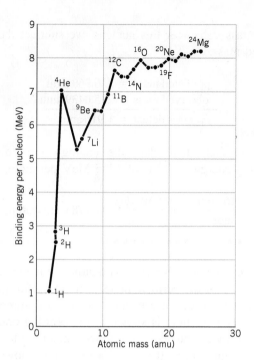

Fig. 15.7 The binding energy per nucleon for most stable nuclei is between 7.8 and 8.8 MeV per nucleon. There is more variability in the binding energy per nucleon for relatively light nuclei. The binding energy per nucleon for ^{4}He is particularly large, which explains why so many of the heavier nuclides undergo α decay.

It also explains why energy is released when relatively heavy nuclei split apart in **fission** (literally, "to split or cleave") **reactions.**

$$^{235}_{92}\text{U} \longrightarrow\ ^{139}_{56}\text{Ba} +\ ^{94}_{36}\text{Kr} + 2\,^{1}_{0}n \qquad \text{(fission)}$$

There are a number of small irregularities in the binding energy curve at the low end of the mass spectrum, as shown in Figure 15.7. The ^{4}He nucleus, for example, is much more stable than its nearest neighbors. The unusual stability of the ^{4}He nucleus explains why α-particle decay is usually much faster than the spontaneous fission of a nuclide into two large fragments.

15.6 The Kinetics of Radioactive Decay

Radioactive nuclei decay by first-order kinetics. The rate of radioactive decay is therefore the product of a rate constant (k) times the number of atoms of the isotope in the sample (N).

$$\text{rate} = -\frac{d(N)}{dt} = k(N)$$

The rate of radioactive decay doesn't depend on the chemical state of the isotope. The rate of decay of ^{238}U, for example, is the same in uranium metal and uranium hexafluoride or any other compound of this element.

The rate at which a radioactive isotope decays is called the **activity** of the isotope. The most common unit of activity is the **curie** (Ci), which was originally defined as the number of disintegrations per second in 1 gram of ^{226}Ra. The curie is now defined as the amount of radioactive isotope necessary to achieve an activity of 3.700×10^{10} disintegrations per second.

Exercise 15.6

Assume that during a cardiac stress test a total of 31.6 millicuries of ^{99m}Tc were injected into a patient to study blood flow through the heart in order to locate possible sites of cardiac infarction (or heart attack). The half-life for the decay of the metastable ^{99m}Tc nuclide to ^{99}Tc by the emission of a gamma ray is 6.04 hours.

$$^{99m}_{43}\text{Tc} \longrightarrow {}^{99}_{43}\text{Tc} + \gamma \quad (\gamma\text{-ray emission})$$

Calculate the number of ^{99m}Tc atoms injected into the patient.

Solution

The rate at which the ^{99m}Tc isotope decays depends on the rate constant for this reaction (k) and the number of technetium atoms in the sample (N).

$$\text{rate} = -\frac{d(N)}{dt} = k(N)$$

The rate constant for the decay of the metastable ^{99m}Tc to ^{99}Tc can be calculated from the half-life of the ^{99m}Tc nuclide as follows.

$$k = \frac{\ln 2}{t_{1/2}} = \frac{0.6931}{t_{1/2}} = \frac{0.6931}{6.04 \text{ hr}} = 0.115 \text{ hr}^{-1}$$

Or

$$k = \frac{0.115}{1 \text{ hr}} \times \frac{1 \text{ hr}}{3600 \text{ s}} = 3.19 \times 10^{-5} \text{ s}^{-1}$$

We can calculate the rate of radioactive decay of the ^{99m}Tc injected into the patient from the activity of the sample and the definition of the curie as a unit of measurement.

$$\text{rate} = 31.6 \times 10^{-3} \text{ Ci} \times \frac{3.700 \times 10^{10} \text{ dis/s}}{1 \text{ Ci}} = 1.17 \times 10^9 \text{ dis/s}$$

We can now calculate the number of ^{99m}Tc atoms in a sample that exhibits this level of activity.

$$\text{rate} = k(N)$$
$$1.17 \times 10^9 \text{ dis/s} = (3.19 \times 10^{-5} \text{ s}^{-1})(N)$$
$$N = 3.67 \times 10^{13}$$

The curie is such a large unit of measurement that even 31.6 millicuries of a radioactive nuclide with a short half-life contain a relatively small number of radioactive nuclei. In this case, the number of nuclei corresponds to about 6×10^{-11} moles.

The relative rates at which radioactive nuclei decay can be described in terms of either the rate constants for the decay or the half-lives of the nuclei. We can conclude that ^{14}C decays more rapidly than ^{238}U, for example, by noting that the rate constant for the decay of ^{14}C is much *larger* than that for ^{238}U.

$$^{14}C: \quad k = 1.210 \times 10^{-4}\,\text{yr}^{-1}$$
$$^{238}U: \quad k = 1.54 \times 10^{-10}\,\text{yr}^{-1}$$

We can reach the same conclusion by noting that the half-life for the decay of ^{14}C is much *shorter* than that for ^{235}U.

$$^{14}C: \quad t_{1/2} = 5730\,\text{yr}$$
$$^{238}U: \quad t_{1/2} = 4.50 \times 10^{9}\,\text{yr}$$

The **half-life** for the decay of a radioactive nuclide is the length of time it takes for exactly half of the nuclei in the sample to decay. In Section 14.12, we noted that the half-life of a first-order process is inversely proportional to the rate constant for this process.

$$t_{1/2} = \frac{\ln 2}{k} = \frac{0.6931}{k}$$

The half-life of a nuclide can be used to estimate the amount of a radioactive isotope left after a given number of half-lives.

Exercise 15.7

Calculate the fraction of the ^{14}C that remains in a sample after eight half-lives.

Solution

Half of the ^{14}C present initially decays during the first half-life, half of what is left decays during the second half-life, and so on. The ^{14}C left after eight half-lives is equal to one-half raised to the eighth power.

$$(1/2)^8 = 0.00391$$

Less than 0.4% of the original ^{14}C is left after eight half-lives.

For more complex calculations, it is easier to convert the half-life of the nuclide into a rate constant and then use the integrated form of the first-order rate law described in Section 14.12.

Exercise 15.8

The half-life for ^{222}Rn is 3.823 days. How long would it take for a sample of ^{222}Rn that weighs 0.750 gram to decay to 0.100 gram?

Solution

We can start by calculating the rate constant for this decay from the half-life.

$$k = \frac{\ln 2}{t_{1/2}} = \frac{0.6931}{3.823 \, d} = 0.1813 \, d^{-1}$$

We then turn to the integrated form of the first-order rate law.

$$\ln\left[\frac{(N)}{(N)_0}\right] = -kt$$

The ratio of the number of atoms that remain in the sample to the number of atoms present initially is the same as the ratio of grams at the end of the time period (0.100 g) to the number of grams present initially (0.750 g).

$$\ln\left[\frac{(0.100)}{(0.750)}\right] = -(0.1813 \, d^{-1})t$$

Solving for t, we find that it takes 11.1 days for 0.750 gram of ^{222}Rn to decay to 0.100 gram of this nuclide.

● ●

15.7 Dating by Radioactive Decay

The Earth is constantly bombarded by various forms of electromagnetic radiation emitted by the Sun. Because cosmic rays are rare, the total energy received in the form of cosmic rays is small—no more than the energy received by the planet from starlight. But the energy of a single cosmic ray is very large, on the order of several billion electron volts ($\approx$200 million kJ/mol). These highly energetic rays react with atoms in the atmosphere to produce neutrons that then react with nitrogen atoms in the atmosphere to produce ^{14}C.

$$^{14}_{7}N + ^{1}_{0}n \longrightarrow ^{14}_{6}C + ^{1}_{1}H$$

The ^{14}C formed in this reaction is a neutron-rich nuclide that decays by electron emission with a half-life of 5730 years.

$$^{14}_{6}C \longrightarrow ^{14}_{7}N + ^{0}_{-1}e$$

Just after World War II, Willard F. Libby proposed a way to use these reactions to estimate the age of carbon-containing substances. The ^{14}C **dating** technique for which Libby received the Nobel Prize was based on the following assumptions.

● ^{14}C is produced in the atmosphere at a more or less constant rate.

● Carbon atoms circulate among the atmosphere, the oceans, and living organisms at a rate very much faster than they decay. As a result, there is a constant concentration of ^{14}C in all living things.

● After death, organisms no longer pick up ^{14}C. Thus, by comparing the activity of a sample with the activity of living tissue, we can estimate how long it has been since the organism died.

This collection of fossil skulls and fragments of jaw bones from the oldest known group of hominids, the Australopithecines, is too old to be dated by ^{14}C. Potassium-argon dating, however, suggests dates from 1.6 to 3.7 million years ago.

The natural abundance of ^{14}C is about 1 part in 10^{12}, and the average activity of living tissue is 15.3 disintegrations per minute per gram of carbon. Samples used for ^{14}C dating can include charcoal, wood, cloth, paper, seashells, limestone, flesh, hair, soil, peat, and bone. Since most iron samples also contain carbon, it is possible to estimate the time since iron was last fired by analyzing for ^{14}C.

Exercise 15.9

The skin, bone, and clothing of "Whiskey Lil," an adult female mummy discovered in Chimney Cave, Lake Winnemucca, Nevada, were dated by radiocarbon analysis. How old is this mummy if the sample exhibited 73.9% of the activity of living tissue?

Solution

Because ^{14}C decays by first-order kinetics, the natural log of the ratio of the ^{14}C in the sample today (N) to the amount that would be present if it were still alive (N_0) is proportional to the rate constant for this decay and the time since death.

$$\ln\left[\frac{(N)}{(N)_0}\right] = -kt$$

The rate constant for this reaction can be calculated from the half-life of ^{14}C, which is 5730 years.

$$k = \frac{\ln 2}{t_{1/2}} = \frac{0.6931}{5730 \text{ yr}} = 1.210 \times 10^{-4} \text{ yr}^{-1}$$

If the sample retained 73.9% of the activity of living tissue, the ratio of the activity today (N) to the original activity (N_0) is 0.739. Substituting what we know into the integrated form of the first-order rate law gives the following result.

$$\ln(0.739) = -(1.210 \times 10^{-4} \text{ yr}^{-1})t$$

Solving this equation for the unknown gives an estimate of the time since death.

$$t = 2.50 \times 10^3 \text{ y}$$

We now know that one of Libby's assumptions is questionable: The amount of ^{14}C in the atmosphere hasn't been constant with time. Because of changes in solar activity and the Earth's magnetic field, it has varied by as much as ±5%. More recently, contamination from the burning of fossil fuels and the testing of nuclear weapons has caused significant changes in the amount of radioactive carbon in the atmosphere. Radiocarbon dates are therefore reported in years before the present era (B.P.). By convention, the present era is assumed to begin in 1950, when ^{14}C dating was introduced.

Studies of bristlecone pines allow us to correct for changes in the abundance of ^{14}C with time. These remarkable trees, which grow in the White Mountains of California, can live up to 5000 years. By studying the ^{14}C activity of samples taken from the annual growth rings in these trees, researchers have developed a calibration curve for ^{14}C dates from the present back to 5145 B.C.

After roughly 46,000 years (eight half-lives), a sample retains only 0.4% of the ^{14}C activity of living tissue. At that point it becomes too old to date by radiocarbon techniques. Other radioactive isotopes, however, can be used to date rocks, soils, or archaeological objects that are much older.

Potassium–argon dating, for example, has been used to date samples up to 4.3 billion years old. Naturally occurring potassium contains 0.0118% by weight of the radioactive ^{40}K isotope. This isotope decays to ^{40}Ar with a half-life of 1.3 billion years. The ^{40}Ar produced after a rock crystallizes is trapped in the crystal lattice. It can be released, however, when the rock is melted at temperatures up to 2000°C. By measuring the amount of ^{40}Ar released when the rock is melted and comparing it with the amount of potassium in the sample, the time since the rock crystallized can be determined.

15.8 Ionizing versus Nonionizing Radiation

We live in a sea of radiation. We are exposed to infrared, ultraviolet, visible, and cosmic rays from the Sun. We are subjected to radio waves from local radio and television transmitters, microwaves from microwave ovens or cellular telephones, and X rays produced by the cathode-ray tubes in our television sets. We are also exposed to both natural sources of radioactivity, including α particles from trace contaminants in brick and clay, β^- emitters in the food chain, and γ rays from soils and rocks.

In recent years, many people have learned to fear the effects of this radiation. They don't want to live near nuclear reactors. They are frightened by reports of links between excess exposure to sunlight and skin cancer. They are afraid of the leakage from microwave ovens or the radiation produced by their television sets.

Several factors combine to heighten the public's anxiety about both the short-range and long-range effects of radiation. Perhaps the most important source of fear is the fact that radiation can't be detected by the average person. Furthermore, the effects of exposure to radiation might not appear for months or years or even decades.

To understand the biological effects of radiation, we must first understand the difference between **ionizing radiation** and **nonionizing radiation.** In general, two things can happen when radiation is absorbed by matter: excitation or ionization. *Excitation* occurs when the radiation excites the translational or vibrational motion of the atoms or molecules, or excites an electron from an occupied orbital into an empty, higher-energy orbital. *Ionization* occurs when the radiation carries enough energy to remove an electron from an atom or molecule.

Because living tissue is 70–90% water by weight, the dividing line between radiation that excites electrons and radiation that forms ions is often assumed to be equal to the energy required to remove an electron from a neutral water molecule: 1216 kJ/mol. Radiation that has less energy can only excite the water molecule. It is therefore called *nonionizing radiation.* Radiation that carries more energy than 1216 kJ/mol can remove an electron from a water molecule and is therefore called *ionizing radiation.*

Table 15.1 contains estimates of the energies of various kinds of radiation. Radio waves, microwaves, infrared radiation, and visible light are all forms of nonionizing radiation. X rays, γ rays, and α and β particles are forms of ionizing radiation. The dividing line between ionizing and nonionizing radiation in the electromagnetic spectrum falls in the ultraviolet portion of the spectrum. It is

Table 15.1
Energies of Ionizing and Nonionizing Forms of Radiation

Radiation	Typical Frequency (s^{-1})	Typical Energy (kJ/mol)	
Particles			
α particles		4.1×10^8	
β particles		1.5×10^7	
Electromagnetic radiation			
Cosmic rays	6×10^{21}	2.4×10^9	Ionizing
γ rays	3×10^{20}	1.2×10^8	Radiation
X rays	3×10^{17}	1.2×10^5	
----Ultraviolet------------------------	--3×10^{15}--	---------1200---------	------
Visible	5×10^{14}	200	
Infrared	3×10^{13}	12	Nonionizing
Microwaves	3×10^9	1.2×10^{-3}	radiation
Radio waves	3×10^7	1.2×10^{-5}	

therefore useful to divide the UV spectrum into two categories: UV_A and UV_B. Radiation at the high-energy end of the UV spectrum can be as dangerous as X rays or γ rays.

When ionizing radiation passes through living tissue, electrons are removed from neutral water molecules to produce H_2O^+ ions. Between three and four water molecules are ionized for every 1.6×10^{-17} joules of energy absorbed in the form of ionizing radiation.

$$H_2O \longrightarrow H_2O^+ + e^-$$

The H_2O^+ ion should not be confused with the H_3O^+ ion produced when acids dissolve in water. The H_2O^+ ion is an example of a *free radical,* which contains an unpaired valence-shell electron. Free radicals are extremely reactive. The radicals formed when ionizing radiation passes through water are among the strongest oxidizing agents that can exist in aqueous solution. At the molecular level, these oxidizing agents destroy biologically active molecules by removing either electrons or hydrogen atoms. This often leads to damage to the membrane, nucleus, chromosomes, or mitochondria of the cell that either inhibits cell division, results in cell death, or produces a malignant cell.

15.9 Biological Effects of Ionizing Radiation

From the time that radioactivity was discovered, it was obvious that it caused damage. Glass containers used to store radium compounds, for example, turned a rich purple and eventually cracked because of radiation damage. As early as 1901, Pierre Curie discovered that a sample of radium placed on his skin produced wounds that were very slow to heal. What some find surprising is the magnitude of the difference between the biological effects of nonionizing radiation, such as light and microwaves, and ionizing radiation, such as high-energy ultraviolet radiation, X rays, γ rays, and α or β particles.

Radiation at the low-energy end of the electromagnetic spectrum, such as radio waves and microwaves, excites the movement of atoms and molecules, which is equivalent to heating the sample. Radiation in or near the visible portion of the spectrum excites electrons into higher-energy orbitals. When the electron eventually falls back to a lower-energy state, the excess energy is given off to neighboring molecules in the form of heat. The principal effect of nonionizing radiation is therefore an increase in the temperature of the system.

We experience the fact that biological systems are sensitive to heat each time we cook with a microwave oven or spend too much time in the Sun. But it takes a great deal of nonionizing radiation to reach dangerous levels. We can assume, for example, that absorption of enough radiation to produce an increase of about 6°C in body temperature would be fatal. Since the average 70-kilogram human is 80% water by weight, we can use the heat capacity of water to calculate that it would take about 1.5 million joules of nonionizing radiation to kill the average human. If this energy were carried by visible light with a frequency of 5×10^{14} s^{-1}, it would correspond to absorption of about seven moles of photons.

Ionizing radiation is much more dangerous. A dose of only 300 joules of X-ray or γ-ray radiation is fatal for the average human, even though this radiation raises the temperature of the body by only 0.001°C. α-particle radiation is even more dangerous; a dose equivalent to only 15 joules is fatal for the average human. Whereas it takes seven moles of photons of visible light to produce a fatal dose of nonionizing radiation, absorption of only 7×10^{-10} moles of the α particles emitted by ^{238}U is fatal.

Ionizing radiation can be measured in three ways:

1. Measure the *activity* of the source in units of disintegrations per second or curies, which is the easiest measurement to make.

2. Measure the radiation to which an object is *exposed* in units of roentgens by measuring the amount of ionization produced when this radiation passes through a sample of air.

3. Measure the radiation *absorbed* by the object in units of radiation absorbed doses or "rads." This is the most useful quantity, but it is the hardest to obtain.

One **radiation absorbed dose,** or **rad,** corresponds to the absorption of 10^{-5} joules of energy per gram of body weight. Because this is equivalent to 0.01 J/kg, one rad produces an increase in body temperature of about 2×10^{-6} °C. At first glance, the rad may seem to be a negligibly small unit of measurement. The destructive power of the radicals produced when water is ionized is so large, however, that cells are inactivated at a dose of 100 rads, and a dose of 400 to 450 rads is fatal for the average human.

Not all forms of radiation have the same efficiency for damaging biological organisms. The faster energy is lost as the radiation passes through the tissue, the more damage it does. To correct for the differences in **radiation biological effectiveness (RBE)** among various forms of radiation, a second unit of absorbed dose has been defined. The **roentgen equivalent man,** or **rem,** is the absorbed dose in rads times the biological effectiveness of the radiation.

$$\text{rems} = \text{rads} \times \text{RBE}$$

Values for the RBE of different forms of radiation are given in Table 15.2.

Estimates of per capita exposure to radiation in the United States are summarized in Table 15.3. These estimates include both external and internal sources of natural background radiation.

Table 15.2

Radiation Biological Effectiveness of Various Forms of Radiation

Radiation	RBE
X rays and γ rays	1
β^- particles	1
Thermal (slow-moving) neutrons	3
Fast-moving neutrons or protons	10
α particles or heavy ions	20

Table 15.3

Average Whole-Body Exposure Levels for Sources of Ionizing Radiation

Source	Per Capita Dose (rems/yr)
Natural background	0.082
Medical X rays	0.077
Nuclear test fallout	0.005
Consumer and industrial products	0.005
Nuclear power industry	0.001
Total	0.170

External sources include cosmic rays from the Sun and α particles or γ rays emitted from rocks and soil. Internal sources include nuclides that enter the body when we breathe (^{14}C, ^{85}Kr, ^{220}Rn, and ^{222}Rn) or through the food chain (^3H, ^{14}C, ^{40}K, ^{90}Sr, ^{131}I, and ^{137}Cs). The actual dose from natural radiation depends on where one lives. People who live in the Rocky Mountains, for example, receive twice as much background radiation as the national average because there is less atmosphere to filter out the cosmic rays from the Sun.

The average dose from medical X rays has decreased in recent years because of advances in the sensitivity of the photographic film used for X rays. Radiation from nuclear test fallout has also decreased as a result of the atmospheric nuclear test ban. The threat of fallout from the testing of nuclear weapons can be appreciated by noting that a Chinese atmospheric test in 1976 led to the contamination of milk in the Harrisburg, Pennsylvania, vicinity at a level of 300 pCi (3.00×10^{-10} Ci) per liter. This was about eight times the level of contamination (41 pCi per liter) that resulted from the accident at Three Mile Island.

In the 1950s and 1960s, an experiment was conducted in St. Louis in which parents were asked to mail in their children's baby teeth when they fell out. Over 300,000 teeth were collected, and it was found that the level of ^{90}Sr in baby teeth increased by 1400% between 1954 and 1964 as a result of contamination of milk by ^{90}Sr in the fallout from atmospheric tests. It was data such as these that convinced the United States to sign the atmospheric test ban.

The contribution to the radiation absorbed dose from consumer and industrial products includes radiation from construction materials, X rays emitted by television sets, and inhaled tobacco smoke. The most recent estimate of the total exposure to radiation emitted from the mining and milling of uranium, the fabrication of reactor fuels, the storage of radioactive wastes, and the operation of nuclear reactors is less than 0.001 rem per year.

Table 15.3 gives an estimate of the total dose from ionizing radiation for the average American of 0.170 rem per year or 170 mrem per year. The Committee on the Biological Effects of Ionizing Radiation of the National Academy of Sciences has estimated that an increase in this dose to a level of 1 rem per year would result in 169 additional deaths from cancer per million people exposed. This would correspond to an increase of 0.1% in the rate of cancer deaths because 170,000 cancer deaths would normally occur in a population this size that was not exposed to this level of radiation.

The principal effect of low doses of ionizing radiation is to induce changes in biological tissue that can lead to cancers that can take up to 20 years to develop. What is the effect of high doses of ionizing radiation? Cells that are actively dividing are more sensitive to radiation than cells that are not. Thus cells in the liver, kidney, muscle, brain, and bone are more resistant to radiation than the cells of bone marrow, the reproductive organs, the epithelium of the intestine, and the skin, which suffer the most damage from radiation. Damage to the bone marrow is the main cause of death at moderately high levels of exposure (200 to 1000 rads). Damage to the gastrointestinal tract is the major cause of death for exposures on the order of 100 to 10,000 rads. Massive damage to the central nervous system is the cause of death from extremely high exposures (over 10,000 rads).

15.10 Natural versus Induced Radioactivity

NATURAL RADIOACTIVITY

The vast majority of the nuclides found in nature are stable. Inasmuch as our planet is 4.6 billion years old, the only radioactive isotopes that should remain are members of three classes.

- Isotopes with half-lives of at least 10^9 years, such as ^{238}U
- Daughter nuclides produced when long-lived radioactive nuclides decay, such as the ^{234}Th ($t_{1/2} = 24.1$ days) produced by the α decay of ^{238}U
- Nuclides such as ^{14}C that are still being synthesized

In Section 15.4, we encountered one factor that influences the stability of a nuclide: the ratio of neutrons to protons. (Nuclei that contain either too many or too few neutrons are unstable.) Another factor that affects the stability of nuclides can be understood by examining patterns in the number of protons and neutrons in stable nuclides.

Half of the elements in the periodic table must have an odd number of protons because atomic numbers that are odd are just as likely to occur as those that are even. In spite of this, about 80% of the stable nuclides have an even number of protons. Very few elements with an odd atomic number have more than one stable isotope. Stable isotopes abound, however, among elements with even atomic numbers. Ten stable isotopes are known for tin ($Z = 50$), for example. It is also interesting to note that 91% of the stable isotopes of elements that have an odd number of protons have an even number of neutrons. These observations suggest that certain combinations of protons and neutrons are particularly stable.

When the structure of the atom was discussed, we found that there are magic numbers of electrons. Electron configurations with 2, 10, 18, 36, 54, and 86 electrons are unusually stable. There also seem to be magic numbers of neutrons and protons. Nuclei with 2, 8, 20, 28, 50, 82, or 126 protons or neutrons are also unusually stable. This observation explains the anomalously large binding energies observed in Figure 15.7 for ^4He, ^{16}O, and ^{20}Ne. In each case, the nuclide has an even number of both protons and neutrons. ^{20}Ne has a magic number of nucleons when both protons and neutrons are counted. ^4He and ^{16}O have magic numbers of both protons and neutrons. The resulting stability of the ^4He nucleus might explain why so many heavy nuclei undergo α-particle decay by ejecting a $^4He^{2+}$ ion or an α particle from the nucleus of the atom.

If nuclei tend to be more stable when they have even numbers of protons and neutrons, it isn't surprising that nuclides with an odd number of both protons and neutrons are unstable. ^{40}K is one of only five naturally occurring nuclides that contain both an odd number of protons and an odd number of neutrons. This nuclide simultaneously undergoes the electron capture and positron emission expected for neutron-poor nuclides and the electron emission observed with neutron-rich nuclides, as we saw in Figure 15.3.

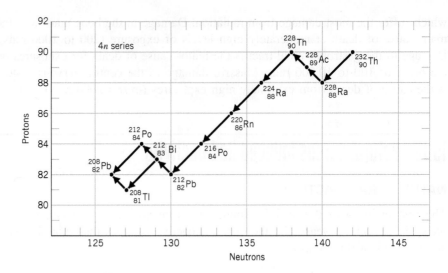

Fig. 15.8 The $4n$ series starts with ^{232}Th and eventually decays to ^{208}Pb.

Only 18 radioactive isotopes with atomic numbers of 80 or less can be found in nature. With the exception of ^{14}C, which is continuously synthesized in the atmosphere, all these elements have lifetimes longer than 10^9 years. Although all these isotopes undergo radioactive decay, they decay so slowly that reasonable quantities are still present, 4.6 billion years after the planet was formed.

Another 45 natural radioactive isotopes have atomic numbers larger than 80. These nuclides fall into three families, one of which is shown in Figure 15.8. The parent nuclide is ^{232}Th, which undergoes α decay to form ^{228}Ra. The product of this reaction decays by β^- emission to form ^{228}Ac, which decays to ^{228}Th, and so on, until the stable ^{208}Pb isotope is formed. This family of radionuclides is called the $4n$ series because all its members have a mass number that can be divided by 4.

A second family of radioactive nuclei starts with ^{238}U and decays to form the stable ^{206}Pb isotope. Every member of this series has a mass number that fits the equation $4n + 2$. The third family, known as the $4n + 3$ series, starts with ^{235}U and decays to ^{207}Pb. A $4n + 1$ series once existed, which started with ^{237}Np and decayed to form the only stable isotope of bismuth, ^{209}Bi. The half-life of every member of this series is less than 2×10^6 years, however, so none of the nuclides produced by the decay of neptunium remain in detectable quantities on the Earth.

INDUCED RADIOACTIVITY

In 1934, Irene Curie, the daughter of Pierre and Marie Curie, and her husband, Frederic Joliot, announced the first synthesis of an artificial radioactive isotope. They bombarded a thin piece of aluminum foil with α particles produced by the decay of polonium and found that the aluminum target became radioactive. Chemical analysis showed that the product of this reaction was an isotope of phosphorus.

$$^{27}_{13}\text{Al} + ^4_2\text{He} \rightarrow ^{30}_{15}\text{P} + ^1_0 n$$

In the next 50 years, more than 2000 other artificial radionuclides were synthesized.

A shorthand notation has been developed for nuclear reactions such as the reaction discovered by Curie and Joliot. The parent (or target) nuclide and the daughter nuclide are separated by parentheses that contain the symbols for the particle that hits the target and the particle or particles released in this reaction.

$$_{13}^{27}\text{Al}(\alpha, n)_{15}^{30}\text{P}$$

The nuclear reactions used to synthesize artificial radionuclides are characterized by enormous activation energies. Three devices have typically been used to overcome these activation energies: linear accelerators, cyclotrons, and nuclear reactors. Linear accelerators or cyclotrons can be used to excite charged particles such as protons, electrons, α particles, or even heavier ions, which are then focused on a stationary target. The following reaction, for example, can be induced by a cyclotron or linear accelerator.

$$_{12}^{24}\text{Mg} + _{1}^{2}\text{H} \longrightarrow _{11}^{22}\text{Na} + _{2}^{4}\text{He}$$

Because these reactions involve the capture of a positively charged particle, they usually produce a *neutron-poor* nuclide.

Artificial radionuclides are also synthesized in nuclear reactors, which are excellent sources of slow-moving or **thermal neutrons.** The absorption of a neutron usually results in a *neutron-rich* nuclide. The following neutron absorption reaction occurs in the cooling systems of nuclear reactors cooled with liquid sodium metal.

$$_{11}^{23}\text{Na} + _{0}^{1}n \longrightarrow _{11}^{24}\text{Na} + \gamma$$

In 1940, absorption of thermal neutrons was used to synthesize the first elements with atomic numbers larger than the heaviest naturally occurring element, uranium. The first of these truly artificial elements were neptunium and plutonium, which were synthesized by Edwin M. McMillan and Philip H. Abelson by irradiating ^{238}U with neutrons to form ^{239}U,

$$_{92}^{238}\text{U} + _{0}^{1}n \longrightarrow _{92}^{239}\text{U} + \gamma \qquad \text{(neutron capture)}$$

which undergoes β^{-} decay to form ^{239}Np and then ^{239}Pu.

$$_{92}^{239}\text{U} \longrightarrow _{93}^{239}\text{Np} + _{-1}^{0}e$$
$$_{93}^{239}\text{Np} \longrightarrow _{94}^{239}\text{Pu} + _{-1}^{0}e$$

Larger bombarding particles were eventually used to produce even heavier transuranium elements.

$$_{99}^{253}\text{Es} + _{2}^{4}\text{He} \longrightarrow _{101}^{256}\text{Md} + _{0}^{1}n$$
$$_{96}^{246}\text{Cm} + _{6}^{12}\text{C} \longrightarrow _{102}^{254}\text{No} + 4_{0}^{1}n$$

The half-lives for α decay and spontaneous fission decrease as the atomic number of the element increases. Element 104, for example, has a half-life for spontaneous fission of 0.3 second. Elements therefore become harder to characterize as the atomic number increases. Recent theoretical work has predicted that a magic number of protons might exist at $Z = 114$ and $Z = 120$. This work suggests that there is an island of stability in the sea of unstable nuclides, as shown in Figure 15.9. If this theory is correct, superheavy elements could be

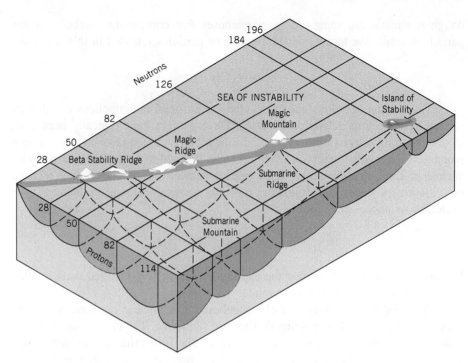

Fig. 15.9 The stable nuclei fall within a narrow band of neutron-to-proton ratios. Glenn Seaborg visualizes this as a ridge of stability in a sea of instability. Synthesis of one of the superheavy elements requires that we jump across the sea of instability in a single step.

formed if we could find a way to cross the gap between elements $Z = 109$ through $Z = 114$.

The synthesis of the element with an atomic number of 114 was recently confirmed in an experiment that involved continuously bombarding a plutonium target with ^{48}Ca ions for eight days. This experiment produced two atoms of element 114 one with an atomic mass of 286 and the other with a mass of 287. One of these atoms disintegrated in 0.3 second; the other underwent α-decay within 0.5 second.

Elements with atomic numbers above 114 have been synthesized in recent years. Element 118, for example, was synthesized in 2006 by bombarding a ^{249}Cf target with ^{40}Ca ions. The search for new elements will continue with attempts to synthesize Element 120 by bombarding a plutonium target with iron isotopes.

Some debate has arisen about the number of neutrons needed to overcome the proton–proton repulsion in a nucleus with 114 protons. The best estimates suggest that at least 184 neutrons, and perhaps as many as 196, would be needed. It is not an easy task to bring together two particles that give both the correct number of total protons and the necessary neutrons to produce a nuclide with a half-life long enough to be detected. If we start with a relatively long-lived parent nuclide, such as ^{251}Cf ($t_{1/2} = 800$ yr) and bombard this nucleus with a heavy ion, such as ^{32}S, we can envision producing a daughter nuclide with the correct atomic number, but the mass number would be too small by at least 16 amu.

$$^{251}_{98}\text{Cf} + ^{32}_{16}\text{S} \longrightarrow ^{282}_{114}\text{X} + ^{1}_{0}n$$

An expanded periodic table for elements up to $Z = 168$ is shown in Figure 15.10. Elements 104 through 112 are transition metals that fill the $6d$ orbitals. Elements 113 through 120 are main-group elements in which the $7p$ and $8s$ orbitals are filled. The next subshell is the $5g$ atomic orbital, which can hold up to 18 electrons. There is reason to believe that the $5g$ and $6f$ orbitals will be filled at the same time. The next 32 elements are therefore grouped into a so-called superactinide series.

H 1																			H 1	He 2
Li 3	Be 4												B 5	C 6	N 7	O 8	F 9	Ne 10	
Na 11	Mg 12												Al 13	Si 14	P 15	S 16	Cl 17	Ar 18	
K 19	Ca 20	Sc 21	Ti 22	V 23	Cr 24	Mn 25	Fe 26	Co 27	Ni 28	Cu 29	Zn 30	Ga 31	Ge 32	As 33	Se 34	Br 35	Kr 36		
Rb 37	Sr 38	Y 39	Zr 40	Nb 41	Mo 42	Tc 43	Ru 44	Rh 45	Pd 46	Ag 47	Cd 48	In 49	Sn 50	Sb 51	Te 52	I 53	Xe 54		
Cs 55	Ba 56	La 57	Hf 72	Ta 73	W 74	Re 75	Os 76	Ir 77	Pt 78	Au 79	Hg 80	Tl 81	Pb 82	Bi 83	Po 84	At 85	Rn 86		
Fr 87	Ra 88	Ac 89	Rf 104	Db 105	Sg 106	Bh 107	Hs 108	Mt 109	(110)	(111)	(112)	(113)	(114)	(115)	(116)	(117)	(118)		
(119)	(120)	(121)	(154)	(155)	(156)	(157)	(158)	(159)	(160)	(161)	(162)	(163)	(164)	(165)	(166)	(167)	(168)		

Lanthanides

Ce 58	Pr 59	Nd 60	Pm 61	Sm 62	Eu 63	Gd 64	Tb 65	Dy 66	Ho 67	Er 68	Tm 69	Yb 70	Lu 71

Actinides

Th 90	Pa 91	U 92	Np 93	Pu 94	Am 95	Cm 96	Bk 97	Cf 98	Es 99	Fm 100	Md 101	No 102	Lr 103

Super-actinides

(122)	(123)	(124)										(153)

Fig. 15.10 An extended version of the periodic table that predicts the positions for elements up to atomic number 168.

15.11 Nuclear Fission

The graph of binding energy per nucleon in Figure 15.6 suggests that nuclides with a mass larger than about 130 amu could spontaneously split apart to form lighter, more stable nuclides. Experimentally, we find that spontaneous fission reactions occur for only the very heaviest nuclides—those with mass numbers of 230 or more. Even when they do occur, these reactions are often very slow. The half-life for the spontaneous fission of ^{238}U, for example, is 10^{16} years, or about 2 million times longer than the age of our planet!

We don't have to wait, however, for slow spontaneous fission reactions to occur. By irradiating samples of heavy nuclides with slow-moving thermal neutrons, it is possible to *induce* fission reactions. When ^{235}U absorbs a thermal neutron, for example, it splits into two particles of uneven mass and releases an average of 2.5 neutrons, as shown in Figure 15.11.

More than 370 daughter nuclides with atomic masses between 72 and 161 amu are formed in the thermal-neutron-induced fission of ^{235}U, including the two products shown below.

$$^{235}_{92}U + ^{1}_{0}n \longrightarrow ^{139}_{56}Ba + ^{94}_{36}Kr + 3^{1}_{0}n$$

Several isotopes of uranium undergo induced fission. But the only naturally occurring isotope in which we can induce fission with thermal neutrons is ^{235}U,

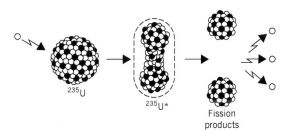

Fig. 15.11 The absorption of a neutron by ^{235}U induces oscillations in the nucleus that deform it until it splits into fragments the way a drop of liquid might break into smaller droplets.

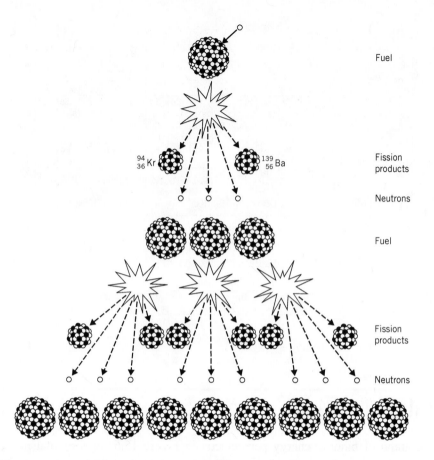

Fig. 15.12 Chain reactions occur when the substance needed to start the reaction is regenerated in the course of the reaction. Induced-fission reactions not only regenerate the neutron needed to continue the reaction but also produce enough neutrons to initiate new reaction chains. Nuclear reactors therefore contain control rods filled with materials that are good neutron absorbers in order to make sure the reaction does not get out of control.

A painting of Fermi's first reactor at the University of Chicago.

which is present at an abundance of only 0.72%. The induced fission of this isotope releases an average of 200 MeV per atom, or 80 million kilojoules per gram of ^{235}U. The attraction of nuclear fission as a source of power can be understood by comparing this value with the 50 kJ/g released when natural gas is burned.

The first artificial nuclear reactor was built by Enrico Fermi and co-workers beneath the University of Chicago's football stadium and brought on line on December 2, 1942. This reactor, which produced several kilowatts of power, consisted of a pile of graphite blocks weighing 385 tons stacked in layers around a cubical array of 40 tons of uranium metal and uranium oxide. Spontaneous fission of ^{238}U or ^{235}U in this reactor produced a very small number of neutrons. But enough uranium was present so that one of these neutrons induced the fission of a ^{235}U nucleus, thereby releasing an average of 2.5 neutrons, which catalyzed the fission of additional ^{235}U nuclei in a chain reaction, as shown in Figure 15.12. The amount of fissionable material necessary for the chain reaction to sustain itself is called the **critical mass.**

The Fermi reactor at Chicago served as a prototype for larger reactors constructed in 1943 at Oak Ridge, Tennessee, and Hanford, Washington, to produce ^{239}Pu for one of the atomic bombs dropped on Japan at the end of World War II. As we have seen, some of the neutrons released in the chain reaction are absorbed by ^{238}U to form ^{239}U, which undergoes decay by the successive loss of two β^- particles to form ^{239}Pu. ^{238}U is an example of a **fertile nuclide.** It doesn't undergo fission with thermal neutrons, but it can be converted to ^{239}Pu, which does undergo thermal-neutron-induced fission.

Fission reactors can be designed to handle naturally abundant ^{235}U, as well as fuels described as slightly enriched (2–5% ^{235}U), highly enriched (20–30% ^{235}U), or fully enriched (>90% ^{235}U). Heat generated in the reactor core is transferred to a cooling agent in a closed system. The cooling agent is then passed through a series of heat exchangers in which water is heated to steam. The steam produced in these exchangers then drives a turbine that generates electrical power. There are two ways of specifying the power of such a plant: the thermal energy produced by the reactor or the electrical energy generated by the turbines. The electrical capacity of the plant is usually about one-third of the thermal power.

It takes 10^{11} fissions per second to produce 1 watt of electrical power. As a result, about 1 gram of fuel is consumed per day per megawatt of electrical energy produced. This means that 1 gram of waste products is produced per megawatt per day, which includes 0.5 gram of ^{239}Pu. These waste products must be either reprocessed to generate more fuel or stored for the tens of thousands of years it takes for the level of radiation to reach a safe limit.

Reactors in which the ratio of the ^{239}Pu or ^{233}U produced to the ^{235}U consumed is greater than 1 generate more fuel than is consumed. Such reactors are known as *breeders,* and commercial breeder reactors are now operating in France.

The key to an efficient breeder reactor is a fuel that gives the largest possible number of neutrons released per neutron absorbed. The breeder reactors being built today use a mixture of PuO_2 and UO_2 as the fuel, and fast neutrons to activate fission. Fast neutrons carry energies of at least several KeV and travel 10,000 or more times faster than thermal neutrons. ^{239}Pu in the fuel assembly absorbs one of these fast neutrons and undergoes fission with the release of three neutrons. ^{238}U in the fuel then captures one of these neutrons to produce additional ^{239}Pu.

The advantage of breeder reactors is obvious—they mean a limitless supply of fuel for nuclear reactors. There are significant disadvantages, however. Breeder reactors are more expensive to build. They are also useless without a subsidiary industry to collect the fuel, process it, and ship the ^{239}Pu to new reactors.

It is the reprocessing of ^{239}Pu that concerns most of the critics of breeder reactors. ^{239}Pu is so dangerous as a carcinogen that the nuclear industry places a limit on exposure to this material that assumes workers inhale no more than *0.2 microgram* of plutonium over their lifetimes. There is also concern that the ^{239}Pu produced by these reactors might be stolen and assembled into bombs by terrorist organizations.

The fate of breeder reactors in the United States is linked to economic considerations. Because of the costs of building these reactors and safely reprocessing the ^{239}Pu produced, the breeder reactor becomes economical only when the scarcity of uranium drives its price so high that the breeder reactor becomes cost-effective by comparison. If nuclear energy is to play a dominant role in the generation of electrical energy in the twenty-first century, breeder reactors eventually may be essential.

Although the "pile" Fermi constructed at the University of Chicago in 1942 was the first artificial nuclear reactor, it was not the first fission reactor to exist on Earth. In 1972, a group of French scientists discovered that uranium ore from a deposit in the Oklo mine in Gabon, West Africa, contained 0.4% ^{235}U instead of the 0.72% abundance found in all other sources of this ore. Analysis of the trace elements in the ore suggested that the amount of ^{235}U in this ore was unusually small because natural fission reactors operated in this deposit for a period of 600,000 to 800,000 years about 2 billion years ago.

15.12 Nuclear Fusion

The graph of binding energy per nucleon in Figure 15.6 suggests another way of obtaining useful energy from nuclear reactions. Fusing two light nuclei can liberate as much energy as the fission of ^{235}U or ^{239}Pu. The fusion of four protons to form a helium nucleus and two positrons, for example, generates 24.7 MeV of energy.

$$4\,{}^{1}_{1}\text{H} \longrightarrow {}^{4}_{2}\text{He} + 2\,{}^{0}_{+1}e$$

Most of the energy radiated from the surface of the Sun is produced by the fusion of protons to form helium atoms within its core.

Fusion reactions have been duplicated in man-made devices. The enormous destructive power of the ^{235}U-fueled atomic bomb dropped on Hiroshima on August 6, 1945, which killed 75,000 people, and the ^{239}Pu-fueled bomb dropped on Nagasaki three days later touched off a violent debate after World War II about the building of the next superweapon—a fusion, or "hydrogen," bomb. Alumni of the Manhattan Project, who had developed the atomic bomb, were divided on the issue. Ernest Lawrence and Edward Teller fought for the construction of the fusion device. J. Robert Oppenheimer and Enrico Fermi argued against it. The decision was made to develop the weapon, and the first artificial fusion reaction occurred when the hydrogen bomb was tested in November 1952.

The history of fusion research is therefore the opposite of fission research. With fission, the reactor came first, and then the bomb was built. With fusion, the bomb was built long before any progress was made toward constructing a controlled fusion reactor. More than 50 years after the first hydrogen bomb was exploded, the feasibility of controlled fusion reactions is still open to debate. The reaction that is most likely to fuel the first fusion reactor is the thermonuclear *d–t*, or deuterium–tritium, reaction. This reaction fuses two isotopes of hydrogen, deuterium (^{2}H) and tritium (^{3}H), to form helium and a neutron.

$$\,{}^{2}_{1}\text{H} + {}^{3}_{1}\text{H} \longrightarrow {}^{4}_{2}\text{He} + {}^{1}_{0}n$$

If we consider the implications of this reaction, we can begin to understand why it is called a **thermonuclear reaction** and why it is so difficult to produce in a controlled manner. The *d–t* reaction requires that we fuse two positively charged particles. This means that we must provide enough energy to overcome the force of repulsion between these particles before fusion can occur. To produce a self-sustaining reaction, we have to provide the particles with enough thermal energy so that they can fuse when they collide.

Each fusion reaction is characterized by a specific *ignition temperature*, which must be surpassed before the reaction can occur. The *d–t* reaction has an ignition temperature above 10^{8} K. In a hydrogen bomb, a fission reaction produced by a small atomic bomb is used to heat the contents to the temperature required to initiate fusion. Obtaining the same result in a controlled reaction is much more challenging.

Any substance at temperatures approaching 10^{8} K will exist as a completely ionized gas, or *plasma*. The goals of fusion research at present include the following.

- To achieve the required temperature to ignite the fusion reaction
- To keep the plasma together at this temperature long enough to get useful amounts of energy out of the thermonuclear fusion reactions
- To obtain more energy from the thermonuclear reactions than is used to heat the plasma to the ignition temperature

These are not trivial goals. The only reasonable container for a plasma at 10^{8} K is a magnetic field. Both doughnut-shaped (toroidal) and linear magnetic bottles have been proposed as fusion reactors. But reactors that produce high enough temperatures for ignition are not the same as the reactors that have produced long enough confinement times for the plasma to provide useful amounts of energy.

A second approach to a controlled fusion reactor involves hitting fuel pellets containing the proper reagents for the thermonuclear reaction with pulsed beams of laser power. If enough power were delivered, the fuel pellets would collapse upon themselves, or implode, to reach densities several orders of magnitude greater than normal. This could produce a plasma both hot enough and dense enough to initiate fusion reactions.

15.13 Nuclear Synthesis

From the most primitive societies to the most complex, people have strived to explain how the world was created. Within the last 60 years, scientists have generated a new story of creation that offers a model for understanding how nuclei are synthesized.

In 1929, Edwin Hubble provided evidence to suggest that our universe is expanding. Between 1946 and 1948, George Gamow and co-workers generated a model which assumed that the primordial substance, or *ylem,* from which all other matter was created was an extraordinarily hot, dense singularity that exploded in a "Big Bang" and has been expanding ever since. This model assumes that neutrons in the ylem were transformed into protons by β^- decay. Neutrons and protons then combined to form ^4He atoms before the temperature and pressure of the fireball decayed to the point at which no further nuclear reactions were possible.

The first-generation stars condensed out of this cloud of hydrogen and helium. As the gas condensed by gravitational attraction, it became warmer. Eventually, the temperature reached 10^7 K, and first-generation, main-sequence stars were born. The temperature at the cores of these stars was high enough to ignite the following thermonuclear reactions.

$$^1_1H + ^1_1H \longrightarrow ^2_1H + ^0_{+1}e$$
$$^1_1H + ^2_1H \longrightarrow ^3_2He$$
$$^3_2He + ^3_2He \longrightarrow ^4_2He + 2\,^1_1H$$

The net result of these reactions is the formation of a helium atom from four protons.

$$4\,^1_1H \longrightarrow ^4_2He + 2\,^0_{+1}e$$

Eventually, the heat generated in this reaction is enough to halt the gravitational collapse of the star, which enters a stable period during which the energy generated by this reaction balances the energy radiated at the surface.

The hydrogen-burning reactions in a main-sequence star are concentrated in the core. When enough hydrogen has been consumed, the core begins to collapse and the temperature of the core rises above 10^8 K. (The larger the star, the more rapidly it radiates energy from its surface and the more rapidly it consumes the hydrogen in the core.) As the core collapses, the hydrogen-containing outer shell expands and the surface of the star cools. Stars that have reached this point in their evolution include the so-called red giants.

Temperatures in the core of these red giants are high enough to ignite other thermonuclear fusion reactions, such as the following.

$$3\,^4_2He \longrightarrow ^{12}_6C$$

This reaction becomes the principal source of energy in a red giant, although there is undoubtedly some burning of hydrogen to helium in the outer shell of the star. As the amount of ^{12}C in the core increases, further reactions occur to form ^{16}O and ^{20}Ne.

$$^{12}_{6}C + {}^{4}_{2}He \longrightarrow {}^{16}_{8}O + \gamma$$
$$^{16}_{8}O + {}^{4}_{2}He \longrightarrow {}^{20}_{10}Ne + \gamma$$

Eventually, the helium in the core is exhausted and the core collapses further, reaching temperatures of about 7×10^8 K. At this point, more complex reactions take place that produce nuclides such as ^{28}Si and ^{32}S.

$$^{12}_{6}C + {}^{16}_{8}O \longrightarrow {}^{28}_{14}Si + \gamma$$
$$^{16}_{8}O + {}^{16}_{8}O \longrightarrow {}^{32}_{16}S + \gamma$$

Further gravitational collapse heats the core to temperatures above 10^9 K, and a complex sequence of reactions takes place to synthesize the very stable nuclei with the highest binding energies, such as Fe and Ni. If the star then explodes as a supernova, its contents are ejected across space. Second-generation stars that condense in this region contain not only hydrogen and helium but also elements with higher atomic number.

The best estimates of the age of the Milky Way suggest that our galaxy is about 15 billion years old. Our Sun and its planets, however, are only 4.6 billion years old. This suggests that the Sun is a second-generation star. In such stars, the transformation of hydrogen to helium can be catalyzed by ^{12}C, as shown in Figure 15.13.

Two processes can synthesize elements with atomic numbers larger than that of iron. One of them is relatively slow (the s-process); the other is very rapid (the r-process). Since the only way to synthesize nuclei with atomic numbers larger than iron is by the absorption of neutrons, both the s-process and the r-process result from (n, γ) reactions.

In the **s-process,** neutrons are captured one at a time to form a neutron-rich nuclide that undergoes α or β^- decay before another neutron can be absorbed. An example of an s-process sequence of reactions starts with ^{120}Sn. The capture of a neutron produces ^{121}Sn, which undergoes β^- decay. If β^- decay occurs before this nuclide captures another neutron, a stable isotope of antimony is formed. Eventually, ^{121}Sb captures a neutron to produce ^{122}Sb, which is transformed into ^{122}Te by β^- decay. ^{122}Te can undergo β^- decay to form ^{122}I, or it can capture a neutron to form ^{123}Te. With ^{123}Te, we encounter a series of stable isotopes of tellurium. Neutrons are slowly absorbed, one at a time, until we reach ^{127}Te, which decays to ^{127}I, the most abundant isotope of iodine.

$$^{120}_{50}Sn(n,\gamma)^{121}_{50}Sn \xrightarrow{\beta^-} {}^{121}_{51}Sb(n,\gamma)^{122}_{51}Sb \xrightarrow{\beta^-} {}^{122}_{52}Te(5n,5\gamma)^{127}_{52}Te \xrightarrow{\beta^-} {}^{127}_{53}I$$

This slow process can't account for very heavy nuclides, such as ^{232}Th and ^{238}U, because the lifetimes of the intermediate nuclei with atomic numbers between 83 and 90 are too short for this step-by-step absorption of neutrons to proceed. Synthesizing appreciable quantities of uranium and thorium requires a rapid process. In the **r-process,** a number of neutrons are captured in rapid succession before there is time for α or β decay to take place. Achieving an r-process reaction, however, requires a very high neutron flux. (These reactions occur during nuclear explosions, for example.) The neutron flux needed to fuel such

$$^{12}_{6}C + {}^{1}_{1}H \longrightarrow {}^{13}_{7}N + \gamma$$

$$^{13}_{7}N \longrightarrow {}^{13}_{6}C + {}^{0}_{+1}e$$

$$^{13}_{6}C + {}^{1}_{1}H \longrightarrow {}^{14}_{7}N + \gamma$$

$$^{14}_{7}N + {}^{1}_{1}H \longrightarrow {}^{15}_{8}O + \gamma$$

$$^{15}_{8}O \longrightarrow {}^{15}_{7}N + {}^{0}_{+1}e$$

$$^{15}_{7}N + {}^{1}_{1}H \longrightarrow {}^{12}_{6}C + {}^{4}_{2}He$$

Fig. 15.13 Second-generation stars use a different mechanism to synthesize helium than first-generation stars. They use a sequence of reactions, such as those shown here, that are catalyzed by isotopes heavier than helium.

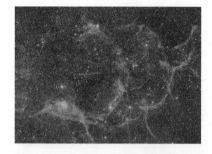

There are two processes for nuclear synthesis. The slow process occurs within a star, such as the Sun, during its lifetime. The fast process occurs when the star goes supernova.

reactions is not likely to occur in a normal star. During the moment when a star explodes as a supernova, however, the conditions are ripe for r-process reactions. The heavier elements on our planet were therefore produced in a series of supernova explosions that occurred in this portion of the galaxy before our solar system condensed.

15.14 Nuclear Medicine

Ever since Roentgen obtained the first X-ray images in 1895, ionizing radiation and radionuclides have played a vital role in medicine. This work has been so fruitful that a separate field known as nuclear medicine has developed. Research in this field focuses on either therapeutic or diagnostic uses of radiation.

There are three standard approaches to fighting cancer: surgery, chemotherapy, and radiation. Surgery, by its very nature, is invasive. Chemotherapy and classic approaches to radiation therapy are not selective. Research in recent years has therefore examined new approaches to radiation therapy that specifically attack tumor cells, without damaging normal tissue. The technique known as *boron neutron capture therapy* provides an example of this work.

Naturally occurring boron consists of two stable isotopes: ^{10}B (19.7%) and ^{11}B (80.3%). ^{10}B absorbs thermal neutrons to form ^{11}B in a nuclear excited state. Although ^{11}B in its nuclear ground state is stable, this excited ^{11}B nuclide undergoes fission to produce ^7Li and an α particle.

$$^{10}_{5}B + {}^{1}_{0}n \longrightarrow {}^{7}_{3}Li + {}^{4}_{2}He$$

Because the energy of a thermal neutron is only about 0.025 eV, the neutrons that are not absorbed do relatively little damage to the normal tissue. The α particle emitted in this reaction has an energy of 2.79 MeV, however, which makes it an extremely lethal form of radiation. The RBE for α-particle radiation is larger than any other particle in Table 15.2 because this relatively massive particle loses energy very efficiently as it collides with matter. Radiation damage from the α particle is therefore restricted to the immediate vicinity of the tissue that absorbed the thermal neutron.

Although other common nuclides in living tissue can absorb thermal neutrons, the ability of ^{10}B to absorb thermal neutrons is 3 orders of magnitude larger than these nuclides. The units with which this measurement is made can be understood by thinking about the area around the nuclide through which the neutron can pass and still be absorbed. The **neutron-capture cross section** for a hydrogen atom corresponds to a circle with a radius of about 2×10^{-13} cm. For a nitrogen atom, the radius of this circle is about 10^{-12} cm. Boron has a neutron-capture cross section that would be described by a circle with a radius of about 2×10^{-9} cm. Virtually all of the neutron capture that occurs is therefore concentrated in the tissue that contains boron.

The potential of boron neutron capture therapy (BNCT) was recognized as early as 1936. Early clinical trials in the late 1950s and early 1960s failed to prolong the lives of patients suffering from brain tumors because the boron compounds available for testing did not concentrate selectively in the tumor cells. Recent research has discovered boron-labeled compounds that are sufficiently selective for BNCT to become useful as an adjunct to surgery—or in place of surgery with inoperable cases.

Key Terms

Activity
Alpha decay
Beta decay
Binding energy
Binding energy per nucleon
^{14}C dating
Critical mass
Curie
Electron capture
Electron (β^-) emission
Fertile nuclide
Fission reactions
Fusion reactions

Gamma emission
Half-life
Induced radioactivity
Ionizing radiation
Isobar
Isotone
Isotope
Mass defect
Metastable
Neutron-capture cross section
Neutron-poor
Neutron-rich

Nonionizing radiation
Nuclide
Positron (β^+) emission
r-process
Radiation absorbed dose (rad)
Radiation biological effectiveness
  (RBE)
Roentgen equivalent man (rem)
s-process
Spontaneous fission
Thermal neutrons
Thermonuclear reaction

Problems

Radioactivity

1. Identify the particle given off when a nucleus undergoes α decay. Identify the particle given off during β decay. Describe the relationship between γ rays and other forms of electromagnetic radiation.

2. Describe an experiment that could be used to determine whether an atom emits α particles, β particles, or γ rays.

The Structure of the Atom

3. Define the terms *atomic number, mass number, nuclide, nucleon,* and *nucleus.*

4. Define the terms *isotopes, isobars,* and *isotones.* Give examples of pairs of nuclides that would be described by each of these terms.

5. Calculate the number of electrons, protons, and neutrons in a neutral ^{90}Sr atom.

6. Calculate the number of electrons, protons, and neutrons in both a ^{40}K$^+$ and an ^{127}I$^-$ ion.

Modes of Radioactive Decay

7. Explain why the three forms of β decay interconvert isobars.

8. In theory, β decay could include reactions in which a positron is captured and the charge on the nucleus increases. Explain why the following positron capture reaction does not occur under "normal" conditions.

$$^{14}_{6}C + \,^{0}_{+1}e \longrightarrow \,^{14}_{7}N + h\nu$$

9. Which of the following reactions interconvert isotopes? Which interconvert isobars? Which interconvert isotones?

  (a) electron emission   (b) electron capture

  (c) positron emission   (d) α emission

  (e) neutron emission   (f) neutron absorption

  (g) α emission followed by two β^- decays

10. Identify the missing particle in each of the following equations and name the form of radioactive decay.

  (a) $^{125}_{53}I + \,^{0}_{-1}e \longrightarrow X$

  (b) $^{240}_{94}Pu \longrightarrow X + \,^{144}_{58}Ce + 2\,^{1}_{0}n$

  (c) $^{90}_{38}Sr \longrightarrow X + \,^{0}_{-1}e$

  (d) $^{40}_{19}K \longrightarrow X + \,^{0}_{+1}e$

  (e) $^{228}_{90}Th \longrightarrow X + \,^{4}_{2}He$

11. Write a balanced equation for the β-particle decay of the ^{99}Mo nuclide.

12. Write a balanced equation for the α-particle decay of the ^{248}Cf nuclide.

13. Write a balanced equation for the electron capture reaction of the ^{62}Cu nuclide.

14. Write a balanced equation for the positron emission reaction of the ^{34}Cl nuclide.

15. Predict the products of the following nuclear reactions.

  (a) electron emission by ^{32}P

  (b) positron emission by ^{11}C

  (c) α decay by ^{212}Rn

  (d) electron capture by ^{125}Xe

16. Predict the products of the following nuclear reactions.

  (a)$^{208}_{82}Pb(^{2}_{1}H,n)$___   (b) $^{95}_{42}Mo(n,\gamma)$___

  (c)$^{202}_{84}Po(^{22}_{10}Ne,4n)$___   (d) $^{14}_{7}N(n,p)$___

17. Identify the missing particle or particles in the following reactions and write a balanced equation for each reaction.

  (a)$^{238}_{92}U(\underline{\quad},3n)^{239}_{94}Pu$ (b) ___$(\alpha,n)^{242}_{96}Cm$

  (c)$^{250}_{98}Cf(^{11}_{5}B,\underline{\quad})^{257}_{103}Lr$ (d) $^{249}_{98}Cf(\underline{\quad},4n)^{257}_{104}Rf$

Neutron-Rich versus Neutron-Poor Nuclides

18. Explain why neutron-rich nuclides become more stable when they undergo decay by electron (β^-) emission.

19. Explain why neutron-poor nuclides become more stable when they undergo decay by either electron capture,

positron (β^+) emission, the emission of an alpha particle, or spontaneous fission.

20. Which of the following nuclides are most likely to be neutron-rich?

    (a) ^{14}C (b) ^{24}Na (c) ^{25}Si (d) ^{27}Al (e) ^{31}P

21. Which of the following nuclides are most likely to be neutron-poor?

    (a) ^3H (b) ^{11}C (c) ^{14}N (d) ^{40}K (e) ^{61}Cu

22. Explain why ^{17}Ne, ^{18}Ne, and ^{19}Ne decay by positron emission but ^{23}Ne and ^{24}Ne decay by electron emission.

23. Which isotope of carbon is most likely to decay by positron emission?

    (a) ^{11}C (b) ^{12}C (c) ^{13}C (d) ^{14}C

24. Which isotope of carbon is most likely to decay by electron emission?

    (a) ^{11}C (b) ^{12}C (c) ^{13}C (d) ^{14}C

Binding Energy Calculations

25. Calculate the binding energy of ^6Li in MeV per atom if the exact mass of this nuclide is 6.01512 amu. Calculate the binding energy per nucleon.

26. Calculate the binding energy of ^{60}Ni in MeV per atom if the exact mass is 59.9332 amu. Calculate the binding energy per nucleon.

27. Calculate the exact mass of ^{238}U if the binding energy per nucleon is 7.570198 MeV.

28. Which nuclide in Problems 25 through 27 has the largest binding energy? Which has the largest binding energy per nucleon?

29. Calculate the energy released in the following reaction.

$$^{10}B(n,\alpha)^7Li$$

Use the following data for the masses of the particles involved in the reaction: ^{10}B = 10.0129 amu; ^7Li = 7.01600 amu; ^4He = 4.00260 amu. Explain why the ability of ^{10}B to release a high-energy α particle after it absorbs a thermal neutron generated considerable interest in getting boron compounds to absorb preferentially into the fast-growing tumor in patients who suffer from brain tumors.

The Kinetics of Radioactive Decay

30. The half-life of ^{32}P is 14.3 days. Calculate how long it would take for a 1.000-gram sample of ^{32}P to decay to each of the following quantities of ^{32}P.

    (a) 0.500 gram   (b) 0.250 gram   (c) 0.125 gram

31. Calculate the half-life for the decay of ^{39}Cl if a 1.000-gram sample decays to 0.125 gram in 165 minutes.

32. Calculate the rate constants in s^{-1} for the decay of the following nuclides from their half-lives.

    (a) ^{18}F, 110 minutes   (b) ^{54}Mn, 312 days

    (c) ^3H, 12.26 years   (d) ^{14}C, 5730 years

    (e) ^{129}I, 1.6×10^7 years

33. A 1.000-gram sample of ^{22}Na decays to 0.20 gram in 6.04 years. Calculate the half-life for this decay, the rate constant, and the time it would take for this sample to decay to 0.075 gram.

34. Calculate the time required for a 2.50-gram sample of ^{51}Cr to decay to 1.00 gram, assuming that the half-life is 27.8 days.

35. A sample of ^{210}Po initially weighed 2.000 grams. After 25 days, 0.125 gram of ^{210}Po remained, the rest of the sample having decayed to the stable ^{206}Pb isotope. Calculate the half-life of ^{210}Po and the mass of ^{206}Pb formed.

36. Forgeries that had been accepted by art authorities as paintings by the Dutch artist Vermeer (1632–1675) have been detected by measuring the activity of the ^{210}Pb isotope in the lead paints. When lead is extracted from its ores, it is separated from ^{226}Ra, which is the source of the ^{210}Pb isotope. The amount of ^{210}Pb in the paint therefore decreases with time. If the half-life for the decay of ^{210}Pb is 21 years, what fraction of the ^{210}Pb would be present in a 300-year-old painting? What fraction would remain in a 10-year-old forgery?

Dating by Radioactive Decay

37. The threat to people's health from radon in the air trapped in their houses has received attention in recent years. If the average level of radon in a house is approximately 1 picocurie (pCi) per liter of air, how many radon atoms are there per liter? (Assume that ^{222}Rn is the principal source of this activity and that the half-life for the decay of this nuclide is 3.823 days.)

38. Calculate the number of disintegrations per minute in a 1.00-mg sample of ^{238}U, assuming that the half-life is 4.47×10^9 years.

39. Calculate the activity in curies for 1.00-mg samples of the following isotopes of uranium.

    (a) ^{228}U, $t_{1/2}$ = 9.3 min   (b) ^{230}U, $t_{1/2}$ = 20.8 d

    (c) ^{236}U, $t_{1/2}$ = 2.39×10^7 yr

40. The ^{14}C in living matter has an activity of 15.3 disintegrations, or "counts," per minute (cpm). What is the age of an artifact that has an activity of 4 cpm? (^{14}C: $t_{1/2}$ = 5730 yr)

41. A skull fragment found in 1936 at Baldwin Hills, California, was dated by radiocarbon analysis. Approximately 100 grams of bone were cleaned and treated with 1 M hydrochloric acid to destroy the mineral content of the bone. The bone protein was collected, dried, and pyrolyzed. The CO_2 produced was collected and purified, and the ratio of ^{14}C to ^{12}C was measured. If this sample contained roughly 5.7% of the ^{14}C present in living tissue, how old is the skeleton? (^{14}C: $t_{1/2}$ = 5730 yr)

42. Measurements on the linen wrappings from the Book of Isaiah in the Dead Sea Scrolls suggest that the scrolls contain about 79.5% of the ^{14}C expected in living tissue. How old are these scrolls? (^{14}C: $t_{1/2}$ = 5730 yr)

43. The Lascaux cave near Montignac in France contains a series of remarkable cave paintings. Radiocarbon dating of charcoal taken from this site suggests an age of 15,520 years. What fraction of the ^{14}C present in living tissue is still present in this sample? (^{14}C: $t_{1/2} = 5730$ yr)

44. Charcoal samples from Stonehenge in England emit 62.3% of the disintegrations per gram of carbon per minute expected for living tissue. What is the age of this charcoal? (^{14}C: $t_{1/2} = 5730$ yr)

45. A lump of beeswax was excavated in England near a collection of Bronze Age objects that are between 2500 and 3000 years old. Radiocarbon analysis of the beeswax suggests an activity roughly 90.3% of that observed for living tissue. Was this beeswax part of the hoard of Bronze Age objects, or did it date from another period? (^{14}C: $t_{1/2} = 5730$ yr)

Ionizing versus Nonionizing Radiation

46. Use Lewis structures to describe the difference between an H_2O^+ ion and an H_3O^+ ion. If a free radical is an ion or molecule that contains one or more unpaired electrons, which of these ions is a free radical?

Biological Effects of Ionizing Radiation

47. Radioactivity is measured in units that describe the amount of radiation given off, the amount of radiation to which an object is exposed, the amount of radiation absorbed, or the effectiveness of the radiation as a threat to biological systems. Sort the following units into these categories.

    (a) curies (b) rads (c) rems (d) roentgens

48. Explain why sources of α particles are intrinsically more dangerous than sources of β^- particles.

Natural versus Induced Radioactivity

49. The first artificial radioactive elements were synthesized by Irene Curie and Frederic Joliot, who bombarded ^{10}B and ^{27}Al with α particles to form ^{13}N and ^{30}P. Write balanced equations for these reactions, identify the particle ejected in each reaction, and predict the mode of decay expected for the products of these reactions.

50. Russell, Soddy, and Fajans predicted that the emission of one α and two β particles by a nuclide would produce an isotope of the parent nuclide. Which isotope of ^{216}Po is produced by such decay? What intermediate nuclides are formed?

51. In the first synthesis of an isotope of mendelevium ($Z = 101$), Ghiorso and co-workers bombarded ^{253}Es with α particles. Starting with less than 10^{-12} gram of einsteinium, they isolated one atom of mendelevium after a period of a few hours. If a neutron was emitted in this reaction, what isotope of Md was produced? Another isotope of mendelevium was produced by bombarding ^{238}U with ^{19}F atoms. If five neutrons were ejected in this reaction, what isotope of Md was produced?

52. How many alpha and beta particles are emitted when ^{232}Th decays to ^{208}Pb?

Nuclear Fission and Nuclear Fusion

53. Explain why relatively light nuclides give off energy when they fuse to form heavier nuclides, whereas relatively heavy nuclides give off energy when they undergo fission.

54. Describe the difference between spontaneous and induced fission reactions. Explain why nuclei undergoing induced fission reactions have much shorter half-lives.

55. Describe the advantages and disadvantages of fusion reactors versus fission reactors.

Nuclear Synthesis

56. What evidence do we have that the Sun is a second-generation star?

57. Describe the difference between the s-process and r-process for the synthesis of nuclides. Explain why the s-process can't synthesize relatively heavy naturally occurring nuclides, such as ^{238}U.

Integrated Problems

58. Calculate the rems of radiation absorbed by the average person from the ^{14}C in his or her body. Assume the activity of the ^{14}C in the average body is 0.08 μCi, the energy of the β^- particles emitted when ^{14}C decays is 0.156 MeV, and about one-third of this energy is captured by the body.

59. Use the following data for ^{48}Cr to calculate the rate constant and the half-life for the decay of this isotope by electron capture.

Mass (g)	Time (min)
100	0
80	444
60	1017
40	1825
20	3205

60. Calculate the weight of 1.00 millicurie (mCi) of ^{14}C, assuming that the half-life of this nuclide is 5730 years.

61. The activity of the ^{14}C in living tissue is 15.3 disintegrations per minute per gram of carbon. The limit for reliable determination of ^{14}C ages is 0.10 disintegration per minute per gram of carbon. Calculate the maximum age of a sample that can be dated accurately by radiocarbon dating if the half-life for the decay of ^{14}C is 5730 years.

62. Use the relationship between the energy and the frequency of a photon to calculate the energy in kilojoules per mole of a photon of blue light that has a frequency of 6.5×10^{14} s^{-1}. Compare the results of this calculation with the ionization energy of water (1216 kJ/mol).

63. Calculate the energy in kilojoules per mole for an X ray that has a frequency of 3×10^{17} s^{-1}. How do the results of this calculation compare with the ionization energy of water (1216 kJ/mol)?

Chapter Sixteen

ORGANIC CHEMISTRY

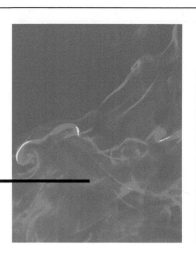

16.1 What Is an Organic Compound?

Suppose that you went to buy gasoline and were faced with three choices of "unleaded" gasoline that had different octane numbers. As you filled the tank, you might wonder, "What is 'unleaded' gasoline, and why would anyone want to add lead to gas?" Or you might wonder, "What would I get for my money if I bought premium gas, with a higher octane number?"

Let's assume that you then stop to buy drugs for a sore back that has been bothering you since you helped a friend move into a new apartment. Once again, you are faced with choices. You could buy aspirin, which has been used for over a hundred years. Or you could buy a more modern painkiller, such as ibuprofen, whose structure is as follows.

Ibuprofen

While you are deciding which drug to buy, you might wonder, "What is the difference between these drugs?" and even, "How do they work?"

You then drive to campus, where you sit in a "plastic" chair to eat a sandwich that has been wrapped in "plastic." You might ask yourself, "What makes one of these plastics flexible while the other is rigid?" While you're eating, a friend stops by and teases you about the effect of your diet on the level of cholesterol in your blood. Which brings up the question, "What is cholesterol?" and "Why do so many people worry about it?"

Answers to each of these questions fall within the realm of a field known as **organic chemistry.** For more than 200 years, chemists have divided materials into two categories. Those isolated from plants and animals were classified as *organic*, while those obtained from minerals were *inorganic*. At one time, chemists believed that organic compounds were fundamentally different from those that were inorganic because organic compounds contained a **vital force** that was only found in living systems.

The first step in the decline of the vital force theory occurred in 1828, when Friedrich Wöhler synthesized urea from inorganic starting materials. Wöhler was trying to make ammonium cyanate (NH_4OCN) from silver cyanate (AgOCN) and ammonium chloride (NH_4Cl). What he expected is described by the following equation.

$$AgOCN(aq) + NH_4Cl(aq) \longrightarrow AgCl(s) + NH_4OCN(aq)$$

The product he isolated from this reaction had none of the properties of cyanate compounds. It was a white, crystalline material that was identical to urea, H_2NCONH_2, which could be isolated from urine.

If the difference between organic and inorganic compounds is not the presence of some mysterious vital force required for their synthesis, what is the basis for distinguishing between these classes of compounds? Most compounds extracted from living organisms contain carbon. It is therefore tempting to identify organic chemistry as the chemistry of carbon. But this definition would include compounds such as calcium carbonate ($CaCO_3$), as well as the elemental forms of carbon—diamond and graphite—that are classified as inorganic. We will therefore define organic chemistry as the chemistry of compounds that contain both carbon and hydrogen.

More than 95% of the compounds that have been isolated from natural sources or synthesized in the laboratory are organic. The special role of carbon in the chemistry of the elements is the result of a combination of factors, including the number of valence electrons on a neutral carbon atom, the electronegativity of carbon, and the atomic radius of carbon atoms (see Table 16.1).

Carbon has four valence electrons—$2s^2 2p^2$—and it might be expected to either gain four electrons or lose four electrons to form an ion. The electronegativity of carbon is too small for carbon to gain electrons from most elements to form C^{4-} ions, and too large for carbon to lose electrons to form C^{4+} ions. Carbon therefore forms covalent bonds with a large number of other elements, including the hydrogen, nitrogen, oxygen, phosphorus, and sulfur found in living systems.

Because they are relatively small, carbon atoms can come close enough together to form strong C=C double bonds or even C≡C triple bonds. Carbon also forms strong double and triple bonds to nitrogen and oxygen. It can even form double bonds to elements such as phosphorus or sulfur that do not form double bonds to themselves.

In the 1970s, the unmanned Viking spacecraft carried out experiments designed to search for evidence of life on Mars. These experiments were based on the assumption that living systems contain carbon, and the absence of any evidence for carbon-based life on that planet was assumed to mean that no life existed. Several factors make carbon essential to life.

- The ease with which carbon atoms form covalent bonds to other carbon atoms

- The strength of covalent C—C single bonds and the covalent bonds carbon forms to other nonmetals, such as N, O, P, and S

- The ability of carbon to form multiple bonds to other nonmetals, including C, N, O, P, and S atoms

These factors provide an almost infinite variety of potential structures for organic compounds such as vitamin C, for which the following structure is drawn.

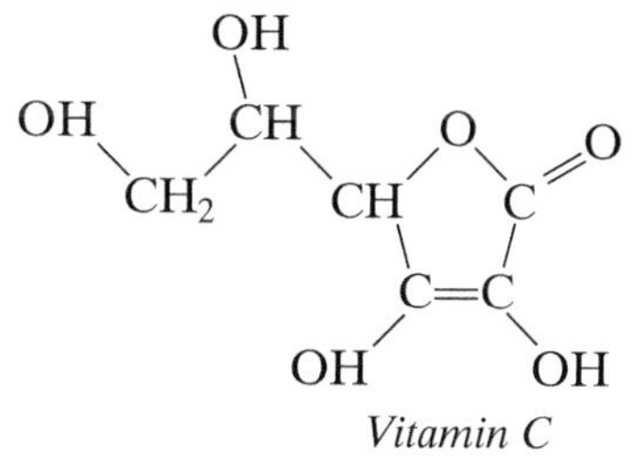

Table 16.1
Physical Properties of Carbon

Electronic configuration	$1s^2 2s^2 2p^2$
Electronegativity	2.54
Covalent radius	0.077 nm

Vitamin C

No other element can provide the variety of combinations and permutations necessary for life to exist.

16.2 The Saturated Hydrocarbons or Alkanes

Compounds that contain only carbon and hydrogen are known as **hydrocarbons.**
Those that contain as many hydrogen atoms as possible are said to be *saturated*.
The saturated hydrocarbons are also known as **alkanes.**

The simplest alkane is methane: CH_4. The Lewis structure of methane can
be generated by combining the four electrons in the valence shell of a neutral carbon atom with four hydrogen atoms to form a compound in which the carbon
atom shares a total of eight valence electrons with the four hydrogen atoms.

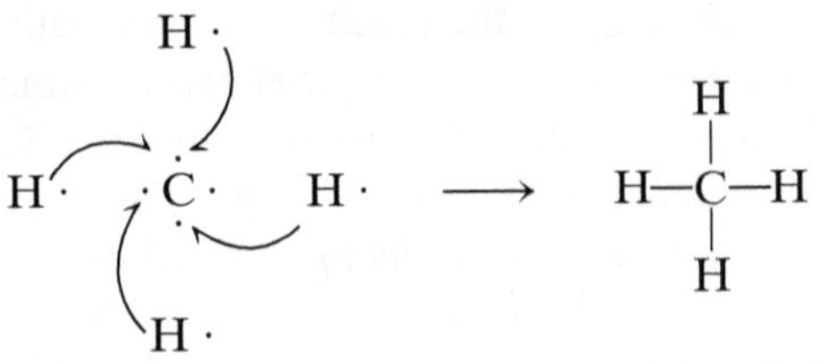

Methane is an example of a general rule that carbon is **tetravalent**; it forms
a total of four bonds in almost all of its compounds. The geometry around the carbon atom is tetrahedral, as shown in the following line drawing, because this is
the most efficient way to arrange the four electron domains in the four C—H
bonds. Solid lines in this figure correspond to bonds that lie in the plane of the
paper on which the diagram is drawn. Dashed lines represent bonds that extend
behind the plane of the paper, and solid wedged lines describe bonds that come
out of the plane of the paper.

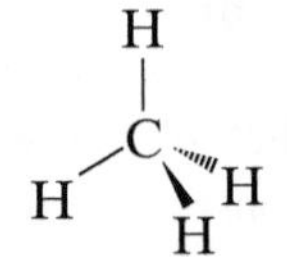

Methane

Exercise 16.1

Use the fact that carbon is usually tetravalent to predict the formula of ethane,
the alkane that contains two carbon atoms.

Solution

As a rule, compounds that contain more than one carbon atom are held together
by covalent C—C single bonds. If we assume that carbon is tetravalent, the
formula of this compound must be C_2H_6.

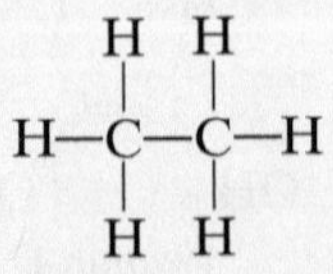

Ethane

The alkane that contains three carbon atoms is known as *propane*, which
has the formula C_3H_8 and the following skeleton structure.

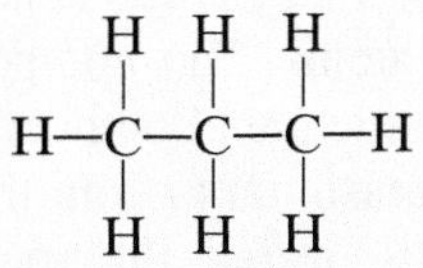

$$H-\underset{\underset{H}{|}}{\overset{\overset{H}{|}}{C}}-\underset{\underset{H}{|}}{\overset{\overset{H}{|}}{C}}-\underset{\underset{H}{|}}{\overset{\overset{H}{|}}{C}}-H \quad \textit{Propane}$$

The four-carbon alkane is *butane,* with the formula C_4H_{10}.

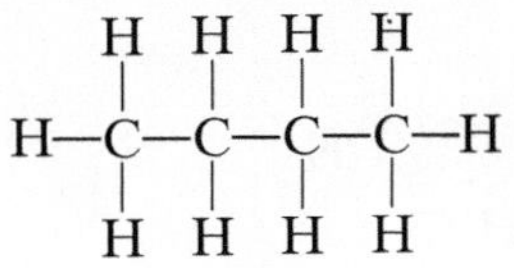

$$H-\underset{\underset{H}{|}}{\overset{\overset{H}{|}}{C}}-\underset{\underset{H}{|}}{\overset{\overset{H}{|}}{C}}-\underset{\underset{H}{|}}{\overset{\overset{H}{|}}{C}}-\underset{\underset{H}{|}}{\overset{\overset{H}{|}}{C}}-H \quad \textit{Butane}$$

• •

The names, formulas, and physical properties for a variety of alkanes with the generic formula C_nH_{2n+2} are given in Table 16.2. The boiling points of the alkanes gradually increase with the molecular weight. At room temperature, the lighter alkanes are gases; the midweight alkanes are liquids; and the heavier alkanes are solids, or tars.

Because the bond angle in a tetrahedron is 109.5°, alkane molecules that contain three or more carbon atoms should not be thought of as "linear," as can be seen in the line drawings of the structures of propane and butane shown below. The alkanes in Table 16.2 are therefore known as **straight-chain hydrocarbons** in which the carbon atoms form a chain that runs from one end of the molecule to the other. The generic formula for these compounds can be understood by assuming that they contain chains of CH_2 groups with an additional hydrogen atom capping either end of the chain. Thus, for every n carbon atoms there must be $2n + 2$ hydrogen atoms: C_nH_{2n+2}.

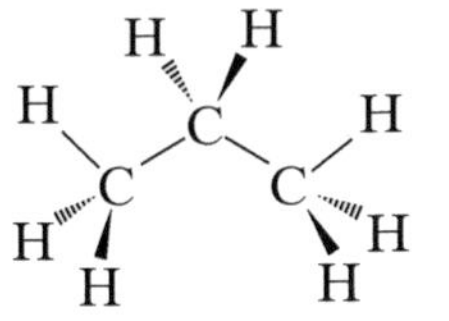

Propane 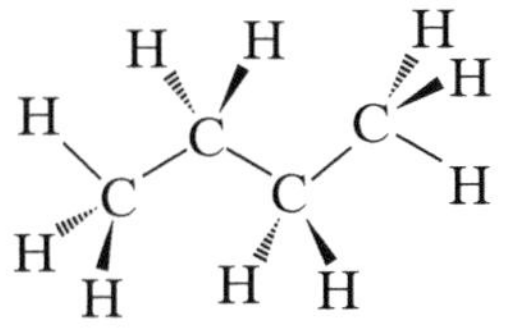 *Butane*

Table 16.2
The Saturated Hydrocarbons, or Alkanes

Name	Molecular Formula	Melting Point (°C)	Boiling Point (°C)	State at 25°C
Methane	CH_4	−182.5	−162	gas
Ethane	C_2H_6	−183.3	−88.6	gas
Propane	C_3H_8	−189.7	−42.1	gas
Butane	C_4H_{10}	−138.4	−0.5	gas
Pentane	C_5H_{12}	−129.7	36.1	liquid
Hexane	C_6H_{14}	−95	69.0	liquid
Heptane	C_7H_{16}	−90.6	98.4	liquid
Octane	C_8H_{18}	−56.8	125.7	liquid
Nonane	C_9H_{20}	−53.5	150.8	liquid
Decane	$C_{10}H_{22}$	−29.7	174.1	liquid
Undecane	$C_{11}H_{24}$	−24.6	195.9	liquid
Dodecane	$C_{12}H_{26}$	−9.6	216.3	liquid
Eicosane	$C_{20}H_{42}$	36.8	343.8	solid
Triacontane	$C_{30}H_{62}$	65.8	449.7	solid

Alkanes also form **branched** structures. The smallest hydrocarbon in which a branch can occur has four carbon atoms. This compound has the same formula as butane (C_4H_{10}), but a different structure. Compounds with the same formula and different structures are known as **isomers** (from the Greek *isos*, "equal," and *meros*, "parts"). When it was first discovered, the branched isomer with the formula C_4H_{10} was therefore given the name *isobutane*.

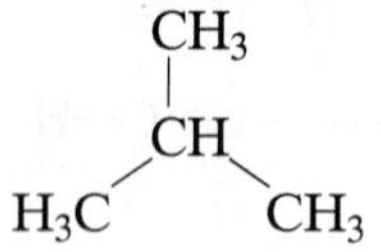

$$\begin{array}{c} CH_3 \\ | \\ CH \\ \diagup \quad \diagdown \\ H_3C \qquad CH_3 \end{array}$$

Isobutane

Isomers, such as butane and isobutane, that differ in the way the atoms are connected are called **constitutional isomers** because they literally differ in their constitution. One contains two CH_3 groups and two CH_2 groups; the other contains three CH_3 groups and one CH group.

There are three constitutional isomers of pentane, C_5H_{12}. The first is "normal" pentane, or *n*-pentane.

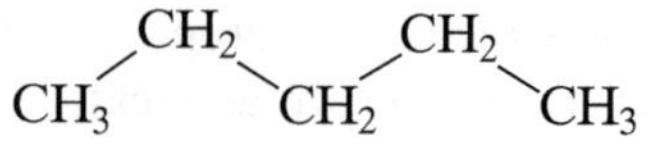

$$\begin{array}{ccccc} & CH_2 & & CH_2 & \\ \diagup & & \diagdown & & \diagdown \\ CH_3 & & CH_2 & & CH_3 \end{array} \quad \textit{n-Pentane}$$

A branched isomer is also possible, which was originally known as isopentane. When a more highly branched isomer was discovered, it was named neopentane (the new isomer of pentane).

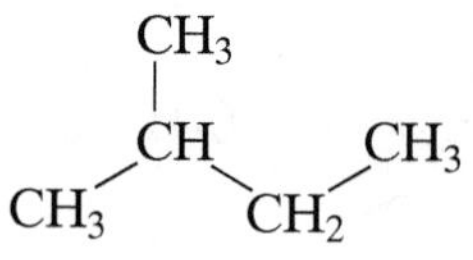

$$\begin{array}{cc} CH_3 & \\ | & \\ CH \quad CH_3 & \\ \diagup \quad \diagdown \diagup & \\ CH_3 \quad CH_2 & \end{array} \quad \textit{Isopentane} \qquad \begin{array}{c} CH_3 \\ | \\ CH_3{-}C{-}CH_3 \\ | \\ CH_3 \end{array} \quad \textit{Neopentane}$$

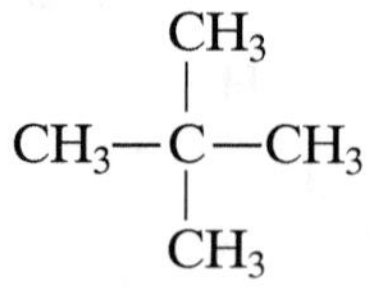

Exercise 16.2

The following structures all have the same molecular formula: C_6H_{14}. Which of these structures represent the same molecule?

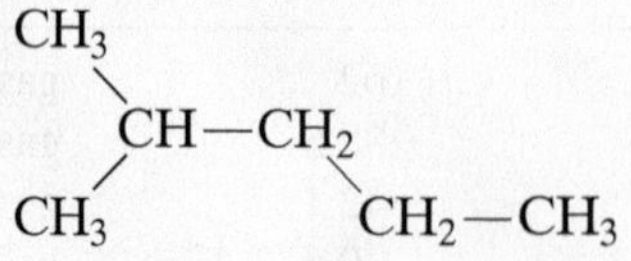

$$\begin{array}{cc} CH_3 & \\ \diagdown & \\ CH{-}CH_2 & \\ \diagup \qquad \diagdown & \\ CH_3 \qquad CH_2{-}CH_3 & \end{array} \qquad \begin{array}{c} CH_3 \diagdown \quad \diagup CH_3 \\ CH \\ | \\ CH_2 \\ | \\ CH_2 \\ | \\ CH_3 \end{array} \qquad \begin{array}{c} CH_3 \\ | \\ CH_3{-}C{-}CH_3 \\ | \\ CH_2 \\ \diagdown \\ CH_3 \end{array}$$

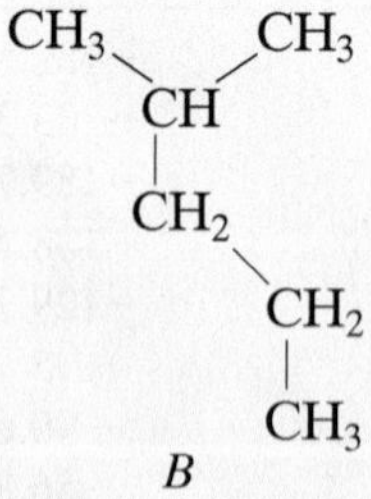

$$\qquad A \qquad\qquad\qquad\qquad B \qquad\qquad\qquad\qquad C$$

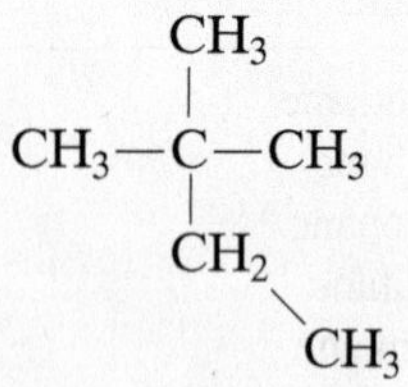

There is no difference between compounds *A* and *B*; they both contain a five-carbon chain with a branch on the second carbon. Compound *C*, on the other hand, contains a four-carbon chain with two branches on the second carbon atom.

Exercise 16.3

Determine the number of constitutional isomers of hexane, C_6H_{14}.

Solution

There are five constitutional isomers of hexane. There is a straight-chain, or normal, isomer.

$$CH_3—CH_2—CH_2—CH_2—CH_2—CH_3$$

There are two isomers with a single carbon branch. One of these compounds has a CH_3_ substituent on the second carbon of a five-carbon chain. The other has a CH_3_ substituent on the third carbon in the chain.

$$
\begin{array}{cc}
\quad\quad CH_3 & \quad\quad\quad\quad CH_3 \\
\quad\quad | & \quad\quad\quad\quad | \\
CH_3—CH—CH_2—CH_2—CH_3 & \quad CH_3—CH_2—CH—CH_2—CH_3
\end{array}
$$

There are another two isomers that have two branches on a four-carbon chain. One of these isomers has CH_3_ substituents on the second and third carbons of the chain. The other isomer has two CH_3_ substituents on the second carbon.

$$
\begin{array}{cc}
\quad\quad CH_3 & \quad\quad CH_3 \\
\quad\quad | & \quad\quad | \\
CH_3—CH—CH—CH_3 & \quad CH_3—C—CH_2—CH_3 \\
\quad\quad\quad\quad | & \quad\quad | \\
\quad\quad\quad\quad CH_3 & \quad\quad CH_3
\end{array}
$$

As we have seen, there are two constitutional isomers with the formula C_4H_{10}; three isomers of C_5H_{12}; and five isomers of C_6H_{14}. The number of isomers of a compound increases rapidly with additional carbon atoms. There are over 4 billion isomers for $C_{30}H_{62}$, for example.

16.3 Rotation Around C—C Bonds

As one looks at the structure of the ethane molecule, it is easy to fall into the trap of thinking about this molecule as if it was static. Nothing could be further from the truth. At room temperature, the average speed of an ethane molecule as it moves through space is about 500 m/s—more than twice the speed of a 747 traveling across the ocean. While the molecule is moving, it tumbles around its center of gravity like an airplane out of control. At the same time, the C—H bonds are vibrating at a rate of almost 10^{14} times per second.

There is another way in which the ethane molecule can move—the CH_3 groups at either end of the molecule can rotate with respect to each other around the C—C bond. When this happens, the molecule passes through an endless number of **conformations** that have slightly different energies. The highest energy conformation corresponds to a structure in which the hydrogen atoms are "eclipsed." If we viewed the molecule along the C—C bond, the hydrogen atoms

> ➤ **CHECKPOINT**
>
> Use a set of molecular models to confirm that the following compounds are isomers,
>
> $$
> \begin{array}{cc}
> \quad\quad CH_3 & CH_3 \\
> \quad\quad | & | \\
> CH_3—CH & —CH—CH_3
> \end{array}
> $$
>
> $$
> \begin{array}{c}
> \quad\quad CH_3 \\
> \quad\quad | \\
> CH_3—C—CH_2—CH_3 \\
> \quad\quad | \\
> \quad\quad CH_3
> \end{array}
> $$
>
> whereas the following are different ways of writing the structure of the same compound.
>
> $$
> \begin{array}{cc}
> \quad\quad CH_3 & CH_3 \\
> \quad\quad | & | \\
> CH_3—CH & —CH—CH_3
> \end{array}
> $$
>
> $$
> \begin{array}{c}
> \quad\quad CH_3 \\
> \quad\quad | \\
> CH_3—CH—CH—CH_3 \\
> \quad\quad\quad\quad | \\
> \quad\quad\quad\quad CH_3
> \end{array}
> $$

on one CH_3 group would obscure those on the other, as shown in the following ball-and-stick model.

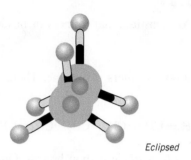

Eclipsed

The lowest energy conformation for ethane is a structure in which the hydrogen atoms are "staggered," as shown in the following ball-and-stick model.

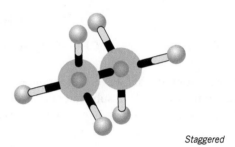

Staggered

The difference between the energies of these conformers is relatively small—only about 12 kJ/mol. But it is large enough that rotation around the C—C bond is not smooth. Although the frequency of this rotation is on the order of 10^{10} revolutions per second, the ethane molecule spends a slightly larger percentage of the time in the staggered conformation.

16.4 The Nomenclature of Alkanes

Common names such as pentane, isopentane, and neopentane are good enough to differentiate between the three isomers with the formula C_5H_{12}. They become less useful, however, as the size of the hydrocarbon chain increases.

The International Union of Pure and Applied Chemistry (IUPAC) has developed a systematic approach to naming alkanes and cycloalkanes based on the following steps.

- Find the longest continuous chain of carbon atoms in the skeleton structure. Name the compound as a derivative of the alkane with this number of carbon atoms. The following compound, for example, is a derivative of pentane because the longest chain contains five carbon atoms.

- Name the substituents on the chain. Substituents derived from alkanes are named by replacing the *-ane* ending with *-yl*. This compound contains a methyl (CH_3-) substituent.

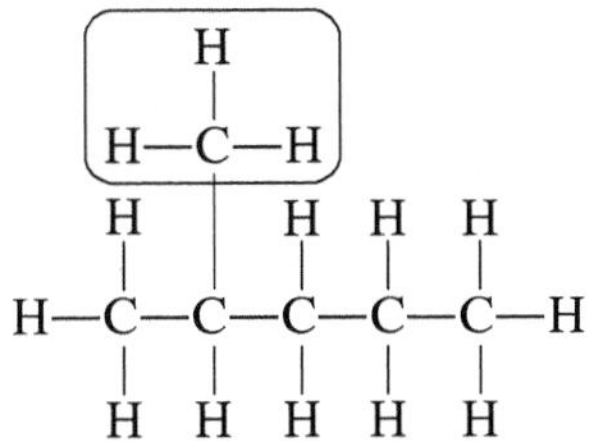

$$
\begin{array}{c}
\boxed{\begin{array}{c} H \\ | \\ H-C-H \end{array}} \\
\begin{array}{ccccc} H & & H & H & H \\ | & & | & | & | \\ H-C-C-C-C-C-H \\ | & | & | & | & | \\ H & H & H & H & H \end{array}
\end{array}
$$

- Number the chain starting at the end nearest the first substituent and specify the carbon atoms on which the substituents are located. Use the lowest possible numbers. This compound, for example, is 2-methylpentane, not 4-methylpentane.

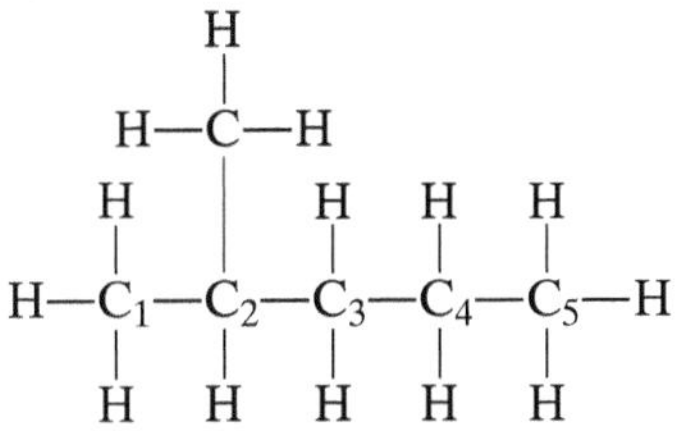

$$
\begin{array}{c}
H \\ | \\ H-C-H \\
\begin{array}{ccccc} H & & H & H & H \\ | & & | & | & | \\ H-C_1-C_2-C_3-C_4-C_5-H \\ | & | & | & | & | \\ H & H & H & H & H \end{array}
\end{array}
$$

- Use the prefixes *di-*, *tri-*, and *tetra-* to describe substituents that are found two, three, or four times, respectively, on the same chain of carbon atoms. The following compound, for example, would be 2,3-dimethylpentane.

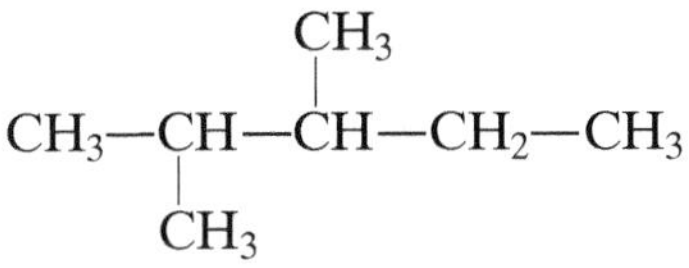

$$
\begin{array}{c}
CH_3 \\ | \\ CH_3-CH-CH-CH_2-CH_3 \\ | \\ CH_3
\end{array}
$$

- Arrange the names of the substituents (not their prefixes) in alphabetical order.

Exercise 16.4

Name the following compound.

$$
\begin{array}{c}
CH_3 \qquad CH_3 \\ | \qquad\quad | \\ CH_3-C-CH_2-CH-CH_3 \\ | \\ CH_3
\end{array}
$$

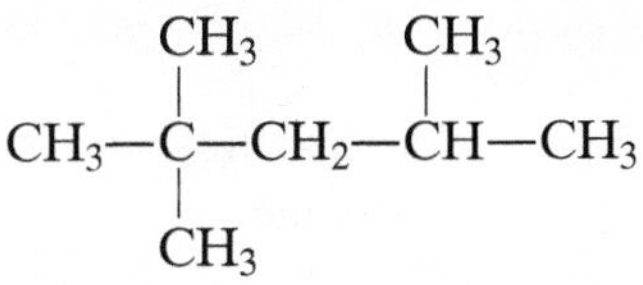

Solution

This compound is a derivative of pentane because the longest continuous chain contains five carbon atoms. There are three identical CH_3- (or methyl) substituents on the backbone. Two of these methyl groups are on the second carbon, and one is on the fourth carbon. This compound is therefore *2,2,4-trimethylpentane*. Since it contains a total of eight carbon atoms, it is also known by the common name *isooctane*. Isooctane is the standard against which fuels are rated. Regular fuel with an "octane number" of 87 burns the way a mixture of 87% isooctane and 13% heptane burns.

Exercise 16.5

Name the following compound.

$$\begin{array}{c} CH_3 \\ | \\ CH_3-CH-CH_2-CH-CH_2-CH_3 \\ | \\ CH_2 \\ | \\ CH_2 \\ | \\ CH_3 \end{array}$$

Solution

The longest continuous chain in the skeleton structure of this compound contains seven carbon atoms. It is therefore named as a derivative of heptane.

$$\begin{array}{c} {}^{1}CH_3 \\ | \\ CH_3-\underset{2}{CH}-\underset{3}{CH_2}-\overset{4}{CH}-CH_2-CH_3 \\ | \\ {}^{5}CH_2 \\ | \\ {}^{6}CH_2 \\ | \\ {}^{7}CH_3 \end{array}$$

The heptane chain contains two substituents—a methyl group on the second carbon atom and an ethyl group on the fourth carbon atom.

Because substituents are listed in alphabetical order, the systematic name for this compound is 4-ethyl-2-methylheptane.

16.5 The Unsaturated Hydrocarbons: Alkenes and Alkynes

Carbon not only forms the strong C—C bonds found in alkanes, but also forms strong C=C double bonds. Compounds that contain C=C double bonds were once known as *olefins* (literally, "to make an oil") because they were hard to crystallize. (They tend to remain oily liquids when cooled.) These compounds are now called **alkenes.**

The connection between alkanes and alkenes can be understood by thinking about a hypothetical reaction in which we break a C—H on both carbon atoms in ethane so that one of the electrons in these bonds ends up on each atom. We then bring the hydrogen atoms together to form an H_2 molecule and allow the electrons on the two carbon atoms to interact to form a double bond between these atoms.

Although this hypothetical reaction does not occur, the opposite reaction is easy to achieve. In the presence of a suitable catalyst, such as platinum metal, we can transform an alkene into the parent alkane.

The generic formula for an alkene with one C=C double bond is C_nH_{2n}.

Alkenes are examples of **unsaturated hydrocarbons** because they have fewer hydrogen atoms than the corresponding alkanes. They were once named by adding the suffix *-ene* to the name of the substituent that carried the same number of carbon atoms.

$$H_2C=CH_2 \qquad \textit{Ethylene}$$

$$CH_2=CH—CH_3 \qquad \textit{Propylene}$$

The IUPAC nomenclature for alkenes names these compounds as derivatives of the parent alkanes. The presence of the C=C double bond is indicated by changing the *-ane* ending on the name of the parent alkane to *-ene.*

$$CH_3—CH_3 \qquad \textit{Ethane} \qquad\qquad CH_2=CH_2 \qquad \textit{Ethene}$$

$$CH_3—CH_2—CH_3 \qquad \textit{Propane} \qquad CH_2=CH—CH_3 \qquad \textit{Propene}$$

The location of the C=C double bond in the skeleton structure of the compound is indicated by specifying the number of the carbon atom at which the C=C bond starts.

$$CH_2=CH-CH_2-CH_3 \qquad CH_3-CH=CH-CH_3$$
1-Butene   2-Butene

The names of substituents are then added as prefixes to the name of the alkene.

Exercise 16.6

Name the following compound.

$$CH_3-CH=\overset{\overset{\textstyle CH_3}{|}}{C}-CH_2-\underset{\underset{\textstyle CH_3}{|}}{CH}-CH_3$$

Solution

This compound is a derivative of hexane because the longest carbon chain contains six carbon atoms. It contains a C=C double bond, which means it is a *hexene*. Because the double bond links the second and third carbon atoms, it is a 2-hexene. Because the CH_3 substituents are on the third and fifth carbon atom, the compound is *3,5-dimethyl-2-hexene.*

Compounds that contain C≡C triple bonds are called **alkynes.** These compounds have four fewer hydrogen atoms than the parent alkanes, so the generic formula for an alkyne with a single C≡C triple bond is C_nH_{2n-2}. The simplest alkyne has the formula C_2H_2 and is known by the common name *acetylene.*

$$H-C\equiv C-H \quad \textit{Acetylene}$$

The IUPAC nomenclature for alkynes names these compounds as derivatives of the parent alkane, with the ending *-yne* replacing *–ane.*

$CH_3C\equiv CCH_2CH_3$ 2-Pentyne

$HC\equiv CCH_2CH_3$ 1-Butyne

$CH_3C\equiv CCH_2\underset{\underset{\textstyle CH_3}{|}}{CH}CH_2CH_3$ 5-Methyl-2-heptyne

In addition to compounds that contain one double bond (*alkenes*) or one triple bond (*alkynes*), we can also envision compounds with two double bonds (*dienes*), three double bonds (*trienes*), or a combination of double and triple bonds.

$$CH_3CH{=}CHCH_2C{\equiv}CH \qquad \textit{4-Hexen-1-yne}$$

$$H_2C{=}CHCH{=}CH_2 \qquad \textit{1, 3-Butadiene}$$

16.6 The Aromatic Hydrocarbons and Their Derivatives

At the turn of the nineteenth century, one of the signs of living the good life was having gas lines connected to your house, so that you could use gas lanterns to light the house after dark. The gas burned in these lanterns was called **coal gas** because it was produced by heating coal in the absence of air. The principal component of coal gas was methane, CH_4.

In 1825, Michael Faraday was asked to analyze an oily liquid with a distinct odor that collected in tanks used to store coal gas at high pressures. Faraday found that this compound had the empirical formula: CH. Ten years later, Eilhardt Mitscherlich produced the same material by heating benzoic acid with lime. Mitscherlich named this substance *benzin*, which became *benzene* when translated into English. He also determined that the molecular formula of this compound is C_6H_6.

Benzene is obviously an unsaturated hydrocarbon because it has far less hydrogen than the equivalent saturated hydrocarbon: C_6H_{14}. Other compounds eventually isolated from coal had similar properties. Their formulas suggested the presence of multiple $C{=}C$ bonds, but these compounds were not reactive enough to be alkenes. Because they often had a distinct odor, or aroma, they became known as **aromatic compounds.**

The structure of benzene was a recurring problem throughout most of the nineteenth century. The first step toward solving this problem was taken by Friedrich August Kekulé in 1865. (Kekulé's interest in the structure of organic compounds may have resulted from the fact that he first enrolled at the University of Giessen as a student of architecture.) One day, while dozing before a fire, Kekulé dreamed of long rows of atoms twisting in a snakelike motion until one of the snakes seized hold of its own tail. This dream led Kekulé to propose that benzene consists of a ring of six carbon atoms with alternating C—C single bonds and C=C double bonds. Because there are two ways in which these bonds can alternate, Kekulé proposed that benzene was a mixture of two compounds in equilibrium.

Kekulé's explanation of the structure of benzene

Kekulé's structure explained the formula of benzene, but it did not explain why benzene failed to behave like an alkene. The unusual stability of benzene was not understood until the development of the theory of resonance. This theory states that molecules for which two or more satisfactory Lewis structures can be drawn are an average, or hybrid, of these structures. Benzene, for example, is a resonance hybrid of the two Kekulé structures.

The resonance structures for benzene

The difference between the equilibrium and resonance descriptions of benzene is subtle, but important. In the equilibrium approach, a pair of arrows is used to describe a reversible reaction, in which the molecule on the left is converted into the one on the right, and vice versa. In the resonance approach, a double-headed arrow is used to suggest that a benzene molecule does not shift back and forth between two different structures; it is a mixture, or hybrid, of these structures.

One way to probe the difference between Kekulé's idea of an equilibrium between two structures and the resonance theory in which benzene is a hybrid mixture of these structures would be to study the lengths of the carbon–carbon bonds in benzene. If Kekulé's idea was correct, we would expect to find a molecule in which the bonds alternate between relatively long C—C single bonds (0.154 nm) and significantly shorter C═C double bonds (0.133 nm). When benzene is cooled until it crystallizes, and the structure of the molecule is studied by X-ray diffraction, we find that the six carbon–carbon bonds in this molecule are all the same length (0.139 nm). The crystal structure of benzene is therefore more consistent with the resonance model of bonding in benzene than the original Kekulé structures.

To emphasize the difference between benzene and a simple alkene, chemists sometimes replace the Kekulé structures for benzene and its derivatives with an aromatic ring in which the circle in the center of the ring indicates that the electrons in the ring are delocalized; they are free to move around the ring.

It is this delocalization of electrons around the aromatic ring that is conveyed by the circle inside the ring.

Because aromatic compounds were extracted from coal tar as early as the 1830s, many of these compounds were given common names that are still in use today. A few of these compounds are as follows.

Toluene   Phenol   Anisole   Aniline   Benzoic acid

There are three ways in which a pair of substituents can be placed on an aromatic ring. (1) In the *ortho* (*o*) isomer, the substituents are in adjacent positions on the ring; (2) in the *meta* (*m*) isomer, they are separated by one carbon atom; and (3) in the *para* (*p*) isomer, they are on opposite ends of the ring. The three isomers of dimethylbenzene, or xylene, follow.

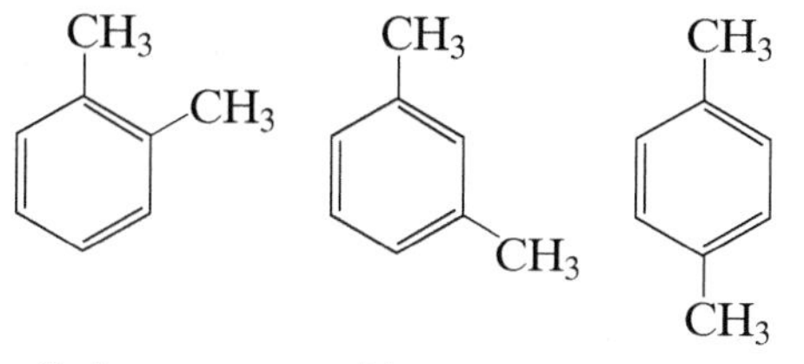

Ortho   *Meta*   *Para*

Exercise 16.7

Predict the structure of *para*-dichlorobenzene, one of the active ingredients in moth balls.

Solution

Para isomers of benzene contain two substituents at opposite positions in the six-membered ring. *Para*-dichlorobenzene therefore has the following structure.

Aromatic compounds can contain more than one six-membered ring. Naphthalene, anthracene, and phenanthrene, shown in Figure 16.1, are examples of aromatic compounds that contain two or more fused benzene rings.

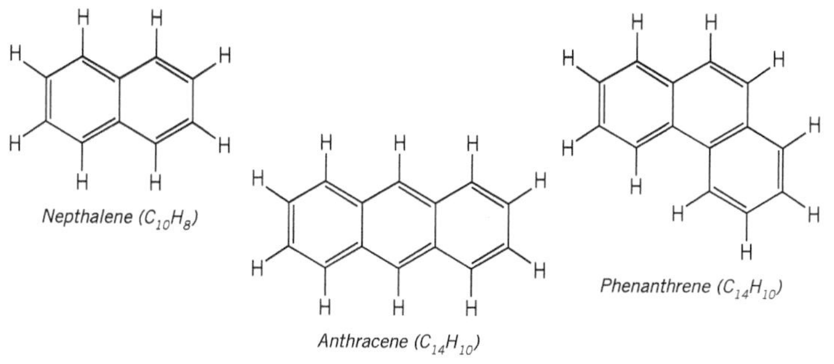

Nepthalene ($C_{10}H_8$)

Anthracene ($C_{14}H_{10}$)

Phenanthrene ($C_{14}H_{10}$)

Fig. 16.1 Examples of fused-ring aromatic compounds.

16.7 The Chemistry of Petroleum Products

The term *petroleum* comes from the Latin stems *petra*, "rock," and *oleum*, "oil." It is used to describe a broad range of hydrocarbons found as gases, liquids, or solids beneath the surface of the Earth. The two most common forms of petroleum are natural gas and crude oil.

Natural gas is a mixture of lightweight alkanes. A typical sample of natural gas when it is collected at its source contains 80% methane (CH_4), 7% ethane (C_2H_6), 6% propane (C_3H_8), 4% butane and isobutane (C_4H_{10}), and 3% pentanes (C_5H_{12}). The C_3, C_4, and C_5 hydrocarbons are removed before the gas is sold. The commercial natural gas delivered to the customer is therefore primarily a mixture of methane and ethane. The propane and butanes removed from natural gas are usually liquefied under pressure and sold as liquefied petroleum gases (*LPG*).

Natural gas did not replace coal as an important source of energy in the United States until after World War II, when a network of gas pipelines was constructed. By 2008, the world's annual consumption of natural gas had grown to more than 100 trillion cubic feet. Slightly less than one-quarter of this volume is used in the United States, where it provides about 24% of the energy consumed.

The first oil well was drilled by Edwin Drake in 1859, in Titusville, Pennsylvenia. It produced up to 800 gallons per day, which far exceeded the demand for this material. One hundred and fifty years later, the total world consumption of oil is more than 3.5 billion gallons per day.

About 650 billion barrels of oil were produced by the petroleum industry between 1859 and 2000.[1] The total proven world reserves of crude oil is slightly more than 1650 billion barrels, which is more than 75% of the 2200 billion barrels of oil estimated to be ultimately recoverable. It took 500 million years for the petroleum beneath the Earth's crust to accumulate. At the present rate of consumption, we might exhaust the world's supply of petroleum by the 200th anniversary of the first oil well.

Crude oil is a complex mixture that is between 50 and 95% hydrocarbon by weight. The first step in refining crude oil involves separating the oil into different hydrocarbon fractions by distillation. A typical set of petroleum fractions is given in Table 16.3. Since a number of factors influence the boiling point of a hydrocarbon, these petroleum fractions are complex mixtures. More than 500 different hydrocarbons have been identified in the gasoline fraction, for example.

About 10% of the product of the distillation of crude oil is a fraction known as **straight-run gasoline,** which served as a satisfactory fuel during the early days of the internal combustion engine. As the automobile engine developed, it was made more powerful by increasing the compression ratio.[2] Straight-run gasoline burns unevenly in high-compression engines, which produces a shock wave that causes the engine to "knock," or "ping." As the petroleum industry matured, it faced two problems: increasing the yield of gasoline from each barrel of crude oil and decreasing the tendency of gasoline to knock when it burned.

Table 16.3
Petroleum Fractions

Fraction	Boiling Range (°C)	Number of Carbon Atoms
Natural gas	< 20	C_1 to C_4
Gasoline	40–200	C_5 to C_{12}, but mostly C_6 to C_8
Kerosene	150–260	mostly C_{12} to C_{13}
Heating oil and diesel fuels	> 260	C_{14} and higher
Lubricants	> 400	C_{20} and above
Asphalt or coke	residue	polycyclic

---

[1]There are 42 gallons of oil in a barrel.

[2]Modern cars run at compression ratios of about 9:1, which means the gasoline–air mixture in the cylinder is compressed by a factor of nine before it is ignited.

The relationship between knocking and the structure of the hydrocarbons in gasoline is summarized in the following general rules.

- Branched alkanes and cycloalkanes burn more evenly than straight-chain alkanes.
- Short alkanes (C_4H_{10}) burn more evenly than long alkanes (C_7H_{16}).
- Alkenes burn more evenly than alkanes.
- Aromatic hydrocarbons burn more evenly than cycloalkanes.

The most commonly used measure of a gasoline's ability to burn without knocking is its **octane number.** Octane numbers compare a gasoline's tendency to knock with the tendency of a blend of two hydrocarbons—heptane and 2,2,4-trimethylpentane, or isooctane—to knock. Gasolines that match a blend of 87% isooctane and 13% heptane are given an octane number of 87.

There are three ways of reporting octane numbers. First, measurements made at high speeds and high temperatures are reported as *motor octane numbers*. Second, measurements taken under relatively mild engine conditions are known as *research octane numbers*. Finally, the *road-index octane numbers* reported on gasoline pumps are an average of these two. Road-index octane numbers for a few pure hydrocarbons are given in Table 16.4.

By 1922 a number of compounds had been discovered that could increase the octane number of gasoline. Adding as little as 6 mL of tetraethyllead to a gallon of gasoline, for example, can increase the octane number by 15 to 20 units. This discovery gave rise to the first "ethyl" gasoline, and enabled the petroleum industry to produce aviation fuels with octane numbers greater than 100. The term *ethyl* was deliberately chosen when marketing this product to avoid the potential adverse reaction to the word "lead." By 1972, however, concerns about the toxicity of pollutants dispersed into the air when tetraethyllead was used as an "antiknock" additive gave rise to an initiative by the EPA to phase out leaded gasoline.

Another way to increase the octane number is **thermal reforming.** At high temperatures (500–600°C) and high pressures (25–50 atm), straight-chain alkanes isomerize to form branched alkanes and cycloalkanes, thereby increasing the octane number of the gasoline. Running this reaction in the presence of hydrogen and a catalyst results in **catalytic reforming,** which can produce a gasoline with even higher octane numbers. Thermal or catalytic reforming and gasoline additives such as tetraethyllead increase the octane number of the straight-run gasoline obtained from the distillation of crude oil, but neither process increases the yield of gasoline from a barrel of oil.

The data in Table 16.4 suggest that we could increase the yield of gasoline by "cracking" the hydrocarbons that end up in the kerosene or fuel oil fractions into smaller pieces. **Thermal cracking** was discovered as early as the 1860s. At high temperatures (500°C) and high pressures (25 atm), long-chain hydrocarbons break into smaller pieces. A saturated C_{12} hydrocarbon in kerosene, for example, might break into two C_6 fragments. Because the total number of carbon and hydrogen atoms remains constant, one of the products of this reaction contains a $C\!=\!C$ double bond.

$$CH_3(CH_2)_{10}CH_3 \longrightarrow CH_3(CH_2)_4CH_3 + CH_2\!=\!CH(CH_2)_3CH_3$$

The presence of alkenes in thermally cracked gasolines increases the octane number (70) relative to that of straight-run gasoline (60), but it also makes thermally cracked gasoline less stable for long-term storage. Thermal cracking has therefore been replaced by **catalytic cracking,** which uses catalysts instead of high temperatures and pressures to crack long-chain hydrocarbons into smaller fragments for use in gasoline.

Table 16.4

Hydrocarbon Octane Numbers

Hydrocarbon	Road Index Octane Number
Heptane	0
2-Methylheptane	23
Hexane	25
2-Methylhexane	44
1-Heptene	60
Pentane	62
1-Pentene	84
Butane	91
Cyclohexane	97
2,2,4-Trimethylpentane (isooctane)	100
Benzene	101
Toluene	112

The quality of diesel fuels is measured in terms of a *cetane number*. Cetane ($C_{16}H_{34}$) is a straight-chain hydrocarbon that contains 16 carbon atoms. The advantage of running an engine on longer hydrocarbons is the increased density of the fuel. The density of diesel fuel (0.85 g/cm^3) is about 18% larger than the density of gasoline (0.72 g/cm^3). Diesel fuels are also easier to refine than gasoline. Diesel fuel and heating oil are obtained in the same fraction when crude oil is refined. Diesel fuel goes through an additional purification step, however, to remove sulfur contaminants.

16.8 The Chemistry of Coal

Coal can be defined as a sedimentary rock that burns. It was formed by the decomposition of plant matter, and it is a complex substance that can be found in many forms. Coal is divided into four classes: anthracite, bituminous, sub-bituminous, and lignite. Elemental analysis gives empirical formulas such as $C_{137}H_{97}O_9NS$ for bituminous coal and $C_{240}H_{90}O_4NS$ for high-grade anthracite. A typical structure for coal is shown in Figure 16.2.

The total energy consumption in the United States has been estimated as 100 quadrillion BTUs or 100×10^{15} kJ/year. Of this total, 41% comes from oil,

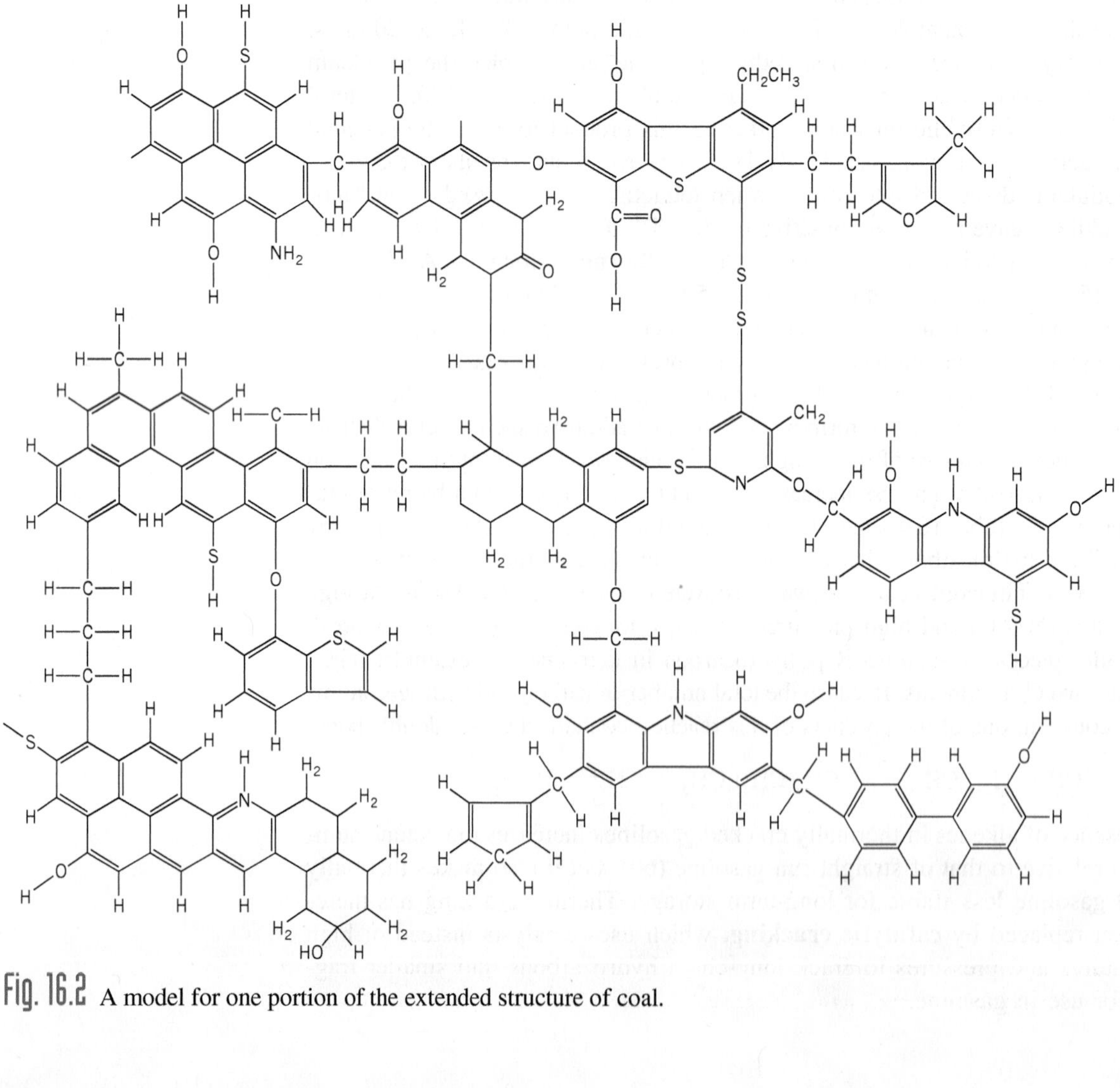

Fig. 16.2 A model for one portion of the extended structure of coal.

20% from natural gas, and 25% from coal. Coal is unique as a source of energy in the United States because none of the 2343 billion pounds used in 2008 was imported. Furthermore, the proven reserves are so large we can continue using coal at this level of consumption for at least 2000 years. Coal is also more cost efficient as a fuel than either natural gas or fuel oil. Coal is more difficult to handle, however. As a result, there has been a long history of efforts to turn coal into either a gaseous or a liquid fuel.

COAL GASIFICATION

As early as 1800, **coal gas** was made by heating coal in the absence of air. Coal gas is rich in CH_4 and gives off up to 20.5 kJ per liter of gas burned. Coal gas— or **town gas,** as it was also known—became so popular that most major cities and many small towns had a local gas house, and gas burners were adjusted to burn a fuel that produced 20.5 kJ/L.

A slightly less efficient fuel known as **water gas** can be made by reacting the carbon in coal with steam.

$$C(s) + H_2O(g) \longrightarrow CO(g) + H_2(g) \qquad \Delta H° = 131.3 \text{ kJ/mol}_{rxn}$$

Water gas burns to give CO_2 and H_2O, releasing roughly 11.2 kJ per liter of gas consumed. Water gas formed by the reaction of coal with oxygen and steam is a mixture of CO, CO_2, and H_2. The ratio of H_2 to CO can be increased by adding water to this mixture, to take advantage of a reaction known as the **water-gas shift reaction.**

$$CO(g) + H_2O(g) \longrightarrow CO_2(g) + H_2(g) \qquad \Delta H° = -41.2 \text{ kJ/mol}_{rxn}$$

Water gas from which the CO_2 has been removed is called **synthesis gas,** because it can be used as a starting material for a variety of organic and inorganic compounds, such as methyl alcohol, or methanol.

$$CO(g) + 2 H_2(g) \longrightarrow CH_3OH(l)$$

Methanol can then be used as a starting material for the synthesis of alkenes, aromatic compounds, acetic acid, formaldehyde, and ethyl alcohol (ethanol). Synthesis gas can also be used to produce methane, or synthetic natural gas (SNG).

$$CO(g) + 3 H_2(g) \longrightarrow CH_4(g) + H_2O(g)$$
$$2 CO(g) + 2 H_2(g) \longrightarrow CH_4(g) + CO_2(g)$$

COAL LIQUEFACTION

The first step toward making liquid fuels from coal involves the manufacture of synthesis gas (CO and H_2) from coal. In 1925, Franz Fischer and Hans Tropsch developed a catalyst that converted CO and H_2 at 1 atm and 250 to 300°C into liquid hydrocarbons. By 1941, Fischer-Tropsch plants produced 740,000 tons of petroleum products per year in Germany.

Fischer-Tropsch technology is based on a complex series of reactions that use H_2 to reduce CO to CH_2 groups linked to form long-chain hydrocarbons.

$$CO(g) + 2 H_2(g) \longrightarrow (-CH_2-)_n(l) + H_2O(g) \qquad \Delta H° = -165 \text{ kJ/mol}_{rxn}$$

The water produced in this reaction combines with CO in the water-gas shift reaction to form H_2 and CO_2.

$$CO(g) + H_2O(g) \longrightarrow CO_2(g) + H_2(g) \qquad \Delta H° = -41.2 \text{ kJ/mol}_{rxn}$$

The overall Fischer-Tropsch reaction is therefore described by the following equation.

$$2\,CO(g) + H_2(g) \longrightarrow (-CH_{2-})_n(l) + CO_2(g) \qquad \Delta H° = -206\,\text{kJ/mol}_{rxn}$$

At the end of World War II, Fischer-Tropsch technology was under study in most industrial nations. The low cost and high availability of crude oil, however, led to a decline in interest in liquid fuels made from coal. Although a synthetic liquid fuels program was run by the U.S. Bureau of Mines as early as 1944, this program was abolished in 1986. One of the few commercial plants using this technology today is in the Sasol complex in South Africa, which uses 30.3 million tons of coal per year.

Another approach to liquid fuels is based on the reaction between CO and H_2 to form methanol, CH_3OH.

$$CO(g) + 2\,H_2(g) \longrightarrow CH_3OH(l)$$

Methanol can be used directly as a fuel, or it can be converted into gasoline with catalysts such as the ZSM-5 zeolite catalyst developed by Mobil Oil Company. In the short run, catalysts such as ZSM-5 will benefit countries such as New Zealand that do not have oil deposits but have an abundant supply of methane that can be converted into methanol for use in the synthesis of gasoline.

As the supply of petroleum becomes smaller and its cost continues to rise, a gradual shift may be observed toward liquid fuels made from coal. Only time will tell whether this takes the form of a return to a modified Fischer-Tropsch technology, the conversion of methanol to gasoline, the generation of biodiesel, or other alternatives.

16.9 Functional Groups

Bromine reacts with 2-butene to form 2,3-dibromobutane.

$$CH_3CH{=}CHCH_3 + Br_2 \longrightarrow CH_3\overset{\displaystyle Br}{\underset{\displaystyle Br}{\overset{|}{\underset{|}{C}}}}HCHCH_3$$

It also reacts with 3-methyl-2-pentene to form 2,3-dibromopentane.

$$CH_3CH{=}\overset{\displaystyle CH_3}{\overset{|}{C}}CH_2CH_3 + Br_2 \longrightarrow CH_3\overset{\displaystyle Br}{\underset{\displaystyle Br}{\overset{|}{\underset{|}{C}}}}H\overset{\displaystyle CH_3}{\overset{|}{C}}CH_2CH_3$$

Instead of trying to memorize both equations, we can build a general rule that bromine reacts with compounds that contain a $C{=}C$ double bond to give the product expected from addition across the double bond. This approach to understanding the chemistry of organic compounds presumes that certain atoms or groups of atoms known as **functional groups** give these compounds their characteristic properties.

Functional groups focus attention on the important aspects of the structure of a molecule. We don't have to worry about the differences between the structures of

Table 16.5

Common Functional Groups

Functional Group	Name	Example
$-\overset{\textstyle \vert}{\underset{\textstyle \vert}{C}}-$	Alkane	$CH_3CH_2CH_3$ (propane)
$C=C$	Alkene	$CH_3CH=CH_2$ (propene)
$C\equiv CH$	Alkyne	$CH_3C\equiv CH$ (propyne)
F, Cl, Br, or I	Alkyl halide	CH_3Br (methyl bromide)
$-OH$	Alcohol	CH_3CH_2OH (ethanol)
$-O-$	Ether	CH_3OCH_3 (dimethyl ether)
$-NH_2$	Amine	CH_3NH_2 (methyl amine)

1-butene and 2-methyl-2-hexene, for example, when these compounds react with hydrogen bromide. We can focus on the fact that both compounds are alkenes that add HBr across the $C=C$ double bond.

$$CH_2=CHCH_2CH_3 + HBr \longrightarrow CH_3\overset{\textstyle \vert}{\underset{\textstyle Br}{C}}HCH_2CH_3$$

$$CH_3\overset{\textstyle CH_3}{\overset{\textstyle \vert}{C}}=CHCH_2CH_2CH_3 \longrightarrow CH_3\overset{\textstyle CH_3}{\overset{\textstyle \vert}{\underset{\textstyle \underset{Br}{\vert}}{C}}}CH_2CH_2CH_2CH_3$$

Some common functional groups are given in Table 16.5.

The $C=O$ group plays a particularly important role in organic chemistry. This group is called a **carbonyl,** and some of the functional groups that include a carbonyl are shown in Table 16.6.

Table 16.6

Functional Groups That Contain a Carbonyl

Functional Group	Name	Example
$-\overset{\textstyle O}{\overset{\textstyle \|}{C}}-H$	Aldehyde	CH_3CHO (acetaldehyde)
$-\overset{\textstyle O}{\overset{\textstyle \|}{C}}-$	Ketone	CH_3COCH_3 (acetone)
$-\overset{\textstyle O}{\overset{\textstyle \|}{C}}-Cl$	Acyl chloride	CH_3COCl (acetyl chloride)
$-\overset{\textstyle O}{\overset{\textstyle \|}{C}}-OH$	Carboxylic acid	CH_3CO_2H (acetic acid)
$-\overset{\textstyle O}{\overset{\textstyle \|}{C}}-O-$	Ester	$CH_3CO_2CH_3$ (methyl acetate)
$-\overset{\textstyle O}{\overset{\textstyle \|}{C}}-NH_2$	Amide	CH_3NH_2 (acetamide)

Exercise 16.8

Root beer hasn't tasted the same since the use of sassafras oil as a food additive was outlawed because sassafras oil is 80% safrole, which has been shown to cause cancer in rats and mice. Identify the functional groups in the structure of safrole.

Solution

Safrole is an aromatic compound because it contains a benzene ring. It is also an alkene because it contains a C=C double bond. The most difficult functional group to recognize in this molecule might be the two ether linkages (—O—).

Exercise 16.9

The following compounds are the active ingredients in over-the-counter drugs used as analgesics (to relieve pain without decreasing sensibility or consciousness), antipyretics (to reduce the body temperature when it is elevated), and/or anti-inflammatory agents (to counteract swelling or inflammation of the joints, skin, and eyes). These so-called nonsteroidal anti-inflammatory drugs (or NSAIDs) act by blocking the synthesis of a family of compounds known as prostaglandins, which have very significant physiological effects in even small quantities. Identify the functional groups in each molecule.

Solution

All three compounds are aromatic. Aspirin is also a carboxylic acid (—CO$_2$H) and an ester (—CO$_2$CH$_3$). Tylenol is also an alcohol (—OH) and an amide (—CONH—). Ibuprofen contains an alkane substituent and a carboxylic acid functional group.

Exercise 16.10

The discovery of penicillin in 1928 marked the beginning of what has been called the "golden age of chemotherapy" in which previously life-threatening bacterial infections were transformed into little more than a source of discomfort. For those who are allergic to penicillin, a variety of antibiotics, including tetracycline, are available. Identify the numerous functional groups in the tetracycline molecule.

Tetracycline

Solution

This compound contains an aromatic ring fused to three six-membered cycloalkane rings. It is also an alcohol (with five —OH groups), a ketone (with C=O groups at the bottom of the second and fourth rings), an amine (the —N(CH$_3$)$_2$ substituent at the top of the fourth ring), and an amide (the —CONH$_2$ group at the bottom right-hand corner of the fourth ring.)

Exercise 16.11

Herbal anesthetics extracted from opium-containing poppies have been known for thousands of years. The problem with their use was summarized by Gabriele Falloppio in the sixteenth century as follows: "When soporifics are weak, they are useless, and when strong, they kill." Cocaine was used as a local anesthetic as early as 1859. By the turn of the twentieth century it had been replaced by procaine, which is commonly known by the trade name novocaine. In recent years, lidocaine has become the local anesthetic of choice for dental work. Identify the functional groups in each of these compounds.

Lidocaine

Cocaine

Procaine (Novocaine)

Solution

All three compounds contain an aromatic ring. Cocaine also contains a seven-membered cycloalkane ring, as well as both amine (—NR$_3$) and ester (—CO$_2$R) functional groups. Lidocaine also contains both amine (—NR$_3$) and amide (—CONHR) functional groups. Procaine also contains both amine (—NR$_2$ and —NH$_2$) and ester (—CO$_2$R) functional groups.

16.10 Oxidation–Reduction Reactions

Focusing on the functional groups in a molecule allows us to recognize patterns in the behavior of related compounds. Consider what we know about the reaction between sodium metal and water, for example.

$$2\,Na(s) + 2\,H_2O(l) \longrightarrow H_2(g) + 2\,Na^+(aq) + 2\,OH^-(aq)$$

We can divide this reaction into two half-reactions. One half-reaction involves the oxidation of sodium metal to form sodium ions.

$$\text{Oxidation:} \qquad Na \longrightarrow Na^+ + e^-$$

The other half-reaction involves the reduction of an H$^+$ ion in water to form a neutral hydrogen atom that combines with another hydrogen atom to form an H$_2$ molecule.

$$\text{Reduction:} \qquad 2\,H^+ + 2\,e^- \longrightarrow H_2$$

Once we recognize that water can be thought of as containing an —OH functional group, we can predict what might happen when sodium metal reacts with an alcohol that contains the same functional group. Sodium metal should react with methanol (CH$_3$OH), for example, to give H$_2$ gas and a solution of the Na$^+$ and CH$_3$O$^-$ ions dissolved in this alcohol.

$$2\,Na(s) + 2\,CH_3OH(l) \longrightarrow H_2(g) + 2\,Na^+(alc) + 2\,CH_3O^-(alc)$$

Because the reactions between sodium metal and either water or an alcohol involve the transfer of electrons, they are examples of oxidation–reduction reactions. But what about the following reaction, in which hydrogen gas reacts with an alkene in the presence of a transition metal catalyst to form an alkane?

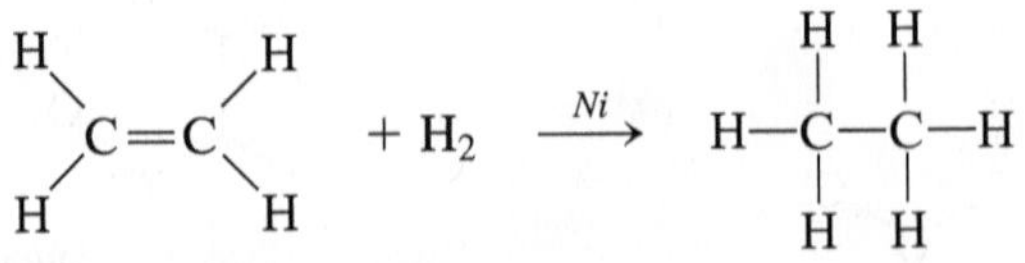

There is no change in the number of valence electrons on any of the atoms in this reaction. Both before and after the reaction, each carbon atom shares a total of eight valence electrons and each hydrogen atom shares two electrons. Instead of electrons, this reaction involves the transfer of atoms—in this case, hydrogen atoms.

There are so many atom-transfer reactions that chemists developed the concept of **oxidation number** (see Section 5.16) to extend the idea of oxidation and reduction to reactions in which electrons aren't necessarily gained or lost.

Oxidation involves an increase in the oxidation number of an atom.

Reduction occurs when the oxidation number of an atom decreases.

During the transformation of ethene into ethane, there is a *decrease* in the oxidation number of the carbon atom. This reaction therefore involves the *reduction* of ethene to ethane. The H_2 is oxidized.

Not all reactions involve a change in oxidation state. Consider the reaction between a carboxylic acid and an amine, for example.

$$CH_3CO_2H + CH_3NH_2 \longrightarrow CH_3CO_2^- + CH_3NH_3^+$$

Or the reaction between an alcohol and a hydrogen bromide.

$$CH_3CH_2OH + HBr \longrightarrow CH_3CH_2Br + H_2O$$

Neither of these reactions would be considered oxidation–reduction reactions because there is no change in the oxidation number of any atom in either reaction.

The oxidation numbers of the carbon atoms in a variety of compounds are given in Table 16.7.

These oxidation numbers can be used to decide whether an organic reaction is an example of an oxidation–reduction reaction.

Table 16.7
Typical Oxidation Numbers of Carbon

Functional Group	Example	Oxidation Number of Carbon in the Example
Alkane	CH_4	−4
Alkyllithium	CH_3Li	−4
Alkene	$H_2C=CH_2$	−2
Alcohol	CH_3OH	−2
Ether	CH_3OCH_3	−2
Alkyl halide	CH_3Cl	−2
Amine	CH_3NH_2	−2
Alkyne	$HC\equiv CH$	−1
Aldehyde	H_2CO	0
Carboxylic acid	HCO_2H	2
Carbon dioxide	CO_2	4

Exercise 16.12

Determine whether each of the following reactions is an oxidation–reduction reaction.

(a) $2\ CH_3OH \xrightarrow{H^+} CH_3OCH_3$

(b) $\overset{O}{\overset{\|}{HCOH}} + CH_3OH \xrightarrow{H^+} \overset{O}{\overset{\|}{HCOCH_3}}$

(c) $CO + H_2 \xrightarrow{catalyst} CH_3OH$

(d) $CH_3Br + 2\ Li \longrightarrow CH_3Li + LiBr$

Solution

(a) This is not an oxidation–reduction reaction because there is no change in the oxidation number of the carbon atoms when an alcohol is converted to an ether.

$$2\ \underset{-2}{CH_3OH} \longrightarrow \underset{-2}{CH_3OCH_3} + H_2O$$

(b) This is not an oxidation–reduction reaction because there is no change in the oxidation number of any of the carbon atoms when a carboxylic acid reacts with an alcohol to form an ester.

$$\underset{+2}{HCO_2H} + \underset{-2}{CH_3OH} \longrightarrow \underset{+2\ \ -2}{HCO_2CH_3} + H_2O$$

(c) This is an oxidation–reduction reaction because the carbon atom is reduced from the $+2$ to -2 oxidation state when CO combines with H_2 to form methanol.

$$\underset{+2}{CO} + H_2 \longrightarrow \underset{-2}{CH_3OH}$$

(d) This is an oxidation–reduction reaction because the carbon atom is reduced from the -2 to the -4 oxidation state when CH_3Br reacts with lithium metal to form CH_3Li.

$$\underset{-2}{CH_3Br} + 2\ Li \longrightarrow \underset{-4}{CH_3Li} + LiBr$$

Because electrons are neither created nor destroyed, oxidation can't occur in the absence of reduction, or vice versa. It is often useful, however, to focus attention on one component of the reaction and ask: Is that substance oxidized or reduced?

 Exercise 16.13

Determine whether the following transformations involve the oxidation or the reduction of the carbon atom.

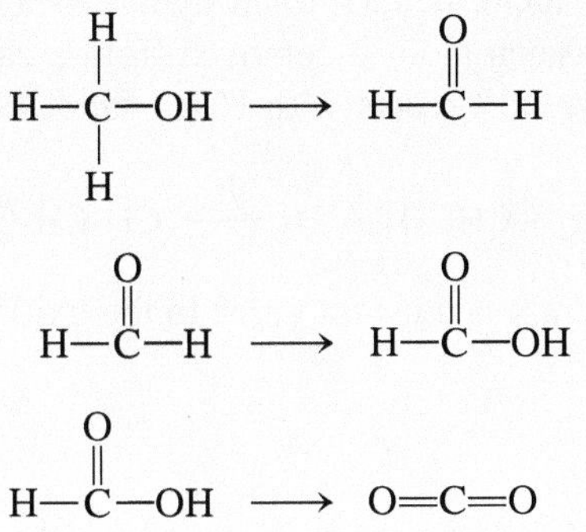

$$
\begin{array}{ccc}
\text{H} & & \text{O} \\
| & & \| \\
\text{H--C--OH} & \longrightarrow & \text{H--C--H} \\
| & & \\
\text{H} & &
\end{array}
$$

$$
\begin{array}{ccc}
\text{O} & & \text{O} \\
\| & & \| \\
\text{H--C--H} & \longrightarrow & \text{H--C--OH}
\end{array}
$$

$$
\begin{array}{ccc}
\text{O} & & \\
\| & & \\
\text{H--C--OH} & \longrightarrow & \text{O=C=O}
\end{array}
$$

Solution

Each of these transformations involves the oxidation of the carbon atom. The first reaction involves oxidation of the carbon atom from the -2 to the 0 oxidation state. In the second reaction, the carbon atom is oxidized to the $+2$ state, and the third reaction involves oxidation of the carbon atom to the $+4$ oxidation state.

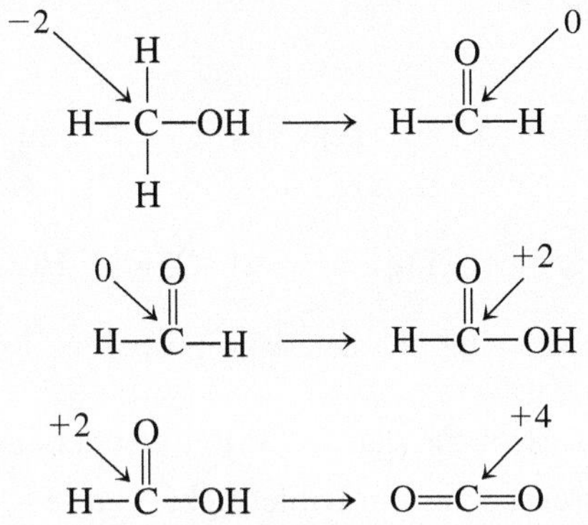

$$
\begin{array}{ccc}
\overset{-2}{\diagdown}\;\;\text{H} & & \text{O}\;\overset{0}{\diagup} \\
| & & \| \\
\text{H--C--OH} & \longrightarrow & \text{H--C--H} \\
| & & \\
\text{H} & &
\end{array}
$$

$$
\begin{array}{ccc}
\overset{0}{\diagdown}\;\text{O} & & \text{O}\;\overset{+2}{\diagup} \\
\| & & \| \\
\text{H--C--H} & \longrightarrow & \text{H--C--OH}
\end{array}
$$

$$
\begin{array}{ccc}
\overset{+2}{\diagdown}\;\text{O} & & \overset{+4}{\diagup} \\
\| & & \\
\text{H--C--OH} & \longrightarrow & \text{O=C=O}
\end{array}
$$

• •

Assigning oxidation numbers to the individual carbon atoms in a complex molecule can be difficult. Fortunately, there is another way to recognize many oxidation–reduction reactions in organic chemistry:

> **Oxidation occurs when hydrogen atoms are removed from a carbon atom or when an oxygen atom is added to a carbon atom.**

> **Reduction occurs when hydrogen atoms are added to a carbon atom or when an oxygen atom is removed from a carbon atom.**

The first reaction in Exercise 16.13 involves oxidation of the carbon atom because a pair of hydrogen atoms are removed from that atom when the alcohol is oxidized to an aldehyde.

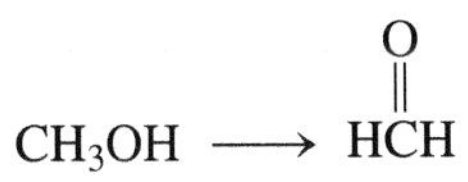

$$
\begin{array}{ccc}
& & \text{O} \\
& & \| \\
\text{CH}_3\text{OH} & \longrightarrow & \text{HCH}
\end{array}
$$

The second reaction in this exercise is an example of oxidation because an oxygen atom is added to the carbon atom when an aldehyde is oxidized to a carboxylic acid.

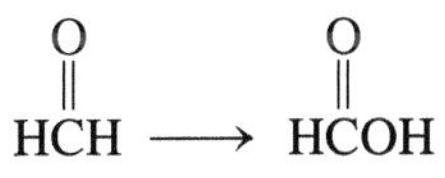

$$
\begin{array}{ccc}
\text{O} & & \text{O} \\
\| & & \| \\
\text{HCH} & \longrightarrow & \text{HCOH}
\end{array}
$$

Reduction, on the other hand, occurs when hydrogen atoms are added to a carbon atom or when an oxygen atom is removed from a carbon atom. An alkene is reduced, for example, when it reacts with H_2 to form the corresponding alkane.

$$CH_2{=}CHCH_3 + H_2 \xrightarrow{\text{Ni}} CH_3CH_2CH_3$$

The following diagram is a useful guide to the oxidation–reduction reactions of organic compounds.

$$CH_4 \xrightarrow{\;\;/\;\;} CH_3OH \longrightarrow \overset{\overset{\textstyle O}{\|}}{HCH} \longrightarrow \overset{\overset{\textstyle O}{\|}}{HCOH} \longrightarrow CO_2$$
$$\;\;-4\qquad\quad -2\qquad\quad 0\qquad\quad\; +2\qquad\quad +4$$

Each of the arrows in this figure involves a two-electron oxidation of a carbon atom along the path toward carbon dioxide. A line is drawn through the first arrow because it is impossible to achieve this transformation in a single step.

16.11 Alkyl Halides

Alkanes react with one of the halogens (F_2, Cl_2, Br_2, or I_2) in the presence of ultraviolet light.

$$CH_4(g) + Cl_2(g) \longrightarrow CH_3Cl(g) + HCl(g)$$

This reaction has the following characteristic properties.

- It doesn't take place in the dark or at low temperatures.
- It occurs in the presence of ultraviolet light or at temperatures above 250°C.
- Once the reaction gets started, it continues after the light is turned off.
- The products of the reaction include CH_2Cl_2 (dichloromethane), $CHCl_3$ (chloroform), and CCl_4 (carbon tetrachloride), as well as CH_3Cl (chloromethane).
- The reaction also produces some C_2H_6.

These facts are consistent with a **chain-reaction** mechanism that involves three processes: chain initiation, chain propagation, and chain termination.

CHAIN INITIATION

A Cl_2 molecule can dissociate into a pair of chlorine atoms by absorbing energy in the form of either ultraviolet light or heat.

$$Cl_2 \longrightarrow 2\,Cl\cdot \qquad \Delta H° = 243.4 \text{ kJ/mol}_{rxn}$$

The chlorine atom produced in this reaction is an example of a **free radical**—an atom or molecule that contains one or more unpaired electrons.

CHAIN PROPAGATION

Free radicals, such as the $Cl\cdot$ atom, are extremely reactive. When a chlorine atom collides with a methane molecule, it can pull a hydrogen atom off the methane molecule to form HCl and a $CH_3\cdot$ radical.

$$CH_4 + Cl\cdot \longrightarrow CH_3\cdot + HCl \qquad \Delta H° = -16 \text{ kJ/mol}_{rxn}$$

If the $CH_3\cdot$ radical then collides with a Cl_2 molecule, it can remove a chlorine atom to form CH_3Cl and a new $Cl\cdot$ radical.

$$CH_3\cdot + Cl_2 \longrightarrow CH_3Cl + Cl\cdot \qquad \Delta H^\circ = -87 \text{ kJ/mol}_{rxn}$$

Because a $Cl\cdot$ atom is generated in the second reaction for every $Cl\cdot$ atom consumed in the first, this reaction continues in a chainlike fashion until the radicals involved in these chain-propagation steps are destroyed.

CHAIN TERMINATION

If a pair of the radicals that keep the chain reaction going collide, they combine in a chain-terminating step. Chain termination can occur in three ways.

$$2\, Cl\cdot \longrightarrow Cl_2 \qquad \Delta H^\circ = -243.4 \text{ kJ/mol}_{rxn}$$
$$CH_3\cdot + Cl\cdot \longrightarrow CH_3Cl \qquad \Delta H^\circ = -330 \text{ kJ/mol}_{rxn}$$
$$2\, CH_3\cdot \longrightarrow CH_3CH_3 \qquad \Delta H^\circ = -350 \text{ kJ/mol}_{rxn}$$

Because the concentration of the radicals is relatively small, these chain-termination reactions are relatively infrequent.

This chain-reaction mechanism for free-radical reactions explains the observations listed for the reaction between CH_4 and Cl_2.

- The reaction doesn't occur in the dark or at low temperatures because energy must be absorbed to generate the free radicals that initiate the reaction.

$$Cl_2 \longrightarrow 2\, Cl\cdot \quad \Delta H^\circ = 243.4 \text{ kJ/mol}_{rxn}$$

- The reaction occurs in the presence of ultraviolet light because a UV photon has enough energy to dissociate a Cl_2 molecule to a pair of $Cl\cdot$ atoms. The reaction occurs at high temperatures because Cl_2 molecules can dissociate to form $Cl\cdot$ atoms by absorbing thermal energy.

- The reaction continues after the light has been turned off because light is only needed to generate the $Cl\cdot$ atoms that start the reaction. The chain reaction then converts CH_4 into CH_3Cl without consuming these $Cl\cdot$ atoms.

$$CH_4 + Cl\cdot \longrightarrow CH_3\cdot + HCl$$
$$CH_3\cdot + Cl_2 \longrightarrow CH_3Cl + Cl\cdot$$

- The reaction doesn't stop at CH_3Cl because the $Cl\cdot$ atoms can abstract additional hydrogen atoms to form CH_2Cl_2, $CHCl_3$, and eventually CCl_4.

$$CH_3Cl + Cl\cdot \longrightarrow CH_2Cl\cdot + HCl$$
$$CH_2Cl\cdot + Cl_2 \longrightarrow CH_2Cl_2 + Cl\cdot, \text{ and so on}$$

- The formation of C_2H_6 is a clear indication that the reaction proceeds through a free-radical mechanism. When two $CH_3\cdot$ radicals collide, they combine to form a ethane molecule.

$$2\, CH_3\cdot \longrightarrow CH_3CH_3$$

Free-radical halogenation of alkanes provides us with another example of the role of atom-transfer reactions in organic chemistry. The net effect of this reaction is to oxidize a carbon atom by removing a hydrogen from this atom.

$$CH_4 + Cl_2 \longrightarrow CH_3Cl + HCl$$
$$\quad -4 \qquad\qquad\qquad -2$$

16.12 Alcohols and Ethers

Alcohols contain an —OH group attached to a saturated carbon. The common names for alcohols are based on the name of the alkyl group.

CH_3OH	*Methyl alcohol*
CH_3CH_2OH	*Ethyl alcohol*
OH | CH_3CHCH_3	*Isopropyl alcohol*

The systematic nomenclature for alcohols adds the ending *-ol* to the name of the parent alkane and uses a number to identify the carbon that carries the —OH group. The systematic name for isopropyl alcohol, for example, is 2-propanol.

CH_3OH	*Methanol*
CH_3CH_2OH	*Ethanol*
OH | CH_3CHCH_3	*2-Propanol*

Exercise 16.14

More than 200 organic compounds have been isolated from the oil that gives rise to the characteristic odor of a rose. One of the most abundant of these compounds is known by the common name *citronellol*. Use the systematic nomenclature to name this alcohol, which has the following structure.

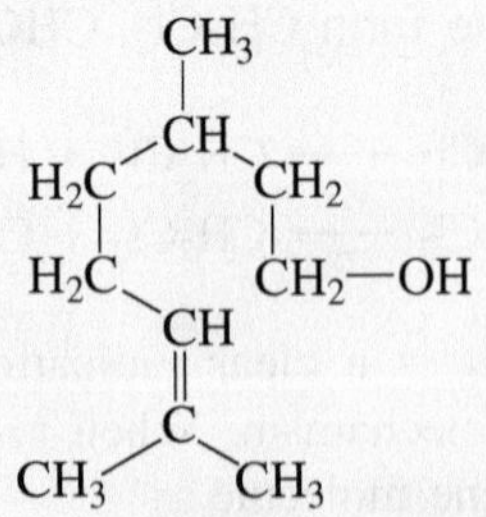

Solution

The longest chain of carbon atoms in this compound including the carbon carrying the —OH group contains eight atoms.

CH₃
|
$\overset{3}{CH}$
$\overset{4}{H_2C}$ $\overset{2}{CH_2}$
$\overset{5}{H_2C}$ $\overset{1}{CH_2}$—OH
$\overset{6}{CH}$
‖
C
CH₃ $\overset{7}{}$ CH₃
$\overset{8}{}$

Because the longest chain contains the —OH group, this compound is named as a derivative of octane. This compound is an alcohol; therefore, it would be tempting to name it as an *octanol*. But it contains a C=C double bond, which means it must be an *octenol*. We now have to indicate that the —OH group is on one end of the chain and the C=C double bond occurs between the sixth and seventh carbon atoms of this chain, which can be done by considering the compound to be a derivative of *oct-6-en-1-ol*. Finally, we have to indicate the presence of a pair of CH₃ groups on the third and seventh carbon atom. This compound is therefore given the systematic name *3,7-dimethyloct-6-en-1-ol*.

● ●

Methanol, or methyl alcohol, is also known as *wood alcohol* because it was originally made by heating wood until a liquid distilled. Methanol is highly toxic, and people have become blind or died from drinking it. Ethanol, or ethyl alcohol, is the alcohol associated with "alcoholic" beverages. It has been made for at least 6000 years by adding yeast to solutions that are rich in either sugars or starches. The yeast cells obtain energy from enzyme-catalyzed reactions that convert sugar or starch to ethanol and CO_2.

$$C_6H_{12}O_6(aq) \longrightarrow 2\ CH_3CH_2OH(aq) + 2\ CO_2(g)$$

When the alcohol reaches a concentration of 10 to 12% by volume, the yeast cells die. Brandy, rum, gin, and the various whiskeys that have a higher concentration of alcohol are prepared by distilling the alcohol produced by this fermentation reaction. Ethanol isn't as toxic as methanol, but it is still dangerous. Most people become intoxicated when their blood alcohol levels reach about 0.1 gram per 100 mL. An increase in the level of alcohol in the blood to between 0.4 and 0.6 g/100 mL can lead to coma or death.

Ethanol is oxidized to CO_2 and H_2O by the alcohol dehydrogenase enzymes in the liver. This reaction gives off 30 kilojoules per gram, which makes ethanol a better source of energy than carbohydrates (17 kJ/g), and almost as good a source of energy as fat (38 kJ/g). An ounce of 80-proof liquor can provide as much as 3% of the average daily caloric intake, and drinking alcohol can therefore contribute to obesity. Many alcoholics are malnourished, however, because of the absence of vitamins in the calories they obtain from alcoholic beverages.

Alcohols are classified as either *primary* (1°), *secondary* (2°), or *tertiary* (3°) on the basis of their structures.

OH   OH
| |
CH₃CH₂OH   CH₃CHCH₃   CH₃CCH₃
|
CH₃

1°   *2°*   *3°*

Ethanol is a primary alcohol because there is only one alkyl group attached to the carbon that carries the —OH substituent. The structure of a primary alcohol can be abbreviated as RCH_2OH, where R stands for an alkyl group. The isopropyl alcohol found in rubbing alcohol is a secondary alcohol, which has two alkyl groups on the carbon atom with the —OH substituent (R_2CHOH). An example of a tertiary alcohol (R_3COH) is *tert*-butyl (or *t*-butyl) alcohol or 2-methyl-2-propanol.

Another class of alcohols are the *phenols*, in which an —OH group is attached to an aromatic ring. Phenols are potent disinfectants. When antiseptic techniques were first introduced in the 1860s by Joseph Lister, it was phenol (or carbolic acid, as it was then known) that was used. Phenol derivatives, such as *o*-phenylphenol, are still used in commercial disinfectants such as Lysol.

Phenol   *o-Phenylphenol*

Alcohols are Brønsted acids in aqueous solution.

$$CH_3CH_2OH(aq) + H_2O(l) \rightleftharpoons H_3O^+(aq) + CH_3CH_2O^-(aq)$$

Alcohols therefore react with sodium metal to produce sodium salts of the corresponding conjugate base.

$$2\,Na(s) + 2\,CH_3OH(l) \longrightarrow 2\,Na^+(alc) + 2\,CH_3O^-(alc) + H_2(g)$$

The conjugate base of an alcohol is known as an **alkoxide.**

$$[Na^+][CH_3O^-] \qquad [Na^+][CH_3CH_2O^-]$$
sodium methoxide   *sodium ethoxide*

Alcohols (ROH) can be thought of as derivatives of water in which one of the hydrogen atoms has been replaced by an alkyl group. If both of the hydrogen atoms are replaced by alkyl groups, we get an **ether** (ROR). These compounds are named by adding the word *ether* to the names of the alkyl groups.

$$CH_3CH_2OCH_2CH_3 \qquad \textit{diethyl ether}$$

Diethyl ether, often known by the generic name "ether," was once used extensively as an anesthetic. Because mixtures of diethyl ether and air explode in the presence of a spark, ether has been replaced by safer anesthetics.

There are important differences between both the physical and chemical properties of alcohols and ethers. Consider diethyl ether and 1-butanol, for example, which are constitutional isomers with the formula $C_4H_{10}O$.

$$CH_3CH_2OCH_2CH_3 \qquad CH_3CH_2CH_2CH_2OH$$
BP = 34.5°C   *BP* = 117.2°C
d = 0.7138 g/mL   *d* = 0.8098 g/mL
insoluble in water   *soluble in water*

The shapes of these molecules are remarkably similar, as shown by the following line structures.

$$CH_3\diagdown_{CH_2}\diagup^{O}\diagdown_{CH_2}\diagup^{CH_3} \qquad CH_3\diagdown_{CH_2}\diagup^{CH_2}\diagdown_{CH_2}\diagup^{OH}$$

Diethyl ether   *1-Butanol*

The fundamental difference between these compounds is the presence of —OH groups in the alcohol that are missing in the ether. Because hydrogen bonds can't form between the molecules in the ether, the boiling point of the ether is more than 80°C lower than the corresponding alcohol. Because there are no hydrogen bonds to organize the structure of the liquid, the ether is also significantly less dense than the corresponding alcohol.

Ethers can act as a hydrogen-bond acceptor, as shown by the above line structure. But they can't act as hydrogen-bond donors. As a result, ethers are less likely to be soluble in water than the alcohol with the same molecular weight. The absence of an —OH group in an ether also has important consequences for its chemical properties. Unlike alcohols, ethers are essentially inert to chemical reactions. They don't react with most oxidizing or reducing agents, and they are stable to most acids and bases, except at high temperatures. They are therefore frequently used as solvents for chemical reactions.

Compounds that are potential sources of an H^+ ion, or proton, are often described as being **protic.** Ethanol, for example, is a *protic* solvent.

$$CH_3CH_2OH(aq) + H_2O(l) \rightleftharpoons H_3O^+(aq) + CH_3CH_2O^-(aq)$$

Substances that can't act as a source of a proton are said to be **aprotic.** Because they don't contain an —OH group, ethers are *aprotic* solvents.

16.13 Aldehydes and Ketones

The connection between the structures of alkenes and alkanes was established in Section 16.5, which noted that we can transform an alkene into an alkane by adding an H_2 molecule across the C=C double bond.

$$\begin{array}{c} H\diagdown \quad \diagup H \\ C=C \\ H\diagup \quad \diagdown H \end{array} + H_2 \xrightarrow{Pt} \begin{array}{c} H \;\; H \\ | \;\;\; | \\ H-C-C-H \\ | \;\;\; | \\ H \;\; H \end{array}$$

The driving force behind this reaction is the difference between the strengths of the bonds that must be broken and the bonds that form in the reaction. In the course of this hydrogenation reaction, a relatively strong H—H bond (435 kJ/mol) and a moderately strong carbon–carbon bond (one of the bonds in the C=C double bond) are broken, but two strong C—H bonds (439 kJ/mol) are formed. The reduction of an alkene to an alkane is therefore an exothermic reaction.

What about the addition of an H_2 molecule across a C=O double bond?

$$\begin{array}{c} H \;\; O \\ | \;\;\; \| \\ H-C-C-H \\ | \\ H \end{array} + H_2 \xrightarrow{Pt} \begin{array}{c} H \;\; OH \\ | \;\;\; | \\ H-C-C-H \\ | \;\;\; | \\ H \;\; H \end{array}$$

Once again, a significant amount of energy has to be invested in this reaction to break the H—H bond (435 kJ/mol) and one of the bonds in the carbon–oxygen double bond ($\approx$375 kJ/mol). The overall reaction is still exothermic, however, because of the strength of the C—H bond (413 kJ/mol) and the O—H bond (467 kJ/mol) that are formed.

The addition of hydrogen across a C=O double bond raises several important points. First, and perhaps foremost, it shows the connection between the chemistry of primary alcohols and aldehydes. But it also helps us understand the origin of the term *aldehyde*. If a reduction reaction in which H_2 is added across a double bond is an example of a *hydrogenation* reaction, then an oxidation reaction in which an H_2 molecule is removed to form a double bond might be called *dehydrogenation*. Thus, using the symbol [O] to represent an oxidizing agent, we see that the product of the oxidation of a primary alcohol is literally an "al-dehyd" or **aldehyde.** It is an *al*cohol that has been *dehyd*rogenated.

$$CH_3CH_2OH \xrightarrow{\text{[O]}} CH_3\overset{\displaystyle O}{\overset{\displaystyle \|}{C}}H$$

This reaction also illustrates the importance of differentiating between primary, secondary, and tertiary alcohols. Consider the oxidation of isopropyl alcohol, or 2-propanol, for example.

$$CH_3\overset{\displaystyle OH}{\overset{\displaystyle |}{C}}HCH_3 \xrightarrow{\text{[O]}} CH_3\overset{\displaystyle O}{\overset{\displaystyle \|}{C}}CH_3$$

The product of this reaction was originally called *aketone,* although the name was eventually softened to *azetone* and finally *acetone.* Thus, it is not surprising that any substance that exhibited chemistry that resembled "aketone" became known as a **ketone.**

Aldehydes can be formed by oxidizing a primary alcohol. Oxidation of a secondary alcohol gives a ketone. What happens when we try to oxidize a tertiary alcohol? The answer is: Nothing happens.

$$CH_3\overset{\displaystyle OH}{\underset{\displaystyle CH_3}{\overset{\displaystyle |}{\underset{\displaystyle |}{C}}}}CH_3 \xrightarrow{\text{[O]}} \!\!\!\!/$$

There aren't any hydrogen atoms that can be removed from the carbon atom carrying the —OH group in a 3° alcohol.

THE NOMENCLATURE OF ALDEHYDES AND KETONES

The common names of aldehydes are derived from the names of the corresponding carboxylic acids.

$$\overset{\displaystyle O}{\overset{\displaystyle \|}{H}}COH \qquad\qquad H\overset{\displaystyle O}{\overset{\displaystyle \|}{C}}H$$
Formic acid   *Formaldehyde*

$$CH_3\overset{\displaystyle O}{\overset{\displaystyle \|}{C}}OH \qquad\qquad CH_3\overset{\displaystyle O}{\overset{\displaystyle \|}{C}}H$$
Acetic acid   *Acetaldehyde*

The systematic names for aldehydes are obtained by adding -*al* to the name of the parent alkane.

$$\underset{\textit{Methanal}}{\overset{\overset{\displaystyle O}{\|}}{HCH}} \qquad \underset{\textit{Ethanal}}{\overset{\overset{\displaystyle O}{\|}}{CH_3CH}}$$

The presence of substituents is indicated by numbering the parent alkane chain from the end of the molecule that carries the —CHO functional group. For example,

$$\overset{\overset{\displaystyle O}{\|}}{BrCH_2CH_2CH} \quad \textit{3-Bromopropanal}$$

The common name for a ketone is constructed by adding *ketone* to the names of the two alkyl groups on the C=O double bond, listed in alphabetical order.

$$\overset{\overset{\displaystyle O}{\|}}{CH_3CCH_2CH_3} \quad \textit{Ethyle methyl ketone}$$

The systematic name is obtained by adding -*one* to the name of the parent alkane and using numbers to indicate the location of the C=O group.

$$\overset{\overset{\displaystyle O}{\|}}{CH_3CCH_2CH_3} \quad \textit{2-Propanone}$$

COMMON ALDEHYDES AND KETONES

Formaldehyde has a sharp, somewhat unpleasant odor. The aromatic aldehydes whose structures are shown below, on the other hand, have a very pleasant "flavor." Benzaldehyde has the characteristic odor of almonds, vanillin is responsible for the flavor of vanilla, and cinnamaldehyde makes an important contribution to the flavor of cinnamon.

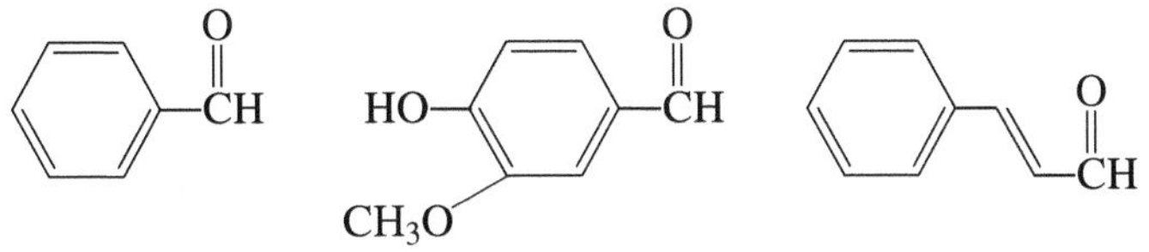

Benzaldehyde   *Vanillin*   *Cinnamaldehyde*

16.14 Reactions at the Carbonyl Group

It is somewhat misleading to write the carbonyl group as a covalent C=O double bond. The difference between the electronegativities of carbon and oxygen is large enough to make the C=O bond moderately polar. As a result, the carbonyl group is best described as a hybrid of the following resonance structures.

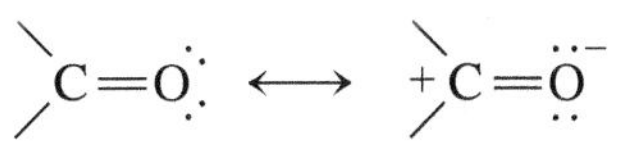

$$\overset{\diagdown}{\underset{\diagup}{C}}=\overset{..}{\underset{..}{O}} \quad \longleftrightarrow \quad \overset{\diagdown}{\underset{\diagup}{C}}{}^{+}=\overset{..}{\underset{..}{O}}{}^{-}$$

We can represent the polar nature of this hybrid by indicating the presence of a slight negative charge on the oxygen (δ^-) and a slight positive charge (δ^+) on the carbon of the C=O double bond.

$$\delta^+ \!\! \diagdown \!\! C = O^{\delta^-}$$

Reagents that react with the electron-rich δ^- end of the C=O bond are called **electrophiles** (literally, "lovers of electrons"). Electrophiles include ions (such as H^+ and Fe^{3+}) and neutral molecules (such as $AlCl_3$ and BF_3) that are *Lewis acids*, or *electron-pair acceptors*. Reagents that attack the electron-poor δ^+ end of this bond are **nucleophiles** (literally, "lovers of nuclei"). Nucleophiles are *Lewis bases* (such as NH_3 or the OH^- ion).

$$\begin{array}{c} \ddot{O}\\ \| \\ C \end{array} \quad \text{Electrophiles} \ (H^+, Fe^{3+}, BF_3, \text{etc.})$$

Nucleophiles
(OH^-, NH_3, etc.)

The polarity of the C=O double bond can be used to explain the reactions of carbonyl compounds. Aldehydes and ketones react with a source of the hydride (H^-) ion because the H^- ion is a Lewis base, or nucleophile, that attacks the δ^+ end of the C=O bond. When this happens, the two valence electrons on the H^- ion form a covalent bond to the carbon atom. Since carbon is tetravalent, one pair of electrons in the C=O bond is displaced onto the oxygen to form an intermediate alkoxide ion with a negative charge on the oxygen atom.

$$\begin{array}{c} \ddot{O} \\ \| \\ C \\ CH_3 \quad CH_3 \\ H:^- \end{array} \longrightarrow \begin{array}{c} :\ddot{O}:^- \\ | \\ CH_3-C-CH_3 \\ | \\ H \end{array}$$

An alkoxide ion

Because it is a strong Brønsted base, the alkoxide ion can then remove an H^+ ion from water to form an alcohol.

$$\begin{array}{c} :\ddot{O}:^- \\ | \\ CH_3-C-CH_3 + H_2O \\ | \\ H \end{array} \longrightarrow \begin{array}{c} :\ddot{O}-H \\ | \\ CH_3-C-CH_3 + OH^- \\ | \\ H \end{array}$$

Common sources of the H^- ion include lithium aluminum hydride ($LiAlH_4$) and sodium borohydride ($NaBH_4$). Both compounds are ionic.

$$LiAlH_4: \quad [Li^+][AlH_4^-]$$
$$NaBH_4: \quad [Na^+][BH_4^-]$$

The aluminum hydride (AlH_4^-) and borohydride (BH_4^-) ions act as if they were complexes between an H^- ion, acting as a Lewis base, and neutral AlH_3 or BH_3 molecules, acting as a Lewis acid.

$$\left[\begin{array}{c} H \\ | \\ H-Al-H \\ | \\ H \end{array} \right]^- \longrightarrow H{:}^- \quad \begin{array}{c} H \\ | \\ Al-H \\ | \\ H \end{array}$$

$LiAlH_4$ is such a good source of the H^- ion that it reacts with the H^+ ions in water or other protic solvents to form H_2 gas. The first step in the reduction of a carbonyl with $LiAlH_4$ is therefore carried out using an ether as the solvent. The product of the hydride reduction reaction is then allowed to react with water in a second step to form the corresponding alcohol.

$$CH_3CH_2\overset{\displaystyle O}{\overset{\displaystyle \|}{C}}H \quad \xrightarrow[\text{2. } H_2O]{\substack{\text{1. } LiAlH_4 \\ \text{in ether}}} \quad CH_3CH_2CH_2OH$$

$NaBH_4$ is less reactive toward protic solvents, which means that borohydride reductions are usually done in a single step, using an alcohol as the solvent.

$$CH_3CH_2\overset{\displaystyle O}{\overset{\displaystyle \|}{C}}CH_3 \quad \xrightarrow[CH_3CH_2OH]{NaBH_4} \quad CH_3CH_2\overset{\displaystyle OH}{\overset{\displaystyle |}{C}}HCH_3$$

16.15 Carboxylic Acids and Carboxylate Ions

When one of the substituents on a carbonyl group is an —OH group, the compound is a **carboxylic acid** with the generic formula RCO_2H. These compounds are acids, as the name suggests, that form **carboxylate ions** (RCO_2^-) by the loss of an H^+ ion.

$$CH_3-C\overset{\displaystyle O}{\underset{\displaystyle OH}{\diagup\!\!\!\diagdown}} \rightleftharpoons CH_3-C\overset{\displaystyle O}{\underset{\displaystyle O^-}{\diagup\!\!\!\diagdown}} + H^+$$

The carboxylate ion formed in this reaction is a hybrid of two resonance structures.

$$\left[CH_3-C\overset{\displaystyle O}{\underset{\displaystyle O^-}{\diagup\!\!\!\diagdown}} \longleftrightarrow CH_3-C\overset{\displaystyle O^-}{\underset{\displaystyle O}{\diagup\!\!\!\diagdown}} \right]$$

Resonance delocalizes the negative charge in the carboxylate ion, which makes this ion more stable than the alkoxide ion formed when an alcohol loses an H^+ ion. By increasing the stability of the conjugate base, resonance increases the acidity of the acid that forms this base. Carboxylic acids are therefore much stronger acids than the analogous alcohols. The value of K_a for a typical carboxylic acid is about 10^{-5}, whereas alcohols have values of K_a of only 10^{-16}.

$$CH_3C\overset{O}{\underset{OH}{}} \rightleftharpoons CH_3C\overset{O}{\underset{O^-}{}} + H^+ \qquad K_a = 1.8 \times 10^{-5}$$

$$CH_3OH \rightleftharpoons CH_3O^- + H^+ \qquad K_a = 8 \times 10^{-16}$$

Carboxylic acids were among the first organic compounds to be discovered. As a result, they have well-established common names that are often derived from the Latin stems of their sources in nature. Formic acid (Latin *formica*, "an ant") and acetic acid (Latin *acetum*, "vinegar") were first obtained by distilling ants and vinegar, respectively. Butyric acid (Latin *butyrum*, "butter") is found in rancid butter, and caproic, caprylic, and capric acids (Latin *caper*, "goat") are all obtained from goat fat. A list of common carboxylic acids is given in Table 16.8.

The systematic nomenclature of carboxylic acids involves adding the ending -*oic acid* to the name of the parent alkane to indicate the presence of the —CO_2H functional group.

HCO_2H	*Methanoic acid*
CH_3CO_2H	*Ethanoic acid*
$CH_3CH_2CO_2H$	*Propanoic acid*

Unfortunately, because of the long history of their importance in biology and biochemistry, you are more likely to encounter these compounds by their common names.

Table 16.8
Common Carboxylic Acids

Common Name	Formula	Solubility in H_2O (g/100 mL)
Saturated carboxylic acids and fatty acids		
Formic acid	HCO_2H	∞
Acetic acid	CH_3CO_2H	∞
Proprionic acid	$CH_3CH_2CO_2H$	∞
Butyric acid	$CH_3(CH_2)_2CO_2H$	∞
Caproic acid	$CH_3(CH_2)_4CO_2H$	0.968
Caprylic acid	$CH_3(CH_2)_6CO_2H$	0.068
Capric acid	$CH_3(CH_2)_8CO_2H$	0.015
Lauric acid	$CH_3(CH_2)_{10}CO_2H$	0.0055
Myristic acid	$CH_3(CH_2)_{12}CO_2H$	0.0020
Palmitic acid	$CH_3(CH_2)_{14}CO_2H$	0.00072
Stearic acid	$CH_3(CH_2)_{16}CO_2H$	0.00029
Unsaturated fatty acids		
Palmitoleic acid	$CH_3(CH_2)_5CH{=}CH(CH_2)_7CO_2H$	
Oleic acid	$CH_3(CH_2)_7CH{=}CH(CH_2)_7CO_2H$	
Linoleic acid	$CH_3(CH_2)_4CH{=}CHCH_2CH{=}CH(CH_2)_7CO_2H$	
Linolenic acid	$CH_3CH_2CH{=}CHCH_2CH{=}CHCH_2CH{=}CH(CH_2)_7CO_2H$	

Table 16.9

Common Dicarboxylic Acids

HO_2CCO_2H	Oxalic acid		
$HO_2CCH_2CO_2H$	Malonic acid	$\underset{\displaystyle HO_2CCH_2CHCO_2H}{\overset{\displaystyle OH}{|}}$	Malic acid
$HO_2CCH_2CH_2CO_2H$	Succinic acid		
$\underset{\displaystyle H}{\overset{\displaystyle HO_2C}{\diagdown}}C{=}C\underset{\displaystyle H}{\overset{\displaystyle CO_2H}{\diagup}}$	Maleic acid	$\underset{\displaystyle HO_2CCHCHCO_2H}{\overset{\displaystyle OH\,OH}{|\ \ |}}$	Tartaric acid
$\underset{\displaystyle H}{\overset{\displaystyle HO_2C}{\diagdown}}C{=}C\underset{\displaystyle CO_2H}{\overset{\displaystyle H}{\diagup}}$	Fumaric acid	$\underset{\displaystyle HO_2CCH_2CCO_2H}{\overset{\displaystyle O}{\|}}$	Oxaloacetic acid

Formic acid and acetic acid have a sharp, pungent odor. As the length of the alkyl chain increases, the odor of carboxylic acids becomes more unpleasant. Butyric acid, for example, is found in sweat, and the odor of rancid meat is due to carboxylic acids released as the meat spoils.

Compounds that contain two $-CO_2H$ functional groups are known as **dicarboxylic acids.** A number of dicarboxylic acids (see Table 16.9) can be isolated from natural sources. Tartaric acid, for example, is a by-product of the fermentation of wine, and succinic, fumaric, malic, and oxaloacetic acid are intermediates in the metabolic pathway used to oxidize sugars to CO_2 and H_2O.

Several tricarboxylic acids also play an important role in the metabolism of sugar. The most important example of this class of compounds is the citric acid that gives so many fruit juices their characteristic acidity.

$$\underset{\displaystyle CH_2CO_2H}{\overset{\displaystyle CH_2CO_2H}{\underset{|}{\overset{|}{H{-}C{-}CO_2H}}}} \quad \textit{Citric acid}$$

16.16 Esters

Carboxylic acids ($-CO_2H$) can react with alcohols (ROH) in the presence of either acid or base to form **esters** ($-CO_2R$). Acetic acid, for example, reacts with ethanol to form ethyl acetate and water.

$$\underset{\displaystyle CH_3COH}{\overset{\displaystyle O}{\|}} + CH_3CH_2OH \underset{\displaystyle}{\overset{\displaystyle H^+}{\rightleftharpoons}} \underset{\displaystyle CH_3COCH_2CH_3}{\overset{\displaystyle O}{\|}} + H_2O$$

This isn't an efficient way of preparing an ester, however, because the equilibrium constant for this reaction is relatively small ($K_c \approx 3$). Chemists tend to synthesize esters in a two-step process. They start by reacting the acid with a chlorinating agent such as thionyl chloride (SOCl$_2$) to form the corresponding **acyl chloride.**

$$\underset{\displaystyle CH_3COH}{\overset{\displaystyle O}{\|}} + SOCl_2 \longrightarrow \underset{\displaystyle CH_3CCl}{\overset{\displaystyle O}{\|}} + SO_2 + HCl$$

They then react the acyl chloride with an alcohol in the presence of base to form the ester.

$$\underset{\displaystyle CH_3CCl}{\overset{\displaystyle O}{\|}} + CH_3CH_2OH \xrightarrow{\;B\;} \underset{\displaystyle CH_3COCH_2CH_3}{\overset{\displaystyle O}{\|}} + BH^+\, Cl^-$$

The base reacts with the HCl given off in this reaction, thereby driving it to completion.

As might be expected, esters are named as if they were derivatives of a carboxylic acid and an alcohol. The ending *-ate* or *-oate* is added to the name of the parent carboxylic acid, and the alcohol is identified using the "alkyl alcohol" convention. The following ester, for example, can be named as a derivative of acetic acid (CH$_3$CO$_2$H) and ethyl alcohol (CH$_3$CH$_2$OH).

$$\underset{\displaystyle CH_3COCH_2CH_3}{\overset{\displaystyle O}{\|}} \quad \textit{Ethyl acetate}$$

Or it can be named as a derivative of ethanoic acid (CH$_3$CO$_2$H) and ethyl alcohol (CH$_3$CH$_2$OH).

$$\underset{\displaystyle CH_3COCH_2CH_3}{\overset{\displaystyle O}{\|}} \quad \textit{Ethyl ethanoate}$$

The term *ester* is commonly used to describe the product of the reaction of any strong acid with an alcohol. Sulfuric acid, for example, reacts with methanol to form a diester known as dimethyl sulfate.

$$HO-\underset{\displaystyle \underset{O}{|}}{\overset{\displaystyle \overset{O}{\|}}{S}}-OH + 2\,CH_3OH \longrightarrow CH_3O-\underset{\displaystyle \underset{O}{|}}{\overset{\displaystyle \overset{O}{\|}}{S}}-OCH_3$$

Phosphoric acid reacts with alcohols to form triesters such as triethyl phosphate.

$$HO-\underset{\displaystyle \underset{OH}{|}}{\overset{\displaystyle \overset{O}{\|}}{P}}-OH + 3\,CH_3OH \longrightarrow CH_3O-\underset{\displaystyle \underset{OCH_3}{|}}{\overset{\displaystyle \overset{O}{\|}}{P}}-OCH_3$$

Compounds that contain the —CO$_2$R functional group might therefore best be called **carboxylic acid esters,** to indicate the acid from which they are formed.

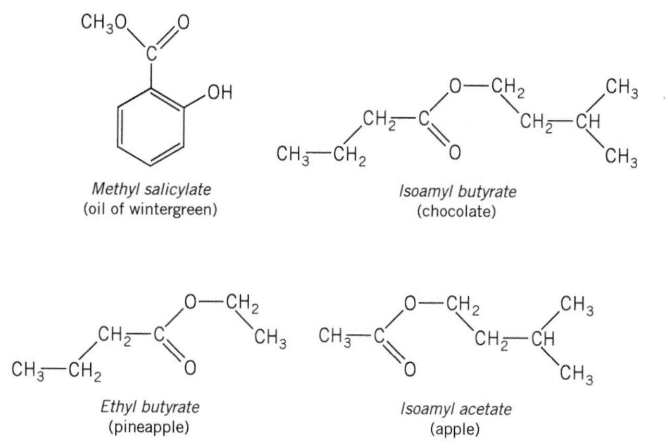

*Methyl salicylate
(oil of wintergreen)*

*Isoamyl butyrate
(chocolate)*

*Ethyl butyrate
(pineapple)*

*Isoamyl acetate
(apple)*

Fig. 16.3 Typical carboxylic acid esters with pleasant odors or flavors.

Carboxylic acid esters with low molecular weights are colorless, volatile liquids that often have a pleasant odor. They are therefore important components of both natural and synthetic flavors (see Figure 16.3).

16.17 Amines; Alkaloids, and Amides

Amines are derivatives of ammonia in which one or more hydrogen atoms are replaced by alkyl groups. We indicate the degree of substitution by labeling the amine as either primary (RNH_2), secondary (R_2NH), or tertiary (R_3N). The common names of these compounds are derived from the names of the alkyl groups.

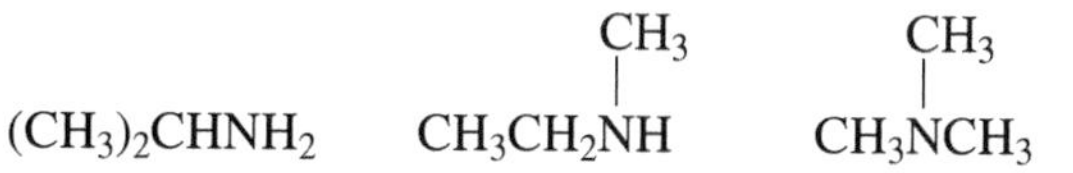

$$(CH_3)_2CHNH_2 \quad CH_3CH_2\overset{\overset{\displaystyle CH_3}{|}}{N}H \quad CH_3\overset{\overset{\displaystyle CH_3}{|}}{N}CH_3$$

Isopropylamine   *Ethylmethylamine*   *Trimethylamine*

The systematic names of primary amines are derived from the name of the parent alkane by adding the prefix *-amino* and a number specifying the carbon that carries the —NH_2 group.

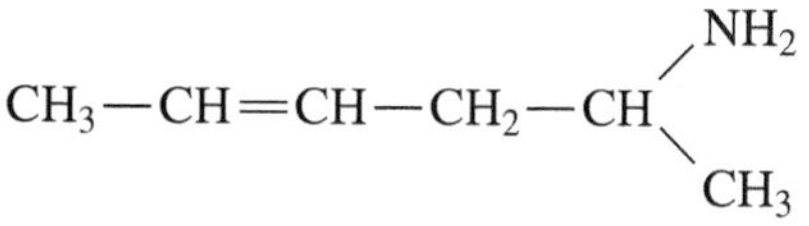

$$CH_3—CH{=}CH—CH_2—CH\overset{\displaystyle NH_2}{\underset{\displaystyle CH_3}{{<}}} \quad \textit{Hex-4-en-2-amine}$$

The chemistry of amines mirrors the chemistry of ammonia. Amines are bases that pick up a proton to form ammonium salts. Trimethylamine, for example, reacts with acid to form the trimethylammonium ion.

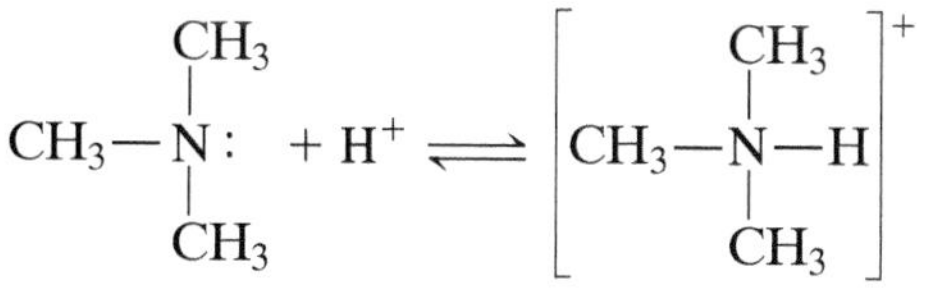

$$CH_3—\overset{\overset{\displaystyle CH_3}{|}}{\underset{\underset{\displaystyle CH_3}{|}}{N}}{:} \; + H^+ \rightleftharpoons \left[CH_3—\overset{\overset{\displaystyle CH_3}{|}}{\underset{\underset{\displaystyle CH_3}{|}}{N}}—H \right]^+$$

These salts are more soluble in water than the corresponding amines, and this reaction can be used to dissolve otherwise insoluble amines in aqueous solution.

The difference between amines and their ammonium salts plays an important role in both over-the-counter and illicit drugs. Cocaine, for example, is commonly

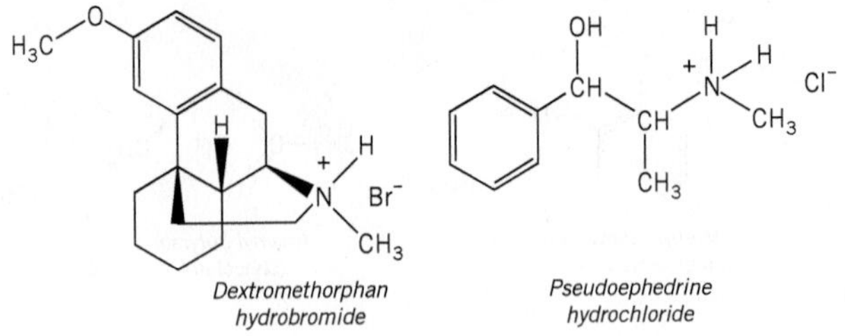

Fig. 16.4 The structures of common ingredients in over-the-counter drugs that are used as ammonium ion salts to increase the solubility of the active ingredient in water

Dextromethorphan hydrobromide

Pseudoephedrine hydrochloride

sold as the hydrochloride salt, which is a white, crystalline solid. By extracting this solid into ether, it is possible to obtain the "free base." A glance at the side panel of almost any over-the-counter medicine will provide examples of ammonium salts that are used to ensure that the drug dissolves in water. Over-the-counter drugs sold as amine hydrochlorides or hydrobromides include dextromethorphan hydrobromide, which is used as a cough suppressant, and pseudoephedrine hydrochloride, which is a decongestant, shown in Figure 16.4.

Amines that are isolated from plants are known as **alkaloids.** They include poisons such as nicotine, coniine, and strychnine shown in Figure 16.5. Nicotine has a pleasant, invigorating effect when taken in minuscule quantities, but is extremely toxic in larger amounts. Coniine is the active ingredient in hemlock, a poison that has been used since the time of Socrates. Strychnine is another toxic alkaloid that has been a popular poison in murder mysteries.

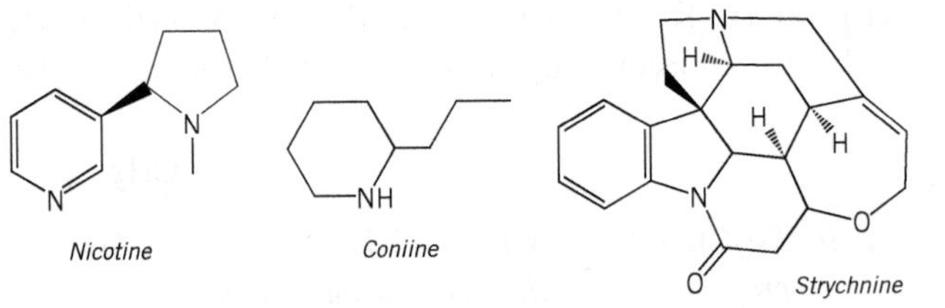

Fig. 16.5 The structures of toxic alkaloids that can be extracted from plants.

Nicotine

Coniine

Strychnine

The alkaloids also include a number of drugs, such as morphine, quinine, and cocaine, shown in Figure 16.6. Morphine is obtained from poppies; quinine can be found in the bark of the chinchona tree; and cocaine is isolated from coca leaves. This family of compounds also includes synthetic analogs of naturally occurring alkaloids, such as heroin and LSD).

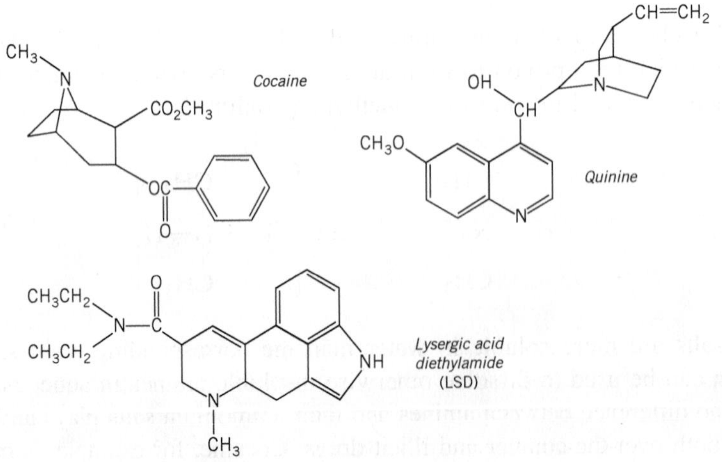

Cocaine

Quinine

Lysergic acid diethylamide (LSD)

Fig. 16.6 The structure of common alkaloids that have been used as drugs.

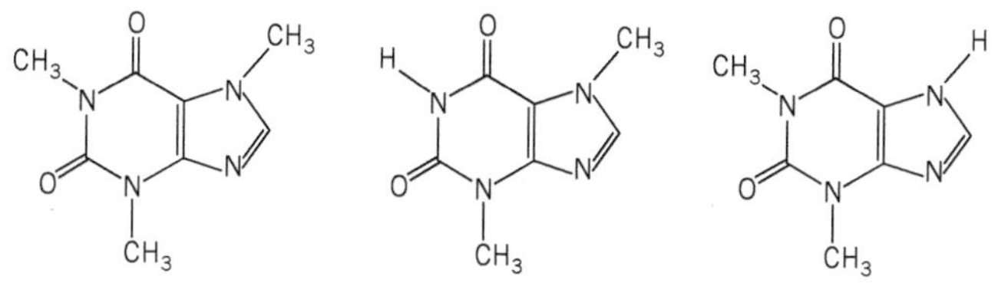

Fig. 16.7 The caffeine in coffee, theobromine in chocolate, and the active ingredient in a common bronchodilator are central nervous system stimulants with very similar structures.

To illustrate the importance of minor changes in the structure of a molecule, three amines with similar structures are shown in Figure 16.7. One of these compounds is caffeine, which is the pleasantly addictive substance that makes a cup of coffee an important part of the day for so many people. Another is theobromine, which is the pleasantly addictive substance that draws so many people to chocolate. The third compound is theophylline, which is a prescription drug used as a bronchodilator by individuals with asthma.

It is tempting to assume that carboxylic acids will react with amines to form the class of compounds known as **amides.** In practice, when aqueous solutions of carboxylic acids and amines are mixed, we get an acid–base reaction.

$$CH_3\overset{O}{\overset{\|}{C}}OH + CH_3NH_2 \rightleftharpoons CH_3\overset{O}{\overset{\|}{C}}O^-\ CH_3NH_3{}^+$$

The best way to prepare an amide is to react the appropriate acyl chloride with an amine.

$$CH_3\overset{O}{\overset{\|}{C}}Cl + 2\ CH_3NH_2 \longrightarrow CH_3\overset{O}{\overset{\|}{C}}NHCH_3 + CH_3NH_3{}^+\ Cl^-$$

Excess amine is used to drive the reaction to completion.

16.18 Alkene Stereoisomers

The geometry around the $C{=}C$ double bond in an alkene plays an important role in the chemistry of these compounds. To understand why, let's return to the hypothetical intermediate in which we have a C_2H_4 molecule with an unpaired electron on each of the carbon atoms.

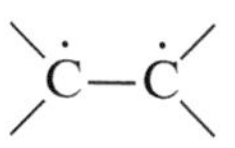

The unpaired electron on one of these carbon atoms interacts with the unpaired electron on the other carbon atom to form a second covalent bond. The geometry around a $C{=}C$ double bond is therefore different from the geometry around a $C{-}C$ single bond. Because of the double bond, the six atoms in a C_2H_4 molecule all lie in the same plane, as shown in Figure 16.8. There is no way to rotate one end of this bond relative to the other without breaking the double bond. Because the double bond is relatively strong (≈ 600 kJ/mol), rotation around the $C{=}C$ double bond cannot occur at room temperature.

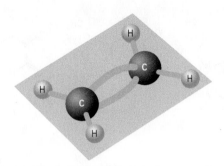

Fig. 16.8 The planar C_2H_4 molecule.

Isomers were defined in Section 16.2 as compounds that have the same molecular formula, but different structures. Isomers can differ in two ways. When they differ in the way the atoms are connected, they are called **constitutional isomers.**[3] Butane and isobutane, for example, are constitutional isomers.

$$CH_3-CH_2-CH_2-CH_3 \qquad\qquad \overset{\displaystyle CH_3}{\underset{\displaystyle }{CH_3-\overset{|}{C}H-CH_3}}$$

Butane   Isobutane

Constitutional isomers have similar chemical properties but different physical properties. Butane, for example, melts at $-138.4°C$ and boils at $-0.5°C$, whereas isobutane melts at $-159.6°C$ and boils at $-11.7°C$.

Isomers in which the atoms are connected in the same way, but differ in how the atoms are arranged in space, are called **stereoisomers.** Alkenes form stereoisomers that differ in the way substituents are arranged around the C=C double bond. The isomer with substituents on the *same side* of the double bond is called **cis,** from a Latin stem meaning "on this side." The isomer in which the substituents are *across from each other*, is called **trans,** from a Latin stem meaning "across." The *cis* isomer of 2-butene, for example, has both CH_3 groups on the top side of the double bond in the structure for this molecule shown below. In the *trans* isomer the CH_3 groups are on opposite sides of the double bond. One CH_3 group is on the top side of the double bond, and the other is on the bottom side of the bond.

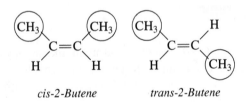

cis-2-Butene   trans-2-Butene

● ●

 Exercise 16.15

Name the straight-chain constitutional and stereoisomers of pentene (C_5H_{10}).

Solution

There are two constitutional isomers of a straight-chain pentene molecule. These isomers differ in the location of the C=C double bond. In one isomer,

---

[3]One of the definitions of constitution in the *Webster's International Dictionary* is "the way in which a person or thing is made up; structure, organization."

the C=C bond is between the first and second carbon atoms in the chain. In the other isomer, it is between the second and third carbon atoms.

$$H_2C=CHCH_2CH_2CH_3 \quad CH_3CH=CHCH_2CH_3$$

1-Pentene   *2-Pentene*

There are no *cis/trans* isomers of l-pentene because there is only one way of arranging the substituents around the double bond.

Cis and *trans* isomers are possible, however, for 2-pentene.

cis-2-Pentene   *trans-2-Pentene*

• •

The limitations of the *cis/trans* convention for describing stereoisomers of alkenes can be appreciated by considering the following compound.

Is this *trans*-3-methyl-2-pentene because the two CH_3 substituents are across the double bond from each other? Or should it be called *cis*-3-methyl-2-pentene because the two "bulky" substituents (CH_3 and CH_3CH_2) are on the same side of the double bond?

An unambiguous system for describing stereoisomers of alkenes has been developed that is based on strict rules for assigning a priority to the substituents on each end of the C=C double bond.

- The highest priority is assigned to the atom that has the largest atomic number. Carbon ($Z = 6$), for example, would have a higher priority than hydrogen ($Z = 1$), and bromine ($Z = 35$) would have a higher priority than chlorine ($Z = 17$).

- When the atoms directly attached to the double bond are the same, compare the substituents on this atom. Consider the CH_3 and CH_2CH_3 substituents, for example. The carbon atom on the CH_2CH_3 substituent would be assigned a higher priority because it is bound to two H atoms and a C atom, whereas the carbon atom in the CH_3 substituent is bound to three H atoms.

- Identify the substituent on each carbon atom that has the highest priority.

- If the substituents with the highest priority on both carbon atoms are on the same side of the double bond, the compound is described as the Z isomer (from the German *zusammen*, which means "together.")

- If the substituents with the highest priority on the carbon atoms are on opposite sides of the double bond, the compound is the *E* isomer (from the German *entgegen*, "opposite.")

Exercise 16.16

Determine whether the following compound should be described as the *E* or *Z* isomer of 3-methyl-2-pentene.

$$CH_3 \quad\quad CH_2CH_3$$
$$\diagdown C = C \diagup$$
$$H \quad\quad CH_3$$

Solution

There is no doubt about the priority of the CH_3 and H substituents on the second carbon atom in this molecule.

Highest priority

$$(CH_3) \quad\quad CH_2CH_3$$
$$\diagdown C = C \diagup$$
$$H \quad\quad CH_3$$

There is no way to assign priority on the basis of the atoms bound to the third carbon in this molecule—they are both carbon atoms. We therefore proceed down the substituent. The next atoms we encounter in the CH_3 group are all hydrogen atoms. But the next atoms in the CH_2CH_3 substituent are two hydrogen atoms and a carbon atom. The CH_2CH_3 group therefore is assigned a higher priority.

Highest priority

$$CH_3 \quad\quad (CH_2CH_3)$$
$$\diagdown C = C \diagup$$
$$H \quad\quad CH_3$$

We therefore conclude that the two substituents with the highest priority are on *the same side* of the C=C double bond. They are both on the top side of the C=C bond in the structure with which this molecule has been drawn.

Highest priority   *Highest priority*

$$(CH_3) \quad\quad (CH_2CH_3)$$
$$\diagdown C = C \diagup$$
$$H \quad\quad CH_3$$

This compound is therefore the *Z* isomer, and it would be named as follows: (*Z*)-3-methyl-2-pentene.

16.19 Stereogenic Atoms

The previous section introduced an example of stereoisomerism: *cis/trans* isomers such as *cis-* and *trans*-2-butene.

$$CH_3 \diagdown C{=}C \diagup CH_3 \qquad CH_3 \diagdown C{=}C \diagup H$$
$$H \diagup \qquad \diagdown H \qquad\qquad H \diagup \qquad \diagdown CH_3$$

cis-2-Butene   *trans*-2-Butene

Cis/trans isomers have similar chemical properties but different physical properties. *Cis*-2-Butene, for example, freezes at −138.9°C, whereas *trans*-2-butene freezes at −105.6°C.

The carbon atoms that form the C=C double bond in 2-butene are called **stereocenters** or **stereogenic atoms.** A stereocenter is an atom for which the interchange of two groups converts one stereoisomer into another. The carbon atoms in the C=C double bond in 2-butene, for example, are stereocenters. Interchanging the CH₃ and H substituents on the carbon atom on either side of the C=C double bond would convert *cis*-2-butene into *trans*-2-butene, and vice versa.

cis-2-Butene   *trans*-2-Butene

CHIRALITY, OR THE "HANDEDNESS" OF A MOLECULE

The *cis/trans* isomers formed by alkenes are not the only example of stereoisomers. To understand the second example of stereoisomers, it might be useful to start by considering a pair of hands. For all practical purposes, they contain the same "substituents"—four fingers and one thumb on each hand. If you clap them together, you will find even more similarities between the two hands. The thumbs are attached at about the same point on the hand, significantly below the point where the fingers start. The second fingers on both hands are usually the longest, then the third fingers, then the first fingers, and finally the "little" fingers.

In spite of their many similarities, there is a fundamental difference between a pair of hands, which can be observed by trying to place your right hand into a typical left-hand glove. Your hands have two important properties: (1) each hand is the *mirror image* of the other, and (2) these mirror images are not *superimposable*. The mirror image of the left hand looks like the right hand, and vice versa, as shown in Figure 16.9.

Some molecules possess a similar "handedness." They exist as stereoisomers whose structures have two characteristic properties: (1) each isomer is the mirror image of the other, and (2) the two isomers are not superimposable. Molecules that possess these properties are said to be **chiral** (literally: "handed"). Molecules that are not chiral are said to be **achiral.** As we have already seen, right- and left-handed gloves are "chiral." (It is difficult, if not impossible, to place the right-hand glove on your left hand or the left-hand glove on your right hand.) Mittens, however, are usually "achiral." (Either mitten can fit on either hand.) Feet and shoes are both chiral, but socks are not.

In 1874 Jacobus van't Hoff and Joseph Le Bel recognized that a compound that contains a single tetrahedral carbon atom with four different substituents could exist in two forms that were mirror images of each other. Consider the CHFClBr molecule, for example, which contains four different substituents on a tetrahedral carbon atom. Figure 16.10 shows one possible arrangement of these substituents and the mirror image of this structure. As noted in Section 16.1, solid lines are

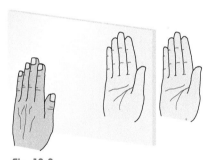

Fig. 16.9 Each hand is the mirror image of the other, but neither hand is superimposable on the other.

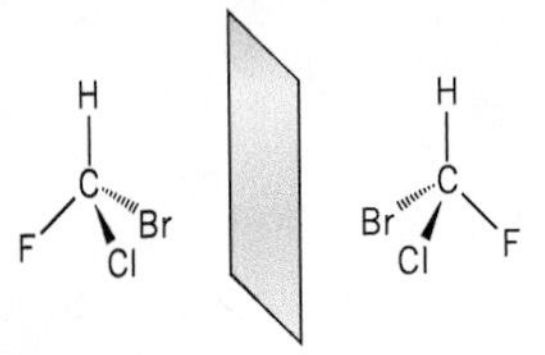

Fig. 16.10 One form of the CHFClBr molecule and its mirror image.

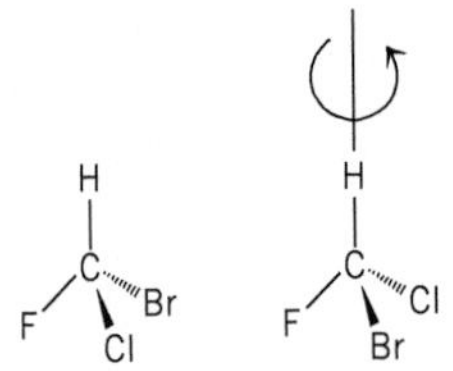

Fig. 16.11 To determine whether a molecule is optically active, we have to decide whether the molecule and its mirror image can be superimposed. We can start by rotating the mirror image by 180° around the C—H.

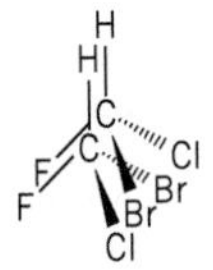

Fig. 16.12 This compound is chiral because these structures are not superimposable.

used to represent bonds that lie in the plane of the paper. Wedges are used for bonds that come out of the plane of the paper toward the viewer; dashed lines describe bonds that go behind the paper.

If we rotate the molecule on the right in Figure 16.10 by 180° around the C—H bond, we get the structures shown in Figure 16.11.

These structures are different because they cannot be superimposed on each other, as shown in Figure 16.12.

CHFClBr is therefore a chiral molecule that exists in the form of a pair of stereoisomers that are mirror images of each other that cannot be superimposed.

As a rule, any tetrahedral atom that carries four different substituents is a stereocenter, or a stereogenic atom. Compounds that contain a single stereocenter are always chiral. Some compounds that contain two or more stereocenters are achiral because of the symmetry of the relationship between the stereocenters.

The prefix "en-" often means "to make, or cause to be," as in "endanger." It is also used to strengthen a term, to make it even more forceful, as in "enliven." Thus, it is not surprising that a pair of stereoisomers that are mirror images of each are called **enantiomers.** They are literally compounds that contain parts that are forced to be across, or anti-, from each other. Stereoisomers that are not mirror images of each other are called **diastereomers.**[4] The *cis/trans* isomers of 2-butene, for example, are stereoisomers, but they are not mirror images of each other. As a result, they are diastereomers.

Exercise 16.17

Which of the following compounds would form enantiomers because the molecule is chiral?

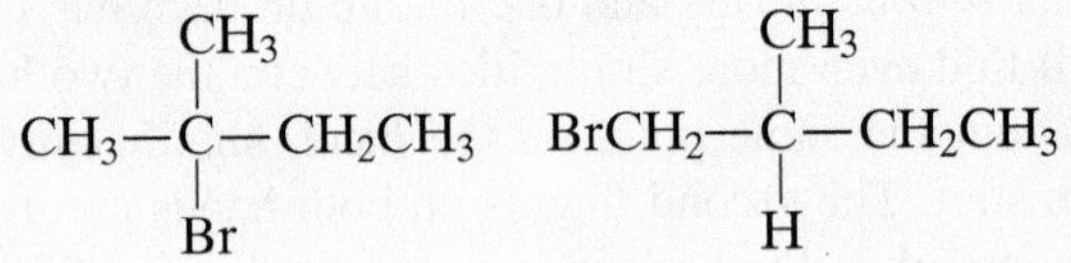

2-Bromo-2-methylbutane   *1-Bromo-2-methylbutane*

Solution

The second carbon atom in 2-bromo-2-methylbutane contains two identical CH_3- substituents. As a result, it is achiral, and this compound does not form enantiomers.

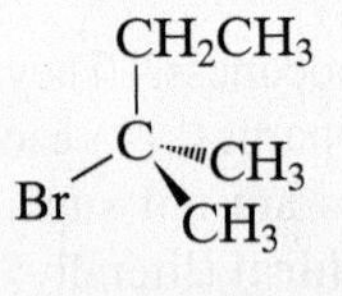

2-Bromo-2-methylbutane

The second carbon atom in 1-bromo-2-methylbutane, on the other hand, carries four different substituents: H, Br, CH_3, and CH_2CH_3. As a result, this molecule is chiral, and it forms enantiomers.

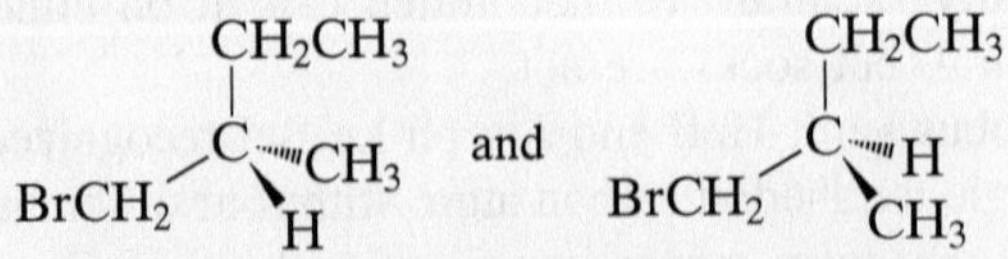

Enantiomers of 1-Bromo-2-methylbutane

---

[4]The prefix "dia-" is often used to indicate "opposite directions," or "across," as in *diagonal*.

Note: Every object has a mirror image. The relevant question is whether the mirror image can be superimposed on the source of the image. If it can, the object is not chiral (*achiral*). If it cannot, the object is chiral, and it can exist as a pair of stereoisomers.

● ●

16.20 Optical Activity

THE DIFFERENCE BETWEEN ENANTIOMERS ON THE MACROSCOPIC SCALE

Light is just one component of the spectrum of electromagnetic radiation that ranges from low-energy radio waves up to very high-energy gamma rays (see Section 3.3). The different forms of radiation in this spectrum all have certain properties in common: They travel at the speed of light, they are composed of waves that have a characteristic frequency (v) and wavelength (λ), and they consist of electric and magnetic fields that oscillate in planes parallel to the axis along which the radiation travels.

If you could analyze the light that travels toward you from a lamp, you would find components of the light oscillating in all of the possible planes parallel to the path of the light. However, when light is passed through a polarizer such as a Nicol prism or the lens of polarized sunglasses, the light that emerges has its oscillations confined to a single plane.

In 1813 Jean Baptiste Biot noticed that plane-polarized light was rotated either to the right or the left when it passed through single crystals of quartz or aqueous solutions of tartaric acid or sugar. Because they interact with light, substances that can rotate plane-polarized light are said to be **optically active.** Those that rotate the plane clockwise (to the right) are said to be **dextrorotatory** (from the Latin *dexter*, "right").[5] Those that rotate the plane counterclockwise (to the left) are called **levorotatory** (from the Latin *laevus*, "left"). The instrument with which optically active compounds are studied is a polarimeter, such as the one shown in Figure 16.13.

Imagine a horizontal line that passes through the zero of a coordinate system. Normally, you would place the negative numbers on the left and the positive

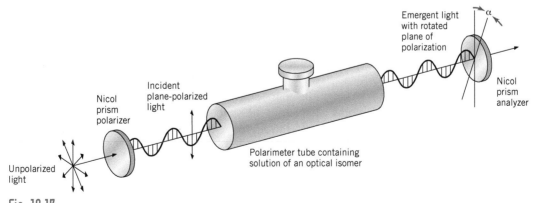

Fig. 16.13 A polarimeter for studying the rotation of plane-polarized light.

---

[5]You might remember that "dextro" means right by noting that the predominantly right-handed world in which we live uses words such as *dextrous* to mean unusually skilled at the use of one's hands—in particular, the "right" hand.

numbers on the right of zero. Thus, it is not surprising that dextrorotatory compounds are indicated with a positive sign (+) and levorotatory compounds with a negative sign (−).

The angle through which an enantiomer rotates plane-polarized light depends on four quantities: (1) the wavelength of the light, (2) the length of the cell through which the light passes, (3) the concentration of the optically active compound in the solution through which the light passes, and (4) the **specific rotation** of the compound, which reflects the relative ability of the compound to rotate plane-polarized light. The specific rotation of the dextrorotatory isomer of glucose is written as follows:

$$[\alpha]_D^{20} = +3.12$$

When the spectrum of sunlight was first analyzed by Joseph von Fraunhofer in 1814, he observed a limited number of dark bands in this spectrum, which he labeled A–H. We now know that the D band in this spectrum is the result of the absorption of light that has a wavelength of 589.6 nm by sodium atoms. The "D" in the symbol for rotation indicates that it is light of this wavelength that was studied. The "20" indicates that the experiment was done at 20°C. The "+" sign indicates that the compound is dextrorotatory; it rotates light clockwise. Finally, the magnitude of this measurement indicates that when a solution of this compound that has a concentration of 1.00 g/mL was studied in a 10-cm cell, it rotated the light by 3.12°.

The magnitude of the rotations observed for a pair of enantiomers is always the same. The only difference between these compounds is the direction in which they rotate plane-polarized light. The specific rotation of the levorotatory isomer of this compound would therefore be −3.12°.

Compounds, such as CHFClBr, that are chiral are optically active. Chirality is a property of a molecule that results from its structure. Optical activity is a macroscopic property of a collection of these molecules that arises from the way they interact with light. To decide whether a compound should be optically active, we look for evidence that the molecules are chiral.

THE DIFFERENCE BETWEEN ENANTIOMERS ON THE MOLECULAR SCALE

Section 16.18 introduced an unambiguous system for differentiating between the stereoisomers of alkenes. A similar strategy, which is based on the Latin terms for left (*sinister*) and right (*rectus*), has been developed for distinguishing between a pair of enantiomers.

- Arrange the four substituents in order of *decreasing* atomic number of the atoms attached to the stereocenter. (The substituent with the highest atomic number gets the highest priority.) The substituents in 2-bromobutane, for example, would be listed in the order: $Br > CH_3$ and $CH_2CH_3 > H$.

- When two or more substituents have the same priority—such as the CH_3 and CH_2CH_3 groups in 2-bromobutane—continue working down the substituent chain until you find a difference. In 2-bromobutane, we would give the CH_2CH_3 group a higher priority than the CH_3 group because the next point down the chain is a carbon atom in the CH_2CH_3 substituent and a hydrogen atom in the CH_3 group. Thus, the four substituents on 2-bromobutane would be listed in the order: $Br > CH_2CH_3 > CH_3 > H$.

- View the enantiomer being named from the direction that places the substituent with the lowest priority as far from the eye as possible. In the following example, this involves rotating counterclockwise around the $C—CH_2CH_3$ bond and tilting the molecule slightly around an axis that lies in the plane of the paper. When this is done, the substituent that has the lowest priority is hidden from the eye.

- Trace a path that links the substituents in *decreasing* order of priority. If the path curves to the right—clockwise—the molecule is the rectus or *R* enantiomer. If it curves to the left—counterclockwise—it is the sinister or *S* enantiomer.

In this example, the path curves to the left, so this enantiomer is the (*S*)-2-bromobutane stereoisomer.

It is important to recognize that the (*R*)/(*S*) system for differentiating between stereoisomers is based on the structure of an individual molecule and the (+)/(−) system is based on the macroscopic behavior of a large collection of molecules. The most complete description of an enantiomer combines aspects of both systems. The enantiomer analyzed in this section is best described as (*S*)-(−)-2-bromobutane. It is the (*S*) enantiomer because of its structure and the (−) enantiomer because samples of the enantiomer with this structure are levorotatory; they rotate plane-polarized light clockwise.

Key Terms

Achiral	Cis	Octane number
Acyl chloride	Coal	Optically active
Alcohol	Coal gas	Organic chemistry
Aldehyde	Conformation	Oxidation number
Alkaloid	Constitutional isomer	Protic
Alkane	Crude oil	Specific rotation
Alkene	Dextrorotatory	Stereo center
Alkoxide	Diastereomer	Stereogenic atom
Alkyne	Dicarboxylic acid	Stereoisomer
Amide	Electrophile	Straight-chain hydrocarbon
Amine	Enantiomer	Straight-run gasoline
Aprotic	Ester	Synthesis gas
Aromatic compound	Ether	Tetravalent
Branched hydrocarbon	Free radical	Thermal cracking
Carbonyl	Functional group	Thermal reforming
Carboxylate ion	Hydrocarbon	Town gas
Carboxylic acid	Isomer	Trans
Carboxylic acid ester	Ketone	Unsaturated hydrocarbon
Catalytic cracking	Levorotatory	Vital force
Catalytic reforming	Natural gas	Water gas
Chain reaction	Nucleophile	Water-gas shift reaction
Chiral		

Problems

What is an Organic Compound?

1. Explain why carbon forms covalent bonds, not ionic bonds, with so many other elements.

2. Explain why carbon forms relatively strong double bonds, not only with itself, but with other nonmetals such as nitrogen, oxygen, phosphorus, and sulfur.

The Saturated Hydrocarbons or Alkanes

3. Use examples to explain the difference between *saturated* and *unsaturated* hydrocarbons and between *straight-chain* and *branched* hydrocarbons.

4. Explain why it is better to describe butane as a "straight-chain hydrocarbon" than as a "linear hydrocarbon."

5. Use the fact that straight-chain alkanes have a CH_3 group at either end and a chain of CH_2 groups down the middle to explain why alkanes have the generic formula C_nH_{2n+2}. Write the generic formulas for cycloalkanes, alkenes, and alkynes.

6. Describe the difference between *n*-pentane, isopentane, and neopentane. Classify these compounds as either stereoisomers or constitutional isomers.

7. Predict the structures of the constitutional isomers of heptane, C_7H_{16}.

8. Write the molecular formula for the saturated hydrocarbon that has the following carbon skeleton and name this compound.

$$C-C-C-C-C$$
$$|\quad\quad|$$
$$C\quad\quad C$$
$$\quad\quad\quad|$$
$$\quad\quad\quad C$$

9. Write the molecular formula for the saturated hydrocarbon that has the following carbon skeleton and name this compound.

$$C-C\quad C-C-C$$
$$\quad\quad|\quad\quad|$$
$$C-C-C-C-C$$
$$\quad\quad|$$
$$\quad\quad C-C$$

10. Explain why it is possible to isolate different constitutional isomers of butane, but not different conformations of butane.

11. Provide the systematic (IUPAC-approved) name for the following compound.

$$CH_3$$
$$|$$
$$CH_3CH_2CHCHCH_2CHCH_3$$
$$\quad\quad\quad|\quad\quad\quad|$$
$$\quad\quad CH_3\quad CH_2CH_3$$

12. Write the formula for the compound known as 1-methyl-1-chlorocyclopentane.

13. Draw the structure of 2,3,4-trimethyl-4-ethyloctane.

14. One way to decide whether a pair of structures represent different compounds is to assign a systematic name to each structure. Use this approach to decide whether the following are isomers or different descriptions of the same compound.

$$CH_3CHCH_2CHCH_3$$

The Unsaturated Hydrocarbons: Alkenes and Alkynes

15. Draw the structures of all the alkenes that have the formula C_6H_{12} and name these compounds.

16. Draw the structures of all the alkynes that have the formula C_5H_8 and name these compounds.

17. Explain why it is a mistake to name a compound 3-pentene. What would be the correct name of this compound?

18. Use the electron domain model to predict the shape of the tetrafluoroethylene (C_2F_4) molecule that is the starting material used to make Teflon.

19. Explain why alkenes can form both constitutional isomers and stereoisomers. What characteristic feature of a pair of alkenes can be used to decide whether they are constitutional isomers or stereoisomers?

20. Explain why alkynes can have constitutional isomers but not stereoisomers.

21. Which of the following compounds does not have the same molecular formula as the others?
    (a) cyclopentane
    (b) methylcyclobutane
    (c) 1-pentene
    (d) pentane
    (e) 1,1-dimethylcyclopropane

22. Which of the following compounds have *cis/trans* isomers?
    (a) $CHCl_3$ (b) $F_2C=CF_2$
    (c) $Cl_2C=CHCH_3$ (d) $FClC=CFCl$
    (e) $H_2C=CHF$

23. Which of the following compounds have *cis/trans* isomers?
    (a) 1-pentene   (b) 2-pentene
    (c) 2-methyl-2-butene   (d) 1-chloro-2-butene
    (e) 2-pentyne

24. Which of the following pairs of compounds would be constitutional isomers?
    (a) $CH_3CH_2OCH_2CH_3$ and $CH_3CH_2CH_2CH_2OH$
    (b) $(CH_3)_2CHCH_3$ and $CH_3CH_2CH_2CH_3$
    (c) $CH_3CH_2CH_2CH_3$ and CH_2CH_2 with CH_3
    (d) structures shown with Cl and H

25. Draw the structure for *cis*-3-ethyl-2-hexene.

26. Draw the structure for 2,6-dimethyl-3-heptyne.

27. Draw the structure for *trans*-6-methyl-2-heptene.

28. Draw the structures for the *cis/trans* isomers of 1,4-dimethylcyclohexane.

29. Draw the structure for 2,3-dimethyl-2-pentene. Explain why this compound can't exist as a pair of *cis/trans* isomers.

30. Draw the structure of (*E*)-2-pentene and determine whether this is the *cis*- or *trans*-isomer of this compound.

31. Only one of the C=C double bonds in *cis*-1,3-pentadiene can be labeled as either *E* or *Z*. Determine whether it is the bond between first and second carbon atoms or the bond between the third and fourth. Classify the geometry around this bond as either *E* or *Z*.

cis-1,3-Pentadiene

The Aromatic Hydrocarbons and Their Derivatives

32. Draw the structures of the following aromatic compounds: aniline, anisole, benzene, and benzoic acid.

33. TNT is an abbreviation for 2,4,6-trinitrotoluene. Toluene is a derivative of benzene in which a methyl (CH_3) group is substituted for one of the hydrogen atoms. Trinitrotoluene is a derivative of toluene in which NO_2 groups have replaced three more hydrogen atoms on the benzene ring. Draw the structures of all the possible isomers of trinitrotoluene. Label the isomer that is 2,4,6-trinitrotoluene.

34. On an exam, a student described benzene as a mixture of two structures that are rapidly being converted from one to the other. What is wrong with this answer?

35. Explain why the 12 atoms in a benzene molecule all lie in the same plane.

The Chemistry of Petroleum Products and Coal

36. Explain why a mixture of CO and H_2 can be used as a fuel. What are the products of the combustion of this mixture, which was once known as "water gas."

37. Natural gas, petroleum ether, gasoline, kerosene, and asphalt are all different forms of hydrocarbons that give off energy when burned. Describe how these substances

differ. What happens to the boiling points of these mixtures as the average length of the hydrocarbon chain increases?

38. Which of the following compounds has the largest octane number?

    (a) *n*-butane

    (b) *n*-pentane

    (c) *n*-hexane

    (d) *n*-octane

39. Which of the following won't increase the octane number of gasoline?

    (a) increasing the concentration of branched-chain alkanes

    (b) increasing the concentration of cycloalkanes

    (c) increasing the concentration of aromatic hydrocarbons

    (d) increasing the average length of the hydrocarbon chains

40. Coal gas can be obtained when coal is heated in the absence of air. Water gas can be obtained when coal reacts with steam. Describe the difference between these gases and explain why much more water gas can be extracted from a ton of coal.

Functional Groups

41. Give examples of compounds that contain each of the following functional groups.

    (a) an alcohol   (b) an aldehyde

    (c) an amine   (d) an amide

    (e) an alkyl halide   (f) an alkene

    (g) an alkyne

42. Describe the difference between the members of each of the following pairs of functional groups.

    (a) an alcohol and an alkoxide ion

    (b) an alcohol and an aldehyde

    (c) an amine and an amide

    (d) an ether and an ester

    (e) an aldehyde and a ketone

43. Classify the following compounds as an alcohol, an aldehyde, an ether, or a ketone.

    (a) CH_3CH_2CH with $\overset{O}{\underset{\|}{}}$ (b) CH_3CH_2OH

    (c) $CH_3CH_2OCH_2CH_3$ (d) $CH_3CH_2CCH_3$ with $\overset{O}{\underset{\|}{}}$

44. Classify the following compounds as primary, secondary, or tertiary alcohols.

    (a) CH_3CH_2OH (b) $CH_3CH_2\overset{CH_3}{\underset{|}{C}}HOH$

    (c) $CH_3CH_2\overset{CH_3}{\underset{|}{\underset{CH_3}{C}}}OH$ (d) $CH_3CH_2\overset{OH}{\underset{|}{C}}HCH_2CH_3$

45. Cortisone is an adrenocortical steroid used as an anti-inflammatory agent. Use the fact that carbon is tetravalent to determine the molecular formula of this compound from the following line drawing. Identify the functional groups present in this molecule.

46. Piperine [(*E,E*)-1-[5-(1,3-benzodioxol-5-yl)-1-oxo-2, 4-pentadienyl]piperidine] can be extracted from black pepper. Identify the functional groups in the structure of this compound.

47. PGE$_2$ is a member of the family of compounds known as prostaglandins, which have very significant physiological effects in even small quantities. They can affect blood pressure, heart rate, body temperature, blood clotting, and conception. Some induce inflammation; others relieve it. Aspirin, which is both an anti-inflammatory and an antipyretic (fever-reducing) drug, acts by blocking the synthesis of prostaglandins. Identify the functional groups in the structure of this molecule and calculate its molecular formula.

48. Pseudoephedrine hydrochloride is the active ingredient in a variety of decongestants, including Sudafed™. Use the structure of this compound shown in Figure 16.4 to identify the functional groups in this molecule.

49. Cocaine was once used in Coca-Cola™. Quinine is still added to tonic water. Use the structures of these

alkaloids shown in Figure 16.6 to identify the functional groups in these compounds.

Oxidation-Reduction Reactions

50. Arrange the following substances in order of increasing oxidation number of the carbon atom.
    (a) C (b) HCHO (c) HCO_2H (d) CO
    (e) CO_2 (f) CH_4 (g) CH_3OH

51. Determine whether or not each of the following is an example of an oxidation–reduction.
    (a) $CH_4 + Cl_2 \longrightarrow CH_3Cl + HCl$
    (b) $CH_3OH \longrightarrow HCHO$
    (c) $HCHO \longrightarrow HCO_2H$
    (d) $CH_3OH + HBr \longrightarrow CH_3Br + H_2O$
    (e) $(CH_3)_2CO \longrightarrow (CH_3)_2CHOH$

52. Classify the following transformations as examples of either oxidation or reduction.
    (a) $CH_3CH_2OH \longrightarrow CH_3CHO$
    (b) $CH_3CHO \longrightarrow CH_3CO_2H$
    (c) $(CH_3)_2C{=}O \longrightarrow (CH_3)_2CHOH$
    (d) $CH_3CH{=}CHCH_3 \longrightarrow CH_3CH_2CH_2CH_3$
    (e) $(CH_3)_2CHC{\equiv}CH \longrightarrow (CH_3)_2CHCH_2CH_3$

53. Which of the following compounds can be oxidized to form an aldehyde?
    (a) CH_3CH_2OH (b) $CH_3CHOHCH_3$
    (c) CH_3OCH_3 (d) $(CH_3)_2C{=}O$

54. Which of the following compounds should be the most difficult to oxidize?
    (a) CH_3CH_2OH (b) $(CH_3)_2CHOH$
    (c) $(CH_3)_3COH$ (d) CH_3CHO

55. Which of the following compounds can be prepared by reducing CH_3CHO?
    (a) CH_3CH_3 (b) CH_3CH_2OH
    (c) CH_3CO_2H (d) $CH_3CH_2CO_2H$

56. Which of the following compounds can be prepared by oxidizing CH_3CHO?
    (a) CH_3CH_3 (b) CH_3CH_2OH
    (c) CH_3CO_2H (e) $CH_3CH_2CO_2H$

57. Predict the product of the oxidation of 2-methyl-3-pentanol.

58. Predict the product of the reduction of 2-methyl-2-pentene with hydrogen gas over a nickel metal catalyst.

Alkyl Halides

59. Use Lewis structures to describe the free-radical chain-reaction mechanism involved in the bromination of methane to form methyl bromide. Clearly label the chain-initiation, chain-propagation, and chain-termination steps.

60. How many different products could be formed by the free-radical chlorination of pentane? If attack at the different hydrogen atoms in this compound were more or less random, what would be the relative abundance of the different isomers of chloropentane formed in this reaction?

61. Consider the reaction between a Cl atom and a pentane molecule. Classify the different intermediates that could be produced in this reaction as either primary, secondary, or tertiary free radicals.

Alcohols and Ethers

62. Describe the differences between the structures of water, methyl alcohol, and dimethyl ether.

63. Draw the structures of the seven constitutional isomers that have the formula $C_4H_{10}O$. Classify these isomers as either alcohols or ethers.

64. Ethyl alcohol and dimethyl ether have the same chemical formula, C_2H_6O. Explain why one of these compounds reacts rapidly with sodium metal but the other does not.

65. Predict the product of the dehydration of ethyl alcohol with sulfuric acid at 130°C.

66. Use examples to explain why it is possible to oxidize either a primary or secondary alcohol, but not a tertiary alcohol.

67. Draw the structure of o-phenylphenol, the active ingredient in Lysol.

68. Explain why alcohols are Brønsted acids in water.

69. Write the structures of the conjugate bases formed when the following alcohols act as a Brønsted acid.
    (a) CH_3CH_2OH (ethyl alcohol)
    (b) C_6H_5OH (phenol)
    (c) $(CH_3)_2CHOH$ (isopropyl alcohol)

70. Use the fact that there are no hydrogen bonds between ether molecules to explain why diethyl ether ($C_4H_{10}O$) boils at 34.5°C, whereas its constitutional isomer—1-butanol ($C_4H_{10}O$)—boils at 118°C.

71. Assign a systematic name to the following alcohols.

A *B*

72. Assign a systematic name to the compound known by the common name *menthol*.

Aldehydes and Ketones

73. At which end of a carbonyl group will each of the following substances react?
    (a) H^+ (b) H^- (c) OH^-
    (d) NH_3 (e) BF_3

74. Explain why mild oxidation of a primary alcohol gives an aldehyde, whereas oxidation of a secondary alcohol gives a ketone.

75. Explain why strong oxidizing agents can't be used to convert a primary alcohol to an aldehyde.

76. Assign both common and systematic names to following compounds.

A   *B*

77. Assign a systematic name to the compound known by the common name *geranial*.

Carboxylic Acids, Carboxylate Ions, and Carboxylic Acid Esters

78. Explain the difference between a carboxylic acid, a carboxylate ion, and a carboxylic acid ester.

79. What major differences between carboxylic acids (such as butyric acid, $CH_3CH_2CH_2CO_2H$) and esters (such as ethyl butyrate, $CH_3CH_2CH_2CO_2CH_2CH_3$) help explain why butyric acid gives rise to the odor of rotten meat but ethyl butyrate gives rise to the pleasant odor of pineapple?

80. Explain the difference between CH_3CH_2OH and CH_3CO_2H that makes one of the C—OH bonds in these compounds more than 10 orders of magnitude more acidic than the other when values of K_a for these compounds are compared.

Amines, Alkaloids, and Amides

81. Classify each of the compounds in Figures 16.5 and 16.6 as either a primary amine, a secondary amine, a tertiary amine, and/or an amide.

82. Explain why reacting a complex amine, such as pseudoephedrine, with an acid makes these compounds more soluble in water.

83. Draw the structure of caffeine and label each of the nitrogen atoms in this compound as either an amine or an amide.

Optical Activity

84. Describe the difference between constitutional isomers and stereoisomers. What characteristic feature of a molecule could be used to distinguish between these two forms of isomers?

85. Objects that cannot be superimposed on their mirror images are said to be *chiral,* and chiral molecules are optically active. Which of the following molecules are optically active?
    (a) CH_4 (b) CH_3Cl (c) $CHCl_3$
    (d) $CHFCl_2$ (e) $CHFClBr$

86. Which of the following compounds are optically active?
    (a) C_2H_4 (b) C_6H_6

    (c) $C_6H_4Cl_2$ (d) $CH_3CH_2CHCH_3$

87. Coniine is the active ingredient in the poison known as hemlock that was given to Socrates in 399 B.C. Use the following diagram of the structure of coniine to predict whether this compound is optically active.

88. Use the structure of caffeine shown here to predict whether this compound is optically active.

89. Determine the number of centers of chirality in the structure of vitamin C shown in Section 16.1.

90. Predict whether the following compound is chiral.

91. Which of the following compounds, which play an important role in the chemistry of biological systems, is chiral?

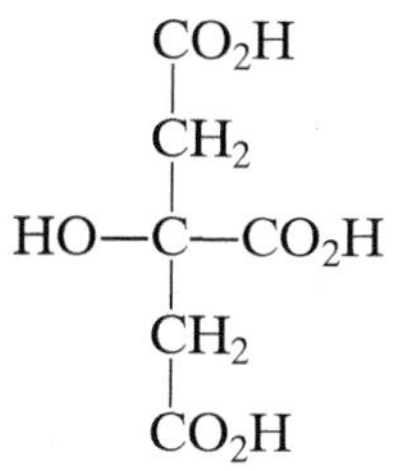

$$CH_2CO_2H$$ citric acid structure

citric acid

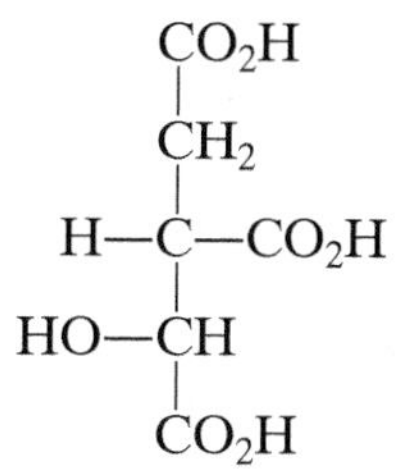

isocitric acid

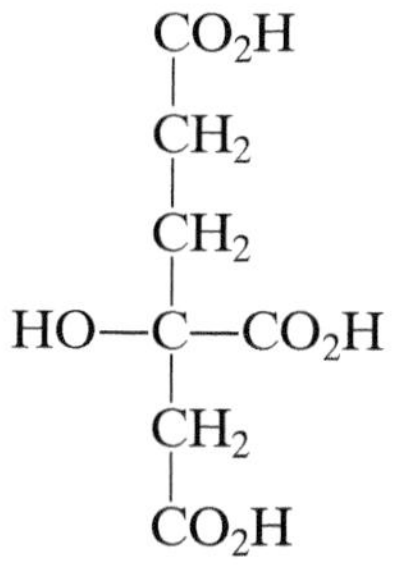

homocitric acid

92. Which (if any) of the carbon atoms in 2-bromo-3-methylbutane are stereocenters?

93. Determine whether the following isomers are enantiomers or diastereomers. What characteristic feature of the molecule can be used to make this decision?

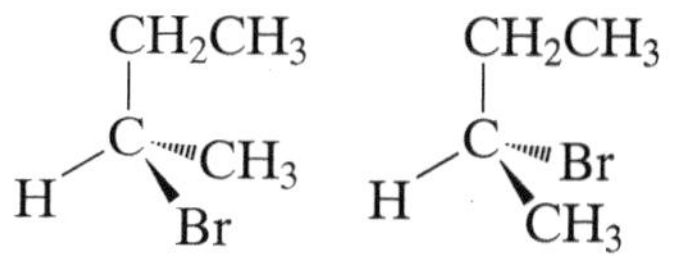

94. Determine whether the following isomers are enantiomers or diastereomers. What characteristic feature of the molecule can be used to make this decision?

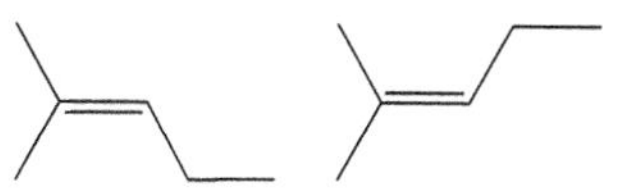

95. Determine whether the following compound is the R or S enantiomer of 2-bromo-butane.

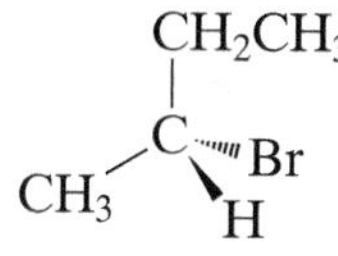

Integrated Problems

96. Describe a way of determining whether a chemical sample is an alkane or an alkene.

97. Describe a way of determining whether a chemical sample is a carboxylic acid or an ester.

98. Describe a way of determining whether a chemical sample is an alcohol or an ether.

99. Describe a way of determining whether a chemical sample is a primary or a tertiary alcohol.

100. Identify the Brønsted acids and the Brønsted bases in the following reaction.

$$CH_3C{\equiv}CH + NH_2^- \longrightarrow CH_3C{\equiv}C^- + NH_3$$

101. Succinic acid plays an important role in the Krebs Cycle, malic acid (*apple acid*) is found in apples, and tartaric acid (*fruit acid*) is found in many fruits. Which of these dicarboxylic acids is chiral?

Appendix A

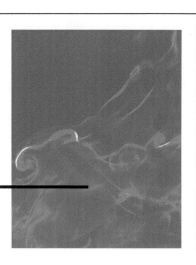

A.1 Systems of Units

All measurements contain a number that indicates the magnitude of the quantity being measured and a set of units that provide a basis for comparing the quantity with a standard reference. There are several systems of units, each containing units for properties such as mass, length, volume, and time.

THE ENGLISH UNITS OF MEASUREMENT

In the **English system** of units in use in the United States, the individual units are defined in an arbitrary way. There are 12 inches in a foot, 3 feet in a yard, and 1760 yards in a mile. There are 2 cups in a pint and 2 pints in a quart but 4 quarts in a gallon.

However, it is not mathematically correct to write:

$$12 \text{ inches} = 1 \text{ foot}$$
$$3 \text{ feet} = 1 \text{ yard}$$
$$1760 \text{ yards} = 1 \text{ mile}$$

The "=" sign used in the above relationships is not the mathematician's equal sign because the relationship must be dimensionally homogeneous (the units must be the same on both sides of the equal sign). The relationships can be correctly written as

$$12 \text{ inches} \equiv 1 \text{ foot}$$
$$3 \text{ feet} \equiv 1 \text{ yard}$$
$$1760 \text{ yards} \equiv 1 \text{ mile}$$

where the symbol $\equiv$ can represent either *physically equivalent* or *defined to be*.

The relationships between some of the common units in the English system are given in Table A.1.

Table A.1
The English System of Units

Length: inch (in.), foot (ft), yard (yd), mile (mi)
12 in. $\equiv$ 1 ft $\qquad$ 5280 ft $\equiv$ 1 mi
3 ft $\equiv$ 1 yd $\qquad$ 1760 yd $\equiv$ 1 mi
Volume: fluid ounce (oz), cup (c), pint (pt), quart (qt), gallon (gal)
2 c $\equiv$ 1 pt $\qquad$ 32 oz $\equiv$ 1 qt
2 pt $\equiv$ 1 qt $\qquad$ 4 qt $\equiv$ 1 gal
Weight: ounce (oz), pound (lb), ton
16 oz $\equiv$ 1 lb $\qquad$ 2000 lb $\equiv$ 1 ton
Time: second (s), minute (min), hour (h), day (d), year (y)
60 s $\equiv$ 1 min $\qquad$ 24 h $\equiv$ 1 d
60 min $\equiv$ 1 h $\qquad$ 365¼ d $\equiv$ 1 yr

More than 300 years ago, the Royal Society of London discussed replacing the irregular English system of units with one based on decimals. It was not until the French Revolution, however, that a decimal-based system of units was adopted. This **metric system** was based on a set of three fundamental quantities. The meter was introduced for measurements of length, the liter for measurements of volume, and the gram for measurements of weight. The relationship between the fundamental units of the metric system and the traditional units of the English system can be found in Table A.2.

The principal advantage of the metric system is the ease with which the base units can be converted into a unit that is more appropriate for the quantity being measured. This is done by adding a prefix to the name of the base unit. The prefix *kilo-* (k), for example, implies multiplication by a factor of 1000. Thus, a kilometer is equal to 1000 meter.

$$1 \text{ km} = 1000 \text{ m}$$

The prefix *milli-* (m), on the other hand, means division by a factor of 1000. A *milliliter* (mL) is therefore equal to 0.001 liter.

$$1 \text{ mL} = 0.001 \text{ L}$$

The common metric prefixes are given in Table A.3.

Table A.2
English and Metric Equivalents

Length: meter (m)	
English	**Metric**
1 m ≡ 1.094 yd	1 yd ≡ 0.9144 m
Volume: liter (L)	
1 L ≡ 1.057 qt	1 qt ≡ 0.9464 L
Mass: gram (g)	
1 lb ≡ 453.6 g	0.002205 lb ≡ 1 g

Table A.3
Metric System Prefixes

Prefix	Symbol	Meaning
femto-	f	$\times$ 1/1,000,000,000,000,000 (10^{-15})
pica-	P	$\times$ 1/1,000,000,000,000 (10^{-12})
nano-	n	$\times$ 1/1,000,000,000 (10^{-9})
micro-	μ	$\times$ 1/1,000,000 (10^{-6})
milli-	m	$\times$ 1/1,000 (10^{-3})
centi-	c	$\times$ 1/100 (10^{-2})
deci-	d	$\times$ 1/10 (10^{-1})
kilo-	k	$\times$ 1,000 (10^{3})
mega-	M	$\times$ 1,000,000 (10^{6})
giga-	G	$\times$ 1,000,000,000 (10^{9})
tera-	T	$\times$ 1,000,000,000,000 (10^{12})

Another advantage of the metric system is the link between the base units of length and volume. By definition, a liter is equal to the volume of a cube exactly 10 cm tall, 10 cm long, and 10 cm wide. Because the volume of a cube with these dimensions is 1000 cubic centimeters (1000 cm^3) and a liter contains 1000 milliliters, 1 mL is equivalent to 1 cm^3.

$$1 \text{ mL} \equiv 1 \text{ cm}^3$$

A third advantage of the metric system is the link between the base units of volume and weight. The gram was originally defined as the mass of 1 mL of water at 4°C. (It is important to specify the temperature because water expands or contracts as the temperature changes.)

SI UNITS OF MEASUREMENT

A series of international conferences on weights and measures has been held periodically since 1875 to refine the metric system. At the 11th conference, in 1960, a new system of units known as the International System of Units (abbreviated

Table A.4
SI Base Units

Physical Quantity	Name of Unit	Symbol
Length	meter	m
Mass	kilogram	kg
Time	second	s
Temperature	Kelvin	K
Electric current	ampere	A
Amount of substance	mole	mol
Luminous intensity	candela	cd

SI in all languages) was proposed as a replacement for the metric system. The seven base units for the **SI system** are given in Table A.4.

DERIVED SI UNITS

The units of every measurement in the SI system are supposed to be derived from one or more of the seven base units. The preferred unit for volume is the cubic meter, for example, because volume has units of length cubed and the SI unit for length is the meter. The preferred unit for speed is meters per second because speed is the distance traveled divided by the time it takes to cover this distance.

$$\text{SI unit of volume: } m^3 \qquad \text{SI unit of speed: } m/s$$

Some of the common derived SI units are given in Table A.5.

Table A.5
Common Derived SI Units in Chemistry

Physical Quantity	Name of Unit	Symbol	SI Unit
Density		ρ rho	kg/m^3
Electric charge	coulomb	C	$A \cdot s$
Electric potential	volt	V	J/C
Energy	joule	J	$kg \cdot m^2/s^2$
Force	newton	N	$kg \cdot m/s^2$
Frequency	hertz	Hz	s^{-1}
Pressure	pascal	Pa	N/m^2
Velocity (speed)	meters per second	v	m/s
Volume	cubic meter	V	m^3

NON-SI UNITS

Strict adherence to SI units would require changing directions such as "add 250 mL of water to a 1-L beaker" to "add 0.00025 cubic meters of water to an 0.001-m^3 container." Because this is not a convenient way to express directions for working in the chemistry laboratory, a number of units that are not strictly acceptable under the SI convention are still in use. Some of the non-SI units are given in Table A.6.

Table A.6
Non-SI Units in Common Use

Physical Quantity	Name of Unit	Symbol	SI Equivalent
Volume	liter	L	1×10^{-3} m^3
Length	angstrom	Å	0.1 nm
Pressure	atmosphere	atm	101.325 kPa
	torr	mmHg	133.32 Pa
Energy	electron volt	eV	1.602×10^{-19} J
Temperature	degree Celsius	°C	$T_K - 273.15$
Concentration	molarity	M	mol/L

CONVERSION FACTORS

Conversion factors are used to convert one unit of measurement to another. A conversion factor is a fraction with its numerator and denominator expressed in different units.

In general, if unit 1 is to be converted to unit 2, the conversion factor is used:

$$\text{Unit 1} \times \text{conversion factor} = \text{Unit 2}$$

Useful conversion factors are given in Table B.2 in Appendix B. The conversion factors between kg and pounds are

$$\frac{1 \text{ kg}}{2.2046 \text{ lb}} \qquad \frac{2.2046 \text{ lb}}{1 \text{ kg}}$$

If we wish to convert 5.34 pounds to kilograms

$$5.34 \text{ lb} \times \frac{1 \text{ kg}}{2.2046 \text{ lb}} = 2.42 \text{ kg}$$

The units in the conversion factor cancel the units on pounds and leave the units of kilograms.

A common unit conversion used in chemistry is the conversion between mass and moles. The molar mass of carbon is 12.011 g/mol. It is tempting to write:

$$1 \text{ mol C} = 12.011 \text{ g C}$$

However, the units on the two sides of the equation are not the same. It is correct to use the molar mass of carbon to form the conversion factors

$$\frac{1 \text{ mol C}}{12.011 \text{ g}} \quad \text{or} \quad \frac{12.011 \text{ g}}{1 \text{ mol C}}$$

These conversion factors can now be used to convert between moles and mass of carbon. The above relationship between moles and mass of carbon can be made dimensionally consistent by using one of the conversion factors.

$$1 \text{ mol C} = 12.011 \text{ g C} \left(\frac{1 \text{ mol C}}{12.011 \text{ g C}} \right) = 1 \text{ mol C}$$

When performing any mathematic operation (addition, subtraction, multiplication, division) the units must be consistent throughout the equation. For example, to convert 34.54 g of carbon to moles.

$$34.54 \text{ g} \left(\frac{1 \text{ mol C}}{12.011 \text{ g C}} \right) = 2.876 \text{ mol C}$$

A.2 Uncertainty in Measurement

As noted in Chapter 1, there is a fundamental difference between stating that there are 12 inches in a foot and stating that the circumference of the Earth at the equator is 24,903.01 miles. The first relationship is based on a *definition*. By convention, there are exactly 12 inches in 1 foot. The second relationship is based on a *measurement*. It reports the circumference of the Earth to within the limits of experimental error in an actual measurement.

Many conversion factors are based on definitions. There are exactly 5280 feet in a mile and 2.54 centimeters in an inch, for example. Conversion factors based on definitions are known with complete certainty. (There is no error or uncertainty associated with the numbers.) Measurements, however, are always accompanied by a finite amount of error or uncertainty, which reflects limitations in the techniques used to make them.

The first measurement of the circumference of the Earth, made in the third century B.C., for example, gave a value of 250,000 stadia, or 29,000 miles. As the quality of the instruments used to make the measurement improved, the amount of error gradually decreased. But it never disappeared. Regardless of how carefully measurements are made, they always contain an element of uncertainty.

SYSTEMATIC AND RANDOM ERRORS

There are two sources of error in a measurement: (1) limitations in the sensitivity of the instruments used, and (2) imperfections in the techniques used to make the measurement. These errors can be divided into two classes: systematic and random.

The idea of systematic error can be understood in terms of the bull's-eye analogy shown in Figure A.1. Imagine what would happen if you aimed at a target with a rifle whose sights were not properly adjusted. Instead of hitting the bull's-eye, you would systematically hit the target at another point. Your results would be influenced by a **systematic error** caused by an imperfection in the equipment being used. Systematic error can also result from mistakes an individual makes while taking the measurement. In the bull's-eye analogy, a systematic error of this kind might occur if you flinched and pulled the rifle toward you each time it was fired.

Fig. A.1 (*a*) Systematic errors give results that are systematically too small or too large. (*b*) Random errors give results that fluctuate between being too small and too large.

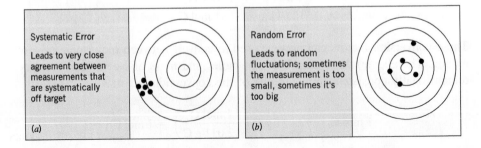

Systematic Error

Leads to very close agreement between measurements that are systematically off target

(a)

Random Error

Leads to random fluctuations; sometimes the measurement is too small, sometimes it's too big

(b)

To understand **random error**, imagine what would happen if you closed your eyes for an instant just before you fired the rifle. The bullets would hit the target more or less randomly. Some would hit too high, others would hit too low. Some would hit too far to the right, others too far to the left. Instead of an error that systematically gives a result too far in one direction, you now have a random error with random fluctuations.

Random errors most often result from limitations in the equipment or techniques used to make a measurement. Suppose, for example, that you wanted to collect 25 mL of a solution. You could use a beaker, a graduated cylinder, or a buret. Volume measurements made with a 50-mL beaker are accurate to within ±5 mL. In other words, you would be just as likely to obtain 20 mL of solution (5 mL too little) as 30 mL (5 mL too much). You could decrease the amount of error by using a graduated cylinder that is capable of measurements to within ±1 mL. The error could be decreased even further by using a buret, which is capable of delivering a volume to within 1 drop, or ±0.05 mL.

Exercise A.1

Which of the following would lead to systematic errors, and which would produce random errors?

(a) Using a 1-qt milk carton to measure 1-L samples of milk.

(b) Using a balance that is sensitive to ±0.1 g to obtain 250-mg samples of vitamin C.

Solution

Procedure (a) would result in a systematic error. The volume would always be too small because a quart is slightly smaller than a liter. Procedure (b) would produce a random error because the equipment used to make the measurement is not sensitive enough.

ACCURACY AND PRECISION

To most people, *accuracy* and *precision* are synonyms, words that have the same or nearly the same meaning. In the physical sciences, however, there is an important difference between the terms. Measurements are *accurate* when they agree with the true value of the quantity being measured. They are *precise* when individual measurements of the same quantity agree.

The difference between accuracy and precision is similar to the difference between two terms from statistics: *validity* and *reliability*. Accuracy means the same thing as validity. A measurement is accurate, or valid, only if we get the correct answer. Precision is the same as reliability. A measurement is precise, or reliable, if we get essentially the same result each time we make the measurement. Measurements are therefore precise when they are reproducible.

To understand how systematic and random errors affect accuracy and precision, let's return to the bull's-eye analogy (see Figure A.2). Systematic errors influence the accuracy, but not the precision, of a measurement. It is possible to get measurements that are consistently the same, but systematically wrong. Random errors influence both the accuracy and the precision of the measurement.

Systematic errors can be reduced by increasing the care and patience of the individual making the measurement or by improving the accuracy of the equipment

Fig. A.2 (*a*) Systematic errors affect the accuracy of a measurement. The measurement may still be precise, but the results are systematically different from the correct answer. (*b*) Random errors can affect both the accuracy and precision of a measurement.

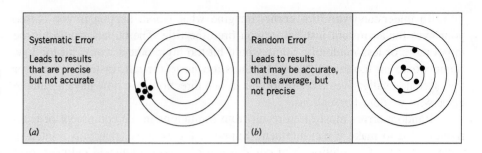

used. Random errors can be reduced by averaging the results of many measurements of the same quantity because the precision of a series of measurements increases with the square root of the number of measurements.

A.3 Significant Figures

The fact that measurements form the basis on which the scientific process is built has an important consequence. It is essential to make sure that the measurement doesn't appear more accurate than the equipment used to make the measurement allows. We can achieve this goal by controlling the number of digits, or **significant figures**, used to report the measurement.

Imagine what would happen if you determined the mass of an old copper penny on a postage scale and then on an analytical balance. The postage scale might give a mass of about 3 g, which means that the penny is closer to 3 g than either 2 g or 4 g. The analytical balance is much more sensitive; it can measure the mass of an object to the nearest ± 0.001 g, in which case you might find that the penny has a mass of 2.531 g.

Postage scale	3 ± 1 g
Analytical balance	2.531 g ± 0.001 g

The postage scale gave a measurement that had only one reliable or significant digit. That measurement is therefore said to be good to only one significant figure. The analytical balance gave four significant figures (2.531). The number of significant figures in a measurement is the number of digits that are known with some degree of confidence—such as 2, 5, and 3 in the measurement with the analytical balance—plus the last digit (in this case, the 1), which is generally an estimate or approximation. This means that some degree of uncertainty is associated with the last reported digit in a measurement. As we improve the sensitivity of the equipment used to make a measurement, the number of significant figures increases.

At first glance, it might seem that we can determine the number of significant figures by simply counting the digits in the measurement. Unfortunately, zeros represent a problem. Zeros in a number can be classified in three ways: leading zeros (such as 0.0035), trailing zeros (such as 350,000), and zeros between two significant figures (such as 3056).

- Leading zeros are never significant. They are only used to set the position of the decimal point. Consider the following numbers: 0.0045, 0.045, 0.45, 4.5, and 45. Each number is known to ± 1 part in 45. Thus, there are two significant figures in each of these numbers.

- Zeros between two significant figures are always significant. In the number 3105 the zero is between 1 and 5, both of which are significant. Therefore, the zero is also significant, giving a total of four significant figures. The numbers 40.05, 0.0102, and 1706.2 have four, three, and five significant figures, respectively.

- Trailing zeros that are not needed to place the decimal point are significant. For example, 4.00 has three significant figures because the trailing zeros are not necessary to show the position of the decimal; they indicate that the number is known to ± 0.01. Trailing zeros where there is no decimal point present a problem. It is often unclear how many zeros are significant. For example, does a measurement of 400 mL represent 400 ± 1 mL, 400 ± 10 mL, or 400 ± 100 mL? The only way to clearly show the number of significant figures in a measurement like this is to express the number in **scientific notation**. The measurement 400 mL can be written as 4.00×10^2 (three significant figures), 4.0×10^2 (two significant figures), or 4×10^2 (one significant figure) depending on the degree to which the measurement is known.

If you are not sure whether a digit is significant, assume that it isn't. If the directions for an experiment read: "Add the sample to 400 mL of water," assume the volume of water is known to one significant figure.

ADDITION AND SUBTRACTION WITH SIGNIFICANT FIGURES

What would be the mass of a solution prepared by adding 0.507 g of salt to 150.0 g of water? If we solved the problem without considering significant figures, we would simply add the two measurements.

$$
\begin{array}{ll}
150.0 \quad \text{g H}_2\text{O} & \text{(without using significant figures)} \\
+0.507 \text{ g salt} \\
\hline
150.507 \text{ g solution}
\end{array}
$$

But this answer doesn't make sense. We only know the mass of the water to the nearest tenth of a gram, so we can only know the total mass of the solution to within ± 0.1 g. Taking significant figures into account, we find that adding 0.507 g of salt to 150.0 g of water gives a solution with a mass of 150.5 g.

$$
\begin{array}{ll}
150.0 \quad \text{g H}_2\text{O} & \text{(using significant figures)} \\
+0.507 \text{ g salt} \\
\hline
150.5 \quad \text{g solution}
\end{array}
$$

Many of the calculations in this book are done by combining measurements with different degrees of accuracy and precision. The guiding principle in carrying out these calculations is easily stated: *The accuracy of the final answer can be no greater than the least accurate measurement.* This principle can be translated into a simple rule for addition and subtraction as follows.

> **When measurements are added or subtracted, the number of significant figures to the right of the decimal in the answer is determined by the measurement with the fewest digits to the right of the decimal.**

Because the least accurate measurement in the example shown above is known to ± 0.1 g, the total mass of the solution can only be known to ± 0.1 g.

MULTIPLICATION AND DIVISION WITH SIGNIFICANT FIGURES

The same principle governs the use of significant figures in multiplication and division: *The final result can be no more accurate than the least accurate measurement*. In this case, however, we count the significant figures in each measurement, not the number of decimal places.

> **When measurements are multiplied or divided, the answer can contain no more total significant figures than the measurement with the fewest total number of significant figures.**

To illustrate this rule, let's calculate the cost of the copper in one of the old pennies that is pure copper. Let's assume that the penny has a mass of 2.531g, that it is essentially pure copper, and that the price of copper is 0.60 cents per gram.

$$2.531 \text{ g} \times \frac{\$0.0060}{\text{g}} = \$0.015$$

There are four significant figures in the mass of the penny (2.531). But there are only two significant figures in the price of copper, so the final answer can only have two significant figures. This calculation helps explain why pennies are no longer made of more or less pure copper because the cost of the metal would be half-again as much as the value of the coin.

ROUNDING OFF

When the answer to a calculation contains too many significant figures, it must be **rounded off**. When we multiply the mass of the copper penny in the calculation shown above by the cost per gram, we get 1.51860000. If we were allowed four significant figures, we might round this off to 1.519 cents. But the price of copper was only known to two significant figures, so the answer in the calculation was rounded off to 1.5 cents.

We can obtain a set of systematic rules for rounding off by considering that 10 digits can occur in the last decimal place in a calculation. One way of rounding off involves underestimating the answer for five of the digits (0, 1, 2, 3, and 4) and overestimating the answer for the other five (5, 6, 7, 8, and 9). This approach to rounding off is summarized as follows.

- If the digit is smaller than 5, drop the digit and leave the remaining number unchanged. Thus, 1.684 becomes 1.68.
- If the digit is 5 or larger, drop the digit and add 1 to the preceding digit. Thus, 1.247 becomes 1.25.

Using these rules ensures that we neither overestimate or underestimate the amount of error in a measurement when we adjust a calculation for the correct number of significant figures.

A.4 Scientific Notation

Chemists routinely work with numbers that are extremely small, such as the mass of a single electron (0.000,000,000,000,000,000,000,000,911 g.) They also work with numbers that are extremely large, such as the number of carbon atoms in a 1-carat diamond (10,030,000,000,000,000,000,000). There isn't a calculator made that will accept either of these numbers as they are written here. Before we can use these numbers, it is necessary to convert them to **scientific notation**—a number between 1 and 10 multiplied by 10 raised to the appropriate exponent. In this notation, the mass of an electron is 9.11×10^{-28} grams and there are 1.003×10^{22} carbon atoms in a 1-carat diamond.

Exponential mathematics can be understood by thinking about the following examples.

- A number raised to the zero power is equal to 1.

$$10^0 = 1$$

- A number raised to the first power is equal to itself.

$$10^1 = 10$$

- A number raised to the nth power is equal to the product of that number times itself $n - 1$ times.

$$10^5 = 10 \times 10 \times 10 \times 10 \times 10 = 100,000$$

- Dividing by a number raised to some exponent is the same as multiplying by that number raised to an exponent of the opposite sign

$$\frac{5}{10^2} = 5 \times 10^{-2} \qquad \frac{5}{10^{-3}} = 5 \times 10^3$$

The following rule can be used to convert numbers into scientific notation:

The exponent in scientific notation is equal to the number of times the decimal point must be moved to produce a number between 1 and 10.

The population of Chicago was recently estimated as $2,833,000 \pm 1000$. Note that the population is only reported to four significant figures because the error in this estimate is ± 1000. To convert the number to scientific notation, we have to move the decimal point *to the left* six times and use the correct number of significant figures.

$$2,833,000 \pm 1000 = 2.833 \times 10^6$$

To convert numbers smaller than 1 into scientific notation, we have to move the decimal point to the right. The decimal point in 0.000985, for example, must be moved *to the right* four times. There are only three significant figures in this number, so the number written in scientific notation can contain only three digits.

$$0.000985 = 9.85 \times 10^{-4}$$

The primary reason for converting numbers into scientific notation is to make calculations with unusually large or small numbers less cumbersome. But there is another important advantage to scientific notation. Because zeros are no longer used to set the decimal point, all of the digits in a number in scientific notation are significant, as shown by the following examples:

1.03×10^{22}	three significant figures
9.852×10^{-5}	four significant figures
2.0×10^{-23}	two significant figures

Exercise A.2

Convert the following numbers into scientific notation.

(a) 0.004694   (b) 1.98   (c) 4,679,000 $\pm$ 100

Solution

(a) 4.694×10^{-3} (b) 1.98×10^0 (c) $4.6790 \pm 0.0001 \times 10^6$

A.5 The Graphical Treatment of Data

If the basis of science is a natural curiosity about the world that surrounds us, an important step in doing science is trying to find patterns in the observations and measurements that result from this curiosity.

Anyone who has played with a prism or seen a rainbow has watched what happens when white light is split into a spectrum of different colors. The difference between the blue and red light in the spectrum is the result of differences in the frequencies and wavelengths of the light, as shown in Table A.7.

Table A.7

Characteristic Wavelengths (λ) and Frequencies (ν) of Light of Different Colors

Color	Wavelength (m)	Frequency (s^{-1})
Violet	4.100×10^{-7}	7.312×10^{14}
Blue	4.700×10^{-7}	6.379×10^{14}
Green	5.200×10^{-7}	5.765×10^{14}
Yellow	5.800×10^{-7}	5.169×10^{14}
Orange	6.000×10^{-7}	4.997×10^{14}
Red	6.500×10^{-7}	4.612×10^{14}

There is an obvious pattern in the data: As the wavelength of the light becomes larger, the frequency becomes smaller. But recognizing this pattern is not enough. It would be even more useful to construct a mathematical equation that fits the data. This would allow us to calculate the frequency of light of any wavelength, such as blue-green light with a wavelength of 5.00×10^{-7} m, or to calculate the wavelength of light of a known frequency, such as blue-violet light with a frequency of 7.00×10^{14} cycles per second.

The first step toward constructing an equation that fits the data involves plotting the data in different ways until we get a straight line. We might decide, for example, to plot the wavelengths on the vertical axis and the frequencies on the horizontal axis, as shown in Figure A.3. When we construct a graph, we should keep the following points in mind.

- The scales of the graph should be chosen so that the data fill as much of the available space as possible.

- It isn't necessary to include the origin (0,0) on the graph. In fact, it may be more efficient to leave out the origin, so that the data fill the available space.

- Once the scales have been chosen and labeled, the data are plotted one point at a time.

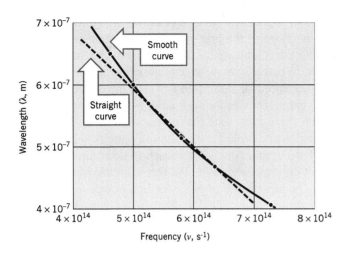

Fig. A.3 Plot of wavelength (λ) in meters versus frequency (ν) in cycles per second for light of different colors. Note that the curve is almost, but not quite, a straight line.

- A straight line or a smooth curve is then drawn through as many points as possible. Because of experimental error, the line or curve may not pass through every data point.

If this process gives a straight line, we can conclude that the quantity plotted on the vertical axis (λ) is **directly proportional** to the quantity on the horizontal axis (ν). We can then fit the graph to the equation for a straight line: $y = mx + b$.

$$\lambda = m\nu + b$$

The graph in Figure A.3 is not quite a straight line. Because the data are known to four significant figures, the deviation from a straight line is not the result of experimental error. We must therefore conclude that the wavelength and frequency of light are not directly proportional. We might therefore test whether the two sets of measurements are inversely proportional. In other words, we might try to fit the data to the following equation.

$$\lambda = m\left(\frac{1}{\nu}\right) + b$$

We can test this relationship by plotting the wavelengths in Table A.4 versus the inverse of the frequencies, as shown in Figure A.4.

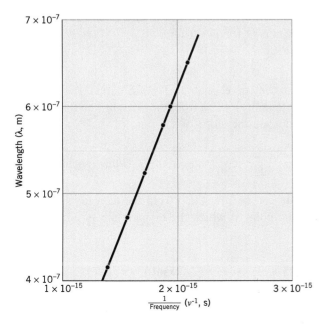

Fig. A.4 A plot of wavelength (λ) versus the inverse of frequency (ν) for light of different colors gives a straight line, within experimental error.

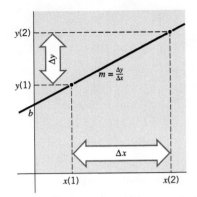

Fig. A.5 To calculate the slope (m) of a straight line, divide the distance between two points on the vertical axis by the corresponding distance between two points along the horizontal axis. The y intercept (b) can be determined by reading the value on the vertical axis that · corresponds to a value of zero on the horizontal axis.

The graph in Figure A.4 gives a beautiful straight-line relationship. We can now calculate the slope of the line (m) and the y intercept (b), as shown in Figure A.5. When doing the calculation, it is important to choose points that are *not* original data points. When this is done, the slope of the line in Figure A.4 is found to be equal to the speed of light: 2.998×10^8 meters per second.

When the graph includes the origin, the y-intercept can be determined by reading the value on the vertical axis that corresponds to a value of zero on the horizontal axis. This isn't possible for the data in Figure A.4, however, because the origin was left off the graph. Another approach to determining the intercept starts by selecting a point on the straight line. We then combine values of y and x read from the plot for this point with the slope of the line to calculate the value of the intercept. When this approach is applied to the data in Figure A.4, we find that the intercept is equal to zero. The data in Figure A.4 therefore fit the following equation.

$$\lambda = (2.998 \times 10^8 \text{ m/s})\left(\frac{1}{\nu}\right)$$

Rearranging the equation, we find that the product of the frequency times the wavelength of light is equal to the speed of light.

$$\nu\lambda = 2.998 \times 10^8 \text{ m/s}$$

Exercise A.3

The following data were obtained from a study of the relationship between the volume (V) and temperature (T) of a gas at constant pressure.

Volume (mL)	273.0	277.4	282.7	287.8	293.1	298.1
Temperature (°C)	0.0	5.0	10.0	15.0	20.0	25.0

Determine the temperature at which the volume of the gas should become equal to zero.

Solution

A plot of the data gives a straight line, as shown in Figure A.6. The equation for the straight line can be written as follows.

$$T = mV + b$$

By choosing any two points on the curve, such as T = 7.5°C and T = 22.5°C, and reading the volumes at those temperatures from the straight line that passes through the points, we can calculate the slope of the line.

$$m = \frac{\Delta T}{\Delta V} = \frac{22.5°C - 7.5°C}{295.5 \text{ mL} - 280.1 \text{ mL}} = 0.974°C/mL$$

The value of b for the equation can be calculated from the data for any point on the straight line, such as the point at which the temperature is 22.5°C and the volume is 295.5 mL.

$$b = T - mV = 22.5°C - \frac{0.974°C}{1 \text{ mL}}(295.5 \text{ mL}) = -265°C$$

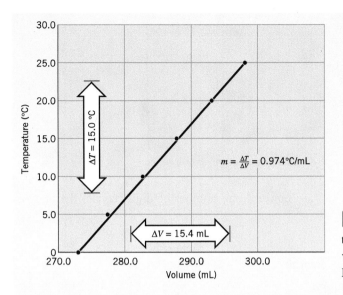

Fig. A.6 Plot of the temperature versus volume data from Exercise A.3.

According to these data, the volume of the gas should become zero when the gas is cooled until the temperature reaches about −265°C.

• •

A.6 Significant figures and Unit Conversion Worksheet

SIGNIFICANT FIGURES

One way in which a scientist can indicate the quality of a measurement or the degree to which a piece of equipment allows a measurement to be made is by the use of significant figures. Consider the measurement 32.45 g. The digits in 32.4 are known with certainty, but there is uncertainty associated with the 0.05. The actual value might be 32.47 g, 32.44 g, or some other value that deviates in the second digit after the decimal. The actual value of the last reported significant figure is uncertain. All other significant figures in a measurement are assumed to be known with certainty. Thus, in 32.45 there are four significant figures and the last digit is uncertain. You should use correct significant figures when recording data and reporting results in the laboratory, as well as when working homework and test questions.

The uncertainty associated with many measurements results from estimating between the smallest divisions on a measuring device. Measurements should always be recorded to show the digits that are known with certainty (those digits corresponding to a division on the measuring device) plus one digit that is uncertain (corresponding to the estimation between the smallest divisions).

1. Figure A.7 shows a diagram of three volumes of water measured in 100 mL graduated cylinders. Record each measurement to the correct number of significant figures. A graduated cylinder is always read from the bottom of the meniscus.

COUNTING SIGNIFICANT FIGURES IN A MEASUREMENT

Expressing numbers in scientific notation can help you count the number of significant figures. The exponential portion of a number expressed in scientific notation is not used in counting significant figures.

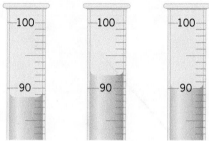

Fig. A.7 Record the volume of liquid in the graduated cylinders.   Always read from the bottom of the meniscus.

If all of the numbers are nonzero, the number of digits equals the number of significant figures.

Measurement	Number of Significant Figures
1426.5	5
32.4561	6

Leading zeros are never significant. They are used only to set the position of the decimal point.

Measurement	Scientific Notation	Significant Figures
0.369	3.69×10^{-1}	3
0.00029	2.9×10^{-4}	2
0.008957	8.957×10^{-3}	4

Zeros between significant figures are always significant.

Measurement	Scientific Notation	Significant Figures
104.56	1.0456×10^{2}	5
0.002305	2.305×10^{-3}	4

Trailing zeros that are *not* needed to set the position of the decimal point are significant. Their function is to show how accurately the measurement is known.

Measurement	Scientific Notation	Significant Figures
12.30	1.230×10^{1}	4
5.00	5.00×10^{0}	3
230.0	2.30×10^{2}	3

Trailing zeros that have no decimal point do not clearly show the number of significant figures. These measurements are best written in scientific notation. For example, the measurement 700 milliliters could represent 7×10^{2} mL, 7.0×10^{2} mL, or 7.00×10^{2} mL (i.e., one, two, or three significant figures, depending on how the measurement was made).

2. Express each of the following measurements in scientific notation and count the number of significant figures.

Measurement	Scientific Notation	Significant Figures
98.30		
104.02		
0.00285		
100.03		
0.04340		

Significant Figures in Calculations. It is important to maintain the proper number of significant figures when doing calculations. The number of digits shown in the result must correctly reflect the significant figures associated with each measurement used in the calculation.

Addition and Subtraction: In addition and subtraction, the answer is restricted to the number of digits *after the decimal.* The measurement with the least number of digits *after the decimal* determines the number of digits after the decimal in the answer.

$$
\begin{array}{ll}
140.15 & \text{(two digits after the decimal)} \\
34.4129 & \text{(four digits after the decimal)} \\
2032.1 & \text{(one digit after the decimal)} \\
\hline
2206.6629 & \text{(calculator answer)}
\end{array}
$$

The answer in correct significant figures is 2206.7 (limited to one digit after the decimal). Note that the last digit needed to be rounded up to 7 because the portion that was dropped (0.0629) was greater than 0.0500.

Multiplication and Division: In multiplication and division the answer can have no more total significant figures than the measurement with the least number of significant figures.

$$12.34 \times 0.0203 \times 25.673 = 6.431137846$$

Significant figures   (4)   (3)   (5)   (calculator answer)

The answer in correct significant figures is 6.43 (limited to three significant figures).

Calculations with Multiple Steps: In calculations with multiple steps, it is necessary to look at the number of significant figures in the answer to each step.

3. Explain why there are only three significant figures in the answer for the following calculation.

$$\frac{(123.4 + 0.42)}{(17.48 - 2.8)} = \frac{123.8}{14.7} = 8.42$$

4. Explain why there are only two significant figures in the answer for the following calculation.

$$\frac{(123.4 + 0.42)}{(17.48 - 17.00)} = \frac{123.8}{0.48} = 2.6 \times 10^2$$

Measurements versus Definitions

5. Use a metric rule to find the diameter of the following circle to two significant figures. (For this measurement, do not estimate between the millimeter marks.)

Diameter = _____   Number of significant figures = ___2___

6. Now use the definition of radius (radius = diameter/2) to find the radius of the circle using the measurement you made of the diameter. Maintain significant figures as described in the previous sections.

   Radius =   _____

   Number of significant figures in numerator =   _____

   Number of significant figures in denominator = _____

   Number of significant figures in radius =   _____

7. Repeat the measurement of the diameter of the circle. This time, record the measurement to three significant figures (estimate between the millimeter marks). Now determine the radius for the circle while maintaining significant figures as described above.

   Radius =   _____

   Number of significant figures in numerator =   _____

   Number of significant figures in denominator = _____

   Number of significant figures in radius =   _____

8. Compare the number of significant figures in the two values of the radius. Does the number of significant figures in the radii that you calculated in steps 6 and 7 differentiate between the number of significant figures in the two measurements that you made?

The purpose of significant figures is to reflect the quality of a *measurement*. The problem with the radii that you calculated is that the 2 in the denominator of the radius formula is not a measurement. Instead it is a definition of radius. The 2 can be assumed to have an infinite number of significant figures. This also applies to counting numbers such as 24 people in a class. There could not be 24.5 or 23.9 people in the class. Numbers in a calculation that are definitions or counting numbers (not measurements) can be considered to have an infinite number of significant figures.

9. Now determine the radius of the circle for both measurements (steps 6 and 7), assuming that the 2 in the denominator has an infinite number of significant figures.

   Measurement 1:

   Radius = _____   Number of significant figures = _____

   Measurement 2:

   Radius = _____   Number of significant figures = _____

UNIT CONVERSIONS

When conversion factors are given (as in Table B.2 in Appendix B), you must determine the appropriate number of significant figures to use in converting a measurement to another unit.

The conversion factors used for converting between pounds and kilograms is the result of a measurement and therefore has a limited number of significant figures. The conversion factor can be found by determining the weight in pounds of a 1.0000 kg mass using a balance or scale. If the mass is found to weigh 2.20 pounds, the conversion factors are:

$$\frac{1.00 \text{ kg}}{2.20 \text{ lb}} \quad \text{or} \quad \frac{2.20 \text{ lb}}{1.00 \text{ kg}}$$

You may assume enough significant figures in the kg mass to equal the significant figures in the measurement of the pounds. If a better balance or scale was used, the conversion factor might be:

$$\frac{1.0000 \text{ kg}}{2.2046 \text{ lb}} \quad \text{or} \quad \frac{2.2046 \text{ lb}}{1.0000 \text{ kg}}$$

Some conversion factors are definitions and therefore have an infinite number of significant figures. Below are examples of conversion factors based on definitions, not measurements.

$$\frac{1.0000 \ldots \text{ ft}}{12.000 \ldots \text{ in.}} \quad \frac{1.0000 \ldots \text{ in.}}{2.5400 \ldots \text{ cm}}$$

To convert 14 lb into kilograms:

$$14 \text{ lb} \times \frac{1 \text{ kg}}{2.2046 \text{ lb}} = 6.4 \text{ kg}$$

To convert 14.0000 lb into kilograms:

$$14.0000 \text{ lb} \times \frac{1 \text{ kg}}{2.2046 \text{ lb}} = 6.3504 \text{ kg}$$

To convert 58 cm into inches:

$$58 \text{ cm} \times \frac{1 \text{ in.}}{2.54 \text{ cm}} = 23 \text{ in.}$$

To convert 58.0000 cm into inches:

$$58.0000 \text{ cm} \times \frac{1 \text{ in.}}{2.54 \text{ cm}} = 22.8346 \text{ in.}$$

10. Explain the difference between the number of significant figures found in each of the previous four conversions.

11. Perform the following calculations and express the result in correct significant figures.
    (a) Convert 18.9 in. into centimeters.
    (b) Convert 47.50 kg into pounds.
    (c) Convert 35.56 in. into centimeters.
    (d) Convert 92.3 cm into feet.

Appendix B

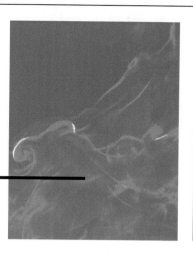

TABLES

Table B.1
Values of Selected Fundamental Constants

Speed of light in a vacuum (c)	$c = 2.99792458 \times 10^8$ m/s
Charge on an electron (q_e)	$q_e = 1.60217733 \times 10^{-19}$ C
Rest mass of an electron (m_e)	$m_e = 9.109389 \times 10^{-28}$ g
	$m_e = 5.485799 \times 10^{-4}$ amu
Rest mass of a proton (m_p)	$m_p = 1.672623 \times 10^{-24}$ g
	$m_p = 1.00727647$ amu
Rest mass of a neutron (m_n)	$m_n = 1.674928 \times 10^{-24}$ g
	$m_n = 1.0086649$ amu
Faraday's constant (F)	$F = 96,485$ C/mol
Planck's constant (h)	$h = 6.626075 \times 10^{-34}$ J $\cdot$ s
Ideal gas constant (R)	$R = 0.0820568$ L $\cdot$ atm/mol $\cdot$ K
	$R = 8.31451$ J/mol $\cdot$ K
Atomic mass unit (amu)	1 amu $= 1.6605402 \times 10^{-24}$ g
Boltzmann's constant (k)	$k = 1.380658 \times 10^{-23}$ J/K
Avogadro's constant (N)	$N = 6.0221367 \times 10^{23}$ mol^{-1}
Rydberg constant (R_H)	$R_H = 1.0973715 \times 10^7$ m^{-1}
	$= 1.0973715 \times 10^{-2}$ nm^{-1}

Table B.2
Selected Conversion Factors

Energy	$\dfrac{1 \text{ J}}{0.2390 \text{ cal}}$	$\dfrac{1 \text{ J}}{1 \times 10^7 \text{ erg}}$	$\dfrac{1 \text{ cal}}{4.184 \text{ J}}$ (by definition)
	$\dfrac{1 \text{ eV/atom}}{1.6021793 \times 10^{-19} \text{ J/atom}}$		$\dfrac{1 \text{ eV/atom}}{96.485 \text{ kJ/mol}}$
Temperature	K = °C + 273.15		
	°C = 5/9(°F − 32)		
	°F = 9/5°C + 32		
Pressure	$\dfrac{1 \text{ atm}}{760 \text{ mm of Hg}}$ (by definition),	$\dfrac{1 \text{ atm}}{760 \text{ torr}}$ (by definition)	
	$\dfrac{1 \text{ atm}}{101.325 \text{ kPa}}$	$\dfrac{1 \text{ atm}}{14.7 \text{ lb/in.}^2}$	
Mass	$\dfrac{1 \text{ kg}}{2.2046 \text{ lb}}$	$\dfrac{1 \text{ lb}}{453.59 \text{ g}}$	
	$\dfrac{1 \text{ oz}}{0.06250 \text{ lb}}$	$\dfrac{1 \text{ oz}}{28.350 \text{ g}}$	
	$\dfrac{1 \text{ ton}}{2000 \text{ lb}}$	$\dfrac{1 \text{ ton}}{907.185 \text{ kg}}$	
	$\dfrac{1 \text{ metric tonne}}{1000 \text{ kg}}$	$\dfrac{1 \text{ metric tonne}}{2204.62 \text{ lb}}$	
Volume	$\dfrac{1 \text{ mL}}{0.001 \text{ L}}$	$\dfrac{1 \text{ mL}}{1 \text{ cm}^3}$ (by definition)	
	$\dfrac{1 \text{ oz (fluid)}}{0.031250 \text{ qt}}$	$\dfrac{1 \text{ oz (fluid)}}{0.029573 \text{ L}}$	
	$\dfrac{1 \text{ qt}}{0.9463529 \text{ L}}$	$\dfrac{1 \text{ L}}{1.05672 \text{ qt}}$	
Length	$\dfrac{1 \text{ m}}{39.370 \text{ in.}}$	$\dfrac{1 \text{ mi}}{1.60934 \text{ km}}$	$\dfrac{1 \text{ in.}}{2.54 \text{ cm}}$ (by defintion)

Table B.3
The Vapor Pressure of Water

Temperature (°C)	Pressure (mmHg)	Temperature (°C)	Pressure (mmHg)	Temperature (°C)	Pressure (mmHg)	Temperature (°C)	Pressure (mmHg)
0	4.6	13	11.2	26	25.2	39	52.4
1	4.9	14	12.0	27	26.7	40	55.3
2	5.3	15	12.8	28	28.3	41	58.3
3	5.7	16	13.6	29	30.0	42	61.5
4	6.1	17	14.5	30	31.8	43	64.8
5	6.5	18	15.5	31	33.7	44	68.3
6	7.0	19	16.5	32	35.7	45	71.9
7	7.5	20	17.5	33	37.7	46	75.7
8	8.0	21	18.7	34	39.9	47	79.6
9	8.6	22	19.8	35	42.2	48	83.7
10	9.2	23	21.1	36	44.6	49	88.0
11	9.8	24	22.4	37	47.1	50	92.5
12	10.5	25	23.8	38	49.7		

Table B.4
Radii of Atoms and Ions

Element	Ionic Radius (nm)	Ionic Charge	Covalent Radius (nm)	Metallic Radius (nm)
Aluminum	0.050	(+3)	0.125	0.1431
Antimony	0.245	(−3)	0.141	
	0.09	(+3)		
	0.062	(+5)		
Arsenic	0.222	(−3)	0.121	0.1248
	0.058	(+3)		
	0.047	(+5)		
Astatine	0.227	(−1)		
	0.051	(+7)		
Barium	0.135	(+2)	0.198	0.2173
Beryllium	0.031	(+2)	0.089	0.1113
Bismuth	0.213	(−3)	0.152	0.1547
	0.096	(+3)		
	0.074	(+5)		
Boron	0.020	(+3)	0.088	0.083
Bromine	0.196	(−1)	0.1142	
	0.039	(+7)		
Cadmium	0.097	(+2)	0.141	0.1489
Calcium	0.099	(+2)	0.174	0.1973
Carbon	0.260	(−4)	0.077	
	0.015	(+4)		
Cesium	0.169	(+1)	0.235	0.2654
Chlorine	0.181	(−1)	0.099	
	0.026	(+7)		
Chromium	0.064	(+3)	0.117	0.1249
	0.052	(+6)		
Cobalt	0.074	(+2)	0.116	0.1253
	0.063	(+3)		
Copper	0.096	(+1)	0.117	0.1278
	0.072	(+2)		
Fluorine	0.136	(−1)	0.064	0.0717
	0.007	(+7)		
Francium	0.176	(+1)		0.27
Gallium	0.062	(+3)	0.125	0.1221
Germanium	0.272	(−4)	0.122	0.1225
	0.053	(+4)		
Gold	0.137	(+1)	0.134	0.1442
	0.091	(+3)		
Hydrogen	0.208	(−1)	0.0371	
	10^{-6}	(+1)		
Indium	0.081	(+3)	0.150	0.1626
Iodine	0.216	(−1)	0.1333	
	0.050	(+7)		
Iron	0.076	(+2)	0.1165	0.1241
	0.064	(+3)		
Lead	0.215	(−4)	0.154	0.1750
	0.120	(+2)		
	0.084	(+4)		
Lithium	0.068	(+1)	0.123	0.152
Magnesium	0.065	(+2)	0.136	0.160

Element	Ionic Radius (nm)	Ionic Charge	Covalent Radius (nm)	Metallic Radius (nm)
Manganese	0.080	(+2)	0.117	0.124
	0.046	(+7)		
Mercury	0.127	(+1)	0.144	0.160
	0.110	(+2)		
Molybdenum	0.062	(+6)	0.129	0.1362
Nickel	0.072	(+2)	0.115	0.1246
Nitrogen	0.171	(−3)	0.070	
	0.013	(+3)		
	0.011	(+5)		
Oxygen	0.140	(−2)	0.066	
	0.176	(−1)		
Phosphorus	0.212	(−3)	0.110	0.108
	0.042	(+3)		
	0.034	(+5)		
Polonium	0.230	(−2)	0.153	0.167
	0.056	(+6)		
Potassium	0.133	(+1)	0.2025	0.2272
Radium	0.140	(+2)		0.220
Rubidium	0.148	(+1)	0.216	0.2475
Scandium	0.081	(+3)	0.144	0.1606
Selenium	0.198	(−2)	0.117	
	0.069	(+4)		
	0.042	(+6)		
Silicon	0.271	(−4)	0.117	
	0.041	(+4)		
Silver	0.126	(+1)	0.134	0.1444
Sodium	0.095	(+1)	0.157	0.186
Strontium	0.113	(+2)	0.192	0.2151
Sulfur	0.184	(−2)	0.104	
	0.037	(+4)		
	0.029	(+6)		
Tellurium	0.221	(−2)	0.137	0.1432
	0.081	(+4)		
	0.056	(+6)		
Thallium	0.095	(+3)	0.155	0.1704
Tin	0.294	(−4)	0.140	0.1405
	0.102	(+2)		
	0.071	(+4)		
Titanium	0.090	(+2)	0.132	0.1448
	0.068	(+4)		
Tungsten	0.065	(+6)	0.130	0.1370
Uranium	0.083	(+6)		0.1385
Vanadium	0.059	(+5)	0.122	0.1321
Xenon			0.130	
Zinc	0.074	(+2)	0.125	0.1332
Zirconium	0.079	(+4)	0.145	0.167

Table B.5
Ionization Energies (kJ/mol)

Atomic Number	Symbol	I	II	III	IV	V	VI	VII
1	H	1,312.0						
2	He	2,372.3	5,250.3					
3	Li	520.2	7,297.9	11,814.6				
4	Be	899.4	1,757.1	14,848.3	21,005.9			
5	B	800.6	2,427.0	3,659.6	25,025.0	32,825.7		
6	C	1,086.4	2,352.6	4,620.4	6,222.5	37,829.4	47,275.6	
7	N	1,402.3	2,856.0	4,578.0	7,474.9	9,444.7	50,370.4	64,358.0
8	O	1,313.9	3,388.2	5,300.3	7,469.1	10,989.2	13,326.1	71,332
9	F	1,681.0	3,374.1	6,050.3	8,407.5	11,022.4	15,163.6	17,867.2
10	Ne	2,080.6	3,952.2	6,122	9,370	12,177	15,238	19,998
11	Na	495.8	4,562.4	6,912	9,543	13,352	16,610	20,114
12	Mg	737.7	1,450.6	7,732.6	10,540	13,629	17,994	21,703
13	Al	577.6	1,816.6	2,744.7	11,577	14,831	18,377	23,294
14	Si	786.4	1,577.0	3,231.5	4,355.4	16,091	19,784	23,785
15	P	1,011.7	1,903.2	2,912	4,956	6,273.7	21,268	25,397
16	S	999.58	2,251	3,361	4,564	7,012	8,495.4	27,105
17	Cl	1,251.1	2,297	3,822	5,158	6,540	9,362	11,017.9
18	Ar	1,520.5	2,665.8	3,931	5,771	7,238	8,780.8	11,994.9
19	K	418.8	3,051.3	4,411	5,877	7,975	9,648.5	11,343
20	Ca	589.8	1,145.4	4,911.8	6,474	8,144	10,496	12,320
21	Sc	631	1,235	2,389	7,099	8,844	10,720	13,310
22	Ti	658	1,310	2,652.5	4,174.5	9,573	11,516	13,590
23	V	650	1,413	2,828.0	4,506.5	6,294	12,362	14,489
24	Cr	652.8	1,592	2,987	4,740	6,690	8,738	15,540
25	Mn	717.4	1,509.0	3,248.3	4,940	6,990	9,220	11,508
26	Fe	759.3	1,561	2,957.3	5,290	7,240	9,600	12,100
27	Co	758	1,646	3,232	4,950	7,670	9,840	12,400
28	Ni	736.7	1,752.9	3,393	5,300	7,280	10,400	12,800
29	Cu	745.4	1,957.9	3,553	5,330	7,710	9,940	13,400
30	Zn	906.4	1,733.2	3,832.6	5,730	7,970	10,400	12,900
31	Ga	578.8	1,979	2,963	6,200			
32	Ge	762.1	1,537.4	3,302	4,410	9,020		
33	As	947	1,797.8	2,735.4	4,837	6,043	12,300	
34	Se	940.9	2,045	2,973.7	4,143.4	6,590	7,883	14,990
35	Br	1,139.9	2,100	3,500	4,560	5,760	8,550	9,938
36	Kr	1,350.7	2,350.3	3,565	5,070	6,240	7,570	10,710
37	Rb	403.0	2,632	3,900	5,070	6,850	8,140	9,570
38	Sr	549.5	1,064.5	4,120	5,500	6,910	8,760	10,200
39	Y	616	1,181	1,980	5,960	7,430	8,970	11,200
40	Zr	660	1,267	2,218	3,313	7,870		
41	Nb	664	1,382	2,416	3,960	4,877	9,899	12,100
42	Mo	684.9	1,558	2,621	4,480	5,910	6,600	12,230
43	Tc	702	1,472	2,850				
44	Ru	711	1,617	2,747				
45	Rh	720	1,744	2,997				
46	Pd	805	1,874	3,177				
47	Ag	731.0	2,073	3,361				
48	Cd	867.7	1,631.4	3,616				
49	In	558.3	1,820.6	2,704	5,200			
50	Sn	708.6	1,411.8	2,943.0	3,930.2	6,974		
51	Sb	833.7	1,595	2,440	4,260	5,400	10,400	

Atomic Number	Symbol	I	II	III	IV	V	VI	VII
52	Te	869.2	1,790	2,698	3,609	5,668	6,820	13,200
53	I	1,008.4	1,845.8	3,200				
54	Xe	1,170.4	2,046	3,100				
55	Cs	375.7	2,440					
56	Ba	502.9	965.23					
57	La	538.1	1,067	1,850.3				
58	Ce	527.8	1,047	1,949	3,547			
59	Pr	523	1,018	2,086	3,761	5,543		
60	Nd	530	1,035	2,130	3,900			
61	Pm	535	1,052	2,150	3,970			
62	Sm	543	1,068	2,260	3,990			
63	Eu	547	1,084	2,400	4,110			
64	Gd	592	1,167	1,990	4,250			
65	Tb	564	1,112	2,110	3,840			
66	Dy	572	1,126	2,200	4,000			
67	Ho	581	1,139	2,203	4,100			
68	Er	589	1,151	2,194	4,120			
69	Tm	596	1,163	2,285	4,120			
70	Yb	603	1,175	2,415	4,216			
71	Lu	524	1,340	2,022	4,360			
72	Hf	642	1,440	2,250	3,210			
73	Ta	761						
74	W	770						
75	Re	760	1,260	2,510	3,640			
76	Os	840						
77	Ir	880						
78	Pt	870	1,791.0					
79	Au	890.1	1,980					
80	Hg	1,007.0	1,809.7	3,300				
81	Tl	589.3	1,971.0	2,878				
82	Pb	715.5	1,450.4	3,081.4	4,083	6,640		
83	Bi	703.3	1,610	2,466	4,370	5,400	8,520	
84	Po	812						
85	At							
86	Rn	1,037.0						
87	Fr							
88	Ra	509.3	979.0					
89	Ac	498.8	1,170					
90	Th	587	1,110	1,930	2,780			
91	Pa	568						
92	U	584						
93	Np	597						
94	Pu	585						
95	Am	578.2						
96	Cm	581						
97	Bk	601						
98	Cf	608						
99	Es	619						
100	Fm	627						
101	Md	635						
102	No	642						

Table B.6
Electron Affinities

Atomic Number	Symbol	Electron Affinity (kJ/mol)	Atomic Number	Symbol	Electron Affinity (kJ/mol)
1	H	72.8	37	Rb	46.89
2	He	*	38	Sr	*
3	Li	59.8	39	Y	0
4	Be	*	40	Zr	50
5	B	27	41	Nb	96
6	C	122.3	42	Mo	96
7	N	−7	43	Tc	70
8	O	141.1	44	Ru	110
9	F	328.0	45	Rh	120
10	Ne	*	46	Pd	60
11	Na	52.7	47	Ag	125.7
12	Mg	*	48	Cd	*
13	Al	45	49	In	29
14	Si	133.6	50	Sn	121
15	P	71.7	51	Sb	101
16	S	200.42	52	Te	190.15
17	Cl	348.8	53	I	295.3
18	Ar	*	54	Xe	*
19	K	48.36	55	Cs	45.49
20	Ca	*	56	Ba	*
21	Sc	*	57–71	La–Lu	50
22	Ti	20	72	Hf	*
23	V	50	73	Ta	60
24	Cr	64	74	W	60
25	Mn	*	75	Re	14
26	Fe	24	76	Os	110
27	Co	70	77	Ir	150
28	Ni	111	78	Pt	205.3
29	Cu	118.3	79	Au	222.74
30	Zn	0	80	Hg	*
31	Ga	29	81	Tl	30
32	Ge	120	82	Pb	110
33	As	77	83	Bi	110
34	Se	194.96	84	Po	180
35	Br	324.6	85	At	270
36	Kr	*	86	Rn	*
			87	Fr	44.0

*These elements have negative electron affinities. Electron affinity is the negative of the enthalpy change for the reaction $X(g) + e^- \rightarrow X^-(g)$.

Table B.7
Electronegativities[a]

Atomic Number	Element	Electro-negativity	Atomic Number	Element	Electro-negativity
1	H	2.300	41	Nb	1.25
2	He	4.157	42	Mo	1.39
3	Li	0.912	43	Tc	1.52
4	Be	1.576	44	Ru	1.66
5	B	2.051	45	Rh	1.79
6	C	2.544	46	Pd	1.91
7	N	3.066	47	Ag	1.98
8	O	3.610	48	Cd	1.52
9	F	4.193	49	In	1.656
10	Ne	4.787	50	Sn	1.824
11	Na	0.869	51	Sb	1.984
12	Mg	1.293	52	Te	2.158
13	Al	1.613	53	I	2.359
14	Si	1.916	54	Xe	2.582
15	P	2.253	55	Cs	0.66
16	S	2.589	56	Ba	0.88
17	Cl	2.869	57–71		
18	Ar	3.242	72		
19	K	0.734	73		
20	Ca	1.034	74		
21	Sc	1.15	75		
22	Ti	1.25	76		
23	V	1.37	77		
24	Cr	1.45	78		
25	Mn	1.55	79		
26	Fe	1.67	80	Hg	1.76
27	Co	1.76	81		
28	Ni	1.86	82		
29	Cu	1.83	83		
30	Zn	1.59	84		
31	Ga	1.756	85		
32	Ge	1.994	86		
33	As	2.211	87		
34	Se	2.424	88		
35	Br	2.685	89		
36	Kr	2.966	90		
37	Rb	0.706	91		
38	Sr	0.963	92		
39	Y	1.00	93–103		
40	Zr	1.12			

[a]Allen electronegativities are taken from L. C. Allen, *Int. J. Quant. Chem.*, **48**, 253–277 (1993); L. C. Allen, *J. Am. Chem. Soc.*, **111**, 9003–9014 (1989).

Table B.8
Acid Dissociation Equilibrium Constants

Compound	Dissociation Reaction	K_a	pK_a
Acetic acid	$CH_3CO_2H + H_2O \rightleftharpoons CH_3CO_2^- + H_3O^+$	1.75×10^{-5}	4.757
Ammonium ion	$NH_4^+ + H_2O \rightleftharpoons NH_3 + H_3O^+$	5.6×10^{-10}	9.25
Arsenic acid	$H_3AsO_4 + H_2O \rightleftharpoons H_2AsO_4^- + H_3O^+$	6.0×10^{-3}	2.22
	$H_2AsO_4^- + H_2O \rightleftharpoons HAsO_4^{2-} + H_3O^+$	1.0×10^{-7}	7.00
	$HAsO_4^{2-} + H_2O \rightleftharpoons AsO_4^{3-} + H_3O^+$	3.0×10^{-12}	11.52
Arsenous acid	$H_3AsO_3 + H_2O \rightleftharpoons H_2AsO_3^- + H_3O^+$	6.3×10^{-10}	9.22
	$H_2AsO_3^- + H_2O \rightleftharpoons HAsO_3^{2-} + H_3O^+$	3.0×10^{-14}	13.52
Benzoic acid	$C_6H_5CO_2H + H_2O \rightleftharpoons C_6H_5CO_2^- + H_3O^+$	6.3×10^{-5}	4.20
Boric acid	$H_3BO_3 + H_2O \rightleftharpoons H_2BO_3^- + H_3O^+$	7.3×10^{-10}	9.14
Carbonic acid	$H_2CO_3 + H_2O \rightleftharpoons HCO_3^- + H_3O^+$	4.5×10^{-7}	6.35
	$HCO_3^- + H_2O \rightleftharpoons CO_3^{2-} + H_3O^+$	4.7×10^{-11}	10.33
Chloric acid	$HClO_3 + H_2O \rightleftharpoons ClO_3^- + H_3O^+$	5.0×10^2	-2.70
Chloroacetic acid	$ClCH_2CO_2H + H_2O \rightleftharpoons ClCH_2CO_2^- + H_3O^+$	1.4×10^{-3}	2.85
Chlorous acid	$HClO_2 + H_2O \rightleftharpoons ClO_2^- + H_3O^+$	1.1×10^{-2}	1.96
Chromic acid	$H_2CrO_4 + H_2O \rightleftharpoons HCrO_4^- + H_3O^+$	9.6	-0.98
	$HCrO_4^- + H_2O \rightleftharpoons CrO_4^{2-} + H_3O^+$	3.2×10^{-7}	6.50
Citric acid	$H_3Cit + H_2O \rightleftharpoons H_2Cit^- + H_3O^+$	7.5×10^{-4}	3.13
	$H_2Cit^- + H_2O \rightleftharpoons HCit^{2-} + H_3O^+$	1.7×10^{-5}	4.77
	$HCit^{2-} + H_2O \rightleftharpoons Cit^{3-} + H_3O^+$	4.0×10^{-7}	6.40
Dichloroacetic acid	$Cl_2CHCO_2H + H_2O \rightleftharpoons Cl_2CHCO_2^- + H_3O^+$	5.1×10^{-2}	1.29
Formic acid	$HCO_2H + H_2O \rightleftharpoons HCO_2^- + H_3O^+$	1.8×10^{-4}	3.75
Glycine	$H_3N^+CH_2CO_2H + H_2O \rightleftharpoons H_3N^+CH_2CO_2^- + H_3O^+$	4.5×10^{-3}	2.35
	$H_3N^+CH_2CO_2^- + H_2O \rightleftharpoons H_2NCH_2CO_2^- + H_3O^+$	2.5×10^{-10}	9.60
Hydrazoic acid	$HN_3 + H_2O \rightleftharpoons N_3^- + H_3O^+$	1.9×10^{-5}	4.72
Hydrobromic acid	$HBr + H_2O \rightleftharpoons Br^- + H_3O^+$	1×10^9	-9
Hydrochloric acid	$HCl + H_2O \rightleftharpoons Cl^- + H_3O^+$	1×10^6	-6
Hydrocyanic acid	$HCN + H_2O \rightleftharpoons CN^- + H_3O^+$	6×10^{-10}	9.22
Hydrofluoric acid	$HF + H_2O \rightleftharpoons F^- + H_3O^+$	7.2×10^{-4}	3.14
Hydroiodic acid	$HI + H_2O \rightleftharpoons I^- + H_3O^+$	3×10^9	-9.5
Hydrogen peroxide	$H_2O_2 + H_2O \rightleftharpoons HO_2^- + H_3O^+$	2.2×10^{-12}	11.66
Hydrogen selenide	$H_2Se + H_2O \rightleftharpoons HSe^- + H_3O^+$	1.0×10^{-4}	4.00
Hydrogen sulfide	$H_2S + H_2O \rightleftharpoons HS^- + H_3O^+$	1.0×10^{-7}	7.00
	$HS^- + H_2O \rightleftharpoons S^{2-} + H_3O^+$	1.3×10^{-13}	12.89
Hypobromous acid	$HOBr + H_2O \rightleftharpoons OBr^- + H_3O^+$	2.4×10^{-9}	8.62
Hypochlorous acid	$HOCl + H_2O \rightleftharpoons OCl^- + H_3O^+$	2.9×10^{-8}	7.54
Hypoiodous acid	$HOI + H_2O \rightleftharpoons OI^- + H_3O^+$	2.3×10^{-11}	10.64
Iodic acid	$HIO_3 + H_2O \rightleftharpoons IO_3^- + H_3O^+$	0.16	0.80
Nitric acid	$HNO_3 + H_2O \rightleftharpoons NO_3^- + H_3O^+$	28	-1.45
Nitrous acid	$HNO_2 + H_2O \rightleftharpoons NO_2^- + H_3O^+$	5.1×10^{-4}	3.29
Oxalic acid	$H_2C_2O_4 + H_2O \rightleftharpoons HC_2O_4^- + H_3O^+$	5.4×10^{-2}	1.27
	$HC_2O_4^- + H_2O \rightleftharpoons C_2O_4^{2-} + H_3O^+$	5.4×10^{-5}	4.27
Perchloric acid	$HOClO_3 + H_2O \rightleftharpoons ClO_4^- + H_3O^+$	1×10^8	-8
Periodic acid	$H_5IO_6 + H_2O \rightleftharpoons H_4IO_6^- + H_3O^+$	2.3×10^{-2}	1.64
Phenol	$C_6H_5OH + H_2O \rightleftharpoons C_6H_5O^- + H_3O^+$	1.0×10^{-10}	10.00
Phosphoric acid	$H_3PO_4 + H_2O \rightleftharpoons H_2PO_4^- + H_3O^+$	7.1×10^{-3}	2.15
	$H_2PO_4^- + H_2O \rightleftharpoons HPO_4^{2-} + H_3O^+$	6.3×10^{-8}	7.20
	$HPO_4^{2-} + H_2O \rightleftharpoons PO_4^{3-} + H_3O^+$	4.2×10^{-13}	12.38
Phosphorous acid	$H_3PO_3 + H_2O \rightleftharpoons H_2PO_3^- + H_3O^+$	1.00×10^{-2}	2.00
	$H_2PO_3^- + H_2O \rightleftharpoons HPO_3^{2-} + H_3O^+$	2.6×10^{-7}	6.59

Compound	Dissociation Reaction	K_a	pK_a
Sulfamic acid	$H_2NSO_3H + H_2O \rightleftharpoons H_2NSO_3^- + H_3O^+$	1.03×10^{-1}	0.987
Sulfuric acid	$H_2SO_4 + H_2O \rightleftharpoons HSO_4^- + H_3O^+$	1×10^3	-3
	$HSO_4^- + H_2O \rightleftharpoons SO_4^{2-} + H_3O^+$	1.2×10^{-2}	1.92
Sulfurous acid	$H_2SO_3 + H_2O \rightleftharpoons HSO_3^- + H_3O^+$	1.7×10^{-2}	1.77
	$HSO_3^- + H_2O \rightleftharpoons SO_3^{2-} + H_3O^+$	6.4×10^{-8}	7.19
Thiocyanic acid	$HSCN + H_2O \rightleftharpoons SCN^- + H_3O^+$	71	-1.85
Trichloroacetic acid	$Cl_3CCO_2H + H_2O \rightleftharpoons Cl_3CCO_2^- + H_3O^+$	0.22	0.66
Water	$H_2O + H_2O \rightleftharpoons OH^- + H_3O^+$	1.8×10^{-16}	15.75

Table B.9

Base Ionization Equilibrium Constants

Compound	Ionization Reaction	K_b	pK_b
Ammonia	$NH_3 + H_2O \rightleftharpoons NH_4^+ + OH^-$	1.8×10^{-5}	4.74
Aniline	$C_6H_5NH_2 + H_2O \rightleftharpoons C_6H_5NH_3^+ + OH^-$	4.0×10^{-10}	9.40
Butylamine	$CH_3(CH_2)_3NH_2 + H_2O \rightleftharpoons CH_3(CH_2)_3NH_3^+ + OH^-$	4.0×10^{-4}	3.40
Dimethylamine	$(CH_3)_2NH + H_2O \rightleftharpoons (CH_3)_2NH_2^+ + OH^-$	5.9×10^{-4}	3.23
Ethanolamine	$HOCH_2CH_2NH_2 + H_2O \rightleftharpoons HOCH_2CH_2NH_3^+ + OH^-$	3.3×10^{-5}	4.50
Ethylamine	$CH_3CH_2NH_2 + H_2O \rightleftharpoons CH_3CH_2NH_3^+ + OH^-$	4.4×10^{-4}	3.37
Hydrazine	$H_2NNH_2 + H_2O \rightleftharpoons H_2NNH_3^+ + OH^-$	1.2×10^{-6}	5.89
Hydroxylamine	$HONH_2 + H_2O \rightleftharpoons HONH_3^+ + OH^-$	1.1×10^{-8}	7.97
Methylamine	$CH_3NH_2 + H_2O \rightleftharpoons CH_3NH_3^+ + OH^-$	4.8×10^{-4}	3.32
Pyridine	$C_5H_5N + H_2O \rightleftharpoons C_5H_5NH^+ + OH^-$	1.7×10^{-9}	8.77
Trimethylamine	$(CH_3)_3N + H_2O \rightleftharpoons (CH_3)_3NH^+ + OH^-$	6.3×10^{-5}	4.20
Urea	$H_2NCONH_2 + H_2O \rightleftharpoons H_2NCONH_3^+ + OH^-$	1.5×10^{-14}	13.82

Table B.10
Solubility Product Equilibrium Constants

Substance	K_{sp}	Substance	K_{sp}	Substance	K_{sp}
AgBr	5.0×10^{-13}	α-CoS	4.0×10^{-21}	$MnCO_3$	1.8×10^{-11}
AgCN	1.2×10^{-16}	β-CoS	2.0×10^{-25}	$Mn(OH)_2$	2×10^{-13}
Ag_2CO_3	8.1×10^{-12}	$Cr(OH)_3$	6.3×10^{-31}	MnS	3×10^{-13}
AgOH	2.0×10^{-8}	CuBr	5.3×10^{-9}	$NiCO_3$	6.6×10^{-9}
$AgC_2H_3O_2$	4.4×10^{-3}	CuCl	1.2×10^{-6}	NiC_2O_4	4×10^{-10}
$Ag_2C_2O_4$	3.4×10^{-11}	CuCN	3.2×10^{-20}	α-NiS	3.2×10^{-19}
AgCl	1.8×10^{-10}	$CuCrO_4$	3.6×10^{-6}	β-NiS	1.0×10^{-24}
Ag_2CrO_4	1.1×10^{-12}	$CuCO_3$	1.4×10^{-10}	γ-NiS	2.0×10^{-26}
AgI	8.3×10^{-17}	$Cu(OH)_2$	2.2×10^{-20}	$PbBr_2$	4.0×10^{-5}
Ag_2S	6.3×10^{-50}	CuI	1.1×10^{-13}	$PbCO_3$	7.4×10^{-14}
AgSCN	1.0×10^{-12}	Cu_2S	2.5×10^{-48}	PbC_2O_4	4.8×10^{-10}
Ag_2SO_4	1.4×10^{-5}	CuS	6.3×10^{-36}	$PbCl_2$	1.6×10^{-5}
$Al(OH)_3$	1.3×10^{-33}	CuSCN	4.8×10^{-15}	$PbCrO_4$	2.8×10^{-13}
AuCl	2.0×10^{-13}	$FeCO_3$	3.2×10^{-11}	PbF_2	2.7×10^{-8}
$AuCl_3$	3.2×10^{-23}	$Fe_2C_2O_4$	3.2×10^{-7}	$Pb(OH)_2$	1.2×10^{-15}
AuI	1.6×10^{-23}	$Fe(OH)_2$	8.0×10^{-16}	PbI_2	7.1×10^{-9}
AuI_3	5.5×10^{-46}	$Fe(OH)_3$	4×10^{-38}	PbS	8.0×10^{-28}
$BaCO_3$	5.1×10^{-9}	FeS	6.3×10^{-18}	$PbSO_4$	1.6×10^{-8}
BaC_2O_4	2.3×10^{-8}	Hg_2Br_2	5.6×10^{-23}	SnS	1.0×10^{-25}
$BaCrO_4$	1.2×10^{-10}	$Hg_2(CN)_2$	5×10^{-40}	$Sn(OH)_2$	1.4×10^{-28}
BaF_2	1.0×10^{-6}	Hg_2CO_3	8.9×10^{-17}	$Sn(OH)_4$	1×10^{-56}
$Ba(OH)_2$	5×10^{-3}	$Hg_2(OAc)_2$	3×10^{-11}	$SrCO_3$	1.1×10^{-10}
$BaSO_4$	1.1×10^{-10}	$Hg_2C_2O_4$	2.0×10^{-13}	SrC_2O_4	1.6×10^{-7}
Bi_2S_3	1×10^{-97}	HgC_2O_4	1×10^{-7}	$SrCrO_4$	2.2×10^{-5}
$CaCO_3$	2.8×10^{-9}	Hg_2Cl_2	1.3×10^{-18}	SrF_2	2.5×10^{-9}
CaC_2O_4	4×10^{-9}	Hg_2CrO_4	2.0×10^{-9}	$SrSO_4$	3.2×10^{-7}
$CaCrO_4$	7.1×10^{-4}	Hg_2I_2	4.5×10^{-29}	TlBr	3.4×10^{-6}
CaF_2	4.0×10^{-11}	Hg_2S	1.0×10^{-47}	TlCl	1.7×10^{-4}
$Ca(OH)_2$	5.5×10^{-6}	HgS	4×10^{-53}	TlI	6.5×10^{-8}
$CdCO_3$	5.2×10^{-12}	$K_2NaCo(NO_2)_6$	2.2×10^{-11}	$Zn(CN)_2$	2.6×10^{-13}
$Cd(CN)_2$	1.0×10^{-8}	$MgCO_3$	3.5×10^{-8}	$ZnCO_3$	1.4×10^{-11}
$Cd(OH)_2$	2.5×10^{-14}	MgC_2O_4	1×10^{-8}	ZnC_2O_4	2.7×10^{-8}
CdS	8×10^{-27}	MgF_2	6.5×10^{-9}	$Zn(OH)_2$	1.2×10^{-17}
$CoCO_3$	1.4×10^{-13}	$Mg(OH)_2$	1.8×10^{-11}	α-ZnS	1.6×10^{-24}
$Co(OH)_3$	1.6×10^{-44}	$MgNH_4PO_4$	2.5×10^{-13}	β-ZnS	2.5×10^{-22}

Table B.11
Complex Formation Equilibrium Constants

Equilibrium	K_f	Equilibrium	K_f
$Ag^+ + 2\,Br^- \rightleftharpoons AgBr_2^-$	2.1×10^7	$Fe^{2+} + 6\,CN^- \rightleftharpoons Fe(CN)_6^{4-}$	1×10^{35}
$Ag^+ + 2\,Cl^- \rightleftharpoons AgCl_2^-$	1.1×10^5	$Fe^{3+} + 6\,CN^- \rightleftharpoons Fe(CN)_6^{3-}$	1×10^{42}
$Ag^+ + 2\,CN^- \rightleftharpoons Ag(CN)_2^-$	1.3×10^{21}	$Fe^{3+} + SCN^- \rightleftharpoons Fe(SCN)^{2+}$	8.9×10^2
$Ag^+ + 2\,I^- \rightleftharpoons AgI_2^-$	5.5×10^{11}	$Fe^{3+} + 2\,SCN^- \rightleftharpoons Fe(SCN)_2^+$	2.3×10^3
$Ag^+ + 2\,NH_3 \rightleftharpoons Ag(NH_3)_2^+$	1.1×10^7	$Hg^{2+} + 4\,Br^- \rightleftharpoons HgBr_4^{2-}$	1×10^{21}
$Ag^+ + 2\,SCN^- \rightleftharpoons Ag(SCN)_2^-$	3.7×10^7	$Hg^{2+} + 4\,Cl^- \rightleftharpoons HgCl_4^{2-}$	1.2×10^{15}
$Ag^+ + 2\,S_2O_3^{2-} \rightleftharpoons Ag(S_2O_3)_2^{3-}$	2.9×10^{13}	$Hg^{2+} + 4\,CN^- \rightleftharpoons Hg(CN)_4^{2-}$	3×10^{41}
$Al^{3+} + 6\,F^- \rightleftharpoons AlF_6^{3-}$	6.9×10^{19}	$Hg^{2+} + 4\,I^- \rightleftharpoons HgI_4^{2-}$	6.8×10^{29}
$Al^{3+} + 4\,OH^- \rightleftharpoons Al(OH)_4^-$	1.1×10^{33}	$I_2 + I^- \rightleftharpoons I_3^-$	7.8×10^2
$Cd^{2+} + 4\,Cl^- \rightleftharpoons CdCl_4^{2-}$	6.3×10^2	$Ni^{2+} + 4\,CN^- \rightleftharpoons Ni(CN)_4^{2-}$	2×10^{31}
$Cd^{2+} + 4\,CN^- \rightleftharpoons Cd(CN)_4^{2-}$	6.0×10^{18}	$Ni^{2+} + 6\,NH_3 \rightleftharpoons Ni(NH_3)_6^{2+}$	5.5×10^8
$Cd^{2+} + 4\,I^- \rightleftharpoons CdI_4^{2-}$	2.6×10^5	$Pb^{2+} + 4\,Cl^- \rightleftharpoons PbCl_4^{2-}$	4×10^1
$Cd^{2+} + 4\,OH^- \rightleftharpoons Cd(OH)_4^{2-}$	4.2×10^8	$Pb^{2+} + 4\,I^- \rightleftharpoons PbI_4^{2-}$	3.0×10^4
$Cd^{2+} + 4\,NH_3 \rightleftharpoons Cd(NH_3)_6^{2+}$	1.3×10^7	$Sb^{3+} + 4\,Cl^- \rightleftharpoons SbCl_4^-$	5.2×10^4
$Co^{2+} + 6\,NH_3 \rightleftharpoons Co(NH_3)_6^{2+}$	1.3×10^5	$Sb^{3+} + 4\,OH^- \rightleftharpoons Sb(OH)_4^-$	2×10^{38}
$Co^{3+} + 6\,NH_3 \rightleftharpoons Co(NH_3)_6^{3+}$	2×10^{35}	$Sn^{2+} + 4\,Cl^- \rightleftharpoons SnCl_4^{2-}$	3.0×10^1
$Co^{2+} + 4\,SCN^- \rightleftharpoons Co(SCN)_4^{2-}$	1×10^3	$Zn^{2+} + 4\,CN^- \rightleftharpoons Zn(CN)_4^{2-}$	5×10^{16}
$Cr^{3+} + 4\,OH^- \rightleftharpoons Cr(OH)_4^-$	8×10^{29}	$Zn^{2+} + 4\,OH^- \rightleftharpoons Zn(OH)_4^{2-}$	4.6×10^{17}
$Cu^{2+} + 4\,OH^- \rightleftharpoons Cu(OH)_4^{2-}$	3×10^{18}	$Zn^{2+} + 4\,NH_3 \rightleftharpoons Zn(NH_3)_4^{2+}$	2.9×10^9
$Cu^{2+} + 4\,NH_3 \rightleftharpoons Cu(NH_3)_4^{2+}$	2.1×10^{13}		

Table B.12
Standard Reduction Potentials

Half-reaction	$E°$ (V)
$3 N_2 + 2 H^+ + 2 e^- \rightleftharpoons 2 HN_3$	-3.1
$Li^+ + e^- \rightleftharpoons Li$	-3.045
$Rb^+ + e^- \rightleftharpoons Rb$	-2.925
$K^+ + e^- \rightleftharpoons K$	-2.924
$Cs^+ + e^- \rightleftharpoons Cs$	-2.923
$Ba^{2+} + 2 e^- \rightleftharpoons Ba$	-2.90
$Sr^{2+} + 2 e^- \rightleftharpoons Sr$	-2.89
$Ca^{2+} + 2 e^- \rightleftharpoons Ca$	-2.76
$Na^+ + e^- \rightleftharpoons Na$	-2.7109
$Mg(OH)_2 + 2 e^- \rightleftharpoons Mg + 2 OH^-$	-2.69
$Mg^{2+} + 2 e^- \rightleftharpoons Mg$	-2.375
$H_2 + 2 e^- \rightleftharpoons 2 H^-$	-2.23
$Al^{3+} + 3 e^- \rightleftharpoons Al$ (0.1 M NaOH)	-1.706
$Be^{2+} + 2 e^- \rightleftharpoons Be$	-1.70
$Ti^{2+} + 2 e^- \rightleftharpoons Ti$	-1.63
$Zn(CN)_4^{2-} + 2 e^- \rightleftharpoons Zn + 4 CN^-$	-1.26
$Mn^{2+} + 2 e^- \rightleftharpoons Mn$	-1.18
$Zn(NH_3)_4^{2+} + 2 e^- \rightleftharpoons Zn + 4 NH_3$	-1.04
$SO_4^{2-} + H_2O + 2 e^- \rightleftharpoons SO_3^{2-} + 2 OH^-$	-0.92
$Cr^{2+} + 2 e^- \rightleftharpoons Cr$	-0.91
$TiO_2 + 4 H^+ + 4 e^- \rightleftharpoons Ti + 2 H_2O$	-0.87
$2 H_2O + 2 e^- \rightleftharpoons H_2 + 2 OH^-$	-0.8277
$Zn^{2+} + 2 e^- \rightleftharpoons Zn$	-0.7628
$Cr^{3+} + 3 e^- \rightleftharpoons Cr$	-0.74
$2 SO_3^{2-} + 3 H_2O + 4 e^- \rightleftharpoons S_2O_3^{2-} + 6 OH^-$	-0.58
$PbO + H_2O + 2 e^- \rightleftharpoons Pb + 2 OH^-$	-0.576
$Ga^{3+} + 3 e^- \rightleftharpoons Ga$	-0.560
$S + 2 e^- \rightleftharpoons S^{2-}$	-0.508
$2 CO_2 + 2 H^+ + 2 e^- \rightleftharpoons H_2C_2O_4$	-0.49
$Ni(NH_3)_6^{2+} + 2 e^- \rightleftharpoons Ni + 6 NH_3$	-0.48
$Co(NH_3)_6^{2+} + 2 e^- \rightleftharpoons Co + 6 NH_3$	-0.422
$Cr^{3+} + e^- \rightleftharpoons Cr^{2+}$	-0.41
$Fe^{2+} + 2 e^- \rightleftharpoons Fe$	-0.409
$Cd^{2+} + 2 e^- \rightleftharpoons Cd$	-0.4026
$PbSO_4 + 2 e^- \rightleftharpoons Pb + SO_4^{2-}$	-0.356
$In^{3+} + 3 e^- \rightleftharpoons In$	-0.338
$Tl^+ + e^- \rightleftharpoons Tl$	-0.3363
$Ag(CN)_2^- + e^- \rightleftharpoons Ag + 2 CN^-$	-0.31
$Co^{2+} + 2 e^- \rightleftharpoons Co$	-0.28
$H_3PO_4 + 2 H^+ + 2 e^- \rightleftharpoons H_3PO_3 + H_2O$	-0.276
$Ni^{2+} + 2 e^- \rightleftharpoons Ni$	-0.23
$2 SO_4^{2-} + 4 H^+ + 2 e^- \rightleftharpoons S_2O_6^{2-} + 2 H_2O$	-0.224
$CO_2 + 2 H^+ + 2 e^- \rightleftharpoons HCO_2H$	-0.20
$O_2 + 2 H_2O + 2 e^- \rightleftharpoons H_2O_2 + 2 OH^-$	-0.146
$Sn^{2+} + 2 e^- \rightleftharpoons Sn$	-0.1364
$Pb^{2+} + 2 e^- \rightleftharpoons Pb$	-0.1263
$CrO_4^{2-} + 4 H_2O + 3 e^- \rightleftharpoons Cr(OH)_3 + 5 OH^-$	-0.12
$WO_3 + 6 H^+ + 6 e^- \rightleftharpoons W + 3 H_2O$	-0.09
$Ru^{3+} + e^- \rightleftharpoons Ru^{2+}$	-0.08
$O_2 + H_2O + 2 e^- \rightleftharpoons HO_2^- + OH^-$	-0.076
$Fe^{3+} + 3 e^- \rightleftharpoons Fe$	-0.036
$2 H^+ + 2 e^- \rightleftharpoons H_2$	$0.0000000\ldots$

Half-reaction	$E°$ (V)
$NO_3^- + H_2O + 2\,e^- \rightleftharpoons NO_2^- + 2\,OH^-$	0.01
$AgBr + e^- \rightleftharpoons Ag + Br^-$	0.0713
$S_4O_6^{2-} + 2\,e^- \rightleftharpoons 2\,S_2O_3^{2-}$	0.0895
$Sn^{4+} + 2\,e^- \rightleftharpoons Sn^{2+}$	0.15
$Cu^{2+} + e^- \rightleftharpoons Cu^+$	0.158
$ClO_4^- + H_2O + 2\,e^- \rightleftharpoons ClO_3^- + 2\,OH^-$	0.17
$SO_4^{2-} + 4\,H^+ + 2\,e^- \rightleftharpoons H_2SO_3 + H_2O$	0.20
$AgCl + e^- \rightleftharpoons Ag + Cl^-$	0.2223
$IO_3^- + 3\,H_2O + 6\,e^- \rightleftharpoons I^- + 6\,OH^-$	0.26
$Hg_2Cl_2 + 2\,e^- \rightleftharpoons 2\,Hg + 2\,Cl^-$	0.2682
$Cu^{2+} + 2\,e^- \rightleftharpoons Cu$	0.3402
$ClO_3^- + H_2O + 2\,e^- \rightleftharpoons ClO_2^- + 2\,OH^-$	0.35
$Ag(NH_3)_2^+ + e^- \rightleftharpoons Ag + 2\,NH_3$	0.373
$O_2 + 2\,H_2O + 4\,e^- \rightleftharpoons 4\,OH^-$	0.401
$H_2SO_3 + 4\,H^+ + 4\,e^- \rightleftharpoons S + 3\,H_2O$	0.45
$HgCl_4^{2-} + 2\,e^- \rightleftharpoons Hg + 4\,Cl^-$	0.48
$Cu^+ + e^- \rightleftharpoons Cu$	0.522
$I_3^- + 2\,e^- \rightleftharpoons 3\,I^-$	0.5338
$I_2 + 2\,e^- \rightleftharpoons 2\,I^-$	0.535
$MnO_4^- + 2\,H_2O + 3\,e^- \rightleftharpoons MnO_2 + 4\,OH^-$	0.588
$ClO_2^- + H_2O + 2\,e^- \rightleftharpoons ClO^- + 2\,OH^-$	0.59
$O_2 + 2\,H^+ + 2\,e^- \rightleftharpoons H_2O_2$	0.682
$Fe^{3+} + e^- \rightleftharpoons Fe^{2+}$	0.770
$Hg_2^{2+} + 2\,e^- \rightleftharpoons 2\,Hg$	0.7961
$Ag^+ + e^- \rightleftharpoons Ag$	0.7996
$Hg^{2+} + 2\,e^- \rightleftharpoons Hg$	0.851
$H_2O_2 + 2\,e^- \rightleftharpoons 2\,OH^-$	0.88
$ClO^- + H_2O + 2\,e^- \rightleftharpoons Cl^- + 2\,OH^-$	0.89
$2\,Hg^{2+} + 2\,e^- \rightleftharpoons Hg_2^{2+}$	0.905
$NO_3^- + 3\,H^+ + 2\,e^- \rightleftharpoons HNO_2 + H_2O$	0.94
$ClO_2 + e^- \rightleftharpoons ClO_2^-$	0.95
$NO_3^- + 4\,H^+ + 3\,e^- \rightleftharpoons NO + 2\,H_2O$	0.96
$Pd^{2+} + 2\,e^- \rightleftharpoons Pd$	0.987
$HNO_2 + H^+ + e^- \rightleftharpoons NO + H_2O$	0.99
$IO_3^- + 6\,H^+ + 6\,e^- \rightleftharpoons I^- + 3\,H_2O$	1.085
$Br_2(aq) + 2\,e^- \rightleftharpoons 2\,Br^-$	1.087
$Cr^{6+} + 3\,e^- \rightleftharpoons Cr^{3+}$	1.10
$ClO_3^- + 2\,H^+ + 2\,e^- \rightleftharpoons ClO_2^- + H_2O$	1.15
$ClO_4^- + 2\,H^+ + 2\,e^- \rightleftharpoons ClO_3^- + H_2O$	1.19
$2\,IO_3^- + 12\,H^+ + 10\,e^- \rightleftharpoons I_2 + 6\,H_2O$	1.19
$HCrO_4^- + 7\,H^+ + 3\,e^- \rightleftharpoons Cr^{3+} + 4\,H_2O$	1.195
$Pt^{2+} + 2\,e^- \rightleftharpoons Pt$	1.2
$MnO_2 + 4\,H^+ + 2\,e^- \rightleftharpoons Mn^{2+} + 2\,H_2O$	1.208
$O_2 + 4\,H^+ + 4\,e^- \rightleftharpoons 2\,H_2O$	1.229
$O_3 + H_2O + 2\,e^- \rightleftharpoons O_2 + 2\,OH^-$	1.24
$Tl^{3+} + 2\,e^- \rightleftharpoons Tl^+$	1.247
$ClO_2 + H^+ + e^- \rightleftharpoons HClO_2$	1.27
$2\,HNO_2 + 4\,H^+ + 4\,e^- \rightleftharpoons N_2O + 3\,H_2O$	1.27
$Au^{3+} + 2\,e^- \rightleftharpoons Au^+$	1.29
$Cr_2O_7^{2-} + 14\,H^+ + 6\,e^- \rightleftharpoons 2\,Cr^{3+} + 7\,H_2O$	1.33
$ClO_4^- + 8\,H^+ + 7\,e^- \rightleftharpoons \tfrac{1}{2}\,Cl_2 + 4\,H_2O$	1.34
$Cl_2 + 2\,e^- \rightleftharpoons 2\,Cl^-$	1.3583
$ClO_4^- + 8\,H^+ + 8\,e^- \rightleftharpoons Cl^- + 4\,H_2O$	1.37
$Au^{3+} + 3\,e^- \rightleftharpoons Au$	1.42

(continued)

Table B.12
Standard Reduction Potentials (continued)

Half-reaction	$E°$ (V)
$ClO_3^- + 6\,H^+ + 6\,e^- \rightleftharpoons Cl^- + 3\,H_2O$	1.45
$PbO_2 + 4\,H^+ + 2\,e^- \rightleftharpoons Pb^{2+} + 2\,H_2O$	1.467
$2\,ClO_3^- + 12\,H^+ + 10\,e^- \rightleftharpoons Cl_2 + 6\,H_2O$	1.47
$HClO + H^+ + 2\,e^- \rightleftharpoons Cl^- + H_2O$	1.49
$MnO_4^- + 8\,H^+ + 5\,e^- \rightleftharpoons Mn^{2+} + 4\,H_2O$	1.491
$HClO_2 + 3\,H^+ + 4\,e^- \rightleftharpoons Cl^- + 2\,H_2O$	1.56
$2\,NO + 2\,H^+ + 2\,e^- \rightleftharpoons N_2O + H_2O$	1.59
$2\,HClO_2 + 6\,H^+ + 6\,e^- \rightleftharpoons Cl_2 + 4\,H_2O$	1.63
$2\,HClO + 2\,H^+ + 2\,e^- \rightleftharpoons Cl_2 + 2\,H_2O$	1.63
$HClO_2 + 2\,H^+ + 2\,e^- \rightleftharpoons HClO + H_2O$	1.64
$MnO_4^- + 4\,H^+ + 3\,e^- \rightleftharpoons MnO_2 + 2\,H_2O$	1.679
$Au^+ + e^- \rightleftharpoons Au$	1.68
$PbO_2 + SO_4^{2-} + 4\,H^+ + 2\,e^- \rightleftharpoons PbSO_4 + 2\,H_2O$	1.685
$N_2O + 2\,H^+ + 2\,e^- \rightleftharpoons N_2 + H_2O$	1.77
$H_2O_2 + 2\,H^+ + 2\,e^- \rightleftharpoons 2\,H_2O$	1.776
$Co^{3+} + e^- \rightleftharpoons Co^{2+}$	1.842
$S_2O_8^{2-} + 2\,e^- \rightleftharpoons 2\,SO_4^{2-}$	2.05
$O_3(g) + 2\,H^+ + 2\,e^- \rightleftharpoons O_2(g) + H_2O$	2.07
$F_2(g) + 2\,H^+ + 2\,e^- \rightleftharpoons 2\,HF(aq)$	3.03

Table B.13

Standard-State Enthalpies, Free Energies, and Entropies of Atom Combination, 298.15 K

Substance	ΔH°_{ac} (kJ/mol$_{rxn}$)	ΔG°_{ac} (kJ/mol$_{rxn}$)	ΔS°_{ac} (J/mol$_{rxn} \cdot$ K)
Aluminum			
Al(s)	−326.4	−285.7	−136.21
Al(g)	0	0	0
Al^{3+}(aq)	−857	−771	−486.2
Al$_2$O$_3$(s)	−3076.0	−2848.9	−761.33
AlCl$_3$(s)	−1395.6	−1231.5	−549.46
AlF$_3$(s)	−2067.5	−1896.4	−574.36
Al$_2$(SO$_4$)$_3$(s)	−7920.1	−7166.9	−2525.9
Barium			
Ba(s)	−180	−146	−107.4
Ba(g)	0	0	0
Ba^{2+}(aq)	−718	−707	−160.6
BaO(s)	−983	−903	−260.88
Ba(OH)$_2 \cdot$ 8 H$_2$O(s)	−9931.6	−8915	−3469
BaCl$_2$(s)	−1282	−1168	−376.96
BaCl$_2$(aq)	−1295	−1181	−378.04
BaSO$_4$(s)	−2929	−2673	−850.1
Ba(NO$_3$)$_2$(s)	−3612	−3217	−1229.4
Ba(NO$_3$)$_2$(aq)	−3573	−3231	−1140.7
Beryllium			
Be(s)	−324.3	−286.6	−126.77
Be(g)	0	0	0
Be^{2+}(aq)	−707.1	−666.3	−266.0
BeO(s)	−1183.1	−1026.6	−283.18
BeCl$_2$(s)	−1058.1	−943.6	−383.99
Bismuth			
Bi(s)	−207.1	−168.2	−130.31
Bi(g)	0	0	0
Bi$_2$O$_3$(s)	−1735.6	−1525.3	−705.8
BiCl$_3$(s)	−951.2	−800.2	−505.6
BiCl$_3$(g)	−837.8	−741.2	−323.79
Bi$_2$S$_3$(s)	−1393.7	−1191.8	−677.2
Boron			
B(s)	−562.7	−518.8	−147.59
B(g)	0	0	0
B$_2$O$_3$(s)	−3145.7	−2926.4	−736.10
B$_2$H$_6$(g)	−2395.7	−2170.4	−763.07
B$_5$H$_9$(l)	−4729.7	−4251.4	−1615.45
B$_5$H$_9$(g)	−4699.2	−4248.2	−1523.75
B$_{10}$H$_{14}$(s)	−8719.3	−7841.2	−2963.92
H$_3$BO$_3$(s)	−3057.5	−2792.7	−891.92
BF$_3$(g)	−1936.7	−1824.9	−375.59
BCl$_3$(l)	−1354.9	−1223.2	−442.7
B$_3$N$_3$H$_6$(l)	−4953.1	−4535.4	−1408.9
B$_3$N$_3$H$_6$(g)	−4923.9	−4532.8	−1319.84

(continued)

Table B.13

Standard-State Enthalpies, Free Energies, and Entropies of Atom Combination, 298.15 K (continued)

Substance	ΔH_{ac}° (kJ/mol$_{rxn}$)	ΔG_{ac}° (kJ/mol$_{rxn}$)	ΔS_{ac}° (J/mol$_{rxn} \cdot$ K)
Bromine			
$Br_2(l)$	−223.768	−164.792	−197.813
$Br_2(g)$	−192.86	−161.68	−104.58
$Br(g)$	0	0	0
$HBr(g)$	−365.93	−339.09	−91.040
$HBr(aq)$	−451.08	−389.60	−207.3
$BrF(g)$	−284.72	−253.49	−104.81
$BrF_3(g)$	−604.45	−497.56	−358.75
$BrF_5(g)$	−935.73	−742.6	−648.60
Calcium			
$Ca(s)$	−178.2	−144.3	−113.46
$Ca(g)$	0	0	0
$Ca^{2+}(aq)$	−721.0	−697.9	−208.0
$CaO(s)$	−1062.5	−980.1	−276.19
$Ca(OH)_2(s)$	−2097.9	−1912.7	−623.03
$CaCl_2(s)$	−1217.4	−1103.8	−380.7
$CaSO_4(s)$	−2887.8	−2631.3	−860.3
$CaSO_4 \cdot 2\ H_2O(s)$	−4256.7	−4383.2	−1553.8
$Ca(NO_3)_2(s)$	−3557.0	−3189.0	−1234.5
$CaCO_3(s)$	−2849.3	−2639.5	−703.2
$Ca_3(PO_4)_2(s)$	−7278.0	−6727.9	−1843.5
Carbon			
$C(graphite)$	−716.682	−671.257	−152.36
$C(diamond)$	−714.787	−668.357	−155.719
$C(g)$	0	0	0
$CO(g)$	−1076.377	−1040.156	−121.477
$CO_2(g)$	−1608.531	−1529.078	−266.47
$COCl_2(g)$	−1428.0	−1318.9	−366.02
$CH_4(g)$	−1662.09	−1534.997	−430.684
$HCHO(g)$	−1509.72	−1412.01	−329.81
$H_2CO_3(aq)$	−2599.14	−2396.02	−683.3
$HCO_3^-(aq)$	−2373.83	−2156.47	−664.8
$CO_3^{2-}(aq)$	−2141.33	−1894.26	−698.16
$CH_3OH(l)$	−2075.11	−1882.25	−651.2
$CH_3OH(g)$	−2037.11	−1877.94	−538.19
$CCl_4(l)$	−1338.84	−1159.19	−602.49
$CCl_4(g)$	−1306.3	−1154.57	−509.04
$CHCl_3(l)$	−1433.84	−1265.20	−567.3
$CHCl_3(g)$	−1402.51	−1261.88	−472.69
$CH_2Cl_2(l)$	−1516.80	−1356.37	−540.1
$CH_2Cl_2(g)$	−1487.81	−1354.98	−447.69
$CH_3Cl(g)$	−1572.15	−1444.08	−432.9
$CS_2(l)$	−1184.59	−1082.49	−342.40
$CS_2(g)$	−1156.93	−1080.64	−255.90
$HCN(g)$	−1271.9	−1205.43	−224.33

Substance	ΔH_{ac}° (kJ/mol$_{rxn}$)	ΔG_{ac}° (kJ/mol$_{rxn}$)	ΔS_{ac}° (J/mol$_{rxn}$ · K)
$CH_3NO_2(l)$	−2453.77	−2214.51	−805.89
$C_2H_2(g)$	−1641.93	−1539.81	−344.68
$C_2H_4(g)$	−2251.70	−2087.35	−555.48
$C_2H_6(g)$	−2823.94	−2594.82	−774.87
$CH_3CHO(l)$	−2745.43	−2515.35	−775.9
$CH_3CO_2H(l)$	−3286.8	−3008.86	−937.4
$CH_3CO_2H(g)$	−3234.55	−2992.96	−814.7
$CH_3CO_2H(aq)$	−3288.06	−3015.42	−918.5
$CH_3CO_2^-(aq)$	−3070.66	−2785.03	−895.8
$CH_3CH_2OH(l)$	−3266.12	−2968.51	−1004.8
$CH_3CH_2OH(g)$	−3223.53	−2962.22	−882.82
$CH_3CH_2OH(aq)$	−3276.7	−2975.37	−1017.0
$C_6H_6(l)$	−5556.96	−5122.52	−1464.1
$C_6H_6(g)$	−5523.07	−5117.36	−1367.7
Chlorine			
$Cl_2(g)$	−243.358	−211.360	−107.330
$Cl(g)$	0	0	0
$Cl^-(aq)$	−288.838	−236.908	−108.7
$ClO_2(g)$	−517.5	−448.6	−230.47
$Cl_2O(g)$	−412.2	−345.2	−225.24
$Cl_2O_7(l)$	−1750	—	—
$HCl(g)$	−431.64	−404.226	−93.003
$HCl(aq)$	−506.49	−440.155	−223.4
$ClF(g)$	−255.15	−223.53	−106.06
Chromium			
$Cr(s)$	−396.6	−351.8	−150.73
$Cr(g)$	0	0	0
$CrO_3(s)$	−1733.6	—	—
$CrO_4^{2-}(aq)$	−2274.4	−2006.47	−768.51
$Cr_2O_3(s)$	−2680.4	−2456.89	−751.0
$Cr_2O_7^{2-}(aq)$	−4027.7	−3626.8	−1214.5
$(NH_4)_2Cr_2O_7$	−7030.7	—	—
$PbCrO_4(s)$	−2519.2	—	—
Cobalt			
$Co(s)$	−424.7	−380.3	−149.475
$Co(g)$	0	0	0
$Co^{2+}(aq)$	−482.9	−434.7	−293
$Co^{3+}(aq)$	−333	−246.3	−485
$CoO(s)$	−911.8	−826.2	−287.60
$Co_3O_4(s)$	−3162	−2842	−1080.3
$Co(NH_3)_6^{3+}(aq)$	−7763.5	−6929.5	−3018
Copper			
$Cu(s)$	−338.32	−298.58	−133.23
$Cu(g)$	0	0	0
$Cu^+(aq)$	−266.65	−248.60	−125.8
$Cu^{2+}(aq)$	−273.55	−233.09	−266.0
$CuO(s)$	−744.8	−660.0	−284.81

(continued)

Table B.13

Standard-State Enthalpies, Free Energies, and Entropies of Atom Combination, 298.15 K (continued)

Substance	ΔH_{ac}° (kJ/mol$_{rxn}$)	ΔG_{ac}° (kJ/mol$_{rxn}$)	ΔS_{ac}° (J/mol$_{rxn}$ · K)
$Cu_2O(s)$	-1094.4	-974.9	-400.68
$CuCl_2(s)$	-807.8	-685.6	-388.71
$CuS(s)$	-670.2	-590.4	-267.7
$Cu_2S(s)$	-1304.9	-921.6	-379.7
$CuSO_4(s)$	-2385.17	-1530.44	-869
$Cu(NH_3)_4^{2+}(aq)$	-5189.4	-4671.13	-1882.5
Fluorine			
$F_2(g)$	-157.98	-123.82	-114.73
$F(g)$	0	0	0
$F^-(aq)$	-411.62	-340.70	-172.6
$HF(g)$	-567.7	-538.4	-99.688
$HF(aq)$	-616.72	-561.98	-184.8
Hydrogen			
$H_2(g)$	-435.30	-406.494	-98.742
$H(g)$	0	0	0
$H^+(aq)$	-217.65	-203.247	-114.713
$OH^-(aq)$	-696.81	-592.222	-286.52
$H_2O(l)$	-970.30	-875.354	-320.57
$H_2O(g)$	-926.29	-866.797	-202.23
$H_2O_2(l)$	-1121.42	-990.31	-441.9
$H_2O_2(aq)$	-1124.81	-1003.99	-407.6
Iodine			
$I_2(s)$	-213.676	-141.00	-245.447
$I_2(g)$	-151.238	-121.67	-100.89
$I(g)$	0	0	0
$HI(g)$	-298.01	-272.05	-88.910
$IF(g)$	-281.48	-250.92	-103.38
$IF_5(g)$	-1324.28	-1131.78	-646.9
$IF_7(g)$	-1603.7	-1322.17	-945.6
$ICl(g)$	-210.74	-181.64	-98.438
$IBr(g)$	-177.88	-149.21	-97.040
Iron			
$Fe(s)$	-416.3	-370.7	-153.21
$Fe(g)$	0	0	0
$Fe^{2+}(aq)$	-505.4	-449.6	-318.2
$Fe^{3+}(aq)$	-464.8	-375.4	-496.4
$Fe_2O_3(s)$	-2404.3	-2178.8	-756.75
$Fe_3O_4(s)$	-3364.0	-3054.4	-1039.3
$Fe(OH)_2(s)$	-1918.9	-1727.2	-644
$Fe(OH)_3(s)$	-2639.8	-2372.1	-901.1
$FeCl_3(s)$	-1180.8	-1021.7	-533.8
$FeS_2(s)$	-1152.1	-1014.1	-463.2
$Fe(CO)_5(l)$	-6019.6	-5590.9	-1438.1
$Fe(CO)_5(g)$	-5979.5	-5582.9	-1330.9

Substance	ΔH°_{ac} (kJ/mol$_{rxn}$)	ΔG°_{ac} (kJ/mol$_{rxn}$)	ΔS°_{ac} (J/mol$_{rxn} \cdot$ K)
Lead			
Pb(s)	-195.0	-161.9	-110.56
Pb(g)	0	0	0
Pb^{2+}(aq)	-196.7	-186.3	-164.9
PbO(s)	-661.5	-581.5	-267.7
PbO$_2$(s)	-970.7	-842.7	-428.9
PbCl$_2$(s)	-797.8	-687.4	-369.8
PbCl$_4$(l)	-1011.0	—	—
PbS(s)	-574.2	-498.9	-252.0
PbSO$_4$(s)	-2390.4	-2140.2	-838.84
Pb(NO$_3$)$_2$(s)	-3087.3	—	—
PbCO$_3$(s)	-2358.3	-2153.8	-685.6
Lithium			
Li(s)	-159.37	-126.66	-109.65
Li(g)	0	0	0
Li$^+$(aq)	-437.86	-419.97	-125.4
LiH(s)	-467.56	-398.26	-233.40
LiOH(s)	-1111.12	-1000.59	-371.74
LiF(s)	-854.33	-784.28	-261.87
LiCl(s)	-689.66	-616.71	-244.64
LiBr(s)	-622.48	-551.06	-239.52
LiI(s)	-536.62	-467.45	-232.78
LiAlH$_4$(s)	-1472.7	-1270.0	-683.42
LiBH$_4$(s)	-1401.9	-1333.4	-675.21
Magnesium			
Mg(s)	-147.70	-113.10	-115.97
Mg(g)	0	0	0
Mg^{2+}(aq)	-614.55	-567.9	-286.8
MgO(s)	-998.57	-914.26	-282.76
MgH$_2$(s)	-658.3	-555.5	-346.99
Mg(OH)$_2$(s)	-2005.88	-1816.64	-637.01
MgCl$_2$(s)	-1032.38	-916.25	-389.43
MgCO$_3$(s)	-2707.7	-2491.7	-724.2
MgSO$_4$(s)	-2708.1	-2448.9	-869.1
Manganese			
Mn(s)	-280.7	-238.5	-141.69
Mn(g)	0	0	0
Mn^{2+}(aq)	-501.5	-466.6	-247.3
MnO(s)	-915.1	-833.1	-275.05
MnO$_2$(s)	-1299.1	-1167.1	-442.76
Mn$_2$O$_3$(s)	-2267.9	-2053.3	-720.1
Mn$_3$O$_4$(s)	-3226.6	-2925.6	-1009.7
KMnO$_4$(s)	-2203.8	-1963.6	-806.50
MnS(s)	-773.7	-695.2	-263.3
Mercury			
Hg(l)	-61.317	-31.820	-98.94
Hg(g)	0	0	0
Hg^{2+}(aq)	$+109.8$	$+132.58$	-207.2

(continued)

Table B.13

Standard-State Enthalpies, Free Energies, and Entropies of Atom Combination, 298.15 K (continued)

Substance	ΔH°_{ac} (kJ/mol$_{rxn}$)	ΔG°_{ac} (kJ/mol$_{rxn}$)	ΔS°_{ac} (J/mol$_{rxn}$ · K)
HgO(s)	−401.32	−322.090	−265.73
HgCl$_2$(s)	−529.0	−421.8	−359.4
Hg$_2$Cl$_2$(s)	−631.21	−485.745	−487.8
HgS(s)	−398.3	−320.67	−260.4
Nitrogen			
N$_2$(g)	−945.408	−911.26	−114.99
N(g)	0	0	0
NO(g)	−631.62	−600.81	−103.592
NO$_2$(g)	−937.86	−867.78	−235.35
N$_2$O(g)	−1112.53	−1038.79	−247.80
N$_2$O$_3$(g)	−1609.20	−1466.99	−477.48
N$_2$O$_4$(g)	−1932.93	−1740.29	−646.53
N$_2$O$_5$(g)	−2179.91	−1954.8	−756.2
NO$_3{}^-$(aq)	−1425.2	−1259.56	−490.1
NOCl(g)	−791.84	−726.96	−217.86
NO$_2$Cl(g)	−1080.12	−970.4	−368.46
HNO$_2$(aq)	−1307.9	−1172.9	−454.5
HNO$_3$(g)	−1572.92	−1428.79	−484.80
HNO$_3$(aq)	−1645.22	−1465.32	−604.8
NH$_3$(g)	−1171.76	−1081.82	−304.99
NH$_3$(aq)	−1205.94	−1091.87	−386.1
NH$_4{}^+$(aq)	−1475.81	−1347.93	−498.8
NH$_4$NO$_3$(s)	−2929.08	−2603.31	−1097.53
NH$_4$NO$_3$(aq)	−2903.39	−2610.00	−988.8
NH$_4$Cl(s)	−1779.41	−1578.17	−682.7
N$_2$H$_4$(l)	−1765.38	−1574.91	−644.24
N$_2$H$_4$(g)	−1720.61	−1564.90	−526.98
HN$_3$(g)	−1341.7	−1242.0	−335.37
Oxygen			
O$_2$(g)	−498.340	−463.462	−116.972
O(g)	0	0	0
O$_3$(g)	−604.8	−532.0	−244.24
Phosphorus			
P(white)	−314.64	−278.25	−122.10
P$_4$(g)	−1199.65	−1088.6	−372.79
P$_2$(g)	−485.0	−452.8	−108.257
P(g)	0	0	0
PH$_3$(g)	−962.2	−874.6	−297.10
P$_4$O$_6$(s)	−4393.7	—	—
P$_4$O$_{10}$(s)	−6734.3	−6128.0	−2034.46
PO$_4{}^{3-}$(aq)	−2588.7	−2223.9	−1029
PF$_3$(g)	−1470.4	−1361.5	−366.22
PF$_5$(g)	−2305.4	—	—
PCl$_3$(l)	−999.4	−867.6	−441.7
PCl$_3$(g)	−966.7	−863.1	−347.01

Substance	ΔH°_{ac} (kJ/mol$_{rxn}$)	ΔG°_{ac} (kJ/mol$_{rxn}$)	ΔS°_{ac} (J/mol$_{rxn}\cdot$K)
$PCl_5(g)$	−1297.9	−1111.6	−624.60
$H_3PO_4(s)$	−3243.3	−2934.0	−1041.05
$H_3PO_4(aq)$	−3241.7	−2833.6	−1374
Potassium			
$K(s)$	−89.24	−60.59	−96.16
$K(g)$	0	0	0
$K^+(aq)$	−341.62	−343.86	−57.8
$KOH(s)$	−980.82	−874.65	−357.2
$KCl(s)$	−647.67	−575.41	−242.94
$KNO_3(s)$	−1804.08	−1606.27	−663.75
$K_2Cr_2O_7(s)$	−4777.4	−4328.7	−1505.9
$KMnO_4(s)$	−2203.8	−1963.6	−806.50
Silicon			
$Si(s)$	−455.6	−411.3	−149.14
$Si(g)$	0	0	0
$SiO_2(s)$	−1864.9	−1731.4	−448.24
$SiH_4(g)$	−1291.9	−1167.4	−422.20
$SiF_4(g)$	−2386.5	−2231.6	−520.50
$SiCl_4(l)$	−1629.3	−1453.9	−589
$SiCl_4(g)$	−1599.3	−1451.0	−498.03
Silver			
$Ag(s)$	−284.55	−245.65	−130.42
$Ag(g)$	0	0	0
$Ag^+(aq)$	−178.97	−168.54	−100.29
$Ag(NH_3)_2^+(aq)$	−2647.15	−2393.51	−922.6
$Ag_2O(s)$	−849.32	−734.23	−385.7
$AgCl(s)$	−533.30	−461.12	−242.0
$AgBr(s)$	−496.80	−424.95	−240.9
$AgI(s)$	−453.23	−382.34	−238.3
Sodium			
$Na(s)$	−107.32	−76.761	−102.50
$Na(g)$	0	0	0
$Na^+(aq)$	−347.45	−338.666	−94.7
$NaH(s)$	−3811.25	−313.47	−228.409
$NaOH(s)$	−999.75	−891.233	−365.025
$NaOH(aq)$	−1044.25	−930.889	−381.4
$NaCl(s)$	−640.15	−566.579	−246.78
$NaCl(g)$	−405.65	−379.10	−89.10
$NaCl(aq)$	−636.27	−575.574	−203.4
$NaNO_3(s)$	−1795.38	−1594.58	−673.66
$Na_3PO_4(s)$	−3550.68	−3224.26	−1094.75
$Na_2SO_3(s)$	−2363.96	−2115.0	−803
$Na_2SO_4(s)$	−2877.20	−2588.86	−969.89
$Na_2CO_3(s)$	−2809.51	−2564.41	−813.70
$NaHCO_3(s)$	−2739.97	−2497.5	−808.0
$NaCH_3CO_2(s)$	−3400.78	−3099.66	−1013.2
$Na_2CrO_4(s)$	−2950.1	−2667.18	−949.53
$Na_2Cr_2O_7(s)$	−4730.6	—	—

(continued)

Table B.13

Standard-State Enthalpies, Free Energies, and Entropies of Atom Combination, 298.15 K (continued)

Substance	ΔH°_{ac} (kJ/mol$_{rxn}$)	ΔG°_{ac} (kJ/mol$_{rxn}$)	ΔS°_{ac} (J/mol$_{rxn} \cdot$ K)
Sulfur			
$S_8(s)$	-2230.440	-1906.00	-1310.77
$S_8(g)$	-2128.14	-1856.37	-911.59
$S(g)$	0	0	0
$S^{2-}(aq)$	-245.7	-152.4	-182.4
$SO_2(g)$	-1073.95	-1001.906	-241.71
$SO_3(s)$	-1480.82	-1307.65	-580.3
$SO_3(l)$	-1467.36	-1307.19	-537.2
$SO_3(g)$	-1422.04	-1304.50	-394.23
$SO_4^{2-}(aq)$	-2184.76	-1909.70	-791.9
$SOCl_2(g)$	-983.8	-879.6	-349.50
$SO_2Cl_2(g)$	-1384.5	-1233.1	-508.39
$H_2S(g)$	-734.74	-678.30	-191.46
$H_2SO_3(aq)$	-2070.43	-1877.75	-648.2
$H_2SO_4(aq)$	-2620.06	-2316.20	-1021.4
$SF_4(g)$	-1369.66	-1217.2	-510.81
$SF_6(g)$	-1962	-1715.0	-828.53
$SCN^-(aq)$	-1391.75	-1272.43	-334.9
Tin			
$Sn(s)$	-302.1	-267.3	-124.35
$Sn(g)$	0	0	0
$SnO(s)$	-837.1	-755.9	-273.0
$SnO_2(s)$	-1381.1	-1250.5	-438.3
$SnCl_2(s)$	-870.6	—	—
$SnCl_4(l)$	-1300.1	-249.8	-570.7
$SnCl_4(g)$	-1260.3	-1122.2	-463.5
Titanium			
$Ti(s)$	-469.9	-425.1	-149.6
$Ti(g)$	0	0	0
$TiO(s)$	-1238.8	-1151.8	-306.5
$TiO_2(s)$	-1913.0	-1778.1	-452.0
$TiCl_4(l)$	-1760.8	-1585.0	-588.7
$TiCl_4(g)$	-1719.8	-1574.6	-486.2
Tungsten			
$W(s)$	-849.4	-807.1	-141.31
$W(g)$	0	0	0
$WO_3(s)$	-2439.8	-2266.4	-581.22
Zinc			
$Zn(s)$	-130.729	-95.145	-119.35
$Zn(g)$	0	0	0
$Zn^{2+}(aq)$	-284.62	-242.21	-273.1
$ZnO(s)$	-728.18	-645.18	-278.40
$ZnCl_2(s)$	-789.14	-675.90	-379.92
$ZnS(s)$	-615.51	-534.69	-271.1
$ZnSO_4(s)$	-2389.0	-2131.8	-862.5

Table B.14
Bond Dissociation Enthalpies

	As	B	Br	C	Cl	F	H	I	N	O	P	S	Si
As	180		255	200	310	485	300	180		330			
B		300	370		445	645		270		525			
Br			195	270	220	240	370	180	250		270	215	330
C				350	330	490	415	210	305	360	265	270	305
Cl					240	250	431	210	190	205	330	270	400
F						160	569		280	215	500	325	600
H							435	300	390	464	325	370	320
I								150		200	180		230
N									160	165			330
O										140	370	423	464
P											210		
S												260	
Si													225

Single-Bond Dissociation Enthalpies (kJ/mol)

Double- and Triple-Bond Dissociation Enthalpies (kJ/mol)

C=C	611		C=S	477
C≡C	837		N=N	418
C=O	745		N≡N	946
C≡O	1075		N=O	594
C=N	615		O=O	498
C≡N	891		S=O	523

Table B.15
Electron Configurations of the First 86 Elements

Atomic Number	Symbol	Electron Configuration
1	H	$1s^1$
2	He	$1s^2 = [\text{He}]$
3	Li	$[\text{He}]\, 2s^1$
4	Be	$[\text{He}]\, 2s^2$
5	B	$[\text{He}]\, 2s^2\, 2p^1$
6	C	$[\text{He}]\, 2s^2\, 2p^2$
7	N	$[\text{He}]\, 2s^2\, 2p^3$
8	O	$[\text{He}]\, 2s^2\, 2p^4$
9	F	$[\text{He}]\, 2s^2\, 2p^5$
10	Ne	$[\text{He}]\, 2s^2\, 2p^6 = [\text{Ne}]$
11	Na	$[\text{Ne}]\, 3s^1$
12	Mg	$[\text{Ne}]\, 3s^2$
13	Al	$[\text{Ne}]\, 3s^2\, 3p^1$
14	Si	$[\text{Ne}]\, 3s^2\, 3p^2$
15	P	$[\text{Ne}]\, 3s^2\, 3p^3$
16	S	$[\text{Ne}]\, 3s^2\, 3p^4$
17	Cl	$[\text{Ne}]\, 3s^2\, 3p^5$
18	Ar	$[\text{Ne}]\, 3s^2\, 3p^6 = [\text{Ar}]$
19	K	$[\text{Ar}]\, 4s^1$
20	Ca	$[\text{Ar}]\, 4s^2$
21	Sc	$[\text{Ar}]\, 4s^2\, 3d^1$
22	Ti	$[\text{Ar}]\, 4s^2\, 3d^2$
23	V	$[\text{Ar}]\, 4s^2\, 3d^3$
24	Cr	$[\text{Ar}]\, 4s^1\, 3d^5$
25	Mn	$[\text{Ar}]\, 4s^2\, 3d^5$
26	Fe	$[\text{Ar}]\, 4s^2\, 3d^6$
27	Co	$[\text{Ar}]\, 4s^2\, 3d^7$
28	Ni	$[\text{Ar}]\, 4s^2\, 3d^8$
29	Cu	$[\text{Ar}]\, 4s^1\, 3d^{10}$
30	Zn	$[\text{Ar}]\, 4s^2\, 3d^{10}$
31	Ga	$[\text{Ar}]\, 4s^2\, 3d^{10}\, 4p^1$
32	Ge	$[\text{Ar}]\, 4s^2\, 3d^{10}\, 4p^2$
33	As	$[\text{Ar}]\, 4s^2\, 3d^{10}\, 4p^3$
34	Se	$[\text{Ar}]\, 4s^2\, 3d^{10}\, 4p^4$
35	Br	$[\text{Ar}]\, 4s^2\, 3d^{10}\, 4p^5$
36	Kr	$[\text{Ar}]\, 4s^2\, 3d^{10}\, 4p^6 = [\text{Kr}]$
37	Rb	$[\text{Kr}]\, 5s^1$
38	Sr	$[\text{Kr}]\, 5s^2$
39	Y	$[\text{Kr}]\, 5s^2\, 4d^1$
40	Zr	$[\text{Kr}]\, 5s^2\, 4d^2$
41	Nb	$[\text{Kr}]\, 5s^1\, 4d^4$
42	Mo	$[\text{Kr}]\, 5s^1\, 4d^5$
43	Tc	$[\text{Kr}]\, 5s^2\, 4d^5$
44	Ru	$[\text{Kr}]\, 5s^1\, 4d^7$
45	Rh	$[\text{Kr}]\, 5s^1\, 4d^8$
46	Pd	$[\text{Kr}]\, 4d^{10}$
47	Ag	$[\text{Kr}]\, 5s^1\, 4d^{10}$
48	Cd	$[\text{Kr}]\, 5s^2\, 4d^{10}$
49	In	$[\text{Kr}]\, 5s^2\, 4d^{10}\, 5p^1$
50	Sn	$[\text{Kr}]\, 5s^2\, 4d^{10}\, 5p^2$
51	Sb	$[\text{Kr}]\, 5s^2\, 4d^{10}\, 5p^3$

Atomic Number	Symbol	Electron Configuration
52	Te	[Kr] $5s^2 4d^{10} 5p^4$
53	I	[Kr] $5s^2 4d^{10} 5p^5$
54	Xe	[Kr] $5s^2 4d^{10} 5p^6$ = [Xe]
55	Cs	[Xe] $6s^1$
56	Ba	[Xe] $6s^2$
57	La	[Xe] $6s^2 5d^1$
58	Ce	[Xe] $6s^2 4f^1 5d^1$
59	Pr	[Xe] $6s^2 4f^3$
60	Nd	[Xe] $6s^2 4f^4$
61	Pm	[Xe] $6s^2 4f^5$
62	Sm	[Xe] $6s^2 4f^6$
63	Eu	[Xe] $6s^2 4f^7$
64	Gd	[Xe] $6s^2 4f^7 5d^1$
65	Tb	[Xe] $6s^2 4f^9$
66	Dy	[Xe] $6s^2 4f^{10}$
67	Ho	[Xe] $6s^2 4f^{11}$
68	Er	[Xe] $6s^2 4f^{12}$
69	Tm	[Xe] $6s^2 4f^{13}$
70	Yb	[Xe] $6s^2 4f^{14}$
71	Lu	[Xe] $6s^2 4f^{14} 5d^1$
72	Hf	[Xe] $6s^2 4f^{14} 5d^2$
73	Ta	[Xe] $6s^2 4f^{14} 5d^3$
74	W	[Xe] $6s^2 4f^{14} 5d^4$
75	Re	[Xe] $6s^2 4f^{14} 5d^5$
76	Os	[Xe] $6s^2 4f^{14} 5d^6$
77	Ir	[Xe] $6s^2 4f^{14} 5d^7$
78	Pt	[Xe] $6s^1 4f^{14} 5d^9$
79	Au	[Xe] $6s^1 4f^{14} 5d^{10}$
80	Hg	[Xe] $6s^2 4f^{14} 5d^{10}$
81	Tl	[Xe] $6s^2 4f^{14} 5d^{10} 6p^1$
82	Pb	[Xe] $6s^2 4f^{14} 5d^{10} 6p^2$
83	Bi	[Xe] $6s^2 4f^{14} 5d^{10} 6p^3$
84	Po	[Xe] $6s^2 4f^{14} 5d^{10} 6p^4$
85	At	[Xe] $6s^2 4f^{14} 5d^{10} 6p^5$
86	Rn	[Xe] $6s^2 4f^{14} 5d^{10} 6p^6$ = [Rn]

Table B.16

Standard-State Enthalpy of Formation, Free Energy of Formation, and Absolute Entropy Data, 298.15 K

Substance	ΔH_f° (kJ/mol$_{rxn}$)	ΔG_f° (kJ/mol$_{rxn}$)	S° (J/mol$_{rxn} \cdot$ K)
Aluminum			
Al(s)	0	0	28.33
Al(g)	326.4	285.7	164.54
Al^{3+}(aq)	-531	-485	-321.7
Al$_2$O$_3$(s)	-1675.7	-1582.3	50.92
AlCl$_3$(s)	-704.2	-628.8	110.67
AlF$_3$(s)	-1504.1	-1425.0	66.44
Al$_2$(SO$_4$)$_3$(s)	-3440.84	-3099.94	239.3
Barium			
Ba(s)	0	0	62.8
Ba(g)	180	146	170.243
Ba^{2+}(aq)	-537.64	-560.77	9.6
BaO(s)	-553.5	-525.1	70.42
Ba(OH)$_2 \cdot$ 8 H$_2$O(s)	-3342.2	-2792.8	427
BaCl$_2$(s)	-858.6	-810.4	123.68
BaCl$_2$(aq)	-871.95	-823.21	122.6
BaSO$_4$(s)	-1473.2	-1362.2	132.2
Ba(NO$_3$)$_2$(s)	-992.07	-769.59	213.8
Ba(NO$_3$)$_2$(aq)	-952.36	-783.28	302.5
Beryllium			
Be(s)	0	0	9.50
Be(g)	324.3	286.6	136.269
Be^{2+}(aq)	-382.8	-379.73	-129.7
BeO(s)	-609.6	-508.3	14.14
BeCl$_2$(s)	-490.4	-445.6	82.68
Bismuth			
Bi(s)	0	0	56.74
Bi(g)	207.1	168.2	187.05
Bi$_2$O$_3$(s)	-573.88	-493.7	151.5
BiCl$_3$(s)	-379.1	-315.0	177.0
BiCl$_3$(g)	-265.7	-256.0	358.85
Bi$_2$S$_3$(s)	-143.1	-140.6	200.4
Boron			
B(s)	0	0	5.86
B(g)	562.7	518.8	153.45
B$_2$O$_3$(s)	-1272.77	-1193.65	53.97
B$_2$H$_6$(g)	35.6	86.7	232.11
B$_5$H$_9$(l)	42.68	171.82	184.22
B$_5$H$_9$(g)	73.2	175.0	275.92
B$_{10}$H$_{14}$(s)	-45.2	192.3	176.56
H$_3$BO$_3$(s)	-1094.33	-968.92	88.83
BF$_3$(g)	-1137.00	-1120.33	254.12
BCl$_3$(l)	-427.2	-387.4	206.3

Substance	ΔH_f° (kJ/mol$_{rxn}$)	ΔG_f° (kJ/mol$_{rxn}$)	S° (J/mol$_{rxn}$ · K)
$B_3N_3H_6(l)$	−541.00	−392.65	199.6
$B_3N_3H_6(g)$	−511.75	−390.00	288.68
Bromine			
$Br_2(l)$	0	0	152.231
$Br_2(g)$	30.907	3.110	245.463
$Br(g)$	111.884	82.396	175.022
$HBr(g)$	−36.40	−53.45	198.695
$HBr(aq)$	−121.55	−103.96	82.4
$BrF(g)$	−93.85	−109.18	228.97
$BrF_3(g)$	−255.60	−229.43	292.53
$BrF_5(g)$	−428.9	−350.6	320.19
Calcium			
$Ca(s)$	0	0	41.42
$Ca(g)$	178.2	144.3	154.884
$Ca^{2+}(aq)$	−542.83	−553.58	−53.1
$CaO(s)$	−635.09	−604.03	39.75
$Ca(OH)_2(s)$	−986.09	−898.49	83.39
$CaCl_2(s)$	−795.8	−748.1	104.6
$CaSO_4(s)$	−1434.11	−1321.79	106.7
$CaSO_4 \cdot 2\,H_2O(s)$	−2022.63	−1797.28	194.1
$Ca(NO_3)_2(s)$	−938.39	−743.07	193.3
$CaCO_3(s)$	−1206.92	−1128.79	92.9
$Ca_3(PO_4)_2$	−4120.8	−3884.7	236.0
Carbon			
$C(graphite)$	0	0	5.74
$C(diamond)$	1.895	2.900	2.377
$C(g)$	716.682	671.257	158.096
$CO(g)$	−110.525	−137.168	197.674
$CO_2(g)$	−393.509	−394.359	213.74
$COCl_2(g)$	−218.8	−204.6	283.53
$CH_4(g)$	−74.81	−50.752	186.264
$HCHO(g)$	−108.57	−102.53	218.77
$H_2CO_3(aq)$	−699.65	−623.08	187.4
$HCO_3^-(aq)$	−691.99	−586.77	91.2
$CO_3^{2-}(aq)$	−677.14	−527.81	−56.9
$CH_3OH(l)$	−238.66	−166.27	126.8
$CH_3OH(g)$	−200.66	−161.96	239.81
$CCl_4(l)$	−135.44	−65.21	216.40
$CCl_4(g)$	−102.9	−60.59	309.85
$CHCl_3(l)$	−134.47	−73.66	201.1
$CHCl_3(g)$	−103.14	−70.34	295.71
$CH_2Cl_2(l)$	−121.46	−67.26	177.8
$CH_2Cl_2(g)$	−92.47	−65.87	270.23
$CH_3Cl(g)$	−80.84	−57.40	234.5
$CS_2(l)$	89.70	65.27	151.34
$CS_2(g)$	117.36	67.12	237.84
$HCN(g)$	135.1	124.7	201.78

(*continued*)

Table B.16

Standard-State Enthalpy of Formation, Free Energy of Formation, and Absolute Entropy Data, 298.15 K (continued)

Substance	ΔH_f° (kJ/mol$_{rxn}$)	ΔG_f° (kJ/mol$_{rxn}$)	S° (J/mol$_{rxn} \cdot$ K)
$CH_3NO_2(l)$	−113.09	−14.42	171.75
$C_2H_2(g)$	226.73	209.20	200.94
$C_2H_4(g)$	52.26	68.15	219.56
$C_2H_6(g)$	−84.68	−32.82	229.60
$CH_3CHO(l)$	−192.30	−128.12	160.2
$CH_3CO_2H(l)$	−484.5	−389.9	159.8
$CH_3CO_2H(g)$	−432.25	−374.0	282.5
$CH_3CO_2H(aq)$	−485.76	−396.46	178.7
$CH_3CO_2^-(aq)$	−486.01	−369.31	86.6
$CH_3CH_2OH(l)$	−277.69	−174.78	160.7
$CH_3CH_2OH(g)$	−235.10	−168.49	282.70
$CH_3CH_2OH(aq)$	−288.3	−181.64	148.5
$C_6H_6(l)$	49.028	124.50	172.8
$C_6H_6(g)$	82.927	129.66	269.2
Chlorine			
$Cl_2(g)$	0	0	223.066
$Cl(g)$	121.679	105.680	165.198
$Cl^-(aq)$	−167.159	−131.228	56.5
$ClO_2(g)$	102.5	120.5	256.84
$Cl_2O(g)$	80.3	97.9	266.21
$Cl_2O_7(l)$	238		
$HCl(g)$	−92.307	−95.299	186.908
$HCl(aq)$	−167.159	−131.228	56.5
$ClF(g)$	−54.48	−55.94	217.89
Chromium			
$Cr(s)$	0	0	23.77
$Cr(g)$	396.6	351.8	174.50
$CrO_3(s)$	−589.5		
$CrO_4^{2-}(aq)$	−881.15	−727.75	50.21
$Cr_2O_3(s)$	−1139.7	−1058.1	81.2
$Cr_2O_7^{2-}(aq)$	−1490.3	−1301.1	261.9
$(NH_4)_2Cr_2O_7(s)$	−1806.7		
$PbCrO_4(s)$	−930.9		
Cobalt			
$Co(s)$	0	0	30.04
$Co(g)$	424.7	380.3	179.515
$Co^{2+}(aq)$	−58.2	−54.4	−113
$Co^{3+}(aq)$	92	134	−305
$CoO(s)$	−237.94	−214.20	52.97
$Co_3O_4(s)$	−891	−774	102.5
$Co(NH_3)_6^{3+}(aq)$	−584.9	−157.0	146
Copper			
$Cu(s)$	0	0	33.150
$Cu(g)$	338.32	298.58	166.38
$Cu^+(aq)$	71.67	49.98	40.6

Substance	ΔH_f° (kJ/mol$_{rxn}$)	ΔG_f° (kJ/mol$_{rxn}$)	S° (J/mol$_{rxn} \cdot$ K)
$Cu^{2+}(aq)$	64.77	65.49	−99.6
$CuO(s)$	−157.3	−129.7	42.63
$Cu_2O(s)$	−168.6	−146.0	93.14
$CuCl_2(s)$	−220.1	−175.7	108.07
$CuS(s)$	−53.1	−53.6	66.5
$Cu_2S(s)$	−79.5	−86.2	120.9
$CuSO_4(s)$	−771.36	−66.69	109
$Cu(NH_3)_4^{2+}(aq)$	−348.5	−111.07	273.6

Fluorine

$F_2(g)$	0	0	202.78
$F(g)$	78.99	61.91	158.754
$F^-(aq)$	−332.63	−278.79	−13.8
$HF(g)$	−271.1	−273.2	173.779
$HF(aq)$	−320.08	−296.82	88.7

Hydrogen

$H_2(g)$	0	0	130.684
$H(g)$	217.65	203.247	114.713
$H^+(aq)$	0	0	0
$OH^-(aq)$	−229.994	−157.244	−10.75
$H_2O(l)$	−285.830	−237.129	69.91
$H_2O(g)$	−241.818	−228.572	188.25
$H_2O_2(l)$	−187.78	−120.35	109.6
$H_2O_2(aq)$	−191.17	−134.03	143.9

Iodine

$I_2(s)$	0	0	116.135
$I_2(g)$	62.438	19.327	260.69
$I(g)$	106.838	70.50	180.791
$HI(g)$	26.48	1.70	206.594
$IF(g)$	−95.65	−118.51	236.17
$IF_5(g)$	−822.49	−751.73	327.7
$IF_7(g)$	−943.9	−818.3	346.5
$ICl(g)$	17.78	−5.46	247.551
$IBr(g)$	40.84	3.69	258.773

Iron

$Fe(s)$	0	0	27.28
$Fe(g)$	416.3	370.7	180.49
$Fe^{2+}(aq)$	−89.1	−78.90	−137.7
$Fe^{3+}(aq)$	−48.5	−4.7	−315.9
$Fe_2O_3(s)$	−824.2	−742.2	87.40
$Fe_3O_4(s)$	−1118.4	−1015.4	146.4
$Fe(OH)_2(s)$	−569.0	−486.5	88
$Fe(OH)_3(s)$	−823.0	−696.5	106.7
$FeCl_3(s)$	−399.49	−334.00	142.3
$FeS_2(s)$	−178.2	−166.9	52.93
$Fe(CO)_5(l)$	−774.0	−705.3	338.1
$Fe(CO)_5(g)$	−733.9	−697.21	445.3

(continued)

Table B.16

Standard-State Enthalpy of Formation, Free Energy of Formation, and Absolute Entropy Data, 298.15 K (continued)

Substance	ΔH_f° (kJ/mol$_{rxn}$)	ΔG_f° (kJ/mol$_{rxn}$)	S° (J/mol$_{rxn} \cdot$ K)
Lead			
Pb(s)	0	0	64.81
Pb(g)	195.0	161.9	175.373
Pb^{2+}(aq)	−1.7	−24.43	10.5
PbO(s)	−217.32	−187.89	68.7
PbO$_2$(s)	−277.4	−217.33	68.6
PbCl$_2$(s)	−359.41	−314.10	136.0
PbCl$_4$(l)	−329.3		
PbS(s)	−100.4	−98.7	91.2
PbSO$_4$(s)	−919.94	−813.14	148.57
Pb(NO$_3$)$_2$(s)	−451.9		
PbCO$_3$(s)	−699.1	−625.5	131.0
Lithium			
Li(s)	0	0	29.12
Li(g)	159.37	126.66	138.77
Li$^+$(aq)	−278.49	−293.31	13.4
LiH(s)	−90.54	−68.35	20.08
LiOH(s)	−484.93	−438.95	42.80
LiF(s)	−615.97	−587.71	35.65
LiCl(s)	−408.61	−384.37	59.33
LiBr(s)	−351.23	−342.00	74.27
LiI(s)	−270.41	−270.29	86.78
LiAlH$_4$(s)	−116.3	−44.7	78.74
LiBH$_4$(s)	190.8	125.0	75.86
Magnesium			
Mg(s)	0	0	32.68
Mg(g)	147.70	113.10	148.650
Mg^{2+}(aq)	−466.85	−454.8	−138.1
MgO(s)	−601.70	−569.43	26.94
MgH$_2$(s)	−75.3	−35.9	31.09
Mg(OH)$_2$(s)	−924.54	−833.58	63.18
MgCl$_2$(s)	−641.32	−591.79	89.62
MgCO$_3$(s)	−1095.8	−1012.1	65.7
MgSO$_4$(s)	−1284.9	−1170.6	91.6
Manganese			
Mn(s)	0	0	32.01
Mn(g)	280.7	238.5	173.70
Mn^{2+}(aq)	−220.75	−228.1	−73.6
MnO(s)	−385.22	−362.90	59.71
MnO$_2$(s)	−520.03	−465.14	53.05
Mn$_2$O$_3$(s)	−959.0	−881.1	110.5
Mn$_3$O$_4$(s)	−1387.8	−1283.2	155.6
KMnO$_4$(s)	−837.2	−737.6	171.76
MnS(s)	−214.2	−218.4	78.2

Substance	ΔH_f° (kJ/mol$_{rxn}$)	ΔG_f° (kJ/mol$_{rxn}$)	S° (J/mol$_{rxn}$ · K)
Mercury			
$Hg(l)$	0	0	76.02
$Hg(g)$	61.317	31.820	174.96
$Hg^{2+}(aq)$	171.1	164.40	−32.2
$HgO(s)$	−90.83	−58.539	70.29
$HgCl_2(s)$	−224.3	−178.6	146.0
$Hg_2Cl_2(s)$	−265.22	−210.745	192.5
$HgS(s)$	−58.2	−50.6	82.4
Nitrogen			
$N_2(g)$	0	0	191.61
$N(g)$	472.704	455.63	153.298
$NO(g)$	90.25	86.55	210.761
$NO_2(g)$	33.18	51.31	240.06
$N_2O(g)$	82.05	104.20	219.85
$N_2O_3(g)$	83.72	139.46	312.28
$N_2O_4(g)$	9.16	97.89	304.29
$N_2O_5(g)$	11.35	115.1	355.7
$NO_3^-(aq)$	−205.0	−108.74	146.4
$NOCl(g)$	51.71	66.08	261.69
$NO_2Cl(g)$	12.6	54.4	272.15
$HNO_2(aq)$	−119.2	−50.6	135.6
$HNO_3(g)$	−135.06	−74.72	266.38
$HNO_3(aq)$	−207.36	−111.25	146.4
$NH_3(g)$	−46.11	−16.45	192.45
$NH_3(aq)$	−80.29	−26.50	111.3
$NH_4^+(aq)$	−132.51	−79.31	113.4
$NH_4NO_3(s)$	−365.56	−183.87	151.08
$NH_4NO_3(aq)$	−339.87	−190.56	259.8
$NH_4Cl(s)$	−314.43	−203.87	94.6
$N_2H_4(l)$	50.63	149.34	121.21
$N_2H_4(g)$	95.40	159.35	238.47
$HN_3(g)$	294.1	328.1	238.97
Oxygen			
$O_2(g)$	0	0	205.138
$O(g)$	249.170	231.731	161.055
$O_3(g)$	142.7	163.2	238.93
Phosphorus			
$P(white)$	0	0	41.09
$P_4(g)$	58.91	24.4	279.98
$P_2(g)$	144.3	103.7	218.129
$P(g)$	314.64	278.25	163.193
$PH_3(g)$	5.4	13.4	210.23
$P_4O_6(s)$	−1640.1		
$P_4O_{10}(s)$	−2984.0	−2697.7	228.86
$PO_4^{3-}(aq)$	−1277.4	−1018.7	−222
$PF_3(g)$	−918.8	−897.5	273.24
$PF_5(g)$	−1595.8		

(continued)

Table B.16

Standard-State Enthalpy of Formation, Free Energy of Formation, and Absolute Entropy Data, 298.15 K (continued)

Substance	ΔH_f° (kJ/mol$_{rxn}$)	ΔG_f° (kJ/mol$_{rxn}$)	S° (J/mol$_{rxn}$ · K)
$PCl_3(l)$	−319.7	−272.3	217.1
$PCl_3(g)$	−287.0	−267.8	311.78
$PCl_5(g)$	−374.9	−305.0	364.58
$H_3PO_4(s)$	−1279.0	−1119.1	110.50
$H_3PO_4(aq)$	−1277.4	−1018.7	−222)
Potassium			
$K(s)$	0	0	64.18
$K(g)$	89.24	60.59	160.336
$K^+(aq)$	−252.38	−283.27	102.5
$KOH(s)$	−424.764	−379.08	78.9
$KCl(s)$	−436.747	−409.14	82.59
$KNO_3(s)$	−494.63	−394.86	133.05
$K_2Cr_2O_7(s)$	−2061.5	−1881.8	291.2
$KMnO_4(s)$	−837.2	−737.6	171.76
Silicon			
$Si(s)$	0	0	18.83
$Si(g)$	455.6	411.3	167.97
$SiO_2(s)$	−910.94	−856.64	41.84
$SiH_4(g)$	34.3	56.9	204.62
$SiF_4(g)$	−1614.94	−1572.65	282.49
$SiCl_4(l)$	−687.0	−619.9	240.
$SiCl_4(g)$	−657.01	−616.98	330.73
Silver			
$Ag(s)$	0	0	42.55
$Ag(g)$	284.55	245.65	172.97
$Ag^+(aq)$	105.579	77.107	72.68
$Ag(NH_3)_2^+(aq)$	−111.29	−17.12	245.2
$Ag_2O(s)$	−31.05	−11.20	−121.3
$AgCl(s)$	−127.068	−109.789	96.2
$AgBr(s)$	−100.37	−96.90	107.1
$AgI(s)$	−61.84	−66.19	−115.5
Sodium			
$Na(s)$	0	0	51.21
$Na(g)$	107.32	76.761	153.712
$Na^+(aq)$	−240.13	−261.905	59.0
$NaH(s)$	−56.275	−33.46	−40.016
$NaOH(s)$	−425.609	−379.494	64.455
$NaOH(aq)$	−470.114	−419.150	48.1
$NaCl(s)$	−411.153	−384.138	72.13
$NaCl(g)$	−176.65	−196.66	229.81
$NaCl(aq)$	−407.27	−393.133	115.5
$NaNO_3(s)$	−467.85	−367.00	116.52

Substance	ΔH_f° (kJ/mol$_{rxn}$)	ΔG_f° (kJ/mol$_{rxn}$)	S° (J/mol$_{rxn} \cdot$ K)
$Na_3PO_4(s)$	−1917.40	−1788.80	173.80
$Na_2SO_3(s)$	−1123.0	−1028.0	155
$Na_2SO_4(s)$	−1387.08	−1270.16	149.58
$Na_2CO_3(s)$	−1130.68	−1044.44	134.98
$NaHCO_3(s)$	−950.81	−851.0	101.7
$NaCH_3CO_2(s)$	−708.81	−607.18	123.0
$Na_2CrO_4(s)$	−1342.2	−1234.93	176.61
$Na_2Cr_2O_7(s)$	−1978.6		
Sulfur			
$S_8(s)$	0	0	31.80
$S_8(g)$	102.30	49.63	430.98
$S(g)$	278.805	238.250	167.821
$S^{2-}(aq)$	33.1	85.8	−14.6
$SO_2(g)$	−296.830	−300.194	248.22
$SO_3(s)$	−454.51	−374.21	70.7
$SO_3(l)$	−441.04	−373.75	113.8
$SO_3(g)$	−395.72	−371.06	256.76
$SO_4^{2-}(aq)$	−909.27	−744.53	20.1
$SOCl_2(g)$	−212.5	−198.3	309.77
$SO_2Cl_2(g)$	−364.0	−320.0	311.94
$H_2S(g)$	−20.63	−33.56	205.79
$H_2SO_3(aq)$	−608.81	−537.81	232.2
$H_2SO_4(aq)$	−909.27	−744.53	20.1
$SF_4(g)$	−774.9	−731.3	292.03
$SF_6(g)$	−1209	−1105.3	291.82
$SCN^-(aq)$	76.44	92.71	144.3
Tin			
$Sn(s)$	0	0	44.14
$Sn(g)$	302.1	267.3	168.486
$SnO(s)$	−285.8	−256.9	56.5
$SnO_2(s)$	−580.7	−519.76	52.3
$SnCl_2(s)$	−325.1		
$SnCl_4(l)$	−511.3	−440.21	258.6
$SnCl_4(g)$	−471.5	−432.2	365.8
Titanium			
$Ti(s)$	0	0	30.63
$Ti(g)$	469.9	425.1	180.2
$TiO(s)$	−519.7	−495.0	34.8
$TiO_2(s)$ rutile	−944.8	−889.5	50.33
$TiCl_4(l)$	−804.2	−737.2	252.3
$TiCl_4(g)$	−763.2	−726.8	354.8
Tungsten			
$W(s)$	0	0	32.64
$W(g)$	849.4	807.1	173.950
$WO_3(s)$	−842.87	−764.083	75.90

(*continued*)

Table B.16

Standard-State Enthalpy of Formation, Free Energy of Formation, and Absolute Entropy Data, 298.15 K (continued)

Substance	ΔH_f° (kJ/mol$_{rxn}$)	ΔG_f° (kJ/mol$_{rxn}$)	S° (J/mol$_{rxn} \cdot$ K)
Zinc			
Zn(s)	0	0	41.63
Zn(g)	130.729	95.145	160.984
Zn^{2+}(aq)	−153.89	−147.06	−112.1
ZnO(s)	−348.28	−318.30	43.64
ZnCl$_2$(s)	−415.05	−369.39	111.46
ZnS(s)	−205.98	−201.29	57.7
ZnSO$_4$(s)	−982.8	−871.5	110.5

Appendix C

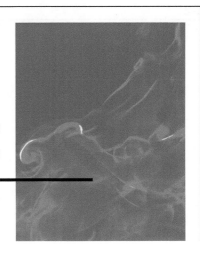

Chapter 1

1-7 Mixture

1-9 The ratio of sulfur to oxygen in the molecule

1-11 (a) sodium (b) magnesium (c) aluminum (d) silicon
(e) phosphorus (f) chlorine (g) argon

1-13 (a) molybdenum (b) tungsten (c) rhodium (d) iridium
(e) palladium (f) platinum (g) silver (h) gold (i) mercury

1-17 The model has been successfully used to predict the outcomes of a multitude of experiments.

1-19 Atoms of different elements will have different weights and chemical properties.

1-21 3.2×10^7 sec

1-23 4×10^{-1} micrometers; 4×10^2 nanometers

1-25 0.1416 m^3/sec

1-27 (a) 3 (b) 4 (c) 1 (d) 5

1-29 (a) 475 (b) 0.0680 (c) 9.46×10^{10} (d) 30.1

1-31 (a) 153.92 (b) 33 (c) 5.10×10^7 (d) 0.0431

1-35 Electron: −1; Proton: +1; Neutron: 0

1-37 Electron

1-41 ^{52}Cr

1-43 53 protons; 74 neutrons; 53 electrons
mass number: 127; atomic number: 53

1-45 ^{79}Se

1-47

	Z	A	e
^{31}P	15	31	15
^{18}O	8	18	8
^{39}K	19	39	19
^{58}Ni	28	58	28

1-49 5.96842 times more massive

1-51 ^{16}O

1-53

Mass (grams)	Z	A	Number of neutrons	Mass (amu)
1.6627×10^{-23}	5	10	5	10.0129
3.9829×10^{-23}	12	24	12	23.9850
2.9888×10^{-23}	8	18	10	17.9992
1.7752×10^{-22}	47	107	60	106.903

1-55 Three

1-57 1.99265×10^{-23} g; 12.0000 amu

1-59 (b)

1-63 $^{52}Cr^{+3}$

1-65 $^{79}Se^{-2}$

1-69 NH_4^+, H_3O^+

1-73 (a) IA (b) IVA (c) IIA (d) VIA (e) IIA (f) VIIIA (g) VIIA

1-75 Seven

1-77 (a) and (d)

1-79 Au

1-81 79.90 amu

1-83 1.2011×10^6 amu; 12.011 amu; No single carbon atom has this mass.

1-85 126.90 amu

1-87 Boron; One isotope has 6 neutrons, 5 protons, and 5 electrons, the other would have 5 neutrons, 5 protons, and 5 electrons.

1-89 (a) 107.87 amu
   (b) Both have 47 protons.
   (c) Both have 47 electrons. ^{107}Ag has 60 neutrons and ^{109}Ag has 62 neutrons.

1-91

	Z	A	e	n	$\%A$
^{6}Li	3	6	3	3	7.42
^{7}Li	3	7	3	4	92.58
^{20}Ne	10	20	10	10	90.51
^{21}Ne	10	21	10	11	0.27
^{22}Ne	10	22	10	12	9.22

1-95 The same

1-103 (a) $4 \text{ Cr}(s) + 3 \text{ O}_2(g) \longrightarrow 2 \text{ Cr}_2\text{O}_3(s)$
   (b) $\text{SiH}_4(g) \longrightarrow \text{Si}(s) + 2 \text{ H}_2(g)$
   (c) $2 \text{ SO}_3(g) \longrightarrow 2 \text{ SO}_2(g) + \text{O}_2(g)$

1-105 (a) $\text{CH}_4(g) + 2 \text{ O}_2(g) \longrightarrow \text{CO}_2(g) + 2 \text{ H}_2\text{O}(g)$
   (b) $2 \text{ H}_2\text{S}(g) + 3 \text{ O}_2(g) \longrightarrow 2 \text{ H}_2\text{O}(g) + 2 \text{ SO}_2(g)$
   (c) $2 \text{ B}_5\text{H}_9(g) + 12 \text{ O}_2(g) \longrightarrow 5 \text{ B}_2\text{O}_3(s) + 9 \text{ H}_2\text{O}(g)$

1-107 (a) $1 \text{ C}_3\text{H}_8(g) + 5 \text{ O}_2(g) \longrightarrow 3 \text{ CO}_2(g) + 4 \text{ H}_2\text{O}(g)$
   (b) $\text{C}_2\text{H}_5\text{OH}(l) + 3 \text{ O}_2(g) \longrightarrow 2 \text{ CO}_2(g) + 3 \text{ H}_2\text{O}(g)$
   (c) $\text{C}_6\text{H}_{12}\text{O}_6(s) + 6 \text{ O}_2(g) \longrightarrow 6 \text{ CO}_2(g) + 6 \text{ H}_2\text{O}(l)$

1-109 The ion has 18 electrons, 20 protons, and 20 neutrons. The chemical symbol for X is Ca.

1-111

classification	group	period	electrons	element
metal	IA	3	11	Na
semimetal	IVA	4	32	Ge
semimetal	IIIA	2	5	B
semimetal	IVA	3	14	Si
nonmetal	VIIA	4	35	Br

Chapter 2

2-1 1.6817 amu

2-3 (a) 6.941 amu and 6.941 g   (b) 12.011 amu and 12.011 g
   (c) 24.305 amu and 24.305 g   (d) 63.546 amu and 63.546 g

2-5 1000 Si atoms

2-7 (a) 40.078 g   (b) 87.62 g   (c) 78.96 g   (d) 72.61 g

2-9 51.996 g

2-11 (a) 12.011 g C   (b) 58.693 g Ni   (c) 200.59 g Hg

2-13 $\dfrac{2.7 \times 10^{26} \text{ atoms}}{\text{new mol}}$

2-15 106.1 g/mol

2-17 8.67×10^{-17} g

2-19 7.306×10^{-23} g

2-21 105 g

2-23 1.2044×10^{24} atoms

2-25 9.2738×10^{-23} g/atom

2-27 12 carbon atoms; 48 hydrogen atoms; 6.022×10^{23} carbon atoms; 24.09×10^{24} hydrogen atoms

2-29 1.20×10^{24} atoms H; 6.02×10^{23} molecules of H_2; 2.02 g H_2

2-31 (d)

2-33 46.025 g/mol; 30.026 g/mol

2-35 (a) 444.556 g/mol   (b) 46.005 g/mol
(c) 97.46 g/mol   (d) 158.032 g/mol

2-37 162.187 g/mol

2-39 (a) 375.937 g/mol   (b) 284.745 g/mol   (c) 444.438 g/mol

2-41 8.95 g CCl_4

2-43 0.212 mol Al

2-45 (a) 21.200%   (b) 13.854%   (c) 16.480%   (d) 46.646%

2-47 31.352% Si

2-49 7.61×10^{-3} mols C

2-51 5.62×10^{21} atoms of Cl

2-53 (a) 36.8 g N and 63.2 g O   (b) 2.63 mols N and 3.95 mols O
(c) 1.50:1   (d) N_2O_3

2-55 Fe_3O_4

2-57 N_2O

2-59 XeF_6

2-61 (c)

2-63 No, only the empirical formula can be determined from percent composition.

2-65 $C_{20}H_{14}O_4$

2-67 $C_{14}H_{18}N_2O_5$

2-71 10 mol CO_2

2-73 12 mol CuO

2-75 2.25 mol O_2

2-77 1. Convert grams of CO reacted to moles of CO.
2. Relate moles of CO reacted to moles of CO_2 produced.
3. Convert moles CO_2 to grams.

2-79 27.4 g CO_2

2-81 9.79 g O_2

2-83 0.511 kg C_2H_5OH

2-85 3.73 g PH_3

2-87 50.0 g N_2

2-89 0.400 mol P_4S_{10}; 0.400 mol P_4S_{10}; 0.500 mol P_4S_{10}

2-91 10.3 g HCl; increase the amount of Cl_2

2-93 0.482 moles of PF_5

2-95 175 g Fe

2-97 The first strip is Pb. The second strip is Al.

2-99 76 g

2-101 (b) and (c)

2-103 5.99 M

2-105 14.8 M

2-107 0.0814 M

2-109 The solution was less than 1.00 M. The solution must be prepared by placing the solute in a container calibrated to hold 1.000 liter. Some water is added to the volumetric flask to completely dissolve the solute. Water is added until the liquid level has been brought to the calibration mark.

2-111 6.3×10^{-8} M

2-113 0.0648 M $AgNO_3$; Either add half as much $AgNO_3$ or use twice as much water.

2-115 0.25 mol

2-117 The 0.25 M solution

2-119 (a) 0.00324 M $AgNO_3$
      (b) 6.48×10^{-5} M $AgNO_3$
      (c) 2.60×10^{-6} M $AgNO_3$

2-121 0.025 M $CuSO_4$

2-123 4.2 mL of 6.0 M HCl would be diluted to 250 mL.

2-125 0.267 L

2-127 172 mL

2-129 Take about 150 mL of distilled water and slowly add 42.1 mL of the HNO_3 with mixing. Bring the solution to a final volume of 200 mL with distilled water.

2-131 36 mL

2-133 0.9808 M NaOH

2-135 0.375 M $C_6H_{12}O_6$

2-137 2

2-139 MnO_2

2-141 39.452 g

2-143 80.0%

2-145 Fe_3O_4

Chapter 3

3-1 The nucleus

3-3 The positive charge and most of the mass

3-7 The wavelength decreases by a factor of 2. The speed is a constant.

3-9 6.0×10^{-7} m

3-11 Red light

3-13 11.92 m; radio-wave

3-15 red; orange

3-17 5.089894×10^{14} cycle/s; 5.084740×10^{14} cycle/s

3-19 580 nm

3-21 3.027×10^{-19} J

3-23 4.914×10^{-7} m; visible

3-25 The Rutherford model of the atom does not specify where electrons might be found in the atom. The Bohr model does.

3-27 As the distance between the electron and nucleus increases the force holding them together decreases according to Coulomb's law.

3-29 (a) (ii) (b) (ii) (c) (i) and (ii)

3-31 Ionized state

3-33 3.025×10^{-19} J; 4.565×10^{14} cycle/s

3-35 $n = 5$

3-37 Na < Li < Be < F

3-39 Na: +1 Mg: +2 Al: +3 Si: +4 P: +5 S: +6 Cl: +7 Ar: +8 K: +1 Ca: +2

3-41 Ar will have a higher ionization energy than Cl$^-$.

3-43 Innermost shell

3-45 The outer-shell electrons in chlorine are farther from the nucleus.

3-47 The ionization energy of F$^-$ would be less than that of Ne.

3-49 Be

3-51 Decrease going from top to bottom

3-53 Because the $1s^1$ electron of hydrogen is closer to the nucleus.

3-57 (b)

3-59 In the first three rows, they are the same.

3-61 No

3-63 No

3-65 800.6 kJ/mol

3-67 Because both elements only have electrons in the 1s subshell.

3-71 Electrons are found in three different subshells, $1s$, $2s$ and $2p$.

3-75 Mg

3-77 Kr

3-81 -1, 0, +1; 0 or +1

3-83 0, 1, 2, 3

3-85 s orbital; p orbital; d orbital; f orbital

3-87 (c) and (e)

3-89 An unpaired electron

3-91 (a) Atoms would be deflected
(b) No deflections
(c) No deflections
(d) Atoms will be deflected into 3 separate areas

3-93 (b)

3-95 5

3-97 $n = 1; l = 0,1; m_l = 0, 1; m_s = +1, -1$
$n = 2; l = 0, 1, 2; m_l = 0, 1; m_s = +1, -1$

3-99 14

3-101 Two

3-103 2, 8, 18, 32

3-105 (d)

3-109 (e)

3-113 row 5, column 5

3-115 Group IA

3-117 [He] $2s^2 2p^3$

3-119 (b)

3-121 (c)

3-123 6

3-125 (d)

3-127 (b)

3-129 (a), (b) and (d) are incorrect.

3-131 (d)

3-133 The size of the atom increases.

3-135 0.1442 nm; 144.2 pm

3-139 The covalent radius decreases when going from left to right. In going down
a single column on the periodic table, the covalent radius increases.

3-143 $Fe^{3+} < Fe^{2+}$

3-145 $F^- < Cl^- < Br^- < I^-$

3-147 (e)

3-149 (c)

3-151 (b)

3-153 (e)

3-155 (d)

3-157 (a)

3-159 Mg < Be < Ne < Na < Li

3-163 (a) Mg < P < Cl   (b) Se < S < O < F   (c) K < P < O

3-165 AVEE will increase from left to right.

3-169 Nonmetals have large values for AVEE.

3-173 Tl < Ga < Al < B

3-177 Increasing Ionization Energy: $O^{2-} < F^- < Ne < Na^+ < Mg^{2+}$
Increasing Radius: $Mg^{2+} < Na^+ < Ne < F^- < O^{2-}$

Increasing Ionization Energy: Na < Mg < O < F < Ne

Increasing Radius: Ne < F < O < Mg < Na

3-179 The order of increasing ionization energy is $S^{2-} < Ar < K^+$. The order of increasing radii would be $K^+ < Ar < S^{2-}$.

3-181 (a) 2   (b) 3   (c) 2   (d) 0   (e) 1

3-185 Ba; BaO and $BaCl_2$

3-187

Atom	Chemical Symbol	Magnetic field behavior	First IE	Atomic Radius	Number of PES Peaks	Core Charge	AVEE	Number of Valence Electrons
X	Ne	not deflected	2.08	0.070	3	+8	2.0	8
Y	Na	deflected	0.50	0.16	4	+1	0.50	1
Z	Mg	not deflected	0.74	0.14	4	+2	0.8	2

3-189 (a) Cl

(b) Cl: $1s^2 2s^2 2p^6 3s^2 3p^5$

(c) Cl^-: $1s^2 2s^2 2p^6 3s^2 3p^6$. The radius of Cl^- would be larger.

(d) It is easier to ionize Cl^- than it is Cl.

(e) AVEE (Cl) < AVEE (Ar)

(f) This element is F.

(g) 7

3-191 (a) (ii)   (b) 2   (c) +2

(d) You are taking an electron from something which is already a positive ion.

Chapter 4

4-3 (a) 8   (b) 4   (c) 5   (d) 7

4-5 All have 8 valence electrons.

4-9 They are the same.

4-21 Each atom should have eight electrons surrounding it.

4-23 (c)

4-25 (a)

4-27 (a) O—S—O   (b) O—S—O   (c) O—C—S
              |
              O

(d) H—N—H   (e) Cl—C—Cl
        | |
        H   Cl
                     (with H on top of C)

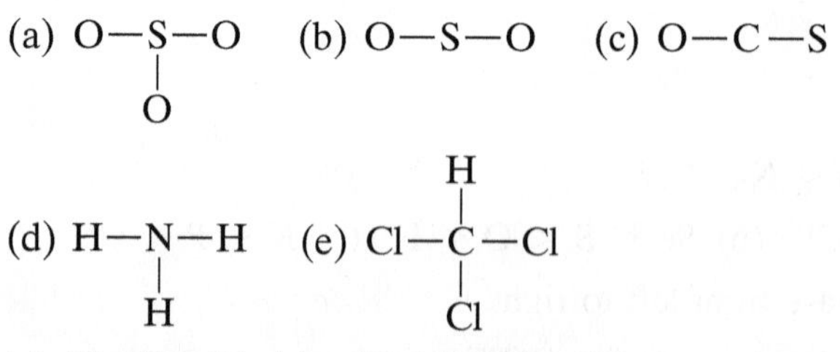

4-29 (a) 22 (b) 34 (c) 48 (d) 32

4-31 (a)   H   (b) O—N—O   (c)   O
              | | |
       H—N—H   O   O—S—O
              | |
              H   O

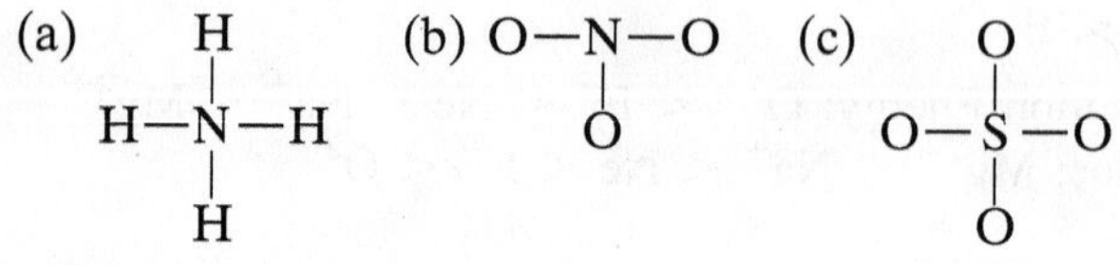

4-33
(a) $\left[:\ddot{O}-N-\ddot{O}: \right]^{-}$ (b) $\left[:\ddot{O}-\ddot{S}-\ddot{O}: \right]^{2-}$ or $\left[:\ddot{O}-\ddot{S}-\ddot{O}: \right]^{2-}$

(c) $\left[\begin{array}{c} :O: \\ \| \\ C \\ :\ddot{O} \quad \ddot{O}: \end{array} \right]^{2-}$ (d) $\left[\ddot{O}=N=\ddot{O} \right]^{+}$

4-35 (a) N_2O :N≡N—$\ddot{O}$: ⟷ $\ddot{N}$=N=$\ddot{O}$

                    I   II

Structure I is preferred.

(b) N_2O_3 :O:
                       ‖
            $\ddot{O}$=$\ddot{N}$—N—$\ddot{O}$:

4-37 $.\ddot{O}.$ $.\ddot{O}.$
        \   /
         N—$\ddot{O}$—N
        /   \
     :$\ddot{O}.$ $.\ddot{O}$:

4-39 (a) :$\ddot{O}$—$\ddot{S}$=O   (b) :$\ddot{O}$—S—$\ddot{O}$:
                                                  ‖
                                                :O:

(c) $\left[:\ddot{O}-\ddot{S}-\ddot{O}: \right]^{2-}$ (d) $\left[\begin{array}{c} :\ddot{O}: \\ | \\ :\ddot{O}-S-\ddot{O}: \\ | \\ :\ddot{O}: \end{array} \right]^{2-}$

4-41 (a), (c) and (d)

4-43 (a), (c) and (d)

4-45 None of these

4-47 N_2 is a triple bond, HNNH is a double bond and H_2NNH_2 is a single bond.

4-49 The bond strengths must be similar.

4-51 $\left[\ddot{S}=C=\ddot{N} \right]^{-} ⟷ \left[:S≡C-\ddot{N}: \right]^{-} ⟷ \left[:\ddot{S}-C≡N: \right]^{-}$

4-53 (b) and (c)

4-57 (b)

4-59 (a) 0   (b) −0.29   (c) −0.19

4-61 −0.34; +0.34

4-63 $\delta_O = -0.35$; $\delta_H = +0.05$

4-65 (a) H   (b) C   (c) B   (d) N   (e) N

4-67 (a) 0   (b) 0   (c) 0   (d) 0

4-69 (a) 0, +1   (b) 0, +1   (c) +1

4-73
          :$\ddot{O}$:⁻   :$\ddot{O}$:⁻   :$\ddot{O}$:⁻
           |⁺ | ‖
⁻:$\ddot{O}$—P—$\ddot{O}$:⁻ ⟷ :$\ddot{O}$—P—$\ddot{O}$:⁻ ⟷ $\ddot{O}$=P=O
           | ‖ |
          :$\ddot{O}$:⁻   :$\ddot{O}$:   :$\ddot{O}$:⁻

           I   II   III

Structure II would be the preferred structure.

4-75 Structure I is best.

4-77 Structure (c) is the best structure.

(a) $\ddot{\ddot{O}} = C = \ddot{N}^{\,-}$ (b) $^{+}:O \equiv C - \ddot{N}:^{\,=}$ (c) $^{-}:\ddot{O} - C \equiv N:$

4-79 (a) 3 domains, trigonal planar, 120°
(b) 4 domains, tetrahedral, 109.5°
(c) 6 domains, octahedral, 90°
(d) 5 domains, trigonal bipyramid, 90° and 120°

4-81 (a) 4 domains, 109.5°   (b) 4 domains, 109.5°
(c) 2 domains, 180°   (d) 3 domains, 120°
(e) 2 domains, 180°

4-83 (a) 3   (b) 4   (c) 8   (d) 4   (e) 1

4-85

Bonding domains	Nonbonding domains	Arrangement	Molecular geometry
2	0	linear	linear
4	0	tetrahedral	tetrahedral
2	2	tetrahedral	bent
5	0	trigonal bipyramidal	trigonal bipyramidal
3	2	trigonal bipyramidal	T-Shaped
5	1	octahedral	square pyramid
4	2	octahedral	square planar

4-87 (a) tetrahedral   (b) tetrahedral   (c) tetrahedral   (d) tetrahedral

4-89 (a) trigonal pyramidal   (b) see-saw
(c) square pyramidal   (d) octahedral

4-91 (a) linear   (b) bent   (c) trigonal planar

4-93 (a)

4-95 (c)

4-97 (a), (c), and (e)

4-99 (a), (b), and (d)

4-103

Bonding Domains	Nonbonding Domains	Bond Angle
2	0	180°
5	0	90° and 120°
3	0	120°
2	1	<120°
2	3	180°
4	0	109.5°
3	1	<109.5°
2	2	<109.5°
2	3	180°
3	2	<90°
4	2	90°

4-105 (a) both $\underline{C}$ 109.5°, $\underline{O}$ <109.5°
(b) 109.5°
(c) $\underline{O}$ <109.5°, $\underline{N}$ <120°
(d) $\underline{H_3C}$ <109.5°, O=$\underline{C}$ 120°
(e) 120°

4-109 The carbon-sulfur bond is a polar bond. In CS_2 the dipoles cancel because the molecule is symmetrical. Therefore CS_2 is not a polar molecule. The dipoles created by the bonds in OCS, however, do not cancel.

4-111 All are polar.

4-117 All of these molecules have an unpaired electron in the Lewis structure.

4-119 (d)

4-121 Group VIIA

4-123 (a)

4-125

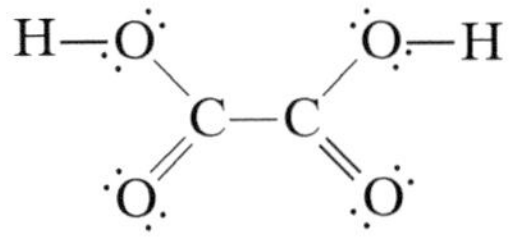

4-131 (a) Nitrogen has an incomplete octet.
     (b) Too many electrons are shown.
     (c) All valence electrons are not shown.
     (d) Hydrogen should only share two electrons.
     (e) Hydrogen should only share two electrons.

Chapter 4 Special Topics

4A-5 (a) sp^3 (b) sp^2 (c) sp^2

4A-7 In ethylene, the hybridization on each carbon is sp^2. Each C—H bond is σ (sp^2-s). One of the C—C bonds is a σ (sp^2-sp^2) and the other is a π ($2p$-$2p$). In acetylene, the hybridization on each carbon is sp. The C—H bonds are σ (sp-s) bonds. One C—C bond is σ (sp-sp) and the other two C—C bonds are π ($2p$-$2p$) bonds.

4A-9 A σ molecular orbital is one that results from the head-on overlap of atomic orbitals, that is the electron density lies along the internuclear axis. A π molecular orbital is one that results from the sideways or parallel overlap of atomic orbitals.

4A-13 (a) 1   (b) 2   (c) 3   (d) 2   (e) 1

4A-17 The peroxide ion is not paramagnetic.

4A-19 (a) diamagnetic   (b) diamagnetic   (c) diamagnetic
      (d) paramagnetic   (e) diamagnetic

Chapter 5

5-1 Decreasing metallic character: Na > Mg > Al > Si > P > S > Cl > Ar
   Increasing nonmetallic character: Na < Mg < Al < Si < P < S < Cl < Ar

5-3 (d)

5-5 (a) Ca   (b) Na   (c) K   (d) Mg

5-7 All the alkali metals have the same valence configuration of xs^1 and very small AVEE values.

5-9 Xenon

5-11 The ionization energy will decrease from Li to Fr.

5-13 The second ionization energy for Li would be much higher than the first.

5-15 Be^{2+}:   [He]
     Mg^{2+}:   [Ne]
     Ca^{2+}:   [Ar]
     Sr^{2+}:   [Kr]
     Ba^{2+}:   [Xe]
     Ra^{2+}:   [Rn]

5-17 Mg

5-19 Atomic size increases from Be to Ra.

5-21 Aluminum

5-23 The ionization energy and AVEE decrease from B to Al.

5-25 Ga: $[Ar]4s^2 3d^{10} 4p^1$
     Ga^+: $[Ar]4s^2 3d^{10}$
     Ga^{+2}: $[Ar]4s^1 3d^{10}$
     Ga^{+3}: $[Ar]3d^{10}$

5-27 (a) 3   (b) −2   (c) −2

5-29 The hydrogen of metal hydrides is a negative ion, H^-.

5-35 Zn: $[Ar]4s^2 3d^{10}$; +2

5-37 Ni: $[Ar]4s^2 3d^8$
     Ni^{2+}: $[Ar]3d^8$

5-39 −1

5-41 (a) $Mg(NO_3)_2$ (b) $Fe_2(SO_4)_3$ (c) Na_2CO_3

5-43 Na_3N; AlN

5-45 6

5-47 (b)

5-49 Group IIIA

5-51 The very high third ionization energy for magnesium

5-53 AlN

5-55 GaAs

5-57 (a) $Ca(s) + H_2(g) \longrightarrow CaH_2(s)$
     (b) $2 Ca(s) + O_2(g) \longrightarrow 2 CaO(s)$
     (c) $8 Ca(s) + S_8(s) \longrightarrow 8 CaS(s)$
     (d) $Ca(s) + F_2(g) \longrightarrow CaF_2(s)$
     (e) $3 Ca(s) + N_2(g) \longrightarrow Ca_3N_2(s)$
     (f) $6 Ca(s) + P_4(s) \longrightarrow 2 Ca_3P_2(s)$

5-63 Coulombic forces

5-71 Electrons of the bonded atoms are localized.

5-79 (e)

5-81 (f)

5-83 (a), (b), (e)

5-85 (e)

5-87 (a) covalent   (b) covalent   (c) covalent   (d) ionic

5-89 No, the compounds would be metallic.

5-91 (a) Covalent   (b) Semi-metal   (c) Ionic   (d) Ionic   (e) Covalent
     (f) Ionic/Metallic   (g) Covalent   (h) Ionic/Covalent   (i) Covalent

5-97 (a) B_2O_3, Mg_3N_2 (b) Mg_3N_2 (c) $BaSi_2$

5-99 (a) -3 (b) $+5$ (c) $+3$ (d) $+2$

5-101 $+3$

5-103 (a) 0 (b) -1 (c) $+1$ (d) $+3$ (e) $+5$ (f) $+7$

5-105 (a) $+2$ (b) $+4$ (c) $+4$ (d) $+6$ (e) $+6$ (f) $+6$ (g) $+8$ (h) $+8$

5-107 (-2) H_2S, ZnS
$(+4)$ SF_4, SO_2, SO_3^{2-}, H_2SO_3
$(+6)$ SF_6, SO_3, SO_4^{2-}, H_2SO_4

5-109 $(+2)$ TiO
$(+3)$ Ti_2O_3, Ti_2S_3, $TiCl_3$
$(+4)$ TiO_2, $TiCl_4$, K_2TiO_3, H_2TiCl_6, $Ti(SO_4)_2$

5-111 (a) $C_{CH_3} = -3$ $C_{central} = +1$ $H = +1$ $Br = -1$
(b) $C = -2$ $H = +1$ $O = -2$
(c) $C_{CH_3} = -3$ $C_{central} = +2$ $H = +1$ $O = -2$
(d) $C_{CH_3} = -3$ $C_{central} = 0$ $H = +1$ $O = -2$

5-113 (a) $C_{CH_3} = -3$ $C_{central} = +3$ $H = +1$ $O = -2$ $Cl = -1$
(b) $C = -2$ $H = +1$ $N = -3$
(c) $C_{CH_3} = -3$ $C_{C=O} = +3$ $C_{CH_2} = -1$ $O = -2$ $H = +1$
(d) $C = -3$ $H = +1$
(e) $C_{CH_2} = -2$ $C_{CH} = -1$ $C_{CH_3} = -3$ $H = +1$
(f) $C = -1$ $H = +1$

5-117 (a) Magnesium is oxidized. Hydrogen is reduced.
(b) Iodine is oxidized. Chlorine is reduced.
(c) This is not an oxidation-reduction reaction.
(d) Sodium is oxidized. Hydrogen is reduced.

5-119 Carbon is oxidized. Sulfur is reduced.

5-123 (a) P_4S_3 (b) SiO_2 (c) CS_2 (d) CCl_4 (e) PF_5

5-125 (a) $SnCl_2$ (b) $Hg(NO_3)_2$ (c) SnS_2 (d) Cr_2O_3 (e) Fe_3P_2

5-127 (a) $Co(NO_3)_3$ (b) $Fe_2(SO_4)_3$ (c) $AuCl_3$ (d) MnO_2 (e) WCl_6

5-129 (a) aluminum chloride
(b) sodium nitride
(c) calcium phosphide
(d) lithium sulfide
(e) magnesium oxide

5-131 (a) antimony(III) sulfide
(b) tin(II) chloride
(c) sulfur tetrafluoride
(d) strontium bromide
(e) silicon tetrachloride

5-133 (a) H_2CO_3 (b) HCN (c) H_3BO_3 (d) H_3PO_3 (e) HNO_2

5-135 $S_2O_3^{2-}$

5-137 (a) calcite, calcium carbonate   (b) barite, barium sulfate

5-139

Compound	ΔEN	$\overline{EN}$	Bond Triangle Classification
FeO	1.94	2.64	Covalent
Fe_2O_3	1.94	2.64	Covalent
$FeCl_2$	1.20	2.27	Covalent
$FeCl_3$	1.20	2.27	Covalent

5-151 (a) C$_3$H$_6$ C$_3$H$_3$N

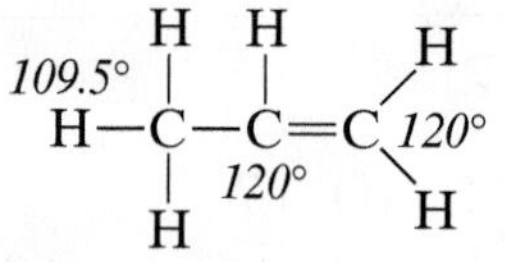

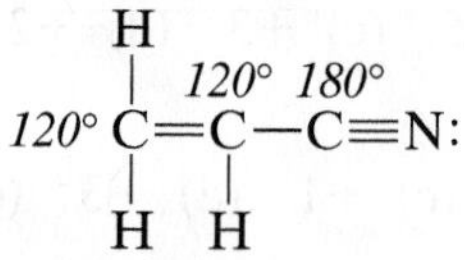

$$109.5° \begin{array}{c} H \ \ H \\ | \ \ \ | \\ H-C-C=C\ 120° \\ | \ \ \ \ 120° \\ H \end{array} \begin{array}{c} H \\ \\ 120° \\ H \end{array} \qquad \begin{array}{c} H \\ | \ \ 120° \ \ 180° \\ 120° \ C=C-C\equiv N: \\ | \ \ \ | \\ H \ \ H \end{array}$$

(b) (iv) is true.

5-147 (a) 2 PbS + 3 O$_2$ $\longrightarrow$ 2 PbO + 2 SO$_2$

(b) It is an oxidation-reduction reaction. The O$_2$ is reduced and the S is oxidized.

5-149 MgO

Chapter 6

6-3 The temperature of a system is a measure of the average kinetic energy of all the particles in the system.

6-5 32°F, 0°C or 273.15 K

6-7 Kinetic energy can be thought of as the energy of motion. The faster something is moving, the greater its kinetic energy.

6-11 Ar, CO, CH$_4$, Cl$_2$

6-15 liquid water: 18 mL, gaseous water: 22.4 L

6-17 Increase

6-19 The pressure of the wind will be the same. The billboard feels a greater force.

6-21 0.9813 atm; 9.941 $\times$ 10^4 Pa

6-23 29.9 in Hg; 1.01 $\times$ 10^5 Pa

6-25 3 $\times$ 10^{-6} atm

6-27 324 mm Hg

6-29 13 L

6-31 5.29 atm

6-33 four-fifths

6-35 0.332 L

6-37 3:1

6-39 18.0 L

6-41 Dry air

6-43 N$_2$O

6-47 (b)

6-49 $82.06 \dfrac{cc \cdot atm}{mol \cdot K}$

6-51 2.5 atm

6-53 0.452 atm

6-55 $\dfrac{\$0.00014}{g\ N_2}$

6-57 253 K

6-59 3.13 atm

6-61 3.30 g/L

6-65 Krypton

6-67 1.3 L

6-69 1.50 atm

6-71 2.3×10^2 K

6-77 (a) 7.0×10^{-21} J/molecule
   (b) NH_3
   (c) No
   (d) Increase the temperature
   (e) If the temperature is kept the same, then the average kinetic energy is the same, but the pressure would be reduced.
   (f) (1) Fix n and V, double the pressure.
       (2) Fix T and V, double the number of molecules present.
       (3) Fix n and T, cut the volume in half.

6-83 The effusion of a gas is the rate at which it will escape though a pinhole into a vacuum.

6-85 $SF_6 < CF_2Cl_2 < Cl_2 < SO_2 < Ar$

6-89 Nitric oxide must be NO and nitrous oxide must be N_2O.

6-91 16.0 g/mol

6-93 234.8 g/mol

6-95 0.542 g Mg

6-97 3.66×10^4 L CO_2

6-99 2.45 L CO_2

6-101 (a), (b), and (d)

6-103 Xenon

6-107 (c)

6-109 (a) Kinetic energy would remain unchanged, but the frequency of collisions would decrease.
   (b) Kinetic energy would decrease as well as the frequency of collisions.
   (c) Kinetic energy and frequency of collisions would both increase.
   (d) Kinetic energy would remain unchanged, but the frequency of collisions would increase.

6-111 58.1 g/mol

6-113 C_3H_6

Chapter 6 Special Topics

6A-1 Smaller

6A-3 At low temperatures or high pressures

6A-7 $\dfrac{He}{Ne} = 1.1$ and $\dfrac{Ar}{Ne} = 1.2$

6A-9 For CO_2 the $\dfrac{an^2}{V^2}$ term is important at this pressure. However for H_2 the nb term is more important.

Chapter 7

7-1 Kinetic energy is energy of motion. Potential energy is stored energy.

7-3 Energy can be transferred from one object to another by mechanical (contact) means.

7-9 The iron atoms in the warm bar would slow down, and the atoms in the cooler bar would speed up until they had the same average kinetic energy. When the average kinetic energies are the same, they have the same temperature.

7-11 As a balloon filled with helium becomes warmer, the particles within become more agitated. This induces a greater pressure on the walls, causing the balloon to expand.

7-15 less than 35°C

7-17 temperature, pressure, volume, enthalpy, and energy of an ideal gas

7-19 (a), (b), (c), (d)

7-27 All chemical reactions involve the breaking and reforming of some kind of bonds between the atoms, ions or molecules.

7-31 (a) 131.29 kJ   (b) 262.58 kJ   (c) 3.93 kJ   (d) 6.56 kJ

7-33 -1.12×10^3 kJ/mol Mg

7-35 Processes (1) and (4) will have the same reaction enthalpy.

7-37 The initial state of the system (the reactants) and the final state of the system (the products)

7-39 25°C and 1 bar of pressure

7-43 In an exothermic process the bond strengths of the products are greater than for the reactants. In an endothermic the reactants have stronger bonds than the products.

7-47 (a) Endothermic   (b) Exothermic   (c) Exothermic

7-49 Yes, to form one mole of $CH_4(g)$ from its atoms, 1662.09 kJ of heat is released. To form two moles, 3324.18 kJ is released.

7-51 (a) Endothermic   (b) Exothermic

7-53 Endothermic

7-55 -1284.4 kJ/mol$_{rxn}$

7-57 -5313.97 kJ/mol$_{rxn}$

7-59 -410.7 kJ/mol$_{rxn}$

7-61 -851.5 kJ/mol$_{rxn}$

7-63 -3310.1 kJ/mol$_{rxn}$

7-65 P_4

7-67 Exothermic

7-71 (a) is the longest and also the weakest bond.
   (c) is the shortest bond with the largest bond strength.

7-75 -189.32 kJ/mol$_{rxn}$

7-77 $+22.8$ kJ/mol$_{rxn}$

7-79 -53.9 kJ/mol$_{rxn}$

7-81 $+178.32$ kJ/mol$_{rxn}$

7-83 -1284.4 kJ/mol$_{rxn}$

7-85 -5313.9 kJ/mol$_{rxn}$

7-87 -410.6 kJ/mol$_{rxn}$

7-89 The reaction with iron (III) oxide evolves more heat per mol aluminum consumed.

7-91 The C—Cl bond length is greater than the C—H bond length and the C—Cl bond enthalpy is predicted to be smaller.

7-93 negative

7-99 (a) $CH_3OCH_3(g) + 3\ O_2(g) \longrightarrow 2\ CO_2(g) + 3\ H_2O(g)$
       $CH_3CH_2OH(g) + 3\ O_2(g) \longrightarrow 2\ CO_2(g) + 3\ H_2O(g)$
   (b) -1329.6 kJ/mol$_{rxn}$; -1277.4 kJ/mol$_{rxn}$
   (c) Dimethyl ether
   (d) Ethyl alcohol

7-101 (b)

7-103 Exothermic

7-109 (a) Mg is metallic, Cl_2 is covalent and $MgCl_2$ is ionic.
   (b) -641.3 kJ/mol
   (c) The product molecule has an ionic bond whereas the reactant molecules have covalent and metallic bonding which are much weaker, therefore it is going to be very exothermic.

Chapter 8

8-5 The level of intermolecular forces that exists between molecules of a substance will determine what phase the substance is found in at room temperature.

8-9 A hydrogen bond will only exist when there is a hydrogen atom that is covalently bound to an electronegative element such as N, O or F.

8-11 Dipole-dipole forces; dispersion forces; and possibly dipole-induced dipole forces and/or hydrogen bonding

8-13 (e)

8-15 The dispersion forces are less in propane.

8-17 The dispersion interactions are weaker in isopentane.

8-19 They will have the same average velocity.

8-23 Vapor pressure is a function of temperature.

8-25 Not all gas particles are moving at exactly the same speed.

8-27 The heat energy required to cause the evaporation of the water in the cloth comes from your forehead.

8-29 The rate of molecules evaporating is the same as the rate of molecules condensing.

8-35 Car wax has a nonpolar surface to it, while water is a very polar molecule. Water beads up because it is more attracted to other water molecules than the surface of the car.

8-37 (a) $CH_3CH_2CH_3$ (b) CH_3CH_3 (c) CH_3CH_2OH
(d) $CH_3CH_2CH_2CH_2CH_3$

8-39 No

8-43 The atmospheric pressure is less in Denver, and water boils at a lower temperature.

8-45 38.1°C

8-47 (c)

8-51 High pressure and low temperature

8-53 There is a phase change at each intersection.

8-55 Water has hydrogen bonding.

8-59 It takes a large amount of energy to disrupt the hydrogen bonds in water.

8-61 Anesthetics and olive oil are relatively nonpolar.

8-63 (e)

8-65 The CCl_4 will favor the reactant. A polar solvent will favor the products.

8-77 (c)

8-79 (e)

8-81 (a) $Mg(s) + 2\ HCl(aq) \longrightarrow MgCl_2(aq) + H_2(g)$
$Mg(s) + 2\ H^+(aq) \longrightarrow Mg^{2-}(aq) + H_2(g)$
(b) $Na_2CO_3(aq) + Ca(NO_3)_2(aq) \longrightarrow 2\ NaNO_3(aq) + CaCO_3(s)$
$CO_3^{2-}(aq) + Ca^{+2}(aq) \longrightarrow CaCO_3(s)$

8-85 The compound on the right (boiling point 35°C) can form a hydrogen bond, the one on the left cannot.

8-91 (b)

8-93 0.433 grams remain and the final pressure will be 325 mm Hg.

8-97 Vapor pressure should increase and surface tension should decrease.

8-99 The intermolecular forces of liquid water are greater than those in liquid methanol.

8-101 (a) *n*-hexane: dispersion; 2-hexanone: dispersion, dipole–induced dipole, and dipole–dipole; 2-hexanol: dispersion, dipole–induced dipole, dipole–dipole and hydrogen bonding
(b) 2-hexanol
(c) 2-hexanol
(d) 2-hexanol

8-103 (a) Boiling point (C) < (A) < (B)
(b) Vapor pressure (B) < (A) < (C)
(c) (B)

8-105 No

Chapter 8 Special Topics

8A-1 Colligative properties are properties of solutions that will vary with the concentration of the solute particles in the solution.

8A-3 (a)

8A-7 262 mm Hg

8A-11 53.8°C

8A-13 78.1°C

8A-17 −0.191°C

8A-19 (c)

8A-21 The salt lowers the melting point of ice, and highways can remain free of ice at temperatures lower than the freezing point of pure water.

8A-23 −0.188°C

8A-25 −1.8°C

Chapter 9

9-1 Crystalline, amorphous, and polycrystalline

9-3 Covalent compounds; metallic compounds; ionic compounds

9-5 (d)

9-9 Diamond, graphite and fullerenes

9-13 Localized electrons

9-17 (c)

9-21 A coordination number is the number of nearest neighbor atoms that "bond" to an atom in a particular crystal structure.

9-25 (b)

9-27 A unit cell is the simplest repeating unit in a crystalline structure.

9-29 An intermetallic compound exists with a defined stoichiometry. An alloy represents a solid solution where the components randomly substitute one for another.

9-33 (a), (d), (e)

9-37 (d)

9-39 (d)

9-41 The lower solubility of NaF reflects its higher lattice energy.

9-45 Ceramics are formed by the fusion of silicon oxides with nonmetal atoms.

9-47 (a) Ionic, but on the border with covalent
     (b) Covalent
     (c) Semiconductor
     (d) Ionic, but on the border with covalent

9-49 The type of the unit cell

9-51 10.50 g/cm^3

9-53 Face-centered cubic

9-55 (d)

9-57 N = 4

9-59 $V_{\text{unit cell}} = 2.40 \times 10^{-23}$ cm^3; $r_{\text{Cr}} = 1.25 \times 10^{-8}$ cm

9-61 4.352×10^{-8} cm

9-63 0.3635 nm; 0.148 nm

9-65 The force that holds the atoms together, the covalent bond, is much stronger than the dispersion forces that hold the molecules together.

9-67
1. (e)
2. (f) or (a)
3. (d) or (b)
4. (f) or (a)
5. (c) or (b)
6. (c)
7. (b)
8. (e)
9. (f), (a), or (c)
10. (d) or (c)
11. (f)

9-69 (a) BaO and MgO (b) MgO

9-71 SiO_2

Chapter 9 Special Topics

9A-3 If a dislocation allows one plane of atoms to slip past another plane more easily, then this would weaken the metal.

9A-11 Metals are good conductors of heat.

9A-13 Concrete

9A-15 Iron

9A-19 (b), (j), and (l) might be ceramics.

9A-21 Engines made from ceramic powders operate at very high temperatures and are therefore more efficient than steel engines.

Chapter 10

10-1 Reactions that go to completion leave no unreacted limiting reagent behind. Reactions that go to equilibrium may leave an appreciable amount of unreacted starting materials.

10-5 If K_c is greater than 1 then the concentration of B must be greater than that of A. Conversely, if K_c is less than 1 then the concentration of A must be greater than that of B.

10-7 The rate of a chemical reaction is equivalent to the measured change in amount of a reactant or product divided by the measured time interval over which the change in amount occurred.

10-9 The rate constant is the proportionality constant between the rate of the reaction and the concentrations of reactants in the rate law. The rate constant is only one part of the entire rate law.

10-13 Equilibrium is when the forward reaction appears to stop and the concentrations of the products and the reactants do not change. A more correct definition of equilibrium is when the rate of the forward reaction is equal to the rate of the reverse reaction.

10-15 (d)

10-17 (a) $K_c = \dfrac{[OF_2]^2}{[O_2][F_2]^2}$ (b) $K_c = \dfrac{[SO_3]^2}{[O_2][SO_2]^2}$ (c) $K_c = \dfrac{[SO_2Cl_2]^2[O_2]}{[SO_3]^2[Cl_2]^2}$

10-19 (a) $K_c = \dfrac{[NO]^2[O_2]}{[NO_2]^2}$ (b) $K_c = \dfrac{[NO_2]^2}{[NO]^2[O_2]}$

The value of K_c for reaction (a) is 1.6×10^{-6}.

10-21 7.8×10^6

10-23 trial I: 59.686
  trial II: 59.45
  trial III: 59.21
  K_c is a constant.

10-25 (c)

10-27 $K_c = 1.5 \times 10^4$; $Q_c = 4.8 \times 10^4$; The system must shift to the left to form reactant.

10-31 $\Delta(N_2) = 0.117$
  $\Delta(H_2) = 0.351$

10-33 (e)

10-35 $[H_2]_{eq} = 0.766\ M$; $[NH_3]_{eq} = 0.156\ M$

10-37 $[N_2O_4]_{eq} = 0.099\ M$; $[NO_2]_{eq} = 2.4 \times 10^{-3}\ M$

10-39 $[NH_3] = 0.01\ M$; $[N_2] = 0.15\ M$; $[H_2] = 0.24\ M$

10-41 $[N_2] = 0.10\ M$; $[O_2] = 0.090\ M$; $[NO] = 1.7 \times 10^{-6}$

10-43 $0.20\ M$ of NOCl can be formed leaving $0.30\ M$ NO.

10-45 $[PCl_5]_{eq} = 0.99\ M$
  $[PCl_3]_{eq} = 6.5 \times 10^{-3}\ M$
  $[Cl_2]_{eq} = 0.21\ M$
  % decomposition: 0.65%

10-47 $[NO_2]_{eq} = 0.10\ M$; $[NO]_{eq} = 2.6 \times 10^{-4}\ M$; $[O_2]_{eq} = 0.050\ M$

10-49 $0.40\ M$

10-55 As the reaction system approaches more closely equilibrium state, the concentrations of all components will tend to change less and less as Q_c gets closer to K_c.

10-59 Equilibrium constants will change with temperature depending upon the nature of the reaction involved.

10-61 Products will be more favored.

10-63 (a) Shift to the right   (b) Shift to the right   (c) Shift to the right

10-65 (a) Shift to the right   (b) Shift to the left   (c) Shift to the right

10-67 To increase the yield of ammonia in the Haber process, increase the pressure, increase the amount of N_2 and/or H_2 in the reaction vessel, remove NH_3 as soon as it forms.

10-69 Shift to the right

10-71 $Ag_2CrO_4(s) \rightleftharpoons 2\ Ag^+(aq) + CrO_4^{2-}(aq)$
  $[Ag^+] = 2\ [CrO_4^{2-}]$ $K_{sp} = [Ag^+]^2[CrO_4^{2-}]$

10-75 (c)

10-77 $Ag_2CO_3(s) \rightleftharpoons 2\ Ag^+(aq) + CO_3^{2-}(aq)$
  $[Ag^+] = 2\ [CO_3^{2-}]$ $K_{sp} = [Ag^+]^2[CO_3^{2-}]$

10-79 $SrF_2(s) \rightleftharpoons Sr^{2+}(aq) + 2\ F^-(aq)$; 2.47×10^{-9}

10-81 (a) $Mg(OH)_2(s) \rightleftharpoons Mg^{2+}(aq) + 2\ OH^-(aq)$
  (b) $K_{sp} = [Mg^{2+}]\ [OH^-]^2$
  (c) 1.7×10^{-4} mol/L
  (d) 9.6×10^{-4} grams

  10-83 (a) 1.4×10^{-15} grams
  (b) 2.4×10^{-17} grams

10-85 1.9×10^{-9} grams

10-87 (e)

10-89 Ag_2S

10-91 Hg_2S

10-93 6.5×10^{-5} mol/L

10-95 5.145×10^{-3}

10-97 1×10^{-10}

10-99 (a) 1.4×10^{-22} grams
(b) 1×10^{-25} grams

10-101 $HgS < Bi_2S_3 < CuS < Ag_2S$

10-103 A precipitate will not form.

10-105 A precipitate will form.

10-107 The silver ion concentration will decrease while the chloride ion concentration will exceed the prior equilibrium concentrations because extra chloride ions have been added.

10-109 1.5×10^{-15} mol/L

10-111 (a)

10-113 (c)

10-115 (a) Graph (3)   (b) Graph (2)   (c) Graph (4)

10-117 Graph (b) is the correct representation.

10-119 (a) $K_{sp} = 2.4 \times 10^{-5}$
(b) $CaSO_4$ will precipitate.
(c) $[Ca^{2+}] = [SO_4^{2-}] = 4.9 \times 10^{-3}$

10-121 Graphs (b) and (e) are possible.

10-123 (a) The rate of the forward reaction equals the rate of the reverse reaction.
(b) 6.44×10^5
(c) The reaction will proceed to the right.

10-125 (e)

Chapter 11

11-1 An acid is a proton donor. A base is a proton acceptor.

11-3 Litmus will turn red in the presence of an acid. Litmus will remain blue in the presence of a base.

11-5 (a) and (b)

11-7 $HBr(g) + H_2O(l) \longrightarrow H_3O^+(aq) + Br^-(aq)$
$H_2O(l) + NH_3(g) \longrightarrow OH^-(aq) + NH_4^+(aq)$

11-15 (a) $\underset{acid}{HCl(aq)} \quad + \underset{base}{H_2O(l)} \quad \longrightarrow \quad \underset{acid}{H_3O^+(aq)} + \underset{base}{Cl^-(aq)}$

(b) $\underset{base}{HCO_3^-(aq)} \quad \underset{acid}{H_2O(l)} \quad \longrightarrow \quad \underset{base}{OH^-(aq)} + \underset{acid}{H_2CO_3(aq)}$

(c) $\underset{base}{NH_3(aq)} \quad + \underset{acid}{H_2O(l)} \quad \longrightarrow \quad \underset{base}{OH^-(aq)} + \underset{acid}{NH_4^+(aq)}$

(d) $\underset{base}{CaCO_3(s)} \quad + \underset{acid}{2\,HCl(aq)} \quad \longrightarrow \quad Ca^{2+}(aq) \quad + \underset{base}{2Cl^-(aq)} + \underset{acid}{H_2CO_3(aq)}$

11-19 (a) H_2O (b) OH^- (c) O^{2-} (d) NH_3

11-21 (a) OH^- (b) H_2O (c) H_3O^+ (d) NH_3

11-23 True

11-29 Addition of a strong acid or base to water increases the concentration of either the H_3O^+ ion or the OH^- ion, respectively.

11-33 0.010 M

11-35 $[H_3O^+] = 1.9 \times 10^{-4} M$; $[OH^-] = 5.3 \times 10^{-11} M$

11-37 The hydronium ion concentration increases. The hydroxide ion concentration decreases. The pH of the solution also decreases.

11-39 (a)

11-41 pH = 5.8; pOH = 8.2

11-43 $[H_3O^+] = 1.6 \times 10^{-4} M$; $[OH^-] = 6.3 \times 10^{-11} M$

11-45 $1.3 \times 10^{-3} M$

11-49 (a) weak (b) weak (c) strong (d) weak (e) strong

11-51 (d)

11-53 $HOAc < HNO_2 < HF < HOClO$

11-55 I^-, ClO_4^-, NO_3^-

11-59 Hydrogen bromide is a stronger acid than water.

11-61 OH^- is a stronger base than HCO_2^-.

11-65 The acid is assumed to be completely dissociated.

11-67 $AsO_4^{-3} < HAsO_4^{-2} < H_2AsO_4^- < H_3AsO_4$

11-69 (d)

11-71 $HOI < HOBr < HOCl$

11-73 $CH_3COOH < CH_2ClCOOH < CCl_3COOH$

11-75 pH = 1.2; pOH = 12.8

11-77 (a) 0 (b) 1.65 (c) 1.44 (d) 1.55

11-79 1.00

11-81 In solutions of very weak acids or bases, the acid or base does not completely dissociate and therefore more fully associated species are present in addition to dissociated species.

11-83 (a)

11-85 $[H_3O^+] = [HCO_2^-] = 4.2 \times 10^{-3} M$; $[HCO_2H] = 9.6 \times 10^{-2} M$

11-87 $1.3 \times 10^{-6} M$

11-89 8.1×10^{-5}

11-93 pH = 4.27; $K_a = 2.91 \times 10^{-8}$

11-95 (c)

11-97 $[OH^-] = [HCO_2H] = 2.1 \times 10^{-6} M$; $[HCO_2^-] = 0.080 M$

11-99 9.3

11-101 5.0×10^{-4}

11-103 11.8

11-105 8.6

11-107 (a)

11-109 (a) $H_2PO_4^-$ (b) H_2CO_3 (c) HSO_4^- (d) HNO_2

11-111 1.72; 3.1

11-113 9.3

11-117 (e)

11-119 (d)

11-121 0.143; 0.00023; 0.014

11-123 10.3

11-125 (a), (c), and (d)

11-127 7.0

11-129 (a) $HOBr(aq) + OH^-(aq) \rightleftharpoons H_2O(l) + OBr^-(aq)$
(b) $HOAc(aq) + NH_3(aq) \rightleftharpoons OAc^-(aq) + NH_4(aq)$
(c) $H_3O^+(aq) + OH^-(aq) \rightleftharpoons 2\,H_2O(l)$
(d) $2\,HCOOH(aq) + Ba(OH)_2(s) \rightleftharpoons 2\,HCOO^-(aq) + 2\,H_2O(l) + Ba^+(aq)$

Reactions (a), (c) and (d) will go to completion.

11-135 (a) methyl orange: yellow
(b) cresol red: red
(c) phenolphthalein: pink
(d) alizarin yellow: yellow

11-137 The rain in Scotland had $[H_3O^+] = 0.004\,M$.
The acetic acid solution had $[H_3O^+] = 0.001\,M$.

11-139 (a) False (b) True (c) True (d) False

11-143 $HBr < HOBrO_2 < HOBrO < NaBr < NaBrO_3 < NaBrO_2$

11-145 (a) Figure (iii) (b) Figure (i) (c) Figure (iv)

11-147 Conjugate base: (c); Conjugate acid: (a)

11-149 (a) $Al_2O_3(s) + 6\,HCl(aq) \longrightarrow 2\,AlCl_3(aq) + 3\,H_2O(l)$
(b) $CaO(s) + H_2SO_4(aq) \longrightarrow CaSO_4(aq) + H_2O(l)$
(c) $Na_2O(s) + H_2O(l) \longrightarrow 2\,Na^+(aq) + 2\,HO^-(aq)$
(d) $MgCO_3(s) + 2\,HCl(aq) \longrightarrow H_2CO_3(aq) + MgCl_2(aq)$
(e) $3\,NaOH(s) + H_3PO_4(aq) \longrightarrow 3\,H_2O(aq) + Na_3PO_4(aq)$

11-151 A. (a) $HBr(aq) + H_2O(l) \longrightarrow H_3O^+(aq) + Br^-(aq)$
(b) $1.0\,M$
(c) $pH = 0$
(d) 1.0×10^{-14}
(e) $Br^-(aq)$
B. (a) $NH_3(aq) + H_2O(l) \longrightarrow NH_4^+(aq) + OH^-(aq)$
(b) 4.2×10^{-3}
(c) $1.0\,M$
(d) $4.2 \times 10^{-3}\,M$
(e) 11.6
C. (a) $1.0\,M$
(b) $Br^-(aq) + H_2O(l) \longrightarrow HBr(aq) + OH^-(aq)$
(c) 7.0

11-153 (a) $HNO_2(aq) + H_2O(l) \longrightarrow NO_2^-(aq) + H_3O^+(aq)$
(b) 1.65
(c) 8.65
(d) 0
(e) Diagram (a)

11-155 (a) 1.0 (b) 7.0 (c) $1 < pH < 7$ (d) $7 < pH < 14$ (e) 13

11-157 $KOH < KClO_3 < KI < NH_4ClO_4 < HClO_3 < HClO_4$

11-159 2.87; 4.57; 8.72; 12.0

11-161 (b)

Chapter 11 Special Topics

11A-1 Assume $[H_2CO_3]$ at equilibrium $= 0.100\ M$.
$[HCO_3^-] = [H_3O^+] = 2.1 \times 10^{-4}\ M$; $[CO_3^{2-}] = 4.7 \times 10^{-11}\ M$

11A-5 (b) and (d)

11A-7 $[H_3O^+] = [HC_2O_4^-] = 0.24\ M$; $[H_2C_2O_4] = 1.01\ M$;
$[C_2O_4^{2-}] = 5.4 \times 10^{-5}\ M$

11A-9 8.4

11A-13 1.6; 4.1; 10.1; 12.7

11A-15 Acidic

Chapter 12

12-3 (a) reduction   (b) oxidation   (c) reduction   (d) oxidation

12-5 $+3$ in all the compounds.

12-7 $CH_4\ (-4) < H_3COH\ (-2) < C$ and $H_2CO\ (0) < CO\ (+2) < CO_2\ (+4)$

12-9 (a) -2 (b) -2 (c) -2 (d) $+6$ (e) $+6$ (f) $+4$ (g) $+6$ (h) $+2$

(a) Na_2S	-2		(e) SF_6	$+6$
(b) MgS	-2		(f) $BaSO_3$	$+4$
(c) CS_2	-2		(g) $NaHSO_4$	$+6$
(d) H_2SO_4	$+6$		(h) SCl_2	$+2$

12-11 As oxidation number increases, the strength of the acid increases.

12-13 (a)

12-15 This is not a redox reaction.

12-25 (a) $Al(s) \rightleftharpoons Al^{3+}(aq) + 3\ e^-$
$3\ e^- + Cr^{3+}(aq) \rightleftharpoons Cr(s)$

(b) $2\ Fe^{2+}(aq) \rightleftharpoons 2\ Fe^{3+}(aq) + 2\ e^-$
$2\ e^- + I_2(aq) \rightleftharpoons 2I^-(aq)$

(c) $Cr(s) \rightleftharpoons Cr^{3+}(aq) + 3\ e^-$
$3\ e^- + 3\ Fe^{3+}(aq) \rightleftharpoons 3\ Fe^{2+}(aq)$

(d) $Zn(s) \rightleftharpoons Zn^{2+}(aq) + 2\ e^-$
$2\ e^- + 2\ H^+(aq) \rightleftharpoons H_2(g)$

12-29 The sign is reversed. The magnitude remains the same.

12-31 (a) The Al is the reducing agent while the Cr^{3+} is the oxidizing agent.
(b) The Cr^{2+} is both the oxidizing and reducing agent.
(c) The Fe is the reducing agent and Cr^{3+} is the oxidizing agent.
(d) The H_2 is the reducing agent and Cr^{3+} is the oxidizing agent.

12-33 Yes

12-35 (a) and (d)

12-37 (a), (b), (d) and (e)

12-39 It cannot take in any more electrons.

12-41 (c)

12-43 (a) $Cr^{3+}(aq)$, $Fe^{3+}(aq)$ (or $Fe^{2+}(aq)$)
     (b) $Mg^{2+}(aq)$, $Cr^{2+}(aq)$
     (c) These ions will not react.

12-45 (d)

12-49 (e)

12-51 (a) K   (b) Sn   (c) Au^+

12-53 $Cl_2 > Br_2 > H_2O_2 > Fe^{3+} > I_2$

12-55 (b)

12-57 -1.67 V; The reaction should not occur spontaneously as written.

12-59 0.364 V; The reaction should occur spontaneously as written.

12-71 $E_{cell} = -1.67 \text{ V} - \dfrac{0.02569}{3} \ln \dfrac{(Fe^{3+})}{(Al^{3+})} = -1.59 \text{ V}$; This reaction will not proceed.

12-73 (a) 1.03 V   (b) 1.00 V   (c) 0.91 V   (d) 0.86 V   (e) 0.0 V

12-75 An increase in temperature would lead to a decrease in E_{cell}.

12-77 2×10^{37}

12-81 96485 s

12-83 0.094 g H_2 and 7.4 g Br_2

12-85 13 g

12-87 1.18 g I_2

12-89 (e)

12-91 $x = 1$, $y = 4$

12-93 $+3$

12-95 Anode: $2 \, F^-(aq) \longrightarrow F_2(g) + 2 \, e^-$
      Cathode: $Ca^{2+}(aq) + 2 \, e^- \longrightarrow Ca(s)$

12-101 The reaction generates hydroxide ions.

12-105 (a) Zinc is the anode and chromium is the cathode.
      (b) Electrons flow from the Zn electrode to the cathode.
      (c) The cathode is positive and the anode is negative.
      (d) anode: $Zn(s) \longrightarrow Zn^{2+}(aq) + 2 \, e^-$
          cathode: $Cr^{3+}(aq) + 3 \, e^- \longrightarrow Cr(s)$
      (e) $3 \, Zn(s) + 2 \, Cr^{3+}(aq) \longrightarrow 3 \, Zn^{2+}(aq) + 2 \, Cr(s)$
      (f) Electrons would flow to the hydrogen electrode from the chromium electrode. The reaction would be $6 \, H^+(aq) + 2 \, Cr(s) \longrightarrow 3 \, H_2(g) + 2 \, Cr^{3+}(aq)$.

12-111 (a) Ag
      (b) 1.56 V
      (c) The Zn cell is the anode and the Ag cell is the cathode.
      (d) Electrons flow from Zn to Ag.
      (e) $2 \, Ag^+(aq) + Zn(s) \longrightarrow 2 \, Ag(s) + Zn^{2+}(aq)$
      (f) 5.8×10^{52}

12-113 0.150 M

12-115 6.1 g $FeCl_2$

Chapter 12 Special Topics

12A-1 This is an oxidation-reduction reaction.

12A-3 (a) $5\ Fe^{2+}(aq) + MnO_4^-(aq) + 8\ H^+(aq) \longrightarrow 5\ Fe^{3+}(aq) + Mn^{+2}(aq)$
$+ 4\ H_2O(l)$

(b) $5\ S_2O_3^{2-}(aq) + 8\ MnO_4^-(aq) + 14\ H^+(aq) \longrightarrow 10\ SO_4^{2-}(aq)$
$+ 8\ Mn^{+2}(aq) + 7\ H_2O(l)$

(c) $5\ PbO_2(s) + 4\ H^+(aq) + 2\ Mn^{2+}(aq) \longrightarrow 5\ Pb^{2+}(aq) + 2\ H_2O(l)$
$+ 2\ MnO_4^-(aq)$

(d) $2\ MnO_4^-(aq) + 6\ H^+(aq) + 5\ SO_2(g) + 2\ H_2O(l) \longrightarrow 2\ Mn^{2+}(aq)$
$+ 5\ H_2SO_4(aq)$

12A-5 (a) $2\ H_2O_2(aq) \longrightarrow O_2(g) + 2\ H_2O(l)$

(b) $3\ NO_2(g) + H_2O(l) \longrightarrow 2\ HNO_3\ (aq) + NO(g)$

12A-7 (a) $6\ HCl(aq) + 2\ HNO_3(aq) \longrightarrow 2\ NO(g) + 3\ Cl_2(g) + 4\ H_2O(l)$

(b) $2\ HBr(aq) + H_2SO_4(aq) \longrightarrow SO_2(g) + Br_2(aq) + 2\ H_2O(l)$

(c) $10\ HCl(aq) + 6\ H^+(aq) + 2\ MnO_4^-(aq) \longrightarrow 5\ Cl_2(g) + 2\ Mn^{2+}(aq)$
$+ 8\ H_2O(l)$

12A-9 $8\ H_2S(g) + 8\ H_2O_2(aq) \longrightarrow S_8(s) + 16\ H_2O(l)$

12A-11 (a) $5\ CH_3OH(aq) + 18\ H^+(aq) + 6\ MnO_4^-(aq) \xrightarrow{H^+} 5\ CO_2(g)$
$+ 6\ Mn^{2+}(aq) + 19\ H_2O(l)$

(b) $3\ CH_3OH(aq) + 4\ MnO_4^-(aq) \longrightarrow 3\ HCO_2H(aq) + H_2O(l)$
$+ 4\ MnO_2(s) + 4\ OH^-(aq)$

12A-13 (a) $2\ SO_3^{2-}(aq) + O_2(g) \xrightarrow{H^+} 2\ SO_4^{2-}(aq)$

(b) $2\ SO_3^{2-}(aq) + O_2(g) \xrightarrow{OH^-} 2\ SO_4^{2-}(aq)$

12A-15 $2\ NH_3(aq) + ClO^-(aq) \longrightarrow N_2H_4(aq) + H_2O(l) + Cl^-(aq)$

12A-17 $CuO(s) + 2\ NH_3(g) \longrightarrow 3\ Cu(g) + 3\ H_2O(l) + N_2(g)$

12A-19 (a) CH_3COH is the oxidized species.

(b) CH_3COCH_3 is the oxidized species.

(c) Both have the same oxidation state.

(d) CO is the oxidized species.

Chapter 13

13-3 Yes, the freezing of liquid water is spontaneous below 0°C but not spontaneous above 0°C.

13-5 Gaseous water has more entropy.

13-9 (d)

13-11 (c)

13-15 All solutions have a concentration of 1.0 M.
All gases have a partial pressure of 1.0 bar (0.9869 atm).

13-17 Zero; Entropy increases

13-19 Gas

13-21 Products; The opposite is true when $\Delta S°$ is negative.

13-23 -1538 J/mol$_{rxn} \cdot$ K; Entropy is most likely not a driving force in the reaction.

13-25 -137.56 J/mol$_{rxn} \cdot$ K; The products of the reaction are more ordered than the reactants.

13-29 (b)

13-31 $\Delta S° < 0$ and $\Delta H° < 0$

13-33 $\Delta H° = -851.5$ kJ/mol$_{rxn}$
$\Delta S° = -38.58$ J/mol$_{rxn} \cdot$ K
The enthalpy change is the driving force behind the reaction.

13-35 $\Delta H°_{rxn} = -602.8$ kJ/mol$_{rxn}$
$\Delta S°_{rxn} = 579.7$ J/mol$_{rxn} \cdot$ K

13-37 $\Delta H° = -220.8$ kJ/mol$_{rxn}$
$\Delta S° = 68$ J/mol$_{rxn} \cdot$ K
All three are driving forces in this reaction.

13-39 $\Delta H° = 24.9$ kJ/mol$_{rxn}$
$\Delta S° = 307.6$ J/mol$_{rxn} \cdot$ K

13-43 $\Delta H° = -34.3$ kJ/mol$_{rxn}$
$\Delta S° = 75.58$ J/mol$_{rxn} \cdot$ K

13-45 (a) $\Delta G°$ will become more positive.
(b) $\Delta G°$ will become more negative.
(c) $\Delta G°$ will become more negative.
(d) $\Delta G°$ will become more positive.

13-47 $\Delta G°$ will increase linearly with increasing temperature.

13-49 $-10°C$ $\Delta G > 0$
$0°C$ $\Delta G = 0$
$+10°C$ $\Delta G < 0$

13-51 The entropy change is positive. The enthalpy change is also positive.

13-55 (a) $\Delta G° = -29.11$ kJ/mol$_{rxn}$
(b) $\Delta G° = 326.4$ kJ/mol$_{rxn}$
(c) $\Delta G° = 130.3$ kJ/mol$_{rxn}$

13-59 (c)

13-61 $P_{COCl_2} = 0.11$ atm; $P_{CO} = P_{Cl_2} = 1.8 \times 10^{-2}$ atm

13-63 $P_{SO_3} = 0.471$ atm
$P_{SO_2} = 1.9 \times 10^{-2}$ atm
$P_{O_2} = 9.7 \times 10^{-3}$ atm

13-67 $[N_2] = 0.40$ M; $[O_2] = 0.60$ M; $[NO] = 3.2 \times 10^{-5}$ M

13-71 The larger the magnitude of $\Delta G°$ the farther the standard state conditions are from equilibrium.

13-73 (a)

13-75 9.3×10^{-5}

13-77 $\Delta H° = -0.25$ kJ/mol$_{rxn}$
$\Delta S° = -92.0$ J/mol$_{rxn} \cdot$ K
$\Delta G° = 27.14$ kJ/mol$_{rxn}$
$K = 1.8 \times 10^{-5}$

13-79 $\Delta G° = -1.5 \times 10^2$ J/mol; $K = 1.9 \times 10^{26}$

13-81 -0.55V

13-83 $\Delta H° = 41.16$ kJ/mol$_{rxn}$
$\Delta S° = 41.51$ J/mol$_{rxn} \cdot$ K
$\Delta G° = 28.62$ kJ/mol$_{rxn}$
$K = 9.6 \times 10^{-6}$

13-85 516 K

13-87

T(K)	K
298	1×10^{-25}
473	9×10^{-13}
673	3×10^{-6}
873	1×10^{-2}

13-89 (a) $\Delta S° = -90.01$ J/mol$_{rxn} \cdot$ K
$\Delta H° = -483.636$ kJ/mol$_{rxn}$
$\Delta G° = -457.144$ kJ/mol$_{rxn}$
  (b) $\Delta S° = -181.226$ J/mol$_{rxn} \cdot$ K
$\Delta H° = -822.306$ kJ/mol$_{rxn}$
$\Delta G° = -768.476$ kJ/mol$_{rxn}$
  (c) $\Delta S° = -198.762$ J/mol$_{rxn} \cdot$ K
$\Delta H° = -92.22$ kJ/mol$_{rxn}$
$\Delta G° = -32.90$ kJ/mol$_{rxn}$

13-91 $NH_3(g) \longrightarrow NH_3(l)$
$\Delta H° = -34.18$ kJ/mol$_{rxn}$
$\Delta S° = -81.15$ J/mol$_{rxn} \cdot$ K

13-95 (a) -114.74 kJ/mol$_{rxn}$
  (b) -146.54 J/mol$_{rxn} \cdot$ K
  (c) -71.1 kJ/mol$_{rxn}$
  (d) products
  (e) 2.86×10^{12}
  (f) to the right
  (g) increase

13-97 $\Delta G°$ will be more positive.

Chapter 14

14-3 The rate is expressed in the form of a change in one of the reactants or products as a function of change in time.

14-7 $\text{rate} = -\dfrac{\Delta x}{\Delta t}$

14-11 time^{-1}

14-13 M^{-2} time^{-1}

14-15 (e)

14-17 $0.0296\ M\ \text{s}^{-1}$

14-19 $-\dfrac{d(\text{NH}_3)}{dt} = 0.800\dfrac{d(\text{O}_2)}{dt}$

$-\dfrac{d(\text{O}_2)}{dt} = 0.833\dfrac{d(\text{H}_2\text{O})}{dt}$

14-21 $0.0114\ M\ \text{s}^{-1}$

14-27 the slowest step in the reaction mechanism

14-31 Rate $= k(\text{NO})^2(\text{H}_2)$

14-33 (a) and (c)

14-37 $k_f/k_r = [\text{products}]/[\text{reactants}] = K_c$

14-39 The rate constant will decrease.

14-41 None

14-43 $k = 1.2 \times 10^2\ M^{-2}\text{s}^{-1}$; Rate $= 15\ M\ \text{s}^{-1}$

14-45 $k = 3.55 \times 10^{-7}\ \text{mmHg}^{-2}\text{s}^{-1}$; Rate $= 5.55\ \text{mmHg s}^{-1}$

14-47 $k = 6.5 \times 10^{-5}\ M^{-1}\text{s}^{-1}$; Rate $= 3.3 \times 10^{-7}\ M\ \text{s}^{-1}$

14-51 (a) $4.0 \times 10^2\ M^{-1}\text{s}^{-1}$ (b) 0.012 sec

14-55 Decrease the half-life by a factor of 2

14-57 1.19×10^{10} yr

14-59 1.90×10^3 yr

14-61 2.4×10^4 yr

14-63 The beeswax did not belong to the Bronze Age.

14-65 Rate $= k(\text{N}_2\text{O})$

14-67 First-order

14-69 A plot of the ln (Ni(CO)$_4$) versus time is linear indicating that the reaction is first-order in Ni(CO)$_4$.

14-71 Second-order reaction; $k = 1.016 \times 10^{-3}\ \text{min}^{-1}\ M^{-1}$; time $= 984$ min;

14-73 zero order; $k = 0.015\ \text{sec}^{-1}$

14-77 rate $= k(\text{Cr}(\text{NH}_3)_5\ \text{Cl}^{2+})(\text{OH}^-)$

14-79 1.1×10^8 s

14-81 That way one can study the order of the reaction with respect to each individual reactant.

14-89 155 kJ/mol$_\text{rxn}$.

14-91 51.6 kJ/mol$_\text{rxn}$

14-93 $0.05\ M^{-1}\text{s}^{-1}$

14-99 (a) IV (b) I (c) III (d) III

14-103 (b), (c), (d), and (e)

14-105 (a) 19.8 min

(b) The third plot is correct.

(c) $0.035\ \text{min}^{-1}$

(d) The half-life would be longer at the lower temperature.

(e) 75 kJ/mol; $K < 1$

14-107 (b)

14-111 (a) Rate $= k(\text{CH}_4)(\text{Cl}_2)^{1/2}$

(b) The data is consistent with the rate law.

Chapter 15

15-5 38 electrons, 38 protons, and 52 neutrons.

15-9 Interconverts isotopes: (e), (f), and (g)
Interconverts isobars, responses (a), (b), and (c)
None of these interconvert isotones.

15-11 $^{99}_{42}Mo \longrightarrow ^{0}_{-1}e + ^{99}_{43}Tc$

15-13 $^{62}_{29}Cu + ^{0}_{-1}e \longrightarrow ^{62}_{28}Ni$

15-15 (a) $^{32}_{16}S$ (b) $^{11}_{5}B$

     (c) $^{208}_{84}Po$ (d) $^{125}_{53}I$

15-17 (a) $^{4}_{2}He$ (b) $^{239}_{94}Pu$

     (c) $4^{1}_{0}n$ (d) $^{12}_{6}C$

15-21 (b)

15-23 (a)

15-25 $32.03 \dfrac{MeV}{atom}$; $5.338 \dfrac{MeV}{nucleon}$

15-27 238.0520 amu

15-29 2.8 MeV

15-31 55 min

15-33 $k = 0.27$ yr^{-1}; $t_{1/2} = 2.6$ yr; $t = 9.6$ yr

15-35 6.3 d; 1.84 g

15-37 $1.76 \times 10^4 \dfrac{atoms}{L}$

15-39 (a) 8.9×10^4 Ci
(b) 27.3 Ci
(c) 6.34×10^{-8} Ci

15-41 2.4×10^4 yr

15-43 15.3%

15-45 The beeswax did not belong to the Bronze Age.

15-49 $^{27}_{13}Al + ^{4}_{2}He \longrightarrow ^{30}_{15}P + ^{1}_{0}n$

     $^{10}_{5}B + ^{4}_{2}He \longrightarrow ^{13}_{7}N + ^{1}_{0}n$

15-51 $^{253}_{99}Es + ^{4}_{2}He \longrightarrow ^{256}_{101}Md + ^{1}_{0}n$

     $^{238}_{92}U + ^{19}_{9}F \longrightarrow ^{252}_{101}Md + 5^{1}_{0}n$

15-59 $k = 5.0 \times 10^{-4}$ min^{-1}; $t_{1/2} = 1.4 \times 10^3$ min

15-61 4.2×10^4 yr

15-63 1×10^5 kJ/(mol photons)

Chapter 16

16-1 Carbon forms covalent bonds with so many other elements because carbon exhibits an electronegativity which is too small for it to gain the needed electrons, but too large for carbon to lose the required number of electrons to form ionic bonds.

16-5 For n carbon atoms, there are $2n$ hydrogen atoms. Two additional hydrogens come from the terminal CH_3 groups. Alkanes, therefore, have the generic formula C_nH_{2n+2}. The generic formula C_nH_{2n} describes a cycloalkane. The generic formula for an alkene with a single $C{=}C$ is C_nH_{2n}. The generic formula for an alkyne with one triple bond is C_nH_{2n-2}.

16-9 $C_{12}H_{26}$; 3,4-diethyl-3-methylheptane

16-11 3,4,6-trimethyloctane

16-13

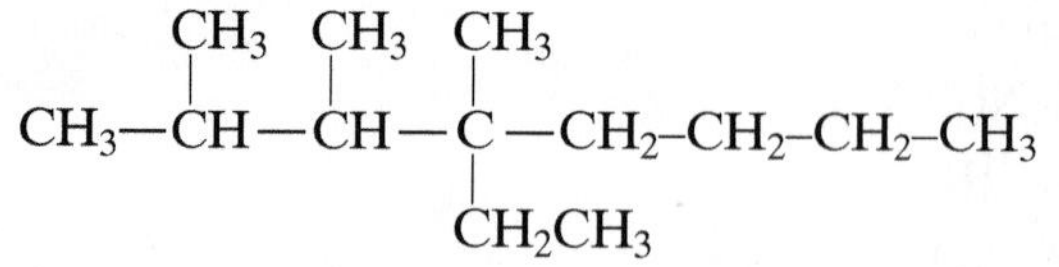

$$CH_3{-}\underset{\underset{}{|}}{CH}{-}\underset{\underset{}{|}}{CH}{-}\underset{\underset{CH_2CH_3}{|}}{\overset{\overset{CH_3}{|}}{C}}{-}CH_2{-}CH_2{-}CH_2{-}CH_3$$

(with CH_3 groups above the first three carbons and CH_2CH_3 below the third)

16-17 The correct name should be 2-pentene.

16-19 Constitutional isomers are associated with changes in attachment of atoms. Stereoisomers retain the same relative attachments, but the component parts of a molecule take up alternate positions relative to a constraint.

16-21 (d)

16-23 (b)

16-25

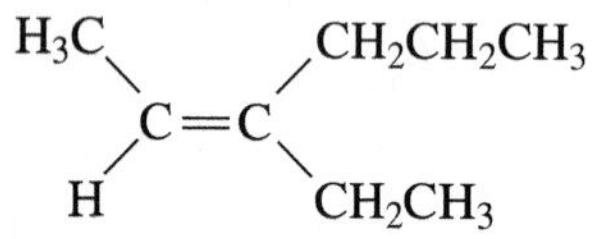

$$\underset{H}{\overset{H_3C}{\diagup}}C{=}C\underset{CH_2CH_3}{\overset{CH_2CH_2CH_3}{\diagdown}}$$

16-27

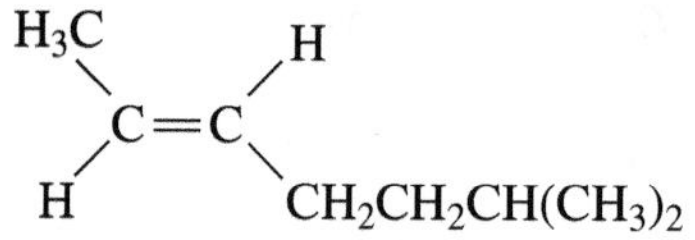

$$\underset{H}{\overset{H_3C}{\diagup}}C{=}C\underset{CH_2CH_2CH(CH_3)_2}{\overset{H}{\diagdown}}$$

16-29

$$\underset{H_3C}{\overset{H_3C}{\diagup}}C{=}C\underset{CH_3}{\overset{CH_2CH_3}{\diagdown}}$$

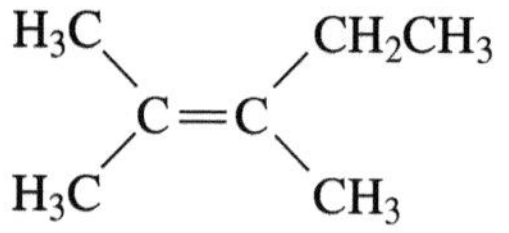

16-31 The bond between the third and fourth carbons is Z.

16-35 Each carbon atom in benzene has three bonding domains with corresponding trigonal planar geometry.

16-39 (d)

16-41 (a) CH_3CH_2OH (b) CH_3CHO (c) CH_3NH_2 (d) CH_3CONH_2
(e) CH_3Br (f) $CH_2{=}CH_2$ (g) $HC{\equiv}CH$

16-43 (a) aldehyde (b) alcohol (c) ether (d) ketone

16-45 Molecular formula of cortisone is $C_{21}H_{28}O_5$. The functional groups include an alkene, two alcohol groups, and three ketone groups.

16-51 (a) Oxidation-reduction reaction   (b) Oxidation-reduction reaction
(c) Oxidation-reduction reaction   (d) Not an oxidation-reduction reaction
(e) Oxidation-reduction reaction

16-53 (a)

16-55 (b)

16-57

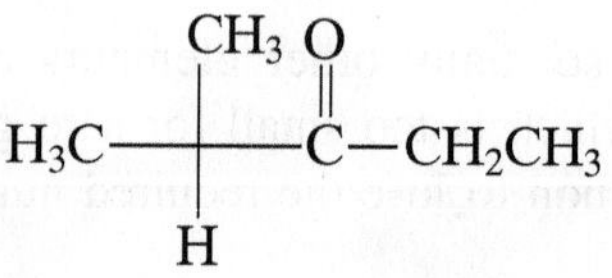

$$H_3C{-}\underset{\underset{H}{|}}{\overset{\overset{CH_3}{|}}{C}}{-}\overset{\overset{O}{\|}}{C}{-}CH_2CH_3$$

16-59 Br_2 + energy $\longrightarrow$ 2Br$^\cdot$ Initiation
CH_4 + Br $\longrightarrow$ $CH_3^\cdot$ + HBr   Propagation
$CH_3^\cdot$ + Br_2 $\longrightarrow$ CH_3Br + Br$^\cdot$ Termination
2Br$^\cdot$ $\longrightarrow$ Br_2
$CH_3^\cdot$ + Br$^\cdot$ $\longrightarrow$ CH_3Br

16-65 $CH_3CH_2OCH_2CH_3(l)$ + $H_2O(l)$

16-71 A: 5-ethyl-3-octanol   B: 5-methyl-1-hepten-4-ol

16-77 3,7-dimethyl-2,6-octadienal

16-85 (e)

16-87 Coniine is optically active.

16-89 Vitamin C has two stereocenters.

16-91 Both isocitric and homocitric acid are chiral.

16-95 R

Appendix D

CHECKPOINT ANSWERS

Chapter 1

- (p. 6) The symbol 8 S represents 8 individual atoms of sulfur. The symbol S_8 represents 8 atoms of sulfur bonded together to form a single molecule.
- (p. 8) 529 mg; 259 cm
- (p. 10) 4.72×10^{-7}; 1.03×10^{22}; 0.0000000754; 3,668,000
- (p. 13) 5; 2; 3; 1; 4
- (p. 13) In 10,000 atoms of lithium you would expect to find approximately 742 ^{6}Li atoms and 9258 atoms of ^{7}Li.
- (p. 14) The ratio of the natural abundance of ^{1}H to ^{2}H is 99.985/0.015 = 6.7×10^3; 2 significant figures
- (p. 16) NaOH: Na^+ and OH^-; K_2SO_4: 2 K^+ and SO_4^{2-}; $BaSO_4$: Ba^{2+} and SO_4^{2-}; $Be_3(PO_4)_2$: 3 Be^{2+} and 2 PO_4^{3-}
- (p. 17) F: 9; Pb: 82; 24: Cr; 74:W
- (p. 22) $H^+(aq) + Cl^-(aq) + Na^+(aq) + OH^-(aq) \longrightarrow$
$$Na^+(aq) + Cl^-(aq) + H_2O(l)$$

Chapter 2

- (p. 33) K atomic weight 39.098 amu; molar mass 39.098 g/mol; U atomic mass 238.03 amu; 238.03 g/mol.
- (p. 35) 2.2×10^{22} Al atoms
- (p. 40) There are two atoms of carbon in one molecule of C_2H_2; two moles of carbon atoms in one mole of C_2H_2; and 1.20×10^{24} atoms of carbon in one mole of C_2H_2.
- (p. 43) CaC_2. The molecular weight must be known to find the molecular formula.
- (p. 48) Four moles of H_2; 2.4×10^{24} molecules of H_2; 4.8×10^{24} atoms of H.
- (p. 49) Two moles of CO require 4.0 moles of H_2. This equals 8.06 grams of H_2.
- (p. 53) In the balanced chemical equation I_2 and Mg react in a one-to-one mole ratio. Four grams of I_2 equals 0.016 moles. Four grams of Mg equals 0.16 moles. There is 10 times more Mg present than I_2. I_2 is the limiting reagent.
- (p. 58) Moles are generally more useful to chemists than mass when relating how much of one chemical substance will react with another.

Chapter 3

- (p. 73) The radius of a typical atom would have to be 10,000 cm. This is equivalent to 100 m.
- (p. 75) The frequency of microwaves covers a range of approximately 3×10^{10} to 3×10^{12} Hz.
- (p. 79) The higher the temperature of an object, the higher the frequency of electromagnetic radiation emitted by the object.
- (p. 85) The first ionization energy of Sr would be greater than that of Rb. Sr has one more proton in its nucleus than Rb and therefore attracts its electrons more strongly than Rb. The first ionization energy of Cl is greater than

that of Br. Bromine has one more shell of electrons than Cl, and therefore its electrons are further from the nucleus and more easily removed.

- (p. 86) Carbon has a core charge of +4, and fluorine has a core charge of +7.

- (p. 87) The first ionization energy of sodium is much less than the first ionization energy of neon. If there were 9 electrons in the second level, we would expect these electrons to be held more tightly by the 11 protons in the nucleus of sodium than the 8 valence electrons held by the 10 protons in the nucleus of neon. This would give sodium a larger first ionization than neon rather than the observed smaller ionization energy.

- (p. 88) Sodium has three shells, whereas lithium only has two. The first ionization energy of sodium is much less than the first ionization energy of lithium. Therefore, the radius of the valence shell of sodium is larger than that of lithium.

- (p. 89) The number of peaks in a photoelectron spectrum is determined by the number of different energies that the electrons in an atom have.

- (p. 90) There is only one peak in the PES of He because both of its electrons have the same energy.

- (p. 92) Neon has 2 electrons in its $n = 1$ shell and 8 electrons in its $n = 2$ shell. Of the 8 electrons in the $n = 2$ shell, 2 are in the s subshell and 6 are in the p subshell.

- (p. 94) The energies and the relative intensities of the PES of Sc are as follows:

Energy (MJ/mol)	Relative Intensity
433	2
48.5	2
39.2	6
5.44	2
3.24	6
0.77	1
0.63	2

Electrons are removed from the subshells of Sc in the following order: 4s, 3d, 3p, 3s, 2p, 2s, 1s.

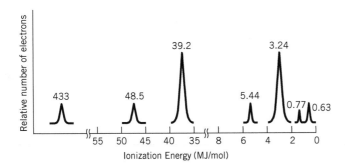

The subshells of Sc are filled with electrons in the following order: 1s, 2s, 2p, 3s, 3p, 4s, 3d. There is a difference in the order in which electrons are added to and taken away from Sc. Electrons are added to the lowest energy subshell (the most stable subshell closest to the nucleus) first and then to subsequent subshells in order of increasing energy until the 3p subshell is filled. After the 3d subshell is filled, the 4s is filled followed by the *lower* energy 3d subshell.

Electrons are removed from the highest energy subshell (the subshell farthest from the nucleus) first and then from lower energy subshells.

- (p. 96) Three quantum numbers are required to describe three dimensions in space. The numbers describe n-size, l-shape, and m-orientation.

- (p. 99) Li, B, and N all have at least one unpaired electron and would therefore be expected to have magnetic properties. Be has no unpaired electrons and would therefore have no magnetic properties and would not be deflected in a Stern-Gerlach experiment.

- (p. 102) C: 2; N: 3; Ne: 0; F: 1; C, N, and F all have at least one unpaired electron and would therefore be expected to have magnetic properties.

- (p. 105) A metallic radius is one-half the distance between two adjacent nuclei of two like atoms measured in a solid crystal lattice. A covalent radius is one-half the distance between two adjacent nuclei of two like atoms of a molecule. The molecule may be in the solid, liquid, or gaseous state.

- (p. 107) The ionic radius of Si^{4+} would be less than the ionic radius of Al^{3+}. Si^{4+} and Al^{3+} have the same number of electrons, but Si^{4+} has one more proton that pulls the electrons more closely to the nucleus.

- (p. 108) Ionization energies obtained from PES are for the removal of only one electron. That electron may be located in any shell or subshell. Second, third, and fourth ionization energies describe the energy to sequentially remove a second, third, and fourth electron from the outermost shells of an atom.

- (p. 110) The maximum positive charge on atoms in Group IVA is +4. For example, Sn has the electron configuration [Kr] $5s^2$ $4d^{10}$ $5p^2$. It can lose four electrons to form a Sn^{4+} ion with the electron configuration: [Kr] $4d^{10}$.

- (p. 111) AVEEs increase from left to right across a period because each subsequent atom in a period has one more positive charge in the nucleus that attracts the electrons more tightly. As you move from the bottom to the top of a group, there are fewer shells containing electrons. The electrons are closer to the nucleus and therefore held tightly by the nucleus. This is the same trend followed by first ionization energy. The trend for AVEEs is the reverse of the trend followed by atomic radii.

Chapter 4

- (p. 125) An atom in Group IVA has 4 valence electrons.

- (p. 127) H_2^+ has only one electron to form the bond between the two hydrogen atoms. A covalent bond is formed by the attraction between the nucleus of one atom and the electron(s) of another atom. Because there is only one electron in H_2^+, both nuclei share the single electron. In addition, there is less electron density to shield the repulsion between the two nuclei in H_2^+. Therefore the nuclei in H_2^+ are not pulled as closely together as the nuclei in H_2.

- (p. 128) In CO_2 the nonbonding electrons are located in the nonbonding electron domains of the oxygen atoms. In CO_2 each oxygen has two nonbonding domains. There are two electrons in each nonbonding domain, but four electrons are located in each of the bonding domains between carbon and oxygen. Four electrons are required to form a double bond.

- (p. 129) Hydrogen needs only two electrons in its outer shell in order to achieve the electron configuration of He. Therefore, hydrogen is capable of forming only one bond and cannot serve as the central atom in a Lewis structure.

- (p. 129) A skeleton structure shows the relative positions of the atoms. A Lewis structure shows the positions of the electrons.
- (p. 131)

$$: \ddot{C}l :$$
$$|$$
$$: \ddot{C}l - \overset{\displaystyle}{\underset{\displaystyle}{C}} - \ddot{C}l : \qquad [: \ddot{O} - H]^-$$
$$|$$
$$: \ddot{C}l :$$

- (p. 133) In order for an atom to expand its octet, it must have free, unoccupied orbitals with an energy close to the energy of its valence electrons (within the same shell). Nitrogen and oxygen have valence electrons in the second shell. The second shell contains one *s* orbital and three *p* orbitals. These four orbitals can contain a maximum of eight total electrons. Because there are no more subshells in the second shell, no more orbitals are available to hold any additional electrons.
- (p. 136) The oxygen–oxygen bond length would be greatest in hydrogen peroxide because single bonds are longer than multiple bonds when comparing bond lengths between the same atoms.
- (p. 138) Isomers have the same molecular formula but differ in the arrangement of the atoms. All of the atoms in the two structures shown for acetate are in the same positions. Only the location of a multiple bond differs in the two Lewis structures. This makes them resonance structures, not isomers.
- (p. 138)

The two resonance Lewis structures of benzene have alternating single and double bonds in two different arrangements. The actual structure of benzene is an average of these two structures. Therefore, we would expect each bond in benzene to behave like a bond and a half (an average of a single bond and a double bond). The bond length of each C—C bond will be greater than the C=C double bond of ethylene but less than the C—C single bond length of ethane.

- (p. 140) The bond in NO would be polar covalent because N and O have different electronegativities. The bond in O_2 would be pure covalent because both oxygen atoms have the same electronegativity.
- (p. 142) The most electronegative atom will have the negative partial charge. If the electronegativity of *B* is greater than that of *A*, *B* would have the negative partial charge.
- (p. 143) The formal charges are: double-bonded oxygen 0, sulfur +1, single-bonded oxygen −1.
- (p. 149) Each carbon in the Lewis structure of benzene has one C—C single bond, one C—C double bond, and one C—H single bond. Therefore, each carbon atom has three electron-pair domains.
- (p. 154) The shape of a molecule can be found using Table 4.3. The shape of a molecule with three bonding domains and one nonbonding domain around the central atom is trigonal pyramidal.

- (p. 155) The C_1—C_2—C_3 is 109°.

  The C_1—C_2—H_g is 109°.

  The H_g—C_1—O is slightly less than 109°.

- (p. 156) There are three bonding electron domains around each carbon. Therefore, the bond angles would be approximately 120°.

- (p. 156) Chloroform would have a dipole moment because the polarity of the C—H bond would be different from that of the three C—Cl bonds. The polarities of the bonds would not cancel and would therefore result in a dipole moment for the molecule.

Chapter 5

- (p. 179) Ca < K < Rb

- (p. 181) Atoms in Group IIA have two electrons in their valence shell that can be more easily lost than their inner-shell electrons. Group IA elements have only one valence shell electron. Therefore, an electron must be removed from an inner shell in order to lose a second electron. These electrons are closer to the nucleus and are more difficult to remove.

- (p. 189) The most common ion of Sr is Sr^{2+}. The peroxide anion is O_2^{2-}. The formula is SrO_2.

- (p. 192) The electrons in metallic bonds are delocalized and therefore free to move from one atom to another and conduct an electric current. The electrons in both covalent and ionic bonds are localized and therefore not free to move to another atom or ion.

- (p. 195) The difference in electronegativity between Zn and S is 0.99. Based only on the difference in electronegativity, we would predict that ZnS is polar covalent.

- (p. 199) Using the bond-type triangle, we find ZnS located directly on the dividing line between covalent and ionic bonding. We would predict that ZnS would have approximately equal covalent and ionic character.

- (p. 201) $BaSi_2$ is primarily metallic, $BaBr_2$ is primarily ionic, and $GaBr_3$ is primarily covalent. $BaBr_2$ would be the best insulator that would also stand up to high heat.

- (p. 207) Formal charge and oxidation numbers are tools that have been developed by chemists. Formal charge treats all bonds as if they are pure covalent bonds. Oxidation numbers treat all bonds as if they were pure ionic bonds. Both of these are useful tools for making predictions about chemical structures and changes that occur during reactions.

- (p. 209) The first reaction is not an oxidation–reduction reaction. None of the atoms undergoes a change in oxidation state. The second reaction is an oxidation–reduction reaction. The oxidation state of N changes from -3 in NH_3 to $+2$ in NO. The oxidation state of O changes from zero in O_2 to -2 in both NO and water. The third reaction is an oxidation–reduction reaction. The oxidation number for carbon in CH_3OH changes from -2 to -3, while the oxidation number of carbon in CO changes from $+2$ to $+3$.

Chapter 6

- (p. 224) If the liquid water and gaseous water are both at the same temperature, then the average kinetic energy of the water molecules should be the same.

- (p. 228) The pressure exerted on the floor by the spike heel will be greater. If the same person wears one shoe with a spike heel and the other with a normal heel, the same total force is exerted on the floor. However, in the case of the spike heel, that force is exerted over a much smaller area of the floor.

- (p. 232) No, the diameter of the tube does not change the relationship between the pressure and volume of the gas.

- (p. 233) $PV \propto T$

- (p. 234) Boyle's law described the relationship between pressure and volume, and Amonton's law described the relationship between pressure and temperature. By combining these laws, we can obtain a relationship between temperature and volume. This is Charles' law.

- (p. 244) If we assume that 1 L of wet air and 1 L of dry air are at the same temperature and pressure, then there must be the same number of moles in both containers. The gas in the container of dry air has an average molecular weight of 29.0 g/mol. The container of wet air contains both air and water vapor. Because it contains both air and water vapor but has the same total number of moles of gas as the dry air, there are fewer moles of dry air (29 g/mol) in this container and more moles of water vapor (18 g/mol) than in the container of dry air. Therefore, the container of wet air is lighter than the container of dry air because the container of wet air has fewer heavy molecules (N_2 and O_2) and more light molecules (H_2O) of gas than the container of dry air.

- (p. 247) As the temperature of a gas increases, the molecules of gas begin to move more rapidly, resulting in their having more kinetic energy.

- (p. 250) The heavier gas molecules will be moving at a slower velocity than the lighter gas molecules. Because kinetic energy is dependent on both mass and velocity, a heavy molecule moving at a slow velocity can have the same kinetic energy as a less massive molecule moving at a greater velocity.

Chapter 7

- (p. 268) No bonds are broken in the reactants. One bond is formed in the products. Therefore, more bonds are made than broken. This reaction will release energy because it is a bond-making process.

- (p. 269) When heat enters a balloon, the temperature of the gas particles increases. This results in an increase in the motion of the gas particles, causing them to strike the sides of the balloon more frequently and with more force. If the balloon is elastic, it will expand, increasing the volume of the gas while the pressure remains constant.

- (p. 271) Specific heats are based on the mass of a substance, and molar heat capacities are based on moles. One mole of iron has a much larger mass than one mole of aluminum. Given two substances with similar structures and properties, the one with the higher molecular weight will have a higher molar heat capacity and a lower specific heat compared to the substance with the lower molecular weight.

- (p. 274) While at rest on the 10th floor, the crate will have only potential energy. During the fall the crate will have both potential and kinetic energy. As the crate falls its velocity increases, thus increasing its kinetic energy. As it moves closer to the ground, its potential energy in relation to the ground decreases. When the crate strikes the ground, it can release its energy in the form of sound and heat or by doing work on the ground (making a hole or indentation).

- (p. 275) According to the first law of thermodynamics, the total energy change must equal zero. This requires that the sum of the change in energy of the system and the change in energy of the surroundings equals zero. If we assume the brick to be our system and the water to be the surroundings, the brick can gain energy only if the surroundings supply that energy. Therefore, the first law does not prevent the brick from becoming hotter and the water from becoming colder. The first law requires only that what is gained by one part of the universe must be lost by another part of the universe.

- (p. 278) The temperature of the gas will increase the most for the gas in a fixed-volume piston and cylinder. All of the heat transferred to this gas is used to increase the temperature of the gas. Some of the heat transferred to the gas in the movable piston and cylinder is used to do work.

- (p. 282) Table sugar can provide or give off energy. This is an exothermic process, and the bond-making process must provide more energy than is required by the bond-breaking process. The products must therefore be more stable than the reactants. The bonds formed are stronger than the bonds broken.

- (p. 283) All of the reactions listed in Table 7.2 release or give off energy. Therefore, these are exothermic reactions, and the sign of the change in enthalpy is negative.

- (p. 288) ΔH_{373}° represents the change in enthalpy for a process that occurs under standard conditions and at 373 K.

- (p. 290)
$$N(g) \longrightarrow N(g)$$
$$H(g) \longrightarrow H(g)$$
$$N(g) + 3\,H(g) \longrightarrow NH_3(g)$$

ΔH_{ac}° for $N(g)$ and $H(g)$ are both zero because both are isolated atoms in the gaseous state. The ΔH_{ac}° for NH_3 is -1171.76 kJ/mol$_{rxn}$. ΔH_{ac}° for NH_3 represents the formation of NH_3 from its gaseous atoms. No bonds are broken in the reactants, but bonds are formed by the products; therefore, ΔH_{ac}° is negative.

- (p. 293)

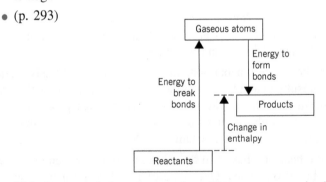

This is an endothermic reaction. The energy required to break bonds is greater than the energy given off in forming bonds. Therefore, the total strength of the bonds is greatest in the reactants.

- (p. 295) Bond breaking: $2\,CO(g)$ $\Delta H^{\circ} = 2(1076.38\,\text{kJ/mol}_{rxn})$
$$= 2152.76\,\text{kJ/mol}_{rxn}$$

$O_2(g)$ $\Delta H^{\circ} = 498.34\,\text{kJ/mol}_{rxn}$

Total bond breaking $\Delta H^{\circ} = 2651.10\,\text{kJ/mol}_{rxn}$

Bond making: $2\,CO_2(g)$ $\Delta H^{\circ} = 2(-1608.53\,\text{kJ/mol}_{rxn}) = -3217.06\,\text{kJ/mol}_{rxn}$

$\Delta H_{rxn}^{\circ} = (2651.10\,\text{kJ/mol}_{rxn}) + (-3217.06\,\text{kJ/mol}_{rxn}) = -565.96\,\text{kJ/mol}_{rxn}$

The reaction is exothermic.

- (p. 297) The enthalpy of atom combination of liquid ethanol is greater than the enthalpy of atom combination of gaseous ethanol. More energy is released in the formation of liquid ethanol than in the formation of gaseous ethanol from its gaseous atoms. The change in state from liquid ethanol to gaseous ethanol requires an input of energy. There are forces of attraction between the molecules of liquid ethanol that are not present in gaseous ethanol.

- (p. 299) The sum of the bond strengths can be estimated using the values from Exercise 7.9 and the discussion that follows this exercise.

$$Si—F = 597 \text{ kJ/mol}$$

$$Si—O = 466 \text{ kJ/mol}$$

$$O—H = 463 \text{ kJ/mol}$$

$$3 \times 597 \text{ kJ/mol} + 466 \text{ kJ/mol} + 463 \text{ kJ/mol} = 2720 \text{ kJ/mol}$$

- (p. 299) When similar atoms are bonded together, the enthalpy of atom combination decreases with increasing bond length. BrF would be expected to have weaker bonds (a smaller enthalpy of atom combination) than ClF because BrF is a larger molecule with a larger bond length.

Chapter 8

- (p. 319) As an atom becomes larger, the electrons farther from the nucleus are more easily induced into a temporary dipole. As the size of an atom increases, there is an increase in the dispersion force between atoms. $Xe > Kr > Ar > Ne > He$.

- (p. 320) Hydrogen bonds between NH_3 molecules are weaker than those between H_2O molecules because oxygen is more electronegative that nitrogen. This gives a larger partial charge to the H in the OH bond.

- (p. 322) (a) These three substances have very different molecular weights. We would predict that the boiling point should increase with an increase in molecular weight due to increasing dispersion forces. Therefore the boiling point of tetrabromobutane (373.6 g/mol) is greater than the boiling point of carbon tetrachloride (154 g/mol), and acetone (58 g/mol) has the lowest boiling point.

    (b) These three substances all have very similar molecular weights (142 to 144 g/mol). When comparing the boiling points of substances with similar molecular weights, it is necessary to examine their intermolecular forces. Octanoic acid is capable of hydrogen bonding and will therefore have the highest boiling point. Nonanal is polar and therefore has dipole–dipole intermolecular forces. Decane is nonpolar and will have only dispersion forces. Therefore, decane is expected to have the lowest boiling point.

- (p. 325) The phase change from liquid to gas requires more enthalpy (enthalpy of vaporization 44 kJ/mol_{rxn}) than the phase change from solid to liquid (enthalpy of fusion 6 kJ/mol_{rxn}). The solid state has the strongest intermolecular forces. Some of these forces must be overcome to melt the solid, and the remaining forces must be overcome to cause the liquid to change to a gas.

- (p. 327) If the volume of the container is decreased, the ideal gas law tells us that the pressure of the vapor in the container should increase. However, if the pressure of the vapor in the container is at its vapor pressure (the maximum pressure of vapor for a given temperature), the pressure cannot increase. Therefore, as the volume is decreased, condensation of the vapor will occur so that the pressure of the vapor remains constant at the vapor pressure.

- (p. 338) In order for a side chain to be hydrophilic, it must be able to interact with water. This means that the side chain should have intermolecular forces similar to the intermolecular forces of water (hydrogen bonding). The side chain on alanine contains carbon and hydrogen atoms that have similar electronegativities. This side chain is nonpolar. The side chain on cysteine is somewhat polar due to the presence of sulfur with its two nonbonding pairs of electrons, but it is not capable of hydrogen bonding. These two side chains are therefore hydrophobic (they do not form hydrogen bonds with water). Lysine contains an NH_3 group, and serine contains an OH group. Both lysine and serine are capable of forming hydrogen bond intermolecular forces. Therefore, both of them will be hydrophilic.

- (p. 342) The bond-breaking process for one mole of $BaCl_2(s)$ is $+1282$ kJ/mol_{rxn}. The formation of attractive forces between one mole of Ba^{2+} ions and water is -718 kJ/mol_{rxn}. The formation of attractive forces between two moles of Cl^- ions and water release $(2) \times (-289$ $kJ/mol_{rxn})$. The change in enthalpy associated with the dissolution of $BaCl_2(s)$ is -14 kJ/mol_{rxn}.

- (p. 343) The bond-breaking process for $AgCl(s)$ is $+533.30$ kJ/mol_{rxn}. The formation of attractive forces between one mole of Ag^+ ions and water is -178.97 kJ/mol_{rxn}. The formation of attractive forces between one mole of Cl^- ions and water is -288.838 kJ/mol_{rxn}. The change in enthalpy associated with the dissolution of $AgCl(s)$ is $+65.49$ kJ/mol_{rxn}.

Chapter 9

- (p. 371) On the atomic scale, molecular solids are composed of individual molecules held together by intermolecular forces. Network covalent solids are composed of a three-dimensional array of atoms held together by covalent bonds. On the macroscopic scale network covalent solids have high melting points, and molecular solids have relatively low melting points.

- (p. 374) Both the strength of the attraction of an atom for its valence electrons and the energy difference between valence subshells within an atom contribute to the AVEE and electronegativity. Atoms that have a relatively small attraction for their valence electrons and have small differences in energy between valence subshells have low AVEE and electronegativity. Atoms with these characteristics can lose their valence electrons to form positive cations. The lost electrons can move easily between available subshells from cation to cation. This results in metals being good conductors of electricity.

- (p. 381) Diffusion of atoms can produce the formation of metallic bonds between different metals.

- (p. 382) On the atomic scale molecular solids are composed of individual molecules held together by intermolecular forces. Ionic solids are composed of cations and anions in a three-dimensional array held together by ionic bonds. Ionic solids have a higher melting point than molecular solids. Ionic solids conduct an electric current when melted. Molecular solids do not.

- (p. 384) The lattice energy of MgF_2 is greater than that of $MgCl_2$ because F^- is a smaller ion than Cl^-. The smaller fluoride ion is held more closely and tightly to the Mg^{2+}.

- (p. 384) Metallic solids are composed of positively charged metal cations in a three-dimensional array. The metal cations are surrounded by a sea of delocalized electrons that form metallic bonds. Ionic solids are composed of cations and anions in a three-dimensional array held together by localized electrons in ionic bonds.

  Metallic solids have a wide range of melting points compared to ionic solids which have relatively high melting points. Metallic solids can conduct electricity in the solid state but ionic solids cannot.

- (p. 384) In order for a substance to be classified as a body-centered cubic cell, all of the positions in the unit cell must be occupied by the same type of atom or ion.

- (p. 385) B_4C has an average EN of 2.3 and a ΔEN of 0.5. This places it in the covalent region of a bond-type triangle. MoC has an average EN of 2.0 and a ΔEN 1.1 and is in the ionic region. Both compounds could serve as insulators because they would not conduct heat or electricity.

- (p. 392) The simple cubic cell has the simplest relationship between the radii of the ions and the length of the edge of the unit cell. In a simple cubic cell the length of the edge of the cell is equal to $r + r$. In a face-centered cube, $2(r + r) = a2^{1/2}$ where a is the length of the edge of the unit cell. For a body-centered cubic cell, the relationship is $2(r + r) = a3^{1/2}$.

Chapter 10

- (p. 411) After the reaction has reached equilibrium, both N_2O_4 and NO_2 will be present in the container.

- (p. 412) Using Table 10.1, we see that at equilibrium the ratio of *trans*-2-butene to *cis*-2-butene is 0.559 to 0.441 which equals 1.27.

- (p. 415) An increase in the rate constant of a reaction will result in an increase in the rate of the reaction. Therefore, the rate of the reaction will increase with temperature.

- (p. 418) If the rate of the reverse reaction is greater than the rate of the forward reaction at a given moment in time, this does not mean that the reverse rate constant is greater than the forward rate constant. The rate of a reaction is dependent on both the rate constant and the concentrations of the reactants. If the rate of the reverse reaction is greater than the forward, this may be due to a higher concentration of products (reactants for the reverse reaction) than reactants for the forward reaction.

- (p. 420) The equilibrium constant can be determined from the division of the forward rate constant by the reverse rate constant. For this reaction the equilibrium constant would be 2.

- (p. 427) The stoichiometry of the reaction is 1:1:1. For every mole of PCl_3 that reacts, one mole of PCl_5 is produced and one mole of Cl_2 is used up. Therefore, if the concentration of PCl_3 decreased by 0.96 mol/L, the concentration of Cl_2 also decreased by 0.96 mol/L, leaving only 0.04 mol/L. The concentration of PCl_5 must increase from zero to 0.96 mol/L.

- (p. 430) The equilibrium concentrations found in Exercise 10.8 give the correct K_c (1.27) within ± 1 in the last significant figure. $0.559/0.441 = 1.27$

- (p. 431) The change in concentrations that takes place during a chemical reaction must follow the stoichiometry of the balanced chemical equation.

- (p. 434) ΔC will be small when the equilibrium constant is significantly smaller than one. ΔC would be small for K values of 1.0×10^{-5} and 1.0×10^{-10}.

- (p. 441) If more P_2 is added to the system at equilibrium, the system will shift to use up the added P_2. The system will shift to the right to produce more P_4. If the pressure increases while the temperature remains constant, a reaction will shift in a direction to relieve the increase in pressure by decreasing the number of moles of gas. For this reaction the equilibrium will shift to the right to produce P_4 and use up P_2.

Chapter 11

- (p. 470) An Arrhenius base must ionize to give OH^- when it dissolves in water. $Mg(OH)_2$ ionizes to give Mg^{2+} and $2\ OH^-$. It is therefore an Arrhenius base. HNO_3 and CH_3CO_2H both ionize to produce H^+ and are therefore Arrhenius acids.

- (p. 473) $H_3PO_4(aq) + H_2O(l) \rightleftharpoons H_2PO_4^-(aq) + H_3O^+(aq)$. The conjugate base is $H_2PO_4^-$. Phosphoric acid is a polyprotic acid, meaning that it can further dissociate to lose an additional proton and produce additional H_3O^+.

$$C_6H_5NH_2(aq) + H_2O(l) \rightleftharpoons C_6H_5NH_3^+(aq) + OH^-$$

The conjugate acid of aniline is $C_6H_5NH_3^+$.

- (p. 478) The K_w will still equal 10^{-14}. The addition of acid will result in a higher overall concentration of H_3O^+, even though some of the hydronium ion from the dissociation of water may shift back to the left. What is most significant about the shift in the equilibrium of water back to the left is the decrease in the concentration of hydroxide ion. This decrease in concentration of the hydroxide ion multiplied by the increase in concentration of the hydronium ion will therefore equal K_w, 10^{-14}.

- (p. 481) pH is $-\log[H_3O^+]$. As the concentration of hydronium ion increases, there will be a decrease in the pH.

- (p. 484)

$$K_w = [OH^-][H_3O^+]$$

$$K_a = \frac{[OH^-][H_3O^+]}{[H_2O]}$$

The K_a expression includes H_2O as a reactant.

- (p. 485) Both solutions have a low concentration of acid. The acid with the larger K_a will be a stronger acid and will dissociate more than the acid with the small K_a. The solution with the larger K_a will produce more hydronium ion and will therefore have a smaller pH.

- (p. 489) A stronger acid produces a weaker conjugate base. HOCl is the stronger acid; therefore its conjugate base, OCl^- will be the weaker base. OBr^- will be the stronger base.

- (p. 491) H_2S is a stronger acid than H_2O. Sulfur is a larger atom than oxygen. The sulfur–hydrogen bond is longer and therefore weaker than the oxygen–hydrogen bond. The sulfur–hydrogen bond is more easily broken than the oxygen–hydrogen bond.

- (p. 493) The only difference between the two acids (HOCl and HOI) is the oxygen bonded to Cl in one acid and to I in the other. Because Cl is more electronegative than I, it will pull electron density away from the oxygen–hydrogen bond more than I. Thus the oxygen–hydrogen bond will be weaker for HOCl than for HOI, and the hydrogen will more easily be lost from HOCl. Therefore, HOCl is a stronger acid than HOI.

- (p. 494) When a strong acid is added to water, it is assumed to completely dissociate to produce hydronium ion. For a monoprotic acid the hydronium concentration will be equal to the initial concentration of strong acid.

- (p. 497) It is not possible to determine which of the two solutions has the higher pH. Solution 1 contains a stronger acid and therefore dissociation is greater than for the acid in solution 2, but the concentration of the acid in solution 2 is greater than the concentration of the acid in solution 1. It is possible that solution 2 could produce more hydronium ion because it is more concentrated, even though it is a weaker acid.

- (p. 499) The $1.0\ M$ solution is more acidic.

- (p. 500) A large K_b indicates a stronger base that would react with water to accept more hydrogen ions and produce more hydroxide ions. Since both bases have the same concentration, the base with the larger K_b is more basic.

- (p. 504) NO_2^- is the conjugate base of HNO_2. Therefore, $K_b = K_w/K_a$.

$$K_b = (1.0 \times 10^{-14})/(5.1 \times 10^{-4}) = 2.0 \times 10^{-11}$$

- (p. 510) If acid is added to a buffer, the pH of the buffer will decrease slightly, but the decrease will not be as great as when the same amount of acid is added to water. If base is added to a buffer, the pH of the buffer will increase slightly but not as much as when the same amount of base is added to water. The resulting pH is dependent on how much acid or base is added.

- (p. 511) Lactic acid produced in the body can react with HCO_3^- by the following reaction.

$$HC_3H_5O_3(aq) + HCO_3^-(aq) \rightleftharpoons C_3H_5O_3^-(aq) + H_2CO_3(aq)$$

- (p. 520) As NaOH is slowly added to a solution of HCl, the pH will gradually increase until just enough NaOH has been added to completely react with the HCl. At this point the pH will increase rapidly for several pH units. As more NaOH is added, the pH will resume a more gradual increase. A plot of pH versus volume of NaOH will yield a typical titration curve.

Chapter 12

- (p. 542) Oxidation is the loss of electrons, whereas reduction is the gain of electrons. One process cannot occur without the other.

- (p. 544) $Fe = +3, Cl = -1; C = -4, H = +1; Mn = +7, O = -2;$ $C = +2, H = +1, O = -2, N = -3$

- (p. 550) The anode is negative, and the cathode is positive.

- (p. 551) The number of atoms is balanced, but the charge is not. There is a charge of $+2$ on the reactant side and a charge of $+3$ on the product side of the chemical equation. The equation is not balanced.

- (p. 553) If the reaction is reversed, the sign of the cell potential is changed.

- (p. 554) Sn(s) is oxidized and nitrogen is reduced. Sn is the reducing agent, and the HNO_3 is oxidizing agent.

- (p. 567) The standard potential corresponds to 1 M concentrations of both Cu^{2+} and Ni^{2+}. A concentration of 1.5 M Cu^{2+} on the reactant side and a concentration of 0.010 M Ni^{2+} on the product side will result in a greater tendency for the reaction to proceed toward the products. This will result in a greater (more positive) value for the potential (E) under these conditions than at standard conditions.

- (p. 587) The student received no credit because the reaction was balanced in base rather than in acid.

Chapter 13

- (p. 596) Dissolving salt in water is a spontaneous process. Even though heat is absorbed during the process, the salt dissolves because the system (water and salt) becomes more disordered as the salt dissolves.

- (p. 600) It is possible for a process to proceed if the entropy of the system decreases as long as the entropy of the surroundings increases even more, causing the entropy of the universe to increase.

- (p. 601) At absolute zero, motion at the molecular and atomic level approaches zero. A substance would have minimum entropy when there is no motion.

- (p. 603) Like cyclopentane being converted to pentene, glucose would have greater freedom of motion (more disorder) in its open-chain form. The sign of ΔS would be positive.

- (p. 604) Both molecules consist of the same kind and number of atoms. Their reactions of atom combination would involve the same gaseous atom reactants. Because the entropy of atom combination of isobutane is more negative than the entropy of atom combination of butane, the formation of isobutane results in the loss of more freedom of motion.

- (p. 609) If ΔH for a spontaneous process, such as dissolving table salt in water, is positive, the entropy of the process must be positive enough to make the change in Gibbs Free energy negative. The increase in disorder must be enough to make up for the endothermic process.

- (p. 611) No, the hardening of cement has a negative ΔS. The cement becomes more ordered. In order for this to be a spontaneous process, it must therefore be exothermic (releasing heat).

- (p. 621) If $Q_p > K_p$ the reaction must proceed toward the reactants. In order for Q_p to decrease to K_p, the numerator (product partial pressures) of the expression must decrease and the denominator (reactant partial pressures) must increase.

- (p. 629) If ΔH is > 0 (endothermic), the reactants are favored. If ΔS is > 0, the products are favored. A change in temperature will change the equilibrium constant. In the relationship between K, ΔH, and ΔS, it is the enthalpy term that is dependent on temperature. As temperature decreases, the term containing enthalpy becomes more negative and therefore K becomes smaller. Thus the equilibrium is shifted toward the reactants.

  If ΔH is > 0 (endothermic), the reactants are favored. If ΔS is < 0, the reactants are favored. These circumstances will have the same result for a decrease in temperature as in the previous question because ΔH is still > 0. Therefore, a decrease in temperature will shift the equilibrium toward the reactants.

Chapter 14

- (p. 641) The kinetics of the reaction is so slow that during a human lifetime there will be no observable change of diamond to graphite.

- (p. 645) If a large value for Δt is used, there will be a significant change in rate during the time interval. The resulting rate would then be more of an average rate during the time period rather than an instantaneous rate at some moment in time.

- (p. 645) As the reaction progresses, the concentration of phenolphthalein decreases as it reacts. The concentration term in the rate law decreases, and therefore the rate of the reaction decreases. As the reaction proceeds, phenolphthalein is used up, and there is a decrease in the probability of a collision between phenolphthalein molecules and base molecules. Thus the reaction rate decreases.

- (p. 646) The rate describes how quickly the reactants are used up and products are produced with respect to time. The rate constant is a proportionality constant used in the rate law. The rate of most reactions decreases with respect to time and concentration of reactants. However, the rate constant remains constant with respect to time and concentration.

- (p. 646) The units of k will be $1/(M \cdot sec) = L/(sec \cdot mol)$.

- (p. 647) The rate of disappearance of HI is twice as fast as the rate of formation of H_2.

$$rate_{HI} = 2(rate_{hydrogen})$$

- (p. 655) The rate of a zero-order reaction is independent of the concentration of the reactants. The rate of reaction of a first-order reaction is dependent on concentration. As a reaction proceeds, the rate of reaction of a zero-order reaction will remain constant, while a first-order reaction will slow down.

- (p. 657) In trials 2 and 3 the concentrations of both NH_4^+ and NO_2^- change. Therefore, there will be two unknowns (m and n) in the equation. One of the concentrations must remain constant between two trials so that its effect will cancel out of the equation. Trials 1 and 2 cannot be used to find the order with respect to NH_4^+ because the concentration of NH_4^+ does not change in these trials. However, the concentration of NO_2^- does change. Any change in the rate will therefore be due to NO_2^- not NH_4^+.

- (p. 657) For trial 1 the following are known: $(NH_4^+) = 5.00 \times 10^{-2}$, $(NO_2^-) = 2.00 \times 10^{-2}$, $m = 1$, $n = 1$, rate $= 2.70 \times 10^{-7}$. Solving the rate law equation for the rate constant, k, we get $2.70 \times 10^{-4}\ M^{-1}sec^{-1}$.

- (p. 662) The rate constant for the decay of ^{15}O is less than the rate constant for ^{19}O. Therefore, the rate of decay of ^{15}O will be slower, and it will have a longer half-life.

- (p. 662) For a zero-order reaction the half-life for the second reaction will be twice as large as that for the first trial. For a first-order reaction the half-life will not change. The half-life of a first-order reaction is independent of concentration. For a second-order reaction, doubling the concentration will decrease the half-life by a factor of two.

- (p. 669) No, this is not true for an endothermic reaction.

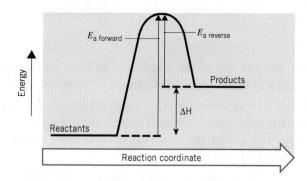

- (p. 671) ΔH is the difference between the energy of the products and the energy of reactants in both diagrams. ΔH will be the same in both diagrams because the enthalpies of products and reactants have not changed.
- (p. 672) The E_a term is a negative in the Arrhenius equation. Therefore, if E_a is large, k will be small. A small k indicates a slow reaction rate. Because a catalyst decreases E_a, the value of k will increase, leading to a faster reaction rate.
- (p. 673) Because the external pressure is less at the top of a mountain, the water will boil at a lower temperature than at sea level. Therefore, the egg will not be as hot in the boiling water. Cooking the egg to make it hard-boiled involves a chemical reaction. According to the Arrhenius equation, the rate constant for this reaction will be smaller at the lower temperature of the boiling water at the top of the mountain.

Chapter 15

- (p. 693) Both mass number and charge must be conserved.
- (p. 695) ^{35}S contains 19 neutrons and 16 protons.

Chapter 16

- (p. 729) In the first structure individual methyl groups (CH_3) are attached to the second and third carbons. In the second structure both methyl groups are attached to the second carbon. Because the methyl groups are bonded to different atoms in the two structures, the structures must be isomers.

    In the second set of structures the methyl groups are bonded to the same carbons (carbon 2 and 3). The only difference is the relative position of the two methyl groups. Because there is free rotation around carbon–carbon single bonds, the two structures represent the same compound.

Photo Credits

Chapter 1-16 Opener:

Chapter 1: *Page 7:* Image originally created by IBM Corporation.

Chapter 2: *Page 42:* Ken Karp/Omni-Photo Communications, Inc.
Page 52: Bob Burch/Bruce Coleman/Photoshot
Page 59: Joshua H. Rickard.

Chapter 3: *Page 72:* Bettmann/© Corbis
Page 73: Richard Megna/Fundamental Photographs
Page 77: Ken Karp

Chapter 4: *Page 144:* © Estate of Irving Geis.
Page 145: Illustration, Irving Geis. Image from Irving Geis Collection/Howard Hughes Medical Institute. Rights owned by HHMI. Not to be reproduced without permission.
Page 175: Ken Karp/Omni-Photo Communications, Inc.

Chapter 5: *Page 179:* Ken Karp/Omni-Photo Communications, Inc.
Page 181(top): Charles D. Winters/Photo Reserachers Inc.
Page 181(bottom): Andy Washnik/Wiley Archive
Page 191: George Bodner
Page 193: Charles D. Winters/Photo Researchers Inc.

Chapter 6: *Page 227:* Ted Kinsman/Photo Researchers
Page 230: Ken Karp/Omni-Photo Communications, Inc.

Chapter 7: *Page 266:* Andrew Lambert Photography/Photo Researchers, Inc.
Page 274: Andy Washnik.
Page 282: Andy Washnik.
Page 283: Charles D. Winters/Photo Researchers, Inc.

Chapter 8: *Page 317:* Perennou Nuridsany/Photo Researchers, Inc.
Page 333: Charles D. Winters/Photo Researchers, Inc.
Page 335: Andy Washnik.
Page 339: Jack Fields/Photo Researchers, Inc.
Page 346: Lawrence Migdale/Photo Researchers, Inc.

Chapter 9: *Page 368:* Astrid & Hanns-Frieder Michler/Photo Researchers, Inc.
Page 373: Scott Camazine/Photo Researchers, Inc.
Page 376-379: George Bodner
Page 387: Lisl Dennis/Getty Images, Inc.

Chapter 10: *Page 409:* Ken Karp/Wiley Archive
*Page 411(top):*Tripos Associates
Page 411(bottom): Tripos Associates

Page 424: Ken Karp/Wiley Archive

Page 444: Courtesy of BASF SE, Photo: BASF Corporate Archives Ludwigshafen.

Page 448: Herve Berthoule/Photo Researchers, Inc.

Chapter 11: *Page 475:* Andy Washnik/Wiley Archive

Page 482: Michael Watson.

Page 495: Ken Karp/Wiley Archive

Page 499: Photodisc/Getty Images, Inc.

Page 500: Art Attack/Photo Researchers, Inc.

Page 510: Copyright of Thermo Fisher Scientific Inc., 2010

Chapter 12: *Page 552:* Ken Karp/Wiley Archive

Page 557: Charles D. Winters/Photo Researchers Inc.

Page 559: Ken Karp/Wiley Archive

Page 564: Andrew Lambert Photography/Photo Researchers, Inc.

Page 575: Ken Karp/Omni-Photo Communications, Inc.

Chapter 13: *Page 627:* Ken Karp/Omni-Photo Communications, Inc.

Chapter 14: *Page 641:* Ken Karp/Wiley Archive.

Page 643: Michael Watson.

Page 652: Wiley Archive

Page 654: Andy Washnik/Wiley Archive

Page 661: ZUMA Press

Page 673: SPL/Photo Researchers, Inc.

Chapter 15: *Page 689:* Science Source/Photo Researchers

Page 703: John Reader/Photo Researchers, Inc.

Page 714: SPL/Photo Researchers, Inc.

Page 718: Courtesy of NASA

INDEX

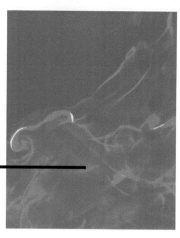

Table of Atomic Number and Average Atomic Weight

Element	Symbol	Atomic Number	Mass (amu)
Actinium	Ac	89	227.03*
Aluminum	Al	13	26.982
Americium	Am	95	241.06*
Antimony	Sb	51	121.76
Argon	Ar	18	39.948
Arsenic	As	33	74.922
Astatine	At	85	209.99*
Barium	Ba	56	137.33
Berkelium	Bk	97	249.08*
Beryllium	Be	4	9.0122
Bismuth	Bi	83	208.98
Bohrium	Bh	107	262.12*
Boron	B	5	10.811
Bromine	Br	35	79.904
Cadmium	Cd	48	112.41
Calcium	Ca	20	40.078
Californium	Cf	98	252.08*
Carbon	C	6	12.011
Cerium	Ce	58	140.12
Cesium	Cs	55	132.91
Chlorine	Cl	17	35.453
Chromium	Cr	24	51.996
Cobalt	Co	27	58.933
Copper	Cu	29	63.546
Curium	Cm	96	244.06*
Dubnium	Db	105	262.11*
Dysprosium	Dy	66	162.50
Einsteinium	Es	99	252.08*
Erbium	Er	68	167.26
Europium	Eu	63	151.97
Fermium	Fm	100	257.10*
Fluorine	F	9	18.998
Francium	Fr	87	223.02*
Gadolinium	Gd	64	157.25
Gallium	Ga	31	69.723
Germanium	Ge	32	72.61
Gold	Au	79	196.97
Hafnium	Hf	72	178.49
Hassium	Hs	108	(265)**
Helium	He	2	4.0026
Holmium	Ho	67	164.93
Hydrogen	H	1	1.0079
Indium	In	49	114.82
Iodine	I	53	126.90
Iridium	Ir	77	192.22
Iron	Fe	26	55.847
Krypton	Kr	36	83.80
Lanthanum	La	57	138.91
Lawrencium	Lr	103	262.11*
Lead	Pb	82	207.2
Lithium	Li	3	6.941
Lutetium	Lu	71	174.97
Magnesium	Mg	12	24.305
Manganese	Mn	25	54.938
Meitnerium	Mt	109	(266)**
Mendelevium	Md	101	258.10*
Mercury	Hg	80	200.59
Molybdenum	Mo	42	95.94
Neodymium	Nd	60	144.24
Neon	Ne	10	20.180
Neptunium	Np	93	237.05*
Nickel	Ni	28	58.693
Niobium	Nb	41	92.906
Nitrogen	N	7	14.007
Nobelium	No	102	259.10*
Osmium	Os	76	190.2
Oxygen	O	8	15.999
Palladium	Pd	46	106.42
Phosphorus	P	15	30.974
Platinum	Pt	78	195.08
Plutonium	Pu	94	239.05*
Polonium	Po	84	209.98*
Potassium	K	19	39.098
Praseodymium	Pr	59	140.91
Promethium	Pm	61	146.92*
Protactinium	Pa	91	231.04
Radium	Ra	88	226.03*
Radon	Rn	86	222.02*
Rhenium	Re	75	186.21
Rhodium	Rh	45	102.91
Rubidium	Rb	37	85.468
Ruthenium	Ru	44	101.07
Rutherfordium	Rf	104	261.11*
Samarium	Sm	62	150.36
Scandium	Sc	21	44.956
Seaborgium	Sg	106	263.12*
Selenium	Se	34	78.96
Silicon	Si	14	28.086
Silver	Ag	47	107.87
Sodium	Na	11	22.990
Strontium	Sr	38	87.62
Sulfur	S	16	32.066
Tantalum	Ta	73	180.95
Technetium	Tc	43	98.906*
Tellurium	Te	52	127.60
Terbium	Tb	65	158.93
Thallium	Tl	81	204.38
Thorium	Th	90	232.04
Thulium	Tm	69	168.93
Tin	Sn	50	118.71
Titanium	Ti	22	47.88
Tungsten	W	74	183.85
Uranium	U	92	238.03
Vanadium	V	23	50.942
Xenon	Xe	54	131.29
Ytterbium	Yb	70	173.04
Yttrium	Y	39	88.906
Zinc	Zn	30	65.39
Zirconium	Zr	40	91.224

Source: These data are based on the definition that the mass of the ^{12}C isotope of carbon is exactly 12 amu. The atomic weight of an element is the weighted average of the masses of the isotopes of that element. Data from "Atomic Weights of the Elements," *Pure and Applied Chemistry*, **63**, 975 (1991).

*Radioactive element has no stable isotope. The mass is for one of the element's common radioisotopes.

**Mass number of the most stable isotope of this radioactive element.